This comprehensive textbook provides a detailed introduction to the basic physics and engineering aspects of lasers, as well as to the design and operational principles of a wide range of optical systems and electro-optic devices. Throughout, full details of important derivations and results are given, as are many practical examples of the design, construction, and performance characteristics of different types of lasers and electro-optic devices.

The first half of the book deals with the fundamentals of laser physics, the characteristics of laser radiation, and discusses individual types of laser, including optically-pumped insulating crystal lasers, atomic gas lasers, molecular gas lasers, and semiconductor lasers. The second half deals with topics such as optical fibers, electro-optic and acousto-optic devices, the fundamentals of nonlinear optics, parametric processes, phase conjugation and optical bistability. The book concludes with chapters on optical detection, coherence theory, and the applications of lasers.

Covering a broad range of topics in modern optical physics and engineering, this book will be invaluable to those taking undergraduate courses in laser physics, optoelectronics, photonics, and optical engineering. It will also act as a useful reference for graduate students and researchers in these fields.

Lasers and Electro-Optics

Lasers and Electro-Optics

Fundamentals and Engineering

CHRISTOPHER C. DAVIS

Professor of
Electrical Engineering,
University of Maryland,
College Park

CAMBRIDGE
UNIVERSITY PRESS

Published by the Press Syndicate of the University of Cambridge
The Pitt Building, Trumpington Street, Cambridge CB2 1RP
40 West 20th Street, New York, NY 10011-4211, USA
10 Stamford Road, Oakleigh, Melbourne 3166, Australia

First published 1996

Printed in Great Britain at the University Press, Cambridge

A catalogue record of this book is available from the British Library

Library of Congress cataloguing in publication data

Davis, Christopher C., 1944–
Lasers and electro-optics : fundamentals and engineering / Christopher C. Davis.
p. cm.
Includes bibliographical references.
ISBN 0-521-30831-3 (hardback.) – ISBN 0-521-48403-0 (pbk.)
1. Lasers. 2. Electrooptics. I. Title.
TA1675.D38 1995
621.36–dc20 94-43230 CIP

TAG

To my beloved wife, Mary, and our children:

Alex, Fiona and Mark.

Concordia res parvae crescunt

Contents

Preface

The author of a text generally feels obligated to explain the reasons for his or her writing. This is a matter of tradition as it provides an opportunity for explaining the development and philosophy of the text, its subject matter and intended audience, and acknowledges the help that the author has received. I hope to accomplish these tasks briefly here.

This text has grown over many years out of notes that I have developed for courses at the senior undergraduate and beginning graduate student level at the University of Manchester, Cornell University, and the University of Maryland, College Park. These courses have covered many aspects of laser physics and engineering, the practical aspects of optics that pertain to an understanding of these subjects, and a discussion of related phenomena and devices whose importance has grown from the invention of the laser in 1960. These include nonlinear optics, electro-optics, acousto-optics, and the devices that take practical advantage of these phenomena. The names given to the fields that encompass such subject matter have included laser physics, optical electronics, optoelectronics, photonics, and quantum electronics.

The last of these names is consistent with a treatment of the laser and associated phenomena as fundamentally quantum mechanical in nature. However, almost all aspects of laser operation and important related phenomena can be explained well classically. Therefore, this text requires no knowledge of quantum mechanics, although a background in electromagnetic theory at the undergraduate level is desirable for a greater understanding. Most electrical engineering majors do not take a course in quantum mechanics until they reach graduate level, and many physics majors will not have acquired sufficient quantum mechanics knowledge at the undergraduate level for this to make a meaningful contribution to better understanding in a study of lasers and electro-optics.

For all the above reasons this text should be suitable for senior undergraduate and beginning graduate students in electrical engineering or physics. It should also prove useful to mechanical engineers or chemists who use lasers and electro-optic devices in their research.

The text is broken up into two principal parts. Chapters 1–13 discuss the basic physics and engineering of lasers of all kinds, beginning with a discussion of the fundamental physics of the stimulated emission process and laser amplifiers. This is followed by chapters on laser resonators and the characteristics of laser radiation and methods for controlling it. There are succeeding chapters that cover optically-pumped insulating crystal lasers, atomic gas lasers, molecular gas lasers of various kinds including gas transport, gas dynamics and chemically-pumped varieties, and tunable lasers. The first section of the book concludes with a chapter on semiconductor lasers that begins with a review of the basic physics necessary for their understanding.

The second part of the text covers various issues of relevance to lasers and electro-optics, including the optics of Gaussian beams, laser resonators and anisotropic crystalline materials. There are chapters on optical fibers, electro-optic and acousto-optic devices, the fundamentals of nonlinear optics, and application of nonlinear optics in harmonic generation, parametric processes, phase conjugation, and optical bistability. The text concludes with chapters on optical detectors and the detection process, coherence theory, and applications of lasers.

I have found that Chapters 1–13 provide sufficient material for a one semester course on lasers, with some applications from Chapter 24 included. Chapters 14–19, with Chapter 23 form the basis of a one semester course on optical design, electro-optic devices and optical detectors. I draw on the somewhat more difficult material in Chapters 20–22 as reference material in both the one semester courses just mentioned, and also as adjunct material in more advanced graduate courses.

I have been an experimentalist in the laser business for almost thirty years. I have always found the laser itself a fascinating device that provides a teaching vehicle for discussing many fundamental concepts and practical aspects of design. I have always tried to introduce practical details of real lasers into my classes as early as a treatment of some of the associated fundamentals permits. This should be apparent in the current text where I digress in Chapter 3 into a fairly detailed practical discussion of two important lasers, even though contextually a fuller discussion of these devices could be left until later. I believe that this makes pedagogical sense as students get a glimpse of where they are headed. Throughout the text I have attempted to provide full details of important derivations, and provide practical examples from the literature on the design, construction, and performance characteristics of lasers and electro-optic devices.

In developing a sound pedagogical approach to teaching the material in this text to many students over more than two decades I have drawn inevitably on the work of many others. There have been many other texts that cover material that is shared in common with the current one. What is different between this and related texts is not so much the analytic treatment of common subject matter, but the specific choice of material presentation sequence, and assorted explanations. As it is said, "there is nothing new under the sun," so it is not the intent of this author to claim that the treatment of particular topics in the current text is necessarily unique. Different authors impart their own slant to the same subject matter: sometimes their treatments converge, particularly when there is one especially good way of explanation that is valid. I have attempted in every case to provide reference to the original literature from which I have benefited in my writing, and I apologize for any inadvertent omissions.

Over the course of many years I have learned much from my contacts in the classroom, office, research laboratory, and at conferences. I am indebted to numerous past and present colleagues and students for their intellectual stimulation, advice, provision of material, and feedback on early versions of the current text that have contributed greatly to the finished product. I would like to thank especially my past and present faculty colleagues Mario Dagenais, Julius Goldhar, Ping-Tong Ho, Urs Hochuli, Terry King, Chi Lee, Ross McFarlane; and finally George Wolga who gave me valuable help at the very beginning of this work. My graduate students and post-doctoral research associates have over several years provided help and advice that have helped me greatly. In particular, I appreciate their forebearance in tolerating a degree of benign neglect, especially during the latter stages of the completion of this work. I am particularly grateful to Walid Atia, Rob Bonney, Simon Bush, Ali Güngör, Pat Mead, Dave Mazzoni, Melody Owens, Saeed Pilevar, and Richard Wagreich for their current help with my research that has provided the time to complete this book. I am most grateful to Joan Hamilton and Nono Kusuma for drawing most of the diagrams and to Dave Mazzoni and Sarah Mulhall for help with additional diagrams.

I appreciate the patience of several editors at Cambridge University Press in waiting for completion of this text, and for accepting ongoing excuses as to why it was not completed earlier. I am most grateful to Patricia Keehn for her expert and careful computer typesetting work using TEX over enormous numbers of revisions of this document.†

Most of all, I am indebted to my family, especially my wife Mary, for their love and their tolerance of irregular hours at home, and "occasional" inattention to family matters.

College Park, Maryland Christopher C. Davis
1995

† In most cases where curves showing the parametric variation of phenomena discussed in the text are given these have been calculated from scratch using Mathcad (©Mathsoft, Inc.) or by the use of Fortran subroutine available on the Unix network at the University of Maryland, Two- and three-dimensional graphics have been produced by Mathcad and Mongo (©John L. Tonry). The entire text has been typeset using Donald Knuth's TEX: *The TEXbook*, Addison-Wesley, Reading, MA, 1984. TEX is a trademark of the American Mathematical Society.

1

Spontaneous and Stimulated Transitions

1.1 Introduction

A laser is an oscillator that operates at very high frequencies. These *optical* frequencies range to values several orders of magnitude higher than can be achieved by the 'conventional' approaches of solid-state electronics or electron tube technology. In common with electronic circuit oscillators, a laser is constructed from an amplifier with an appropriate amount of positive feedback. The acronym LASER, which stands for *light amplification by stimulated emission of radiation*, is in reality therefore a slight misnomer†.

In this chapter we shall consider the fundamental processes whereby amplification at optical frequencies can be obtained. These processes involve the fundamental atomic nature of matter. At the atomic level matter is not a continuum, it is composed of discrete particles – atoms, molecules or ions. These particles have energies that can have only certain discrete values. This discreteness or *quantization*, of energy is intimately connected with the duality that exists in nature. Light sometimes behaves as if it were a wave and in other circumstances it behaves as if it were composed of particles. These particles, called *photons*, carry the discrete packets of energy associated with the wave. For light of frequency ν the energy of each photon is $h\nu$, where h is Planck's constant – 6.6×10^{-34} J s. The energy $h\nu$ is the *quantum* of energy associated with the frequency ν. At the microscopic level the amplification of light within a laser involves the emission of these quanta. Thus, the term *quantum electronics* is often used to describe the branch of science that has grown from the development of the maser in 1954 and the laser in 1960.

The widespread practical use of lasers and optical devices in applications such as communications, and increasingly in areas such as signal processing and image analysis has lead to the use of the term *photonics.* Whereas, electronics uses electrons in various devices to perform analog and digital functions, photonics aims to replace the electrons with photons. Because photons have zero mass, do not interact with each other to any significant extent, and travel at the speed of light photonic devices promise small size and high speed.

1.2 Why 'Quantum' Electronics?

In 'conventional' electronics, where by the word 'conventional' for the present purposes we mean frequencies where solid-state devices such as transistors or diodes will operate, say below 10^{11} Hz, an oscillator is conveniently constructed by applying an appropriate amount of positive feedback to an amplifier. Such an

† The more truthful acronym LOSER was long ago deemed inappropriate.

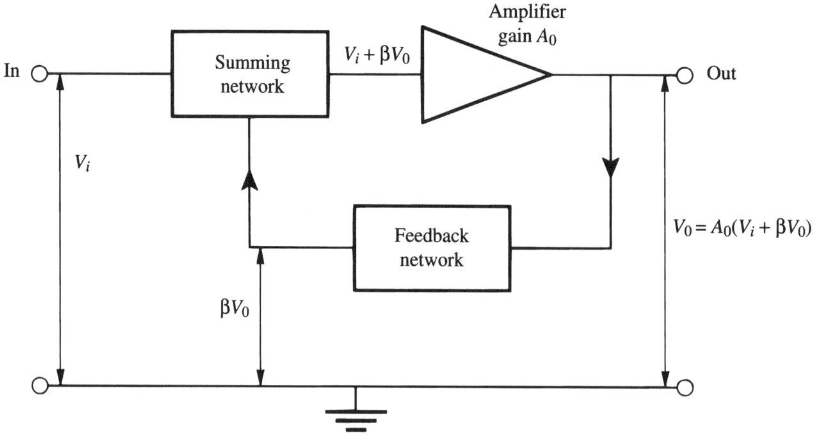

Fig. 1.1. Circuit diagram of a simple amplifier with feedback.

arrangement is shown schematically in Fig. (1.1). The input and output voltages of the amplifier are V_i and V_0 respectively. The overall gain of the system is A, where $A = V_0/V_i$. Now,

$$V_0 = A_0(V_i + \beta V_0)$$

so

$$V_0 = \frac{A_0 V_i}{1 - \beta A_0}$$

and

$$A = \frac{A_0}{1 - \beta A_0}. \tag{1.1}$$

If $\beta A_0 = +1$ then the gain of the circuit would apparently become infinite, and the circuit would generate a finite output without any input. In practice electrical 'noise', which is a random oscillatory voltage generated to a greater or lesser extent in all electrical components in any amplifier system, provides a finite input. Because βA_0 is generally a function of frequency the condition $\beta A_0 = +1$ is generally satisfied only at one frequency. The circuit oscillates at this frequency by amplifying noise at this same frequency that appears at its input. However, the output does not grow infinitely large, because as the signal grows, A_0 falls – this process is called saturation. This phenomenon is fundamental to all oscillator systems. A laser (or maser) is an optical (microwave) frequency oscillator constructed from an optical (microwave) frequency amplifier with positive feedback, shown schematically in Fig. (1.2). Light waves which become amplified on traversing the amplifier are returned through the amplifier by the reflectors and grow in intensity, but this intensity growth does not continue indefinitely because the amplifier saturates. The arrangement of mirrors (and sometimes other components) that provides the feedback is generally referred to as the laser cavity or resonator.

We shall deal with the full characteristics of the device consisting of amplifying medium and resonator later, for the moment we must concern ourselves with the problem of how to construct an amplifier at optical frequencies. The frequencies involved are very high, for example lasers have been built which operate from

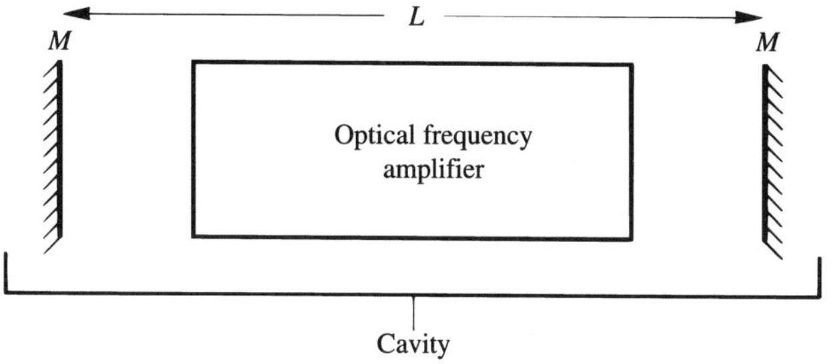

Fig. 1.2. Schematic diagram of a basic laser structure incorporating an amplifying medium and two feedback mirrors, *M*.

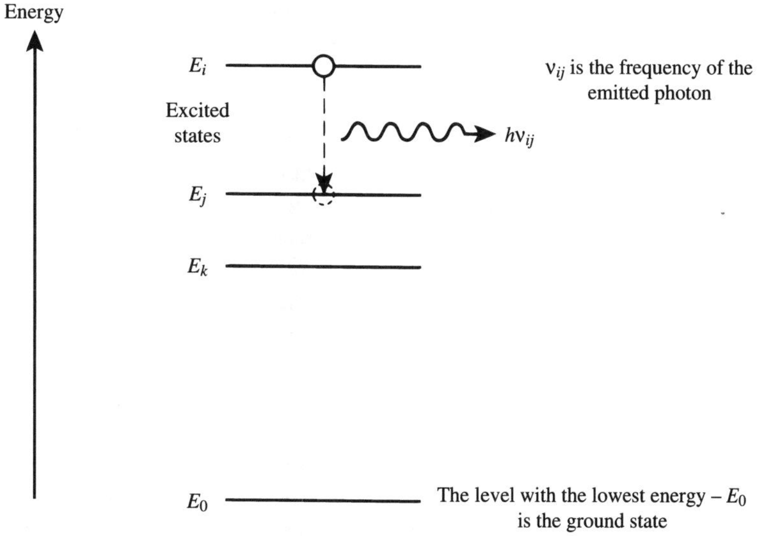

Fig. 1.3. Simple schematic energy level diagram for a particle.

very short wavelengths, for example 109.8 nm, using para-hydrogen gas as the amplifying medium, to 2650 μm using methyl bromide as the amplifying medium. This is a frequency range from 2.73×10^{15} Hz down to 1.13×10^{11} Hz. The operating frequencies of masers overlap this frequency range at the low frequency end, the fundamental difference between the two devices is essentially only one of scale. If the length of the resonant cavity which provides feedback is L, then for $L \gg \lambda$, where λ is the wavelength at which oscillation occurs, we have a laser: for $L \sim \lambda$ we have a maser.

1.3 Amplification at Optical Frequencies

How do we build an amplifier at such high frequencies? We use the energy delivered as the particles that consitute the amplifying medium make jumps between their different energy levels. The medium may be gaseous, liquid, a crystalline or glassy solid, an insulating material or a semiconductor. The electrons that are bound within the particles of the amplifying medium, whether these are atoms, molecules or ions, can occupy only certain discrete energy levels. Consider such a system of energy levels, shown schematically in Fig. (1.3). Electrons can make jumps between these levels in three ways.

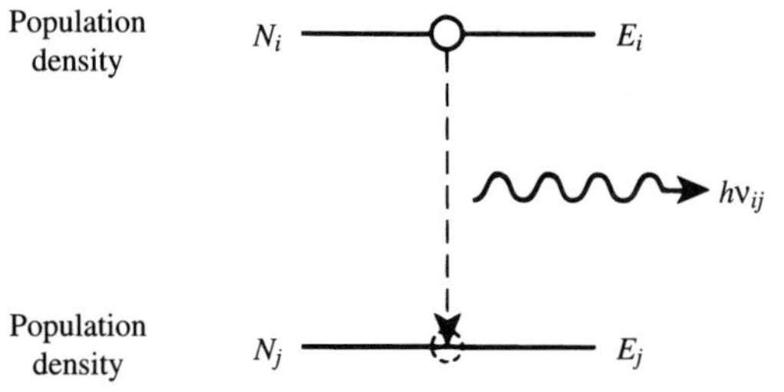

Fig. 1.4. Schematic representation of spontaneous emission between two levels of energy E_i and E_j.

1.3.1 Spontaneous Emission

An electron spontaneously falls from a higher energy level to a lower one as shown in Fig. (1.4), the emitted photon has frequency

$$\nu_{ij} = \frac{E_i - E_j}{h}. \tag{1.2}$$

This photon is emitted in a random direction with arbitrary polarization (except in the presence of magnetic fields, but this need not concern us here). The photon carries away momentum $h/\lambda = h\nu/c$ and the emitting particle (atom, molecule or ion) recoils in the opposite direction. The probability of such a spontaneous jump is given quantitatively by the Einstein A coefficient defined as A_{ij} = 'probability' per second of a spontaneous jump from level i to level j.

For example, if there are N_i particles per unit volume in level i then $N_i A_{ij}$ per second make jumps to level j. The total rate at which jumps are made between the two levels is

$$\frac{dN_{ij}}{dt} = -N_i A_{ij}. \tag{1.3}$$

There is a negative sign because the population of level i is decreasing.

Generally an electron can make jumps to more than one lower level, unless it is in the first (lowest) excited level. The total probability that the electron will make a spontaneous jump to any lower level is A_i s^{-1} where

$$A_i = \sum_j A_{ij}. \tag{1.4}$$

The summation runs over all levels j lower in energy than level i and the total rate at which the population of level i changes by spontaneous emission is

$$\frac{dN_i}{dt} = -N_i A_i,$$

which has the solution

$$N_i = \text{constant} \times e^{-A_i t}. \tag{1.5}$$

If at time $t = 0, N_i = N_i^0$ then

$$N_i = N_i^0 e^{-A_i t}, \tag{1.6}$$

so the population of level i falls exponentially with time as electrons leave by spontaneous emission. The time in which the population falls to $1/e$ of its initial value is called the natural lifetime of level i, τ_i, where $\tau_i = 1/A_i$. The magnitude of this lifetime is determined by the actual probabilities of jumps from level i

by spontaneous emission. Jumps which are likely to occur are called *allowed* transitions, those which are unlikely are said to be *forbidden. Allowed* transitions in the visible region typically have A_{ij} coefficients in the range 10^6–10^8 s^{-1}. *Forbidden* transitions in this region have A_{ij} coefficients below 10^4 s^{-1}. These probabilities decrease as the wavelength of the jump increases. Consequently, levels that can decay by allowed transitions in the visible have lifetimes generally shorter than 1 μs, similar forbidden transitions have lifetimes in excess of 10–100 μs. Although no jump turns out to be absolutely forbidden, some jumps are so unlikely that levels whose electrons can only fall to lower levels by such jumps are very long lived. Levels with lifetimes in excess of 1 hour have been observed under laboratory conditions. Levels which can only decay slowly, and usually only by forbidden transitions, are said to be *metastable.*

When a particle changes its energy spontaneously the emitted radiation is not, as might perhaps be expected, all at the same frequency. Real energy levels are not infinitely sharp, they are smeared out or *broadened.* A particle in a given energy level can actually have any energy within a finite range. The frequency spectrum of the spontaneously emitted radiation is described by the *lineshape function*, $g(\nu)$. This function is usually normalized so that

$$\int_{-\infty}^{\infty} g(\nu)d\nu = 1. \tag{1.7}$$

$g(\nu)d\nu$ represents the probability that a photon will be emitted spontaneously in the frequency range $\nu + d\nu$. The lineshape function $g(\nu)$ is a true probability function for the spectrum of emitted radiation and is usually sharply peaked near some frequency ν_0, as shown in Fig. (1.5). For this reason the function is frequently written $g(\nu_0, \nu)$ to highlight this. Since negative frequencies do not exist in reality the question might properly be asked: 'Why does the integral have a lower limit of minus infinity?' This is done because $g(\nu)$ can be viewed as the Fourier transform of a real function of time, so negative frequencies have to be permitted mathematically. In practice $g(\nu)$ is only of significant value for some generally small range of positive frequencies so

$$\int_{0}^{\infty} g(\nu)d\nu \simeq 1. \tag{1.8}$$

The amount of radiation emitted spontaneously by a collection of particles can be described quantitatively by their *radiant intensity* $I_e(\nu)$. The units of radiant intensity are watts per steradian.† The total power (watts) emitted in a given frequency interval $d\nu$ is

$$W(\nu)d\nu = \int_S I_e(\nu)d\nu d\Omega, \tag{1.9}$$

where the integral is taken over a closed surface S surrounding the emitting particles.

The total power emitted is

$$W_0 = \int_{-\infty}^{\infty} W(\nu)d\nu. \tag{1.10}$$

† The steradian is the unit of solid angle, Ω. The surface of a sphere encompasses a solid angle of 4π steradians.

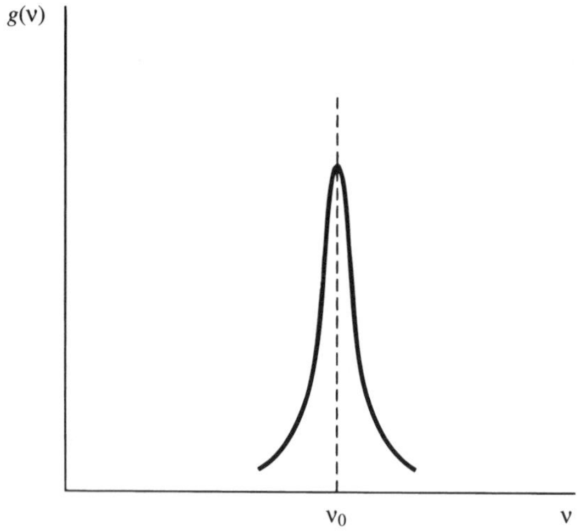

Fig. 1.5. A lineshape function $g(\nu_0, \nu)$.

$W(\nu)$ is closely related to the lineshape function

$$W(\nu) = W_0 g(\nu). \tag{1.11}$$

For a collection of N_i identical particles the total spontaneously emitted power per frequency interval is

$$W(\nu) = N_i A_i h\nu g(\nu). \tag{1.12}$$

Clearly this power decreases with time if the number of excited particles decreases.

For a plane electromagnetic wave we can introduce the concept of *intensity*, which has units of W m^{-2}. The intensity is the average amount of energy per second transported across unit area in the direction of travel of the wave. The spectral distribution of intensity, $I(\nu)$, is related to the total intensity, I_0, by

$$I(\nu) = I_0 g(\nu). \tag{1.13}$$

It is worth pointing out that in reality perfect plane waves do not exist, such waves would have a unique propagation direction and infinite radiant intensity. However, to a very good degree of approximation we can treat the light from a small source as a plane wave if we are far enough away from the source. The light coming from a star represents a very good example of this.

1.3.2 Stimulated Emission

Besides being able to make transitions from a higher level to a lower one by spontaneous emission, electrons can also be stimulated to make these jumps by the action of an external radiation field, as shown in Fig. (1.6).

Let the energy density of the externally applied radiation field at frequency ν be $\rho(\nu)$ (energy per unit volume per unit frequency interval; i.e., J m^{-3} Hz^{-1}). If ν is the same frequency as a transition between two levels labelled 2 and 1, the rate at which stimulated emissions occur is $N_2 B'_{21}(\nu)$ s^{-1} Hz^{-1} m^{-3} where $B'_{21}(\nu)$ is a function specific to the electron jump between the two levels and N_2 is the number of particles per unit volume in the upper level of the transition. The frequency

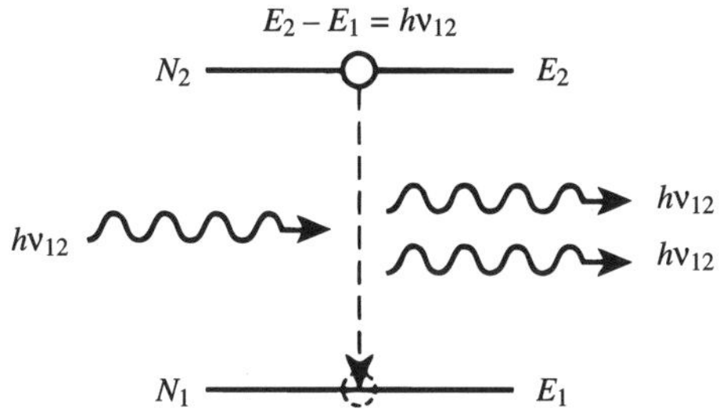

Fig. 1.6. Schematic representation of stimulated emission between two levels of energy E_2 and E_1.

dependence of $B'_{21}(\nu)$ is the same as the lineshape function

$$B'_{21}(\nu) = B_{21}g(\nu_0, \nu). \tag{1.14}$$

B_{21} is called the Einstein coefficient for stimulated emission. The total rate of change of population density by stimulated emission is

$$\begin{aligned} \frac{dN_2}{dt} &= -N_2 \int_{-\infty}^{\infty} B'_{21}(\nu)\rho(\nu)d\nu \\ &= -N_2 B_{21} \int_{-\infty}^{\infty} g(\nu_0, \nu)\rho(\nu)d\nu. \end{aligned} \tag{1.15}$$

Note, that for the dimensions of both sides of Eq. (1.15) to balance B_{21} must have units m^3 J^{-1} s^{-2}. To evaluate the integral in Eq. (1.15) we must consider how energy density is related to intensity and how these quantitites might actually vary with frequency.

1.4 The Relation Between Energy Density and Intensity

The energy density of a radiation field $\rho(\nu)$ (energy per unit volume per unit frequency interval) can be simply related to the intensity of a plane electromagnetic wave. If the intensity of the wave is $I(\nu)$ (for example, watts per unit area per frequency interval). Then

$$\rho(\nu) = \frac{I(\nu)}{c}, \tag{1.16}$$

where c is the velocity of light in the medium.† This is illustrated in Fig. (1.7). All the energy stored in the volume of length c passes across the plane A in one second so

$$\rho(\nu) \times A \times c = I(\nu) \times A. \tag{1.17}$$

The energy density in a general radiation field $\rho(\nu)$ is a function of frequency ν. If $\rho(\nu)$ is independent of frequency the radiation field is said to be *white*, as shown in Fig. (1.8). If the radiation field is *monochromatic* at frequency ν_{21}, its spectrum is as shown in Fig. (1.9). The ideal *monchromatic* radiation field has an infinitely narrow energy density profile at frequency ν_{21}. This type of profile is called a δ-function. The properties of this function are described in more detail in Appendix 1.

† $c = c_0/n$, where c_0 is the velocity of light in a vacuum and n is the *refractive index*.

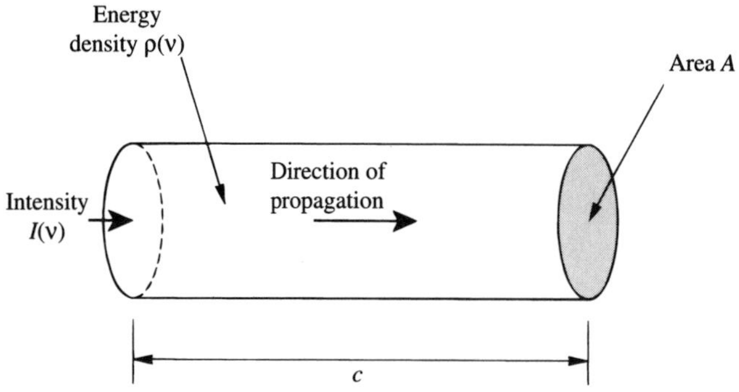

Fig. 1.7. A volume of space swept through per second by part of a plane wave.

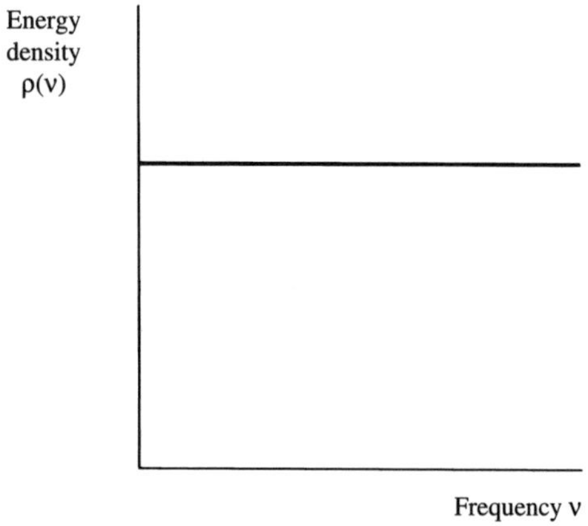

Fig. 1.8. A 'white' energy density spectrum.

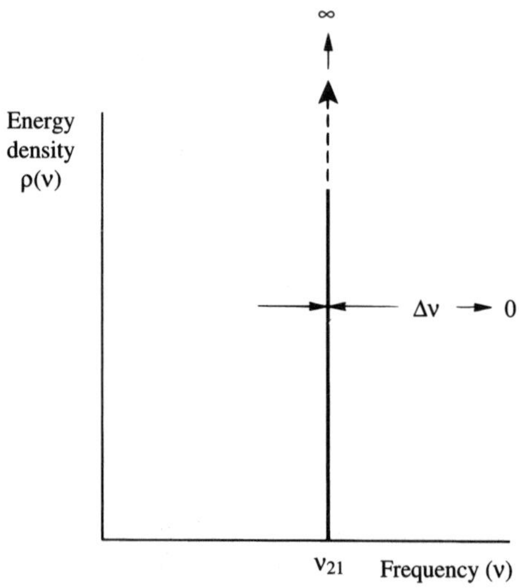

Fig. 1.9. A monochromatic energy density spectrum.

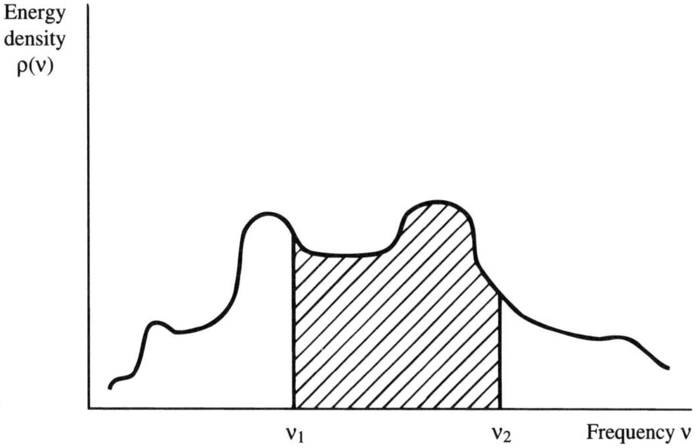

Fig. 1.10. A generalized energy density spectrum.

For a general radiation field the total energy stored per unit volume between frequencies ν_1 and ν_2 is

$$\int_{\nu_1}^{\nu} \rho(\nu) d\nu.$$

The energy density of a general radiation field is a function of frequency, for example as shown in Fig (1.10). For a monochromatic radiation field at frequency ν_{21}

$$\rho(\nu) = \rho_{21}\delta(\nu - \nu_0). \tag{1.18}$$

The δ-function has the property

$$\delta(\nu - \nu_{21}) = 0 \text{ for } \nu \neq \nu_{21}, \tag{1.19}$$

and

$$\int_{-\infty}^{\infty} \delta(\nu - \nu_{21}) d\nu = 1. \tag{1.20}$$

So for a monochromatic radiation field the total stored energy per unit volume is

$$\int_{-\infty}^{\infty} \rho(\nu) d\nu = \int_{-\infty}^{\infty} \rho_{21}\delta(\nu - \nu_{21}) d\nu = \rho_{21}. \tag{1.21}$$

For such a monochromatic radiation field Eq. (1.15) can be written as

$$\begin{aligned} \frac{dN_2}{dt} &= -N_2 B_{21} \int_{-\infty}^{\infty} g(\nu_0, \nu)\rho_{21}\delta(\nu - \nu_{21}) d\nu \\ &= -N_2 B_{21} g(\nu_0, \nu_{21})\rho_{21}. \end{aligned} \tag{1.22}$$

It is very important to note that the rate of stimulated emissions produced by this input monochromatic radiation is directly proportional to the value of the lineshape function at the input frequency. The maximum rate of stimulated emission is produced, all else being equal, if the input radiation is at the line center frequency ν_0.

If the stimulating radiation field has a spectrum that is broad, we can assume that the energy density $\rho(\nu)$ is constant over the narrow range of frequencies where

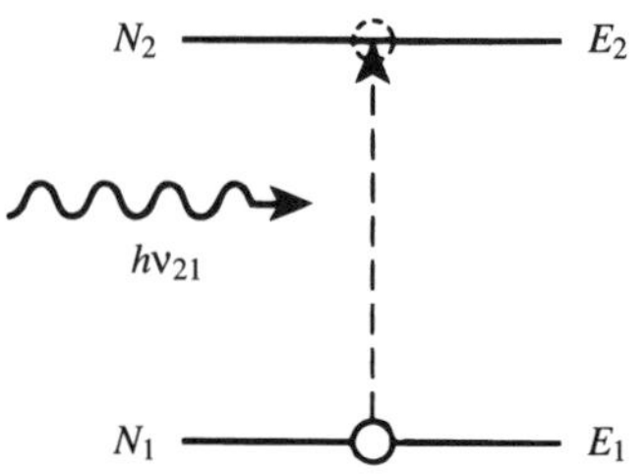

Fig. 1.11. Schematic representation of stimulated absorption between two levels of energy E_1 and E_2.

$g(\nu_0, \nu)$ is significant. In this case Eq. (1.15) gives

$$\frac{dN_2}{dt} = -N_2 B_{21} \rho(\nu), \tag{1.23}$$

where $\rho(\nu) \simeq \rho(\nu_0)$ is the energy density in the frequency range where transitions take place.

1.4.1 Stimulated Absorption

As well as making stimulated jumps in a downward direction, electrons may make transitions in an upward direction between energy levels of a particle by absorbing energy from an electromagnetic field, as shown in Fig. (1.11). The rate of such absorptions and the rate at which electrons leave the lower level is,

$$N_1 \rho(\nu) B_{12} g(\nu_0, \nu) \text{ s}^{-1} \text{ Hz}^{-1} \text{ m}^{-3},$$

which yields a result similar to Eq. (1.15)

$$\frac{dN_1}{dt} = -N_1 B_{12} \int_{-\infty}^{\infty} g(\nu_0, \nu) \rho(\nu) d\nu. \tag{1.24}$$

Once again B_{12} is a constant specific to the jump and is called the Einstein coefficient for stimulated absorption. Here again $\rho(\nu)$ is the energy density of the stimulating field. There is no analog in the absorption process to spontaneous emission. A particle cannot spontaneously *gain* energy without an external energy supply. Thus, it is unnecessary for us to continue to describe the absorption process as stimulated absorption.

It is interesting to view both stimulated emission and absorption as photon–particle collision processes. In *stimulated* emission the incident photon produces an identical photon by 'colliding' with the electron in an excited level, as shown in Fig. (1.12a.) After the stimulated emission process, both photons are travelling in the same direction and with the same polarization as the incident photon originally had. When light is described in particle terms, polarization can be viewed as describing the angular motion or spin of individual photons. Left and right hand circularly polarized light corresponds in this particle picture to beams of photons that spin clockwise and counterclockwise, respectively, about their direction of propagation. Linearly polarized light corresponds to a beam of photons that have no net angular momentum about an axis parallel to their direction of propagation. In stimulated emission the stimulated photon has exactly the same frequency as the stimulating photon. In absorption the incident photon disappears, as shown in Fig. (1.12b). In both stimulated emission and absorption the atom recoils to conserve linear momentum.

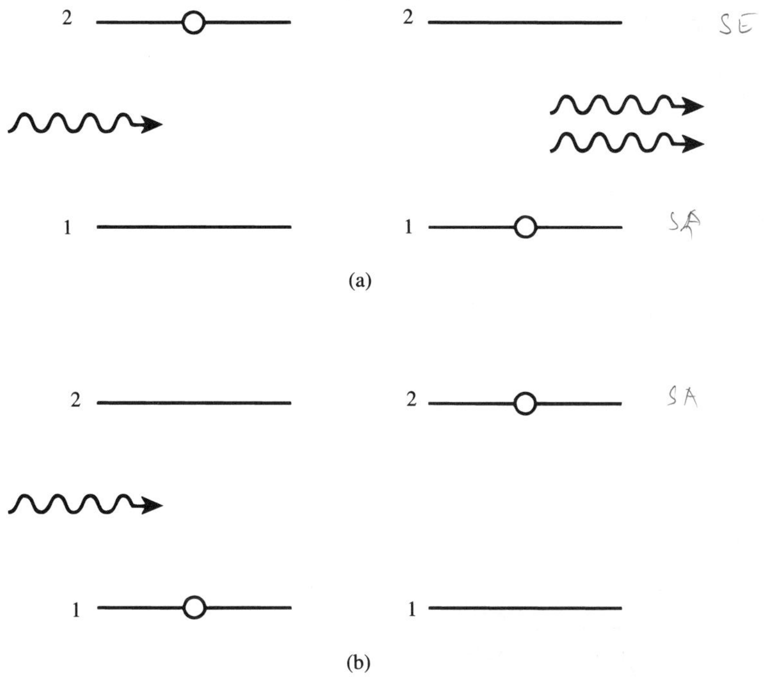

Fig. 1.12. A photon–particle 'collision' picture of the stimulated emission and absorption processes: (a) stimulated emission, (b) absorption.

1.5 Intensity of a Beam of Electromagnetic Radiation in Terms of Photon Flux

If the intensity of a beam of light is $I(\nu)$ (W m^{-2} per frequency interval) then the number of photons in the beam crossing unit area per unit time is

$$N_{photons} = \frac{I(\nu)}{h\nu}(\text{photons s}^{-1}\text{ m}^{-1}\text{ per frequency interval}). \tag{1.25}$$

If the beam is monochromatic and has total intensity $I(\nu_{21})$ then

$$N_{photons} = \frac{I(\nu_{21})}{h\nu}\text{ photons s}^{-1}\text{ m}^{-2}. \tag{1.26}$$

1.6 Black-Body Radiation

A particularly important kind of radiation field is that emitted by a *black body*. Such a body absorbs with 100% efficiency all the radiation falling on it, irrespective of the radiation frequency. Black-body radiation has a spectral distribution of the kind shown in Fig. (1.13). A close approximation to a black body (absorber and emitter) is an enclosed cavity containing a small hole, as shown in Fig. (1.14). Radiation that enters the hole has very little chance of escaping. If the inside of this cavity is in thermal equilibrium it must lose as much energy as it absorbs and the emission from the hole is therefore characteristic of the equilibrium temperature T inside the cavity. Thus this type of radiation is often called 'thermal' or 'cavity' radiation.

In the early days of the quantum theory, the problem of describing theoretically the spectral profile of the emission from a black body was a crucial one. In the latter part of the nineteenth century experimental measurements of this spectral profile

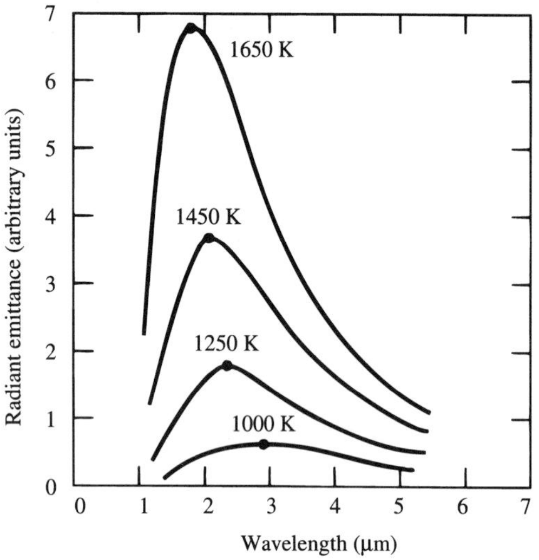

Fig. 1.13. Spectral distribution of black-body radiation at different temperatures.

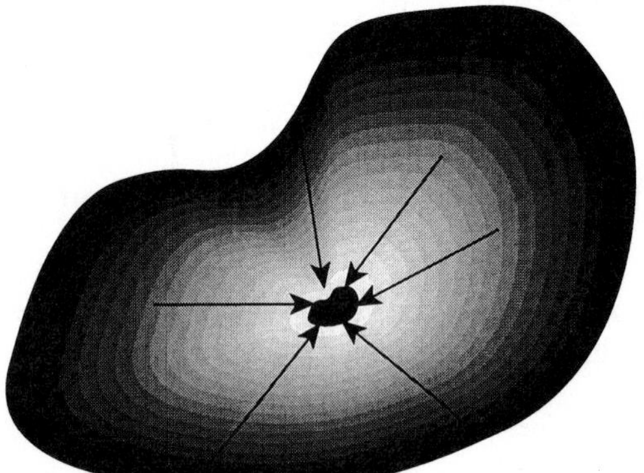

Fig. 1.14. Simple model of a black-body absorber/emitter – an enclosed cavity containing a small hole.

had already been obtained and the data had even been fitted to an empirical formula. Attempts to explain the form of the data were based on treating the electromagnetic radiation as a collection of oscillators, each oscillator with its own characteristic frequency. The problem was to determine how many oscillations at a given frequency could be fitted inside a cavity.

Thermodynamically, the shape of the cavity for which the calculation is performed is arbitrary (provided it is much larger than the wavelength of any of the oscillations) otherwise we could make a heat engine by connecting together cavities of different shapes. If, for example, two cavities of different shapes, but at the same temperature, were connected together with a reflective hollow pipe, we could imagine placing filters having different narrow frequency bandpass characteristics in the pipe. Unless the radiation emitted in each elemental frequency band from both cavities was identical, one cavity could be made to heat up and the other cool down, thereby violating the second law of thermodynamics. For convenience purposes, we choose a cubical cavity with sides of length L. A plane electromagnetic wave will 'fit' inside this cavity if it satisfies appropriate *periodic* boundary

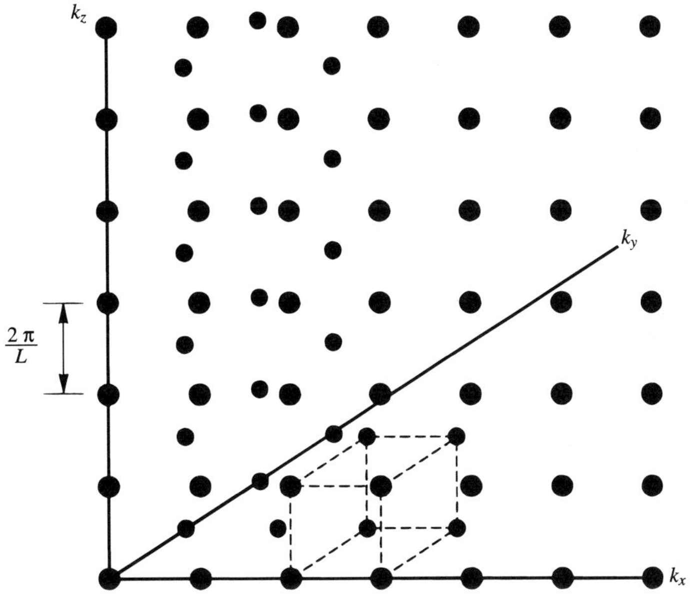

Fig. 1.15. Allowed values of k_x, k_y, k_z in **k**-space for a cubical cavity of side L.

conditions. If the wave has a spatial variation written in complex notation as $e^{i\mathbf{k}\cdot\mathbf{r}}$ these boundary conditions can be written as

$$e^{ik_x x} = e^{ik_x(x+L)} \tag{1.27}$$

with similar equations for the y and z directed components of the wave. Equations like (1.27) are satisfied if

$$k_x = \frac{2\pi\ell}{L}, k_y = \frac{2\pi m}{L}, k_z = \frac{2\pi n}{L}, \tag{1.28}$$

where k_x, k_y, k_z are the components of the wave vector **k** of the oscillation and ℓ, m, n are integers.

$$\mathbf{k} = \frac{\omega}{c}\hat{\mathbf{k}}, \quad |\mathbf{k}| = 2\pi/\lambda, \tag{1.29}$$

where ω is the angular frequency of the wave, λ its wavelength, c is the velocity of light and $\hat{\mathbf{k}}$ is a unit vector in the direction of the wave vector.

In a three-dimensional space whose axes are k_x, k_y, and k_z (called **k**-space) the possible k values that are periodic inside the cube form a lattice as shown in Fig. (1.15). The size of a unit cell of this lattice is $(2\pi/L)^3$. Each cell corresponds to one possible *mode*, characterized by its own values of k_x, k_y, and k_z which are periodic inside the cube. Note that the spacing of adjacent modes, say in the k_x direction, is $2\pi/L$. Thus the permitted values of $k_x, 0, 2\pi/L, 4\pi/L$ etc., correspond to oscillation wavelengths $\infty, L, L/2, L/3$ etc. Thus in counting the permitted values of k_x all the intuitive values of λ which 'fit' between the walls of the cavity are included. The total number of modes of oscillation with $|\mathbf{k}| \leq k$ is

$$N_k = \frac{\text{total volume of } k\text{-space with } |\mathbf{k}| \leq k}{\text{volume of unit cell}} \times 2$$
$$= 2 \times \frac{4}{3}\pi k^3/(2\pi/L)^3. \tag{1.30}$$

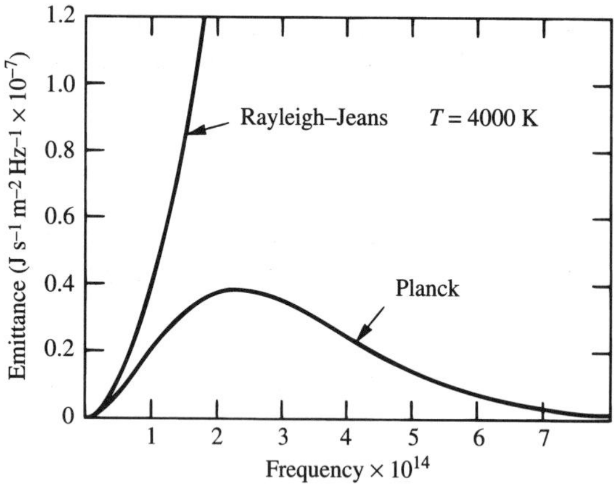

Fig. 1.16. The Rayleigh–Jeans prediction of the spectral intensity of a black body compared to the Planck formula.

The factor of 2 enters because we must take account of the two distinct polarizations of the radiation field. Therefore,

$$N_k = \frac{k^3 L^3}{3\pi^2}. \tag{1.31}$$

Now since $k = 2\pi\nu/c$, the number of modes with frequency $\leq \nu$ is

$$N_\nu = \frac{8\pi\nu^3}{3c^3} L^3. \tag{1.32}$$

Since L^3 is the volume of the enclosure, V,

$$\frac{N_\nu}{V} = \frac{8\pi\nu^3}{3c^3}. \tag{1.33}$$

The mode-density (per unit volume per unit frequency interval) is

$$p(\nu) = \frac{1}{V}\frac{dN_\nu(\nu)}{d\nu} = \frac{8\pi\nu^2}{c^3}. \tag{1.34}$$

Rayleigh and Jeans attempted to use this type of frequency dependence to describe the spectral composition of black-body radiation. They assumed that the law of equipartition of energy held for the distribution of energy among the various modes of the radiation field. Consequently, they assigned an equal energy kT to each mode of oscillation and predicted that the intensity distribution of the black-body radiation would be

$$I(\nu) \propto \frac{8\pi\nu^2}{c^3}.$$

This did not fit experimental observations of black-body radiation except at relatively low frequencies (in the red and infrared regions of the spectrum). The predicted large increase in $I(\nu)$ at high frequencies was in dramatic conflict with observations made of black-body emission intensities in the ultraviolet. This conflict between theory and experiment was called the *ultraviolet catastrophe*, as shown in Fig. (1.16). Planck resolved these difficulties with his quantum hypothesis. He proposed that each oscillation mode could only take certain quantized energies

$$E_{n_\nu} = (n + \tfrac{1}{2})h\nu, \qquad n = 0, 1, 2, 3..., \tag{1.35}$$

where the contribution $\frac{1}{2}h\nu$ is called the zero point energy. The probability of finding energy E_n in a particular mode of oscillation is given by classical Maxwell–Boltzmann statistics, i.e.,

$$\frac{P(n)}{P(0)} = \frac{e^{-E_n/kT}}{e^{-E_0/kT}} = e^{-nh\nu/kT}, \tag{1.36}$$

where h is Planck's constant, k is Boltzmann's constant, T is the absolute temperature, and $P(0)$ is the probability of finding the lowest energy in the mode. Consequently the average energy of a mode is:

$$\overline{E}_\nu = \sum_{n=0}^{\infty} \text{energy of excitation} \times \text{probability} = \sum_{n=0}^{\infty} P(n)E_n, \tag{1.37}$$

giving

$$\overline{E}_\nu = \sum_{n=0}^{\infty} P(0)e^{-nh\nu/kT}(n + \tfrac{1}{2})h\nu. \tag{1.38}$$

Now, if a particular oscillation is excited it must be in one of the quantized states, therefore

$$\sum_{n=0}^{\infty} P(n) = 1,$$

so

$$\sum_{n=0}^{\infty} P(0)e^{-nh\nu/kT} = 1. \tag{1.39}$$

Thus

$$P(0) = \frac{1}{\sum_{n=0}^{\infty} e^{-nh\nu/kT}}, \tag{1.40}$$

and

$$E_\nu = \frac{\sum_{n=0}^{\infty}(n + \frac{1}{2})h\nu e^{-nh\nu/kT}}{\sum_{n=0}^{\infty} e^{-nh\nu/kT}} \tag{1.41}$$

$$= \frac{1}{2}h\nu + \frac{(h\nu e^{-h\nu/kT} + 2h\nu e^{-2h\nu/kT} + ...nh\nu e^{-nh\nu/kT})}{1 + e^{-h\nu/kT} + e^{-2h\nu/kT} + ...e^{-nh\nu/kT}} \tag{1.42}$$

$$= \frac{1}{2}h\nu - \frac{\frac{d}{d(1/kT)}(e^{-h\nu/kT} + e^{-2h\nu/kT} + ...)}{1/(1 - e^{-h\nu/kT})} \tag{1.43}$$

$$= \frac{1}{2}h\nu - \frac{\frac{d}{d(1/kT)}[(1/(1 - e^{-h\nu/kT})]}{1/(1 - e^{-h\nu/kT})} \tag{1.44}$$

$$= \frac{1}{2}h\nu + \frac{h\nu e^{-h\nu/kT}}{(1 - e^{-h\nu/kT})^2}(1 - e^{-h\nu/kT}). \tag{1.45}$$

We have used a mathematical 'trick' here to sum the series in the numerator by recognizing that it is the derivative of a geometric series that we can sum easily. So

$$\overline{E}_\nu = \frac{1}{2}h\nu + \frac{h\nu}{(e^{h\nu/kT} - 1)}. \tag{1.46}$$

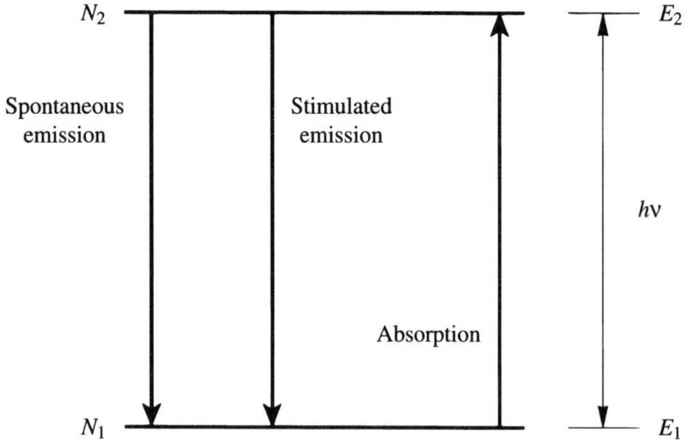

Fig. 1.17. Radiative processes connecting two energy levels in thermal equilibrium at temperature T.

This is the average energy per mode. Consequently, the stored energy in the black-body radiation field treated as a collection of quantized oscillators is

$$\rho(\nu) = p(\nu)\overline{E}_\nu = \frac{8\pi h\nu^3}{c^3}\left(\frac{1}{2} + \frac{1}{e^{h\nu/kT} - 1}\right). \qquad (1.47)$$

The $\frac{1}{2}$ factor comes from zero point energy that cannot be released, so the available stored energy in the field is

$$\rho(\nu) = \frac{8\pi h\nu^3}{c^3}\left(\frac{1}{e^{h\nu/kT} - 1}\right) \qquad (1.48)$$

This formula predicts exactly the observed spectral character of black-body radiation and was the first spectacular success of quantum theory. Note that

$$\rho(\nu) = \frac{8\pi h\nu^3}{c^3}\left(\frac{1}{e^{h\nu/kT} - 1}\right) = \frac{8\pi\nu^2}{c^3} \times h\nu \times \left(\frac{1}{e^{h\nu/kT} - 1}\right), \qquad (1.49)$$

which is (the number of modes per volume per frequency interval) $\times$ photon energy $\times 1/(e^{h\nu/kT} - 1)$. The quantity $1/(e^{h\nu/kT} - 1)$ represents the average number of photons in each mode, this is called the *occupation number* of the modes of the field.

1.7 Relation Between the Einstein A and B Coefficients

We can derive a useful relationship between Einstein's A and B coefficients by considering a collection of atoms in thermal equilibrium inside a cavity at temperature T. The energy density of the radiation within the cavity is given by

$$\rho(\nu) = \frac{8\pi h\nu^3}{c^3}\left(\frac{1}{e^{h\nu/kT} - 1}\right), \qquad (1.50)$$

since the radiation in thermal equilibrium in the cavity will be black-body radiation. Although real atoms in such a cavity possess many energy levels, we can restrict ourselves to considering the dynamic equilibrium between any two of them, as shown in Fig. (1.17). The jumps which occur between two such levels as a result of interaction with radiation essentially occur independently of the energy levels of the system which are not themselves involved in the jump.

In thermal equilibrium the populations N_2 and N_1 of these two levels are constant so

$$\frac{dN_2}{dt} = \frac{dN_1}{dt} = 0, \tag{1.51}$$

and the rates of transfer between the levels are equal. Since the energy density of a black-body radiation field varies very little over the range of frequencies where transitions between levels 2 and 1 take place we can use Eqs. (1.3) and (1.23) and write

$$\frac{dN_2}{dt} = -N_2 B_{21}\rho(\nu) - A_{21}N_2 + N_1 B_{12}\rho(\nu). \tag{1.52}$$

Therefore, substituting from Eq. (1.48)

$$N_2\left[B_{21}\frac{8\pi h\nu^3}{c^3(e^{h\nu/kT}-1)} + A_{21}\right] = N_1\left[B_{12}\frac{8\pi h\nu^3}{c^3(e^{h\nu/kT}-1)}\right]. \tag{1.54}$$

For a collection of particles that obeys Maxwell–Boltzmann statistics, in thermal equilibrium energy levels of high energy are less likely to be occupied then levels of low energy. In exact terms the ratio of the population densities of two levels whose energy difference is $h\nu$ is

$$\frac{N_2}{N_1} = e^{-h\nu/kT}. \tag{1.54}$$

So,

$$\frac{8\pi h\nu^3}{c^3(e^{h\nu/kT}-1)} = \frac{A_{21}}{B_{12}e^{h\nu/kT} - B_{21}}. \tag{1.55}$$

This equality can only be satisfied if

$$B_{12} = B_{21} \tag{1.56}$$

and

$$\frac{A_{21}}{B_{21}} = \frac{8\pi h\nu^3}{c^3}, \tag{1.57}$$

so a single coefficient A_{21} (say) will describe both stimulated emission and absorption. Eqs. (1.56) and (1.57) are called the *Einstein relations*. The stimulated emission rate is W_{21}, where

$$W_{21} = B_{21}\rho(\nu) = \frac{c^3 A_{21}}{8\pi h\nu^3}\rho(\nu), \tag{1.58}$$

which is proportional to energy density. The spontaneous emission rate is A_{21}, which is independent of external radiation.

If we ignored A_{21} in describing the equilibration of the two energy levels previously considered, then in equilibrium we would have

$$N_2 B_{21} = N_1 B_{12}, \tag{1.59}$$

and if $B_{12} = B_{21}$, this would imply $N_1 = N_2$, which we know cannot be so.

Although spontaneous emission would appear to be a different kind of radiative process from stimulated emission, in fact it is not. The modes of the radiation field contain photons. Even a mode of the radiation field containing no photons has a zero point energy of $\frac{1}{2}h\nu$. This zero point energy cannot be extracted from the mode, that is detected, whereas photons in the mode can be. Spontaneous emission corresponds to stimulated emission resulting from this zero point energy of the radiation field.

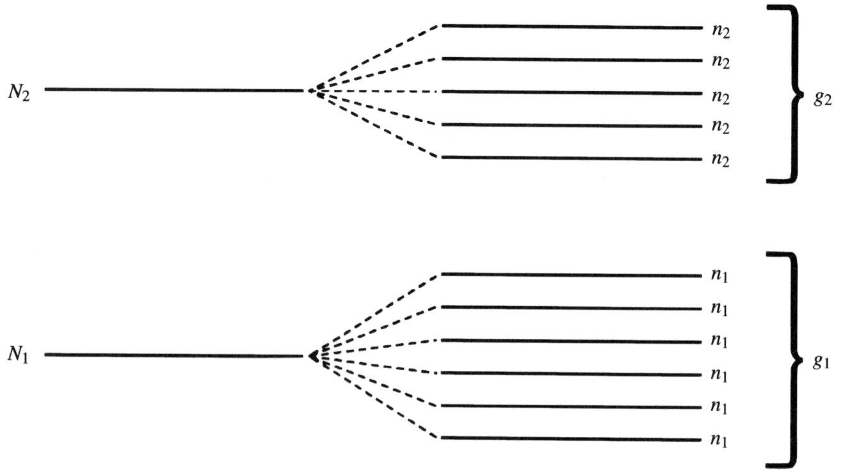

Fig. 1.18. Two energy levels, each of which has a number of sub-levels of the same energy.

1.8 The Effect of Level Degeneracy

In real systems containing atoms, molecules or ions, it frequently happens that different configurations of the system can have exactly the same energy. If a given energy level corresponds to a number of different arrangements specified by an integer g, we call g the *degeneracy* of the level. We call the separate states of the system with the same energy *sub-levels*. The levels 2 and 1 that we have been considering may consist of a number of degenerate sub-levels, where each sub-level has the same energy, as shown in Fig. (1.18), with g_2 sub-levels making up level 2 and g_1 sub-levels making up level 1. For each of the sub-levels of levels 1 and 2 with population n_1, n_2 respectively, the ratio of populations is

$$\frac{n_2}{n_1} = e^{-h\nu/kT}, \tag{1.60}$$

and

$$N_1 = g_1 n_1, \quad N_2 = g_2 n_2. \tag{1.61}$$

Therefore

$$\frac{n_2}{n_1} = \frac{g_1 N_2}{g_1 N_2}, \tag{1.62}$$

and

$$\frac{N_2}{N_1} = \frac{g_2}{g_1} e^{-h\nu/kt}. \tag{1.63}$$

From Eqs. (1.53) and (1.63) it follows that in this case, where degenerate levels are involved, that the Einstein relations become

$$g_1 B_{12} = g_2 B_{21}, \tag{1.64}$$

and as before

$$\frac{A_{21}}{B_{21}} = \frac{8\pi h\nu^3}{c^3}. \tag{1.65}$$

Note that

$$A_{21} = B_{21}\frac{8\pi h\nu^3}{c^3} = B_{21}\frac{8\pi\nu^2}{c^3}h\nu, \tag{1.66}$$

which can be described as

$B_{21}\times$ no. of modes per unit volume per frequency interval
$\times$ photon energy.

If there were only one photon in each mode of the radiation field, then the resulting energy density would be

$$\rho(\nu) = p(\nu)\overline{E}_\nu = \frac{8\pi\nu^2}{c^3}h\nu. \tag{1.67}$$

The resulting number of stimulated transitions would be

$$W_{21} = B_{21}\frac{8\pi\nu^2}{c^3}h\nu = A_{21}, \tag{1.68}$$

thus, the number of spontaneous transitions per second is equal to the number of stimulated transitions per second that would take place if there was just one photon excited in each mode.

1.9 Ratio of Spontaneous and Stimulated Transitions

It is instructive to examine the relative rates at which spontaneous and stimulated processes occur in a system in equilibrium at temperature T. This ratio is

$$R = \frac{A_{21}}{B_{21}\rho(\nu)}. \tag{1.69}$$

We choose the $\rho(\nu)$ appropriate to a black-body radiation field, since such radiation is always present to interact with an excited atom that is contained within an enclosure at temperature T.

$$R = \frac{A}{B\rho(\nu)} = (e^{h\nu/kT} - 1). \tag{1.70}$$

If we use $T = 300$ K and examine the *microwave region*, $\nu = 10^{10}$ (say), then

$$\frac{h\nu}{kT} = \frac{6.626\times 10^{-34}\times 10^{10}}{1.38\times 10^{-23}\times 300} = 1.6\times 10^{-3}$$

so

$$R = e^{0.0016} - 1 \approx 0.0016$$

and stimulated emission dominates over spontaneous. Particularly, in any microwave laboratory experiment

$$\rho(\nu)_{laboratory\ created} > \rho(\nu)_{black-body}$$

and spontaneous emission is negligible. However, spontaneous emission is still observable as a source of noise – the randomly varying component of the optical signal.

In the *visible region*

$$\nu \approx 10^{15} \quad \frac{h\nu}{kT} \approx 160 \text{ and } A >> B\rho(\nu),$$

So, in the visible and near-infrared region spontaneous emission generally dominates unless we can arrange for there to be several photons in a mode. The average number of photons in a mode in the case of black-body radiation is

$$\overline{n}(\nu) = \frac{1}{e^{h\nu/kT} - 1}, \qquad (1.71)$$

which is very small in the visible and infrared.

1.10 Problems

(1.1) In a *dispersive* medium the refractive index varies with wavelength. We can define *a group* refractive index by the relation

$$n_g = n - \lambda \frac{dn}{d\lambda}.$$

(i) Prove that $n_g = n + \nu dn/d\nu$.

(ii) Prove that if a black-body cavity is filled with such a dispersive material then the radiation mode density, p_ν, satisfies

$$p_\nu = \frac{8\pi\nu^2 n^2 n_g}{c_0^3}.$$

(iii) Prove also that in such a situation

$$\frac{A_{21}}{B_{21}} = \frac{8\pi h\nu^3 n^2 n_g}{c_0^3}.$$

(1.2) Calculate the photon flux (photons m^{-2} s^{-1}) in a plane monochromatic wave of intensity 100 W m^{-2} at a wavelength of (a) 100 nm, and (b) 100 μm.

(1.3) What is the total number of modes per unit volume for visible light?

(1.4) Prove Wien's displacement law for black-body radiation, namely

$$\lambda_m T = 2898 \ \mu\text{m K}.$$

λ_m is the wavelength of peak emission of the black-body at absolute temperature T K

(1.5) What fraction of the total emission of a normalized Lorentzian lineshape of FWHM $\Delta\nu$ occurs between $\nu_0 - \Delta\nu$ and $\nu_0 + \Delta\nu$?

(1.6) Estimate the total force produced by the photon pressure of the sun on an aluminum sheet of area 10^6 m^2 situated on the surface of the earth. Use sensible values for the parameters of the problem in order to produce a numerical result.

(1.7) Calculate the total stored energy in a 1 m^3 box that lies between the wavelengths 10.5 μm and 10.7 μm at a temperature of 3000 K.

(1.8) What would the spectral energy density distribution be if 'black-body' radiation could only occupy *two-photon* states? Namely, the only allowed energies for the various modes of a cavity would be

$$E_n = (2n + 1)h\nu.$$

Hint: First calculate mode density, and then average energy per mode.

(1.9) At what temperature would the stimulated and spontaneous emission rates be equal for particles in a cavity and a transition at a wavelength of 1 μm.

(1.10) A point source emits a Lorentzian line of FWHM 1 GHz at λ_0= 500 nm and total radiated power of 1 W. A square bandpass optical filter with transmittance 0.8 between $\nu_0 - \Delta\nu$ and ν_0 covers an optical detector of active surface area 10 mm^2 placed 1 m from the source. What light intensity reaches the detector?

Hint: $\int dx/(1+x^2) = \arctan x$

2

Optical Frequency Amplifiers

2.1 Introduction

We saw in Chapter 1 how the stimulated emission and absorption processes can cause the intensity of a wave propagating through a medium to change. The overall intensity will increase if the number of stimulated emissions can be made larger than the number of absorptions. If this situation occurs then we have built an amplifier that operates through the mechanism of stimulated emission. This *laser amplifier*, in common with electronic amplifiers, only has useful gain over a particular frequency bandwidth. The operating frequency range will be determined by the lineshape of the transition, and, roughly speaking, has the same order of frequency width as the width of the lineshape. To consider in more detail how laser amplifiers operate, we need to examine the various mechanisms by which transitions between different energy states of a particle turn out to cover a range of frequencies. This *line broadening* affects in a fundamental way not only the frequency bandwidth of the amplifier, but also its gain.

To turn a laser amplifier into an oscillator we need to supply an appropriate amount of positive feedback. The level of oscillation will stabilize because the amplifier saturates. We shall see that laser amplifiers fall into two categories that saturate in different ways. The *homogeneously* broadened amplifier consists of a number of amplifying particles that are essentially equivalent while the *inhomogeneously* broadened amplifier contains amplifying particles with a distribution of amplification characteristics. Finally we shall introduce the concept of the complex susceptibility for absorbing transitions by considering the classical electron oscillator model for interaction of an electron with an external electromagnetic field.

2.2 Homogeneous Line Broadening

So far we have been treating the energy levels of atoms, molecules or ions as if they were sharp distinct states of clearly defined energy. In practice this is not so, all energy states are *smeared out* over a finite (although usually very small) range of energies. At the fundamental level this smearing out of the energy is caused by the uncertainty involved in the energy measurement process. This gives rise to an intrinsic and unavoidable amount of line broadening called *natural broadening*.

2.2.1 Natural Broadening

This most fundamental source of line broadening arises, as just mentioned, because of uncertainty in the exact energy of the states involved. This uncertainty in

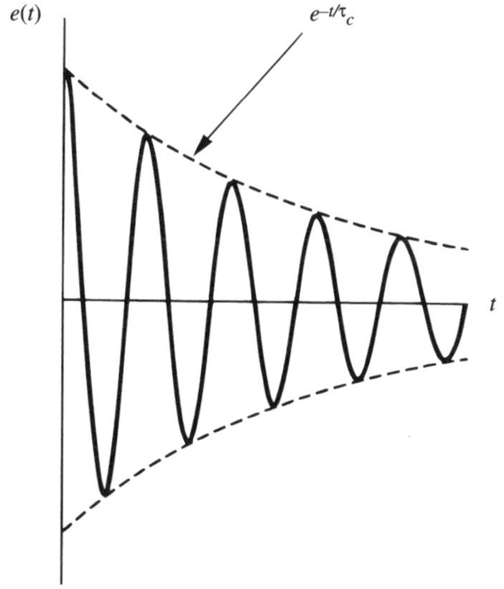

Fig. 2.1. The damped oscillation of the electric field produced by an excited particle as it decays.

measured energy, ΔE, arises from the time, Δt, involved in making such a measurement. Heisenberg's uncertainty principle[2.1]–[2.3] tells us that the product of these *uncertainties* $\Delta E \Delta t \sim \hbar$.† Now, since an excited particle can only be observed for a time that is of the order of its lifetime, the measurement time Δt is roughly the same as the lifetime, so

$$\Delta E \sim \hbar/\tau = A\hbar. \tag{2.1}$$

The uncertainly in emitted frequency $\Delta \nu$ is $\Delta E/h$, so

$$\Delta \nu \sim A/2\pi. \tag{2.2}$$

When the decay of the excited atom is viewed as a photon emission process we can think of the atom, initially placed in the excited state at time $t' = 0$, emitting a photon at time t. The distribution of these times t among many such atoms varies as $e^{-t/\tau}$. Our knowledge of when the photon is likely to be emitted with respect to the time origin restricts our ability to be sure of its frequency. For example, if a photon is observed at time t, and is known to have come from a state with lifetime τ, we know that the *probable* time t' at which the atom became excited was $t - \tau < t' \leq t$. The longer the lifetime of the state the greater the uncertainty about when the atom acquired its original excitation. In the limit as $\tau \rightarrow \infty$, our knowledge of the time of excitation becomes infinitely uncertain and we can ascribe a very well defined frequency to the emitted photon, in this limit the electromagnetic waveform emitted by the atom approaches infinite length and is undamped.

We can put the above approximate determination of $\Delta \nu$ on a more exact basis by considering the exponential intensity decay of a group of excited atoms: The decay of each individual excited atom is modelled as an exponentially decaying (damped) cosinusoidal oscillation, as shown schematically in Fig. (2.1). It must be stressed, however, that this is only a convenient way of *picturing* how an excited particle decays and emits electromagnetic radiation. It would not be possible

† $\hbar = h/2\pi$.

in practice to observe such an electromagnetic field by watching a single excited particle decay. We can only observe a classical field by watching many excited particles simultaneously. However, within the framework of our model we can represent the electric field of a decaying excited particle as

$$e(t) = E_0 e^{-t/\tau_c} \cos \omega_0 t. \tag{2.3}$$

What time constant τ_c applies to this damped oscillation? The instantaneous intensity $i(t)$ emitted by an individual excited atom is

$$i(t) \propto |e(t)|^2 = E_0^2 e^{-2t/\tau_c} \cos^2 \omega_0 t. \tag{2.4}$$

If we observe many such atoms the total observed intensity is

$$\begin{aligned} I(t) &= \sum_{particles} i(t) = \sum_i E_0^2 e^{-2t/\tau_c} \cos^2(\omega_0 t + \epsilon_i) \\ &= \sum_i \frac{E_0^2}{2} e^{-2t/\tau_c} [1 + \cos 2(\omega_0 t + \epsilon_i)], \end{aligned} \tag{2.5}$$

where ϵ_i is the phase of the wave emitted by atom i. In the summation the cosine term gets smeared out because individual atoms are emitting with random phases. So, $I(t) \propto e^{-2t/\tau_c}$ However, we know that $I(t) \propto e^{-t/\tau}$, where τ is the lifetime of the emitting state, so the time constant τ_c is in fact $= 2\tau$. Thus:

$$e(t) = E_0 e^{-t/2\tau} \cos \omega_0 t. \tag{2.6}$$

To find the frequency distribution of this signal we take its Fourier transform

$$E(\omega) = \frac{1}{2\pi} \int_{-\infty}^{\infty} e(t) e^{-i\omega t} dt, \tag{2.7}$$

where

$$\begin{aligned} e(t) &= \frac{E_0}{2}(e^{i(\omega_0 + i/2\tau)t} + e^{-i(\omega_0 - i/2\tau)t}) \text{ for } t > 0 \\ &= 0 \text{ for } t < 0. \end{aligned} \tag{2.8}$$

The start of the period of observation at $t = 0$, taken for example at an instant when all the particles are pushed into the excited state, allows the lower limit of integration to be changed to 0, so

$$\begin{aligned} E(\omega) &= \frac{1}{2\pi} \int_0^{\infty} e(t) e^{-i\omega t} dt \\ &= \frac{E_0}{4\pi} \left[\frac{i}{(\omega_0 - \omega + i/2\tau)} - \frac{i}{(\omega_0 + \omega - i/2\tau)} \right]. \end{aligned} \tag{2.9}$$

The intensity of emitted radiation is

$$I(\omega) \propto | E(\omega) |^2 = E(\omega) E^*(\omega) \propto \frac{1}{(\omega - \omega_0)^2 + (1/2\tau)^2}. \tag{2.10}$$

Or, in terms of ordinary frequency

$$I(\nu) \propto \frac{1}{(\nu - \nu_0)^2 + (1/4\pi\tau)^2}. \tag{2.11}$$

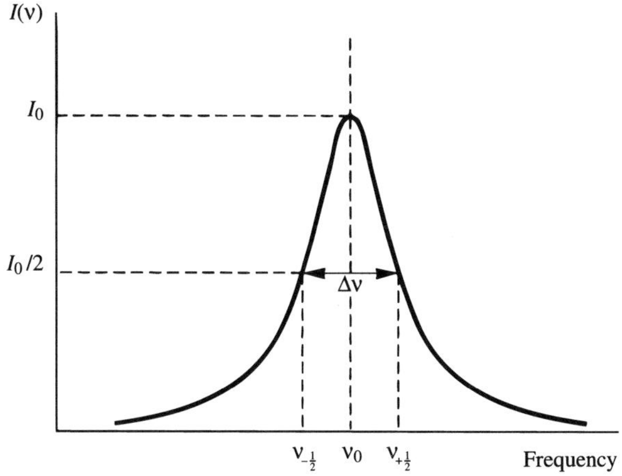

Fig. 2.2. Lorentzian lineshape function for natural broadening.

The full width at half maximum height (FWHM) of this function is found from the half intensity points of $I(\nu)$ that occur at frequencies $\nu_{\pm\frac{1}{2}}$ as shown in Fig. (2.2). This occurs where

$$\left(\frac{1}{4\pi\tau}\right)^2 = (\nu_{\pm\frac{1}{2}} - \nu_0)^2. \tag{2.12}$$

The FWHM is $\Delta\nu = \nu_{+\frac{1}{2}} - \nu_{-\frac{1}{2}}$, which gives†

$$\Delta\nu = \frac{1}{2\pi\tau} = \frac{A}{2\pi}. \tag{2.13}$$

So from Eq. (2.11)

$$I(\nu) \propto \frac{1}{(\nu - \nu_0)^2 + (\Delta\nu/2)^2}. \tag{2.14}$$

The normalized form of this function is the lineshape function for natural broadening

$$g(\nu)_N = \frac{(2/\pi\Delta\nu)}{1 + [2(\nu - \nu_0)/\Delta\nu]^2}. \tag{2.15}$$

This type of function is called a Lorentzian. Since natural broadening is the same for each particle it is said to be a *homogeneous* broadening mechanism. Other such mechanisms of homogeneous broadening exist, for example:

(i) In a crystal the constituent particles of the lattice are in constant vibrational motion. This collective vibration can be treated as equivalent to sound waves bouncing around inside the crystal. These sound waves, just like electromagnetic waves, can only carry energy in discrete amounts. The packets of acoustic energy are called *phonons*, and are analogous in many ways to photons, the principal differences between them being that the phonons only travel at the speed of sound and can only exist in a material medium. Collisions of phonons with the particles of the lattice perturb the phase of any excited, emitting particles present. This type of

† A more exact treatment gives $\Delta\nu = (A_1 + A_2)/2\pi$ where A_2 and A_1 are the Einstein coefficients of the upper and lower levels of the transition.

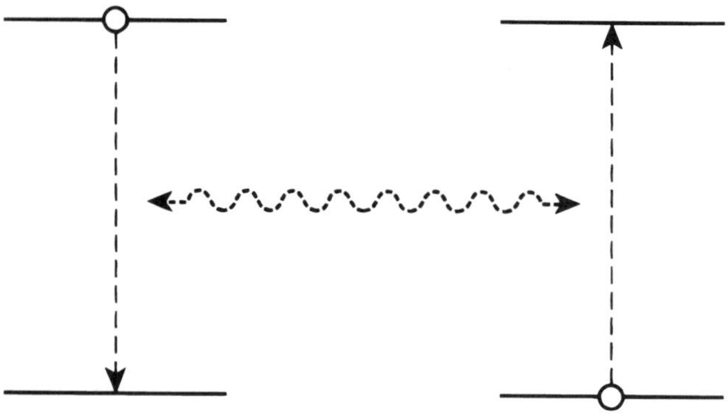

Fig. 2.3. Schematic illustration of how resonance and Van der Waals broadening arise.

collision, which does not abruptly terminate the lifetime of the particle in its emitting state, is called a *soft* collision.

(ii) By *pressure* broadening, particularly in the gasous and liquid phase: interaction of an emitting particle with its neighbors causes perturbation of its emitting frequency and subsequent broadening of the transition. This interaction may arise in a number of ways:

(a) Collisions with neutral particles, which may be *soft* or *hard*. A hard collision causes abrupt decay of the emitting species.

(b) Collisions with charged particles. These collisions need not be very direct, but may involve a very small interaction that occurs when the charged particle passes relatively near, but perhaps as far as several tens of atomic diameters away from, the excited particle. In any case the relative motion of the charged and excited particles leads to a time-varying electric field at the excited particle that perturbs its energy states. This general effect in which an external electric field perturbs the energy levels of an atom (molecule or ion) is called the *Stark* effect; hence line broadening caused by charged particles (ions or electrons) is called Stark broadening.

(c) By Van der Waals and resonance interactions (usually small effects). Resonance interactions occur when an excited particle can easily exchange energy with like neighbors, the effect is most important for transitions involving the ground state since in this case there are generally many particles near an excited particle for which the possibility of energy exchange exists. Such a process is shown schematically in Fig. (2.3). Broadening occurs because the *possibility* of energy exchange exists, not because an actual emission/reabsorption process occurs.

2.3 Inhomogeneous Broadening

When the environment or properties of particles in an emitting sample are non-identical, *inhomogeneous* broadening can occur. In this type of broadening the shifts and perturbations of emission frequencies differ from particle to particle.

For example, in a real crystal the presence of imperfections and impurities in the crystal structure alters the physical environment of atoms from one lattice site to another. The random distribution of lattice point environments leads to a distribution of particles whose center frequencies are shifted in a random way throughout the crystal.

2.3.1 Doppler Broadening

In a gas the random distribution of particle velocities leads to a distribution in the emission center frequencies of different emitting particles seen by a stationary observer. For an atom whose component of velocity towards the observer is v_x the observed frequency of the transition, whose stationary center frequency is ν_0, is

$$\nu = \nu_0 + \frac{v_x}{c}\nu_0, \tag{2.16}$$

where c is the velocity of light in the gas.

The Maxwell–Boltzman distribution of atomic velocities for particles of mass M at absolute temperature T is[2.4],[2.5]

$$f(v_x, v_y, v_z) = \left(\frac{M}{2\pi kT}\right)^{3/2} \exp\left[-\frac{M}{2kT}\left(v_x^2 + v_y^2 + v_z^2\right)\right]. \tag{2.17}$$

The number of atoms per unit volume that have velocities simultaneously in the range $v_x \to v_x + dv_x, v_y \to v_y + dv_y, v_z \to v_z + dv_z$ is $Nf(v_z, v_y, v_z)dv_z dv_y dv_z$ where N is the total number of atoms per unit volume. The $(M/2\pi kT)^{3/2}$ factor is a normalization constant that ensures that the integral of $f(v_x, v_y, v_z)$ over all velocities is equal to unity, i.e.,

$$\iiint_{-\infty}^{\infty} f(v_x, v_y, v_z) dv_x dv_d dv_z = 1. \tag{2.18}$$

The normalized one-dimensional velocity distribution is

$$f(v_x) = \sqrt{\frac{M}{2\pi kT}} e^{\frac{Mv_x^2}{2kT}}. \tag{2.19}$$

This Gaussian-shaped function is shown in Fig. (2.4). It represents the probability that the velocity of a particle towards an observer is in the range $v_x \to v_x + dv_x$. This is the same as the probability that the frequency be in the range

$$\nu_0 + \frac{v_x}{c}\nu_0 \to \nu_0 + \left(\frac{v_x + dv_x}{c}\right) = \nu_0 + \frac{v_x}{c}\nu_0 + \frac{dv_x}{c}\nu_0. \tag{2.20}$$

The probability that the frequency lies in the range $\nu \to \nu + d\nu$ is the same as the probability of finding the velocity in the range $(\nu - \nu_0)c/\nu_0 \to (\nu - \nu_0)c/\nu_0 + (c/\nu_0)d\nu$, so the distribution of the emitted frequencies is

$$g(\nu) = \frac{c}{\nu_0}\sqrt{\frac{M}{2\pi kT}} \exp\left[\left(-\frac{M}{2kT}\right)\left(\frac{c^2}{\nu_0^2}\right)(\nu - \nu_0)^2\right]. \tag{2.21}$$

This is already normalized (since $f(v_x)$ was normalized). It is called the normalized Doppler-broadened lineshape function. Its FWHM is

$$\Delta\nu_D = 2\nu_0\sqrt{\frac{2kT\ln 2}{Mc^2}}. \tag{2.22}$$

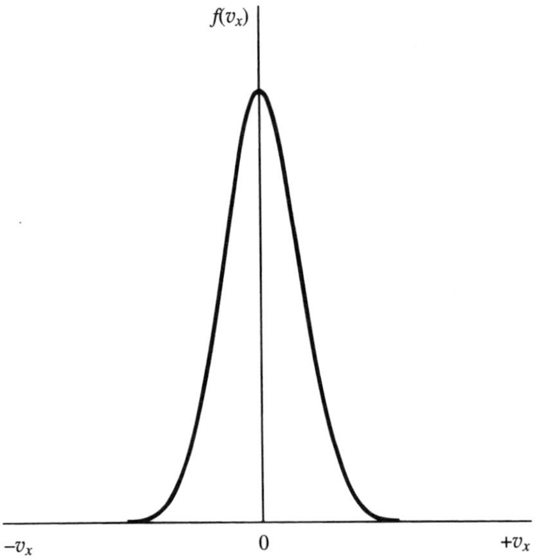

Fig. 2.4.
The one-dimensional Gaussian distribution of velocities for the particles in a gas.

It can be seen that this increases with $\sqrt{T}$ and falls with atomic mass as $1/\sqrt{M}$. In terms of the Doppler-broadened linewidth $\Delta\nu_D$ the normalized Doppler lineshape function is

$$g(\nu) = \frac{2}{\Delta\nu_D}\sqrt{\frac{\ln 2}{\pi}}\, e^{-[2(\nu-\nu_0)/\Delta\nu_D]^2 \ln 2}. \tag{2.23}$$

This is a Gaussian function. It is shown in Fig. (2.5) compared with a Lorentzian function of the same FWHM. The Gaussian function is much more sharply peaked while the Lorentzian has considerable intensity far away from its center frequency, in its *wings*.

2.3.1.1 Example.

The 632.8 nm transition of neon is the most important transition to show laser oscillation in the helium–neon laser. The atomic mass of neon is 20. Therefore, using

$$M = 20 \times 1.67 \times 10^{-27} \text{ kg},$$
$$\nu_0 = 3 \times 10^8/632.8 \times 10^{-9} \text{ Hz},$$
$$T = 400 \text{ K},$$

which we shall see is an appropriate temperature to use for the gas in a He–Ne laser, the Doppler width is $\Delta\nu_D \sim 1.5$ GHz.† Doppler broadening usually dominates over all other sources of broadening in gaseous systems, except occasionally in very heavy gases at high pressures and in highly ionized plasmas of light gases, in the latter case Stark broadening frequently dominates.

Doppler broadening gives rise to a Gaussian lineshape and is a form of inhomogeneous broadening. The inhomogeneous lineshape covers a range of frequencies because many *different* particles are being observed. The particles are *different*

† The gigahertz (GHz) is a unit of frequency $\equiv 10^9$ Hz, another designation of high frequency is the terahertz (THz) $\equiv 10^{12}$ Hz.

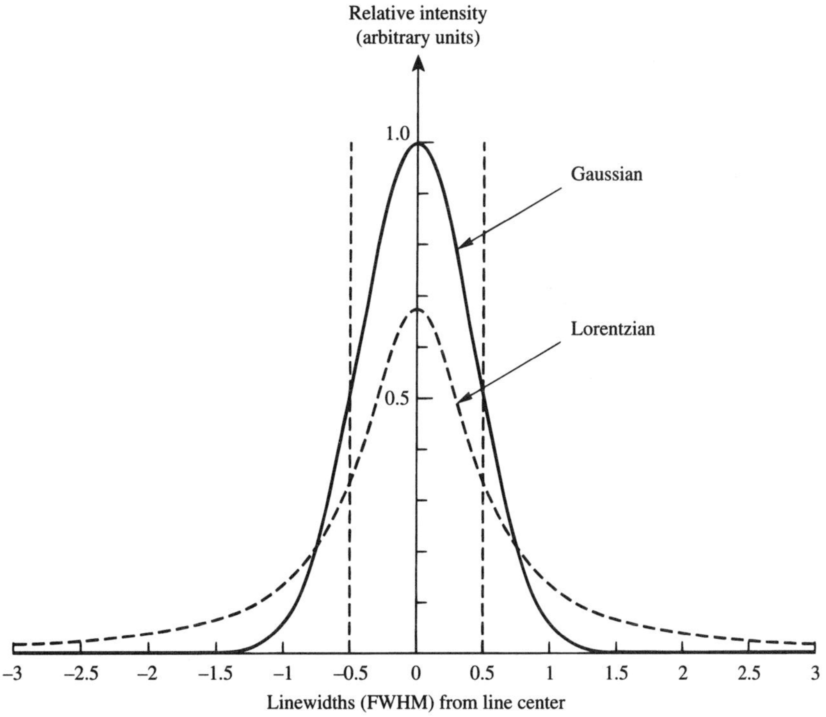

Fig. 2.5. Comparison of normalized Lorentzian and Gaussian lineshapes.

in the sense that they have different velocities, and consequently different center frequencies. Homogeneous broadening also always occurs at the same time as inhomogeneous broadening, to a greater or lesser degree. To illustrate this, imagine a hypothical experiment in which only those particles in a gas within a certain narrow velocity range are observed. The center frequencies of these particles are confined to a narrow frequency band and in this sense there is no inhomogeneous broadening – all the *observed* particles are the same. However, a broadened lineshape would still be observed – the homogeneous lineshape resulting from natural and pressure broadening. This is illustrated in Fig. (2.6). When all particles are observed, irrespective of their velocity, an *overall* lineshape called a *Voigt* profile will be observed. This overall lineshape will be considered in more detail later, but results from the superposition of Lorentzian lineshapes spread across the Gaussian distribution of Doppler-shifted center frequencies, as shown in Fig. (2.7). If the constituent Lorentzians have FWHM $\Delta\nu_L \ll \Delta\nu_D$ then the overall lineshape remains Gaussian and the system is properly said to be *inhomogeneously* broadened. On the other hand if all observed particles are identical, or almost identical, so that $\Delta\nu_D \ll \Delta\nu_L$ then the system as a whole is *homogeneously* broadened.

In solid materials inhomogeneous broadening, when it is important, results from lattice imperfections and impurities that cause the local environments of individual excited particles to differ in a random way. We shall assume that the broadening that thereby results also gives rise to a Gaussian lineshape of appropriate FWHM, which we shall also designate as $\Delta\nu_D$.†

† The designation $\Delta\nu_D$ originates from Doppler broadening, but is used generally to designate inhomogeneous linewidth.

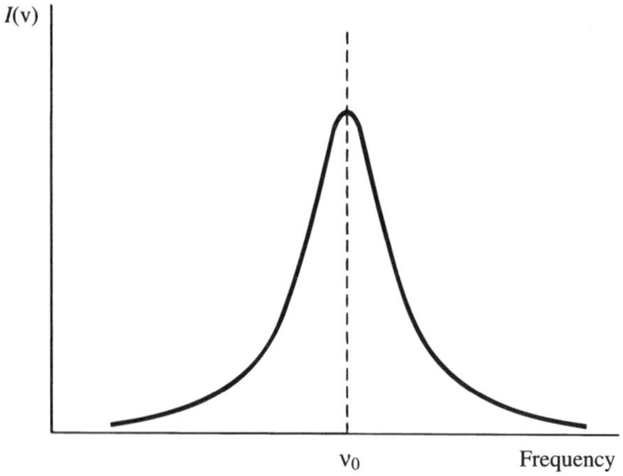

Fig. 2.6. Homogeneous broadening of a group of particles in a gas that have the same velocity.

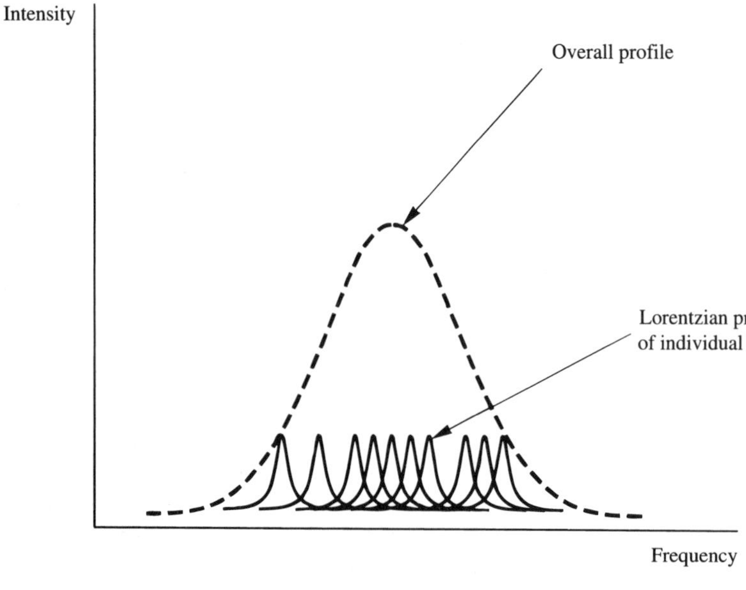

Fig. 2.7. A Doppler-broadened distribution of Lorentzian lineshapes.

2.4 Optical Frequency Amplification with a Homogeneously Broadened Transition

In an optical frequency amplifier we are generally concerned with the interaction of a monochromatic radiation field with a transition between two energy states whose center frequency is at, or near, the frequency of the monochromatic field. The magnitude of this interaction with each particle is controlled by the homogeneous lineshape function of the transition.

In the general case, the monochromatic radiation field and the center frequency of the transition are not the same. This situation is shown schematically in Fig. (2.8). The stimulating radiation field is taken to be at frequency ν whilst the center frequency of the transition is at ν'. The closer ν is to ν', the greater the number of transitions that can be stimulated. The stimulated transitions occur at frequency ν, since this is the frequency of the stimulating radiation. The number

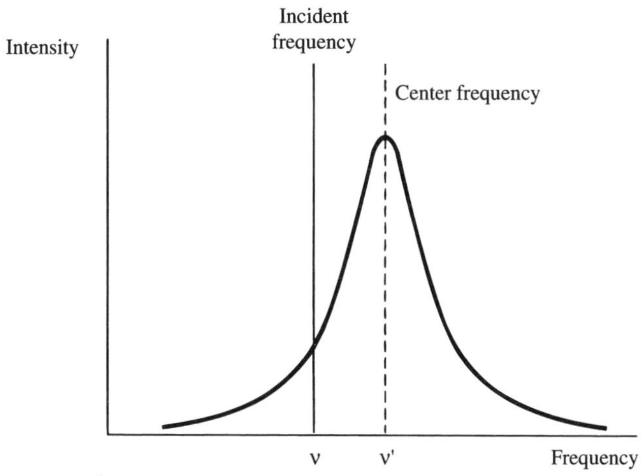

Fig. 2.8. A monochromatic field interacting with a homogeneously broadened lineshape.

of stimulated transitions is proportional to the homogeneous lineshape function $g(\nu', \nu)$ and can be written as

$$N_S = N_2 B_{21} \rho(\nu) g(\nu', \nu). \tag{2.24}$$

We have written $g(\nu', \nu)$ to indicate that this lineshape function has its center frequency at ν' but is being evaluated at frequency ν. For example, for a Lorentzian broadened line

$$g(\nu', \nu) = \frac{(2/\pi\Delta\nu)}{1 + [2(\nu - \nu')/\Delta\nu]^2}. \tag{2.25}$$

$g(\nu', \nu)$ has its maximum value when $\nu' = \nu$. It is important to stress that this lineshape function $g(\nu', \nu)$ that is used here is the *homogeneous* lineshape function of the individual particles in the system, even though the contribution of homogeneous broadening to the overall broadening in the system may be small, for example when the overall broadening is predominantly inhomogeneous. The important point about the interaction of a particle with radiation is that an excited atom, molecule or ion can only interact with a monochromatic radiation field that overlaps its homogeneous (usually Lorentzian-shaped) lineshape profile. For example, consider the case of two excited atoms with different center frequencies, these may be the different center frequencies of atoms with different velocities relative to a fixed observer. The homogeneous lineshapes of these two atoms are shown in Fig. (2.9) together with a monochromatic radiation field at frequency ν. Particle A with center frequency ν_A and *homogeneous width* $\Delta\nu$ can interact strongly with the field while the interaction of particle B is negligible.

We can analyze the interaction between a plane monochromatic case and a collection of homogeneously broadened particles with reference to Fig. (2.10). As the wave passes through the medium it grows in intensity if the number of stimulated emissions exceeds the number of absorptions. The change in intensity of the wave in travelling a small distance dz through the medium is

$$\begin{aligned} dI_\nu &= (\text{number of stimulated emissions} - \text{number of absorptions})/\text{vol} \\ &\quad \times h\nu \times dz \\ &= \left(N_2 B_{21} g(\nu', \nu) \frac{I_\nu}{c} - N_1 B_{12} g(\nu', \nu) \frac{I_\nu}{c} \right) h\nu dz. \end{aligned} \tag{2.26}$$

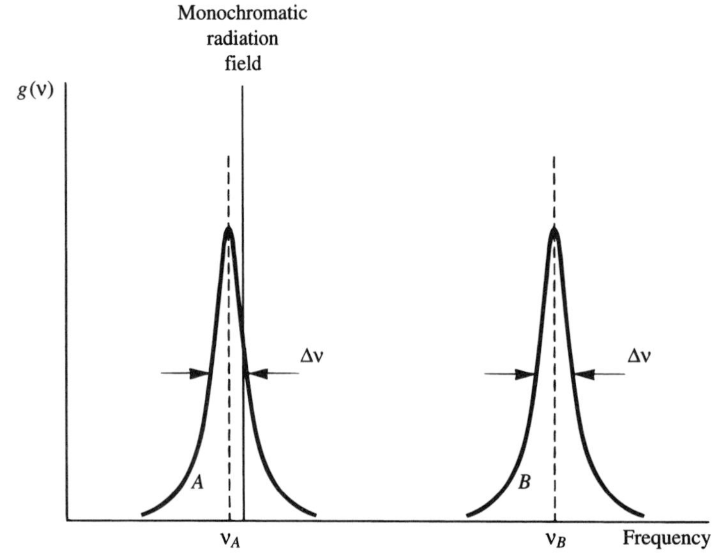

Fig. 2.9. A monochromatic radiation field interacting with two homogeneously broadened lineshapes whose center frequencies are different.

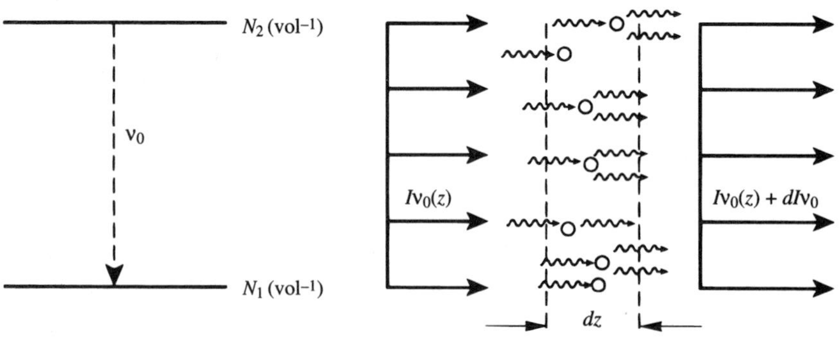

Fig. 2.10. A plane wave interacting with a collection of homogeneously broadened particles.

Use of the Einstein relations, Eqs. (1.64) and (1.65), gives

$$dI_\nu = \frac{I_\nu}{c}\left(N_2 - \frac{g_2}{g_1}N_1\right)\frac{c^3 A_{21}}{8\pi h\nu^3}h\nu g(\nu',\nu)dz. \tag{2.27}$$

Therefore,

$$\frac{dI_\nu}{dz} = \left(N_2 - \frac{g_2}{g_1}N_1\right)\frac{c^2 A_{21}}{8\pi\nu^2}g(\nu',\nu)I_\nu, \tag{2.28}$$

which has the solution

$$I_\nu = I_\nu(0)e^{\gamma(\nu)z}, \tag{2.29}$$

where $I_\nu(0)$ is the initial intensity at $z = 0$, and

$$\gamma(\nu) = \left(N_2 - \frac{g_2}{g_1}N_1\right)\frac{c^2 A_{21}}{8\pi\nu^2}g(\nu',\nu). \tag{2.30}$$

$\gamma(\nu)$ is called the gain coefficient of the medium and has the same frequency dependence as $g(\nu', \nu)$. If

$$N_2 > \frac{g_2}{g_1} N_1$$

then $\gamma(\nu) > 0$ and we have an optical frequency amplifier. If

$$N_2 < \frac{g_2}{g_1} N_1$$

then $\gamma(\nu) < 0$ and net absorption of the incident radiation occurs.

For a system in thermal equilibrium

$$\frac{N_2}{N_1} = \frac{g_2}{g_1} e^{-h\nu/kT} \tag{2.31}$$

and for $T > 0$, $e^{-h\nu/kT} < 1$, which implies, that in thermal equilibrium at positive temperatures, we cannot have positive gain. However, if we allow the formal existence of a negative temperature, at least for our system of two levels, then we can have

$$N_2 > \frac{g_2}{g_1} N_1.$$

Such a situation, which is essential for the construction of an optical frequency amplifier, is called a state of *population inversion* or *negative temperature*. This is not a true state of thermal equilibrium and can only be maintained by feeding energy into the system.

In the above discussion we have neglected the occurrence of spontaneous emission: this is reasonable for a truly plane wave as the total number of spontaneous emissions into the zero solid angle subtended by the wave is zero. If the wave being amplified were diverging into a small solid angle $\delta\omega$, then $N_2 A_{21} \delta\omega / 4\pi$ spontaneous emissions per second per unit volume would contribute to the increase in intensity of the wave. However, such emissions, being independent of the incident wave, are not in a constant phase relationship with this wave as are the stimulated emissions.

2.4.1 The Stimulated Emission Rate in a Homogeneously Broadened System

The stimulated emission rate $W_{21}(\nu)$ is the number of stimulated emissions per particle per second per unit volume caused by a monochromatic imput wave at frequency ν:

$$W_{21}(\nu) = B_{21} g(\nu', \nu) \rho_\nu. \tag{2.32}$$

ρ_ν(J m^{-3}) is the energy density of the stimulating radiation. Eq. (2.32) can be rewritten in terms of more practical parameters as

$$W_{21}(\nu) = \frac{A_{21} c^2 I_\nu}{8\pi h \nu^3} g(\nu', \nu). \tag{2.33}$$

$W_{21}(\nu)$ has units m^{-3} s^{-1} per particle. Note that the frequency variation of $W_{21}(\nu)$ follows the lineshape function. The total number of stimulated emissions is

$$N_s = N_2 W_{21}(\nu). \tag{2.34}$$

2.5 Optical Frequency Amplification with Inhomogeneous Broadening Included

Although in this discussion so far we have been restricting our attention to homogeneous systems we can show that Eqs. (2.26)–(2.30) hold generally, even in a system with inhomogeneous broadening, if we take $g(\nu', \nu)$ as the total lineshape function.

In an inhomogeneously broadened system we can divide the atoms up into classes, each class consisting of atoms with a certain range of center emission frequencies and the same homogeneous lineshape. For example, the class with center frequency ν'' in the frequency range $d\nu''$ has $Ng_D(\nu', \nu'')d\nu''$ atoms in it, where $g_D(\nu', \nu'')$ is the normalized inhomogeneous distribution of center frequencies – the inhomogeneous lineshape function centered at ν'. This class of atoms contributes to the change in intensity of a monochromatic wave at frequency ν as

$$\begin{aligned}&\Delta(dI_\nu)(\text{from the group of particles in the band } d\nu'') = \\ &\qquad (N_2 B_{21} g_D(\nu', \nu'')d\nu'' g_L(\nu'', \nu)\frac{I_\nu}{c} \\ &\qquad - N_1 B_{12} g_D(\nu', \nu'')d\nu'' g_L(\nu'', \nu)\frac{I_\nu}{c})h\nu dz,\end{aligned} \tag{2.35}$$

where $g_L(\nu'', \nu)$ is the homogeneous lineshape function of an atom at center frequency ν''. Eq. (2.35) is equivalent to Eq. (2.26). The increase in intensity from all the classes of atoms is found by integrating over these classes, that is, over the range of center frequencies ν'', so Eq. (2.35) becomes

$$dI_\nu = \frac{I_\nu}{c}(N_2 B_{21} - N_1 B_{12})\left[\int_{-\infty}^{\infty} g_D(\nu', \nu'')g_L(\nu'', \nu)d\nu''\right]\nu dz, \tag{2.36}$$

which in a similar fashion to Eqs. (2.26)–(2.30) gives

$$\gamma(\nu) = \left(N_2 - \frac{g_2}{g_1}N_1\right)\frac{c^2 A_{21}}{8\pi\nu^2}g(\nu', \nu), \tag{2.37}$$

where $g(\nu', \nu)$ is now the overall lineshape function defined by the equation

$$g(\nu', \nu) = \int_{-\infty}^{\infty} g_D(\nu', \nu'')g_L(\nu'', \nu)d\nu''. \tag{2.38}$$

In other words the overall lineshape function is the convolution[2.6] of the homogeneous and inhomogeneous lineshape functions. The convolution integral in Eq. (2.38) can be put in more familiar form if we measure frequency relative to the center frequency of the overall lineshape, that is, we put $\nu' = 0$, and Eq. (2.38) becomes

$$\begin{aligned}g(0, \nu) &= \int_{-\infty}^{\infty} g_D(0, \nu'')g_L(\nu'', \nu)d\nu'' \\ &= \int_{-\infty}^{\infty} g_D(0, \nu'')g_L(0, \nu - \nu'')d\nu'',\end{aligned} \tag{2.39}$$

which can be written in simple form as

$$g(\nu) = \int_{-\infty}^{\infty} g_D(\nu'')g_L(\nu - \nu'')d\nu''. \tag{2.40}$$

This is recognizable as the standard convolution integral of two functions $g_D(\nu)$ and $g_L(\nu)$.

If $g_D(\nu', \nu'')$ is indeed a normalized Gaussian lineshape as in Eq. (2.23) and $g_L(\nu'', \nu)$ is a Lorentzian, then Eq. (2.38) can be written in the form

$$g(\nu', \nu) = \frac{2}{\Delta\nu_D}\sqrt{\frac{\ln 2}{\pi}}\frac{y}{\pi}\int_{-\infty}^{\infty}\frac{e^{t^2}}{y^2 + (x-t)^2}dt, \tag{2.41}$$

where $y = \Delta\nu_D\sqrt{\ln 2}/\Delta\nu_D$ and $x = 2(\nu - \nu')\sqrt{\ln 2}/\Delta\nu_D$.

This is one way of writing a normalized Voigt profile[2.7],[2.8]. The integral in Eq. (2.40) cannot be evaluated analytically but must be evaluated numerically. For this purpose the Voigt profile is often written in terms of the error function for complex argument $W(z)$, which is available in tabulated form[2.9]

$$g(\nu', \nu) = \frac{2}{\Delta\nu_D}\sqrt{\frac{\ln 2}{\pi}}\mathscr{R}[W(z)], \tag{2.42}$$

where $z = x + iy$ and $\mathscr{R}[w(z)]$ denotes the real part of the function.

2.6 Optical Frequency Oscillation – Saturation

If we can force a medium into a state of population inversion for a pair of its energy levels, then the transition between these levels forms an optical frequency amplifier. To turn this amplifier into an oscillator we need to apply appropriate feedback by inserting the amplifying medium between a pair of suitable mirrors. If the overall gain of the medium exceeds the losses of the mirror cavity and ancilliary optics then oscillation will result. The level at which this oscillation stabilizes is set by the way in which the amplifier saturates. How does this happen?

2.6.1 Homogeneous Systems

Consider an amplifying transition at center frequency ν_0 between two energy levels of an atom. Suppose we maintain this pair of levels in population inversion by feeding in energy. In equilibrium in the absence of an external radiation field, the rates R_2 and R_1 at which atoms are fed into these levels must be balanced by spontaneous emission and non-radiative loss processes (such as collisions). The population densities and effective lifetimes of the two levels are N_2, τ_2 and N_1, τ_1 respectively as shown in Fig. (2.11). These effective lifetimes include the effect of non-radiative deactivation. If X_{2j} is the rate per particle per unit volume by which collisions depopulate level 2 and cause the particle to end up in a lower state j we can write

$$\frac{1}{\tau_2} = \sum_j (A_{2j} + X_{2j}). \tag{2.43}$$

In equilibrium

$$\frac{dN_2^o}{dt} = R_2 - \frac{N_2^o}{\tau_2} = 0, \tag{2.44}$$

where the term N_2^o/τ_2 is the total loss rate per unit volume from spontaneous emission and other deactivation processes, so

$$N_2^o = R_2\tau_2, \tag{2.45}$$

where the superscript o indicates that the population is being calculated in the absence of a radiation field. Similarly, for the lower level of the transition, in

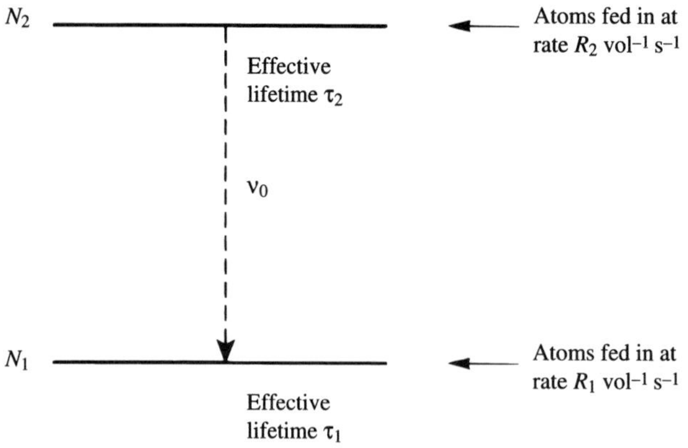

Fig. 2.11. Simple energy level system used in a discussion of amplifier saturation.

equilibrium

$$\frac{dN_1^o}{dt} = R_1 + N_2^o A_{21} - \frac{N_1^o}{\tau_1} = 0. \tag{2.46}$$

The term N_1^o/τ_1 is the total loss rate per unit volume from the level by spontaneous emission and other deactivation processes, while the term $N_2^o A_{21}$ is the rate at which atoms are feeding into level 1 by spontaneous emission from level 2. So from Eqs. (2.45) and (2.46)

$$N_1^o = (R_1 + N_2^o A_{21})\tau_1 = (R_1 + R_2\tau_2 A_{21})\tau_1. \tag{2.47}$$

The population inversion is

$$\left(N_2^o - \frac{g_2}{g_1}N_1^o\right) = \Delta N^o = R_2\tau_2 - \frac{g_2}{g_1}\tau_1(R_1 + R_2\tau_2 A_{21}), \tag{2.48}$$

or when the level degeneracies are equal

$$\Delta N^o = R_2\tau_2 - \tau_1(R_1 + R_2\tau_2 A_{21}). \tag{2.49}$$

If we now feed in a monochromatic (or other) signal then stimulated emission and absorption processes will occur. We take the energy density of this signal at some point within the medium to be $\rho(\nu) = I(\nu)/c$. The rate at which this signal causes stimulated emissions is

$$W_{21}(\nu) = \int_{-\infty}^{\infty} B_{21} g(\nu_0, \nu)\rho(\nu)ds \text{ per particle per second,} \tag{2.50}$$

where $g(\nu_0, \nu)$ is the *homogeneous* lineshape function. If the input radiation were *white*, which in this context means that $\rho(\nu)$ is a constant over the range of frequencies spanned by the lineshape function, then

$$W_{21}(\nu) = B_{21}\rho(\nu)\int_{-\infty}^{\infty} g(\nu_0, \nu)d\nu = B_{21}\rho(\nu). \tag{2.51}$$

The total rate at which a monochromatic plane wave causes stimulated transitions is

$$W_{21}(\nu) = \int_{-\infty}^{\infty} B_{21} g(\nu_0, \nu)\rho(\nu)d\nu = \frac{I_\nu}{c} B_{21} g(\nu_0, \nu)\int_{-\infty}^{\infty} \delta(\nu - \nu'')d\nu''$$
$$= B_{21} g(\nu_0, \nu)\frac{I_\nu}{c}. \tag{2.52}$$

Returning once more to a consideration of the pair of energy levels shown in Fig. (2.11), in equilibrium, in the presence of a radiation field,

$$\frac{dN_2}{dt} = R_2 - \frac{N_2}{\tau_2} - N_2 B_{21} g(\nu_0, \nu)\rho(\nu) + N_1 B_{12} g(\nu_0, \nu)\rho(\nu) = 0 \tag{2.53}$$

and

$$\frac{dN_1}{dt} = R_1 + N_2 A_{21} - \frac{N_1}{\tau_1} + N_2 B_{21} g(\nu_0, \nu)\rho(\nu) - N_1 B_{12} g(\nu_0, \nu)\rho(\nu) = 0. \tag{2.54}$$

If we write $B_{21} g(\nu_0, \nu)\rho(\nu) = W_{12}(\nu)$, the stimulated emission rate at frequency ν per particle, and neglect degeneracy factors so that we can assume that $W_{12}(\nu) = W_{21}(\nu) = W$ (equivalent to $B_{12} = B_{21}$), then we have

$$R_2 - \frac{N_2}{\tau_2} - N_2 W + N_1 W = 0, \tag{2.55}$$

and

$$R_1 + N_2 A_{21} - \frac{N_1}{\tau_1} + N_2 W - N_1 W = 0. \tag{2.56}$$

From Eq. (2.55)

$$N_2 = \frac{N_1 W + R_2}{1/\tau_2 + W} \tag{2.57}$$

and from Eq. (2.56)

$$N_2 = \frac{N_1/\tau_1 + N_1 W - R_1}{A_{21} + W}, \tag{2.58}$$

so

$$N_1 = \frac{R_1/\tau_2 + R_2 A_{21} + W(R_1 + R_2)}{1/\tau_1\tau_2 + W(1/\tau_1 + 1/\tau_2 - A_{21})}. \tag{2.59}$$

The population inversion in the system is now

$$N_2 - N_1 = \frac{N_1 W + R_2}{1/\tau_2 + W} - N_1 = \frac{R_2 - N_1/\tau_2}{1/\tau_2 + W}. \tag{2.60}$$

From Eq. (2.59) this gives

$$\begin{aligned} N_2 - N_1 &= \frac{1/\tau_2(R_2/\tau_1 - R_2 A_{21} - R_1/\tau_2) + W(R_2/\tau_1 - R_2 A_{21} - R_1/\tau_2)}{(1/\tau_2 + W)(1/\tau_1\tau_2 + W/\tau_2 + W/\tau_1 - W A_{21})} \\ &= \frac{R_2/\tau_1 - R_1/\tau_2 - R_2 A_{21}}{1/\tau_1\tau_2 + W(1/\tau_2 + 1/\tau_1 - A_{21})}. \end{aligned} \tag{2.61}$$

Multiplying the numerator and denominator of Eq. (2.61) by $\tau_1\tau_2$ we get

$$N_2 - N_1 = \frac{R_2\tau_2 - R_1\tau_1 - R_2\tau_1\tau_2 A_{21}}{1 + W\tau_2(1 + \tau_1/\tau_2 - A_{21}\tau_1)}. \tag{2.62}$$

The numerator of this expression is just the population inversion in the absence of any light signal, ΔN^o. Thus,

$$N_2 - N_1 = \frac{\Delta N^o}{1 + W\tau_2(1 + \tau_1/\tau_2 - A_{21}\tau_1)} \tag{2.63}$$

or with the substitution

$$\phi = A_{21}\tau_2\left[1 + (1 - A_{21}\tau_2)\frac{\tau_1}{\tau_2}\right] \tag{2.64}$$

$$N_2 - N_1 = \frac{\Delta N^o}{1 + \phi W/A_{21}}. \tag{2.65}$$

Now from Eq. (2.33)

$$W = \frac{c^2 A_{21}}{8\pi h\nu^3} I_\nu g(\nu_0, \nu). \tag{2.66}$$

If we define

$$I_s(\nu) = \frac{8\pi h\nu^3}{c^2\phi g(\nu_0, \nu)} \tag{2.67}$$

then

$$N_2 - N_1 = \frac{\Delta N^o}{1 + I_\nu/I_s(\nu)} \tag{2.68}$$

and $I_s(\nu)$, called the saturation intensity, is the intensity of an incident light signal (power area^{-1}) that reduces the population inversion to half its value when no signal is present. Note that the value of the saturation intensity depends on the frequency of the input signal relative to the line center.

Returning to our expression for the gain constant of a laser amplifier

$$\gamma(\nu) = (N_2 - N_1)\frac{c^2 A_{21}}{8\pi\nu^2} g(\nu_0, \nu) \tag{2.69}$$

the gain as a function of intensity is, in a *homogeneously* broadened system,

$$\gamma(\nu) = \frac{\Delta N^o}{[1 + I_\nu/I_s(\nu)]}\frac{c^2 A_{21}}{8\pi\nu^2} g(\nu_0, \nu), \tag{2.70}$$

which is reduced, that is, *saturates* as the strength of the amplified signal increases. A good optical amplifier should have a large value of saturation intensity, from Eq. (2.67) this implies that ϕ should be a minimum. In such systems often $A_{21} \simeq 1/\tau_2$ so $\phi \simeq 1$.

2.6.2 Inhomogeneous Systems

The problem of gain saturation in inhomogeneous media is more complex. For example, in a gas a plane monochromatic wave at frequency ν interacts with a medium whose individual particles have Lorentzian homogeneous lineshapes with FWHM $\Delta\nu_N$, but whose center frequencies are distributed over an inhomogeneous (Doppler) broadened profile of width (FWHM) $\Delta\nu_D$. The Lorentzian contribution to the overall lineshape is

$$g_L(\nu', \nu) = \frac{(2/\pi\Delta\nu_N)}{1 + [2(\nu - \nu')/\Delta\nu_N]^2}, \tag{2.71}$$

where ν' is the center frequency of a particle set by its velocity relative to the observer. The Doppler broadened profile of all the particles is

$$g_D(\nu_0, \nu') = \frac{2}{\Delta\nu_D}\sqrt{\frac{\ln 2}{\pi}}e^{-[(2(\nu'-\nu_0)/\Delta\nu_D)^2\ln 2]}, \tag{2.72}$$

where ν_0 is the center frequency of a particle at rest. The overall lineshape (from all the particles) is a sum of Lorentzian profiles spread across the particle velocity distribution, as shown in Fig. (2.7).

If $\Delta\nu_N \gg \Delta\nu_D$ the overall profile remains approximately Lorentzian and the observed behavior of the system will correspond to *homogeneous* broadening. Such a situation is likely to arise for long wavelength transitions in a gas, particularly if this has a high atomic or molecular weight, at pressures where pressure broadening (which is a homogeneous process) is important, and frequently in solid materials. If $\Delta\nu_D \gg \Delta\nu_N$ (as is often the case in gases) the overall lineshape remains Gaussian and the system is *inhomogeneously* broadened.

Once again we reduce the problem to a consideration of the interaction of a plane electromagnetic wave with a two-state system as shown in Fig. (2.11). As the wave passes through the system its intensity changes according to whether the medium is amplifying or absorbing. We take the intensity of the monochromatic wave to be $I(\nu, z)$ at plane z within the medium. The individual particles of the medium have a distribution of emission center frequencies (or absorption center frequencies) because of their random velocities (or, for example, their different crystal environments). We take the population density functions (atoms vol^{-1} Hz^{-1}) in the upper and lower levels whose center frequency is at ν' to be $N_2(\nu', z)$ and $N_1(\nu', z)$, respectively, at plane z. Atoms are fed into levels 2 and 1 at rates $R_2(\nu')$ and $R_1(\nu')$. These rates are assumed uniform throughout the medium. $N_2(\nu', z), N_1(\nu', z), R_2(\nu')$, and $R_1(\nu')$ are assumed to follow the Gaussian frequency dependence set by the particle velocity distribution, and are normalized so that, for example, the total pumping rate of level 2 is

$$R_2 = \int_{-\infty}^{\infty} R_2(\nu')d\nu' = R_{20} \int_{-\infty}^{\infty} e^{[2(\nu'-\nu_0)/\Delta\nu_D]^2 \ln 2} d\nu'. \tag{2.73}$$

In practice, the primary pumping process may not have this Gaussian dependence, but even when this is the case, the effect of collisions among particles that have been excited will be to *smear out* any non-Gaussian pumping process into a near-Gaussian form. This conclusion is justified by observations of Doppler-broadened lines under various excitation conditions where deviations from a true Gaussian lineshape are found to be minimal. From Eq. (2.73)

$$R_2 = \sqrt{\frac{\pi}{\ln 2}} \frac{\Delta\nu_D}{2} R_{20}, \tag{2.74}$$

where R_{20} is a pumping rate constant and the total population density of level 2 at plane z is

$$N_2 = \int_{-\infty}^{\infty} N_2(\nu', z)d\nu'. \tag{2.75}$$

The rate equations for the atoms whose center frequencies are at ν' are

$$\frac{dN_2}{dt}(\nu', z) = R_2(\nu') - N_2(\nu', z)\left[\frac{1}{\tau_2} + B'_{21}(\nu', \nu)\frac{I(\nu, z)}{c}\right] + N_1(\nu', z)B'_{12}(\nu', \nu)\frac{I(\nu, z)}{c} \tag{2.76}$$

and

$$\frac{dN_1}{dt}(\nu',z) = R_1(\nu') + N_2(\nu',z)\left[A_{21} + B'_{21}(\nu',\nu)\frac{I(\nu,z)}{c}\right] - N_1(\nu',z)\left[\frac{1}{\tau_1} + B'_{12}(\nu',\nu)\frac{I(\nu,z)}{c}\right]. \qquad (2.77)$$

Here we have used the modified Einstein coefficients $B'_{21}(\nu',\nu)$, $B'_{12}(\nu',\nu)$ that describe stimulated emission processes when the stimulating radiation is at frequency ν and the particle's center emission frequency is at ν'. Written out in full

$$B'_{21}(\nu',\nu) = B_{21}g(\nu',\nu), \qquad (2.78)$$

where $g(\nu',\nu)$ is the *homogeneous* lineshape function. The rate of change of intensity of the incident wave due to atoms with center frequencies in a small range $d\nu'$ at ν' is

$$\left[\frac{dI}{dz}(\nu,z)\right]_{d\nu'} = h\nu\frac{I(\nu,z)}{c}[B'_{21}(\nu',\nu)N_2(\nu',z) - B'_{12}(\nu',\nu)N_1(\nu',z)]d\nu'. \quad (2.79)$$

The total rate of change of intensity due to all the atoms, that is from all possible center frequencies ν', is

$$\frac{dI(\nu,z)}{dz} = \frac{h\nu I(\nu,z)}{c}\int_{-\infty}^{\infty}[B'_{21}(\nu',\nu)N_2(\nu',z) - B'_{12}(\nu',\nu)N_1(\nu',z)]d\nu'. \quad (2.80)$$

In the steady state

$$\frac{dN_2}{dt}(\nu',z) = \frac{dN_1}{dt}(\nu',z) = 0, \qquad (2.81)$$

so from Eqs. (2.76) and (2.77)

$$B'_{21}(\nu',\nu)N_2(\nu') - B'_{12}(\nu',\nu)N_1(\nu') = \frac{B'_{21}(\nu',\nu)\left[\dfrac{R_2(\nu')}{1/\tau_2} - \left(\dfrac{g_2}{g_1}\right)\dfrac{(R_2(\nu')A_{21} + R_1(\nu')/\tau_2)}{1/\tau_1\tau_2}\right]}{1 + \left[\dfrac{(g_1/g_2)(1/\tau_2 - A_{21})}{1/\tau_1\tau_2} + \tau_2\right]B'_{21}(\nu',\nu)\dfrac{I(\nu,z)}{c}}. \qquad (2.82)$$

We note that

$$R_2(\nu') = R_{20}e^{-[2(\nu'-\nu_0)/\Delta\nu_D]^2\ln 2}. \qquad (2.83)$$

Substituting in Eq. (2.80) from (2.82) and (2.83) and bearing in mind that $B'_{21}(\nu',\nu)$ has a Lorentzian form,

$$B'_{21}(\nu',\nu) = \frac{B_{21}(2/\pi\Delta\nu_N)}{1 + [2(\nu-\nu')/\Delta\nu_N]^2}, \qquad (2.84)$$

where $\Delta\nu_N$ is the *homogeneous* FWHM of the transition, gives t

$$\frac{1}{I(\nu,z)}\frac{dI(\nu,z)}{dz} = \gamma(\nu) = \frac{\gamma_0\int_{-\infty}^{\infty}d\nu'(\frac{2}{\pi\Delta\nu_N})\{1 + [2(\nu-\nu')/\Delta\nu_N]^2\}^{-1}e^{-[\frac{2(\nu'-\nu)}{\Delta\nu_D}]^2\ln 2}}{1 + \eta I(\nu,z)(\frac{2}{\pi\Delta\nu_N})[1 + [2(\nu-\nu')/\Delta\nu_N]^2]^{-1}}. \qquad (2.85)$$

We have made the substitutions

$$\gamma_0 = \frac{h\nu}{c} B_{21} \left[R_{20}\tau_2 - \frac{g_2}{g_1} \left(\frac{R_{20}A_{21} + R_{10}A_2}{1/\tau_1\tau_2} \right) \right], \tag{2.86}$$

$$\eta = \left[\frac{g_2}{g_1} \frac{(1/\tau_2 - A_{21})}{1/\tau_1\tau_2} + \tau_2 \right] \frac{B_{21}}{c}. \tag{2.87}$$

Eq. (2.85) can be written

$$\gamma(\nu) = \frac{2\gamma_0 \Delta\nu_N}{\pi} \int_{-\infty}^{\infty} \frac{e^{-[2(\nu'-\nu_0)/\Delta\nu_D]^2 \ln 2} d\nu'}{4(\nu - \nu')^2 + \Delta\nu_N^2 [1 + 2\eta I(\nu, z)/\pi\Delta\nu_N]}. \tag{2.88}$$

Although it is fairly clear from Eq. (2.86) that the gain of the amplifier falls as $I(\nu, z)$ increases, it is not easy to see from the integral exactly how this occurs. If the intensity is small the gain approaches its *unsaturated* value

$$\gamma_0(\nu) = \frac{2\gamma_0 \Delta\nu_N}{\pi} \int_{-\infty}^{\infty} \frac{e^{-[2(\nu'-\nu_0)\Delta\nu_D]^2 \ln 2} d\nu'}{4(\nu - \nu')^2 + \Delta\nu_N^2}. \tag{2.89}$$

If Eq. (2.88) is examined closely, for frequencies ν' close to the input frequency ν the integrand can be written approximately as

$$\frac{e^{-[2(\nu'-\nu_0)/\Delta\nu_D]^2 \ln 2}}{\Delta\nu_N^2 [1 + 2\eta I(\nu, z)/\pi\Delta\nu_N]}.$$

On the other hand for frequencies ν' far from the input frequency, the integrand can be written approximately as

$$\frac{e\{-[2(\nu' - \nu_0)/\Delta\nu_D]^2 \ln 2\}}{4(\nu - \nu')^2 + \Delta\nu_N^2},$$

which can be seen to be identical to the integrand in the unsaturated gain expression, Eq. (2.89). Thus, we conclude that in making their contribution to the overall gain, particles whose frequencies are far from the input frequency are relatively unaffected by the input radiation, whereas particles whose frequencies are close to that of the input show strong saturation effects. The gain in the system comes largely from those particles whose frequencies are within (roughly) a homogeneous linewidth of the input radiation frequency. The consequences of this are best illustrated by considering a hypothetical experiment, shown schematically in Fig. (2.12), in which the small-signal gain of a predominantly inhomogeneously broadened amplifier is measured with and without a strong saturating signal simultaneously present. Without the presence of a strong signal at a fixed frequency ν_S the observed (small-signal) gain follows the Gaussian curve of the overall line profile of the amplifier, as shown in Fig. (2.13a). However if we perform this experiment again when a strong fixed frequency field is also present, which causes saturation of the gain, we find the gain is reduced locally by the saturating effect of the strong field as shown in Fig. (2.13b). This phenomenon is called *hole burning*[2.10]. The width of the *hole* that is thus produced is determined by the

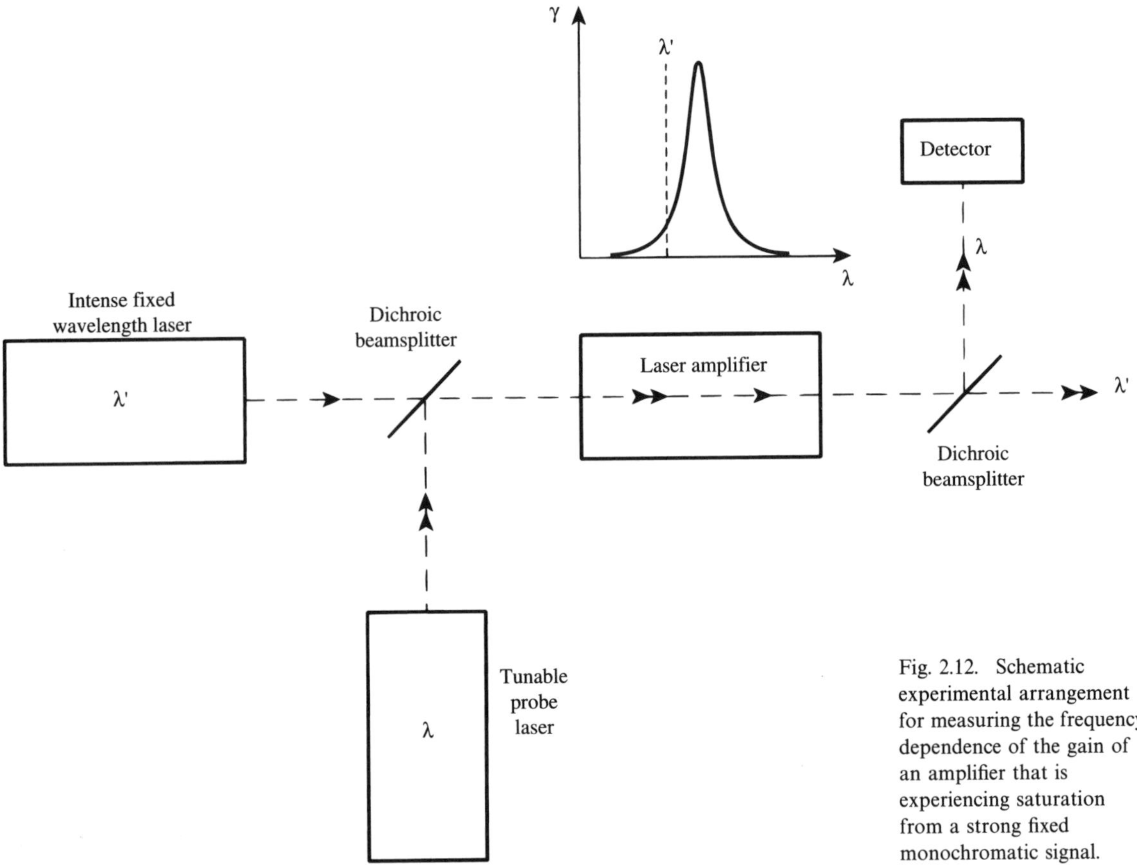

Fig. 2.12. Schematic experimental arrangement for measuring the frequency dependence of the gain of an amplifier that is experiencing saturation from a strong fixed monochromatic signal.

quantity

$$\Delta\nu_N^2\left[1+\frac{2\eta I(\nu,z)}{\pi\Delta\nu_N}\right].$$

If

$$\Delta\nu_N\sqrt{1+\frac{2\eta I(\nu,z)}{\pi\Delta\nu_N}}\ll\Delta\nu_D,$$

for example, in a gaseous system where Doppler broadening is the largest contribution to the total observed line broadening, Eq. (2.88) can be integrated by bringing the much less sharply peaked exponential factor outside the integral. In this case, the sharply peaked Lorentzian lineshape makes the integrand largest for frequencies ν' near to ν; over a small range of frequencies ν' near ν the exponential factor remains approximately constant so Eq. (2.88) can be written

$$\gamma(\nu)=\frac{2\gamma_0\Delta\nu_N}{\pi}e^{-[2(\nu-\nu_0)\Delta\nu_D]^2\ln 2}\times\int_{-\infty}^{\infty}\frac{d\nu'}{4(\nu-\nu')^2+\Delta\nu_N^2[1+2\eta I(\nu,z)/\pi\Delta\nu_N]}. \tag{2.90}$$

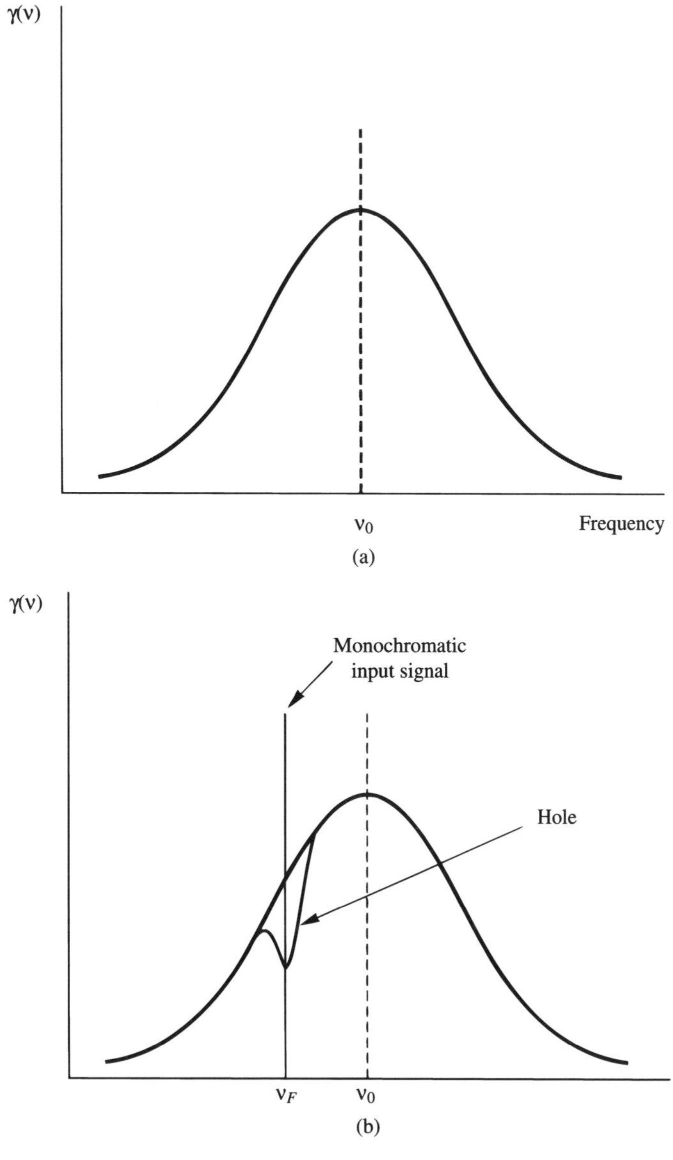

Fig. 2.13. Gain as a function of frequency in an inhomogeneously broadened amplifier. (a) Small-signal situation when no saturation has occurred. (b) Showing the production of a 'hole' in the gain curve by a strong monochromatic input at frequency ν_F.

Now the integral can be evaluated to give

$$\gamma(\nu) = \gamma_0 \left[1 + \frac{2\eta I(\nu, z)}{\pi \Delta\nu_N}\right]^{-\frac{1}{2}} e^{-[2(\nu-\nu_0)/\Delta\nu_D]^2 \ln 2}, \tag{2.91}$$

which gives

$$\gamma(\nu) = \gamma_0 \left[1 + \frac{I(\nu, z)}{I_s'(\nu)}\right]^{-\frac{1}{2}} e^{-[2(\nu-\nu_0)/\Delta\nu_D]^2 \ln 2} \tag{2.92}$$

where $I_s'(\nu) = \pi\Delta\nu_N/2\eta$ is called the saturation intensity for inhomogeneous broadening. Note that γ_0 is the small-signal gain at line center of the inhomogeneously broadened line. It is left as an exercise to the reader to show that γ_0 can be written in the form

$$\gamma_0 = \frac{1}{4\pi}\sqrt{\frac{\ln 2}{\pi}}\frac{\lambda^2 A_{21}}{\Delta\nu_D}\left(N_2 - \frac{g_2}{g_1}N_1\right). \tag{2.93}$$

When Doppler broadening dominates in a system, incident radiation at frequency ν cannot interact with those atoms whose Doppler-shifted frequency is different from ν by much more than $\Delta\nu_N$.

If the amplifier is homogeneously broadened, that is, if

$$\Delta\nu_N\sqrt{1 + \frac{\eta I(\nu, z)}{\pi\Delta\nu_N}} \gg \Delta\nu_D,$$

Eq. (2.88) can be integrated by bringing the less-sharply peaked Lorentzian factor outside the integral to give

$$\gamma(\nu) = \frac{\gamma_0\Delta\nu_D}{\Delta\nu_N\sqrt{\pi\ln 2}}\left\{\left[\frac{2(\nu-\nu_0)}{\Delta\nu_N}\right]^2 + 1 + \frac{I(\nu, z)}{I_s'(\nu)}\right\}^{-1}$$

$$= \frac{\Delta\nu_D}{2}\sqrt{\frac{\pi}{\ln 2}}\frac{\gamma_0 g(\nu_0, \nu)}{\left[1 + I\nu, z)/I_s(\nu)\right]}, \tag{2.94}$$

where $g(\nu_0, \nu)$ is the homogeneous lineshape function

$$g(\nu_0, \nu) = \frac{(2/\pi\Delta\nu_N)}{1 + [2(\nu - \nu_0)\Delta\nu_N]^2} \tag{2.95}$$

and $I_s(\nu)$ is the saturation intensity for homogeneous broadening given by

$$I_s(\nu) = \frac{2I_s'(\nu)}{\pi\Delta\nu_N g(\nu_0, \nu)} = \frac{1}{\eta g(\nu_0, \nu)}. \tag{2.96}$$

It can be seen from Eq. (2.87) that for $g_2 = g_1$, η reduces to the expression

$$\eta = \left[\left(\frac{1/\tau_2 - A_{21}}{1/\tau_1\tau_2}\right) + \tau_2\right]\frac{B_{21}}{c} \tag{2.97}$$

and $I_s(\nu)$ reduces to the expression obtained previously as the saturation intensity for homogeneous broadening. Namely,

$$I_s(\nu) = \frac{8\pi h\nu^3}{c^2\phi g(\nu_0, \nu)}, \tag{2.98}$$

where

$$\phi = A_{21}\tau_2\left[1 + (1 - A_{21}\tau_2)\frac{\tau_1}{\tau_2}\right]. \tag{2.99}$$

2.7 Power Output from a Laser Amplifier

For a laser amplifier of length ℓ and gain coefficient $\gamma(\nu)$ the output intensity for a monochromatic input intensity of I_0 (W m^{-2}) at frequency ν is

$$I = I_0 e^{\gamma(\nu)\ell} \tag{2.100}$$

if saturation effects are neglected. If saturation effects cannot be neglected then the differential equation that describes how intensity increases must be re-examined. This is

$$\gamma(\nu) = \frac{1}{I}\frac{dI}{dz}. \tag{2.101}$$

Table 2.1. *Iterative Solution of Eq. (2.104).*

LHS	RHS
10	2.57
8	5
7	6.98
6.9	7.21
6.99	7.00
6.993	6.994
6.9934	6.9934

For a homogeneously broadened amplifier with saturation an explicit solution to this equation can be found. In this case, if $\gamma_0(\nu)$ is the small-signal gain,

$$\gamma(\nu) = \frac{\gamma_0(\nu)}{1 + I/I_s(\nu)} = \frac{1}{I}\frac{dI}{dz}, \tag{2.102}$$

which can be rewritten in the form

$$\frac{dI}{I} + \frac{dI}{I_s(\nu)} = \gamma_0(\nu)dz. \tag{2.103}$$

The solution to this equation is

$$I = I_0 e^{\gamma_0(\nu)\ell - (I - I_0)/I_s(\nu)}. \tag{2.104}$$

We can best illustrate how this equation can be used to find the output of a saturated amplifier with a numerical example.

2.7.0.1 Example.
A homogeneously broadened laser has a saturation intensity of 3 W m^{-2} and a small-signal gain at line center of 0.5 m^{-1}. Calculate the output intensity if a 5 Wm^{-2} monochromatic signal at line center enters an amplifier 2 m long.

If saturation were negligible the output intensity would be

$$I = I_0 e^{\gamma_0 \ell} = 5e^{0.5\times 2} = 13.59 \text{ W m}^{-2}.$$

Eq. (2.104) must be solved iteratively, we know the solution will be somewhere between the input intensity and the output when saturation is neglected, i.e., $5 < I < 13.59$. We make an initial guess for I (LHS) and evaluate the RHS of Eq. (2.104). The successive calculation of the LHS and RHS for difficult guesses for I is shown in Table (2.1). The answer, obtained when the LHS and RHS agree, is 6.9934 W m^{-2}.

2.8 The Electron Oscillator Model of a Radiative Transition

When a particle decays from an excited state into a lower state, we can model the resultant electric field as a damped oscillation. This behavior leads us to point out the analogy between the decay of an excited particle and the damped oscillation of an electric circuit. For example, for the *RLC* circuit shown in Fig. (2.14) the resonant frequency is $\nu_0 = 1/2\pi\sqrt{LC}$. If the sinusoidal driving voltage is disconnected from the circuit, then the oscillation of the circuit decays exponentially – provided the circuit is underdamped, as shown in detail in Appendix 4. The power spectrum

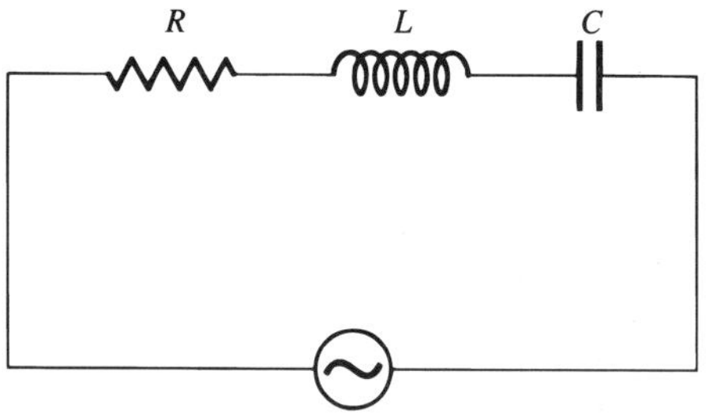

Fig. 2.14. *RLC* circuit.

of the decaying electric current is Lorentzian, just as it is for a spontaneous transition. The FWHM of the circuit resonance, is ν_0/Q, where Q is called the *quality factor* of the circuit, analogous to the homogeneously broadened linewidth $\Delta\nu_N$. A transition between states almost always has $\Delta\nu_N \ll \nu_0$ so clearly has a very high Q.

In the classical theory of how a particle responds to electromagnetic radiation each of the n electrons attached to the particle is treated as a damped harmonic oscillator. For example, when an electric field acts on an atom, the nucleus, which is positively charged, moves in the direction of the field while the electron cloud, which is negatively charged, moves in the opposite direction to the field. The resultant separation of the centers of positive and negative charge causes the atom to become an elemental dipole. If the separation of the nucleus and electron cloud is d, then the resultant dipole has magnitude ed and points from the negative towards the positive charge.† As the frequency of the electric field that acts on the atom increases, the amount of nuclear motion decreases much more rapidly than that of the electrons. At optical frequencies we generally neglect the motion of the nucleus, its great inertia compared to the electron cloud prevents it following the rapidly oscillating applied electric field. If the vector displacement of the ith electron on the atom from its equilibrium position is $\mathbf{x}_i$ then at any instant the atom has acquired a dipole moment

$$\boldsymbol{\mu} = -\sum_{i=1}^{n} e\mathbf{x}_i, \tag{2.105}$$

where the summation runs over all the n electrons on the atom. The magnitude of the displacement of each electron depends on the value of the electric field $\mathbf{E}_i$ at the electron

$$k_i\mathbf{x_i} = -e\mathbf{E_i}, \tag{2.106}$$

where k_i is a force constant. A time-varying field $\mathbf{E}$ leads to a time-varying dipole moment. This dipole moment can become large if there is a resonance between the applied field and a particular electron on the atom. This happens if the frequency of the field is near the natural oscillation frequency of a particular electron. Classically, the resonance frequency of electron i is, by analogy with a mass attached to a spring, $\omega_i = \sqrt{k_i/m}$. If the applied electric field is near this frequency, one electron in the summation in Eq. (2.106) makes a dominant contribution to the dipole moment and we can treat the atom as a single electron oscillator. The physical significance of the resonant frequencies of the electrons is that they correspond to the frequencies

† The magnitude of the electronic charge is $e \simeq 1.6 \times 10^{-19}$C, the charge on an electron is $-e$.

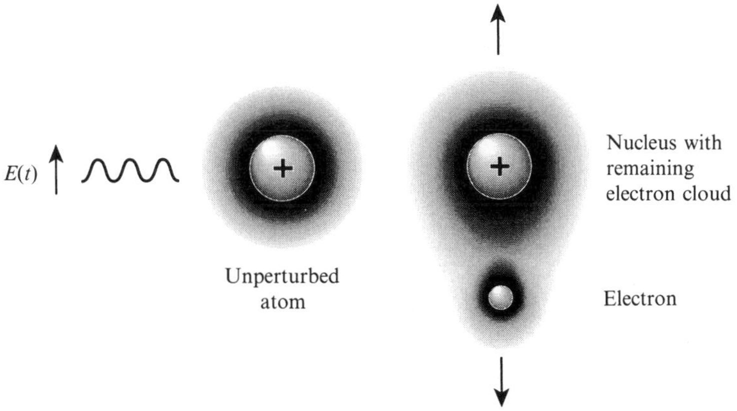

Fig. 2.15. Electron cloud and nucleus are displaced in opposite directions by an applied field.

of transitions that the electrons of the atom can make from one energy state to another. If we confine our attention to one of these resonances then we can treat an n-electron atom as a one-electron classical oscillator. A time-varying electric field $E(t)$, which we assume *a priori* to be at a frequency near to an atomic resonance, perturbs only a single electron to a significant degree, thereby inducing a dipole moment which varies at the same frequency as the applied field. The electron and nucleus are perturbed in opposite directions by the field as shown in Fig. (2.15). We shall see that classically the atom absorbs energy from this field and that the maximum absorption of energy leads to the maximum induced dipole moment (polarization).

The motion of the electrons on each of the particles in the medium, which for simplicity can be assumed to be identical, obeys the differential equation

$$\frac{d^2x}{dt^2} + 2\Gamma\frac{dx}{dt} + \frac{k}{m}x = -\frac{e}{m}E(t). \tag{2.107}$$

The terms on the left hand side of the equation represent, reading from left to right: the acceleration of the electron, a damping term proportional to the electron velocity, and a restoring force. These terms are balanced by the effect of the applied electric field $E(t)$. The restoring force is analogous to the restoring force acting on a mass suspended from a spring and given a small displacement from its equilibrium position. The damping can be regarded as a viscous drag that the moving electron experiences because of its interaction with the other electrons on the particle.

We take $E(t) = \mathcal{R}(Ee^{i\omega t})$ and $x(t) = \mathcal{R}[X(\omega)e^{i\omega t}]$. The possibility that $E(t)$ and $x(t)$ are not in phase is taken into account by allowing the function $X(\omega)$ to include a phase factor. If we define a resonant frequency by $\omega_0 = \sqrt{k/m}$ then the differential equation (2.107) above becomes

$$(\omega_0^2 - \omega^2)X + 2i\omega\Gamma X = -\frac{e}{m}E, \tag{2.108}$$

giving

$$X(\omega) = \frac{-(e/m)E}{\omega_0^2 - \omega^2 + 2i\omega\Gamma}. \tag{2.109}$$

This is the amplitude of the displacement of the electron from its equilibrium position as a function of the frequency of the applied field.

Near resonance $\omega \simeq \omega_0$ so

$$X(\omega \simeq \omega_0) = \frac{-(e/m)E}{2\omega_0(\omega_0 - \omega) + 2i\omega_0\Gamma}. \tag{2.110}$$

Now, the dipole moment of a single electron is $\mu(t) = -e[x(t)]$. This dipole moment arises from the separation of the electron charge cloud and the nucleus.

If there are N electron oscillators per unit volume there results a net polarization (dipole moment per unit volume) of:

$$p(t) = -Nex(t) = \mathscr{R}[P(\omega)e^{i\omega t}], \tag{2.111}$$

where $P(\omega)$ is the complex amplitude of the polarization

$$\begin{aligned} P(\omega) = -NeX(\omega) &= \frac{(Ne^2/m)E_0}{2\omega_0(\omega_0 - \omega) + 2i\omega_0\Gamma} \\ &= \frac{-i(Ne^2/(2m\omega_0\Gamma)}{1 + i(\omega - \omega_0)/\Gamma}E_0. \end{aligned} \tag{2.112}$$

The electronic susceptibility $\chi(\omega)$ is defined by the equation

$$P(\omega) = \epsilon_0\chi(\omega)E_0, \tag{2.113}$$

where ϵ_0 is the permittivity of free space. $\chi(\omega)$ is complex and can be written in terms of its real and imaginary parts as

$$\chi(\omega) = \chi'(\omega) - i\chi''(\omega), \tag{2.114}$$

so

$$p(t) = \mathscr{R}[\epsilon_0\chi(\omega)E_0e^{i\omega t}] = \epsilon_0E_0\chi'(\omega)\cos\omega t + \epsilon_0E_0\chi''\sin\omega t. \tag{2.115}$$

It is a commonly used convention that Eq. (2.114) has the negative sign.

Therefore $\chi'(\omega)$, the real part of the susceptibility, is related to the in-phase polarization, while $\chi''(\omega)$, the complex part, is related to the out of phase (quadrature) component. From Eqs. (2.112) and (2.113)

$$\chi(\omega) = -i\left(\frac{Ne^2}{2m\omega_0\Gamma\epsilon_0}\right)\frac{1}{1 + i(\omega - \omega_0)/\Gamma} \tag{2.116}$$

so

$$\chi'(\omega) = \left(\frac{Ne^2}{2m\omega_0\Gamma\epsilon_0}\right)\frac{(\omega_0 - \omega)/\Gamma}{1 + (\omega - \omega_0)^2/\Gamma^2}, \tag{2.117}$$

$$\chi''(\omega) = \left(\frac{Ne^2}{2m\omega_0\Gamma\epsilon_0}\right)\frac{1}{1 + (\omega - \omega_0)^2/\Gamma^2}. \tag{2.118}$$

Changing to conventional frequency, $\nu = \omega/2\pi$, and putting $\Delta\nu = \Gamma/\pi$, which is the FWHM of the Lorentzian shape that describes $\chi''(\omega)$, we obtain

$$\chi''(\nu) = \left(\frac{Ne^2}{16\pi^2m\nu_0\epsilon_0}\right)\frac{\Delta\nu}{(\Delta\nu/2)^2 + (\nu - \nu_0)^2}, \tag{2.119}$$

$$\chi'(\nu) = \frac{2(\nu_0 - \nu)}{\Delta\nu}\chi''(\nu) = \left(\frac{Ne^2}{8\pi^2m\nu_0\epsilon_0}\right)\frac{\nu_0 - \nu}{(\Delta\nu/2)^2 + (\nu - \nu_0)^2}, \tag{2.120}$$

Fig. (2.16) is a graph of χ'' and χ' normalized to the peak value χ_0'' of χ''. Note that χ'' has the characteristic Lorentzian shape common to the frequency response of *RLC* circuits and homogeneously broadened spectral lines.†

† It can be shown that although Eqs. (2.119) and (2.120) have the correct frequency dependence for the susceptibility of a real atom, they have incorrect magnitude. This arises because of inadequacies in the classical

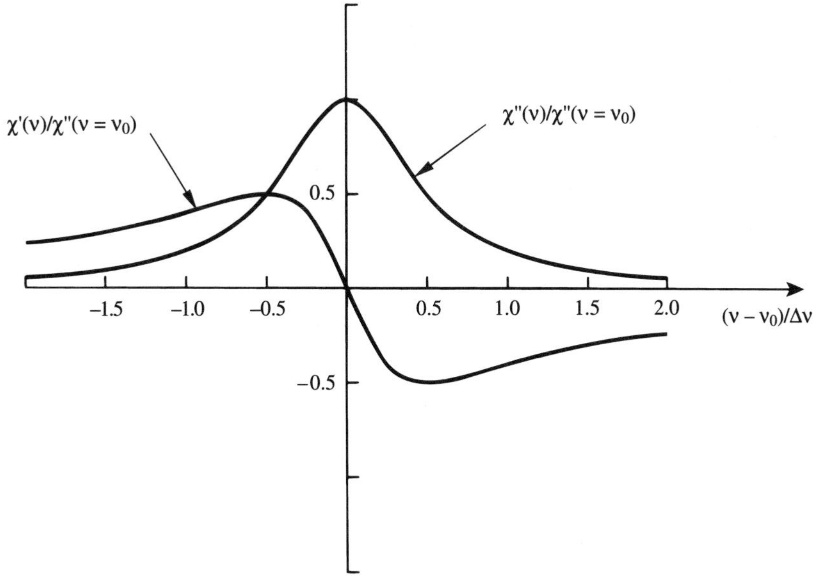

Fig. 2.16. Frequency variation of the real, $\chi'(\nu)$, and imaginary, $\chi''(\nu)$, parts of the susceptibility calculated using the electron oscillator model.

2.9 What Are the Physical Significances of χ' and χ''?

To understand the physical significance of a complex susceptibility it will be helpful if we briefly review some basic electromagnetic theory.

The relationship between the applied electic field **E** and the electron displacement vector **D** is

$$\begin{aligned}\mathbf{D} &= \epsilon_0\mathbf{E} + \mathbf{P} \\ &= \epsilon_0(1+\chi)\mathbf{E},\end{aligned} \tag{2.121}$$

which by introducing the dielectric constant $\epsilon_r = 1 + \chi$ can be written as

$$\mathbf{D} = \epsilon_0\epsilon_r\mathbf{E}. \tag{2.122}$$

The refractive index n of the medium is related to ϵ_r by $n = \sqrt{\epsilon_r}$.

When an external electric field interacts with a group of particles there are two contributions to the induced polarization, a macroscopic contribution $\mathbf{P}_m$ from the collective properties of the particles, for example their arrangement in a crystal lattice, and a contribution from the polarization $\mathbf{P}_t$ associated with transitions in the medium, so, in general

$$\mathbf{P} = \mathbf{P}_m + \mathbf{P}_t. \tag{2.123}$$

Usually there are many transitions possible for the particles of the medium but only one will be in near resonance with the frequency of an applied field. $\mathbf{P}_t$ is dominated by the contribution of this single transition near resonance. Far from any such resonance $\mathbf{P}_t$ is negligible and $\mathbf{P} = \mathbf{P}_m$, which allows us to define the macroscopic dielectric constant ϵ_r (far from resonance) from

$$\mathbf{P} = \epsilon_0\mathbf{E} + \mathbf{P}_m = \epsilon_0\mathbf{E} + \chi_m\epsilon_0\mathbf{E} = \epsilon_r\epsilon_0\mathbf{E}. \tag{2.124}$$

where χ_m is the macroscopic, nonresonant, susceptibility. However, if the frequency

electron oscillator model and these inadequacies should be borne in mind before the classical model is used to make detailed predictions about the behavior of an atom.

of the electric field is near the frequency of a possible transition within the medium then there is a significant contribution to **P** from this transition. Other possible transitions far from resonance do not contribute and we can write

$$\mathbf{D} = \epsilon_0 \mathbf{E} + \mathbf{P}_m + \mathbf{P}_t = \epsilon_r \epsilon_0 \mathbf{E} + \mathbf{P}_t. \tag{2.125}$$

$\mathbf{P}_t$ is related to the complex susceptibility that results from the transition according to $\mathbf{P}_t = \epsilon_0 \chi(\omega)\mathbf{E}$. Therefore we can rewrite Eq. (2.120) as

$$\mathbf{D} = \epsilon_0 \left[\epsilon_r + \chi(\omega)\right] \mathbf{E} = \epsilon_0 \epsilon_r^* \mathbf{E}, \tag{2.126}$$

so the complex susceptibility modifies the effective dielectric constant from ϵ_r to ϵ_r^*.

When an electromagnetic wave propagates through a medium with a complex susceptibility, both the amplitude and phase velocity of the wave are affected. This can be illustrated easily for a plane wave propagating in the z direction with a field variation $\sim e^{i(\omega t - kz)}$. k is the propagation constant, given by the expression

$$k = \omega\sqrt{\mu\epsilon} = \omega\sqrt{\mu_r \mu_0 \epsilon_r \epsilon_0}, \tag{2.127}$$

where for optical materials the relative permeability μ_r is usually unity. For a complex dielectric constant Eq. (2.127) can be rewritten as

$$k' = \omega\sqrt{\mu_0 \epsilon_r \epsilon_0}\sqrt{1 + \frac{\chi(\omega)}{\epsilon_r}} = k\sqrt{1 + \frac{\chi(\omega)}{\epsilon_r}}, \tag{2.128}$$

where k' is now the new propagation constant, which differs from the nonresonant propagation constant k because of the complex susceptibility resulting from a transition. If $|\chi(\omega)| \ll \epsilon_r$ Eq. (2.128) can be simplified by the use of the binomial theorem to give

$$k' = k\left[1 + \frac{\chi(\omega)}{2\epsilon_r}\right] = k\left[1 + \frac{\chi'(\omega)}{2\epsilon_r} - \frac{i\chi''(\omega)}{2\epsilon_r}\right]. \tag{2.129}$$

The wave now propagates through the medium as $e^{-ik'z}$. Written out in full the electric field varies as

$$\begin{aligned} E &= E_0 \exp\left(i\left\{\omega t - k\left[1 + \frac{\chi'(\omega)}{2\epsilon_r} - \frac{i\chi''(\omega)}{2\epsilon_r}\right] z\right\}\right) \\ &= E_0 \exp\left(i\left\{\omega t - k\left[1 + \frac{\chi'(\omega)}{2\epsilon_r}\right] z\right\}\right) \exp\left[-\frac{k\chi''(\omega)}{2\epsilon_r} z\right]. \end{aligned} \tag{2.130}$$

Clearly, this is a wave whose phase velocity is

$$c' = \frac{\omega}{k[1 + \chi'(\omega)/2\epsilon_r]} = \frac{\omega}{k + \Delta k} \tag{2.131}$$

and whose field amplitude changes exponentially with distance. If we write

$$\gamma = -\frac{k\chi''(\omega)}{\epsilon_r} \tag{2.132}$$

then the wave changes its electric field amplitude with distance according to $e^{(\gamma/2)z}$.

Now, the intensity of the wave is $I \propto E(z,t)E^*(z,t)$, so the intensity of the wave changes as it passes through the medium as $I \propto e^{\gamma z}$. We can identify γ as the familiar gain coefficient of the medium, which was calculated previously by considering the spontaneous and stimulated radiative jumps between two energy levels.

Now from Eq. (2.132),

$$\gamma(\nu) = -\frac{k\chi''(\nu)}{n^2} = -\left(\frac{2\pi\nu_0 n}{c_0}\right)\frac{\chi''(\nu)}{n^2}, \tag{2.133}$$

which from Eq. (2.118), obtained by consideration of the system as a collection of classical oscillators, is

$$\gamma(\nu) = -\left(\frac{2\pi\nu}{nc_0}\right)\left(\frac{Ne^2}{16\pi^2 m\nu_0\epsilon_0}\right)\frac{\Delta\nu}{(\Delta\nu/2)^2 + (\nu-\nu_0)^2}, \tag{2.134}$$

which is *always negative.* Clearly, this is an incorrect result since we have previously shown that

$$\gamma(\nu) = \left(N_2 - \frac{g_2}{g_1}N_1\right)\frac{c^2 A_{21}}{8\pi\nu^2}g(\nu_0,\nu), \tag{2.135}$$

which can be positive or negative depending on the sign of $N_2 - (g_2/g_1)N_1$. Thus, the classical electron oscillator model appears to predict only absorption of incident radiation. It is possible, however, within the framework of the classical electron oscillator model to show that in certain conditions stimulated emission can occur. This will be explained in the last section of this chapter. Although the classical electron oscillator model is instructive, it is not entirely adequate in describing the interaction between particles and radiation. It is better to accept that $\gamma(\nu) = -k\chi''(\nu)/n^2$ and use Eq. (2.135) as the expression for $\gamma(\nu)$. In this case we find that the imaginary part of the complex susceptibility of the medium is

$$\chi''(\nu) = -\frac{n^2\gamma(\nu)}{k} = -\left(\frac{n^2 c}{2\pi\nu}\right)\gamma(\nu), \tag{2.136}$$

which from Eq. (2.132) gives

$$\chi''(\nu) = -\frac{\left[N_2 - (g_2/g_1)N_1\right]n^2c^3A_{21}}{8\pi^3\nu^3\Delta\nu}\frac{1}{1+[2(\nu-\nu_0)/\Delta\nu]^2}. \tag{2.137}$$

This quantum mechanical susceptibility is negative or positive depending on whether $N_2 - (g_2/g_1)N_1$ is positive or not. A *negative* value of $\chi''(\nu)$ corresponds to a system in population inversion.

Now it is shown in Appendix 3 that the power expended in inducing polarization in a medium is $\mathbf{E}\cdot d\mathbf{P}/dt$. If we write

$$|\mathbf{P}| = P(t) = \mathscr{R}(Pe^{i\omega t}], \tag{2.138}$$

which in an isotropic medium is parallel to **E**, where

$$|\mathbf{E}| = \mathscr{R}(Ee^{i\omega t}), \tag{2.139}$$

then the rate of expenditure of energy is

$$\frac{\text{power}}{\text{volume}} = \mathscr{R}(Ee^{i\omega t})\frac{\partial}{\partial t}\mathscr{R}(Pe^{i\omega t}). \tag{2.140}$$

Remembering that

$$\mathbf{P} = \epsilon_0\chi\mathbf{E}, \tag{2.141}$$

and averaging we have

$$\frac{\overline{\text{power}}}{\text{volume}} = \overline{\mathcal{R}(Ee^{i\omega t})\mathcal{R}(i\omega\epsilon_0\chi Ee^{i\omega t})} \tag{2.142}$$

$$= \frac{1}{2}\mathcal{R}(i\omega\epsilon_0\chi EE^*) \tag{2.143}$$

$$= \frac{\omega}{2}\epsilon_0 \mid E \mid^2 \mathcal{R}(i\chi). \tag{2.144}$$

Since $\chi = \chi' - i\chi''$

$$\frac{\overline{\text{power}}}{\text{volume}} = \frac{\omega\epsilon_0\chi''}{2} \mid E \mid^2 . \tag{2.145}$$

If our system is in a state of population inversion, that is with a negative χ'', then from Eq. (2.145) the average power per volume used in inducing polarization in the medium is *negative* – and the system provides us with more energy than is put in by the electromagnetic wave.

2.10 The Classical Oscillator Explanation for Stimulated Emission

If there were no applied electric field acting on the electron, then from Eq. (2.107) the position of the electron would satisfy the equation

$$x(t) = x_0 e^{-\Gamma t}\cos(\omega_0 t + \phi_0), \tag{2.146}$$

where ω_0 is the resonant frequency and ϕ_0 is a phase factor set by the initial conditions. If at time $t = 0$ the position and velocity of the electron are a_0, v_0, respectively, then

$$\begin{aligned} a_0 &= x_0 \cos\phi_0, \\ v_0 &= -\Gamma x_0 \cos\phi_0 - \omega_0 x_0 \sin\phi_0. \end{aligned} \tag{2.147}$$

When the electric field is applied we have already seen that in the steady state energy is apparently only absorbed. However, this impression is erroneous. It neglects that in reality no electromagnetic field interacts indefinitely with an electron. Therefore, we must consider what happens when an electron, which can already be regarded as oscillating if it is in an excited state, is suddenly subjected to the additional perturbation of an applied field. We are going to be interested in the behavior of the electron over the first few cycles of the applied field so we neglect damping and write

$$\frac{d^2x}{dt^2} + \omega_0^2 = -\frac{eE_0}{2m}(e^{i\omega t} + e^{-i\omega t}), \tag{2.148}$$

where the applied field is $E_0 \cos\omega t$ and has been written in its complex exponential form. By introducing a new variable $z = \dot{x} + i\omega_0 x$, where the dot indicates differentiation with respect to time, Eq. (2.148) can be rewritten in the form

$$\frac{dz}{dt} - i\omega_0 z = -\frac{eE_0}{2m}(e^{i\omega t} + e^{-i\omega t}). \tag{2.149}$$

This equation can be solved by multiplying each term by $e^{-i\omega_0 t}$ and then integrating to give

$$ze^{-i\omega_0 t} = -\frac{eE_0}{2m}\int (e^{i(\omega-\omega_0)t} + e^{-i(\omega+\omega_0)t})dt. \tag{2.150}$$

This gives

$$\frac{dx}{dt} + i\omega_0 x = -\frac{eE_0}{2m}\left[-\frac{ie^{i\omega t}}{(\omega-\omega_0)} + \frac{ie^{-i\omega t}}{(\omega+\omega_0)} + Ae_0^{i\omega t}\right], \tag{2.151}$$

where A is a constant of integration. By integrating a second time in a similar manner and introducing the initial values of position and velocity it is left as an exercise to the reader to show that the final solution is

$$x(t) = -\frac{eE_0}{m}\left[\frac{\cos\omega_0 t - \cos\omega t}{(\omega^2-\omega_0^2)}\right] + \sqrt{\left(\frac{v_0}{\omega_0}\right)^2 + x_0^2}\cos(\omega_0 t + \phi), \tag{2.152}$$

where $\tan\phi = -v_0/\omega_0 x_0$.

By the use of the trigonometrical identity

$$\cos X - \cos Y = -2\sin\left(\frac{X+Y}{2}\right)\sin\left(\frac{X-Y}{2}\right) \tag{2.153}$$

and assuming that the applied frequency is close to resonance Eq. (2.152) can be written

$$x(t) = -\frac{eE_0}{2m\omega_0}t\sin\omega_0 t + \sqrt{\left(\frac{v_0}{\omega_0}\right)^2 + x_0^2}\cos(\omega_0 t + \phi). \tag{2.154}$$

Thus, near resonance the amplitude of oscillation will increase linearly with time, which is, of course, a consequence of our neglect of damping. However, it is interesting to use Eq. (2.154) to calculate the work done during the first n cycles of the applied field. This work is calculated as the work done by the electric field in polarizing the medium: the polarization $\mathbf{P}$ is proportional to electron displacement. The work done is $\mathbf{E}\cdot\partial\mathbf{P}/\partial t$ (see Appendix 3). During the first n cycles the total work done by the field is

$$W = \int_0^{2n\pi/\omega_0} \mathbf{E}\cdot\frac{\partial\mathbf{P}}{\partial t}dt = -NeE_0\int_0^{2n\pi/\omega_0}(\cos\omega_0 t)\dot{x}(t)dt, \tag{2.155}$$

where N is the total number of particles per unit volume. Writing $(v_0/\omega_0)^2 + x_0^2 = a^2$ and substituting from Eq. (2.154) gives

$$\begin{aligned} W = &- NeE_0\int_0^{2n\pi/\omega_0}\left[-\frac{eE_0}{m\omega_0}\sin 2\omega_0 t - \frac{eE_0 t}{4m}(1+\cos 2\omega_0 t)\right. \\ &\left. - \frac{a\omega_0}{2}\sin\phi + \frac{a\omega_0}{2}\sin(2\omega_0 t + \phi)\right]dt \end{aligned} \tag{2.156}$$

Clearly, the first and last terms of the integrand average to zero over a whole number of cycles. The remaining terms can be integrated to give

$$W = \frac{Ne^2E_0^2}{m}\left(\frac{n^2\pi^2}{2\omega_0^2} + \frac{n\pi ma}{eE_0}\sin\phi\right). \tag{2.157}$$

This work done by the applied field is negative, implying that the oscillating electrons supply energy to the field if $\sin\phi < 0$ and $|\sin\phi| > eE_0 n\pi/2ma\omega_0^2$. This is the condition set by classical theory for stimulated emission to occur. Because the charge on the electron is negative, stimulated emission can only result if the electron velocity when the applied field is turned on is in the direction of the field. If the electron velocity is in the same direction as the field the electron is decelerated by the field and consequently radiates energy. If the electron were accelerated by the field then absorption of energy from the field would occur.

There is a maximum number of cycles of the applied field after which the oscillating electrons start, and continue indefinitely, to absorb energy. This is set by the condition

$$n < 2ma\omega_0^2/eE_0\pi. \tag{2.158}$$

After a long enough time the motion of the electron is dominated by the first term in Eq. (2.151) and can be written

$$x(t) = -\frac{eE_0}{2m\omega_0} t \sin \omega_0 t \tag{2.159}$$

and the electron velocity is, for large enough t

$$\dot{x}(t) \simeq -\frac{eE_0}{2m} t \cos \omega_0 t. \tag{2.160}$$

The electron now has a velocity that is oppositely directed from the applied field, the electron is being accelerated and absorbs energy from the field.

We conclude by saying that when an electric field near resonance is applied to an already oscillating electron, stimulated emission can occur at early times provided the initial velocity of the electron is in the same direction as the field.

2.11 Problems

(2.1) A homogeneously broadened laser amplifier has a small-signal gain at line center of 0.6 m^{-1} and is at line center 3 m long. The saturation intensity of the medium at line center is 3 W m^{-2}. An input signal of intensity 2.5 W m^{-2} enters the amplifier. What is the output intensity if: (a) The input is at line center and gain saturation is neglected? (b) The input is one linewidth (FWHM) from line center and gain saturation is neglected? (c) As in (a) but gain saturation is included, (d) As in (b) but gain saturation is included.
Hint: (c) and (d) need to be solved numerically.

(2.2) A laser amplifier contains two groups of particles centered at frequencies $\nu_0 - a/2$, $\nu_0 + a/2$. Each group is individually homogenously broadened with homogeneous FWHM $\Delta\nu_L$. (a) At what two frequencies $\nu_0 \pm b$ is the gain of the amplifier a maximum? Consider all three cases: (i) $a \gg \Delta\nu_L$, (ii) $a \sim \Delta\nu_L$, (iii) $a \ll \Delta\nu_L$. (b) What is the relative gain at frequency ν_0 as a function of $a/\Delta\nu_L$?

(2.3) What is the overall lineshape function if

$$g_D(\nu_0, \nu) = \begin{cases} 1/\Delta\nu_D & \text{for } \nu_0 - \Delta\nu_D/2 \le \nu \le \nu_0 + \Delta\nu_D/2 \\ 0, & \text{elsewhere} \end{cases}$$

$$g_L(\nu', \nu'') = \frac{(2/\pi\Delta\nu_L)}{1 + \left[2(\nu' - \nu'')/\Delta\nu_L\right]^2} ?$$

(2.4) What is the stimulated emission rate if: $A_{21} = 10^8$ s^{-1}, the line is homogeneously broadened with $\Delta\nu = 1$ GHz, the input radiation is 'white' with $\rho(\nu) = 1$ J m^{-3} Hz^{-1}? What would be the stimulated emission rate if the input radiation were monochromatic, one FWHM from line center, and had intensity 1 W m^{-2}? Take the line center frequency to be 10^{14} Hz.

(2.5) A homogeneously broadened amplifier with FWHM=1 GHz and a small-signal gain at line center of 1 m^{-1} has a saturation intensity at line center of 1 W m^{-2}. An input signal of intensity 2 W m^{-2} enters the amplifier, which is 0.5 m long. Calculate: (a) The output intensity neglecting saturation if the input is at 500 MHz above line center. (b) The output intensity including gain saturation if the input signal is 500 MHz above line center. (c) The output intensity including gain saturation if the input signal is at line center (d) The output intensity 500 MHz above line center if an additional 1 W m^{-2} signal is also injected 500 MHz below line center.

(2.6) What is the saturation intensity one FWHM from line center in a naturally broadened amplifier if the amplifier has the following parameters: A_{21}=10^8 s^{-1}, τ_1=1 ns, τ_2=5 ns, λ_0=1 μm.

(2.7) The lifetime of a particular excited argon level varies with pressure as $\tau = 10^{-8}/(1+P)$ s at 2000 K, where P is pressure measured in atmospheres. At what pressure will the collisional broadening be as great as the Doppler broadening? Neglect the effect of the lower level lifetime. Take $\lambda_0 = 488$ nm.

(2.8) A homogeneously broadened gas laser amplifier with $\Delta\nu = 1$ GHz working at 300 nm has a small-signal gain coefficient at line center of 1 m^{-1}. Its susceptibility as a function of frequency varies as

$$\chi''(\nu) \propto \frac{1}{1 + [2(\nu - \nu_0)/\Delta\nu)]^2}$$

and

$$\chi'(\nu) = \frac{2(\nu - \nu_0)}{\Delta\nu}\chi''(\nu).$$

Both χ' and $\chi'' \ll 1$. The refractive index of the gas is $n = 1$. Calculate the slowest phase velocity that will be observed for waves travelling through the amplifier.

(2.9) A gaseous medium has a Lorentzian broadened lineshape of FWHM 1 GHz. Other parameters of the medium are: $N_2 = 5 \times 10^{16}$ m^{-3}, $N_1 = 5 \times 10^{15}$ m^{-3}, $g_2 = g_1$, $A_{21} = 10^8$ s^{-1}, $\tau_2 = 5$ ns, $\tau_1 = 1$ ns, $\lambda_0 = 1$ μm.

Calculate: (i) $I_S(\nu_0)$, (ii) $\chi = \chi' - i\chi''$, (iii) the change in phase velocity of a weak signal propagating 1 FWHM from line center relative to c_0, and (iv) what would happen if $\tau_1 = 1$ μs and the medium was pumped from $N_2 = 0$, $N_1 = 0$ at time 0 at a rate $R_2 = 10^{20}$ m^{-3} s^{-1}, $R_1 = 0$?

(2.10) How is the result given in Eq. (2.126) modified if $|\chi(\omega)|$ is not much smaller than ϵ_r? Derive new expressions for the phase velocity and gain coefficient.

(2.11) Write a computer program to plot the variation in phase velocity as a function of gain one FWHM on the low frequency side of line center as the gain increases from 0 to $1/\lambda$ m^{-1} for a medium with $\epsilon_r = 1.5$, and $\lambda = 1\mu m$.

References

[2.1] R.H. Dicke and J.P. Wittke, *Introduction to Quantum Mechanics*, Addison-Wesley, Reading, MA, 1960.

[2.2] L. Liboff, *Introductory Quantum Mechanics*, Holden-Day, Inc., San Francisco, 1980.

[2.3] R.L. White, *Basic Quantum Mechanics*, McGraw-Hill, New York, 1966.

[2.4] Sir J.H. Jeans, *An Introduction to the Kinetic Theory of Gases*, Cambridge University Press, Cambruidge, 1940.

[2.5] R.D. Present, *Kinetic Theory of Gases*, McGraw-Hill, New York, 1958.
[2.6] D.C. Champeney, *Fourier Transforms and Their Physical Applications*, Academic Press, London, 1973.
[2.7] A.C.G. Mitchell and M.W. Zemansky, *Resonance Radiation and Excited Atoms*, Cambridge University Press, Cambridge, 1971.
[2.8] P.W. Milonni and J.H. Eberly, *Lasers*, Wiley, New York, 1988.
[2.9] M. Abramowitz and I.A. Stegun, *Handbook of Mathematical Functions*, Dover, New York, 1968.
[2.10] W.R. Bennett, Jr., 'Hole-burning effects in a He–Ne optical maser,' *Phys. Rev.*, **126**, 580–593, 1962.

3

Introduction to Two Practical Laser Systems

3.1 Introduction

To give a little more practical emphasis to some of the ideas we have dealt with so far, let us consider some of the details of two real systems where population inversion and laser oscillation can be obtained. One of these lasers uses an amplifying medium that is a crystalline solid, the other a gas. In each case, the amplifying medium is pumped into a state of population inversion by feeding energy into it in an appropriate way. Laser oscillation occurs when the amplifying medium is placed between a pair of suitable aligned mirrors that provide the necessary optical feedback to cause oscillation to occur. The first of these two systems is the ruby laser. This was the first laser, originally made to work in July 1960 by Theodore Maiman of the Hughes Aircraft Company in Malibu, California.

3.1.1 The Ruby Laser

The amplifying medium of this laser is crystalline aluminum oxide (sapphire) doped with Cr^{3+} ions, typically in concentrations of about 0.05% by weight. The crystal structure of the undoped sapphire is shown in Fig. (3.1). When the crystal is doped with Cr^{3+} ions some of the aluminum sites become occupied by chromium. It is these Cr^{3+} ions that are the active particles in the amplification process. A schematic energy level diagram for the Cr^{3+} ions in such material is shown in Fig. (3.2). Population inversion is produced by irradiating such a ruby crystal with light that causes absorption transitions from the ground state into the broad absorption bands labelled $^4F_1$ and $^4F_2$. The absorption spectrum of a typical ruby laser crystal is shown in Fig. (3.3), which clearly shows these two broad absorption bands, one in the violet–blue and the other in the green–yellow region, that lead to the ^4F levels. These absorption bands are very broad because the very many sub-levels of the ^4F levels of slightly different energy are "smeared" together by line broadening. The sharper absorption peak due to the laser transition itself is also visible as a small peak around 700 nm in Fig. (3.3). Fig. (3.4) shows a more detailed absorption spectrum of the laser transition that illustrates that the laser transition consists of two closely spaced lines, labelled R_2 and R_1. In both Figs. (3.3) and (3.4) the amount of absorption is seen to depend on the relative linear polarization of the incident light and the orientation of the crystal.†

† A fuller discussion of the effect of crystal symmetry on the propagation of light is given in Chapter 18.

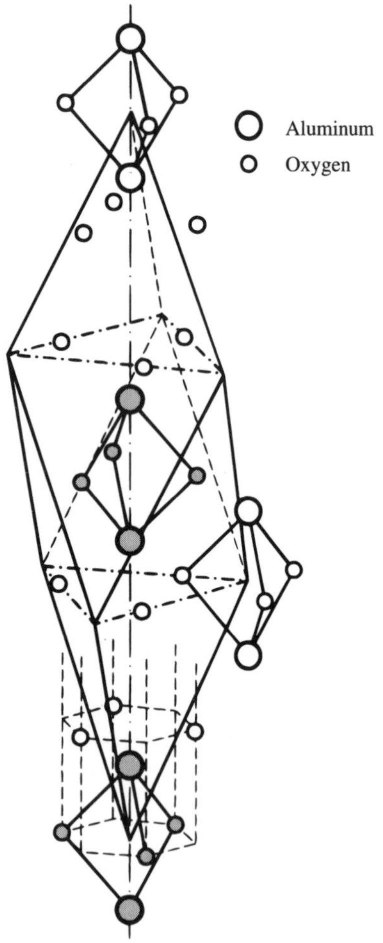

Fig. 3.1. Crystal structure of sapphire: α-Al_2O_3 (aluminum oxide). The shaded atoms make up a unit cell of the structure. The aluminum atom inside the dashed hexagonal prism experiences an almost cubic field symmetry from the oxygen atoms on the prism[3.1].

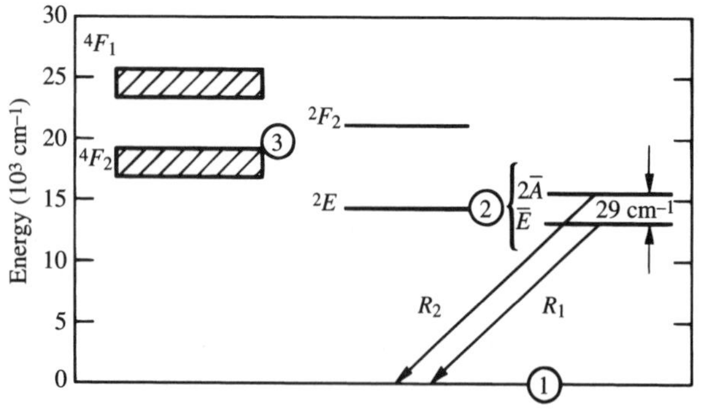

Fig. 3.2. Schematic energy level diagram for ruby – Cr^{3+} ions in sapphire[3.2].

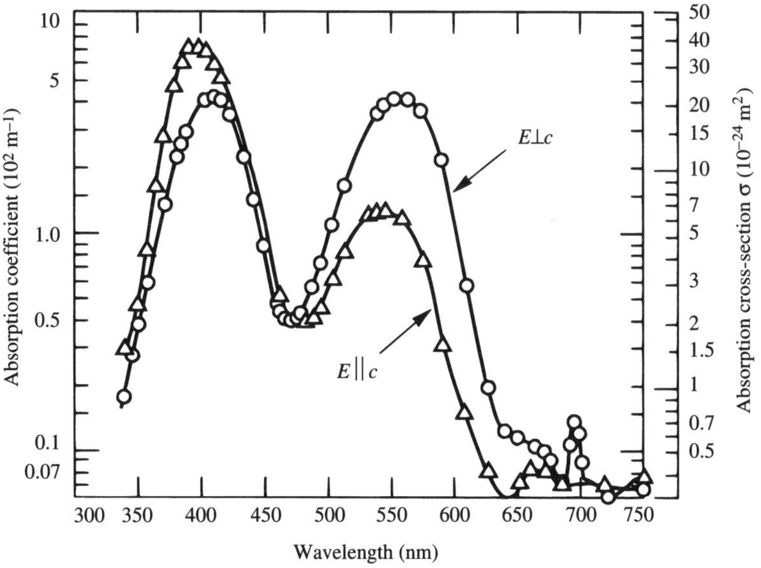

Fig. 3.3. Absorption coefficient and absorption cross-section as a function of wavelength for pink ruby. These absorption spectra are slightly different depending on whether the incident polarized light being absorbed is linearly polarized with its electric vector parallel, or perpendicular, to the c symmetry axis of the crystal[3.3].

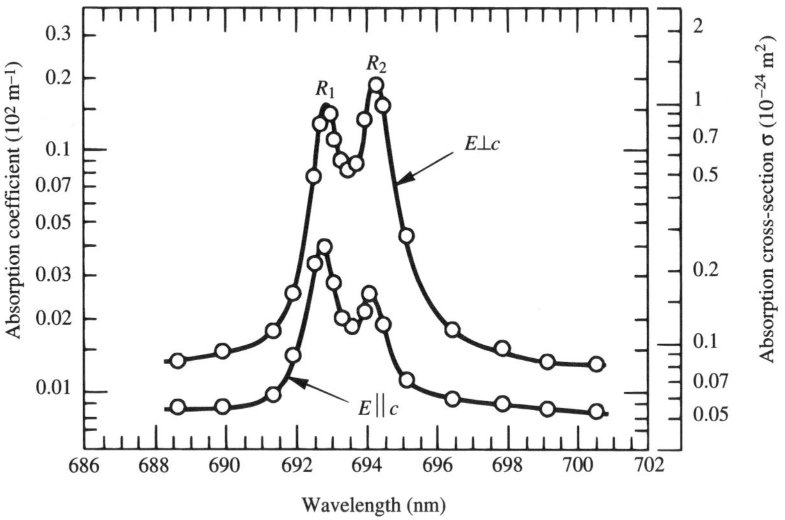

Fig. 3.4. Detailed absorption spectrum of pink ruby in the 686–702 nm region showing the absorption peaks corresponding to the R_1 and R_2 components of the ruby laser transition[3.3].

The $^4F_1$ and $^4F_2$ levels decay (with lifetimes $\sim$ 50 ns) preferentially into the upper laser level ^2E. This level actually consists of a pair of closely spaced levels separated by 29 cm^{-1}.‡

‡The unit cm^{-1} is frequently used as a unit of energy. If two energy levels are 1 cm^{-1} apart, this implies that the wavelength of the transition between them is 1 cm, and its frequency is 3×10^{10} Hz i.e.,

$$\Delta E = h\nu = \frac{hc}{\lambda} = 3 \times 10^{10} h.$$

If two levels are k cm^{-1} apart, then

$$\Delta E = \frac{hc}{\lambda} = \frac{hc}{1/k} = (k \times 3 \times 10^{10})h.$$

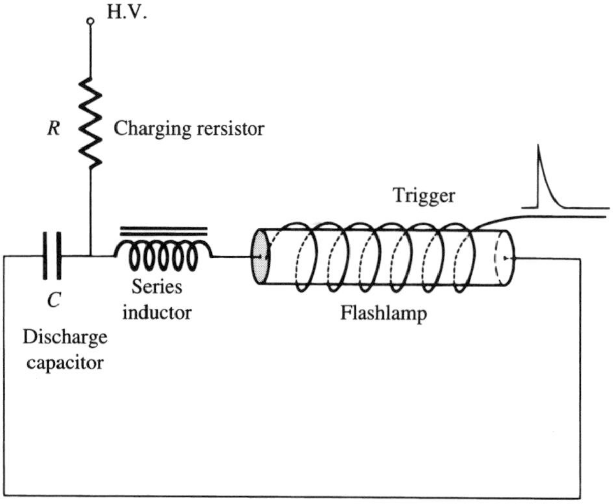

Fig. 3.5. Simple electrical circuit for driving a flashlamp.

The splitting of this energy level arises because of the symmetry of the crystal lattice. The Cr^{3+} ions in the aluminum oxide lattice have energy levels which are affected by the spatial arrangement of aluminum and oxygen ions around them. If this arrangement had exact octahedral symmetry around each Cr^{3+} ion no splitting would occur. In practice the symmetry of the arrangement is rhombohedral and the small energy splitting results.

In practice, to produce a state of population inversion in the crystal, it is necessary to irradiate the crystal with a very intense light source. This is most easily accomplished with a flashlamp, which uses a tube, generally made of quartz and filled with a noble gas, through which energy stored in a capacitor charged to high voltage is discharged. Such a lamp can, for the short period of time during which current flows through it, produce a very high light intensity. A simple electrical circuit for the operation of such a lamp is shown in Fig. (3.5). The spectral characteristics of the lamp will depend on a number of parameters: for example, capacitor size, voltage, tube diameter, the filling gas and its pressure. Generally if the discharge is of short duration $\lesssim$ 10 μs and high energy, the hot gas in the lamp will emit approximately as a black body, perhaps at an effective temperature of 20,000 K, where much of its emission would be in the blue and ultraviolet. For ruby laser excitation the emission of the lamp is optimized in the region of the crystal's absorption bands by using slower flashlamp pulses (100 μs–10 ms) and appropriate gas mixtures. If the flashlamp intensity is great enough, then sufficient ground state $^4A_2$ C_r^{3+} ions can be excited to the $2\overline{A}$ and $\overline{E}$ levels via the broad absorption bands $^4F_2$ and $^4F_1$ that a population inversion exists between the $2\overline{A}$ and $\overline{E}$ levels and the split ground state. An efficient transfer occurs from the broad absorption bands to these upper laser levels because the radiative decay from these bands to the ground state has a lifetime of about 3 μs, against a value of about 50 ns into the upper laser levels. Consequently, about 99% of the ions that reach these bands transfer into the upper laser levels. These levels have an overall lifetime $\tau_2 \simeq$ 3 ms and lose their energy almost exclusively by spontaneous emission to the ground state so that $A_{21} \simeq 1/\tau_2$. Because these levels are relatively long lived, if the exciting flash is shorter than

Fig. 3.6. Schematic energy level diagram of three- and four-level lasers.

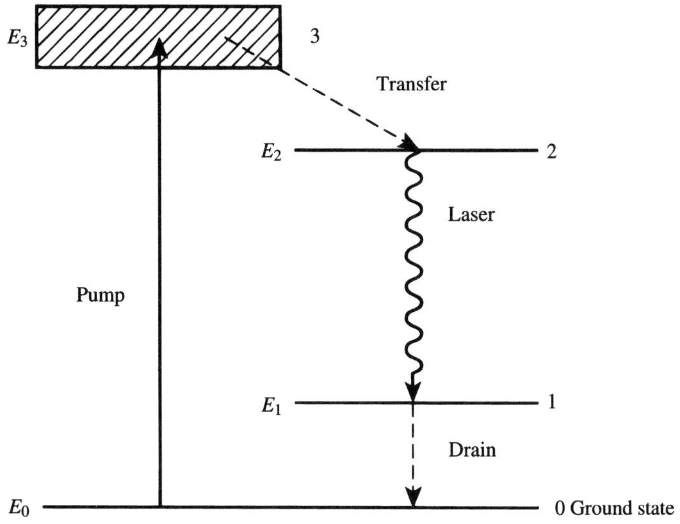

their overall lifetime then most of the ions transferred into them will still be there when the flash is finished. The main laser transition that occurs is R_1, between level $\overline{E}$ and the ground state, its wavelength is 694.3 nm (a visible dark red color). Laser oscillation can also be obtained on the transition R_2, but only by taking special measures to reduce the gain of R_1, which otherwise operates preferentially.

Because of the arrangement of its energy levels and the way it is excited, the ruby laser falls into the class of what are called three-level lasers. Most efficient lasers fall into the class of four-level lasers (two-level lasers are also possible). The difference between three- and four-level systems is best illustrated with reference to the schematic energy level diagram in Fig. (3.6). If $E_1 \ll kT$ the system is a three-level laser system, if $E_1 \gg kT$ it is a four-level system. Since in the ruby laser $E_1 \sim 0$, it is a three-level laser. We shall see later that it is intrinsically more difficult to obtain population inversion in a three- rather than a four-level system. With this fact in mind it is perhaps surprising that the first operational laser was of the three-level type.

Typical ruby lasers employ cylindrical crystals 1–20 mm in diameter, 20–200 mm in length. Usually the resonant cavity consists of the parallel polished faces of the crystal, one of which is coated to make it 100% reflecting. The Fresnel reflection from the other end is often sufficient to provide the optical feedback necessary for oscillation, although it may be made to have a higher reflection coefficient by giving it a suitable coating. Theodore Maiman's original ruby laser employed a very small crystal surrounded by a helical flashlamp, the whole being contained inside a coaxial cylindrical metal reflector, as shown in Fig. (3.7)[3.4]. In this arrangement the optical pumping of the laser is axisymmetric, which offers certain advantages in many applications, since the population inversion produced in the crystal is then itself axisymmetric. Prior to obtaining laser oscillation in the ruby system, Maiman and his colleagues had measured the absorption spectrum and the relative decay rate of the important energy levels[3.3]. It was these measurements

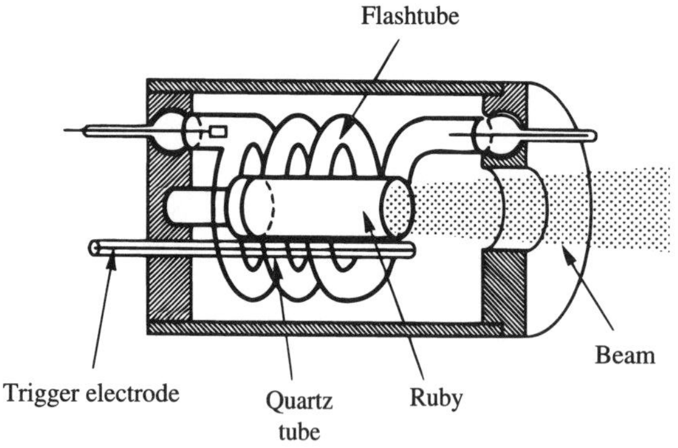

Fig. 3.7. Schematic arrangement of Maiman's original ruby laser.

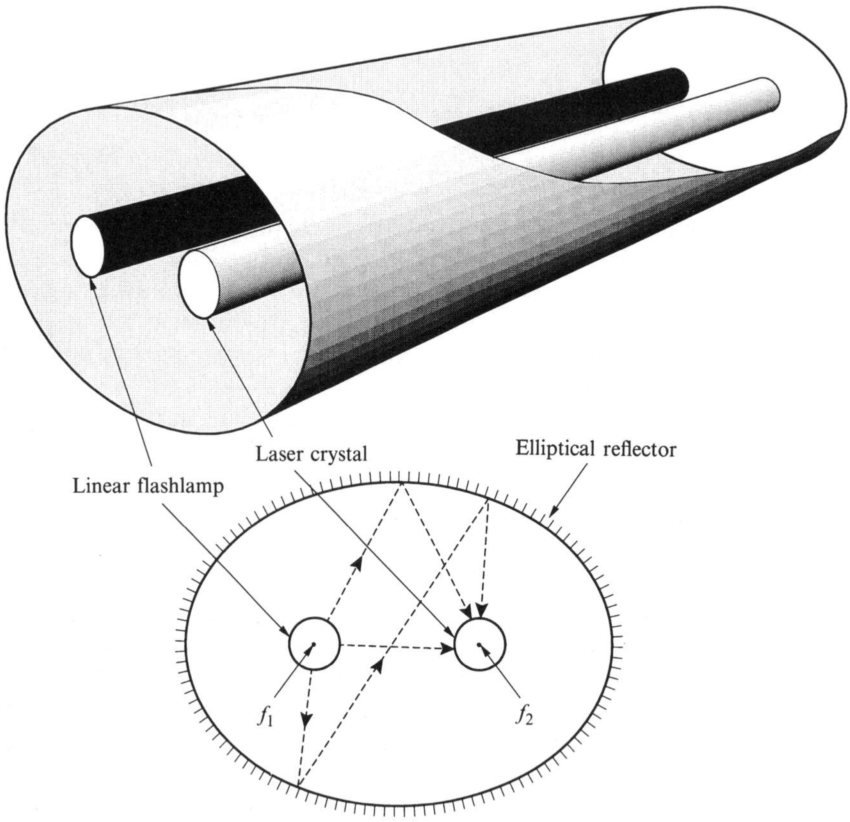

Fig. 3.8. Elliptical reflector arrangement for optical pumping of a laser crystal by a linear flashlamp.

which made them believe that it would be possible to obtain population inversion in the ruby system.†

Although helical flashlamp excitation is still sometimes used, other methods of optical pumping are more common. The most usual arrangement employs a cylindrical flashlamp and crystal placed along the two focal lines of an elliptical reflector, as shown in Fig. (3.8). It is a property of such an elliptical reflector that

† Surprisingly enough, when the first observations of laser action were made and Maiman submitted the article describing the results to *Physical Review Letters*, his article was rejected. Consequently, the first article describing successful operation of a laser appeared in the periodical *Nature*.

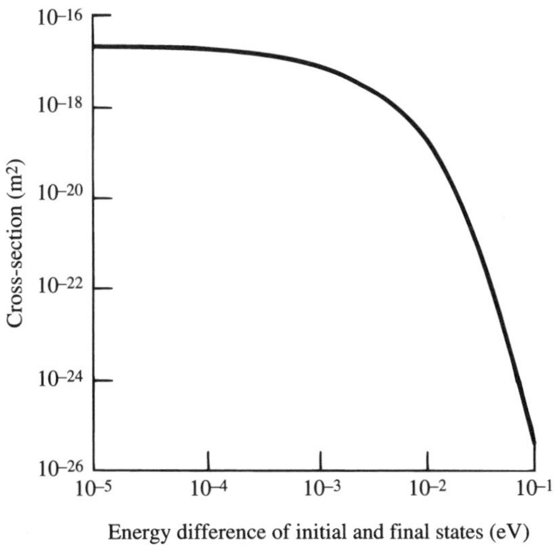

Fig. 3.9. Calculated variation of energy transfer cross-section for a collision between two atomic species as a function of the energy discrepancy ΔE_∞[3.3]. The probability of excitation transfer is linearly dependent on the cross-section.

rays of light from one focus, after reflection, pass through the other focus. Thus, light from the flashlamp is efficiently reflected towards the laser crystal.

3.2 The Helium–Neon Laser

The helium–neon laser was the first gas laser to be operated, and also the first CW (continuous wave) laser. The production of a population inversion in this system occurs as a result of energy transfer processes between metastable helium atoms and ground state neon atoms, which are as a result excited to various upper laser levels.

In a collisional excitation transfer process of the kind

$$A + B^* \overset{\leftarrow}{\rightarrow} A^* + B \pm \Delta E_\infty,$$

where the colliding particles are not molecular and an asterisk indicates that the particle is in an excited state, the probability of the reaction occurring becomes extremely small if the energy discrepancy ΔE_∞ of the reaction exceeds kT[3.5], as shown in Fig. (3.9). ΔE_∞ is the energy difference between the excited states A^* and B^* at large (theoretically infinite) separation of the reacting particles (ΔE is a function of the separation of the two atoms in an A^* and a B state respectively because of forces between them; the effect of this force goes to zero as the two atoms become infinitely far apart).

If either of the colliding particles is molecular, then the probability of reaction can remain large even if ΔE_∞ exceeds kT, since one or the other of the particles may be able to release or store energy in internal degrees of freedom, such as vibration.

In the helium–neon laser the excitation transfer processes occur in a glow discharge in a helium–neon mixture at a total pressure of about 1 torr.† The mixture is usually in the range 5:1–10:1 helium to neon. The discharge is produced

† The torr, named for the seventeeth century Italian mathematician Torricelli, who invented the mercury manometer, is a unit of pressure equal to 1 mm of mercury. Standard atmospheric pressure is 760 torr. 1 torr $\equiv$ 133 Pascal (1 Pa $\equiv$ 1 N m^{-2}).

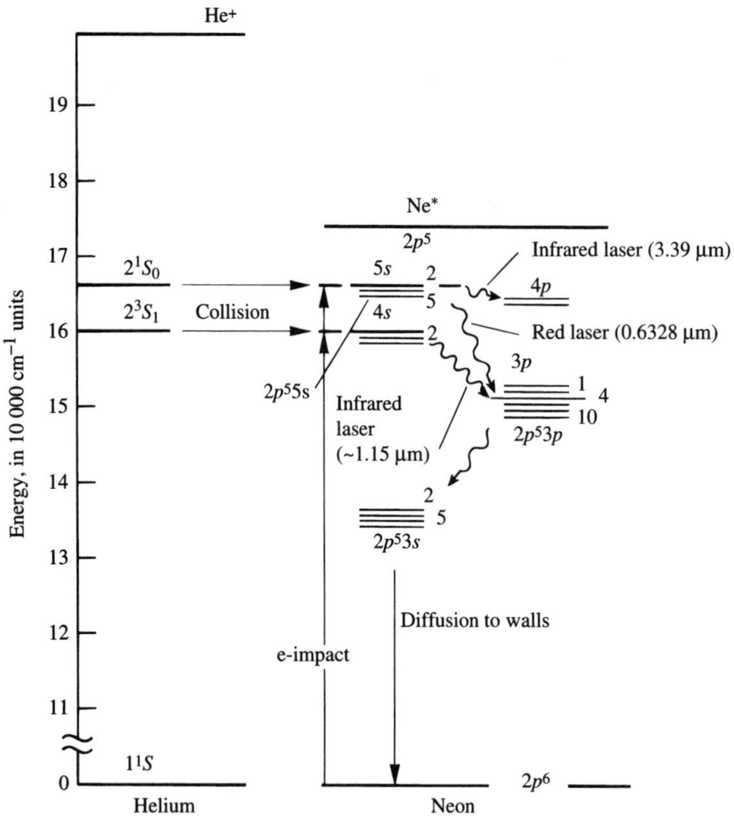

Fig. 3.10. Energy level diagram for the helium–neon laser.

either by DC or RF excitation in tubes usually of 1–20 mm inside diameter, and carries a current density from 0.1 to 1 A cm^{-2}. By various collisional and radiative processes electrons excite ground state helium atoms to the metastable 2^1S_0 and 2^3S_1 levels, as shown in Fig. (3.10)

$$\text{He}(^1S_0 = \text{ground state}) + \text{e} \rightarrow \text{He}(2^1S_0 \text{ or } 2^3S_1)$$

or

$$\text{He}(^1S_0) + \text{e} \rightarrow \text{He}^*(\text{higher nonmetastable state})$$
$$\text{He}^* \rightarrow \text{He}(2^1S_0)(\text{or } 2^3S_1) + h\nu.$$

As the 2^1S_0 and 2^3S_1 states are metastable they do not lose energy readily by undergoing spontaneous emission (the radiative lifetime of the 2^3S_1 state is well in excess of one hour). However, these metastable states can lose energy in collisions, and in collisions with ground state neon atoms high probabilities exist for energy transfer in reactions such as

$$\text{He}(2^1S_0) + \text{Ne}(^1S_0) \rightarrow \text{He}(^1S_0) + \text{Ne}(2p^55s) + \Delta E_1,$$
$$\text{He}(2^3S_1) + \text{Ne}(^1S_0) \rightarrow \text{He}(^1S_0) + \text{Ne}(2p^54s) + \Delta E_2.$$

The energy discrepancies ΔE_1, ΔE_2 are small, ΔE_1 is $\sim$ 400 cm^{-1}, comparable with, or less than kT for the hot atoms in the discharge. The neon 4s and 5s levels that are excited in this way, under appropriate conditions, exhibit population inversion with respect to the 3p and 4p levels. The existence of such a population inversion

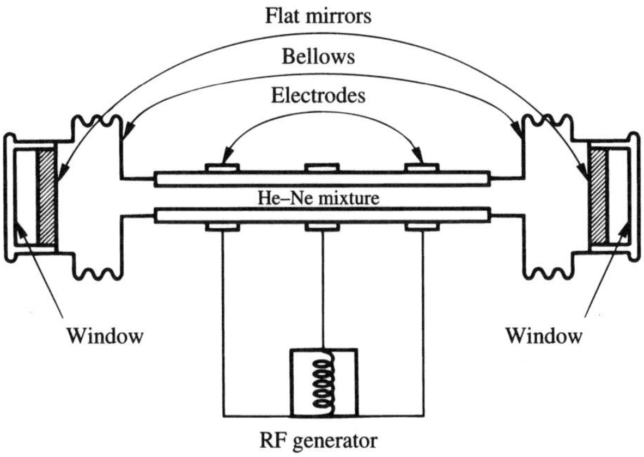

Fig. 3.11. Schematic arrangement of the first gas laser.

does, however, require that the lower laser levels not be efficiently populated. This could occur either because neon atoms are excited by electron collisions or because population builds up in them if they become long lived.

One interesting laser transition that results because of population inversion is between one of the 4*s* and one of the 3*p* levels and occurs at 1.15 μm. This transition (in the near-infrared and consequently invisible to the human eye) was the first transition to be observed in the He–Ne laser. When Javan, Bennett, and Herriott first obtained oscillation on the 1.15 μm transition in 1961[3.6], their successs was essentially a result of investigating a system where, theoretically, solid evidence existed for the possible obtainment of a population inversion. Their laser, the first CW and gas laser is shown in Fig. (3.11). Many of the features of this system are of interest, particularly since most of the technological features of the device have been superseded in more recent He–Ne gas lasers. The glow discharge was produced in this first gas laser by coupling RF power to the tube with external electrodes: in most modern CW gas lasers direct DC input to the gas is preferred. The laser mirrors situated at each end of the amplifier tube were plane, and were fixed inside metal assemblies to allow them to be aligned parallel while retaining the vacuum integrity of the structure. Nowadays, Brewster angle window assemblies allow the laser mirrors to be situated outside the tube. Such windows allow a plane-polarized laser oscillation to occur without the Fresnel reflection loss normally associated with in-cavity windows. It also turns out that plane parallel mirror resonators are extremely sensitive to misalignment (misalignment of a few seconds of arc will prevent oscillation being obtained). In order to be sure that their laser mirrors would be aligned, Javan and his coworkers fixed one mirror to an electromechanical device that wobbled the mirror through a range of angular positions near the expected position where parallelism with the other mirror would occur. By using an image converter tube, which made the infrared laser beam visible, they were able to detect the burst of oscillation as the laser mirror passed through the alignment position. It was subsequently discovered that the use of spherical mirrors of appropriate radius of curvature made the alignment tolerances of the cavity much less severe (oscillation could be sustained over perhaps a $\frac{1}{2}$ degree range near the coaxial position). A schematic diagram of a gas laser system with DC excitation, Brewster windows and external spherical mirrors is shown in Fig. (3.12).

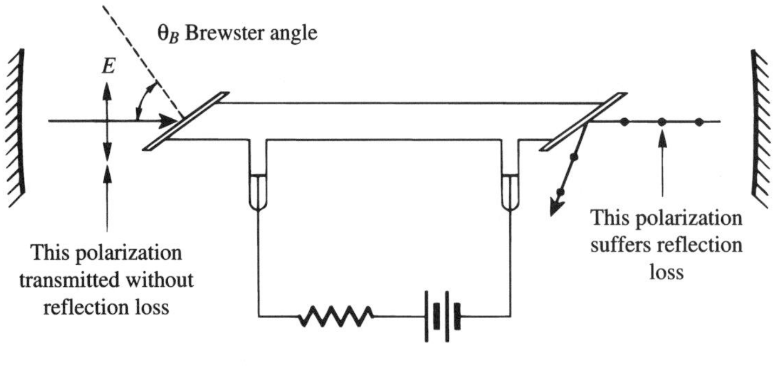

Fig. 3.12. Typical schematic design of a modern gas laser.

The 1.15 μm transition first observed in the helium–neon laser is now mainly of historical importance. Many other laser transitions between the (5*s*, 4*s*) and (4*p*, 3*p*), levels have been observed since, the most important of which are between the 5*s* and 3*p* and the 5*s* and 4*p* levels. Two of these transitions are particularly noteworthy, those at 632.8 nm and 3.39 μm. The red laser transition at 632.8 nm is the one most familiar to anyone who has ever seen one of the many thousands of commercial helium–neon lasers which are in widespread use. Population inversion on this transition results from the energy transfer pumping from helium 2^1S metastables. The lower level of the transition has a radiative lifetime of about 10 ns, about a factor of ten times shorter than that of the upper level. The desirability of such a favorable lifetime ratio in a laser system will become more apparent later. It can be seen from Fig. (3.10) that the lower level of the 632.8 nm transition decays to one of the 3*s* levels. This level is metastable and loses its energy mostly in collisions with the discharge tube walls. If the population of this level rises too high, repopulation of the lower laser level from it may destroy the population inversion. This repopulation may result from electron collisional excitation of the 3*p* levels from the 3*s* levels, or by the reabsorption of photons emitted by atoms in the lower laser level (this latter phenomenon is called *radiation trapping*[3.7],[3.8]). If the rate of collisional deactivation of the 3*s* levels is increased, for example, by reducing the diameter of the laser tube, then the gain of the 632.8 nm transition (and also the 1.15 μm transition) is increased. The gain increases inversely with tube diameter, but cannot be increased too far by this means or the available volume of inverted gas falls, diffraction losses increase, and the laser's power output decreases.

The 3.39 μm laser transition shares a common upper level with the 632.8 μm transition but has a very much higher small-signal gain (as high as 50 dB m^{-1}). This high gain is partly due to a very favorable lifetime ratio for the 3.39 μm transition and also to its longer wavelength. In gas lasers of this kind the gain is proportional to at least λ^2 (λ^3 if the transition is predominantly Doppler-broadened). Because of its higher gain the 3.39 μm transition has a lower oscillation threshold under conditions of comparable feedback than the 632.8 nm transition. Under such conditions it will oscillate preferentially and prevent oscillation at 632.8 nm. This occurs because it builds up in oscillation first and "clamps" the population inversion (because of gain saturation) at a threshold level for its oscillation. This will be discussed in more detail in Chapter 5. The population inversion may be clamped at a level too low for laser oscillation at 632.8 nm. To prevent this occurring, the oscillation threshold at 3.39 μm must be raised. In short lasers with quartz or glass windows these windows introduce sufficient loss for this to occur, also the resonator mirrors will generally have lower reflectivity at 3.39 μm than at 632.8 nm. In long helium–neon lasers ($\gtrsim$ 1 m) the loss introduced by the windows and mirrors may not prevent parasitic oscillation at 3.39 μm. There are two common ways to lower the gain at 3.39 μm in such lasers. An in-cavity methane cell (which absorbs 3.39 μm strongly but not 632.8 nm) can be used or a spatially inhomogeneous magnetic field can be applied to the discharge tube. Such a field introduces a Zeeman splitting† of the 3.39 μm laser transition that is different at different points along the discharge: this effectively increases the linewidth of the laser transition and reduces its gain. The effect of the magnetic field is much smaller at 632.8 nm because its Doppler-broadened linewidth is already large. The spatially inhomogeneous magnetic field can be supplied by a series of bar magnets placed outside the tube and spaced along its length.

References

[3.1] V. Evtuhov and J.K. Neeland, 'Pulsed ruby lasers,' in *Laser: A Series of Advances*, Vol. I, A.K. Levine, Ed., Marcel Dekker, New York, 1966.

[3.2] T.H. Maiman, R.H. Hoskins, I.J. D'Haenens, C.K. Asawa, and V. Evtuhov, 'Stimulated emission in fluorescent solids II spectroscopy and stimulated emission in ruby,' *Phys. Rev.* **123**, 1151–1157, 1961.

[3.3] D.C. Cronemeyer, 'Optical absorption characteristics of pink ruby,' *J. Opt. Soc. Am.* **56**, 1703–1706, 1966.

[3.4] T.H. Maiman, 'Stimulated optical radiation in ruby,' *Nature*, **187**, 493–494, 1960.

[3.5] N.F. Mott and H.S.W. Massey, *The Theory of Atomic Collisions*, Oxford University Press, Oxford, 1965.

[3.6] A. Javan, W.R. Bennett, Jr., and D.R. Herriott, 'Population inversion and continuous optical maser oscillation in a gas discharge containing a He-Ne mixture,' *Phys. Rev. Lett.*, **6**, 106–110, 1961.

[3.7] T. Holstein, 'Imprisonment of resonance radiation in gases,' *Phys. Rev.*, **72**, 1212–1233, 1947.

[3.8] T. Holstein, 'Imprisonment of resonance radiation in gases II,' *Phys. Rev.*, **83**, 1159–1168, 1951.

† An effect where a magnetic field separates an energy level into a series of closely-spaced sub-levels of slightly different energy. The transition changes from a single transition at a unique center frequency to a group of closely-spaced transitions at center frequencies that depend on the strength of the applied magnetic field.

4

Passive Optical Resonators

4.1 Introduction

In this chapter we shall examine the passive properties of optical resonators consisting of two plane-parallel, flat mirrors placed a distance apart. At first, the properties of standing electromagnetic waves in such a system and the way in which their stored energy is lost if the mirrors are not totally reflecting will be considered. Then we shall analyze an optical device, called the Fabry–Perot etalon or interferometer, which represents the archetypal passive resonant structure that is used in a laser. We shall see that this device has a series of equally spaced resonant frequencies and in transmission acts as a *comb* filter. The filter properties of this device allow it to be used as a high resolution instrument for analysis of the spectral content of light.

4.2 Preliminary Consideration of Optical Resonators

Before considering some of the fine detail involved in the process of laser oscillation, it will be useful to consider some preliminary aspects of the optical resonators used to provide the feedback in laser systems. Oscillation in these devices occurs because the amplifying medium is placed between suitable aligned mirrors: usually just two conormal facing mirrors are used. The passive properties of this pair of mirrors, the *optical resonator* or *cavity* of the laser, affect the way in which the oscillation occurs. The resonator has resonance frequencies of its own that interact with the resonance (line center) frequency of the amplifying medium and control the output oscillation frequency of the laser. Before exploring this further, let us consider at what frequency a laser would oscillate if the resonator did not interact with the gain profile of the amplifying medium in any way. Suppose the amplifying medium has a gain profile (gain/frequency response) of a Gaussian form, as shown in Fig. (4.1). Such a gain profile occurs in a gaseous amplifying medium where the individual homogeneous lineshapes of the atoms are significantly narrower than the overall Doppler width of the spontaneous transition.

The maximum gain of the medium is at frequency ν_0, the line center, so it is perhaps logical to expect that oscillation will buildup at this frequency rather than at any other. If we view the build-up of oscillation as a process triggered by spontaneous emission we can see why this is so. A photon travelling in a direction that keeps it bouncing back and forth within the resonator is more likely to be emitted in a narrow band of frequencies $\Delta\nu_0$ near ν_0 than in some other band $\Delta\nu_1$ at frequency ν_1. As oscillation builds up, one can imagine photons spontaneously emitted at all points of the lineshape being amplified to some extent, but oscillation at ν_0 builds up fastest. As its intensity grows it depletes the atomic population

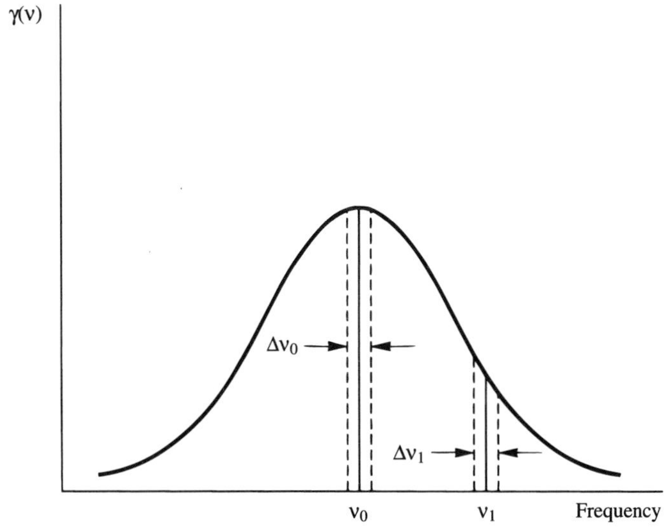

Fig. 4.1. Build-up of laser oscillation at two different frequencies in a noninteractive laser cavity.

by causing sufficient stimulated emission that the medium ceases to be amplifying at frequencies near ν_0 (within a few homogeneous widths, say). If the medium has a homogeneous (Lorentzian) gain profile, then since photons oscillating at ν_0 can stimulate emission from all the atoms in the medium, it is easy to see that oscillation at frequency ν_0 can suppress oscillation at any other frequency under the gain profile. The possibility of additional oscillation at frequencies far away from ν_0 in an inhomogeneously broadened gaseous amplifier is not precluded by this discussion, this in fact often happens, as we shall see later.

The monochromatic character of the oscillation can be predicted by a simple consideration of the shape of the gain profile of the amplifier. In the early stages of oscillation, photons with a frequency distribution $g(\nu_0, \nu)$ (the total lineshape) are being amplified in a material whose gain/frequency response is $\gamma(\nu)$ (proportional to $g(\nu_0, \nu)$). The amplification process changes the lineshape of the emitted photons circulating in the cavity by a process that is dependent on the product of $g(\nu_0, \nu)$ and $\gamma(\nu)$, that is on $[g(\nu_0, \nu)]^2$. The resulting profile of the laser radiation is dependent on higher powers of $[g(\nu_0, \nu)]^2$ as the oscillation is dependent on many passes of photons back and forth through the amplifying medium. For Gaussian lineshapes like $e^{-[2(\nu-\nu_0)/\Delta\nu_D]^2 \ln 2}$, which is like e^{-x^2/σ^2}, the product of two lineshapes produces a narrower profile, for example

$$[e^{-x^2/\sigma^2}]^2 = e^{-2x^2/\sigma^2} = e^{-x^2/(\sigma/\sqrt{2})^2}, \tag{4.1}$$

a function that has a width $1\sqrt{2}$ less than the original. The same can also be shown to be true for Lorentzian profiles. In both cases the gain of the medium causes a narrowing of the original spontaneously emitted lineshape. Thus, we can see that in a noninteractive laser cavity, the laser oscillation will be highly monochromatic and at the line center.

Before considering the interaction that occurs when an amplifying medium is placed inside a resonator, let us review certain passive aspects of the resonator itself. Although, as we shall see later, most practical laser systems use spherical mirrors it is easier to treat the simpler plane-parallel case. The purpose of the

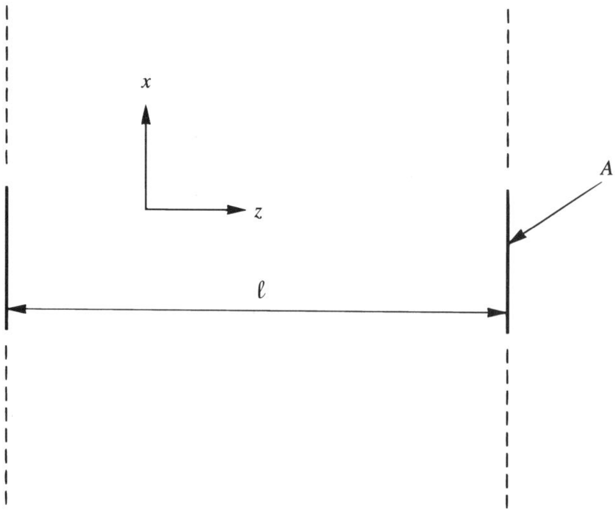

Fig. 4.2. Resonant system consisting of two perfectly-conducting, parallel, infinite planes.

resonator is to provide the feedback necessary to cause oscillation. The oscillation occurs as a result of spontaneous emission into those modes that keep radiation within the resonator after multiple reflections between the cavity mirrors. Emission into other modes escapes from the resonator. There are very many such modes: for example, at a wavelength of 1 μm the number of modes in 1 m^3 that lie in a frequency range of the order of a typical spontaneous emission linewidth, 10^9 Hz, (say) is, from Eq. (1.34)

$$\frac{8\pi\nu^2}{c^3}\Delta\nu = \frac{8\pi \times (3 \times 10^{14})^2 \times 10^9}{(3 \times 10^8)^3} \sim 8 \times 10^{13}\text{modes.}$$

Most of these modes will not lie on, or near, the normals to the resonator mirrors and will not undergo feedback. In common with a conventional electronic resonant circuit, the optical resonator has a quality factor Q that varies from one spontaneous emission mode to another. Those spontaneous emission modes that lie perpendicular, or nearly so, to the parallel resonator mirror surfaces have the highest Q. The Q is defined by the relation

$$Q = \frac{2\pi\nu_0 U}{P}, \tag{4.2}$$

where ν_0 is the resonant frequency under consideration, U is the field energy stored in the resonator and P is the power dissipated by the resonator.

4.3 Calculation of the Energy Stored in an Optical Resonator

Consider the case of a standing electromagnetic wave between two perfectly conducting infinite planes of separation ℓ, as shown in Fig. (4.2). Such a wave corresponds to a mode of this resonant system that lies normal to its end planes (reflectors). Any energy stored in such a mode suffers no losses and remains stored indefinitely.

We shall calculate the electromagnetic energy stored between area A of these plates, in a volume $V = A\ell$. Take the electric field of the standing wave inside the resonator as

$$E(z,t) = E_x \sin \omega t \sin kz, \tag{4.3}$$

where in order for the electric field to be zero on each reflector, $k = n\pi/\ell$, where n is an integer.† The total average stored energy per unit volume is

$$\frac{U}{V} = \overline{\frac{1}{2}(\epsilon E_x^2 + \mu H_y^2)}, \tag{4.4}$$

where ϵ, μ are the permittivity and permeability, respectively, of the medium filling the resonator, and the bar indicates averaging over the whole resonator. Since the magnetic field is

$$H_y = \frac{E_x}{Z} = E_x \sqrt{\frac{\epsilon}{\mu}}, \tag{4.5}$$

where Z is the impedance of the medium between the plates, the total average stored energy per unit volume is

$$\frac{U}{V} = \epsilon \overline{E_x^2}. \tag{4.6}$$

The total energy stored is

$$U = \frac{A}{T}\epsilon \int_0^\ell \int_0^T E^2(z,t)dzdt = \frac{1}{4}\epsilon E^2 V, \tag{4.7}$$

where T is the oscillation period of the field. If the power input to the resonator is P, then

$$Q = \frac{2\pi\nu_0 U}{P} = \frac{2\pi\nu_0 \epsilon E^2 V}{4P}, \tag{4.8}$$

and

$$E = \sqrt{\frac{4QP}{2\pi\nu_0 \epsilon V}} \tag{4.9}$$

So high electric field amplitudes are obtained, for a given power input, in a resonator with a high Q.

If we tried to build a laser with a closed resonator, for example by having an amplifying medium filling a sphere, then there would be no preferred direction in the system. Any of the $8\pi\nu^2 V d\nu/c^3$ modes of the resonator in the frequency range $d\nu$ might have similar Q factors and a confusing situation would arise, since the possibility of simultaneous oscillation on many modes having arbitrary direction and frequency would exist. This problem is avoided by using an open resonator, as then only a few of the total number of possible modes have a high Q.

† The spacing of successive standing wave values of $k, \pi/\ell$, might appear at first glance to be in contradiction with the mode spacing $2\pi/\ell$ used earlier in our treatment of black-body radiation. However, we were counting travelling waves there and k could take both positive and negative values. If we want to count standing waves, spaced by π/ℓ, we must only include the positive octant of the **k**-space shown in Fig. (1.15).

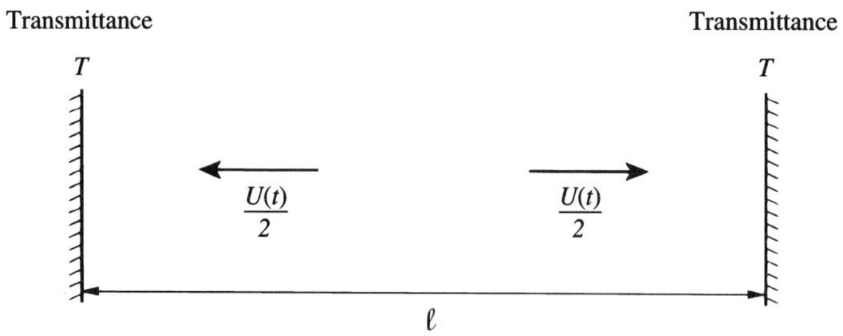

Fig. 4.3. Stored energy within a resonator at time t. The resonator is symmetrical, with both mirrors having identical transmittance T, so equal amounts of energy are propagating towards each mirror.

4.4 Quality Factor of a Resonator in Terms of the Transmission of its End Reflectors

There is an alternative way of looking at the quality factor Q in an open resonator that has two reflectors with equal transmittance T (their fractional intensity, or energy transmission). To simplify the analysis we will assume that $T << 1$.

Consider the decay of stored energy in such a resonator of length ℓ. The initial stored energy is U_0, but at a later time this has been reduced to $U(t)$ because of transmission through the end reflectors. At any given time, equal amounts of this energy are travelling in both directions within the resonator, as shown in Fig. (4.3). In a short time dt, the energy lost from the resonator is

$$-dU = U(t)\left(\frac{Tc}{\ell}\right)dt, \tag{4.10}$$

where c is the velocity of light in the medium between the mirrors. So,

$$-\frac{dU}{U} = \left(\frac{c}{\ell}T\right)dt; \tag{4.11}$$

and with the boundary condition at $t = 0$, $U(0) = U_0$:

$$U = U_0 e^{-(cT/\ell)t}. \tag{4.12}$$

The energy stored in the resonator decays exponentially with a time constant $\tau_0 = \ell/cT$. Highly reflective ends on the resonator imply a long time constant τ_0.

The rate at which energy is dissipated in the resonator is†

$$-\frac{dU}{dt} = \left(\frac{cT}{\ell}\right)U_0 e^{-(cT/\ell)t}. \tag{4.13}$$

The Q of the resonator is

$$Q = \frac{2\pi\nu_0 \times \text{stored energy}}{\text{rate at which energy is dissipated}} = \frac{2\pi\nu_0 U_0 e^{-(cT/\ell)t}}{(cT/\ell)U_0 e^{-(cT/\ell)t}}, \tag{4.14}$$

giving

$$Q = \frac{2\pi\nu_0\ell}{cT} = 2\pi\nu_0\tau_0. \tag{4.15}$$

So a long time constant for decay of energy stored in the resonator implies a high quality factor. If the resonator contains an amplifying medium, and the gain of this medium is great enough, it prevents the decay of energy and sustains an oscillation.

† When the mirror transmittances are not equal and/or are not significantly smaller than unity, the treatment given here for the rate of loss of energy from the resonator is not correct. It can be shown that in these circumstances energy is not lost from the resonator in an exponential fashion.

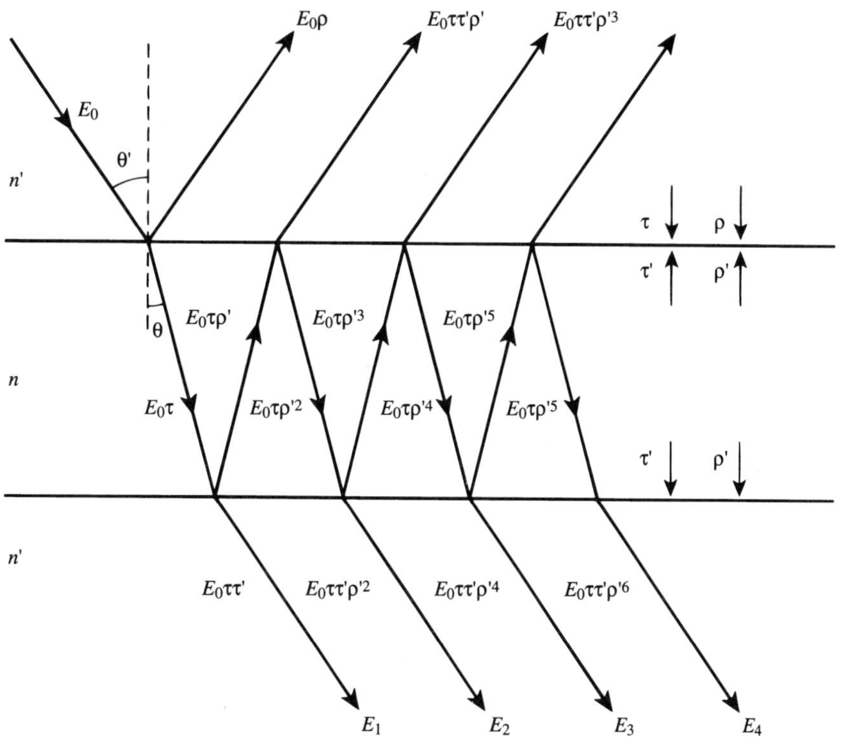

Fig. 4.4. Electric field amplitudes of the various singly and multiply reflected and transmitted waves in a Fabry–Perot etalon.

4.5 Fabry–Perot Etalons and Interferometers

In the nineteenth century the French physicists Fabry and Perot[4.1] developed and analyzed an optical instrument that, although it took a slightly different form, is essentially identical to the structure that serves as the resonator in most laser systems. In the Fabry–Perot *etalon*, which consists of a pair of plane, parallel, optical interfaces or reflectors of constant separation, interference occurs between the beams of light that are multiply reflected between the two interfaces or reflectors. If the optical spacing of these interfaces or reflectors can be changed, the device is called a Fabry–Perot interferometer.

In an idealized device the plane-parallel interfaces are considered to be of infinite extent. We shall see later that in practical devices that have finite interfaces or reflectors, diffraction effects occur that lead to the loss of energy sideways and not just through the interfaces. The simplest kind of etalon consists of just a piece of plane, parallel-sided material of refractive index n immersed in a medium of refractive index n'. Ideally, the parallel faces of the device should be extremely flat, to $1/100$ of a wavelength or less at the wavelength of operation. We consider what happens when a plane wave of frequency ν is incident upon this slab at angle of incident θ', as shown in Fig. (4.4).

The reflection coefficient (the ratio of the reflected to incident field amplitudes at the interface) is taken as ρ for waves travelling from n' to n and ρ' for waves travelling from n to n'. The transmission coefficients are correspondingly τ and τ'; ρ, ρ', τ, and τ' are most easily determined by the use of impedance techniques, discussed in detail in Chapter 14. If the amplitude of the incident electric vector

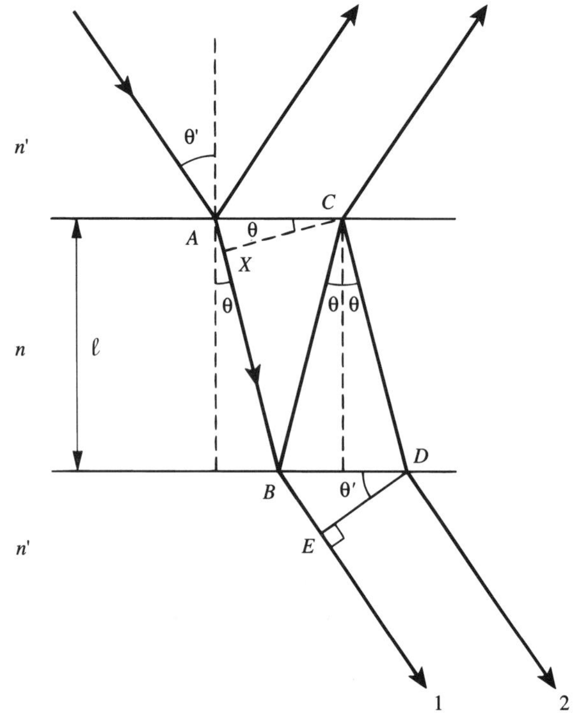

Fig. 4.5. To illustrate the path difference between successive transmitted waves in a Fabry–Perot etalon.

is E_0, then the resultant amplitudes of the various reflected and transmitted waves are as shown in Fig. (4.4). Other types of plane-parallel etalons and interferometers can be analyzed in a similar manner: for example, the etalon might consist of a pair of plane reflectors held a fixed distance apart with the interspace filled with air or some other gas. In this case the medium between the plates has $n \simeq 1$.

The phase difference between two successive transmitted waves in Fig. (4.4) can be found by reference to Fig. (4.5). In this figure

$$CD = BC = \frac{\ell}{\cos\theta},$$
$$BD = 2\ell \ \tan\theta,$$
$$BE = BD\sin\theta' = BD\sin\theta\ \frac{n}{n'}, \tag{4.16}$$

where in the last of these three relations we have used Snell's law. The additional optical distance travelled by wave 2 over wave 1 is

$$\begin{aligned} n(BC + CD) - n'(BE) &= \frac{2n\ell}{\cos\theta} - 2n\ell\tan\theta\sin\theta \\ &= \frac{2n\ell}{\cos\theta} - \frac{2n\ell\sin^2\theta}{\cos\theta} \\ &= 2\ell n\cos\theta, \end{aligned} \tag{4.17}$$

so the phase difference δ between successive transmitted waves is

$$\delta = \frac{4\pi n\ell\cos\theta}{\lambda_0} + 2\epsilon, \tag{4.18}$$

where ϵ is the phase change (if any) that occurs on reflection and λ_0 is the wavelength of the wave *in vacuo*.

If the incident wave is of the form $E = |E_0|e^{i(\omega t - kz)}$, or, in terms of its complex amplitude, $E = E_0 e^{i\omega t}$ (where $E_0 = |E_0|e^{-ikz}$), in Fig. (4.4)

$$\begin{aligned} E_1 &= E_0 e^{-i\delta_0} \tau\tau', \\ E_2 &= E_0 e^{-i\delta_0} \tau\tau' \rho'^2 e^{-i\delta}, \\ E_3 &= E_0 e^{-i\delta_0} \tau\tau' \rho'^4 e^{-2i\delta}, \end{aligned} \tag{4.19}$$

and so on, where δ_0 is the phase difference introduced by the optical path AB.

The total complex amplitude of the transmitted beam is

$$E_t = E_0 e^{-i\delta_0} \tau\tau'(1 + \rho'^2 e^{-i\delta} + \rho'^4 e^{-2i\delta} + ...). \tag{4.20}$$

This is a geometric series with ratio $\rho'^2 e^{-i\delta}$, and first term 1. Its sum to n terms is

$$(E_t)_n = E_0 e^{-i\delta_0} \tau\tau' \frac{[1 - (\rho'^2)^n e^{-in\delta}]}{(1 - \rho'^2 e^{-i\delta})}; \tag{4.21}$$

and since $|\rho'| < 1$, the sum to infinity is

$$E_t = \frac{E_0 e^{-i\delta_0} \tau\tau'}{1 - \rho'^2 e^{-i\delta}}. \tag{4.22}$$

If the interfaces between the media with refractive indices n and n' are not made specially reflecting (by, for example, having reflective coatings placed on them) then in the case of normal incidence†

$$\rho = \frac{n' - n}{n' + n}, \tag{4.23}$$

$$\rho' = \frac{n - n'}{n + n'}, \tag{4.24}$$

so $\rho = -\rho'$ and there is a phase change on reflection from the interface if $n' > n$. Furthermore,

$$\tau = \frac{2n'}{n' + n}, \tag{4.25}$$

$$\tau' = \frac{2n}{n' + n}. \tag{4.26}$$

Note that $|\rho^2| = |\rho'|^2 = R$ which is called the *reflectance* of the interface: it relates the reflected and incident intensities since these are proportional to $|$ electric field $|^2$:

$$R = \left| \frac{E_{\text{reflected}}}{E_{\text{incident}}} \right|^2. \tag{4.27}$$

By a similar procedure as for the transmitted wave we can see that

$$\begin{aligned} E_r &= E_0\rho + E_0\tau\tau'\rho' e^{-i\delta} + E_0\tau\tau'\rho'^3 e^{-2i\delta} + ... \\ &= E_0[\rho + \tau\tau'\rho' e^{-i\delta}(1 + \rho'^2 e^{-i\delta} + \rho'^4 e^{-2i\delta} + ...)], \end{aligned} \tag{4.28}$$

Summing Eq. (4.28) to infinity gives

$$E_r = \frac{E_0(\rho - \rho\rho'^2 e^{-i\delta} + \tau\tau'\rho' e^{-i\delta})}{1 - \rho'^2 e^{-i\delta}}. \tag{4.29}$$

† See Appendix 4.

As far as the transmitted *intensity* through one interface is concerned we can define a transmittance T

$$T = \frac{I_{transmitted}}{I_{incident}} = \left(\frac{|E^2_{transmitted}|}{Z_{in\ transmitted\ medium}}\right)\left(\frac{Z_{in\ incident\ medium}}{|E_{incident^2}|}\right)$$

$$= \left|\begin{matrix}\text{transmission}\\ \text{coefficient}\end{matrix}\right|^2 \left(\frac{Z_{in\ incident\ medium}}{Z_{in\ transmitted\ medium}}\right). \tag{4.30}$$

For the $n' \to n$ interface

$$Z' = \sqrt{\frac{\mu'\mu_0}{\epsilon'\epsilon_0}}, \tag{4.31}$$

where ϵ', μ' are the dielectric constant and relative magnetic permeability, respectively, of the material. If $\mu' = 1$, we can write $\sqrt{\epsilon'} = n'$. So

$$Z' = \sqrt{\frac{\mu_0}{\epsilon_0}}\frac{1}{n'}. \tag{4.32}$$

In normal incidence

$$T = \frac{\tau' Z'}{Z} = \left(\frac{2n'}{n'+n}\right)^2 \frac{n}{n'} = \frac{4nn'}{(n'+n)^2}. \tag{4.33}$$

For the $n \to n'$ interface at normal incidence

$$T = \frac{\tau'^2 Z}{Z'} = \frac{4nn'}{(n'+n)^2}. \tag{4.34}$$

So the intensity transmission coefficient T is independent of which way the wave travels through the interface and

$$R + T = \frac{(n-n')^2}{(n+n')^2} + \frac{4nn'}{(n'+n)^2} = 1. \tag{4.35}$$

Eq. (4.35) is not surprising because of energy conservation at the interface. Now

$$\tau\tau' = \frac{4nn'}{(n'+n)^2} = T, \tag{4.36}$$

so in (4.22)

$$E_t = \frac{E_0 T e^{-i\delta_0}}{1 - Re^{-i\delta}}, \tag{4.37}$$

and in (4.29)

$$E_r = \frac{E_0 r(1 - e^{-i\delta})}{1 - Re^{-i\delta}} = \frac{E_0\sqrt{R}(1 - e^{-i\delta})}{1 - Re^{-i\delta}}. \tag{4.38}$$

These formulae apply in general when R and T are the reflectance and transmittance at the appropriate angle of incidence of the plane-parallel reflecting faces that constitute the etalon. Although we have shown that the intensity reflectance coefficient of an interface between two media of different refractive indices is the same in both directions of wave propagation only in the special case of normal incidence, it can be shown to be generally true (see Appendix 4). The relations

$$\begin{aligned} R &= \rho^2 = \rho'^2, \\ T &= \tau\tau', \end{aligned} \tag{4.39}$$

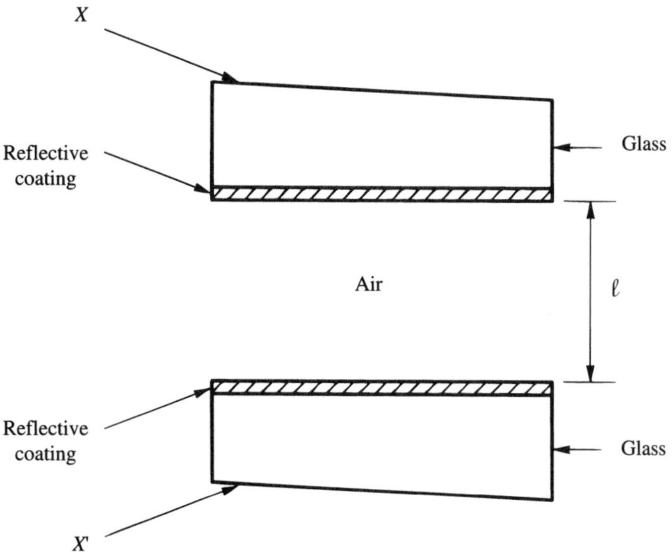

Fig. 4.6. A practical air-spaced Fabry–Perot etalon. The outer faces X, X' are generally antireflection coated and make a small wedge angle with the inner reflecting faces.

are correct results independent of the angle of incidence.

A very popular form of Fabry–Perot etalon or interferometer consists of a pair of parallel, flat, transparent substrates to each of which a reflective coating has been applied, as shown in Fig. (4.6). Although strictly this is a composite device, containing more than two parallel flat interfaces, it can still be treated in the manner detailed above, where R and T are taken as the reflectance and transmittance of the reflective coating applied to each substrate. If the outer faces of the device X and X' are also taken into account, the system consists of a number of etalons in series and its analysis should be modified accordingly. However, it is common practice to coat faces X and X' with an antireflecting layer and have the substrates slightly wedge-shaped to prevent this difficulty from arising.

Strictly, and particularly if the reflecting faces of an air-spaced etalon are metal coated, the absorption of light in passing through the reflecting film should be taken into account. In any practical instrument it is desirable for this absorption to be kept as small as possible. If the fractional energy absorption in passing through the reflective film is A', then

$$R + T + A' = 1; \tag{4.40}$$

if A' is kept very small we can still use $R + T = 1$.

The overall intensity reflection characteristic of the etalon is

$$\frac{I_r}{I_i} = \frac{E_r E_r^*}{E_0 E_0^*} = \frac{E_0 E_0^* R(1 - e^{-i\delta})(1 - e^{i\delta})}{E_0 E_0^*(1 - Re^{-i\delta})(1 - Re^{i\delta})}, \tag{4.41}$$

where I_i is the incident and I_r the reflected intensity. Eq. (4.41) reduces to

$$\frac{I_r}{I_i} = \frac{R[2 - \left(e^{i\delta} + e^{-i\delta}\right)]}{[1 + R^2 - R(e^{i\delta} + e^{-i\delta})]} = \frac{R(2 - 2\cos\delta)}{1 + R^2 - R(2\cos\delta)} \tag{4.42}$$

and writing $\cos\delta = \cos(\delta/2+\delta/2) = \cos^2(\delta/2) - \sin^2(\delta/2) = 1 - 2\sin^2(\delta/2)$

$$\frac{I_r}{I_i} = \frac{4R\sin^2(\delta/2)}{(1-R)^2 + 4R\sin^2(\delta/2)}. \tag{4.43}$$

In a similar way

$$\frac{I_t}{I_i} = \frac{(1-R)^2}{(1-R)^2 + 4R\sin^2(\delta/2)} = \frac{1}{1+4R/(1-R)^2\sin^2(\delta/2)}. \tag{4.44}$$

The transmission characteristics of the device are interesting. The transmittance is unity whenever $\sin(\delta/2) = 0$, i.e. when $\delta/2 = m\pi$ where m is an integer, i.e., when

$$\frac{4\pi n\ell\cos\theta}{\lambda_0} + 2\epsilon = 2m\pi. \tag{4.45}$$

If the phase change on reflection is 0 or π then for maximum transmittance

$$\ell = \frac{m\lambda_0}{2n\cos\theta} = \frac{m\lambda}{2\cos\theta}, \tag{4.46}$$

where λ is the wavelength in the material between the reflectors. In normal incidence $\ell = m\lambda/2$, a perhaps intuitively obvious result. The frequencies of maximum transmission satisfy

$$\nu_m = \frac{mc_0}{2n\ell\cos\theta}, \tag{4.47}$$

where c_0 is the velocity of light *in vacuo*. Adjacent frequencies at which the etalon shows maximum transmission are separated by a frequency

$$\Delta\nu = \frac{c_0}{2n\ell\cos\theta}. \tag{4.48}$$

This is called the *free spectral range* of the etalon. These frequencies of maximum transmission are equally spaced. A device that has this characteristic is a *comb filter*. If we allow for losses in the etalon we find that the peak transmission falls to

$$\frac{I_t}{I_0} = \frac{(1-R)^2A}{(1-RA)^2}, \tag{4.49}$$

where the intensity of the wave on one pass through the etalon changes by a factor A.

If the transmission characteristics of a Fabry–Perot etalon are plotted for various values of R using Eq. (4.44) the curves shown in Fig. (4.7) are obtained. As R increases the sharpness of the transmission peaks increases. The quantity

$$F = \frac{\pi\sqrt{R}}{1-R}, \tag{4.50}$$

gives a measure of this sharpness and is called the *finesse* of the instrument.

Near a transmission maximum we can write $\delta = 2m\pi + \Delta$, where Δ is a small angle. In this case Eq. (4.44) becomes

$$\frac{I_t}{I_i} = \frac{1}{1+F^2\Delta^2/\pi^2}. \tag{4.51}$$

This is a Lorentzian function of Δ with FWHM $2\pi/F$. Now, since

$$\delta = \frac{4\pi n\ell\nu\cos\theta}{c_0}, \tag{4.52}$$

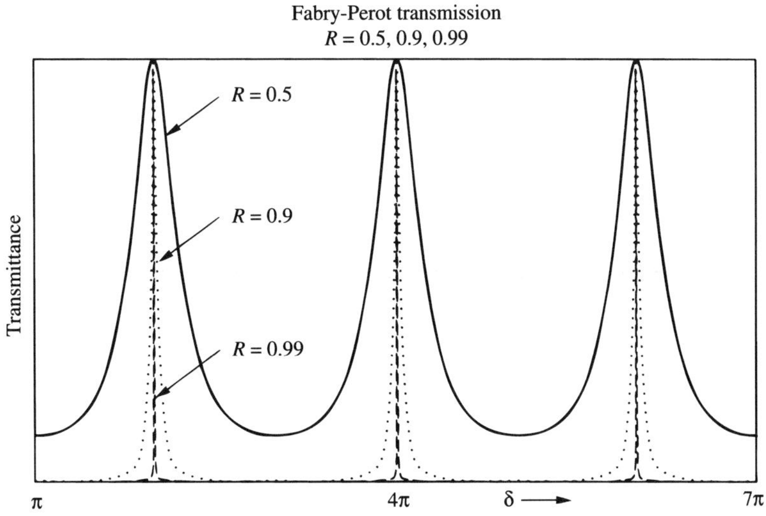

Fig. 4.7. Theoretical transmission characteristic of a Fabry–Perot etalon calculated from Eq. (4.44) for different values of mirror reflectance R. The theoretical reflection characteristic that corresponds to these curves can be viewed by turning the diagram upside down.

it follows from Eq. (4.47) that

$$\Delta = \frac{4\pi n\ell}{c_0}(\nu - \nu_m)\cos\theta, \tag{4.53}$$

so the transmission peaks are Lorentzian functions of the frequency spacing from the transmission maximum ν_m, provided the angular deviation from the center of the peak remains a small angle. The frequency FWHM of these peaks is, therefore:

$$\Delta\nu_{1/2} = \frac{c_0}{(2n\ell\cos\theta)F} = \frac{\Delta_\nu}{F}, \tag{4.54}$$

where $\Delta\nu$ is the free spectral range. The higher the finesse the narrower become the transmission peaks.

4.6 Internal Field Strength

Since a laser consists essentially of a Fabry–Perot etalon with an amplifying medium between its reflectors, it is important to see how the external and internal electric fields of a wave passing through such an etalon are related. It is simplest to do this for a wave in normal incidence. Fig. (4.8) shows schematically the electric field amplitudes inside the etalon of the various singly and multiply reflected components of the input wave. For convenience these various components are shown displaced sideways from each other although in reality they all overlap.

The phase shift between components such as 1 and 3 or 2 and 4 is δ where as before

$$\delta = \frac{4\pi n\ell}{\lambda_0} + 2\epsilon.$$

Bearing in mind that the waves travelling from top to bottom are of the form $E_1 \propto e^{i(\omega t - kz_1)}$ and those travelling from bottom to top are of the form $E_2 \propto e^{i(\omega t - kz_2)}$, the total electric field amplitude between the etalon reflectors, for example at

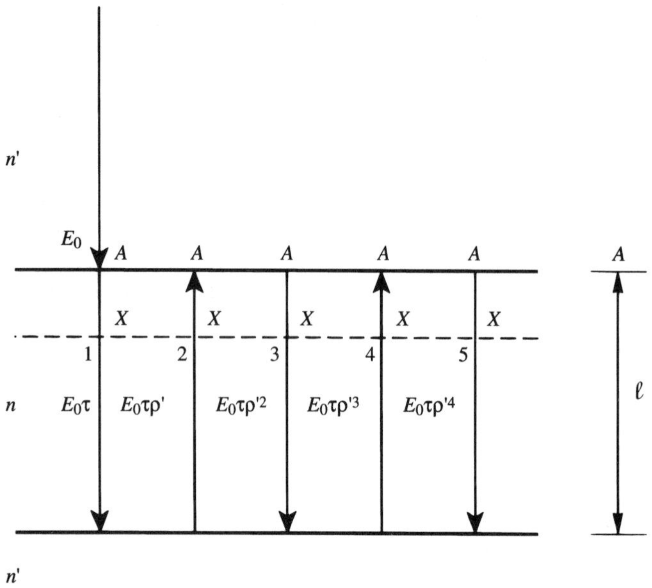

Fig. 4.8. Field amplitudes of the multiply reflected waves inside a Fabry–Perot etalon in normal incidence.

point X, is

$$E_{int} = E_0 e^{-i\delta_0}\tau + E_0 e^{-i(\delta-\delta_0-\epsilon)}\tau\rho' + E_0 e^{-i(\delta_0+\delta)}\tau\rho'^2 + \\ + E_0 e^{-i(2\delta-\delta_0-\epsilon)}\tau\rho'^3 + E_0 e^{-i(\delta_0+2\delta)}\tau\rho'^4 + ..., \tag{4.55}$$

where δ_0 is the phase shift corresponding to the path AX and ϵ is the phase change on reflection

$$\begin{aligned} E_{int} =& E_0\tau e^{-i\delta_0}\left(1 + \rho'^2 e^{-i\delta} + \rho'^4 e^{-2i\delta} + \dots \right. \\ & \left. + \rho' e^{-i(\delta-2\Delta_0-\epsilon)} + \rho'^3 e^{-i(2\delta-2\delta_0-\epsilon)} + \dots\right) \\ =& E_0\tau e^{-i\delta_0}\left(\frac{1}{1-\rho'^2 e^{-i\delta}} + \frac{\rho' e^{-i(\delta-2\delta_0-\epsilon}}{1-\rho'^2 e^{-i\delta}}\right) \\ =& \frac{E_0\tau e^{-i\delta_0}\left(1 + \rho' e^{-i(\delta-2\delta_0-\epsilon)}\right)}{1 - Re^{-i\delta}}. \end{aligned} \tag{4.56}$$

The intracavity stored energy density depends on

$$E_{int}E^*_{int} = \frac{|E_0|^2\tau^2\left[1 + R + 2\rho'\cos(\delta - 2\delta_0 - \epsilon)\right]}{1 + R^2 - 2R\cos\delta}. \tag{4.57}$$

To study the variation of intracavity stored energy with phase shift δ it is simpler if we choose a location within the system where $2\delta_0 + \epsilon = 2p\pi$, where p is any positive or negative integer, or zero. In this case

$$E_{int}E^*_{int} = \frac{|E_0|^2\tau^2\left(1 + R + 2\sqrt{R}\cos\delta\right)}{(1 + R^2 - 2R\cos\delta)}. \tag{4.58}$$

The variation of stored energy with phase shift depends on

$$\frac{\partial}{\partial\delta}(E_{int}E^*_{int}) = \frac{-2|E_0|^2\tau^2\sqrt{R}\sin\delta\left[1 + R^2 + \sqrt{R}(1+R)\right]}{(1 + R^2 - 2R\cos\delta)^2}. \tag{4.59}$$

Therefore, the turning points of $E_{int}E^*_{int}$ are determined by the condition

$$\sin\delta = 0. \tag{4.60}$$

So, $E_{int}E^*_{int}$ is a maximum or a minimum when $\delta = m\pi$, where m is any integer.
When $\delta = (2m+1)\pi$, from Eq. (4.58)

$$E_{int}E^*_{int} = \frac{|E_0|^2\tau^2(1+R-2\sqrt{R})}{(1+R)^2}, \tag{4.61}$$

whereas when $\delta = 2m\pi$.

$$E_{int}E^*_{int} = \frac{|E_0|^2\tau^2(1+R+2\sqrt{R})}{(1-R)^2}. \tag{4.62}$$

Clearly, the maxima of $E_{int}E^*_{int}$, where $E_{int}E^*_{int}$ is proportional to the standing electric field energy density inside the etalon, occur when $\delta = 2m\pi$ and thus correspond to the transmission maxima of the system.

It is worth noting from (4.62) that for $\delta = 2m\pi, \epsilon = 0$, $E_{int}E^*_{int}$ appears to go to infinity as $R \to 1$. This curious result arises because of the way we dealt with Eq. (4.55). Since when $R \to 1, \tau \to 0$ it is clear from (4.55) that the internal field in this case goes to zero. No energy is transmitted through the etalon in this case even though Eq. (4.44) predicts the existence of apparent transmission maxima for $R \to 1, \delta = 2m\pi$. These infinitely narrow, finite height, transmission maxima are an artifact of our mathematical treatment since, in the limiting case when $R \to 1, \tau \to 0$, each term in the infinite series (4.20) is zero.

4.7 Fabry–Perot Interferometers as Optical Spectrum Analyzers

Since a Fabry–Perot etalon or interferometer has a transmission characteristic that is a function of frequency, it can be used to analyze the spectral output of a source of light. If we illuminate the etalon with white light, so that angle θ is a constant, as shown in Fig. (4.9) then for a given spacing ℓ the frequency distribution of transmitted light only shows large intensities for frequencies that satisfy

$$\nu_m = \frac{mc_0}{2n\ell\cos\theta}. \tag{4.63}$$

If we illuminate the etalon normally with a monochromatic source, then a signal will only be transmitted provided the frequency of the monochromatic source satisfies

$$\nu_0 = \frac{mc_0}{2n\ell} \tag{4.64}$$

for some integer m, as shown in Fig. (4.10). In the Fabry–Perot interferometer, in which one of the plates can be moved, the transmitted intensity as a function of plate separation, for illumination with a monochromatic source of frequency ν_0 is also shown in Fig. (4.10). Transmission maxima occur for plate separations that satisfy

$$\ell = \frac{mc_0}{2n\nu_0\cos\theta}. \tag{4.65}$$

$\Delta\ell$, the plate movement between successive maxima, is equal to

$$\frac{c_0}{2n\nu_0\cos\theta} = \frac{\lambda_0}{2n\cos\theta} = \frac{\lambda}{2} \quad \text{in normal incidence,} \tag{4.66}$$

where λ is the wavelength of the monochromatic signal *in the medium* between the interferometer plates. In normal incidence plate separations for maximum transmission are separated by half wavelength intervals. This half wavelength movement of one interferometer plate relative to the other is equivalent to introducing one

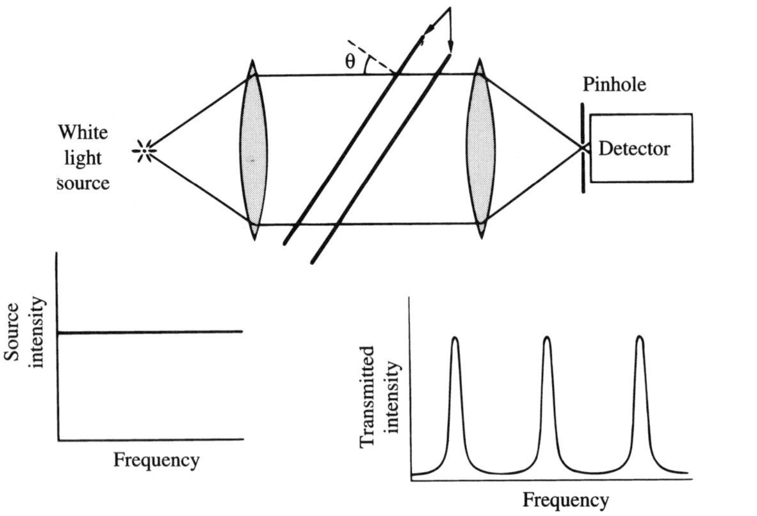

Fig. 4.9. Schematic arrangement in which a Fabry–Perot etalon is illuminated at angle θ with parallel light from a white light source.

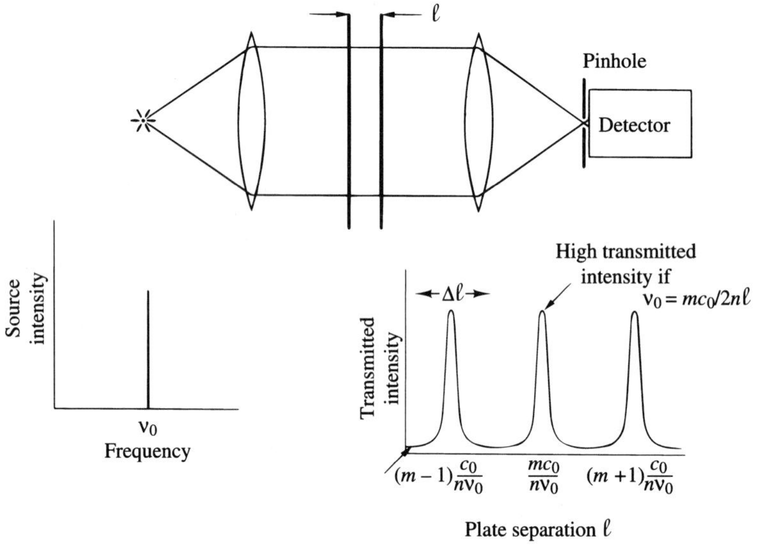

Fig. 4.10. Schematic arrangement in which a Fabry–Perot interferometer of adjustable spacing is illuminated normally with monochromatic parallel light.

whole wavelength additional path difference between successive transmitted rays – this is the constructive interference condition.

If it is not practical to construct a Fabry–Perot interferometer for the spectral analysis of a light signal, for example, when the source being analyzed is of a transient nature of sufficiently short duration that movement of one interferometer plate is not possible, it is still possible to use a Fabry–Perot etalon and utilize the angular discrimination between transmitted beams at different frequencies. For example, consider the case where an etalon is illuminated with a monochromatic point source, as shown in Fig. (4.11). The angles at which transmission maxima occur satisfy

$$\cos\theta_m = \frac{mc_0}{2n\ell\nu_m}. \tag{4.67}$$

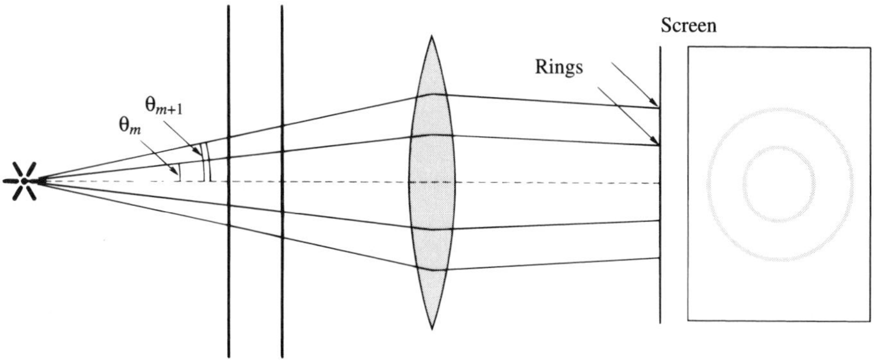

Fig. 4.11. Schematic arrangement in which a Fabry–Perot etalon is illuminated with a monochromatic point source and a ring pattern of transmitted light is observed.

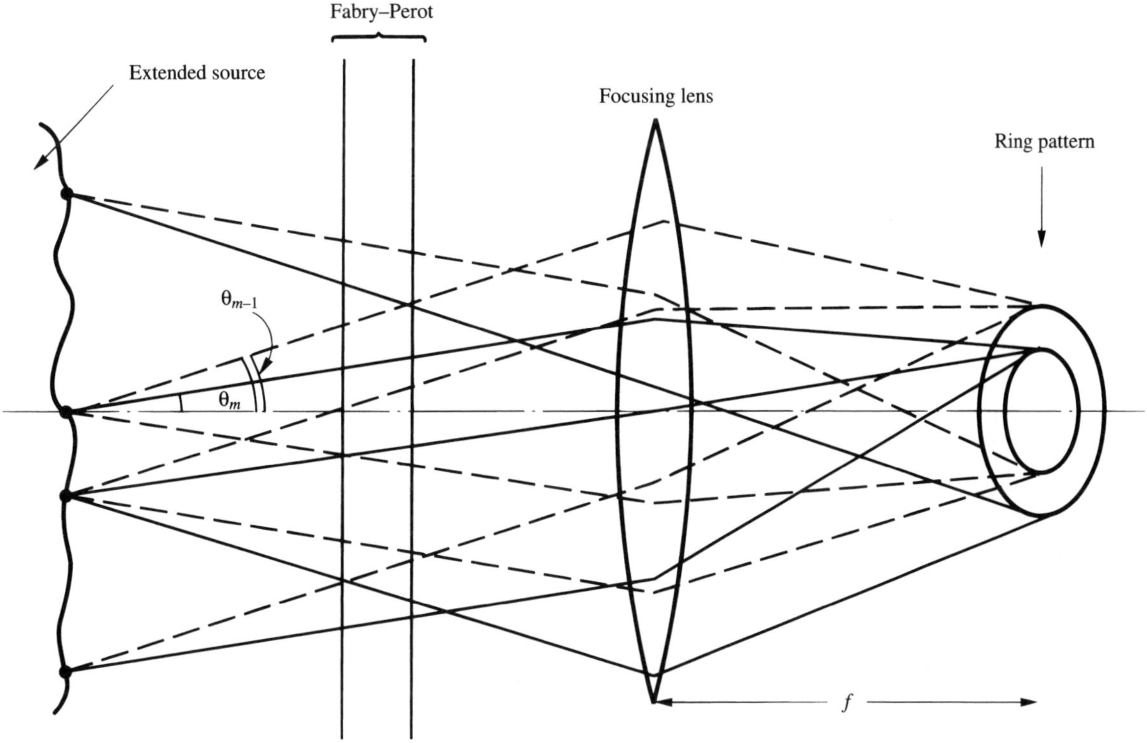

Fig. 4.12. Schematic arrangement used for viewing Fabry–Perot rings with a diffuse monochromatic source and an air-spaced etalon.

The loci of rays of maximum transmission lie along the surfaces of cones with semivertical angles θ_m. The intersection of these cones with a screen produces a series of bright rings. These rings can, if necessary, be focused by a lens. This mode of usage is most useful if the source under study is diffuse, as illustrated in Fig. (4.12). In this case the lens brings all the transmitted intensity maxima at angle θ_m into focus in a ring in the focal plane. If there is a bright point at the center of the focused ring pattern this implies $\theta_m = 0$, therefore

$$m\lambda_0 = 2n\ell. \tag{4.68}$$

For the next order transmission maximum

$$\cos\theta_{m-1} = (m-1)\frac{\lambda_0}{2n\ell}. \tag{4.69}$$

If θ_{m-1} is small, this gives

$$2n\ell\left(1-\frac{\theta_{m-1}^2}{2}\right) = (m-1)\lambda_0. \tag{4.70}$$

Subtracting Eq. (4.68) from Eq. (4.70) gives

$$2n\ell\frac{\theta_{m-1}^2}{2} = \lambda_0, \tag{4.71}$$

and finally

$$\theta_{m-1} = \sqrt{\frac{\lambda_0}{n\ell}}. \tag{4.72}$$

4.7.1 Example:

For λ_0 = 500nm, n = 1, ℓ = 100mm, $\theta \simeq 2.24\times10^{-3}$ radian. If the focal length of the lens is f, the physical separation of adjacent rings near the center of the pattern is

$$\rho \simeq f\theta = f\sqrt{\frac{\lambda_0}{n\ell}}. \tag{4.73}$$

For $f = 500$ mm, $f\theta \simeq 1$ mm.

Although the finesse of a Fabry–Perot interferometer has been defined as

$$F = \frac{\pi\sqrt{R}}{1-R}, \tag{4.74}$$

and is the factor that determines the sharpness of transmission maxima, the true experimental finesse is also dependent on the surface flatness of the mirrors and on the angular spread of the beam incident on the spectrometer[4.2]–[4.6]. For finite-size end reflectors diffraction losses also occur. Diffraction occurs whenever a plane wave front is restricted spatially,† and leads to an effective range of propagation directions in the resultant wave front and a further reduction in finesse.

If a Fabry–Perot system is to be used to study a source with two closely spaced monochromatic frequencies, as shown in Fig. (4.13), then these frequencies can be considered resolved if the transmitted maximum of one is a frequency $\Delta\nu_{1/2}$ from the other, that is the resolution of the instrument becomes $\Delta\nu_{1/2}$, where $\Delta\nu_{1/2} = \Delta\nu/F$, so a high finesse implies a high resolving power.

Consider the following three Fabry–Perot systems with mirrors of different reflectance.

For

$$R = 0.9, \quad F = \frac{\pi\sqrt{0.9}}{0.1} \simeq 30,$$

$$R = 0.95, \quad F = \frac{\pi\sqrt{0.95}}{0.05} = 61,$$

$$R = 0.99, \quad F = \frac{\pi}{0.01} = 314.$$

† See Chapter 6.

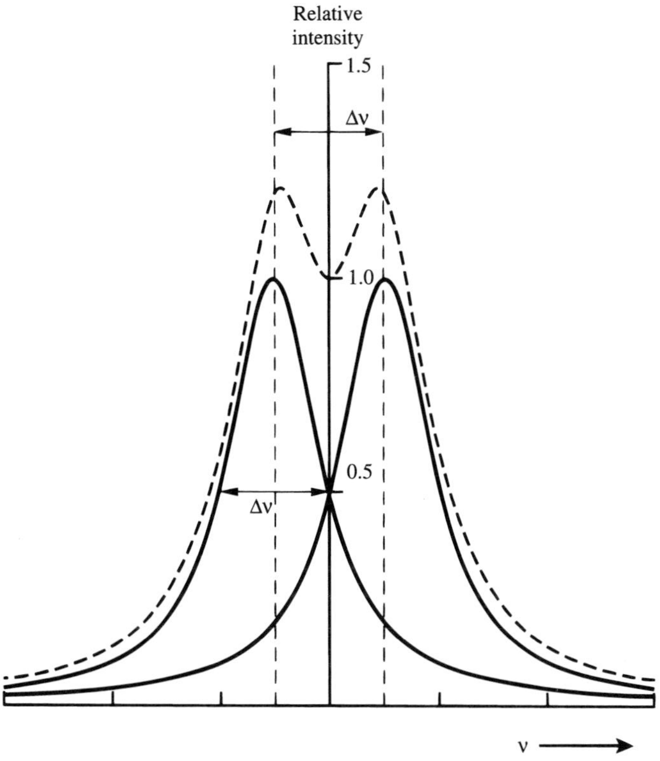

Fig. 4.13. Transmitted intensity (dashed line) as a function of plate separation when a Fabry–Perot interferometer is illuminated with two monochromatic lines (solid lines) closely spaced in frequency that the interferometer is *just* able to resolve.

Table 4.1. *Characteristics of Fabry–Perot interferometers with different reflectances and spacings.*

ℓ(mm)	$\Delta\nu$(Hz)	R	F	$\Delta\nu_{1/2}$(Hz)	$\nu/\Delta\nu_{1/2}$ = Resolving power
10	1.5×10^{10}	0.9	30	5×10^{8}	1.2×10^{6}
100	1.5×10^{9}	0.9	30	5×10^{7}	1.2×10^{7}
10	1.5×10^{10}	0.95	61	2.46×10^{8}	2.44×10^{6}
100	1.5×10^{9}	0.95	61	2.46×10^{7}	2.44×10^{7}

With finesses of this order, the phase difference between the transmission maximum and its half intensity points is indeed a small angle, as we had assumed previously. Table (4.1) shows the free spectral range of two of the above etalons and their resolving power for two different mirror spacings ℓ, under conditions of normal incidence operation at $\lambda = 500$ nm. We take $n = 1$ so $\Delta\nu = c_0/2\ell$.

It is clear that we can increase the ability of the instrument to resolve closely-spaced wavelengths by increasing the separation of the mirrors and increasing their reflectance. However, this procedure also brings neighboring transmission peaks closer together.

Problems can arise with a Fabry–Perot system when a light source under study is emitting different frequencies that satisfy (or almost satisfy) the transmission

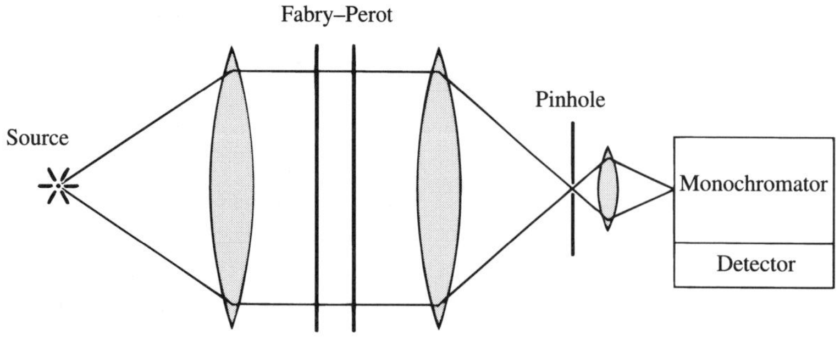

Fig. 4.14. Schematic arrangement for performing high resolution spectroscopy in which a Fabry–Perot interferometer is used in conjunction with a monochromator.

maximum criterion for different integers m, m'

$$\nu_m = \frac{mc_0}{2n\ell}, \quad \nu'_m = \frac{m'c_0}{2n\ell}. \tag{4.75}$$

In order to keep the instrument useful when this happens, it is often necessary to use it in series with some other wavelength selective device. For example, with a grating or prism monochromator as shown in Fig. (4.14). This technique is particularly successful if the ν_m, ν'_m are reasonably well separated in frequency. When it is necessary to study the spectral emission of any light source, conventional or laser, to extremely high precision, the use of a Fabry–Perot interferometer in this way is often the only feasible method of approach.

Some important physical applications of such interferometers include:

(1) High resolution spectroscopy, including measurements of the Zeeman effect and hyperfine structure.

(2) Lineshape studies that allow determination of local temperature and other parameters of the system including some excited state lifetimes.

(3) By measuring the width of spectral lines emitted by a plasma the charged particle density in the plasma can be measured. By scanning the plate spacing very fast it is possible to make such measurements even when the plasma is transient.

4.8 Problems

(4.1) Prove Eq. (4.49).

(4.2) A Fabry–Perot etalon is 3 mm thick and is made of glass (n=1.5). Diverging laser radiation of wavelength 510.6 nm is incident in a range of angles about the normal to the etalon. Calculate the radius of the first three rings observed on a screen 1 m away from the etalon. Calculate also the minimum amount by which the thickness of etalon must be changed to obtain a bright spot at the center of the pattern

(4.3) A Fabry–Perot etalon is illuminated with monochromatic radiation at a wavelength of 488.79 nm (*in vacuo*). The etalon has n=1.55 and is 7.4 mm thick. (i) Calculate the minimum change in temperature necessary to produce a bright spot at the center of the ring pattern. Take the coefficient of expansion of the etalon as 3×10^{-6} K^{-1}. (ii) What is the maximum divergence of the input light if only one ring is seen? (iii) A second monochromatic signal is present at 489.32 nm: (a) Do the transmitted rings have the same order within a spectral range? (b) What minimum equal reflectance of the plates of the etalon is needed to just resolve the two wavelengths?

(4.4) A Fabry–Perot etalon made of glass (n=1.5) of thickness 2mm is illuminated with diverging radiation of wavelength $\lambda_0 = 510.554$ nm from an extended source. A lens of focal length 0.5 m is used to focus the rings onto a screen. A circular aperture of radius 30 mm is cut in the screen concentric with the ring pattern. How many rings go through the aperture? What is the minimum change in refractive index required to get a bright spot at the center of the pattern?

(4.5) Analyze a Fabry–Perot etalon of index n that is bounded on one side with a medium of index n' and on the other by index of refraction n''. Calculate the transmittance as a function of the thickness of the medium. How is the definition of the finesse changed in this case?

(4.6) The *contrast ratio* of a Fabry–Perot interferometer is defined as

$$C = \frac{(I_t/I_0)_{max}}{(I_t/I_0)_{min}}.$$

Prove that

$$C = 1 + \frac{4F^2}{\pi^2}.$$

(4.7) For a real mirror $R+T+A' = 1$ where A' is the fractional energy absorption of the mirror. Derive the new version of Eq. (4.44) that holds in this case and give an equation for the new finesse.

References

[4.1] Max Born and Emil Wolf, *Principles of Optics Electromagnetic Theory of Propagation, Interference and Diffraction of light*, 6th Edition, Pergamon Press, Oxford, 1980.

[4.2] M. Françon, *Optical Interferometry*, Academic Press, New York, 1966.

[4.3] W.H. Steel, *Interferometry*, 2nd Edition Cambridge University Press, Cambridge, 1983.

[4.4] P. Hariharan, *Optical Interferometry*, Academic Press, Sydney, 1985.

[4.5] J.H. Moore, C.C. Davis, and M.A. Coplan, *Building Scientific Apparatus*, 2nd Edition Addison-Wesley, Redwood City, CA, 1988.

5

Optical Resonators Containing Amplifying Media

5.1 Introduction

In this chapter we shall combine what we have learned about optical frequency amplification and the resonant, or feedback, characteristics of Fabry–Perot systems, in order to study laser oscillators. When a Fabry–Perot resonator is filled with an amplifying medium, laser oscillation will occur at specific frequencies if the gain of the medium is large enough to overcome the loss of energy through the mirrors and by other mechanisms within the laser medium. The onset of laser oscillation and the frequency, or frequencies, at which it occurs is governed by threshold amplitude and phase conditions, which will be derived. Once laser oscillation is estalished, it stabilizes at a level that depends on the saturation intensity of the amplifying medium and the reflectance of the laser mirrors. We shall conclude the chapter by investigating how these factors affect the output power that can be obtained from a laser, and how this can be optimized.

5.2 Fabry–Perot Resonator Containing an Amplifying Medium

Fig. (5.1) represents a Fabry-Perot resonator, whose interior is filled with an amplfying medium and which has plane mirrors. We consider the complex amplitudes of the waves bouncing backwards and forwards normally between the resonator mirrors. These waves result from an incident beam with electric vector E_0 at the first mirror as shown in Fig. (5.1), where E_0 is the complex amplitude at some reference point. The reflection and transmission coefficients in the various directions at the mirrors are as shown in the figure. We include in the absorption coefficients A, A_1, A_2 at the two mirrors any reflection losses or scattering that send energy out of the resonator: we do not at this stage include diffraction losses, which result from the finite lateral dimensions of the mirrors or medium. If there were nothing inside the resonator then a wave propagating between the mirrors would propagate as $E_0e^{i(\omega t-kz)}$ to the right and $E_0e^{i(\omega t+kz)}$ to the left. The presence of a gain medium changes the otherwise passive propagation factor k to

$$k'(\omega) = k + \Delta k = k + \frac{k\chi'(\omega)}{2n^2} \tag{5.1}$$

and the gain coefficient $\gamma(\omega) = -k\chi''(\omega)/n^2$ causes the complex amplitude of the wave to change with distance as $e^{(\gamma/2)z}$. We allow for the possible existence of a distributed loss per pass given by an absorption coefficient α. Such absorption causes a fractional change in intensity for a single pass through the medium

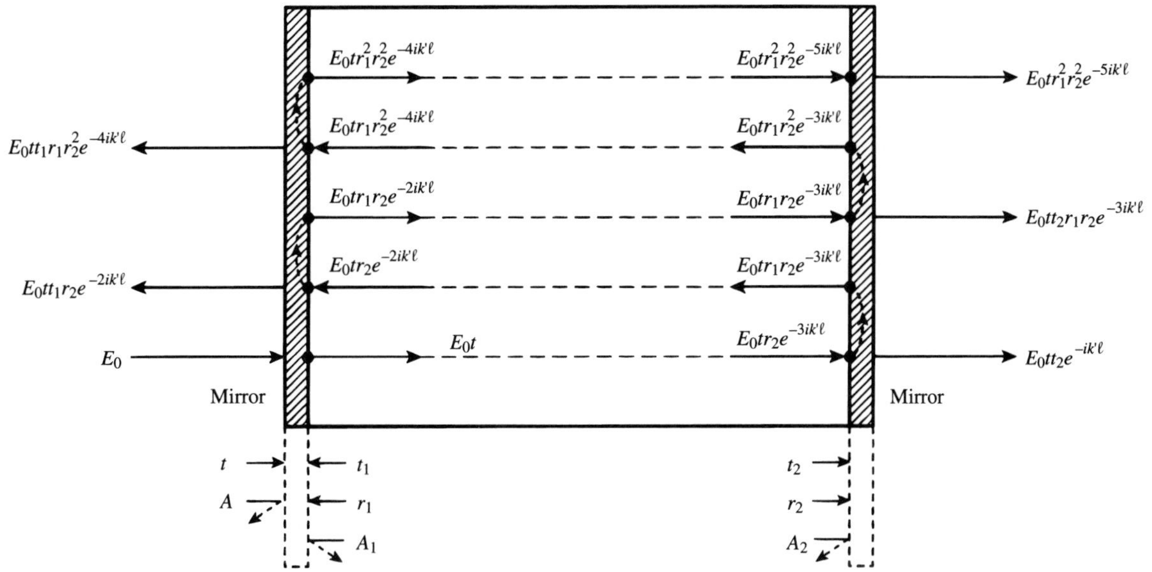

Fig. 5.1. The amplitude of the electric field vectors of the successively transmitted, amplified, and reflected waves in a Fabry–Perot system containing an amplifying (or absorbing) medium. The absorption factors A, A_1, and A_2 are not used explicitly in the analysis given in the text, but they modify the values of t, t_1 and, t_2.

of $e^{-\alpha\ell}$. Such a distributed loss could, for example, arise from scattering by crystal imperfections in a laser rod. This distributed loss modifies the complex amplitude by a factor $e^{-i\alpha\ell/2}$ per pass. Therefore, the full propagation constant of the wave in the presence of both gain and loss is

$$k'(\omega) = k + k\frac{\chi'(\omega)}{2n^2} - \frac{ik\chi''(\omega)}{2n^2} - \frac{i\alpha}{2} \tag{5.2}$$

and the wave propagates as $e^{i(\omega t \pm k'z)}$.

A wave travelling to the right with complex amplitude E_0 at plane $z = 0$ in the resonator, the left hand mirror, has at plane ℓ, the right hand mirror, become

$$E = E_0 e^{i(\omega t - k\ell)} = E_0 e^{-ik\ell} e^{i\omega t} = E_0' e^{i\omega t}.$$

This wave then begins to propagate to the left as

$$E = E_0' e^{i(\omega t + kz)}.$$

At plane $-\ell$, the left hand mirror, with the right hand mirror now taken as the origin, it has become once more a wave travelling to the right

$$E = E_0 e^{ik\ell} e^{i(\omega t - k\ell)} = E_0 e^{-2ik\ell} e^{i\omega t}.$$

In this way we can write down the complex amplitudes of successive rays travelling at normal incidence between the two reflectors, as shown in Fig. (5.1).

The output beam through the right hand mirror arises from the transmission of waves travelling to the right: its total electric field amplitude is,

$$\begin{aligned} E_t &= E_0 t t_2 e^{-ik'\ell} + E_0 t t_2 r_1 r_2 e^{-3ik'\ell} + \dots \\ &= E_0 t t_2 e^{-ik'\ell}(1 + r_1 r_2 e^{-2ik'\ell} + r_1^2 r_2^2 e^{-4ik'\ell} + \dots) \\ &= \frac{E_0 t t_2 e^{-ik'\ell}}{1 - r_1 r_2 e^{-2ik'\ell}} \\ &= \frac{E_0 t t_2 e^{-i(k+\Delta k)\ell} e^{(\gamma-\alpha)\ell/2}}{1 - r_1 r_2 e^{-2i(k+\Delta k)\ell} e^{(\gamma-\alpha)\ell}}, \end{aligned} \tag{5.3}$$

where

$$\gamma(\nu) = \left[N_2 - \left(\frac{g_2}{g_1}\right) N_1\right] \left(\frac{c^2 A_{21}}{8\pi\nu^2}\right) g(\nu_0, \nu).$$

To make a further distinction between the characteristics of this active system bounded by two reflective interfaces we have replaced the ρ and τ coefficients of the last chapter with r and t, respectively. The ratio of input to output intensities is

$$\frac{E_t}{E_0} = \frac{I_t}{I_0} = \frac{t^2 t_2^2 e^{(\gamma-\alpha)\ell}}{(1 - r_1 r_2 e^{-2i(k+\Delta k)\ell} e^{(\gamma-\alpha)\ell})(1 - r_1 r_2 e^{2i(k+\Delta k)\ell} e^{(\gamma-\alpha)\ell})}, \tag{5.4}$$

which becomes

$$\frac{I_t}{I_0} = \frac{t^2 t_2^2 e^{(\gamma-\alpha)\ell}}{1 + r_1^2 r_2^2 e^{2(\gamma-\alpha)\ell} - 2 r_1 r_2 e^{(\gamma-\alpha)\ell}[\cos 2(k + \Delta k)\ell]}. \tag{5.5}$$

In a passive resonator, which has no gain γ or loss α, $\Delta k = 0$, and if $r_1 = r_2 = R$

$$\frac{I_t}{I_0} = \frac{T^2}{1 + R^2 - 2R \cos 2k\ell}.$$

This is the same result as we had before, since $2k\ell = \delta$ (compare with Eq. (4.44).

In a resonator containing an active medium, as $\gamma - \alpha$ increases from zero, the denominator of Eq. (5.3) approaches zero and the whole expression blows up when

$$r_1 r_2 e^{-2i(k+\Delta k)\ell} e^{(\gamma-\alpha)\ell} = 1. \tag{5.6}$$

When this happens we have an infinite amplitude transmitted wave for a finite amplitude incident wave. In other words, a finite amplitude transmitted wave for zero incident wave – *oscillation*. Physically, Eq. (5.6) is the condition that must be satisfied for a wave to make a complete round trip inside the resonator and return to its starting point with the same amplitude and, apart from a multiple of 2π, the same phase.

Eq. (5.6) provides an amplitude condition for oscillation that gives an expression for the threshold gain constant, $\gamma_t(\nu)$,

$$r_1 r_2 e^{[\gamma_t(\omega)-\alpha]\ell} = 1. \tag{5.7}$$

To satisfy Eq. (5.6), $e^{-2i(k+\Delta k)\ell}$ must be real, which provides us with the phase condition

$$2[k + \Delta k(\nu)]\ell = 2\pi m, m = 1, 2, 3\ldots. \tag{5.8}$$

The threshold gain coefficient can be written

$$\gamma_t(\nu) = \alpha - \frac{1}{\ell} \ln r_1 r_2, \tag{5.9}$$

which from the gain equation (2.68) gives the population inversion needed for oscillation

$$\left(N_2 - \frac{g_2}{g_1} N_1\right)_t = \frac{8\pi}{g(\nu_0, \nu) A_{21} \lambda^2} \left(\alpha - \frac{1}{\ell} \ln r_1 r_2\right). \tag{5.10}$$

For a homogeneously broadened transition the parametric variation of Eq. (5.10) that depends on the gain medium can be written as

$$\left(N_2 - \frac{g_2}{g_1} N_1\right)_t \propto \frac{\Delta\nu}{A_{21} \lambda^2}.$$

Whereas, for an inhomogeneously broadened transition since $\Delta\nu_D \propto 1/\lambda$

$$\left(N_2 - \frac{g_2}{g_1}N_1\right)_t \propto \frac{1}{A_{21}\lambda^3}.$$

Clearly, lower inversions are needed to achieve laser oscillation at longer wavelengths. It is much easier to build lasers that oscillate in the infrared than at visible, ultraviolet or X-wavelengths. For example, in an inhomogeneously broadened laser, a population inversion 10^6 times greater would be required for oscillation at 200 nm than at 20 μm (all other factors such as A_{21} being equal). In practice, since A_{21} factors generally increase at shorter wavelengths the difference in population inversion may not need to be as great as this.

In a resonator such as is shown in Fig. (5.1), if $R_1 = r_1^2 \simeq 1, R_2 = r_2^2 \simeq 1$ and distributed losses are small, a wave starting with intensity I inside the resonator will, after one complete round trip, have intensity $IR_1R_2e^{-2\alpha\ell}$, the change in intracavity intensity after one round trip is

$$dI = (R_1R_2e^{-2\alpha\ell} - 1)I. \tag{5.11}$$

This loss occurs in a time $dt = 2\ell/c$. So,

$$\frac{dI}{dt} = cI[R_1R_2e^{-2\alpha\ell} - 1]/2\ell. \tag{5.12}$$

This equation has the solution

$$I = I_0 \exp\{-[1 - R_1R_2e^{-2\alpha\ell}]ct/2\ell\}, \tag{5.13}$$

where I_0 is the intensity at time $t = 0$. The time constant for intensity (energy) loss is

$$\tau_0 = \frac{2\ell}{c(1 - R_1R_2e^{-2\alpha\ell})}. \tag{5.14}$$

Now if $R_1R_2e^{-2\alpha\ell} \simeq 1$, with α small as we have assumed here, then

$$(1 - R_1R_2e^{-2\alpha\ell}) \simeq -\ln(R_1R_2e^{-2\alpha\ell}) = -\ln(R_1R_2) + 2\alpha\ell \tag{5.15}$$

and we get

$$\tau_0 = \frac{2\ell}{c(2\alpha\ell - \ln R_1R_2)} = \frac{1}{c[\alpha - (1/2\ell)\ln r_1r_2]} \tag{5.16}$$

Thus, the threshold population inversion can be written

$$N_t = \frac{8\pi}{A_{21}\lambda^2 g(\nu)c\tau_0}. \tag{5.17}$$

5.2.1 Threshold Population Inversion – Numerical Example

For the 488 nm transition in the argon ion laser (discussed in Chapter 9)

$$\lambda = 488 \text{ nm}; c = 3 \times 10^8 \text{ m s}^{-1}; A_{21} \simeq 10^9 \text{ s}^{-1}; \Delta\nu_D \sim 3 \text{ GHz}.$$

Take $\ell = 1$ m, $R_1 = 100\%$, $R_2 = 90\%$ (typical values for a practical device). Since this is a gas laser internal losses are easily kept small so $\alpha \simeq 0$. In this case

$$\tau_0 = 2\ell/c(1 - R_1R_2)$$
$$= 66.67 \text{ ns}.$$

For oscillation at, or near line center

$$g(\nu_0, \nu_0) = \frac{2}{\Delta\nu_D}\sqrt{\frac{\ln 2}{\pi}} = \frac{0.94}{\Delta\nu_D} \sim \frac{1}{\Delta\nu_D}.$$

The threshold inversion is, from Eq. (5.17),

$$N_t = \frac{8\pi \times 3 \times 10^9}{10^9 \times (488 \times 10^{-9})^2 \times 3 \times 10^8 \times 66.67 \times 10^{-9}} = 1.58 \times 10^{13}\ \mathrm{m}^{-3}.$$

5.3 The Oscillation Frequency

To determine the frequency at which laser oscillation can occur we return to the phase condition, Eq. (5.8). This phase condition was

$$(k + \Delta k)\ell = m\pi, \tag{5.18}$$

which from Eq. (5.1) gives

$$k\ell\left[1 + \frac{\chi'(\nu)}{2n^2}\right] = m\pi. \tag{5.19}$$

Now, from Eq. (2.118)

$$\chi'(\nu) = \frac{2(\nu_0 - \nu)}{\Delta\nu}\chi''(\nu), \tag{5.20}$$

where ν_0 is the line center frequency and $\Delta\nu$ is its *homogeneous* FWHM, and

$$\gamma(\nu) = -\frac{k\chi''(\nu)}{n^2}. \tag{5.21}$$

So we must have

$$\frac{2\pi\nu\ell}{c}\left[1 - \frac{(\nu_0 - \nu)}{\Delta\nu}\frac{\gamma(\nu)}{k}\right] = m\pi, \tag{5.22}$$

and rearranging,

$$\nu\left[1 - \frac{(\nu_0 - \nu)}{\Delta\nu}\frac{\gamma(\nu)}{k}\right] = \frac{mc}{2\ell} = \nu_m, \tag{5.23}$$

where ν_m is the mth resonance of the passive laser resonator in normal incidence as calculated previously. Eq. (5.23) can be rewritten as

$$\nu = \nu_m - (\nu - \nu_0)\frac{\gamma(\nu)c}{2\pi\Delta\nu}. \tag{5.24}$$

We expect the actual oscillation frequency ν to be close to ν_m so we can write $(\nu - \nu_0) \simeq (\nu_m - \nu_0)$ and $\gamma(\nu) \simeq \gamma(\nu_m)$, to give

$$\nu = \nu_m - (\nu_m - \nu_0)\frac{\gamma(\nu_m)c}{2\pi\Delta\nu}. \tag{5.25}$$

At threshold

$$\gamma_t(\nu_m) = \alpha - \frac{1}{\ell}\ln r_1 r_2,$$

and if $\alpha \simeq 0, r_1 = r_2 = \sqrt{R}$

$$\gamma_t(\nu_m) = \frac{1 - R}{\ell}. \tag{5.26}$$

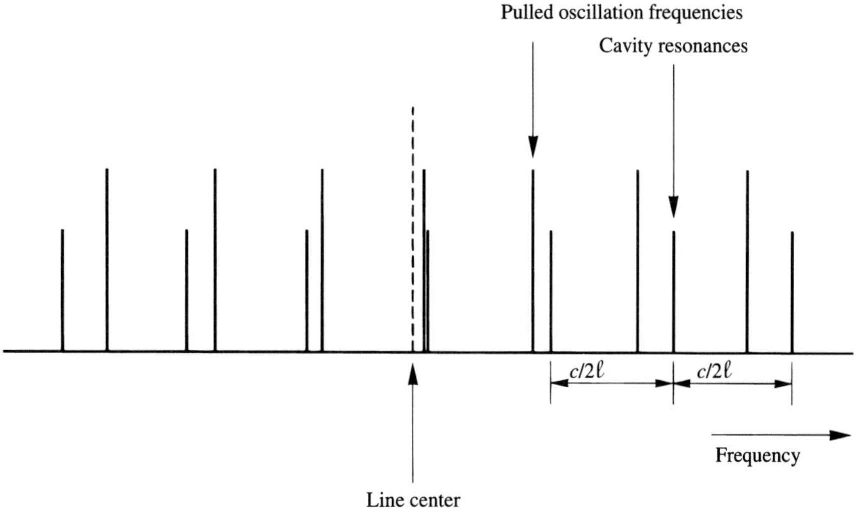

Fig. 5.2. Relative position of line center, Fabry–Perot resonances, and pulled oscillation frequencies that satisfy the phase condition (5.18).

Now the FWHM of the passive resonances (the transmission intensity maxima of the Fabry–Perot) is

$$\Delta\nu_{1/2} = \frac{\Delta\nu_{FSR}}{F} = \frac{c(1-R)}{2\pi\ell\sqrt{R}}, \tag{5.27}$$

which with $R \simeq 1$ gives

$$\Delta\nu_{1/2} = \frac{c(1-R)}{2\pi\ell}, \tag{5.28}$$

and finally,

$$\nu = \nu_m - (\nu_m - \nu_0)\frac{\Delta\nu_{1/2}}{\Delta\nu}. \tag{5.29}$$

Thus, if ν_m coincides with the line center, oscillation occurs at the line center. If $\nu_m \neq \nu_0$, oscillation takes place near ν_m but is shifted slightly towards ν_0. This phenomena is called 'mode-pulling' and is illustrated in Fig. (5.2).

5.4 Multimode Laser Oscillation

We have seen that for oscillation to occur in a laser system the gain must reach a threshold value $\gamma_t(\nu) = \alpha - (1/\ell)\ln r_1 r_2$. For gain coefficients greater than this, oscillation can occur at, or near (because of mode-pulling effects), one or more of the passive resonance frequencies of the Fabry–Perot laser cavity. The resulting oscillations of the system are called longitudinal modes. As oscillation at a particular one of these mode frequencies builds up, the growing intracavity energy density depletes the inverted population and gain saturation sets in. The reduction in gain continues until

$$\gamma(\nu) = \gamma_t(\nu) = \alpha - \frac{1}{\ell}\ln r_1 r_2. \tag{5.30}$$

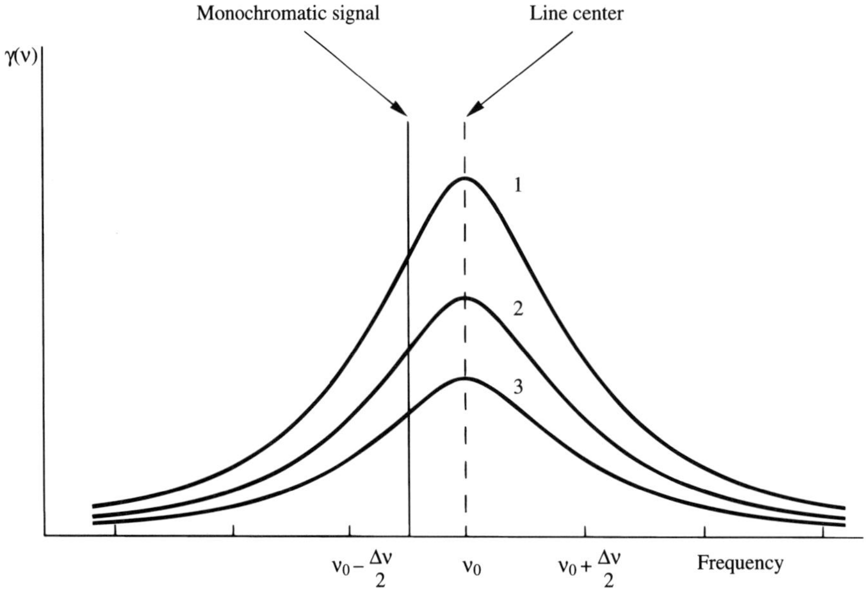

Fig. 5.3. Saturation of gain of a homogeneously broadened transition by a monochromatic signal whose intensity increases from 1→2→3.

Further reduction of $\gamma(\nu)$ below $\gamma_t(\nu)$ does not occur, otherwise the oscillation would cease. Therefore, the gain is stabilized at the loss

$$\alpha - \frac{1}{\ell}\ln r_1 r_2.$$

Usually α, r_1, and r_2 are nearly constant over the frequency range covered by typical amplifying transitions, so over such moderate frequency ranges, 10^{11} Hz say, $\alpha-(1/\ell)\ln r_1 r_2$ as a function of frequency is a straight line parallel to the frequency axis. This line is called the *loss line.* At markedly different frequencies α, r_1, and r_2 can be expected to change: for example, a laser mirror with high reflectivity in the red region of the spectrum could have quite low reflectivity in the blue.

In a homogeneously broadened laser, because the reduction in gain caused by a monochromatic field is uniform across the whole gain profile, the clamping of the gain at $\gamma_t(\nu)$ leads to final oscillation at only one of the cavity resonance frequencies, the one where the original unsaturated gain was highest. We can show this schematically by plotting $\gamma(\nu)$ at various stages as oscillation builds up. Remember first the effect on $\gamma(\nu)$ produced by a monochromatic light signal of increasing intensity as shown in Fig. (5.3). Note that the gain profile is depressed uniformly even though the saturating signal is not at the line center, as predicted by Eq. (2.69).

In a laser, as oscillation begins, several such monochromatic fields start to build up at those cavity resonances where gain exceeds loss, as shown in Fig. (5.4). The oscillation stabilizes when the highest (small-signal) gain has been reduced to the loss line by saturation as shown in Figs. (5.5) and (5.6). Thus, in a *homogeneously* broadened laser, oscillation only occurs at *one* longitudinal mode frequency.

In an inhomogeneously broadened laser, the onset of gain saturation due to a monochromatic signal only reduces the gain locally over a region which is of the order of a homogeneous width. Only particles whose velocities (or environments in a crystal) make their center emission frequencies lie within a homogeneous width of the monochromatic field can interact strongly with it. Schematically, the effect

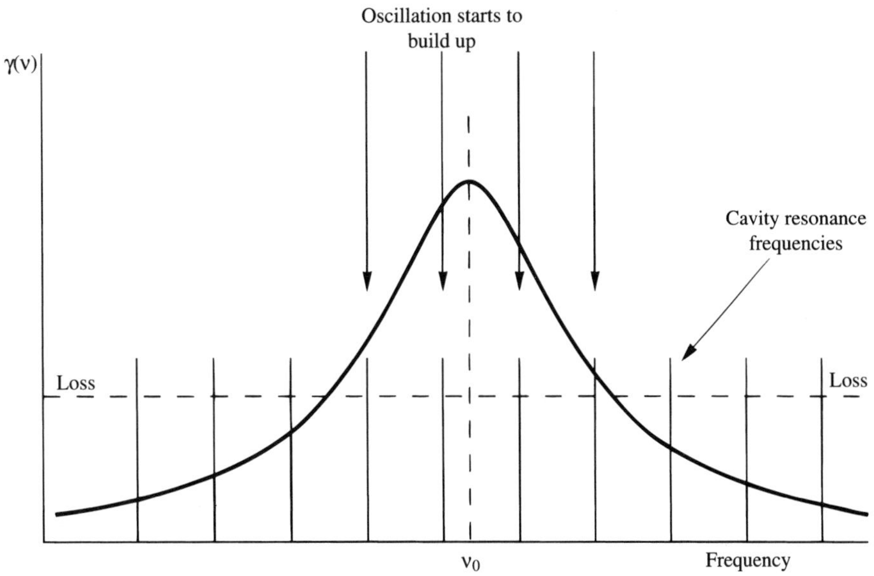

Fig. 5.4. Schematic illustration of the onset of oscillation at cavity resonances that lie above the loss line in a homogeneously broadened laser.

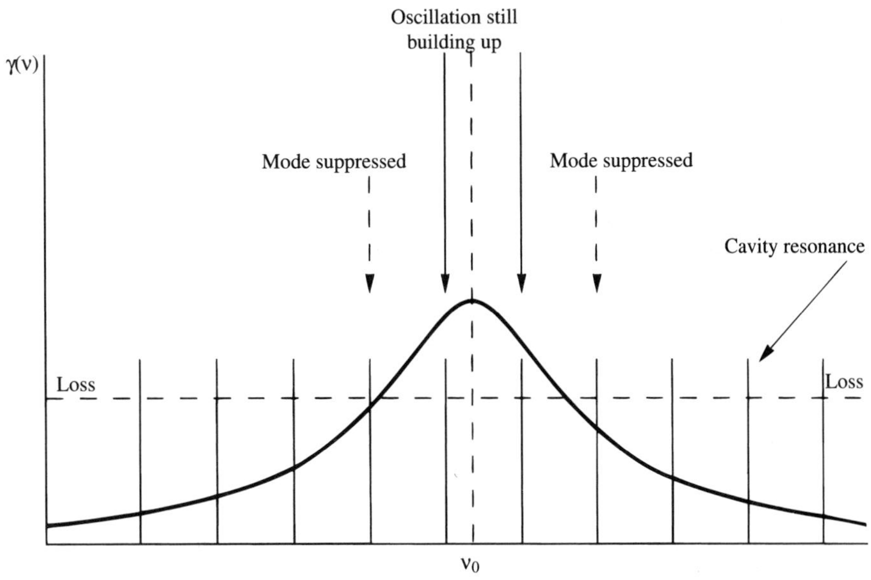

Fig. 5.5. Oscillation building up in a homogeneously broadened laser. Gain saturation has already suppressed oscillation at two of the cavity modes that were above the loss line in Fig. (5.4).

of an increasing intensity monochromatic field on the gain profile is as shown in Fig. (5.7). A localized dip, or *hole*, in the gain profile occurs. If only one cavity resonance has a small-signal gain above the loss line then only this longitudinal mode oscillates. The stabilization of the oscillation might be expected to occur schematically as shown in Fig. (5.8). However, the situation is not quite as simple as this! Oscillation at this single longitudinal mode frequency implies waves travelling in both directions inside the laser cavity. These waves can be represented by

(a) the wave travelling to the right $\sim E_0 e^{i(\omega t - kz)}$,

(b) the wave travelling to the left $\sim E_0 e^{i(\omega t + kz)}$,

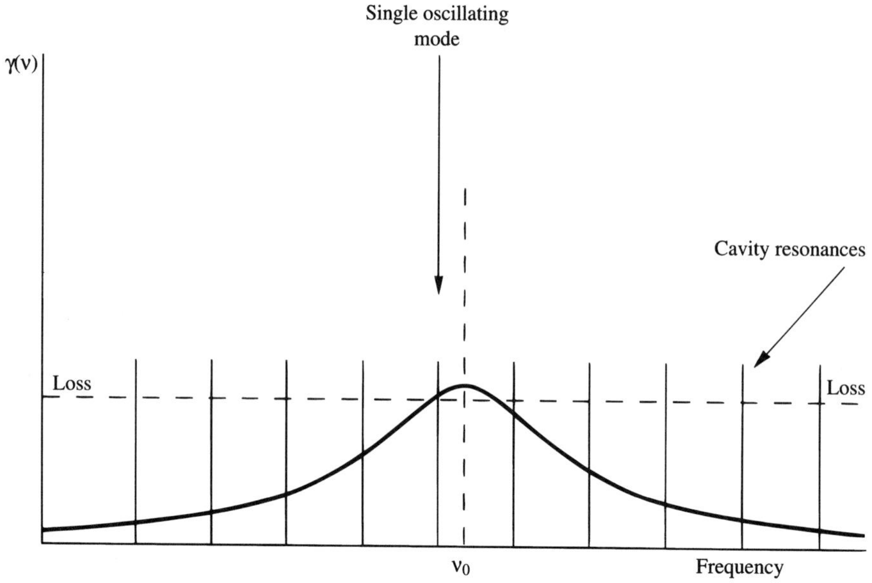

Fig. 5.6. Oscillation stabilized in a homogeneously broadened laser. The gain has been uniformly saturated until only one mode remains at the loss line.

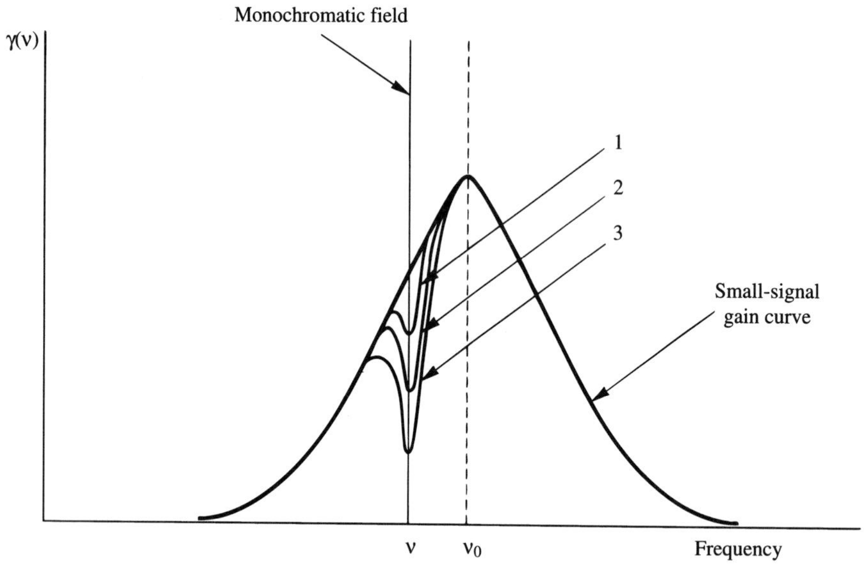

Fig. 5.7. Localized gain saturation in an inhomogeneously broadened amplifier produced by a monochromatic signal whose intensity increases from 1→2→3.

where we choose for convenience that $\omega = 2\pi\nu < 2\pi\nu_0$. Wave (a) can interact with particles whose center frequency is near ν. These particles are, as far as their Doppler shifts are concerned, moving away from an observer looking into the laser from right to left. Their center frequencies satisfy $\nu = \nu_0 - | v | \nu_0/c$, where positive atom velocities correspond to particles moving from left to right. Wave (b) which is travelling in the opposite direction (to the left) and is monitored, still at frequency $\nu (< \nu_0)$ by a second observer looking into the laser from left to right cannot interact with the same velocity group of particles as wave (a). The particles

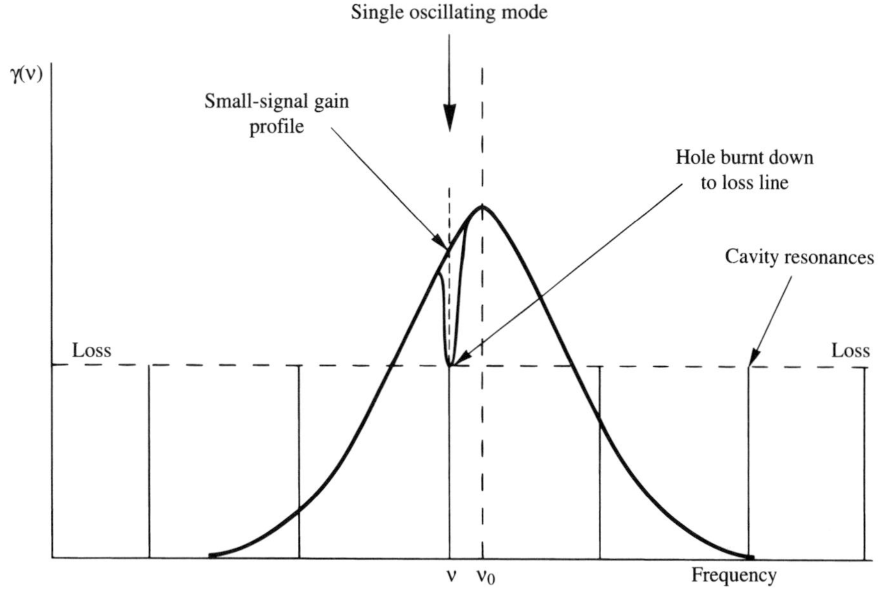

Fig. 5.8. Simplified illustration of stablization of oscillation at a single longitudinal mode in an inhomogeneously broadened laser.

which interacted with wave (a) were moving away from the first observer and were Doppler shifted to lower frequencies so as to satisfy

$$\nu = \nu_0 - \frac{|v|}{c}\nu_0. \tag{5.31}$$

The second observer sees these particles approaching and their center frequency as

$$\nu = \nu_0 + \frac{|v|}{c}\nu_0, \tag{5.32}$$

so they cannot interact with wave (b). Wave (b) interacts with particles moving away from the second observer so that their velocity would be the solution of

$$\nu = \nu_0 - \frac{|v|}{c}\nu_0. \tag{5.33}$$

These particles would be monitored by the first observer at center frequency

$$\nu = \nu_0 + \frac{|v|}{c}\nu_0. \tag{5.34}$$

So the oscillating waves interact with two velocity groups of particles as shown in Fig. (5.9). This leads to saturation of the gain by a single laser mode in an inhomogeneously broadened laser both at the frequency of the mode ν and at a frequency $\nu_0 + (\nu_0 - \nu)$, which is equally spaced on the opposite side of the line center, as shown in Fig. (5.10). The power output of the laser (strictly the intracavity power) comes from those groups of particles that have gone into stimulated emission and left the two holes. The combined area of these two holes gives a measure of the laser power.

If the frequency of the oscillating mode is moved in towards the line center, the main hole and image hole begin to overlap. This corresponds physically to the left and right travelling waves within the laser cavity beginning to interact with the same velocity group of particles. As the oscillating mode moves in towards the line

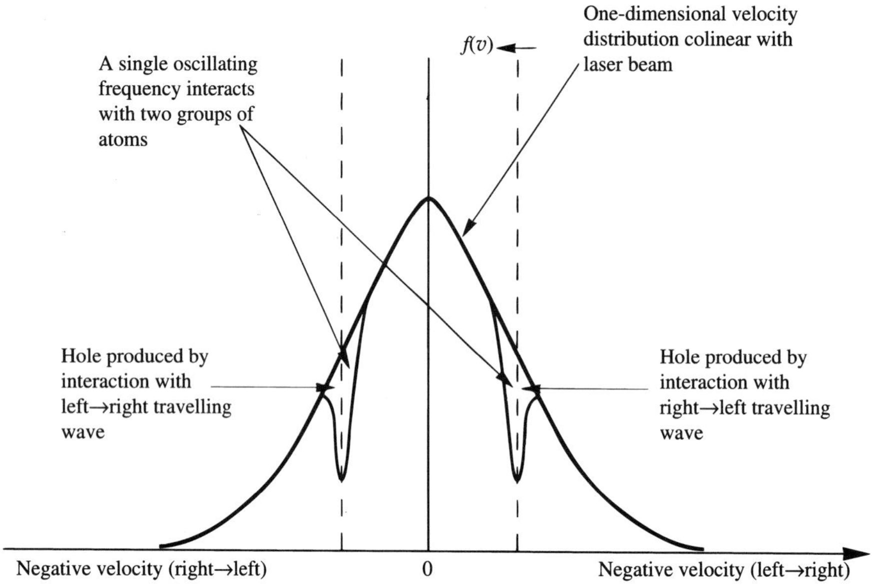

Fig. 5.9. Production of two holes in the velocity distribution of a collection of amplifying particles by a single cavity mode.

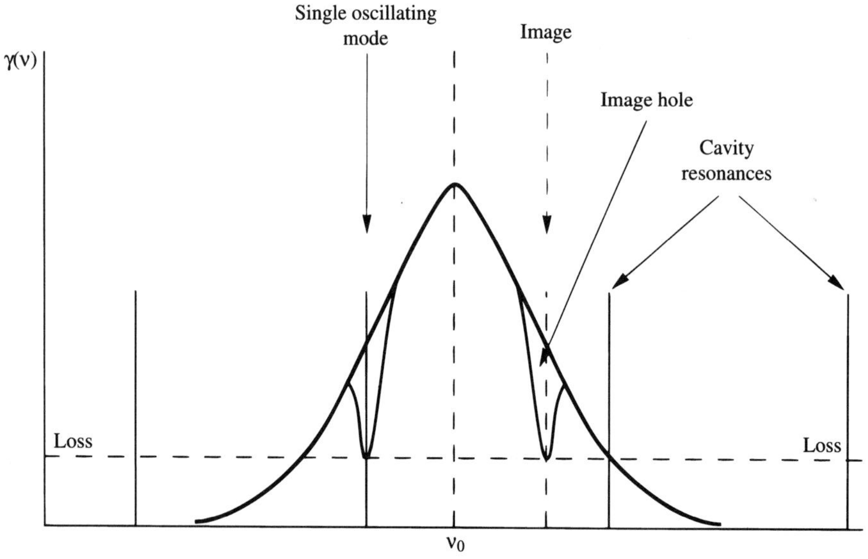

Fig. 5.10. Stabilization of a single longitudinal mode in an inhomogeneously broadened laser.

center, the holes overlap further, the combined area decreases and the laser output power falls, reaching a minimum at the line center. This phenomena is called the Lamb dip[5.1], named after Willis E. Lamb, Jr, who first predicted the effect, and is illustrated in Fig. (5.11). When the cavity resonance is at the line center frequency ν_0, both travelling waves are interacting with the same group of atoms – those with near-zero directed velocity along the laser resonator axis.

Because hole-burning in gain saturation in inhomogeneously broadened lasers is localized near the frequency of a cavity mode, one oscillating mode does not reduce

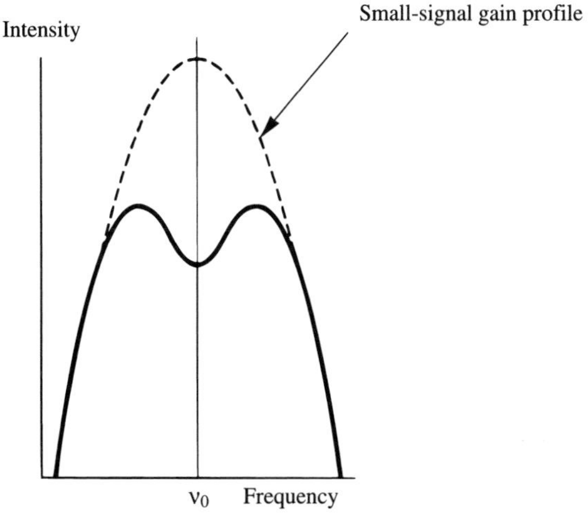

Fig. 5.11. The Lamb dip – a reduction in the intensity of a single oscillating longitudinal mode in an inhomogeneously broadened laser as its frequency is scanned through line center.

the gain at other cavity modes, so simultaneous oscillation at several longitudinal modes is possible. If several such modes have small-signal gains above the loss line the oscillation stabilizes in the manner shown in Fig. (5.12a). The output frequency spectrum from the laser would appear as is shown in Fig. (5.12b). This simultaneous oscillation at several closely spaced frequencies ($c/2\ell$ apart) can be observed with a high resolution spectrometer – for example a scanning Fabry–Perot interferometer as shown in Fig. (5.13). The multiple modes are almost exactly $c/2\ell$ in frequency apart, but are not exactly equally spaced because of mode-pulling. This effect can be observed in the beat spectrum observed with a square-law optical detector (which means most optical detectors). Such a detector responds to the intensity, not the electric field of an incident light signal.

5.5 Mode-Beating

Suppose we shine the light from a two-mode laser on a square-law detector. The incident electric field is

$$E_i = \mathscr{R}(E_1 e^{i\omega t} + E_2 e^{i(\omega+\Delta\omega)t}), \tag{5.35}$$

where E_1 and E_2 are the complex amplitudes of the two modes and $\Delta\omega$ is the frequency spacing between them. Using real notation for these fields the output current i from the detector is

$$\begin{aligned} i &\propto \{|E_1|\cos[(\omega t + \phi_1) + |E_2|\cos[(\omega+\Delta\omega)t + \phi_2]\}^2 \\ &\propto |E_1|^2\cos^2(\omega t + \phi_1) + |E_2|^2\cos^2[(\omega+\Delta\omega)t + \phi_2] \\ &\quad + 2|E_1||E_2|\cos[(\omega t + \phi_1)\cos[(\omega+\Delta\omega)t + \phi_2] \\ &\propto |E_1|^2\cos^2(\omega t + \phi_1) + |E_2|^2\cos^2[(\omega+\Delta\omega)t + \phi_2] \\ &\quad + |E_1||E_2|\cos[(2\omega+\Delta\omega)t + \phi_1 + \phi_2] \\ &\quad + |E_1||E_2|\cos(\Delta\omega t + \phi_2 - \phi_1). \end{aligned} \tag{5.36}$$

And since, for example,

$$|E_1|^2\cos^2(\omega t + \phi_1) = \tfrac{1}{2}|E_1|^2[1 + \cos 2(\omega t + \phi_1)], \tag{5.37}$$

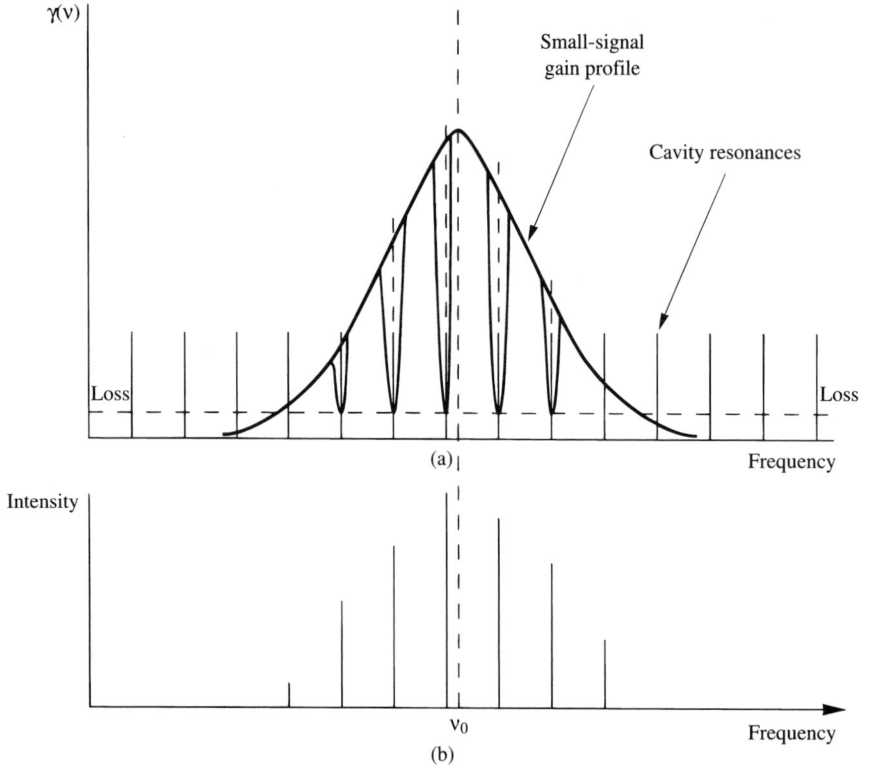

Fig. 5.12. Multi-longitudinal-mode oscillation in an inhomogeneously broadened laser. (a) Only the primary holes are shown burnt down to the loss line. The image holes are not shown. (b) Schematic laser output spectrum.

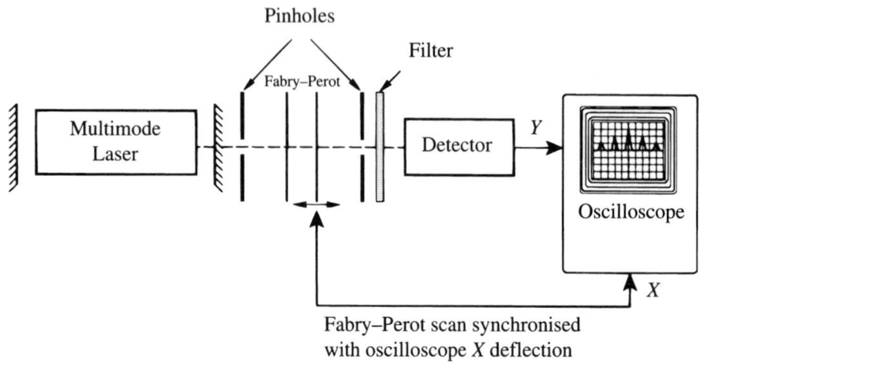

Fig. 5.13. Experimental arrangement with a scanning Fabry–Perot interferometer for observing multi-mode laser oscillation.

the output frequency spectrum of the detector appears to contain the frequencies $2\omega, 2(\omega + \Delta\omega), 2\omega + \Delta\omega$ and $\Delta\omega$. However, the first three of these frequencies are very high, particularly for light in the visible and infrared regions of the spectrum, and do not appear in the output of the detector. It is as if the high frequency terms are averaged to zero by the detector time response to give

$$i \propto \frac{|E_1|^2}{2} + \frac{|E_2|^2}{2} + |E_1||E_2|\cos(\Delta\omega t + \phi_2 - \phi_1), \tag{5.38}$$

so only the difference frequency beat $\Delta\omega$ is observed. This result can be derived directly using the *analytic* signal of the incident electric field. The *analytic* signal of a field which is $E(t) = \mathscr{R}(Ee^{i\omega t})$ is

$$V(t) = Ee^{i\omega t}, \tag{5.39}$$

so that $E(t) = \mathscr{R}[V(t)]$. The response of a square-law detector can be found directly from $i \propto V(t)V^*(t)$. The significance of this is discussed further in Chapter 23.

If the output from the square-law detector is analyzed with a radio-frequency spectrum analyzer (because it is in this frequency range where the difference frequencies between longitudinal laser modes are usually observed) different displays are obtained according to how many longitudinal modes of a multimode laser are simultaneously oscillating. Fig. (5.14) gives some examples. Because Eq. (5.29) is not quite exact, the beat frequencies can split as shown because of nonlinear mode-pulling. This splitting will only be observed if nonlinear mode-pulling is large and the spectrum analyzer that analyzes the output of the photo-dector has high resolution.

If a predominantly inhomogeneously broadened laser also has a significant amount of homogeneous broadening, the holes burnt in the gain curve can start to overlap, for example, when $\Delta\nu \gtrsim c/2\ell$. If $\Delta\nu$ is large enough this causes neighboring oscillating modes to compete, and may lead to oscillation on a strong mode suppressing its weaker neighbors, as shown in Fig. (5.15). This effect has been observed in several laser systems, for example in the argon ion laser, where an increase in the strength of the oscillation can lead to the successive disappearance, first of every other mode, then two modes out of every three, and so on.

5.6 The Power Output of a Laser

When a laser oscillates, the intracavity field grows in amplitude until saturation reduces the gain to the loss line for each oscillating mode. What this means in practice can be best illustrated with reference to Fig. (5.16).

For an asymmetrical resonator, whose mirror reflectances are not equal, the distribution of standing wave energy within the resonator is not symmetrical. For example, in Fig. (5.16), if $R_2 > R_1$ the distribution of intracavity travelling wave intensity will be schematically as shown, and

$$\frac{I_3}{I_2} = R_2; \qquad \frac{I_1}{I_4} = R_1. \tag{5.40}$$

The left travelling wave, of intensity I_-, grows in intensity from I_3 to I_4 on a single pass. The right travelling wave, of intensity I_+, grows in intensity from I_1 to I_2 on a single pass. The total output intensity is

$$I_{out} = T_2 I_2 + T_1 I_4. \tag{5.41}$$

However, calculation of I_2 and I_4 is not straightforward in the general case. I_2 grows from I_1 through a gain process that depends in a complex way on $I_+ + I_-$ as does the growth of I_3 to I_4. We can identify at least three scenarios in which the calculation proceeds differently:

(a) A homogeneously broadened amplifier and single mode operation.

(b) An inhomogeneously broadened amplifier and single mode operation.

(c) An inhomogeneously broadened amplifier and multimode operation.

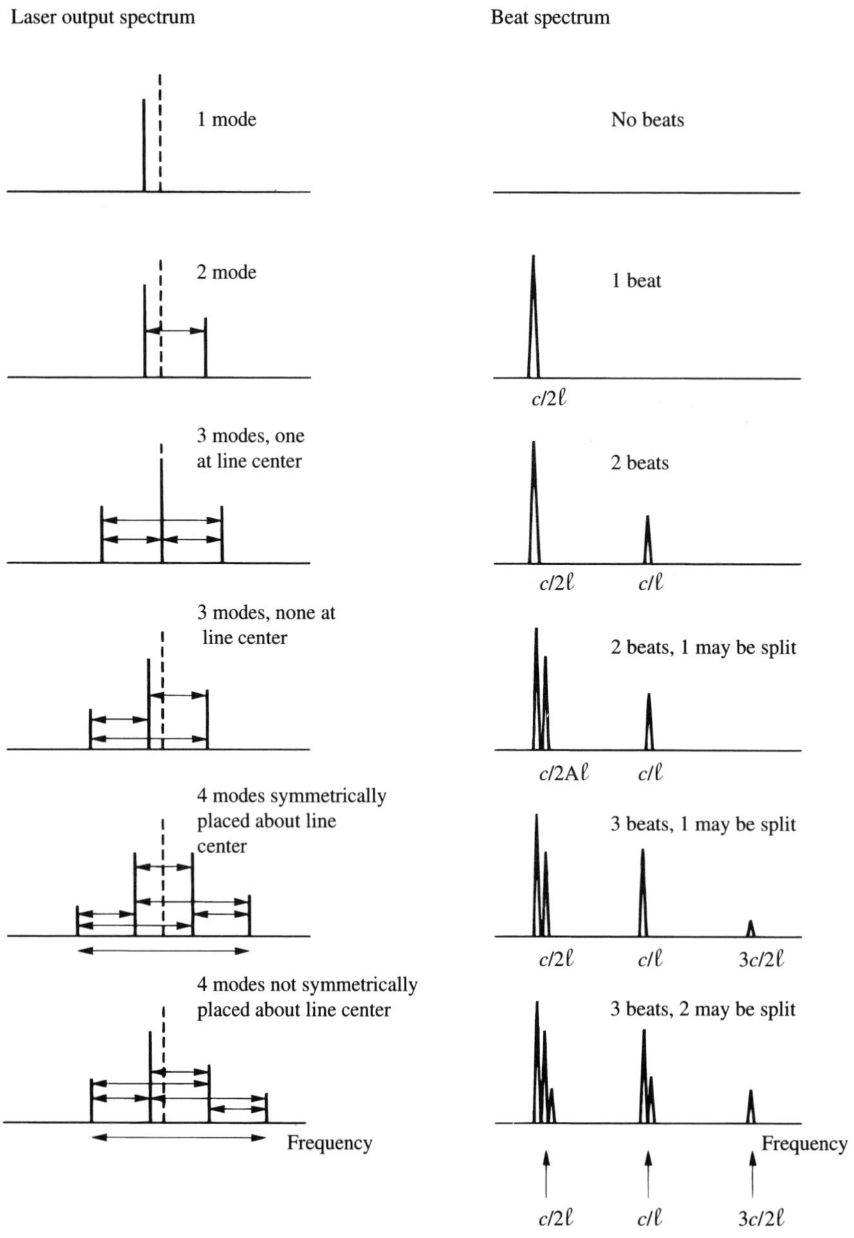

Fig. 5.14. Schematic mode-beating spectra observed with a square-law detector and a multimode laser.

In both cases (b) and (c) the calculation of the output power becomes more complicated as the homogeneous contribution to the broadening grows more significant compared to $\Delta\nu_D$. This additional complexity arises because each oscillating mode burns both a primary and an image hole in the gain curve. The resultant distribution of overlapping holes makes the gain for each mode dependent not only on its own intensity but also on the intensity of the other simultaneously oscillating modes. The presence of distributed intracavity loss presents additional complications. We shall not attempt to deal with these complex situations here

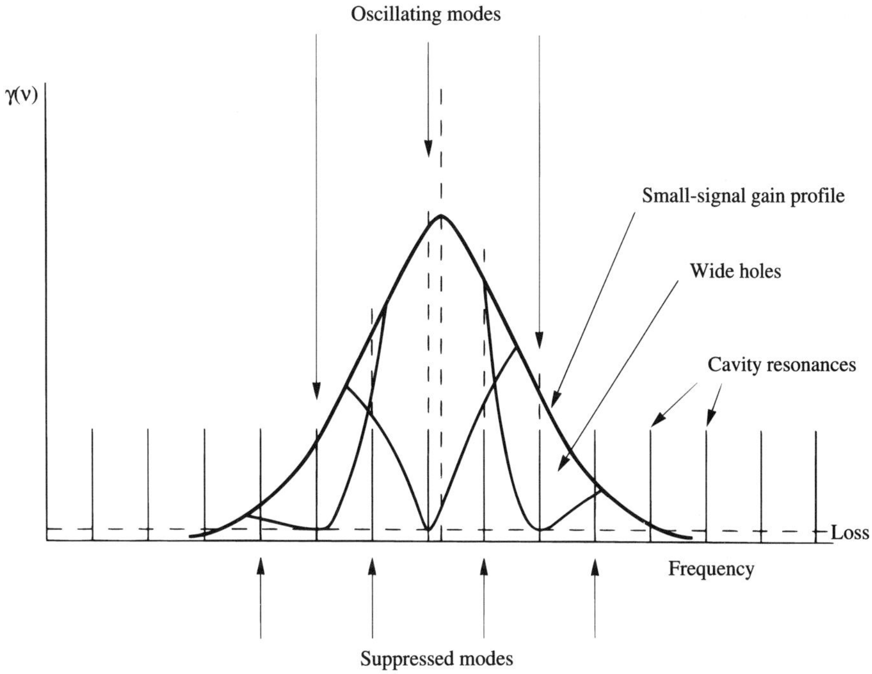

Fig. 5.15. Schematic illustration of mode competition in an inhomogeneously broadened laser in which there is significant homogeneous broadening.

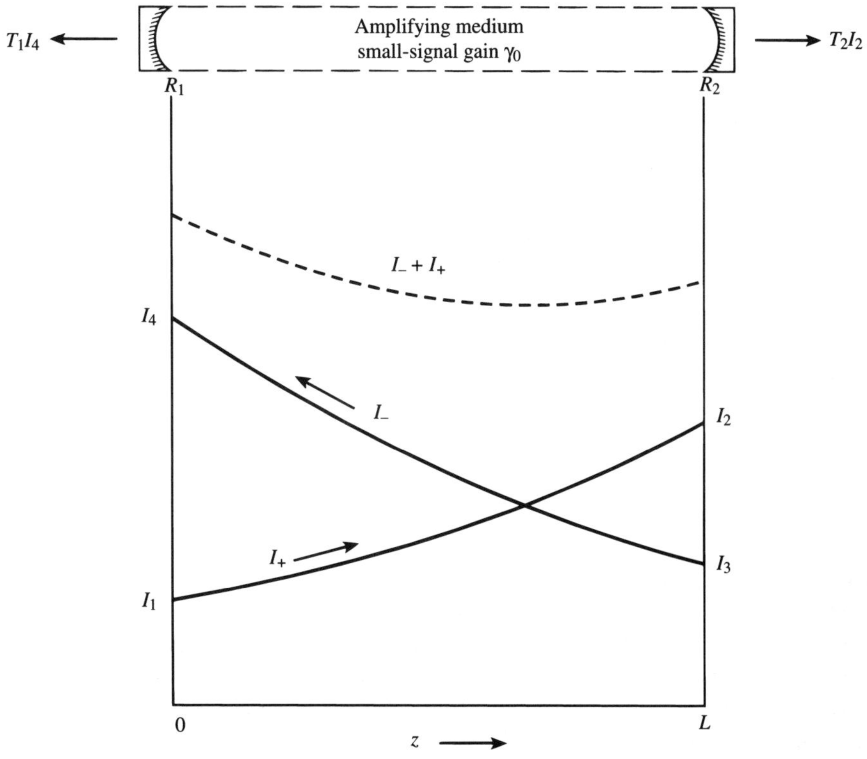

Fig. 5.16. Distribution of wave intensities in an oscillating laser cavity with unequal laser intensities.

but will follow Rigrod[5.3] in dealing with a homogeneously broadened amplifier in which the primary intensity loss occurs at the mirrors. Inhomogeneously broadened systems and multimode operation have been discussed elsewhere by Smith.[5.4]

In a purely homogeneously broadened system the saturated gain in Fig. (5.16) is

$$\gamma(z) = \frac{\gamma_0}{1+(I_+ + I_-)/I_s}. \tag{5.42}$$

Both I_- and I_+ grow according to $\gamma(z)$

$$\frac{1}{I_+}\frac{dI_+}{dz} = -\frac{1}{I_-}\frac{dI_-}{dz} = \gamma(z). \tag{5.43}$$

Consequently,

$$I_+ I_- = \text{constant} = C. \tag{5.44}$$

From Eq. (5.39)

$$I_4 I_1 = I_2 I_3 = C, \tag{5.45}$$

and therefore from Eq. (5.40)

$$I_2/I_4 = \sqrt{R_1/R_2}. \tag{5.46}$$

For the right travelling wave, using Eqs. (5.43) and (5.44) gives

$$\frac{1}{I_+}\frac{dI_+}{dz} = \frac{\gamma_0}{1+(I_+ + C/I_+)/I_s}, \tag{5.47}$$

which can be integrated to give

$$\gamma_0 L = \ln\left(\frac{I_2}{I_1}\right) + \frac{(I_2 - I_1)}{I_s} - \frac{C}{I_s}\left(\frac{1}{I_2} - \frac{1}{I_1}\right). \tag{5.48}$$

In a similar way, for the left travelling wave

$$\gamma_0 L = \ln\left(\frac{I_4}{I_3}\right) + \frac{(I_4 - I_3)}{I_s} - \frac{C}{I_s}\left(\frac{1}{I_4} - \frac{1}{I_3}\right). \tag{5.49}$$

Adding Eqs. (5.48) and (5.49) and using Eqs. (5.40), (5.45) , and (5.46) gives

$$I_2 = \frac{I_s\sqrt{R_1}(\gamma_0 L + \ln\sqrt{R_1 R_2})}{(\sqrt{R_1}+\sqrt{R_2})(1-\sqrt{R_1 R_2})}. \tag{5.50}$$

From Eq. (5.46)

$$I_4 = I_2\sqrt{\frac{R_2}{R_1}}. \tag{5.51}$$

Now

$$T_1 = 1 - R_1 - A_1, \tag{5.52}$$

$$T_2 = 1 - R_2 - A_2, \tag{5.53}$$

so from Eqs. (5.41) and (5.50), if $A_1 = A_2 = A$

$$I_{out} = I_s\frac{(1 - A - \sqrt{R_1 R_2})}{1-\sqrt{R_1 R_2}}(\gamma_0 L + \ln\sqrt{R_1 R_2}). \tag{5.54}$$

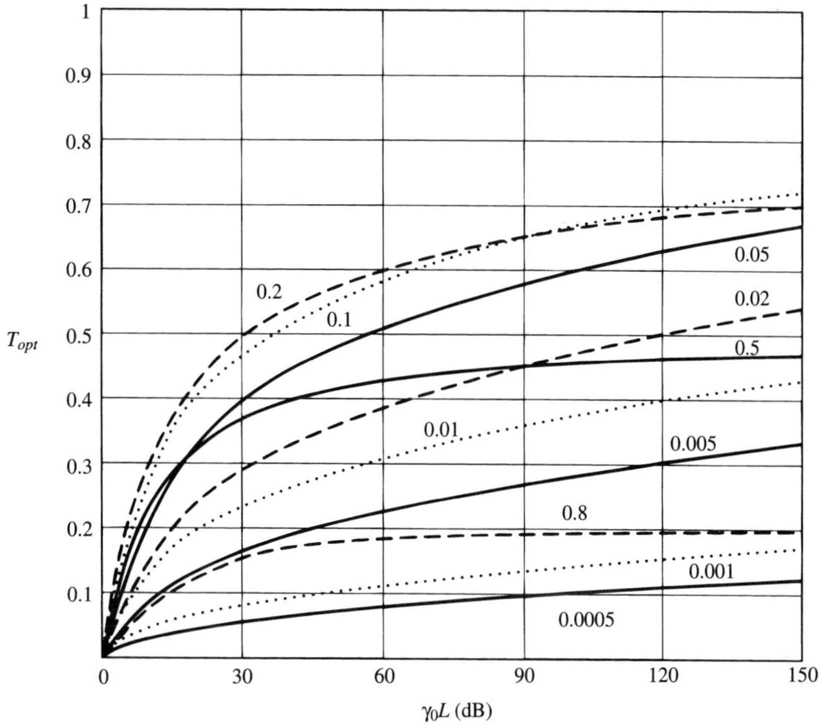

Fig. 5.17. Calculated optimum coupling for a symmetrical resonator for various values of the loss paramater A and the unsaturated gain.

If one mirror is made perfectly reflecting, say $T_1 = 0,\ R_1 = 1$, then

$$I_{out} = T_2 I_2 = \frac{T_2 I_s[\gamma_0 L + \frac{1}{2}\ln(1 - A_2 - T_2)]}{(A_2 + T_2)}. \tag{5.55}$$

For a symmetrical resonator, defined by

$$R_1 R_2 = R^2,$$
$$R = 1 - A - T, \tag{5.56}$$

the output intensity at each mirror is

$$\frac{I_{out}}{2} = \frac{I_s}{2}\frac{(1 - A - R)}{1 - R}(\gamma_0 L + \ln R). \tag{5.57}$$

5.7 Optimum Coupling

To maximize the output intensity from the symmetrical resonator we must find the value of R such that $\partial I_{out}/\partial R = 0$ which gives

$$\frac{T_{opt}}{A} = \left(\frac{1 - A - T_{opt}}{A + T_{opt}}\right)[\gamma_0 L + \ln(1 - A - T_{opt})]. \tag{5.58}$$

For small losses, such that $A + T_{opt} \ll 1$ Eq. (5.58) gives

$$\frac{T_{opt}}{A} = \sqrt{\frac{\gamma_0 L}{A}} - 1. \tag{5.59}$$

Fig. (5.17) shows the calculated optimum coupling for various values of the loss parameter A and the unsaturated gain in dB (4.343 $\gamma_0 L$).

In practice, it should be pointed out, the optimum mirror transmittance in a laser system is generally determined empirically. For example, for the CW CO_2 laser, whose unsaturated gain varies roughly inversely with the tube diameter d, the optimum mirror transmittance has been determined to be $T \simeq L/500d$.

5.8 Problems

(5.1) A four-level laser is pumped into its pump band at a rate 10^{24} m^{-3} s^{-1}, the transfer efficiency to the upper level is 0.5. The lifetime of the upper laser level is 7×10^{-4} s. For the laser transition $A_{21} = 10^3$ s^{-1}, $\lambda_0 = 1$ μm. The laser is homogeneously broadened with $\Delta\nu = 1$ GHz. Assume $n = 1.6$. The amplifying medium is 20 mm long. Neglect lower laser level population. (a) What is the gain at line center? (b) What minimum value of R_2 is needed to get oscillation if mirror 1 has $R_1 = 1$? Assume $\alpha = 0$.

(5.2) How many longitudinal modes will oscillate in an inhomogenously broadened gas laser with ℓ=1 m, $\gamma(\nu_0)$=1 m^{-1}, R_1=R_2=99%, $\alpha_{distributed\ loss}$=0.001 m^{-1}, $\lambda_0 = 500$ nm, $\Delta\nu_D = 3$ GHz.

(5.3) A gas laser is operating simultaneously on five modes, none of which is at line center. The laser beam illuminates a square-law detector. Draw the RF beat spectrum that will be observed. Is the beat signal near $c/2\ell$ split?

(5.4) Derive an expression for the peak transmittance of a Fabry–Perot filter that has a round-trip in-cavity absorption that causes the intensity to change from I_0 to AI_0 on a round trip between the two mirrors.

(5.5) Derive the amplitude condition and the laser oscillation frequencies for a laser whose cavity is of length L and whose amplifying medium is of length ℓ, where $\ell < L$.

(5.6) A laser is exactly 1 m long and has a wavelength $\lambda_0 = 632.8$ nm. The mirrors of the laser have $R = 99\%$. The index of refraction of the amplifying medium is exactly 1.0001, $\Delta\nu = 100.000$ MHz. The laser operates on only the two modes nearest to the line center. The laser output illuminates a photodiode whose output is mixed with a 150 MHz local oscillator. What is the frequency of the lowest beat signal observed? Take the velocity of light in free space to be 2.997×10^8 m s^{-1}.

(5.7) A gas laser with $\lambda_0 = 325$ nm has $\gamma_0(\nu_0) = 0.1$ m^{-1}, and $\Delta\nu_D = 3$ GHz. The laser medium fills the space between two mirrors 500 mm apart. The refractive index of the laser medium can be assumed to be 1, $\Delta\nu_{homogeneous} = 10$ MHz. The distributed loss parameter is $\alpha_{distributed\ loss} =$ 0.01 m^{-1}. What minimum equal mirror reflectance is needed to allow ten longitudinal modes to oscillate. What would happen if $\Delta\nu_{homogeneous} =$ 500 MHz?

(5.8) A Fabry–Perot interferometer has two mirrors of reflectance R spaced by a distance ℓ and the space between its plates is filled with a material of absorption coefficient α, refractive index n. Calculate its contrast ratio $C = (I_t/I_0)_{max}/(I_t/I_0)_{min}$.

(5.9) A single mode homogeneously broadened laser system with a small signal gain at the oscillation frequency of 0.001 m^{-1} has total power output of 1 mW when operated in a symmetrical resonator with $R_1 = R_2 = 0.99$,

L=0.5 m. The loss in each mirror is 10^{-4}. Calculate the saturation intensity.

(5.10) Write a computer program to solve Eq. (5.58) for the optimum mirror transmittance for arbitrary values of L, γ_0, and A. Find the value of T_{opt} for $L = 1$ m, $\gamma_0 = 1$ m^{-1}, $A = 10^{-4}$.

References

[5.1] W.E. Lamb, 'Theory of an optical maser,' *Phys. Rev.* **134A**, 1429–1450, 1964.

[5.2] A. Szoke and A. Javan, 'Isotope shift and saturation behavior of the 1.15 μm transition of neon,' *Phys. Rev. Lett*, **10**, 521–524, 1963.

[5.3] W.W. Rigrod, 'Saturation effects in high-gain lasers,' *J. Appl. Phys.* **36**, 2487–2490, 1965; see also W.W. Rigrod, 'Gain saturation and output power of optical masers,' *J. Appl. Phys.* **34**, 2602–2609, 1963 and W.W. Rigrod, 'Homogeneously broadened CW lasers with uniform distributed loss,' *IEEE J. Quant. Electron.* **QE-14**, 377–381, 1978.

[5.4] P.W. Smith, 'The output power of a 6328 Å He–Ne gas laser,' *IEEE J. Quant. Electron.* **QE-2**, 62–68, 1966.

6

Laser Radiation

6.1 Introduction

In this chapter we shall examine some of the characteristics of laser radiation that distinguish it from ordinary light. Our discussion will include the monochromaticity and directionality of laser beams, and a preliminary discussion of their *coherence* properties. Coherence is a measure of the temporal and spatial phase relationships that exist for the fields associated with laser radiation.

The special nature of laser radiation is graphically illustrated by the ease with which the important optical phenomena of interference and diffraction are demonstrated using it. This chapter includes a brief discussion of these two phenomena with some examples of how they can be observed with lasers. Interference effects demonstrate the coherence properties of laser radiation, while diffraction effects are intimately connected with the beam-like properties that make this radiation special.

6.2 Diffraction

Diffraction of light results whenever a plane wave has its lateral extent restricted by an obstacle. By definition, a plane wave travelling in the z direction has no field variations in planes orthogonal to the z axis, so the derivatives $\partial/\partial x$, or $\partial/\partial y$ operating on any field component give zero. Clearly this condition cannot be satisfied if the wave strikes an obstacle: at the edge of the obstacle the wave is obstructed and there must be variations in field amplitude in the lateral direction. In other words, the derivative operations $\partial/\partial x$, $\partial/\partial y$ do not give zero and the wave after passing the obstacle is no longer a plane wave. When the wave ceases to be a plane wave, its phase fronts are no longer planes and there is no unique propagation direction associated with the wave. The range of $\mathbf{k}$ vectors associated with the new wave gives rise to lateral variations in intensity observed behind the obstacle – the *diffraction pattern.*

We can show that at sufficiently great distances from the obstacle, or aperture, the diffraction pattern can be obtained as a Fourier transform of the amplitude distribution at the obstacle or aperture. There is a close parallel between this phenomenon, where light travels within a range of angular directions determined by the size of an obstacle, and the frequency spread associated with a waveform that is restricted in time. We saw in Chapter 2 that the frequency spectrum of a waveform is given by the Fourier transform of its waveform. It is perhaps not surprising to learn that the angular spectrum of a plane wave transmitted through an aperture is a Fourier transform over the shape and transmission of the aperture.

Consider two plane waves whose field variations are of the general form $e^{i(\omega t-\mathbf{k}\cdot\mathbf{r})}$

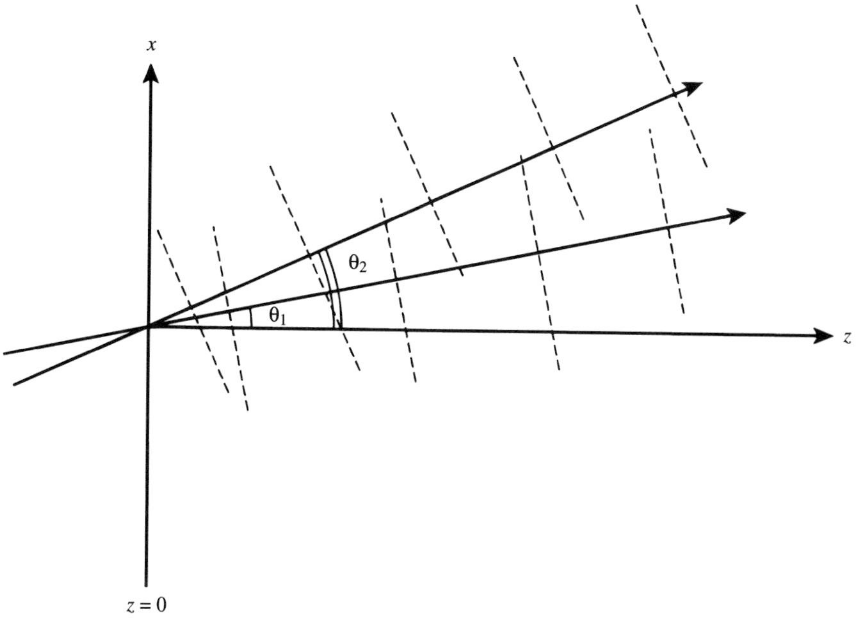

Fig. 6.1. Superposition of two plane waves travelling at different angles in the xz plane.

travelling at slightly different small angles and giving rise to a resultant field in the $z = 0$ plane, as shown in Fig. (6.1).

If the first wave has field amplitudes of the form

$$V_1 = V_0 e^{i(\omega t - k\sin\theta_1 x - k\cos\theta_1 z)} \tag{6.1}$$

and the second has

$$V_2 = V_0 e^{i(\omega t - k\sin\theta_2 x - k\cos\theta_2 z)}, \tag{6.2}$$

the resultant disturbance at $z = 0$ is

$$V_1 + V_2 = V_0 e^{i\omega t}\left(e^{-ik_1 x} + e^{-ik_2 x}\right). \tag{6.3}$$

Because θ is a small angle we can assume that the V_1 and V_2 vectors are parallel, and we have written

$$k_1 = k\sin\theta_1; k_2 = k\sin\theta_2, k = 2\pi/\lambda. \tag{6.4}$$

The assumption that the field vectors are parallel becomes more valid the further from the obstacle the resultant field amplitude is being calculated.

The resultant disturbance due to many such waves can be written as

$$V = V_0 e^{i\omega t}\sum_n e^{-ik_n x} \tag{6.5}$$

In the limit of infinitely many waves, this summation can be written as an integral over a continuous distribution of waves for which the amplitude distribution is $a(k_x)$. For example, the total amplitude of the group of waves travelling in the small range of angles corresponding to a range dk_x at k_x is $a(k_x)dk_x$. The total disturbance in the plane $z = 0$ is

$$V_0(x) = \int_{-\infty}^{\infty} a(k_x)e^{-ik_x x}dk_x. \tag{6.6}$$

The limits on the integral in Eq. (6.6) are set to $\pm\infty$. Implicit in this is the physical reality that $a(k_x) = 0$ for any $|k_x| > |\mathbf{k}|$.

Recognizing Eq. (6.6) as a Fourier transform we can write

$$a(k_x) = \frac{1}{2\pi}\int_{-\infty}^{\infty} V_0(x)e^{ik_x x}dx. \tag{6.7}$$

Thus, if we know the field distribution at the aperture or obstacle, given by $V(x)$, we can calculate $a(k_x)$, which gives a measure of the contribution of various plane waves in the diffraction pattern. This is because an observer cannot distinguish between light coming from an aperture and a collection of plane waves that combine in space to give nonzero field amplitudes only in a region corresponding to the aperture.

6.3 Two Parallel Narrow Slits

As an example, let us look at the case of a pair of slits, which we will represent as a pair of δ-function sources. If the slits are at $x = \pm a$ in the plane $z = 0$ and are parallel to the y axis,

$$V_0(x) = \delta(x - a) + \delta(x + a). \tag{6.8}$$

So

$$\begin{aligned} a(k_x) &= \frac{1}{2\pi}\int_{-\infty}^{\infty} [\delta(x - a) + \delta(x + a)]e^{ik_x x}dx \\ &= \frac{1}{2\pi}(e^{ik_x a} + e^{-ik_x a}) = \frac{\cos k_x a}{\pi}. \end{aligned} \tag{6.9}$$

Note that

$$|a(k_x)|^2 \propto \cos^2 k_x a = \cos^2\left(\frac{2\pi a \sin\theta}{\lambda}\right). \tag{6.10}$$

In the center of the diffraction pattern $|a(k_0)|^2 \propto 1$. Thus:

$$\frac{|a(k_x)|^2}{|a(k_0)|^2} = \cos^2\left(\frac{2\pi a \sin\theta}{\lambda}\right). \tag{6.11}$$

We can interpret $|a(k_x)|^2$ as the intensity in the diffraction pattern at angle θ relative to the intensity in the center of the pattern. The conditions under which Eq. (6.7) can be used to calculate the relative intensity in the diffraction pattern can be roughly stated as $z \gg D^2/\lambda$; $\theta^2 \ll z\lambda/D^2$ where D is the maximum lateral dimension of the obstacle or aperture. When these conditions are satisfied the diffraction pattern is referred to as a *Fraunhofer* diffraction pattern. Diffraction patterns observed too close to an obstacle or aperture for the above conditions to be satisfied are called *Fresnel* diffraction patterns[6.1]–[6.7].

6.4 Single Slit

The diffraction pattern for a single slit of width $2d$ can be calculated in the same way as above. The Fourier transform of the slit is

$$a(k_x) = \frac{1}{2\pi}\int_{-d}^{d} e^{ik_x x}dx = \frac{\sin k_x d}{\pi k_x} \tag{6.12}$$

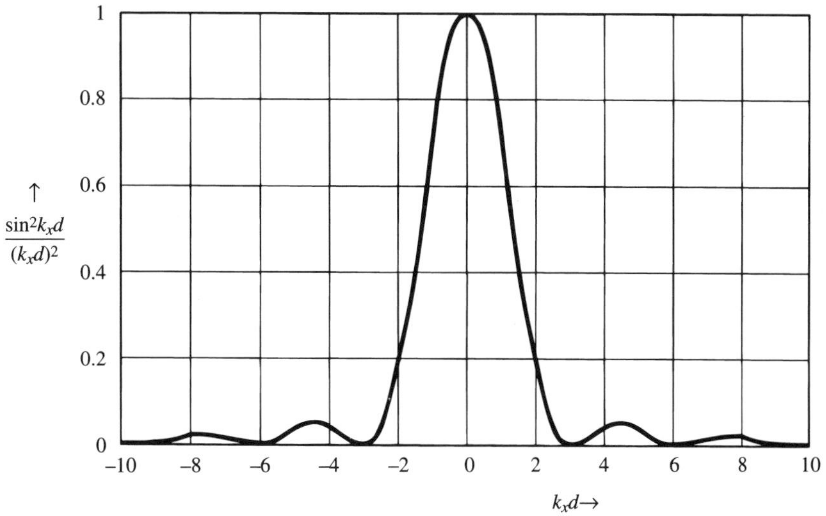

Fig. 6.2. Diffraction pattern of a single slit of width $2d$.

and the relative intensity at angle θ is

$$I = \frac{|a(k_x)|^2}{|a(k_0)|^2} = \frac{\sin^2 k_x d}{(k_x d)^2} = \frac{\sin^2(\frac{2\pi}{\lambda} d \sin\theta)}{(\frac{2\pi d \sin\theta}{\lambda})^2} \tag{6.13}$$

which is shown graphically in Fig. (6.2). The center of the diffraction pattern is an intensity maximum. The first minimum occurs when

$$\frac{2\pi d \sin\theta}{\lambda} = \pi, \tag{6.14}$$

which gives

$$\theta_{min} = \frac{\lambda}{2d} = \frac{\lambda}{w}, \tag{6.15}$$

where w is the width of the slit.

6.5 Two-Dimensional Apertures

We can generalize Eq. (6.7) for the case of a two-dimensional aperture by writing

$$a(k_x, k_y) = \frac{1}{4\pi^2} \int_{-\infty}^{\infty} \xi(x, y) e^{i(k_x x + k_y y)} dx dy, \tag{6.16}$$

where $\xi(x, y)$ is the aperture function, $k_x = |\mathbf{k}| \sin\theta$ and $k_y = |\mathbf{k}| \sin\phi$, where θ and ϕ are the angles the wave vector of the contributing plane waves make with the z axis when projected onto the xz and zy planes, respectively.

6.5.1 Circular Aperture

For a circular aperture of radius R illuminated normally with a coherent plane wave (so that all points in the aperture have the same phase) we can write

$$\begin{aligned} a(k_x, k_y) = &\frac{1}{4\pi^2} \int_{-R}^{R} \int_{-\sqrt{R^2-x^2}}^{\sqrt{R^2-x^2}} [\cos(k_x x + k_y y) \\ &+ i \sin(k_x x + k_y y)] dx dy, \end{aligned} \tag{6.17}$$

which can be written

$$a(k_x, k_y) = \frac{1}{4\pi^2} \int_{-R}^{R} \int_{-\sqrt{R^2-x^2}}^{\sqrt{R^2-x^2}} (\cos k_x x \cos k_y y - sink_x x \sin k_y y$$
$$+ i \sin k_x x \cos k_y y + i \cos k_x \sin k_y y) dxdy. \tag{6.18}$$

All the terms in the integrand which contain sin( ) give zero because sine is an odd function: $\sin k_x x = -\sin -(k_x x)$. Thus:

$$a(k_x, k_y) = \frac{1}{4\pi^2} \int_{-R}^{R} \int_{-\sqrt{R^2-x^2}}^{\sqrt{R^2-x^2}} \cos k_x x \cos k_y y dxdy. \tag{6.19}$$

Because the aperture is rotationally symmetric we can choose $k_y = 0$, which gives

$$a(k_x, k_y) = \frac{1}{2\pi^2} \int_{-R}^{R} \sqrt{R^2 - x^2} \cos(k_x x) dx. \tag{6.20}$$

Put $x = R\cos\chi$, $\rho = k_x R$ to give

$$a(k_x, k_y) = \frac{R^2}{2\pi^2} \int_0^{\pi} \sin^2\chi \cos(\rho\cos\chi) d\chi. \tag{6.21}$$

The Bessel function of order 1 is defined by[6.8]

$$J_1(\rho) = \frac{\rho}{\pi} \int_0^{\pi} \sin^2\chi \cos(\rho\cos\chi) d\chi, \tag{6.22}$$

so

$$a(k_x, k_y) = \frac{R^2}{2\pi} \frac{J_1(\rho)}{\rho}. \tag{6.23}$$

The relative intensity corresponding to the direction k_x, k_y is

$$I = \frac{|a(k_x, k_y)|^2}{|a(k_0, k_0)|^2} = 4 \left[\frac{J_1(\rho)}{\rho}\right]^2, \tag{6.24}$$

where we have used the fact that $\lim_{\rho\to 0}[J_1(\rho)/\rho]^2 = 1/2$[6.8]. Remembering that $\rho = k_x R = |\mathbf{k}| R \sin\theta$:

$$I = 4\frac{J_1(2\pi R\sin\theta/\lambda)^2}{(2\pi R\sin\theta/\lambda)^2}. \tag{6.25}$$

This diffraction pattern is shown in Fig. (6.3). The central disc is called Airy's disc, after Sir G.B. Airy (1801–92), who first solved this problem. The minima occur when $J_1(\rho) = 0$, which is satisfied for $\rho = 1.220\pi$, 2.635π, 2.233π, 2.679π, 3.238π, 3.699π, etc. The first minimum satisfies $\rho = 1.220\pi$ which gives

$$\frac{2\pi R\sin\theta}{\lambda} = 1.22\pi \tag{6.26}$$

and

$$\sin\theta = \frac{1.22\lambda}{2R} = \frac{1.22\lambda}{D}, \tag{6.27}$$

where D is the diameter of the circular aperture.

For small angles the diffraction angle in Eq. (6.27) can be written as $\theta \simeq 1.22\lambda/D$. This angle provides a measure of when diffraction effects are important. Diffraction effects becomes negligible when $D >> \lambda$. The value of θ sets limits to the ability

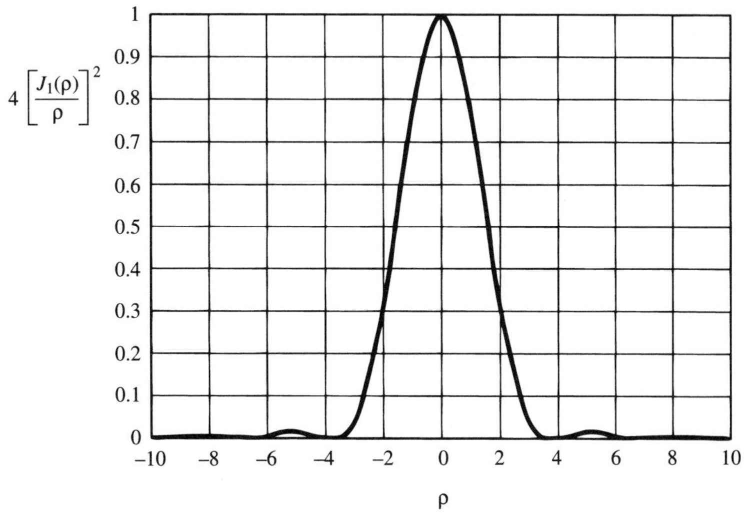

Fig. 6.3. The Airy diffraction pattern of a circular aperture.

of optical systems to produce images, limits the ability of the focusing lens to focus plane waves to small spots, and represents a fundamental deviation of the performance of an optical system from that represented by geometrical optics. We will examine these points in further detail in Chapters 14, 15, and 16. A fundamental observation (see Chapter 14.6) is that the smallest focal spot that can be produced with a plane wave and lens, and also the smallest size object that can be imaged without excessive diffraction, are both about the size of the wavelength of the light used.

6.6 Laser Modes

In the last chapter we saw that when a laser oscillates it emits radiation at one or more frequencies that lie close to passive resonant frequencies of the cavity. These frequencies are called *longitudinal* modes. In our initial discussion of these modes we treated them as plane waves reflecting back and forth between two plane laser mirrors. In practice, laser mirrors are not always plane. Usually at least one of the laser mirrors will have concave spherical curvature. The use of spherical mirrors relaxes the alignment tolerance that must be maintained for adequate feedback to be achieved. Even if the laser mirrors are plane, the waves reflecting between them cannot be plane, as true plane waves can only exist if there is no lateral restriction of the wave fronts. Practical laser mirrors are of finite size so any wave reflecting from them will spread out because of diffraction. We have just shown that this diffractive spreading of a plane wave results when the plane wave is restricted laterally, for example by passing the wave through an aperture. Reflection from a finite size mirror produces equivalent effects. We can explain this phenomenon qualitatively by introducing the concept of Huygens secondary wavelets.† If a plane mirror is illuminated by a plane wave then each point on the mirror can be treated as a source of a spherical wave called a secondary wavelet. The overall reflected wave is the envelope of the sum total of secondary wavelets originating

† Christian Huygens (1629–1695) was a Dutch astronomer who first suggested the concept of secondary wavelets.

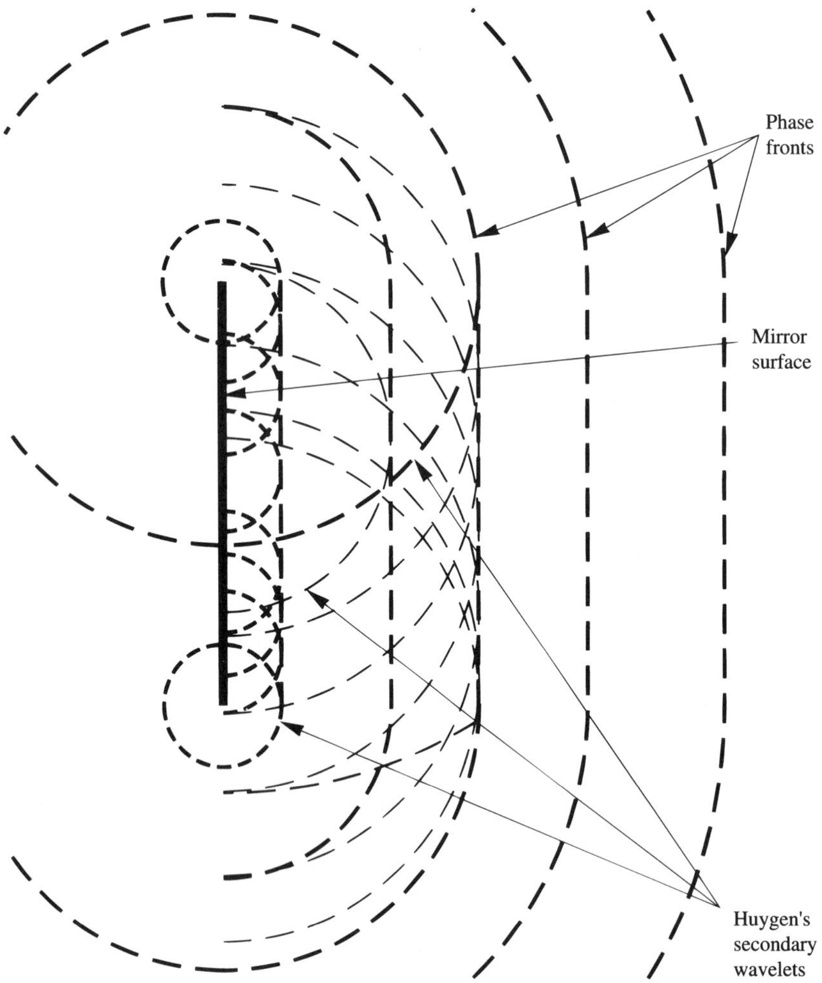

Fig. 6.4. Secondary wavelets originating from a finite plane mirror.

from every point on the mirror surface, as shown in Fig. (6.4). This construction shows that the reflected wave from a finite size mirror is not a plane wave.

The existence of diffraction in a laser resonator places restrictions on the minimum size of mirrors that can be used at a given wavelength λ and spacing ℓ. According to Eq. (6.27) a plane wave that reflects from a mirror of diameter d will spread out into a range of directions characterized by a half angle θ, where $\theta = \lambda/d$.

Consider the resonator shown in Fig. (6.5), whose reflectors have diameter d_1, d_2, respectively and spacing ℓ. A wave diffracting from mirror M_1 towards mirror M_2 will lose substantial energy past the edges of M_2 if the diffraction angle $\theta \leq d_2/\ell$. This gives us the Fresnel condition

$$\frac{d_2 d_2}{\lambda \ell} \geq 1. \tag{6.28}$$

The actual waves that reflect back and forth between the mirrors of a laser resonator are not plane waves but they do have characteristic spatial patterns of electric (and magnetic) field amplitude; they are called *transverse modes.* To be

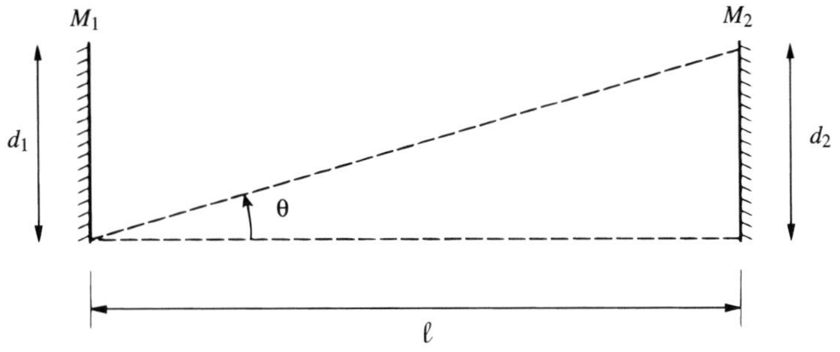

Fig. 6.5. Geometry for calculating the magnitude of diffraction effects in a laser resonator.

amplified effectively such modes must correspond to rays which make substantial numbers of specular reflections before being lost from the cavity. A transverse mode is a field configuration on the surface of one reflector that propagates to the other reflector and back, returning in the same pattern, apart from a complex amplitude factor that gives the total phase shift and loss of the round trip. To each of these transverse modes there corresponds a set of longitudinal modes spaced by approximately, $c/2\ell$.

A more detailed treatment of these transverse modes is given in Chapter 16; however, a few of their important properties will be briefly reviewed here.

(a) The nature of these transverse modes is a function of the reflector sizes, their radii of curvature, the presence of additional limiting apertures between the mirrors, and the resonator length.

(b) These modes are affected by the existence of spatial variations of gain in the laser medium.

(c) Only certain configurations of laser mirrors allow propagating transverse modes to exist that do not suffer substantial diffraction loss from the laser cavity, as illustrated in Fig. (6.6).

(d) The transverse modes, because they are essentially propagating beam solutions of Maxwell's equations, are analogous to the confined beam TEM modes in waveguides and are labelled accordingly. A laser mode of order m, n would be labelled TEM_{mn}.

(e) The transverse modes can have cartesian (rectangular) or polar (circular) symmetry. Cartesian symmetry usually arises when some element in the laser cavity imposes a preferred direction on the direction of the electric and magnetic field vectors (for example, Brewster windows in the laser cavity). If distance along the resonator axis is measured by the coordinate z then a transverse mode with its electric field in the x direction would be a function of the form $E^x_{mn}(x, y, z)e^{i(\omega t \pm kz)}$ with a magnetic field of the form $H^y_{mn}(x, y, z)e^{i(\omega t \pm kz)}$

(f) Dependent on the radii of the resonator mirrors and their separation, the longitudinal modes associated with each transverse mode may have the same or different frequencies. Each individual longitudinal mode associated with a transverse mode TEM_{mn} is labelled TEM_{mnq}. In the general case the longitudinal mode frequencies of two different TEM modes will be different, as shown in Fig. (6.7).

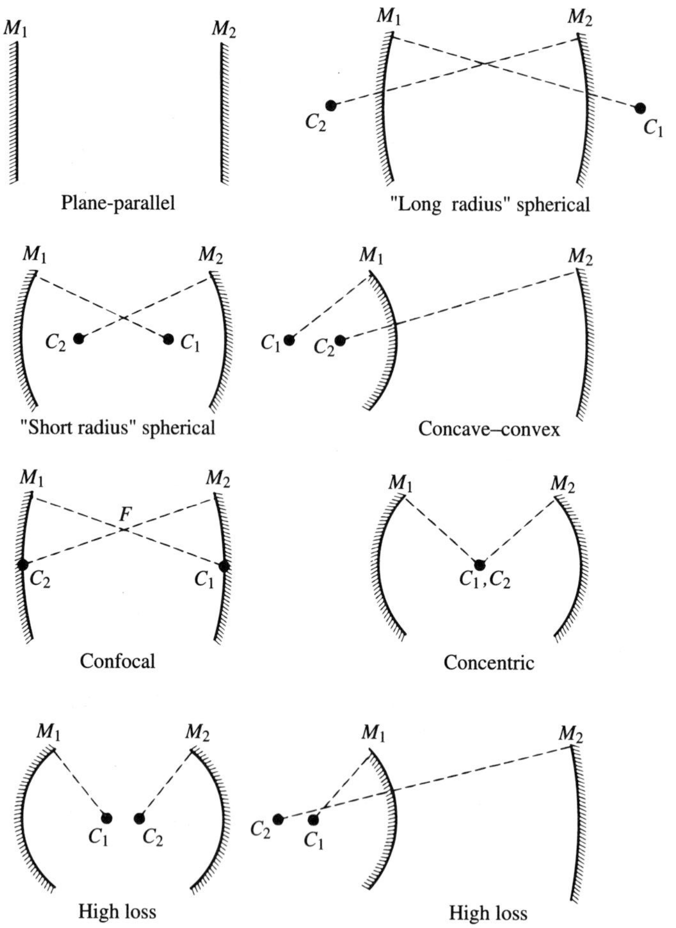

Fig. 6.6. Laser resonator configurations that use two spherical mirrors. C_1 and C_2 are the centers of curvature of mirrors M_1 and M_2, respectively.

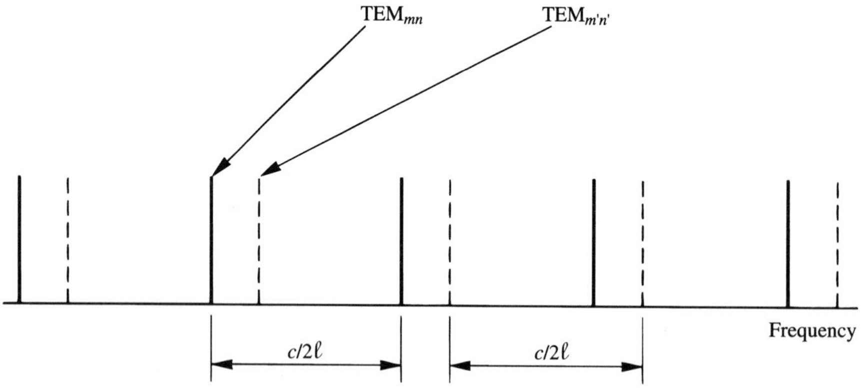

Fig. 6.7. Schematic frequencies of longitudinal modes belonging to two different transverse modes TEM_{mn} and $TEM_{m'n'}$.

(g) Since the field distribution of the transverse modes is propagating in both directions inside the laser resonator, this field distribution is maintained in the output beam from the laser, and the resultant intensity pattern shows an xy spatial dependence in a plane perpendicular to the direction

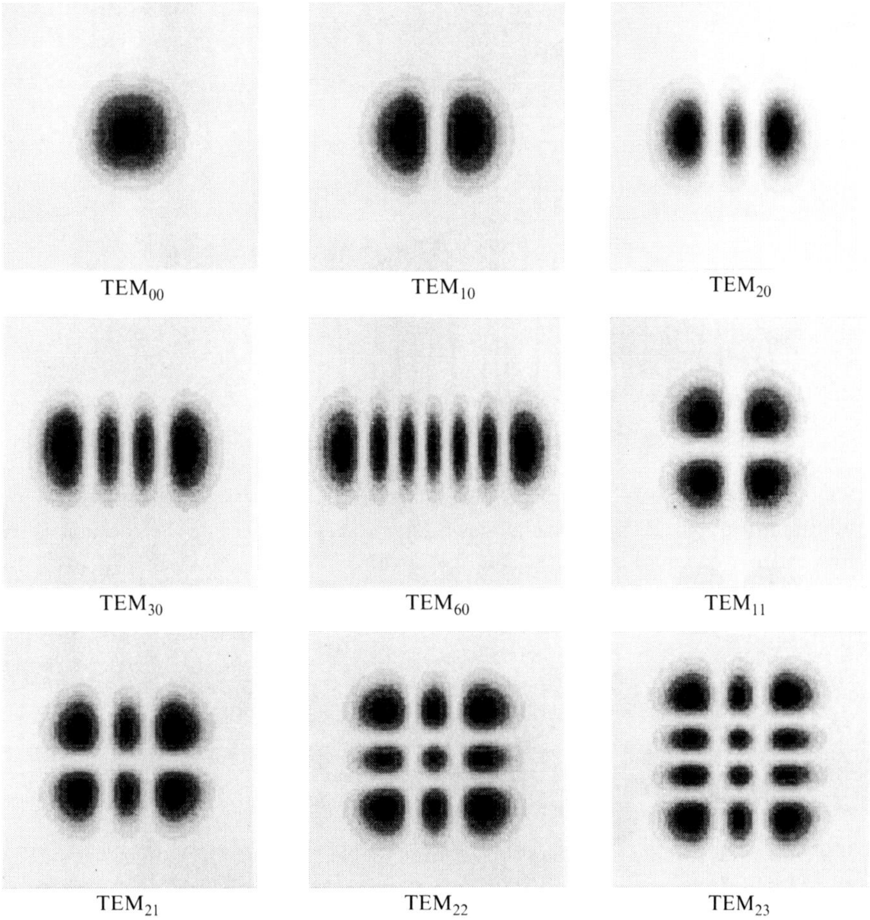

Fig. 6.8. Intensity patterns for different TEM modes possessing cartesian symmetry.

of propagation of the laser beam. These patterns have an intensity distribution

$$I(x, y, z) \propto [E^x_{mn}(x, y, z)]^2 \tag{6.29}$$

and are called mode patterns. Some examples of simple transverse mode patterns having cartesian symmetry are shown in Fig. (6.8). It can be seen that the number of xy nodal lines in the intensity pattern determines the designation TEM$_{mn}$.

6.7 Beam Divergence

Since the oscillating field distributions inside a laser are not plane waves, when they propagate through the mirrors as output beams they spread by diffraction.

The semivertical angle of the cone into which the output beam diverges is†

$$\theta_{beam} = \tan^{-1}\left(\frac{\lambda}{\pi w_0}\right) \approx \frac{\lambda}{\pi w_0}. \tag{6.30}$$

† For a derivation of this result see Chapter 16.

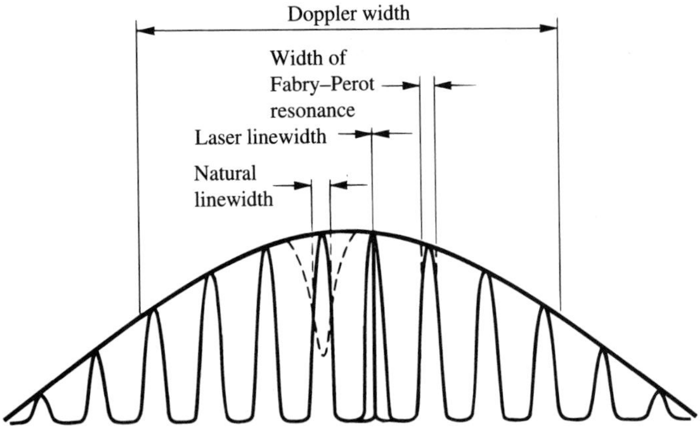

Fig. 6.9. Linewidth factors in a laser.

where λ is the wavelength of the output beam and w_0 is a parameter called the 'minimum spot size' that characterizes the transverse mode. For the special case of two mirrors of equal radii in a confocal arrangement, as shown in Fig. (6.6), the value of w_0 is

$$(w_0)_{conf.} = \sqrt{\frac{\lambda\ell}{2\pi}}, \tag{6.31}$$

so

$$\theta_{beam(conf.)} = \tan^{-1}\frac{\lambda}{\pi}\sqrt{\frac{2\pi}{\lambda\ell}} = \tan^{-1}\sqrt{\frac{2\lambda}{\pi\ell}}. \tag{6.32}$$

If we take the specific example of 632.8 nm He–Ne laser with a symmetric confocal resonator 0.3 m long, w_0=0.17 mm and θ_{beam}=1.2 mrad=0.66°. This is a highly directional beam, but the beam does become wider the further it goes away from the laser. Such a beam is, however, highly useful in providing the perfect straight line reference. For this reason lasers are finding increasing use in construction – for tunneling, leveling, and surveying. Over a 100 m distance the laser beam just described would have expanded to a diameter of 230 mm. After travelling the distance to the moon (~390 000 km) the beam would be ≃900 km in diameter.

6.8 Linewidth of Laser Radiation

A single longitudinal mode of a laser is an oscillation resulting from the interaction of a broadened gain curve with a passive resonance of the Fabry–Perot laser cavity. The frequency width of the gain curve is $\Delta\nu$, the frequency width of the passive cavity resonance is $\Delta\nu_{1/2} = \Delta\nu_{FSR}/F$. We expect the linewidth of the resulting oscillation to be narrower than either of these widths, as shown schematically in Fig. (6.9). It can be shown that the frequency width of the laser oscillation itself is[6.9]–[6.12]

$$\Delta\nu_{laser} = \frac{\pi h\nu_0(\Delta\nu_{1/2})^2}{P}\frac{N_2}{[N_2 - (g_2/g_1)N_1]_{threshold}}, \tag{6.33}$$

where P is the output power.

Eq. (6.9) predicts very low linewidths for many lasers. For a typical He–Ne laser with 99% reflectance mirrors and a cavity 30 cm long

$$\Delta\nu_{1/2} = \frac{c(1-R)}{2\pi\ell} = 1.59 \text{ MHz}.$$

The factor $N_2/[N_2 - (g_2/g_1)N_1]_{threshold}$ is close to unity for a typical low power, say 1 mW, laser. Consequently,

$$\Delta\nu_{laser} = \frac{\pi \times 6.626 \times 10^{-34} \times 3 \times 10^8 \times 1.59^2 \times 10^{12}}{10^{-3} \times 632.8 \times 10^{-9}} = 2.5 \times 10^{-3} \text{ Hz}.$$

Such a small linewidth is never observed in practice because thermal instabilities and acoustic vibrations lead to variations in resonator length that further broaden the output radiation lineshape. The best observed minimum linewidths for highly stablized gas lasers operating in the visible region of the spectrum are around 10^3 Hz. Even if macroscopic thermal and acoustic vibrations could be eliminated from the system, a fundamental limit to the resonator length stability would be set by Brownian motion of the mirror assemblies. For example, consider two laser mirrors mounted on a rigid bar. The mean stored energy in the Brownian motion of the whole bar is $\overline{E} = kT$.

The frequency spread of the laser output that thereby results is

$$\Delta\nu_{Brownian} = \nu\sqrt{\frac{2kT}{YV}}, \tag{6.34}$$

where Y is the Young's modulus of the bar material and V is the volume of the mounting bar. Typical values of $\Delta\nu_{Brownian}$ are ~ 2 Hz.

6.9 Coherence Properties

Because of its extremely narrow output linewidth the output beam from a laser exhibits considerable *temporal coherence* (longitudinal coherence). To illustrate this concept, consider two points A and B a distance L apart in the direction of propagation of a laser beam, as shown in Fig. (6.10). If a definite and fixed phase relationship exists between the wave amplitudes at A and B, then the wave shows temporal coherence for a time c/L. The further apart A and B can be, while still maintaining a fixed phase relation with each other, the greater is the temporal coherence of the output beam. The maximum separation at which the fixed phase relationship is retained is called the *coherence length*, L_c, which is a measure of the length of the continuous uninterrupted wave trains emitted by the laser. The coherence length is related to the *coherence time* τ_c by $L_c = c\tau_c$. The coherence time itself is a direct measure of the monochromaticity of the laser, since by Fourier transformation, done in an analogous manner to the treatment of natural broadening,

$$\tau_c \simeq \frac{1}{\Delta\nu_L} \text{ and } L_c \simeq \frac{c}{\Delta\nu_L}. \tag{6.35}$$

The coherence length and time of a laser source are considerably better than a conventional monochromatic source (a spontaneously emitted line source). The greatly increased coherence can be demonstrated in a Michelson interferometer

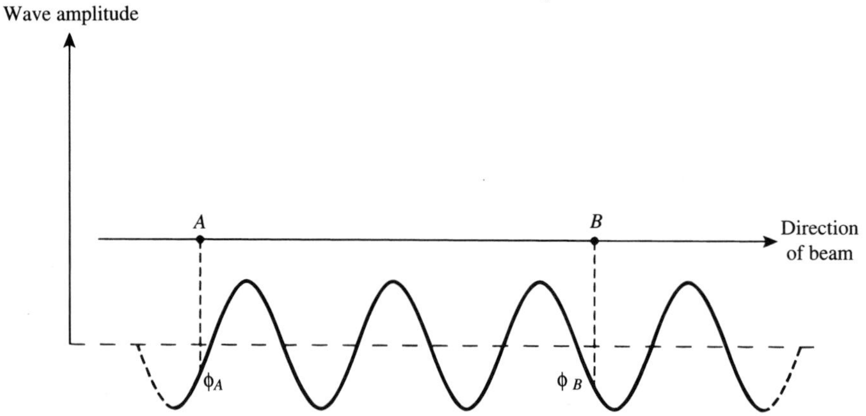

Fig. 6.10. To illustrate the concept of temporal coherence. If an unbroken wave train connects the points A and B then the phase difference $(\phi_B - \phi_A)$ will have a constant value.

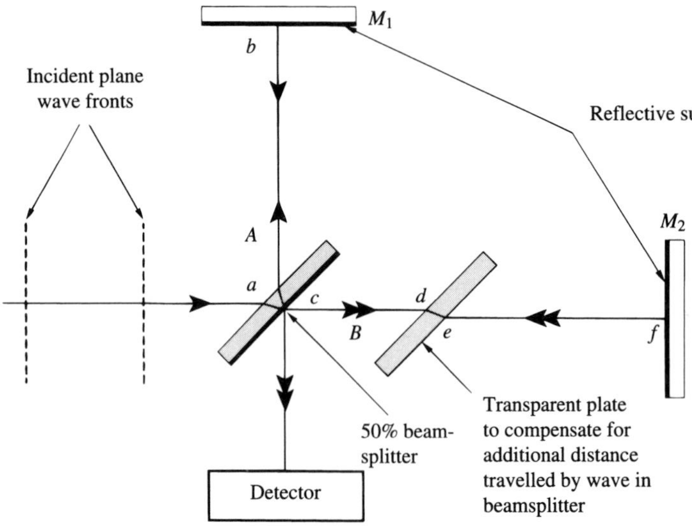

Fig. 6.11. Michelson interferometer. Only the primary reflections are shown. Weak reflections from one surface of the beamsplitter and both faces of the compensating plate also occur unless these faces are antireflection coated. (See Chapter 14.)

experiment, which allows interference between waves at longitudinally different positions in a wavefront to be studied, as shown in Fig. (6.11).

The operation of this instrument can be described as follows: Incident waves divide at the input beamsplitter. One part of the wave, A, takes the path $c \to a \to b \to a \to c \to$ detector. The other part of the incident wave, B, takes the path $c \to d \to e \to f \to e \to d \to c \to$ detector. We can write the electric field of wave A, at the detector, as

$$E_A = E_0 e^{i(\omega t + \phi_A)}, \tag{6.36}$$

where ϕ_A is the phase shift experienced along the path $c \to a \to b \to a \to c \to$ detector. Similarly, for wave B

$$E_B = E_0 e^{i(\omega t + \phi_B)}. \tag{6.37}$$

The signal from the detector can be written as

$$i(t) \propto |E_A + E_B|^2, \tag{6.38}$$

since the detector responds to the intensity of the light, which gives

$$i(t) \propto 2E_0^2[1+\cos(\phi_B-\phi_A)]. \tag{6.39}$$

We expect maximum signal (corresponding to maximum observed illumination) if waves A and B are in phase, that is if $\phi_B - \phi_A = 2n\pi$. Minimum signal results if $\phi_B - \phi_A = (2n+1)\pi$, when the two waves are out of phase. We can write

$$\phi_B - \phi_A = \frac{2\pi L}{\lambda}, \tag{6.40}$$

where L is the difference in path length for the two waves A and B. If L is altered, for example by moving mirror M_2 to the right in Fig. (6.11), we expect the detector signal to go up and down between its maximum and minimum values. However, this interference phenomenon will only be observed provided $L \lesssim L_c$. For a laser with $\Delta\nu \sim 1$ kHz this would require $L \lesssim 300$ km. In practice, it is not feasible to build laboratory interferometers with path differences as large as this, although large path differences can be obtained by incorporating a long optical fiber into one of the paths in Fig. (6.11). If the linewidth of the source is large then it is quite easy to demonstrate the disappearance of any interference effects of $L > L_c$. For example, with a 'white' light source that covers the spectrum from 400–700 nm.

$$\Delta\nu = 3 \times 10^8 \left(\frac{1}{400 \times 10^{-9}} - \frac{1}{700 \times 10^{-a}}\right) = 3.2 \times 10^{14} \text{ Hz}$$

and therefore $L_c \sim 10^{-3}$ mm. In this case interference effects will only be observed by careful adjustment of the interferometer so both paths are almost equal. Interference fringes will be detected due to interference between the two parts of the split wave, A and B, if the coherence length of the incident wave is greater than the distance $2(cd+ef)-2(ab)$.

It should be noted that the nonobservation of fringes in a Michelson interferometer does not imply that interference between waves is not occurring, merely that the continuing change in phase relationship between these waves shifts the position of intensity maxima and minima around in a time $\sim \tau_c$, so apparent uniform illumination results.

A laser also possesses *spatial* (lateral) coherence, which implies a definite fixed phase relationship between points separated by a distance L transverse to the direction of beam propagation. The transverse coherence length, which has similar physical meaning to the longitudinal coherence length, is

$$L_{tc} \sim \frac{\lambda}{\theta_{beam}} \sim \pi\omega_0 \tag{6.41}$$

for a laser source.

6.10 Interference

The existence of spatial coherence in a wavefront and the limit of its extent can be demonstrated in a classic Young's slits interference experiment. In this experiment a pair of thin, parallel slits or a pair of pinholes is illuminated normally with a spatially coherent monochromatic plane wave, as illustrated in Fig. (6.12). A laser beam operating in a *uniphase* mode, TEM_{00}, and expanded so that its central portion illuminates both slits or pinholes serves as an ideal source to illustrate this

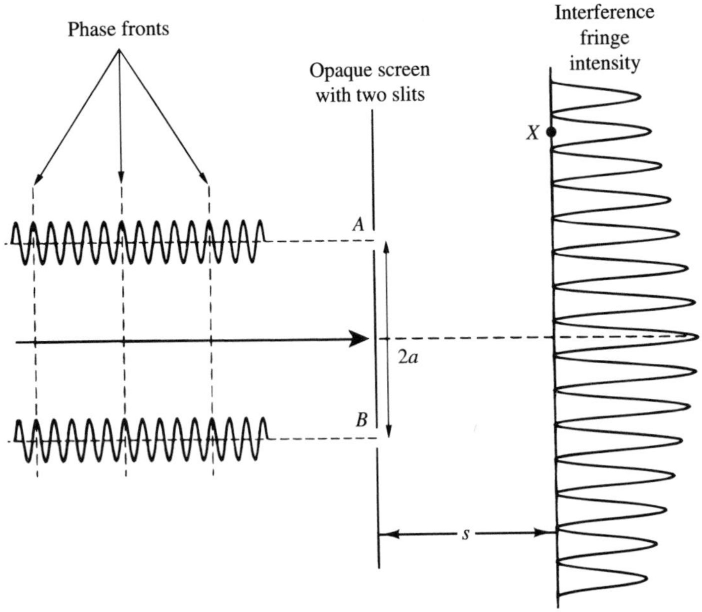

Fig. 6.12. Young's slits interference experiment to demonstrate the existence of spatial coherence in a phase front.

classic interference phenomenon, and at the same time test the spatial coherence properties of the laser beam.

The field amplitudes produced by the plane wave at the slits are equal and in phase and can be represented as

$$V = V_0 \cos(\omega t + \phi), \tag{6.42}$$

where ϕ is a phase factor.

Each slit acts as a source of secondary wavelets that propagate towards the observation screen. The resultant field amplitude at a point such as X will result from the superposition of the waves that have propagated from A and B. The field amplitude at X from A is

$$V_1 = b_1 V_0 \cos(\omega t + \phi - kR_1), \tag{6.43}$$

where

$$R_1 = AX = \sqrt{(x-a)^2 + s^2}; \tag{6.44}$$

$k_1 = 2\pi/\lambda$ is the wavelength of the plane wave, and b_1 is a constant that depends on the geometry of the experiment. The field amplitude at X due to B is

$$V_2 = b_2 V_0 \cos(\omega t + \phi - kR_2), \tag{6.45}$$

where

$$R_2 = BX = \sqrt{(x+a)^2 + s^2}. \tag{6.46}$$

If $s \gg a, x$ then the magnitudes of V_1 and V_2 can be taken as equal, but they differ in phase. The total field at X is

$$V_x = V_1 + V_2 = bV_0[\cos(\omega t + \phi - kR_1) + \cos(\omega t + \phi - kR_2)], \tag{6.47}$$

where b is a constant. If $s \gg a, x$ then θ is a small angle and V_1, V_2 can be taken as parallel, independent of the polarization of the input wave. From (6.47) we get

$$V_x = 2bV_0 \cos\left[\omega t + \phi - \frac{k(R_1 + R_2)}{2}\right] \cos\left[\frac{k(R_2 - R_1)}{2}\right] \tag{6.48}$$

and the intensity at X is

$$I \propto V_x^2 \propto \cos^2\left\{\omega t + \phi - \left[\frac{k(R_1 + R_2)}{2}\right]\right\} \cos^2\left[\frac{k(R_2 - R_2)}{2}\right], \tag{6.49}$$

which gives

$$I \propto \frac{1}{2}\left\{1 + \cos 2\left[\omega t + \phi - \frac{k(R_1 + R_2)}{2}\right]\right\} \cos^2\left[\frac{k(R_2 - R_1)}{2}\right]. \tag{6.50}$$

Because ω is a very high frequency ($\sim 10^{15}$ Hz in the visible region) the detector does not see the term at frequency 2ω. It is as if the detector averages this term, which is as often positive as it is negative, to zero.† The observed intensity is therefore

$$I \propto \cos^2 \frac{k(R_2 - R_1)}{2}. \tag{6.51}$$

Now, if $s \gg a, x$,

$$R_2 = s\sqrt{1 + \left(\frac{x+a}{s}\right)^2} = s\left[1 + \frac{1}{2}\left(\frac{x+a}{s}\right)^2\right], \tag{6.52}$$

$$R_1 = s\left[1 + \frac{1}{2}\left(\frac{x-a}{s}\right)^2\right], \tag{6.53}$$

so

$$R_2 - R_1 = 2ax/s, \tag{6.54}$$

which gives

$$I \propto \cos^2\left(\frac{kax}{s}\right). \tag{6.55}$$

We can write $\theta = x/s$ for $x \ll s$ giving

$$I \propto \cos^2(ka\theta) = \cos^2\left(\frac{2\pi a\theta}{\lambda}\right). \tag{6.56}$$

Note that this is the same result as Eq. (6.10). The intensity is a maximum whenever

$$\frac{2\pi a\theta}{\lambda} = m\pi, \tag{6.57}$$

giving

$$\theta_{max} = \frac{m\lambda}{2a}, \tag{6.58}$$

whereas, for the minima

$$\frac{2\pi a\theta}{\lambda} = \frac{(2m+1)\pi}{2}, \tag{6.59}$$

† For a further discussion of this point see Chapter 23.

giving

$$\theta_{\min} = \left(\frac{2m+1}{4a}\right)\lambda. \tag{6.60}$$

The interference pattern on the screen appears as a series of equally spaced alternate bright and dark bands.

The appearance of these bright and dark bands does, however, depend crucially on the spatial coherence of the waves illuminating the two slits. If the slit variation were increased to beyond the lateral coherence length, so that $2a > L_{tc}$, then the fringe pattern would disappear.

The classic diffraction pattern of a circular aperture can also be easily observed by illuminating a small circular hole with the central portion of a TEM_{00} mode laser beam. The cleanest patterns can be obtained by focusing the laser beam with a lens and placing a small circular aperture (of size smaller than the focused laser beam) in the focal plane of the lens.

6.11 Problems

(6.1) A pair of narrow slits is illuminated with a monochromatic plane wave with $\lambda_0 = 488$ nm. At t a distance of 500 nm behind the slits the dark band spacing is 5 nm. What is the spacing of the slits?

(6.2) Calculate the interference pattern produced by four narrow slits, equally spaced by $2a$. Extend your analysis to N equally spaced slits. What happends as $N \to \infty$?

(6.3) Calculate the Fraunhofer diffraction pattern of a rectangular aperture of dimension $a \times b$.

(6.4) A Michelson interferometer with identical arms shows sharp interference effects when illuminated by a point source. A glass slab of thickness 10 mm, refractive index 1.6 is placed in one arm. No interference effects are then observed. What can you say about the coherence properties of the source?

(6.5) A gas laser with $\lambda_0 = 446$ nm has a resonator 0.5 m long. One of the laser mirrors is randomly vibrating with an amplitude of 10 nm. Estimate how much will this effectively broaden the linewidth of the emitted radiation?

References

[6.1] M. Born and E. Wolf, *Principles of Optics*, 6th Edition, Pergamon Press, Oxford, 1980.

[6.2] R.W. Ditchburn, *Light*, 3rd Edition Academic Press, 1976.

[6.3] E. Hecht and A. Zajac, *Optics*, 2nd Edition, Addison-Wesley, Reading, MA, 1987.

[6.4] F.A. Jenkins and H.E. White, *Fundamentals of Optics*, 3rd Edition McGraw-Hill, New York, 1957.

[6.5] M.V. Klein and T.E. Furtak, *Optics*, 2nd Edition Wiley, New York, 1970.

[6.6] R.S. Conghurst, *Geometrical and Physical Optics*, 3rd Edition, Longman, London, 1973.

[6.7] F.G. Smith and J.H. Thompson, *Optics*, Wiley, London, 1971.

[6.8] A. Abramowitz, and I.A. Stegun, Eds., *Handbook of Mathematical Functions*, Dover Publications, New York, 1968.

[6.9] P.W. Milonni and J.H. Eberly, *Lasers*, Wiley, New York, 1988.

[6.10] A.E. Siegman, *Lasers*, University Science Books, Mill Valley, CA, 1986.
[6.11] A. Yariv, *Introduction to Optical Electronics*, 4th Edition, Holt, Rinehart and Winston, New York, 1991.
[6.12] A. Yariv, *Quantum Electronics*, 3rd Edition, Wiley, New York, 1989.

7

Control of Laser Oscillators

7.1 Introduction

During laser oscillation one or more distinct mode frequencies can be emitted, which in the general case behave as independent monochromatic oscillations. When more than one mode oscillates this multimode operation usually corresponds to simultaneous oscillation on more than one longitudinal mode of a single transverse mode of the cavity. It is also possible for more than one transverse mode to oscillate simultaneously, but this is not of great importance or interest to us here. In this chapter we will examine how multimode operation can be suppressed in favor of single mode operation. We shall also see how many simultaneously oscillating modes can become *locked* together in phase to produce a form of laser oscillation in which the laser output becomes a train of very short pulses. This *mode-locked* operation is of considerable fundamental and practical interest and we shall examine the various ways in which the phenomenon can be induced.

7.2 Multimode Operation

We have seen that in an inhomogeneously broadened laser simultaneous oscillation is possible on more than one longitudinal mode. These oscillations can be coupled to each other by applying an external perturbation, as we shall see shortly. However, in the absence of such perturbation, each mode oscillates independently of the others. We can represent the total electric field of the laser beam as

$$E(t) = \sum_{n=1}^{N} E_n(t)\cos[\omega_n t + \phi_n(t)], \tag{7.1}$$

where ω_n is the frequency of the nth of N modes and $\phi_n(t)$ is its phase. The frequences ω_n, ω_{n+1} are separated by approximately $\pi c/\ell$, but successive adjacent modes are not exactly evenly spaced because of mode-pulling. The phase term fluctuates with time in a random way because the coherence time of the beam is finite. These fluctuations can be attributed to three main causes:

(i) spontaneous emission, of which there is always some even into the narrow solid angle occupied by the laser beam,

(ii) fluctuation in the index of refraction of the amplifying medium,

(iii) vibrations of the mechanical structure of the laser resonator.

The amplitude $E_n(t)$ changes with time if the frequency ω_n drifts relative to the center of the gain profile. Such drift can be made very small by stabilizing the

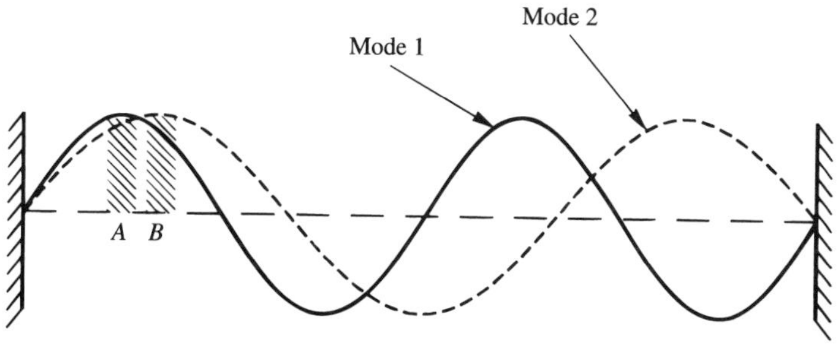

Fig. 7.1. Standing wave patterns of two adjacent oscillating longitudinal modes in a laser resonator.

frequency of the laser. This is most simply done by stabilizing the optical length L of the resonator. The optical length can be defined in the most general way as

$$L = \int_0^{\ell} n(x)dx, \tag{7.2}$$

where $n(x)$ is the refractive index of the medium (or media) between the resonator mirrors placed at the geometric spatial locations 0 and ℓ.

If the laser intensity is monitored with a fast optical detector, it will be observed to fluctuate with time because of the beating together of the various longitudinal modes. We saw in Chapter 5 that these beats allow the number of oscillating modes to be determined. Although these beats might generally be expected to give only high frequency fluctuations this is not so. The beat signals from different pairs of modes also beat with each other to give fluctuations down to quite low frequencies. The laser intensity also fluctuates because of variations in the mode amplitudes $E_n(t)$ caused by fluctuations in the gain of the amplifying medium.

Fluctuations in amplitude and laser frequency can be controlled in several ways. The simplest approach to ideal behavior is to cause the laser to oscillate in only a single longitudinal and transverse mode. The transverse mode most desirable in this context is the so-called *fundamental* mode in which the distribution of intensity across the laser beam is Gaussian.† Oscillation in the fundamental transverse mode can be controlled by appropriate selection of the radii of curvature of the laser mirrors and the size of apertures in the laser cavity – for example, the apertures provided by the lateral size of the amplifying medium itself.

7.3 Single Longitudinal Mode Operation

Single longitudinal mode operation in a homogeneously broadened laser should be automatic. Surprisingly, sometimes this is not so. The strongest mode should suppress all its neighbors, but because of the standing wave character of the in-cavity field this may not occur if the maxima of field amplitudes of two different modes do not overlap in space. This phenomenon is called *spatial hole-burning.* Fig. (7.1) shows schematically how it happens. Mode 1 draws its gain from, for example, particles located near a field maximum such as A. Mode 2 draws its gain from particles located near field maximum B. The two modes will not compete strongly unless particles can move rapidly from A to B. This phenomena is called *spatial cross-relaxation.* Clearly, such effects are more important in a gas than a solid-state laser.

† To be discussed in detail in Chapter 16.

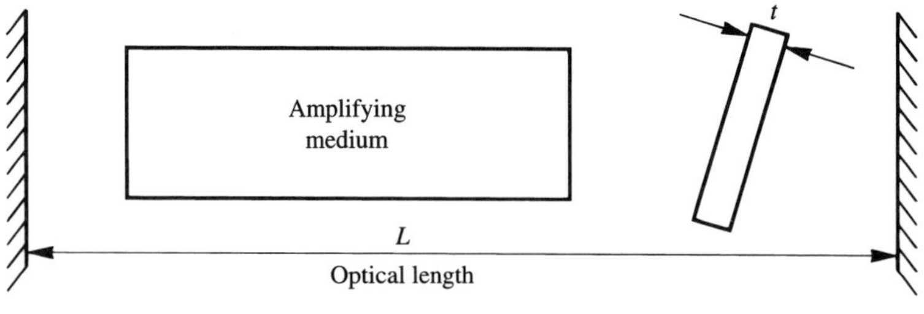

Fig. 7.2. Use of a tilted intracavity etalon to obtain single longitudinal mode laser oscillation. The optical length includes both amplifying medium and etalon.

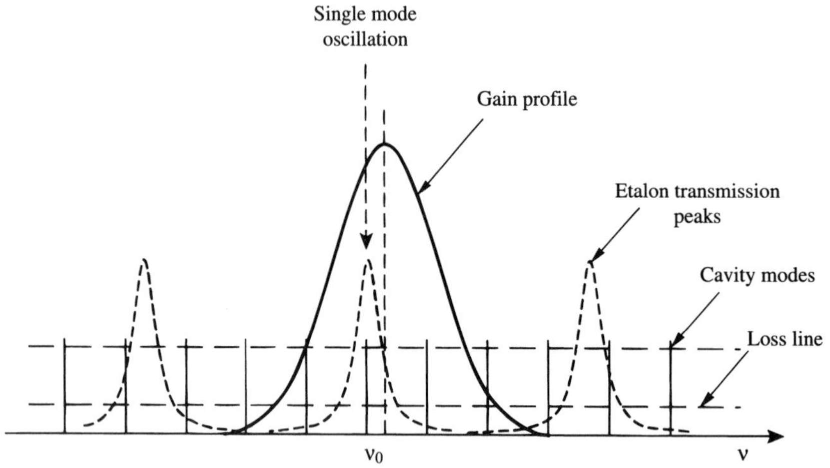

Fig. 7.3. Intracavity etalon transmission peaks superimposed on laser gain profile. Main cavity resonances span the gain curve. In this figure four such resonances lie above the loss line when the etalon is absent.

To force a laser into single longitudinal mode operation several methods have been used:

(i) By making the laser sufficiently short that only one cavity resonance lies under the gain profile and above the loss line. This generally requires $c/2L > \Delta\nu_D$.

(ii) With an intracavity etalon of thickness $t << L$, as shown in Fig. (7.2). The etalon is usually tilted slightly to prevent complications caused by reflections between the etalon faces and laser mirrors. The transmission maxima of such an etalon are relatively wide in frequency compared with the transmission maxima of the main cavity, but they are separated by a much greater frequency, $\Delta\nu_{etalon} >> \Delta\nu_{cavity}$,

$$\Delta\nu_{etalon} = \frac{c_0}{2nd}, \tag{7.3}$$

where n is the refractive index of the etalon.

For example, for a 10 mm thick fused silica etalon $\Delta\nu_{etalon} \simeq 10$ GHz. If this etalon is used inside the cavity of a laser with a gain linewidth of 1 GHz (say), then it is likely that only one of the main cavity resonances will be within a transmission maximum of the in-cavity etalon. This situation is shown schematically in Fig. (7.3).

(iii) By placing a very thin absorbing film inside the cavity. The position of the film, which could be a thin layer of metal evaporated onto a glass substrate, is adjusted to be at a node of the intracavity standing wave of a strong longitudinal mode. Other modes suffer loss at the film and can be suppressed.

(iv) By making the laser homogeneously broadened. This is sometimes possible in the case of gas lasers by making them operate at high gas pressure.

(v) With a Fox–Smith interferometer as shown in Fig. (7.4), which acts as a frequency-dependent[7.1] cavity mirror. Laser oscillation can use either mirror A or mirror B, the combination of which behaves like a Michelson interferometer. A longitudinal mode that divides into two parts at the beamsplitter will suffer high loss from the cavity unless both parts reflect in phase towards the amplifying medium.

(vi) If one of the main cavity mirrors is an etalon, both faces of which are reflective, then a frequency-dependent cavity mirror results and can discriminate between longitudinal modes.

(vii) In homogeneously broadened lasers, multimode oscillation frequently results because the standing waves corresponding to different modes deplete the population inversion at different spatial locations. This is the spatial hole-burning already mentioned. Single longitudinal mode operation can be obtained if the periodicity of the intracavity field can be smoothed out. This requires the wave inside the cavity to look like a travelling rather than a standing wave, which can be accomplished by building the laser in the form of a ring. In Fig. (7.5) waves can travel around the three-mirror cavity in both directions, if one of these waves can be suppressed, then a travelling wave exists inside the laser medium and no spatial hole-burning results. To explain how this might be done, it is actually simplest to describe a clever scheme that uses a conventional cavity with two mirrors but with elements inside the cavity that change the polarization state of the wave. The laser configuration shown in Fig. (7.6) has a Brewster window in the laser cavity and two quarter-wave plates. These optical components, which will be described in detail in Chapter 18, change linearly polarized wave light passing through them into circularly polarized light. At point A inside the cavity the wave that oscillates with minimum loss will be linearly polarized in the direction shown. With respect to x and y axes oriented at 45° to the plane of the diagram we can represent a wave travelling to the right as

$$\begin{aligned}(E_x)_A &= E_0 \cos(\omega t - kz),\\ (E_y)_A &= E_0 \cos(\omega t - kz).\end{aligned} \tag{7.4}$$

After passage through quarter-wave plate L_1, which is oriented appropriately, the electric fields at point B are

$$\begin{aligned}(E_x)_B &= E_0 \cos(\omega t - kz + \pi/2),\\ (E_y)_B &= E_0 \cos(\omega t - kz).\end{aligned} \tag{7.5}$$

This is left hand circularly polarized light (LHCP). The second quarter-wave plate L_2 is oriented to produce a $\pi/2$ phase shift of the y-directed electric vector, so at

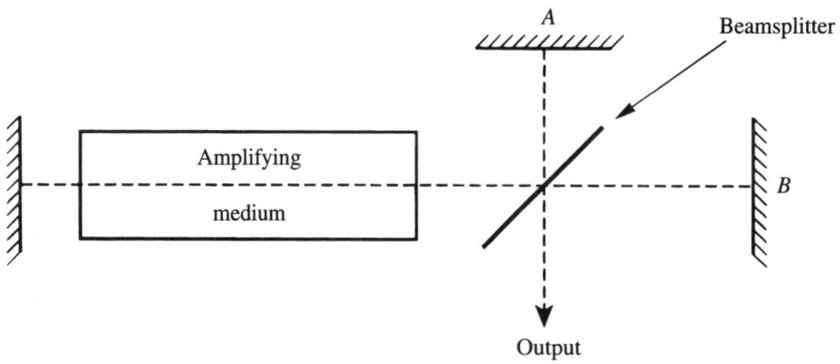

Fig. 7.4. Laser resonator incorporating a Fox–Smith interferometer for encouraging single longitudinal mode laser oscillation.

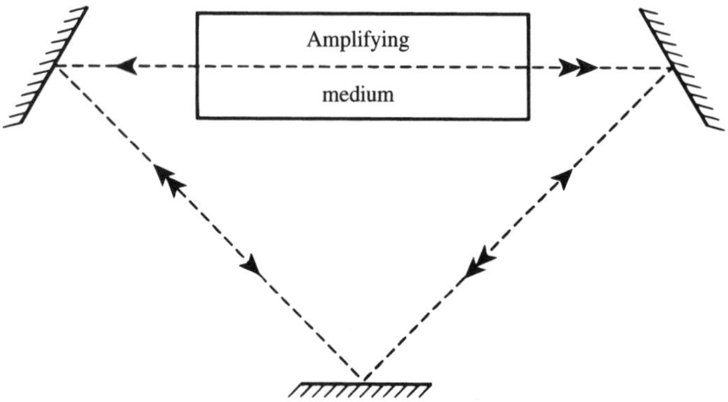

Fig. 7.5. Ring laser configuration with counterpropagating travelling wave fields. Unidirectional laser oscillation in such a cavity suppresses spatial hole-burning and will encourage single mode oscillation.

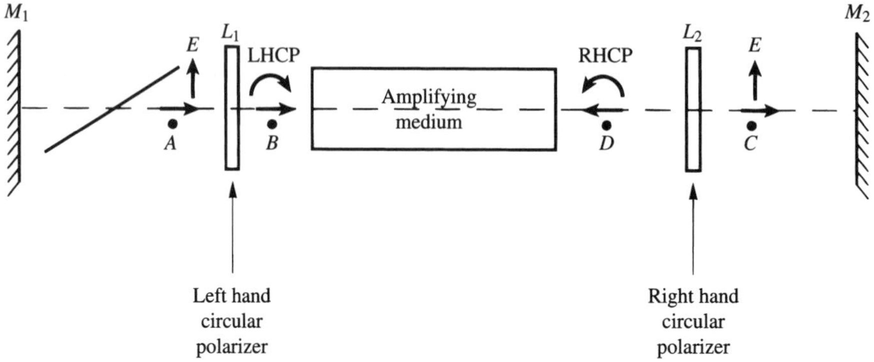

Fig. 7.6. 'Twisted-mode' arrangement for obtaining single mode operation of a homogeneously broadened laser by eliminating standing waves in the amplifying medium.

point C the right travelling wave can be written as

$$\begin{aligned}(E_x)_C &= E_0\cos(\omega t - kz + \pi/2),\\ (E_y)_C &= E_0\cos(\omega t - kz + \pi/2),\end{aligned} \tag{7.6}$$

which is once again linearly polarized. This wave reflects from mirror M_2 and returns through L_2. At point D this left travelling wave can be written as

$$\begin{aligned}(E_x)_D &= E_0\cos(\omega t + kz + \pi/2),\\ (E_y)_D &= E_0\cos(\omega t + kz + \pi),\end{aligned} \tag{7.7}$$

which is right hand circularly polarized light (RHCP). After a second pass through L_1 this wave becomes linearly polarized in its original direction once more.

Inside the amplifying medium the total electric field resulting from the right and left travelling waves is

$$E_x = E_0 \cos(\omega t - kz + \pi/2) + E_0 \cos(\omega t + kz + \pi/2)$$
$$= -2E_0 \sin \omega t \cos kz, \tag{7.8}$$

$$E_y = E_0 \cos(\omega t - kz) + E_0 \cos \omega t + kz + \pi$$
$$= 2E_0 \sin \omega t \sin kz. \tag{7.9}$$

The intracavity intensity is proportional to

$$E_x^2 + E_y^2 = 4E_0^2 \sin^2 \omega t, \tag{7.10}$$

which is independent of z. No standing wave distribution of energy density exists in the cavity and single longitudinal mode operation can be obtained.

7.4 Mode-Locking

So far in this chapter we have discussed ways in which to cause a laser to generate one or more CW oscillating frequencies corresponding to cavity modes. However, it is also possible to cause the laser to generate a train of regularly spaced, generally very short, pulses. A laser that operates in this way is said to be *mode-locked.* This kind of behavior often occurs spontaneously and is then referred to as *self-mode-locking*. A fundamental mode-locked pulse train consists of a series of pulses separated by the cavity round-trip time $2\ell/c$. That this kind of behavior could occur should not seem surprising. Early in the development of the laser, Fleck[7.2] showed that as a laser oscillation built up from spontaneous emission there was a natural tendency for the laser output to show strong fluctuations on a time scale that corresponded to the round-trip time in the cavity. A single spontaneous photon emitted into a high Q mode of the cavity will bounce back and forth in the resonator, and be amplified leading to a pulse train of spacing $2\ell/c$. Whether such a group of photons will succeed in competing with other growing oscillations will determine whether the laser ends up operating CW or mode-locked.†

There are two ways of looking at mode-locking. The more common, but less physically satisfying, is to treat the independent oscillating cavity modes whose electric fields are given in Eq. (7.1) as modes that have had their phases locked together, so that each mode has $\phi_n(t) = \phi_0$. By a shift of our choice of time origin we can arbitrarily set $\phi_0 = 0$. The total electric field of the laser can then be represented as

$$E(t) = \mathscr{R}\left[\sum_{n=1}^{N} E_n(t) e^{i\omega_n t}\right]. \tag{7.11}$$

To simplify the analysis we take $E_n(t) = 1$ and

$$\omega_n = \omega_0 + \left[n - \frac{(N+1)}{2}\right] \Delta\omega_c, \tag{7.12}$$

where $\Delta\omega_c = \pi c/\ell$.

† In this context the term CW is used to include lasers in which the laser excitation is pulsed, but in which the output radiation consists of one or more waves at longitudinal mode frequencies that last for the period of excitation.

Substituting Eq. (7.12) into Eq. (7.11) and summing the resultant geometric series gives

$$E(t) = \cos \omega_0 t \frac{\sin(N\Delta\omega_c t/2)}{\sin(\Delta\omega_c t/2)}. \tag{7.13}$$

This is a pure oscillation at frequency ω_0 modulated with the *envelope* function

$$f(t) = \frac{\sin(N\Delta\omega_c t/2)}{\sin(\Delta\omega_c t/2)}. \tag{7.14}$$

The average power corresponding to this envelope is

$$P(t) \propto \frac{\sin^2(N\Delta\omega_c t/2)}{\sin^2(\Delta\omega_c t/2)}. \tag{7.15}$$

Eq. (7.15) represents a periodic train of pulses that have the following properties:

(a) the pulse spacing is $\Delta T = 2\pi/\Delta\omega_c = 2\ell/c$;

(b) the peak power in the train is N times the average power, and the peak field is N times the average for a single mode;

(c) the pulses within the train become narrower as N increases, and for large N approach a value $\tau = \Delta T/N$.

These characteristics are evident from Fig. (7.7), which is a plot of Eq. (7.15). For an inhomogeneously broadened laser with linewidth $\Delta\nu_D$, the number of independent oscillating modes will be

$$N \sim \frac{2\pi\Delta\nu_D}{\Delta\omega_c}, \tag{7.16}$$

so the mode-locked pulse width will approach a value $\tau \simeq 1/\Delta\nu_D$.

For a homogeneously broadened laser, which ideally oscillates in a single longitudinal mode, the frequency explanation of the generation of short 'mode-locked' pulses seems less than satisfactory. This brings us to a more internally self-consistent model of mode-locking as a mode of operation that the laser finds *energy advantageous*, and which does not require the *a priori* assumption of independent oscillating cavity modes that become locked. This argument is illuminated by considering one of the most common ways in which mode-locking is accomplished – passive locking with a saturable dye cell in the laser cavity as shown in Fig. (7.8). The dye cell is frequently placed in close proximity to one of the cavity mirrors.

The oscillation threshold for an individual cavity mode in this arrangement will be high because of the intracavity loss presented by the absorbing dye. However, if the laser oscillates in a 'bouncing-pulse' mode in which the intracavity intensity is concentrated, then this pulse can 'bleach' the dye.†

In this situation a bouncing pulse is more energy efficient in overcoming the intracavity loss presented by the dye. The actual shape of the pulse will be determined by a self-consistency condition. The pulse train $f(t)$ will have a Fourier transform ($\mathscr{FT}$) $F(w)$ that describes the frequency content of the bouncing pulse. Each of the frequency components of this effective pulse train will be amplified by the gain medium according to a saturated gain profile $g_s(w)$. It is clear that the *narrowest* width that the bouncing pulse can have is $\sim 1/\Delta\nu$, where $\Delta\nu$ is the width of the gain profile. In common with any amplifier, frequencies outside the

† A dilute dye becomes less absorbing at high light intensities, as stimulated absorption and emission of dye molecules become balanced as the ground state of the dye is depleted by excitation.

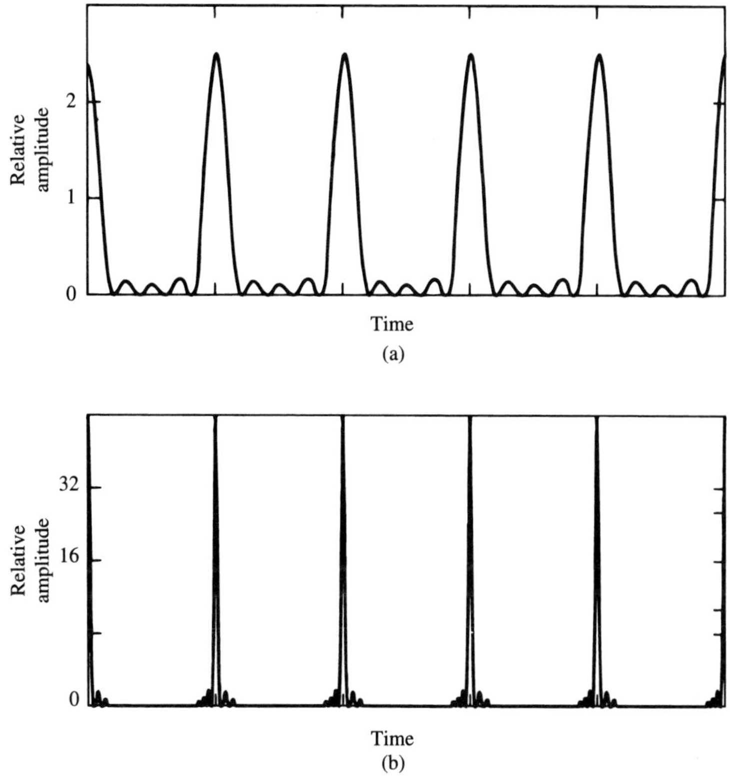

Fig. 7.7. Mode-locked pulse trains calculated from Eq. (7.15) (a) $N = 5$ and (b) $N = 20$.

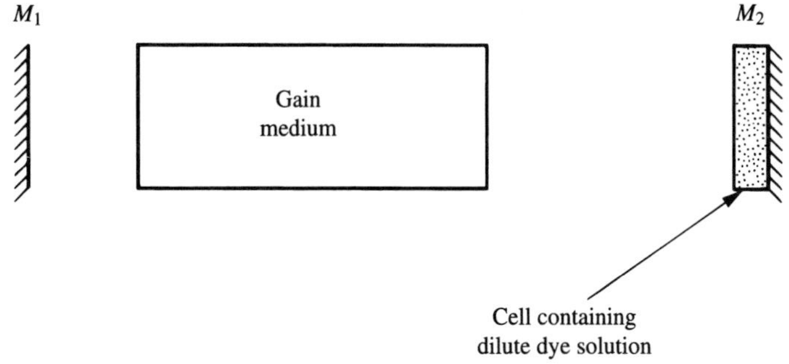

Fig. 7.8. Schematic diagram for passive mode-locking with a saturable dye.

gain bandwidth are not amplified. Schematically, we could write a self-consistency equation

$$\mathscr{F}\mathscr{T}[f(t)] = F(\omega)g_s(\omega)\alpha_d(\omega), \tag{7.17}$$

where $\alpha_d(\omega)$ is the effective absorption spectrum of the bleached dye. This explanation does not impose an inhomogeneously broadened description of the gain medium. Any broadband gain profile can, in principle, amplify a short pulse. Of course, once we accept the bouncing-pulse model of mode-locking we can examine $F(\omega)$ and will find that the principal components of the Fourier transform will correspond to longitudinal mode frequencies.†

† In a sense this is an example of what comes first – *the chicken or the egg?*

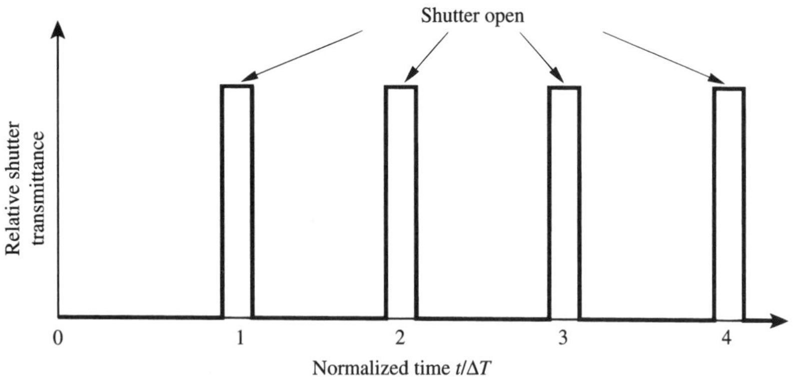

Fig. 7.9. Time sequence of opening of an intracavity shutter used to produce AM mode-locking.

7.5 Methods of Mode-Locking

In general, mode-locked behavior is caused to occur by modulating the gain or loss of the laser cavity in a periodic way, usually at a frequency $f_m = c/2L$. In amplitude modulation (AM) mode-locking the magnitude of cavity loss (or gain) is modulated; in phase (or frequency) modulation (FM) mode-locking only the complex part of the gain is modulated. In *active* mode-locking the modulation is introduced by external means, for example, with an intracavity modulator. In *passive* or *self*-mode-locking the modulation is created by the bouncing mode-locked pulse itself.

7.5.1 *Active Mode-Locking*

In the simplest form of AM mode-locking the intracavity loss is switched periodically with an intracavity shutter as shown schematically in Fig. (7.9). The precise time variation of the periodic intracavity loss is not very important, except that its period should equal the round-trip time, ΔT, for an optical pulse bouncing back and forth in the cavity. It is energy advantageous for a bouncing pulse to develop, rather than one or more CW longitudinal modes, as the pulse will adjust its arrival time at the intracavity modulator to correspond to the time of maximum transmission. An alternative way of viewing what happens is to realize that the modulator generates new frequency components in a longitudinal mode of frequency ν_m passing through it. These new frequencies are called *side-bands*† and have values

$$\nu_k = \nu_m + kf_m \qquad k = 0, \pm1, \pm2 \ldots . \tag{7.18}$$

Each of these side-bands also corresponds to another longitudinal cavity mode with frequency $\nu_{(m+k)}$.

The interaction between the fundamental longitudinal modes and side-bands causes the phases of the longitudinal modes to lock together and generate a mode-locked pulse train. In an inhomogeneously broadened laser the existence of several independent longitudinal modes whose phases can become locked is clear. In a homogeneously broadened laser in which only one such mode *should* oscillate, the side-bands of the dominant mode are not suppressed by gain-saturation: they compete for gain and a self-consistent bouncing pulse results. Almost all the lasers that

† For a more detailed discussion see Chapter 19.

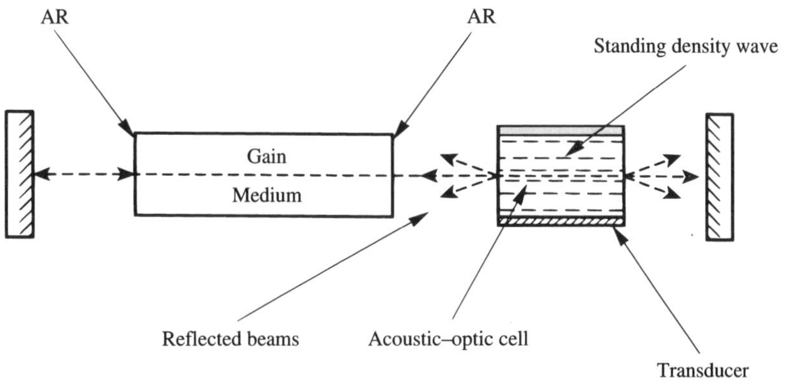

Fig. 7.10. Arrangement for mode-locking with an intracavity acousto-optic modulator (AOM) AR – antireflection coated face.

are used to generate very short mode-locked pulses are (ideally) homogeneously broadened: for example Nd:YAG, Nd:glass, dye and semiconductor lasers.

A common method of intracavity AM mode-locking is to use an acousto-optic modulator.† This was the method used in the first experimental observation of mode-locking using a helium–neon laser[7.3]. The acousto-optic modulator can be used to deflect light from the cavity. The undeflected beam is thereby amplitude modulated.

In mode-locking applications the acousto-optic modulator is used in a standing sound wave configuration; the acousto-optic modulator material is designed to resonate for sound waves at frequency f_s.‡ For an acousto-optic medium used in this way the variation of refractive index is

$$n(z,t) = n_0 + \Delta n \cos 2\pi f_s t \cos k_s z. \tag{7.19}$$

This standing wave acts like a phase diffraction grating that deflects light. The light travelling orthogonally to the sound wave is deflected at frequency $2f_s$, because the maximum index variation in Eq. (7.19) occurs twice per cycle of the standing sound wave. A typical experimental arrangement for mode-locking with an intracavity acoustic cell is shown in Fig. (7.10).

Mode-locking can also be induced through periodic variation of the gain of the amplifying medium. In a semiconductor laser (see Chapter 13) this can be simply done by modulating the drive current to the laser at frequency f_m. Alternatively, one periodically modulated laser can be used to mode-lock another. This approach is called *synchronous* pumping. For example, an argon laser that is itself mode-locked, but does not generate very short mode-locked pulses, can be used to periodically pump a dye laser (an optically excited laser that uses a dilute organic dye solution as its gain medium – see Chapter 12). The dye laser generates much shorter mode-locked pulses than were injected from the pump laser. The success of this scheme requires that the cavities of pump and sychronously mode-locked lasers be matched so that

$$f_m(\text{pump}) = \frac{1}{T}, \tag{7.20}$$

† A full discussion of these devices is given in Chapter 19.

‡ This is in contrast to acousto-optic frequency shifter applications where one end of the crystal is terminated in a matched acoustic load or absorber so that only a unidirectional travelling sound wave results.

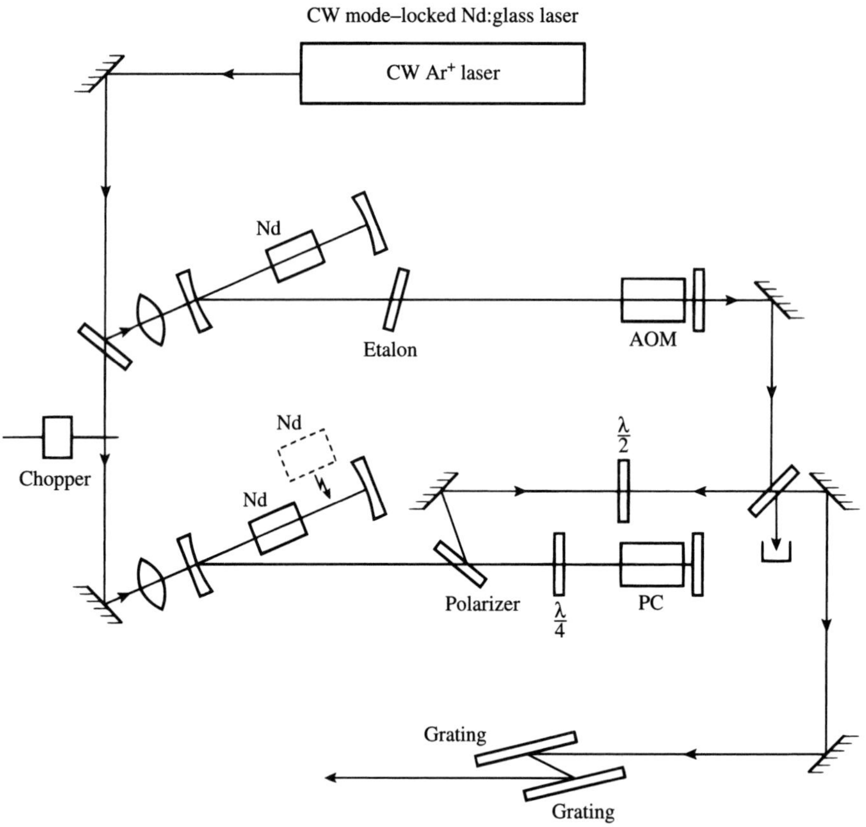

Fig. 7.11. System for generating ultrashort pulses from a neodymium glass laser system[7.7]. The CW Ar^+ laser pumps a Nd:phosphate glass laser that is mode-locked with an acousto-optic modulator (AOM). The mode-locked pulses are further amplified in a *regenerative* amplifier where they may make at least 60 round trips before being switched out of the cavity with the Pockels cells (PC) (an electro-optic crystal that can switch the polarization state of the beam). (Courtesy of Professor Chi H. Lee.)

where T is the cavity round-trip time in the pumped laser. Synchronous pumping allows the use of a convenient mode-locked pump, such as a Nd:YAG or argon ion laser, which do not themselves intrinsically generate the shortest mode-locked pulses, to pump dyes or doped glasses and crystals that support a large gain-bandwidth.

In the ideal case the length of the mode-locked pulses generated by any laser, whether it be homogeneously or inhomogeneously broadened, can approach a value $\tau \sim 1/\Delta\nu$, where $\Delta\nu$ is the width of the gain profile. Consequently gain media with large values of $\Delta\nu$ are desirable for the shortest mode-locked pulse generation, these include dye lasers,[7.4],[7.5] Nd:glass lasers[7.6],[7.7] and titanium–sapphire lasers[7.8]–[7.11]. Typical pulse lengths generated by such lasers are in the picosecond range, although special configurations achieve subpicosecond values. Fig. (7.11) shows a good example of an advanced mode-locked laser system that can generate sub-picosecond pulses. In this scheme a CW argon ion laser pumps as acousto-optically mode-locked Nd:phosphate glass oscillator. The mode-locked pulses are then further amplified by causing them to make multiple passes inside a Nd:phophate glass *regenerative* amplifier[7.7].†

It is usual in a mode-locked system using a linear resonator to place the intracavity modulator or absorbing dye close to one end mirror to ensure that

† Synchronous mode-locking can also be achieved if the optical lengths of the *pump* laser cavity and the *pumped* laser cavity have a ratio that is of the form $L_{pump}/\,L_{pumped} = p/p'$, where p and p' are small integers.

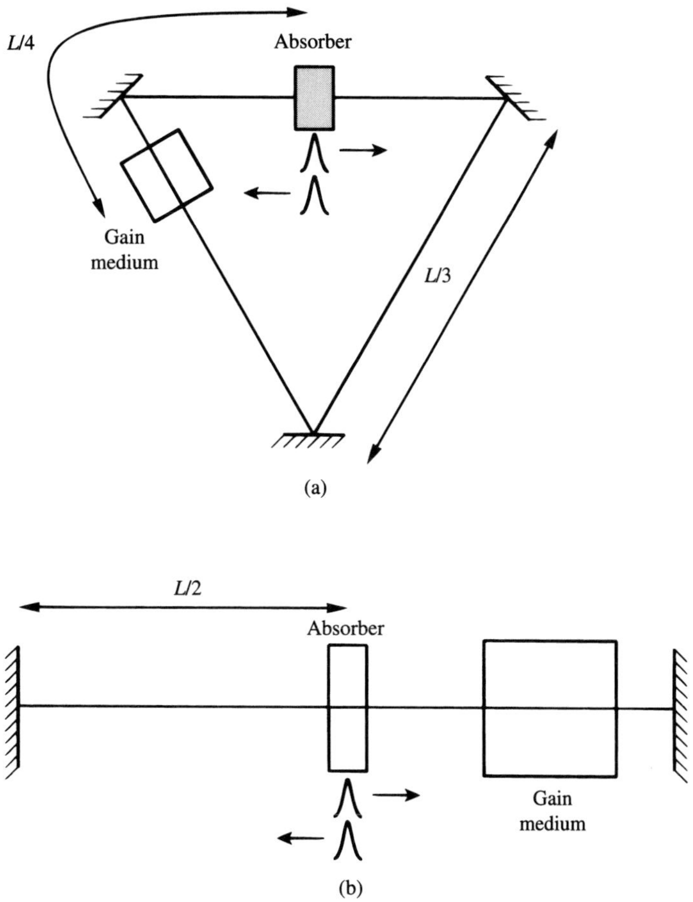

Fig. 7.12. Schematic placement of saturable absorber for CPM mode-locking: (a) ring cavity, (b) linear cavity.

the laser pulse passes the modulator only once in every round trip. If this is not done, then more than one bouncing pulse can develop within the cavity and the output will not be a regular pulse train of spacing $2\ell/c$. In laser cavities configured in the form of a ring, the placement of the intracavity modulator or absorber is not so important. In this case the possibility exists for two counterpropagating pulses to develop in the ring. In ring and linear resonators where two bouncing pulses develop, it is possible to cause the pulses to overlap or *collide* as they pass through an intracavity saturable absorber (dye). Because of the nonlinear variation of absorption of the dye with intensity the superimposed counterpropagating short pulses actually become shorter in order to maximize bleaching of the dye. In this scheme, called colliding pulse mode-locking (CPM), the typical placement of the absorber is at the center of a linear cavity or one quarter perimeter away from the gain medium in a ring cavity as shown in Fig. (7.12). CPM lasers can directly generate mode-locked pulses of sub-picosecond generation[7.5],[7.12].

The shortest pulses generated to date in ring cavity CPM are 28 fs long[7.13]. There is also an approach called additive pulse mode-locking (APM). The basic idea involves two coherent pulses with a relative phase shift interfering so that their wings tend to cancel out[7.14],[7.15].

7.6 Pulse Compression

Any short pulse can be represented by its Fourier transform

$$F(\omega) = \mathscr{F}\mathscr{T}\{f(t)\}. \tag{7.21}$$

The Fourier transform describes the spectral composition of the pulse $f(t)$ in both amplitude and phase. For example,

$$F(\omega) = A(\omega)e^{i\phi(\omega)}. \tag{7.22}$$

If we pass our original short pulse through a system in which the optical delay varies with frequency as $\Delta\phi(\omega)$, then the emerging pulse will have a Fourier transform

$$F'(\omega) = F(\omega)e^{i\Delta\phi(\omega)}. \tag{7.23}$$

If the lower frequency components suffer a greater phase shift than the higher frequency components, this is called a negative group velocity delay. In this case the output pulse, $g(t)$ will be

$$g(t) = \mathscr{F}\mathscr{T}\{F'(\omega)\}. \tag{7.24}$$

If the original pulse is written as

$$f(t) = a(t)e^{-i[\omega_0 t + \phi(t)]}, \tag{7.25}$$

where $a(t)$ is a smooth, slowly varying function of time (the pulse envelope), then the *instantaneous* frequency of the pulse is

$$\omega(t) = \omega_0 + \frac{d\phi}{dt}. \tag{7.26}$$

If $\phi(t)$ is nonlinear, the pulse is said to be *chirped.* If

$$\phi(t) = bE^2, \tag{7.27}$$

the frequency chirping is linear, and with $b > 0$, is positive. The leading edge of the pulse has a lower instantaneous frequency than the trailing edge – it is red-shifted relative to the blue-shifted trailing edge. In a medium with a negative group velocity delay characteristic, the lower frequencies in the leading edge of the pulse travel more slowly than the higher frequencies in the trailing edge. The back of the pulse catches up with the front and the pulse is compressed, as shown schematically in Fig. (7.13).

A simple pulse compressor can be constructed with two diffraction gratings as shown in Fig. (7.14)[7.16]. The first order diffraction angle θ satisfies

$$\sin\theta = \frac{\lambda}{d} - \sin\theta_{in}. \tag{7.28}$$

The longer of two wavelengths diffracts at a larger angle and takes a longer path between the two gratings – thereby acquiring a larger phase shift and satisfying the condition for negative group delay.

An optical fiber can also be used to produce pulse compression. Because of the Kerr effect the refractive index of the fiber varies with intensity as

$$n = n_0 + \gamma I. \tag{7.29}$$

Accordingly, the peak of a pulse is slowed relative to its tail. There are additional effects on the pulse because of the dispersion in the fiber $n(\lambda)$. If $n(\lambda)$ lies in a

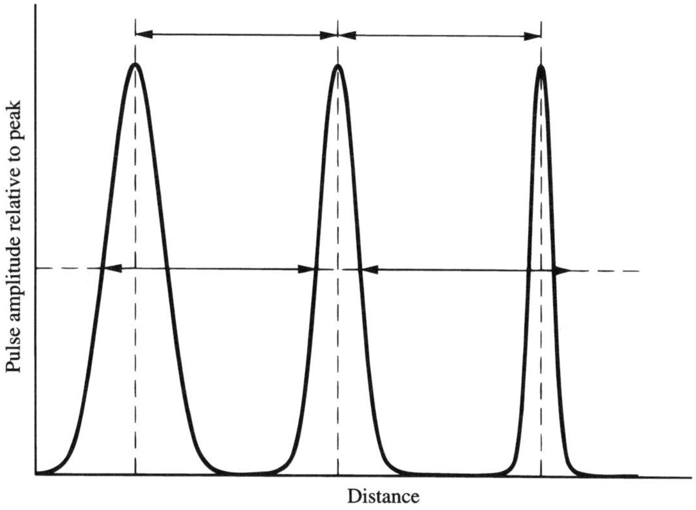

Fig. 7.13. Pulse compression resulting from the travelling edge of a pulse catching up with its leading edge. The arrows in the figure correspond to equal time intervals.

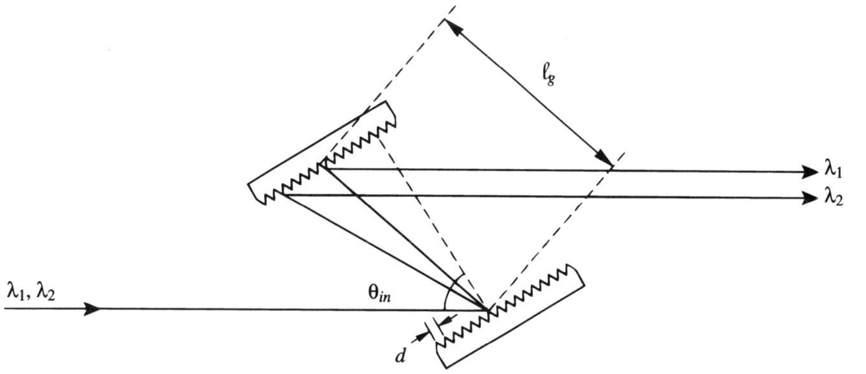

Fig. 7.14. Diffraction grating pair used to produce pulse compression of a 'chirped' pulse.

region of negative group velocity dispersion (which for silica fibers lies beyond 1.3 μm) then the index nonlinearity and dispersion work together to make the pulse shorter.† To date pulses have been compressed to only 6 fs in length[7.17].

References

[7.1] P.W. Smith, 'Mode selection in lasers,' *Proc. IEEE*, **60**, 422–440, 1972. See also A.G. Fox and P.W. Smith, *Phys. Rev. Lett.*, **18**, 826–828, 1967.

[7.2] J.A. Fleck, 'Mode-locked pulse generation in passively switched lasers,' *Appl. Phys. Lett.*, **12**, 178–181, 1968.

[7.3] L.E. Hargrove, R.L. Fork and M.A. Pollack, 'Locking of He-Ne laser modes induced by synchronous intracavity modulation,' *Appl. Phys. Lett.*, **5**, 4–5, 1964.

[7.4] C.V. Shank, and E.P. Ippen, 'Subpicosecond kilowatt pulses from a mode-locked CW dye laser,' *Appl. Phys. Lett.* **24**, 373–375, 1974.

[7.5] R.L. Fork, B.I. Greene, and C.V. Shank, 'Generation of optical pulses shorter than 0.1 psec by colliding pulse mode-locking,' *Appl. Phys. Lett.*, **38**, 671–672, 1981.

[7.6] D.J. Bradley and W. Sibbett, 'Streak-camera studies of picosecond pulses from a mode-locked Nd:glass laser,' *Opt. Commun.*, **9**, 17–20, 1973.

† An optical soliton (see Chapter 17) develops for similar reasons.

[7.7] L. Yan, P-T. Ho, C.H. Lee, and G.L. Burdge, 'Generation of ultrashort pulses from a neodymiun glass laser system,' *IEEE J. Quantum Electron.*, **QE-26**, 2431–2440, 1989.

[7.8 J.D. Kafka, M.L. Watts, D.J. Roach, M.S. Keirstead, H.W. Schaaf, and T. Baer, in *Ultrafast Phenomena VII*, C.B. Harris, E.R. Ippen, G.A. Mourou, and A.H. Zewail, Eds., Springer-Verlag, New York, 1990, pp. 66–68.

[7.9] N. Sarnkara, Y. Ishida, and H. Nakano, 'Generation of 50-fsec pulses from a pulse-compressed, CW, passively mode-locked Ti:sapphire laser,' *Opt. Lett.*, **16**, 153–155, 1991.

[7.10] D.E. Spence, P.N. Kean, and W. Sibbett, '60-fsec pulse generation from a self-mode-locked Ti:sapphire laser,' *Opt. Lett.*, **16**, 42–44, (1991).

[7.11] J. Squier, F. Salin, and G. Mourou, '100-fs pulse generation and amplification in Ti:Al_2O_3,' *Opt. Lett.* **16**, 324–326, 1991.

[7.12] P.M.W. French and J.R. Taylor, 'Generation of sub-100 fsec pulses tunable near 497nm from a colliding-pulse mode-locked ring dye laser,' *Opt. Lett.*, **13**, 470–472, 1988.

[7.13] J.A. Valdmanis, R.L. Fork, and J.P. Gordon, 'Generation of optical pulses as short as 28 femtoseconds directly from a laser balancing self-phase modulation, group-velocity dispersion, saturable absorption, and saturable gain,' *Opt. Lett.*, **10**, 131–133, 1985.

[7.14] E.P. Ippen, H.A. Haus, and C.Y. Liu, 'Additive pulse mode-locking,' *J. Opt. Soc. Am.*, **B6**, 1736–1745, 1989.

[7.15] K.J. Blow and D. Wood, 'Mode-locked lasers with nonlinear external cavities,' *J. Opt. Soc. Amer.*, **B5**, 629–632, 1988.

[7.16] E.B. Treacy, 'Optical pulse compression with diffraction gratings,' *IEEE J. Quantum Electron.*, **QE-5**, 454–458, 1969.

[7.17] R.L. Fork, C.H. Brito Cruz, P.C. Becker, and C.V. Shank, 'Compression of optical pulses to six femtoseconds by using cubic phase compensation,' *Opt. Lett.*, **12**, 483–485, 1987.

8

Optically Pumped Solid-State Lasers

8.1 Introduction

In this chapter we shall discuss in some detail the operating principles, characteristics, and design features of solid-state lasers in which the laser medium is an insulating or glassy solid. In many of these lasers the active particles are impurity ions doped into a host matrix. These lasers are pumped optically, generally with a pulsed or continuous lamp, although they can also be pumped by another laser. Our discussion will build on the brief introduction to one of this class of laser, the ruby laser, given in Chapter 3. The chapter will conclude with a discussion of the characteristics of the radiation emitted by such lasers and how this radiation can be modified and controlled in time.

8.2 Optical Pumping in Three- and Four-Level Lasers

The optical pumping process in an insulating solid-state laser can be illustrated schematically with reference to Fig. (8.1). Light from the pumping lamp(s) excites ground state particles into an absorption band, labelled 3 in the figure. Ideally, particles that reach this state should transfer rapidly into the upper laser level, level 2. If this transfer process is preferentially to level 2 rather than to level 1 a population inversion will result between levels 2 and 1 and laser action can be obtained. The drain transition from level 1 back to the ground state should be fast, to keep level 1 from becoming a 'bottleneck.' The performance of the laser will be influenced by several factors.

8.2.1 Effective Lifetime of the Levels Involved

The length of time a particle can remain in an excited level is governed by its *effective* lifetime. This lifetime is influenced by both radiative and nonradiative processes. The transfer transition from level 3 to level 2 in Fig. (8.1) is generally nonradiative. Particles make the transition by dumping their excess energy into the lattice – they literally heat up the medium. Such a nonradiative process from level i to level j can be described by a rate coefficient X_{ij}, similar to a spontaneous emission coefficient A_{ij}. The overall rate at which particles leave level i is

$$\frac{dN_i}{dt} = -\sum_j N_i A_{ij} - \sum_j N_i X_{ij}, \tag{8.1}$$

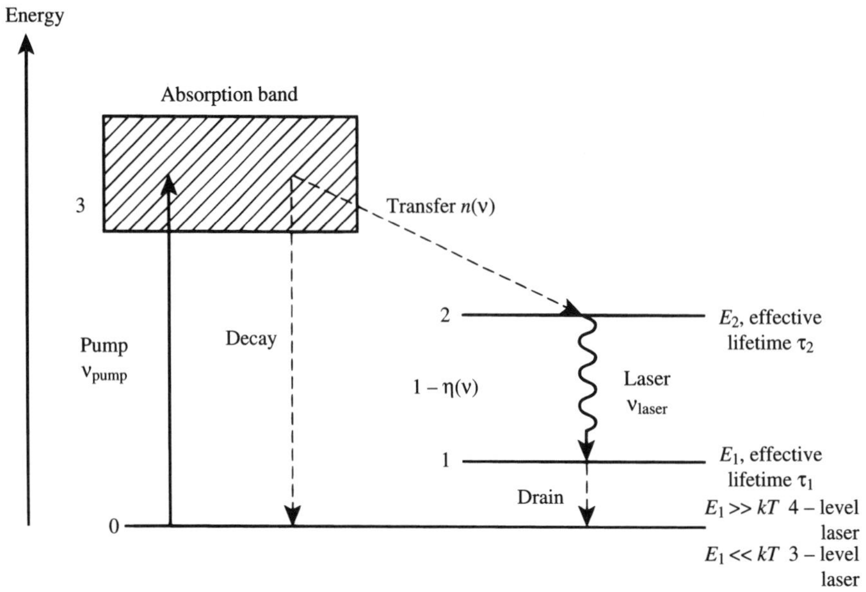

Fig. 8.1. Schematic diagram of a general optically pumped laser system. In a good laser system the transfer coefficient should be close to unity and transfer should occur rapidly.

where the summation in j runs over all levels below level i in energy. The effective lifetime of level i is

$$\tau_i = 1/\sum_j (A_{ij} + X_{ij}). \tag{8.2}$$

The nonradiative decay of an excited state can be thought of as a collisional process, in a similar way to stimulated emission, in which a quantized packet of acoustic energy within the solid, called a *phonon*, collides with the excited particle. These quantized acoustic energy packets always exist in a solid, even at absolute zero: they correspond closely to the waves we considered in our discussion of black-body radiation, except that the modes being counted are now vibrational motions of the particles constituting the solid. The perturbation of excited states by these phonons leads to line broadening, which can be substantial. Broad absorption bands in a solid result from the line-broadened 'smearing' together of levels that would be sharp and distinct in the vapor phase. The vastly increased numbers of neighboring particles in a solid (or liquid) compared to a gas causes large amounts of line broadening, although the extent of the broadening can vary greatly from level to level.

8.2.2 Threshold Inversion in Three- and Four-Level Lasers

If the lower laser level is very close to ($E_1 \ll kT$), or is, the ground state, the system is a three-level laser. In such a laser the threshold inversion is substantially higher than in a four-level system ($E_1 \gg kT$). If the total number of particles per unit volume participating in the pumping and lasing process is N we can write

$$N = N_3 + N_2 + N_1 + N_{gs}, \tag{8.3}$$

where the subscripts indicate the level involved, and N_{gs} is the population density in the ground state. Eq. (8.3) assumes that other levels of the system are not actively involved and that their populations are negligible.

In a three-level laser level 1 is essentially the ground state so we can write

$$N = N_3 + N_2 + N_1. \tag{8.4}$$

In a good solid-state laser, level 3 transfers its excitation very rapidly to level 2 so its population can be neglected. In this case

$$N = N_2 + N_1. \tag{8.5}$$

The threshold inversion is

$$N_t = N_2 - \frac{g_2}{g_1} N_1. \tag{8.6}$$

From Eqs. (8.5) and (8.6)

$$(N_2)_{\text{3-level}} = \frac{(g_2/g_1)N + N_t}{\left[(g_2/g_1 + 1)\right]}, \tag{8.7}$$

which if $g_2 = g_1$ gives $N_2 \simeq N/2$.

In a good four-level laser in which the lower laser level remains relatively depopulated we can write

$$(N_2)_{\text{4-level}} \simeq N_t.$$

Consequently, the relative rate at which level 2 must be excited to produce an inversion (all other factors such as decay rates being comparable) is

$$\frac{(N_2)_{\text{3-level}}}{(N_2)_{\text{4-level}}} \simeq \frac{(g_2/g_1)N + N_t}{\left[(g_2/g_1) + 1\right] N_t}. \tag{8.8}$$

Usually, $N \gg N_t$, and if we take the simple case $g_2 = g_1$, the ratio of pumping rates becomes $N/2N_t$. This number can easily be 10^4 or more, demonstrating once again how surprising it was that a three-level laser was the first ever to be operated.

8.2.3 Quantum Efficiency

If the average energy of photons from the pumping lamp is $h\overline{\nu}_{pump}$ and the laser photon is $h\nu$ the intrinsic quantum efficiency of the pumping process is

$$\eta_i = \frac{h\nu}{h\overline{\nu}_{pump}}. \tag{8.9}$$

8.2.4 Pumping Power

The rate at which particles must be excited to the upper laser level to sustain a population inversion is

$$R_2 = N_2/\tau_2 (\text{particles m}^{-3}\ \text{s}^{-1}). \tag{8.10}$$

If the average probability factor for a particle in level 3 transferring to level 2 rather than elsewhere is $\overline{\eta}$ (called the branching factor) then the rate at which level 3 must be pumped is

$$R_3 = \frac{N_2}{\overline{\eta}\tau_2}. \tag{8.11}$$

The corresponding absorbed pump power is

$$'P_A = \frac{N_2 h\overline{\nu}_{pump}}{\overline{\eta}\tau_2}. \tag{8.12}$$

8.2.5 *Threshold Lamp Power*

To determine the power and spectral characteristics of the lamp(s) needed to create an inversion, we must relate the threshold pumping rate R_2 to the absorption coefficient in the pump band, and electrical and geometrical factors that determine how efficiently the lamp generates pump light and couples this into the laser medium.

The absorption band has a lineshape function $g_3(\nu)$. If the energy density of the pump radiation in the laser medium is $\rho_P(\nu)$ then the rate at which level 3 is excited is

$$R_3 = \int N_0 B_{03} g_3(\nu)\rho_P(\nu)d\nu. \tag{8.13}$$

If we assume plane wave illumination of the laser medium then we can write $\rho(\nu) = I(\nu)/c$ and Eq. (8.13) becomes

$$R_3 = \int N_0 \frac{c^2 A_{30}}{8\pi\nu^2}\frac{I(\nu)}{h\nu} g_3(\nu)d\nu. \tag{8.14}$$

From Eq.(8.14) we can recognize $N_0 c^2 A_{30} g_3(\nu)/8\pi\nu^2$ as the absorption (negative gain) coefficient of the laser medium $\alpha(\nu)$. So

$$R_3 = \int \frac{I(\nu)\alpha(\nu)d\nu}{h\nu}. \tag{8.15}$$

The integration covers the range of frequencies encompassed by the absorption band.

If we allow for the possibility of a frequency-dependent branching factor, the rate of excitation of level 2 is

$$R_2 = \int \frac{I(\nu)\alpha(\nu)\eta(\nu)}{h\nu} d\nu. \tag{8.16}$$

If the absorption band over which the integration is carried out is narrow, of width $\Delta\nu_3$, we can replace the quantities in (8.16) by averages and write

$$R_2 = \overline{I}(\nu)\overline{\alpha}(\nu)\overline{\eta(\nu)}\Delta\nu_3/h\overline{\nu}_{pump}. \tag{8.17}$$

8.3 Pulsed Versus CW Operation

If the pumping lamp is a flashlamp of duration $t_p \ll \tau_2$, then at the end of the flash, all the excited particles will still be in level 2. In this case

$$N_2 = t_p\overline{I(\nu)}\ \overline{\alpha(\nu)}\ \overline{\eta(\nu)}\Delta\nu_3/h\overline{\nu}_{pump}. \tag{8.18}$$

The flashlamp energy needed to achieve $N_2 \geq N_t$ depends additionally on the following factors: The electrical efficiency e of the lamp defined as

$$e = \frac{\text{joules of light energy out}}{\text{capacitor joules in}}. \tag{8.19}$$

(For xenon flashlamps this efficiency factor can be as high as 80%)[8.1]. The fraction f of this light in the spectral region that will pump the absorption band is

$$f = \frac{\text{light energy within absorption band}}{\text{total light energy}}, \tag{8.20}$$

and the geometrical efficiency g with which the light is coupled to the laser medium, which usually takes the form of a cylindrical rod or flat slab

$$g = \frac{\text{light energy within absorption band reaching laser medium}}{\text{total light energy within absorption band}}. \tag{8.21}$$

The total energy that must reach the laser crystal, in the right spectral region, to reach threshold can be written as

$$U_t = t_p\overline{I(\nu)} = \frac{N_t h\overline{\nu}_{pump}}{\overline{\alpha(\nu)}\overline{\eta(\nu)}}. \tag{8.22}$$

8.3.1 Threshold for Pulsed Operation of a Ruby Laser

In the ruby laser the typical Cr^{3+} ion concentration is $\sim$ 0.05% by weight, equivalent to about 10^{25} Cr^{3+} m^{-3}. Since this is a three-level laser, roughly half these ions must be excited to the upper laser level to reach threshold. Therefore, we can take $N_t = 5 \times 10^{24}$ m^{-3}. Fig.(3.3) shows that the approximate average absorption coefficient of ruby in the 350–600 nm region is about 100 m^{-1}. Assuming efficient transfer from the pump bands we take $\overline{\eta(\nu)} = 1$. The average pump wavelength is $\simeq$ 475 nm. Therefore,

$$U_t \simeq \frac{5 \times 10^{24} \times 6.66 \times 10^{-34} \times 3 \times 10^8}{100 \times 475 \times 10^{-9}} = 21 \text{ kJ m}^{-2}.$$

For a cylindrical ruby crystal 20 mm long $\times$ 10 mm diameter, if the lamp energy reaches the crystal uniformly over its curved surface the threshold input energy reaching the crystal in the appropriate spectral band is about 13 J. For a modern flashlamp the electrical efficiency e can be 50% or more. The spectral output of the lamp will approximate a black body with superimposed spectral features, as shown in Fig. (8.2). It is reasonable to assume that 25% of the emitted light is in the ruby pump bands. If the optical arrangement of lamp(s) and crystal is efficient the geometrical factor g should be 50% or better. The electrical input to reach threshold will therefore be about 200 J. Small ruby crystals in an efficient pumping arrangement such as the axisymmetric ellipsoidal cavity shown in Fig. (8.3) can have thresholds as low as 34 J.

To achieve CW operation in a solid-state insulating laser the pumping rate of the upper laser level must be sufficient to maintain $N_2 = N_t$ in the face of all the spontaneous relaxation processes. The absorbed power per unit volume must be

$$U_t = \frac{N_t h\overline{\nu}_{pump}}{\tau_2\overline{\eta(\nu)}}. \tag{8.23}$$

8.3.2 Threshold for CW Operation of a Ruby Laser

Although the ruby laser is not commonly operated in a CW mode it is interesting to calculate its threshold pump rate. We use the same values for the parameters in

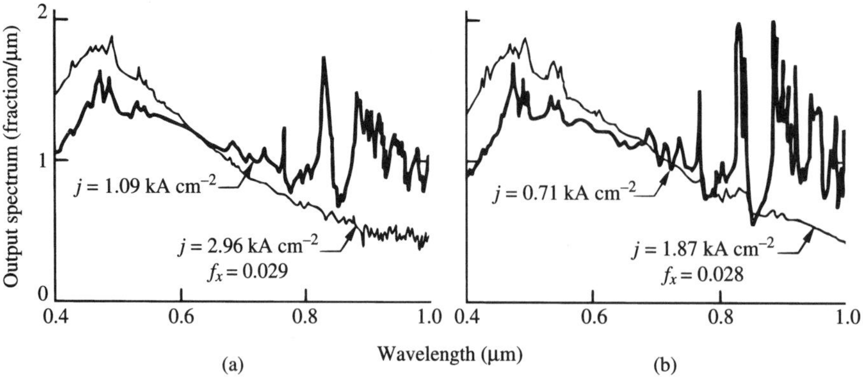

Fig. 8.2. Examples of output spectra from various flashlamps at different flashlamps current densities; f_x is the fraction of explosion energy at which each lamp was operated: (a) 1.5 cm-bore lamp; (b) 4.2 cm-bore lamp[8.2].

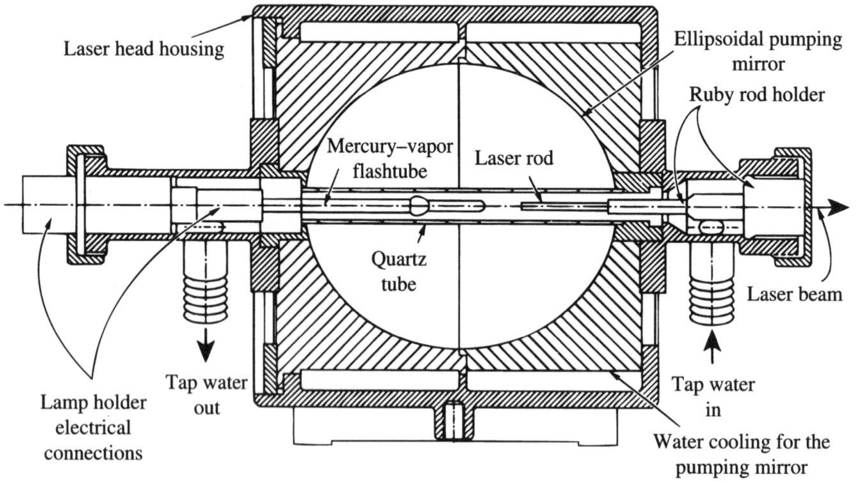

Fig. 8.3. Ellipsoidal cavity for solid-state laser pumping that provides axisymmetric illumination of a cylindrical laser rod by a cylindrical lamp[8.5],[8.6].

Eq. (8.23) as before, with the addition of the value for τ_2, 3 ms, to get

$$U_t = \frac{5 \times 10^{24} \times 6.626 \times 10^{-34} \times 3 \times 10^8}{3 \times 10^{-3} \times 475 \times 10^{-9}} \simeq 7 \times 10^8 \text{ W m}^{-3}.$$

In the first CW ruby laser experiments, Nelson and Boyle[8.3] obtained threshold pump rates of this order. They had to use a 850 W mercury xenon lamp focused onto a ruby crystal 11.5 mm long by 0.61 mm in diameter kept at liquid nitrogen temperature inside a dewar to achieve 300 mW of power actually absorbed in the crystal.

8.4 Threshold Population Inversion and Stimulated Emission Cross-Section

The small-signal gain at the center of the line can be written in the form

$$\gamma_0 = \left(N_2 - \frac{g_2}{g_1} N_1 \right) \sigma_0, \tag{8.24}$$

where σ_0 is called the stimulated emission cross-section. It serves as a useful parameter for comparing different laser media. From Eq. (2.30) it is clear that for a homogeneously broadened line, for which

$$g(\nu_0, \nu_0) = \frac{2}{\pi \Delta \nu}, \tag{8.25}$$

$$\sigma_0 = \frac{c^2 A_{21}}{4\pi^2 \nu^2 \Delta \nu}. \tag{8.26}$$

For the ruby laser $\sigma_0 \simeq 1.2 \times 10^{-24}$ m^2.

To achieve threshold the minimum inversion, N_t is

$$N_t = \left(N_2 - \frac{g_2}{g_1} N_1 \right)_t = \frac{1}{\sigma_0} \left(\alpha - \frac{1}{\ell} \ln r_1 r_2 \right). \tag{8.27}$$

Therefore, the stimulated emission cross-section together with parameters of the laser cavity allow quick estimation of the threshold inversion. We shall meet some examples of this later.

8.5 Paramagnetic Ion Solid-State Lasers

A large number of paramagnetic ions from the iron, rare-earth (lanthanide) and actinide groups of the periodic table exhibit laser action when doped into a large number of host crystals or glasses. In a separate class of solid-state lasers – the so-called *stoichiometric* lasers – the active ion is not an impurity dopant but is an intrinsic part of the lattice. These lasers have not to date achieved significant importance and we will not discuss them further. The large diversity of doped crystalline lasers is demonstrated by Table (8.1), which lists some of the most important such lasers, together with some of their relevant operating parameters. These are all optically pumped lasers; most are four-level systems. By far the most important is the Nd^{3+} laser, which is operated successfully in a variety of host materials, for example in yttrium aluminum garnet ($Y_3Al_5O_{12}$,YAG), calcium tungstate ($CaWO_4$), lithium yttrium fluoride ($LiYF_4$,YLF), $Ca_5(PO_4)_3F$(FAP), $YAlO_3$(YALO), gadolinium gallium garnet (GGG), scandium-substituted GGG (GSGG), and in various glasses. Another important laser of this type is the holmium (Ho^{3+}) laser in $LiYF_4$ (YLF), frequently with erbium ions (Er^{3+}) or thulium ions (Tm^{3+}) added to the host to assist in the absorption of flashlamp energy and transfer into the Ho^{3+} upper laser level.

The chromium (Cr^{3+}) in $BeAl_2O_4$ (alexandrite) laser, is interesting as its laser wavelength can be tuned continuously from 700 to 800 nm.

The Ti^{3+} in Al_2O_3, the titanium–sapphire laser, is a very attractive source for the generation of tunable near-infrared radiation and has largely replaced dye lasers in this application. Fig. (8.4) shows the absorption and emission spectroscopic properties of the Ti^{3+}:Al_2O_3 material. When pumped by a high power argon ion laser, a power output in excess of 1 W can be obtained[8.4].

Although specific details of their operation differ from one doped insulating crystal or glass laser to another, many of the general features of their construction and operation are similar. The Nd:YAG laser serves as a benchmark for this discussion.

8.6 The Nd:YAG Laser

YAG has a combination of desirable properties as a host medium for Nd^{3+} ions: it has relatively high thermal conductivity, which allows it to disperse the waste heat

Table 8.1. *Characteristics of Some Optically Pumped Solid State Laser Systems.*[a]

Ion	Host	Laser wavelength (μm)	Lower laser level energy (cm^{-1})	Laser transition	Laser ion concentration %	τ_2 (ms)	$\Delta\nu$ (GHz)	Pulsed P or CW	Operating temperature (K)	Method of pumping
Cerium, Ce^{3+}	$LiYF_4$(YLF)	0.286	Variable: this is a tunable laser	$5d \rightarrow {}^2F_{7/2}$	0.05	18 ns		P	300	KrF
Neodymium, Nd^{3+}	$Bi_4Ge_3O_{12}$	1.0638	~2000	${}^4F_{3/2} \rightarrow {}^4I_{11/2}$	0.3			P low threshold	77	Xe
	CaF_2	1.0448	2034		0.4–0.6	1.1	270	P	120	Xe
	CaF_2–YF_3	1.0461	2032		0.2–12	0.48	900	P	300	Xe
	CaF_2–YF_3	1.0632	2170		0.2–12	0.48	2700	P	300	Xe
	$CaMoO_4$	1.061	~2000		1.8	0.12	–	P or CW	300	Xe
	$Ca_5(PO_4)_3F$	1.063	1900		1	0.25	180	P or CW	300	Xe
	$CaWO_4$	0.9145	471	${}^4F_{3/2} \rightarrow {}^4I_{9/2}$	0.14–3	0.18	450	P	77	Xe
		1.0649	2016	${}^4F_{3/2} \rightarrow {}^4I_{11/2}$	0.14–3	0.18	210	P or CW	85	Hg
		1.3340	~3970	${}^4F_{3/2} \rightarrow {}^4I_{13/2}$	0.14–3	0.18	600	P	300	Xe
		1.3370	~3925		0.14–3	0.18	690	CW	300	Xe
	$CaY_2Mg_2Ge_3O_{12}$	1.05986	2008	${}^4F_{3/2} \rightarrow {}^4F_{11/2}$	2–6	0.305	1109	CW (high efficiency)	300	Ar^+ laser
	CeF_3	1.0638	2189	${}^4F_{3/2} \rightarrow {}^4I_{11/2}$	~4	~0.27	~1050	P	300	Xe
	$GdGa_5O_{12}$ (GGG)	1.0621	2064		1	0.27	216	P	300	Xe
	$Gd_3Sc_2Al_3O_{12}$	1.06	1978		1	0.275	345	CW	300	W
	$Gd_3Sc_2Ga_3O_{12}$	1.0612	2070		1	0.26	420	P	300	Xe
	LaF_3	1.04065	1983		1–2	0.6–0.7	750	P or CW	300	Xe
		1.3310	4070	${}^4F_{3/2} \rightarrow {}^4I_{13/2}$	2	0.6–0.7	630	P	300	Xe
	LaF_3	1.3310	4070	${}^4F_{3/2} \rightarrow {}^4I_{13/2}$	2	0.6–0.7	630	P	300	Xe
	$LiYF_4$ (YLF)	1.0471	2042	${}^4F_{3/2} \rightarrow {}^4I_{11/2}$		0.5	360	P	300	Xe
		1.0530			2	0.5	375	P	300	Xe

Table 8.1. *Characteristics of Some Optically Pumped Solid State Laser Systems (continued).*

Ion	Host	Laser wave-length (μm)	Lower laser level energy (cm^{-1})	Laser trans-ition	Laser Ion concen-tration %	τ_2 (ms)	$\Delta\nu$ (GHz)	Pulsed P or CW	Operating temp-erature (K)	Method of pumping
Nd^{3+}	$Lu_3Al_5O_{12}$	1.06425	2099		~0.6	0.245	160	P or CW	300	Xe
	SrF_2	1.0370	2008		0.8	1.1	600	P	300	Xe
	$YAlO_3$	0.930	670	$^4F_{3/2} \rightarrow ^4I_{9/2}$	1	0.18	900	P	300	XeF laser
		1.0795	2157	$^4F_{3/2} \rightarrow ^4I_{11/2}$	1–3	0.18	330	P or CW (low threshold)	300	Xe
		1.0645	2026		1–3	0.18	285	P or CW	300	Xe
	$Y_3Al_5O_{12}$ (YAG)	0.8910	200	$^4F_{3/2} \rightarrow ^4I_{9/2}$	1	0.255		P	300	Ar^+ laser
		0.8999	311		1	0.255		P	300	Ar^+ laser
		0.9385	852		1	0.255		P	300	Ar^+ laser
		0.09460	852		1	0.255	270	P	300	Xe
		1.06415	2110	$^4F_{3/2} \rightarrow ^4I_{11/2}$	~1	0.255	195	P or CW (low threshold)	300	Xe
		1.0682	2146		~1	0.255	~300	P	300	Xe
		1.3188	3222	$^4F_{3/2} \rightarrow ^4I_{13/2}$	~1	0.255	195	CW	300	Kr
		1.3382	4034		~1	0.255	231	CW	300	Kr
		1.3564	~4000		~1	0.255	–	CW	300	Kr
		1.4140	~4000		~1	0.255	–	CW	300	Kr
	Y_2O_3	~1.0746	1895	$^4F_{3/2} \rightarrow ^4I_{11/2}$	1.5	0.34	240	CW (very low threshold)	300	Kr^+ laser
		~1.358	3837	$^4F_{3/2} \rightarrow ^4I_{13/2}$	1.5	0.34	360	CW (very low threshold)	300	Kr^+ laser
	$Y_3Sc_2Ga_3O_{12}$	1.0583	~2000	$^4F_{3/2} \rightarrow ^4I_{11/2}$	1		240	P	300	Xe
		1.3310	~4000	$^4F_{3/2} \rightarrow ^4I_{13/2}$	1		420	P	300	Xe
Nd^{3+}	YVO_4	1.0625	1964		~1	0.092		P	300	Xe
		1.0634	~2000		~1	0.092	210	CW	300	Ar^+ laser
		1.3425	3913		0.1	0.033	240	P	300	Xe

Table 8.1. *Characteristics of Some Optically Pumped Solid State Laser Systems (continued).*

Ion	Host	Laser wave-length (μm)	Lower laser level energy (cm^{-1})	Laser trans-ition	Laser ion concen-tration %	τ_2 (ms)	Δν (GHz)	Pulsed P or CW	Operating temp-erature (K)	Method of pumping
Holmium Ho^{3+}	CaF_2-ErF_3(3%)	2.06	~ 250	$^4I_7 \rightarrow ^5I_8$	0.5–1			P	298	Xe
	$LiYF_4$ (YLF)	0.7505	~5300	$^5S_2 \rightarrow ^5I_7$	2	0.09	360	P	300	Xe
		0.9794	~5300	$^5F_5 \rightarrow ^5I_7$	2	0.1	360	P	90	Xe
		1.3960	~11300	$^5S_2 \rightarrow ^5I_5$	2	0.05		P	300	Xe
	$LaYF_4$ (+5% Er)	2.066	300	$^5I_7 \rightarrow ^5I_8$	2			P	300	Xe
	Li(Y,Er)F_4 (+6.7% Tm)	2.0654	~ 300	$^5I_7 \rightarrow ^5I_8$	1.7	12		P	300	Xe
	$Lu_3Al_5O_{12}$ (+10% Yb, 0.3% Cr)	2.9460	~5400	$^5I_6 \rightarrow ^5I_7$	10		270	P	300	Xe
	$Y_3Al_5O_{12}$ (YAG)	2.0914	532	$^5I_7 \rightarrow ^5I_8$	~4	4–5		P	77	Xe
		2.9403	~5400	$^5I_6 \rightarrow ^5I_7$	10		270	P	300	Xe
	YAG + 50% Er	~2.13	532	$^5I_7 \rightarrow ^5I_8$	1.65	5		P	300	Xe
	6–7% Tm							CW	77	W
Erbium Er^{3+}	CaF_2	2.7307	~6500	$^4I_{11/2} \rightarrow ^4I_{13/2}$	4			P	300	Xe
	$LiYF_4$	0.85	6714	$^4S_{3/2} \rightarrow ^4I_{13/2}$	2	0.2	300	P	300	Xe
	(YLF)	~1.2308	10313	$^4S_{3/2} \rightarrow ^4I_{11/2}$	2	0.2		P	300	Xe
		1.7320	12572	$^4S_{3/2} \rightarrow ^4I_{9/2}$	2	0.2		P	300	Xe
		2.870	6738	$^4I_{11/2} \rightarrow ^4I_{13/2}$	2	10		P	300	Xe
Er^{3+}	$Lu_3Al_5O_{12}$	0.86325	6818	$^4S_{3/2} \rightarrow ^4I_{13/2}$	1.5	0.13	240	P	300	Xe
		1.6525	534	$^4I_{13/2} \rightarrow ^4I_{15/2}$	2–5	6–7	195	P	77	Xe
		1.7762	12772	$^4S_{3/2} \rightarrow ^4I_{9/2}$	1.5	0.11		P	300	Xe
		2.8298	6885	$^4I_{11/2} \rightarrow ^4I_{13/2}$	~33	0.12		P	300	Xe
	+5%Yb, 0.3%Cr	2.8298	~6885	$^4I_{11/2} \rightarrow ^4I_{13/2}$	33	0.12		P (low threshold)	300	Xe
	$Y_3Al_5O_{12}$ (YAG)	0.8627	~6800	$^4S_{3/2} \rightarrow ^4I_{13/2}$	2.5	0.12		P	300	Xe
		2.8302	~6800	$^4I_{11/2} \rightarrow ^4I_{13/2}$	33	0.09		P (low threshold)	300	Xe

Table 8.1. *Characteristics of Some Optically Pumped Solid State Laser Systems (continued).*

Ion	Host	Laser wave-length (μm)	Lower laser level energy (cm^{-1})	Laser trans-ition	Laser ion concen-tration %	τ_2(ms)	$\Delta\nu$ (GHz)	Pulsed P or CW	Operating temp-erature (K)	Method of pumping
Thulium Tm^{3+}	$LiYF_4$ (YLF)	0.4526	5969	$^1D_2 \rightarrow ^3F_4$	10	0.001	8990	P	300	XeF laser
	$YAlO_3$ (+ 1% Cr)	2.34	8250	$^3F_4 \rightarrow ^3H_5$	1	~0.1		P	300	Xe
	Y_3AlO_{12} (YAG) + 0.1% Cr	2.324	~8300	$^3F_4 \rightarrow ^3H_5$	1			CW	300	W
Chromium Cr^{3+}	Al_2O_3 (sapphire)	0.6943(R_1)	0	$^2E(\overline{E}) \rightarrow ^4A_2$	~0.05	3	~300	P	300	Xe
		0.6929(R_2)	0	$^2E(2\overline{A}) \rightarrow ^4A_2$	~0.05	3	225	P	300	Xe
		0.6943(R_1)	0	$^2E(2\overline{A}) \rightarrow ^4A_2$	~0.05	3	~300	CW	300	Hg
	$BeAl_2O_4$ (alexandrite)	0.701→0.818	varies	$^4T_2 \rightarrow ^4A_2$		0.26		P	300	Xe
		0.744 → 0.788	varies	$^4T_2 \rightarrow ^4A_2$	0.27	0.26		CW	300	Hg
Cobalt (Co^{2+})	MgF_2	1.63→2.11 tunable	varies	$^4T_2 \rightarrow ^4T_1$	0.9			CW	80	Nd:YAG laser
Nickel (N_i^{2+}		1.674→1.676	varies	$^3T_2 \rightarrow ^3A_2$	0.5	11.5		P	82→100	
	MgF_2	1.731→1.756	varies						100→192	
		1.785→1.797	varies						198→240	
Ti^{3+}	Al_2O_3	0.66–1.1	varies					CW	300	Ar^+ laser
		0.66–1.38						pulsed		frequency doubled Nd:YAG
								CW		laser

[a]Abstracted from a more comprehensive listing in *Handbook of Laser Science and Technology*, M.J. Weber, Ed., Vol. I, *Lasers and Masers*, CRC Press, Boca Raton, FL, 1982.

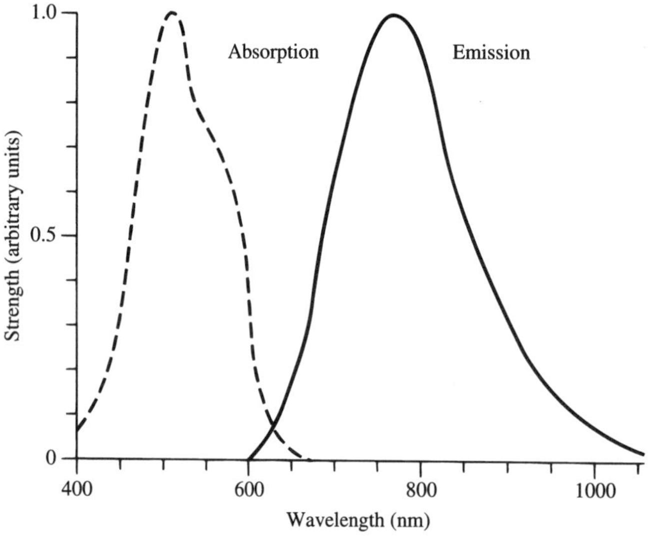

Fig. 8.4. Absorption and emission properties of titanium doped sapphire.

from the optical pumping process; it has high mechanical strength, and can be grown as crystals of large size with good optical quality. The Nd^{3+} ions substitute within the YAG lattice in a single site so the emission and absorption lines are homogeneously broadened. Typical Nd^{3+} doping densities range up to 1%.

An energy level diagram for the Nd^{3+} ions in YAG is shown in Fig. (8.5). The primary pumping process is absorption of lamp energy from the ground state $^4I_{9/2}$ into the $^4F_{5/2}$ level followed by transfer to the upper laser level $^4F_{3/2}$. This consists of a number of closely spaced levels. Two of these, labelled R_2 and R_1 in Fig. (8.5), serve as the upper levels of the closely spaced transitions ℓ_2 and ℓ_1. The stronger of these, ℓ_2, has $\lambda_2 = 1.06415$ μm, the other, ℓ_1, has $\lambda_1 = 1.0646$ μm. Each of these transitions is substantially homogeneously broadened, with $\Delta\nu \sim 2 \times 10^{11}$ Hz, and because they are so close in frequency, $\sim 10^{11}$ Hz apart, they form a single asymmetric lineshape. This lineshape, shown in Fig. (8.6) contributes to the effective overall gain of the laser, which sees its peak gain near 1.06415 μm. It is interesting in a case like this to compute the effective spontaneous emission coefficient describing the transition.

8.6.1 *Effective Spontaneous Emission Coefficient*

At ambient temperature the populations of the two levels R_2 and R_1 are in thermal equilibrium. The ratio of their populations is the ratio of their degeneracy factors.

$$\frac{N_{R_2}}{N_{R_1}} = \frac{g_{R_2}}{g_{R_1}} = \frac{2}{3}.$$

The effective spontaneous emission coefficient A_{21} is related to the coefficients for ℓ_2 and ℓ_1 separately according to

$$A_{21} = A_{\ell_1}\frac{N_{R_1}}{N_{R_2}} + A_{\ell_2}. \tag{8.28}$$

For values $A_{\ell_2} = 1440$ s^{-1}, $A_{\ell_1} \simeq 250$ s^{-1}; $A_{21} = 1815$ s^{-1}. With these parameters the stimulated emission cross-section for the Nd:YAG laser is $\simeq 9 \times 10^{-23}$ m^2.

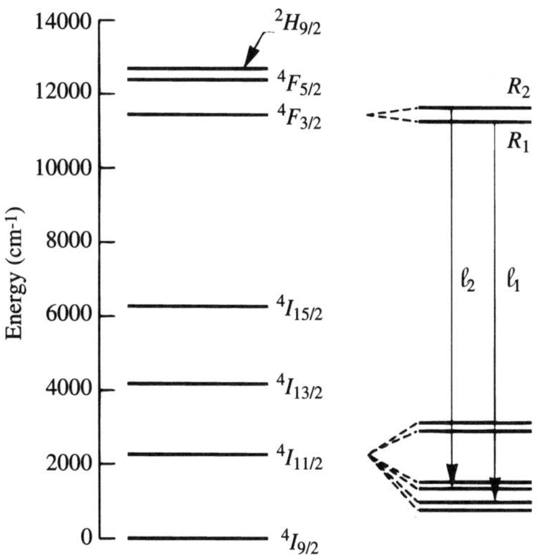

Fig. 8.5. Energy level diagram for Nd^{3+} : YAG, emissions ℓ_1 and ℓ_2 correspond to the two transitions that make up the laser line[8.6].

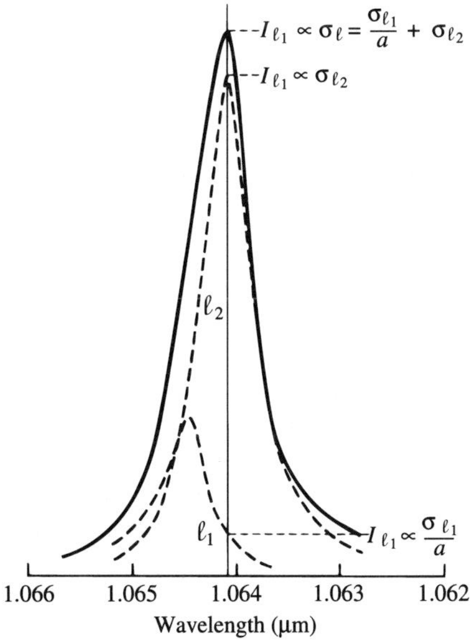

Fig. 8.6. Spontaneous emission spectrum of the 1.06 μm transition in Nd^{3+}:YAG at room temperature. The two Lorentzian lineshapes that contribute to the laser transition are shown by dashed lines[8.6]. The factor a is a lineshape factor that gives the contribution to the intensity at the center of ℓ_2 from the line ℓ_1.

8.6.2 Example – Threshold Pump Energy of a Pulsed Nd:YAG Laser

We take for an example a crystal 50 mm long by 5 mm in diameter. One end of the crystal has a reflectance of 100%, the other a reflectance of 92%. For a good quality crystal $\alpha \simeq 0$. The threshold gain is

$$\gamma_t = \alpha - \frac{1}{\ell}\ln r_1 r_2 = 0.833 \text{ m}^{-1}.$$

Therefore, from Eq. (8.27) the threshold inversion is $N_t \simeq 9.3 \times 10^{21}$ m^{-3} – more than 500 times smaller than for the ruby laser. The energy that must be absorbed to reach threshold is

$$U_t = \frac{N_t h\overline{\nu}_{pump}}{\overline{\eta(\nu)}}, \tag{8.29}$$

which with $\overline{\eta(\nu)} \simeq 1$ and a pump wavelength of 810 nm gives $U_t = 2280$ J m^{-3}. For the crystal specified this gives a threshold absorbed energy of 2.24 mJ.

Overall conversion efficiencies from electrical energy to radiant emission in the Nd^{3+} absorption bands may range as high as 15% so the necessary electrical input to reach threshold may be as small as 15 mJ. With these small thresholds it is not surprising that high energy pulsed operation of Nd:YAG lasers is routine. Single transverse mode outputs up to 50 J are readily available. Although 1.06 μm is the commonest wavelength produced by these lasers, a laser transition at 1.3188 μm is also easy to obtain that has the $^4I_{13/2}$ level in Fig. (8.5) as its lower level.

8.7 CW Operation of the Nd:YAG Laser

CW laser operation of Nd:YAG lasers is possible because of their low threshold pump requirements. Outputs of tens of watts are readily obtained in the TEM_{00} mode. To operate the Nd:YAG crystal described above in a CW mode requires a steady pumping power (W m^{-3}) of

$$P_t = \frac{N_t h\overline{\nu}_{pump}}{\tau_2 \overline{\eta(\nu)}}, \qquad (8.30)$$

where if spontaneous emission is the dominant loss process from the upper laser level $\tau_2 = 1/A_{21}$. With the parameter used above

$$P_t = 4.15 \times 10^6 \text{ J m}^{-3}$$

and for the 50 mm × 5 mm diameter crystal $P_t = 4.07$ W.

Because CW arc lamps are not as efficient at converting electrical energy to light as are flashlamps, the factors e, f, and g can be taken to be on the order of 10%, 10%, and 50% respectively. The lamp power required to reach threshold would be $\simeq 800$ W. The overall efficiency for producing laser power is on the order of 1–3%, as can be seen from Fig. (8.7), which shows the actual performance characteristics of some CW Nd:YAG lasers pumped by krypton arc lamps.

8.8 The Nd^{3+} Glass Laser

The construction of doped crystalline lasers of high power or energy output, particularly in configurations where the output of a laser oscillator is amplified by successive crystals, requires the growth of crystals of substantial size. This must be accomplished without the introduction of optical inhomogeneities into the crystal. In practice, most such crystals are grown by using a small seed crystal and pulling a larger crystal from the melt – the so-called Czochralski method, which is illustrated in Fig. (8.8). Such methods become difficult to use if very large crystals are needed. However, the use of glasses to serve as host media, particularly for Nd^{3+} ions, circumvents this difficulty as extremely large pieces of doped glass are easily fabricated.

In a glass the environments of the Nd^{3+} ions vary much more than in a crystalline material because of the random structural character of the glass matrix. Therefore, not only is the laser substantially inhomogeneously broadened, but it has a much broader linewidth than, for example, Nd:YAG. The laser transition linewidth is typically on the order of 300 cm^{-1}, $\sim 10^{13}$ Hz, and its actual shape

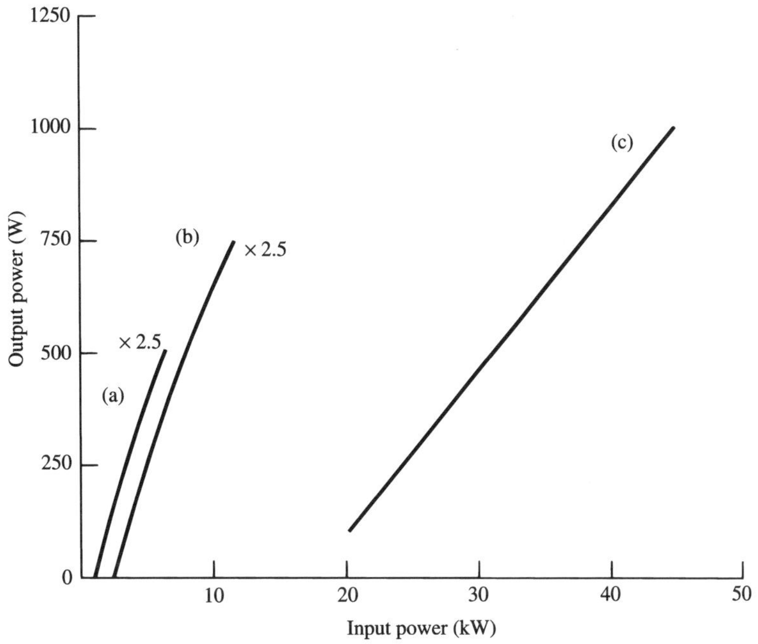

Fig. 8.7. Performance of CW Nd:YAG lasers at 1.06 μm pumped by krypton arc lamps: (a) 6.3×100 mm rod in a single-elliptical pump cavity, (b) 6.3×100 mm rod in a double-elliptical cavity, (c) eight 6 mm×75 mm rods in series, each pumped with an individual lamp[8.7].

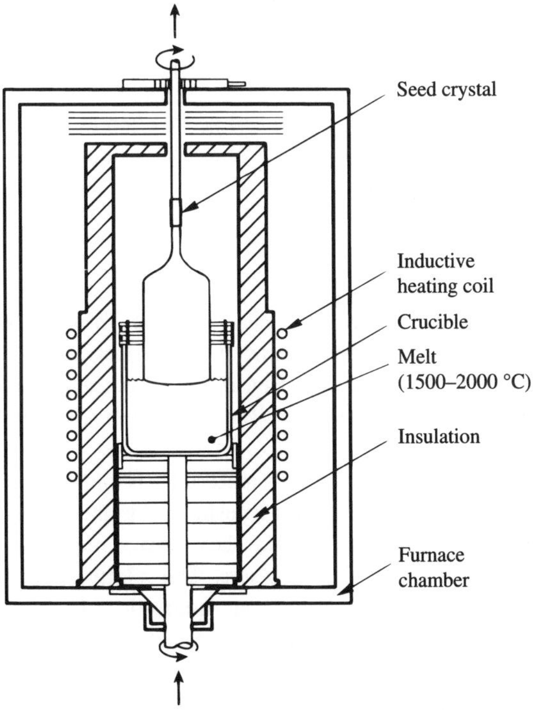

Fig. 8.8. Schematic arrangement of an apparatus for pulling a crystal boule from the melt by the Czochralski method.

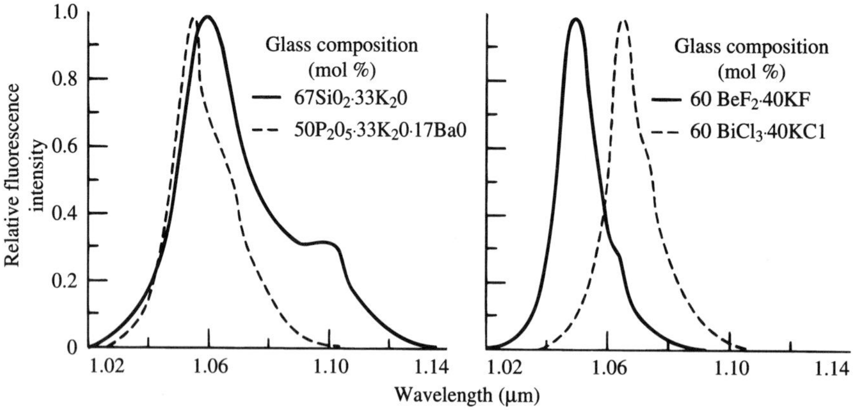

Fig. 8.9. Spontaneous emission lineshape of the 1.06 μm neodymium laser transition in various glasses at 295 K[8.8].

Table 8.2. *Spectroscopic properties of the $^4F_{3/2} \rightarrow {}^4I_{11/2}$ transition of Nd^{3+} ions in different glasses at 295K[8.8].*

Glass	Refractive index n	Cross-section (10^{24} m^2)	Wavelength λ_p (nm)	Line–width Δ_{eff}(nm)	Lifetime τ_R (μs)
Oxides					
Silicates	1.46 to 1.75	0.9 to 3.6	1057 to 1088	34 to 55	170 to 1090
Phosphates	1.49 to 1.63	2.0 to 4.8	1052 to 1057	22 to 35	280 to 530
Borates	1.51 to 1.69	2.1 to 3.2	1054 to 1063	34 to 38	270 to 450
Germanates	1.61 to 1.71	1.7 to 2.5	1060 to 1063	36 to 43	300 to 460
Tellurites	2.0 to 2.1	3.0 to 5.1	1056 to 1063	26 to 31	140 to 240
Halides					
Fluoroberyllates	1.28 to 1.38	1.6 to 4.0	1046 to 1050	19 to 29	460 to 1030
Fluoroaluminates	1.41 to 1.48	2.2 to 2.9	1049 to 1051	30 to 33	420 to 570
Fluorozirconates	1.52 to 1.56	2.9 to 3.0	1049	26 to 27	430 to 450
Chlorides	1.67 to 1.91	6.0 to 6.3	1062 to 1064	19 to 20	180 to 220

varies from one glass to another, as can be seen from Fig. (8.9) and Table (8.2). The lifetime of the upper laser level also depends on the type of glass and the Nd^{3+} concentration as can be seen from Fig. (8.10). Fig. (8.11) shows a typical energy level diagram for Nd^{3+} ions in glass together with the absorption spectrum of the material.

The world's most powerful lasers are based on this material. Because their linewidths are much larger than in YAG, their inversion threshold is higher; however, once they are pumped sufficiently above threshold they offer comparable (energy-out)/(energy-in) performance to Nd:crystalline lasers. The relatively long upper laser level lifetime ~100 μs, allows them to store larger quantities of energy – up to 0.5 MJ m^3. The best example of their capability is in their use in inertial confinement fusion (ICF) research, which will be discussed in more detail in Chapter 24. In such high energy applications the output of a well controlled Nd:YAG crystal laser oscillator will be amplified up to a million times by a series of laser amplifiers of progressively larger aperture, as shown schematically in Fig. (8.12).

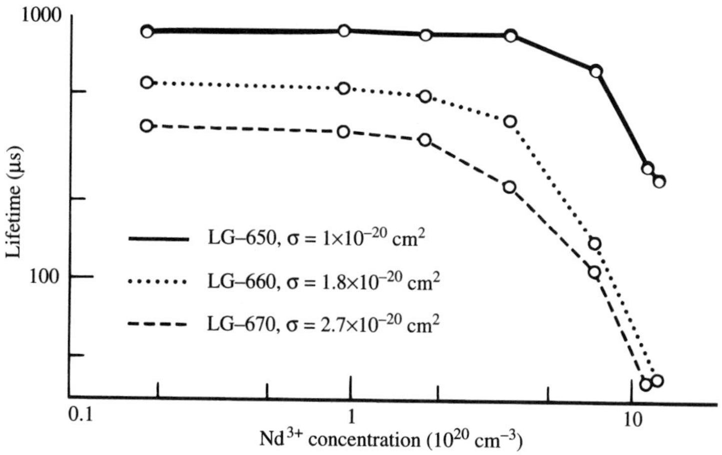

Fig. 8.10. Lifetime of the upper laser level as a function of Nd^{3+} concentration for different silicate glass laser materials. The σ values quoted are the peak values of the stimulated emission cross sections of the ${}^4F_{3/2} \rightarrow {}^4I_{11/2}$ transition[8.9].

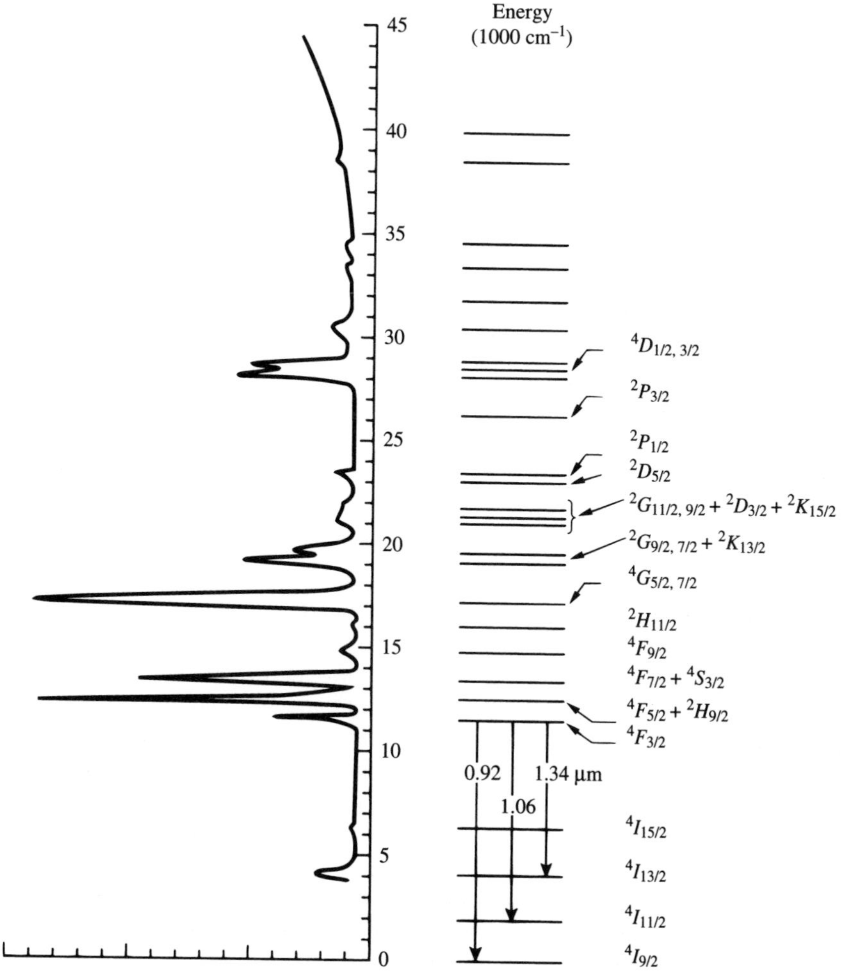

Fig. 8.11. Absorption spectrum and energy levels of Nd^{3+} in glass[8.10].

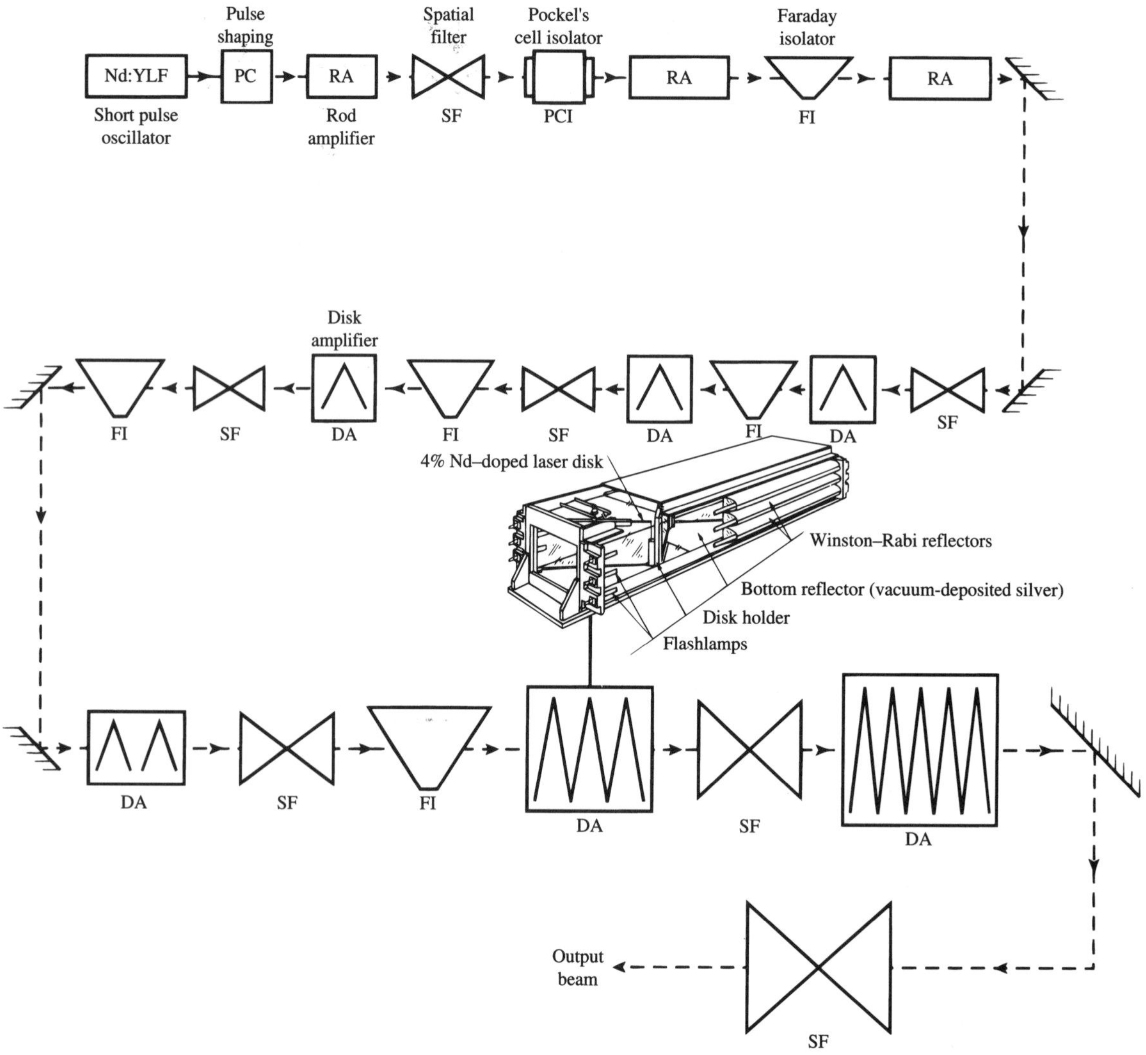

Fig. 8.12. Schematic arrangement of oscillator/amplifier chain for production of very high energy laser pulses from a Nd^{3+} laser. The isolators allow radiation to pass only in one direction. Spatial filters clean up the radial intensity profile of the beam. As the beam increases in intensity down this chain the size of the beam increases as does the size of the successive rod amplifiers (RA) and disk amplifiers (DA).

These amplifiers utilize large slabs of Nd:glass material placed at Brewster's angle and pumped through their flat faces with arrays of linear flashlamps. A large oscillator/amplifier(s) chain of this kind will contain many additional components over and above the oscillator, its mirror feedback system, and the successively larger amplifiers. There will be Faraday isolators (see Chapter 14), which allow light to pass in one direction but not the other, beam expanders, spatial filters for smoothing the transverse spatial profile of the laser beam (see Chapter 16), electro-optic switches (see Chapter 19), and nonlinear crystals for harmonic generation (see Chapter 21).

The large Nova laser at the Lawrence Livermore National Laboratory delivers a final amplified pulse energy of approximately 100 kJ at 1.06 μm in 1 ns – corresponding to a peak power of 100 TW (terawatts). By the use of nonlinear techniques a substantial part (25 kJ) of this laser energy can be converted to its third harmonic (0.35 μm) for irradiation of nuclear fusion targets. In such multi-megajoule laser systems the advantages of Nd:glass becomes apparent: it has a high damage threshold so can amplify high intensity pulses; and because of its large linewidth its gain coefficient is smaller than for Nd:YAG, which helps to eliminate the occurrence of spontaneous parasitic oscillations in amplifier sections.

Amplification of very short optical pulses can be achieved with a glass laser amplifier, because of the broad linewidth $\sim 10^{13}$ Hz, the amplification of pulses ~ 1 ps in length is feasible. In mode-locked operation the broad linewidth of the Nd:glass laser allows it to generate pulses in the 2–20 ps range.

8.9 Geometrical Arrangements for Optical Pumping

In Chapter 3 we have already briefly discussed the optical pumping arrangement in which the laser crystal was pumped by a helical lamp wrapped around the crystal, and the elliptical reflector scheme in which flashlamp and crystal lie along the two focal lines of an elliptical cross-section cylindrical reflector. Other optical pumping arrangements are also commonly used.

In the elliptical reflector arrangement, all the light from the lamp (along one focal line) passes through the crystal (along the other focal line). In this arrangement linear lamp and laser rod are quite separate, which makes access and replacement of either a simple matter. High purity, deionized water coolant can be circulated directly inside the reflector to keep both lamp and laser rod cool. This arrangement is very efficient and convenient but the illumination it provides is not axi-symmetric. The side of the laser crystal nearest to the flashlamp receives its illumination directly, and the greater intensity of this direct illumination produces a population inversion in the system which is greater on the side of the crystal nearest to the lamp. The resultant laser beam has an asymmetric intensity distribution that is skewed towards the side of the laser rod facing the lamp.

The close-coupling optical pumping arrangements shown in Fig. (8.13) are simple and convenient to manufacture, and provide comparable efficiency to elliptical cavities, although again without axial symmetry (unless very many close coupled lamps are used). However, the close proximity of flashlamp and laser can lead to thermal dissipation problems. If the laser crystal is heated by the pumping lamp its oscillation threshold will rise. This can occur for several reasons: because of an increase in the linewidth of the laser transition, thermal population of the lower laser level, or particularly, the production of refractive index inhomogeneities in the laser crystal. In very high power solid-state laser systems, especially Nd^{3+}/glass, the laser medium is frequently fabricated as a series of flat slabs placed at Brewster's angle and optically pumped through their flat faces by arrays of linear flashlamps. Special reflector arrangements have been developed for efficiently coupling light from these flashlamp arrays into the laser slabs, as shown in Fig.(8.14).

8.9.1 Axisymmetric Optical Pumping of a Cylindrical Rod

An ingenious method for obtaining axisymmetric irradiation of the laser crystal is to use an ellipsoidal mirror (an ellipsoid of revolution) as shown in Fig. (8.3). Practically all the light from the flashlamp passes through the laser crystal. However, even, in this case where the illumination of the laser crystal is axisymmetric, the volume illumination of the crystal is not uniform. This can be illustrated by considering a simple two-dimensional model where a cylindrical transparent rod is

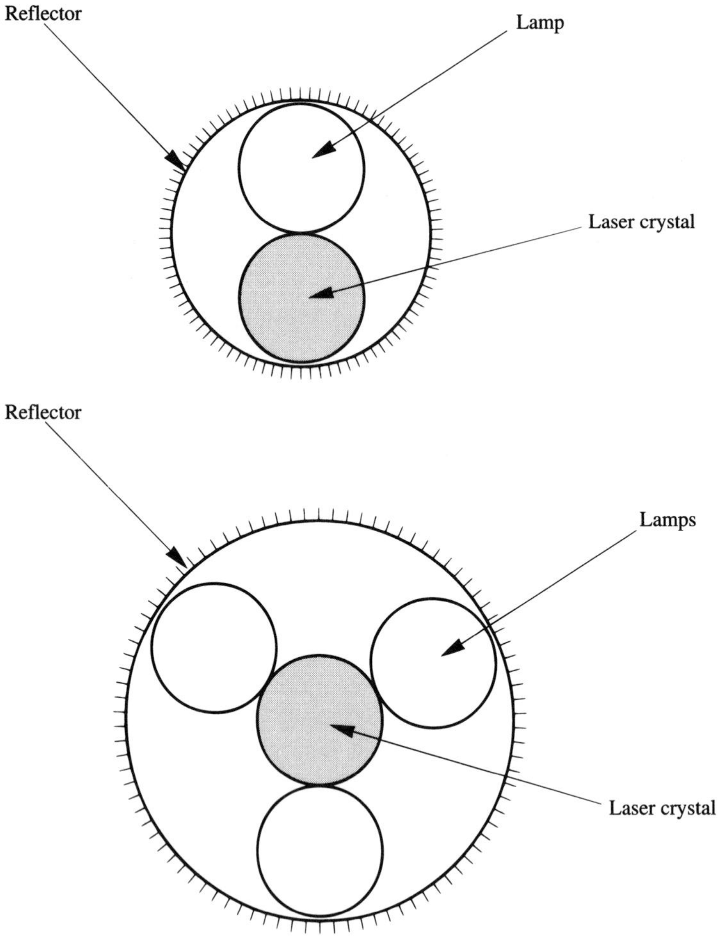

Fig. 8.13. Lamp/reflector configurations used in close optical coupling arrangements for solid state laser pumping.

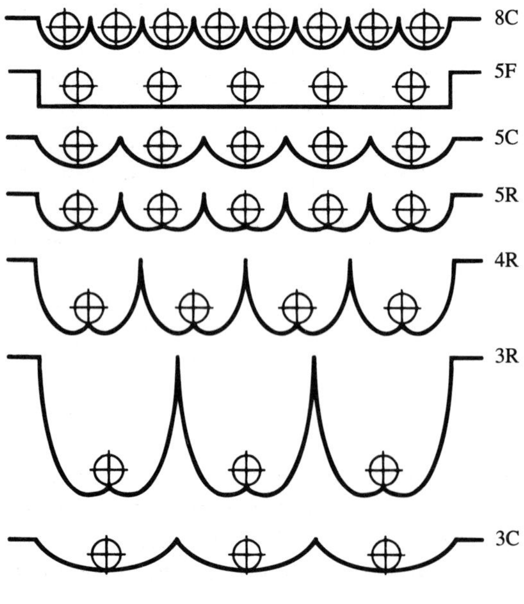

Fig. 8.14. Reflector configurations used in face pumping of slab amplifiers. The classification numbers indicate the number of lamps and reflector type used by the Lawrence Livermore National Laboratory C – cylindrical reflector, F – flat, R – Rabi[8.11].

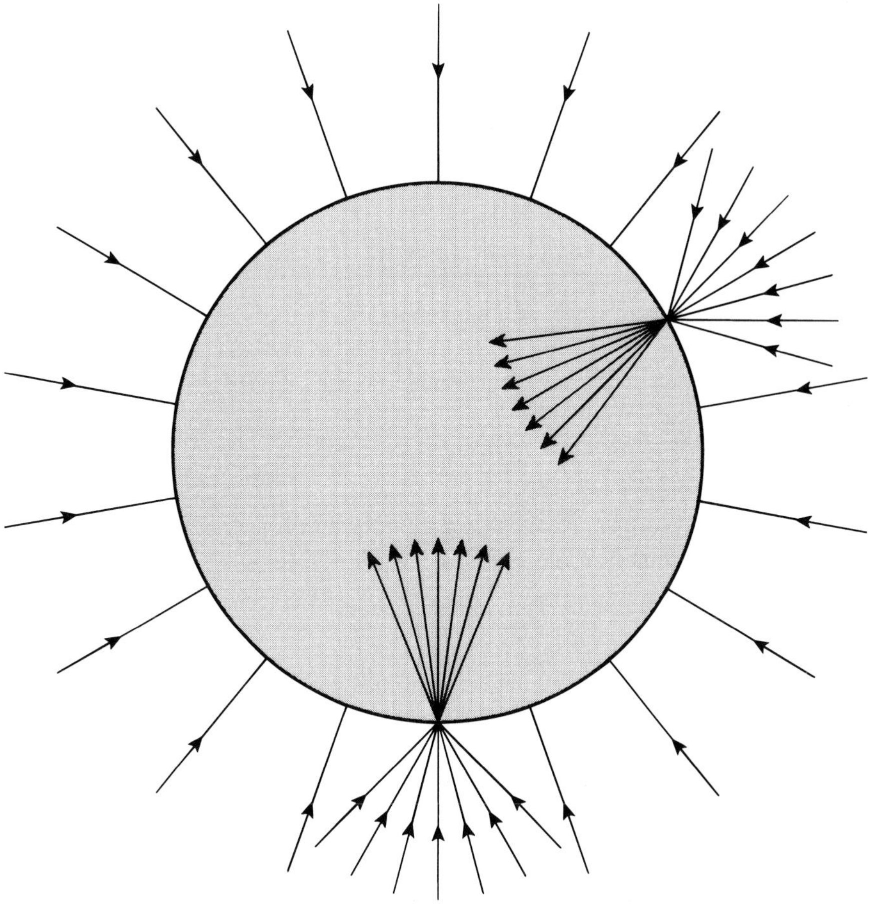

Fig. 8.15. Circularly symmetric illumination of a laser rod showing the refraction of rays through the cylindrical surface.

uniformly illuminated around its circumference as shown in Fig. (8.15). Note that groups of rays that do not strike the surface of the rod normally will be focused through the curved surface by refraction. We assume that the illumination consists of many, incoherent plane waves of equal intensity which have their electric vectors parallel to the axis of the cylinder[8.12] (it can be shown that the final result is the same if the electric vectors of the incident waves are perpendicular to the cylinder axis). If we assume that interference effects between these waves are 'smeared out,' then the energy density in the rod due to all the illuminating waves can be taken as the sum of their individual energy densities.

We consider the group of waves whose propagation vectors lie within a small range of angles $d\alpha$. Their contribution to the total average energy density ρ_0, at a point outside the rod, is

$$\Delta\rho_0 = \frac{\rho_0}{2\pi} d\alpha, \tag{8.31}$$

since the total energy density comes from waves covering the total circular angle of 2π. Now, if there are m waves altogether (where in the limit we will let m become infinitely large) the energy density associated with each wave is ρ_0/m. If the electric field of each wave is of the form $E\cos(\omega t + \phi)$ (where E becomes infinitely small as $m \to 0$) then the time-averaged energy density outside the rod due to an individual wave is

$$d\rho_0 = \overline{\epsilon_0 E^2 \cos^2(\omega t + \phi)}, \tag{8.32}$$

where the bar indicates time averaging, which gives

$$d\rho_0 = \frac{1}{2}\epsilon_0 E^2. \tag{8.33}$$

The total energy density due to the m waves is

$$\rho_0 = \sum_m d\rho_0 = \frac{1}{2}\epsilon_0 \sum_m E^2 = \frac{1}{2}\epsilon_0 E_0^2, \tag{8.34}$$

where E_0 is the root mean square value of the total electric field due to all the waves. Thus, from (8.31) and (8.34)

$$\Delta\rho_0 = \frac{\epsilon_0 E_0^2}{4\pi} d\alpha. \tag{8.35}$$

The group of waves which make this contribution to the energy density, strike the outer wall of the cylindrical rod at angle α, as shown in Fig. (8.16). Their incident intensity is

$$I = c_0 \Delta\rho \tag{8.36}$$

and the intensity transmitted through the surface, from Eq. (4.30) with $n' = 1$, is

$$I_t = nt^2 I, \tag{8.37}$$

where t is the transmission coefficient, which in this case is given by †

$$t = \frac{2\cos\alpha}{\cos\alpha + n\cos\beta}. \tag{8.38}$$

The angle of refraction β is related to the angle of incidence α by Snell's law

$$\frac{\sin\alpha}{\sin\beta} = n, \tag{8.39}$$

where n is the refractive index of the cylinder. The total internal flux transmitted through a small area of surface dA of unit length along the cylinder axis is, from (8.37)

$$F_{int}^0 = nt^2 I \, dA \cos\beta, \tag{8.40}$$

since $dA \cos\beta$ is the effective area of the element of surface normal to the direction of propagation of the wave.

Once this group of waves has entered the cylinder they reflect around inside, maintaining internal angles of incidence and reflection of β, as shown in Fig. (8.16). Each time they strike the wall their intensity is diminished by a factor r^2, where r is the reflection coefficient given by‡

$$r = \frac{n\cos\beta - \cos\alpha}{n\cos\beta + \cos\alpha}. \tag{8.41}$$

We consider the contribution that these waves make to the energy density in a small annulus of the cylinder of radius R and thickness dR. On any given pass

† See Appendix 4.
‡ See Appendix 4.

Fig. 8.16. Geometry of irradiation of a cylindrical laser rod in a cylinder cross-section.

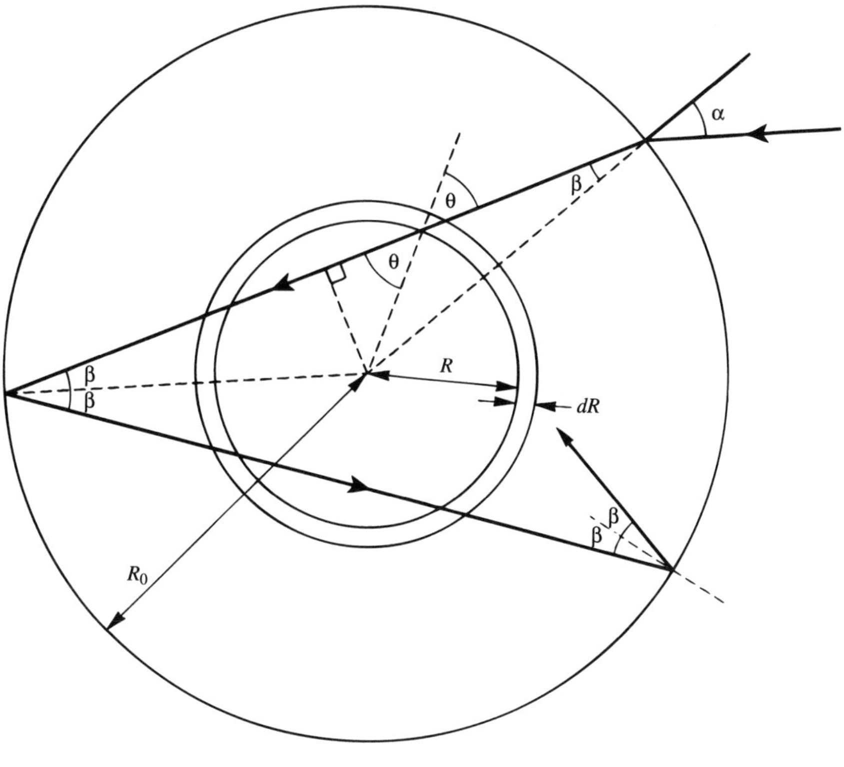

across the cylinder, between internal reflections, the waves strike this region at an angle θ where

$$\theta = \text{arc } \sin(R_0 \sin\beta / R), \tag{8.42}$$

where, of course, if $R < R_0 \sin\beta$, the particular group of waves does not pass through the part of the cylinder of radius R. If the flux that strikes this annular region is F^j_{int}, where F^j_{int} is the internal flux after j internal reflections, then the stored energy within the annulus due to this passage of the internally reflected wave is

$$du^j(r) = \frac{2F^j_{int} n dR}{c_0 \cos\theta}, \tag{8.43}$$

which gives

$$du^j(r) = \frac{2r^{2j} F^0_{int} nRdR}{c_0\sqrt{R^2 - R_0^2 \sin^2\beta}}. \tag{8.44}$$

The total stored energy in the annulus is

$$du(r) = \sum_{j=0}^{\infty} \frac{2r^{2j} F^0_{int} nRdR}{c_0\sqrt{R^2 - R_0^2 \sin^2\beta}}, \tag{8.45}$$

which substituting from (8.40) and summing the geometric series gives

$$du(r) = \frac{2n^2t^2I\cos\beta dARdR}{c_0\sqrt{R^2 - R_0^2\sin^2\beta(1-r^2)}}. \tag{8.46}$$

The energy density in the annulus, since its area is $2\pi RdR$, is

$$d\rho(r) = \frac{n^2t^2I\cos\beta dA}{\pi c_0\sqrt{R^2 - R_0^2\sin^2\beta(1-r^2)}}. \tag{8.47}$$

The total energy density at radius r due to all the waves that strike an area $2\pi R_0$ of the outer cylinder in a small angular range $d\alpha$ at α is

$$\Delta\rho(r) = \frac{2n^2t^2IR_0\cos\beta}{c_0\sqrt{R^2 - R_0^2\sin^2\beta(1-r^2)}}. \tag{8.48}$$

Now, from (8.38) and (8.41)

$$\frac{t^2}{1-r^2} = \frac{\cos\alpha}{n\cos\beta}, \tag{8.49}$$

and since $I = c_0\Delta\rho_0$ and $\Delta\rho_0 = (\rho_0/2\pi)\, d\alpha$,

$$\Delta\rho(r) = \frac{n\rho_0 R_0}{\pi}\frac{\cos\alpha d\alpha}{\sqrt{R^2 - R_0^2\sin^2\alpha/n}}. \tag{8.50}$$

The total energy density at radius R, $\rho(R)$, is obtained by integrating (8.50) over the range of angles of incidence α that allow waves to reach radius R

$$\rho(R) = \frac{n\rho_0 R_0}{\pi}\int_{-\alpha_0}^{\alpha_0}\frac{\cos\alpha d\alpha}{\sqrt{R^2 - R_0^z\sin^2\alpha/n}}, \tag{8.51}$$

where $\sin\alpha_0$ is the smaller of 1 and nR/R_0. The solution of (8.51) is

$$\begin{aligned} \frac{\rho(R)}{\rho_0} &= n^2 && \text{for } 0 \le R \le R_0/n \\ &= \left(\frac{2n^2}{\pi}\right)\text{arc sin}\left(\frac{R_0}{nR}\right) && \text{for } R_0/n \le R \le R_0, \end{aligned} \tag{8.52}$$

which is shown in Fig. (8.17) for the case $n = 1.76$.

The energy density within the cylinder is a maximum, and constant, for all radii $\le R_0/n$. Fig. (8.17) also shows the energy density as a function of radius inside a cylindrical rod when the incident radiation consists of all possible polarizations in three dimensions. If the cylinder absorbs the incident radiation then these energy density functions are modified as illustrated in Fig. (8.18), which shows both the two- and three-dimensional calculated energy density functions for different values of d where $d = 2R_0\times$ absorption coefficient.

It is clear from Figs. (8.17) and (8.18) that if a cylindrical laser crystal is made up as a composite rod with an inner absorbing core, for example Cr^{3+} doped α-Al_2O_3 (pink ruby) of radius R_0/n, and an outer sheath of transparent material of the same refractive index, for example α-Al_2O_3 (sapphire) as shown in Fig. (8.19), then very close to uniform illumination of the amplifying medium will be achieved. Such composite laser rods do, in fact, have lower oscillation thresholds than comparable uniform rods although manufacture of these composite rods is more difficult.

Fig. 8.17. Normalized energy density as a function of radius in a transparent cylindrical rod illuminated uniformly in a cylinder cross-section. In the two-dimensional model all illuminating rays travel in planes orthogonal to the cylinder axis. In the three-dimensional model the illuminating rays take all possible directions[8.12].

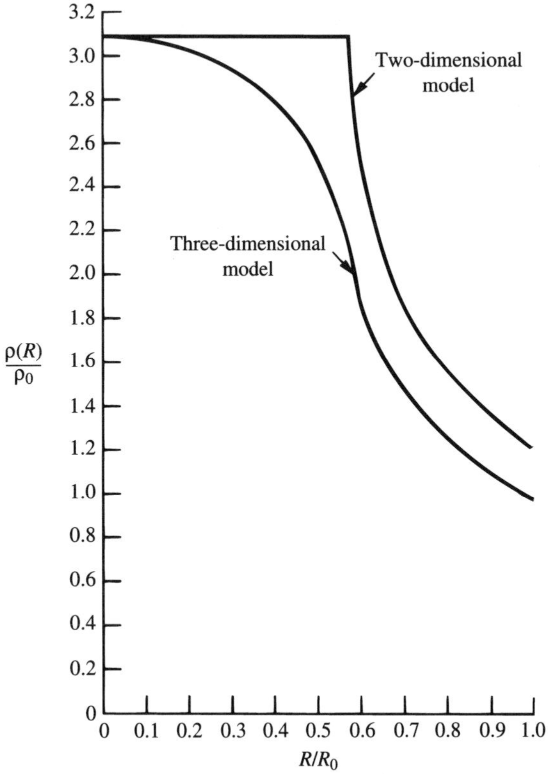

Fig. 8.18. Normalized energy density inside a cylindrical laser rod uniformly illuminated with linearly polarized light polarized along the cylinder axis. The energy density distributions are shown for different amounts of rod absorption where d=cylinder diameter × absorption coefficient[8.12].

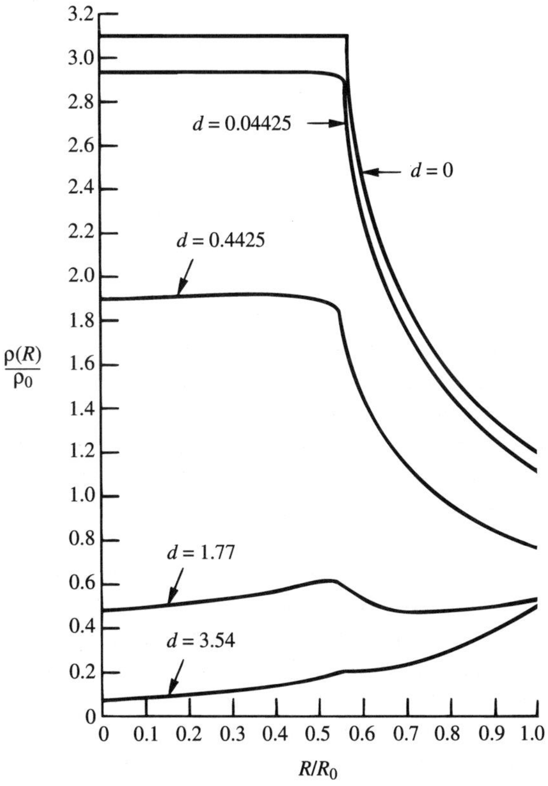

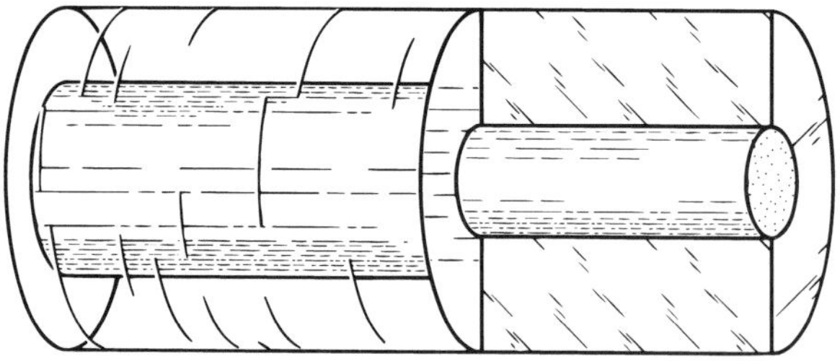

Fig. 8.19. Composite cylindrical laser rod with doped center[8.13].

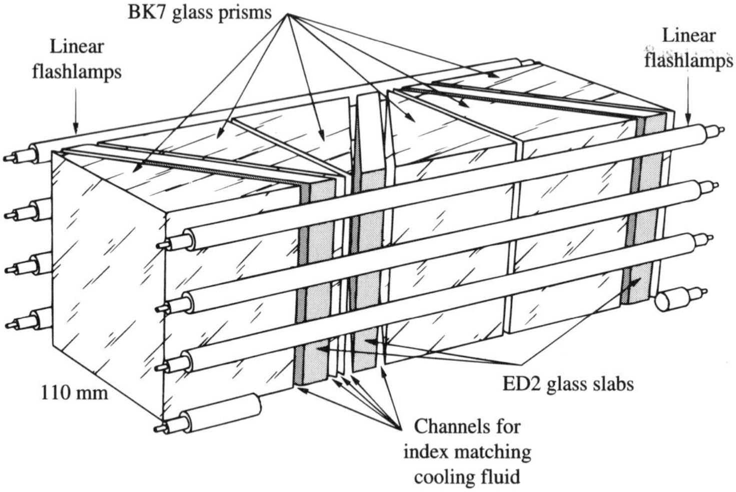

Fig. 8.20. Schematic arrangement of a 'zig-zag' face-pumped laser amplifier module that incorporates prisms and index matching coolant for reducing Fresnel reflection losses at the neodymium glass (ED2) slabs[8.14].

8.10 High Power Pulsed Solid-State Lasers

The input energies to the flashlamps used in pumping high energy solid state lasers can be very large, ranging up to several thousand joules for each lamp. Typical flash durations in these systems are in the 100 μs–1 ms range. These large energy inputs can cause severe heating of the laser rod and consequently restricts operation to pulsed operation, unless special cooling arrangements are incorporated.

For this reason, high power solid state lasers cannot be built just by increasing the diameter and length of a cylindrical laser rod. Large diameter rods cannot eliminate waste heat effectively, and are also difficult to fabricate.

The highest available energy outputs are obtained from Nd^{3+} glass lasers, which at high power levels are efficiently pumped in a face-pumped slab configuration, one example of such a system, which contains some interesting features, is shown in Fig. (8.20). In this arrangement the thermal and optical pumping inhomogeneity problems associated with large diameter cylindrical laser crystals are considerably reduced. Irradiation of the crystal is efficient over a large surface/volume ratio and large size glass slabs, placed at Brewster's angle, can be used.

Development of high power Nd^{3+} glass systems of this kind is continuing in many laboratories as part of a program to obtain laser-induced fusion. Current attention is focused on the development of lasers with output energies up to 1 MJ to be delivered to pellets of nuclear fuel on time scales from 100 ps–1 ns. This corresponds to peak powers of 10^{14}–10^{15} W.

Table 8.3. *Diode-Pumped Solid State Lasers.*

Ion	Transition	Wavelength (μm)	Operating temperature (K)
Nd^{3+}	$^4F_{3/2} \rightarrow {}^4I_{11/2}$	1.06	300
	$^4F_{3/2} \rightarrow {}^4I_{13/2}$	1.32	300
	$^4F_{3/2} \rightarrow {}^4I_{9/2}$	0.95	
Ho^{3+}	$^5I_7 \rightarrow {}^5I_8$	2.1	300
Er^{3+}	$^4I_{13/2} \rightarrow {}^4I_{13/2}$	2.8	300
	$^4I_{13/2} \rightarrow {}^4I_{9/2}$	1.6	300
Tm^{3+}	$^3F_4 \rightarrow {}^3H_5$	2.3	300
U^{3+}	$^4I_{11/2} \rightarrow {}^4I_{9/2}$	2.61	4.2
Dy^{2+}	$^5I_7 \rightarrow {}^5I_8$	2.36	1.9
Yb^{3+}	$^2F_{7/2} \rightarrow {}^2F_{5/2}$	1.03	77

8.11 Diode-Pumped Solid-State Lasers

The development of efficient, high power semiconductor lasers using GaAlAs operating near 800 nm has created an important new class of optically pumped lasers in which a crystalline laser is optically pumped by a semiconductor laser. We shall discuss the characteristics of the latter in detail in Chapter 13. The most common laser of this kind is the Nd:YAG operating at 1.06 μm or 1.32 μm, although other hosts for the Nd^{3+} ions, such a YLF, $La_2Be_2O_5$ (BEL) and YVO_4, have been used. A fortuitous concidence between available high power GaAlAs lasers at ~809 nm and the absorption of Nd^{3+} ions into the $4F_{5/2}$ pump band, see Fig. (8.4), provides for very efficient conversion from 809 nm to laser oscillation at 1.06 μm, over 10% conversion is easily obtained with pump powers of ~200 mW.

Table (8.3) summarizes the diode-pumped solid-state laser transitions that have been obtained to date[8.15]. One laser in this host that is worthy of note is the 2.1 μm holmium YAG laser, which has potential optical radar (LIDAR) applications because its wavelength makes it far less of an eye hazard than lasers operating at 1.06 μm and shorter wavelengths.

Diode-pumped lasers in which the semiconductor lasers are arranged to inject their pump radiation through the cylindrical faces of a laser rod are in many ways akin to lamp-pumped lasers of the same kind. However, because of the high electrical efficiency of GaAlAs laser diodes (> 10%) the overall electrical to optical conversion efficiency of these lasers is very high, approaching 10%.

If the diode laser pump radiation is injected along the axis and matched to the transverse mode geometry of the solid state laser then very stable, narrow linewidth laser oscillation can be obtained: Fig. (8.21) shows the clever monolithic Nd:YAG laser design of Kane and Byer[8.16]. In this laser the magneto-optical properties of YAG are used to obtain unidirectional amplification of a travelling wave in the laser cavity. This prevents the spatial hole-burning that can occur

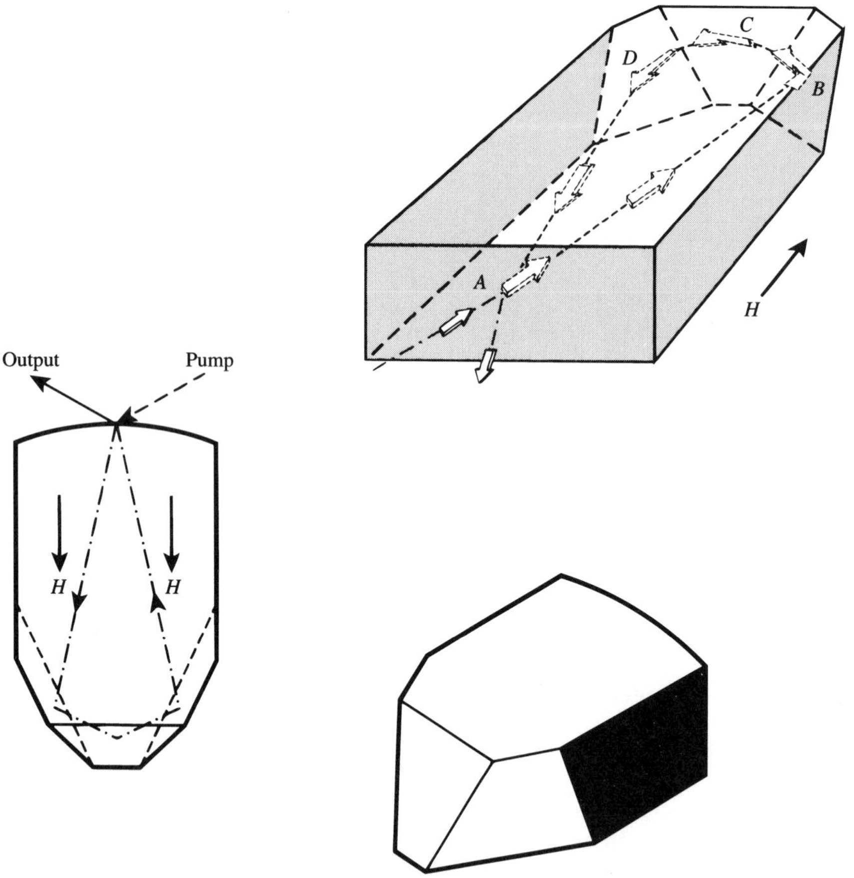

Fig. 8.21. Monolithic integrated diode-pumped ND:YAG laser – the *MISER* laser[8.15]. Faraday rotation within the laser crystal, produced by integral permanent magnet, leads to unidirectional oscillation and single longitudinal mode oscillation as described in Chapter 7. The curved face is partially transmitting and selects a specific polarization state for oscillation. Total internal reflection occurs at points *B*, *C*, and *D*. Faraday radiation takes place along the paths *AB* and *DA*.

in a standing wave homogeneously broadened laser, which can allow multiple longitudinal modes to oscillate.

8.12 Relaxation Oscillations (Spiking)

When a laser system is pushed into a state of population inversion, as soon as its gain passes the threshold value the oscillation grows in amplitude and, by reducing the gain through stimulated emission, stabilizes the gain at the loss line. The approach to a stable oscillation may occur smoothly, as if the approach to the equilibrium situation were damped. However, oscillation about the equilibrium position is frequently observed. This oscillation appears as a time-varying output laser intensity following the attainment of a population inversion sufficient to start the oscillation. It is called 'spiking.' It results from the attempt of the intracavity field and the population inversion to become balanced and can be visualized as an oscillatory energy exchange process between the intracavity intensity and the population inversion. As the intracavity intensity increases, the upper level population is reduced so the gain falls – this reduces the intracavity intensity. This reduction of the intracavity intensity causes the gain to increase again. The intracavity intensity then increases again, and so on.

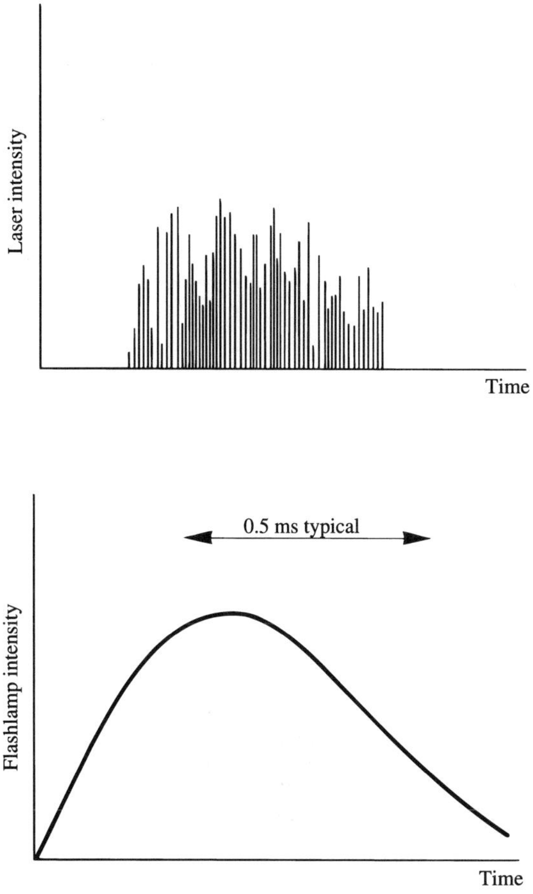

Fig. 8.22. Schematic illustration of the 'spiking' phenomenon in the output of a pulsed solid state laser.

If we observe the output intensity from, for example, a pulsed ruby laser, we generally see the sort of behavior illustrated in Fig. (8.22). The laser emission in successive spikes tends to shift between different longitudinal and transverse modes. Suppose the emission in one spike is in a longitudinal mode of order m so that the oscillation frequency is

$$\nu = \frac{mc}{2\ell}\left(\lambda = \frac{2\ell}{m}\right),$$

then the standing wave field in the laser cavity has nodes and antinodes of intensity in a certain pattern, as shown in Fig. (8.23). Following laser emission in the spike the inverted population is depleted near the antinodes of the field relative to the nodes; this causes a longitudinal mode of different order to oscillate in the next spike. This mode will have higher gain because its field antinodes will be near the field nodes of the previously oscillating mode as shown schematically in Fig. (8.23).

One phenomenon which can suppress this longitudinal mode jumping is called *spatial cross-relaxation*, where an excited particle at position A in Fig. (8.23) can rapidly move to a position B. This smears out any spatial sinusoidal variation of gain within the laser medium and sustains oscillation on a particular high gain longitudinal mode. In a crystal lattice, because the excited particles are fixed, it

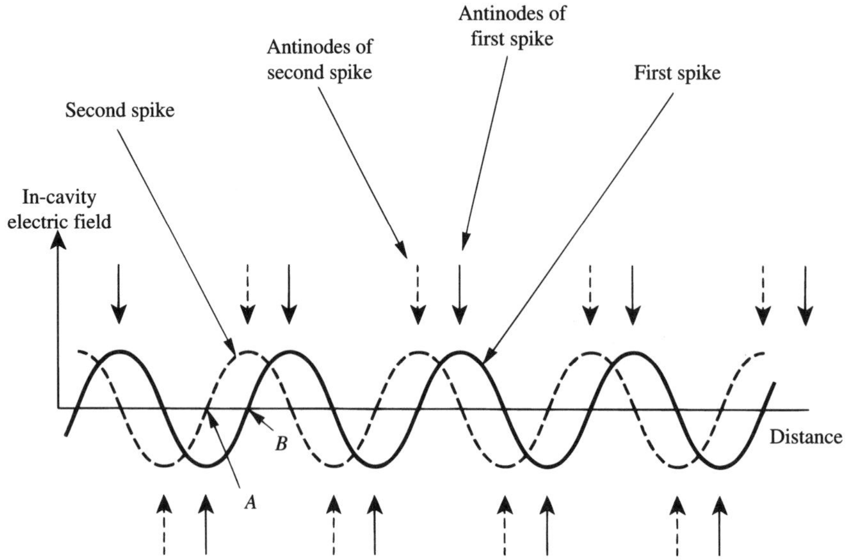

Fig. 8.23. Illustration of how oscillation can jump from one longitudinal mode to another in successive 'spikes' because of population depletion.

is relatively difficult for them to transfer excitation from one position to another. However, the transfer can still occur radiatively, the relative efficiency of this process depending on the Einstein coefficient A_{21} for the laser transition, and the absorption coefficient of the laser crystal on this transition (which, of course, depends on the atom concentration and the linewidth of the transition).

Jumping from one transverse mode to another in successive spikes is understandable in much the same way. Oscillation in one TEM mode depletes a certain region of the crystal preferentially. This favors subsequent oscillation on a different TEM mode, which has its maxima of intracavity intensity in the least depleted spatial region of the laser medium.

Frequency and mode jumping of the above kinds also tend to be enhanced by nonuniformity of optical pumping, and heating of the crystal during the flash, which produces spatial refractive index variations. Simultaneous oscillation on different longitudinal and transverse modes may occur even though the laser transition is homogeneously broadened. The simultaneously oscillating modes arrange to use different spatial regions of the crystal, so although in theory they can compete for the same group of atoms, in fact they do not. A simple illustration of how this can happen is given in Fig. (8.24). In gaseous lasers spatial cross-relaxation is fast and in *homogeneously* broadened gas lasers the above type of process, which would lead to multi-mode oscillation, is generally suppressed.

8.13 Rate Equations for Relaxation Oscillation

In describing relaxation oscillation in a laser we follow the approach first given by Statz and deMars[8.17] . We assume that laser oscillation builds up in a single mode of the optical resonator.

Let q be the number of photons in the mode per unit volume. The mode corresponds to a monochromatic radiation field at frequency ν' whose energy

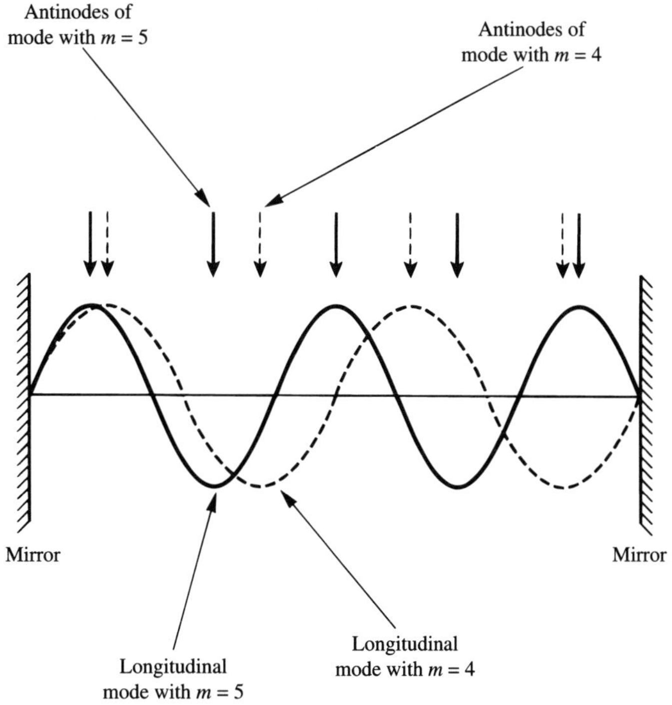

Fig. 8.24. Illustration of how two different longitudinal modes partially avoid competition with each other by using different spatial regions of inverted population in a gain medium.

density is a δ-function

$$\rho(\nu) = qh\nu'\delta(\nu - \nu'). \tag{8.53}$$

The upper and lower laser levels have population densities N, N_1, and effective lifetimes τ, τ_1, respectively. We assume that the laser is a *good* one such that $\tau \gg \tau_1$, and $N \gg N_1$. We neglect N_1. The population inversion in this case is the same as N and obeys the differential equation

$$\frac{dN}{dt} = R - NW - \frac{N}{\tau}, \tag{8.54}$$

where R is the rate per unit volume at which particles are fed into the upper laser level, and W is the stimulated emission rate. In this case

$$W = \int qh\nu'\delta(\nu - \nu')B_{21}g(\nu_0, \nu)d\nu, \tag{8.55}$$

which gives

$$W = qh\nu' B_{21}g(\nu_0, \nu') = qB. \tag{8.56}$$

The constant B is

$$B = h\nu' B_{21}g(\nu_0, \nu') = \frac{c^3}{8\pi\nu'^2}A_{21}g(\nu_0, \nu'), \tag{8.57}$$

which is the ratio of the spontaneous emission rate per frequency interval to the mode density. Eq. (8.54) therefore becomes

$$\frac{dN}{dt} = R - qBN - \frac{N}{\tau}. \tag{8.58}$$

The equation describing the rate of change of the number of photons in the mode is

$$\frac{dq}{dt} = qBN - \frac{q}{\tau_0} + N\epsilon A_{21}, \tag{8.59}$$

where τ_0 is the time constant of the laser cavity, for example given by Eq. (4.11), and ϵ is the probability that a photon will be emitted spontaneously into the particular mode we are considering. Given the very large number of available modes for spontaneous emission we neglect the term containing ϵ in Eq. (8.59). We cannot solve Eqs. (8.58) and (8.59), which are called the *Statz–deMars* equations, exactly, although it is easy to examine their solution numerically using a computer. To find an analytic solution we consider small departures from the equilibrium situation in which $dN/dt = 0$, and $dq/dt = 0$.

In equilibrium, $N = N_0$, where from Eq. (8.59)

$$N_0 = \frac{1}{B\tau_0} \tag{8.60}$$

and from Eqs. (8.58), and (8.60)

$$q_0 = R\tau_0 - \frac{1}{B\tau}. \tag{8.61}$$

There are no photons in the mode when $q_0 = 0$. This occurs when the pumping rate is at its threshold value, R_t, where

$$R_t = \frac{1}{B\tau\tau_0}. \tag{8.62}$$

We define a pumping factor r, where

$$r = \frac{R}{R_t}, \tag{8.63}$$

so Eq. (8.61) can be written

$$q_0 = \frac{r-1}{B\tau}. \tag{8.64}$$

We solve Eqs. (8.58) and (8.59) approximately by considering small oscillations about the equilibrium position. We write

$$\begin{aligned} N(t) &= N_0 + N'(t), \qquad N' \ll N_0, \\ q(t) &= q_0 + q'(t), \qquad q' \ll q_0. \end{aligned} \tag{8.65}$$

Substitution in Eqs. (8.58) and (8.59) gives

$$\frac{dN'}{dt} = R - B(q_0 + q')(N_0 + N') - \left(\frac{N_0 + N'}{\tau}\right), \tag{8.66}$$

$$\frac{dq'}{dt} = B(q_0 + q')(N_0 + N') - \left(\frac{q_0 + q'}{\tau_0}\right). \tag{8.67}$$

Neglecting the product of small quantities gives

$$\frac{dN'}{dt} = \left(R - Bq_0N_0 - \frac{N_0}{\tau}\right) - N'\left(Bq_0 + \frac{1}{\tau}\right) - Bq'N_0, \tag{8.68}$$

$$\frac{dq'}{dt} = \left(Bq_0N_0 - \frac{q_0}{\tau_0}\right) + Bq_0N' + q'\left(BN_0 - \frac{1}{\tau_0}\right). \tag{8.69}$$

By virtue of Eqs. (8.58) and (8.59) the first term in Eq. (8.68) and the first and third terms in Eq. (8.69) are zero. That is

$$\frac{dN'}{dt} = -N\left(Bq_0 + \frac{1}{\tau}\right) - q'BN_0 \tag{8.70}$$

and

$$\frac{dq'}{dt} = NBq_0. \tag{8.71}$$

Differentiation of Eq. (8.70) followed by substitution from Eq. (8.71) gives

$$\frac{d^2N'}{dt^2} + \frac{dN'}{dt}\left(Bq_0 + \frac{1}{\tau}\right) + \frac{N'q_0B}{\tau_0} = 0. \tag{8.72}$$

Use of Eq. (8.61) gives us

$$\frac{d^2N'}{dt^2} + \frac{dN'}{dt}(RB\tau_0) + N'\left(RB - \frac{1}{\tau\tau_0}\right) = 0. \tag{8.73}$$

Similarly, from Eq. (8.71)

$$\frac{d^2q'}{dt^2} - Bq_0\frac{dN'}{dt} = 0. \tag{8.74}$$

Substitution from (8.70) gives

$$\frac{d^2q'}{dt^2} + Bq_0\left(Bq_0 + \frac{1}{\tau}\right)N' + q'Bq_0BN_0 = 0, \tag{8.75}$$

which becomes

$$\frac{d^2q'}{dt^2} + \frac{dq'}{dt}\left(Bq_0 + \frac{1}{\tau}\right) + q'\left(\frac{q_0B}{\tau_0}\right) = 0. \tag{8.76}$$

Finally, from eq. (8.61)

$$\frac{d^2q'}{dt^2} + \frac{dq'}{dt}\left(RB\tau_0\right) + q'\left(RB - \frac{1}{\tau\tau_0}\right) = 0. \tag{8.77}$$

So, q' and N' both obey the same differential equation, which is logical since the fluctuations in photon density are closely coupled to fluctuations in population inversion.

For a trial solution of Eq. (8.77) we try $q' = q_0'e^{-(\eta - i\xi)t}$. Remembering that

$$q_0 = \left(\frac{RB\tau_0 - 1/\tau}{B}\right)$$

we find

$$\eta = \frac{1}{2}\left(q_0B + \frac{1}{\tau}\right) = \frac{RB\tau_0}{2} = \frac{r}{2\tau} \tag{8.78}$$

and

$$\xi = \sqrt{\frac{q_0B}{\tau_0} - \frac{q_0^2B^2}{4} - \frac{q_0B}{2\tau} - \frac{1}{4\tau^2}} = \sqrt{\frac{(r-1)}{\tau\tau_0} - \left(\frac{q_0B}{2} + \frac{1}{2\tau}\right)^2}. \tag{8.79}$$

Eq. (8.79) can be simplied by using Eqs. (8.61)–(8.63) to give

$$\xi = \sqrt{\frac{(r-1)}{\tau\tau_0} - \left(\frac{r}{2\tau}\right)^2}. \tag{8.80}$$

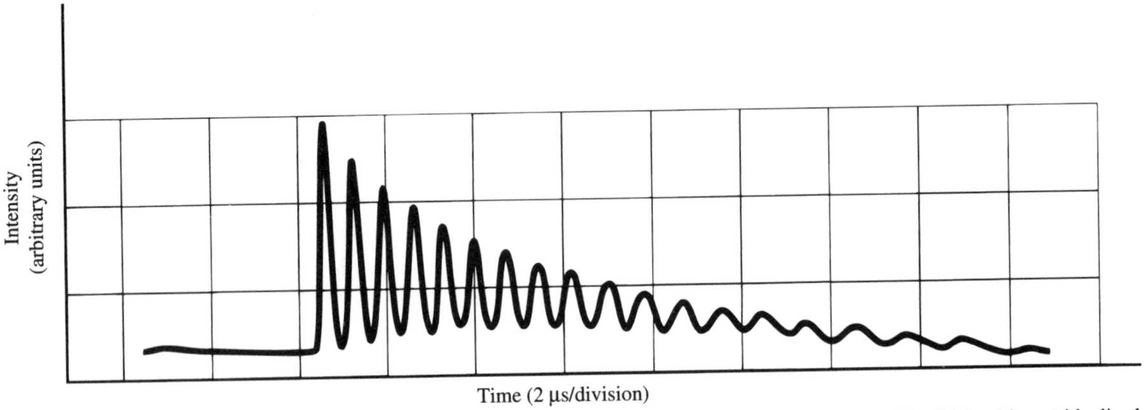

Fig. 8.25. Almost idealized relaxation oscillations obtained from a Nd^{3+}:$CaWO_4$ laser.

The photon density is a damped oscillatory function of angular frequency ξ, provided

$$\left(\frac{r-1}{\tau\tau_0}\right) > \left(\frac{r}{2\tau}\right)^2,$$

of the form $q' = q_0 e^{-\eta t} e^{i\xi t}$.

If we take the specific example of a ruby laser with

$$\tau = 3 \times 10^{-3} \text{ s},$$

$$\tau_0 = 10^{-8} \text{ s},$$

$$r, = 2,$$

Eq. (8.79) gives the frequency of the relaxation oscillation as 1.82×10^5 s^{-1} and its period ($= 2\pi/\xi$) as 3.45×10^{-5} s. Such a damped oscillation in the output of a Nd^{3+}: calcium tungstate laser is shown in Fig. (8.25). It would appear from Eq. (8.80) that for pumping far enough above threshold, when $r \gg 1$ and $r > 4\tau/\tau_0$, the output laser intensity would cease to be oscillatory. Such an assumption cannot be made with any certainty, however, because under these circumstances, q' and N' no longer satisfy the conditions $q' \ll q_0$, $N' \ll N_0$, and our solution of the Statz–deMars equations breaks down.

8.14 Undamped Relaxation Oscillations

Under normal circumstances relaxation oscillations are damped, although the relaxation oscillation may become undamped if the pumping rate is varying with time. Suppose, $R = R_0 + r'(t)$, in which case Eq. (8.68) becomes

$$\frac{dN'}{dt} = r' - R_0 B\tau_0 N' - \frac{q'}{\tau_0} \tag{8.81}$$

and Eq. (8.77) takes the form

$$\frac{d^2q'}{dt^2} + \frac{r}{\tau}\frac{dq'}{dt} + \frac{1}{\tau\tau_0}(r-1)q' = -\frac{1}{\tau}(r-1)r'. \tag{8.82}$$

This is now the differential equation describing a damped oscillator that is being driven with a 'force' $-(1/\tau)(r-1)r'$, and as such can have an oscillatory solution of large amplitude if the driving force has a frequency component near the natural frequency ξ.

Suppose that $r'(t)$ is oscillatory and can be written as

$$r'(t) = r_0 e^{i\xi t},$$

then Eq. (8.82) becomes

$$\frac{d^2q'}{dt} + \frac{r}{\tau}\frac{dq'}{dt} + \frac{1}{\tau\tau_0}(r-1)q' = -\frac{1}{\tau}(r-1)r_0 e^{i\xi' t}. \tag{8.83}$$

We try as a solution $q' = Qe^{i\xi' t}$.

Substitution in Eq. (8.83) gives

$$-\xi'^2 Q + i\xi'\frac{r}{\tau}Q + \frac{1}{\tau\tau_0}(r-1)Q = -\frac{1}{\tau}(r-1)r_0 \tag{8.84}$$

and

$$Q = \frac{(r-1)r_0}{\tau\xi'^2 - ir\xi' - (r-1)/\tau_0}. \tag{8.85}$$

From Eqs. (8.78) and (8.80), Eq. (8.85) can be written as

$$Q = \frac{(r-1)r_0/\tau}{(\xi' - \xi - i\eta)(\xi' + \xi - i\eta)}. \tag{8.86}$$

This amplitude becomes large as $\xi' \to \xi$, particularly if the damping factor η is small.

Undamped relaxation oscillations are quite common in real lasers. A component in the pumping process that fluctuates at a natural resonance frequency of the laser system will tend to drive a relaxation oscillation at, or near, the resonance frequency. Fig. (8.26) shows just such an oscillation, and its frequency spectrum observed in the output of a Nd:YAG laser pumped by a semiconductor laser diode. Such relaxation oscillations can be suppressed by minimizing fluctuations in the pump through feedback control, by increasing the damping factor η, or by channeling the fluctuations in the pump into a frequency region far from a natural resonance frequency of the laser.

8.15 Giant Pulse (Q-Switched) Lasers

In a pulsed solid-state laser the time dependence of the optical pumping leads naturally to 'spiking' in the laser output. The pseudo-random nature of this light output is generally undesirable but can be controlled by a technique called *Q-switching*[8.7]–[8.9]. If the feedback process within the oscillator is blocked, for example by preventing reflection from one of the laser mirrors, then the cavity has low Q. If the lamp is fired inversion builds up within the laser, but oscillation does not start. If the obstacle to feedback is suddenly removed, the Q of the cavity switches to a higher value and oscillation begins. Because the laser now finds itself with an inversion much above threshold the intracavity intensity grows very rapidly and the laser delivers a giant or *Q-switched* pulse of high intensity and short duration. This duration is typically on the order of the natural lifetime τ_0 of the laser cavity described in Section 4.4

$$\tau_0 = \frac{\ell}{c(1-R)}. \tag{8.87}$$

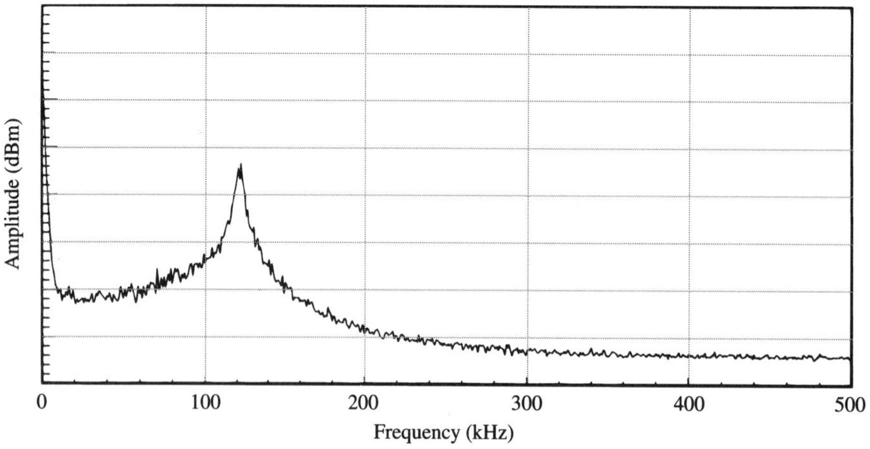

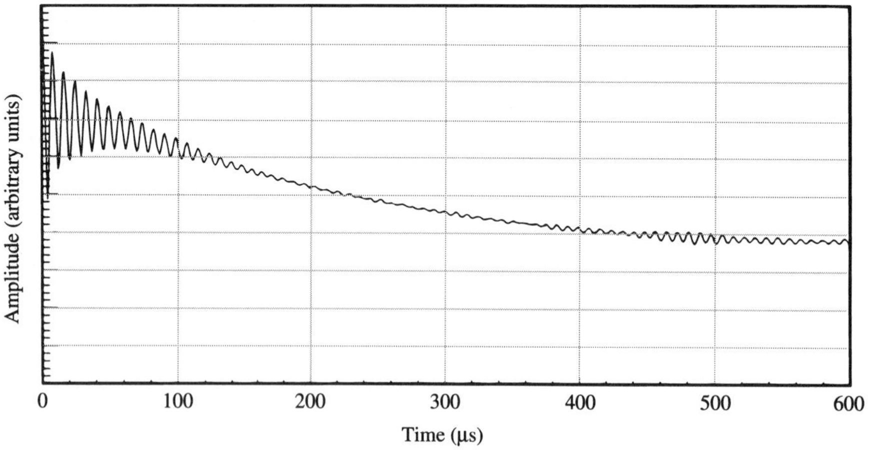

Fig. 8.26. Relaxation oscillations in the output of a semiconductor diode laser pumped Nd:YAG laser. (Courtesy of Dr Simon P. Bush.)

This giant pulse may so deplete the inversion in the laser that no subsequent laser emission occurs for that particular firing of the flashlamp. There are several ways of Q-switching[8.19]–[8.21]. They all rely on removal of an obstacle to reflection from one laser mirror after a substantial population has been stored in the upper laser level.

(a) A spinning wheel with a hole or slot placed inside the laser cavity[8.22]. This method is largely of historic interest, but it is the easiest to understand. The lamp is synchronized to fire before the spinning wheel brings the hole into alignment. To provide a rapid transition from low to high Q, the hole in the wheel should be of small size, with a pair of lenses to focus the intracavity laser beam, as shown in Fig. (8.27).

(b) One of the end reflectors is spun about an axis orthogonal to the alignment axis of the laser cavity. This is most easily accomplished if one of the laser mirrors is replaced with a spinning roof prism, as shown in Fig. (8.28a). A roof prism has the property of reflecting any ray incident orthogonal to the ridge line of the roof back parallel to itself. Therefore, if such a prism is spun about an axis perpendicular to the ridge, alignment of the

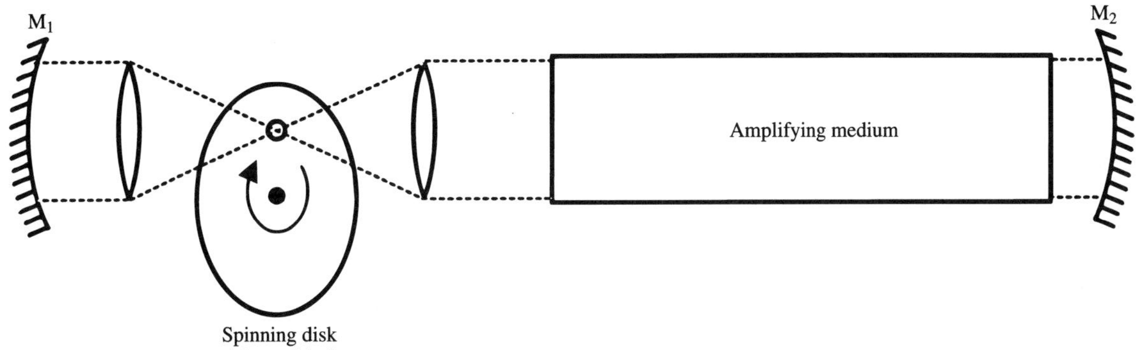

Fig. 8.27. Rotating intracavity wheel with a hole for Q-switching.

laser cavity is guaranteed twice per rotation. Alternatively, the roof prism can remain fixed, or be fabricated directly onto the end of the laser rod. A conventional laser mirror spun about its diameter perpendicular to the ridge of the roof prism as shown in Fig. (8.28b) is guaranteed to align twice per rotation.

(c) If a glass cell containing a dilute solution of absorbing dye such as cryptocyanine is placed inside the cavity then the oscillation threshold predicted by Eq. (5.9) is raised to

$$\gamma_t = \alpha - \frac{1}{\ell}(\alpha_d d + \ln r_1 r_2), \tag{8.88}$$

where α_d is the absorption coefficient of the dye and d its thickness. The laser will not start to oscillate until this threshold is reached. When oscillation does start the increasing intracavity intensity 'bleaches' the dye by exciting ground state dye molecules into excited states. Solid materials with absorbing impurities can be used in a similar way[8.23]. This is the reverse of the gain saturation process described in Section 2.6. In describing this process theoretically, provided the Q-switched pulse is of short duration, spontaneous emission of the dye molecules can be neglected[8.24].

(d) With an optically active element in the laser cavity that changes the polarization state of the light travelling through it by either the Kerr effect†, Faraday effect‡ or by electro-optic activity in a crystal§. In a Kerr-active medium or electro-optic crystal a transverse electric field of sufficient magnitude will change linearly polarized light to circularly polarized or rotate its plane of linear polarization by 90°. In a Faraday-active medium an axial magnitude field causes the plane of linear polarization of a wave travelling along the field to rotate. The general way in which these phenomena can be used to build a Q-switch is best illustrated with Fig. (8.29), which shows an electro-optic crystal used with a transverse electric field as the switching element. The crystal has sufficient voltage applied to it that the electric field causes input linearly polarized light

† See Chapter 20.
‡ See Chapter 14.
§ See Chapter 19.

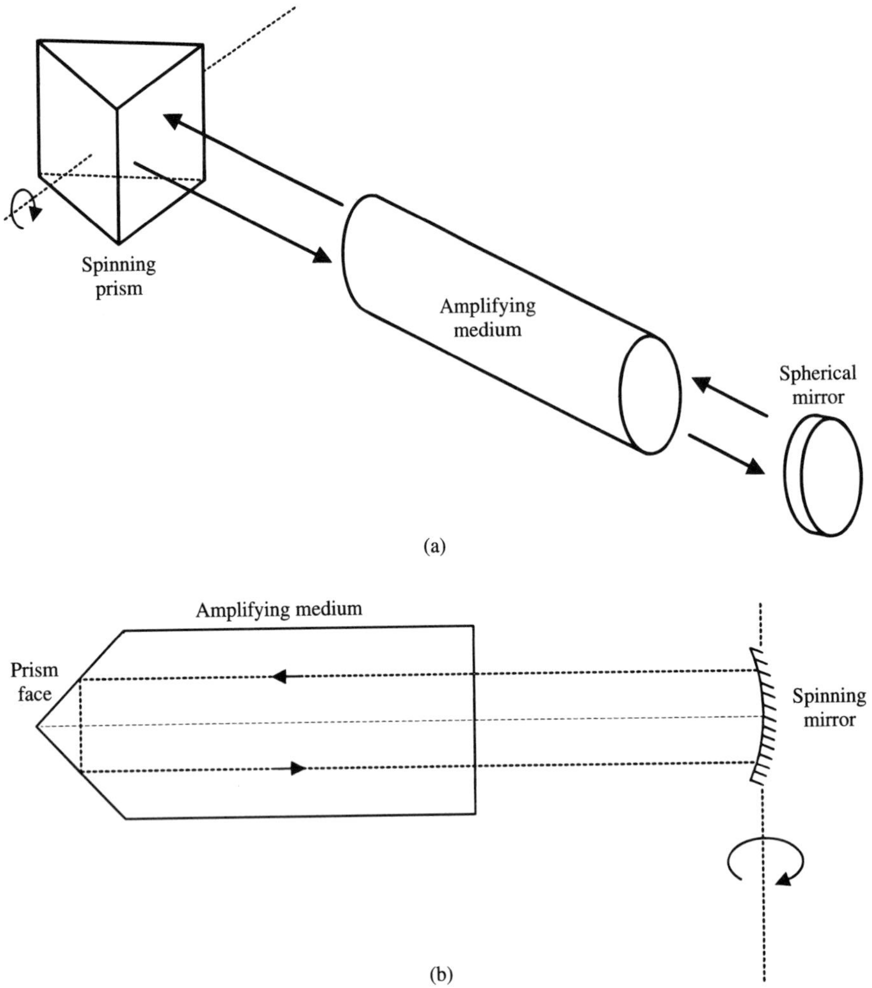

Fig. 8.28. (a) Q-switching with one fixed cavity mirror and a rotating roof prism. (b) Laser rod with roof prism fashioned on one end where Q-switching is accomplished by spinning the other cavity reflector.

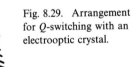

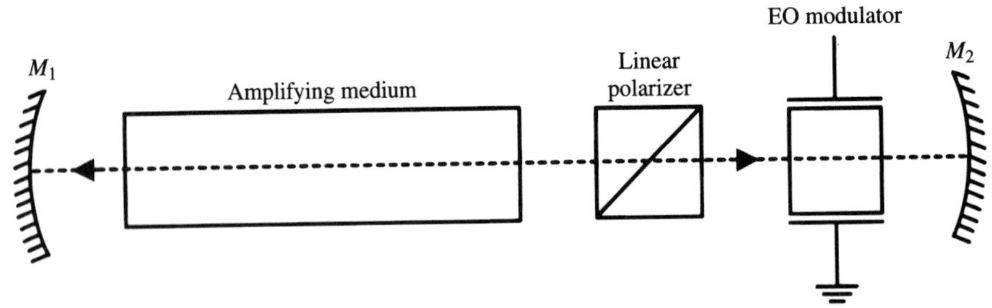

Fig. 8.29. Arrangement for Q-switching with an electrooptic crystal.

to be converted to circularly polarized light on output. Exactly why this occurs in certain materials is explained in detail in Chapters 18 and 19. The circularly polarized light strikes mirror M_2, and returns through the crystal. In this second pass through the crystal the circularly polarized light is converted back to linearly polarized, but with its polarization direction orthogonal to its original direction. This light cannot pass the linear polarizator, and reflection from mirror M_2 is effectively blocked. Fast removal of the voltage applied to the crystal opens up mirror M_2 for reflection and laser oscillation occurs.

8.16 Theoretical Description of the Q-Switching Process

In a simple model of the Q-switching process the Q-switched pulse is assumed to be short enough that spontaneous emission processes during the pulse can be neglected and the changeover from low Q to high Q is instantaneous. For a resonator of volume V the total number of photons is

$$\phi = Vq \tag{8.89}$$

and the total inversion is

$$n = \left(N_2 - \frac{g_2}{g_1}N_1\right)V. \tag{8.90}$$

The gain of the amplifying medium is γ. We neglect the additional complexity of gain saturation. Radiation of intensity I passing through the amplifier grows according to

$$\frac{dI}{dz} = \gamma I \tag{8.91}$$

and as a function of time

$$\frac{dI}{dt} = \frac{dI}{dz}\frac{dz}{dt} = c\gamma I, \tag{8.92}$$

where c is the velocity of light within the laser medium. For a laser cavity of length ℓ containing an amplifying medium of length L, only a fraction L/ℓ of all the light is being amplified. We can describe the average intensity increase within the cavity by Eq. (8.92) with a scale factor of L/ℓ. The total number of photons changes according to

$$\frac{d\phi}{dt} = \phi\left(\frac{c\gamma L}{\ell} - \frac{1}{\tau_0}\right), \tag{8.93}$$

where the cavity time constant τ_0 includes all passive loss effects such as mirror transmission and scattering within the amplifying medium. If we introduce a normalized time unit $\tau = t/\tau_0$, Eq. (8.93) becomes

$$\frac{d\phi}{d\tau} = \phi\left(\frac{\gamma}{\ell/cLT_0} - 1\right). \tag{8.94}$$

Oscillation will not occur if $\gamma < \gamma_t = \ell/(cL\tau_0)$; in this case $d\phi/d\tau = 0$. Therefore we can write

$$\frac{d\phi}{d\tau} = \phi\left(\frac{\gamma}{\gamma_t} - 1\right) \tag{8.95}$$

and since gain is proportional to population inversion

$$\frac{d\phi}{d\tau} = \phi\left(\frac{n}{n_t} - 1\right). \tag{8.96}$$

Each stimulated emission reduces the upper laser level population by one and increases the lower level by the same number. Consequently,

$$\frac{dn}{d\tau} = -2\phi\frac{n}{n_t}. \tag{8.97}$$

Eqs. (8.96) and 8.97) can be solved numerically very easily, for example by a Runge–Kutta[8.26] method. Before giving examples of such solutions we can learn quite a lot about the way the population inversion and photon density behave by examining the equations. If we divide Eq. (8.96) by Eq. (8.97) we get

$$\frac{d\phi}{dn} = \frac{1}{2}\left(\frac{n_t}{n} - 1\right), \tag{8.98}$$

which has the solution

$$\phi = \frac{1}{2}(n_t \ln n - n) + \text{ constant}. \tag{8.99}$$

If the initial photon density and inversion are ϕ_0, n_0, respectively, then

$$\phi_0 = \frac{1}{2}(n_t \ln n_0 - n_0) + \text{ constant}. \tag{8.100}$$

Combining Eqs. (8.99) and (8.100)

$$\phi - \phi = \frac{1}{2}\left[n_t \ln\left(\frac{n}{n_0}\right) - (n - n_0)\right]. \tag{8.101}$$

Since there is negligible photon density before laser action starts, we set $\phi_0 = 0$. Therefore,

$$\phi = \frac{1}{2}\left[n_t \ln\left(\frac{n}{n_0}\right) - (n - n_0)\right]. \tag{8.102}$$

After the Q-switched pulse is over $\phi \to 0$ and the final inversion n_f satisfies

$$n_t \ln\left(\frac{n_f}{n_0}\right) - (n_f - n_0) = 0, \tag{8.103}$$

or

$$\frac{n_f}{n_0} = e^{-(n_0 - n_f)/n_t}. \tag{8.104}$$

The quantity $(n_0 - n_f)/n_0$ represents the fraction of the original inversion that is converted to laser energy, which is

$$\left(\frac{n_0 - n_f}{n_0}\right) = 1 - e^{-(n_0 - n_f)/n_t}. \tag{8.105}$$

This approaches unity as $n_0/n_t \to \infty$.

If n_f is very small this means the Q-switched pulse has very efficiently depleted the inversion. The value of n_f can swing below the threshold value n_t, analogous to the underdamped discharge of a capacitor.

The output power of the laser is

$$P = \frac{\phi h\nu}{\tau_0}, \tag{8.106}$$

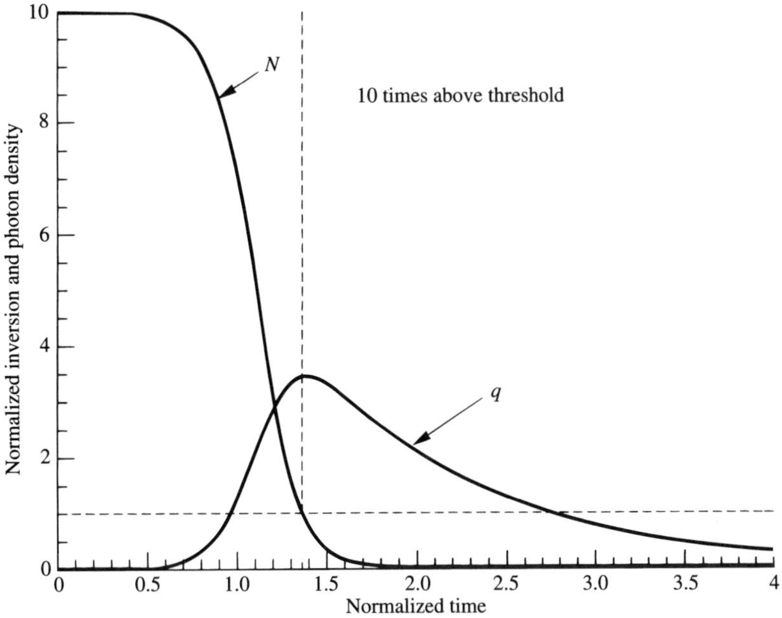

Fig. 8.30. Numerical simulation of Q-switch behavior from Eqs. (8.105) and (8.106) for a system that is pumped initially to 10 times above threshold.

since the rate at which instantaneous energy stored is being lost from the cavity depends only on the cavity lifetime. Therefore from Eq. (8.102)

$$P = \frac{h\nu}{2\tau_0}\left[n_t \ln\left(\frac{n}{n_0}\right) - (n - n_0)\right]. \tag{8.107}$$

Maximum power output results when $d\phi/d\tau = 0$, which from Eq. (8.96) occurs when $n = n_t$. The maximum power output is

$$P_{max} = \frac{h\nu}{2\tau_0}\left[n_t \ln\left(\frac{n_t}{n_0}\right) - \left(n_t - n_0\right)\right]. \tag{8.108}$$

If the laser is Q-switched from far above threshold then $n_0 \gg n_t$ and

$$P_{max} \simeq \frac{n_0 h\nu}{2\tau_0} \tag{8.109}$$

and the maximum number of photons stored in the cavity is

$$\phi_{max} = \frac{P_{max} h\nu}{\tau_0} = \frac{n_0}{2}. \tag{8.110}$$

All these predictions are borne out by actual numerical solution of Eqs. (8.96) and (8.97), as can be seen from Figs. (8.30) and (8.31). To summarize: these simulations demonstrate that:

(i) The Q-switched pulse has higher amplitude and turns on faster as the initial inversion increases from the threshold value – see Fig. (8.30).

(ii) The inversion falls more rapidly at the onset of Q-switching as the initial inversion increases from the threshold value – see Fig. (8.30).

(iii) The peak photon density occurs when the inversion passes through the threshold value – see Figs. (8.29) and (8.30).

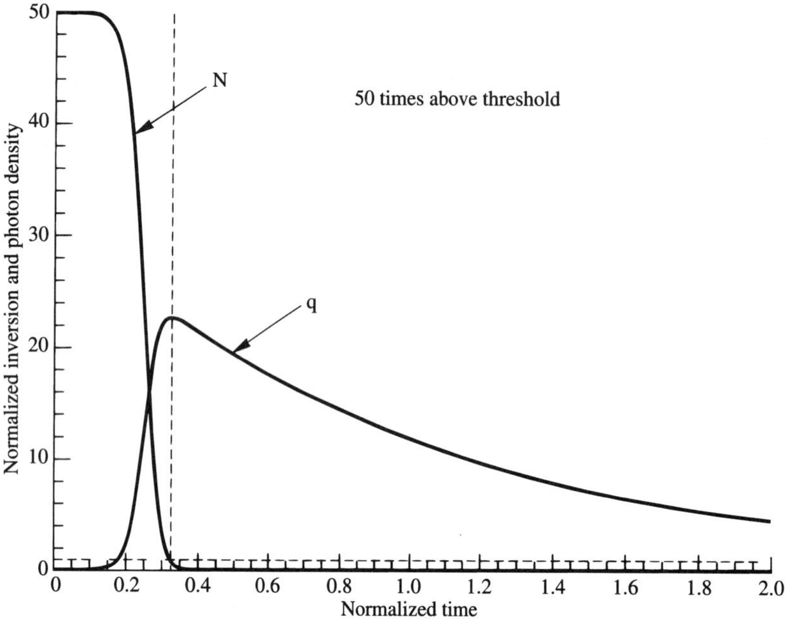

Fig. 8.31. Numerical simulation of Q-switched behavior from Eqs. (8.105) and (8.106) for a system that is pumped initially to 50 times above threshold.

(iv) After the inversion has fallen to a low value the remaining photon density decays with the characteristic lifetime of the cavity.

(v) The final inversion falls increasingly below threshold as the initial inversion rises from the threshold value – see Fig. (8.30).

8.16.1 *Example Calculation of Q-Switched Pulse Characteristics*

We take as an example a Nd:YAG laser with a 1% Nd^{3+} doping density. The density of YAG is 4550 kg m^{-3} so a 1% by weight doping level of Nd^{3+} corresponds to 1.9×10^{26} ions m^{-3}. For the Nd:YAG laser crystal considered earlier in this chapter the threshold inversion was 9.3×10^{21} ions m^{-3}, which corresponds to an excited Nd^{3+} ion population within the crystal volume ($\simeq 10^{-6}$ m^3) of 9.3×10^{15}. For this laser, with $R_{avg} = 0.96$ the cavity lifetime is

$$t_0 = \frac{\ell}{c(1 - R_{avg})} = \frac{50 \times 10^{-3}}{2.997 \times 10^8 \times 0.04} = 4.17 \text{ ns.}$$

If we can pump this crystal to 100 times above threshold then the small signal gain will be 88.3 m^{-1}. The peak Q-switched power will be, from Eq. (8.118)

$$P_{max} \simeq \frac{9.3 \times 10^{17} \times 6.626 \times 10^{-34} \times 2.947 \times 10^8}{2 \times 4.17 \times 10^{-9} \times 1.06 \times 10^{-6}} = 2.05 \times 10^7 \text{W.}$$

The total output pulse energy is

$$U \simeq \frac{n_0 h\nu}{2} \simeq 87 \text{ mJ,}$$

where the factor of 2 in the denominator results from the 'bottleneck' imposed by the lower laser level. Since the Q-switched pulse is very fast, ions that reach the lower laser level can be assumed to remain there during the pulse so the inversion will be lost when half the initially excited ions have emitted.

8.17 Problems

(8.1) Consider a ruby laser rod whose pump band absorption characteristics correspond to the data in Fig. 3.3, i.e., the concentration of Cr^{3+} is 1.88×10^{25} m^{-3}. Use the data given for $E\perp c$. Assume that the laser rod is rectangular (10 mm$\times$ 10 mm$\times$ 100 mm) and is closed-coupled to the flashlamp so that 25% of the flashlamp radiation enters the rod. Calculate the optical flux requirement from the lamp (W m^{-2} Hz^{-1}) to exceed laser threshold by 100% on the R_1 transition.

(a) The two cavity mirror reflectances are 90% and 100%.

(b) The flashlamp pulse is a square pulse of 10^{-4} s duration with a flat spectral output from 300 nm to 700 nm.

(c) The efficiency η for transfer of energy to the upper laser level is 50% for both pump bands.

(d) For the laser transition $\tau_2 \simeq 1/A_{21} = 3\times 10^{-3}$ s, $\Delta\nu = 200$ GHz chomogeneously broadened).

(e) $T = 300$ K.

Take both the $2\overline{A}$ and $\overline{E}$ upper ruby laser states into consideration. The ratio of degeneracy factors for the upper/lower levels is 1/2. Make a suitable approximation in evaluating the intergral for the population of the upper laser levels.

References

[8.1] Laser Program Annual Report, C.D. Hendricks, Ed., Lawrence Livermore National Laboratory, Livermore, CA, UCRL-50021-83, 1984, 6–19.

[8.2] Laser Program Annual Report, C.D. Hendricks, Ed., Lawrence Livermore National Laboratory, UCRL-500021-83, 1984, 6–22, 1983.

[8.3] D.F. Nelson and W.S. Boyle, 'A continuously operating ruby optical maser,' *Appl. Opt.*, **1**, 181–183, 1962; see also *Appl. Opt. Suppl.*, **1**, 99–101, 1962.

[8.4] A. Sanchez, R.E. Fahey, A.J. Strauss, and R.L. Aggrawal, 'Room temperature continuous-wave operation of a Ti:Al_2O_3 laser,' *Opt. Lett.*, **11**, 363–364, 1986.

[8.5] D. Röss, *Lasers, Light Amplifiers and Oscillators*, Academic Press, New York, 1968.

[8.6] D. Röss and G.'Seidlov, 'Pumping new life into ruby lasers,' *Electronics*, Sept. 5, 1966.

[8.7] T. Kushida, H.M. Marcos, and J.E. Geusic, 'Laser transition cross section and fluorescence branching ratio for Nd^{+3} in yttrium aluminium garnet,' *Phys. Rev.*, **167**, 289–291, 1968.

[8.8] 1981 Laser Program Annual Report, Lawrence Livermore National Laboratory, Livermore, CA, E.V. George, Ed., UCRL-50021-81, 1982.

[8.9] Laser Program Annual Report 83, C.D. Hendricks, Ed., Lawrence Livermore National Laboratory, Livermore, CA, UCRL-50021-83, 1984.

[8.10] S.E. Stokowski, 'Glass Lasers,' in *Handbook of Laser Science and Technology*, Vol. I *Lasers and Masers*, M.J. Weber, Ed., CRC Press, Boca Raton, FL, 1982.

[8.11] 1985 Laser Program Annual Report, Lawrence Livermore National Laboratory, M.L. Rufer and P.W. Murphy, Eds., UCRL-50021-85, 1986.

[8.12] J. McKenna, 'The focusing of light by a dielectric rod,' *Appl. Opt.*, **2**, 303–310, 1963; see also C.H. Cooke, J. McKenna, and J.G. Skinner, 'Distribution of absorbed power in a side-pumped ruby rod,' *Appl. Opt.*, **3**, 957–961, 1964.

[8.13] G.E. Devlin, J. McKenna, A.D. May, and A.L. Schawlor, 'Composite rod optical masers,' *Appl. Opt.*, **1**, 11–15, 1962.

[8.14] M.J. Lubin, J.M. Soures, and L.M. Gelman, 'A larger aperture Nd-glass face pumped laser amplifier for high peak power applications,' General Electric Report No. 72CRD143, May 1972.

[8.15] T.Y. Fan and R.C. Byer, 'Diode Laser-Pumped Solid State Lasers,' *IEEE J. Quant. Electron.*, **QE-24**, 895–912, 1988.

[8.16] T.J. Kane and R.C. Byer, 'Monolithic, unidirectional single-mode Nd:YAG ring laser,' *Opt. Lett.*, **10**, 65-67, 1985.

[8.17] H. Statz and G. deMars, 'Transients and oscillation pulses in masers,' in *Quantum Electronics*, Columbia University Press, Cambridge, 1960, pp. 530–537.

[8.18] L.F. Johnson, 'Optically pumped pulses crystal lasers other than ruby,' in *Lasers: A Series of Advances*, Vol. 1, A.K. Levine, Ed., Marcel Dekker, New York, 1966.

[8.19] R.W. Hellworth, 'Control of fluorescent pulsations,' in *Advances in Quantum Electronics*, J.R. Singer, Ed., Columbia University Press, New York, 1961, pp. 334–341.

[8.20] F.J. McClung and R.W. Hellworth, 'Giant optical pulsations from ruby,' *J. Appl. Phys.*, **33**, 828–829, 1962.

[8.21] R.W. Hellworth, 'Q modulation of lasers,' in *Lasers: A Series of Advances* Vol. 1, A.K. Levine, Ed., Marcel Dekker, New York, 1966, pp. 253–294.

[8.22] R.J. Collins and P. Kisliuk, 'Control of population inversion in pulsed optical masers by feedback modulation,' *J. Appl. Phys.*, **33**, 2009–2011, 1962.

[8.23] G. Bret and F. Gines, 'Giant-pulse laser and light amplifier using variable transmission coefficient glasses as light switches,' *Appl. Phys. Lett.* ,**4**, 175–176, 1964.

[8.24] A. Szabo and R.A. Stern, 'Theory of laser, giant pulsing by a saturable absorber,' *J. Appl. Phys.*, **36**, 1562–1566, 1965.

[8.25] W.G. Wagner and B.A. Lengyel, 'Evolution of the giant pulse in a laser,' *J. Appl. Phys.*, **34**, 2042–2046, 1963.

[8.26] W.H. Press, S.A. Teukolsky, W.T. Vetterling, and B.P. Flannery, *Numerical Recipes, The Art of Scientific Computing*, 2nd Edition, Cambridge University Press, Cambridge, 1992.

9

Gas Lasers

9.1 Introduction

In this chapter we shall consider some of the fundamental processes that are used to produce population inversion in gases. We shall see that the technological features of gas lasers, and the efficiency with which they can be made to operate, are intimately connected with the particular mechanism(s) used to excite the upper laser level. Most of our attention will be devoted toward a consideration of gas lasers in which the laser action involves energy levels of an atom. Consideration of most of the features of molecular gas lasers will be delayed until the next chapter.

Most gas lasers are excited by electron collisions in various types of gas discharge and electron beam apparatus. (In the context of 'gas lasers' at this point we exclude those gaseous lasers where the pumping occurs as a result of chemical or photochemical processes.) However, there are several gas laser systems, particularly operating in the far-infrared, where optical pumping (with auxiliary lasers) has been successful.

9.2 Optical Pumping

Because gaseous systems do not in general have broad absorption bands, optical pumping of laser action in such systems with nonlaser sources is not generally feasible. An early laser system which was an exception to this general rule was the pure cesium laser first operated by Rabinowitz, Jacobs, and Gould[9.1],[9.2], and illustrated in Fig. (9.1). This system is now largely of academic interest, but it does provide a good example of where conventional optical pumping of gases can be successful. Cesium vapor has a strong absorption on the transition $6S_{1/2} \rightarrow 8P^0_{1/2}$ at 388.8616 nm which is in very good coincidence with the strong helium line at 388.864 nm from the transition $3p\ ^3P_{2,1} \rightarrow 2s^3S_1$, as shown in Fig. (9.1). The line broadening of the Cs atoms is sufficiently large that significant absorption of the helium transition occurs. Laser action from the Cs atoms was obtained at 7.18 μm, 3.20 μm, 1.38 μm and 0.8919 μm. Population inversion results on these transitions because their lower levels decay rapidly with lifetimes shorter than that of their upper level. The power output at 7.18 μm was about 25 μW for an input of 800 W to the helium lamp.

Rather more power output and efficiency has been obtained from the optically pumped mercury laser, first developed by Djeu and Burnham[9.3], where pumping of the mercury upper laser level is accomplished by successive absorption of two photons from a mercury pumping lamp, as illustrated in Fig. (9.2). Artusy, Holmes, and Siegman[9.4] have obtained $\sim$ 3 mW output at 546.1 nm with 300 W input to the pumping lamp.

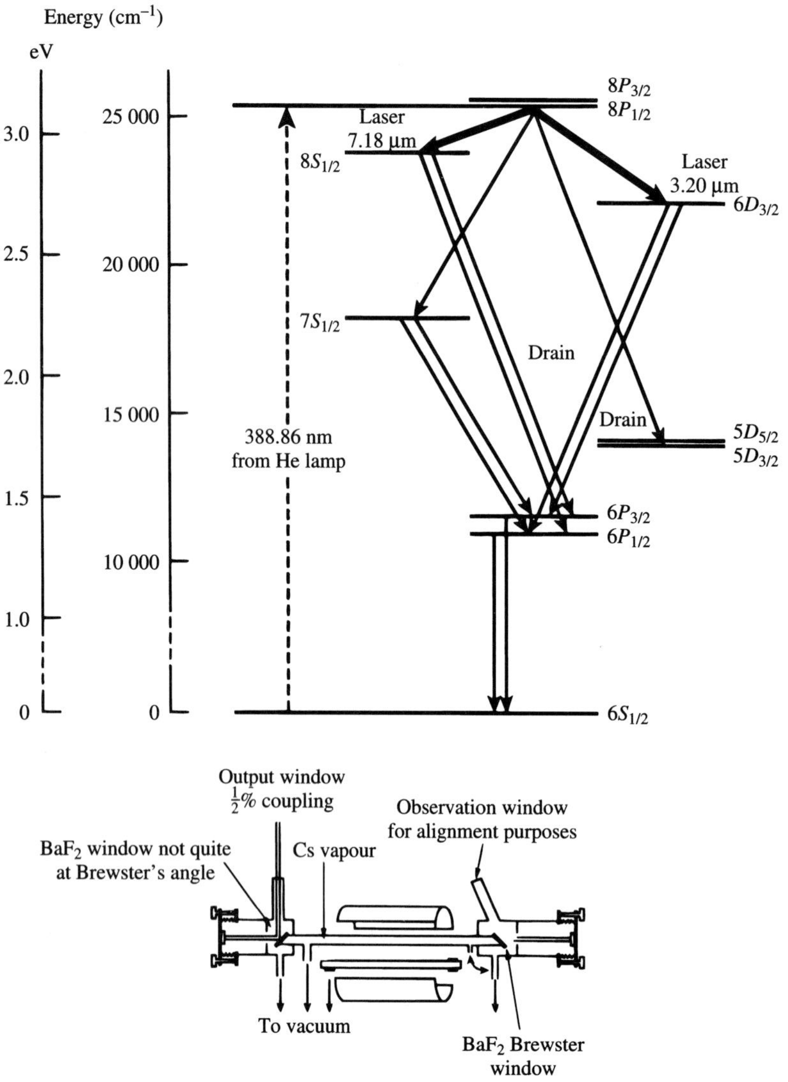

Fig. 9.1. The optically pumped cesium laser of Jacobs *et al.*[9.1], showing the construction of the laser system and the coincidence between the helium pump line and cesium absorption. Laser radiation is coupled from the cavity of one of the laser tube end windows, which is not quite set at Brewster's angle and provides about 0.5% output coupling.

Optical pumping of laser transitions in molecular gases by auxiliary lasers has, however, had much more widespread success than optical pumping with conventional light sources as in the two systems just described. The availability of powerful pump lasers with either discrete tunability, for example the many vibration–rotation transitions in the CO_2 laser†, and continuous tunability, for example from dye lasers,‡ allow a laser to be operated in very close coincidence with an absorption of a molecular vapor, particularly since molecules possess many closely spaced absorption lines. The resultant selectively pumped level serves as the upper level of the laser transition. In this way visible and infrared laser oscillation (0.544 μm – 1.355 μm) has been obtained from I_2 vapor pumped with a frequency doubled (0.53 μm) Nd:YAG laser. However, of more importance are the very many far-infrared

† See Chapter 10.
‡ See Chapter 12.

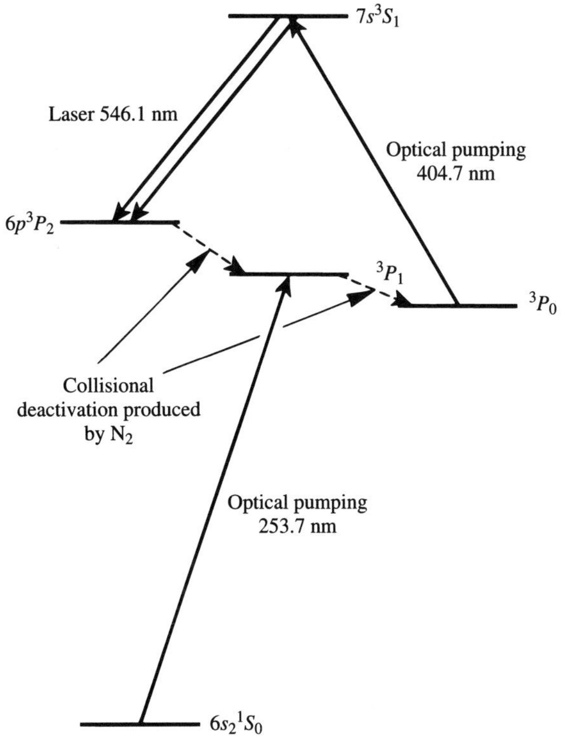

Fig. 9.2. Energy level scheme and excitation pathways in the optically pumped mercury laser.

laser transitions, spanning from the far-infrared up into the millimeter wave region, which can be excited by optically pumping various molecules such as CD_3OH, C_2H_5OH, CH_3F, C_2H_5F, $C_2H_2F_2$, $C_2H_4F_4$, HCOOH, NH_3, CH_3NH_2, CH_3CN, CH_2CHCN, CH_3CCH, CH_3Cl, C_2H_5Cl, CH_2Cl_2, CH_2CHCl, CH_3Br and CH_3I, with CO_2 lasers[9.5]–[9.7]. The excitation scheme of a typical far-infrared laser of this type is shown schematically in Fig. (9.3). Further discussion of some of the technological aspects of these lasers appears later in Section 11.8 on 'Far-Infrared Lasers'.

9.3 Electron Impact Excitation

Most successful excitation of laser action in gases is accomplished by electron impact processes. Excitation of the upper laser level can occur: directly, as in the pure neutral argon laser:

$$\mathrm{Ar} + \mathrm{e} \rightarrow \mathrm{Ar}_u^* + \mathrm{e},$$

where Ar_u^* denotes the upper laser level; and by electron impact followed by radiative cascade:

$$\mathrm{Ar} + \mathrm{e} \rightarrow \mathrm{Ar}^* + \mathrm{e},$$

$$\mathrm{Ar}^* \rightarrow \mathrm{Ar}_u^* + h\nu.$$

The upper laser level can be excited by excitation transfer from some species itself excited by electron impact, as, for example, in the He–Ne laser:

$$\mathrm{He} + \mathrm{e} \rightarrow \mathrm{He}_M^* (\text{metastable helium atom})$$

$$\mathrm{He}_M^* + \mathrm{Ne} \rightarrow \mathrm{He} + \mathrm{Ne}^*.$$

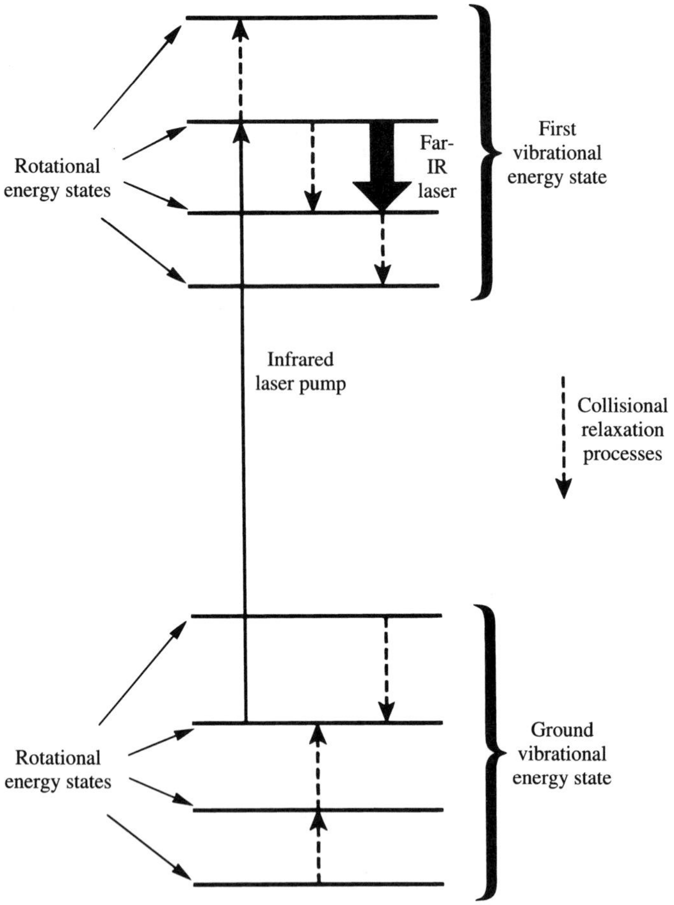

Fig. 9.3. Schematic energy level diagram and excitation scheme in a laser-pumped far-infrared laser.

This process requires a close energy coincidence, unless one of the colliding species is molecular and can take up excess energy, as for example in the Ne–O_2 system:

$$Ne + e \rightarrow Ne^*,$$

$$Ne^* + O_2 \rightarrow O_u^* + O + Ne,$$

or via a repulsive state of a molecule

$$M^* + AB \rightarrow AB^* + M,$$

$$AB^* \rightarrow A_u^* + B,$$

or

$$AB + e \rightarrow AB^* + e,$$

$$AB^* \rightarrow A_u^* + B.$$

9.4 The Argon Ion Laser

An extremely important gas laser system where the excitation is by electron impact is the argon ion laser. In this system the excitation follows a number of paths, illustrated in Fig. (9.4), depending for their relative importance on the exact

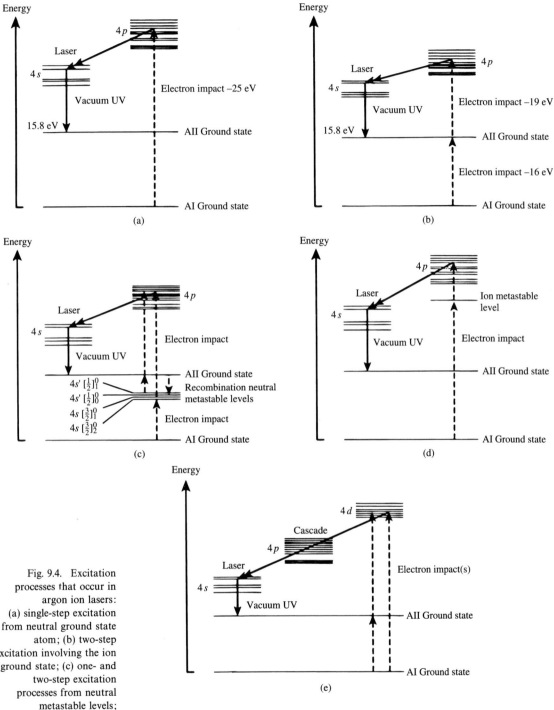

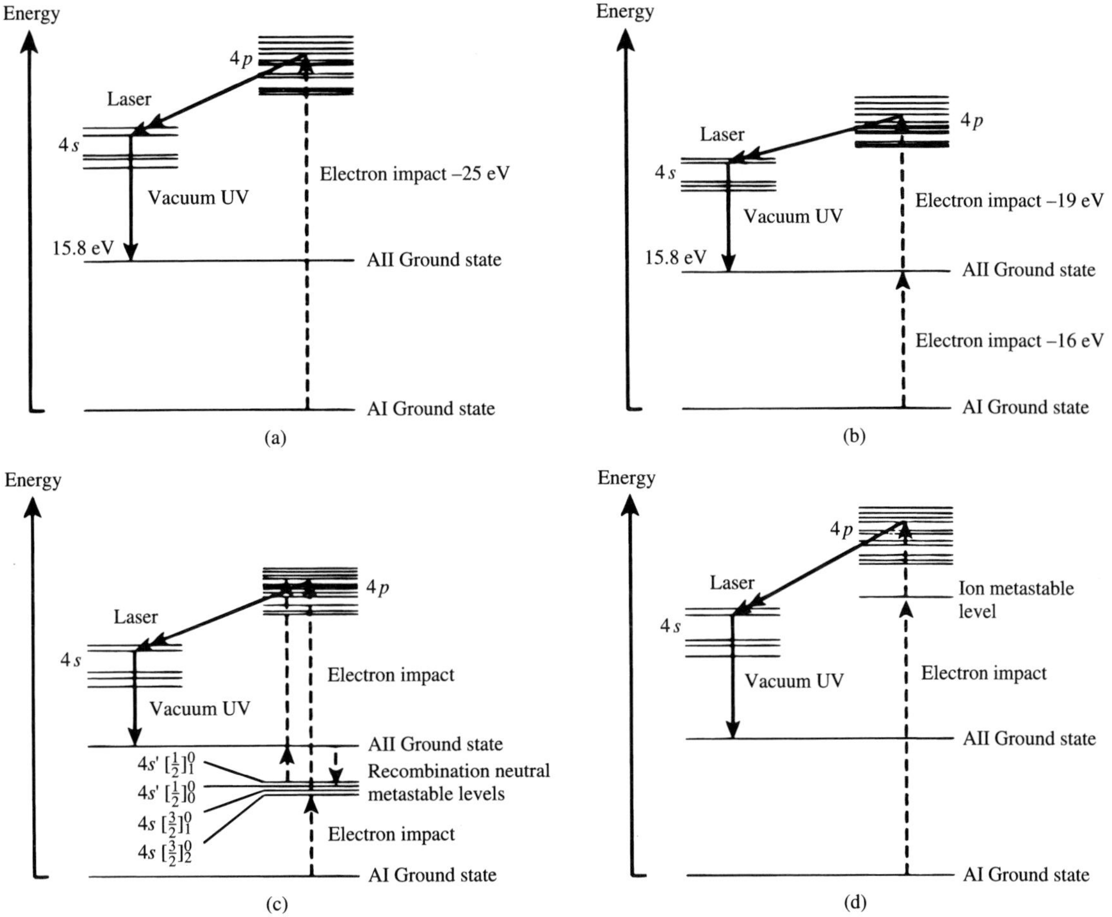

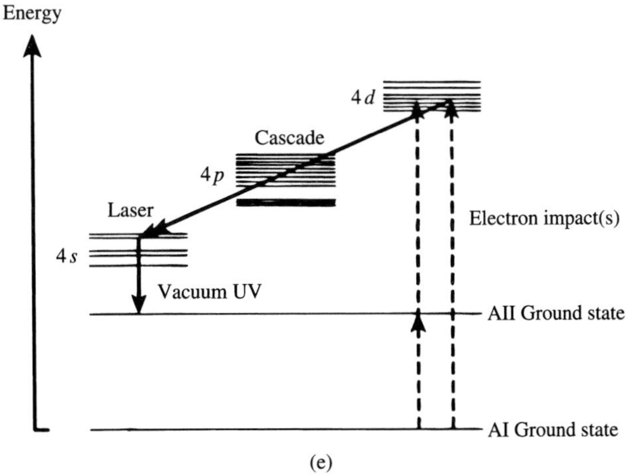

Fig. 9.4. Excitation processes that occur in argon ion lasers: (a) single-step excitation from neutral ground state atom; (b) two-step excitation involving the ion ground state; (c) one- and two-step excitation processes from neutral metastable levels; (d) two-step excitation involving ion metastable levels; (e) one- and two-step excitation involving cascades.

conditions in the system

$$\mathrm{Ar} + \mathrm{e} \rightarrow \mathrm{Ar}_{\mathrm{u}}^{+*} + 2\mathrm{e}, \quad \text{one-step process}$$

$$\left.\begin{array}{l} \mathrm{Ar} + \mathrm{e} \rightarrow \mathrm{Ar}^{+} + 2\mathrm{e} \\ \mathrm{Ar}^{+} + \mathrm{e} \rightarrow \mathrm{Ar}_{\mathrm{u}}^{+*} + \mathrm{e}, \end{array}\right\} \quad \text{two-step process}$$

$$\left.\begin{array}{l} \mathrm{Ar} + \mathrm{e} \rightarrow \mathrm{Ar}^{+} + 2\mathrm{e} \\ \mathrm{Ar}^{+} + \mathrm{e} \rightarrow \mathrm{Ar}^{+*} + \mathrm{e} \\ \mathrm{Ar}^{+*} \rightarrow \mathrm{Ar}_{\mathrm{u}}^{+*} + h\nu \end{array}\right\} \quad \text{cascade process}$$

and other similar processes involving metastable argon species. Similar processes prevail in the excitation of other noble gas and similar ion lasers. The efficiency of direct electron-pumped lasers of this type depends on the relative cross-sections for the excitation of various laser levels. In a gas discharge, where the excitation is generally performed, the electrons have a range of energies. The *cross-section*, $\sigma_j(E)$, for excitation of level j from the ground state by collision with an electron of energy E is essentially a probability factor that indicates how likely this excitation is to take place. We often assume that the distribution of electron energies $f(E)$ follows a Maxwellian distribution appropriate to some electron temperature T_e:

$$f(E) = 2\sqrt{\frac{E}{\pi k^3 T_e^3}}\, e^{-E/kT_e}, \tag{9.1}$$

where $f(E)dE$ is the fraction of electrons with energy between E and $E + dE$. The cross-section $\sigma_j(E)$ for a collisional excitation to a state j when the electron has energy E is defined from the rate of such processes:

$$R_j(E) = NN_e\sigma_j(E)v, \tag{9.2}$$

where N is the density of the particles being excited, N_e is the electron density and v the electron velocity; $N_e v$ is the flux of colliding elections. The total excitation rate is

$$R_j = NN_e \int_{E_j}^{\infty} f(E)\sigma_j(E)\sqrt{\frac{2E}{m}}\,dE. \tag{9.3}$$

Electron impact excitation of the laser system is effective when the rate of excitation of the upper level sufficiently exceeds the lower in relation to the ratio of their effective lifetimes.

The argon ion laser and a number of other gaseous ion lasers fall into a category of gas laser that offers a significant advantage over gas lasers of the same general type as the helium–neon laser. This advantage is in connection with the saturation of the pumping process.

9.5 Pumping Saturation in Gas Laser Systems

In gas lasers such as the helium–neon laser, and many other neutral gas lasers, the excitation of the upper laser level involves either energy transfer processes involving the collision of neutral species or electron collision with neutral species leading to the direct excitation of the upper laser level. When the excitation involves energy transfer from excited neutrals the efficiency of excitation of the upper laser level

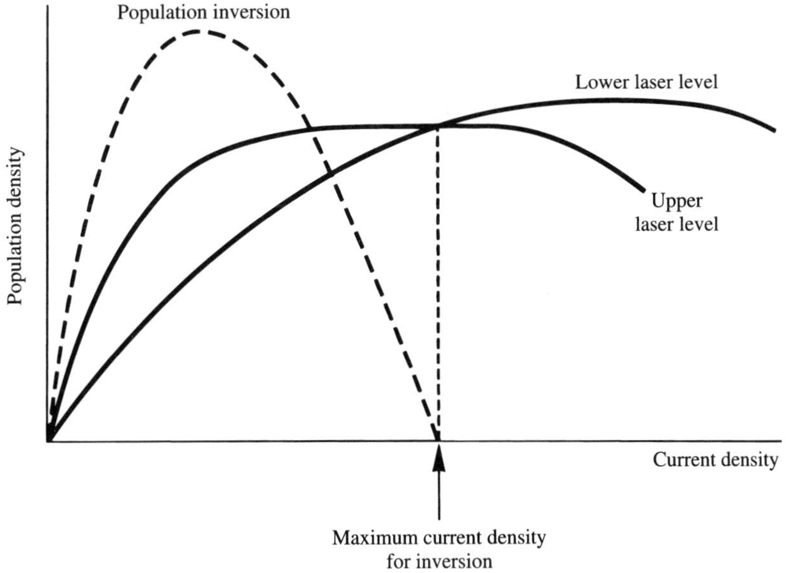

Fig. 9.5. Schematic variation of upper and lower laser population densities and population inversion in a laser where pumping saturation of the upper laser level occurs before pumping saturation of the lower.

depends on the concentration of these excited neutrals. When the concentration of these excited neutrals is maximized the rate of excitation of the upper laser level is also maximized. In general the rate of excitation of these neutral species in a gas discharge is a saturating function of the current density: a point is reached when the population of the neutral species reaches a value essentially independent of the current density. This situation arises because the neutrals are both created and destroyed by electron collisions. Thus, in either an excitation-transfer or direct-electron-impact-excited gas laser the population of the upper laser level cannot be increased beyond a certain point by further increasing the discharge current. In such systems the same saturation also occurs for the lower laser level but may, and frequently does, occur only when the lower level has already reached a population that quenches laser action, as illustrated in Fig. (9.5). The discharge current at which this saturation occurs is relatively low in neutral gas laser systems (and in some ion laser systems where the excitation involves energy transfer from neutral species). For example saturation of the upper laser level population (and hence the power output) occurs in helium – neon lasers at a current density ~ 1 A cm^{-2}.

However, in laser systems where the excited levels are ionic, this saturation does not occur until very much higher current densities are reached; for example, in the pure noble gas ion lasers not until the current density reaches hundreds of A cm^{-2}. Since the power output is a function of current density (typically it is $\propto I^n$, where n is between 1 and 6), the saturated power output of ion lasers can be much larger than from neutral lasers. For example, an argon ion laser 1 m long can under appropriate conditions deliver a power output of tens of watts whereas an equivalent length helium – neon laser might deliver 100 mW at 632.8 nm at best.

9.6 Pulsed Ion Lasers

Because ion lasers, particularly the noble gas ion lasers, depend for their efficient operation on the presence of significant numbers of ions in the discharge plasma,

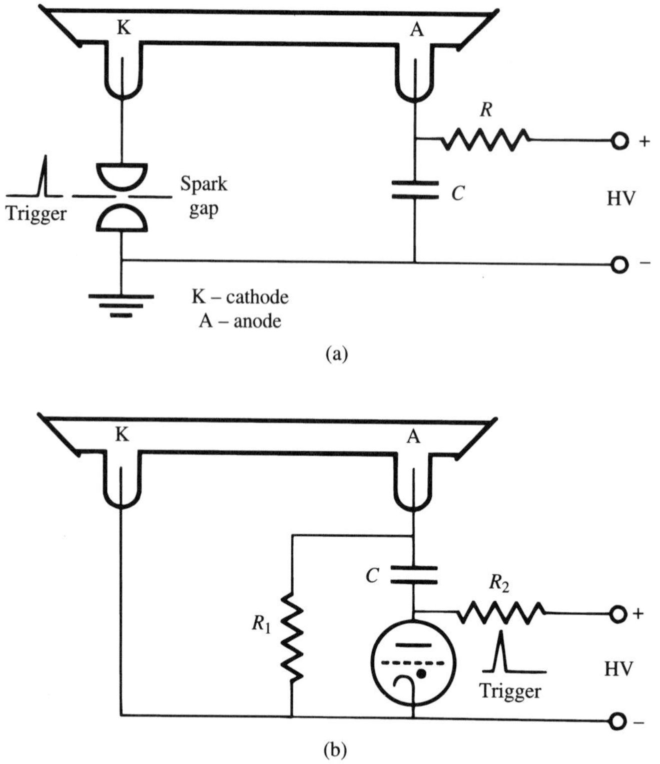

Fig. 9.6. (a)Schematic arrangement for pulsed excitation of a gas laser tube in which the switching element is a triggered spark gap. The charging time constant RC is chosen to be less than the inverse of the operating pulse repetition frequency (prf). (b) As (a) but with a thyratron as the switching element. The charging time constant R_2C is chosen less than $1/\text{prf}$. R_1 provides virtual grounding of the cathode during each charging cycle: $R_1 < CR_2$. However, R_1 is chosen to be larger than the impedance of the operating discharge. In high prf operation the charging resistors can be replaced by chokes (inductors).

they operate best under discharge conditions of current and pressure very different to those found in neutral gas lasers. By operating under conditions of extreme ionization, laser action between levels of multiply ionized species can been obtained, for example in oxygen O^{4+}(OV). Laser action can be obtained most easily in such systems by pulsed discharge techniques, where a capacitor charged to high voltage is switched through the laser tube with a thyratron or spark gap as illustrated in Fig. (9.6). In this mode of operation, because the overall duty cycle of the discharge tube is small, thermal dissipation problems are small and the technology of the laser tube is simple, requiring only pyrex glass, or perhaps more refractive quartz tubulation, with ambient cooling. Current densities of several tens of thousands of A cm^{-2} are easily handled. Some interesting laser transitions that can be excited in this way are listed in Table (9.1).

9.7 CW Ion Lasers

The most important lasers in this category are the argon, krypton, xenon, and neon ion lasers, a list of whose most important lines and typical relative power outputs are given in Table (9.2). Because these lasers operate best at very high current densities, usually in excess of 100 A cm^{-2} and have very low overall electrical efficiency $\lesssim 0.1\%$ the amount of heat that is dissipated in the laser tube is very high and requires selection of special refractory materials for the laser tube, as well as careful attention to thermal stress and cooling problems. In the first CW ion lasers, water cooled fused quartz tubes were found adequate

Table 9.1. *Some interesting pulsed ion laser transitions. These data were obtained using a 7 mm bore by 1.5 m long discharge tube*[9.8].

Lasing species	Wavelength (nm)	Current density (A cm^{-2})	Optimum gas pressure (mTorr)	Peak output power (W)
Ne III	247.34	5700	30	140
Ne II	332.37	3000	13	23
Ne II	337.83	3400	13	21
Ar IV	211.4	>8000	12	20
Ar III	351.11	~7200	10	1000
Ar III	363.79	5700	8	240
Ar II	457.94			
Ar II	476.49			
Ar II	487.99			
Ar II	496.51	500	6	~10
Ar II	501.72			
Ar II	514.53			
Kr IV	219.19	>9800	8	600
Kr IV	225.46	>7800	6	80
Kr III	350.74	~6200	8	130
Kr II	743.58	>7800		400
Xe III or IV	231.54	~5700	5	1400
Xe IV	324.69	>8500	5	260
Xe IV	333.09	>8000	5	330
Xe IV	364.54	>8500	7	3600
Xe IV	430.58	>8000	11	1000
Xe IV	495.42			
Xe IV	500.78			
Xe IV	515.91			
Xe IV	526.02	~9400	11	~1000
Xe IV	535.29			
Xe IV	539.46			
O II	374.95	~3100	15	135
O III	375.99	4800	25	170
O III	559.24	4800	30	180
N IV	347.88	4800	16	400
N III	409.73	3900	11	230

but subsequently demand for higher laser power outputs, and hence power inputs, showed that other materials had greater strength, thermal shock resistance, could withstand greater thermal loading and the internal erosion caused by the gaseous discharge plasma itself, and were easier to cool. The various materials that have been used in modern ion lasers include beryllium oxide (BeO), graphite, discharge-confining structures made up of separate spaced metal disks, and to a lesser degree familiar refractory ceramics based on alumina (Al_2O_3).

Table 9.2. *Some important CW ion laser transitions with examples of their operating characteristics*[9.9],[9.10].

Lasing species	Wavelength (nm)	Operating conditions	Power output (W)
Ne II	332.4	1.7 mm bore by 0.34 m	0.1
Ne II	337.8	long tube, 18 A,	0.03
Ne II	339.3	1100 G, 0.7 Torr	0.003
Ar III	351.1	6 mm bore by 1.85 m	7.2
Ar III	363.8	long tube, 480 A	8.8
		0 G, 0.7–1.4 Torr	
Ar II	457.9	3 mm bore tube by 1 m	0.11
Ar II	465.8	long tube 250 mTorr, 20 A,	0.05
Ar II	472.7	500 G	0.65
Ar II	476.5		0.225
Ar II	488.0	3 mm bore by 1 m long tube 250 mTorr, 20 A 500 G	1.0
		10 cm bore by 1 m long tube 50 mTorr, 150 A, 0 G	100
Ar II	496.5	3 mm bore by 1 m long	0.18
Ar II	501.70	tube, 250 mTorr, 20 A	0.11
		500 G	
Ar II	514.5	similar to 488 nm line above	
Kr II	350.7	6 mm bore by	5.36
Kr II	356.4	1.85 m long tube,	1.34
		440 A, 0 G, 0.6–1.2 Torr	
Kr II	406.7	1.7 mm bore by 0.34 m long tube, 20 A, 1200 G, 0.35 Torr	0.045
Kr II	476.2	3 mm bore by 1 m long tube,	0.04
Kr II	482.5	500 mTorr, 20 A, 500 G	0.025
Kr II	520.8		0.055
Kr II	530.9		0.15
Kr II	568.2		0.11
Kr II	647.1		0.38
Kr II	676.4		0.09
Kr II	752.5		0.075
Xe III	374.6	6 mm bore by 1.85 m long	0.18
Xe III	378.1	tube, 340 A, 0 G, 0.8–1.1	1.62
		Torr	

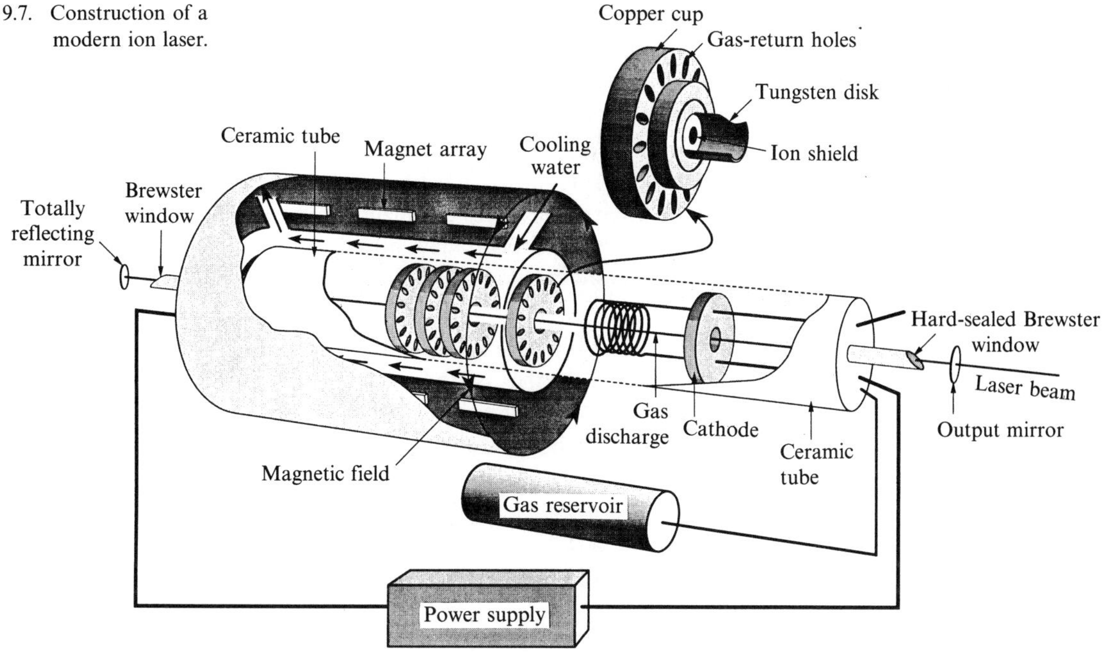

Fig. 9.7. Construction of a modern ion laser.

Over time there have been many variations in plasma tube design in these lasers, but Fig. (9.7) represents what at present has become the most reliable and accepted structure for high current density operation. The discharge is confined by a series of tungsten disks that cool radiatively inside a water cooled ceramic outer gas envelope. Some additional features of the design are worthy of comment:

(i) The tube has Brewster windows, familiar in almost all gas laser systems. Systems have been constructed with antireflection coated perpendicular windows. Such windows do not impose a preferred polarization direction on the laser output, which as a consequence tends to be unpolarized, or sometimes circularly polarized. The power output of the laser is not generally increased by the use of perpendicular windows instead of Brewster windows.

(ii) Inversion occurs in a fairly dense gas discharge plasma produced by passing a high current through the central holes in a line of spaced tungsten disks. Provided the central hole diameter to disk thickness is large enough the discharge passes preferentially through the hole rather than jumping from disk to disk. Structures have also been built in which the gas discharge tube forms one arm of a current loop that acts as the secondary winding of a high frequency transformer[9.10].

(iii) The current density in the capillary region is increased by the application of an axial magnetic field whose magnitude does not generally exceed 0.1 T. The optimum field strength is governed by other factors such as gas pressure, discharge current, tube diameter and the magnitude of Zeeman splitting of the laser transition induced by the field.

(iv) The high current flowing through the gas heats it considerably, so that it expands from the capillary region into the cooler parts of the system, leaving an active particle density in the capillary region much lower than the static filling pressure of the system would indicate.

(v) The high discharge current drives neutral particles towards the anode and creates a pressure differential across the tube that must be prevented as much as possible by the use of an external pressure-equalizing flow path (through which the main discharge current should show no tendency to go). This is formed in this case by internal bypass holes built into the disk structure of the segmented laser tube.

(vi) The high discharge currents (>10 A) for operation are generally supplied by a heated thermonic emitting cathode, either of the oxide coated or dispenser type (a cathode with a porous tungsten matrix, which is impregnated with emitting material that continually diffuses to and replenishes the emitting surface).

(vii) The main discharge capillary is water-cooled indirectly, by radiational cooling of the hot confining disks through an outer water-cooled ceramic envelope. Lower power devices can operate with forced air cooling.

(viii) The hot discharge region should be thermally isolated from the resonator system, otherwise thermal instability of the resonator length, particularly during the warm up period just after the laser has been turned on, will produce unstable operation.

(ix) Efficient laser action, at the 100 W level in the blue–green region, from tubes a little over a meter long, can be obtained from argon lasers operated in relatively wide-bore water-cooled quartz tubes at very high currents (>100 A). Krypton offers slightly less powerful operation in the yellow and red. A mixture of Kr and Ar can be used to provide a 'white' light laser output that is useful for TV and wide-screen display purposes.

(x) CW ultraviolet laser action can be obtained at higher current densities than for visible operation, for example, from excited singly ionized neon or doubly ionized argon and krypton.

(xi) Reactive gases can be excited in a radio frequency external electrode fused quartz system. Internal thermonic cathodes cannot withstand the chemical reactivity of gases such as chlorine.

It should be pointed out that there are significant physical differences between the low-current-density, glow-discharge plasmas used to excite helium–neon, neutral noble gas, carbon dioxide, carbon monoxide and other similar neutral species lasers, and the high-current, low-pressure, quasi-arc discharge used to excite noble gas and other similar ion lasers. The different properties of these gas laser discharge plasmas are discussed in a later section.

9.8 'Metal' Vapor Ion Lasers

These are ion lasers where the laser transitions occur between singly ionized states of (usually low ionization potential) metallic or semimetallic elements. They are excited by collisional processes involving helium or neon metastable atoms and ions in a discharge consisting largely of helium (or neon) with admixture of a

Table 9.3. *Some important metal vapor ion laser transitions with their operating characteristics*[9.10].

Lasing species	Wavelength (nm)	Excitation mechanism	Typical operating characteristics	Power output (mW)
Cd II	325.0	Penning	3.4 Torr He, ∼2	20
Cd II	441.6	ionization	mTorr Cd 2.4 mm	200
		(PI)	bore by 143 cm long	
			tube, 110 mA	
			cataphoresis	
			discharge	
Zn II	491.2	Charge	16 Torr He, 200 mTorr	28
Zn II	492.4	transfer (CT)	Zn 6.3 mm bore by	
			85 cm long hollow	
			cathode	
			discharge, 4.8 A	
Zn II	589.4	PI	1.5 mm by 50 cm long tube	4
Zn II	747.9	PI	∼ 10 mTorr Zn, ∼2	15
			Torr He, 70 m A	
			cataphoresis	
			discharge	
Se II	497.6	CT	7 Torr He, 5 mTorr Se	30
Se II	499.3	CT	3 mm bore by 2 m long tube	30
Se II	506.9	CT	400 m A cataphoresis	50
Se II	517.6	CT	discharge	50
Se II	522.7	CT		50
Se II	530.5	CT		30

small amount of the metal vapor. The excitation processes of greatest importance are, for example, in the case of helium/metal vapor discharges

$$M + He^*_m \rightarrow M^{+*} + He + e,$$

where He_m is a metastable helium atom and M^{+*} is the upper laser level. This process is called Penning ionization[9.11]. Charge transfer is also an important excitation mechanism:

$$M + He^+ \rightarrow M^{+*} + He.$$

Large cross-sections exist for these processes when the energy of the final state M^{+*} is close ($< kT$) to the energy of the exciting particle He^*_m or He^+ (or the corresponding neon species[9.12],[9.13]).

Materials that have been excited into laser action in this way include cadmium, selenium, tellurium, iodine, arsenic, zinc, lead, tin, mercury copper, silver, and gold[9.14]. The charge transfer process is the more common of the two above excitation mechanisms. As an excitation process it offers the significant advantage over Penning ionization of having a much higher current saturation level since the exciting species are ions, rather than neutral metastables. The latter, for reasons previously discussed, have populations that saturate at low current levels.

Some of the more important metal vapor laser transitions are listed in Table (9.3) together with the discharge in which they are observed and their dominant excita-

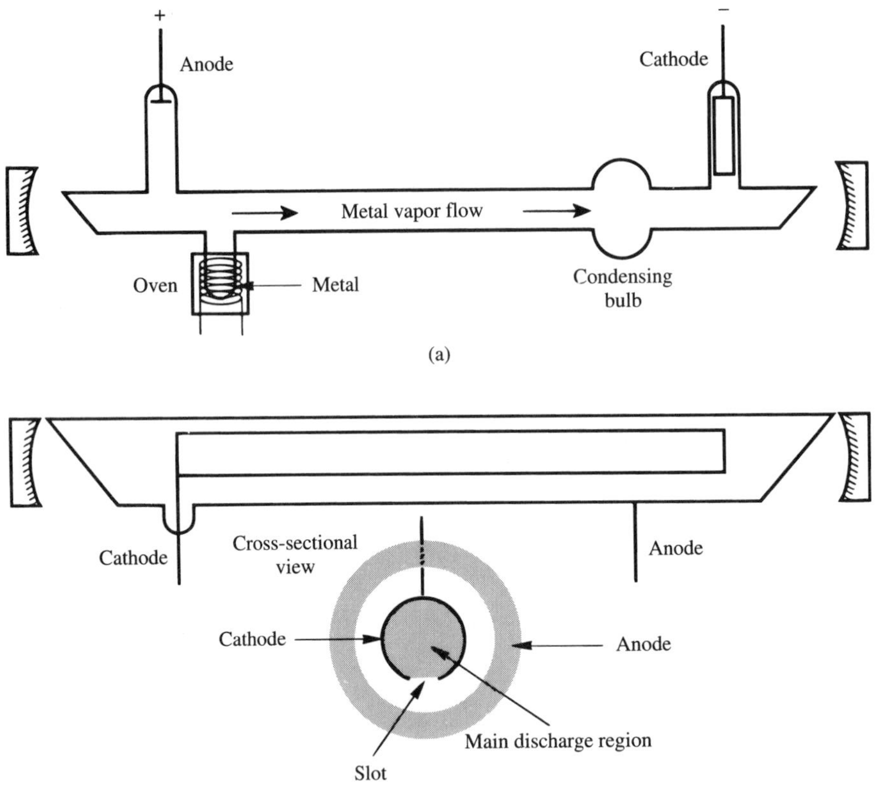

Fig. 9.8. Metal vapor laser structures: (a) cataphoretic flow metal vapor laser; (b) slotted hollow cathode metal vapor laser.

tion mechanism. In many respects these lasers are comparable with the helium-neon laser: they operate in relatively low current density glow discharges ($\lesssim$ few A cm^{-2}) and offer comparable power outputs and efficiencies. The He–Cd laser does however offer the two very attractive short wavelengths 441.6 nm ($\sim$ 100 mW from a 1 m tube) and 325 nm ($\sim$ 3 mW from a 1 m tube) while the He–Se laser offers slightly greater efficiency for the production of blue–green laser radiation than the argon-ion laser at low power levels ($\lesssim$ 100 mW).

The discharge structures in which laser action is excited in metal vapor lasers are of two main types: (a) cataphoretic flow and (b) hollow cathode, which are shown in Fig. (9.8). In the cataphoretic flow discharge, metal atoms are evaporated from an oven at the anode end of a helium or neon gas discharge at a few torr pressure and are swept along the discharge towards the cathode by cataphoresis (a process where heavy atoms in a glow discharge at relatively low current experience a net cathode directed force). If the conditions of operation are optimum, a uniform distribution of a few mTorr of metal atoms occurs throughout the positive column of the discharge. When they reach the cathode end of the positive column the metal atoms largely recondense, so it is possible to use a thermionic cathode, although a large cold cathode (usually aluminum) is adequate to provide the currents necessary for laser oscillation.

In the hollow cathode structure the excitation region consists largely of the extended negative glow inside the cathode. This is a region containing large numbers of energetic electrons, which generally makes it more suitable for the

Fig. 9.9. Schematic I–V characteristic of a gas discharge over several orders of current density.

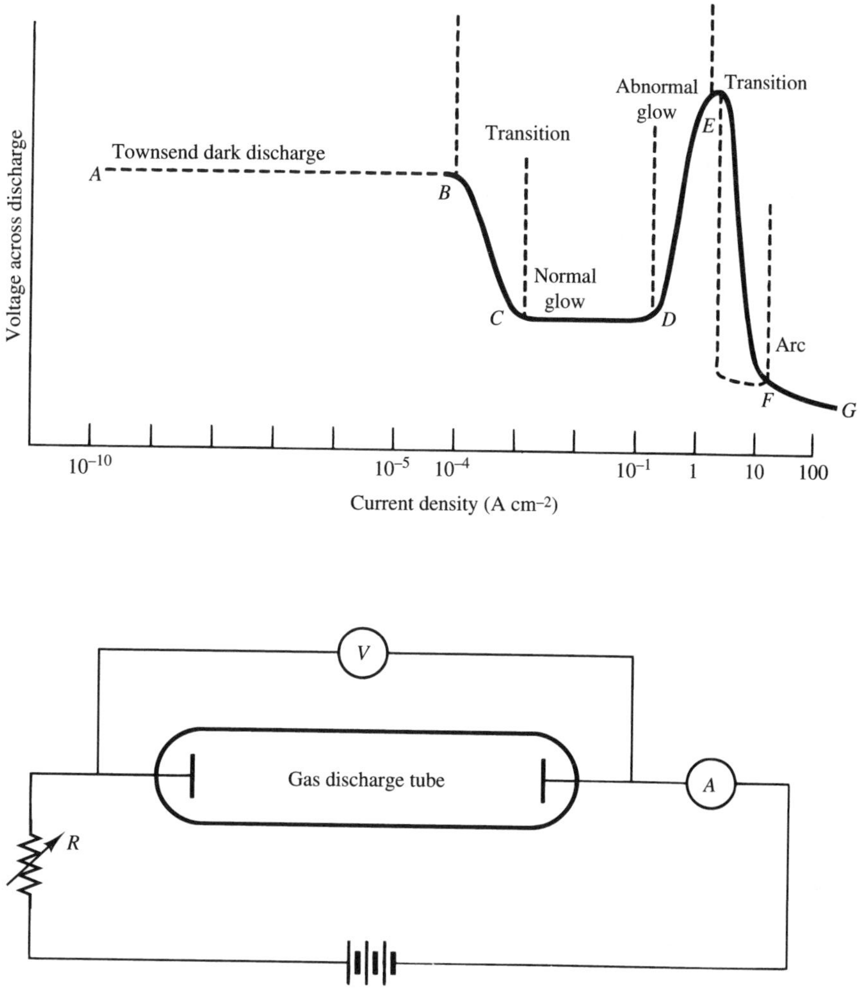

operation of charge transfer excited lasers in which the production of helium ions is necessary.

9.9 Gas Discharges for Exciting Gas Lasers

If one steadily increases the current through a gas discharge and plots the resulting I–V characteristic of the tube the schematic behavior illustrated in Fig. (9.9) is observed. The general experimental arrangement for conducting such an experiment is also shown in the figure. In the normal glow and arc regions the I–V characteristic is almost flat, or has a negative slope, so the series ballast resistor is very important to prevent current runaway. The various regions of operation of the discharge illustrated in Fig. (9.9) are characterized by particular physical phenomena:

(i) $A \rightarrow B$: The Townsend dark discharge occurs when the number of electrons flowing through the gas is insufficient, and their mean energies are too low, for significant secondary ionization, and subsequent cumulative current growth, to occur. The current flow will consist mainly of the drift of residual free charges in the gas, for example, those produced by cosmic radiation and radioactivity of the discharge tube and its gas fill.

(ii) $B \rightarrow C$: Cumulative secondary ionization grows and increases the current flow and the necessary operating voltage falls.

(iii) $C \rightarrow D$: The normal glow region. The discharge is now self-sustaining by secondary ionization; the current is liberated from the cathode by ion bombardment (plus thermionic emission if the discharge is being operated with a thermionic cathode). The normal glow is characterized by a constant current density at the cathode surface (and in the discharge tube itself if this allows expansion of the column of conducting gas with increase of current). Because the normal glow has such a flat I–V characteristic it is essential to stabilize the current with a ballast resistor.

(iv) $D \rightarrow E$: When the whole cathode surface has been covered in the operation of a glow discharge the substaining voltage must now be increased to provide additional current flow.

(v) $E \rightarrow F$: The transition to arc operation can occur abruptly and region EF may not be observed very readily in practice.

(vi) $F \rightarrow G$: The arc is characterized by high current flow and low sustaining voltage. Emission from the cathode, if this is not provided by external heating of thermionic materials, can occur because of the production, by ion bombardment, of local thermionic emitting hot spots. The gas discharge plasma of the arc is becoming highly ionized.

In the region of glow and arc discharge operation, the voltage/length characteristic of the gas discharge is schematically of the form shown in Fig. (9.10). The bulk of the region between anode and cathode is filled by the luminous positive column which is a quasi-neutral plasma. In glow discharge excited lasers the charged particle density in this region is typically 10^{22}–10^{24} particles m^{-3}. In the arc region the charged particle density is typically $> 10^{25}$ particles m^{-3}. Although the arc is a more highly ionized plasma than the glow discharge, the electron temperature in the plasma is not necessarily higher.

In the region of glow discharge operation the discharge is characterized by a number of distinct luminous and nonluminous regions, as illustrated in Fig. (9.11). Electrons that leave the cathode pick up energy in the cathode-fall until they acquire sufficient energy to excite visible radiation from the particles of the gas, this produces the negative glow. The region near the cathode where they have insufficient energy to excite such radiation is the cathode dark space. In the negative glow the electrons lose energy by exciting gas particles until they no longer have sufficient energy to excite visible radiation, so another dark space, the Faraday dark space, results before the electrons have picked up sufficient energy to excite visible radiation again at the head of the positive column. This successive acceleration and deceleration of the discharge electrons can cause the positive column to break up into bright and dark segments called *striations.*

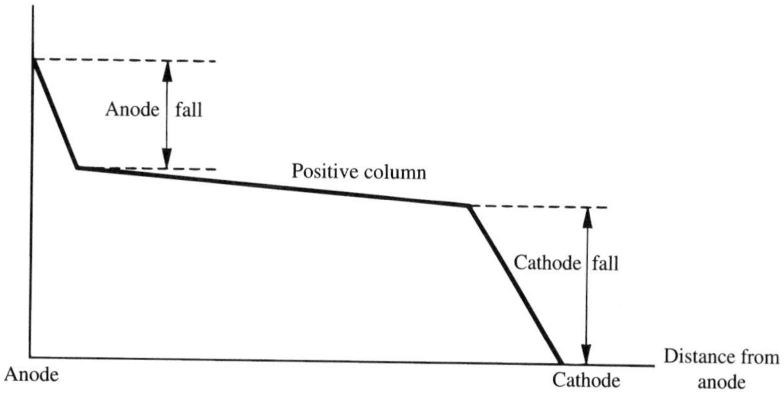

Fig. 9.10. Schematic voltage drop with position along a gas discharge tube in the glow or arc region.

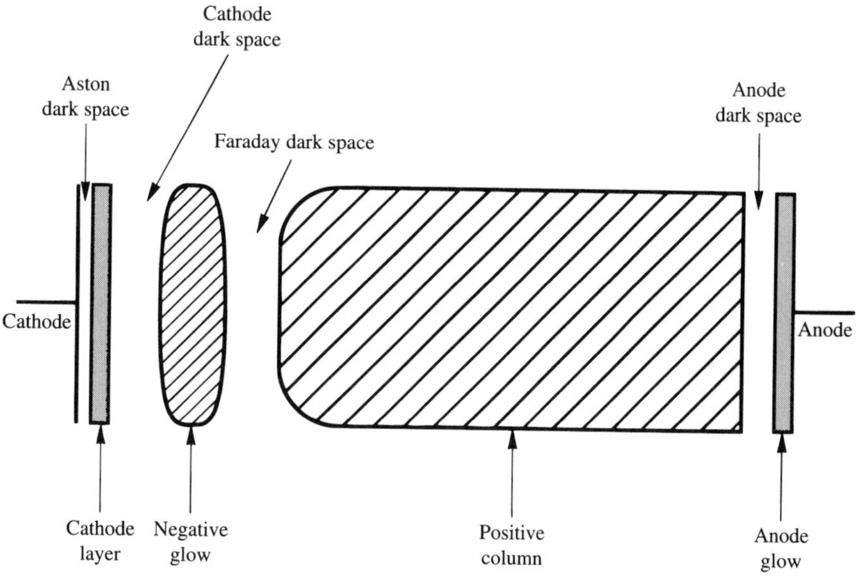

Fig. 9.11. Schematic appearance of a glow discharge in a cylindrical tube showing the various regions of the discharge.

9.10 Rate Equations for Gas Discharge Lasers

The discharge plasma of an electrically excited gas laser is characterized by a number of parameters that vary from one type of laser to another and crucially affect the way in which such different lasers operate.

Since the electrons of the discharge are fundamentally responsible, directly or indirectly, for the excitation of both the upper and lower laser levels, the energy distribution of these electrons and their density are crucial in determining whether a population inversion can be produced in the gas discharge. It is possible to formulate simple rate equations that describe the excitation of the upper and lower levels of a gaseous laser system. Let us consider the simple system shown in Fig. (9.12) where both upper and lower levels of the laser transition are populated by electron collision; this will allow us to see how the operation of the system is affected by changes in the electron energy distribution and density.

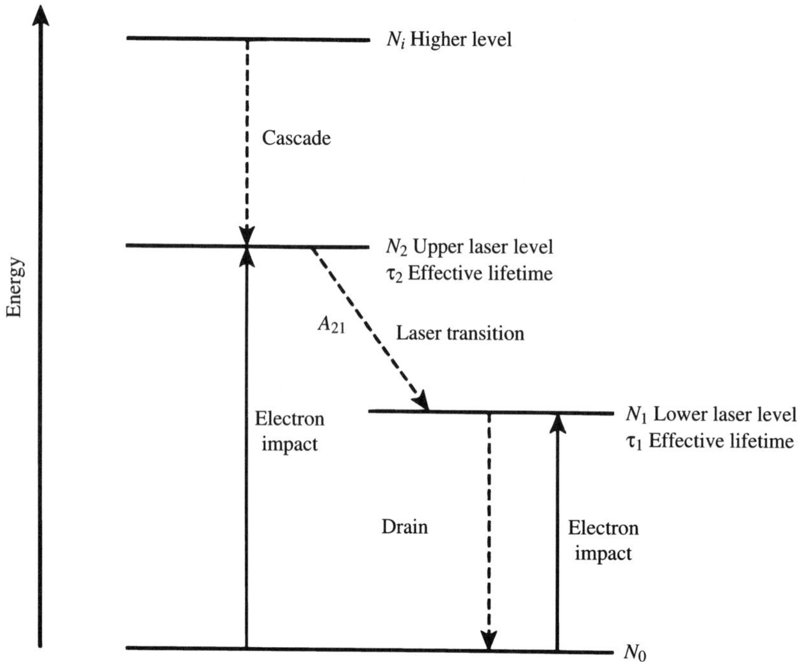

Fig. 9.12. Schematic energy level diagram of a gas laser system in which levels are excited by electron collisions.

Let the ground state density in the system be N_0, of the upper and lower laser levels N_2 and N_1, respectively, and of a general higher lying level N_i. The electron density is N_e. The rate of change of population of the upper laser level is:

$$\frac{dN_2}{dt} = N_0 N_e \langle \sigma_{01} v \rangle + \sum_i N_i A_{i1} - \frac{N_1}{\tau_1}, \tag{9.4}$$

where the summation over i here includes the upper laser level and $\langle \sigma_{01} v \rangle$ is the average of $\sigma_{01} v$ over the electron velocity distribution. In equilibrium

$$\frac{dN_2}{dt}, \frac{dN_1}{dt} = 0. \tag{9.5}$$

The condition for population inversion is that

$$\frac{N_2}{g_2} - \frac{N_1}{g_1} > 0. \tag{9.6}$$

Therefore,

$$\frac{\tau_2}{g_2}\left[N_0 N_e \langle \sigma_{02} v \rangle + \sum_i N_i A_{i2}\right] > \frac{\tau_1}{g_1}\left[N_0 N_e \langle \sigma_{01} v \rangle + \sum_i N_i A_{i1}\right]. \tag{9.7}$$

If we neglect the radiative cascade processes, this reduces to

$$\langle \sigma_{02} v \rangle > \frac{\tau_1}{\tau_2}\frac{g_2}{g_1}\langle \sigma_{01} v \rangle. \tag{9.8}$$

Thus the ease of obtaining population inversion is assisted by (i) a favorable cross-section ratio and (ii) a favorably small lifetime ratio, τ_1/τ_2, and ratio of degeneracy

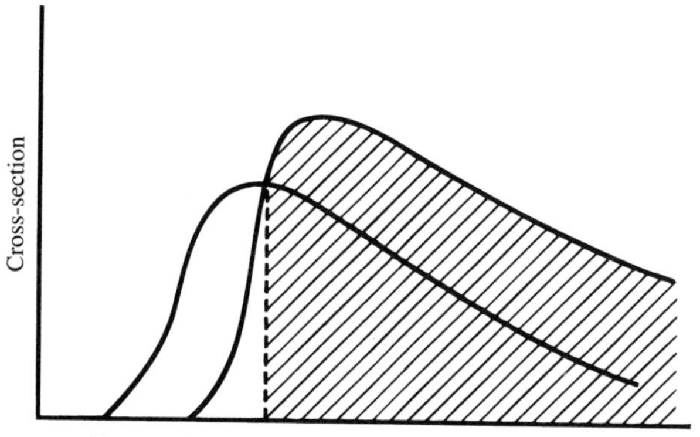

Fig. 9.13. Schematic variation of the collision cross-section for excitation of upper and lower laser levels as a function of electron energy. Electrons with energy above E_m are more efficient at exciting the upper laser level than the lower. In this case inversion is not reduced by an increase in electron energy.

factors, g_2/g_1. The magnitude of the achieved population inversion is

$$\left(N_2 - \frac{g_2}{g_1}N_1\right) = N_0 N_e \left(\langle\sigma_{02}v\rangle\tau_2 - \langle\sigma_{01}v\rangle\frac{g_2}{g_1}\tau_1\right), \tag{9.9}$$

which increases with electron density.

Let us examine the cross-section ratio in a little more detail. The cross-section for the upper and lower laser levels may be of the form shown schematically in Fig. (9.13), in which case a favorable cross-section ratio exists for electron energies above E_m and an optimum electron energy distribution exists for the velocity averaged cross-section difference given by $\langle\sigma_{02}v\rangle - \langle\sigma_{01}v\rangle$ to be as large as possible. If the electron energy distribution is Maxwellian and can be characterized by an electron temperature T_e, then in such a system as the above, increase of the electron temperature will not automatically destroy the inversion, but it will eventually reduce its magnitude. An optimum value of electron temperature exists for the production of the maximum inversion.

In some laser systems the cross-sections may instead be of the form shown schematically in Fig. (9.14). A situation such as shown in Fig. (9.14) can arise when the electron impact process exciting the upper laser level corresponds to an 'optically forbidden' transition and where the cross-section for the lower laser level corresponds to an 'optically allowed' transition. For example, in the two-step excitation of the singly-ionized noble gas ion laser levels from the ion ground state, the excitation of the upper laser levels is 'optically forbidden', since these levels do not make electric dipole transitions to the ground state. The excitation of the lower laser levels from the ion ground state is allowed, as one would expect, since the lower laser levels make very strong, short life-time optical transitions to the ground state. In systems of this kind an inversion is only possible (in our simplified model) for a small range of electron temperatures and continued increase of electron temperature will ultimately destroy the inversion.

The electron temperature that prevails in a gas discharge is mainly a function of the electric field gradient E and the pressure p. The electric field influences the rate at which electrons pick up energy between collisions while the pressure controls

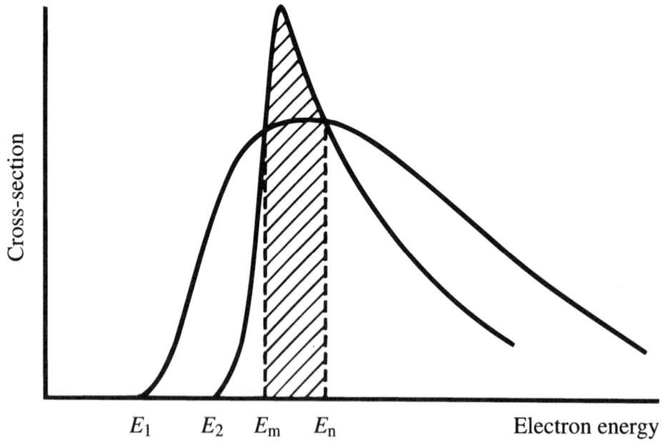

Fig. 9.14. Schematic variation of excitation cross-section with electron energy for upper and lower laser levels where inversion can only be achieved over a specific range of electron energies.

the mean free path between these collisions. It is clear that

$$T_e = f\left(\frac{E}{p_0}\right), \tag{9.10}$$

where p_0 is the pressure that corresponds to the (hot) particle density in the tube reduced to standard temperature and pressure (STP).

In CW argon ion lasers the optimum electron temperature is in the range 2–6 eV; typical operating conditions involve axial fields $\sim$ 1 kV m^{-1} and pressures $p_0 \sim 50$ mTorr, i.e., $E/p_0 \sim 200$ kV m^{-1} Torr^{-1} (the actual filling pressure of the laser is much higher than p_0 but gas heating drives gas from the hot discharge capillary).

Gas discharges used to excite lasers are quasi-neutral, the mean electron and ion densities at any point in the plasma are equal, and unless the gas is highly ionized the density of neutral atoms exceeds that of the charged particles. Some of the detailed aspects of laser operation are controlled by the temperature (and consequent Doppler broadening) of these particles in the plasma and their spatial distribution and velocities.

Although some gas lasers are not excited in cylindrical structures, most gas lasers utilize discharges in cylindrical tubes for their operation. Generally speaking the longer the wavelength of the laser transition the wider the bore of the discharge tube that will be used to excite it. This is particularly a consequence of the Fresnel diffraction condition for low loss transverse mode operation, i.e., that

$$\frac{D}{\lambda \ell} \gtrsim 1, \tag{9.11}$$

where D is the diameter of the discharge tube and ℓ is its length. A larger tube bore is also consistent with the generally lower electron densities needed to achieve population inversion in lasers operating at longer wavelengths. This allows wider bore tubes to be used at the same overall current, and for a greater volume of gas to be excited.

9.11 Problems

(9.1) In the helium–neon laser helium metastables (2^1S) transfer their excitation to neon atoms producing a population of neon atoms in the 5s$'$ $\left[\frac{1}{2}\right]_1^o$ state.†

† In the notation for the neon states the subscript is the total angular momentum J of the state, so $g = 2J + 1$.

Two important laser transitions originate from this state.

$$5s'\left[\tfrac{1}{2}\right]_1^o \rightarrow 3p'\left[\tfrac{3}{2}\right]_2 \text{ at } 632.8 \text{ nm}$$

$$5s'\left[\tfrac{1}{2}\right]_1^o \rightarrow 4p'\left[\tfrac{3}{2}\right]_2 \text{ at } 3.39\ \mu\text{m}.$$

(a) Calculate the minimum length of laser tube required to produce laser oscillation at each of these two wavelengths. (b) What would happen if a small quantity of methane that produced an absorption of 4 m^{-1} at 3.39 μm were added to the helium–neon gas mixture. Consider the longitudinal mode at the line centre only. You may assume: Helium metastables are only destroyed by transfer of their energy to neon atoms and by collisions with the discharge tube walls. The only important collisional process involving neon atoms is the energy transfer process from helium (2^1S) metastables. The lower levels of both laser transitions are unpopulated. Given:

discharge tube diameter = 7 mm,
helium pressure = 1 Torr,
gas temperature = 400 K,
mirror reflectances = 98% at 632.8 nm), 28% (at 3.39 μm)
rate of production of helium (2^1S) metastables = 10^{23} m^{-3} s^{-1}
rate coefficient for transfer process 4.8×10^{-18} m^3 s^{-1}
(He(2^1S) + Ne (ground state) → Ne ($5s'\left[\frac{1}{2}\right]_1^o$ + He (ground state))
lifetime of $5s'\left[\frac{1}{2}\right]_1^o$ state = 62.5 ns,
Einstein A coefficients for the transitions: 632.8 nm : 1.4×10^6 s^{-1}, 3.39 μm: 9.6×10^5 s^{-1}.

Hint: Use kinetic theory to estimate the average column destruction rate of helium metastables at the discharge tube wall.

The rate coefficient k_T for the collisional transfer reaction (a) is defined from

$$\text{Rate of production of Ne } 5s'\left[\tfrac{1}{2}\right]_1^o \text{ states vol}^{-1} =$$

$$k_T\left[\text{He }(2^1S)\right]\text{[Ne(ground state)]}$$

where [He(2^1S)] and [Ne(ground state)] are the population densities (atoms vol^{-1}) in the helium metastable and neon ground states respectively.

References

[9.1] S. Jacobs, P. Rabinowitz, and G. Gould, 'Coherent light amplification in optically pumped Cs vapor,' *Phys. Rev. Lett.*, **7**, 415–417, 1961.

[9.2] P. Rabinowitz, S. Jacobs, and G. Gould, 'Continuously optically pumped Cs laser,' *Appl. Opt.*, **1**, 511–516, 1962.

[9.3] N. Djeu and R. Burnham, 'Optically pumped CW Hg laser at 546.1 nm,' *Appl. Phys. Lett.*, **25**, 350–351, 1974.

[9.4] M. Artusy, N. Holmes, and A.E. Siegman, 'D.C.-excited and sealed-off operation of the optically pumped 546.1 nm Hg laser,' *Appl. Phys. Lett.*, **28**, 1331–1334, 1976..

[9.5] T.Y. Chang and T. Bridges, 'Laser action at 452, 496, and 541 μm in optically pumped CH_3F,' *Opt. Commun.*, **1**, 1117–1118, 1970.

[9.6] T.Y. Chang, 'Optically pumped submillimeter sources,' *IEEE Trans. Microwave Theory Tech.*, **MTT-22**, 983–988, 1974.

[9.7] D.T. Hodges, 'A review of advances in optically pumped far IR lasers,' *Infrared Phys.*, **18**, 375–384, 1978.

[9.8] J.B. Marling, 'Ultraviolet ion laser performance and Spectroscopy-Part I: New Strong noble gas transitions below 2500Å,' *IEEE J. Quant. Electron.*, **QE-11**, 822–834, 1975.

[9.9] T.K. Tio, H.H. Luo, and S.C. Lin, *Appl. Phys. Lett.*, **29**, 795–797, 1976.

[9.10] C.C. Davis and T.A. King, 'Gaseous Ion Lasers,' in *Advances in Quantum Electronics*, D.W. Goodwin, Ed., Vol. 3, Academic Press, London, 1975.

[9.11] C.E. Webb, A.R. Turner-Smith, and J.M. Green, 'Optical excitation in charge transfer and Penning ionization,' *J. Phys. B.*, **3**, L134–L138, 1970.

[9.12] D.S. Duffendack and K.T. Thomson, 'Some factors affecting action cross section for collisions of the second kind between atoms and ions,' *Phys. Rev.*, **43**, 106–111, 1933.

[9.13] O.S. Duffendack and W.H. Gran, 'Regularity along a series in the variation of the action cross section with energy discrepancy in impacts of the second kind,' *Phys. Rev.*, **51**, 804–809, 1937.

[9.14] W.B. Bridges, 'Ionized Gas Lasers,' in *Handbook of Laser Science and Technology*, Vol. II, M.J.Weber, Ed., *Gas Laser*, CRC Press, Boca Raton, FL, 1982.

10

Molecular Gas Lasers I

10.1 Introduction

Many important laser systems operate using molecular species. The laser transition occurs between energy levels of the molecule, which may be in the gaseous, liquid, or solid state. To understand more about molecular lasers it is important to consider the additional complexity of the energy level structure of a molecule compared with an atom. In this chapter we will explain the three different kinds of energy level – electronic, vibrational, and rotational, that occur for molecular species and then continue to explain how such a complex energy level structure allows the possibility of laser oscillation over a very broad wavelength range.

10.2 The Energy Levels of Molecules

Electronic Energy States.

In an atom, the orbiting electrons move in the spherically symmetric potential of the nucleus, and the various energy levels of the system correspond, in a simple sense, to different orbital arrangements of these electrons. For example, an excited electron frequently moves into an orbit that takes it further from the nucleus. In a molecule the electrons travel in orbits that surround all the nuclei of the molecule, although quite often there will be considerable localization of some electrons near a particular nucleus. The *electronic* energy states of the molecule result from different arrangements of the orbiting electrons about the nuclei. Electrons that move from one electronic energy level of a molecule to another experience changes in energy that are broadly comparable to such jumps in atoms. The emitted radiation is typically in the ultraviolet–near-infrared region, that is the energy changes involved are on the order of 10 000 cm^{-1}. However, the nuclei of a molecule can also move in ways that give rise to *vibrational* and *rotational* energy states.

Vibrational Energy States.

The nuclei of a stable molecule are held together by forces that make it more desirable for the molecule to stay together rather than to fly apart. The molecule minimizes its energy by remaining as a molecule. To illustrate this, let us consider the simple saturation in which two atoms A and B are brought together. Fig. (10.1) shows the potential energy of the composite $(A + B)$ molecule as the internuclear separation varies. At large internuclear separation the atoms A and B are essentially independent. As A and B approach, two alternatives exist. If the reduction in overall potential energy that results from electrons initially on A and B now orbiting both nuclei offsets the Coulomb repulsion between the A and B nuclei, a *bound* molecule results. This is shown as the potential curve I with the energy

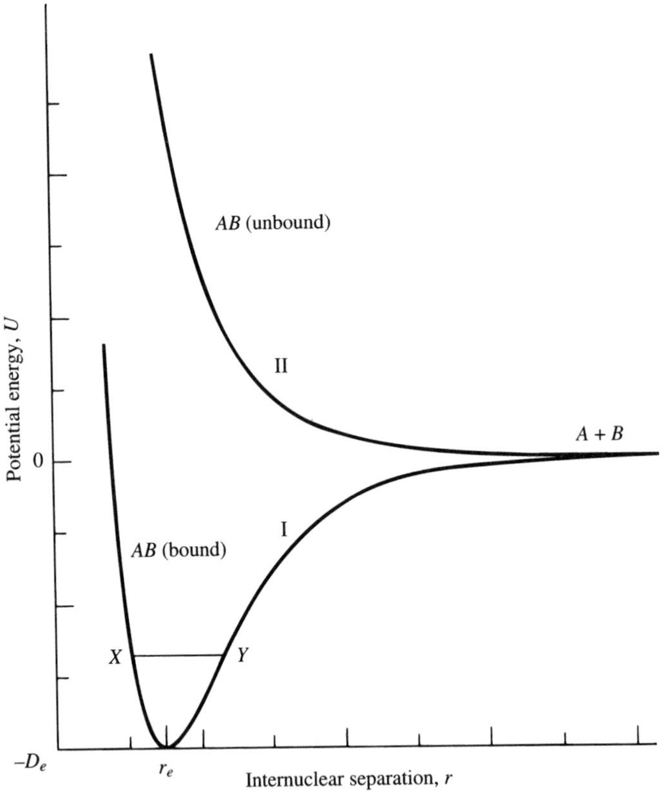

Fig. 10.1. Schematic potential energy level diagram of a diatomic molecule AB.

minimum at internuclear separation r_e. If the nuclei A and B are brought closer still, the potential energy rises again as Coulombic repulsion becomes important. If the depth of the potential dip or *well*, D_e, on curve I is great enough, roughly $> kT$, the molecule will be stable enough to resist a tendency to fly apart because of thermal excitation.

If, as A and B are brought closer from large internuclear separations, the potential energy rises continually, as in curve II, then no *bound* state of the molecule occurs. The nuclei A and B will find it advantageous, in an energy sense, to fly apart. Curve II is referred to an *unbound* state of the molecule AB.

The molecule AB can also be formed by joining together atoms that are themselves in excited states. For example,

$$A + B^* \rightarrow (AB)^*,$$
$$A^* + B^* \rightarrow (AB)^{**}.$$

In this way we can explain the existence of excited electronic states, which may be bound or unbound, as shown in Fig. (10.2).

When a bound molecule, such as AB, sits in the potential minimum on curve I in Fig. (10.1) its nuclei do not remain stationary – they vibrate with respect to each other. We can think of any molecule as a collection of masses M_i held together by springs. Each spring has an equilibrium length r_{ij} and a spring constant k_{ij}. Any displacement of any of the masses from their equilibrium positions will cause the structure to vibrate. The form of these vibrations can be complex for molecules

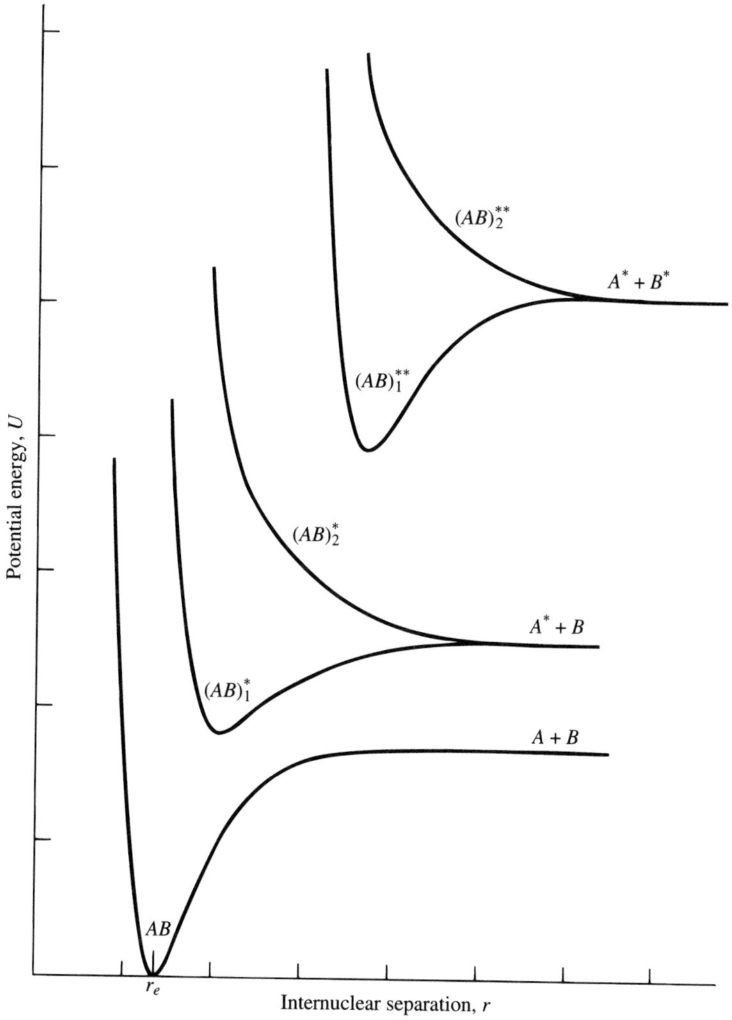

Fig. 10.2. Potential energy level diagram showing the ground and excited electronic states.

with many nuclei. However, for a simple diatomic molecule AB the only possible vibration of the molecule involves a change in the internuclear separation. So, for example, a molecular vibration would correspond to motion along a line such as XY inside the potential well in Fig. (10.1). The line is horizontal because as the molecule vibrates its energy does not change. We refer to a line such as XY as a *vibrational* energy level. We can analyze such vibrational motion with reference to Fig. (10.3).

In equilibrium the spacing of the two nuclei is r_e. At any instant during vibration the spacing is $r(t)$. In the simplest model of the force between the two nuclei it is assumed that the force acting on each nucleus is proportional to the change in internuclear separation. During the vibration the center of mass of the molecule does not move, so we measure the position of each nucleus relative to the center of mass. For nucleus A the equation of motion is

$$m_A \frac{d^2 r_A}{dt^2} = -k[r(t) - r_e], \tag{10.1}$$

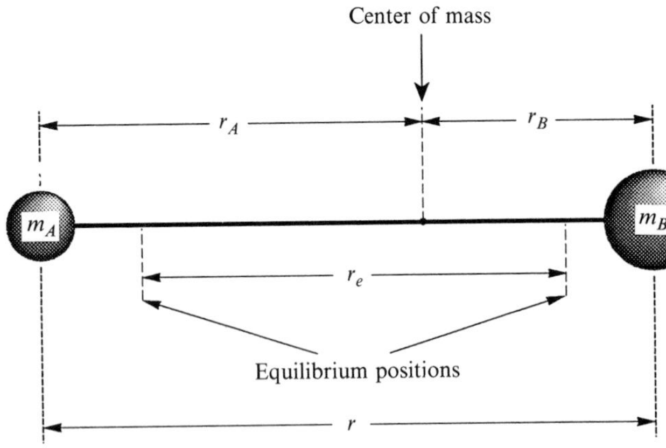

Fig. 10.3. Simple model for a vibrating diatomic molecule.

where k is the *force constant* of the bond between the two nuclei (analogous to the spring constant of a spring connecting two vibrating masses). For nucleus B

$$m_B \frac{d^2 r_B}{dt^2} = -k[r(t) - r_e]. \tag{10.2}$$

The distances to the center of mass must satisfy the following equations

$$r_A + r_B = r, \tag{10.3}$$

$$m_A r_A = m_B r_B, \tag{10.4}$$

which gives

$$r_A = \left(\frac{m_B}{m_A + m_B}\right) r, \qquad r_B = \left(\frac{m_A}{m_A + m_B}\right) r. \tag{10.5}$$

From either Eq. (10.1) or Eq. (10.2)

$$\left(\frac{m_A m_B}{m_A + m_B}\right) \frac{d^2 r}{dt^2} = -k[r(t) - r_e], \tag{10.6}$$

which is generally written in the form

$$\mu \frac{d^2 r}{dt^2} = -k[r(t) - r_e], \tag{10.7}$$

where $\mu = m_A m_B/(m_A + m_B)$ is called the *reduced mass.*

Eq. (10.7) is the equation of a simple harmonic oscillator. The solution to the equation is

$$r = r_0 \sin(\nu_{vib} t + \phi) + r_e, \tag{10.8}$$

where the vibrational frequency $\nu_{vib} = \sqrt{k/\mu}$. This motion is quantized, just as were the oscillations of the radiation field considered in Chapter 1. The only permitted energies of the vibrational motion are

$$E_{vib} = h\nu_{vib}(v + \tfrac{1}{2}). \tag{10.9}$$

v is an integer called the *vibrational quantum number*. These energy levels are equally spaced and are shown inside the simple harmonic potential well in Fig. (10.4). In

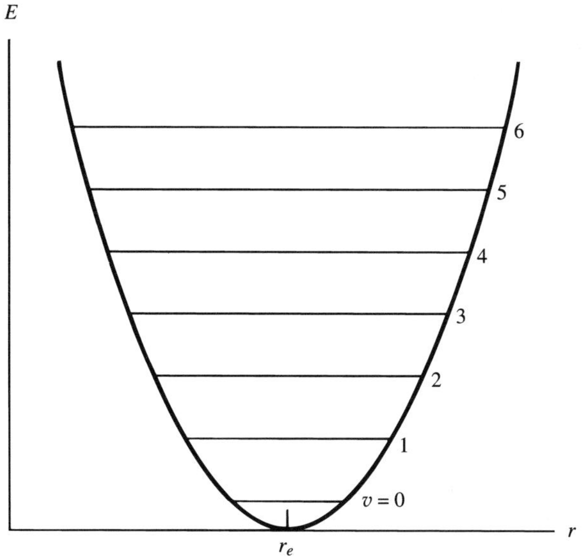

Fig. 10.4. Equally-spaced vibrational energy levels in a simple harmonic potential well.

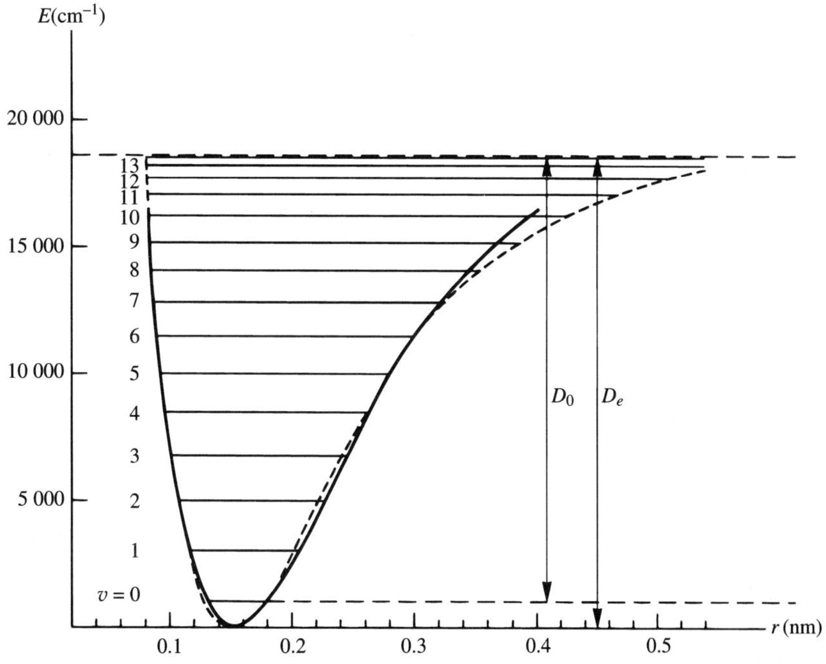

Fig. 10.5. Vibrational energy levels in an anharmonic potential well.

real diatomic molecules the potential well shown in Fig. (10.1) is not of the symmetrical shape shown in Fig. (10.4). In particular, the well has only finite height in the direction of increasing nuclear separation. This reflects the fact that if the molecule vibrates energetically enough it will fly apart. In a real potential well of the shape shown in Fig. (10.5) the discrete vibrational levels are almost equally spaced near the bottom of the well, but become closer together in energy as the vibrational quantum number v increases. A molecule can jump from one

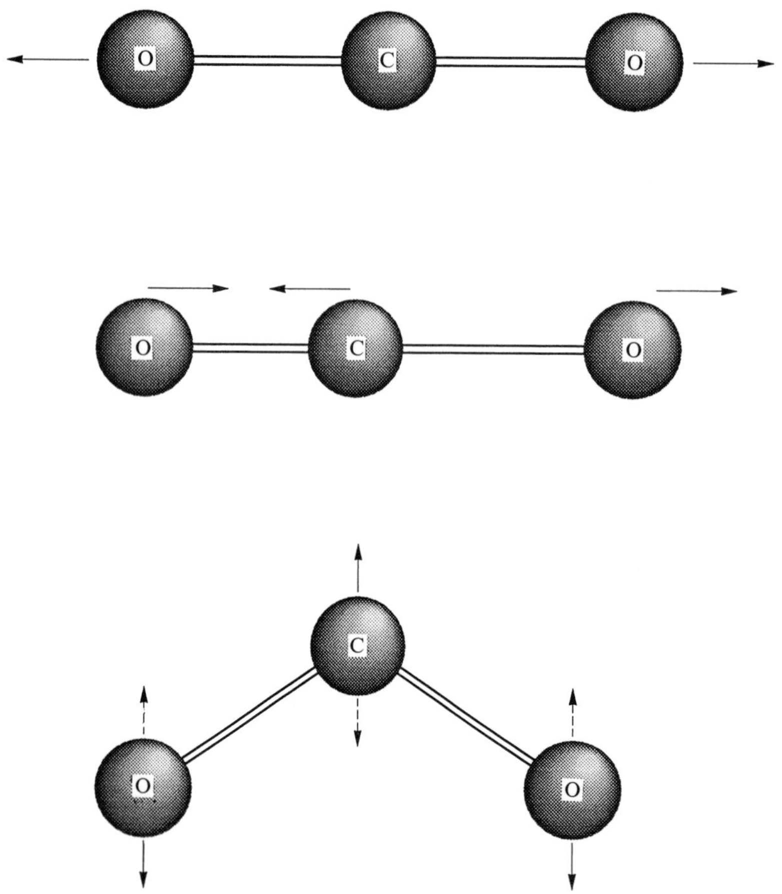

Fig. 10.6. Normal modes of vibration of the CO_2 molecule.

vibrational energy level in the well to another. Usually this process is relatively unlikely except for $\Delta v = \pm 1$. If $\Delta v = +1$ the molecule absorbs a vibrational quantum, if $\Delta v = -1$ it emits a vibrational quantum. The energies of such vibrational quanta are on the order of 1000 cm^{-1}, so the emitted or absorbed radiation is in the infrared region.

10.3 Vibrations of a Polyatomic Molecule

When a molecule with more than two nuclei vibrates, more than one kind of vibration can occur. However, it is always possible to treat any complex vibrational motion of such a polyatomic molecule as a superposition of motions in a finite number of characteristic vibrations called *normal modes.* For example, in the molecule CO_2 shown in Fig. (10.6) where the oxygen, carbon, and oxygen nuclei are in a straight line in the equilibrium position, there are three such normal modes of vibration, also shown in Fig. (10.6). The mode with the highest vibrational frequency is called the *antisymmetric stretch.* In this characteristic motion the length of one C–O bond increases as the other decreases: the molecule remains linear. In the mode of next highest vibrational frequency, the *symmetric stretch*, the vibration is symmetric in the sense that the two oxygen atoms vibrate in antiphase

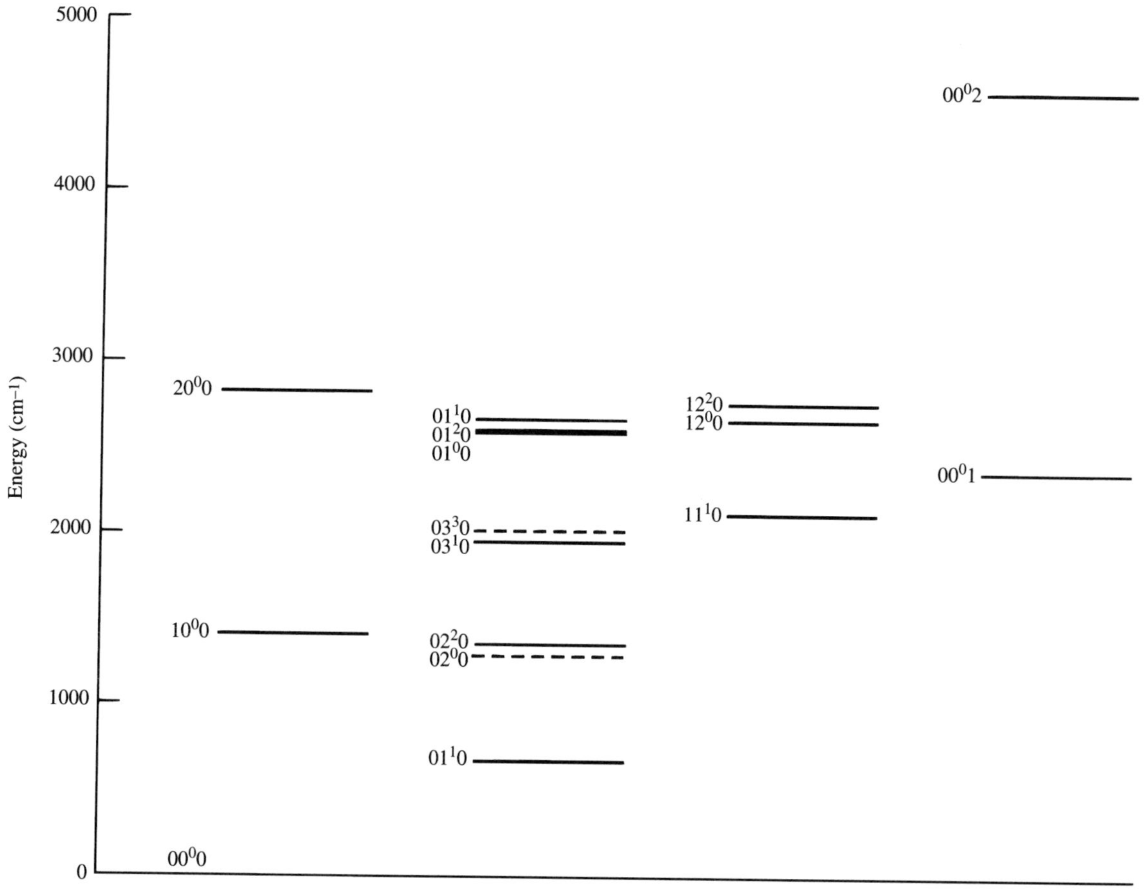

Fig. 10.7. Vibrational energy level diagram of the carbon dioxide molecule in its lowest electronic state.

and the carbon nucleus remains stationary. The lowest vibrational frequency mode is the *bending* vibration in which the molecule deviates from linearity. To describe the energy levels of the CO_2 molecule, it is necessary to specify the number of vibrational quanta present in each mode. For example, a level designated (v_1, v_2, v_3) has v_1 quanta in the symmetric stretching vibration, v_2 in the bending vibration, and v_3 in the asymmetric stretch. Its total vibrational energy is

$$E_{vib}(v_1, v_2, v_3) = h\nu_1(v_1 + \tfrac{1}{2}) + h\nu_2(v_2 + \tfrac{1}{2}) + h\nu_3(v_3 + \tfrac{1}{2}). \tag{10.10}$$

Fig. (10.7) shows the actual vibrational energy levels of the CO_2 molecule in its lowest electronic state. The superscript ° in the designation (10°0), for example, indicates whether the molecule has also acquired angular momentum about its axis as a result of the whirling motion that results if two bending vibrations that are not in phase occur at right angles to each other. A very important series of laser transitions occurs between the (00°1) → (02°0) and (00°1) → (10°0) vibrational levels as we shall discuss in more detail shortly.

In general, a molecule containing N nuclei has $3N - 6$ vibrational normal modes, unless the molecule is linear, in which case there are $3N - 5$: in CO_2, $3N - 5 = 4$: 1 antisymmetric stretch, 1 symmetric stretch and 2 bending modes. When a polyatomic molecule jumps from one vibrational level to another, large simultaneous changes in v_1, v_2, v_3, etc., are usually unlikely.

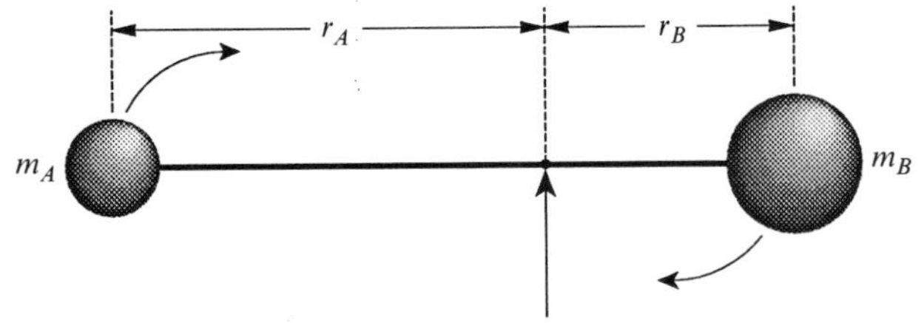

Fig. 10.8. Diatomic molecule AB used to illustrate the concept of rotational energy states.

10.4 Rotational Energy States

Just to complicate the situation further, molecules can also possess rotational energy – imagine twirling a bar-bell about its center of mass! This rotational motion is simplest to understand for the rigid bar-bell diatomic molecule AB shown in Fig. (10.8). If the angular velocity of the molecule about its center of mass is ω then the rotational energy is

$$E_{rot} = \tfrac{1}{2} I \omega^2, \tag{10.11}$$

where I is the moment of inertia of the molecule about the axis of rotation. Eq. (10.11) can also be written in the form

$$E_{rot} = \frac{L^2}{2\omega}, \tag{10.12}$$

where $L = I\omega$ is the angular momentum.

The angular momentum is

$$\begin{aligned} L &= m_A r_A^2 \omega + m_B r_B^2 \omega \\ &= \mu r^2 \omega. \end{aligned} \tag{10.13}$$

The rotational energy is also quantized, because for rotational motion quantum theory shows that angular momentum can only take certain discrete values. The resultant quantized rotational energy levels are

$$E_{rot} = BJ(J+1) - DJ^2(J+1)^2 + ..., \tag{10.14}$$

where J is an integer and usually $D \ll B$. These energies are usually quoted in cm^{-1} units so B is also usually measured in cm^{-1} units. Therefore, a typical set of rotational levels increases in spacing as J increases, as shown in Fig. (10.9).

A molecule can change its rotational state with the subsequent absorption or emission of radiation. Because typical rotational energy levels are spaced by just a few cm^{-1} these transitions occur in the far-infrared region, roughly from 50–1000 μm.

10.5 Rotational Populations

When a molecular gas is in thermal equilibrium, the number of molecules in a given rotational level per unit volume depends statistically on the total number of molecules and the temperature. If N_v is the population density of vibrational level

Fig. 10.9. Rotational energy level diagram associated with a particular vibrational energy state.

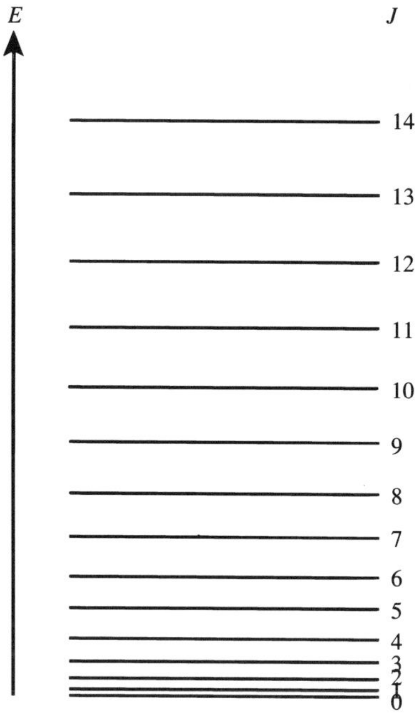

v then we can write

$$N_v = \sum_{J=0}^{\infty} N_J(v), \tag{10.15}$$

where $N_J(v)$ is the population density of the Jth rotational level of vibrational level v. In thermal equilibrium the $N_J(v)$ obey Maxwell–Boltzmann statistics, so that

$$N_J(v) = N_0(v) g(J) e^{E_J/kT_{rot}}, \tag{10.16}$$

where $N_0(v)$ is the population density of the lowest rotational level ($J = 0$) of vibrational level v. The parameter T_{rot} is called the rotational temperature, which is normally the same as the macroscopic temperature of the gas.† $g(J)$ is the degeneracy factor of rotational level J, where

$$g(J) = 2J + 1. \tag{10.17}$$

So from Eqs. (10.14) and (10.16)

$$N_J(v) = N_0(v)(2J + 1)e^{-hcBJ(J+1)/kT_{rot}} \tag{10.18}$$

The factor of hc ($c = 3 \times 10^8$ m s^{-1}) in the exponential of Eq. (10.18) is included to correct the B constant from cm^{-1} to joules. Fig. (10.10) shows the schematic variation of the rotational population of the different rotational levels of a vibrational level. Note that the population peaks for a certain value of J: $N_J(v)$ increases because of the degeneracy factor, but eventually decreases because of the exponential factor.

† The macroscopic temperature, more properly called the *translational* temperature is a measure of the average energy and velocity distribution of the molecules.

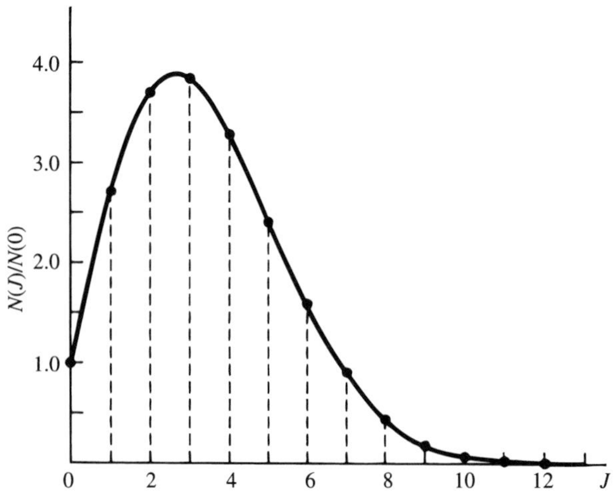

Fig. 10.10. Variation of population in thermal equilibrium among the various rotational levels of a particular vibrational level.

If we examine the possibility of population inversion on a transition between rotational levels of two different vibrational levels we can write the gain as

$$\gamma \propto (N_2 - \frac{g_2}{g_1} N_1)$$

$$\propto N_0(v')(2J'+1)e^{-AJ'(J'+1)} - \left(\frac{2J'+1}{2J''+1}\right) N_0(v'')(2J''+1)e^{-AJ''(J''+1)}$$

$$\propto N_0(v')(2J'+1)e^{-AJ'(J'+1)} \left[1 - \frac{N_0(v'')}{N_0(v')} e^{-A[J''(J''+1)-J'(J'+1)]}\right] \quad (10.19)$$

where A is a constant.† If Eq. (10.19) is examined carefully, it can be seen that even if $N_0(v') \leq N_0(v'')$, if $J'' > J'$, i.e., for a P-branch transition, it is still possible for the transition to exhibit gain. This is called a *partial* population inversion. It is quite common for molecular lasers to operate this way. The vibrational levels as a whole do not have inverted populations, but an individual pair of rotational levels does.

10.6 The Overall Energy State of a Molecule

A molecule simultaneously possesses rotational, vibrational, and electronic energy. A jump from one energy level can involve a change in one, two or all three of these types of excitation. For example, Fig. (10.11) shows schematically the vibrational–rotational energy levels of a molecule in a particular electronic state. An emission would correspond to a change from upper level (v', J') to lower level (v'', J''). Transitions where $J' - J'' = +1$ are called R-branch transitions and are labelled by the rotational quantum number of the lower level. Transitions where $J' - J'' = -1$ are called P-branch transitions and are also labelled by the rotational quantum number of the lower level. Transitions where $J' - J'' = 0$ (Q-branch transitions) also sometimes occur. Changes of J greater than 1 are sometimes seen but are usually much less likely. Fig. (10.12) shows a transition from a vibrational–rotational level of one electronic state to a vibrational–rotational level of a lower

† In these molecular systems it is customary to use single primes to denote the upper level and double primes to denote the lower.

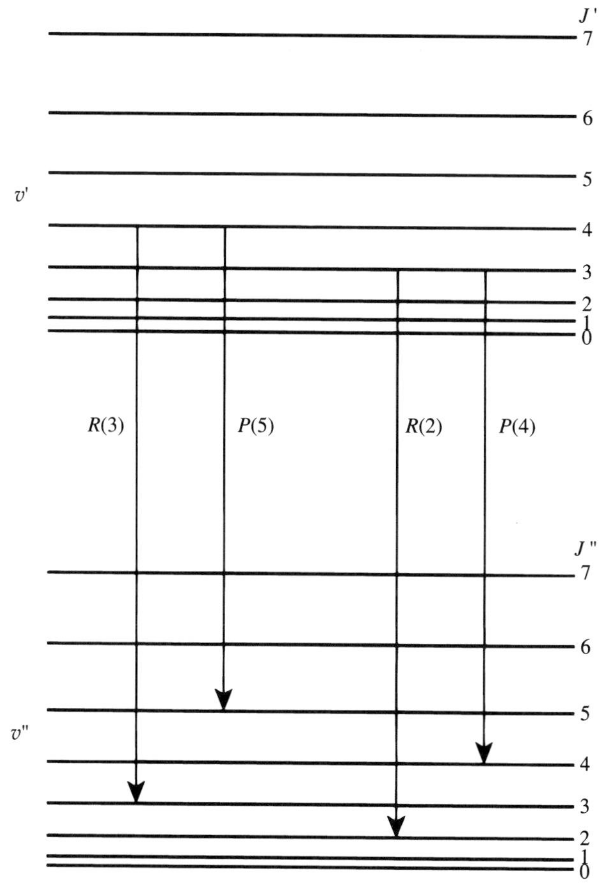

Fig. 10.11. Schematic vibrational–rotational energy level diagram of a molecule in a particular electronic state.

electronic state. Since the emission of the photon is essentially instantaneous on the time scale of molecular vibration, the internuclear separation does not change during the transition, so the jump is always a vertical line. This is very important in determining whether given changes of electronic state are likely or not.

Armed with this brief introduction to the energy levels of molecules we are now in a position to discuss the fascinating variety of molecular lasers and the methods used to excite them.

10.7 The Carbon Dioxide Laser

CO_2 lasers can be excited in a number of ways: they can be operated in a CW mode using both low and high pressure gas mixtures containing CO_2. They can be operated in a pulsed mode, expecially at pressures on the order of atmospheric pressure, in a pulsed mode when excited by high energy electron beams, and they can be excited as a result of chemical reactions. We shall touch briefly on all these methods of excitation, but in the present chapter will concern ourselves principally with low pressure, gas discharge operation.

A typical low pressure CW CO_2 laser operates using a gas mixture containing CO_2, nitrogen, and helium in the approximate ratio CO_2: N_2:He = 1:1:8. The

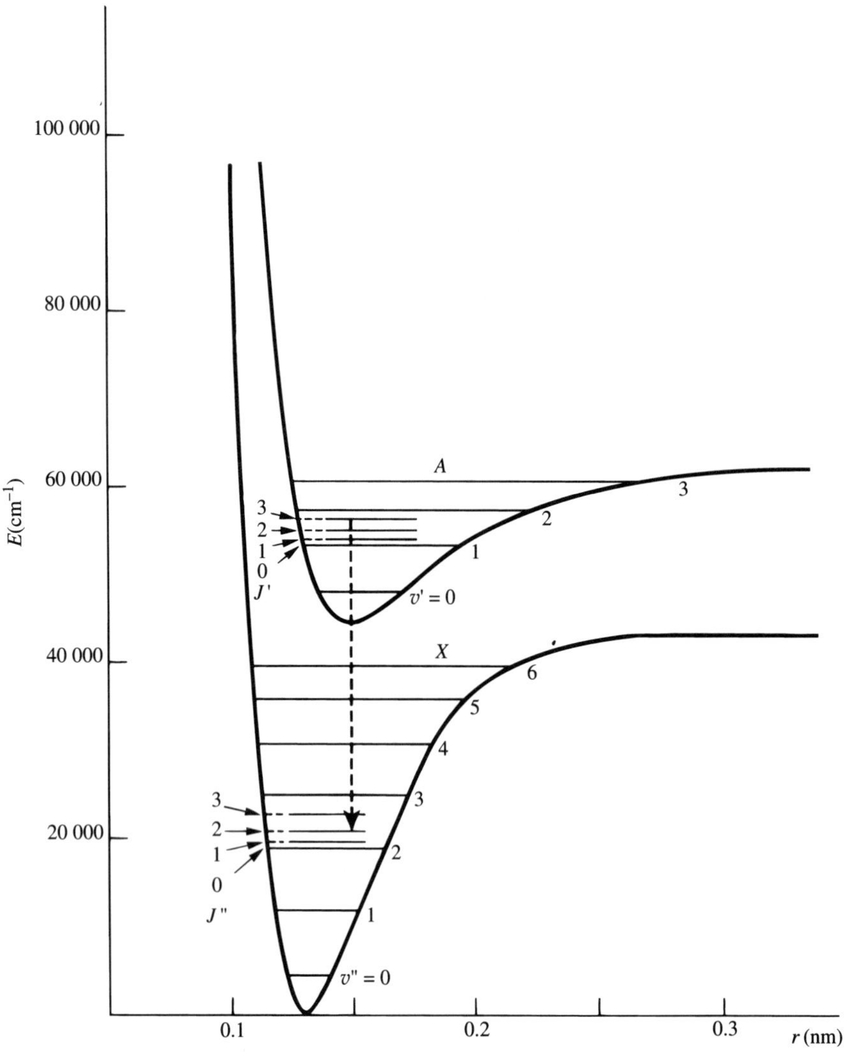

Fig. 10.12. Schematic diagram showing a transition between vibrational–rotational levels of two different electronic states.

optimum total gas pressure varies depending on the diameter of the discharge tube but is typically on the order of 20 Torr. Operating currents are low, for example 50 mA in a 10 mm diameter tube, so these lasers are easy to construct. If the laser has Brewster windows they must be made of a material that is transparent in the 10 μm range: sodium chloride, potassium chloride, zinc selenide, and germanium are commonly used. The laser tube is usually water cooled. Fig. (10.13) shows some of the essential features of a modern laser of this kind. The CO_2 laser was originally reported in 1964 by C.K.N. Patel at Bell Telephone Laboratories[10.1]. He initially reported low power operation using a discharge through CO_2 alone, but soon discovered that nitrogen, in particular, enhanced the power output enormously[10.2]. The effect of nitrogen is nothing short of spectacular: a laser 1 m long with an internal diameter of 10 mm will deliver a power output of several milliwatts when operated with pure CO_2, but can easily provide more than 10 W when nitrogen is added and the mixture, pressure, and current optimized. The role of nitrogen in

Fig. 10.13. The essential features of a modern CW low pressure CO_2 laser.

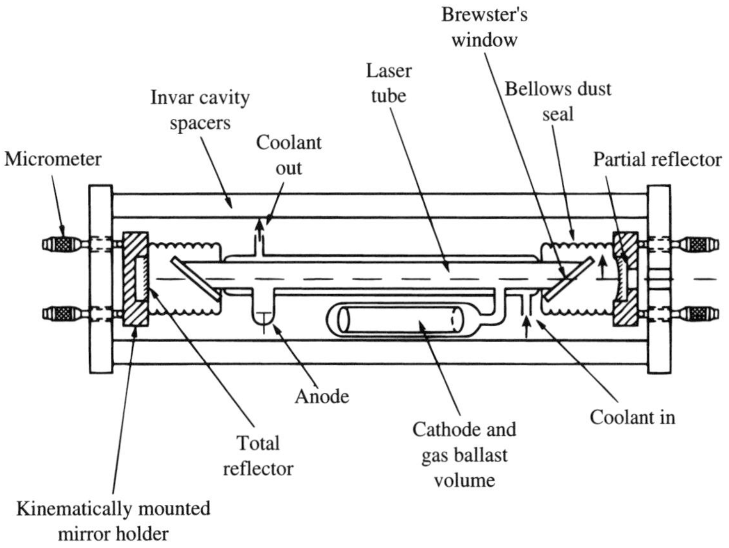

Fig. 10.14. Excitation processes and vibrational energy levels important in the CO_2 laser where vibrational energy is transferred from nitrogen molecules.

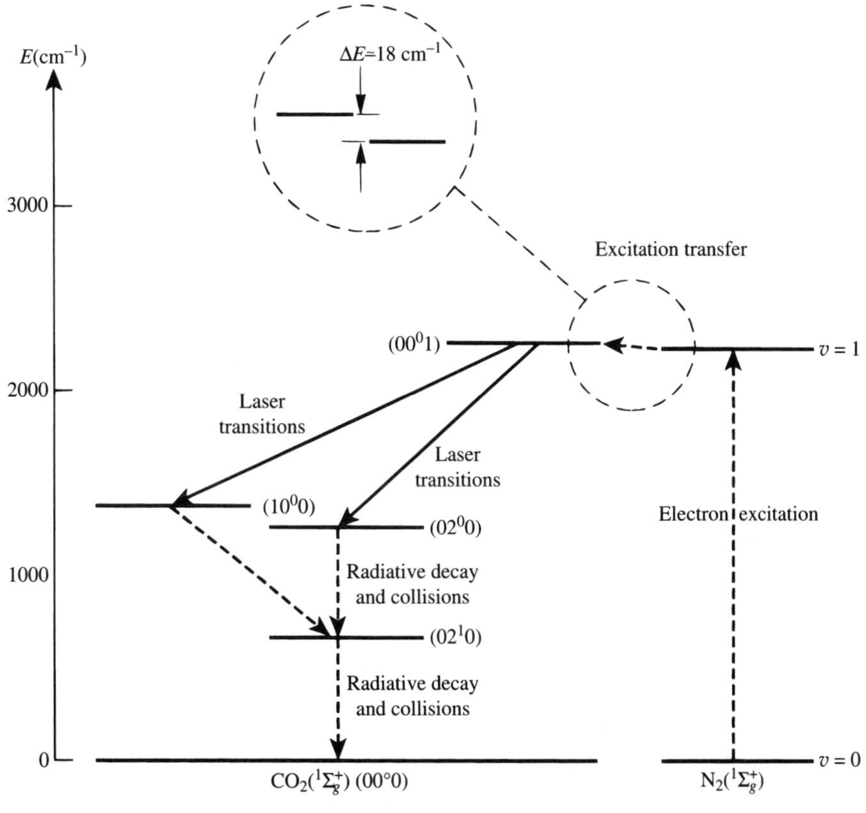

the excitation of the laser can be illustrated with reference to Fig. (10.14), which shows some of the low-lying vibrational energy levels of the CO_2 molecule in its ground electronic state. Electron collisions excite ground state nitrogen molecules to their first vibrational level (v=1). Once in this level, the nitrogen molecules are metastable, they cannot readily lose energy radiatively and can only return to the ground state as a result of collisions. Because N_2 is what is called a *homonuclear* diatomic,† as it vibrates it has no dipole moment, and a dipole moment is needed for spontaneous emission to occur. The population of $N_2(v = 1)$ molecules builds up and these molecules transfer their excitation to CO_2 molecules, thereby populating the upper CO_2 laser level (00°1).

$$N_2(v = 1, E = 2330.7 \text{ cm}^{-1}) + CO_2(00°0) \rightarrow N_2(v = 0)$$
$$+ CO_2(00°1)(E = 2349.2 \text{ cm}^{-1}) - 18.5 \text{ cm}^{-1}.$$

This excitation transfer reaction is efficient because of a close energy coincidence between $N_2(v = 1)$ and CO_2 in the (00°1) vibrational level (which has one quantum of excitation in the antisymmetric vibration). The energy discrepancy of the excitation transfer reaction (18.5 cm^{-1}), is much less than kT at ambient temperature (kT=208 cm^{-1} at 300 K). The highly selective excitation of the (00°1) level produces population inversion between this level and the (10°0) and (02°0) levels. The upper laser level has a radiative lifetime of 4 ms. The lower laser levels have long radiative lifetimes $\simeq$ 4.5 s) but are depopulated efficiently by collisions with nitrogen molecules, helium atoms, and to a lesser extent by other CO_2 molecules. Collisional relaxation of the lower laser level is the principal role of the helium, although its presence also makes operation of a uniform gas discharge much easier. The (10°0) and (02°0) lower laser levels are close enough in energy that they, too, constantly exchange energy through collisions.

The optimum electron temperature for direct excitation of the $N_2(v = 1)$ level is very low (since the threshold energy is $\sim \frac{1}{4}$ eV). Most coaxial gas discharge CO_2 lasers operate with high electron temperatures than this and excite higher levels of the N_2 molecule that subsequently reach the $v = 1$ level of the N_2 molecule by cascade. Typical axial fields and pressures are ~10 kV m^{-1}, $p_0 \simeq$ 10 Torr, i.e., $E/p_0 \simeq 1$ kV cm^{-1} Torr^{-1}.

Because the upper and lower laser levels are vibrational levels, they each have associated rotational levels. Consequently a very large number of laser transitions between different rotational levels of the upper and lower vibrational levels are possible, as can be seen from Table (10.1). The laser transitions between (00°1) and (10°0) are centered around 10.6 μm, those between (00°1) and (02°0) around 9.6 μm. The individual laser transitions are labelled as P- or R-branch as described previously. The strongest transition is the 10.6 μm $P(20)$ transition of the (00°1)→(10°0) *band*‡

$$CO_2(00°1)(J' = 19) \rightarrow CO_2(10°0)(J'' = 20) + h\nu(\lambda = 10.591 \ \mu\text{m}).$$

Because the lower laser level of the CO_2 laser is $\simeq$ 1390 cm^{-1} above the ground state, this is a four-level laser. However, if the laser gas is allowed to become too hot, this ceases to be the case. This is why cooling of the laser discharge is necessary. The gain of the laser will actually decrease as the temperature of

† Its nuclei are identical, except for the possibility of different isotopic species of nitrogen.
‡ A *band* is a group of transitions between the rotational levels of two vibrational levels.

Table 10.1. *Laser transitions from $^{12}C^{16}O_2$ lasers in the 9–11 μm region. Wavelengths and energies are calculated from accurate molecular constants*[10.3].

Wavelength (μm)	Energy (cm^{-1})	Transition	Wavelength (μm)	Energy (cm^{-1})	Transition
9.09349	1099.6872	*R*(62)	9.41471	1062.165965	*P*(2)
9.09976	1098.9301	*R*(60)	9.42889	1060.570666	*P*(4)
9.10623	1098.14940	*R*(58)	9.444333	1058.948714	*P*(6)
9.11291	1097.344886	*R*(56)	9.455805	1057.300161	*P*(8)
9.11979	1096.516356	*R*(54)	9.47306	1055.625068	*P*(10)
9.12689	1095.663612	*R*(52)	9.48835	1053.923503	*P*(12)
9.13420	1094.786462	*R*(50)	9.50394	1052.195545	*P*(14)
9.14173	1093.884721	*R*(48)	9.51981	1050.441282	*P*(16)
9.14948	1092.958211	*R*(46)	9.53597	1048.660810	*P*(18)
9.15745	1092.006758	*R*(44)	9.55243	1046.854234	*P*(20)
9.16565	1091.030196	*R*(42)	9.56918	1045.021670	*P*(22)
9.17407	1090.028367	*R*(40)	9.58623	1043.163239	*P*(24)
9.18273	1089.001119	*R*(38)	9.60357	1041.279074	*P*(26)
9.19161	1087.948306	*R*(36)	9.62122	1039.36915	*P*(28)
9.20073	1086.869791	*R*(34)	9.63917	1037.434110	*P*(30)
9.21009	1085.765445	*R*(32)	9.65742	1035.473616	*P*(32)
9.21969	1084.635145	*R*(30)	9.67597	1033.487000	*P*(34)
9.22953	1083.478778	*R*(28)	9.69483	1031.477430	*P*(36)
9.23961	1082.296237	*R*(26)	9.71400	1029.442092	*P*(38)
9.24995	1081.087426	*R*(24)	9.73348	1027.382171	*P*(40)
9.26053	1079.852255	*R*(22)	9.75326	1025.297865	*P*(42)
9.27136	1078.590644	*R*(20)	9.77336	1023.189375	*P*(44)
9.28244	1077.302520	*R*(18)	9.79377	1021.056912	*P*(46)
9.29379	1075.987820	*R*(16)	9.81450	1018.900693	*P*(48)
9.30539	1074.646490	*R*(14)	9.83554	1016.720942	*P*(50)
9.31725	1073.278484	*R*(12)	9.85690	1014.517888	*P*(52)
9.32937	1071.883766	*R*(10)	9.87858	1012.291767	*P*(54)
9.34176	1070.462308	*R*(8)	9.90057	1010.042823	*P*(56)
9.35441	1069.014093	*R*(6)	9.92289	1007.771302	*P*(58)
9.36734	1067.539110	*R*(4)	9.94552	1005.47746	*P*(60)
9.38053	1066.037360	*R*(2)	9.96849	1003.1615	*P*(62)
9.39400	1064.508853	*R*(0)	9.99177	1000.8238	*P*(64)
			10.01538	998.4646	*P*(66)
10.02591	997.41550	*R*(62)	10.44059	957.800537	*P*(4)
10.03347	996.66441	*R*(60)	10.45823	956.184982	*P*(6)
10.04132	995.884686	*R*(58)	10.47619	954.545087	(*P* 8)
10.04948	995.076610	*R*(56)	10.49449	952.880850	*P*(10)
10.05793	994.240442	*R*(54)	10.51312	951.192264	*P*(12)
10.06668	993.376427	*R*(52)	10.53209	949.479314	*P*(14)
10.07572	992.484803	*R*(50)	10.55140	947.741979	*P*(16)
10.08506	991.565748	*R*(48)	10.57105	945.980230	*P*(18)
10.09469	990.619630	*R*(46)	10.59104	944.194030	*P*(20)
10.10462	989.646506	*R*(44)	10.61139	942.383336	*P*(22)
10.11484	988.646626	*R*(42)	10.63210	940.548098	*P*(24)
10.12535	987.620181	*R*(40)	10.65316	938.688257	*P*(26)
10.13616	986.567352	*R*(38)	10.67459	936.803747	*P*(28)
10.14725	985.488312	*R*(36)	10.69639	934.894496	*P*(30)
10.15865	984.383226	*R*(34)	10.71857	932.960421	*P*(32)
10.17033	983.252249	*R*(32)	10.74112	931.001434	*P*(34)

Table 10.1. *Laser transitions from $^{12}C^{16}O_2$ lasers in the 9–11 μm region. Wavelengths and energies are calculated from accurate molecular constants (continued).*

Wavelength (μm)	Energy (cm^{-1})	Transition	Wavelength (μm)	Energy (cm^{-1})	Transition
10.18231	982.095531	*R*(30)	10.76406	929.017437	*P*(36)
10.19458	980.913211	*R*(28)	10.78739	927.008325	*P*(38)
10.20725	979.705421	*R*(26)	10.81111	924.973985	*P*(40)
10.22001	978.472286	*R*(24)	10.83524	922.914294	*P*(42)
10.23317	977.213922	*R*(22)	10.85978	920.829123	*P*(44)
10.24663	975.930439	*R*(20)	10.88473	918.718331	*P*(46)
10.26039	974.621939	*R*(18)	10.91010	916.581770	*P*(48)
10.27445	973.288517	*R*(16)	10.93590	914.419283	*P*(50)
10.28880	971.930258	*R*(14)	10.96214	912.230703	*P*(52)
10.30347	970.547244	*R*(12)	10.98882	910.015853	*P*(54)
10.31843	969.139547	*R*(10)	11.01595	907.774549	*P*(56)
10.33370	967.707233	*R*(8)	11.04354	905.50659	*P*(58)
10.34928	966.250361	*R*(6)	11.07160	903.21177	*P*(60)
10.36518	964.768982	*R*(4)	11.10014	900.88992	*P*(62)
10.38138	963.263140	*R*(2)	11.12915	898.54082	*P*(64)
10.39790	961.732874	*R*(0)	11.15867	896.1643	*P*(66)
			11.18868	893.7602	*P*(68)
10.42327	959.391745	*P*(2)			

the discharge gas rises. With water cooling to keep the walls of the discharge tube at ambient temperature (300 K) the gain of a CO_2 laser is much higher than, for example, a helium-neon laser. This is reflected in the optimum mirror reflectance for maximum power output, which for a CO_2 laser roughly depends on its length/diameter ratio, L/d, according to the formula[10.4]

$$R \simeq 1 - L/500d. \qquad (10.20).$$

For $L/d = 100$ the optimum reflectance is 80%: for a helium-neon laser of comparable dimensions the optimum reflectance is $\simeq$ 99%. The CO_2 laser is a very efficient laser, its quantum efficiency is $\simeq$ 40%, and overall electrical efficiencies close to this value have been achieved in practice. The maximum achievable power output from these CW low pressure lasers is about 300 W m^{-1} of discharge length, although much higher output powers can be achieved by using different excitation schemes, as we shall see in the next chapter.

10.8 The Carbon Monoxide Laser

Carbon monoxide, CO, is a diatomic molecule with the simple vibrational energy level structure shown in Fig. (10.15). Because the vibration of the molecule is anharmonic, the vibrational energy levels become slightly closer together as the vibrational quantum number increases. If pure CO is excited in a gas discharge, laser oscillation can occur between the rotational levels of adjacent vibrational levels, as shown in Fig. (10.15). Usually the laser transition occurs to the next lowest vibrational level, but can also occur to a level two levels lower. In theory

Fig. 10.15. Vibrational energy level diagram of CO in the ground electronic state.

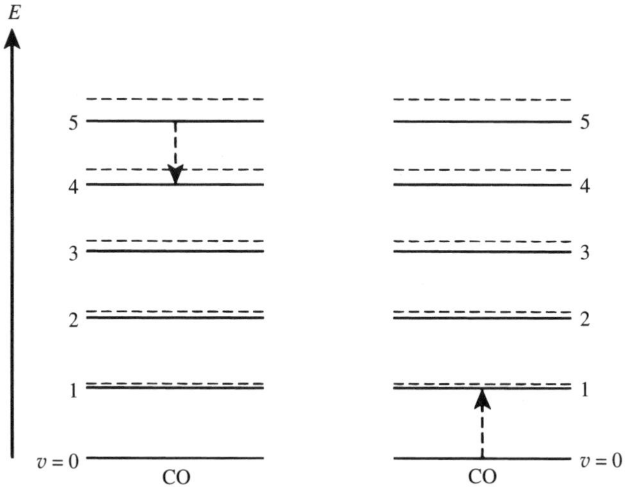

the laser process should be 100% efficient. The lower level of one laser transition can serve as the upper level of another laser transition. Further, a process called *laddering* or *Treanor* pumping redistributes vibrational energy so none is lost except by laser emission. For example, if laser oscillation occurs from level v to level $v-1$, the following sort of process can redistribute the energy back into level v

$$CO(v) \rightarrow CO(v-1) + h\nu,$$

$$CO(v-1) + CO(v=0) \rightleftarrows CO(v-2) + CO(v=1) - \Delta E_{v-1,0},$$

$$CO(v-1) + CO(v=1) \rightleftarrows CO(v) + CO(v=0) + \Delta E_{v-1,1}.$$

For every CO molecule that loses a quantum of vibrational energy in a collision, another CO molecule gains one. The small energy discrepancy ΔE arises because the vibrational energies that are exchanged are not quite equal. The larger the value of v, the greater the energy discrepancy, so the exchange process becomes more efficient in pumping higher v molecules up and lower v molecules down. This process is conducive to maintaining population in the higher vibrational levels. The strongest laser transitions occur in the 5 μm region in vibrational jumps from $7 \rightarrow 6$, $6 \rightarrow 5$, $5 \rightarrow 4$, although laser action all the way from $37 \rightarrow 36$ down to $1 \rightarrow 0$ has been observed.

The CO laser is technologically important as it is a very powerful laser. The construction of a typical low pressure device would be very similar to Fig. (10.12), although for maximum power output the laser tube must be kept colder than is the case for a CO_2 laser. Water cooling can be used but liquid nitrogen cooling enhances power output considerably. It has been determined that the addition of other gases also improves performance. N_2 stores vibrational energy that it can transfer to CO molecules, and it also helps maintain the electron energy distribution in the discharge in a favorable range for laser pumping. The addition of a small quantity of xenon also helps: it lowers the E/p_0 ratio† in the discharge because of the low ionization potential of the xenon. Some added oxygen helps to prevent the decomposition of CO and the formation of carbon deposits. Helium is added as an inert buffer to help maintain a smooth, uniform gas discharge.

† See Chapter 9.

Table 10.2. *Some molecular gas discharge lasers.*

Molecule	Laser wavelength range
N_2O	10 – 11 μm
H_2O	119 μm
HCN	337 μm
DCN	195 μm

Typical operating parameters described for a laser 1.25 m long with a 21 mm bore are: CO:0.5 Torr; He:16 Torr; N_2:1.5 Torr; O_2:0.01 Torr; Xe:0.3 Torr[10.5]. The discharge required 12.4 kV at 12 mA. This laser provided a CW power of 70 W with electrical efficiency of 47%.

10.9 Other Gas Discharge Molecular Lasers

Many small molecules will lase when excited in low pressure gas discharges. Table (10.2) lists some molecules that have been found useful in special applications. The three far-infrared lasers listed in Table (10.2) fall into a category that will be discussed further in the next chapter. These lasers use transitions between adjacent rotational levels of the *same* vibrational level.

References

[10.1] C.K.N. Patel, 'Continuous-wave laser action on vibrational-rotational transitions of CO_2,' *Phys. Rev.*, **136A**, 1187–1193, 1964.

[10.2] C.K.N. Patel, 'CW high power N_2-CO_2 laser,' *Appl. Phys. Lett.*, **7**, 15–17, 1965.

[10.3] T.Y. Chang, 'Vibrational Transition Lasers,' in *Handbook of Laser Science and Technology*, M.J. Weber, Ed., CRC Press, Boca Raton, FL, 1982.

[10.4] D.C. Tyte, 'Carbon dioxide laser,' in *Advances in Quantum Electronics,*" Vol. 1, D.W. Goodwin, Ed., Academic Press, London, 1970.

[10.5] M.L. Bhaumik, W.B. Lacina, and M.M. Mann, 'Enhancement of CO laser efficiency by addition of xenon,' *IEEE J. Quant. Electron.*, **QE-6**, 575–576, 1970.

11

Molecular Gas Lasers II

11.1 Introduction

In this chapter we shall continue our discussion of molecular gas lasers. Many of the lasers to be discussed here provide substantial CW and pulsed power output, or have unusual and innovative technical features. Some of the lasers to be discussed have already been encountered in another context in earlier chapters. For many lasers a change in the method of excitation enhances some important aspects of laser performance, for example in providing higher power output or operation in a new wavelength range. Radical departures from traditional methods of gas discharge excitation have been particularly important in allowing the development of many of the laser systems to be described in the present chapter.

11.2 Gas Transport Lasers

In many laser systems a fundamental limit to the average output power is set by the build-up of waste heat that results from inefficient laser operation. Even in the relatively high efficiency CW CO_2 and CO lasers, collisions that destroy vibrationally excited molecules, rather than just leading to energy exchange from one molecule to another, cause the temperature of the laser medium to rise. In these lasers the temperature rise reduces the population inversion through thermal excitation of the lower laser level. The rise in temperature also reduces the gain through an increase in the Doppler width. Waste heat can also produce changes in the optical properties of the laser medium in a spatially inhomogeneous way. This leads to a phenomenon called *thermal lensing*. We have already seen in Chapter 8 that this can be a serious problem in optically-pumped solid-state lasers. The change from rod-shaped amplifiers to large disks placed at Brewster's angle was driven by the need to provide a better surface-area-to-volume ratio to enhance heat removal.

In a cylindrical, gas discharge laser, heat must be removed by transport of hot particles to the cooled tube walls. Except at very low pressures diffusion is the process that controls the way hot particles reach the tube walls. The characteristic time for this process is the *thermal conduction time*,

$$T_c \simeq \frac{D^2}{\bar{c}\lambda}, \tag{11.1}$$

where D is the diameter of the tube, $\bar{c}$ is the mean velocity of the gas particles, and λ is the *mean free path* of these particles.† When diffusive thermal conduction is the only means for removing waste heat, there is an upper limit of about 1 kW m^{-1} for the input power, regardless of the tube diameter. Consequently, CW

† The mean free path is the average distance travelled by a gas particle between collisions[11.1].

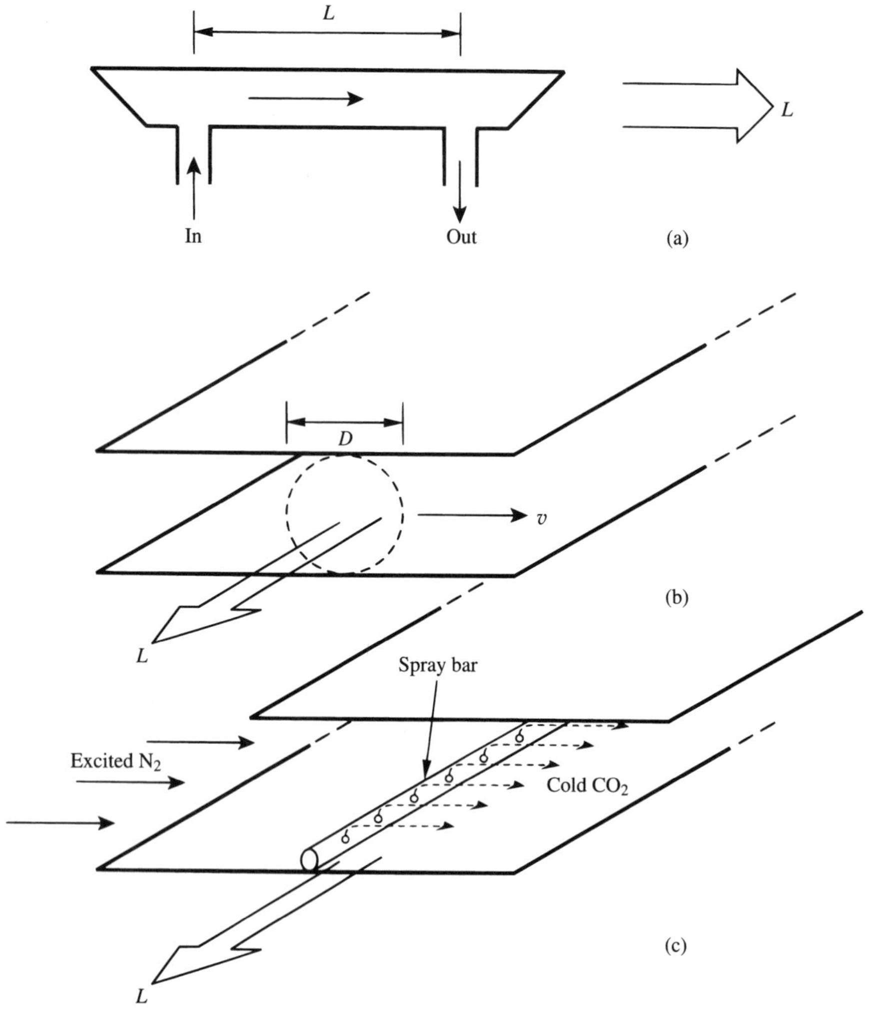

Fig. 11.1. Some examples of gas transport laser geometries: (a) axial flow laser $(LIv)_x$ geometry–resonator axis, current direction and gas velocity are in common axial direction; (b) transverse flow laser; (c) flow-mixing laser.

gas discharge CO_2 lasers operating at 30% efficiency have a fundamental power output limit of around 300 W m^{-1}. A further limit to the laser power is set by the saturation intensity, which as we saw in Chapter 2 is related to the rate of removal of particles from the lower laser level. For the CO_2 lasers we have discussed so far, which are *static* lasers, the saturation intensity is $\sim$ 60 W cm^2.

In a *gas transport* laser by flowing gas through the system we can improve laser operation in a number of ways:

(i) The gas flow provides convective cooling. Waste heat is removed at a rate determined by the residence time of particles in the system, which is a function of the flow velocity v and the dimension over which particles flow through the system. The least efficient flow geometry is shown in Fig. (11.1a). In this arrangement the laser axis is coincident with the flow direction. The characteristic time for heat removal is $\tau_{flow} = L/v$. For a transverse flow geometry, such as shown in Fig. (11.1b) $\tau_{flow} = D/v$ and the removal of waste heat is much more efficient.

Fig. 11.2. $(L)_x\ (Iv)_z$ gas transport laser geometry, where L is the laser beam direction, I the current direction and V the gas flow velocity direction. The gas flow passes through the grid or mesh electrodes.

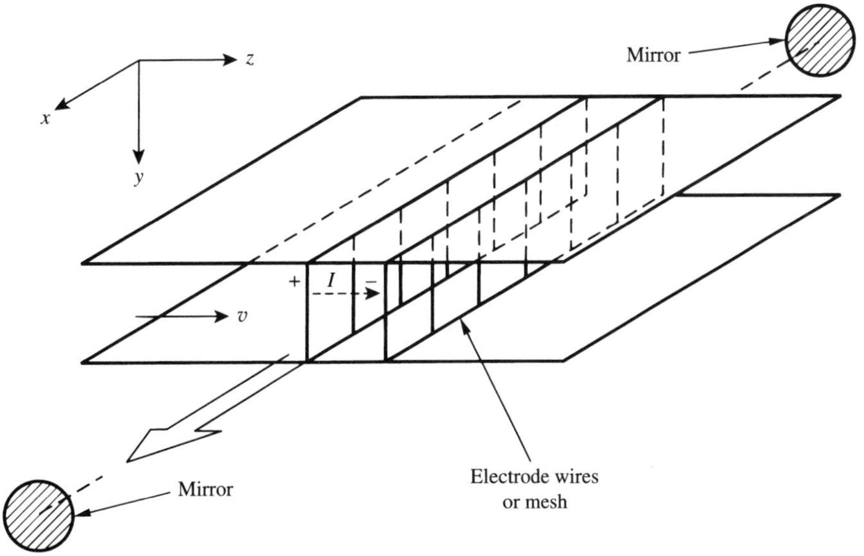

(ii) The gas flow removes particles in the lower laser level and other undesirable metastables from the amplifying region. This can enhance the power output capability even of atomic gas lasers in which build-up of metastables can destroy the inversion or lead to a small saturation intensity. Examples that fall within this category are the copper vapor and helium–xenon lasers[11.2] where high pulse-repetition-frequency operation can produce average power outputs of tens of watts.

(iii) The gas flow removes decomposition products of the discharge. In chemical lasers, which we shall discuss later in this chapter, the flow removes undesirable reaction products from the laser.

(iv) The gas flow can be used as a vehicle for mixing the laser gases, or in the case of a chemical laser, the reacting species. For example, in a CO_2 laser, nitrogen molecules can be excited in a discharge region downstream and CO_2 molecules injected upstream, where laser action can then occur. This is an example of a *flow-mixing* laser, as shown in Fig. (11.1c).

Gas transport lasers can be classified according to the relative directions of the laser axis (L), current flow (I) and gas velocity (v). For example Fig. (11.1a) is $(LIv)_x$. Fig. (11.1b) could be $(LI)_x(v)_z$, or $(L)_x(Iv)_z$, or $(L)_x(I)_y(v)_z$ dependent on whether the electrodes were placed for longitudinal current flow along the laser axis, for current flow transverse to the laser axis in the direction of gas flow, or for current flow transverse to both laser axis and gas flow. Fig. (11.2) gives an example of a $(L)_x(Iv)_z$ system. Geometries involving transverse current flow are generally to be preferred, because lower operating voltages are required to excite a shorter distance between electrodes.

When gas transport techniques are used in a laser oscillator the output transverse mode pattern can become unsymmetrical because the system lacks cylindrical symmetry and can be influenced by flow distortion. For some industrial applications such as cutting and welding the resultant laser beam can still be focused sufficiently well to be useful. However, if improved beam geometry is desired, the gas transport

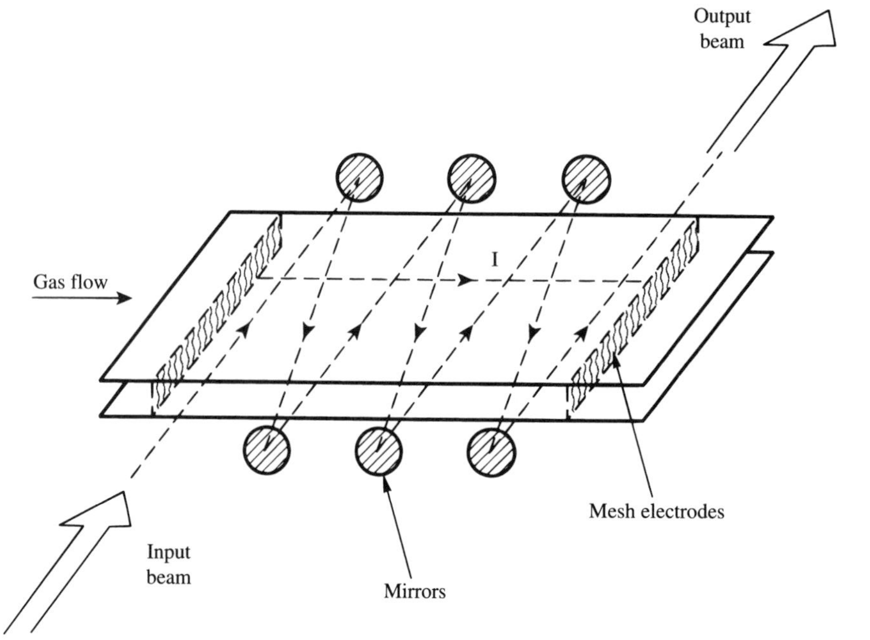

Fig. 11.3. Schematic arrangement of gas transport laser amplifier.

device can be used as an amplifier for the beam from a oscillator giving a controlled beam profile. To enhance its efficiency when used in this mode, the beam to be amplified can be multipassed as shown in Fig. (11.3).

Gas transport CO_2 lasers provide large power output from relatively small packages. Fig. (11.4) shows a schematic of how such a device could be operated in practice. The hot gases emerging from the gain region are cooled in a heat exchanger and then recirculated. The power output increases with the rate of flow of gas through the laser cavity, as shown in Fig. (11.5). A power output of 1 kW CW can be obtained from a CO_2 laser of this kind only about 0.5 m in length. Larger devices can provide CW power outputs in excess of 10 kW.

11.3 Gas Dynamic Lasers

In the gas transport lasers described above, the main purpose of the gas flow is to remove waste heat generated from the primary electrical (or chemical) excitation. In gas dynamic lasers the population inversion itself is produced as a result of the flow of gas from a high to a low pressure region. The inversion is produced, "gas dynamically" by the rapid expansion of the gas, the subsequent gas flow also removes waste heat and reaction products from the gain region.

All gases cool when they expand, even slowly, from high to low pressure, provided they are above a characteristic temperature called the *Joule–Thomson* temperature, which for most gases is at cryogenic temperatures. If the expansion of the gas is rapid, through a nozzle, cooling always occurs, and is very efficient. By 'cooling' in this context we mean a decrease in the translational temperature of the molecules. Expansion of a hot molecular gas through a supersonic nozzle channels the random kinetic motion of the hot molecules into directed motion in the direction of the flow. Since energy must be conserved, the random component of

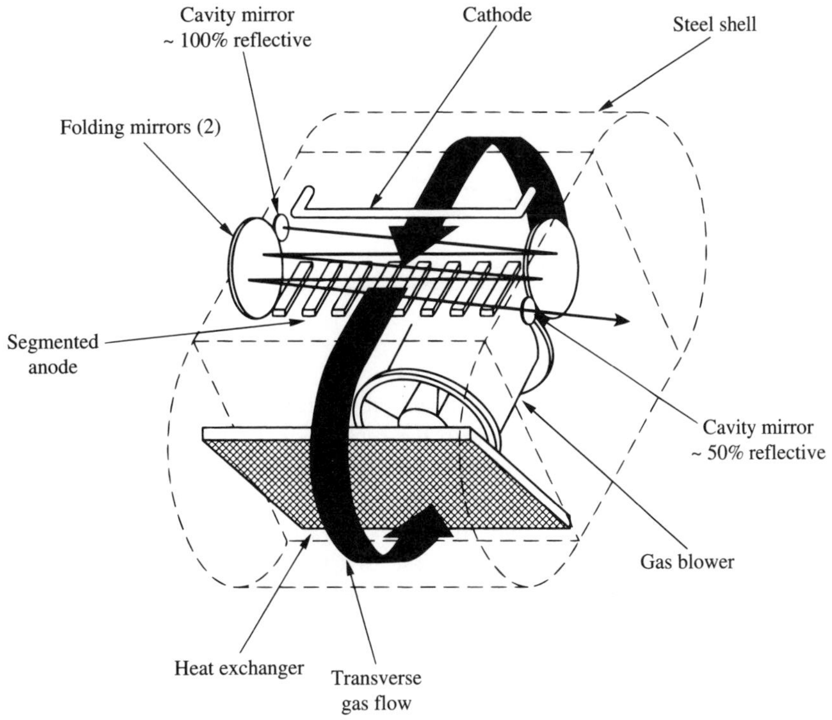

Fig. 11.4. Schematic diagram of GTE/Sylvania gas transport laser.[11.3]

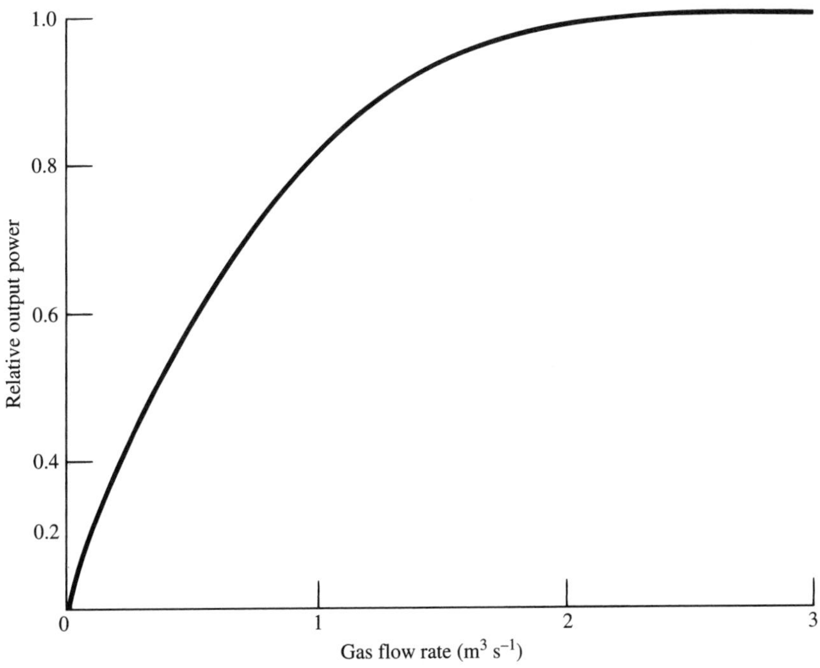

Fig. 11.5. Relative output power versus gas flow rate for a GTE/Sylvania gas transport laser.

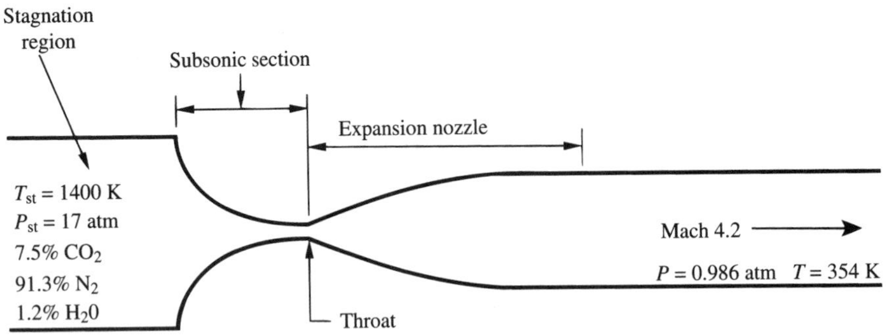

Fig. 11.6. A simple gas dynamic laser expansion nozzle.

molecular velocity must decrease, and the gas cools translationally. The rotational temperature remains in equilibrium with the translational temperature, so it also falls. However, the gas may not equilibrate as far at its vibrational populations are concerned. A population inversion can result if excitation remains *frozen* in higher vibrational levels while at the same time lower vibrational levels lose population through collisions. For example, in the case of CO_2, in thermal equilibrium at temperature T, the population densities of the upper (00°1) and lower (10°0, 02°0) levels N_2, N_1, respectively satisfy the relation

$$\frac{N_2}{N_1} = \frac{g_2}{g_1} e^{-h\nu/kT_{trans}}. \tag{11.2}$$

After expansion of the CO_2 gas through a supersonic nozzle the population density N_1 falls rapidly through collision. The lower laser level comes rapidly into thermal equilibrium with the ground state. However, the upper laser level is not as efficiently relaxed by collisions, and downstream of the nozzle a population inversion can result. Fig. (11.6) shows a schematic diagram of a gas dynamic CO_2 laser nozzle. The gas flows from left to right. The high pressure, high temperature region to the left of the nozzle is called the *stagnation region.* A typical operating mixture might be 7.5% CO_2, 1.2% H_2O and 91.3% N_2 at 17 atm, and 1400 K. After expansion through the nozzle the directed gas velocity is Mach 4.2,† the pressure has fallen to 0.086 atm and the translational temperature has fallen to 354 K. The water vapor is added to the gas mix because it is particularly efficient at relaxing the CO_2 lower laser levels. Vibrationally excited nitrogen also persists downstream of the nozzle and assists in population of the CO_2 upper laser level. The reason why population inversion results downstream of the nozzle is best illustrated with the aid of Fig. (11.7), which shows the conversion of energy initially stored in translation and rotation in the stagnation region into directed motion downstream of the nozzle. The figure also shows the reduction of lower laser level population downstream and the *freezing-in* of upper level population. To extract laser power from the system the laser axis is setup to run orthogonal to the gas flow a few centimeters down from the nozzle, where the maximum inversion exists. In practice an arrangement of many nozzles in parallel would be used, as shown in Fig. (11.8). The energy theoretically available from such a laser is essentially that energy stored in the vibrationally excited hot nitrogen and in the asymmetric stretching mode of CO_2. This energy is 35 kJ per kg of gas with a composition 10% CO_2, 90% N_2 at a stagnation temperature of 1400 K. To achieve the high temperatures and pressures

† The Mach number is the ratio of the gas flow velocity to the velocity of sound in the gas.

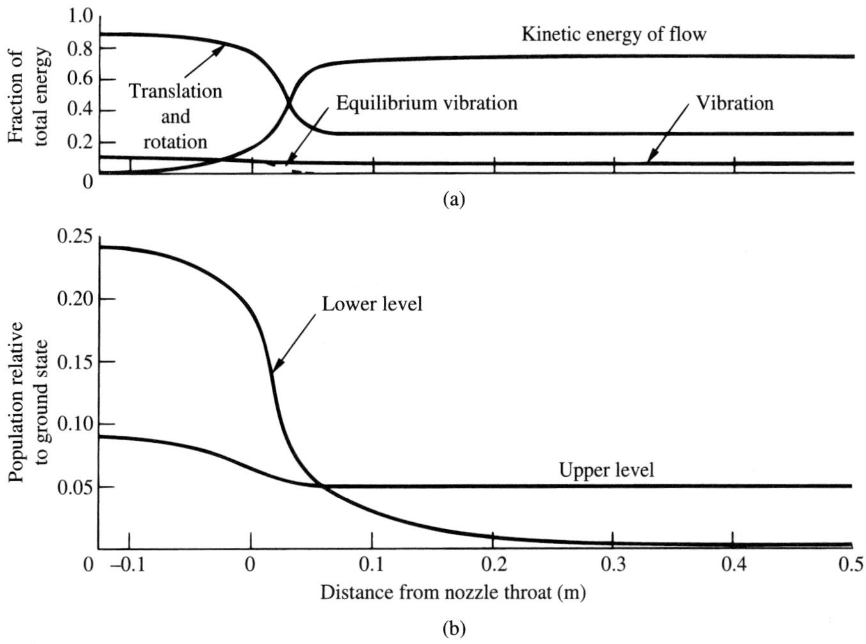

Fig. 11.7. (a) Partition of energy in the gas dynamic laser arrangement shown in Fig. (11.6), (b) relative population of upper and lower CO_2 laser levels.

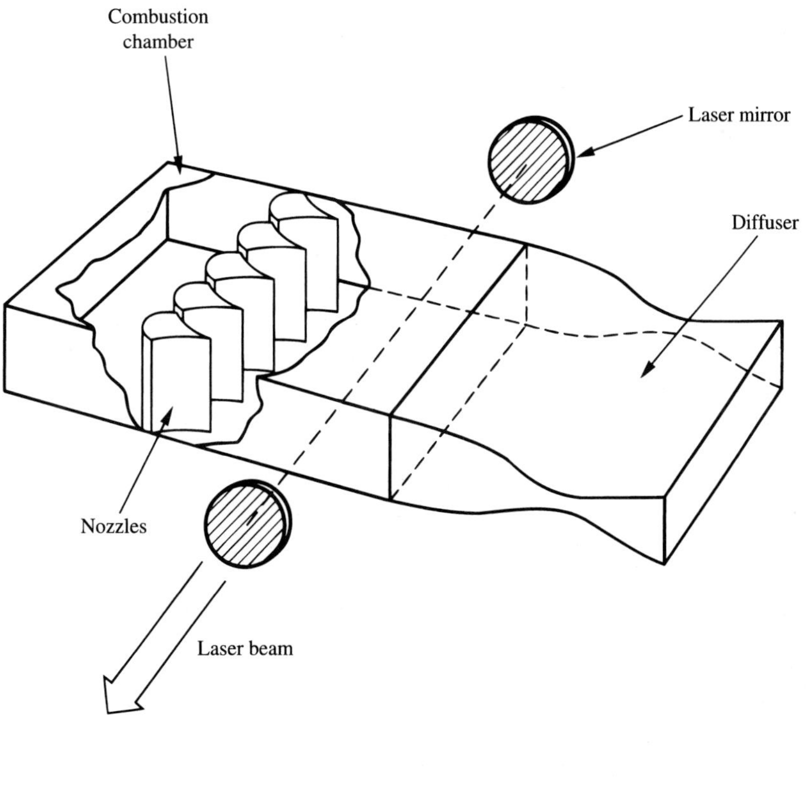

Fig. 11.8. Schematic arrangement of a high power, multinozzle gas dynamic laser[11.4].

needed in the stagnation region electrical heating can be used. However, it is fortuitous that the burning of appropriate hydrocarbon fuels with air will produce just the hot high pressure gases needed. Continuous burning of fuel will provide CW laser operation. Pulsed ignition of the fuel will give explosive production of gas in the stagnation region followed by pulsed laser operation. Some of the fuel mixtures that have been used in gas dynamic CO_2 lasers include: CO (or C_2N_2), air, CH_4 and N_2 (the methane provides the hydrogen needed for water production); CO_2, N_2, O_2, H_2 (pulsed operation). Other hydrocarbons such as acetylene and propane can also be used, or more exotic nitrogen-containing hydrocarbons, which when burnt provide CO_2, H_2O, and N_2 in the correct proportions. Gas dynamic laser action in carbon monoxide can be obtained in similar ways but the available power outputs are much smaller.

The gas dynamic CO_2 laser is the highest power CW laser currently operating. This laser falls into the category of potential "Star Wars" weaponry. Pilotless drone aircraft have been shot down with lasers of this kind. CW power outputs of several MW have been reported. It is worth briefly commenting on some of the technological problems associated with such high power lasers. The windows and mirrors of the laser system must withstand very great energy flux without damage. This requires windows of the lowest absorbtivity possible and mirrors of solid, water-cooled metal such as copper or copper alloys whose intrinsic infrared reflectance is high. Because of the violent excitation conditions in these lasers it is difficult to obtain good beam quality. This is not surprising when one considers the fact that the optical cavity runs right through the exhaust of an arrangement that is not much different from a rocket or jet engine. However, oscillator/amplifier configurations, unstable resonators,† and special beam profiling technique are available to tackle the problem of output beam quality.

11.4 High Pressure Pulsed Gas Lasers

These lasers, of which the pulsed high pressure CO_2 laser is the best example, are frequently called TEA lasers – *transversely excited atmospheric pressure lasers.* Their development arose from a desire to obtain larger power outputs from the CO_2 laser. Conventional discharge excited lasers generally operate at relatively low pressure ($<$ 50 Torr) since it is only at such pressures that a stable, uniform glow discharge can be excited in a longitudinally-excited discharge. Increase of pressure in the discharge leads to the formation of narrow current-bearing filaments (called sparks, arcs or lightning steamers). Consequently, the laser medium is not uniformly excited. Furthermore, at high pressures the E/p_0 value of the discharge may become too low for efficient excitation of the laser levels (see Chapter 9). Transverse excitation overcomes this problem as the applied voltage acts across a smaller dimension of the laser structure. If uniform transverse excitation of a high pressure discharge can be achieved, the possibility of large laser powers exists as there are many more laser molecules available in the gas – if they can be excited. The problem of filament formation can be overcome in the case of transverse pulsed excitation. The power input requirements would be prohibitive for CW operation in this way – even if such excitation were achieved. To prevent filament formation in pulsed excitation several methods have been used, but they all have

† See Chapter 16.

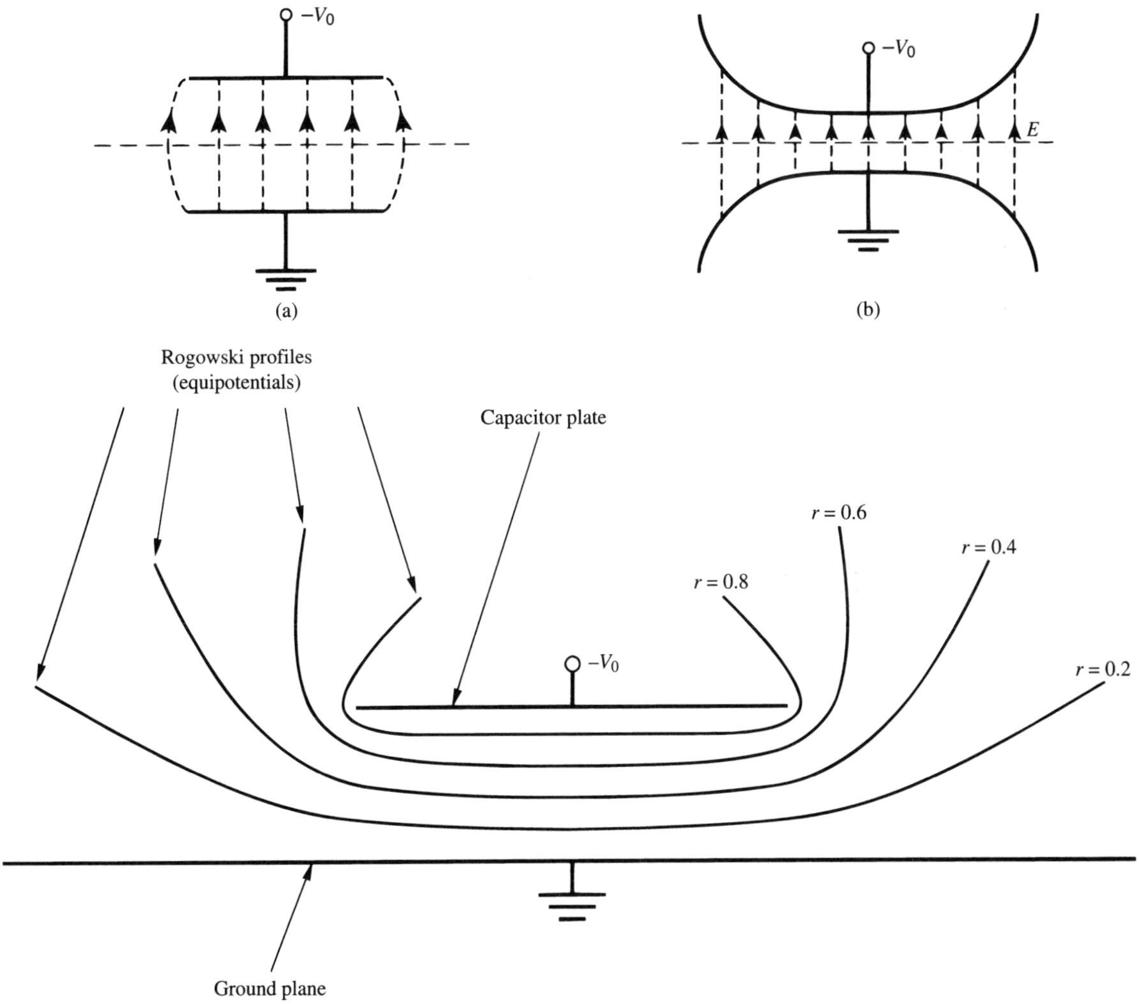

Fig. 11.9.
(a) Plane-parallel electrodes have a high field at their edges; (b) a double Rogowski profile has a uniform field between the electrodes; (c) Rogowski profiles – the equipotentials of a two-dimensional parallel plate capacitor. The potential of each profile satisfies $V = rV_0$.

one feature in common: they excite the gas in such a way as to encourage uniform glow discharge excitation and discourage the formation of filaments.

(i) *Rogowski profile electrodes.* If a transversely excited laser structure is constructed like a large plane-parallel capacitor, sparks will form at the edges of the electrodes because edge effects cause the electric field to be higher there. To prevent this the electrodes must be curved away at the edges in such a way as to ensure the field is never higher at the edges than in the center of the structure. This is done with electrodes formed in a *Rogowski* profile. Two such electrodes can be used as shown in Fig. (11.9b), or one such electrode and a much larger plane electrode placed where the equipotential of the double Rogowski structure would be. The Rogowski profiles follow the equipotential surfaces of the parallel-plate capacitor structure shown

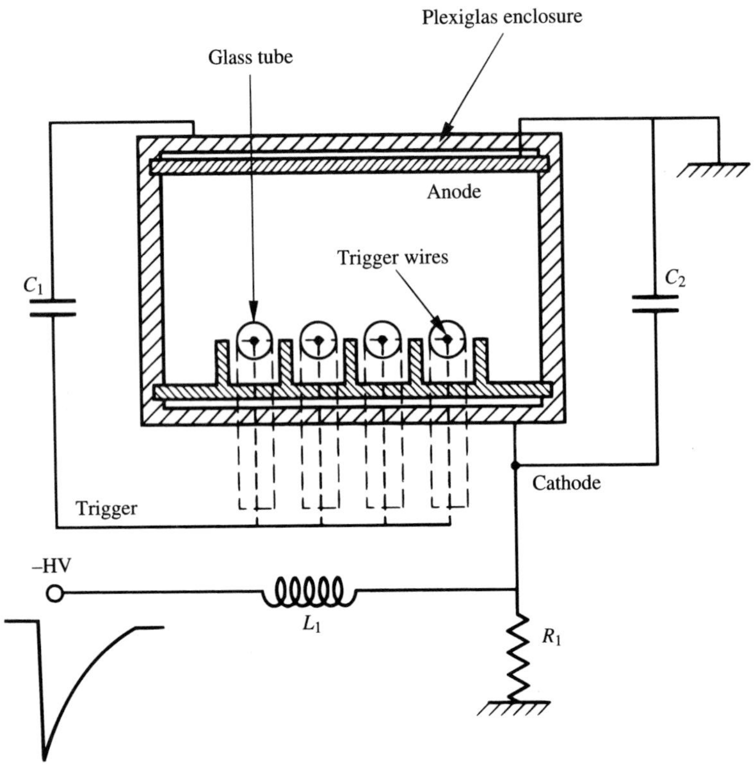

Fig. 11.10. Head-on view of a double discharge TEA laser structure. Preionization is created by striking a discharge with the trigger wires.

in Fig. (11.9c). The parametric equations of the Rogowski profile are[11.5]

$$x = \frac{a}{\pi}[\phi + e^{\phi}\cos(r\pi)],$$

$$y = \frac{a}{\pi}[r\pi + e^{\phi}\sin(r\pi)].$$

For electrodes fabricated so that $r\pi = \pi/2$, or $r\pi = 2\pi/3$ it has been found experimentally that sparking does not occur at the edge of the electrodes. However, even in structures with $\pi/2$ or $2\pi/3$ electrodes, fine adjustment of the operating voltage and gas mixture is still necessary to prevent filamentary behavior. These devices use large operating voltages – typically 20–50 kV.

(ii) *Preionization techniques.* In a transversely-excited structure, even one with Rogowski electrodes, a uniform discharge can be encouraged to form by the use of preionization techniques. Some charge carriers are injected into the space between the main electrodes just prior to application of the main excitation voltage pulse. Preionization can be provided by arrays of insulated trigger wires placed near the cathode as shown in Fig. (11.10) which are excited by a high voltage pulse and lead to production of a corona discharge near the surface of the electrode. Alternatively, as shown in Fig. (11.11) preionization can be provided by firing an array of sparks below a mesh anode or at the side of the structure. These sparks feed energetic ultraviolet photons into the laser gas causing photoionization. Pulsed CO_2 lasers of this sort have been constructed with volumetric

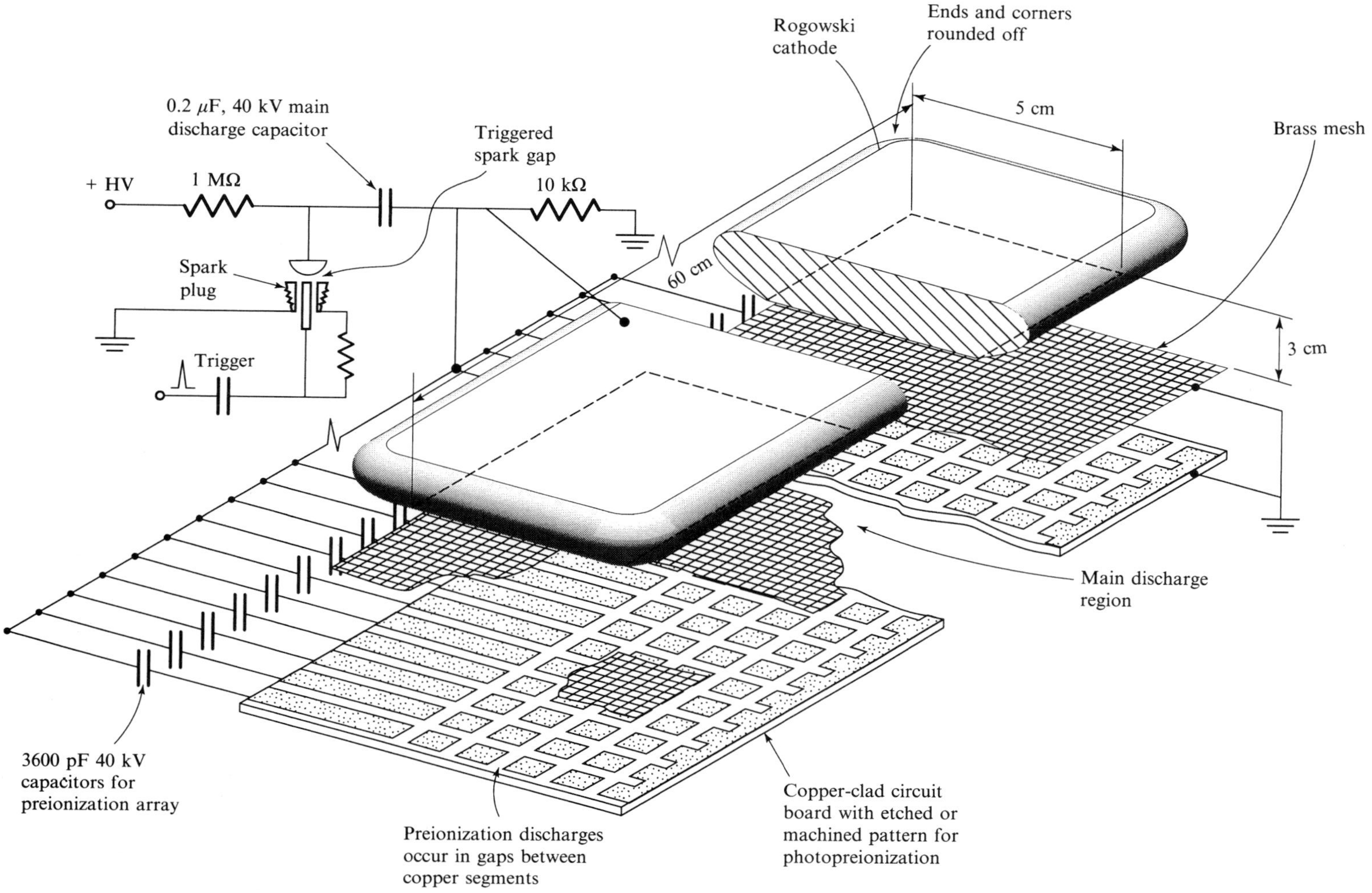

Fig. 11.11 TEA CO_2 laser structure using ultraviolet photo-preionization with a spark array. Typical dimensions are shown.

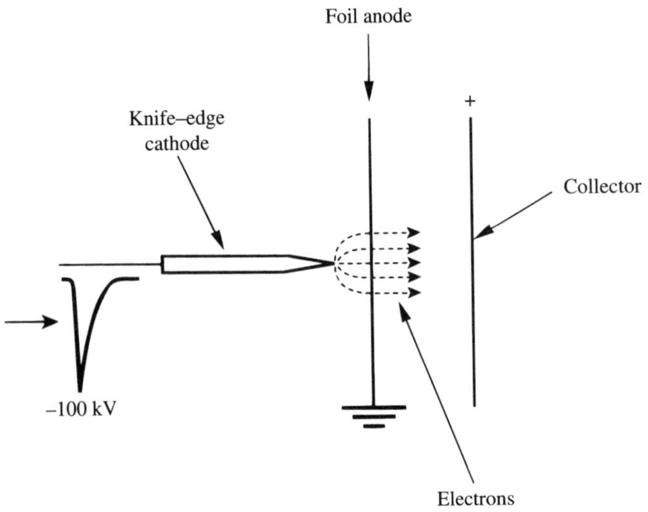

Fig. 11.12. Schematic diagram of a simple e-beam ionizer for a TEA laser.

power outputs of 35 J l^{-1}. The typical laser pulse from these devices lasts about 100 μs so peak power outputs are on the order of 10 MW J^{-1}. Shorter pulses can also be obtained. Outputs of many kilojoules in 1 ns can be produced by large oscillator/amplifier configurations.

(iii) *E-beam preionization and excitation.* High energy electron beams can be used both as a means for preionizing a large volume of high pressure gas and as the primary excitation source in a high pressure gas laser. In its simplest form such an electron beam is generated by applying a large voltage pulse, typically >100 kV, between a knife-edge cathode and a thin metal foil anode in a vacuum. Electrons are drawn from the cathode by field emission and accelerate towards the anode. Most of the high energy electrons penetrate the foil and are then available to excite, or preionize, a volume of high pressure gas on the other side of the foil, as shown in Fig. (11.12).

In TEA lasers it is uncommon for the voltage and pressure condition necessary for uniform, and therefore efficient, discharge operation to be the same as the voltage and pressure conditions which provide the optimum E/p_0 ratio for excitation of the upper laser level. To solve this problem an e-beam pulser/sustainer arrangement can be used, as shown schematically in Fig. (11.13). The electron beam energy is chosen to provide uniform penetration of electrons throughout the volume to be excited. The current needed from the e-beam for preionization is relatively low. Once preionization has occurred the main discharge capacitor is able to provide current to the discharge. The voltage, V, to which this capacitor is charged, is usually insufficient to cause gas breakdown without preionization, but is chosen to provide the desired E/p_0 ratio for laser excitation. The bulk of the excitation energy is provided by this *sustainer* capacitor, whose capacitance can be chosen to provide the length of excitation pulse required.

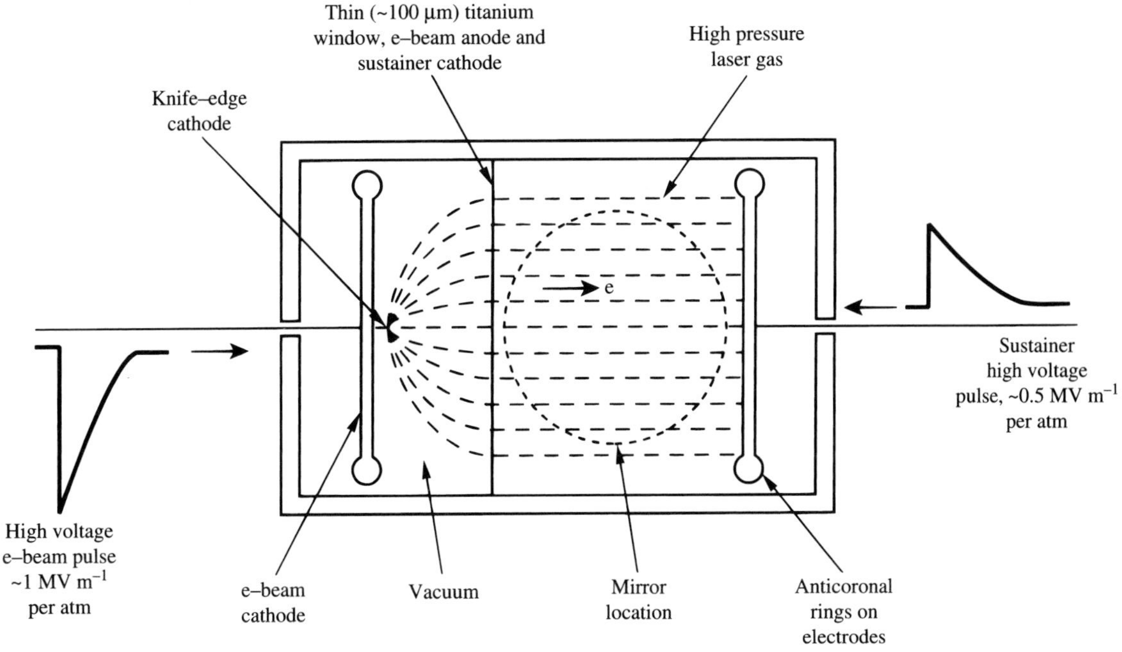

Fig. 11.13. Transverse cross-section of an e-beam pulser/sustainer laser.

Fig. 11.14. Energy level diagram for the N_2 molecule showing an excitation pathway and a laser transition from the upper level ($C^3\Pi_u(v' = 0)$) to the lower level ($B^3\Pi_g(v'' = 0)$) at 337 nm.

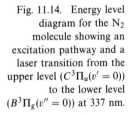

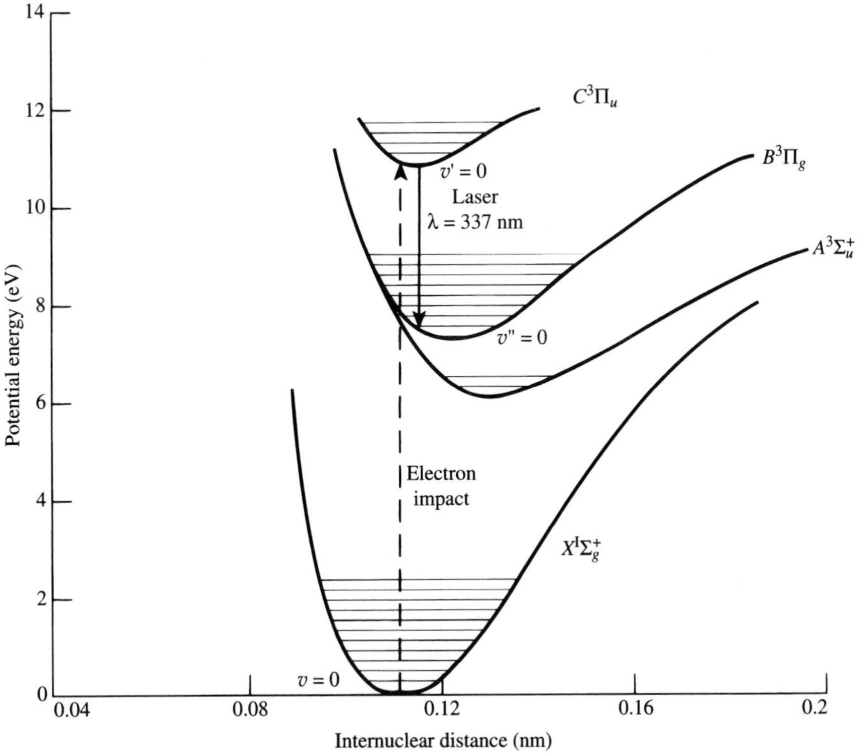

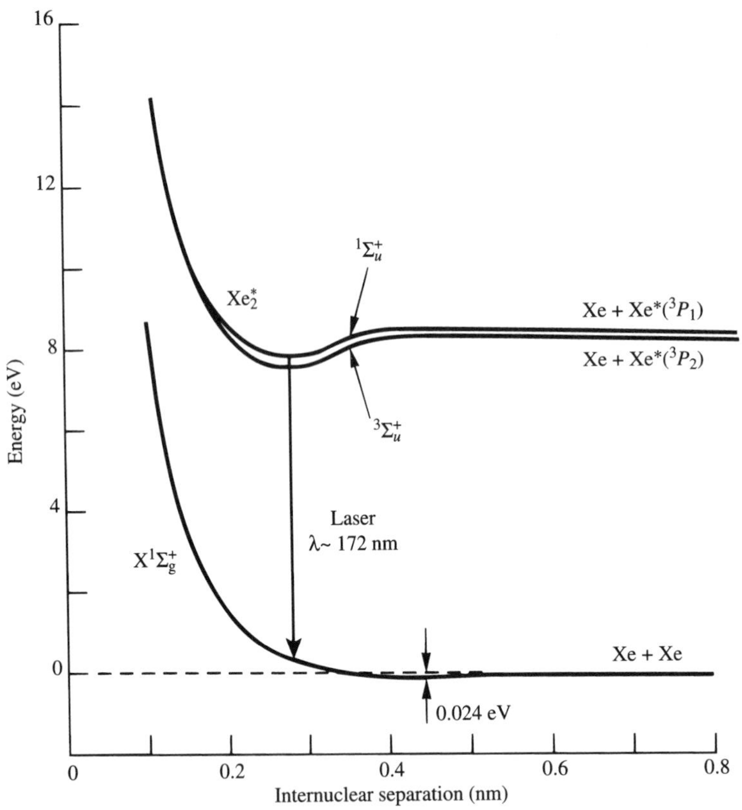

Fig. 11.15. Partial energy level diagrams of the Xe_2 molecule showing some stable bound states of Xe_2^* and the laser transition. There is a very weak Van der Waals binding force between the two Xe atoms in the ground state.

11.5 Ultraviolet Molecular Gas Lasers

Several important molecular gas lasers use fast pulse electrical discharge or e-beam pumping to excite laser action on electronic transitions in the ultraviolet. Fast pulse, high power excitation is required because: (a) the upper laser is short lived; (b) the lower laser level may be longer lived than the upper so the system is self-terminating; (c) a large upper laser level population is required to achieve threshold gain at short wavelengths. Fig. (11.14) shows an energy level diagram for the nitrogen molecule showing the electronic states that are important in this laser. Electrons excite N_2 molecules from the lowest vibrational level of the $^1\Sigma_g^+$ ground electron state to various vibrational levels of the $C^3\Pi_u$ state. Laser emission then occurs to various vibrational levels of the $B^3\Pi_g$ state. This state cannot lose energy readily so the laser is self-terminating. The most important laser transitions are around 337.1 nm. Lasers of this kind can generate MW peak powers in pulses 10–20 ns long. Broadly similar laser action can be obtained using CO ($\lambda \sim 190$ nm) and hydrogen ($\lambda \sim 160$ nm, 116 nm)[11.6].

A much more important series of lasers exists where the laser transition occurs between two electronic states, the lower of which is unbound. The lower laser level produced by stimulated emission flies apart on the time scale of a molecular vibration ($\sim 10^{-13}$ s). There could not be a more efficient way to depopulate the laser level and provide a perfect population inversion! Electronic transitions of this sort have been known for many years to occur in high pressure, rare gas discharges, and indeed serve as practical ultraviolet high intensity sources. Although the rare gases do not form stable molecules such as He_2, Ne_2,... etc. called *dimers* in the ground state, these molecules are known to exist in excited electronic states. These molecules are called *excimers*, a contraction from *excited dimer*. The Xe_2^* excimer laser was the first of this type to be operated, its energy level diagram is shown in Fig. (11.15). This laser is excited by high energy e-beam using high pressure xenon (1 – 30 atm)[11.7]. Although there are very many different collisional reactions operative in the excitation process, it can be represented in a simple way as

$$\begin{aligned} &\mathrm{Xe} + e \rightarrow \mathrm{Xe}^*, \\ &\mathrm{Xe}^* + \mathrm{Xe} + \mathrm{Xe} \rightarrow \mathrm{Xe}_2^* + \mathrm{Xe}, \\ &\mathrm{Xe}_2^* \rightarrow \mathrm{Xe}_2 \ + \ h\nu \\ &\qquad\quad \rightarrow \mathrm{Xe} + \mathrm{Xe}. \end{aligned} \tag{11.3}$$

It is clear from Fig. (11.15) that the laser transition can occur over a range of energies, so the laser is tunable over a small range near 171.6 nm.

If spontaneous emission on the $({}^1\Sigma_u^+ \rightarrow {}^1\Sigma_g^+)$ transition occurs, the observed line profile is very broad. In consequence of this and the short operating wavelength of the laser, the population inversion needed for oscillation is very large. This is the principal reason why high power pulsed excitation is needed to reach threshold. Such required pumping powers are too great to be realistic on a CW basis.

Most ultraviolet molecular gas lasers are the so-called *excimer* lasers. These lasers use electronic transitions between an excited bound state of a molecule formed from two different atoms, frequently a rare gas† and a halogen, for example ArF*, and the unbound (repulsive) ground state of the molecule, as shown in Fig. (11.16). Since these molecular lasers utilize molecules that are not strictly dimers, they should really be called *exciplex* lasers. This word was introduced as a contraction of *excited complex*. There are several important rare gas halide and other halide lasers, listed in Table (11.1). These lasers provide not only ultraviolet radiation but also intense visible radiation, for example, from HgBr. These lasers are attractive because of their large energy outputs and potentially great electrical efficiency. For example, e-beam excitation of a mixture containing 5 Torr of F_2 in 2 atm of argon has generated 55 ns pulses of energy 92 J and a peak power of 1.6 GW[11.8]. Most of these lasers provide greatest power and efficiency when e-beam excited. However, most of them can also be excited in TEA laser structures, provided care is taken to minimize the inductance of the discharge circuit so that fast rise time electrical discharges can be obtained. Typical operating conditions of the rare gas halide lasers require a large pressure of the rare gas with a much smaller amount of the halogen, or a halogen-bearing molecule. For example, the ArF* laser operates well with an Ar+NF_3 mixture. All these lasers have broad spontaneous emission profiles and need large population inversion to reach threshold. However, they store a great deal of energy in the excited molecules and once laser action can be achieved very large output energies are possible. These lasers are being seriously investigated as drivers for laser-induced thermonuclear fusion.

† Not all the *rare* gases are particularly rare, however, the alternative name *noble* gas, which was coined to describe their lack of reactivity is not particularly appropriate either, since some of the rare gas halides are actually stable in the ground state, for example XeF_6.

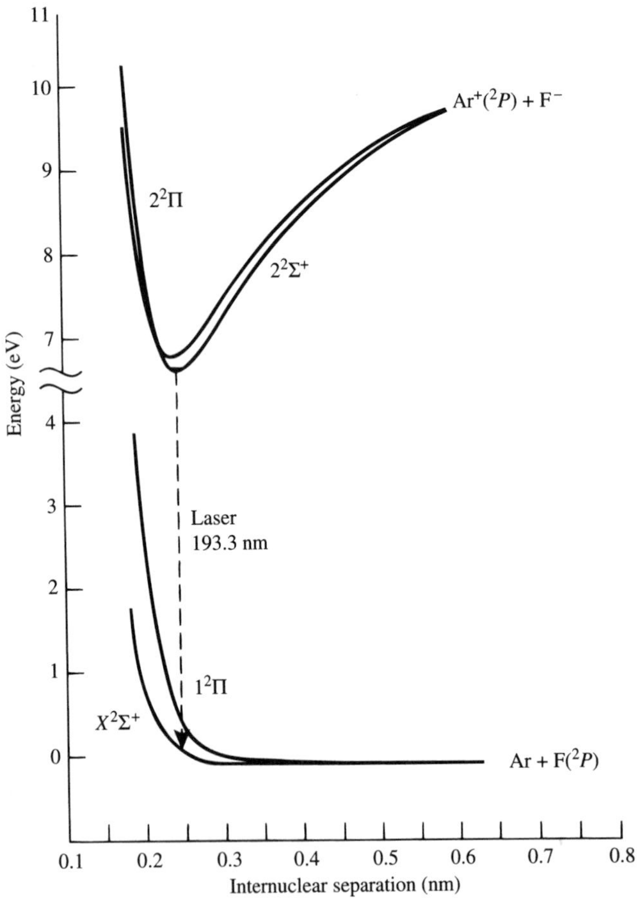

Fig. 11.16. Partial energy level diagram of the Ar–F system showing stable states of ArF^* and the important laser transition at 193 nm.

Table 11.1. *Exciplex Lasers.*

Exciplex molecule	Principal laser wavelengths (nm)
ArCl	169, 175
ArF	193.3
ArO	558
HgBr	499–504.6
HgCl	551.6–559
KrCl	222
KrF	248.4–249.1
KrO	557.81
XeBr	281.8
XeCl	307–308.43
XeF	348.8–354.0, 483, 486
XeO	537.6, 544.2
Kr_2F	430
Xe_2Cl	518

11.6 Photodissociation Lasers

These are lasers in which an ultraviolet flashlamp or the sun is used to decompose (photolyse) a molecule into fragments, one of which is produced in an excited state. The best example is the atomic iodine photodissociation laser. Several fluoro-organic iodides when photolysed produce atomic iodine in an excited atomic state. One of the best molecules for this is i-C_3F_2I (iso-perfluoropropyl iodide), which is a liquid at room temperature but has a vapor pressure of more than 200 Torr. An ultraviolet photon at ~270 nm decomposes this molecule as follows:

$$\text{i-}C_3F_7I + h\nu(\lambda \sim 270 \text{ nm}) \rightarrow \text{i-}C_3F_7 + I^*(^2P_{1/2}). \quad (11.4)$$

A rare gas is usually added to the laser mixture. Laser oscillation occurs at 1.315 μm

$$I^*(^2P_{1/2}) \rightarrow I(^2P_{3/2}, \text{ ground state}) + h\nu(\lambda = 1.315\ \mu\text{m}). \quad (11.5)$$

This might appear to be a three-level laser level since the lower laser level is the atomic ground state, however, the ground state atomic iodine is quite rapidly removed to reform the parent molecule (and, unfortunately, some molecular iodine)

$$I + \text{i-}C_3F_7 + M \rightarrow \text{i-}C_3F_7I + M \quad (11.6)$$

(M is any other molecule, or rare gas atom present)

$$I + I + M \rightarrow I_2 + M. \quad (11.7)$$

Molecular iodine makes subsequent laser action less efficient so it must be removed.

When operated in a pulsed mode, large oscillator/amplifier(s) iodine laser systems have yielded output energies of 1 kJ in 1 ns - a peak power of 10^{12} W[11.9]. These systems are similar to optically-pumped solid-state lasers, except that a quartz tube full of the laser gas mixture replaces the crystalline laser rod. These lasers can also be operated CW with a continuous ultraviolet lamp as the pump, provided the laser gas is recirculated and the iodine removed[11.10].

11.7 Chemical Lasers

A chemical laser system is one in which the reaction of two or more molecules leads to the generation of reaction products in excited states. An important laser of this kind is the hydrogen fluoride laser. In this laser, electrical initiation of a reaction between hydrogen and fluorine leads to the following reactions

$$H_2 + e \rightarrow 2H + e \quad (11.8a)$$

$$H + F_2 \rightarrow HF(v^{**}) + F + 4.24 \text{ eV} \quad (11.8b)$$

$$F + H_2 \rightarrow HF(v^*) + H + 1.38 \text{ eV} \quad (11.8c)$$

where the asterists indicate vibrationally excited molecules. Reactions (11.8b) and (11.8c) cyclically repeat, and constitute what is called a *chain reaction.* Each of these reactions liberates energy which can emerge in part as vibrational energy of the hydrogen fluoride molecule. Sufficient energy is liberated by reaction (11.8b) to excite the HF to vibrational levels up to v=6. Laser oscillation is possible on vibrational transitions $6 \rightarrow 5$, $5 \rightarrow 4$, $4 \rightarrow 3$, $3 \rightarrow 2$, $2 \rightarrow 1$, and $1 \rightarrow 0$. The wavelength range runs from 2.41–3.38 μm. Reaction (11.8c) will excite HF molecules up to $v = 3$. The DF laser operates in a similar fashion but has laser

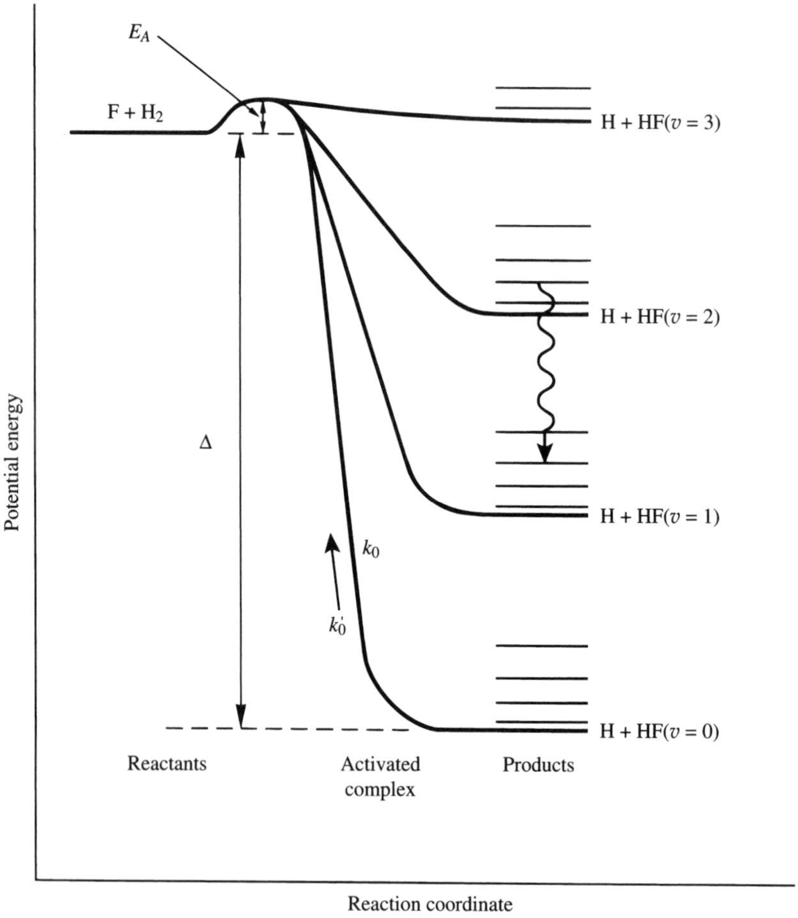

Fig. 11.17. Schematic representation of the progress of the reaction between fluorine atoms and molecular hydrogen to generate vibrationally excited HF. E_A is the activation energy of the reaction[11.11].

transitions $4 \to 3$, $3 \to 2$, $2 \to 1$, $1 \to 0$ in the 3.5–4.06 μm range. We can illustrate schematically the progress of a reaction such as (11.8c) with reference to Fig. (11.17). The energy of the various molecules involved is plotted as a function of a variable called the *reaction coordinate*, which is a parameter that indicates in a qualitative manner the progress of the reaction from starting material (reactants) to reaction products. The activation energy needed to cause the reaction to start is generally supplied by the kinetic energy of a colliding hydrogen atom and fluorine molecule. The energy ΔE liberated by the reaction allows the possibility of the excitation of several vibrational levels of the product.

In the case of the HF or DF chemical laser, the starting material can be a mixture of $H_2(D_2)$ and F_2, but often is a mixture of $H_2(D_2)$ and some other fluorinated molecule such as NF_3, SF_6, IF_5 or SO_2F_2. The reaction between H_2 and F_2 is very violent, and the laser can constitute a virtual "bomb". Very high energy systems of this kind use high pressure starting mixtures and are initiated by e-beam excitation to generate a uniform reaction throughout the laser volume. Pulses in excess of 4 kJ in 25 ns have been obtained. Very large CW powers are also possible. Fig. (11.18) shows an interesting system in which a mixture of ethylene (C_2H_4), nitrogen trifluoride (NF_3), and helium is burnt to generate fluorine atoms that then react with

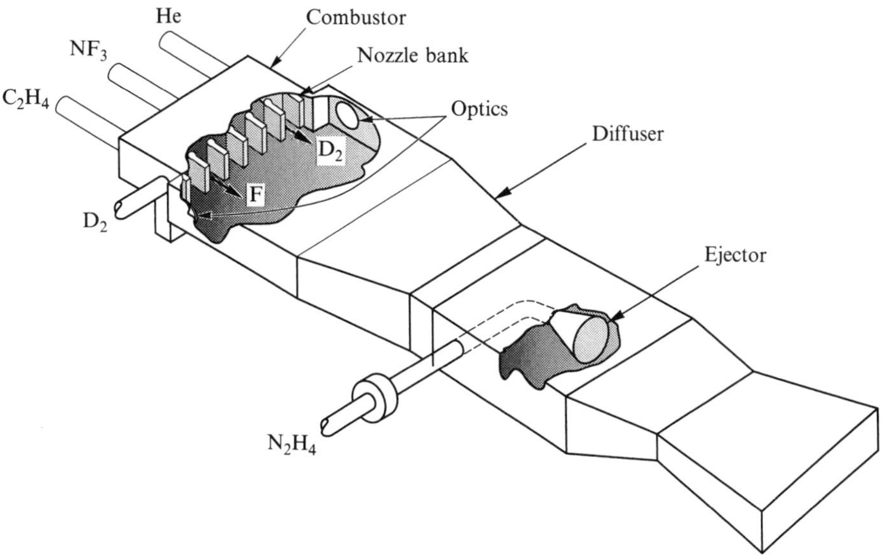

Fig. 11.18. Schematic diagram of high power CW DF chemical laser.

deuterium to generate vibrationally excited DF in a supersonic flow from a bank of nozzles. The removal of gases downstream of the nozzles is assisted by high pressure hydrazine (N_2H_4) injected to provide momentum to the gas flow. This scheme is the basis for the highest power CW chemical laser currently in operation, known as MIRACL (Middle Infrared Advanced Chemical Laser), at White Sands, New Mexico, which can generate output powers in excess of 2 MW from vibrationally excited DF. This laser has proved capable of shooting down small missiles.

There are several other important chemical lasers. The CO chemical laser generally uses the following reaction sequence

$$CS_2 + O \rightarrow CS + SO, \tag{11.9}$$

$$CS + O \rightarrow CO(v^*) + S + 3.6\, eV. \tag{11.10}$$

This reaction is a "flame" reaction and continues as long as oxygen and carbon disulphide are supplied. Continuous powers of 700 W have been obtained from lasers of this kind[11.12].

The hydrogen chloride laser uses a reaction scheme

$$H + Cl_2 \rightarrow HCl(v^{**}) + Cl,$$

$$Cl + H_2 \rightarrow HCl(v^*) + H$$

and lases in the 3.57–4.11 μm region.

One final interesting system is the chemical iodine laser, which involves the generation of *electronically* excited iodine atoms as a result of the reactions

$$O_2(^1\Delta) + O_2(^1\Delta) \rightarrow O_2(^1\Sigma) + O_2(^3\Sigma, \text{ ground state}),$$

$$O_2(^1\Sigma) + I_2 \rightarrow O_2(^3\Sigma) + 2I(^2P_{3/2}, \text{ ground state}),$$

$$O_2(^1\Delta) + I(^2P_{3/2}, \text{ ground state}) \rightarrow O_2(^3\Sigma) + I^*(^2P_{1/2}).$$

Laser emission occurs at 1.315 μm, the same wavelength as in the previously described iodine photodissociation laser. The electronically excited molecular oxygen $O_2(^1\Sigma)$ is generated by a chemical reaction between gaseous chlorine and liquid hydrogen peroxide

$$Cl_2 + H_2O_2 + 2NaOH \rightarrow O_2(^1\Delta) + 2NaCl + 2H_2O.$$

This laser system is potentially both powerful and efficient[11.13]. CW powers of hundreds of watts have been obtained[11.14].

11.8 Far-Infrared Lasers

Far-infared (FIR) lasers are generally molecular gas lasers in which the laser transition occurs between two rotational levels associated with the same vibrational level. As we have already mentioned briefly in Chapters 9 and 10, these lasers can be optically pumped by an auxiliary laser or excited by an electrical discharge. A few technological features of these devices make them different from the gas lasers we have discussed so far. Because of their long operating wavelength, these lasers have narrow gain linewidths. Consequently, only one axial model of a resonator lies under the gain profile. The length, L, of the laser resonator must be physically adjustable so that $mc/2L = \nu_0$, where ν_0 is the center frequency of the gain profile. The schematic variation of power output as the resonator length is varied is shown in Fig. (11.19). In order to minimize diffraction losses, characterized by the Fresnel number $D^2/\lambda L$, where D is the diameter of the tube, relatively large diameter laser tubes are generally used. This is particularly true in the design shown in Fig. (11.20a). This is a traditional laser resonator in which the optical pumping is done by a CO_2 laser beam that enters through one of the resonator mirrors. A simple way for the CO_2 laser beam to enter is for it to be focused through a small hole in the center of one laser mirror. The FIR laser output can be extracted through a hole in the other resonator mirror, or through a *dichroic* or *hybrid* laser mirror.† It is more common these days, however, to build the laser tube as an overmoded waveguide. The laser tube can be made of polished copper or of glass coated with gold on its interior, as shown schematically in Fig. (11.20b). The CO_2 beam reflects off the interior walls, and makes multiple passes inside the resonator so as to pump the absorbing molecules efficiently. The FIR laser oscillation occurs in a transverse mode of the waveguide structure. Diffraction losses are reduced because the oscillating wavelength cannot leak sideways from the resonator: it is guided by the tube walls.

Some important FIR lasers are listed in Table (11.2).

11.9 Problems

(11.1) From the data in Fig. (11.7) calculate the population inversion and gain as a function of distance from the nozzle.

† A dichroic mirror has different reflectance characteristics in two wavelength regions, in this case it would have high reflectance in the 10 μm region and the desired output coupling reflectance at the FIR wavelength. A hybrid mirror performs a dichroic function but uses a metal grid deposited on a dielectric internal to provide the required reflectance characteristics. Such metal grids can provide a linearly polarized output beam. If the grid consists of a series of parallel wires the **E**-field polarization parallel to the wires induces a current in the wires and suffers high loss. The polarization perpendicular to the wire suffers much less loss and is transmitted.

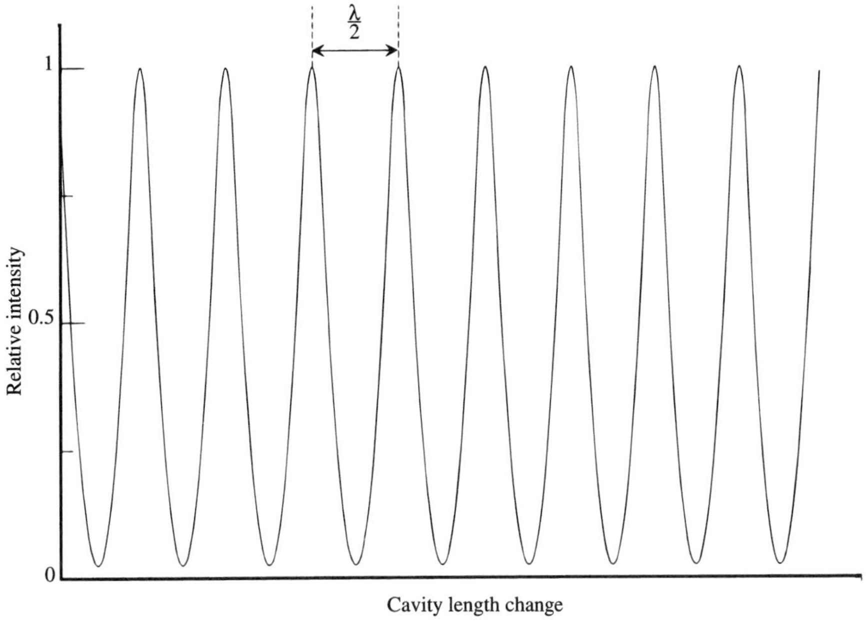

Fig. 11.19. Schematic variation of the power output of a far-infrared laser as the length of the resonator is adjusted.

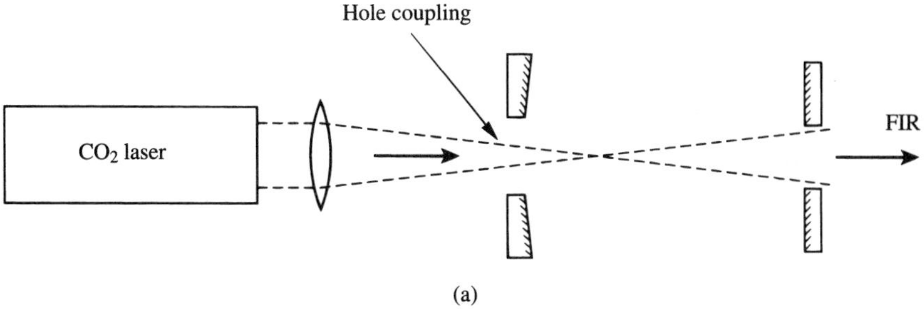

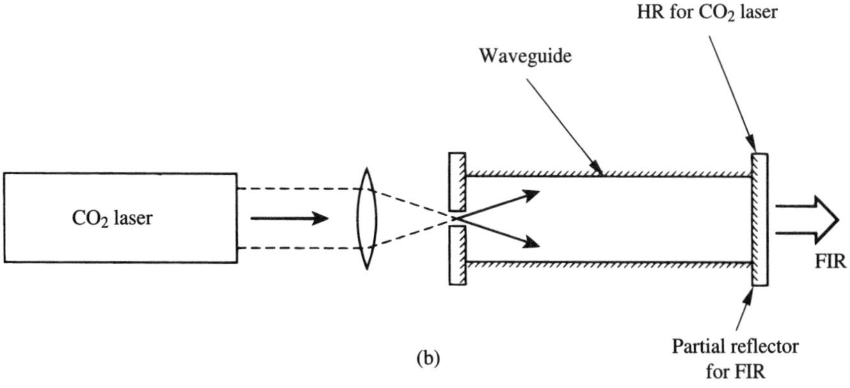

Fig. 11.20. (a) Far-infrared laser pumped by a CO_2 laser that enters through a hole in one of the resonator mirrors. (b) Far-infrared waveguide laser design.

Table 11.2. *Important optically pumped far infrared lasers.*

Molecule	Wavelength (μm)	Typical power output (mW)	Pump laser and wavelength (μm)
NH_3	81.5	50	N_2O 10.78
$^{15}NH_3$	152.9	200	$^{13}CO_2$ 10.78
$^{10}BCl_3$	19.4	50	CO_2
	19.1	50	CO_2
	18.3	50	CO_2
$^{11}BCl_3$	20.6	100	CO_2
	20.2	100	CO_2
CH_3F(methyl fluoride)	372.68	10	CO_2 10.16
	192.78	10	CO_2 10.17
HCOOH (formic acid)	513.0157	5	CO_2 9.23
	302.2781	5	CO_2 9.27
HCOOD	926.2087	10	CO_2 10.29
	919.9355	10	CO_2 10.17
DCOOH	380.5654	10	CO_2 10.30
CH_3Cl (methyl chloride)	334.0	50	CO_2 9.75
CH_2F_2	214.5791	500	CO_2 9.20
(difluoromethane)	184.30590	500	CO_2 9.21
$CH_3{}^{79}Br$	414.98	20	CO_2 10.38
CH_3I (methyl iodide)	1253.738	10	CO_2 10.72
CH_3OH (methanol)	170.57637	10	CO_2 9.69
	164.7832	10	CO_2 9.33
	118.8349	20	CO_2 9.69
	96.52239	20	CO_2 9.33

References

[11.1] Sir J.H. Jeans, *An Introduction to the Kinetic Theory of Gases*, Cambridge University Press, Cambridge, 1962.

[11.2] C.C. Davis, 'Neutral gas lasers,' in *Handbook of Laser Science and Technology,"* Vol. I, *Lasers and Masers*, M.J. Weber, Ed., CRC Press, Boca Raton, FL, 1982; and references therein.

[11.3] W.B. Tiffany, R. Targ, and J.D. Foster, 'Kilowatt CO_2 gas-transport laser,' *Appl. Phys. Lett.*, **15**, 91–93, 1969.

[11.4] J.D. Anderson, Jr., *Gasdynamic Lasers: An Introduction*, Academic Press, New York, 1976.

[11.5] J.D. Cobine, *Gaseous Conductors*, Dover, New York, 1958.

[11.6] R.S. Davis and C.K. Rhodes, 'Electronic transition lasers,' in *Handbook of Laser Science and Technology*, Vol. I, *Lasers and Masers,* M.J. Weber, Ed., CRC Press, Boca Raton, FL, 1982; and references therein.

[11.7] M.H.R. Hutchinson, 'Excimer lasers,' in *Tunable Lasers*, L.F. Lollenauer and J.C. While, Eds., Topics in Applied Physics, Vol. 59, Springer-Verlag, Berlin, 1987.

[11.8] J.M. Hoffman, A.K. Hays, and G.C. Tisone, 'High-power noble-gas-halide lasers,' *Appl. Phys. Lett.*, **28**, 538–539, 1976.

[11.9] G. Brederlow, E. Fill, and K.J. White, *The High-Power Iodine Laser*, Springer-Verlag, Berlin, 1983.

[11.10] L.A. Schlie and R.D. Rathge, 'Long operating time CW atomic iodine probe laser at 1.315 μm,' *IEEE J. Quant. Electron.*, **QE-20**, 1187–1196, 1984.

[11.11] K.L. Kompa, *Chemical Lasers*, Topics in Current Chemistry, Vol. 37, Springer-Verlag, Berlin, 1973.

[11.12] R.J. Richardson, H.Y. Agero, H.V. Lilenfield, T.J. Menne, J.A. Smith, and C.E. Wiswall, 'CO chemical laser utilizing combustor-generated reactants.' *J. Appl. Phys.*, **50**, 7939–7947, 1979.

[11.13] G.N. Hays and G.A. Fisk, 'Chemically pumped iodine laser as a fusion driver,' *IEEE J. Quant. Electron.*, **QE-17**, 1823–1827, 1981.

[11.14] K. Shimizu, T. Sawano, T. Tokuda, S. Yoshida, and I. Tanaka, 'High power stable chemical oxygen iodine lasers,' *J. Appl. Phys.*, **69**, 79–83, 1991.

12

Tunable Lasers

12.1 Introduction

All lasers are to some extent *tunable*. Their output frequency can be varied continuously without discontinuous changes in output power by moving the position of the oscillating modes under the gain profile. However, if the gain profile is not very wide then this range of tunability is limited. For any atomic gas laser, for example, where gain profiles typically have Doppler widths on the order of 1 GHz, tunability of a single axial mode over about a 1 GHz range can be accomplished by changing the optical length $n\ell$ of the cavity. This can be done by moving one of the mirrors with a piezoelectric transducer† and thereby varying the geometric length ℓ, or by adjusting the laser pressure so as to adjust the index n. Although a tunable frequency range of 1 GHz might seem large in absolute terms, it represents a very small fraction of the operating frequency of the laser, $10^{14} - 10^{15}$ Hz, say. Discontinuous tunability in the infrared can be obtained by using a molecular gas laser where several vibrational–rotational transitions have gain. Systems, such as the CO_2 laser, can offer many lines over a relatively broad wavelength region, but a graph of output power versus frequency is not continuous.

To achieve continuous tunable laser operation over a broad wavelength region we must use an amplifying medium with a broad gain profile. The gain of such a laser is inversely proportional to the width of this gain profile so, all else being equal, we must pump the system harder, and more efficiently, to reach threshold. We must also provide wavelength selective feedback in the laser cavity to encourage laser oscillation at the desired wavelength.

12.2 Organic Dye Lasers

Although they were not, chronologically, the first tunable lasers, lasers using a liquid gain medium consisting of organic dye molecules in solution are the most important and widely used tunable lasers. *Dye* molecules are so called because they impart characteristic and generally bright colors to otherwise colorless media. They do this by selectively absorbing radiation, so that, for example, a dye looks red because it absorbs incident blue and green light.

12.2.1 Energy Level Structure

From a structural standpoint organic dyes are hydrocarbon molecules that contain carbon–carbon double bonds and, in particular, sequences of alternate single

† See Chapter 19.

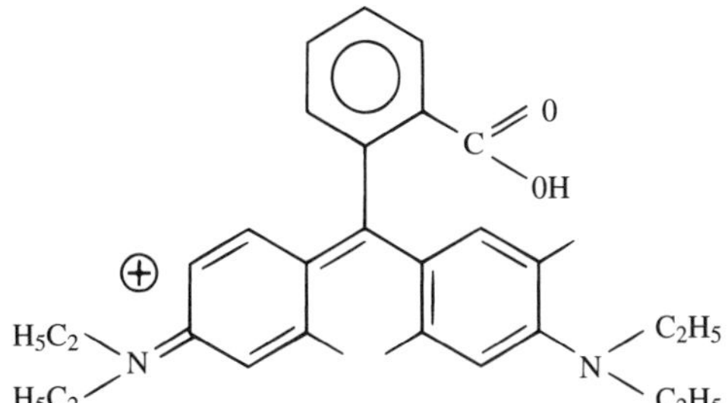

Rhodamine B

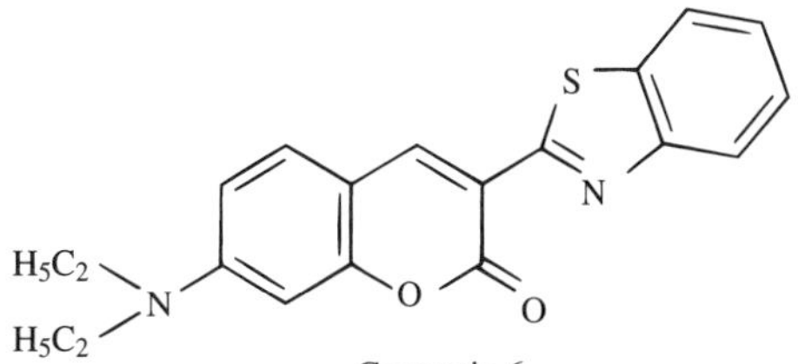

Coumarin 6

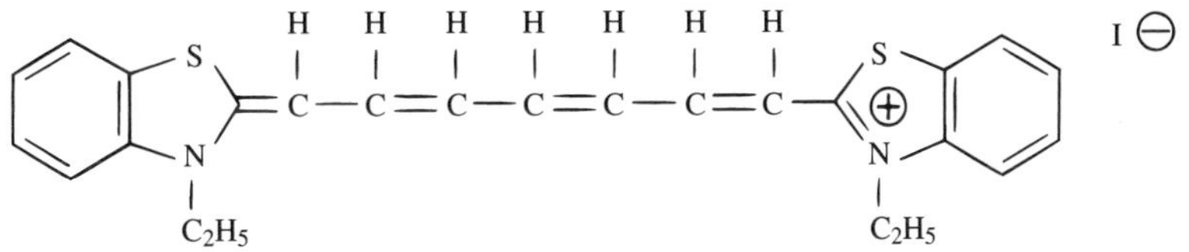

3,3 Diethyl thiatricarbocyanine iodide

Fig. 12.1. Some aromatic organic dye molecules.

and double bonds. Such structures are said to be *conjugated*. Fig. (12.1) shows three examples. All these molecules contain the hexagonal carbon ring structures (benzene rings) that characterize *aromatic* hydrocarbons. It is the relatively loosely-bound electrons associated with these benzene rings that lead to the interesting electronic energy level structure of organic dyes. These loosely-bound electrons actually travel freely from nucleus to nucleus within the plane of the ring. In this orbital motion in the ground state there are an even number of electrons that can be grouped together in pairs. Each pair consists of one electron with the direction of its spin momentum vector pointing up and one electron with its spin momentum vector pointing down so the overall angular momentum is zero. In a single benzene ring we can group these electrons into three pairs above the plane of the ring, and three pairs below. A single electron from one of these pairs can be excited into a range of states, as shown in Fig. (12.2). The states marked S_1, S_2, are called *singlet* states, S_0 is the ground state. The states T_1, T_2 are called *triplet* states. The distinction between these two kinds of states is that in the singlet state the excited electron has moved to a higher energy but retained its original spin orientation so the overall angular momentum of the electrons remains zero. In the triplet state the excited electron has moved to a higher energy state but its spin direction has flipped over, say from up to down so the electrons now have net angular momentum. Each singlet and triplet state has associated with it the vibrational and rotational energy states we have discussed previously. At ambient temperature most dye molecules will be in the lowest vibrational level of S_0. Incident light of appropriate energy will excite these molecules to higher singlet states. Excitation

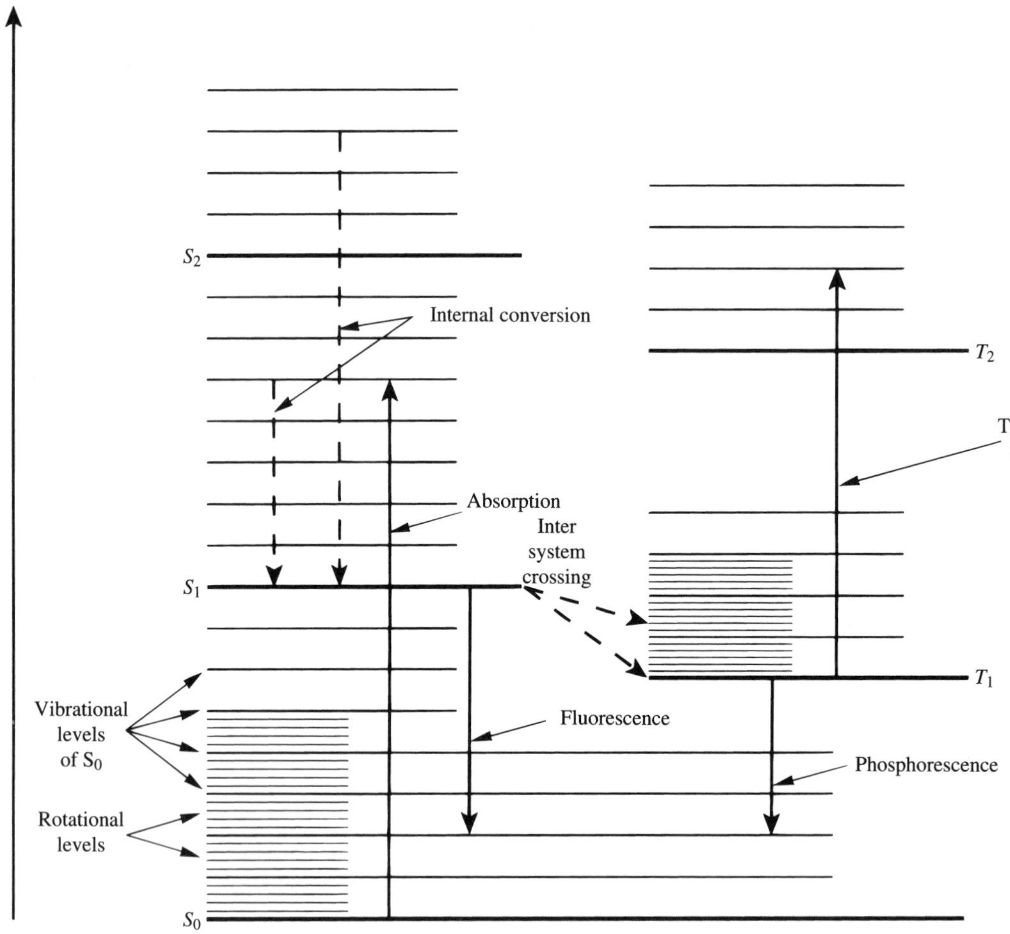

Fig. 12.2. Schematic energy level diagram for an aromatic organic dye molecule showing various important radiative and nonradiative processes.

of the vibrational levels of S_1 usually involves visible or ultraviolet photons. Thus a characteristic absorption spectrum results, which *may* show vibrational structure, although in solution this structure will likely be smeared out. Once a vibrational level of S_1 has been excited, the molecule, generally on about a 1 ps time scale, *internally converts* to the lowest vibrational level of S_1. *Fluorescence* is spontaneous emission from this lowest vibrational level of S_1 to the various vibrational levels of S_0, as shown in Fig. (12.2)†. In a simple model the fluorescence and absorption spectra should be mirror images, as shown on Fig. (12.3a). This is not quite the case in reality as the real spectra in Fig. (12.3b) show. Molecules can also move nonradiatively from $S_1 \rightarrow T_1$ (or $S_2 \rightarrow T_2$ etc.). This process, called *intersystem crossing*, is relatively slow compared to internal conversion but can occur as fast as about 10 ns. Spontaneous emission from $T_1 \rightarrow S_0$ is called *phosphorescence*. Because this process involves the electron losing energy *and* flipping its spin it is a relatively unlikely process and can have a lifetime from 1 μs to many milliseconds.

† Only one aromatic molecule, azulene (and its derivatives) fluoresces readily from $S_2 \rightarrow S_0$ [12.1].

If molecules reach T_0, then absorption to higher triplet states $T_1, T_2, \ldots$ is an efficient process. All these processes are summarized in Fig. (12.2).

12.2.2 Pulsed Laser Excitation

The earliest dye lasers were excited by the pulses from Q-switched ruby lasers. Dye solutions of concentration varying from 10^{-6} to 10^{-3} moles l^{-1} were placed in small optical curvettes inside a laser cavity and the pump laser radiation was directed into the curvette either from the side of the curvette, or through one of the dichroic laser mirrors, as shown in Fig. (12.4). The pumping process involves the following steps: (i) rapid excitation of a vibrational–rotational level of S_1; (ii) internal conversion to the lowest vibrational level of S_1; (iii) laser emission to the vibrational-rotational levels of S_0 sufficiently far above the ground state that their populations are low. A large population density in S_1 is required to achieve threshold because of the large $\Delta\nu$ of the fluorescence spectrum. The threshold will increase even further if the distributed loss, α, within the resonator is not minimized. Distributed loss occurs for various reasons: in particular, intersystem crossing populates T_1 and triplet–triplet absorption occurs in the same wavelength region as fluorescence. It has been discovered, however, that the T_1 population can be kept small by several means. These include using dyes where intersystem crossing is inefficient, and by adding oxygen or detergent to the dye solution, both of which rapidly relax T_1 molecules to S_0. Absorption from $S_1 \rightarrow S_2$ can also occur at the laser wavelength, which is an important loss mechanism in some dyes. Regions of unpumped dye can contribute to α through absorption of the laser wavelength, particularly if this is near the wavelength λ_m in Fig. (12.3b).

Pulsed dye laser operation can also be achieved with fast flashlamp excitation in geometrical arrangements similar to those used to excite solid-state crystalline lasers. Fig. (12.5) shows schematically a coaxial flashlamp arrangement which has been used to obtain dye laser pulses up to 400 J in energy[12.2].

Although a dye laser is predominantly homogeneously broadened, this does not preclude the simultaneous oscillation of several simultaneous modes. With such a broad gain profile, many modes within a relatively broad wavelength region will have almost identical gain. There is insufficient time for the strongest mode to suppress its neighbors when the laser emits only short pulses. Furthermore, spatial inhomogeneity of the optical pumping and thermal and acoustic fluctuations of the liquid gain medium add an effective inhomogeneous contribution to the broadening. The output of a pulsed dye laser can be several nm wide unless additional frequency selectivity is incorporated into the resonant cavity. If one of the cavity mirrors is replaced by a reflective diffraction grating, as shown in Fig. (12.6), only the wavelength that satisfies the diffraction condition will experience reflection back along the resonator axis. It is common to include an intracavity beam expanding telescope so that a large area of the diffraction grating is illuminated. For a reflective diffraction grating with groove spacing d the condition for retroreflection is

$$m\lambda = 2d\sin\theta, \qquad (12.1)$$

where m is an integer, and θ is the angle between the resonator axis and the normal to the grating. The output wavelength of the laser can be tuned by altering the angle θ. The output linewidth will also be reduced because only a small range of

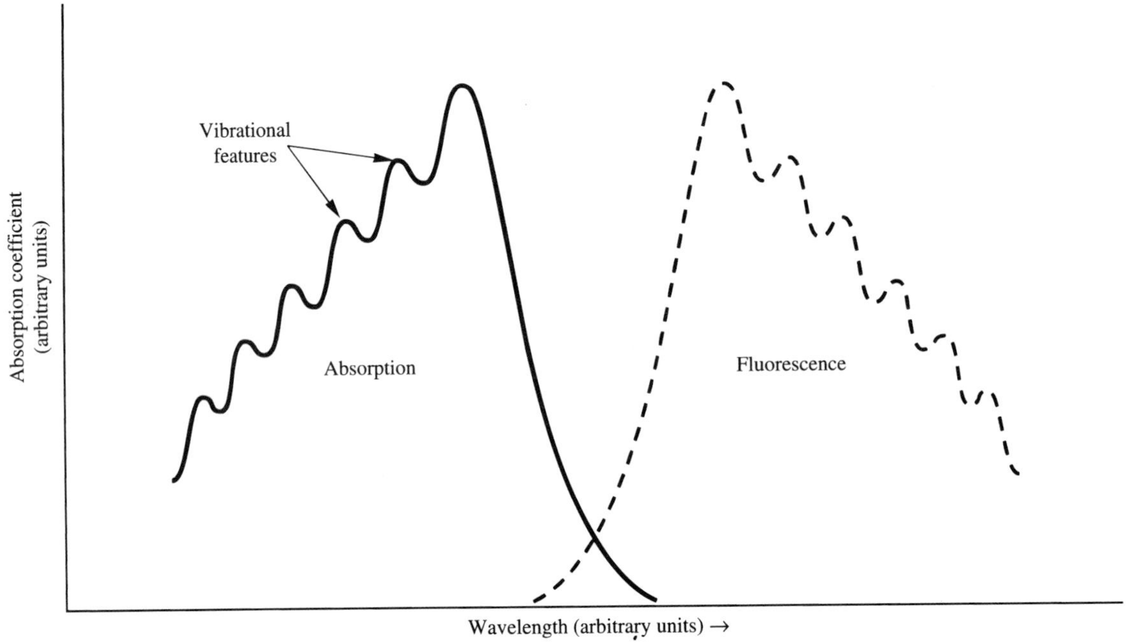

(a)

Fig. 12.3. (a) Idealized absorption and fluorescence spectra of an aromatic organic dye. The vibrational structure is usually smeared out in solution.

wavelengths will diffract within an angular range $\Delta\theta$ that keeps them travelling back-and-forth within the resonator. A reduction in the output linewidth of about a factor 100 can be achieved with a grating having a small groove spacing, say $d = 0.5$ μm. The diffraction gratings used for this purpose are generally fabricated with a groove profile to maximize their efficiency for diffracting at the angle for which m in Eq. (12.1) is equal to unity. For additional reduction in the output linewidth intracavity Fabry–Perot etalons prisms, or birefringent crystal waveplates† can be used. The interested reader should consult Reference [12.4] for more details. One popular simple arrangement is shown in Fig. (12.7). Line-narrowing of the output radiation in this case is accomplished by using a diffraction grating at a large angle of incidence (so-called *grazing* incidence). The pump laser is focused into the dye curvette, often with a cylindrical lens so as to produce a line image. For high power operation the dye would be continuously recirculated through a filter from a reservoir of dye solution.

12.2.3 CW Dye Laser Operation

The pump intensity required to reach threshold in a CW dye laser is very high, typically 50 kW cm^{-2}. It is achieved by focusing a high power visible CW laser into a small volume of dye solution ($\sim$ 50 μl). The dye solution is recirculated by spraying it from a slit nozzle in the form of a thin, planar jet, collecting the jet

† See Chapter 18 for a detailed discussion of waveplates – components that affect the polarization state of a light beam in a way that depends on the wavelength.

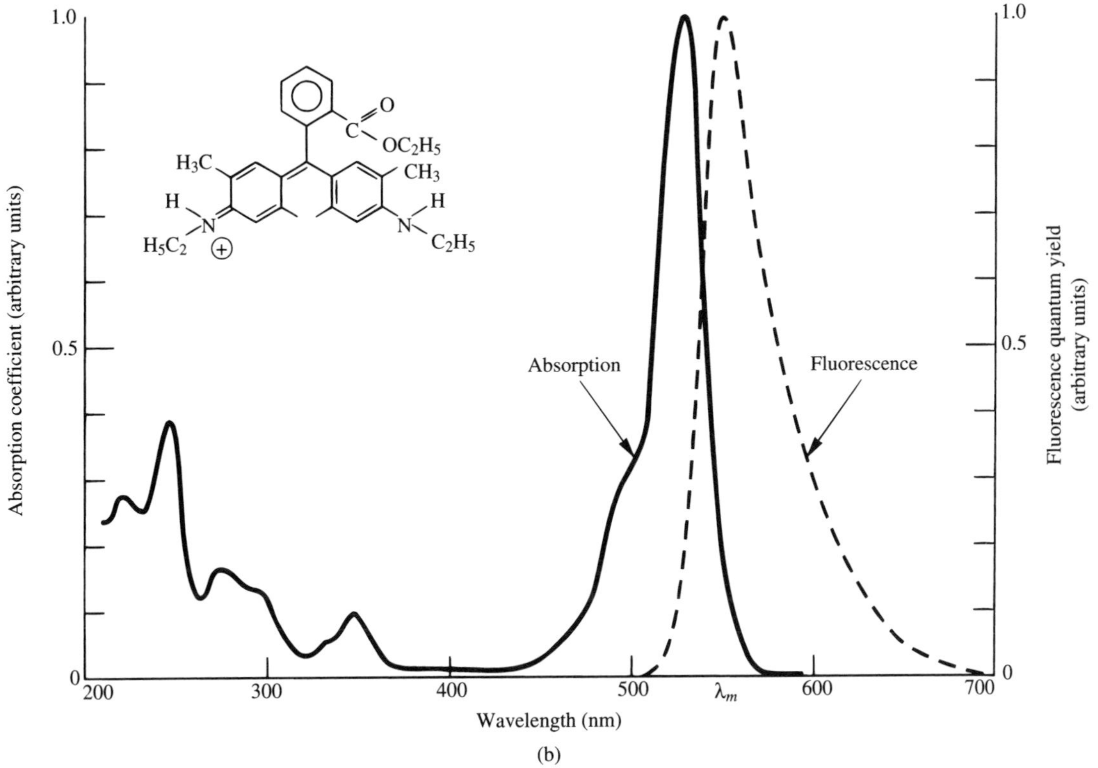

Fig. 12.3. (*cont.*) (b) Absorption and emission spectra of the organic dye Rhodamine 6G dissolved in ethanol.

and returning it to the dye reservoir. Frequently, the plane of the dye jet is set at Brewster's angle. Because of its desirable solvent and viscosity properties ethylene glycol is frequently used as the solvent for the dye. Tuning can be accomplished with either an intracavity prism, or a reflective grating. Fig. (12.8) shows such an arrangement in which the dispersion of the tuning prism allows the pump laser beam to enter without having to pass through one of the dye laser mirrors. Because this three-mirror resonator involves oblique incidence on one of the cavity mirrors it can introduce astigmatism into the resultant laser beam. However, it is possible to place a Brewster-angle oriented dye cell or dye jet so that its astigmatism compensates for that of the resonator[12.7]. The pump lasers most commonly used are argon and krypton ion lasers, which operate in the ultraviolet–red part of the spectrum. Fig. (12.9) shows the tuning capability of various dyes pumped by various argon and krypton ion laser lines. The use of different dye/solvent combinations provides tunability over a broad region.

Because the pump laser is focused to a small spot (10–20 μm in diameter) in the dye solution it is essential to design a dye laser cavity with a small spot size, and to locate its beamwaist at the focus of the pump beam.

12.3 Calculation of Threshold Pump Power in Dye Lasers

The calculation of the threshold pump power in a dye laser follows similar lines to the calculation performed for optically pumped solid-state lasers in Chapter

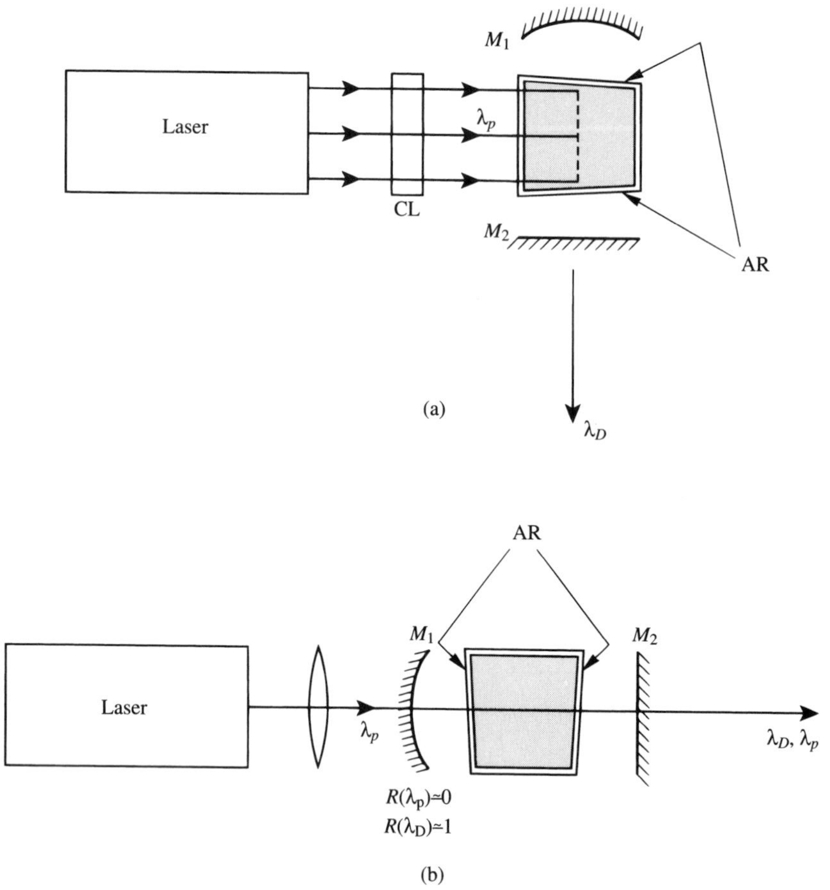

Fig. 12.4. Simple arrangements for (a) side-pumping and (b) end-pumping of simple pulsed dye lasers. In each case the dye solution is contained in an optical cuvette with wedged, antireflection coated (AR) faces. In (a) the laser beam is focused with a cylindrical lens (CL) to a line image in the cuvette. In (b) the pump laser is focused with a spherical lens (L). M_1 and M_2 are the dye laser cavity mirrors.

8. However, we must allow for the possible build-up of triplet state molecules, particularly in the T_1 state, since these constitute a pump-dependent distributed absorption loss. The various important pathways in the system are shown schematically in Fig. (12.10).

Pump radiation populates a vibrational level of S_1 but very rapid internal conversion transfers these molecules to the lowest vibrational level of S_1. If the ground state dye concentration is N, then the absorption coefficient for the pump radiation can be written as

$$\alpha_p = \sigma_p N, \tag{12.2}$$

where σ_p is the absorption cross-section at the pump wavelength. If all the pump radiation of intensity I_p is absorbed by the dye solution of thickness d then we can write the pumping rate as

$$R = \frac{I_p}{h\nu_p d}. \tag{12.3}$$

Molecules leave S_0 by spontaneous emission (fluorescence) with lifetime τ_f and by transferring to T_1 with rate constant k_{ST}. Clearly the upper laser level population

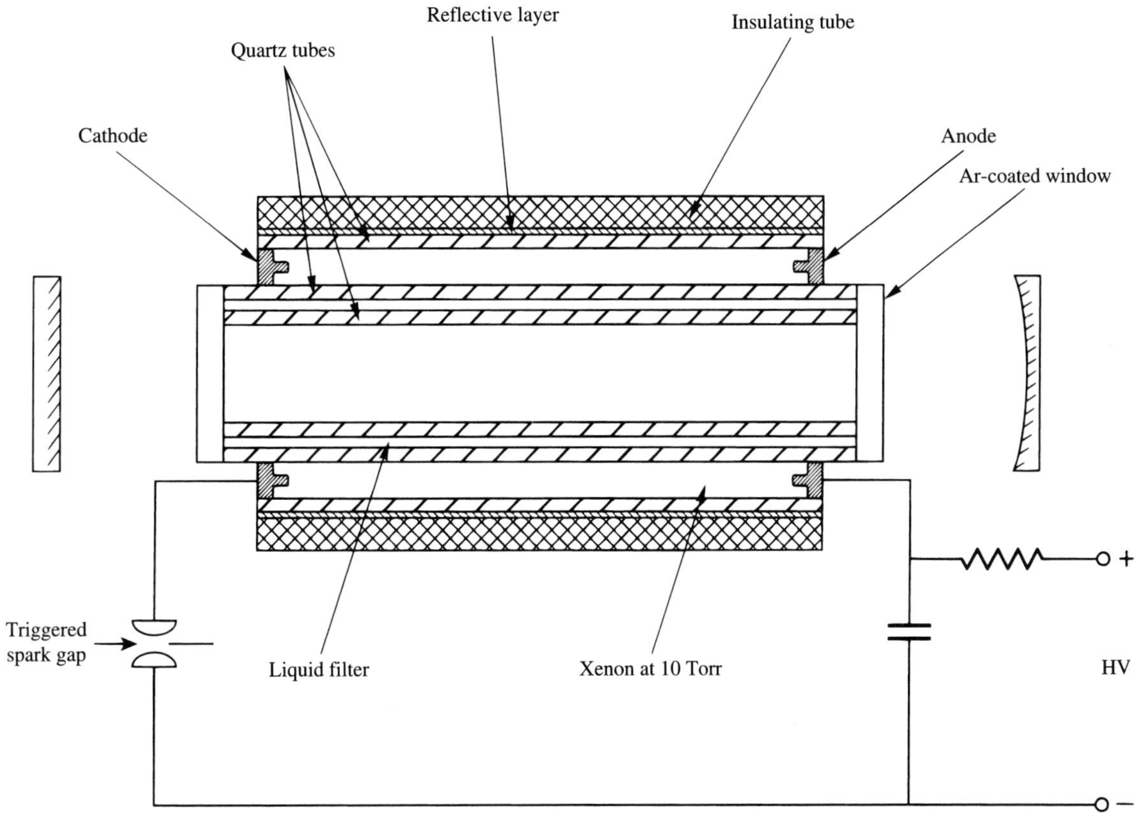

Fig. 12.5. Schematic diagram of a high energy pulsed dye laser using a coaxial flashlamp. The liquid filter absorbs short wavelength ultraviolet that can photodissociate dye molecules.

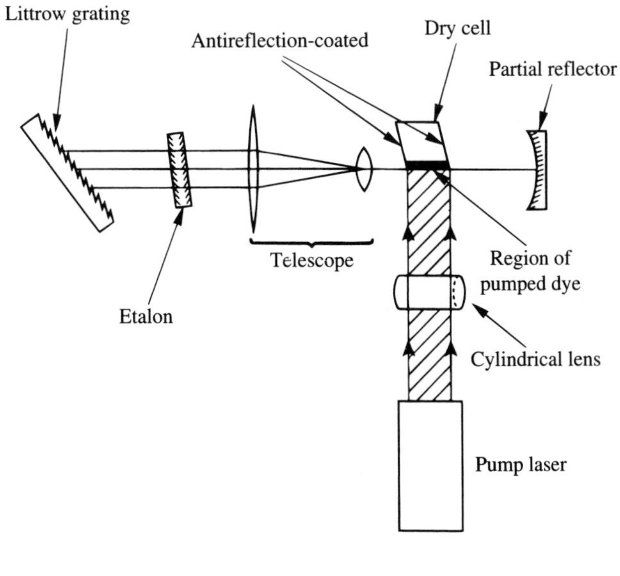

Fig. 12.6. A pulsed dye laser design that uses a reflective diffraction grating for wavelength tuning and an intracavity etalon for line-narrowing. This is referred to as a Hansch-type dye laser[12.3].

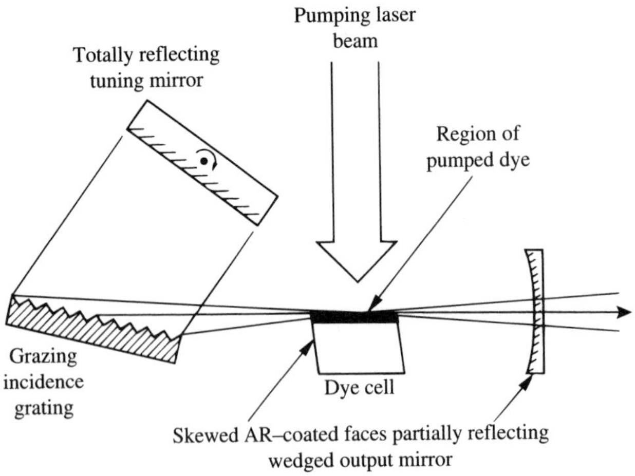

Fig. 12.7. Littman-type pulsed dye laser[12.5],[12.6] that uses a grazing incidence grating for simultaneous tuning and line-narrowing.

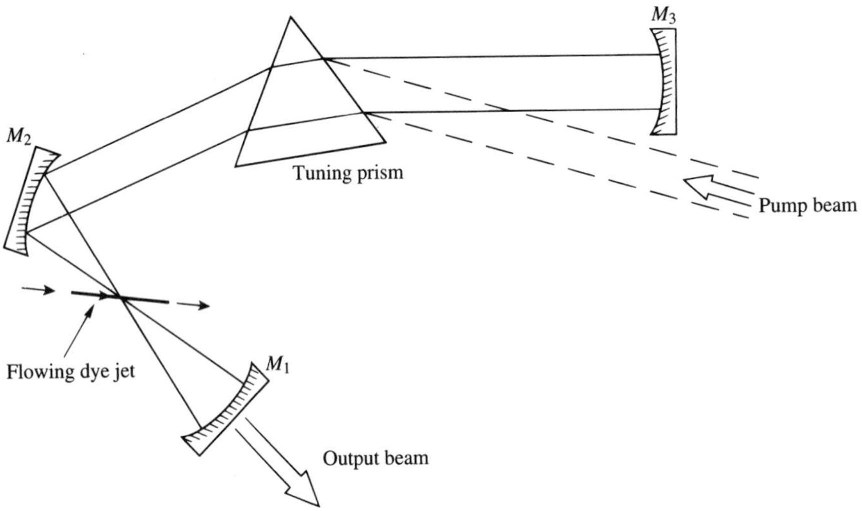

Fig. 12.8. Schematic of a CW dye laser design incorporating an astigmatically compensated laser cavity and an intracavity flowing dye jet at Brewster's angle.

density N_2 satisfies the equation

$$\frac{d}{dt}N_2 = R - \frac{N_2}{\tau_f} - k_{ST}N_2. \tag{12.4}$$

12.3.1 Pulsed Operation

In short pulse operation with an excitation pulse faster than $1/k_{ST}$ (typically $\simeq$ 100 ns) we can neglect transfer to the triplet state. The pump intensity needed to achieve an upper level population density then satisfies

$$I_p = N_2 h\nu_p d/\tau_f. \tag{12.5}$$

The lower laser level population is close to zero, so to reach threshold $N_2 \simeq N_t$. If we neglect distributed loss, which is reasonable in pulsed operation, then from

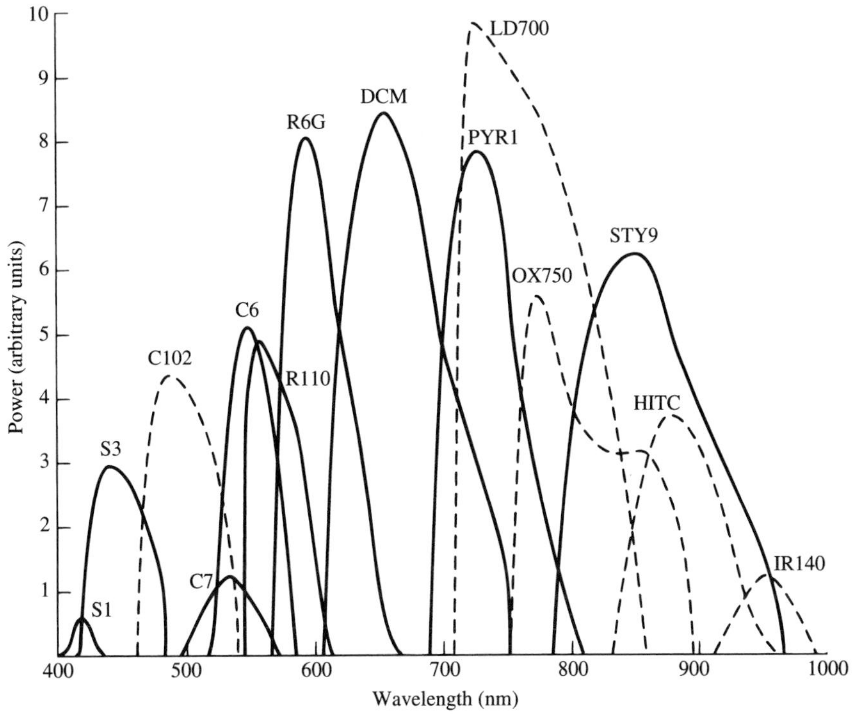

Fig. 12.9. Typical tuning curves of various dyes pumped by argon and krypton ion laser lines. The dyes and pump wavelenth are as follows: S1 (Stilbene 1, Ar^+ 351–364 nm); S3 (Stilbene 3, Ar^+ 351–364 nm); C102 (Coumarin 102, Kr^+ 407–423 nm); C7 (Coumarin 7, Ar^+ 476 nm); C6 (Coumarin 6, Ar^+ 488 nm); R110 (Rhodamine 110; Ar^+ 458–514 nm); R6G (Rhodamine 6G, Ar^+ 458–514 nm); DCM (Ar^+ 458–514 nm); PYR1 (Pyridene1, Ar^+ 458–514 nm); LD700 (Kr^+ 647 and 676 nm); 0×750 (Oxazine, 750 Kr^+ 647 and 676 nm); STY (Styryl 9 Ar^+ 458–514 nm); HITC (HITC-P Kr^+ 647 and 676 nm); IR140 (Kr^+ 752 and 799 nm) (Tuning curves courtesy of Spectra-Physics, Inc.).

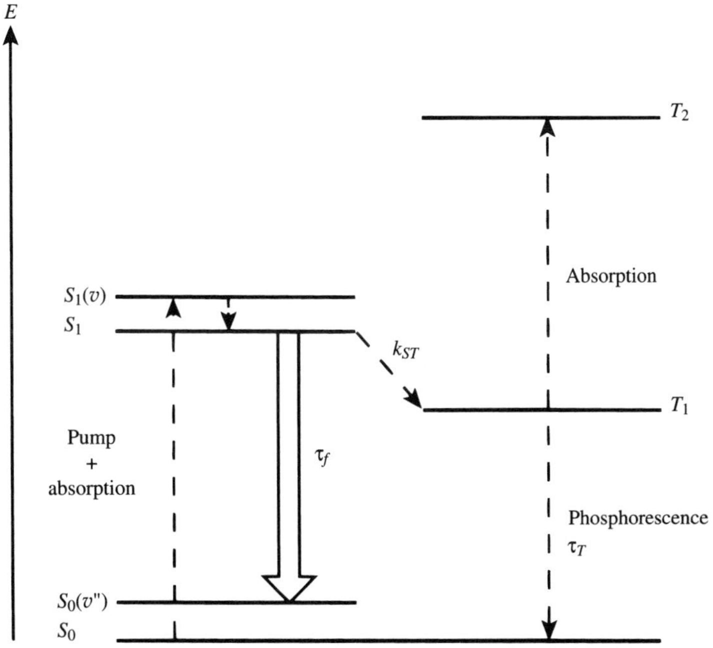

Fig. 12.10. Excitation and relaxation pathways in a dye laser.

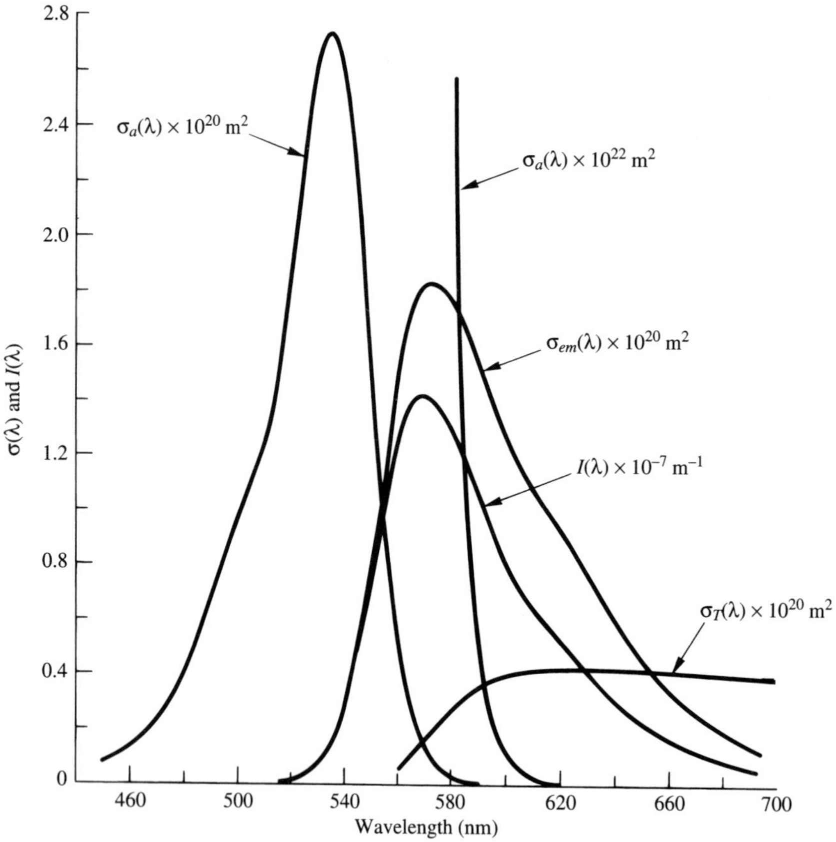

Fig. 12.11. Absorption and emission cross-sections for the dye Rhodamine 6G. Data from Snaveley[12.8]. The emission spectrum has been normalized so that $\int I(\lambda)d\lambda = 0.92$, the measured fluorescence quantum efficiency.

Eq. (5.17) the threshold inversion density can be written as

$$N_t = -\frac{8\pi\tau_f \ln r_1 r_2}{\lambda^2 g(\nu_0, \nu)\ell}. \tag{12.6}$$

For a homogeneously broadened fluorescence spectrum $g(\nu_0, \nu) = 2/\pi\Delta\nu$. If we design a laser cavity such that $R_1 = 1$, $R_2 = (1 - T)$ then $-\ln r_1 r_2 = T/2$. Combining Eqs. (12.5) and (12.6) gives

$$I_p = 2\pi^2 h\nu_p \Delta\nu T d/\lambda^2 \ell. \tag{12.7}$$

In a transversely pumped geometry, such as shown in Fig. (12.4), some reasonable values for the parameters are: $d = 1$ mm, $\ell = 10$ mm, $T = 0.01$. For Rhodamine 6G in aqueous solution a frequency–doubled† Nd:YAG laser operating at 530 nm is a suitable pump. The peak of the fluorescence spectrum is $\lambda = 580$ nm with a FWHM of about 60 nm, equivalent to 5.4×10^{13} Hz, as shown in Fig. (12.11). The calculated threshold pump intensity is 2.1×10^4 W cm^{-2}. Such intensities are easily obtained with focused pump beams of modest energy. For example, with a line image of the pump beam in the dye solution with a width of 1 mm the total pump peak is only 2.1 kW. With a Q-switched laser pulse 10 ns in length the threshold pump energy is only 21 μJ.

† See Chapter 21.

12.3.2 CW Operation

In a CW dye laser, we can no longer neglect the distributed loss that results from transfer of molecules to the triplet state. If the phosphorescence lifetime in Fig. (12.10) is τ_T the triplet state obeys the equation

$$\frac{d}{dt}N_T = N_2 k_{ST} - \frac{N_T}{\tau_T}. \tag{12.8}$$

The $S \rightarrow T$ transfer rate can be estimated from the fluorescence quantum efficiency ϕ by assuming that any molecule that does not fluoresce to the ground state crosses to T_1. Clearly,

$$\phi = \frac{1/\tau_f}{1/\tau_f + k_{ST}}, \tag{12.9}$$

so

$$k_{ST} = \frac{1}{\tau_f}(1-\phi). \tag{12.10}$$

In equilibrium, from Eq. (12.8),

$$N_T = N_2 k_{ST} \tau_T. \tag{12.11}$$

The threshold gain is

$$\gamma_t(\lambda) = \alpha(\lambda) - \frac{1}{2\ell}\ln R_1(\lambda)R_2(\lambda), \tag{12.12}$$

where the explicit dependence of all the quantities on wavelength has been noted, particularly because laser oscillation potentially can occur over a broad wavelength region. The distributed loss comes from $S_0 \rightarrow S_1$ and $T_1 \rightarrow T_2$ absorption so

$$\alpha(\lambda) = N_0\sigma_a(\lambda) + N_T\sigma_T(\lambda), \tag{12.13}$$

where $\sigma_a(\lambda)$, $\sigma_T(\lambda)$ are the absorption cross-sections for absorption (at the laser wavelength, λ) from the ground state and from T_1, respectively. If we write

$$\gamma_t(\lambda) = N_2\sigma_L(\lambda), \tag{12.14}$$

where σ_L is the stimulated emission cross-section, then Eq. (12.12) becomes

$$N_2\sigma_L(\lambda) = N_0\sigma_a(\lambda) + N_T\sigma_T(\lambda) - \frac{1}{2\ell}\ln R_1(\lambda)R_2(\lambda). \tag{12.15}$$

Now the total dye concentration N must satisfy

$$N = N_0 + N_2 + N_T, \tag{12.16}$$

so from Eq. (12.15)

$$N_2(\lambda) = \frac{N\sigma_2(\lambda) - (1/2\ell)\ln R_1(\lambda)R_2(\lambda)}{\sigma_L(\lambda) + \sigma_a(\lambda) - k_{ST}\tau_T(\sigma_T - \sigma_a)}. \tag{12.17}$$

Values for the various cross-sections used in Eq. (12.17) are shown for the dye Rhodamine 6G in Fig. (12.11). If the laser cavity does not provide wavelength-selective feedback, the laser will oscillate at the wavelength for which $N_2(\lambda)$ is a minimum – the *intrinsic* wavelength. The threshold inversion increases as the laser wavelength is tuned to values below the intrinsic value because $\sigma_a(\lambda)$ increases. The ideal dye is one for which $\sigma_a(\lambda) \simeq 0$, $k_{ST}\tau_T = 0$. This latter condition can be obtained, approximately, by adding a triplet state quencher such as a detergent to the solution. In this case

$$N_2(\lambda) = \frac{-\frac{1}{2}\ln R_1(\lambda)R_2(\lambda)}{\sigma_L(\lambda)}. \tag{12.18}$$

This inversion density is equivalent to the one considered previously in pulsed operation. The threshold pump intensities are comparable. In practice, desirable CW pump intensities exceed 100 kW cm^{-2}. A 10 W argon ion laser focused to a Gaussian spot of spotsize $w_0 = 5\ \mu m$ has a peak intensity of 25 MW cm^{-2}.

12.4 Inorganic Liquid Lasers

Several rare earth ions will exhibit laser action when a molecule containing the ion is dissolved in an appropriate solvent. Some examples are given in Table (12.1). These lasers have characteristics in some ways intermediate between crystalline solid-state lasers and glass lasers containing the active ion. For example, Fig. (12.12) shows the fluorescence spectra of neodymium ions in two different inorganic solvents, selenium oxychloride and phosphorous oxychloride†, compared to their fluorescence spectrum in glass. The emission of the ions is broader in the glass, which can be interpreted as evidence of the broad range of local environments experienced by Nd^{3+} ions in the glass. In the liquid system some local structure exists as each Nd^{3+} ion bonds to molecules of the inorganic solvent in a specific way. Nd^{3+} liquid lasers can provide substantial power outputs, pulsed outputs of several hundred joules, and peak powers of several GW have been obtained. Because the laser emission wavelength of Nd^{3+} ions in liquid solution is still 1.06 μm hybrid high energy oscillator/amplifier systems can be constructed where Nd:YAG and Nd:glass components are also used. One important advantage and one important disadvantage of the liquid system in this regard are worth mentioning. A liquid gain medium is self-repairing. If optical damage occurs and the liquid is circulated a homogeneous medium is presented for the next firing of the laser. However, the index of refraction of a liquid changes 10^2–10^3 times faster with temperature than in a solid so liquid gain media excited by flashlamps can show very strong thermal lensing effects.

12.5 Free Electron Lasers

An electron travelling undisturbed will not spontaneously emit energy. However, if the electron accelerates it must absorb energy, if it decelerates it must radiate it. For example, if an electron increases its velocity in a particle accelerator it does so by absorbing energy from the electromagnetic field supplied by the accelerating electrode. If an electron is constrained to move in a circle by a combination of electric and magnetic fields then it emits *synchrotron* radiation. Since the electron is experiencing a central acceleration it should absorb energy. However, if the electron is constrained to a circular path it cannot increase its kinetic energy so the energy absorbed must be reradiated. The spontaneous radiation emitted by an electron as it decelerates is called *Bremsstrahlung*. A special kind of radiation called *Cerenkov*

† The laser medium is usually made by dissolving Nd_2O_3 in the solvent; typical Nd^{3+} ion concentrations are in the range 6×10^{24}–6×10^{26} m^{-3}.

Table 12.1. *Some liquid lasers.*

Ion	Solvent	Laser wavelength (μm)
Nd^{3+}	$SeOCl_2$–$SnCl_4$	1.056, 1.33
	$POCl_3$–$SnCl_4$	1.052–1.06
	$POCl_3$–$ZrCl_4$	1.056
Eu^{3+} (in a complex with phosphorous ions and benzoylacetonate)	Ethanol	0.6129
Eu^{+} (in a complex with dimethyl ammonium ions and trifluoro-acetylacetonate)	Acetonitrile	0.6119
Tb^{3+} (in a complex with phosphorous ions and thenoyltrifluoracetonate)	Polymethylmethacrylate (plexiglas)	0.545

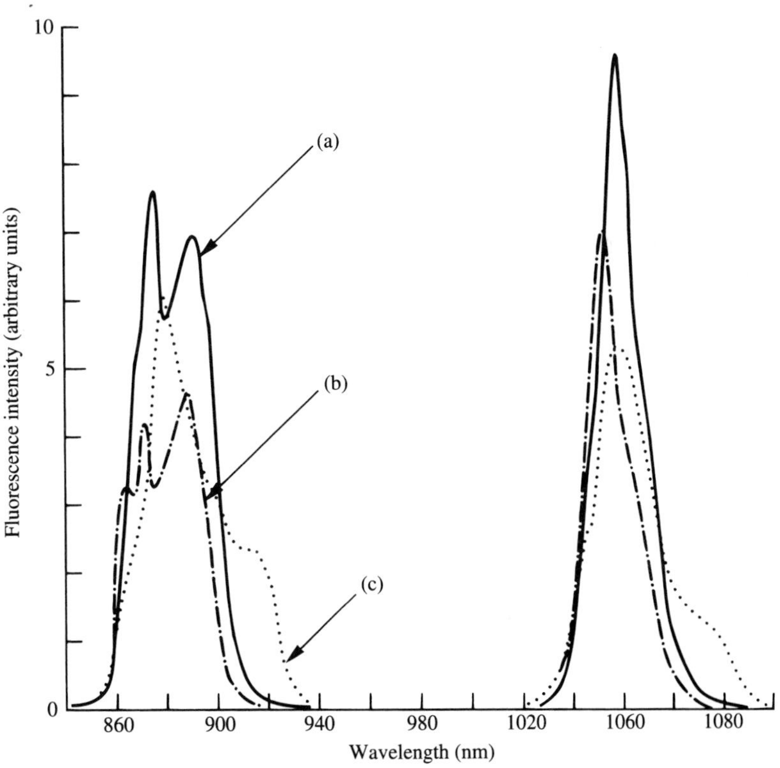

Fig. 12.12. Fluorescence spectra of Nd^{3+} ions in: (a) $SeOCl_2$ solvent; (b) $POCl_3$ solvent; (c) glass.

radiation is emitted when an electron with velocity v, where $v > c_0/n$, enters a medium of refractive index n. This radiation is emitted because the electron cannot have a velocity in the medium greater than the velocity of light in the medium.

The general form of the electromagnetic field produced by an accelerating or decelerating electron is complex, but a couple of special cases will serve to illustrate some features of the field. If the electron is non-relativistic (its velocity is much less than the velocity of light) then the electric and magnetic field vectors of the radiation produced by the electrons are[12.9]

$$\mathbf{E} = \frac{e}{4\pi\epsilon_0 c^2 r^3}\mathbf{r} \times (\mathbf{r} \times \dot{\mathbf{v}}), \tag{12.19}$$

$$\mathbf{B} = \frac{e}{4\pi\epsilon_0 c^3 r^2}\dot{\mathbf{v}} \times \mathbf{r}, \tag{12.20}$$

where $\mathbf{v}$ is the velocity vector of the electron and $\mathbf{r}$ is the vector from the electron to the point of observation.†. Eqs. (12.19) and (12.20) are similar to the field of a small radiating electric dipole and have the form shown in Fig. (12.13a). If the velocity of the electron is comparable to the velocity of light then the form of the electromagnetic field takes a relatively simple form if $\dot{\mathbf{v}}$ and $\mathbf{v}$ are parallel. In this case

$$\mathbf{E} = \frac{e}{4\pi\epsilon_0 c^2 s^3}\mathbf{r} \times (\mathbf{r} \times \dot{\mathbf{v}}), \tag{12.21}$$

$$\mathbf{B} = \frac{er}{4\pi\epsilon_0 c^3 s^3}\dot{\mathbf{v}} \times \mathbf{r}, \tag{12.22}$$

where $s = r$ - $(\{\mathbf{v} \cdot \mathbf{v}\})/c$. The fields associated with a decelerating electron are in this case directed much more in the forward direction, as shown in Fig. (12.13b).

The above line of discussion is intended to show that for an accelerated or decelerated electron there is a spontaneous emission process. In the free electron laser the spontaneous emission of the electron is channelled into a particular mode (or set of modes) of the radiation field by selecting the velocity of the beam in such a way that the radiation emitted by each electron adds coherently to the radiation from the other electrons in the beam. In such *free-electron* lasers (FELs) an electron beam is perturbed in such a way as to encourage its interaction with a photon beam and stimulated emission into a particular direction and frequency band. Free electrons are not restricted to specific energies as are bound electrons, they can have any energy. The electrons make what is called *free–free* transitions – they jump between two energy states in the continuum of free-electron states.

In a Bremsstrahlung FEL a relativistic‡ electron beam is fired through a periodic static magnetic field (a *wiggler* or *undulator*), which forces the electrons to oscillate and radiate energy in a narrow frequency band. For example, in Fig. (12.14) an electron beam travelling in the z direction has its velocity in the y direction modulated by the Lorentz force, $\mathbf{F} = -e(\mathbf{v} \times \mathbf{B})$, of the alternating magnetic flux $\mathbf{B}$ provided by the north and south poles of the wiggler magnet. The electrons in the beam see the wiggler field as a pump electromagnetic wave. These FELs can operate in two regimes depending on whether the electron beam is sufficiently dense for particle interactions within the beam to be important or not. When the electrons in the beam do not interact the laser operates in the Compton regime, if interactions are important they operate in the Raman regime. FELs operating in the Compton regime use very high energy, low density beams, typically with beam energies in the 20 MeV–3 GeV range and currents of tens of mA and provide stimulated emission at short wavelengths (near-infrared, visible, and ultraviolet).

† Strictly $\mathbf{r}$ is the *retarded* position vector of the electron—the position where the electron was when it generated the effect that propagates with the velocity of light to the point where the field is observed. For velocities much smaller than the velocity of light this distinction is unimportant.

‡ So called because the velocity of the electrons in the beam is close to the velocity of light.

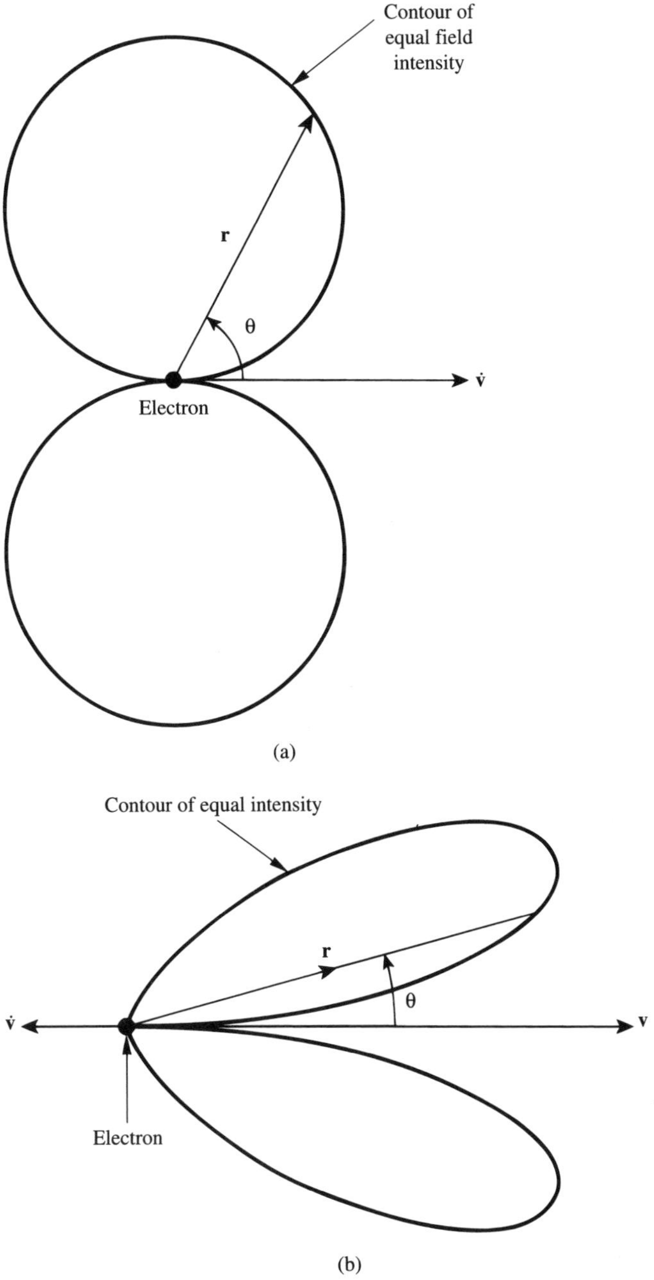

Fig. 12.13. (a) Angular distribution of the electromagnetic field produced by a low velocity electron. The field pattern is symmetrical about the direction of electron motion[12.9]. (b) Radiation field pattern of a relativistic electron that is being decelerated along its direction of motion[12.9].

FELs operating in the Raman regime utilize the interactions between particles in the beam to enhance the gain at specific frequencies. These devices typically use pulse-line accelerators with energies up to about 10 MeV and currents up to 3 kA and provide substantial gain and power output in the middle-infrared to millimeter wave range. The wiggler fields usually have periods on the order of a few centimeters and use fields of about 0.1 T.

If the wiggler field has a period ℓ_0 and the electron beam velocity along the axis

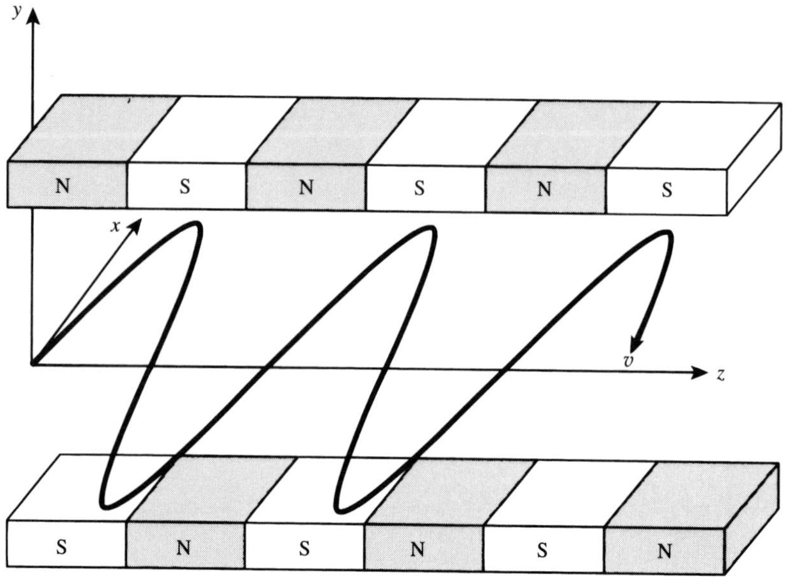

Fig. 12.14. Geometry of a Bremsstrahlung FEL in which an electron beam passes through the alternating north–south magnetic fields of a 'wiggler'[12.11].

is v then the laser wavelength satisfies the approximate relationship

$$\lambda = \frac{\ell_0}{2\gamma^2}, \tag{12.23}$$

where γ is the relativistic factor, $(1 - v^2/c^2)^{-1/2}$. Clearly if λ_0 is on the order of centimeters, large values of γ are needed to achieve short wavelength operation. We can represent the axial variation of the wiggler magnetic field in Fig. (12.14) as

$$B_x(z) = B_0 \sin k_0 z, \tag{12.24}$$

where $k_0 = 2\pi/\ell_0$, with ℓ_0 the wiggler period. The motion of an electron is governed by the Lorentz force acting on it. Because an electron must be assumed to be travelling near the velocity of light its mass, measured in the rest frame of the magnet, has increased to γm. Consequently, it obeys the equation of motion

$$\frac{d}{dt}(\gamma m\mathbf{v}) = \frac{e}{c}(\mathbf{v} \times \mathbf{B}). \tag{12.25}$$

If we assume that γ does not change, which is only true if we neglect radiation, then

$$\frac{d\mathbf{v}}{dt} = \frac{e}{\gamma m}(\mathbf{v} \times \mathbf{B}). \tag{12.26}$$

Eq. (12.26) can be written in the form

$$\frac{dv_x}{dt} = -\frac{\Omega v_z}{\gamma} \sin k_0 z, \tag{12.27}$$

$$\frac{dv_z}{dt} = \frac{\Omega v_x}{\gamma} \sin k_0 z, \tag{12.28}$$

where $\Omega = eB_0/mc$. If $v_x << v_z$ we can put $z = v_z t$ and solve Eqs. (12.27) and

(12.28) to give

$$v_x = \frac{v_0\Omega}{\gamma\omega_0}\cos\omega_0 t, \tag{12.29}$$

$$v_z = v_0 - \frac{2v_0\Omega^2}{4\gamma^2\omega_0^2}\cos 2\omega_0 t, \tag{12.30}$$

where v_0 is the velocity of the electron beam before it enters the wiggler field and $\omega_0 = k_0 v_z$. The position of the electron satisfies the equations

$$x = \frac{v_0\Omega}{\gamma\omega_0^2}\sin\omega_0 t, \tag{12.31}$$

$$z = v_0 t - \frac{v_0\Omega^2}{4\gamma^2\omega_0^3}\sin 2\omega_0 t. \tag{12.32}$$

Therefore, the electron undergoes a lateral simple-harmonic motion as it traverses the wiggler field. The oscillating electron emits radiation at a wavelength that must be calculated by considering the Doppler shift resulting from the electron motion.

As far as the electron is concerned, the wiggler field appears as a time varying magnetic (and electric) field, that is as an electromagnetic wave. In this field the electron oscillates with a frequency

$$\omega_0' = \gamma k_0 v_z = \gamma k_0 v_0, \tag{12.33}$$

the relativistic factor γ appears because the electrons see many more undulations of the field within their unit of time. To the observer at rest in the laboratory this corresponds to the relativistic phenomenon of a moving clock appearing to run slow. The electron can interact with radiation that has frequency ω_0' (within the frame of reference travelling with the electron). This radiation is observed in the laboratory at frequency ω_s where

$$\omega_0' = \gamma(\omega_s - \mathbf{k}_s \cdot \mathbf{v}). \tag{12.34}$$

ω_s is greater than ω_0' if the radiating electron is travelling toward the observer. In Eq. (12.34) $k_s = \omega_s/c$, is the wave vector observed in the laboratory frame. If $\mathbf{k}$ is nearly parallel to $\mathbf{v}$ then

$$\omega_0' = \gamma\omega_s\left(1 - \frac{v}{c}\cos\theta\right), \tag{12.35}$$

where θ is the angle between $\mathbf{k}$ and $\mathbf{v}$ ($\simeq v_0$). For θ equal to a small angle, $\cos\theta = 1 - \theta^2/2$ and Eq. (12.35) gives

$$\gamma k_0 v_0 = \gamma\omega_s\left(1 - \frac{v_0}{c}\right). \tag{12.36}$$

Since $1 - (v_0/c) \simeq 1/2\gamma^2, k_0 = 2\pi/\ell_0, \omega_s = 2\pi c/\lambda_s$ we get

$$\lambda_s \simeq \frac{\ell_0}{2\gamma^2}. \tag{12.37}$$

This radiation is emitted in the forward direction, parallel to the electron velocity.

The Doppler shifting of the emitted radiation wavelength is quite significant. For example, if $\ell_0 \sim 10$ mm and the electron beam energy is 2 MeV then λ_s is in the submillimeter region, if the beam energy increases to 50 MeV the emitted radiation moves into the near-infrared. Practical FELs can take many forms. Fig. (12.15)

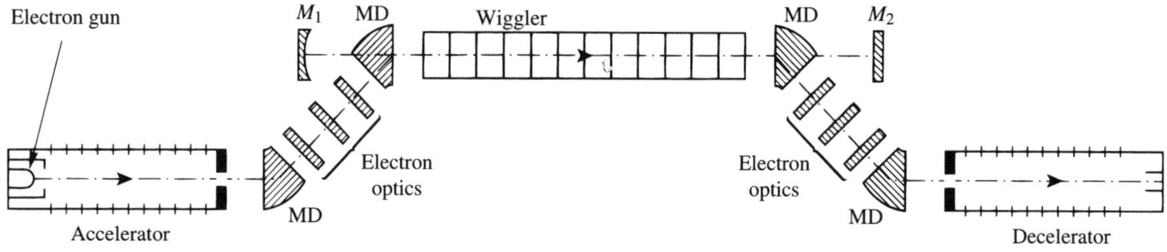

Fig. 12.15. Schematic arrangement of an FEL in which a beam of electrons from an accelerator is deflected into and out of a Fabry–Perot resonator.

shows a schematic diagram of how pulses of electrons from an accelerator are deflected into, and out of an optical Fabry–Perot resonator. Inside the resonator the electron beam interacts with the wiggler field. The theory of operation of such devices is not quite so straightforward as the simple discussion here suggests[12.10] and additional effects must be considered when the electron beam interacts with itself so as to form 'bunches' or when the electron beam has a high particle density so that plasma-like phenomena must be considered.

12.6 Problems

(12.1) Calculate the spectral width of the dye laser output when wavelength selection is provided by a diffraction grating if the laser beam divergence is θ_{beam}. Hint: consider the *angular dispersion* of the grating $d\theta/d\lambda$ calculated from Eq. (12.1).

(12.2) (Computer project) Use the data in Fig. (12.11) to calculate the intrinsic wavelength of a Rhodamine 6G dye laser as a function of concentration N.

References

[12.1] J.B. Birks, *Photophysics of Aromatic Molecules*, Wiley-Interscience, New York, 1970.

[12.2] F.N. Baltakov, B.A. Barikhin, and L.V. Sukhanov, '400-J pulsed laser using a solution of rhodamine-6G in ethanol,' *JETP Lett.*, **19**, 174–175, 1974.

[12.3] T.W. Hänsch, 'Repetitively pulsed tunable dye laser for high resolution spectroscopy,' *Appl. Opt.*, **11**, 895–898, 1972.

[12.4] F.P. Schäfer, Ed., *Dye Lasers*, 2nd revised edition, Topics in Applied Physics, Vol. 1, Springer-Verlag, Berlin, 1977.

[12.5] M.G. Littman and H.J. Metcalf, 'Spectrally narrow dye laser without beam expander,' *Appl. Opt.*, **17**, 2224–2227, 1978.

[12.6] M. Littman and J. Montgomery, 'Grazing-incidence designs improve pulsed dye lasers,' *Laser Focus*, **24**, 70–86, 1988.

[12.7] H.W. Kogelnik, E.P. Ippen, A. Dienes, and C.V. Shank, 'Astigmatically compensated cavities for CW dye lasers,' *IEEE J. Quantum Electron.*, **QE-8**, 373–379, 1972.

[12.8] B.B. Snavely, 'Continuous-wave dye lasers,' in *Dye Lasers*, F.P. Schäfer, Ed., 2nd revised edition, Topics in Applied Physics, Vol. 1, Springer-Verlag, Berlin 1977.

[12.9] W.K.H. Panofsky and M. Phillips, *Classical Electricity and Magnetism*, Addison-Wesley, Reading, MA, 1955.

[12.10] T.C. Marshall, *Free Electron Lasers*, Macmillan, New York, 1985.

[12.11] D. Prosnitz, 'Free electron lasers,' in *Handbook of Laser Science and Technology*, Vol. I, *Lasers and Masers*, M.J. Weber, Ed., CRC Press, Boca Raton, FL, 1982.

13
Semiconductor Lasers

13.1 Introduction

The semiconductor laser, in various forms, is the most widely used of all lasers, it is manufactured in the largest quantities, and is of the greatest practical importance. Every compact disc (CD) player contains one. Much of the world's long and medium distance communication takes place over optical fibers along which propagate the beams from semiconductor lasers.

These lasers operate by using the jumps in energy that can occur when electrons travel between semiconductors containing different types and levels of controlled impurities (called *dopants*). In this chapter we will discuss the basic semiconductor physics that is necessary to understand how these lasers work, and how various aspects of their operation can be controlled and improved. Central to this discussion will be what goes on at the junction between p- and n-type semiconductors. The ability to grow precisely-doped single- and multi-layer semiconductor materials and fabricate devices of various forms – at a level that could be called molecular engineering – has allowed the development of many types of structures that make efficient semiconductor lasers. In some respects the radiation from semiconductor lasers is far from ideal, its coherence properties are far from perfect, being intermediate between those of a low pressure gas laser and an incoherent line source. We will discuss how these coherence properties can be controlled in semiconductor lasers.

13.2 Semiconductor Physics Background

When atoms combine to form a solid the sharply defined energy states that we associate with free atoms, such as in a gas, broaden considerably because of the large number of interactions between an atom and its neighbors. The result is a series of broad energy states, called *bands*. The energy states of the outermost electrons of the atoms constituting the solid broaden and combine to produce a band of filled energy states called the *valence band*. Excited energy states of the atoms broaden and combine to give rise to a series of excited energy bands, which at absolute zero are empty. The lowest, empty energy band is called the *conduction band*. The relative disposition of the valence and conduction bands determines whether the solid will be an insulator, a metal, or a semiconductor.

The band structure can be drawn in a one-dimensional energy representation as shown in Fig. (13.1). The energy spacing, E_g, between the top of the highest filled energy band and the bottom of the lowest unfilled energy band is called the *band gap*. In an insulator, $E_g >> kT$, so virtually no electrons can be thermally excited into the conduction band to give rise to electrical conductivity. Electrical

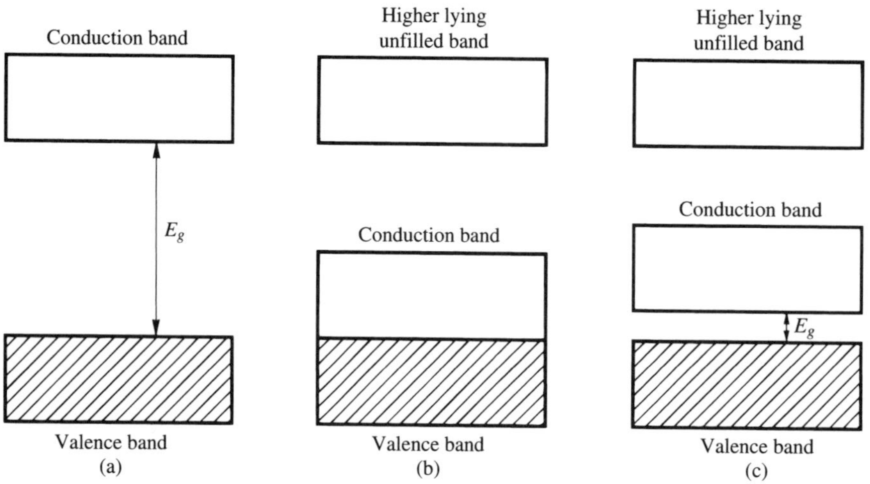

Fig. 13.1. Simple band diagrams for (a) an insulator, (b) a metal, and (c) a semiconductor, where $E_g \sim kT$.

conductivity can only result if there are unfilled states within an energy band. In a semiconductor E_g is a few times larger than kT; for a very small energy gap the material can be referred to as a semimetal – such as bismuth.

The most important semiconductor material in electronics is silicon (in the early days it was germanium). However, the most important semiconductor materials in photonics are binary materials, such as the III–V† semiconductors GaAs, InSb, ternary materials such as $Al_xGa_{1-x}As$ (where x is a fractional factor ≤ 1 that determines the precise stochiometry of the material), and quaternary materials such as $In_xGa_{1-x}\,As_yP_{1-y}$. The ternary and quaternary materials can be viewed as binary materials in which one or both principal constituents have been replaced to some extent by atoms from the same group in the periodic table. Although a broad range of multicomponent crystals can be grown in this way, arbitrary compositions are not all possible.[13.1] The most important parameter that changes from one semiconductor system to the next is the band gap, as can be seen in Table (13.1). We shall see that the primary process that leads to light generation in semiconductors involves an electron falling from the conduction band to the valence band and releasing a photon. The characteristic wavelength associated with the bandgap energy is $\lambda = ch/E_g$.

When an electric field is applied to a semiconductor conduction results because electrons in the conduction band acquire directed velocities. These velocities are in the opposite direction to the applied field because the electrons are negatively charged.‡ Each electron moving in a material has momentum $p = m^*v$. The mass is written as m^* to indicate that this is the *effective* mass of an electron moving in the lattice, which because of the wave/particle dual character of the electron and its interaction with the lattice is not the same as the free electron mass, and may be negative. The momentum can be related to the wavelength of the electron, viewed as a wave, by the famous de Broglie relation

$$p = h/\lambda, \tag{13.1}$$

† The roman numerals indicate the group in the periodic table to which each atomic constituent belongs.

‡ It sometimes happens that because of the interaction between the electrons and the lattice the electrons do not move in this intuitive way, they behave as if their mass had becomed *negative*. For further information on the interesting way in which charge carriers move in semiconductors consult Kittel[13.2].

Table 13.1. *Band gap characteristics of important semiconductor materials.*

Material	Energy gap at 300 K (eV)	$\lambda = ch/E_g(\mu m)$
Diamond[a]	5.4	0.23
Germanium	0.66	1.88
Silicon	0.66	1.88
InSb	0.17	7.3
InAs	0.36	3.5
InP	1.35	0.92
GaP	2.25	0.55
GaP	2.25	0.55
GaAs	1.42	0.87
$In_{0.8}Ga_{0.2}As_{0.34}P_{0.65}$	1.1	1.13
$In_{0.5}Ga_{0.5}P$	2.0	0.62
GaSb	0.68	0.73
AlSb	1.6	1.58
$Al_xGa_{1-x}As$ $(0 \leq x < 0.37)$	1.42–1.92	0.65–0.87
$In_{0.53}Ga_{0.47}As$	0.74	1.67

[a] Although diamond is an insulator it can be made conductive by irradiation with ultraviolet light.

where h is Planck's constant. Since the wave vector $\mathbf{k}$ of the electron has magnitude $|\mathbf{k}| = 2\pi/\lambda$ we can relate momentum, kinetic energy, E, and wave vector

$$E = \frac{1}{2}m^*v^2 = \frac{p^2}{2m^*}, \tag{13.2}$$

$$p = \hbar k, \tag{13.3}$$

where $\hbar = h/2\pi$, and

$$E = \frac{\hbar^2 k^2}{2m^*}. \tag{13.4}$$

In this simple picture the energy bands can be drawn as parabolas on an E–k diagram. Fig. (13.2) gives an example.

For the conduction band, E in Eq. (13.2) would be measured from the bottom of the band at energy E_c; in the valence band it would be measured from the top of the band at energy E_v. Clearly in Fig. (13.2) the effective mass of electrons in the valence band is negative. To explain how electrons move within a band we must allow for the existence of at least one vacant state in the band. We must also remember that because electrons behave like waves, like the modes of the black-body radiation field they are quantized. Consequently, in a cubical crystal of side L the allowed wave vectors of the electron can only take values

$$k_i = \frac{2\pi n}{L}(i = x, y, z), \tag{13.5}$$

where n is a positive or negative integer, or zero. When this feature of the energy bands is included there are discrete electron states available in the valence and conduction bands as shown in Fig. (13.3).

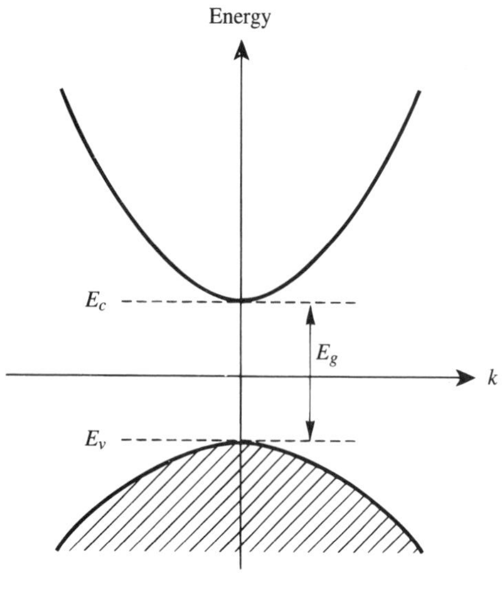

Fig. 13.2. Parabolic energy band diagram for an insulator or semiconductor crystal where no carriers have been thermally excited into the conduction band, cross-hatching indicates filled states at 0 K.

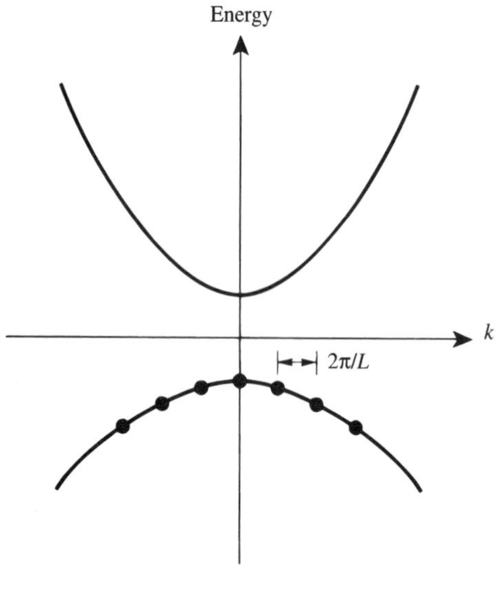

Fig. 13.3. Parabolic energy band diagram showing the filled discrete states in the valence band at 0 K.

For a filled energy band, such as the valence band shown in Fig. (13.3), the vector sum of the momenta of all the electrons is zero, that is

$$\sum_i \mathbf{k}_i = 0. \tag{13.6}$$

If an electron is excited from valence band to conduction band by absorbing a photon, then for a *direct* band gap transition the electron does not change its wave vector significantly† in crossing the gap, as shown in Fig. (13.4). If the energy minimum of the conduction band is not at the same k value as the energy maximum, then we have an *indirect* band gap semiconductor.

The vacancy left in the valence band behaves very much like a positive charge,

† This is a very small change in momentum because of the momentum of the absorbed photon.

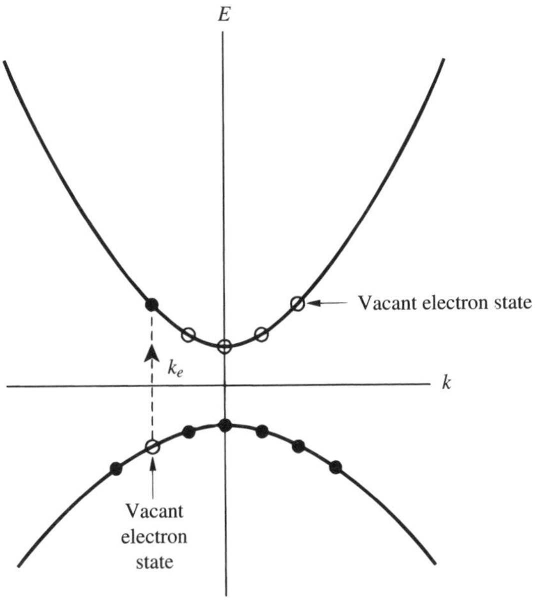

Fig. 13.4. Parabolic energy band diagram of a direct band gap semiconductor in which an electron has been excited from the valence band to a vacant state in the conduction band leaving a vacancy in the valence band.

called a *hole*. The wave vector of the electron is $\mathbf{k}_e$, so to conserve total momentum of all the electrons (in both valence and conduction bands), the total momentum of the electron in the valence band is

$$\mathbf{k}_h = -\mathbf{k}_e. \tag{13.7}$$

Therefore, the hole is not really located on the E–k diagram at the location indicated by the electron vacancy. If an electric field is applied to a semiconductor with a conduction band electron and valence band hole, then electrical conduction occurs through motion of the conduction band electron in the opposite direction to the applied field, and the positively charged valence band hole in the same direction as the field. The effective mass of the hole is positive, opposite in sign to the valence band electrons, but equal to them in magnitude

$$m_h^* = -m_e^* \text{ (valence band)}. \tag{13.8}$$

Under the action of an applied field the vacancy in the valence band does not move in the simple fashion that would be expected from Fig. (13.4). Motion of electrons into and out of the vacancy leads to motion of the vacancy in the *opposite* direction to the applied field because the equivalent hole is moving in the *same* direction as the field.

Fortunately, in describing the basic properties of semiconductors that allow them to be both detectors and generators of light we do not need to worry about the somewhat complex behavior of vacancies, but treat the material as containing negatively charged electrons and positively charged holes.

13.3 Carrier Concentrations

In a very pure (*intrinsic*) semiconductor, except at absolute zero, there are some electrons in the conduction band, and, of course, corresponding holes in the

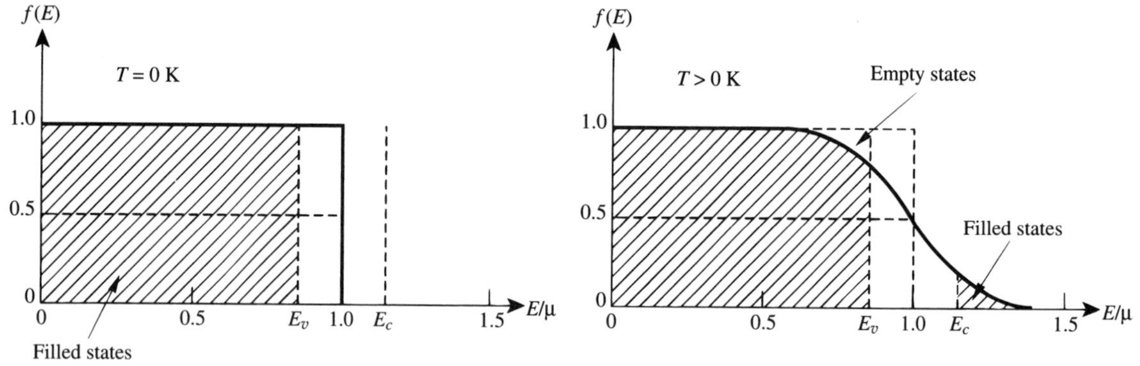

Fig. 13.5.
The Fermi–Dirac distribution shown relative to the band gap in an intrinsic semiconductor.

valence band. To determine their concentrations at temperature T we must find the probability that an electron is thermally excited across the band gap. This probability is not $\sim e^{-E_g/kT}$, as might be expected from Maxwell–Boltzmann statistics, because electrons distribute themselves among available energy states according to Fermi–Dirac statistics[13.3]. Electrons avoid each other to the extent that no two electrons can occupy exactly the same energy state. Consequently, the probability an electron will occupy an energy state at energy E_k is given by the Fermi-Dirac distribution

$$f(E_k) = \frac{1}{e^{(E_k-\mu)/kT}+1}. \tag{13.9}$$

The reference energy μ is called the Fermi level.† For $E_k - \mu >> kT$ this can be approximated as

$$f(E_k) = e^{(\mu-E_k)/kT}. \tag{13.10}$$

Fig. (13.5) shows the shape of this distribution relative to the band gap in an intrinsic semiconductor both at absolute zero and at a finite temperature.

A few significant features of this figure exist. The Fermi level is situated in the center of the band gap. At the Fermi level $f(E) = \frac{1}{2}$. This is the probability of a state *at that energy* being occupied. Since there are no energy states in the band gap there will be no electrons in a state at energy μ. All states up to E_v, the top of the valence band, are filled.

At $T > 0$ K some states above E_c will become filled, while vacancies will exist in the top of the valence band.

To calculate the electron concentration we need, in addition to the probability of occupation of each mode, the actual number of states involved. This calculation is very similar to the one we performed earlier for black-body radiation. The number of states per unit volume for which the electron waves have wave vectors within a small range dk at $k = |\mathbf{k}|$ is (cf. Eq. (1.31))

$$\rho(k)dk = \frac{k^2}{\pi^2}dk. \tag{13.11}$$

† For those readers with an interest in chemical thermodynamics, μ is the *chemical potential* of the electrons in the semiconductor.

The electrons also have two different kinds of state analogous to the two polarization states of electromagnetic waves. These two states are called the *spin* states of the electrons.

For an electron in the conduction band the total energy is

$$E_k = E_c + \frac{\hbar k^2}{2m_c^*}, \tag{13.12}$$

where m_c^* is the electron effective mass in the conduction band. Therefore, from Eq. (13.12)

$$k = \sqrt{\frac{2m_e^*}{\hbar^2}}\sqrt{E_k - E_c} \tag{13.13}$$

and

$$dk = \frac{1}{2}\sqrt{\frac{2m_e^*}{\hbar^2}}\frac{dE_k}{\sqrt{E_k - E_c}}. \tag{13.14}$$

The density of states $\rho(E_k)$, which tells us how many electron states exist in a range of energies between E_k and $E_k + dE_k$, is found from

$$\rho(E_k)dE_k = \rho(k)dk. \tag{13.15}$$

It follows that

$$\rho(E_k)dE_k = \frac{k^2}{\pi^2}dk \tag{13.16}$$

and

$$\rho(E_k)dE_k = \frac{1}{2\pi^2}\left(\frac{2m_e^*}{\hbar^2}\right)^{3/2}\sqrt{E_k - E_c}dE_k. \tag{13.17}$$

The total number of electrons in the conduction band is

$$n = \int_{E_c}^{\infty} \rho(E_k)f(E_k)dE_k. \tag{13.18}$$

Substitutions from Eq. (13.10) and Eq. (13.17) gives

$$n = \frac{1}{2\pi^2}\left(\frac{2m_e^*}{\hbar^2}\right)^{3/2} e^{\mu/kT}\int_{E_c}^{\infty} \sqrt{E_k - E_c}e^{-(E_k/kT)}dE_k, \tag{13.19}$$

which can be integrated to give

$$n = 2\left(\frac{m_e^* kT}{2\pi\hbar^2}\right)^{3/2} e^{(\mu - E_c)/kT}. \tag{13.20}$$

We can derive a similar result for the concentration of holes in the valence band if we note that the Fermi–Dirac distribution for the holes is

$$f_h(E) = 1 - f(E), \tag{13.21}$$

since a hole is equivalent to a ‘missing’ electron. Therefore,

$$f_h(E) = \frac{1}{e^{(\mu - E_k)/kT} + 1}. \tag{13.22}$$

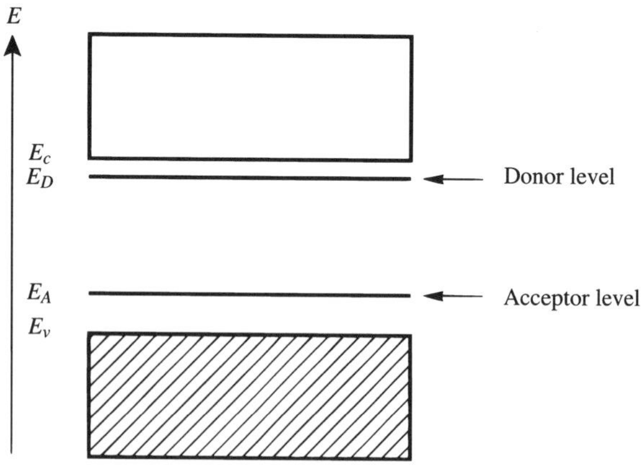

Fig. 13.6. Schematic band diagram showing position of donor and acceptor levels in the band gap of a doped semiconductor.

The corresponding hole concentration is

$$p = 2\left(\frac{m_h^* kT}{2\pi\hbar^2}\right)^{3/2} e^{(E_v-\mu)/kT} \tag{13.23}$$

and, noting that $E_c - E_v = E_g$,

$$np = 4\left(\frac{kT}{2\pi\hbar^2}\right)^3 \left(m_e^* m_h^*\right)^{3/2} e^{-E_g/kT} = K^2 e^{-E_g/kT}. \tag{13.24}$$

13.4 Intrinsic and Extrinsic Semiconductors

The conductivity of a semiconductor is determined by the concentration of conduction band electrons and valence band holes, and by the ability of these charges to move through the solid under the action of an electric field, characterized by their mobilities μ_e and μ_h. The conductivity is

$$\sigma = ne\mu_e + pe\mu_h, \tag{13.25}$$

where e is the magnitude of the electronic charge. The conductivity of semiconductors can be controlled by doping. It generally lies between the values for metals and insulators, at ambient temperatures typically in the range from 10^{-7} S m^{-1} to 10^4 S m^{-1},† but depends strongly on temperature. A true insulator would have $\sigma < 10^{-9}$ S m^{-1}, while a good conductor typically has $\sigma > 10^6$ S m^{-1}. Very pure semiconductors (in which $n = p$ in Eq. (13.24) generally have lower conductivity than *impurity* semiconductors in which impurity dopants are added to provide *extrinsic* conductivity. In an n-type semiconductor an impurity is added that can release electrons. For example, GaAs can be made n-type by doping with silicon or germanium in gallium sites. The impurity group IV atom has an additional electron than the group III atom it replaces, which it can *donate* to the conduction band. GaAs can be made p-type by doping with zinc or cadmium (group II elements) in gallium sites, where the group II atoms have one too few electrons and would like to *accept* another. Conversely, GaAs can also be made p-type by replacing arsenic (group V) with a group IV atom (Si or Ge). In an extrinsic semiconductor the impurity atoms occupy energy levels within the band gap, as shown in Fig. (13.6).

† The siemen (S) is the SI unit of conductance, equivalent to the unit Ω^{-1}.

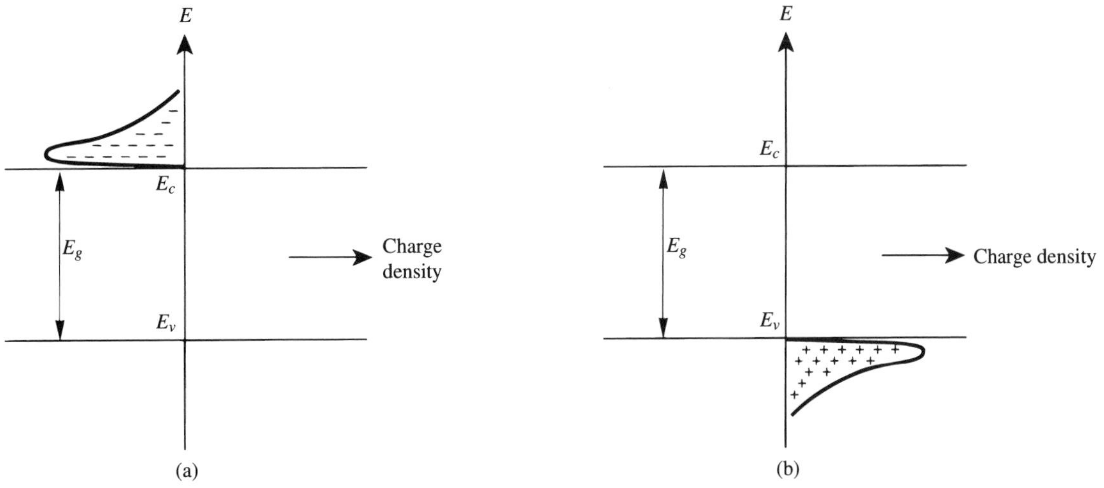

Fig. 13.7. Schematic variation of carrier concentration in degenerately doped (a) *n*- and (b) *p*-type semiconductors.

In *n*-type material the energy gap that electrons must cross to reach the conduction band is lower than in intrinsic material, so $n > p$. However, if the doping level is not too great Eq. (13.24) still holds. In *p*-type material $p > n$.

When an intrinsic semiconductor is doped the Fermi level moves from the center of the band gap. It moves up toward the conduction band with increasing *n*-doping and down towards the valence band with *p*-doping. Its precise location is a function of the doping level. At high enough doping levels the Fermi level moves into the conduction band in *n*-type material and into the valence band in *p*-type material. When this happens the material is said to be *degenerately* doped. This happens because at high doping levels, $\sim 10^{18}$ m^{-3} in GaAs, the impurity atoms begin to interact with each other and their once sharp energy levels start to spread and merge with the neighboring band. The schematic charge distribution with energy in degenerately doped *n*- and *p*-type materials is shown in Fig. (13.7).

There are a few holes in the valence band in the *n*-type material and a few electrons in the conduction band in the *p*-type material, but their numbers are small and are not shown in Fig. (13.7). If we could somehow superimpose the two halves of Fig. (13.7) we would have a condition of population inversion – filled electron energy states above empty ones. In essence, we do this by forward biasing a *p–n* junction. However, before describing precisely how this is done and how a semiconductor laser is constructed, we must first describe a little of the physics that goes on at the junction between two differently doped semiconductors.

13.5 The *p–n* Junction

In its simplest form the *p–n* junction is the interface region between two identical semiconductors, one of which is *n*-doped, the other *p*-doped. This is called a *p-n homojunction.*

The Fermi level in a semiconductor corresponds to its chemical potential, so that when a *p-n* junction is formed the Fermi levels line up with each other leading to the electron energy versus position diagram shown in Fig. (13.8). The low resistance conduction direction of the diode is shown running from $p \rightarrow n$. In the

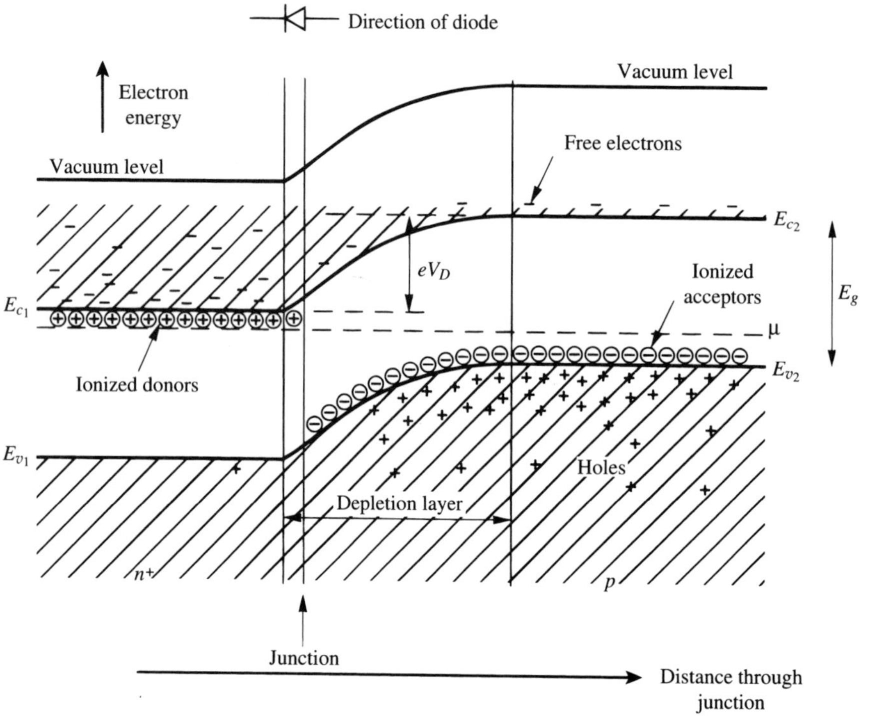

Fig. 13.8. Band diagram of a $p-n$ Junction in equilibrium showing relative position of electrons and holes.

figure the n-type material is more heavily doped then the p-type material so its Fermi level is closer to the conduction band. The materials are designated p and n^+ to indicate this. Typical doping levels might result from a donor concentration n_D of $\sim 10^{24}$ m^{-3} and acceptor concentration $n_A \sim 10^{22}$ m^{-3}. The vacuum level is the energy to which an electron would need to be raised to escape completely from the semiconductor. The electrons in the n-type material come largely from ionized donors whose energy is near the conduction band edge E_{c_1}. The holes in the p-type material come largely from ionized acceptors whose energy is near the valence band edge E_{v_2}. In thermal equilibrium an electron that travels from n^+ to p must climb an energy hill, shown in the diagram as eV_D, where V_D is called the diffusion potential. Note that

$$eV_d = E_g - (E_{c_1} - \mu) - (\mu - E_{v_2}). \tag{13.26}$$

In a region called the *depletion layer*, which spans the junction, electrons fall down the potential hill, while holes 'fall' up the hill (because they are positively charged their energy is lower on the p-side of the junction). However, across the junction the product pn remains constant, according to Eq. (13.24). The hill constitutes an energy barrier to the passage of electrons from $n^+ \rightarrow p$ or holes from $p \rightarrow n^+$. The schematic variation of electron and hole concentration across the junction is shown in Fig. (13.9).

The concentration of electrons and holes far from the junction is respectively n_{10}, p_{10} in the n^+ material, n_{20}, p_{20} in the p-type material. From Eq. (13.24)

$$n_{10}p_{10} = n_{20}p_{20} = n_i^2, \tag{13.27}$$

Fig. 13.9. Schematic charge density variation of electrons and holes across a *p–n* junction in equilibrium.

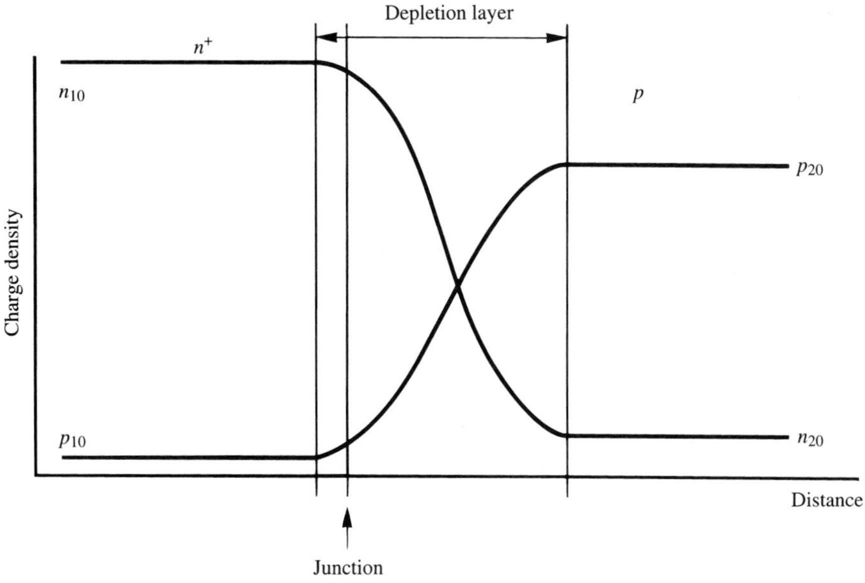

where n_i is the equilibrium electron (and hole) concentration in undoped (intrinsic) material.

The probability factor that describes the likelihood of a charge carrier being able to 'climb' the hill is $e^{-e(V_D-V)/kT}$. It should not be too surprising therefore that the current/voltage relation for the diode is[13.2]

$$I = I_S(e^{eV/kT} - 1). \tag{13.28}$$

When the *p-n* junction is forward biased the charge concentration across the junction changes to the schematic form shown in Fig. (13.10).

Current flow injects extra electrons into the *p*-type material, and a relatively smaller number of extra holes into the *n*-type material (because the *p*-type is less heavily doped). Note, that to preserve electrical neutrality across the junction, the concentration of free electrons in the n^+ material and holes in the *p*-type material† rises to balance the injected excess carriers. In each material these excess carriers must disappear within a short distance of the junction as they represent a deviation from the thermal equilibrium carrier distribution.‡

The excess carriers disappear by recombination as electrons and holes recombine. Of course, all this means is that an electron at a given energy falls into an empty electron state of lower energy. If this recombination involves a band-to-band radiative transition as shown schematically in Fig. (13.11), then a photon is produced, which may escape from the semiconductor and be observed as *injection luminescence*. This process, in semiconductors, is the spontaneous emission process with which we are already familiar. The efficiency with which excess carriers recombine radiatively strongly determines the utility of the junction as a *light*

† The *majority* carriers in each material, respectively.

‡ The characteristic distance over which these excess carriers disappear is closely related to the so-called *diffusion length*, which for electrons in *p*-type material is $L_p = \sqrt{D_e \tau_p}$ where D_e is the electron diffusion coefficient ($m^2\ s^{-1}$) and τ_p is the lifetime of free electrons in *p*-type material.

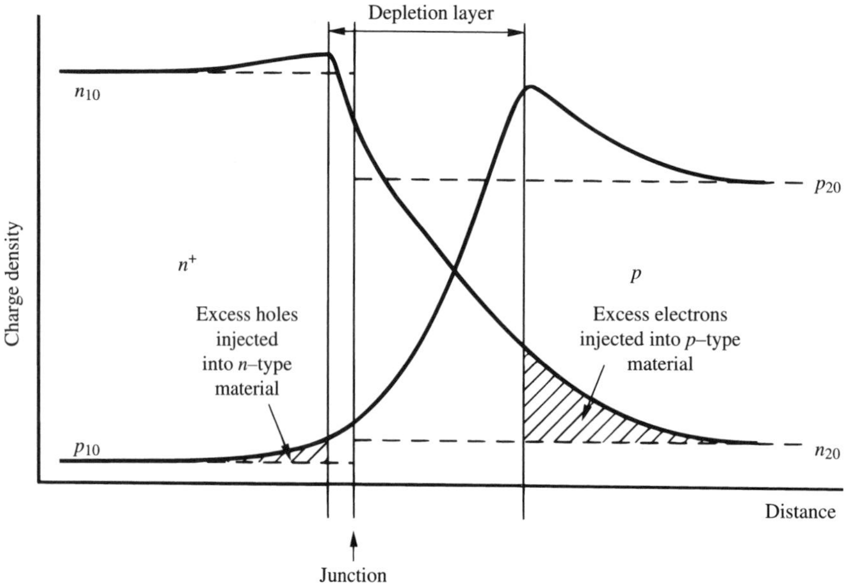

Fig. 13.10. Charge density variation of electrons and holes across a forward biased *p*-*n* junction.

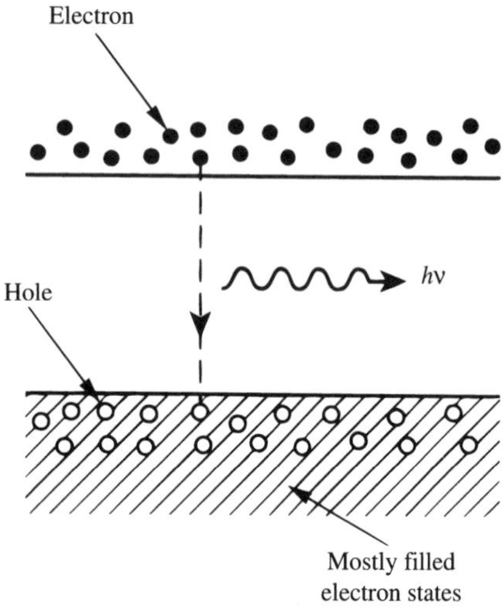

Fig. 13.11. Band-to-band radiative recombination.

emitting diode – LED. Early semiconductor diode lasers were constructed by using a $p-n$ junction with degenerately doped materials as shown in Fig. (13.12). Under forward bias the energy band diagram would distort near the junction, as shown in Fig. (13.13) to provide a region where filled electron states in the valence band find themselves directly above empty electron states (holes) in the valence band. The resulting population inversion could be large enough to permit laser oscillation at high bias currents and/or low temperatures. We will not spend further time in discussing these devices here: further discussion can be found elsewhere[13.4]–[13.6].

Fig. 13.12. Schematic band structure at the junction between two degenerately doped semiconductors at low temperature.

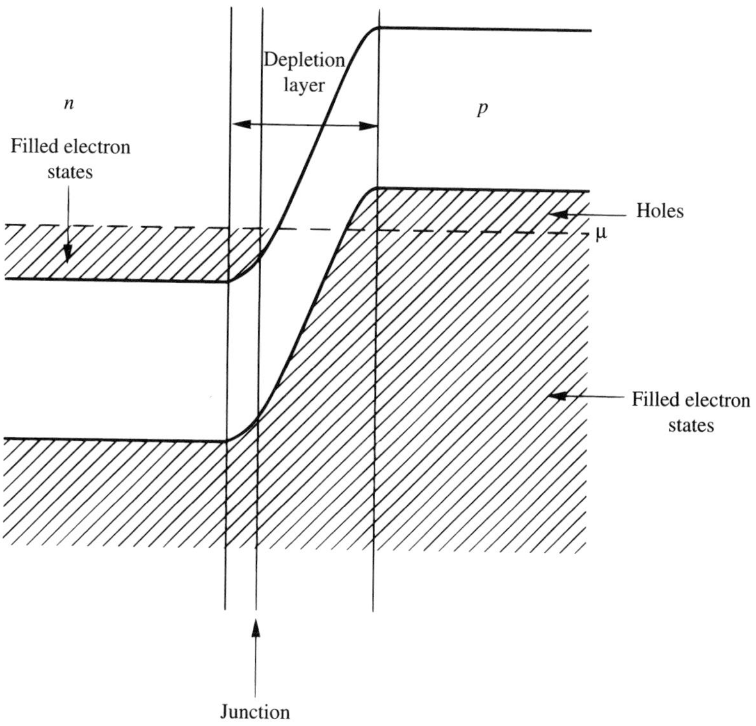

Fig. 13.13. Forward-biased *p-n* junction between degenerately doped semiconductors showing schematically how a population inversion can be obtained.

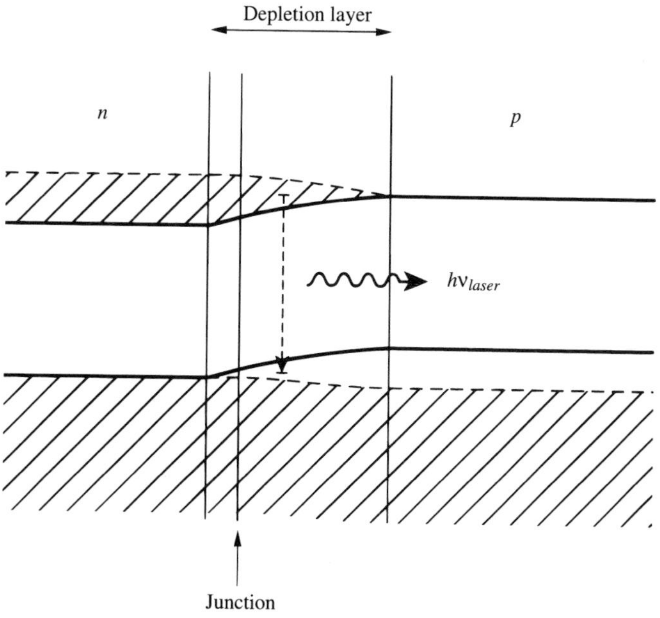

13.6 Recombination and Luminescence

The rate at which electrons and holes recombine near a p–n junction is determined by the excess volume density of these carriers† and their lifetimes. For example, for excess electrons recombining in p-type material‡

$$\frac{d(\Delta n)}{dt} = -\frac{\Delta n}{\tau_p}, \tag{13.29}$$

where Δn is the excess electron volume density injected into the p-region and τ_p is the lifetime of these electrons in the p-type material. A similar relation holds for holes injected into the n-region.

Eq. (13.29) is analogous to Eq. (1.3), the characteristic equation describing population loss from an excited state by spontaneous emission. The actual rate of recombination depends on the local electron n and hole p concentrations. If both radiative and nonradiative recombination processes occur then the recombination rate can be written as

$$\frac{d(\Delta n)}{dt} = -\Delta n\left(\frac{1}{\tau_{rr}} + \frac{1}{\tau_{nr}}\right), \tag{13.30}$$

where τ_{rr}, τ_{nr} are lifetimes for radiative and nonradiative recombination, respectively. The fraction of recombinations that occurs radiatively is called the *internal quantum efficiency*, η_{int}, where

$$\eta_{int} = \frac{1}{1 + \tau_{rr}/\tau_{nr}}. \tag{13.31}$$

The overall recombination rate R_{rr} is

$$R_{rr} = rnp, \tag{13.32}$$

where r is the recombination rate constant ($m^3\ s^{-1}$) that is characteristic of the material. Consequently, the generated optical power per unit volume is

$$\phi = rnp\eta_{int}(\overline{h\nu}), \tag{13.33}$$

where $\overline{h\nu}$ is the average photon energy of radiative recombinations.

In an efficient LED (or semiconductor laser) most of the carriers that recombine do so radiatively: other recombination processes in which electron energy is lost in producing phonons, or in recombination with ionized impurities are undesirable. In most of these devices much of the current through the junction is carried by electrons, which generally have higher mobility than the holes: for example in GaAs, $\mu_e = 0.8\ m^2\ V^{-1}\ s^{-1}$; $\mu_h = 0.03\ m^2\ V^{-1}\ s^{-1}$. We define the *injection efficiency*, η_i, as

$$\eta_i = \frac{\text{electron current}}{\text{total current}}. \tag{13.34}$$

† Densities above their equilibrium values with no forward bias applied.

‡ Structures in which excess holes recombine in n-type material behave similarly and their analysis is similar to the analysis for structures in which electrons carry most of the current.

The optical power generated by an injected current I is therefore

$$\phi = \eta_{int} \frac{I}{e} \overline{h\nu}. \tag{13.35}$$

In a good LED using direct band gap material $\eta_{int} \simeq 0.5$. Most of the radiation is generated within a few diffusion lengths of the junction; in typical devices this is a few tens of micrometers, but depends on temperature and doping level.

The probability of band-to-band recombination, which is the most desirable process for generating high energy photons, is reduced in indirect band gap semiconductors such as silicon, germanium or gallium phosphide (GaP). As these materials have relatively large band gaps this is unfortunate if short wavelength LEDs are to be constructed. Fortunately, the probability of radiative recombination can be increased by appropriate doping, for example by doping GaP with nitrogen or ZnO.

13.6.1 The Spectrum of Recombination Radiation

We expect the photons generated by band-to-band recombination to contain a range of energies rising from a minimum value $\lambda_g = ch/E_g$. To calculate the shape of the recombination radiation we must consider the electron and hole population densities as a function of energy. For the electrons the population density with energy is, cf. Eq. (13.18),

$$n(E_2) = \rho_c(E_2) f(E_2), \tag{13.36}$$

where E_2 is energy within the conduction band and $\rho_c(E_2)$ is the density of states within the band.

For the holes in the valence band the population density as a function of the energy E_1 within the band† is

$$p(E_1) = \rho_v(E_1)[1 - f(E_1)]. \tag{13.37}$$

From Eq. (13.17)

$$\rho_c(E_2) = 4\pi \left(\frac{2m_e^*}{h^2} \right)^{3/2} \sqrt{E_2 - E_c}, \tag{13.38}$$

and similarly

$$\rho_v(E_1) = 4\pi \left(\frac{2m_h^*}{h^2} \right)^{3/2} \sqrt{E_v - E_1}. \tag{13.39}$$

The rate of spontaneous emission resulting from recombination will depend on several factors: (i) the number of electron states within each small range of energy, dE_2, in the conduction band, (ii) the probability that these states are occupied, $f(E_2)$, (iii) the number of electron states within each small range of energies, dE_1, in the valence band, and (iv) the probability that these states are empty, $1 - f(E_1)$. Therefore, we write the total rate per unit volume at which recombinations

† E_1 is the energy of the vacant electron state in the valence band.

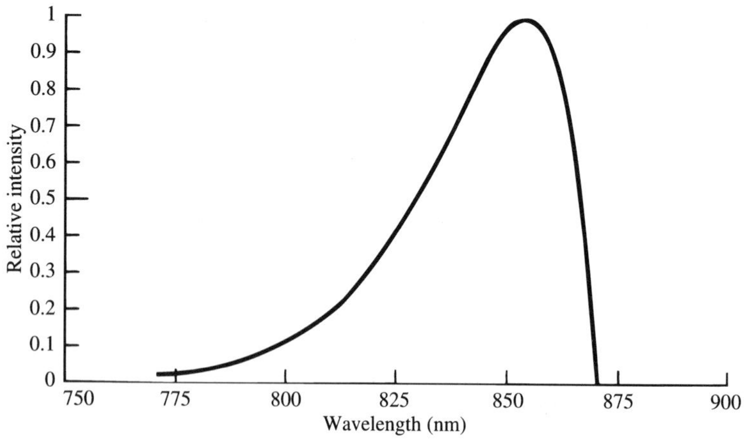

Fig. 13.14. Recombination spectrum of GaAs calculated from the simple model given by Eq. (13.45).

occur as

$$R_{spont} = \int\int A\rho_c(E_2)f_2(E_2)\rho_v(E_1)[1 - f_1(E_1)]dE_1dE_2, \tag{13.40}$$

where A is an Einstein coefficient for spontaneous emission and the integrals run over the conduction and valence bands, respectively. If A does not depend on the energies E_2 and E_1, then

$$\begin{aligned} R_{spont} &= A\int \rho_c(E_2)f_2(E_2)dE_2 \int \rho_v(E_1)[1 - f_1(E_1)]dE_1 \\ &= Anp, \end{aligned} \tag{13.41}$$

where n and p are the total electron and hole densities, respectively. A typical value for A in the case of GaAs is 10^{-16} m^3 s^{-1}.

Unfortunately, in calculating the spectral distribution of recombination radiation we need to use Eqs. (13.36) and (13.37), which are complicated. Consequently, the simplifying assumption is made that charge particle density decreases exponentially from the band edge; for example:

$$n(E_2) = n_c e^{-(E_2 - E_c)/kT}, \tag{13.42}$$

and

$$p(E_1) = p_v e^{-(E_v - E_1)/kT}. \tag{13.43}$$

The luminescence spectrum will have a lineshape

$$g(\nu) \propto \int_{E_c}^{E_v + h\nu} n(E_2)p(E_1)dE_2, \tag{13.44}$$

subject to the constraint $E_2 - E_1 = h\nu$. Substitution from Eqs. (13.42) and (13.43) and integration gives

$$g(\nu) \propto (h\nu - E_g)e^{-(h\nu - E_g)/kT}. \tag{13.45}$$

This lineshape function is shown in Fig. (13.14).

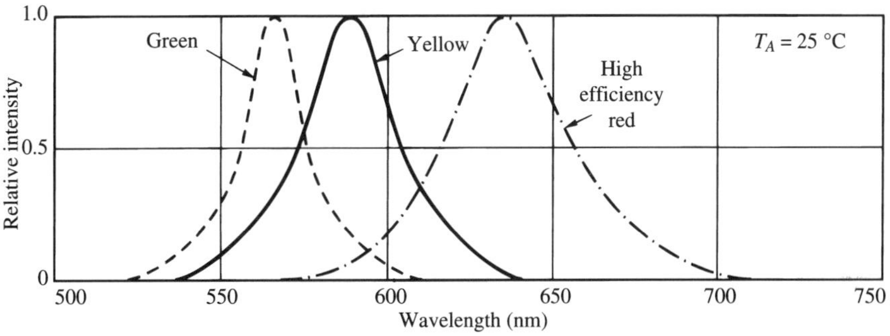

Fig. 13.15. Spectra of Hewlett-Packard red, yellow, and green LEDs.

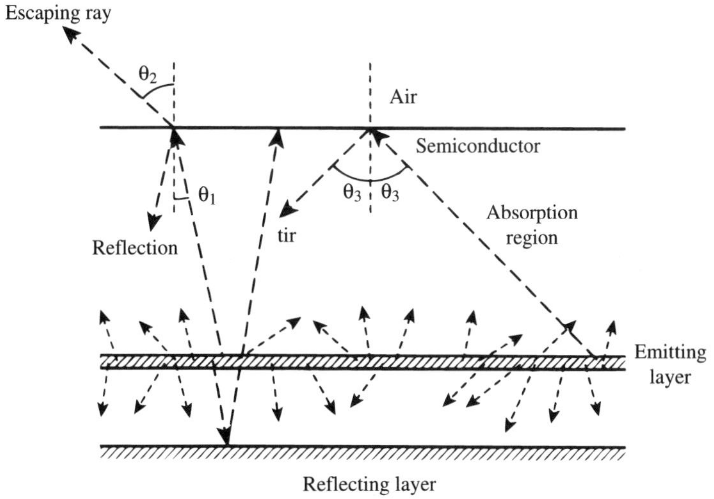

Fig. 13.16. Schematic illustration of factors that affect the external quantum efficiency. The totally internally reflected (tir) ray is incident on the interface at an angle $\theta_3 > \theta_c$.

Actual recombination spectra are more symmetric than Fig. (13.14), as shown in Fig. (13.15), because of energy band smearing that occurs near the band edge of real semiconductors.†

13.6.2 External Quantum Efficiency

It is not sufficient to build an LED by using an active region where η_{int} is high. This will guarantee efficient photon generation, but unless these photons escape from the device useful light will not be generated. There are four principal ways in which generated photons fail to escape from the active region, shown in Fig. (13.16). Of potentially greatest importance is the reabsorption of photons in bulk material between the active layer and the surface. Reflectance losses at the outer semiconductor/air interface can be reduced by coating this interface appropriately. The half of the radiation that is not emitted towards the exit surface can be redirected with a reflective layer.

† Called band-tailing.

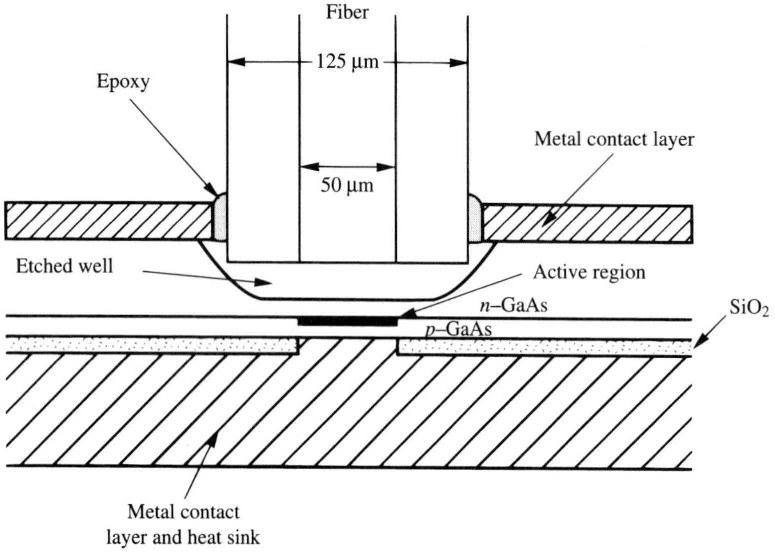

Fig. 13.17. Burrus-type LED with attached optical fiber.

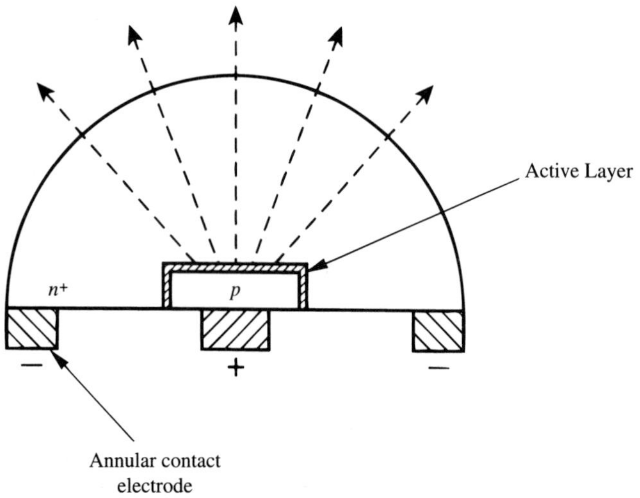

Fig. 13.18. Simple hemispherical LED.

The *external quantum efficiency* η_{ext} is the ratio

$$\eta_{ext} = \frac{\text{photons leaving exit interface}}{\text{number of carriers crossing junction}}. \tag{13.46}$$

The most efficient way to maximize η_{ext} is to build a planar structure in which the active layer is very close to the surface. This is exemplified by the Burrus-type surface emitting LED[13.7] shown in Fig. (13.17) with an attached optical fiber.

A structure of this kind is fabricated in a series of steps involving the deposition of one layer of material on another with selective etching to produce for example the 'etched-well' that allows the end of the optical fiber to be placed close to the active region. Fig. (13.18) shows a simple structure, widely used in simple solid-state lamps, in which the hemispherical (or ellipsoidal) shape permits a greater fraction of the emitted light to escape through the outer surface.

In all these structures a potentially important loss mechanism is the reabsorption of emitted photons within the unexcited regions between the active layer and the outer surface. This occurs because the emitted photons have $h\nu > E_g$, so they can easily be reabsorbed. However, it is possible to build composite structures in which the band gaps of the semiconductor regions near the active layer are larger so that reabsorption cannot occur. This is made possible through the incorporation of *heterojunctions*.†

13.7 Heterojunctions

Although it is possible to build good LEDs and obtain semiconductor laser oscillation by using a forward biased homojunction, such devices are not optimal. Their efficiency can be characterized by the use of the internal quantum efficiency, η_{int}, considered as

$$\eta_{int} = \frac{\text{number of band-to-band radiative recombinations}}{\text{number of carriers crossing junction}}. \tag{13.47}$$

If both radiative and nonradiative recombinations occur, then for electrons injected into p-type material

$$\frac{1}{\tau_p} = A + X, \tag{13.48}$$

where A is an Einstein coefficient describing radiative recombination and X is a coefficient for nonradiative recombination.

Clearly,

$$\eta_{int} = \frac{1}{1 + (X/A)}, \tag{13.49}$$

so we want $A >> X$ for efficient photon generation. Unfortunately, this is not the whole story. If we want a bright, efficient, photon emitter we need to guarantee that as many carriers as possible recombine soon after crossing the junction, and do not escape to travel into the bulk material far from the junction. In an ideal device electrons and holes would be 'trapped' in the region where recombination is desired. This trapping can be effected by the use of layered semiconductor structures that use heterojunctions.

13.7.1 Ternary and Quaternary Lattice-Matched Materials

It is possible to replace Ga atoms in GaAs with Al to make the ternary material $Ga_{1-x}Al_xAs$ without any significant change in the atomic arrangement of the crystal lattice. For compositions containing up to 37% Al, the ternary material remains a direct band gap semiconductor. At the same time the band gap changes from 1.42 eV(GaAs) to 1.92 eV($Ga_{0.63}Al_{0.37}As$). Because the lattice spacing of the GaAlAs is very close to that of GaAs, it is possible to grow a layer of GaAlAs on GaAs, or vice versa, without introducing strain at the interface between the

† A junction between two different semiconductors, for example GaAs and GaAlAs.

two materials. The two materials are said to be *lattice-matched*. The junction between the GaAs and GaAlAs is called a heterojunction: either material can additionally be *n*- or *p*-doped. So, for example, we could have *n–N*, *n–P*, *P–N*, or *p–P* heterojunctions.† Other important lattice-matched heterojunctions are $GaAs/In_xGa_{1-x}As_{1-y}P_y$, $InP/In_rGa_{1-r}As_{1-s}P_s$, $GaSb/In_tGa_{1-t}As_uSb_{1-u}$. Not all compositional ranges (x, y, r, s, t, u) lead to lattice-matched, direct band gap semiconductors, and some compositions cannot be grown[13.2]. Some specific examples of important ternary and quaternary materials are listed in Table (13.1).

13.7.2 Energy Barriers and Rectification

We will use the system GaAl/GaAlAs as the archetype for explaining the properties of heterojunctions that make them desirable for the fabrication of LEDs and lasers. Because these two materials have different energy gaps, when a junction between them is formed, energy barriers to the flow of electrons and holes result. The Fermi levels in the two materials line up across the junction so that deformation of the band edge occurs near the junction as shown schematically in Fig. (13.19).

The *n–P* heterojunction has rectification properties like a traditional diode. Moreover, even the *n–N* and *p–P* heterojunctions can show diode-like behavior because of the energy barriers in Figs. (13.19b) and (13.19c)[13.8],[13.9]. For example, there is a significant probability that an electron in the conduction band moving from $n \rightarrow N$ in Fig. (13.19b) or from $p \rightarrow P$ in Fig. (13.19c) will reflect because of the energy discontinuity at the junction.

13.7.3 The Double Heterostructure

To maximize the recombination efficiency, and therefore potential light output, at a *p-n* junction it is desirable to place a heterojunction on each side of the region where electrons and holes recombine. A schematic diagram of such a double heterostructure is given in Fig. (13.20).

The active GaAs layer is sandwiched between two layers of GaAlAs whose larger energy gaps reflect both electrons and holes back into the active layer, as shown schematically in Fig. (13.21). The contact layers are highly doped *n*- and *p*-type layers of GaAs. The + superscript is used to denote a high doping level. The metal contact layer can be deposited most easily on a small band gap material with a relatively large conductivity. Otherwise, a rectifying metal/semiconductor junction will result.‡

One additional advantage of this type of structure has already been mentioned, namely the transparency of the confining layers to photons emitted from the active layer because of the larger band gap of the material in these layers. But, there is a second advantage: the confining layers have a lower refractive index than the active layer. Consequently, photons emitted parallel, or nearly so, to the junctions can be guided by total internal reflection and this leads to enhanced efficiency of both edge emitting LEDs and lasers. Fig. (13.22) shows an edge emitting LED design. The insulating SiO_2 layer confines the current to a specific lateral region. The emerging optical beam has an approximately elliptical shape whose semimajor axis

† The upper case letters refer to the semiconductor with the larger band gap, in this case the GaAlAs.
‡ Called a Schottky diode.

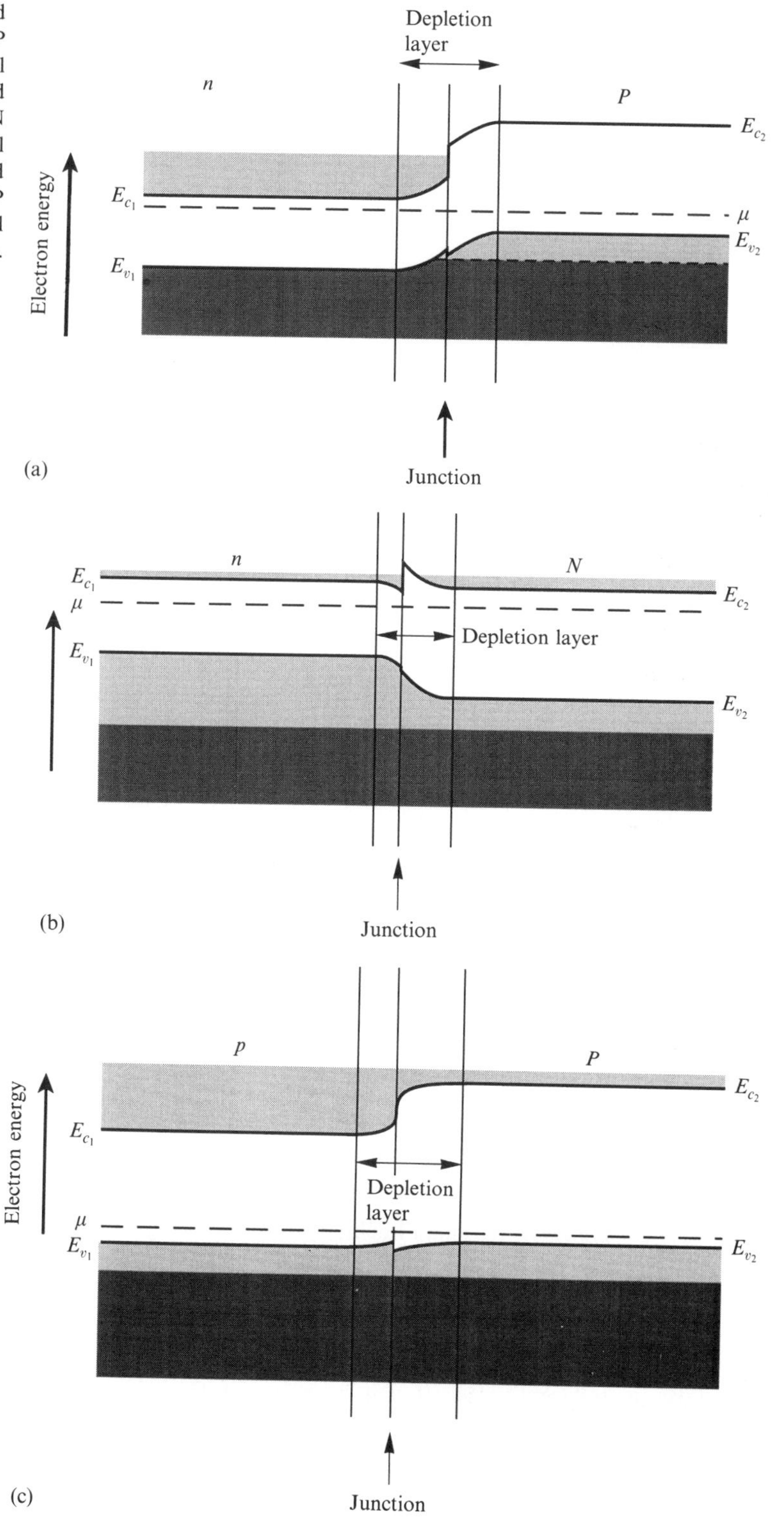

Fig. 13.19. (a) Band energy diagram of a n–P heterojunction in thermal equilibrium. (b) Band energy diagram of a n–N heterojunction in thermal equilibrium. (c) Band energy diagram of a p–P heterojunction in thermal equilibrium.

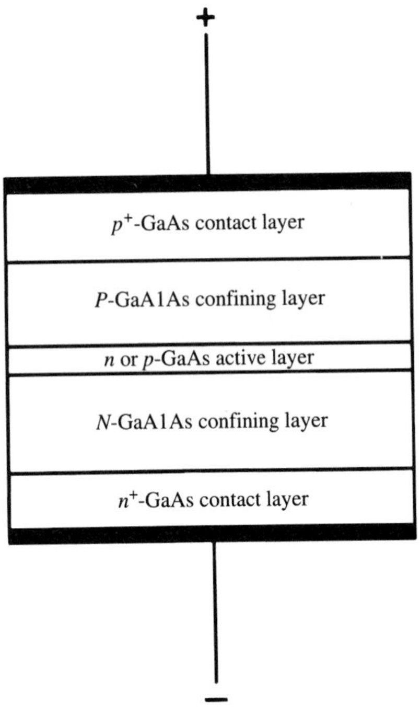

Fig. 13.20. Schematic diagram of a layered double heterostructure

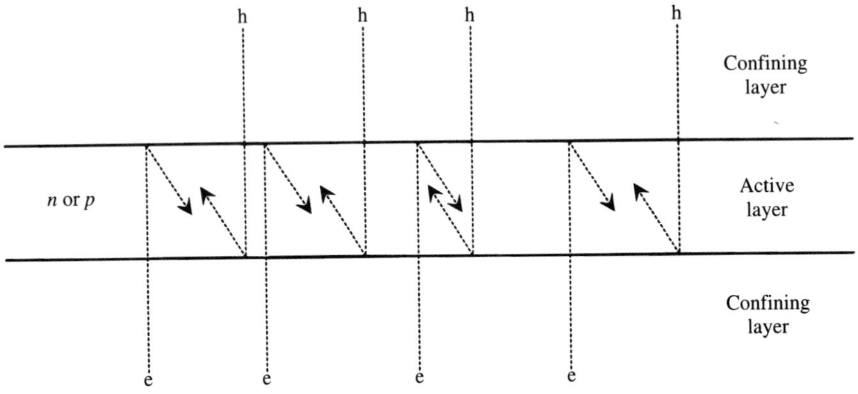

Fig. 13.21. Schematic illustration of how electrons (e) and holes (h) reflect at the edges of the confining layer in a double heterostructure.

is parallel to the junction. However, the small thickness of the active layer leads to diffraction such that this elliptical beam becomes elliptical with its semimajor axis perpendicular to the junction as the beam travels away from the device.

The double heterostructure shown in Fig. (13.22) is biased so that the p–N (or n–P) junction at the edge of the active layer is conducting.

If the active layer is sufficiently thin it can be assumed that the electron concentration n does not vary much within it. The recombination rate per unit volume within this layer is

$$-\frac{dn}{dt} = \frac{n}{\tau_{rr}} + \frac{n}{\tau_{nr}}, \tag{13.50}$$

Fig. 13.22. Edge emitting LED design.

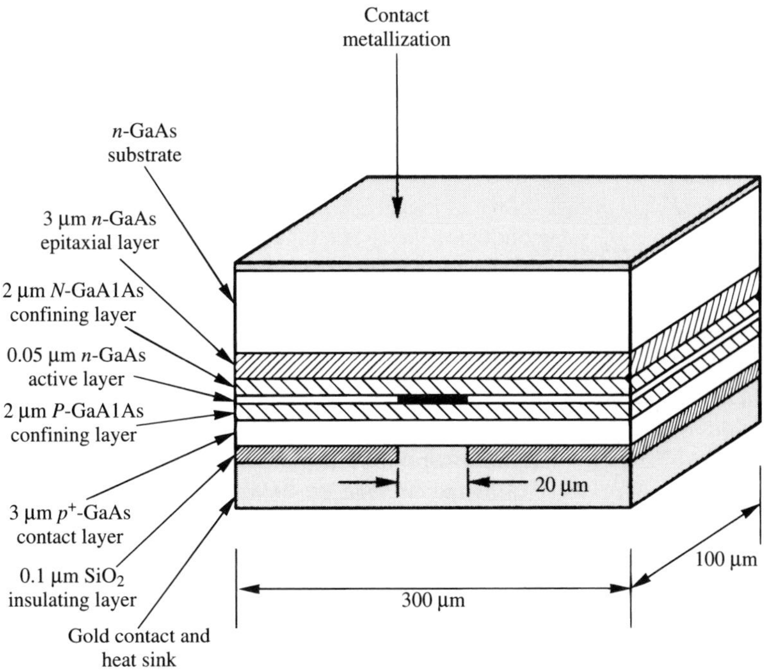

where τ_{rr}, τ_{nr} are the recombination times for radiative and non- radiative recombination, respectively, within the active layer.

However, an additional term should be added to Eq. (13.44) to include *surface-recombination* effects that occur at the heterojunction boundaries of the layer. This surface recombination occurs at a rate per unit area of

$$-\frac{dn}{dt} = nv_S, \tag{13.51}$$

where v_S is called the *recombination velocity*. Surface recombination is enhanced by any lattice mismatch at the heterojunction: it can also occur at the outside bounding faces of the active layer but this effect is not significant except in a very small area device. Therefore for an active layer of thickness d the overall recombination time τ satisfies

$$\frac{1}{\tau} = \frac{1}{\tau_{rr}} + \frac{1}{\tau_{nr}} + \frac{2v_S}{d}. \tag{13.52}$$

The factor of 2 comes from the two interfaces of the active layer and we have assumed that v_S is the same for both interfaces. A typical value for v_S at a GaAs/GaAlAs heterojunction is 10 m s^{-1}. The internal quantum efficiency within the active layer is clearly

$$\eta_{\text{int}} = \frac{\tau}{\tau_{rr}}. \tag{13.53}$$

The radiative recombination time in the active layer will vary depending on doping level and also depends on the injected current, but values in the range 1–100 ns are typical. The nonradiative lifetime τ_{nr} is decreased by the presence of impurities and lattice defects, but a value of 100 ns might be typical. Consequently for an active layer of thickness 0.05 μm and for $\tau = 1$ ns, $\eta_{int} \simeq 0.7$.

13.8 Semiconductor Lasers

The edge emitting LED design shown in Fig. (13.22) can, in the right circumstances, become an edge emitting semiconductor laser. All that is required is a condition of population inversion within the active region that provides sufficient gain to overcome cavity losses. In many such lasers the cavity is formed by the plane parallel facets at opposite ends of the active region. However, before discussing these and other semiconductor laser designs in more detail we need to return to some fundamental considerations.

In common with all lasers, gain results in the active region if the rate of stimulated emission exceeds that of absorption. The probability of a band-to-band stimulated emission depends on four factors: the Einstein coefficient, B_{21}, for the transition, the energy density of the stimulating radiation, $\rho(\nu_{21})$, the probability that the upper energy level is occupied by an electron, f_2, *and* the probability $1-f_1$ that the lower energy level is *not* occupied by an electron; namely it is occupied by a hole. Formally, we can write

$$\text{gain} \propto [B_{21}f_2(1-f_1) - B_{12}f_1(1-f_2)]\rho(\nu_{21}), \tag{13.54}$$

which if we assume that $B_{12} = B_{21}$, gives

$$\text{gain} \propto f_2 - f_1. \tag{13.55}$$

The existence of gain is a nonequilibrium situation that can be represented schematically as the existence of a region of filled electron states above empty electron states, as shown in Fig. (13.23).

We can represent the energies that bound the filled electron states in the conduction and valence bands as μ_c, μ_v, respectively, referred to as quasi-Fermi levels. From Eq. (13.9)

$$f_2(E_2) = \frac{1}{e^{(E_2-\mu_c)/kT} + 1}, \tag{13.56}$$

$$f_1(E_1) = \frac{1}{e^{(E_1-\mu_v)/kT} + 1}. \tag{13.57}$$

The condition $f_1 > f_2$ results if

$$(\mu_c - \mu_v) > h\nu_{21}, \tag{13.58}$$

a condition first derived by Bernard and Duraffourg[13.10] and, of course, $h\nu_{21} > E_g$.

In a heavily doped p-type semiconductor the Fermi level can be forced down into the valence band. If such a material is used as the active region in an N-p-P double heterostructure then the energy levels in thermal equilibrium appear as shown in Fig. (13.24). However, under sufficient forward bias the situation shown in Fig. (13.25) will result. Within the active layer the situation shown in Fig. (13.23) has been created: filled electron states above a quasi-Fermi level μ_c in the conduction band and μ_v in the valence band.

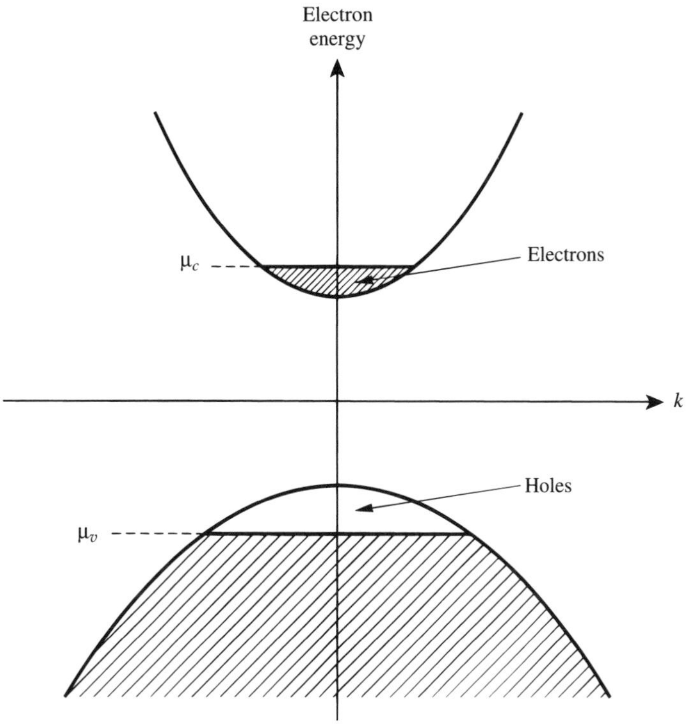

Fig. 13.23. Population inversion shown in an E–k diagram.

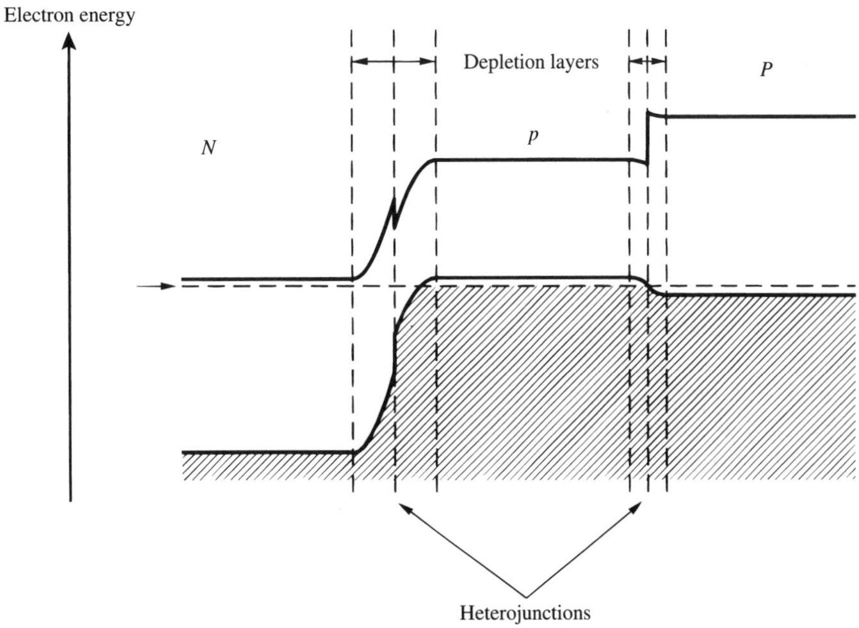

Fig. 13.24. Schematic band diagram in thermal equilibrium of an N–p–P double heterostructure in which the p region is heavily doped. Note that the Fermi level in the p region has moved down into the valence band.

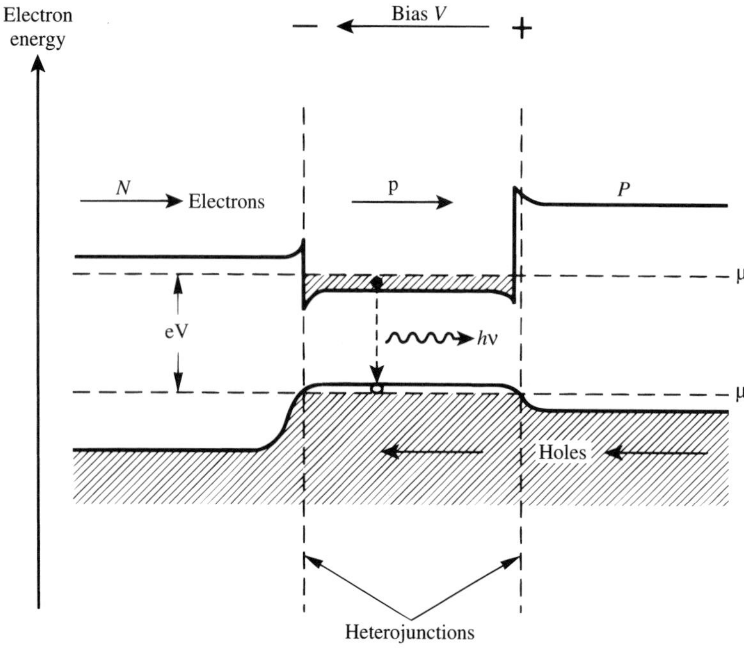

Fig. 13.25. Schematic band diagram of a forward biased N–p–P double heterostructure in which the p region is heavily doped. Electrons flowing from the N region and holes flowing from the P region are trapped in the p region potential well creating a population inversion.

13.9 The Gain Coefficient of a Semiconductor Laser

To derive an expression for the gain coefficient of a semiconductor laser it is best to return to the fundamentals discussed in Chapter 2. In that chapter we saw that the increase in intensity, dI_ν (Wm^{-2}) of a small band of frequencies near ν on passing a distance dz through a medium could be written as

$$dI_\nu = \left(\text{no. of stimulated emission/vol—no. of absorptions/vol.}\right) \times h\nu dz,$$
$$= \left[B_{21}N_2 g(\nu',\nu)\frac{I_\nu}{c} - B_{12}N_1 g(\nu',\nu)\frac{I_\nu}{c}\right] h\nu dz, \qquad (13.59)$$

where $N_2 g(\nu',\nu)$ is the density of filled upper level states per frequency interval, and $N_1 g(\nu',\nu)$ is the density of filled lower level states per frequency interval. I_ν ($\mathrm{W\ m}^{-2}$) is the intensity of input monochromatic radiation.

In a semiconductor laser the probability of stimulated emission depends on several factors: (i) an Einstein coefficient B, which we shall assume applies also to absorption processes; (ii) the density of states in the upper level, $\rho_c(E_2)$ and the probability that these states are occupied, $f_2(E_2)$; and (iii) the density of states in the lower level, $\rho_v(E_1)$, and the probability that these states are empty, $1 - f_1(E_1)$. Consequently, Eq. (13.52) becomes

$$dI_\nu = \left\{\rho_c(E_2)\rho_\nu(E_1)f_2(E_2)[1 - f_1(E_1)]Bh\frac{I_\nu}{c} - \rho_c(E_2)\rho_\nu(E_1)f_1(E_1)[1 - f_2(E_2)]Bh\frac{I_\nu}{c}\right\} h\nu dz. \qquad (13.60)$$

The additional factor h in Eq. (13.60) enters because we will be integrating over energy rather than frequency.

If the dispersive nature of the medium is taken into account then† the relation between the Einstein A and B coefficients in this case is

$$\frac{A}{B} = \frac{8\pi h\nu^3 n^2 n_g}{c_0^3}. \tag{13.61}$$

Therefore,

$$dI_\nu = \frac{hc_0^2 A}{8\pi n n_g \nu^2} \rho_c(E_2)\rho_\nu(E_1)[f_2(E_2) - f_1(E_1)]I_\nu dz, \tag{13.62}$$

where dI_ν now has dimensions $\mathrm{Wm^{-2}\ J^{-1}}$. B, the coefficient for stimulated emission for states distributed over an energy range dE has units $\mathrm{J^{-1}\ m^6\ s^{-2}}$. Eq. (13.62) can be integrated to give

$$\gamma(\nu) = \frac{hc_0^2}{8\pi n n_g \nu^2} \int_{E_c}^{E_v + h\nu} A\rho_c(E_2)\rho_v(E_2 - h\nu)[f_2(E_2) - f_1(E_2 - h\nu)]dE_2. \tag{13.63}$$

The limits of integration are the same as applied in our previous calculation of the spectrum of recombination radiation.

How does the gain $\gamma(\nu)$ depend on the current flowing through the junction? For a current I crossing a p–n junction of width w and height h the injected current density is

$$J = \frac{I}{hw}. \tag{13.64}$$

In a double heterojunction laser, ideally, all the injected carriers recombine in the region between the two heterojunctions, of spacing d. Therefore, the total spontaneous emission rate is

$$R_{\text{spont}} = \frac{\eta_{int} J}{ed}. \tag{13.65}$$

So, from Eq. (13.41)

$$Anp = \frac{\eta_{int} J}{ed}. \tag{13.66}$$

To calculate the gain from Eq. (13.63) is not straightforward because the Fermi–Dirac functions $f_2(E_2)$, $f_1(E_1)$ depend on the location of the quasi-Fermi levels μ_c, μ_v in the conduction and valence bands, respectively, in the active region. But, we can estimate typical values that might be expected:

13.9.1 Estimation of Semiconductor Laser Gain

We consider the case of an n-doped GaAs active region with a donor concentration of $n_D = 10^{24}\ \mathrm{m^{-3}}$, an internal quantum efficiency of 0.6 and an injected current of 10 mA that enters a 10 μm wide by 10 μm high active region whose

† See Problem (1.1).

thickness between heterojunctions in 0.5 μm. From Eqs. (13.66) and (13.64), taking $A = 10^{-16}$ m^3 s^{-1}

$$np = \frac{0.6 \times 0.01}{1.6 \times 10^{-19} \times 100 \times 10 \times 0.5 \times 10^{-18} \times 10^{-16}}$$
$$= 7.5 \times 10^{47}\ \text{m}^{-6}. \tag{13.67}$$

Furthermore, in the active region $n - p = n_D = 10^{24}$ m^{-3}, which gives $n = 1.5 \times 10^{24}$ m^{-3}, $p = 0.5 \times 10^{24}$ m^{-3}.

The electron density satisfies

$$n = \int_{E_c}^{\infty} \rho_c(E_2) f_2(E_2) dE_2. \tag{13.68}$$

If the temperature is sufficiently low we can approximate $f_2(E_2) = 1$ for $E_2 \le \mu_c$ and $f_2(E_2) = 0$ for $E_2 > \mu_c$, with similar relations for the holes. Consequently, using Eq. (13.38)

$$n = \int_{E_c}^{\mu_c} \rho_c(E_2) dE_2 = \frac{1}{2\pi^2}\left(\frac{2m_e^*}{\hbar^2}\right)^{3/2} \int_{E_c}^{\mu_c} \sqrt{E_2 - E_c} dE_2 \tag{13.69}$$

and

$$p = \int_{\mu_v}^{E_v} \rho_v(E_1) dE_1 = \frac{1}{2\pi^2}\left(\frac{2m_h^*}{\hbar^2}\right)^{3/2} \int_{\mu_v}^{E_v} \sqrt{E_v - E_1} dE_1. \tag{13.70}$$

In GaAs $m_e^* = 0.067\ m_e, m_h^* = 0.48\ m_e$, so evaluating the constants and performing the integration gives

$$n = 1.23 \times 10^{54}(\mu_c - E_c)^{3/2}, \tag{13.71}$$

$$p = 2.35 \times 10^{55}(E_v - \mu_v)^{3/2}. \tag{13.72}$$

With the electron and hole concentrations calculated previously

$$\mu_c - E_c = 1.14 \times 10^{-20}\ \text{J} = 0.071\ \text{eV}, \tag{13.73}$$

$$E_v - \mu_v = 7.88 \times 10^{-22}\ \text{J} = 0.0048\ \text{eV}. \tag{13.74}$$

The band gap energy in GaAs is 1.42 eV, so this calculation predicts that gain will result for photon energies from $\mu_c - \mu_v$ to E_g, i.e., wavelengths from 832 nm to 873 nm. With the parameters calculated so far we can calculate the gain at a wavelength within this range, at, say, 838 nm≡1.48 eV. In this case

$$\gamma(\nu) = \frac{hc_0^2}{8\pi n n_g \nu^2} \int_{E_a}^{E_b} A \times 6.5 \times 10^{109} \times \sqrt{E_2 - E_c}\sqrt{E_v - E_2 + h\nu} dE_2. \tag{13.75}$$

The limits on the integral are set by the condition $f_2(E_2) > f_1(E_1)$ and $E_2 - E_1 = h\nu$. Clearly,

$$E_a = E_c + 0.055(\text{eV}), \tag{13.76}$$
$$E_b = E_c + 0.06(\text{eV}). \tag{13.77}$$

If we assume that A is independent of energy and has a value of 10^{-16} m^3 s^{-1}, and using the values $n = 3.6$, $n_g = 4$ appropriate to GaAs, Eq. (13.75) becomes

$$\gamma(\nu) = 2.52 \times 10^7 \int_{E_c+0.055}^{E_c+0.06} \sqrt{E_2 - E_c}\sqrt{E_v - E_2 + 1.48}dE_2, \tag{13.78}$$

where the intergand is expressed in eV units. Now $E_v - E_2 + 1.48 = 0.06 - (E_2 - E_c)$, so writing $E_2 - E_c = 0.06\cos^2\theta$ the integral is transformed to give

$$\begin{aligned}\gamma(\nu) &= 1.62 \times 10^5 \int_0^{0.293} \sin^2\theta\cos^2\theta d\theta \\ &= 1.06 \times 10^4 \text{ m}^{-1} = 10.6 \text{ mm}^{-1}.\end{aligned} \tag{13.79}$$

Such a high gain figure is very typical for modern semiconductor lasers. The effective gain in the laser is, of course, reduced by the distributed loss coefficient α, which in semiconductor lasers results from scattering. This scattering increases with doping level, but will be in the 10^3–10^4 m^{-1} range. If $\alpha = 10^3$ m^{-1}, the effective γ is 9.6 mm^{-1}. In a laser 500 μm long the single pass intensity increase would be $e^{4.8} = 122$.

13.10 Threshold Current and Power–Voltáge Characteristics

The active region in a semiconductor laser is typically very thin, ranging from a few micrometers down to a few nanometers in so-called single quantum well (SQW) lasers, which we will discuss briefly later. The laser cavity is typically a few hundred micrometers long, although shorter devices are also made. Longer cavity semiconductor lasers are also fabricated in which additional structures are incorporated into the cavity region to provide single frequency operation or allow for intracavity modulation. Because the gain can be very high the natural Fresnel reflectance at the cleaved facet ends of the laser provides more than sufficient feedback to produce laser oscillation. In fact, in applications in which a semiconductor laser structure is to be used purely as an amplifier, extraordinary steps must be taken to reduce the facet reflectance, perhaps below 10^{-4} [13.11] to prevent spurious oscillation.

The performance of a semiconductor laser is defined by several parameters: its threshold current density J_{th}, power versus voltage (P–I) characteristic, and its effective current density J_{eff}, which is defined as (cf. Eq. (13.66))

$$J_{eff} = \frac{\eta_{int} J}{d}. \tag{13.80}$$

The threshold current density J_{th} is the value at which the gain exceeds the cavity losses, which we expect to be

$$\gamma_{th} = \alpha - \frac{1}{\ell}\ln\sqrt{R_1 R_2}, \tag{13.81}$$

where R_1, R_2 are the end facet reflectancess. However, an additional factor must be included in Eq. (13.81). As has already been mentioned, there is a refractive index discontinuity on both sides of the active layer in a double heterostructure laser, as shown schematically in Fig. (13.26).

This gives rise to a *wave guiding* effect or confinement of the transverse mode of the laser to a region in, and on both sides of, the active region. The confinement

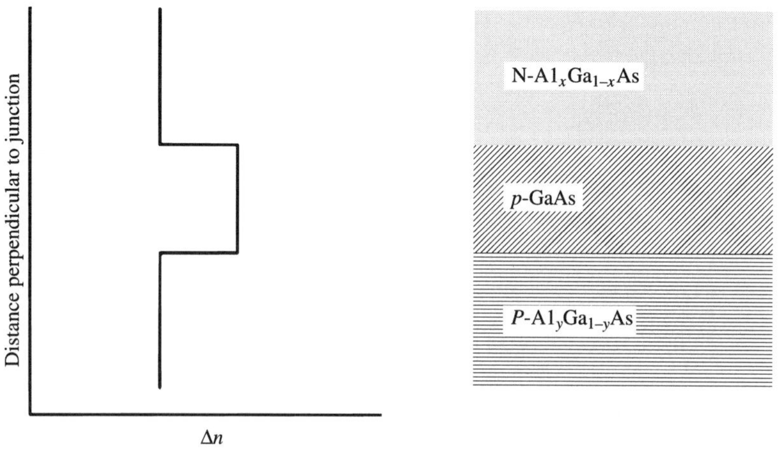

Fig. 13.26. Schematic variation of refractive index across a N–p–P double heterojunction.

factor, Γ, is a measure of the fraction of the oscillating field distribution that experiences gain within the active layer. The threshold gain is therefore

$$\gamma_{th} = \frac{1}{\Gamma}\left(\alpha - \frac{1}{\ell}\ln\sqrt{R_1 R_2}\right). \tag{13.82}$$

In many semiconductor lasers, so long as they oscillate in a single transverse mode[13.12] the gain is directly proportional to the excess current density above threshold, namely,

$$\gamma = \beta[J_{eff} - (J_{eff})_{th}]. \tag{13.83}$$

The ideal $P-I$ characteristic is of the general form shown in Fig. (13.27), although very different looking curves are experimentally observed.

13.11 Longitudinal and Transverse Modes

The active region is generally a rectangular parallelepiped with an end aperture of dimension *wd* with waveguiding refractive index discontinuities in both the direction perpendicular to the junction, and sometimes additionally in the direction parallel to the junction – in so-called *index-guided* stripe geometry lasers. The dimension *wd*, and the extent to which the laser is above threshold leads to the possibility of different transverse emitted field distributions as the current is increased, as shown in Fig. (13.28).

If the stripe width *w* is maintained sufficiently small, then a single laser field distribution can be maintained to relatively high power levels. The conditions under which this will occur depends on the index discontinuities involved, as will be discussed later in Chapter 17.

Below threshold a semiconductor laser emits a broad, continuous spectrum like an LED. However, as the current density rises above threshold distinct longitudinal modes will be observed to oscillate. Particularly in index-guided lasers, one or two modes may begin to dominate as the current rises far above threshold, as shown in Fig. (13.29)

Because the index of the active region, and the confining layer, varies with wavelength, the longitudinal modes of a semiconductor laser are not equally

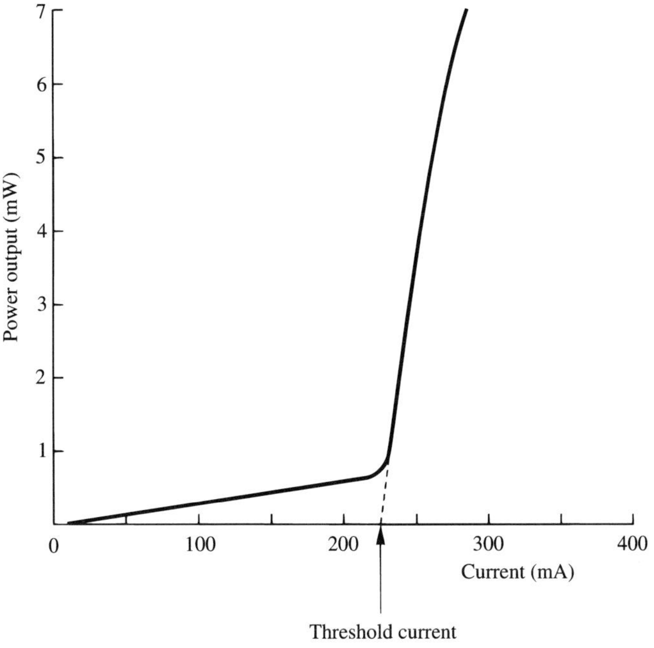

Fig. 13.27. *P–I* characteristic of a InGaAsP/InP double heterojunction laser[13.13].

spaced. Schematically, we can write

$$n_m \nu_m = \frac{m c_0}{2\ell}, \tag{13.84}$$

where n_m is the refractive index that applies at the frequency of the cavity resonance of index m, and

$$\Delta(n\nu) = n\delta\nu + \nu\frac{dn}{d\nu}\delta\nu. \tag{13.85}$$

13.12 Semiconductor Laser Structures

The majority of modern semiconductor lasers use some variant on the double heterostructure configuration shown schematically in Fig. (13.22). The principal difference between many semiconductor laser structures and Fig. (13.22) is that generally the structure is fabricated so as to confine the current to a specific region within the device. Such *stripe-geometry* lasers can be fabricated in several ways, as shown schematically in Fig. (13.30).

If the region of current flow is defined by a conductive stripe on one of the electrodes, or by making the confining layer resistive by proton bombardment or oxygen implantation (which creates defects in the lattice) then the active region will be laterally restricted to the region where current flows as shown in Figs. (13.30b), (13.30c), and (13.30f). If there is no additional modification of the active region in the lateral direction, then the laser is described as *gain-guided* as in Figs. (13.30a) and (13.30f). On the other hand, if the fabrication process involves surrounding an active region of GaAs by GaAlAs on all four sides as in Fig. (13.30b), then the laser is described as *index-guided*. Index-guided structures provide the greatest control of the emitted transverse mode, at the expense of fabrication complexity.

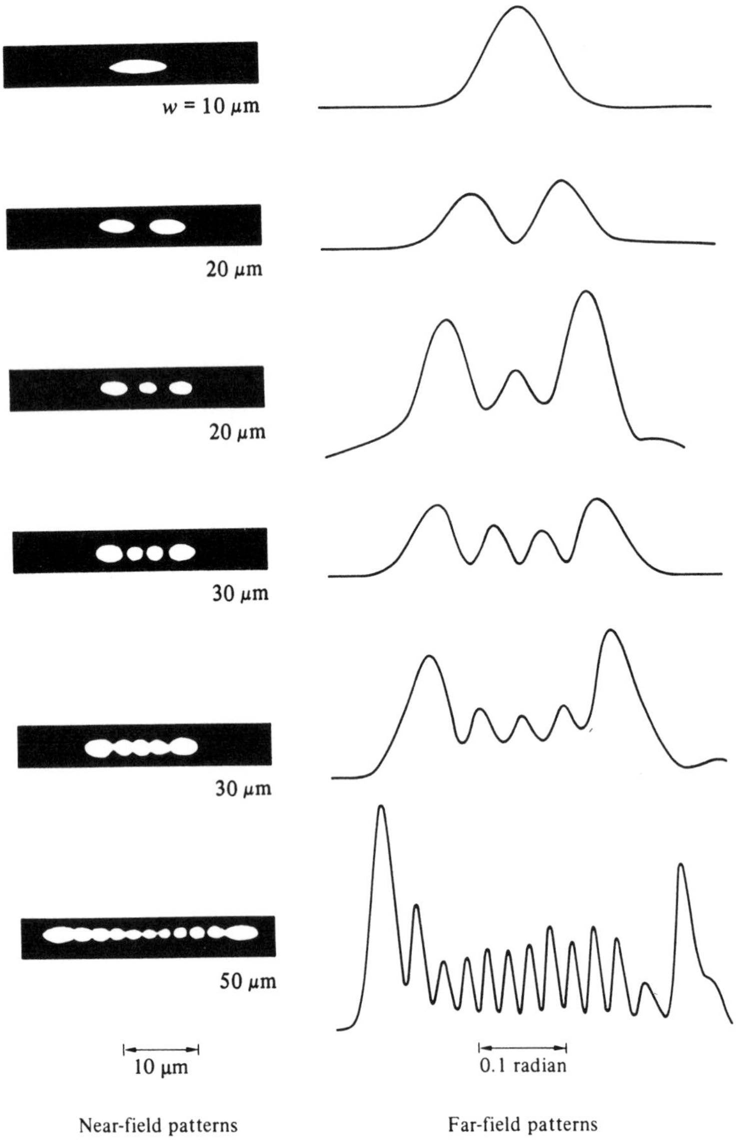

Fig. 13.28. Near-field and far-field intensity patterns for different stripe widths and currents in a double heterojunction planar stripe laser[13.14] (reproduced with the permission of the publisher).

There are many other structures that incorporate the essential fabrication schemes shown in Fig. (13.30). A currently popular semiconductor laser design is the GRINSCH (graded-index separate confinement heterostructure), an example of which is shown in Fig. (13.31). The active layer in this laser is very thin – a single quantum well (SQW)† with graded layers of GaAlAs on each side to confine the mode optically. The active region could also be a multiple quantum well of, for example, four 0.013 μm thick layers of $Ga_{0.94}Al_{0.06}As$ separated by three 0.009 μm layers of $Ga_{0.8}Al_{0.2}As$[13.16]. For additional technical details the reader should consult the specialized literature[13.15],[13.19].

† See Section 13.15.

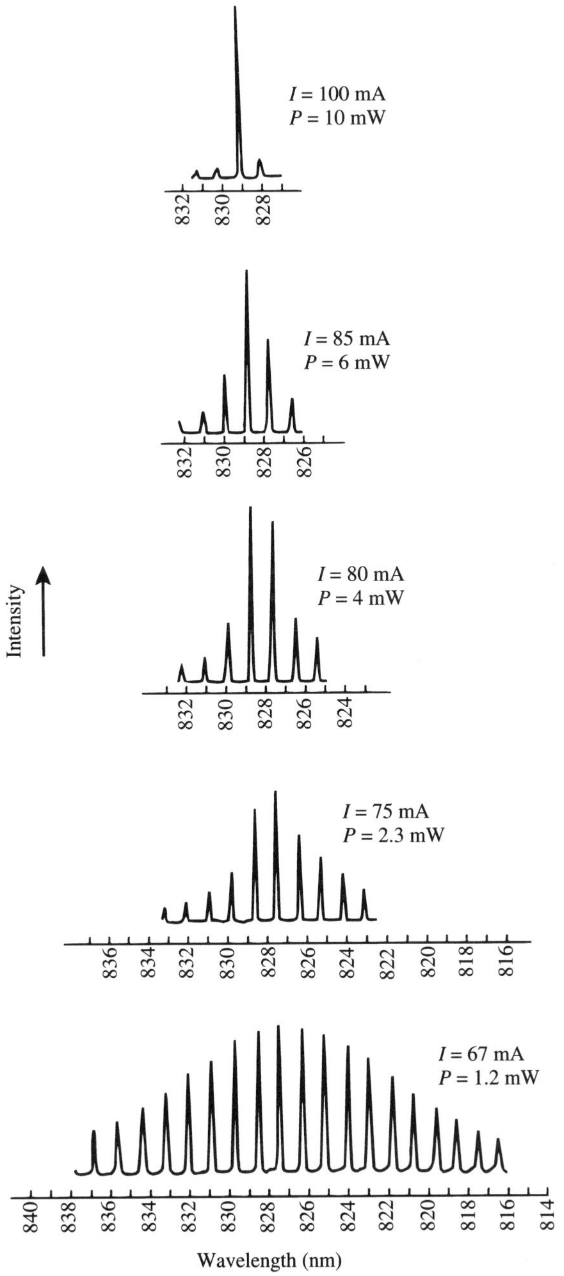

Fig. 13.29. Output frequency spectra of a fairly 'well-behaved' CW laser diode at different currents[13.15].

13.12.1 *Distributed Feedback (DFB) and Distributed Bragg Reflection (DBR) Lasers*

Most edge emitting semiconductor lasers that we have discussed so far have a tendency to oscillate simultaneously on several cavity modes. Generally, special features must be incorporated into the laser structure to encourage single mode operation. This can be done by including a periodic structure or *grating* in the

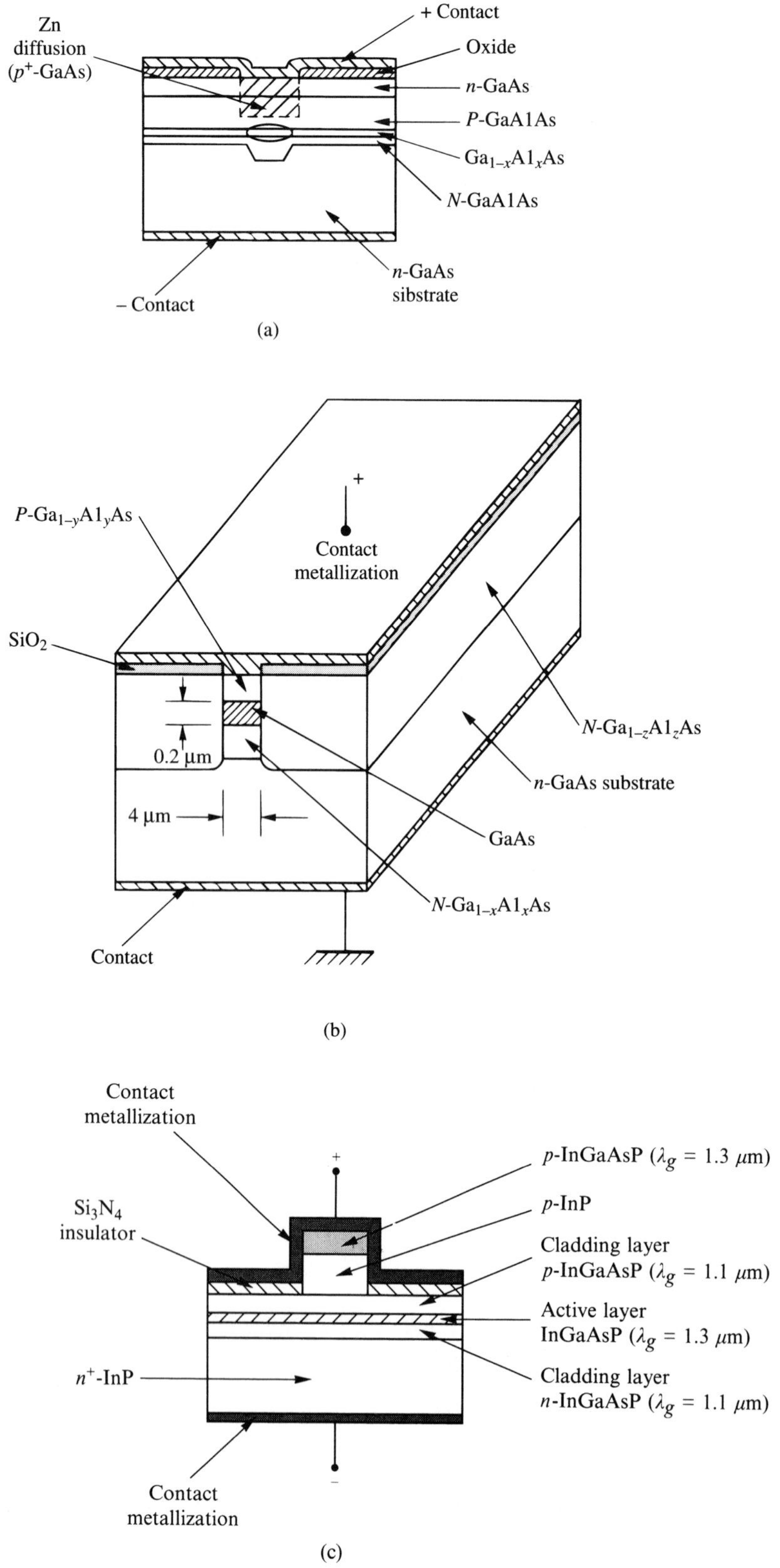

Fig. 13.30. Schematic construction of various types of stripe-geometry, double heterostructure lasers: (a) stripe confinement by oxide layer and deep zinc diffusion; (b) buried stripe; (c) cladded ridge[13.17].

Fig. 13.30. (*cont.*) (d) etched mesa; (e) constricted double heterostructure (CDH) (will give single transverse mode operation[13.18]; (f) stripe definition by oxygen implantation or photon bombardment.

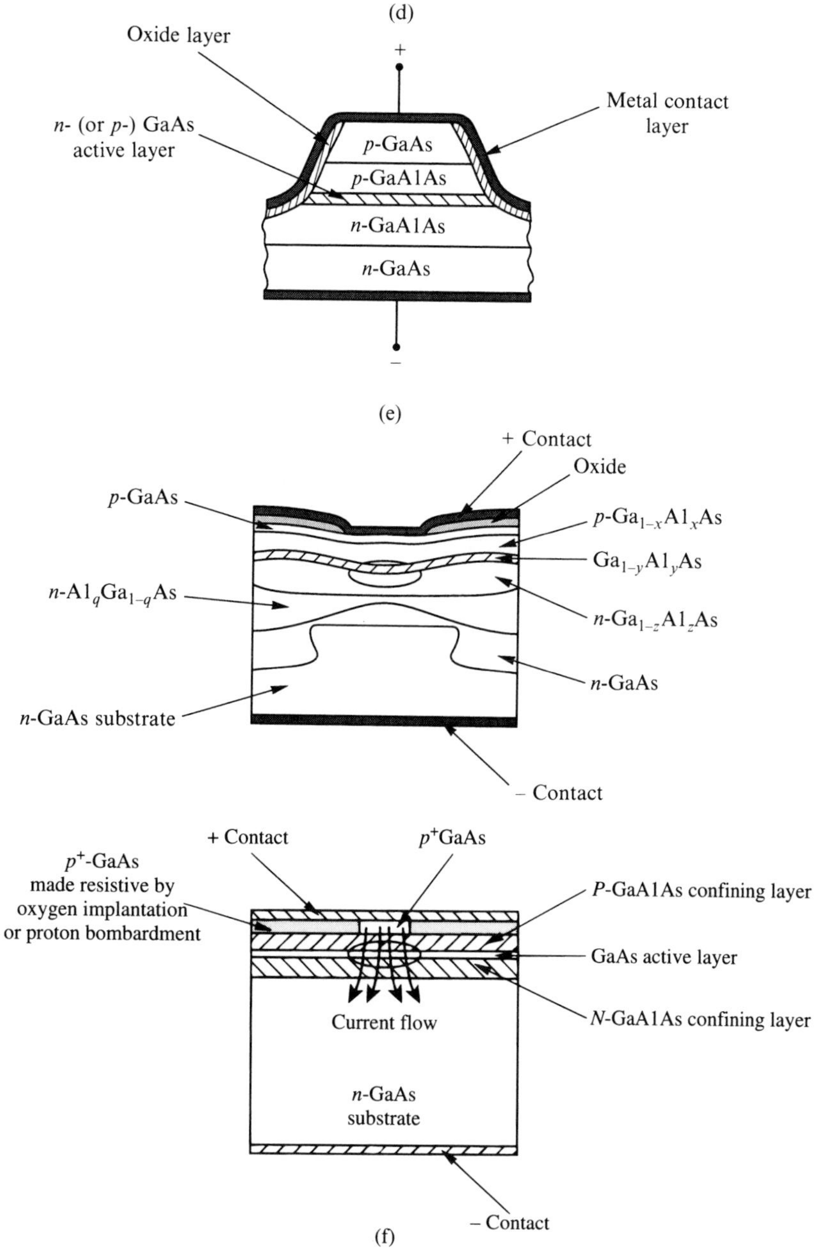

device near the active layer where the oscillating transverse mode will interact with it. Fig. (13.32) shows a planar waveguiding structure that includes a surface corrugation of periodicity d. Light that satisfies the Bragg condition will be reflected efficiently back on itself. This condition is set by waves scattering off successive corrugations reflecting and constructively interferring. If the index of the guiding layer is n, the Bragg condition can be written as

$$2d \sin\theta = m\lambda_0/n. \tag{13.86}$$

The angle θ is generally close to 90°, its effect can be included in an *effective*

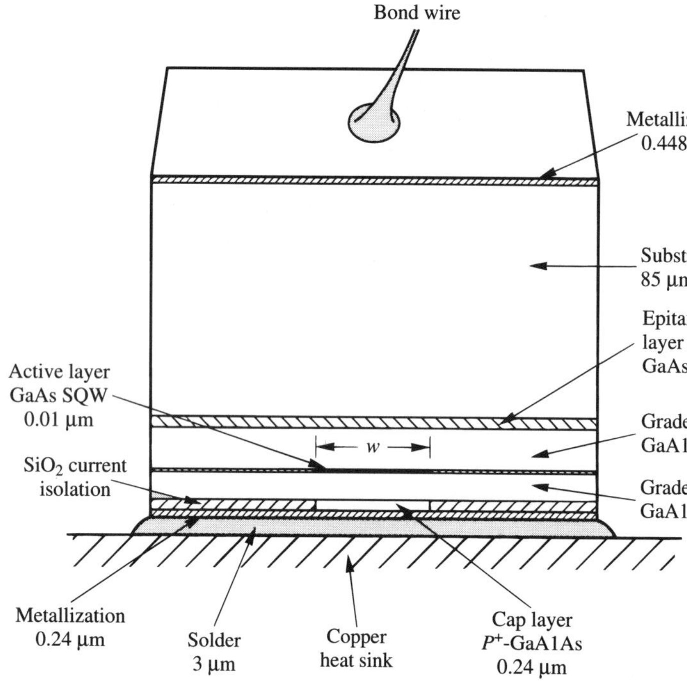

Fig. 13.31. GRINSCH SQW design used by McDonnell–Douglas Electronics Systems Co. Typical thicknesses of the various layers are shown. For a narrow stripe laser $w \sim 5\ \mu m$, for a broad stripe laser $w \sim 60\ \mu m$.

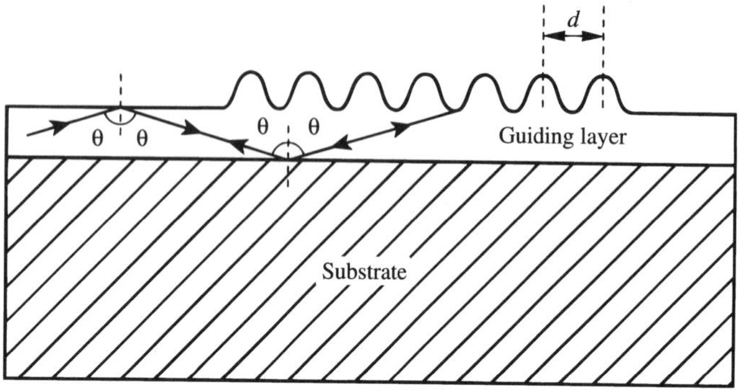

Fig. 13.32. Bragg reflection in a periodic waveguide structure.

refractive index, n_{eff}, for the guided wave so that the grating period d required for operation at wavelength λ_0 is

$$d = \frac{m\lambda_0}{2n_{eff}}. \tag{13.87}$$

A laser in which a periodic structure is placed in close proximity to the active region over a substantial length of the gain region is called a *distributed feedback* DFB laser. It oscillates at a wavelength determined by the Bragg reflection condition given in Eq. (13.87) above. Fig. (13.33a) shows how such a DFB laser is fabricated. The corrugations are separated from the active region and are formed between an $Al_{0.17}Ga_{0.83}As$ layer and an $Al_{0.07}Ga_{0.93}As$ layer. The thin p-type $Al_{0.17}Ga_{0.83}As$ layer and an n-type $Al_{0.3}Ga_{0.7}As$ layer confine the optical field to the active region. However, sufficient field amplitudes leak through the p-type $Al_{0.17}Ga_{0.83}As$ to allow interaction with the corrugation.

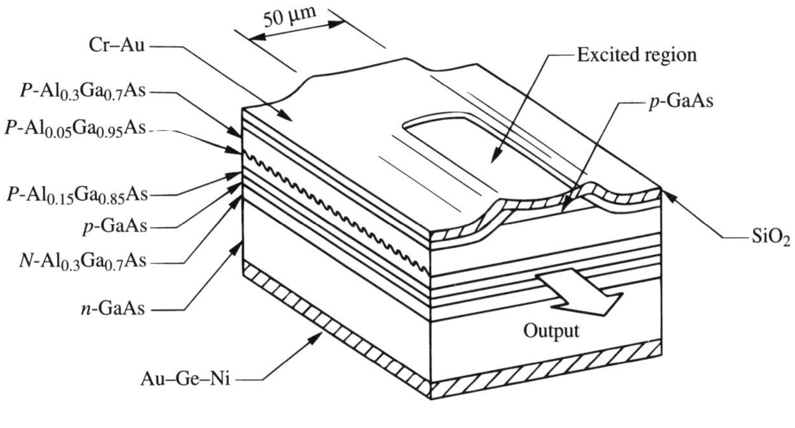

(a)

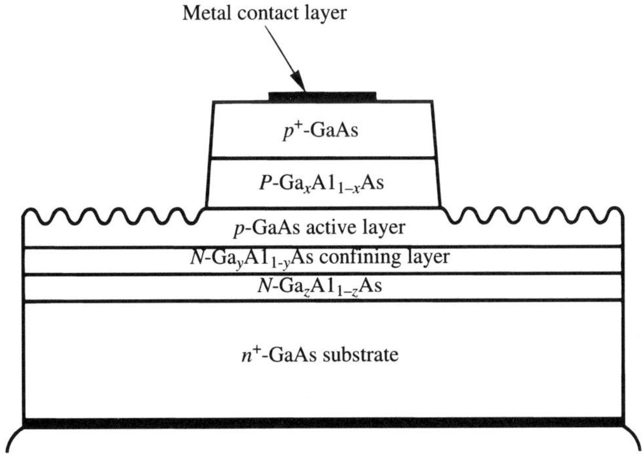

(b)

Fig. 13.33. (a) A GaAlAs double heterojunction DFB laser structure[13.20]. (b) DBR laser structure.

The corrugations on the confining layer are produced by photolithography in which holographic techniques are used and a periodic interference pattern is projected onto the surface to be etched. If a laser beam of wavelength λ is split into two parts and then the two beams reinterfere at an angle 2θ to each other, and each at angle θ to the surface normal, a grating of period $d = \lambda/2\sin\theta$ results. Thus, for example, a layer of photoresist can be periodically exposed in preparation for etching. In a laser incorporating a periodic structure a detailed analysis can provide specific information about the threshold gain, and will predict the precise resonance frequencies at which laser oscillation will occur[13.4],[13.19].

Selective cavity reflectance can also be provided in a semiconductor laser structure in which a grating is used at each end of the gain region to provide feedback at wavelength that satisfy the Bragg reflection condition. Such lasers are called *distributed Bragg reflection* DBR lasers. Fig. (13.33b) shows an example of such a laser.

A grating structure, which can be fabricated at one or both ends of the cavity, provides selective cavity reflection to keep the left and right travelling waves in phase. DBR lasers can provide stable, single-frequency operation and tunability.

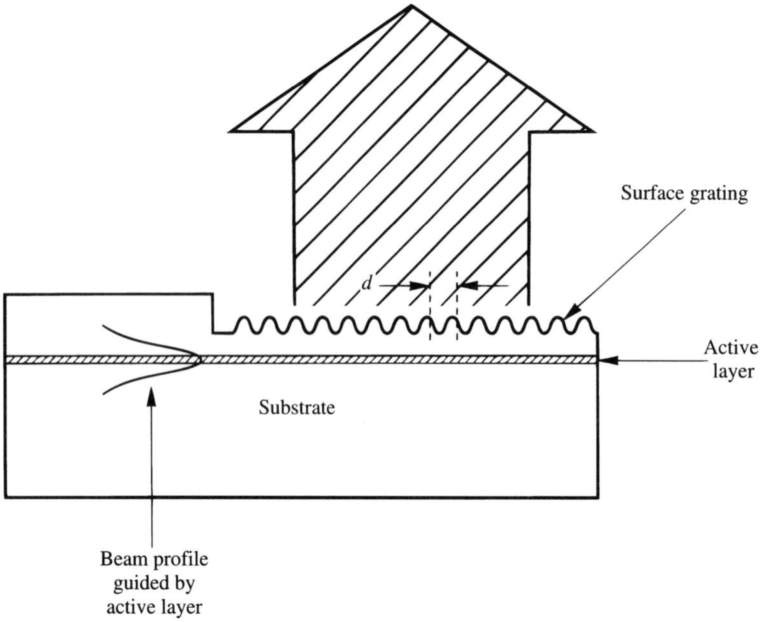

Fig. 13.34.
Grating-coupled surface emitting laser.

The gain and Bragg reflection regions are separate, which allow independent control of the carrier density in the grating region. Changes in carrier density in this region change the effective index of the cavity mode and tune the laser frequency. DBR lasers can also provide narrower oscillating linewidth (<1 nm) than conventional double heterojunction lasers. An additional advantage of both DFB and DBR lasers is that their lasing frequency is less dependent on temperature than in a Fabry–Perot cavity semiconductor laser. The emission from the latter follows the temperature dependence of the energy gap, whereas the DFB and DBR structures follow the smaller temperature dependence of the refractive index and grating spacing.

13.13 Surface Emitting Lasers

In recent years a new class of semiconductor laser, the surface emitting laser, has emerged in which the laser beam is emitted perpendicular to the wafer. These new structures can be fabricated by conventional planar integrated circuit techniques and can provide emission over a broad area. In contrast to the edge emitting devices we have considered so far, they also can be scaled down to *very* small size. Literally, many of these miniature lasers can be fabricated in an area smaller than the head of a pin.

One version of surface emitting laser is shown in Fig. (13.34). In this *grating-coupled* surface emitting laser a periodic structure is etched onto the surface of the confining layer on one side of the junction. If the grating period, d, is chosen appropriately, then a diffraction maximum exists in the vertical direction for waves travelling in the active layer. This results if

$$m\lambda = d. \tag{13.88}$$

In principle, a laser of this kind can emit over a large surface area, and provide

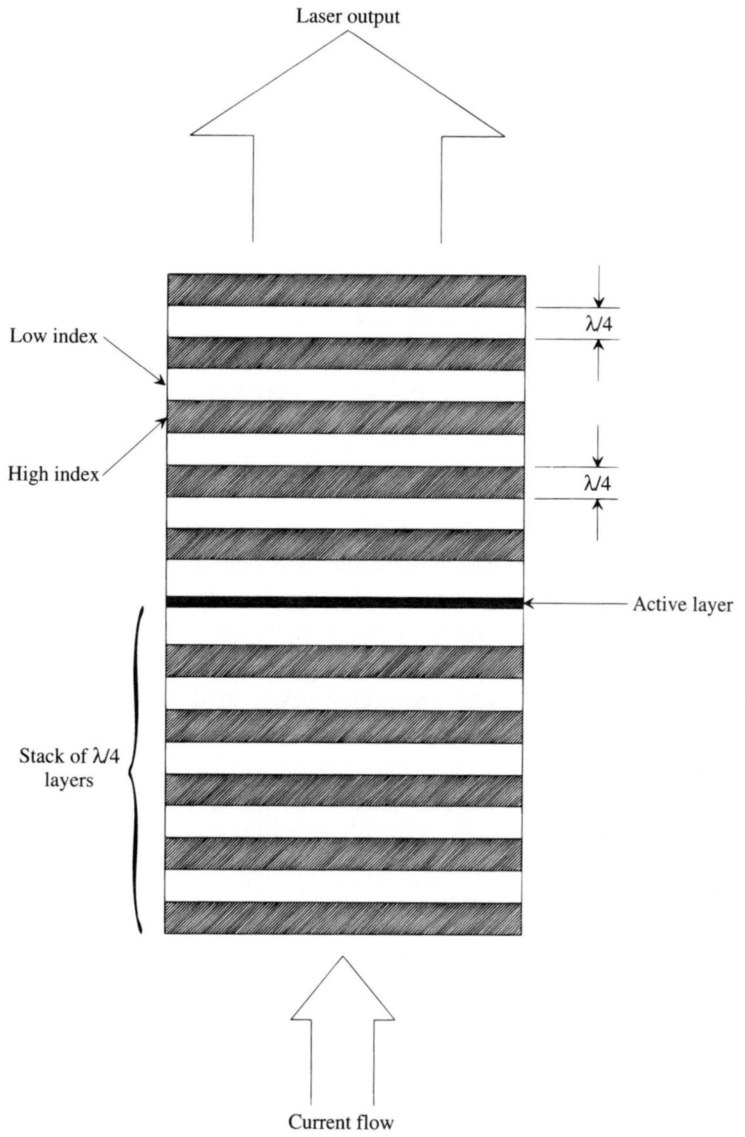

Fig. 13.35. Schematic structure of a vertical cavity surface emitting laser.

a beam of high power and low beam divergence. Pulsed powers in excess of 4 W have been achieved[13.21] and high CW powers[13.22].

However, the most technologically important surface emitting lasers are *vertical cavity surface emitting lasers*, VCSELs. These lasers are schematically of the form shown in Fig. (13.35). The active layer is bounded on each side in the vertical direction with alternate high and low index layers each of which is designed to be a quarter wavelength thick at the laser wavelength. Each of these stacks of $\lambda/4$ layers acts as a narrow band, high reflectivity mirror, thereby forming a vertical cavity. There are several advantages to this type of laser structure:

(a) A very large number of small lasers can be fabricated to occupy a given area. A collaboration between AT&T Bell Laboratories and Bellcore has

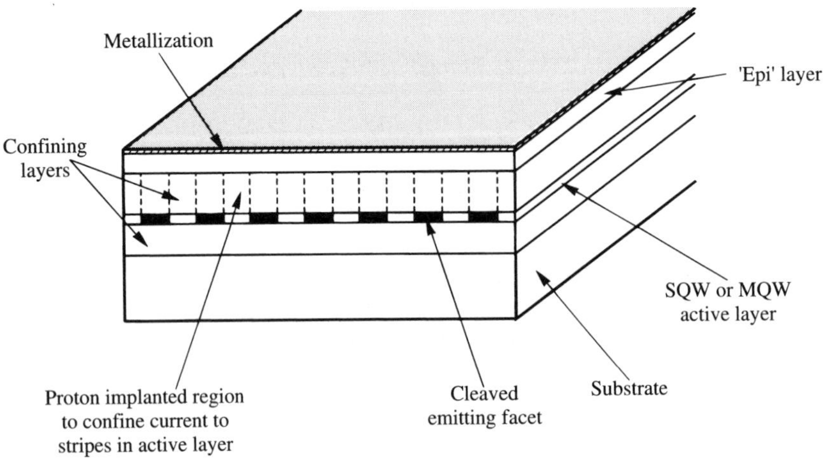

Fig. 13.36. Edge emitting laser diode array.

produced VCSEL microlasers at densities of over twenty thousand per square millimeter[13.23]. These 'microlasers' are only about 5.5 μm high and have diameters in the region of 1-5 μm. The active layer is only 24 nm thick. The threshold currents for lasing are only about 1 mA and with improved microfabrication techniques could go as low as 10 μA.

(b) Because these lasers are very short their cavity mode spacing is very large, $\sim$180 nm for a 5.5 μm laser. This ensures that they operate in a single mode.

(c) The small diameter and cylindrical shape of these lasers provides a circular output beam.

(d) The matrix of optical emitters that results has potential in many optical computing and optical interconnect applications.

13.14 Laser Diode Arrays and Broad Area Lasers

There is a limit to the maximum optical power that can be extracted from an edge emitting, stripe-geometry laser. A primary limit is set by the intracavity power density, which at high enough levels will lead to optical damage to the laser facets. An oscillating laser mode that is 4 μm wide by 500 nm high with an output power of 100 mW will be close to the optical damage threshold, since its intracavity peak power density is on the order of 1 MW mm^{-2}. To produce higher output power than this the most successful approach has involved the use of laser diode *arrays*. In these devices, many stripe lasers are fabricated in close proximity to each other, as shown in Fig. (13.36).

The current is confined to a series of stripe regions by producing regions of semi-insulating semiconductor through proton implantation, which produces defects in the semiconductor lattice. Alternatively, stripe metal contacts on the top surface can be used to delineate the regions of current flow. Typically, ten stripe lasers would be fabricated with 4 μm wide active regions on 10 μm centers. The proximity of the individual stripes to each other causes overlap of adjacent oscillating intracavity field distributions. A phase-locking of the fields in the separate stripes occurs leading to a combined, coherent wavefront being emitted from the whole facet. The optical

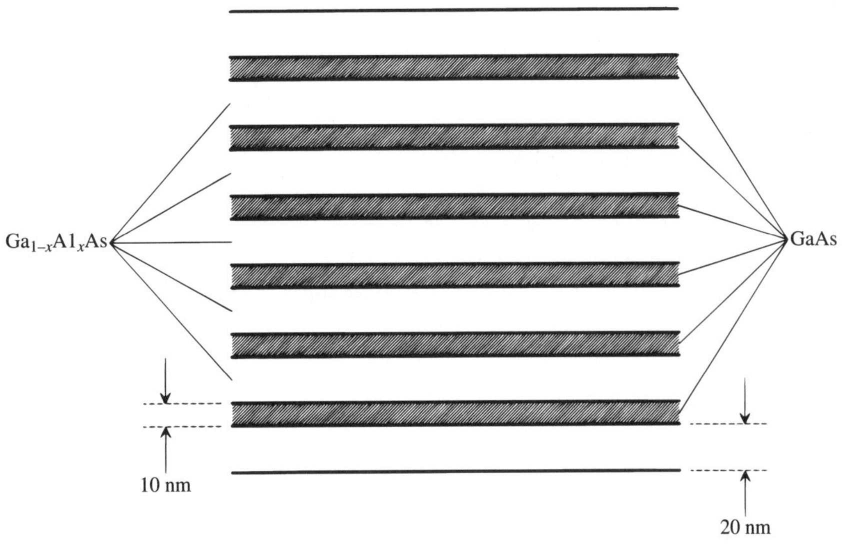

Fig. 13.37. Layered structure of the active region of a multiple quantum well semiconductor låser.

quality of this wavefront is not generally high as it has several maxima and minima – very much like the diffraction pattern that would result from a series of regularly spaced small apertures. The total power output from an array of ten elements can be as high as 1 W. Arrays of up to 40 elements have been fabricated that produce 2.5 W. If a larger number of stripe lasing elements is included in a single line then parasitic laser oscillation in the transverse direction can occur. Even higher powers can be produced by two-dimensional arrays. For example a CW output of 38 W has been produced from twenty 10-stripe phased arrays spaced on 500 μm centers.

A broad area laser (BAL) is an alternative approach to generating a high power output. In these devices the active region is made large, but is not divided into stripes. For example, a 100 μm wide SQW structure has produced a CW power of 3.8 W at a current of 5 A. The most important application of these high power arrays and BALs is in the pumping of solid-state lasers, which have already been described in Chapter 8.

13.15 Quantum Well Lasers

Quantum well semiconductor lasers use a very thin active region, for example a 20 nm thick layer of GaAs bounded on each side by GaAlAs. If the laser has only one such thin layer it is called a single quantum well (SQW) laser. On the other hand if the device is fabricated as a multilayer sandwich of thin layers of alternating GaAs and GaAlAs, as shown schematically in Fig. (13.37), then it is called a multiple quantum well (MQW) laser. In these lasers the total number of injected carriers necessary to achieve inversion is dramatically reduced compared to conventional heterostructure devices.

A detailed treatment of the operation of these devices is beyond our scope here, but it is still possible to explain semiquantitatively why they have attractive properties. In the energy band diagram of a SQW laser shown in Fig. (13.38) the quantum well is a region of minimum potential energy for electrons in the conduction band and holes in the valence band. The reduced threshold injection

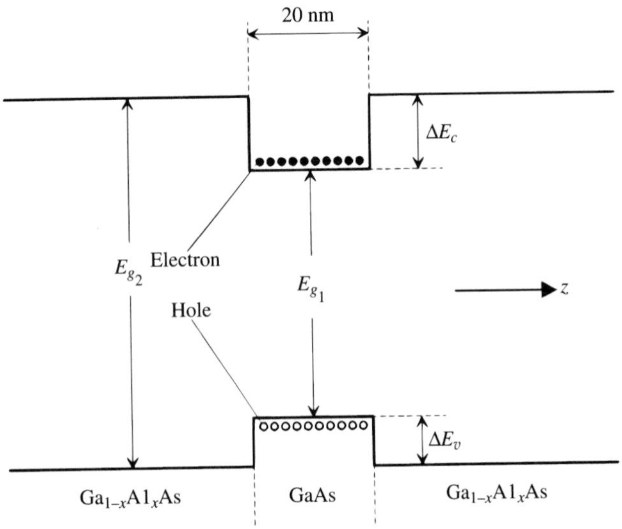

Fig. 13.38. Energy band diagram of a SQW laser.

current in these devices results largely because the very thin quantum well modifies the density of states available to electrons and holes. Recall that these particles have wave-like properties and must satisfy periodic boundary conditions within a box (just like black-body radiation). The number of modes that exist in the two-dimensional region parallel to the layers is easily calculated as

$$N_{2D}(k_{||}) = \frac{k_{||}^2 L_x L_y}{2\pi} = \frac{k_{||}^2 A}{2\pi}, \tag{13.89}$$

where $A = L_x L_y$ is the area of the junction region.

Therefore, the two-dimensional density of states is

$$\rho(k_{||})dk_{||} = \frac{dN_{2D}(k_{||})}{dk_{||}}dk_{||} = \frac{Ak_{||}}{\pi}dk_{||}, \tag{13.90}$$

or in terms of energy per unit area

$$\rho_{QW}(E) = \frac{1}{A}\frac{dN_{2D}}{dE} = \frac{k_{||}}{\pi}\frac{dk_{||}}{dE}. \tag{13.91}$$

The number of modes that exist in the z (perpendicular) direction is severely reduced because of the small size in this direction. Qualitatively, there may be only one such mode, corresponding roughly to a half-wavelength of the electron wave fitting in the energy minimum in the conduction band in Fig. (13.38). The total energy of an electron in the quantum well can be written as

$$E = \frac{\hbar^2 k_{||}^2}{2m_c} + E_Q, \tag{13.92}$$

where E_Q is the energy associated with the restricted motion of the electrons perpendicular to the junction.

From Eq. (13.92)

$$k_{||} = \sqrt{\frac{2m_c}{\hbar^2}(E - E_Q)}. \tag{13.93}$$

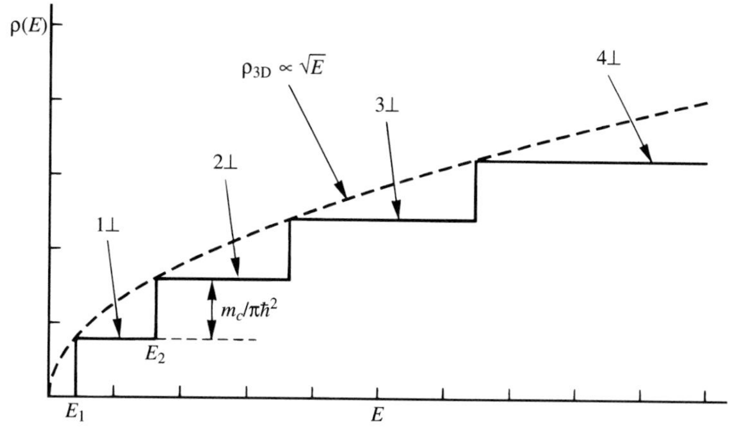

Fig. 13.39. Density of states in a SQW compared to the full three-dimensional density of the states function, E_1 is the minimum electron energy for one mode perpendicular to the layers, E_2 for two modes, and so on. The number 2⊥ on the curve, for example, indicates that there are two permitted modes in the perpendicular direction.

Therefore,

$$\frac{dk_{\|}}{dE} = \frac{m_c/\hbar^2}{k_{\|}} \tag{13.94}$$

and

$$\rho_{QW}(E) = \frac{m_c}{\pi\hbar^2}. \tag{13.95}$$

This result holds when only a single mode exists perpendicular to the thin layer. At critical electron energies the electron wavelength becomes short enough that additional modes perpendicular to the layers become possible. The consequence of this is a step-like density of states function, as shown in Fig. (13.39). Each step corresponds to an increase in the QW density of states $= m_c/\pi\hbar^2$.

QW lasers have lower thresholds, higher gains, and generate narrower linewidth laser radiation than more conventional heterostructure lasers. A SQW laser generates the narrowest linewidth and has the lowest threshold current but at current densities above threshold the effective gain of a MQW laser is higher. This is a consequence of the ratio Γ between the width of the active region and the width of the oscillating intracavity mode. To provide modal confinement, the region on each side of the quantum well is frequently a region of graded index, as shown in Fig. (13.40).

Semiconductor lasers do not generate laser radiation whose linewidth is narrow compared to low pressure gas lasers. They can often be encouraged to oscillate in a single transverse mode, but will generally produce several longitudinal modes, of typical spacing 0.1 nm (for a 200 μm long GaAs laser). The gain profile is generally Lorentzian, of a width determined by the phonon collisions within the semiconductor. The calculated Schawlow–Townes linewidths of small narrow stripe lasers are typically a few MHz, but values as narrow as this are quite difficult to achieve in practice. Unless special measures are taken typical linewidths are tens of MHz. Lasers with extended cavities, either external or incorporated into the semiconductor structure, have generated laser linewidths as narrow as 10 kHz.

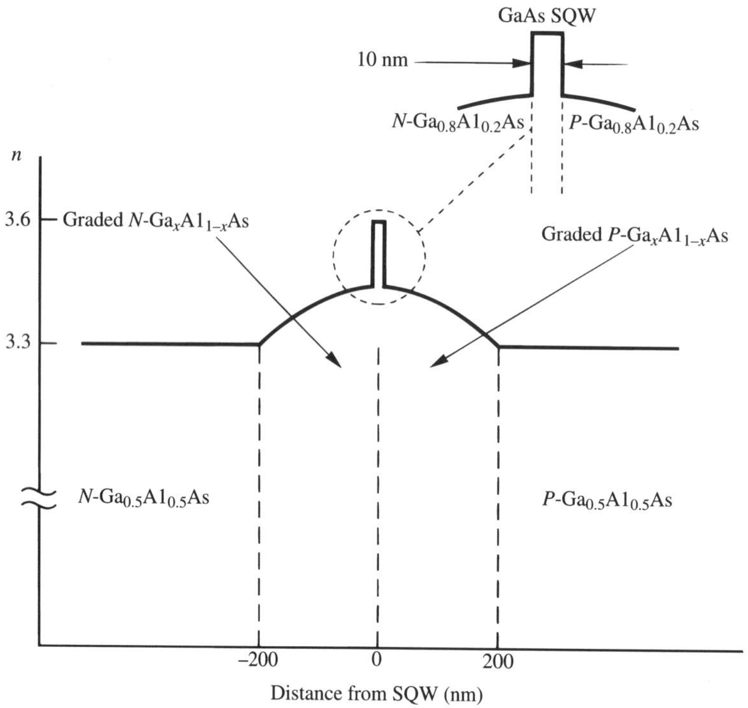

Fig. 13.40. Schematic variation of the refractive index change across a QW laser with a graded layer of $Ga_{0.5}Al_{0.5}As \rightarrow Ga_{0.8}Al_{0.2}As$ on each side of the QW (GaAs) region. The Ga concentration increases towards the QW region in the graded layers on either side.[13.24]

13.16 Problems

(13.1) Prove that the maximum fraction of the emitted radiation from the active layer in Fig. (13.16) that might escape from the semiconductor/air interface if the reflectance R at the outer semiconductor/air surface is assumed to be independent of angle is $f = \frac{1}{2}n_s^2$, where n_s is the refractive index of the semiconductor.

(13.2) An edge emitting LED with a center wavelength of 810 nm has an active region 100 μm wide by 1 μm thick. Estimate the distance from the output facet at which the output beam will have become circular.

(13.3) The recombination time in a GaAs/GaAlAs laser is $\sim$ 4 ns, the cavity length between end facets of the diode is 500 μm, the reflectance at each end is $R =$ 0.31, the distributed loss is 1000 m^{-1}, and the oscillating beam width is 4 μm. If all the electrons injected into the junction region combine there, calculate the threshold current density for oscillation. Take the gain (m^{-1}) of the laser to be $N/10^{23}$, where N is the carrier density (m^{-3}) in the junction region.

(13.4) A GaAs laser has a cavity length of 0.3 mm and a refractive index of 4. Its threshold gain coefficient is 6 mm^{-1}. If the small signal gain is increased to 12 mm^{-1} how many longitudinal modes will oscillate if the gain curve has a Gaussian frequency variation with a full width at half maximum height of 0.1 μm at 1.3 μm.

(13.5) Calculate the wavelength of peak intensity for the recombination spectrum given by Eq. (13.45). Plot the variation of the wavelength of peak emission for GaAs from 0 to 500 K.

References

[13.1] J. Gowar, *Optical Communication Systems*, Prentice Hall, London, 1984.

[13.2] C. Kittel, *Introduction to Solid State Physics*, 6th Edition, Wiley, New York, 1986.

[13.3] F. Reif, *Fundamentals of Statistical and Thermal Physics*, McGraw Hill, New York, 1965.

[13.4] A. Yariv, *Introduction to Optical Electronics*, 1st Edition, Holt Rinehart and Winston Inc., New York, 1971.

[13.5] M.I. Nathan, 'Semiconductor Lasers,' *Appl. Opt.*, **5**, 1514–1528, 1966.

[13.6] H.C. Casey, Jr, and M.B. Panish, *Heterostructure Lasers*, Parts A and B, Academic Press, New York, 1978; see also M.B. Panish, 'Heterostructure Injection Lasers,' *Proc. IEEE*, **64**, 1512–1540, 1976.

[13.7] C.A. Burrus and B.I. Miller, 'Small-area, double-heterostructure, aluminum-gallium arsenide electroluminescent diode source for optical-fiber transmission lines,' *Opt. Commun.*, **4**, 307–309, 1971.

[13.8] A. Chandra and L.F. Eastman, 'Rectification at $n-n$ $GaAs$: $(Ga, Al)As$ heterojunctions,' *Electron. Lett.*, **15**, 90–91, 1979.

[13.9] J.F. Womac and R.H. Rediker, 'The graded-gap $Al_xGa_{1-x}As$-GaAs heterojunction,' *J. Appl. Phys.*, **43**, 4129–4133, 1972.

[13.10] M.G. Bernard and G. Duraffourg, 'Laser conditions in semiconductors,' *Phys. Stat. Solids*, **1**, 699–703, 1961.

[13.11] I.F. Wu, I. Riant, J-M. Verdiell, and M. Dagenais, 'Real-time in situ monitoring of antireflection coatings for semiconductor laser amplifiers by ellipsometry,' *IEEE Photonics Technology Lett.*, **4**, 991–993, 1992.

[13.12] R.L. Hartmann and R.W. Dixon, 'Reliability of DH GaAs lasers at elevated temperatures,' *Appl. Phys. Lett.*, **26**, 239–240, 1975.

[13.13] G.H. Olsen, C.J. Nuese, and M. Ettenberg, *Appl. Phys. Lett.*, **34**, 262–264, 1979.

[13.14] H. Yonezu, I. Sakuma, K. Kobayashi, T. Kamejima, M. Unno, and Y. Nannichi, 'A GaAs-$Al_xGa_{1-x}As$ double heterostructure planar stripe laser,' *Japan J. Appl. Phys.*, **12**, 1585–1592, 1973.

[13.15] H. Kressel, M. Effenberg, J.P. Wittke, and I. Ladany, 'Laser diodes and LEDs for optical fiber communication,' in *Semiconductor Devices for Optical Communication*, H. Kressel, Ed., Springer-Verlag, New York, 1982.

[13.16] D.R. Scifres, R.D. Burnham and W. Streifer, 'High Power Coupled Multiple Stripe Quantum Well Injection Lasers,' *Appl. Phys. Lett.*, **41**, 118–120 (1982).

[13.17] I.P. Kaminow, L.W. Stulz, J.S. Ko, A.G. Dentai, R.E. Nahory, J.C. DeWinter, and R.L. Hartman, 'Low threshold InGaAsP ridge waveguide lasers at 1.3 μm,' *IEEE J. Quant. Electron.*, **QE-19**, 1312–1319, 1983.

[13.18] D. Botez, 'CW high-pressure single-mode operation of constricted double-heterojunction AlGaAs lasers with a large optical cavity,' *Appl. Phys. Lett.*, **36**, 190–192, 1980.

[13.19] P.K. Cheo, *Fiber Optics and Optoelectronics*, 2nd Edition, Prentice Hall, Englewood Cliffs, New Jersey 1990; A. Yariv, *Optical Electronics*, 3rd Edition, Holt, Rinehart and Winston, New York, 1985.

[13.20] K. Aiki, M. Nakamura, J. Umeda, A. Yariv, A. Katziv, and H.W. Yen, 'GaAs-GaAlAs distributed feedback laser with separate optical and carrier confinement,' *Appl. Phys. Lett.*, **27**, 145–146, 1975.

[13.21] D.F. Welch, R. Parke, A. Hardy, R. Waaits, W. Striefer, and D.R. Scifres. 'High-power, 4W pulsed, grating-coupled surface emitting laser,' *Electron. Lett.*, **25**, 1038–1039, 1989.

[13.22] J.S. Mott and S.H. Macomber, 'Two-dimensional surface emitting distributed feedback laser array,' *IEEE Photon. Technal. Lett.*, **1**, 202–204, 2989.

[13.23] J.L. Jewel, 'Microlasers,' *Sci. Am.*, Nov. 1991, 86–94.

[13.24] A. Yariv, *Quantum Electronics*, 3rd Edition, Wiley, New York, 1989.

14

Analysis of Optical Systems I

14.1 Introduction

In this chapter we shall look at some very useful, and quite simple, techniques for analyzing optical systems. Most of our attention will be directed towards optical systems that have axial symmetry. So, for example, we shall examine various properties of lenses, flat and spherical mirrors, plane-parallel slabs of transparent material, and polarizers. In our analysis we will be confronted with the problem of how the components of the system affect the passage of light. To describe the behavior of the system in detail we need to know how a wave passing through is modified in terms of intensity, phase shift, direction of propagation, and polarization state. Our discussion will be simplified by assuming that we are dealing with light that is monochromatic, or at least contains only a small range of wavelengths. This allows us to neglect the effects of *dispersion*, the variation of the refractive index with wavelength. Dispersion, defined quantitatively as $dn/d\lambda$, causes the properties of an optical system containing transmissive components, such as lenses and optical fibers, to vary with wavelength.

14.2 The Propagation of Rays and Waves through Isotropic Media

An isotropic optical material is one whose properties are independent of the direction of propagation or polarization state of an electromagnetic wave passing through the material. Several simple relations hold for such a material:

(i) The polarization **P** induced by the electric field **E** of the wave is parallel to **E**, so

$$\mathbf{P} = \epsilon_0 \chi \mathbf{E}. \tag{14.1}$$

χ, the electric susceptibility, is a scalar.

(ii) It follows that the electric displacement vector **D** of the wave is parallel to **E**

$$\mathbf{D} = \epsilon_0 \mathbf{E} + \mathbf{P}, \tag{14.2}$$

which from Eq. (14.1) gives

$$\mathbf{D} = \epsilon_0 (1 + \chi) \mathbf{E} \tag{14.3}$$

or

$$\mathbf{D} = \epsilon_0 \epsilon_r \mathbf{E}. \tag{14.4}$$

ϵ_r is called the dielectric constant of the material. For an absorbing (or amplifying material) ϵ_r is complex, $\epsilon_r = \epsilon' - i\epsilon''$. The refractive index of

Fig. 14.1. Angular relationships between the vectors **D, E, B, H, k**, and **S** in an isotropic medium.

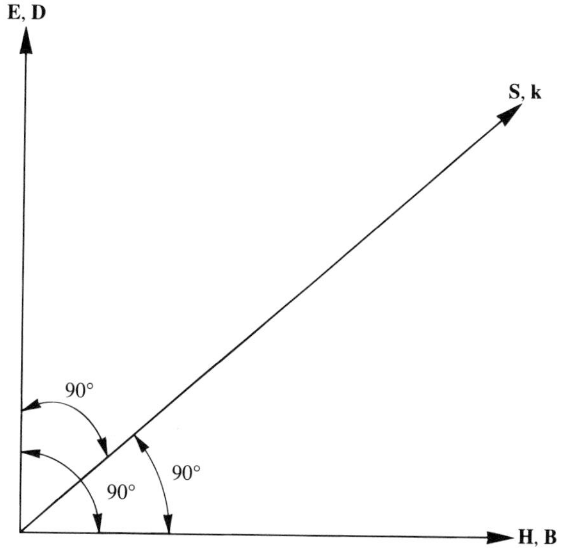

the material is $n = \sqrt{\epsilon_r}$, or in a lossy material $n = \sqrt{\epsilon'}$. The dielectric constant, ϵ_r, and consequently the refractive index, n, depend on the frequency of the wave. This phenomenon is called dispersion and is the reason why a prism separates white light into its constituent colors.

(iii) The Poynting vector of the wave, $\mathbf{S} = \mathbf{E} \times \mathbf{H}$, which represents the magnitude and direction of energy flow, is parallel to **k**, the wave vector. The wave vector **k** is perpendicular to the surfaces of constant phase (phasefronts) and is by definition perpendicular to the vectors **D** and **B** as shown in Fig. (14.1). The time averaged magnitude of **S** is called the intensity of the wave

$$I = < |\mathbf{S}| >_{AV} . \tag{14.5}$$

The local direction of **S** is called the *ray* direction. Only for plane waves is the ray direction the same at all points on the phasefront. Many properties of an optical system can be best understood by considering what happens to a ray of light entering the system. How does its direction change as it encounters mirrors and lenses, and what happens to the intensity of the wave to which the ray belongs as it crosses interfaces between different components of the system?

14.3 Simple Reflection and Refraction Analysis

The phenomena of reflection and refraction are most easily understood in terms of plane electromagnetic waves – those sorts of waves where there is a unique direction of energy flow (the ray direction). Other types of wave, such as spherical waves and Gaussian beams, are also important; however, the part of their wavefront which strikes an optical component can frequently be approximated as a plane wave, so plane wave considerations of reflection and refraction still hold true. The ray direction is the same as the direction of the Poynting vector $\mathbf{S} = \mathbf{E} \times \mathbf{H}$. In isotropic media (gases, liquids, glasses, and crystals of cubic symmetry) the ray

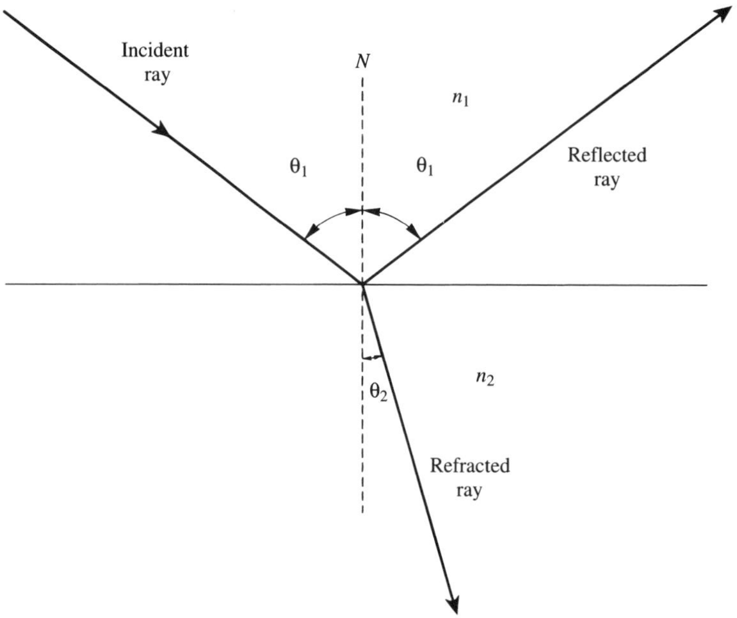

Fig. 14.2. Incident, reflected, and refracted ray directions at the planar boundary between two media.

direction and the direction of **k**, the propagation vector, are the same. Except for plane waves the direction of **k** varies from point to point on the phase front.

When light strikes a plane mirror, or the planar boundary between two media of different refractive index, the *angle of incidence* is always equal to the *angle of reflection*, as shown in Fig. (14.2). This is the fundamental *law of reflection.*

When a light ray crosses the boundary between two media of different refractive index, the *angle of refraction* θ_2, shown in Fig. (14.2) is related to the angle of incidence θ_1, by *Snell's law*

$$\frac{\sin\theta_1}{\sin\theta_2} = \frac{n_2}{n_1}. \tag{14.6}$$

This result does not hold true in general unless both media are isotropic.

Since $\sin\theta_2$ cannot be greater than unity, if $n_2 < n_1$ the maximum angle of incidence for which there can be a refracted wave is called the *critical angle* θ_c where

$$\sin\theta_c = n_2/n_1 \tag{14.7}$$

as illustrated in Fig. (14.3a).

If θ_1 exceeds θ_c the boundary acts as a very good mirror, as illustrated in Fig. (14.3b). This phenomenon is called *total internal reflection.* Several types of reflecting prisms operate this way. When total internal reflection occurs, there is no transmission of energy through the boundary. However, the fields of the wave do not abruptly go to zero at the boundary. There is said to be an *evanescent wave* on the other side of the boundary, the field amplitudes of which decay exponentially with distance. Because of the existence of this evanescent wave, other optical components should not be brought too close to a totally reflecting surface, or energy will be coupled to them via the evanescent wave and the efficiency of total internal reflection will be reduced. With extreme care this effect

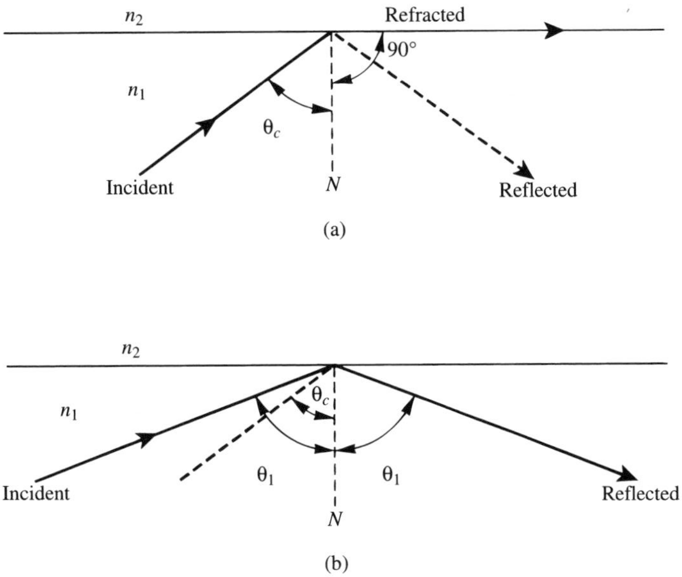

Fig. 14.3. (a) Incident ray at the critical angle striking the boundary between a dense and a less dense medium. (b) Total internal reflection at the boundary between a dense and a less dense medium.

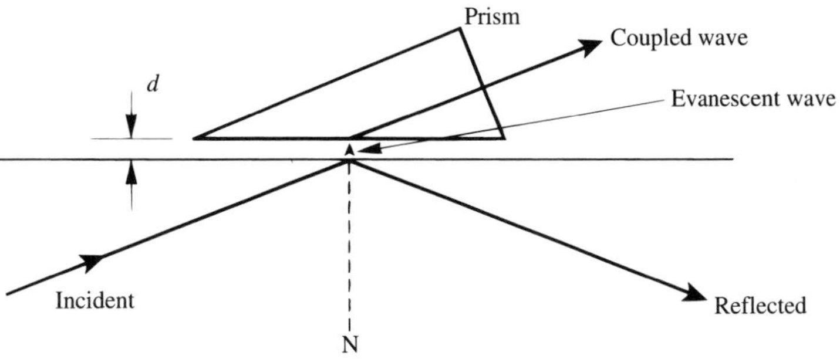

Fig. 14.4. Schematic arrangement with which evanescent coupling could be observed. The spacing between the boundary and the coupling prism must be very small ($d \sim$ wavelength) for significant coupling to occur.

can be used to produce a variable-reflectivity, totally-internally-reflecting surface, as shown in Fig. (14.4). If one or both of the media in Fig. (14.2) are anisotropic, like calcite, crystalline quartz, ammonium dihydrogen phosphate (ADP), potassium dihydrogen phosphate (KDP) or tellurium, in general the incident wave will split into two components, one of which obeys Snell's law (the *ordinary wave*) and one which does not (the *extraordinary wave*). This phenomenon is called *double refraction*, and is discussed in detail in Chapter 18.

If an optical system contains only planar interfaces, the path of a ray of light through the system can be easily calculated using only the law of reflection and Snell's law. This simple approach neglects diffraction effects, which become significant unless the lateral dimensions (*apertures*) of the system are *all* much larger than the wavelength (say 10^4 times larger). The behavior of light rays in more complex systems containing nonplanar components, but where diffraction effects are negligible, is better described with the aid of *paraxial ray* analysis[14.1]. Transmitted and reflected intensities and polarization states cannot be determined by the above methods and are most easily determined by the *method of impedances*.

14.4 Paraxial Ray Analysis

A plane wave is characterized by a unique propagation direction given by the wave vector **k**. All fields associated with the wave are, at a given time, equal at all points in infinite planes orthogonal to the propagation direction. In real optical systems such plane waves do not exist, as the finite size of the elements of the system restricts the lateral extent of the waves. Nonplanar optical components will cause further deviations of the wave from planarity. Consequently, the wave acquires a ray direction which varies from point to point on the phase front. The behavior of the optical system must be characterized in terms of the deviations its elements cause to the bundle of rays that comprise the propagating, laterally-restricted wave. This is most easily done in terms of paraxial rays. In a cylindrically-symmetric optical system, for example, a coaxial system of spherical lenses or mirrors, *paraxial rays* are those rays whose directions of propagation occur at sufficiently small angles θ to the symmetry axis of the system that it is possible to replace $\sin\theta$ or $\tan\theta$ by θ – in other words paraxial rays obey the small angle approximation.

14.4.1 Matrix Formulation

In an optical system whose symmetry axis is in the z direction, a paraxial ray in a given cross-section (z = constant) is characterized by its distance r from the z axis and the angle r' it makes with that axis. If the values of these parameters at two planes of the system (an *input* and an *output* plane) are $r_1 r_1'$ and $r_2 r_2'$ respectively, as shown in Fig. (14.5a), in the paraxial ray approximation there is a linear relation between them of the form

$$\begin{aligned} r_2 &= A r_1 + B r_1', \\ r_2' &= C r_1 + D r_1' \end{aligned} \tag{14.8}$$

or, in matrix notation

$$\begin{pmatrix} r_2 \\ r_2' \end{pmatrix} = \begin{pmatrix} A & B \\ C & D \end{pmatrix} \begin{pmatrix} r_1 \\ r_1' \end{pmatrix}. \tag{14.9}$$

$\begin{pmatrix} A & B \\ C & D \end{pmatrix}$ is called the ray transfer matrix, **M**; its determinant is usually unity,† i.e., $AD - BC = 1$.

Optical systems made of isotropic material are generally reversible - a ray which travels from right to left with input parameters r_2, r_2' will leave the system with parameters r_1, r_1' – thus:

$$\begin{pmatrix} r_1 \\ r_1' \end{pmatrix} = \begin{pmatrix} A' & B' \\ C' & D' \end{pmatrix} \begin{pmatrix} r_2 \\ r_2' \end{pmatrix}, \tag{14.10}$$

where the reverse ray transfer matrix satisfies

$$\begin{pmatrix} A' & B' \\ C' & D' \end{pmatrix} = \begin{pmatrix} A & B \\ C & D \end{pmatrix}^{-1}.$$

The ray transfer matrix allows the properties of an optical system to be described in general terms by the location of its *focal points* and *principal planes*, whose

† When the media to the left of the input plane and to the right of the output plane have the same refractive index.

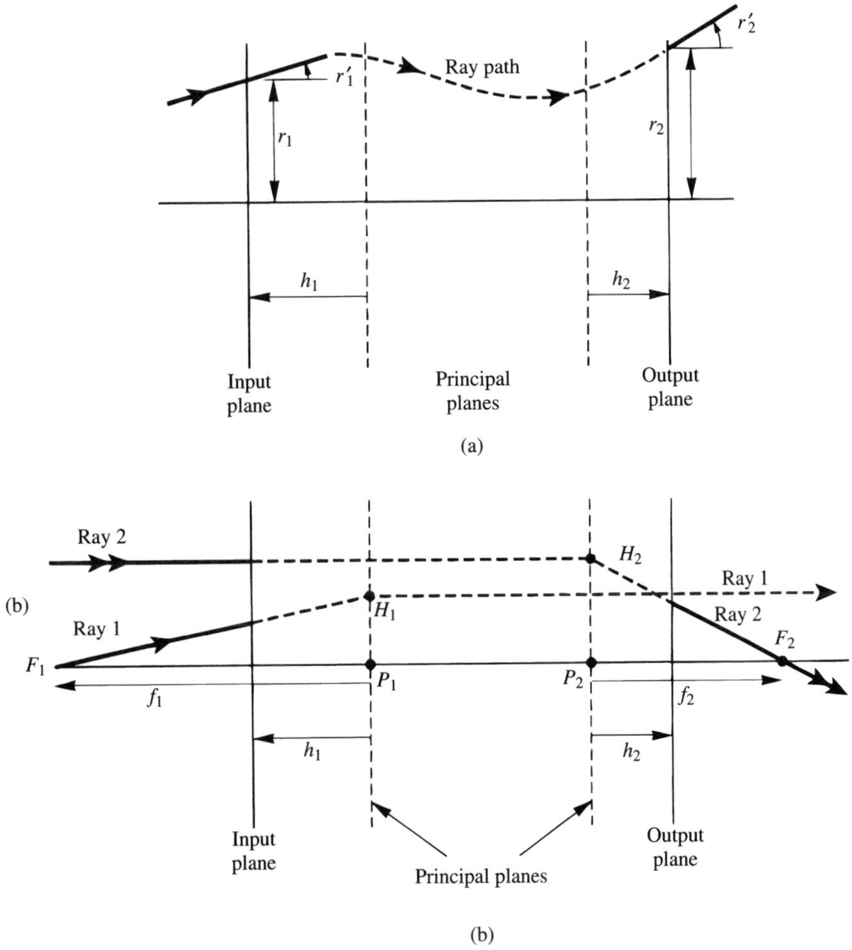

Fig. 14.5. (a) Paraxial ray path from input to output plane. (b) Principal planes, points, and rays of a paraxial optical system.

location is determined from the elements of the matrix. The significance of these features of the system can be illustrated with the aid of Fig. (14.5b). An input ray which passes through the *first focal point*, F_1 (or would pass through this point if it did not first enter the system) emerges travelling parallel to the axis. The intersection point of the extended input and output rays, point H_1 in Fig. (14.5b), defines the location of the *first principal plane*. Conversely, an input ray travelling parallel to the axis will emerge at the output plane and pass through the second focal point, F_2 (or appear to have come from this point). The intersection of the extension of these rays, point H_2, defines the location of the *second principal plane*. Rays 1 and 2 in Fig. (14.5b) are called the *principal rays* of the system. The location of the principal planes allows the corresponding emergent ray paths to be determined as shown in Fig. (14.5b). The dashed lines in this figure, which permit geometric construction of the location of output rays 1 and 2, are called *virtual ray paths*. Both F_1 and F_2 lie on the axis of the system. The axis of the system intersects the principal planes at the *principal points*, P_1 and P_2, in Fig. (14.5b).

The distance f_1 from the first principal plane to the first focal point is called the *first focal length*; f_2 is the *second focal length*.

In most practical situations, the refractive indices of the media to the left of the input plane (the *object space*) and to the right of the output plane (the *image space*) are equal. In this case we can derive simple relations between the focal lengths f_1 and f_2 and h_1 and h_2, the distances of the input and output planes from the principal planes, measured in the sense shown in Fig. (14.5b).

We can break up the system shown in Fig. (14.5b) into three parts, the region from the input plane to the first principal plane, the region between the two principal planes, and the region from second principal plane to output plane.

If we write the transfer matrix from the left to the right principal planes as $\begin{pmatrix} A'' & B'' \\ C'' & D'' \end{pmatrix}$, then the overall transfer matrix is

$$\begin{pmatrix} A & B \\ C & D \end{pmatrix} = \begin{pmatrix} 1 & h_2 \\ 0 & 1 \end{pmatrix} \begin{pmatrix} A'' & B'' \\ C'' & D'' \end{pmatrix} \begin{pmatrix} 1 & h_1 \\ 0 & 1 \end{pmatrix}, \tag{14.11}$$

which gives

$$\begin{pmatrix} A'' & B'' \\ C'' & D'' \end{pmatrix} = \begin{pmatrix} 1 & -h_2 \\ 0 & 1 \end{pmatrix} \begin{pmatrix} A & B \\ C & D \end{pmatrix} \begin{pmatrix} 1 & -h_1 \\ 0 & 1 \end{pmatrix} \tag{14.12}$$

and therefore

$$\begin{pmatrix} A'' & B'' \\ C'' & D'' \end{pmatrix} = \begin{pmatrix} A - h_2C & B - h_2A - h_2(D - h_1C) \\ C & D - h_1C \end{pmatrix}. \tag{14.13}$$

Clearly, $C'' = C$ for the second principal ray in Fig. (14.5b), the distance of the ray from the axis, r, does not change, therefore $A'' = 1$ giving

$$r = (A - h_2C)r, \tag{14.14}$$

so

$$h_2 = \frac{A-1}{C}. \tag{14.15}$$

Furthermore, for the first principal ray, whose input angle is r'

$$r = A''r + B''r' \tag{14.16}$$

so $B'' = 0$.

Consequently, if the media to the left of the input plane and the right of the output plane are the same, using $\det(M) = 1$, gives

$$h_1 = \frac{D-1}{C}. \tag{14.17}$$

Note that both A'' and D'' are equal to unity.

For the second principal ray in Fig. (14.5b)

$$\begin{pmatrix} r_2 \\ r_2' \end{pmatrix} = \begin{pmatrix} A'' & B'' \\ C'' & D'' \end{pmatrix} \begin{pmatrix} r_1 \\ 0 \end{pmatrix} \tag{14.18}$$

so

$$r_2' = C''r_1. \tag{14.19}$$

Furthermore, from Fig. (14.5b), it is easy to see that

$$-r_2' = \frac{r_1}{f_2}. \tag{14.20}$$

Combining Eqs. (14.19) and (14.20) gives

$$C'' = -\frac{1}{f_2}. \tag{14.21}$$

By a similar procedure using the first principal ray it can be shown that

$$C'' = -\frac{1}{f_1}. \tag{14.22}$$

We have already seen from Eq. (14.13) that $C'' = C$, therefore,

$$f_1 = f_2 = f. \tag{14.23}$$

If the media to the left and right of the input plane are the same, both principal focal lengths are equal. If the elements of the transfer matrix are known, the locations of the focal points and principal planes are determined. Graphical construction of ray paths through the system using the methods of *ray tracing* is then straightforward (see Section 14.4.2).

In using matrix methods for optical analysis, a consistent sign convention must be employed. In the present discussion: a ray is assumed to travel in the positive z direction from left to right through the system. The distance from the first principal plane to an object is measured positive from right to left – in the negative z direction. The distance from the second principal plane to an image is measured positive from left to right – in the positive z direction. The lateral distance of the ray from the axis is positive in the upward direction, negative in the downward direction. The acute angle between the system axis direction and the ray, say r_1' in Fig. (14.5a), is positive if a counterclockwise motion is necessary to go from the positive z direction to the ray direction. When the ray crosses a spherical interface the radius of curvature is positive if the interface is convex to the input ray. The use of ray transfer matrices in optical system analysis can be illustrated with some specific examples.

(i) *Uniform optical medium.* In a uniform optical medium of length d no change in ray angle occurs, as illustrated in Fig. (14.6a), so

$$\begin{aligned} r_2' &= r_1' \\ r_2 &= r_1 + d r_1'. \end{aligned} \tag{14.24}$$

Therefore,

$$\mathbf{M} = \begin{pmatrix} 1 & d \\ 0 & 1 \end{pmatrix}. \tag{14.25}$$

The focal length of this system is infinite and it has no specific principal planes.

(ii) *Planar interface between two different media.* At the interface, as shown in Fig. (14.6b), $r_1 = r_2$ and from Snell's law, using the approximation $\sin\theta = \theta$

$$r_2' = \frac{n_1}{n_2} r_1'. \tag{14.26}$$

Therefore

$$\mathbf{M} = \begin{pmatrix} 1 & 0 \\ 0 & n_1/n_2 \end{pmatrix}. \tag{14.27}$$

(iii) *A parallel-sided slab of refractive index n bounded on both sides with media of refractive index 1* (Fig. 14.6c).

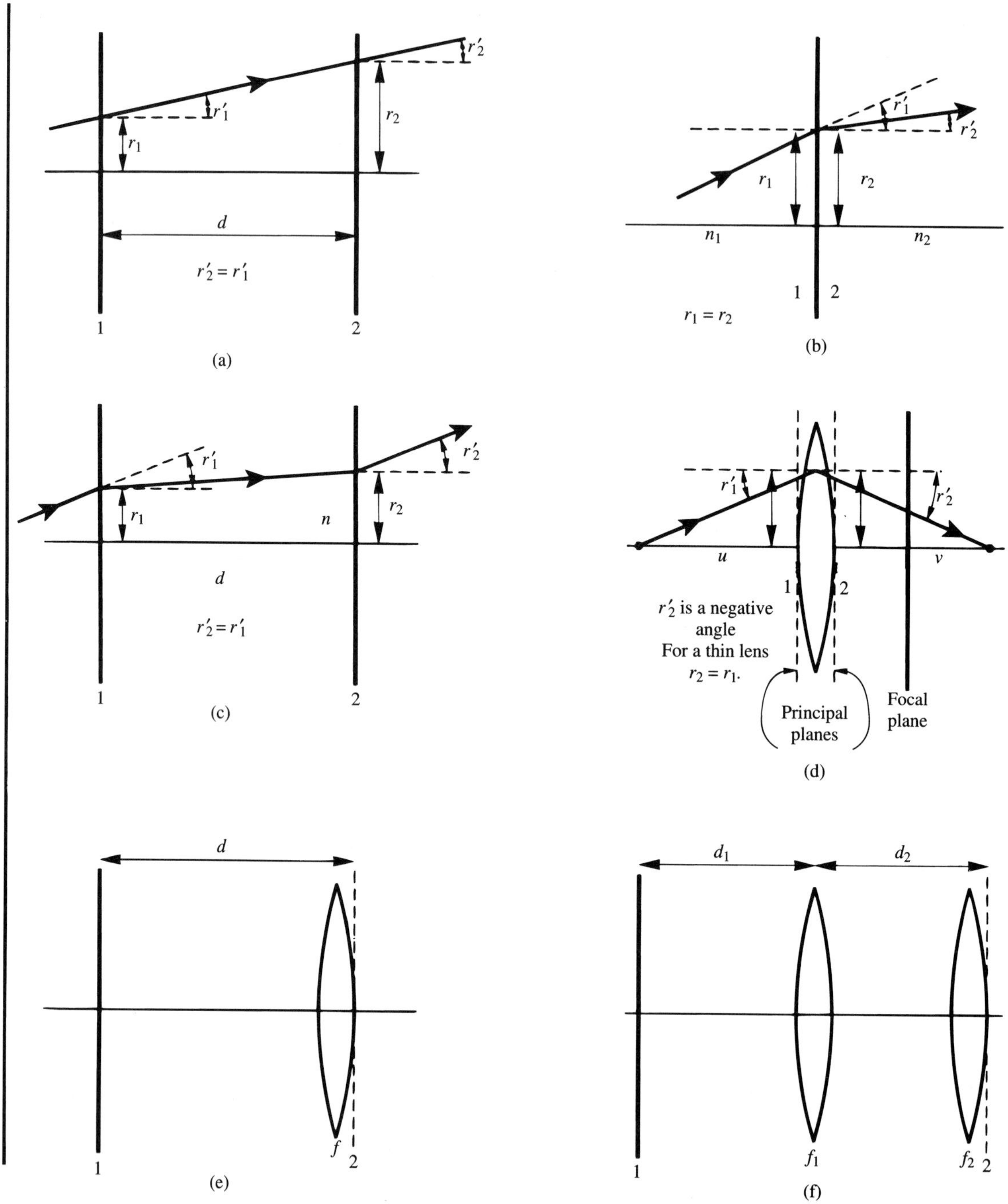

Fig. 14.6. Simple optical systems: (a) a length of uniform medium; (b) the planar interface between two media;
(c) a parallel-sided slab;
(d) thin lens; (e) a length of uniform medium plus a thin lens; (f) two spaced thin lenses.

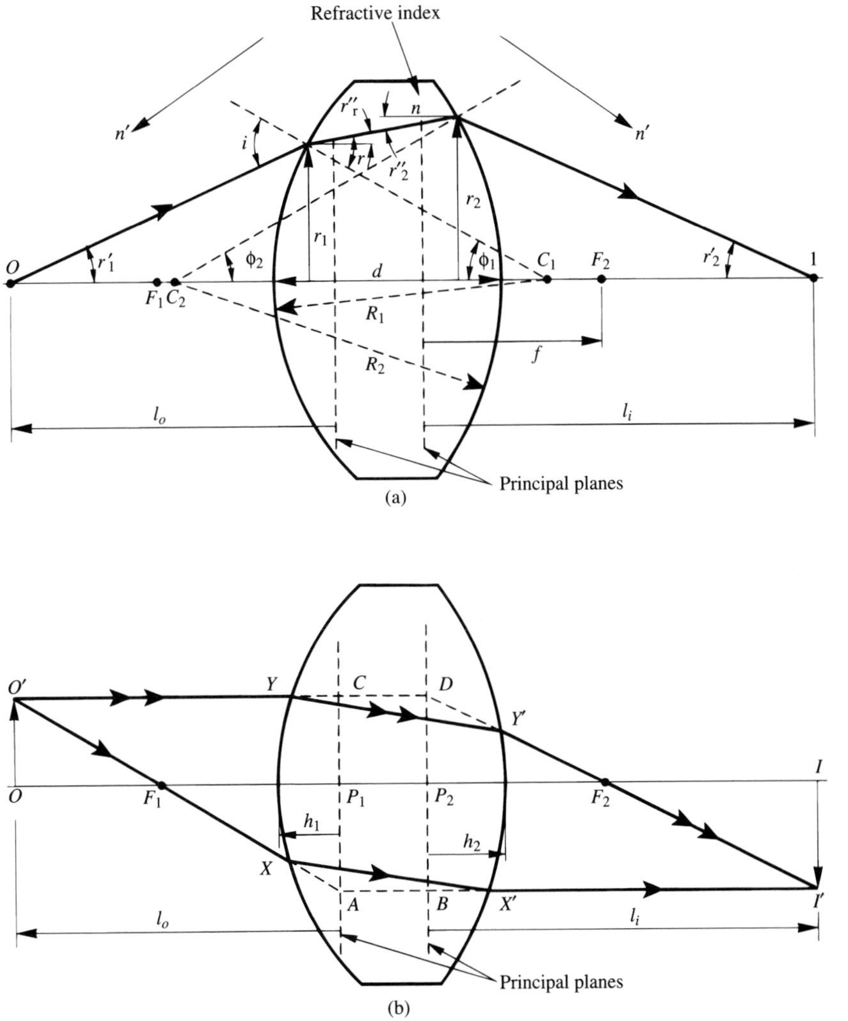

Fig. 14.7. (a) Ray path through a thick lens. C_1 and C_2 are the centers of curvature of the two spherical surfaces of which the faces of the lens are a part. (b) Principal rays and planes for a thick lens. The dashed lines show the use of the principal planes for determing the exit trajectories of the input principal rays. The actual ray paths through the lens are shown by the solid lines.

In this case,

$$\mathbf{M} = \begin{pmatrix} 1 & d/n \\ 0 & 1 \end{pmatrix}. \tag{14.28}$$

The principal planes of this system are the boundary faces of the optically dense slab.

(iv) *Thick lens* The ray transfer matrix of the thick lens shown in Fig. (14.7a) is the product of the three transfer matrices:

$$\mathbf{M}_3 \cdot \mathbf{M}_2 \cdot \mathbf{M}_1 = \begin{pmatrix} \text{matrix for} \\ \text{second spherical} \\ \text{interface} \end{pmatrix} \begin{pmatrix} \text{matrix for} \\ \text{medium of} \\ \text{length } d \end{pmatrix} \begin{pmatrix} \text{matrix for} \\ \text{first spherical} \\ \text{of interface} \end{pmatrix}. \tag{14.29}$$

Note the order of these three matrices; $\mathbf{M}_1$ comes on the right because it operates first on the column vector which describes the input ray.

At the first spherical surface

$$n'(r_1' + \phi_1) = nr = n(r_2'' + \phi_1), \tag{14.30}$$

which, since $\phi_1 = r_1/R_1$, can be written as

$$r_2'' = \frac{n'r_1'}{n} + \frac{(n'-n)r_1}{nR_1}. \tag{14.31}$$

The transfer matrix at the first spherical surface is

$$\mathbf{M}_1 = \begin{pmatrix} 1 & 0 \\ (n'-n)/nR_1 & n'/n \end{pmatrix} = \begin{pmatrix} 1 & 0 \\ -D_1/n & n'/n \end{pmatrix}, \tag{14.32}$$

where $D_1 = (n-n')/R_1$ is called the power of the surface. If R_1 is measured in meters the units of D_1 are *diopters.*

In the paraxial approximation all the rays passing through the lens travel the same distance d in the lens. Thus,

$$\mathbf{M}_2 = \begin{pmatrix} 1 & d \\ 0 & 1 \end{pmatrix}. \tag{14.33}$$

The ray transfer matrix at the second interface is

$$\mathbf{M}_3 = \begin{pmatrix} 1 & 0 \\ (n-n')/n'R_2 & n/n' \end{pmatrix} = \begin{pmatrix} 1 & 0 \\ -D_2/n' & n/n' \end{pmatrix}, \tag{14.34}$$

which is identical in form to $\mathbf{M}_1$. Note that in this case both r_2' and R_2 are negative. The overall transfer matrix of the thick lens is

$$\mathbf{M} = \mathbf{M}_3\mathbf{M}_2\mathbf{M}_1 = \begin{pmatrix} 1 - dD_1/n & dn'/n \\ dD_1D_2/nn' - D_1/n' - D_2/n' & 1 - dD_2/n \end{pmatrix}. \tag{14.35}$$

If Eq. (14.35) is compared with Eqs. (14.15) and (14.17), it is clear that the locations of the principal planes of the thick lens are

$$h_1 = \frac{d}{(n/n')\left(1 + D_1/D_2 - dD_1/n\right)}, \tag{14.36}$$

$$h_2 = \frac{d}{(n/n')\left(1 + D_2/D_1 - dD_2/n\right)}, \tag{14.37}$$

A numerical example will best illustrate the location of the principal planes for a biconvex thick lens.

Suppose:

$$n' = 1(\text{air}), n = 1.5(\text{ glass}),$$

$$R_1 = -R_2 = 50 \text{ mm},$$

$$d = 10 \text{ mm}.$$

In this case $D_1 = D_2 = 0.01 \text{ mm}^{-1}$ and, from Eqs. (14.36) and (14.37), $h_1 = h_2 = 3.448$ mm. These principal planes are symmetrically placed inside the lens. Fig. (14.7b) shows how the principal planes can be used to trace the principal ray paths through a thick lens.

From Eqs. (14.18) and (14.11)

$$r_2' = \frac{-r_1}{f} + \left(1 - \frac{dD_2}{n}\right) r_1', \tag{14.38}$$

where the focal length is

$$f = \left(\frac{D_1 + D_2}{n'} - \frac{dD_1 D_2}{nn'} \right)^{-1} . \tag{14.39}$$

If ℓ_o is the distance from the object O to the first principal plane, and ℓ_i the distance from the second principal plane to the image I in Fig. (14.7b), then from the similar triangles $OO'F_1$, P_1AF_1 and P_2DF_2, $II'F_2$

$$\frac{OO'}{II'} = \frac{OF_1}{F_1P_1} = \frac{\ell_o - f_1}{f_1},$$

$$\frac{OO'}{II'} = \frac{P_2F_2}{F_2I} = \frac{f_2}{\ell_i - f_2}. \tag{14.40}$$

If the media on both sides of the lens are the same, then $f_1 = f_2 = f$ and it immediately follows that

$$\frac{1}{\ell_o} + \frac{1}{\ell_i} = \frac{1}{f}. \tag{14.41}$$

This is the fundamental imaging equation.

(v) *Thin lens.* If a lens is sufficiently thin that to a good approximation $d = 0$ the transfer matrix is

$$\mathbf{M} = \begin{pmatrix} 1 & 0 \\ -(D_1 + D_2)/n' & 1 \end{pmatrix} . \tag{14.42}$$

As shown in Fig. (14.6d) the principal planes of such a thin lens are at the lens. The focal length of the thin lens is f, where

$$\frac{1}{f} = \frac{D_1 + D_2}{n'} = \left(\frac{n}{n'} - 1 \right) \left(\frac{1}{R_1} - \frac{1}{R_2} \right), \tag{14.43}$$

so the transfer matrix can be written very simply as

$$\mathbf{M} = \begin{pmatrix} 1 & 0 \\ -1/f & 1 \end{pmatrix} . \tag{14.44}$$

The focal length of the lens depends on the refractive index of the lens material and the refractive index of the medium with which it is immersed. In air

$$\frac{1}{f} = (n-1) \left(\frac{1}{R_1} - \frac{1}{R_2} \right) . \tag{14.45}$$

For a biconvex lens R_2 is negative and

$$\frac{1}{f} = (n-1) \left(\frac{1}{\mid R_1 \mid} + \frac{1}{\mid R_2 \mid} \right) . \tag{14.46}$$

For a biconcave lens

$$\frac{1}{f} = -(n-1) \left(\frac{1}{\mid R_1 \mid} + \frac{1}{\mid R_2 \mid} \right) . \tag{14.47}$$

The focal length of any diverging lens is negative. For a thin lens object and image distances are measured to a common point. It is common practice to rename the distances ℓ_o and ℓ_i in this case so that the imaging Eq. (14.41) reduces to its familiar form

$$\frac{1}{v} + \frac{1}{u} = \frac{1}{f}, \tag{14.48}$$

u is called the *object distance* and v the *image distance*.

(vi) *A length of uniform medium plus a thin lens* (Fig. (14.6e)). This is a combination of systems (i) + (v); its overall transfer matrix is found from Eqs. (14.25) and (14.44) as

$$\mathbf{M} = \begin{pmatrix} 1 & 0 \\ -1/f & 1 \end{pmatrix} \begin{pmatrix} 1 & d \\ 0 & 1 \end{pmatrix} = \begin{pmatrix} 1 & d \\ -1/f & 1-d/f \end{pmatrix}. \tag{14.49}$$

(vii) *Two thin lenses.* As a final example of the use of ray transfer matrices consider the combination of two thin lenses shown in (Fig. (14.6f).) The transfer matrix of this combination is

$$\mathbf{M} = \begin{pmatrix} \text{matrix for} \\ \text{second lens}, f_2 \end{pmatrix} \begin{pmatrix} \text{matrix for} \\ \text{uniform medium}, d_2 \end{pmatrix} \begin{pmatrix} \text{matrix for} \\ \text{first lens}, f_1 \end{pmatrix} \times \begin{pmatrix} \text{matrix for} \\ \text{uniform medium}, d_1 \end{pmatrix}, \tag{14.50}$$

which can be shown to be

$$\mathbf{M} = \begin{pmatrix} 1 - \dfrac{d_2}{f_1} & d_1 + d_2 - \dfrac{d_1 d_2}{f_1} \\ -\dfrac{1}{f_1} - \dfrac{1}{f_2} + \dfrac{d_2}{f_1 f_2} & 1 - \dfrac{d_1}{f_1} - \dfrac{d_1}{f_2} - \dfrac{d_2}{f_2} + \dfrac{d_1 d_2}{f_1 f_2} \end{pmatrix}. \tag{14.51}$$

The focal length of the combination is

$$f = \frac{f_1 f_2}{d_2 - (f_1 + f_2)}. \tag{14.52}$$

The optical system consisting of two thin lenses is the archetypal system used in analyses of the stability of lens waveguides and optical resonators[14.1].

(viii) *Spherical mirrors.* The object and image distances from a spherical mirror also obey Eq. (14.48) where the focal length of the mirror is $R/2$; f is positive for a concave mirror, negative for a convex mirror. Positive object and image distances for a mirror are measured positive in the normal direction away from its surface. If a negative image distance results from Eq. (14.48) this implies a *virtual image* (behind the mirror).

14.4.2 Ray Tracing

Practical implementation of paraxial ray analysis in optical system design can be very conveniently carried out graphically by *ray tracing*. In ray tracing a few simple rules allow geometrical construction of the principal ray paths from an object point, although these constructions do not take into account the nonideal behavior, or *aberrations*, of real lenses. The first principal ray from a point on the object passes through (or its projection passes through) the first focal point. From the point where this ray, or its projection, intersects the first principal plane the output ray is drawn parallel to the axis. The actual ray path between input and output planes can be found in simple cases, for example, the path XX' in the thick lens shown in Fig. (14.7b). The second principal ray is directed parallel to the axis; from the intersection of this ray, or its projection, with the second principal plane the output ray passes through (or appears to have come from) the second focal point. The actual ray path between input and output planes can be found in simple cases, for example, the path YY' in the thick lens shown in Fig. (14.7b). The intersection of the two principal rays in the image space produces the image point which corresponds to the same point on the object. If only the back-projections of the output principal rays appear to intersect, this intersection point lies on a *virtual* image.

In the majority of applications of ray tracing a quick analysis of the system is desired. In this case, if all lenses in the system are treated as thin lenses, the position of the principal planes need not be calculated beforehand and ray tracing becomes particularly easy. For a thin lens a third principal ray is useful for determining image location. The ray from a point on the object which passes through the center of the lens is not deviated by the lens. The use of ray tracing to determine the size and position of the real image produced by a convex lens and the virtual image produced by a concave lens is shown in Fig. (14.8). Fig. (14.8a) shows a converging lens where the input principal ray parallel to the axis actually passes through the focal point. Fig. (14.8b) shows a diverging lens where this ray emerges from the lens appearing to have come from the focal point. More complex systems of lenses can be analyzed in the same way. Once the image location has been determined by the use of the principal rays, the path of any group of rays, a ray *pencil*, can be found; see, for example, the cross-hatched ray pencil shown in Fig. (14.8a).

The use of ray tracing rules to analyze spherical mirror systems is similar to that described above, although the ray striking the center of the mirror in this case reflects so that the angle of incidence equals the angle of reflection. Figs. (14.8c) and (14.8d) illustrate the use of ray tracing to locate the images produced by concave and convex mirrors. In the case of a convex mirror the ray tracing construction is modified in the same way as it was for a defocusing lens. A ray parallel to the axis that would pass through the focal point *if it did not first encounter the mirror surface* reflects parallel to the axis. A ray parallel to the axis reflects at the mirror surface so as to appear in projection to have come from the focal point.

14.4.3 Imaging and Magnification

In Fig. (14.8a) the ratio of the height of the image to the height of the object is called the *magnification*, *m*. In the case of a thin lens

$$m = \frac{b}{a} = \frac{v}{u}. \tag{14.53}$$

For a more general system the magnification can be obtained from the ray transfer matrix equation

$$\begin{pmatrix} b \\ b' \end{pmatrix} = \begin{pmatrix} A & B \\ C & D \end{pmatrix} \begin{pmatrix} a \\ a' \end{pmatrix} = \mathbf{M} \begin{pmatrix} a \\ a' \end{pmatrix}, \tag{14.54}$$

where a' is the angle of a ray through a point on the object and b' through the corresponding point on the image. The matrix $\mathbf{M}$ in this case includes the entire system from object O to image I. So, in Fig. (14.8a)

$$\mathbf{M} = \begin{pmatrix} \text{matrix for} \\ \text{uniform medium} \\ \text{of length v} \end{pmatrix} \begin{pmatrix} \text{matrix for} \\ \text{lens} \end{pmatrix} \begin{pmatrix} \text{matrix for} \\ \text{uniform medium} \\ \text{of length u} \end{pmatrix}. \tag{14.55}$$

For imaging, b must be independent of a', so $B = 0$.

The magnification is

$$m = \frac{b}{a} = A \tag{14.56}$$

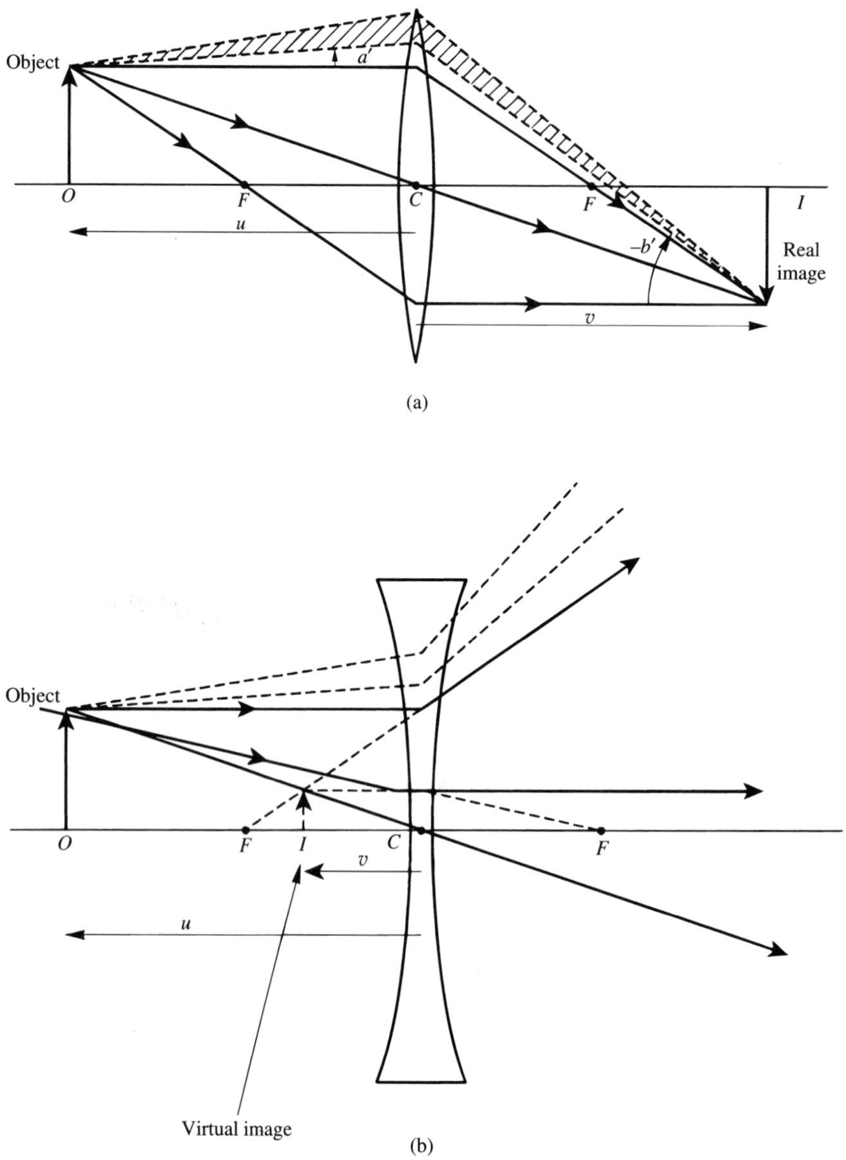

Fig. 14.8. Ray tracing diagrams: (a) real image production by a converging lens; (b) virtual image production by a diverging lens.

so the ray transfer matrix of the imaging system can be written

$$\mathbf{M} = \begin{pmatrix} m & 0 \\ -1/f & 1/m \end{pmatrix}, \tag{14.57}$$

where it should be noted the result, $\det(M) = 1$, has been used.

The *angular magnification* of the system is defined as

$$m' = \left(\frac{b'}{a'}\right)_{a=0}, \tag{14.58}$$

which gives $m' = 1/m$.

Note that $mm' = 1$, a useful general result.

Fig. 14.8. (*cont.*) (c) real image production by a concave mirror; (d) virtual image production by a convex mirror.

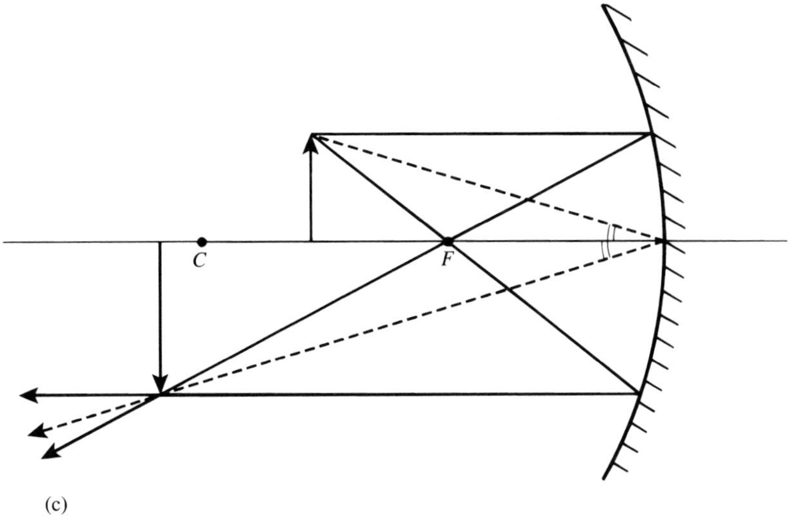

(c)

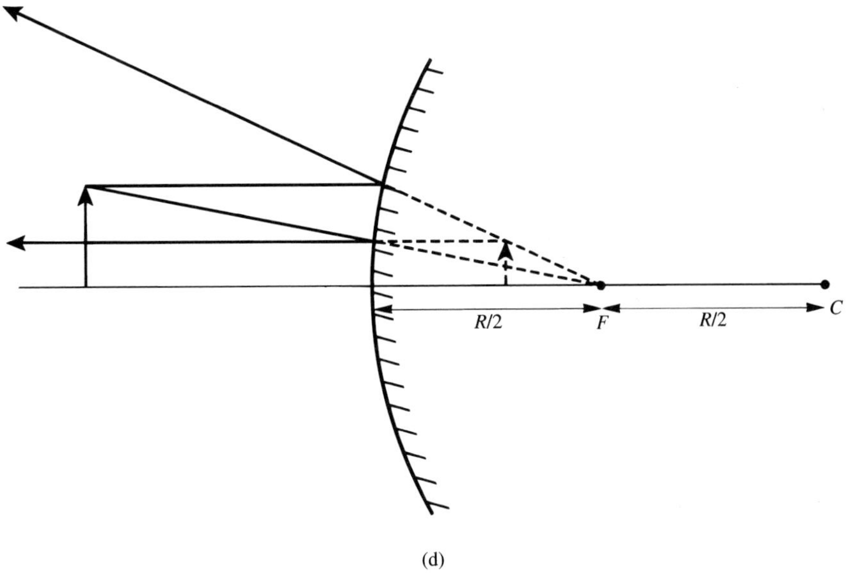

(d)

14.5 The Use of Impedances in Optics

The method of impedances is the easiest way to calculate the fraction of incident intensity transmitted and reflected in an optical system. It is also a good way to follow the changes in polarization state which result when light passes through an optical system.

The impedance of a plane wave travelling in a medium of relative permeability μ_r and dielectric constant ϵ_r is

$$Z = \sqrt{\frac{\mu_r \mu_0}{\epsilon_r \epsilon_0}} = Z_0 \sqrt{\frac{\mu_r}{\epsilon_r}}. \tag{14.59}$$

If $\mu_r = 1$, as is usually the case for optical media, the impedance can be written

$$Z = Z_0/n. \tag{14.60}$$

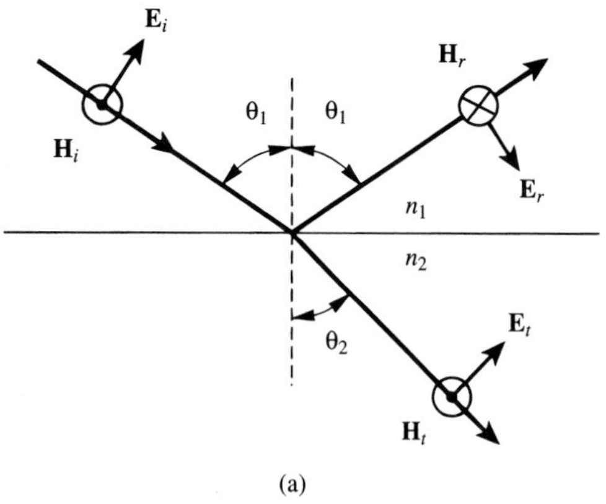

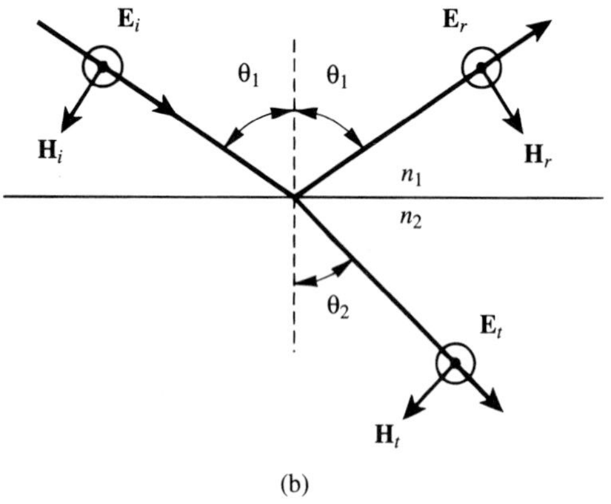

Fig. 14.9. Field vectors of a plane wave striking a planar boundary: (a) an electric field in the plane of incidence – a P-wave or TM wave; (b) an electric field perpendicular to the plane of incidence – an S-wave or TE wave.

This impedance relates the transverse **E** and **H** fields of the wave

$$Z = \frac{E_{tr}}{H_{tr}}. \tag{14.61}$$

When a plane wave crosses a planar boundary between two different media the components of both **E** and **H** parallel to the boundary have to be continuous across that boundary. Fig. (14.9a) illustrates a plane wave polarized in the plane of incidence striking a planar boundary between two media of refractive index n_1 and n_2, respectively. In terms of the magnitudes of the vectors involved

$$E_i \cos\theta_1 + E_r \cos\theta_1 = E_t \cos\theta_2, \tag{14.62}$$

$$H_i - H_r = H_t. \tag{14.63}$$

Eq. (14.63) can be written

$$\frac{E_i}{Z_1} - \frac{E_r}{Z_1} = \frac{E_t}{Z_2}. \tag{14.64}$$

It is easy to eliminate E_t between Eqs. (14.62) and (14.64) to give

$$\rho = \frac{E_r}{E_i} = \frac{Z_2 \cos\theta_2 - Z_1 \cos\theta_1}{Z_2 \cos\theta_2 + Z_1 \cos\theta_1}, \tag{14.65}$$

where ρ is the *reflection coefficient* of the surface. The fraction of the incident energy reflected from the surface is called the *reflectance*, R, where

$$R = \rho^2. \tag{14.66}$$

Similarly, the *transmission coefficient* of the boundary is

$$\tau = \frac{E_t \cos\theta_2}{E_i \cos\theta_1} = \frac{2Z_2 \cos\theta_2}{Z_2 \cos\theta_2 + Z_1 \cos\theta_1}. \tag{14.67}$$

By a similar treatment applied to the geometry shown in Fig. (14.9b) it can be shown that for a plane wave polarized perpendicular to the plane of incidence

$$\rho = \frac{Z_2 \sec\theta_2 - Z_1 \sec\theta_1}{Z_2 \sec\theta_2 + Z_1 \sec\theta_1}, \tag{14.68}$$

$$\tau = \frac{2Z_2 \sec\theta_2}{Z_2 \sec\theta_2 + Z_1 \sec\theta_1}. \tag{14.69}$$

If the effective impedance for a plane wave polarized in the plane of incidence (*P polarization*) and incident on a boundary at angle θ is defined as

$$Z' = Z \cos\theta, \tag{14.70}$$

and for a wave polarized perpendicular to the plane of incidence (*S polarization*) as

$$Z' = Z \sec\theta, \tag{14.71}$$

then a universal pair of formulae for ρ and τ results:

$$\rho = \frac{Z_2' - Z_1'}{Z_1' + Z_2'}, \tag{14.72}$$

$$\tau = \frac{2Z_2'}{Z_1' + Z_2'}. \tag{14.73}$$

It will be apparent from an inspection of Fig. (14.9) that Z' is just the ratio of the electric field component parallel to the boundary and the magnetic field component parallel to the boundary. For reflection from an ideal mirror $Z_2' = 0$. Note that from Eqs. (14.72) and (14.73)

$$\tau = 1 + \rho. \tag{14.74}$$

In normal incidence Eqs. (14.65) and (14.68) become identical and can be written as

$$\rho = \frac{Z_2 - Z_1}{Z_2 + Z_1} = \frac{n_1 - n_2}{n_1 + n_2}. \tag{14.75}$$

Note that there is a change of phase of π in the reflected field relative to the incident field when $n_1 < n_2$.

Since intensity $\propto$ (electric field)^2, the fraction of the incident energy that is reflected is

$$R = \rho^2 = \left(\frac{n_1 - n_2}{n_1 + n_2}\right)^2. \tag{14.76}$$

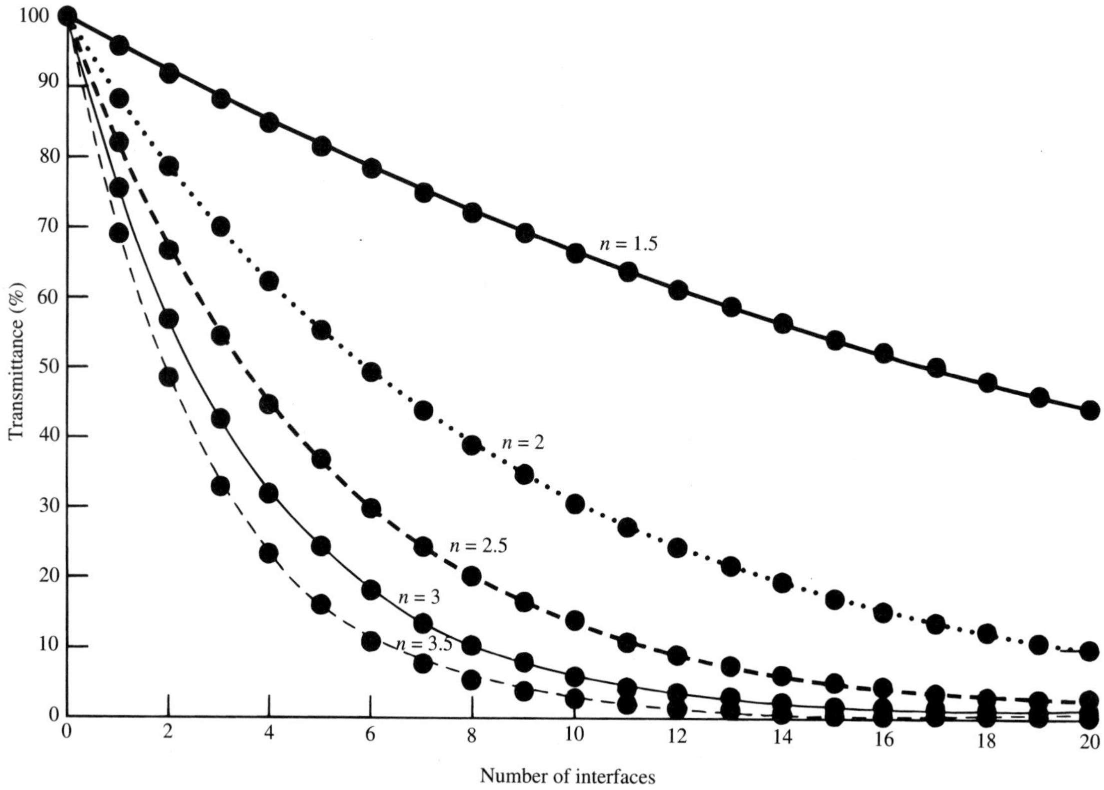

Fig. 14.10. Change in the reflectance at the boundary between two media as the index mismatch increases and as the number of independent boundaries crossed increases.

R increases with the index mismatch between the two media, and with the number of interfaces as shown in Fig. (14.10). However, it should be noted that if a series of boundaries are parallel and the incident light is fairly monochromatic, then Fabry – Perot effects occur and the transmittance of the structure must be calculated differently – see Section 14.5.2.

It is important to note that although the energy reflectance at a boundary is $|\rho|^2$ the energy transmitance is *not* $|\tau|^2$. This is because the impedances are different on the two sides of the boundary. For example, if a wave is incident from medium 1 to medium 2 the intensity is $I_1 = |E_i|^2/2Z_1$, and the transmitted intensity is $I_2 = |E_t|^2/2Z_2$. The transmittance of the boundary is

$$T = \frac{I_2}{I_1} = \tau^2 \frac{Z_1}{Z_2}. \qquad (14.77)$$

If the wave were travelling in the reverse direction across the boundary then its transmission coefficient, as in Eq. (14.73) would be

$$\tau\prime = \frac{2Z_1'}{Z_1' + Z_2'} \qquad (14.78)$$

and

$$\tau\tau\prime = \frac{4Z_1'Z_2'}{(Z_1' + Z_2')^2} = \frac{\tau^2 Z_1}{Z_2} = T. \qquad (14.79)$$

If there is no absorption of energy at the boundary the fraction of energy transmitted, called the *transmittance*, is

$$T = 1 - R = \frac{4n_1 n_2}{(n_1 + n_2)^2}. \tag{14.80}$$

14.5.1 Reflectance for Waves Incident on an Interface at Oblique Angles

If the wave is not incident normally, it must be decomposed into two linearly polarized components, one polarized in the plane of incidence, and one polarized perpendicular to the plane of incidence.

For example, consider a plane polarized wave incident on an air/glass ($n = 1.5$) interface at an angle of incidence of 30° with a polarization state exactly intermediate between the S and P polarizations.

The angle of refraction at the boundary is found from Snell's law

$$\sin\theta_2 = \sin 30°/1.5,$$

so $\theta_2 = 19.47°$.

The effective impedance of the P component in the air is

$$Z'_{P1} = 376.7\cos\theta_1 = 326.23\ \Omega$$

and in the glass

$$Z'_{P2} = \frac{376.7}{1.5}\cos\theta_2 = 236.7\ \Omega.$$

Thus, from Eq. (14.65), the reflection coefficient for the P component is

$$\rho_P = \frac{236.77 - 326.33}{236.77 + 326.23} = -0.159.$$

The fraction of the intensity associated with the P component that is reflected is $\rho_P^2 = 0.0253$.

For the S component of the input wave

$$Z'_{S1} = 376.7/\cos\theta_1 = 434.98\ \Omega$$
$$Z'_{S2} = 376.7/(1.5\cos\theta_2) = 266.37\ \Omega$$
$$\rho_S = \frac{266.37 - 434.98}{266.37 + 434.98} = -0.240.$$

The fraction of the intensity associated with the S polarization component that is reflected is $\rho_S^2 = 0.0578$. Since the input wave contains equal amounts of S and P polarization, the overall reflectance in this case is

$$R = <\rho^2>_{avg} = 0.0416 \simeq 4\%.$$

Note that the reflected wave now contains more S polarization than P, so the polarization state of the reflected wave has been changed.

14.5.2 *Brewster's Angle*

Returning to Eq. (14.65) it might be asked whether the reflectance is ever zero. It is clear that ρ will be zero if

$$n_1 \cos\theta_2 = n_2 \cos\theta_1, \tag{14.81}$$

which from Snell's law, Eq. (14.6), gives

$$\cos\theta_1 = \frac{n_1}{n_2}\sqrt{1-\left(\frac{n_1^2}{n_2^2}\right)\sin^2\theta_1}, \tag{14.82}$$

giving the solution

$$\theta_1 = \theta_B = \text{arc }\sin\sqrt{\frac{n_2^2}{n_1^2+n_2^2}} = \text{arc }\tan\frac{n_2}{n_1}. \tag{14.83}$$

The angle θ_B is called *Brewster's angle*, a wave polarized in the plane of incidence and incident on a boundary at this angle is totally transmitted. This fact is put to good use in the design of low-reflection-loss windows in laser systems, as we have seen peviously. If Eq. (14.68) is inspected carefully it will be seen that there is no angle of incidence that yields zero reflection for a wave polarized perpendicular to the plane of incidence.

14.5.3 *Transformation of Impedance through Multilayer Optical Systems*

The impedance concept allows the reflection and transmission characteristics of multilayer optical systems to be evaluated very simply. If the incident light is incoherent then the overall transmission of a multilayer structure is just the product of the transmittances of its various interfaces. For example, an air/glass interface transmits about 96% of the light in normal incidence. The transmittance of a parallel-sided slab is $0.96 \times 0.96 \equiv 92\%$. This simple result ignores the possibility of interference effects between reflected and transmitted waves at the two faces of the slab. If the faces of the slab are very flat and parallel, and the light is coherent, such effects cannot be ignored. In this case, the method of transformed impedances is useful.

Consider the three-layer structure shown in Fig. (14.11a). The path of a ray of light through the structure is shown. The angles $\theta_1, \theta_2, \theta_3$ can be calculated from Snell's law. As an example consider a wave polarized in the plane of incidence. The effective impedances of media 1, 2, and 3 are:

$$\begin{aligned} Z_1' &= Z_1\cos\theta_1 = Z_0\cos\theta_1/n_1, \\ Z_2' &= Z_2\cos\theta_2 = Z_0\cos\theta_2/n_2, \\ Z_3' &= Z_3\cos\theta_3 = Z_0\cos\theta_3/n_3. \end{aligned} \tag{14.84}$$

It can be shown that the reflection and transmission coefficients of the structure are exactly the same as the equivalent structure in Fig. (14.11b) for normal incidence, where the effective thickness of layer 2 is now

$$d' = d\cos\theta_2. \tag{14.85}$$

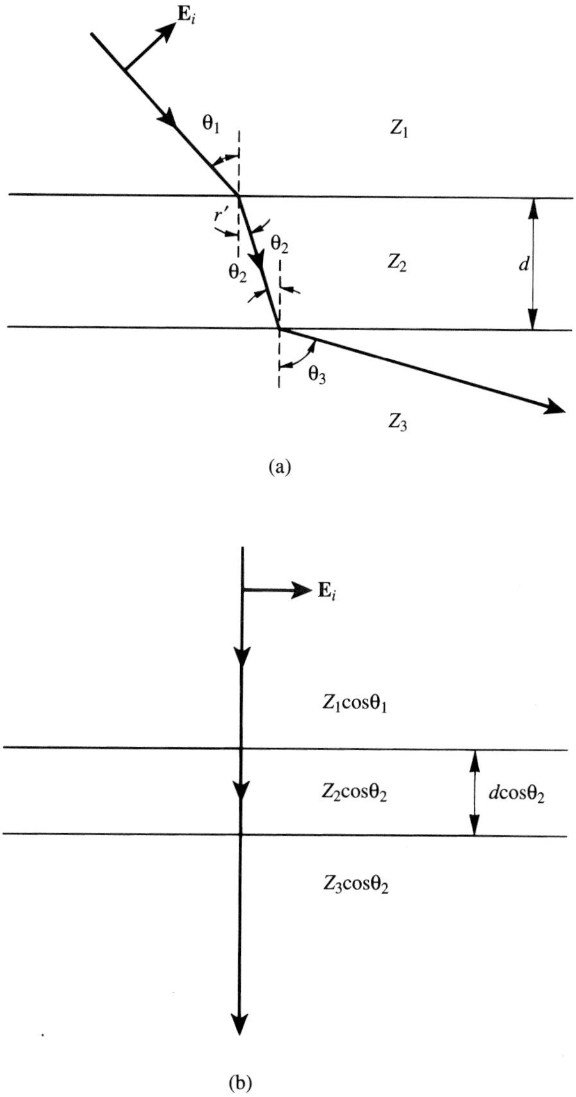

Fig. 14.11. Waves passing through a dielectric slab: (a) waves incident at an angle being refracted and transmitted; (b) normal incidence structure equivalent to (a).

The reflection and transmission coefficients of the structure can be calculated from its equivalent structure using the transformed impedance concept[14.2].

The transformed impedance of medium 3 at the boundary between media 1 and 2 is

$$Z_3'' = Z_2' \left(\frac{Z_3' \cos k_2 d' + i Z_2' \sin k_2 d'}{Z_2' \cos k_2 d' + i Z_3' \sin k_2 d'} \right), \tag{14.86}$$

where $k_2 = 2\pi/\lambda_2 = 2\pi n_2/\lambda_0$.

The reflection coefficient of the whole structure is now just

$$\rho = \frac{Z_3'' - Z_1'}{Z_3'' + Z_1'} \tag{14.87}$$

and its transmission coefficient

$$\tau = \frac{2Z_3''}{Z_3'' + Z_1'}. \tag{14.88}$$

In a structure with more layers, the transformed impedance formula, Eq. (14.86) can be used sequentially, starting at the last optical surface and working back to the first.

14.5.4 Polarization Changes

A change in polarization state can occur in two principal ways: (a) the wave can reflect off an interface for which the reflection coefficient is different for light polarized in the plane of incidence (a *P-wave*), or perpendicular to the plane of incidence (an *S-wave*), and (b) the wave can be transmitted through an anisotropic, or optically active medium that alters the polarization state of the wave. These effects are considered in further detail in Chapters 18 and 21. Optically active media rotate the plane of polarization of a linearly polarized wave passing through them. This can occur in solids or liquids in which the constituent particles are *chiral* in character, that is they have distinct left or right handed symmetry. A good example would be an aqueous sugar solution. A solution of the sugar dextrose, for example, rotates the plane of polarization of a linearly polarized wave in a clockwise direction viewing in the direction of propagation while its *enantiomorph*,† laevulose, rotates the plane of polarization in a counter-clockwise direction. One way to account for the phenomenon of optical activity is for left and right hand circularly polarized light to have different phase velocities in the medium. Optical activity can also be induced by the action of an external field. For example, in the liquid nitrobenzene, application of a transverse electric field causes the liquid to become optically active. This is the Kerr effect and can be used to make shutters for Q-switching. Many materials become optically active if they are placed in an axial magnetic field. This is the Faraday effect. The actual amount of rotation is described by the *Verdet* constant α for the material. For a slab of material of length ℓ and an applied axial flux density $\mathbf{B}$

$$\theta_{rot} = \alpha\ell \mid \mathbf{B} \mid . \tag{14.89}$$

If α is positive the direction of rotation is clockwise if the light propagates in the same direction as $\mathbf{B}$, it is counterclockwise if the direction of propagation is antiparallel to $\mathbf{B}$. Thus, a to-and-fro pass of light through the field doubles the angle of rotation.‡ The Faraday effect can be used to make an optical isolator; the operation of which is shown schematically in Fig. (14.12). A plane polarized wave is rotated by 45° on passing through the Faraday-active medium from left to right. It then passes through a second polarizer oriented at 45° to the input polarizer. A wave passing from right to left also experiences a 45° rotation but cannot then pass the entrance linear polarizer. Thus, the source of radiation is protected from undesirable feedback from optical components further down the chain. The use of isolators is extremely important in high power oscillator/amplifier(s) systems. Without their use a reflected wave of initially weak intensity could be amplified on its return passage through the system with disastrous consequences for the oscillator.

† The molecule with the same atomic constituents, but mirror image structure.

‡ This is in contrast to natural optical activity where a to-and-fro pass leads to no net rotation.

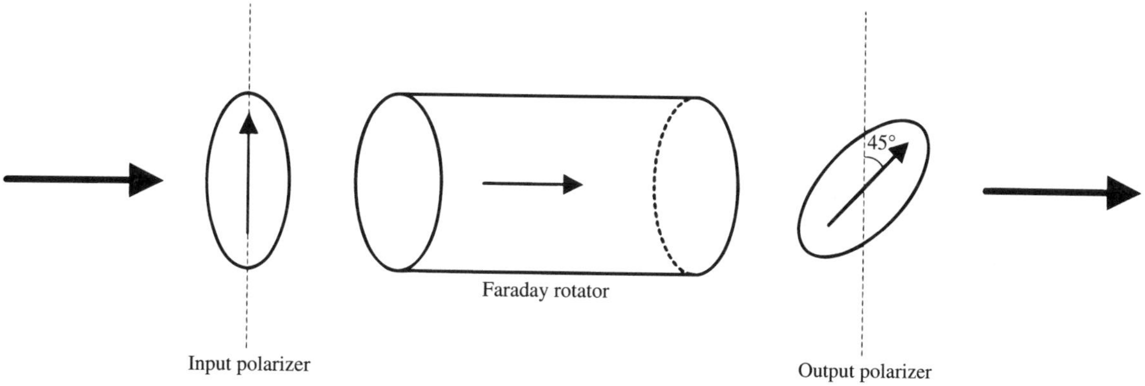

Fig. 14.12. Schematic construction of a Faraday isolator. A section of Faraday material producing a 45° rotation of the plane of linear polarization is placed between two polarizers oriented at 45° with respect to each other.

14.6 Problems

(14.1) A slab of glass with $n = 1.55$ is exactly 3.5 mm thick and is bounded on both sides by air, $n = 1$. Coherent light of wavelength 1 μm is incident on the slab. Calculate: (a) The fractional intensity transmission for P-waves incident at an angle of 30°. (b) The fractional intensity transmission for S-waves incident at an angle of 30°. (c) The rotation of the plane of polarization of the transmittted and reflected waves for a wave initially composed of 50% S-wave and 50% P-wave. (d) The rotation of the plane of polarization of the transmitted and reflected waves for a wave initially composed of 30% S-wave and 70% P-wave.

(14.2) How would the answers to Problems (14.1a), and (14.1b) be modified if the light were incoherent?

(14.3) A thick lens with $n = 1.55$ has $R_1 = 60$ mm, $R_2 = -40$ mm with $d = 15$ mm. Calculate the positions, of the principal planes and the focal length of the lens. Draw a sketch to illustrate your answer.

(14.4) An object is placed 500 mm from the R_1 face of the thick lens in Problem (14.3). Calculate where the image is.

(14.5) For a thin lens with $u = 50$ mm, $f = 20$ mm draw a ray tracing diagram to illustrate the location of the image and its magnification. State the image distance, magnification, and angular magnification found from the diagram. Repeat for $u = 20$ mm, $f = 50$ mm and $u = 50$ mm, $f = -20$ mm. Mention any important differences among these three situations.

(14.6) Develop a matrix approach for dealing with flat and spherical mirrors in paraxial ray analysis. Hint: A mirror folds a ray path back on itself.

(14.7) What is the overall transmittance through a glass parallel-sided slab of thickness 3 mm, refractive index 1.55, when light of wavelength 632.8 nm is incident at an angle of 40°? if the incident light is coherent and is an equal mixture of S and P polarizations.

(14.8) (Computer project) Write a computer program to plot the variation in reflectance for P- and S-waves incident on an air/glass boundary at angles from 0 to $\pi/2$. Take $n_1 = 1$, $n_2 = 1.5$. Calculate the S-wave reflectance at Brewster's angle.

(14.9) (Computer project) Write a computer program to calculate the reflectance as a function of wavelength in normal incidence for a single $\lambda_s/4$ antireflection layer, where λ_s is the wavelength in the layer. Plot R from $\lambda_s/2$ to $1.5\lambda_s$.

(14.10) (Computer project) Write a computer program to calculate and plot the reflectance of a single $\lambda_s/4$ antireflection layer designed for normal incidence as the incidence angle varies from 0 to 90°.

(14.11) For a spherical mirror of aperture diameter D and focal length f calculate the blurring along the axis, caused by spherical aberration, of the image of an object at a distance u from the mirror, What is the magnitude of the blurring if $u = 3$ m, $D = 3$ m, $f = 2.5$ m. Hint: Calculate the path of the extreme ray – one that just hits the edge of the mirror – by exact trigonometry.

(14.12) (Computer project) A triangular prism has index n=1.5 and apex angle $\theta = 30°$. Plot the angular deviation of a ray of light passing through the prism (and undergoing two refractions) as a function of the angle of incidence. For what angle of incidence is this angle a minimum? Use this result as a clue in deriving the minimum deviation angle theoretically.

References

[14.1] H. Kogelnik and T. Li, "Laser beams and resonators," *Proc. IEEE*, **54**, 1312–1329, 1966.

[14.2] S. Ramo, J.R. Whinnery, and R. Van Duzer, *Fields and Waves in Communication Electronics*, 3rd Edition, Wiley, New York, 1993.

15

Analysis of Optical Systems II

15.1 Introduction

In this chapter we shall use the paraxial ray transfer matrix techniques introduced in the last chapter to analyze periodic optical systems. These are optical systems in which a ray passes through a sequence of optical components that repeat in a periodic fashion along the axis. We shall find that this approach can also be used to study the paths of light rays bouncing back and forth between pairs of mirrors. This will allow us to deduce the *stability* condition for an optical resonator. In such a system a ray never deviates more than a certain finite distance from the axis and remains confined.

15.2 Periodic Optical Systems

If we increase the number of lenses in an optical system we can produce arrangements that periodically refocus a ray of light propagating paraxially through the system. When this periodic refocusing prevents the ray being lost from the lens sequence, then we have constructed a stable lens waveguide. As an example let us analyze the biperiodic lens sequence shown in Fig. (15.1). The 'unit cell' of this periodic structure consists of the two lenses of focal lengths f_1 and f_2. Now in Fig. (15.1)

$$\begin{aligned} r_2 &= Ar_1 + Br_1', & r_3 &= Ar_2 + Br_2', \\ r_2' &= Cr_1 + Dr_1', & r_3' &= Cr_2 + Dr_2', \end{aligned} \tag{15.1}$$

or in general for transmission through n sections (each consisting in this case of two lenses)

$$\begin{pmatrix} r_{n+1} \\ r_{n+1}' \end{pmatrix} = \begin{pmatrix} A & B \\ C & D \end{pmatrix}^n \begin{pmatrix} r_1 \\ r_1' \end{pmatrix}. \tag{15.2}$$

We can evaluate such a matrix product by Sylvester's theorem

$$\begin{pmatrix} A & B \\ C & D \end{pmatrix}^n = \frac{1}{\sin\phi} \begin{pmatrix} A\sin n\phi - \sin(n-1)\phi & B\sin n\phi \\ C\sin n\phi & D\sin n\phi - \sin(n-1)\phi \end{pmatrix}, \tag{15.3}$$

where $\cos\phi = \frac{1}{2}(A+D)$ and

$$\sin\phi = \sqrt{1 - \frac{1}{4}(A+D)^2}. \tag{15.4}$$

Clearly,

$$r_{n+1} = \{[A\sin n\phi - \sin(n-1)\phi]r_1 + B(\sin n\phi)r_1'\}/\sin\phi. \tag{15.5}$$

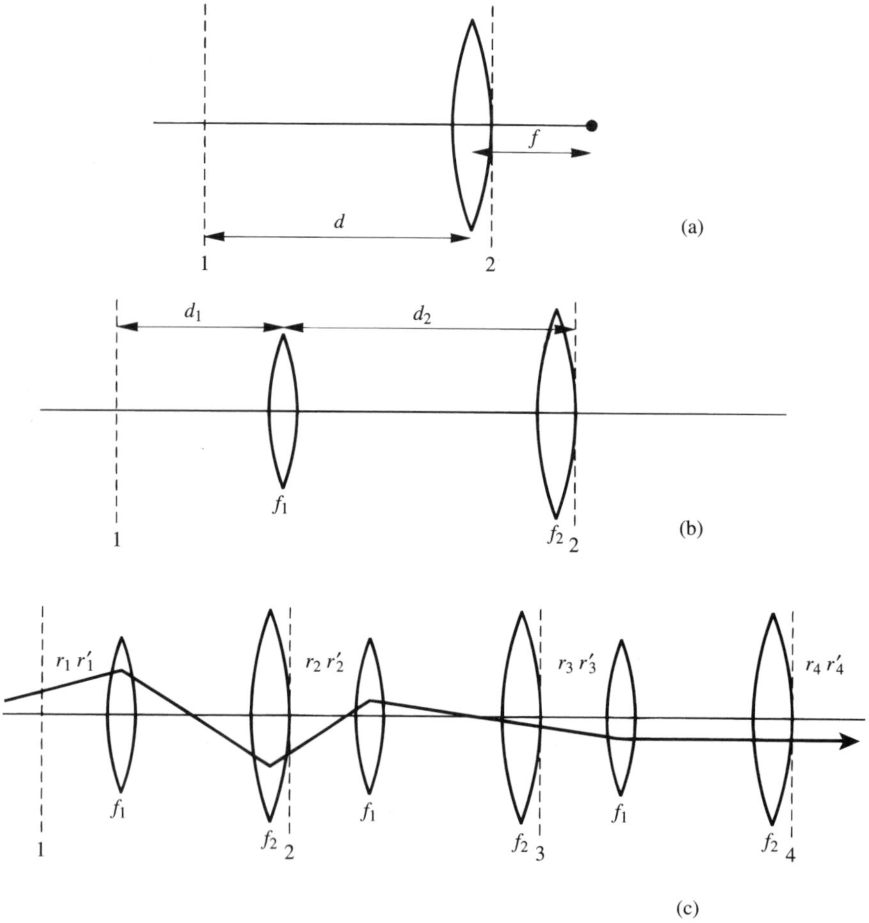

Fig. 15.1. A biperiodic lens sequence: (a) basic unit on which the sequence is based: a length of uniform medium plus a lens; (b) unit cell of sequence consisting of two lens of different focal lengths; (c) three unit cells of the sequence: the paraxial ray parameters are indicated at the input and output planes of each unit cell.

If the angular factors in (15.5) remain real, then r_{n+1} will oscillate as a function of n (the number of pairs of lenses traversed) and the structure has periodic refocusing properties. However, if ϕ becomes imaginary, then we can write

$$\sin\phi = \sin i\psi = \frac{1}{2i}(e^{-\psi} - e^{\psi}) = i\sinh(\psi). \tag{15.6}$$

Now the dependence of r_{n+1} on n has become a hyperbolic function and the ray diverges more and more from the axis as it passes through the system.

The condition for ϕ to be real and $\sin\phi$ to remain oscillatory is

$$|\cos\phi| \leq 1, \tag{15.7}$$

which from (15.4) gives

$$\left|\frac{1}{2}(A + D)\right| \leq 1. \tag{15.8}$$

If this condition is satisfied the lens or other optical sequence is said to be stable.

15.3 The Identical Thin Lens Waveguide

If $f_1 = f_2 = f$ and $d_1 = d_2 = d$ for thin lenses in the lens sequence considered above, then the unit cell of the structure becomes a single lens, whose transfer matrix is

$$\begin{pmatrix} A & B \\ C & D \end{pmatrix} = \begin{pmatrix} 1 & d \\ -1/f & 1-d/f \end{pmatrix}, \tag{15.9}$$

and the stability condition becomes

$$\left|1 - \frac{d}{2f}\right| \leq 1, \tag{15.10}$$

which is satisfied if

$$0 \leq d \leq 4f. \tag{15.11}$$

So a sequence of diverging lenses ($f < 0$) is of course unstable. If the Eq. (15.10) is not satisfied a ray rapidly escapes from the lens waveguide.

If a complete cone of optical rays is propagating through the system (a family of rays that lie inside a cone of semivertical angle r_1' at the input plane of the system) then the family of rays constitutes a beam whose radius at the nth lens is

$$r_{n+1} = \frac{1}{\sin\phi}\left\{[A\sin n\phi - \sin(n-1)\phi]r_1 + B(\sin n\phi)r_1'\right\}, \tag{15.12}$$

where $A = 1, B = d$. In this case

$$\cos\phi = 1 - \frac{d}{2f},$$

and

$$\sin\phi = \sqrt{\frac{d}{f} - \frac{d^2}{4f^2}} = \frac{d}{2f}\sqrt{\frac{4f}{d} - 1}, \tag{15.13}$$

so

$$\begin{aligned} r_{n+1} &= \frac{1}{\sin\phi}[(\sin n\phi - \sin n\phi\cos\phi + \cos n\phi\sin\phi)r_1 + d(\sin n\phi)r_1'] \\ &= \left(\frac{d}{2f}r_1 + dr_1'\right)\frac{\sin n\phi}{\sin\phi} + r_1\cos n\phi. \end{aligned} \tag{15.14}$$

We can write this in the form

$$r_{n+1} = r_{max}\sin(n\phi + \alpha), \tag{15.15}$$

where

$$r_{max}\cos\alpha = \frac{(dr_1/2f + dr_1')}{\sin\phi}; \quad r_{max}\sin\alpha = r_1. \tag{15.16}$$

Therefore,

$$r_{max}^2 = \frac{(dr_1/2f + dr_1')^2}{\sin^2\phi} + r_1^2, \tag{15.17}$$

which from (15.9) gives

$$r_{max}^2 = \left(\frac{4f}{4f-d}\right)(r_1^2 + dr_1r_1' + dfr_1'^2), \tag{15.18}$$

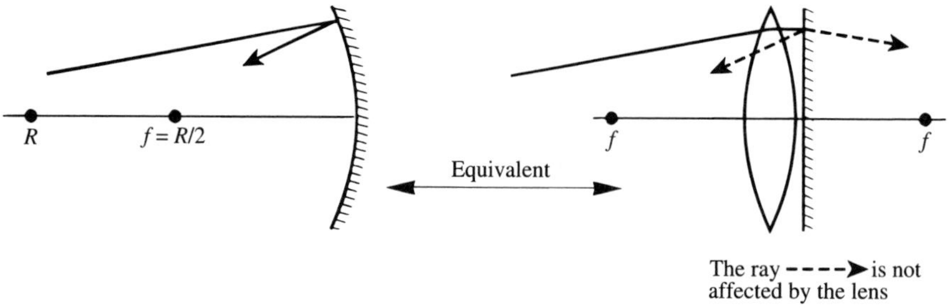

Fig. 15.2. A ray striking a concave mirror is equivalent to a thin lens/flat mirror combination in which the ray only interacts with the lens once.

and

$$\tan\alpha = \frac{r_1 \sin\phi}{dr_1/2f + dr_1'} = \frac{\sqrt{4f/d - 1}}{1 + 2fr_1'/r_1}. \tag{15.19}$$

Note that the stability condition is effectively restated by Eqs. (15.18) and (15.19): if $d > 4f$ then r_{max} and $\tan\alpha$ become imaginary.

15.4 The Propagation of Rays in Mirror Resonators

When a ray of light strikes a spherical mirror of radius of curvature R, the path of the reflected ray is the same, apart from being in the mirror image direction, as the same ray traversing a thin lens of focal length $R/2$, i.e., the spherical mirror is equivalent to a superimposed 'one-way' thin lens and plane mirror *when the reflected ray is considered to be unaffected by the lens* as illustrated in Fig. (15.2). So a two-mirror system with mirror radii R_1, R_2 spaced d apart, is equivalent to a biperiodic lens sequence with lenses of focal length $f_1 = R_1/2$ and $f_2 = R_2/2$ spaced the same distance d apart. The only difference between the two structures is that the ray path in the mirror system is folded back on itself. By considering this duality between curved mirrors and biperiodic lens sequences we can arrive at some important considerations regarding the stability of such curved mirror resonators.

The curved mirror resonator shown in Fig. (15.3) is equivalent to the biperiodic lens sequence shown in Fig. (15.3). For a symmetrical mirror resonator the equivalent biperiodic lens sequence becomes an identical lens waveguide as shown in Fig. (15.4). We can resolve the paraxial propagation of a ray in such a resonator into independent motions in the x and y planes. Following Eq. (15.15), after n reflections we write the coordinates of the ray as

$$x_n = x_{max}\sin(n\phi + \alpha_x), \quad y_n = y_{max}\sin(n\phi + \alpha_y), \tag{15.20}$$

where α_x and α_y are constants.

The locus of points x_n, y_n on the surface of one of the mirrors is an ellipse.

It must be remembered that these considerations only apply to rays propagating sufficiently near parallel to the axis of the system that the paraxial ray approximation still holds.

If in (15.20) ϕ satisfies

$$2p\phi = 2q\pi, \tag{15.21}$$

where p and q are any two integers, then after p round trips the ray will retrace its pattern between the mirrors.

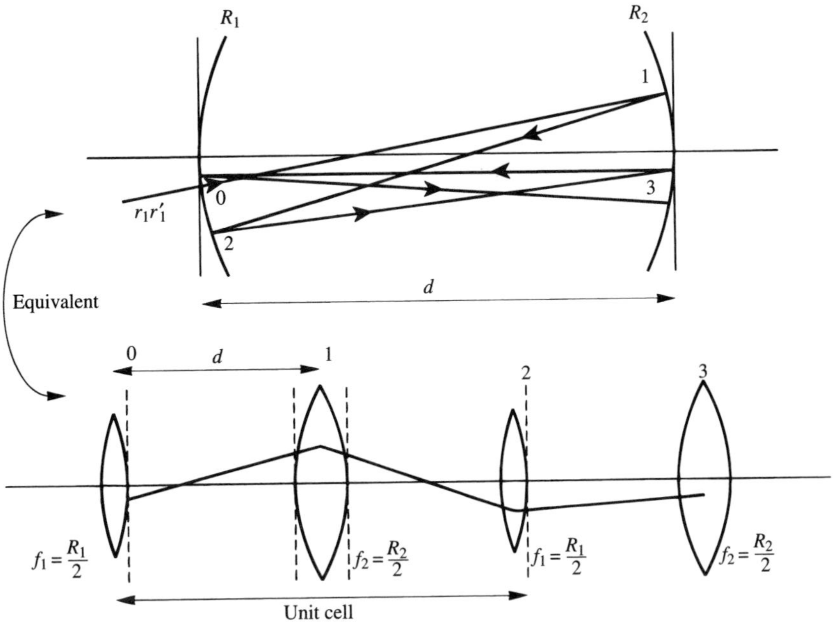

Fig. 15.3. Curved mirror resonator and its equivalent biperiodic lens sequence.

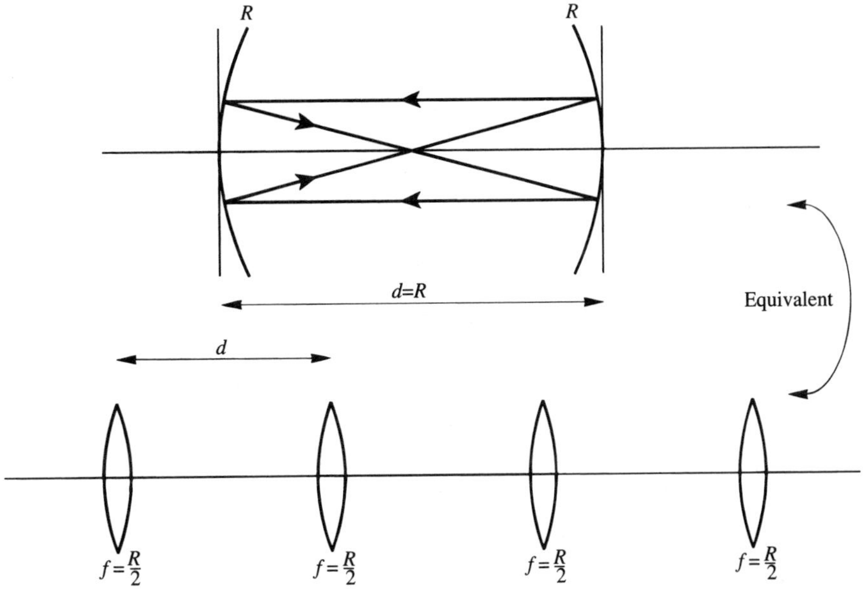

Fig. 15.4. Confocal symmetric resonator and its equivalent lens sequence.

For example if $q = 1$, $p = 2$, so that $\phi = \pi/2$ then $d = 2f = R$ (from $\cos\phi = 1 - d/2f$) and the ray retraces its path after two round trips as illustrated in Fig. (15.4). When this situation exists the resonator is said to be *symmetrical confocal* as the two mirrors have a common focal point.

For a general resonator with mirrors of radii R_1, R_2 spaced a distance d apart, the stability condition, i.e., the condition that a paraxial ray remains so and eventually retraces its pattern between the two mirrors, is the same as the stability condition for the equivalent biperiodic lens sequence. Now for the equivalent biperiodic lens sequence, the unit cell transfer matrix is from (14.51)

$$\begin{pmatrix} A & B \\ C & D \end{pmatrix} \equiv \begin{pmatrix} 1-d/f_1 & 2d-d^2/f_1 \\ -1/f_1-1/f_2+d/f_1f_2 & 1-d/f_1-2d/f_2+d^2/f_1f_2 \end{pmatrix} \tag{15.22}$$

and the stability condition is $\mid \frac{1}{2}(A+D) \mid \leq 1$, which gives

$$\left| \frac{1}{2}\left(2-\frac{2d}{f_1}-\frac{2d}{f_2}+\frac{d^2}{f_1f_2}\right) \right| \leq 1. \tag{15.23}$$

This can be simplified to the stability condition

$$0 \leq 1-\frac{d}{2f_2}-\frac{d}{2f_1}+\frac{d^2}{4f_1f_2} \leq 1. \tag{15.24}$$

This stability condition for the biperiodic lens sequence can be rewritten as

$$0 \leq \left(1-\frac{d}{2f_2}\right)\left(1-\frac{d}{2f_1}\right) \leq 1. \tag{15.25}$$

For the equivalent spherical mirror resonator, the stability condition is

$$0 \leq \left(1-\frac{d}{R_2}\right)\left(1-\frac{d}{R_1}\right) \leq 1, \tag{15.26}$$

where d is the spacing between the mirrors. The mirror curvatures and spacings that satisfy this equation can be represented on a resonator stability diagram, as shown in Fig. (15.5). It can be seen from Fig. (15.5) that both the plane-parallel, confocal and concentric cavities are right on the margin of stability – in unstable equilibrium. In these situations a slight perturbation of the cavity, for example, either by increasing the mirror spacing in the concentric case or by any convex distortion of one of the plane mirrors in the plane-parallel case, will move the resonator into the unstable region.

Although we have considered the stability properties of these resonators by a geometric optics approach, by considering the paths of rays between the reflectors, we shall see in the next chapter that the same stability conditions apply to the real types of wave that propagate between the reflectors of such resonators. However, before considering these waves in a little more detail, we shall deal with a ray propagation problem that arises frequently in laser systems and in the propagation of waves through thin film and fiber optic waveguides.

15.5 The Propagation of Rays in Isotropic Media with Refractive Index Gradients

In this case by 'isotropic' we continue to imply that in the medium **P** and **E** are parallel, and that the direction of the wave normal is the same as the direction of energy propagation. Energy propagation occurs in the direction of the Poynting vector **S**, which is the direction of the geometric-optical light ray. As far as the (scalar) refractive index is concerned the medium is not isotropic.

The light wave in the medium, at any point will have a field dependence of the general form

$$U_0 = U(x,y,z)e^{i(\omega t-\mathbf{k}\cdot\mathbf{r})}, \tag{15.27}$$

Fig. 15.5. Stability diagram for optical resonators[15.1].

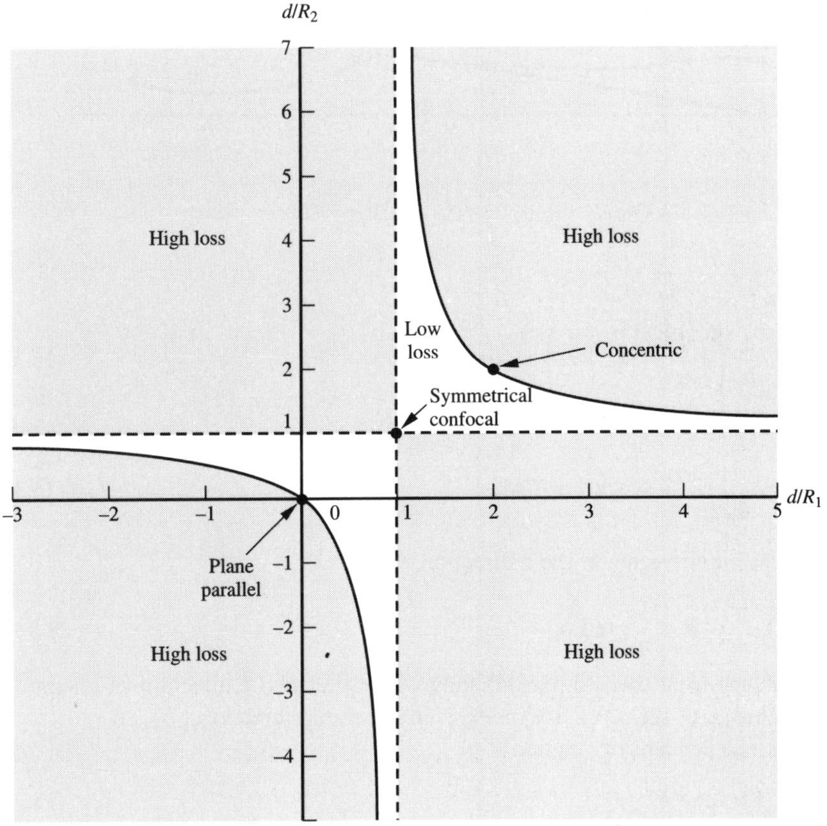

where the wave vector is

$$\mathbf{k} = \frac{2\pi}{\lambda}\hat{\mathbf{s}} = \frac{2\pi}{\lambda_0}n\hat{\mathbf{s}} = k_0 n\hat{\mathbf{s}} \tag{15.28}$$

and $\hat{\mathbf{s}}$ is a unit vector in the direction of energy flow. So

$$U_0 = U(x, y, z, t)e^{ik_0 n(\hat{\mathbf{s}}\cdot\mathbf{r})} \tag{15.29}$$

We call $n(\hat{\mathbf{s}} \cdot \mathbf{r})$ the optical path $\zeta(r)$. It can be shown[15.2] that

$$\text{grad } \zeta = n\hat{\mathbf{s}}. \tag{15.30}$$

The surface where $\zeta(\mathbf{r})$ is constant is the geometric wave front. It can also be shown[15.2] that†

$$\frac{d}{ds}\left(n\frac{d\mathbf{r}}{ds}\right) = \text{grad } n. \tag{15.31}$$

This is the differential equation that describes ray propagation in an optically inhomogeneous medium. For paraxial rays $\mathbf{r}(z) = r(z)\hat{\rho} + \mathbf{z}$ where $\hat{\rho}$ is a unit vector in the direction of r and $d/ds = d/dz$ as shown in Fig. (15.6) so (15.31)

† See Appendix 7.

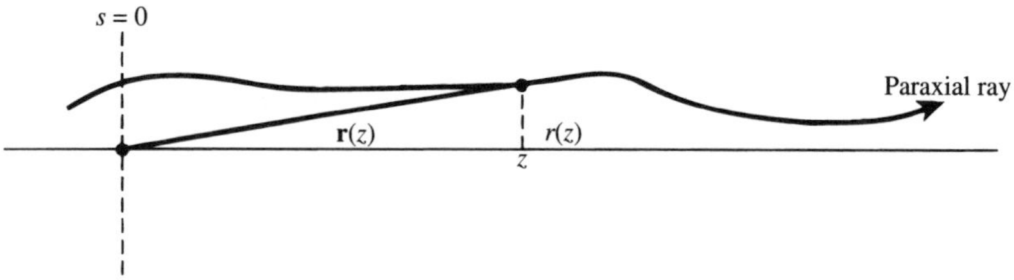

Fig. 15.6. Path of a paraxial ray in a system where the refractive index varies with position.

becomes

$$\frac{d}{dz}\left\{n\frac{d}{dz}\left[r(z)\hat{\rho}+\mathbf{z}\right]\right\} = \text{grad } n, \tag{15.32}$$

which if n does not vary in the z direction gives

$$n\frac{d}{dz}\left[\frac{dr(z)}{dz}\hat{\rho}+\hat{\mathbf{k}}\right] = \text{grad } n, \tag{15.33}$$

where $\hat{\mathbf{k}}$ is a unit vector in the z direction, thus

$$\frac{d^2r(z)}{dz^2}\hat{\rho} = \frac{1}{n}\text{ grad } n. \tag{15.34}$$

It is not surprising that Eq. (15.34) shows that grad n, the direction of maximum rate of change of the refractive index, is in the radial direction.

If n is independent of position

$$r = az + b, \tag{15.35}$$

where a and b are constants of integration. The path of propagation of the ray is a straight line, as expected. Frequently, however, we encounter a refractive index that varies quadratically, or approximately so, as a function of distance r from the axis of symmetry

$$n = n_0 - \frac{1}{2}n_2r^2. \tag{15.36}$$

Such a quadratic index profile can arise in a number of ways:

(i) Propagation of a laser beam with a transverse Gaussian-like intensity profile in a slightly absorbing medium. The heating caused by absorption of the beam is largest on axis, and if $dn/dT < 0$, as is the case for most materials, then the index is smallest on axis. We will see that this leads to spreading and defocusing of the beam. If, however, $dn/dT > 0$, as is the case in certain glasses, then thermal self-focusing of the beam can occur.

(ii) In cylindrical laser crystals the pumping radiation heats up the crystal most on its outer surface, which generally leads to a refractive index that increases towards the center of the rod (for $dn/dT < 0$).

(iii) Gas lenses with a quadratic index profile can be made for focusing and guiding laser beams.

(iv) In optical waveguides made from composite materials having diffrent refractive indices: either as a planar sandwich, as in a heterostructure laser, or in a cylindrical configuration, such as a fiber optic light guide.

(v) Small gradient index lenses, which take the form of small cylinders with plane ends, are now in widespread use in optical systems. These lenses are fabricated by introducing a radial concentration of dopant into the glass. They are frequently referred to as GRIN or SELFOC lenses.

For a quadratic refractive index variation

$$n = n_0 - \tfrac{1}{2}n_2 r^2 = n_0 - \tfrac{1}{2}n_2(x^2 + y^2), \tag{15.37}$$

where $\mathbf{r} = x\hat{\mathbf{i}} + y\hat{\mathbf{j}}$;

$$\text{grad } n = -n_2 x\,\hat{\mathbf{i}} - n_2 y\,\hat{\mathbf{j}}, \tag{15.38}$$

so from equation (15.31)

$$\left(\frac{d^2r}{dz^2}\right)\hat{\rho} = \frac{-n_2 x\hat{\mathbf{i}} - n_2 y\hat{\mathbf{j}}}{n_0 - \frac{1}{2}n_2(x^2 + y^2)}, \tag{15.39}$$

and

$$\left[\frac{d^2r}{dz^2} - \frac{1}{2}\frac{n_2}{n_0}(x^2 + y^2)\frac{d^2r}{dz^2}\right]\hat{\rho} + \frac{n_2}{n_0}x\hat{\mathbf{i}} + \frac{n_2}{n_0}y\hat{\mathbf{j}} = 0 \tag{15.40}$$

For paraxial rays r is small, so $r^2 = x^2 + y^2$ is negligible and we neglect the second term in (15.40), which gives

$$\frac{d^2r}{dz^2} + \left(\frac{n_2}{n_0}\right)r = 0. \tag{15.41}$$

If the initial parameters of the ray are r_0, r_0' at plane $z = 0$, then we know that $r(0) = r_0, dr/dz(r = 0) = r_0'$ and the solution of (15.41) can be easily shown to be

$$r(z) = \cos\left(\sqrt{\frac{n_2}{n_0}}\right) z r_0 + \sqrt{\frac{n_0}{n_2}} \sin\left(\sqrt{\frac{n_2}{n_0}}\right) z r_0', \tag{15.42}$$

which can be verified by writing down

$$r'(z) = \frac{dr}{dz} = -\sqrt{\frac{n_2}{n_0}} \sin\left(\sqrt{\frac{n_2}{n_0}} z\right) r_0 + \cos\left(\sqrt{\frac{n_2}{n_0}}\right) z r_0' \tag{15.43}$$

$$\frac{d^2r}{dz^2} = -\sqrt{\frac{n_2}{n_0}} \cos\left(\sqrt{\frac{n_2}{n_0}} z\right) r_0 - \sqrt{\frac{n_2}{n_0}} \sin\left(\sqrt{\frac{n_2}{n_0}} z\right) r_0'. \tag{15.44}$$

It can be seen from Eq. (15.42) that the ray oscillates back and forth around the axis, and the medium acts like a lens provided $n_2 > 0$. If the medium has $n_2 < 0$, then the solution of (15.41) is

$$r(z) = \cosh\left(\sqrt{\frac{|n_2|}{n}} z\right) r_0 + \sqrt{\frac{n_0}{|n_2|}} \sinh\left(\sqrt{\frac{|n_2|}{n}} z\right) r_0' \tag{15.45}$$

and

$$r'(z) = \sqrt{\frac{|n_2|}{n_0}} \sinh\left(\sqrt{\frac{|n_2|}{n_0}} z\right) r_0 + \cosh\left(\sqrt{\frac{|n_2|}{n_0}} z\right) r_0'. \tag{15.46}$$

In this case a propagating ray will diverge from the axis and not experience periodic focusing within the medium.

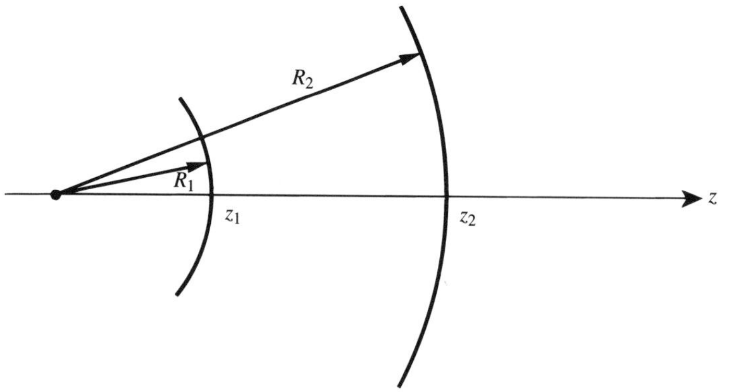

Fig. 15.7. Change in radius of curvature of a spherical wave as it propagates away from its source.

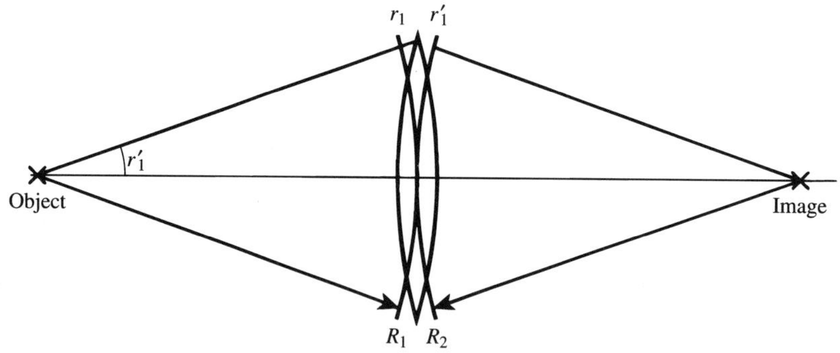

Fig. 15.8. Focusing of a spherical wave by a lens.

15.6 The Propagation of Spherical Waves

Although the transverse modes that propagate inside a laser cavity are not plane waves, in many cases we can derive useful information about them by considering them as such, for example, in the calculation of longitudinal mode frequencies. Also, although these transverse modes are not spherical waves, they frequently exhibit similar properties to spherical waves and it is worthwhile reviewing briefly some of the propagation characteristics of spherical waves.

Spherical waves are waves that originate from a point source in an isotropic medium. For these waves **P** is parallel to **E** and n must be uniform or spherically symmetric. When they propagate their radius of curvature increases as shown in Fig. (15.7), according to

$$R_2(z_2) = R_1(z_1) + z_2 - z_1. \tag{15.47}$$

We use the sign convention that the spherical wave has positive radius of curvature when it is convex in the positive z direction (direction of propagation). When such a wave encounters a thin lens it is focused (if the lens has a positive focal length) as shown in Fig. (15.8). In this case

$$\frac{1}{R_1} - \frac{1}{R_2} = \frac{1}{f}, \text{ so } \frac{1}{R_2} = \frac{1}{R_1} - \frac{1}{f}, \tag{15.48}$$

(note that R_2 is negative). If instead we consider one ray OX that helps to make up this spherical wave, then if r_1, r_1' are the input parameters of this ray

$$R_1 = r_1/r_1',$$

or, in general, at any plane

$$R(z) = \frac{r(z)}{r'(z)}. \tag{15.49}$$

For propagation of this ray through any optical system with transfer matrix $\begin{pmatrix} A & B \\ C & D \end{pmatrix}$ we know that

$$\begin{pmatrix} r_2 \\ r_2' \end{pmatrix} = \begin{pmatrix} A & B \\ C & D \end{pmatrix} \begin{pmatrix} r_1 \\ r_1' \end{pmatrix},$$

so

$$\begin{aligned} r_2 &= Ar_1 + Br_1' \\ r_2' &= Cr_1 + Dr_1'. \end{aligned} \tag{15.50}$$

The radius of curvature of the outgoing spherical wave (of which the ray is a part) is

$$R_2 = \frac{r_2}{r_2'} = \frac{Ar_1 + Br_1'}{Cr_1 + Dr_1'} = \frac{A(r_1/r_1') + B}{C(r_1/r_1') + D}, \tag{15.51}$$

which gives the simple final result

$$R_2 = \frac{AR_1 + B}{CR_1 + D}. \tag{15.52}$$

Therefore, we can calculate the change in the radius of curvature of the spherical wave as it propagates by considering the transfer matrix of the system. As a spherical wave comes closer to plane, its radius approaches infinity.

15.7 Problems

(15.1) An input ray enters the following sequence with ray parameters r_{in}=0.1 m, r'_{in}=0.01 rad: a thin lens of focal length 3 m, an air path of length 0.5 m, and a rectangular glass slab of length 1 m and refractive index =1.5. The final flat face of the glass slab is coated to reflect light. What are the ray parameters when the ray returns to the input plane?

(15.2) A Galilean telescope, consisting of a converging lens and diverging lens, is being used to expand a laser beam. If the two lenses of the telescope have focal lengths f_1, f_2 and are placed a distance d apart, use ray transfer matrices to show what the configuration should be to minimize the beam divergence of the output beam.

(15.3) Find a proof for Sylvester's theorem.

(15.4) A spherical wave of radius of curvature 100 mm enters a planar, parallel sided slab with a graded index of the form $n(r) = 1.5 - \frac{1}{2}n_2r^2$, where $n_2 = 10^2$ m^{-2}. How thick must the slab be for the central part of the spherical wavefront to emerge as a plane wave?

References

[15.1] H. Kogelnik and T. Li, 'Laser beams and resonators,' *Proc. IEEE*, **54**, 1312–1329, 1966.

[15.2] M. Born and E. Wolf, *Principles of Optics*, 6th Edition, Pergamon Press, Oxford, 1980.

16

Optics of Gaussian Beams

16.1 Introduction

In this chapter we shall look from a wave standpoint at how narrow beams of light travel through optical systems. We shall see that special solutions to the electromagnetic wave equation exist that take the form of narrow beams – called *Gaussian beams*. These beams of light have a characteristic radial intensity profile whose width varies along the beam. Because these Gaussian beams behave somewhat like spherical waves, we can match them to the curvature of the mirror of an optical resonator to find exactly what form of beam will result from a particular resonator geometry.

16.2 Beam-Like Solutions of the Wave Equation

We expect intuitively that the transverse modes of a laser system will take the form of narrow beams of light that propagate between the mirrors of the laser resonator and maintain a field distribution that remains distributed around and near the axis of the system. We shall therefore need to find solutions of the wave equation that take the form of narrow beams and then see how we can make these solutions compatible with a given laser cavity.

Now, the wave equation is, for any field or potential component U_0 of an electromagnetic wave

$$\nabla^2 U_0 - \mu\epsilon_r\epsilon_0 \frac{\partial^2 U_0}{\partial t^2} = 0, \tag{16.1}$$

where ϵ_r is the dielectric constant, which may be a function of position. The non-plane-wave solutions that we are looking for are of the form

$$U_0 = U(x, y, z)e^{i[\omega t - \mathbf{k}(\mathbf{r})\cdot\mathbf{r}]}. \tag{16.2}$$

We allow the wave vector $\mathbf{k}(\mathbf{r})$ to be a function of $\mathbf{r}$ to include situations where the medium has a nonuniform refractive index. From Eqs. (16.1) and (16.2)

$$\nabla^2 U + \mu\, \epsilon_r\epsilon_0\omega^2 U = 0, \tag{16.3}$$

where ϵ_r and μ may be functions of $\mathbf{r}$. We have seen previously that the propagation constant in the medium is $k = \omega\sqrt{\epsilon_r\epsilon_0\mu}$, so

$$\nabla^2 U + k(\mathbf{r})^2 U = 0. \tag{16.4}$$

This is the time-independent form of the wave equation, frequently referred to as the *Helmholtz* equation.

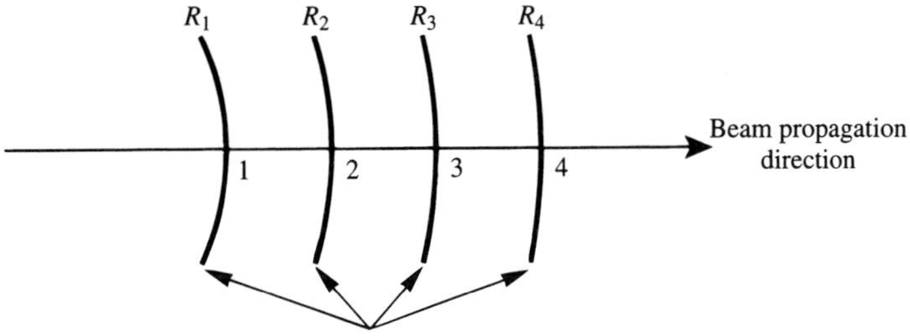

Fig. 16.1. The changing curvature of a spherical wave as it propagates in the z direction.

In general, if the medium is absorbing, or exhibits gain, then its dielectric constant ϵ_r has real and imaginary parts.

$$\epsilon_r = 1 + \chi(\omega) = 1 + \chi'(\omega) - i\chi''(\omega) \tag{16.5}$$

and

$$k = k_0\sqrt{\epsilon_r}, \tag{16.6}$$

where $k_0 = \omega\sqrt{\epsilon_0\mu}$

If the medium were conducting, with conductivity σ, then its complex propagation vector would obey

$$k^2 = \omega^2\mu\ \epsilon_r\epsilon_0\left(1 - i\frac{\sigma}{\omega\epsilon_r\epsilon_0}\right), \tag{16.7}$$

so conductivity can be included as a contribution to the imaginary part of the dielectric constant[16.1].

We know that simple solutions of the time-independent wave equation above are transverse plane waves. However, these simple solutions are not adequate to describe the field distributions of transverse modes in laser systems. Let us look for solutions to the wave equation which are related to plane waves, but whose amplitude varies in some way transverse to their direction of propagation. Such solutions will be of the form $U = \psi(x, y, z)e^{-ikz}$ for waves propagating in the positive z direction. For functions $\psi(x, y, z)$ which are localized near the z axis the propagating wave takes the form of a narrow beam. Further, because $\psi(x, y, z)$ is not uniform, the surfaces of constant phase in the wave will no longer necessarily be plane. If we can find solutions of the wave equation where $\psi(x, y, z)$ gives phase fronts that are spherical (or approximately so over a small region) then we can make the propagating beam solution $U = \psi(x, y, z)e^{-ikz}$ satisfy the boundary conditions in a resonator with spherical reflectors, provided the mirrors are placed at the position of phase fronts whose curvature equals the mirror curvature. Thus the propagating beam solution becomes a satisfactory transverse mode of the resonator. For example in Fig. (16.1) if the propagating beam is to be a satisfactory transverse mode, then a spherical mirror of radius R_1 must be placed at position 1 or one of radius R_2 at 2, etc. Mirrors placed in this way lead to reflection of the wave back on itself.

Substituting $U = \psi(x, y, z)e^{-ikz}$ in Eq. (16.4) we get

$$\left(\frac{\partial^2\psi}{\partial x^2}\right)e^{-ikz} + \left(\frac{\partial^2\psi}{\partial y^2}\right)e^{-ikz}$$
$$+\left(\frac{\partial^2\psi}{\partial z^2}\right)e^{-ikz} - 2ik\frac{\partial\psi}{\partial z}e^{-ikz} - k^2\psi(x, y, z)e^{-ikz}$$
$$+k^2\psi(x, y, z)e^{-ikz} = 0, \tag{16.8}$$

which reduces to

$$\frac{\partial^2\psi}{\partial x^2} + \frac{\partial^2\psi}{\partial y^2} - 2ik\frac{\partial\psi}{\partial z} + \frac{\partial^2\psi}{\partial z^2} = 0. \tag{16.9}$$

If the beam-like solution we are looking for remains paraxial then ψ will only vary slowly with z, so we can neglect $\partial^2\psi/\partial z^2$ and get

$$\frac{\partial^2\psi}{\partial x^2} + \frac{\partial^2\psi}{\partial y^2} - 2ik\frac{\partial\psi}{\partial z} = 0. \tag{16.10}$$

We try as a solution

$$\psi(x, y, z) = \exp\left\{-i\left[P(z) + \frac{k}{2q(z)}r^2\right]\right\}, \tag{16.11}$$

where $r^2 = x^2 + y^2$ is the square of the distance of the point x, y from the axis of propagation. $P(z)$ represents a phase shift factor and $q(z)$ is called the beam parameter. We shall see the significance of these parameters shortly.

Substituting in Eq. (16.10) and using the relations below that follow from Eq. (16.11)

$$\frac{\partial\psi}{\partial x} = -\frac{ik}{2q(z)}\exp\left\{-i\left[P(z) + \frac{k}{2q(z)}r^2\right]\right\}\cdot 2x \tag{16.12}$$

$$\frac{\partial^2\psi}{\partial x^2} = -\frac{ik}{q(z)}\exp\left\{-i\left[P(z) + \frac{k}{2q(z)}r^2\right]\right\} - \frac{k^2}{4q^2(z)}\exp\left\{-i\left[P(z) + \frac{k}{2q(z)}r^2\right]\right\}\cdot 4x^2, \tag{16.13}$$

$$\frac{\partial\psi}{\partial z} = \exp\left\{-i\left[P(z) + \frac{k}{2q(z)}r^2\right]\right\}\left[-i\left(\frac{dP}{dz} - \frac{k}{2q^2}\frac{dq}{dz}r^2\right)\right] \tag{16.14}$$

we get

$$-2k\left[\frac{dP}{dz} + \frac{i}{q(z)}\right] - \left[\frac{k^2}{q^2(z)} - \frac{k^2}{q^2(z)}\frac{dq}{dz}\right](r^2) = 0. \tag{16.15}$$

Since this equation must be true for all values of r the coefficients of different powers of r must be independently equal to zero so

$$\frac{dq}{dz} = 1 \tag{16.16}$$

and

$$\frac{dP}{dz} = \frac{-i}{q(z)}. \tag{16.17}$$

The solution $\psi(x, y, z) = \exp\{-i\left[P(z) + kr^2/2q(z)\right]\}$ is called the *fundamental*

Gaussian beam solution of the time-independent wave equation since its 'intensity' as a function of x and y is

$$UU^* = \psi\psi^* = \exp\left\{-i\left[P(z) + \frac{k}{2q(z)}r^2\right]\right\}\exp\left\{i\left[P^*(z) + \frac{k}{2q^*(z)}r^2\right]\right\}, \tag{16.18}$$

where $P^*(z)$ and $q^*(z)$ are the complex conjugates of $P(z)$ and $q(z)$, respectively, so

$$UU^* = \exp\left\{-i\left[P(z) - P^*(z)\right]\right\}\exp\left\{\frac{-ikr^2}{2}\left[\frac{1}{q(z)} - \frac{1}{q^*(z)}\right]\right\}. \tag{16.19}$$

For convenience we introduce two real beam parameters $R(z)$ and $w(z)$ that are related to $q(z)$ by

$$\frac{1}{q} = \frac{1}{R} - \frac{i\lambda}{\pi w^2}, \tag{16.20}$$

where both R and w depend on z. It is important to note that $\lambda = \lambda_0/n$ is the wavelength *in the medium.* From Eqs. (16.19) and (16.20) above we can see that

$$\begin{aligned} UU^* &\propto \exp\frac{-ikr^2}{2}\left(-\frac{i\lambda}{\pi w^2} - \frac{i\lambda}{\pi w^2}\right) \\ &\propto \exp\frac{-2r^2}{w^2}. \end{aligned} \tag{16.21}$$

Thus, the beam intensity shows a Gaussian dependence on r, the physical significance of $w(z)$ is that it is the distance from the axis at point z where the intensity of the beam has fallen to $1/e^2$ of its peak value on axis and its amplitude to $1/e$ of its axial value: $w(z)$ is called the *spot size.* With these parameters

$$U = \exp\left\{-i\left[kz + P(z) + \frac{kr^2}{2}\left(\frac{1}{R} - \frac{i\lambda}{\pi w^2}\right)\right]\right\}. \tag{16.22}$$

We can integrate Eqs. (16.16) and (16.17). From Eq. (16.16)

$$q = q_0 + z, \tag{16.23}$$

where q_0 is a constant of integration that gives the value of the beam parameter at plane $z = 0$. From Eq. (16.17)

$$\frac{dP(z)}{dz} = \frac{-i}{q(z)} = \frac{-i}{q_0 + z}, \tag{16.24}$$

so

$$P(z) = -i[\ln(z + q_0)] + (\theta + i\ln q_0), \tag{16.25}$$

where the constant of integration, $(\theta + i\ln q_0)$, is written in this way so that substituting from Eqs. (16.23) and (16.25) in Eq. (16.22) we get

$$U = \exp\left\{-i\left[kz - i\ln\left(1 + \frac{z}{q_0}\right) + \theta + \frac{kr^2}{2}\left(\frac{1}{R} - \frac{i\lambda}{\pi w^2}\right)\right]\right\}. \tag{16.26}$$

The factor $e^{-i\theta}$ is only a constant phase factor that we can arbitrarily set to zero and get

$$U = \exp\left\{-i\left[kz - i\ln\left(1 + \frac{z}{q_0}\right) + \frac{kr^2}{2}\left(\frac{1}{R} - \frac{i\lambda}{\pi w^2}\right)\right]\right\}. \tag{16.27}$$

The radial variation in phase of this field for a particular value of z is

$$\phi(z) = kz + \frac{kr^2}{2R} - \mathscr{R}\left[i\ln\left(1 + \frac{z}{q_o}\right)\right]. \tag{16.28}$$

There is no radial phase variation if $R \to \infty$. We choose the value of z where this occurs as our origin $z = 0$. We call this location the *beamwaist*; the complex beam parameter here has the value

$$\frac{1}{q_0} = -i\frac{\lambda}{\pi w_0^2}, \tag{16.29}$$

where $w_0 = w(0)$ is called the *minimum spot size.* At an arbitrary point z

$$q = \frac{i\pi w_0^2}{\lambda} + z, \tag{16.30}$$

and using $1/q = 1/R - i\lambda/\pi w^2$ we have the relationship for the *spot size* $w(z)$ at position z

$$w^2(z) = w_0^2\left[1 + \left(\frac{\lambda z}{\pi w_0^2}\right)^2\right]. \tag{16.31}$$

So w_0 is clearly the *minimum* spot size. The value of R at position z is

$$R(z) = z\left[1 + \left(\frac{\pi w_0^2}{\lambda z}\right)^2\right]. \tag{16.32}$$

We shall see shortly that $R(z)$ can be identified as the radius of curvature of the phase front at z. From Eq. (16.31) we can see that $w(z)$ expands with distance z along the axis along a hyperbola that is inclined to the axis at an angle $\theta_{beam} = \tan^{-1}(\lambda/\pi w_0)$. This is a small angle if the beam is to have small divergence, in which case

$$\theta_{beam} \simeq \lambda/\pi w_0, \tag{16.33}$$

where θ_{beam} is the half angle of the diverging beam shown in Fig. (16.2). From Eq. (16.27) it is clear that the surfaces of constant phase are those surfaces which satisfy

$$kz + \mathscr{R}\left[-i\ln\left(1 + \frac{z}{q_0}\right)\right] + \frac{kr^2}{2}\left[\frac{1}{R}\right] = \text{constant}, \tag{16.34}$$

which gives

$$k\left[z + \frac{r^2}{2R}\right] = \text{constant} - \mathscr{R}\left[-i\ln\left(1 + \frac{z}{q_0}\right)\right]. \tag{16.35}$$

Now,

$$\ln\left(1 + \frac{z}{q_0}\right) = \ln\left(1 - \frac{iz\lambda}{\pi w_0^2}\right) = \ln\sqrt{1 + \left(\frac{\lambda z}{\pi w_0^2}\right)^2} - i\tan^{-1}\left(\frac{\lambda z}{\pi w_0^2}\right), \tag{16.36}$$

where we have used the relation

$$\ln(x + iy) = \ln\sqrt{x^2 + y^2} + i\tan^{-1}(y/x). \tag{16.37}$$

Fig. 16.2. Characteristics of a Gaussian beam: (a) Gaussian field distribution along a phase front; (b) beam contour showing the hyperbolic envelope of beam expansion contour and asymptotic angle θ.

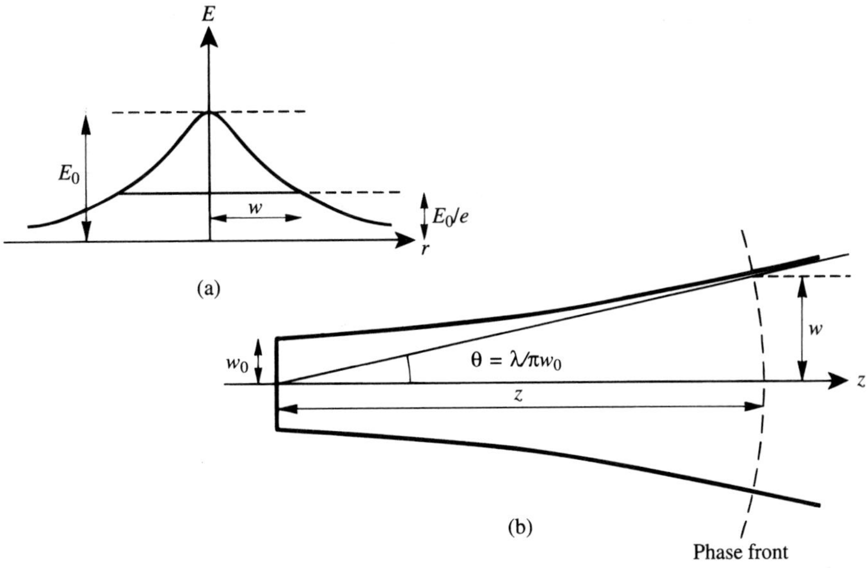

So, from Eq. (16.37)

$$\mathscr{R}\left[-i\ln\left(1+\frac{z}{q_0}\right)\right]=\tan^{-1}\left(\frac{\lambda z}{\pi w_0^2}\right). \tag{16.38}$$

The term $kz = 2\pi z/\lambda$ in Eq. (16.35) is very large, except very close to the beamwaist and for long wavelengths, it generally dominates the z dependence of Eq. (16.35). Also, $\tan^{-1}(\lambda z/\pi w_0^2)$ at most approaches a value of $\pi/2$, so we can write

$$k\left[z+\frac{r^2}{2R}\right]=\text{constant} \tag{16.39}$$

as the equation of the surface of constant phase at z, which can be rewritten

$$z=\frac{\text{constant}}{\text{k}}-\frac{r^2}{2R}. \tag{16.40}$$

This is a parabola, which for $r^2 \ll z^2$ is a very close approximation to a spherical surface of which $R(z)$ is the radius of curvature.

The complex phase shift a distance z from the point where q is purely imaginary, the point called the beamwaist, is found from Eq. (16.17),

$$\frac{dP(z)}{dz}=\frac{-i}{q}=\frac{-i}{z+i\pi w_0^2/\lambda}. \tag{16.41}$$

Thus,

$$P(z)=-i\ln\left[1-i\left(\frac{\lambda z}{\pi w_0^2}\right)\right]=-i\ln\sqrt{1+\left(\frac{\lambda z}{\pi w_0^2}\right)^2}-\tan^{-1}\left(\frac{\lambda z}{\pi w_0^2}\right), \tag{16.42}$$

where the constant of integration has been chosen so that $P(z) = 0$ at $z = 0$.

The real part of this phase shift can be written as

$$\Phi=\tan^{-1}\left(\frac{\lambda z}{\pi w_0^2}\right), \tag{16.43}$$

which marks a distinction between this Gaussian beam and a plane wave. The imaginary part is, from Eqs. (16.31) and (16.42),

$$\mathscr{I}[P(z)] = -i\ln\left[w(z)/w_0\right] \tag{16.44}$$

This imaginary part gives a real intensity dependence of the form $w_0^2/w^2(z)$ on axis, which we would expect because of the expansion of the beam. For example, part of the 'phase factor' in Eq. (16.11) gives

$$e^{-i\mathscr{I}[P(z)]} = e^{-\ln(w/w_0)} = \frac{w_0}{w}. \tag{16.45}$$

Thus, from Eqs. (16.11), (16.20), and (16.44) we can write the spatial dependence of our Gaussian beam

$$U = \psi(x, y, z)e^{-ikz} \tag{16.46}$$

as

$$U(r, z) = \frac{w_0}{w}\exp\left[-i(kz - \Phi) - r^2\left(\frac{1}{w^2} + \frac{ik}{2R}\right)\right]. \tag{16.47}$$

This field distribution is called the *fundamental* or TEM_{00} Gaussian mode. Its amplitude distribution and beam contour in planes of constant z and planes lying in the z axis, respectively, are shown in Fig. (16.2).

In any plane the radial intensity distribution of the beam can be written as

$$I(r) = I_0 e^{-2r^2/w^2}, \tag{16.48}$$

where I_0 is the axial intensity.

16.3 Higher Order Modes

In the preceding discussion, only one possible solution for $\psi(x, y, z)$ has been considered, namely the simplest solution where $\psi(x, y, z)$ gives a beam with a Gaussian dependence of intensity as a function of its distance from the axis: the width of this Gaussian beam changes as the beam propagates along its axis. However, higher order beam solutions of Eq. (16.10) are possible.

16.3.1 Beam Modes with Cartesian Symmetry

For a system with Cartesian symmetry we can try for a solution of the form

$$\psi(x, y, z) = g\left(\frac{x}{w}\right)h\left(\frac{y}{w}\right)\exp\left\{-i\left[P(z) + \frac{k}{2q(z)}(x^2 + y^2)\right]\right\}, \tag{16.49}$$

where g and h are functions of x and z, and y and z, respectively, $w(z)$ is a Gaussian beam parameter and $P(z)$, $q(z)$ are the beam parameters used previously.

For real functions g and h, $\psi(x, y, z)$ represents a beam whose intensity patterns scale according to a Gaussian beam parameter $w(z)$. If we substitute this higher order solution into Eq. (16.10) then we find that the solutions have g and h obeying the following relation.

$$gh = H_m\left(\sqrt{2}\frac{x}{w}\right)H_n\left(\sqrt{2}\frac{y}{w}\right), \tag{16.50}$$

where H_m, H_n are solutions of the differential equation

$$\frac{d^2H_m}{dx^2} - 2x\frac{dH_m}{dx} + 2mH_m = 0. \tag{16.51}$$

The solution to Eq. (16.51) is the Hermite polynomial H_m of order m: m and n are called the transverse mode numbers. Some of the low order Hermite polynomials are:

$$H_0(x) = 1;\ H_1(x) = x;\ H_2(x) = 4x^2 - 2;\ H_3(x) = 8x^3 - 12x. \tag{16.52}$$

The overall amplitude behavior of this higher order beam is

$$\begin{aligned} U_{m,n}(r,z) =& \frac{w_0}{w} H_m\left(\sqrt{2}\frac{x}{w}\right) H_m\left(\sqrt{2}\frac{y}{w}\right) \exp\Big[-i(kz - \Phi) \\ & -r^2\left(\frac{1}{w^2} + \frac{ik}{2R}\right)\Big]. \end{aligned} \tag{16.53}$$

The parameters $R(z)$ and $w(z)$ are the same for all these solutions. However, Φ depends on m, n, and z as

$$\Phi(m,n,z) = (m+n+1)\tan^{-1}\left(\frac{\lambda z}{\pi w_0^2}\right). \tag{16.54}$$

Since this Gaussian beam involves a product of Hermite and Gaussian functions it is called a *Hermite–Gaussian* (HG) mode and has the familiar intensity pattern observed in the output of many lasers, as shown in Fig. (16.3). The mode is designated a TEM_{mn} HG mode. For example, in the plane $z = 0$ the electric field distribution of the TEM_{mn} mode is, if the wave is polarized in the x direction,

$$E_{m,n}(x,y) = E_0 H_m\left(\sqrt{2}\frac{x}{w_0}\right) H_n\left(\sqrt{2}\frac{y}{w_0}\right) \exp\left(-\frac{x^2+y^2}{w_0^2}\right). \tag{16.55}$$

The lateral intensity variation of the mode is

$$I(x,y) \propto |E_{m,n}(x,z)|^2 \tag{16.56}$$

and in surfaces of constant phase

$$E_{m,n}(x,y) = E_0 \frac{w_0}{w} H_m\left(\sqrt{2}\frac{x}{w}\right) H_n\left(\sqrt{2}\frac{y}{w}\right) \exp\left(-\frac{x^2+y^2}{w^2}\right), \tag{16.57}$$

where it is important to remember that w is a function of z. Some examples of the field and intensity variations described by Eqs. (16.55) and (16.56) are shown in Figs. (16.4) and (16.5). The phase variation

$$\Phi(m,n,z) = (m+n+1)\tan^{-1}\left(\frac{\lambda z}{\pi w_0^2}\right) \tag{16.58}$$

means that the phase velocity increases with mode number so that in a resonator different transverse modes have different resonant frequencies.

16.3.2 Cylindrically Symmetric Higher Order Beams

In this case we try for a solution of Eq. (16.10) in the form

$$\psi = g\left(\frac{r}{w}\right) \exp\left\{-i\left(P(z) + \frac{k}{2q(z)}r^2 + \ell\phi\right)\right\} \tag{16.59}$$

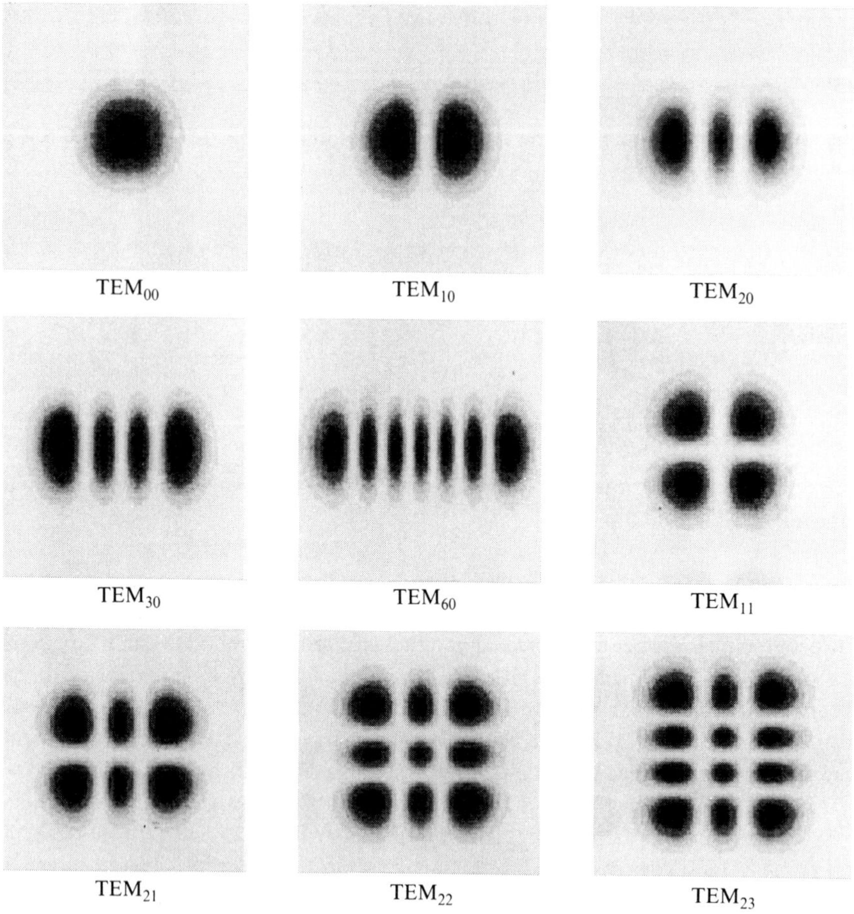

Fig. 16.3. Hermite–Gaussian laser mode intensity patterns.

and find that

$$g = \left(\sqrt{2}\frac{r}{w}\right)^{\ell} L_p^{\ell}\left(2\frac{r^2}{w^2}\right), \tag{16.60}$$

where L_p^{ℓ} is an associated Laguerre polynomial, p and ℓ are the radial and angular mode numbers and P_p^{ℓ} obeys the differential equation

$$x\frac{d^2L_p^{\ell}}{dx^2} + (\ell + 1 - x)\frac{dL_p^{\ell}}{dx} + pL_p^{\ell} = 0. \tag{16.61}$$

Some of the associated Laguerre polynomials of low order are:

$$L_0^{\ell}(x) = 1, \tag{16.62}$$

$$L_1^{\ell}(x) = \ell + 1 - x, \tag{16.63}$$

$$L_2^{\ell}(x) = \frac{1}{2}(\ell + 1)(\ell + 2) - (\ell + 2)x + \frac{1}{2}x^2. \tag{16.64}$$

These are called Laguerre–Gaussian (LG) modes. Some examples of their intensity mode patterns are given in Fig. (16.6). These 'axisymmetric' modes are designated

TEM_{pl}. Figs. (16.7) and (16.8) give some three-dimensional representations of their field and intensity distributions. The angular variation of these patterns is shown explicitly by looking at the magnitude of the electric field, which from Eq. (16.59) is

$$|E(r,\phi)| = E_0 \left(\sqrt{2}\frac{r}{w}\right)^{\ell} L_p^{\ell}\left(2\frac{r^2}{w^2}\right) e^{-r^2/w^2}\cos(\ell\phi). \tag{16.65}$$

A true axisymmetric intensity distribution frequently results in practice because the choice of origin for ϕ is arbitrary. The modes with $\phi = \phi_0$ or $\phi = \phi_0 + \pi/2$ give identical patterns, which superimposed have cylindrical symmetry. The best example of this is the TEM_{01}^* mode, called the *donut* mode. This mode can be regarded as the superposition of two LG modes or the superposition of the two HG modes TEM_{01} and TEM_{10}. As is the case for beams with cartesian symmetry the beam parameters $w(z)$ and $R(z)$ are the same for all cylindrical modes. The phase factor Φ for these cylindrically symmetric modes is

$$\Phi(p,\ell,z) = (2p+\ell+1)\tan^{-1}\left(\frac{\lambda z}{\pi w_0^2}\right). \tag{16.66}$$

The question could be asked – why should lasers that frequently have apparently cylindrical symmetry both in construction and excitation geometry generate transverse modes with cartesian rather than radial symmetry? The answer is that this usually results because there is indeed some feature of laser construction or method of excitation that removes an apparent equivalence of all radial directions. For example, in a laser with Brewster windows the preferred polarization orientation imposes a directional constraint that forms modes of cartesian (HG) rather than radial (LG) symmetry. The simplest way to observe the HG modes in the laboratory is to place a thin wire inside the laser cavity. The laser will then choose to operate in a transverse mode that has a nodal line in the location of the intracavity obstruction. Adjustment of the laser mirrors to slightly different orientations will usually change the output transverse mode, although if the mirrors are adjusted too far from optimum the laser will go out.

16.4 The Transformation of a Gaussian Beam by a Lens

A lens can be used to focus a laser beam to a small spot, or systems of lenses may be used to expand the beam and recollimate it. An ideal thin lens in such an application will not change the transverse mode intensity pattern measured at the lens but it will alter the radius of curvature of the phase fronts of the beam.

We have seen that a Gaussian laser beam of whatever order is characterized by a complex beam parameter $q(z)$ which changes as the beam propagates in an isotropic, homogeneous material according to $q(z) = q_0 + z$, where

$$q_0 = \frac{i\pi w_0^2}{\lambda} \quad \text{and} \quad \frac{1}{q(z)} = \frac{1}{R(z)} - \frac{i\lambda}{\pi w^2}. \tag{16.67}$$

$R(z)$ is the radius of curvature of the (approximately) spherical phase front at z and is given by

$$R(z) = z\left[1+\left(\frac{\pi w_0^2}{\lambda z}\right)^2\right]. \tag{16.68}$$

This Gaussian beam becomes a true spherical wave as $w \to \infty$. Now, a spherical

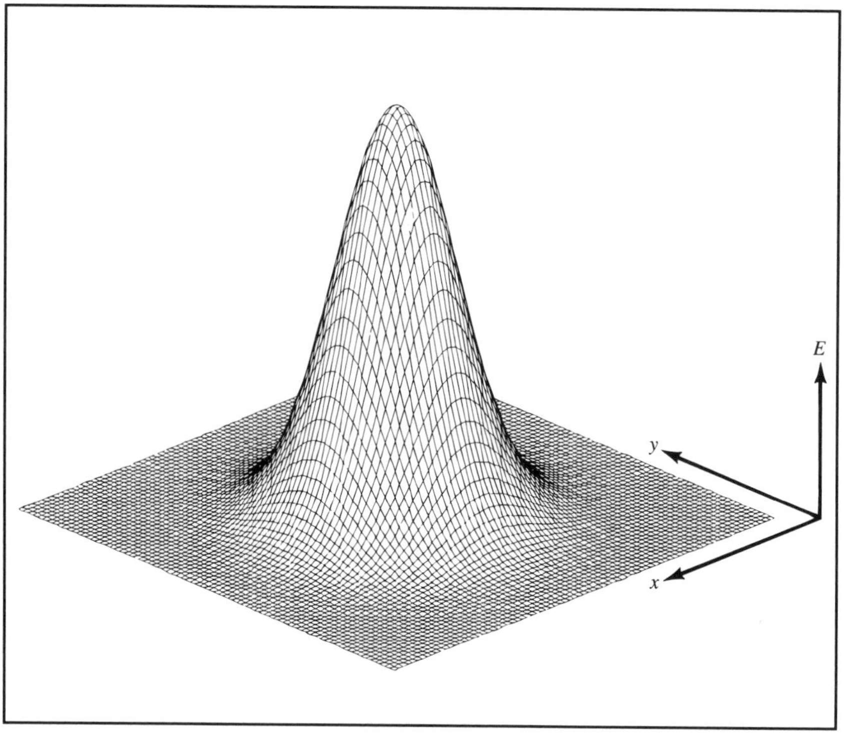

(a)

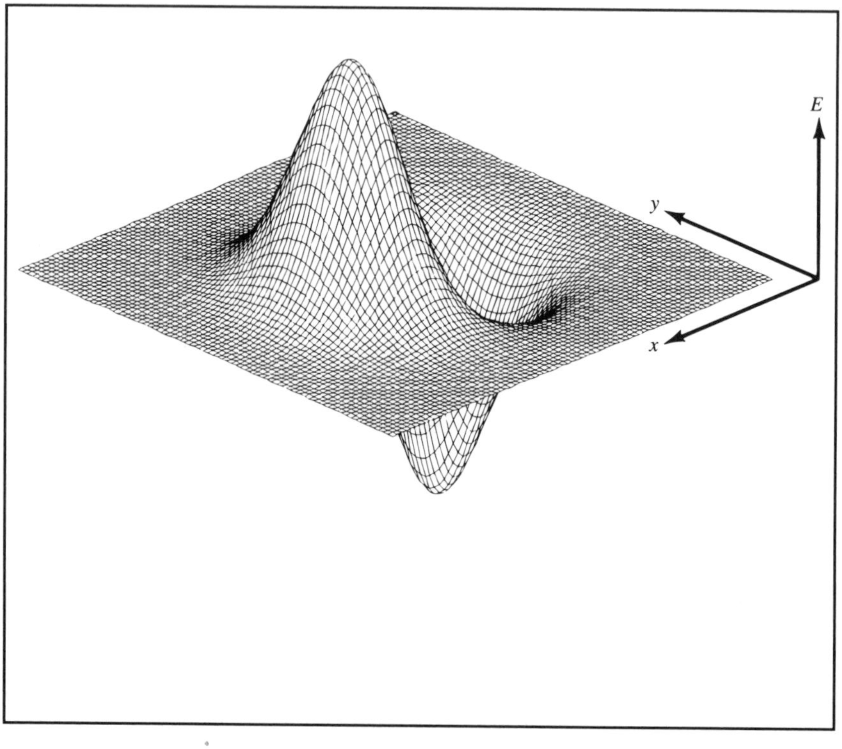

(b)

Fig. 16.4.
Three-dimensional plots of the field distribution of some Hermite–Gaussian laser modes.

Fig. 16.4. (*cont.*)

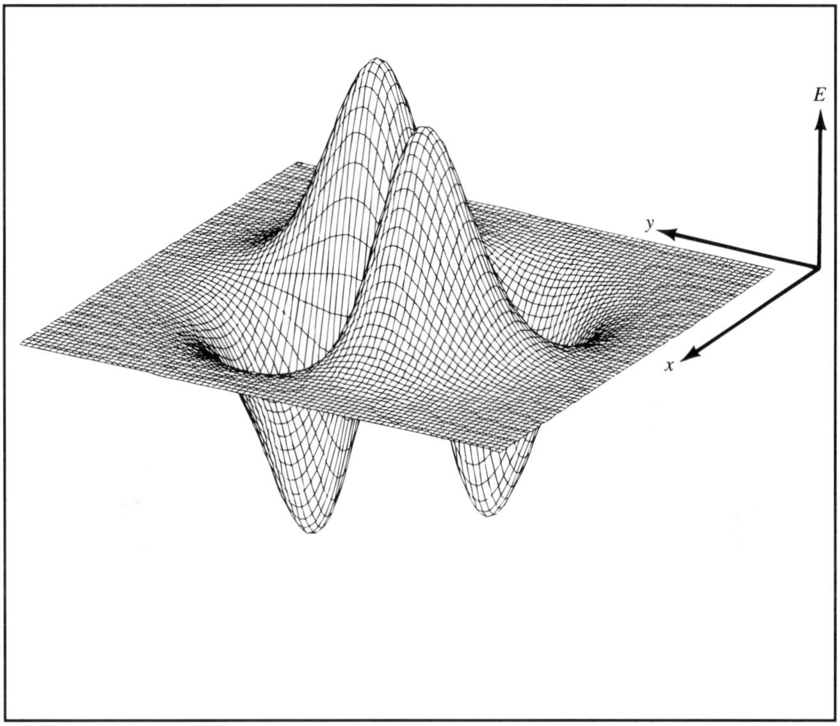

(c)

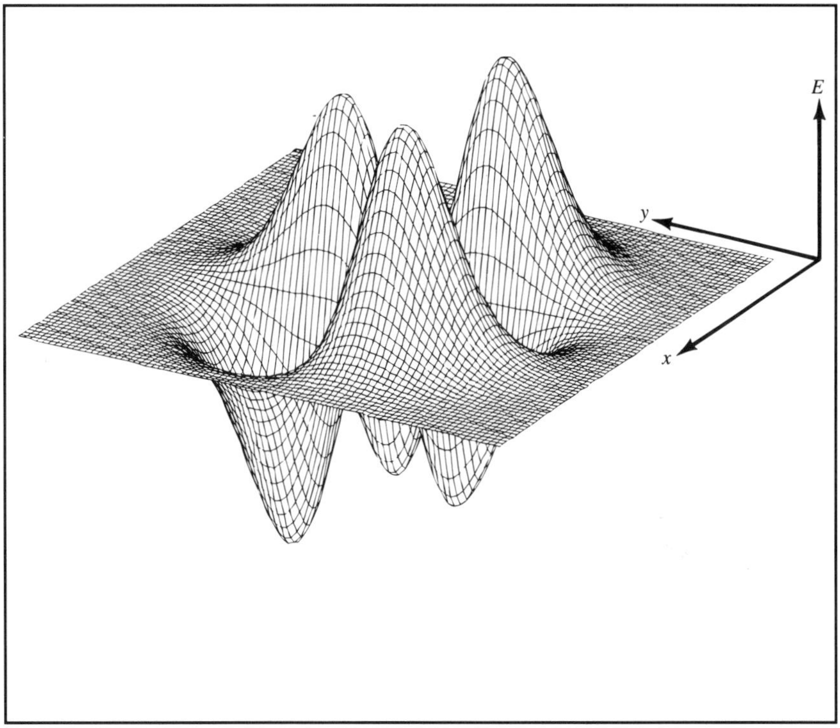

(d)

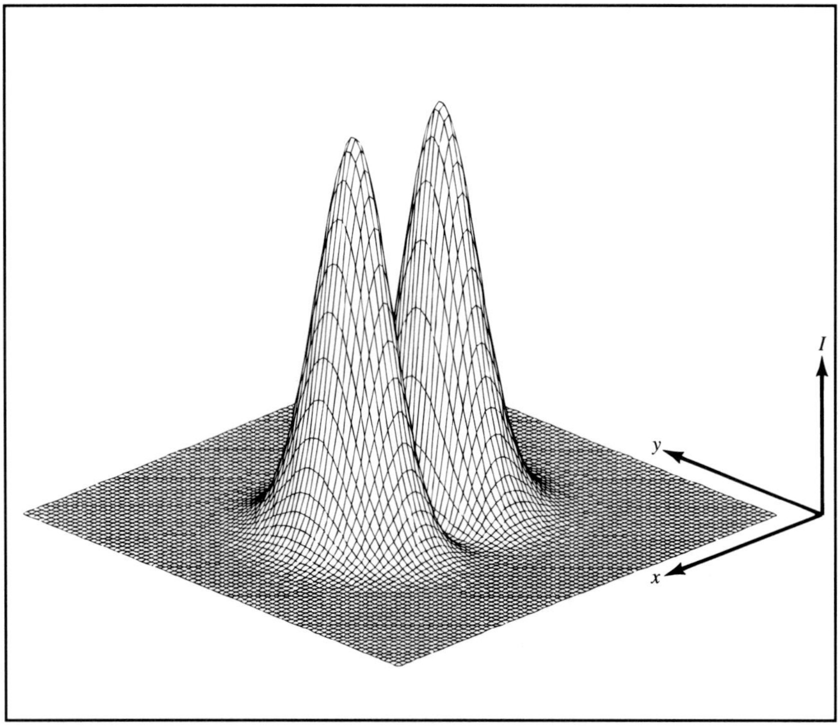

(a)

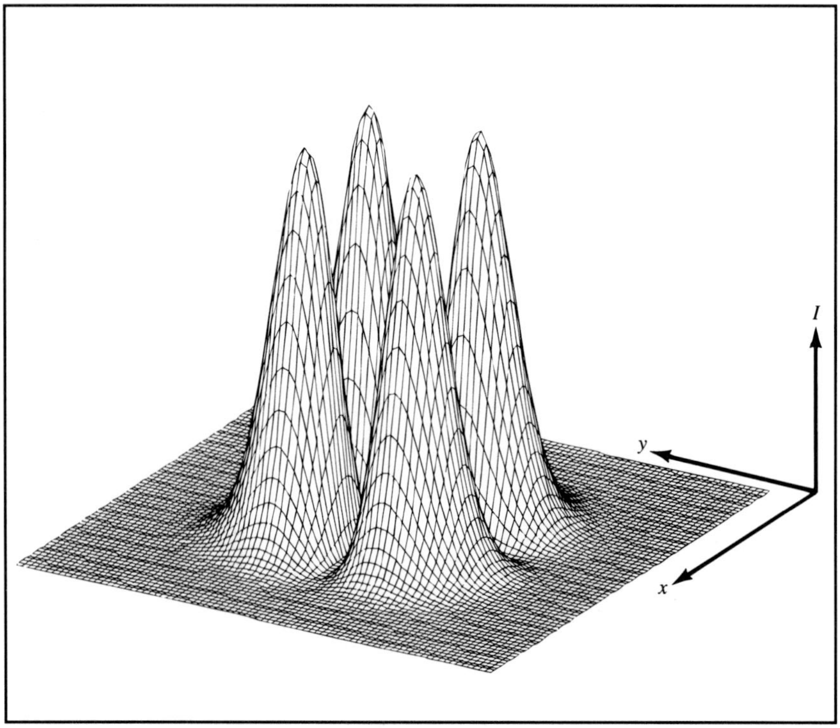

(b)

Fig. 16.5.
Three-dimensional plots of the intensity distribution of some Hermite–Gaussian laser modes.

Fig. 16.5. (*cont.*)

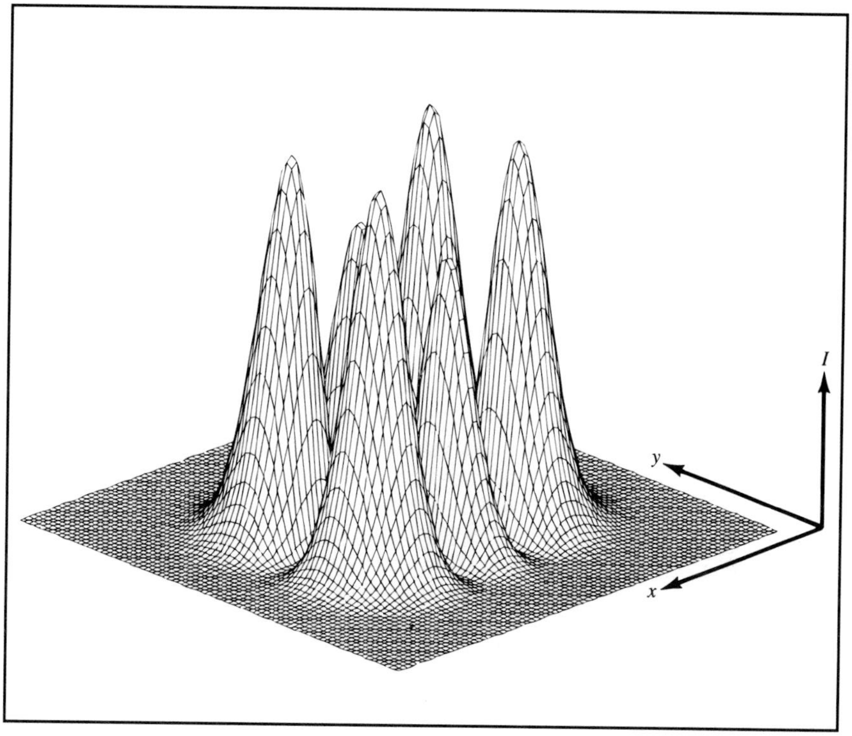

(c)

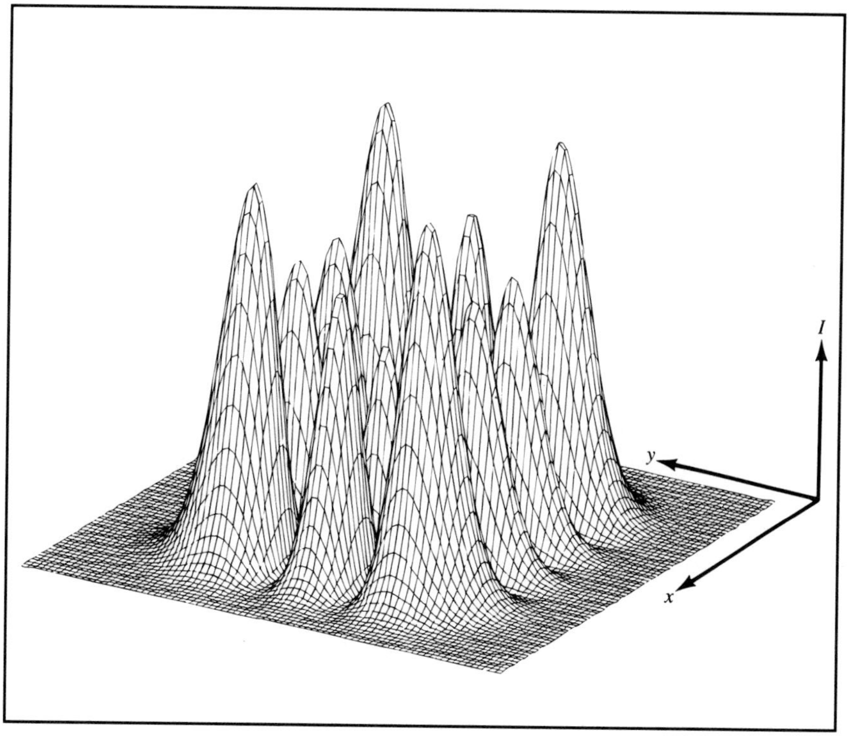

(d)

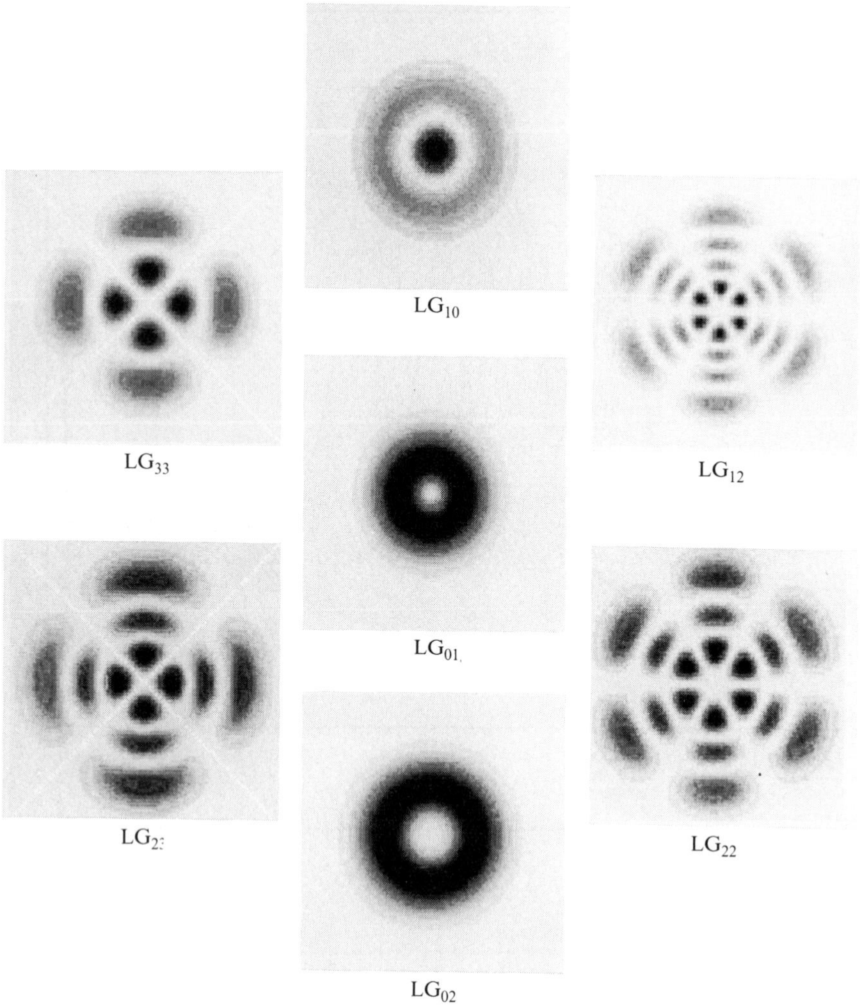

Fig. 16.6.
Laguerre–Gaussian laser mode intensity patterns.

wave changes its radius of curvature as it propagates according to $R(z) = R_0 + z$ where R_0 is its radius of curvature at $z = 0$. So the complex beam parameter of a Gaussian wave changes in just the same way as it propagates as does the radius of curvature of a spherical wave.

When a Gaussian beam strikes a thin lens the spot size, which measures the transverse width of the beam intensity distribution, is unchanged at the lens. However, the radius of curvature of its wavefront is altered in just the same way as a spherical wave. If R_1 and R_2 are the radii of curvature of the incoming and outgoing waves measured at the lens, as shown in Fig. (16.9), then as in the case of a true spherical wave

$$\frac{1}{R_2} = \frac{1}{R_1} - \frac{1}{f}. \tag{16.69}$$

So, for the change of the overall beam parameter, since w is unchanged at the lens, we have the following relationship between the beam parameters, *measured at the lens*

$$\frac{1}{q_2} = \frac{1}{q_1} - \frac{1}{f}. \tag{16.70}$$

If instead q_1 and q_2 are measured at distances d_1 and d_2 from the lens as shown in Fig. (16.10) then at the lens

$$\frac{1}{(q_2)_L} = \frac{1}{(q_1)_L} - \frac{1}{f} \tag{16.71}$$

and since $(q_1)_L = q_1 + d_1$ and $(q_2)_L = q_2 - d_2$ we have

$$\frac{1}{q_2 - d_2} = \frac{1}{q_1 + d_1} - \frac{1}{f}, \tag{16.72}$$

which gives

$$q_2 = \frac{(1 - d_2/f)q_1 + (d_1 + d_2 - d_1 d_2/f)}{(-q_1/f) + (1 - d_1/f)}. \tag{16.73}$$

If the lens is placed at the beamwaist of the input beam then

$$\frac{1}{(q_1)_L} = -i\frac{\lambda}{\pi w_0^2}, \tag{16.74}$$

where w_0 is the spot size of the input beam. The beam parameter immediately after the lens is

$$\frac{1}{(q_2)_L} = -\frac{i\lambda}{\pi w_0^2} - \frac{1}{f}, \tag{16.75}$$

which can be rewritten as

$$q_{2L} = \frac{-\pi w_0^2 f}{i\lambda f + \pi w_0^2}. \tag{16.76}$$

At a distance d_2 after the lens

$$q_2 = \frac{-\pi w_0^2 f}{i\lambda f + \pi w_0^2} + d_2, \tag{16.77}$$

which can be rearranged to give

$$\frac{1}{q_2} = \frac{(d_2 - f) + (\lambda f/\pi w_0^2)^2 d_2 - i(\lambda f^2/\pi w_0^2)}{(d_2 - f)^2 + (\lambda f d_2)^2/\pi w_0^2}. \tag{16.78}$$

The location of the new beamwaist (which is where the beam will be focused to its new minimum spot size) is determined by the condition $\mathscr{R}(1/q_2) = 0$: namely

$$(d_2 - f) + \left(\frac{\lambda f}{\pi w_0^2}\right)^2 d_2 = 0, \tag{16.79}$$

which gives

$$d_2 = \frac{f}{1 + (\lambda f/\pi w_0^2)^2}. \tag{16.80}$$

Almost always $(\lambda f/\pi w_0^2) \ll 1$ so the new beamwaist is very close to the focal point of the lens.

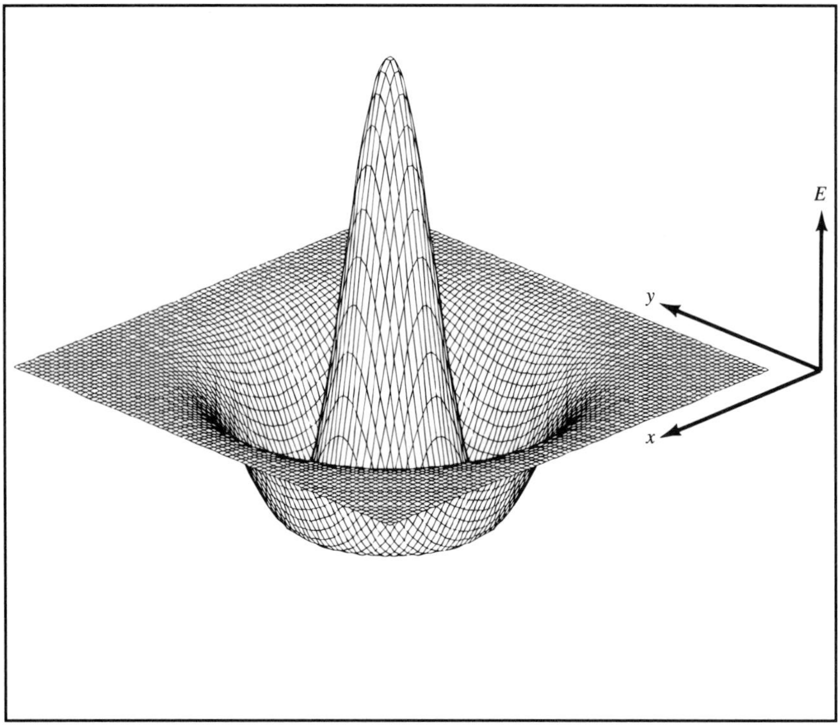

(a)

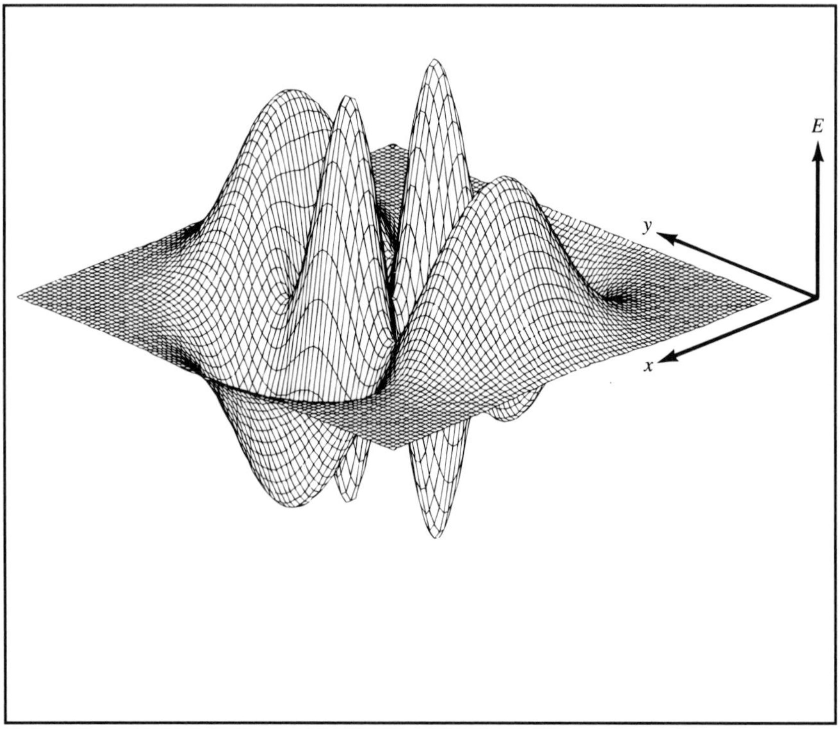

(b)

Fig. 16.7.
Three-dimensional plots of the field distribution of some Laguerre–Gaussian laser modes.

Fig. 16.7. (*cont.*)

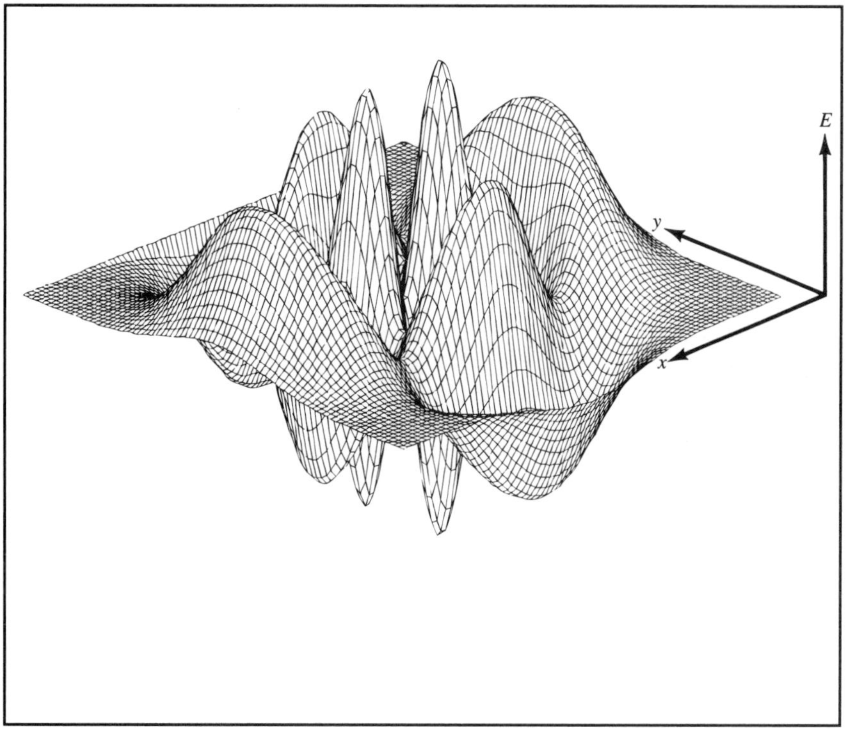

(c)

Examination of the imaginary part of the right hand side of Eq. (16.78) reveals that the spot size of the focused beam is

$$w_2 = \frac{f\lambda/\pi w_0}{\sqrt{1+(\lambda f/\pi w_0^2)^2}}, \tag{16.81}$$

which provided $\lambda f/\pi w_0^2 << 1$, as is frequently the case, gives

$$w_2 \simeq f\theta, \tag{16.82}$$

where $\theta = \lambda/\pi w_0$ is the half angle of divergence of the input beam.

It is straightforward to show that in general the minimum spot size of a TEM_{00} Gaussian beam focused by a lens is

$$w_f = \frac{f\lambda}{\pi w_1}\left[\left(1-\frac{f}{R_1}\right)^2 + \left(\frac{\lambda f}{\pi w_1^2}\right)^2\right]^{-1/2}, \tag{16.83}$$

where w_1 and R_1 are the laser-beam spot size and radius of curvature at the input face of the lens. If the lens is placed very close to or very far from the beamwaist of the beam being focused, Eq. (16.83) reduces to

$$w_f = f\theta_B, \tag{16.84}$$

where θ_B is the beam divergence at the input face of the lens.

Thus, if the focusing lens is placed a great distance from, or very close to, the

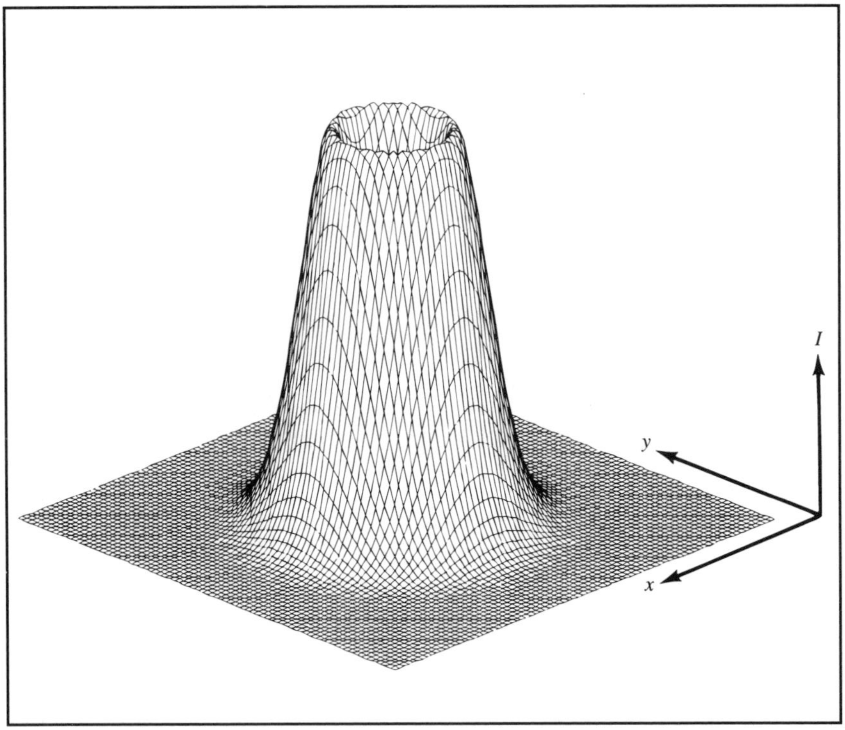

(a)

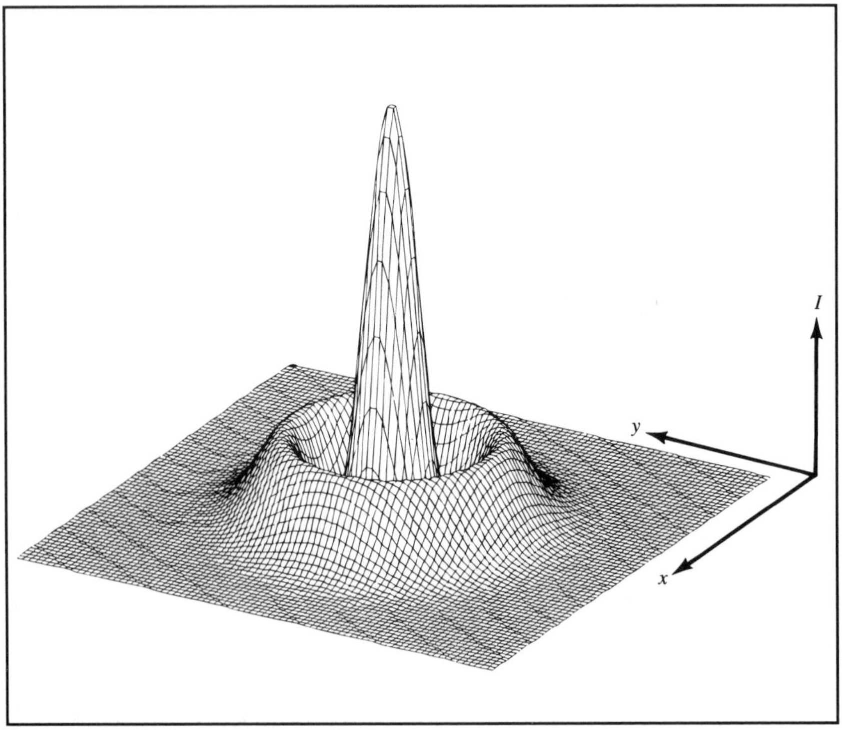

(b)

Fig. 16.8.
Three-dimensional plots of the intensity distribution of some Laguerre–Gaussian laser modes.

Fig. 16.8. (*cont.*)

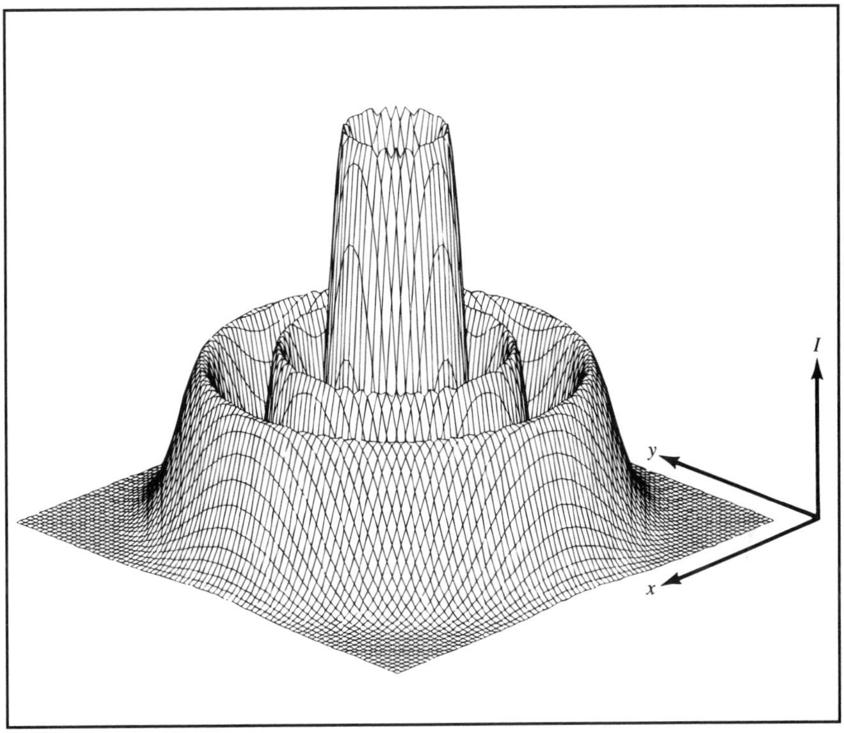

(c)

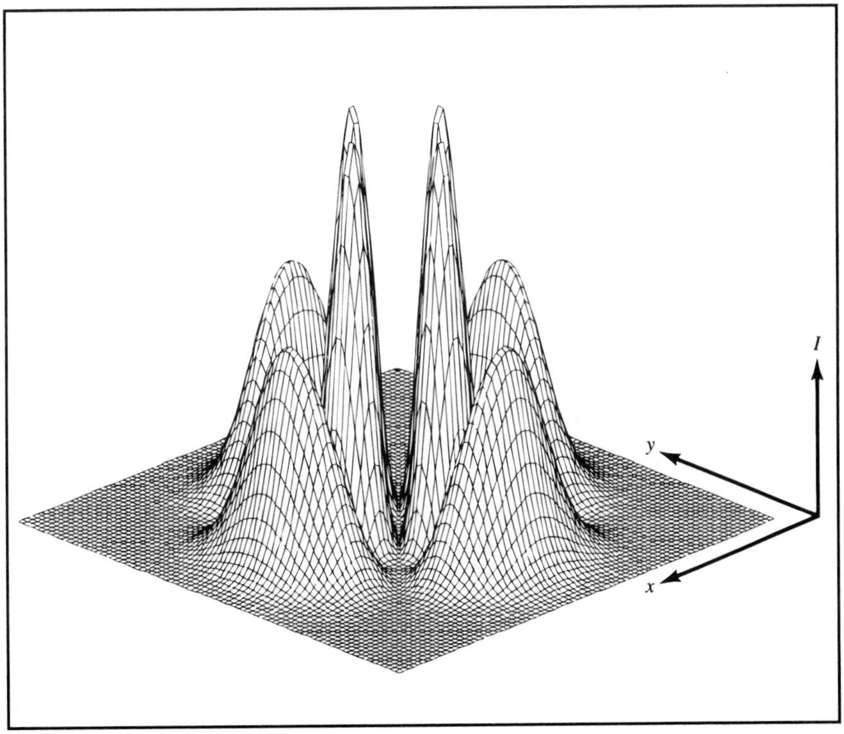

(d)

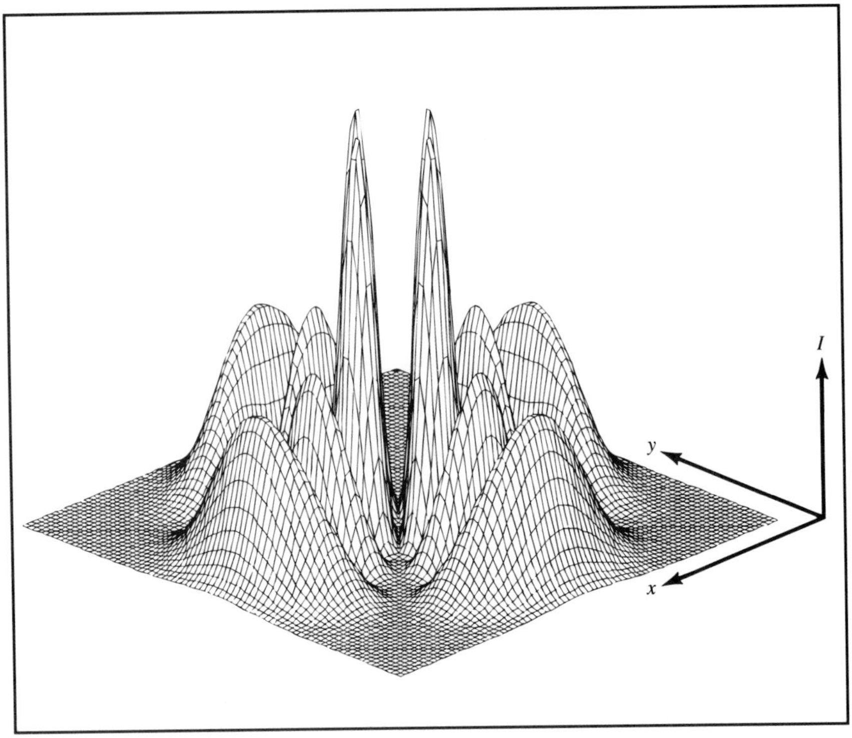

(e)

Fig. 16.8. (*cont.*)

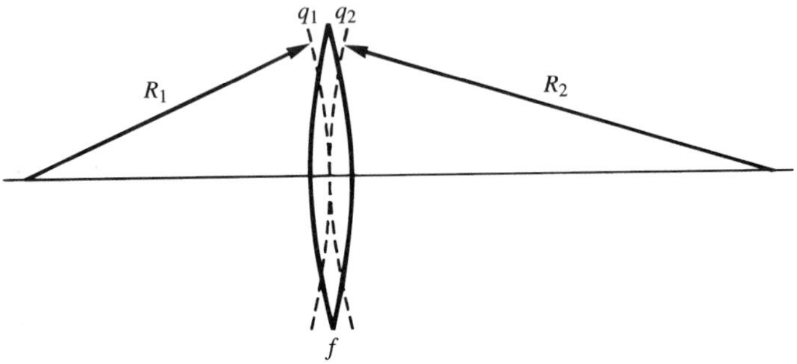

Fig. 16.9. Transformation of a Gaussian beam by a lens. The input and output phase front curvatures and beam parameters are indicated.

input beamwaist then the size of the focused spot is always close to $f\theta$. If, in fact, the beam incident on the lens were a plane wave, then the finite size of the lens (radius r) would be the dominant factor in determining the size of the focused spot. We can take the 'spot size' of the plane wave as approximately the radius of the lens, and from Eq. (16.67), setting $w_0 = r$, the radius of the lens, we get

$$w_2 = \frac{\lambda f}{\pi r}. \tag{16.85}$$

This focused spot cannot be smaller than a certain size since for any lens the value of r clearly has to satisfy the condition $r \leq f$

Thus, the minimum focal spot size that can result when a plane wave is focused

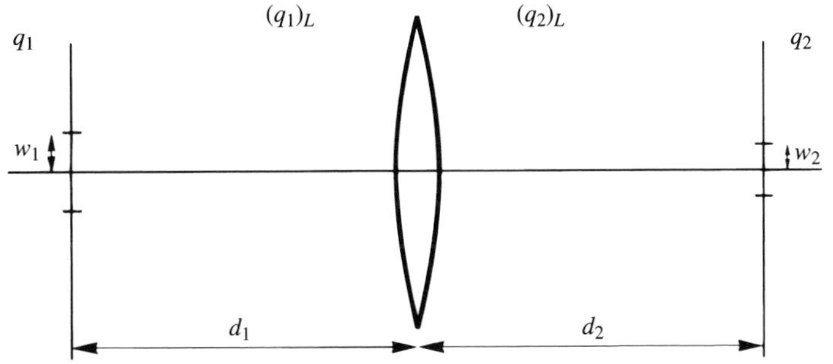

Fig. 16.10. Optical system diagram for transformation of a Gaussian beam from an input to an output plane.

by a lens is

$$(w_2)_{min} \sim \frac{\lambda}{\pi}. \tag{16.86}$$

We do not have an equality sign in Eq. (16.86) because a lens for which $r = f$ does not qualify as a thin lens, so Eq. (16.85) does not really hold exactly.

We can see from Eq. (16.67) that in order to focus a laser beam to a small spot we must either use a lens of very short focal length or a beam of small beam divergence. We cannot, however, reduce the focal length of the focusing lens indefinitely, as when the focal length does not satisfy $f_1 \ll w_0, w_1$, the lens ceases to satisfy our definition of it as a thin lens ($f \ll r$). To obtain a laser beam of small divergence we must expand and recollimate the beam; there are two simple ways of doing this:

(i) With a Galilean telescope as shown in Fig. (16.11). The expansion ratio for this arrangement is $-f_2/f_1$, where it should be noted that the focal length f_1 of the diverging input lens is negative. This type of arrangement has the advantage that the laser beam is not brought to a focus within the telescope, so the arrangement is very suitable for the expansion of high power laser beams. Very high power beams can cause air breakdown if brought to a focus, which considerably reduces the energy transmission through the system.

(ii) With an astronomical telescope, as shown in Fig. (16.12). The expansion ratio for this arrangement is f_2/f_1. The beam is brought to a focus within the telescope, which can be a disadvantage when expanding high intensity laser beams because air breakdown at the common focal point can occur. The telescope can be evacuated or filled to high pressure to help prevent such breakdown occurring. An advantage of this system is that by placing a small circular aperture at the common focal point it is possible to obtain an output beam with a smoother radial intensity profile than the input beam. The aperture should be chosen to have a radius about the same size, or slightly larger than, the spot size of the focused Gaussian beam at the focal point. This process is called spatial filtering and is illustrated in Fig. (16.13). In both the astronomical and Galilean telescopes, spherical aberration† is reduced by the use of bispherical lenses. This distributes the focusing power over the maximum number of surfaces.

† The lens aberration in which rays of light travelling parallel to the axis that strike a lens at different distances from the axis are brought to different focal points.

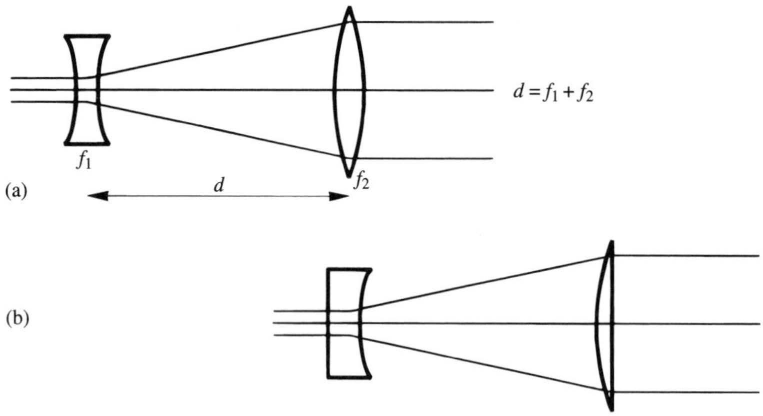

Fig. 16.11. Galilean telescope laser beam expanders: (a) using bispherical lenses; (b) using plano-spherical lenses.

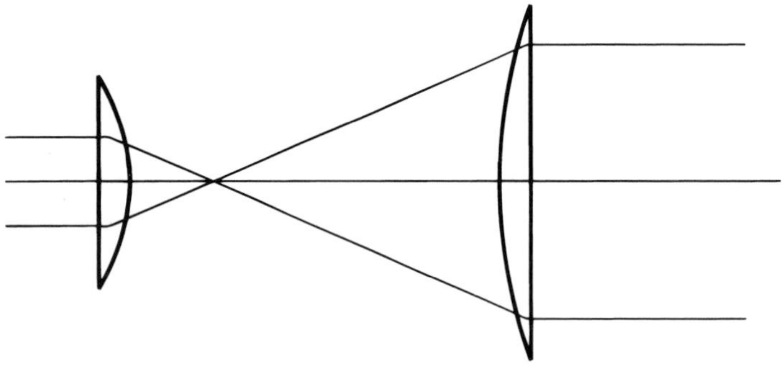

Fig. 16.12. Astronomical telescope laser-beam expander in which the beam is brought to an internal focus.

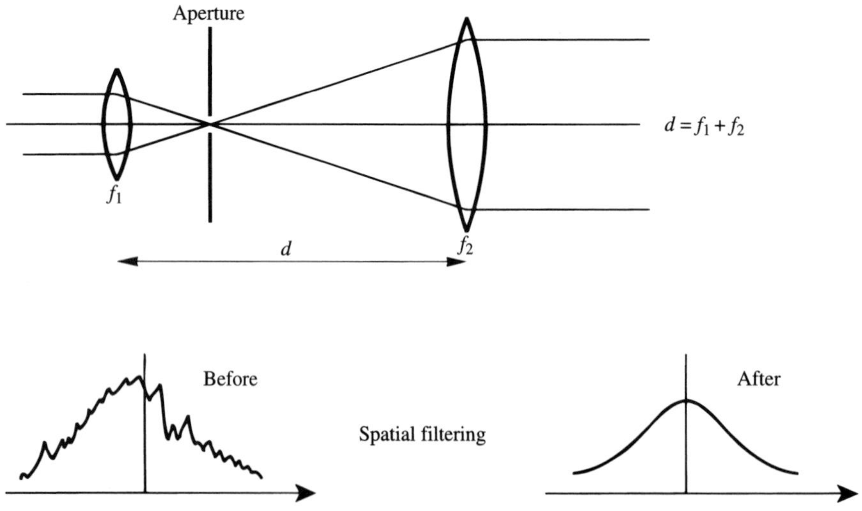

Fig. 16.13. Beam expander incorporating a spatial filter for smoothing the intensity profile.

16.5 Transformation of Gaussian Beams by General Optical Systems

As we have seen, the complex beam parameter of a Gaussian beam is transformed by a thin lens in just the same way as the radius of curvature of a spherical wave. Now, the transformation of the radius of curvature of a spherical wave by an optical system with transfer matrix $\begin{pmatrix} A & B \\ C & D \end{pmatrix}$ obeys

$$R_2 = \frac{AR_1 + B}{CR_1 + D}. \tag{16.87}$$

Thus, by continuing to draw a parallel between the q of a Gaussian beam and the R of a spherical wave, we can postulate that the transformation of the complex beam parameter obeys a similar relation, i.e.,

$$q_2 = \frac{Aq_1 + B}{Cq_1 + D}, \tag{16.88}$$

where $\begin{pmatrix} A & B \\ C & D \end{pmatrix}$ is the transfer matrix for paraxial rays. Eq. (16.88) can be justified for a general optical system by considering that such a system could, in principle, be replaced by an arrangement of spaced thin lenses. Each thin lens, or length of uniform medium, obeys Eq. (16.88). We can illustrate the use of the transfer matrix in this way by following the propagation of a Gaussian beam in a lens waveguide.

16.6 Gaussian Beams in Lens Waveguides

In a biperiodic lens sequence containing equally spaced lenses of focal lengths f_1 and f_2 the transfer matrix for n unit cells of the sequence is, from Eq. (15.3),

$$\begin{pmatrix} A & B \\ C & D \end{pmatrix}^n = \frac{1}{\sin\phi}\begin{pmatrix} A\sin n\phi - \sin(n-1)\phi & B\sin n\phi \\ C\sin n\phi & D\sin n\phi - \sin(n-1)\phi \end{pmatrix}, \tag{16.89}$$

so from Eq. (16.88) and writing $q_2 = q_{n+1}$, since we are interested in propagation through n unit cells of the sequence,

$$q_{n+1} = (1/\sin\phi)\frac{[A\sin n\phi - \sin(n-1)\phi]q_1 + B\sin n\phi}{C\sin(n\phi)q_1 + D\sin n\phi - \sin(n-1)\phi}. \tag{16.90}$$

The condition for stable confinement of the Gaussian beam by the lens sequence is the same as in the case of paraxial rays. This is the condition that ϕ remains real, i.e., $|\cos\phi| \leq 1$, where

$$\cos\phi = \tfrac{1}{2}(A + D) = 1 - \frac{d}{f_1} - \frac{d}{f_2} + \frac{d^2}{2f_1f_2} \tag{16.91}$$

which gives

$$0 \leq \left(1 - \frac{d}{2f_1}\right)\left(1 - \frac{d}{2f_2}\right) \leq 1. \tag{16.92}$$

16.7 The Propagation of a Gaussian Beam in a Medium with a Quadratic Refractive Index Profile

We can most simply deduce what happens to a Gaussian beam in such a medium, whose refractive index variation is given by $n(r) = n_0 - \frac{1}{2}n_2r^2$ by using Eq. (16.88) and the transfer matrix of the medium, which from Eqs. (15.42) and (15.43) is

$$\begin{pmatrix} \cos\sqrt{n_2/n_0}z & \sqrt{n_0/n_2}\sin\sqrt{n_2/n_0}z \\ -\sqrt{n_2/n_0}\sin\sqrt{n_2/n_0}z) & \cos\sqrt{n_2/n_0}z \end{pmatrix}. \tag{16.93}$$

Thus if the input beam parameter is q_0 we have

$$q_{out}(z) = \frac{\cos\left(z\sqrt{n_2/n_0}\right)q_{in} + \sqrt{\frac{n_0}{n_2}}\sin\left(z\sqrt{n_2/n_0}\right)}{-\sqrt{n_2/n_0}\sin\left(z\sqrt{\frac{n_2}{n_0}}\right)q_{in} + \cos\left(z\sqrt{n_2/n_0}\right)} \tag{16.94}$$

where $1/q_{in} = 1/R_0 - i\lambda/\pi w_0^2$. The condition for stable propagation of this beam is that $q_{out}(z) = q_{in}$, which from Eq. (16.94) gives

$$-\sqrt{\frac{n_2}{n_0}}\sin z\sqrt{n_2/n_0}q^2 + \cos z\sqrt{n_2/n_0}q =$$

$$\cos z\sqrt{n_2/n_0}q + \sqrt{n_0/n_2}\sin z\sqrt{n_2/n_0}, \tag{16.95}$$

which has the solution

$$q^2 = -\frac{n_0}{n_2} \text{ and } q = i\sqrt{\frac{n_0}{n_2}}. \tag{16.96}$$

This implies the propagation of a Gaussian beam with planar phase fronts and constant spot size

$$w = \left(\frac{n_o}{n_2}\right)^{1/4}\sqrt{\frac{\lambda}{\pi}}. \tag{16.97}$$

Once again we stress that $\lambda = \lambda_0/n$ is the wavelength in the medium.†

Propagation of such a Gaussian beam without spreading is clearly very desirable in the transmission of laser beams over long distances. In most modern optical communication systems laser beams are guided inside optical fibers that have a maximum refractive index on their axis. These fibers, although they may not have an exact parabolic radial index profile, achieve the same continuous refocusing effect. We shall consider the properties of such fibers in greater detail in Chapter 17.

16.8 The Propagation of Gaussian Beams in Media with Spatial Gain or Absorption Variations

We can describe the effect of gain or absorption on the field amplitudes of a wave propagating through an amplifying or lossy medium by using a complex propagation constant k', where

$$k' = k + i\gamma/2. \tag{16.98}$$

† In the relationship between beam parameter q and the radius and spot size of the Gaussian beam a uniform refactive index is assumed. In taking $n = n_0$ in a medium with a quadratic index profile we are assuming only a small variation of the index over the width of the beam. So, for example, $n_2w^2 \ll n_0$.

This result is exact at line center, otherwise the value of k would be written as $k + \Delta k$ as in Eq. (5.1).

For a wave travelling in the positive z direction the variation with field amplitude with distance is $e^{-ik'z}$.

Consequently, γ can be identified as the intensity gain coefficient of the medium and k as the effective magnitude of the wave vector, satisfying

$$k = \frac{2\pi}{\lambda}, \tag{16.99}$$

where λ is the wavelength in the medium.

If we replace k by k' in Eq. (16.47) we can represent the field of the Gaussian beam as

$$U(r,z) = \frac{w_o}{w} \exp\left[-i(k'z - \Phi) - r^2\left(\frac{1}{w^2} + \frac{ik'}{2R}\right)\right], \tag{16.99}$$

which gives

$$U(r,z) = \frac{w_0}{w} e^{\gamma z/z} e^{-i(kz-\phi)} e^{-r^2(1/w^2 - \gamma/4R)}, \tag{16.100}$$

with intensity variation

$$I(r,z) = \left(\frac{w_0}{w}\right)^2 e^{\gamma z} e^{-2r^2(1/w^2 - \gamma/4R)}. \tag{16.101}$$

The intensity grows with distance as $e^{\gamma z}$, as usual, but there is a modification of the 'apparent' spot size to

$$w' = \frac{w}{\sqrt{1 - \gamma w^2/4R}}. \tag{16.102}$$

This increase in spot size is negligible even for an extremely high gain laser. For example, the xenon laser at 3.5 μm has shown[16.2] a small signal gain of 400 dB m^{-1}. In this case $10 \log_{10}(I/I_0) = 400$ and $I/I_0 = e^{\gamma}$, giving $\gamma = 92.1$ m^{-1}.

The smallest phase front curature that this laser beam could exhibit is, from Eq. (16.32)

$$R_{min} = \frac{2\pi w_0^2}{\lambda} \tag{16.103}$$

where the spot size is $\sqrt{2}w_0$. Therefore,

$$\frac{\gamma w^2}{4R} = \frac{\gamma\lambda}{4\pi} = 2.6 \times 10^{-5}, \tag{16.104}$$

so there is a negligible change in the spot size.

In practice, the occurrence of gain saturation leads to much greater modification of a propagating Gaussian beam than the subtle effect described above. This is an example of a situation in which the gain medium affects the Gaussian beam because the gain is spatially nonuniform. To illustrate this let us consider a medium where the gain varies quadratically with distance from the axis.

16.9 Propagation in a Medium with a Parabolic Gain Profile

It is convenient in this case to include the gain as part of a complex refractive index. If the modified propagation constant is k', we can write

$$k' = n'k_0 \tag{16.105}$$

where k_0 is the propagation constant *in vacuo*.

Therefore,

$$n' = \frac{k}{k_0} + \frac{i\gamma}{2k_0}, \tag{16.106}$$

where k is the real part of k'.

Consequently, if the gain varies parabolically with distance from the axis

$$\gamma(r) = \gamma_0 - \gamma_2 r^2 \tag{16.107}$$

The modified propagation constant is

$$k'(r) = k + \frac{1}{2} i(\gamma_0 - \gamma_2 r^2), \tag{16.108}$$

and the complex refractive index is

$$n(r) = \frac{k + \frac{1}{2} i(\gamma_0 - \gamma_2 r^2)}{k_0} \tag{16.109}$$

giving

$$n(r) = n_0 + \frac{i\gamma_0}{2k_0} - \frac{i\gamma_2}{2k_0} r^2. \tag{16.110}$$

The transfer matrix **M** for a length z of this medium is similar to Eq. (16.93), namely

$$\mathbf{M} = \begin{pmatrix} \cos(z\sqrt{n_2/n'}) & \sqrt{n'/n_2}\sin(z\sqrt{n_2/n'}) \\ -\sqrt{n_2/n'}\sin(z\sqrt{n_2/n'}) & \cos(z\sqrt{n_2/n'}) \end{pmatrix} \tag{16.111}$$

where $n_2 = i\gamma_2/k_0, n' = n_0 + i\gamma_0/2k_0$.

The beam parameter of a Gaussian beam propagating through such a medium will be constant, as was the case for Eq. (16.95), if

$$q^2 = -n'/n_2 \text{ and } q = i\sqrt{\frac{n'}{n_2}}. \tag{16.112}$$

Thus,

$$q = \sqrt{\frac{\gamma_2}{2n_0k_0}}\left[\left(1 - \frac{\gamma_0}{4k_0}\right) - i\left(1 + \frac{\gamma_0}{4k_0}\right)\right]; \tag{16.113}$$

$k_0 = 2\pi/\lambda_0$ is much larger than achievable values of γ_0, even for millimeter wave lasers, so

$$q = \sqrt{\frac{\gamma_2}{2n_0k_0}}(1-i), \tag{16.114}$$

which from Eq. (16.17) gives

$$R = \sqrt{\frac{2n_0k_0}{\gamma_2}} = 2\sqrt{\frac{\pi}{\lambda\gamma_2}}, \tag{16.115}$$

$$w^2 = 2\sqrt{\frac{\lambda}{\pi\gamma_2}}. \tag{16.116}$$

Remember that λ is the wavelength in the medium of refractive index n_0. Thus, we have a Gaussian beam that does not spread yet has spherical wavefronts.

We can understand this phenomenon qualitatively, as follows. The Gaussian

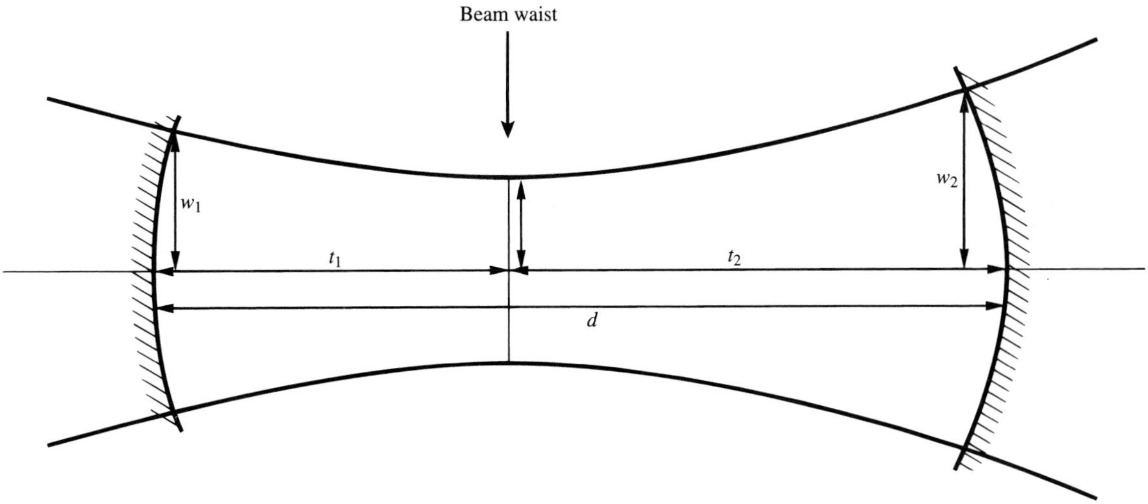

Fig. 16.14. Resonator using two concave mirrors showing the location of the beamwaist and the contour of the Gaussian beam.

beam, in the absence of gain, expands as it propagates. However, in the presence of gain that is highest on axis there is a compensatory effect where the amplification of the beam makes it wish to decrease in width as it propagates: the peak grows more than the wings of the profile.

A radial distribution of gain occurs in most gas lasers because of radial variations in the electron, or metastable, density. This affects radially the rate of excitation of laser levels, either those excited directly by electrons or by transfer of energy from metastables. The radial variation is close to quadratic in lasers where the excitation is by single-collision direct electron impact, but where the radial electron density is parabolic as a function of radius. The radial electron density is nearly parabolic in gas discharges at moderate to high pressures where the flow of charged particles to the walls occurs by ambipolar diffusion.† Quasi-parabolic gain profiles can also result in optically-pumped cylindrical solid state lasers.

16.10 Gaussian Beams in Plane and Spherical Mirror Resonators

We have already mentioned that a beam-like solution of Maxwell's equations will be a satisfactory transverse mode of a plane or spherical mirror resonator provided we place the resonator mirrors at points where their radii of curvature match the radii of curvature of the phase fronts of the beam. So for a Gaussian beam a double-concave mirror resonator matches the phase fronts as shown in Fig. (16.14) and plano-spherical and concave-convex resonators as shown in Figs. (16.15) and (16.16).

We can consider these resonators in terms of their equivalent biperiodic lens sequences as shown in Fig. (16.17). Propagation of a Gaussian beam from plane 1 to plane 3 in the biperiodic lens sequence is equivalent to one complete round trip

† A situation in which positive ions and electrons diffuse relative to each other at a rate set by the concentration gradient of each. The discharge usually remains electrically neutral because the charge densities of positive and negative charge are spatially equal.

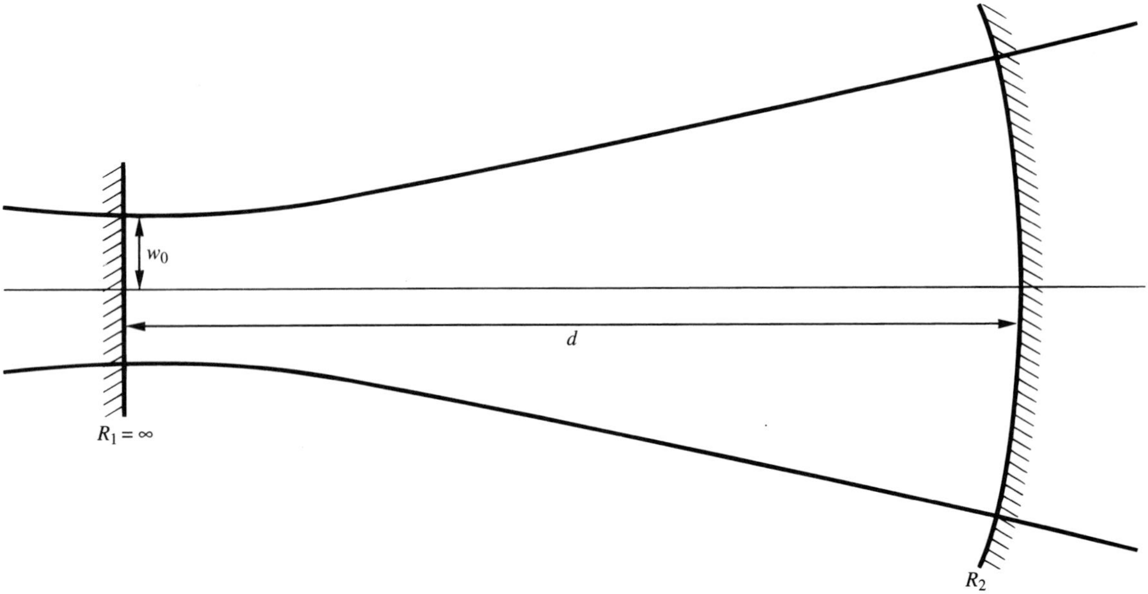

Fig. 16.15. Laser resonator that uses a plane mirror and a concave mirror.

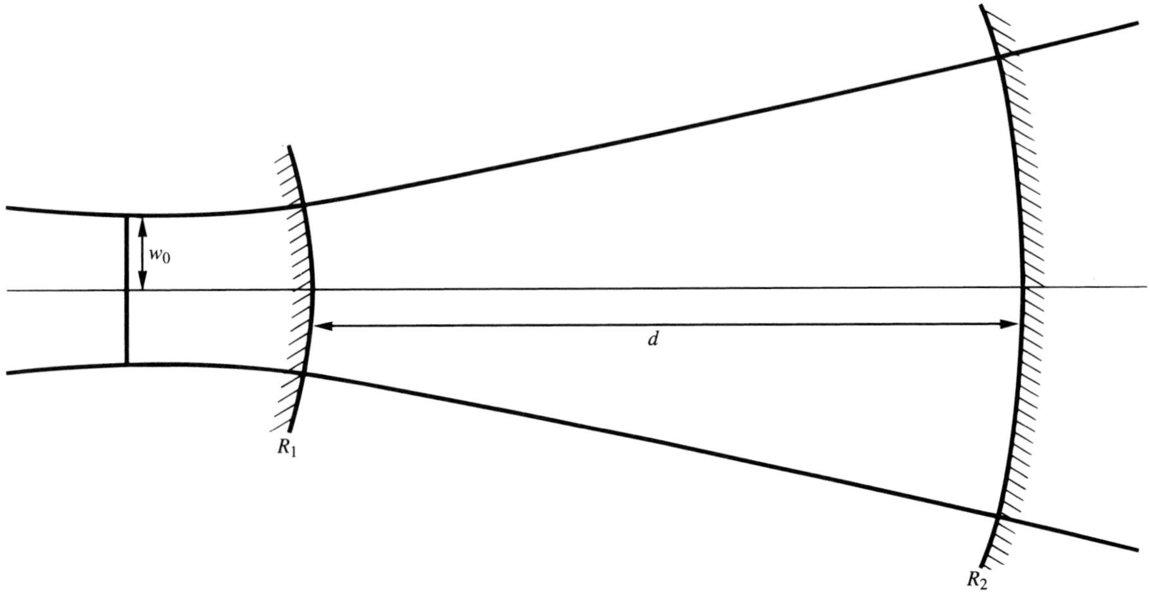

Fig. 16.16. Convex–concave laser resonator.

Fig. 16.17. Biperiodic lens sequence equivalent to the resonator in Fig. (16.14).

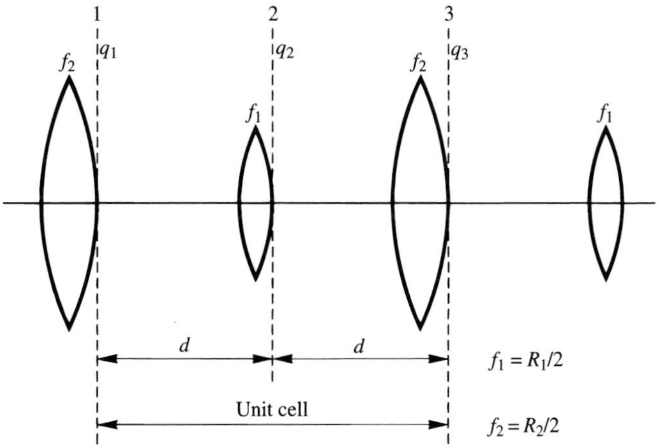

inside the equivalent spherical mirror resonator. If the complex beam parameters at planes 1, 2, and 3 are q_1, q_2, and q_3, then

$$\frac{1}{q_1 + d} - \frac{1}{f_1} = \frac{1}{q_2}, \tag{16.117}$$

which gives

$$q_2 = \frac{(q_1 + d)f_1}{f_1 - q_1 - d} \tag{16.118}$$

and

$$\frac{1}{q_2 + d} - \frac{1}{f_2} = \frac{1}{q_3}. \tag{16.119}$$

If the Gaussian beam is to be a real transverse mode of the cavity, then we want it to repeat itself after a complete round trip – at least as far as spot size and radius of curvature are concerned – that is we want

$$q_3 = q_1 = q. \tag{16.120}$$

16.11 Symmetrical Resonators

If both mirrors of the resonator have equal radii of curvature $R = 2f$, the condition that the beam be a transverse mode is

$$q_3 = q_2 = q_1 = q. \tag{16.121}$$

From Eq. (16.119) above

$$fq - (q + d)q = (q + d)f, \tag{16.122}$$

which gives

$$\frac{1}{q^2} + \frac{1}{df} + \frac{1}{fq} = 0, \tag{16.123}$$

which has roots

$$\frac{1}{q} = \frac{-1}{2f} \pm \frac{1}{2}\sqrt{\frac{1}{f^2} - \frac{4}{fd}} = -\frac{1}{2f} \pm i\sqrt{\frac{1}{fd} - \frac{1}{4f^2}}. \tag{16.124}$$

Now, $1/q = 1/R - i\lambda/\pi w^2$, so for a real spot size, i.e., w^2 positive, we must take the negative sign above. We do not get a real spot size if the resonator is geometrically unstable ($d > 4f$).

The spot size at either of the two (equal curvature) mirrors is found from

$$\frac{\lambda}{\pi w^2} = \sqrt{\frac{1}{fd} - \frac{1}{4f^2}}, \tag{16.125}$$

$$\frac{\lambda}{\pi w^2} = \sqrt{\frac{2}{Rd} - \frac{1}{R^2}}, \tag{16.126}$$

which gives

$$w^2 = \frac{\lambda R/\pi}{\sqrt{2R/d - 1}}. \tag{16.127}$$

The radius of curvature of the phase fronts at the mirrors is the same as the radii of curvature of the mirrors.

Because this resonator is symmetrical, we expect the beamwaist to be in the center of the resonator, at $z = d/2$ if we measure from one of the mirrors. So since $q_0 = q + z$ and $q_0 = i\pi w_0^2/\lambda$ from Eq. (16.124) we get

$$w_0^2 = \frac{\lambda}{2\pi}\sqrt{d(2R - d)}. \tag{16.128}$$

Note that the spot size gets larger as the mirrors become closer and closer to plane.

Now we know that a TEM mode, the TEM_{mn} cartesian mode, for example, propagates as

$$U_{m,n}(r, z) = \frac{w_0}{w} H_m\left(\sqrt{2}\frac{x}{w} H_m\right)\left(\sqrt{2}\frac{y}{w}\right) \times \exp\left[-i(kz - \Phi) - r^2\left(\frac{1}{w^2} + \frac{ik}{2R}\right)\right], \tag{16.129}$$

where the phase factor associated with the mode is

$$\Phi(m, n, z) = (m + n + 1)\tan^{-1}\left(\frac{\lambda z}{\pi w_0^2}\right). \tag{16.130}$$

On axis, $r = 0$, the overall phase of the mode is $kz - \Phi$.

In order for standing wave resonance to occur in the cavity, this phase shift from one mirror to the other must correspond to an integral number of half wavelengths, i.e.,

$$kd - 2(m + n + 1)\tan^{-1}\left(\frac{\lambda d}{2\pi w_0^2}\right) = (q + 1)\pi, \tag{16.131}$$

where the cavity in this case holds $q + 1$ half wavelengths. For cylindrical $\text{TEM}_{p\ell}$ modes the resonance condition would be

$$kd - 2(2p + \ell + 1)\tan^{-1}\left(\frac{\lambda d}{2\pi w_0^2}\right) = (q + 1)\pi. \tag{16.132}$$

If we write $\Delta\nu = c/2d$ (the frequency spacing between modes), then since $k = 2\pi\nu/c$

$$kd = 2\pi\nu d/c = \pi\nu/\Delta\nu, \tag{16.133}$$

we get

$$\frac{\nu}{\Delta\nu} = (q+1) + \frac{2}{\pi}(m+n+1)\tan^{-1}\left(\frac{1}{\sqrt{2R/d-1}}\right). \tag{16.134}$$

By writing $2\tan^{-1}\left(1/\sqrt{2R/d-1}\right) = x$ and using $\sin^2(x/2) = \frac{1}{2}(1-\cos x)$, $\cos^2(x/2) = \frac{1}{2}(1+\cos x)$, we get the formula

$$\frac{\nu}{\Delta\nu} = (q+1) + \frac{1}{\pi}(m+n+1)\cos^{-1}(1-d/R) \tag{16.135}$$

for the resonant frequencies of the longitudinal modes of the TEM_{mn} cartesian mode in a symmetrical mirror cavity.

From inspection of the above formula we can see that the resonant frequencies of longitudinal modes of the same order (the same integer q) depend on m and n. When the cavity is plane-plane, in which case $R = \infty$, $\nu/\Delta\nu = (q+1)$, the familar result for a plane–parallel Fabry–Perot cavity results.

When $d = R$ the cavity is said to be confocal and we have the special relations

$$w^2 = \frac{\lambda d}{\pi}, \quad w_0^2 = \frac{\lambda d}{2\pi}, \tag{16.136}$$

$$\nu/\Delta\nu = (q+1) + \frac{1}{2}(m+n+1). \tag{16.137}$$

The beam spot size increases by a factor of $\sqrt{2}$ between the center and the mirrors.

16.12 An Example of Resonator Design

We consider the problem of designing a symmetrical (double concave) resonator with $R_1 = R_2 = R$ and mirror spacing $d = 1$ m for $\lambda = 488$ nm.

If the system were made confocal, $R_1 = R_2 = R = d$ then the minimum spot size would be

$$(w_0)_{confocal} = \sqrt{\frac{\lambda d}{2\pi}} = 0.28 \text{ mm}.$$

At the mirrors $(w_{1,2})_{confocal} = w_0\sqrt{2} \simeq 0.39$ mm.

If we wanted to increase the mirror spot size to 2 mm, say, we would need to use longer radius mirrors $R \gg d$ such that

$$w^2 = (\lambda R/\pi)/\sqrt{\frac{2R}{d} - 1} = 4 \times 10^{-6}$$

which using

$$\frac{w}{\sqrt{\lambda d/2\pi}} \simeq \left(\frac{2R}{d}\right)^{1/4}$$

gives $R \simeq 346d = 346$ m (so R is $\gg d$).

Thus, it can be seen that to increase the spot size even by the small factor above we need to go to very large radius mirrors. Even in this large radius spherical mirror resonator the TEM_{00} transverse mode is still a very narrow beam.

In the general case the resonator mirrors do not have equal radii of curvature, so we need to solve Eqs. (16.117) and (16.119) for $q_1 = q_3 = q$. To find the position of the mirrors relative to the position of the beamwaist it is easiest to proceed directly from the equation describing the variation of the curvature of the wavefront with distance inside the resonator, i.e.,

$$R(z) = z\left[1 + \left(\frac{\pi w_0^2}{\lambda z}\right)^2\right], \tag{16.138}$$

where z is measured from the beamwaist.† So for the general case shown in Fig. (16.14), the curvature of the Gaussian beam at mirror 1 is

$$R_1 = -t_1\left[1 + \left(\frac{\pi w_0^2}{\lambda t_1}\right)^2\right], \tag{16.139}$$

which gives

$$R_1 t_1 + t_1^2 + \left(\frac{\pi w_0^2}{\lambda}\right)^2 = 0. \tag{16.140}$$

Similarly, the curvature of the Gaussian beam at mirror 2 is

$$R_2 = t_2\left[1 + \left(\frac{\pi w_0^2}{\lambda t_2}\right)^2\right], \tag{16.141}$$

which gives

$$R_2 t_2 - t_2^2 - \left(\frac{\pi w_0^2}{\lambda}\right)^2 = 0. \tag{16.142}$$

Therefore, we get

$$t_1 = -\frac{R_1}{2} \pm \frac{1}{2}\sqrt{R_1^2 - 4\left(\frac{\pi w_0^2}{\lambda}\right)^2}, \tag{16.143}$$

$$t_2 = \frac{R_2}{2} \pm \frac{1}{2}\sqrt{R_2^2 - 4\left(\frac{\pi w_0^2}{\lambda}\right)^2}. \tag{16.144}$$

If we choose the minimum spot size that we want in the resonator the above two equations will tell us where to place our mirrors.

If we write $d = t_2 + t_1$ then we can solve Eqs. (16.143) and (16.144) for w_0.

If we choose the positive sign in Eqs. (16.143) and (16.141), then

$$t_1 = -\frac{R_1}{2} + \frac{1}{2}\sqrt{R_1^2 - 4\left(\frac{\pi w_0^2}{\lambda}\right)^2}, \tag{16.145}$$

and

$$t_2 = \frac{R_2}{2} + \frac{1}{2}\sqrt{R_2^2 - 4\left(\frac{\pi w_0^2}{\lambda}\right)^2}. \tag{16.146}$$

† The radius of the Gaussian beam is negative for $z < 0$, positive for $z > 0$, so there is a difference in sign convention between the curvature of the Gaussian beam phase fronts and the curvature of the two mirrors that constitute the optical resonator.

Therefore

$$d = \left(\frac{R_2 - R_1}{2}\right) + \frac{1}{2}\sqrt{R_2^2 - 4\left(\frac{\pi w_0^2}{\lambda}\right)^2} + \frac{1}{2}\sqrt{R_1^2 - 4\left(\frac{\pi w_0^2}{\lambda}\right)^2}, \quad (16.147)$$

so

$$(2d - R_2 + R_1)^2 = R_2^2 - 4\left(\frac{\pi w_0^2}{\lambda}\right)^2 + R_1^2 - 4\left(\frac{\pi w_0^2}{\lambda}\right)^2 + 2\sqrt{\left[R_1^2 - 4\left(\frac{\pi w_0^2}{\lambda}\right)^2\right]\left[R_2^2 - 4\left(\frac{\pi w_0^2}{\lambda}\right)^2\right]}. \quad (16.148)$$

Eq. (16.148) can be solved for w_0 to give

$$w_0^4 = \left(\frac{\lambda}{\pi}\right)^2 \frac{d(-R_1 - d)(R_2 - d)(R_2 - R_1 - d)}{(R_2 - R_1 - 2d)^2}. \quad (16.149)$$

The spot sizes at the two mirrors are found from

$$w_1^4 = \left(\frac{\lambda R_1}{\pi}\right)^2 \frac{(R_2 - d)}{(-R_1 - d)} \frac{d}{(R_2 - R_1 - d)}, \quad (16.150)$$

$$w_2^4 = \left(\frac{\lambda R_2}{\pi}\right)^2 \frac{(-R_1 - d)}{(R_2 - d)} \frac{d}{(R_2 - R_1 - d)}, \quad (16.151)$$

where we must continue to remember that for the double-concave mirror resonator we are considering, R_1, the radius of curvature of the Gaussian beam at mirror 1, is negative.

The distances t_1 and t_2 between the beamwaist and the mirrors are

$$t_1 = \frac{-d(R_2 - d)}{R_1 - R_2 + 2d}, \quad (16.152)$$

$$t_2 = \frac{d(R_1 + d)}{R_1 - R_2 + 2d}. \quad (16.153)$$

For this resonator the resonant condition for the longitudinal modes of the TEM_{mn} cartesian mode is

$$\frac{\nu}{\Delta\nu} = (q + 1) + \frac{1}{\pi}(m + n + 1)\cos^{-1}\sqrt{\left(1 + \frac{d}{R_1}\right)\left(1 - \frac{d}{R_2}\right)}, \quad (16.154)$$

where the sign of the square root is taken the same as the sign of $1 + d/R_1$, which is the same as the sign of $1 - d/R_2$ for a stable resonator.

16.13 Diffraction Losses

In our discussions of transverse modes we have not taken into account the finite size of the resonator mirrors, so the discussion is only strictly valid for mirrors whose physical size makes them much larger than the spot size of the mode at the mirror in question. If this condition is not satisfied, some of the energy in the Gaussian beam leaks around the edge of the mirrors. This causes modes of large spot size to suffer high diffraction losses. So, since plane parallel resonators in theory have infinite spot size, these resonators suffer enormous diffraction losses.

Because higher order TEM_{mn} modes have larger spot sizes than low order modes, we can prevent them from operating in a laser cavity by increasing their diffraction loss. This is done by placing an aperture between the mirrors whose size is less than the spot size of the high order mode at the position of the aperture, yet greater than the spot size of a lower order mode (usually TEM_{00}) that it is desired to have operate in the cavity.

The diffraction loss of a resonator is usually described in terms of its Fresnel number $a_1a_2/\lambda d$, where a_1, a_2 are the radii of the mirrors and d is their spacing. Very roughly the fractional diffraction loss at a mirror is $\approx w/a$. When the diffraction losses of a resonator are significant then the real field distribution of the modes can be found using Fresnel–Kirchoff diffraction theory. If the field distribution at the first mirror of the resonator is U_{11}, then the field U_{21} on a second mirror is found from the sum of all the Huygen's secondary wavelets originating on the first mirror as[16.3]:

$$U_{21}(x, y) = \frac{ik}{4\pi}\int_1 U_{11}(x, y)\frac{e^{-ikr}}{r}(1 + \cos\theta)dS. \tag{16.155}$$

The integral is taken over the surface of mirror 1: r is the distance of a point on mirror 1 to the point of interest on mirror 2; θ is the angle which this direction makes with the axis of the system; dS is an element of area. Having found $U_{21}(x, y)$ by the above method we can recalculate the field on the first mirror as

$$U_{12}(x, y) = \frac{ik}{4\pi}\int_2 U_{21}(x, y)\frac{e^{ikr}}{r}(1 + \cos\theta)dS \tag{16.156}$$

and if we require $U_{12}(x, y)$ to be equivalent to $U_{11}(x, y)$, apart from an arbitrary phase factor, then our transverse field distribution becomes self-consistent and is a solution for the transverse mode distribution of a resonator with finite aperture mirrors. This is the approach that was first used by Fox and Li[16.4] to analyze the transverse field distributions that would result in optical resonators.

16.14 Unstable Resonators

Many high power lasers have sufficiently large gain that they do not require the amount of feedback provided by a stable optical resonator with high reflectance mirrors. To achieve maximum power output from such lasers, one resonator mirror must have relatively low reflectance and the laser beam should have a large enough spot size so as to extract energy from a large volume of the medium. However, as the spot size of the oscillating transverse mode becomes larger, an increasing portion of the energy lost from the cavity 'leaks' out past the edge of the mirrors. As the configuration of the cavity approaches the unstable regions in Fig. (15.5) the spot size grows infinitely large. At the same time, the divergence of the output beam falls, making focusing and control of the output easier. Operation of the resonator in the unstable region provides efficient energy extraction from a high gain laser, and provides a low divergence beam of generally quite high transverse mode quality.

Schawlow first suggested the use of an unstable resonator, with its associated intracavity beam defocusing properties, to counteract the tendency of some lasers to focus internally into high intensity filaments. This property of unstable resonators is valuable in reducing the occurrence of 'hot spots' in the output beam. These

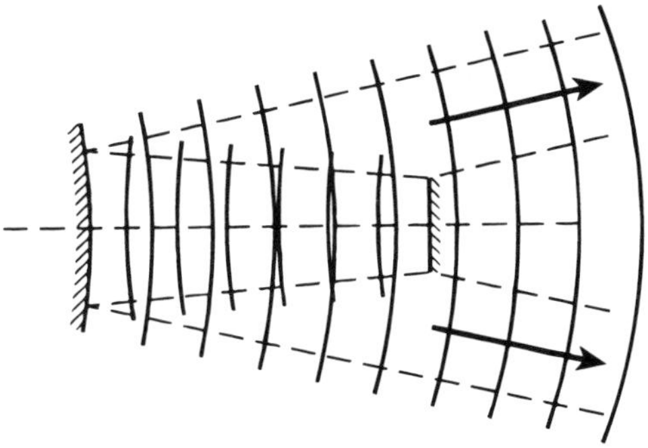

Fig. 16.18. Cassegrain unstable resonator.

can result if a high gain laser with spatial variations in its refractive index profile (produced, for example, by nonhomogeneous pumping) is operated in a stable resonator arrangement. A simple unstable resonator configuration that accomplishes this is shown in Fig. (16.18). The output laser beam will have a characteristic 'donut' shaped intensity pattern in the near field.

The loss per pass of an unstable resonator can be determined approximately by a simple geometric optics analysis first presented by Siegman[16.5]. The geometry used in the analysis is shown in Fig. (16.19). The wave originating from mirror M_1 is assumed to be a spherical wave whose center of curvature is at point P_1 (not in general the same point as the center of curvature of mirror M_1). The right-travelling wave is partially reflected at mirror M_2, and is now assumed to be a spherical wave that has apparently originated at point P_2. This wave now reflects from M_1, again, as if it had come from point P_1 and another round-trip cycle commences. The positions of P_1 and P_2 must be self-consistent for this to occur, each must be the virtual image of the other in the appropriate spherical mirror. The distances of P_1 and P_2 from M_1 and M_2 are written as r_1L, r_2L, respectively, where L is the resonator length. r_1 and r_2 are measured in the positive sense in the direction of the arrows in Fig. (16.19). If, for example, P_2 were to the left of M_2, then r_2 would be negative. We also use our previous sign convention that the radii of curvature of M_1, and M_2, R_1, R_2, respectively, are both negative.

From the fundamental lens equation we have

$$\frac{1}{r_1} - \frac{1}{r_2+1} = -\frac{2L}{R_1} = 2(g_1 - 1), \tag{16.157}$$

$$\frac{1}{r_2} - \frac{1}{r_1+1} = -\frac{2L}{R_2} = 2(g_2 - 1), \tag{16.158}$$

where g_1 and g_2 are dimensionless parameters.

So

$$r_1 = \pm\frac{\sqrt{1 - 1/(g_1g_2)} - 1 + 1/g_1}{2 - 1/g_1 - 1/g_2}, \tag{16.159}$$

$$r_2 = \pm\frac{\sqrt{1 - 1/(g_1g_2)} - 1 + 1/g_2}{2 - 1/g_1 - 1/g_2}. \tag{16.160}$$

These equations determine the location of the virtual centers of the *spherical* waves in the resonator.

A wave of unit intensity that leaves mirror M_1 will partially spill around the

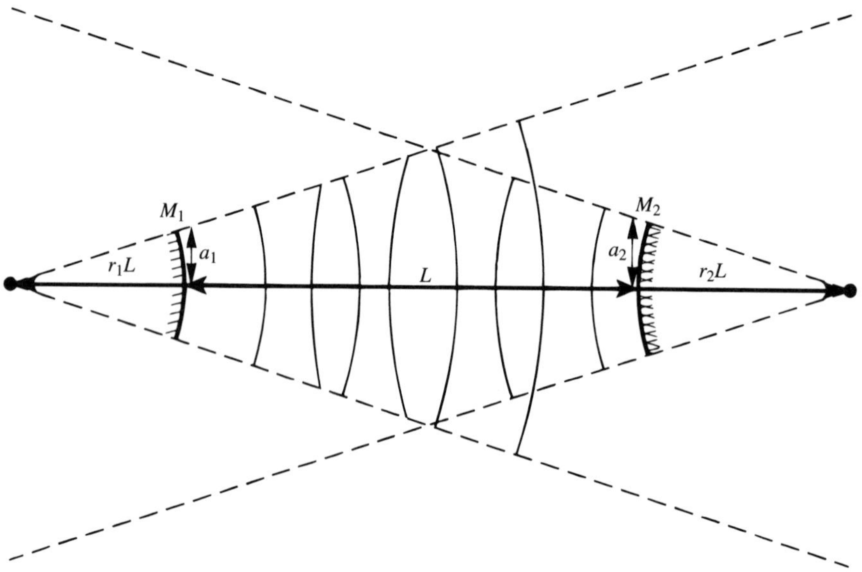

Fig. 16.19. Geometry for unstable resonator calculation[16.5].

edge of mirror M_2 – this is sometimes loosely referred to as a *diffraction loss*, but it is not. A fraction Γ_1 of the energy is reflected at M_1 so a fraction $1-\Gamma_1$ is lost from the resonator. A fraction Γ_2 of this reflected wave reflects from mirror M_2. Thus, after one complete round trip our initial intensity has been reduced to $\Gamma^2 \equiv \Gamma_1\Gamma_2$.

We can get some feeling for how the loss per round trip actually depends on the geometry of the resonator by considering the resonator in Fig. (16.19) as a purely two-dimensional one. This is equivalent to having infinitely long strip-shaped mirrors that only curve in the lateral direction.

From the geometry of Fig. (16.19) it is easy to see that

$$\Gamma_1 = \frac{r_1 a_2}{(r_1+1)a_1}, \tag{16.161}$$

$$\Gamma_2 = \frac{r_2 a_1}{(r_2+1)a_2}. \tag{16.162}$$

Thus,

$$\Gamma_{2D}^2 = \Gamma_1\Gamma_2 = \pm\frac{r_1 r_2}{(r_1+1)(r_2+1)}. \tag{16.163}$$

Surprisingly, the round-trip energy lost is *independent* of the dimensions of the two mirrors.

16.15 Problems

(16.1) Compute and plot the fraction of laser power in a TEM_{00} mode of spot size w_0 that passes through an aperture of radius a as a function of the ratio a/w_0.

(16.2) A Gaussian beam has a minimum spot size $w_0 = 10\ \mu m$ and a wavelength of 880 nm. Calculate: (a) The radius of curvature 1 m from the beamwaist. (b) The spot size 1 m from the beamwaist. Where does the beam have its minimum radius of curvature?

(16.3) The beam in Problem (16.2) is focused by a lens with focal length 10 mm placed at the beamwaist. Calculate the position of the new beamwaist, and the new minimum spotsize.

(16.4) Design a GRIN lens system of reasonable parameters to collimate the laser beam in Problem (16.2).

(16.5) The laser beam in Problem (16.2) is allowed to enter a medium with a quadratic index profile, with $n_0 = 1.5$, and $n_2 = 0.1\ \mathrm{m}^{-2}$. The beam enters a section of this medium 2 m long with its beamwaist at the front face. The beam emerges from the medium and travels 2 m further. What are its spot size and radius of curvature at this point?

(16.6) Prove that when a laser beam of spot size w_0 is focused by a thin lens placed at the beamwaist, there is a maximum possible distance at which a focus can be achieved. What is this maximum distance if w_0=30 mm? What happens to the Gaussian beam when it is not brought to a focus by the lens? Where has the beamwaist moved to?

(16.7) Prove that to produce the smallest spot size at range d with a thin lens placed at the beamwaist the optimum thin lens has $f' = d$.

(16.8) A 1.55 μm semiconductor laser emits an elliptical Gaussian beam with minimum spot sizes in two orthogonal directions at the output facet of 100 μm, and 10 μm, respectively. At what distance from the facet has the output beam become circular? What is the spot size at this point?

(16.9) A cylindrical lens (one with two different focal lengths in orthogonal planes) is placed at the circular beam location in Problem (16.8). What two focal lengths are required to bring the semiconductor laser to a circular focus 20 mm from this point?

(16.10) For a resonator with two concave mirrors with $R_1 = 1$ m, $R_2 = 2$ m, for what value of d and specific ray parameters does the intracavity ray retrace its path after two round trips. Draw ray-tracing diagrams to illustrate these situations.

(16.11) A Gaussian beam with a minimum spot size of 20 μm is focused by a thin lens with a focal length of 5 mm that is placed 10 mm from the beamwaist. Calculate the location and size of the new beamwaist. Take the wavelength as 1.55 μm.

(16.12) (Computer project) A 1.3 μm laser has a minimum spot size of 0.5 mm. Design a GRIN lens choosing reasonable values for its length, diameter, and n_2 parameter for focusing this beam to a 10 μm spotsize. Place the GRIN lens at 20 mm from the original beamwaist.

(16.13) A 633 nm laser has a resonator with two concave mirrors of radii 10 m and 5 m, respectively, placed 0.5 m apart. Calculate: (a) the spotsize of the resultant Gaussian beam; (b) the location of its beamwaist.

(16.14) Design a symmetrical biconcave resonator ($R_1 = R_2$) with a minimum spot size of 1 mm for 1.06 μm and a spacing of 100 mm.

(16.15) Prove Eq. (16.83).

(16.16) Explain what the different resonator arrangements are that correspond to the different choices of sign in Eqs. (16.143) and (16.144). Do the different choices affect the spot size?

References

[16.1] S. Ramo, J.R. Whinnery, and T. Van Duzer, *Fields and Waves in Communication Electronics*, 3rd Edition, Wiley, New York, 1993.

[16.2] J.W. Kluver, 'Laser amplifier noise at 3.5 microns in helium-xenon,' *J. Appl. Phys.*, **37**, 2987–2999, 1966.

[16.3] M. Born and E. Wolf, *Principles of Optics*, 6th Edition, Pergamon Press, Oxford, 1980.

[16.4] A.G. Fox and T. Li, 'Resonant Modes in a Maser Interferometer,' *Bell Sys. Tech. J.*, **40**, 453–488, 1961.

[16.5] A.E. Siegman, 'Unstable optical resonators,' *Appl. Opt.*, **13**, 353–367, 1974.

17

Optical Fibers and Waveguides

In this chapter we shall discuss from both a ray and wave standpoint how light can be guided along by planar and cylindrical dielectric waveguides. We shall explain why optical fibers are important and useful in optical communication systems and discuss briefly how these fibers are manufactured. Some practical details of how fibers are used and how they are integrated with other important optical components will conclude the chapter.

17.1 Introduction

We saw in the previous chapter that a Gaussian beam can propagate without beam expansion in an optical medium whose refractive index varies in an appropriate manner in the radial direction. This is a rather specific example of how an optical medium can guide light energy. However, we can discuss this phenomenon in more general terms. By specifying the spatial variation in the refractive index, and through the use of the wave equation with appropriate boundary conditions, we can show that dielectric waveguides will support certain 'modes' of propagation. However, it is helpful initially to see what can be learned about this phenomenon from ray optics.

17.2 Ray Theory of Cylindrical Optical Fibers

17.2.1 Meridional Rays in a Step-Index Fiber

A step-index fiber has a central core of index n_1 surrounded by cladding of index n_2 where $n_2 < n_1$. When a ray of light enters such a optical fiber, as shown in Fig. (17.1), it will be guided along inside the core of the fiber if the angle of incidence between core and cladding is greater than the critical angle. Two distinct types of rays can travel along inside the fiber in this way: *meridional* rays travel in a plane that contains the fiber axis, *skew* rays travel in a nonplanar zig-zag path and never cross the fiber axis, as illustrated in Fig. (17.2).

For the meridional ray in Fig. (17.1), total internal reflection (TIR) occurs within the core if $\theta_i > \theta_c$, or $\sin\theta_i > n_2/n_1$. From Snell's law, applied to the ray entering the fiber,

$$\sin\theta = \sin\theta_0/n_1. \tag{17.1}$$

Since $\theta + \theta_i = 90°$, the condition for total internal reflection is

$$\sin\theta_0 < \sqrt{n_1^2 - n_2^2}. \tag{17.2}$$

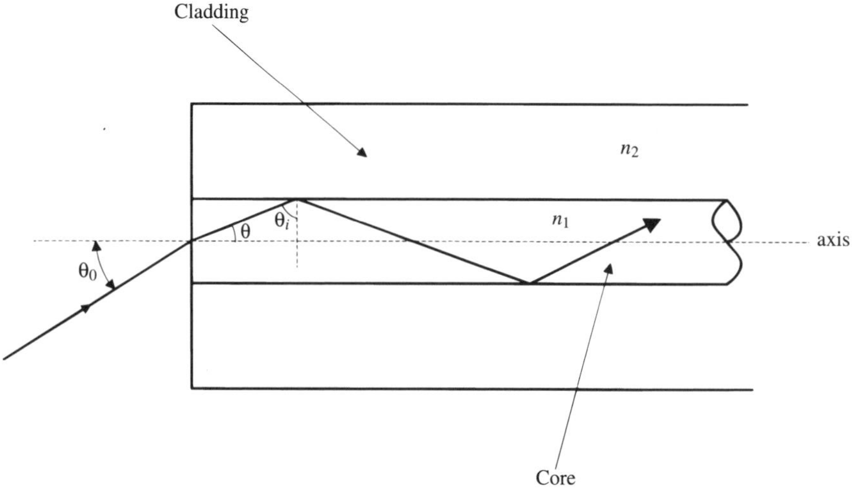

Fig. 17.1. Meridional ray entering fiber and being guided by total internal reflection

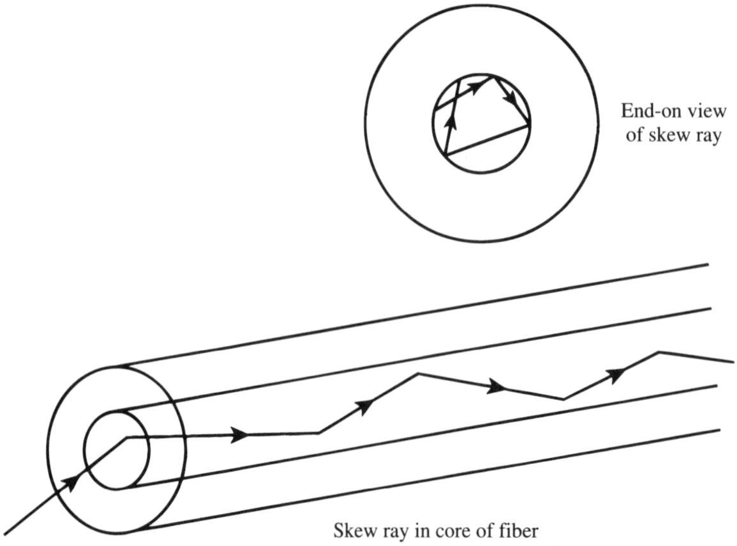

Fig. 17.2. Skew rays

For most optical fibers the relative difference between the index of core and cladding $\Delta = (n_1 - n_2)/n_1$ is small, so Eq. (17.2) can be written

$$\sin\theta_0 < \sqrt{(n_1 - n_2)(n_1 + n_2)}, \tag{17.3}$$

which since $n_1 \simeq n_2$ can be written

$$\sin\theta_0 < n_1\sqrt{2\Delta}. \tag{17.4}$$

If a lens is used to focus light from a point source into a fiber, as shown in Fig. (17.3), then there is a maximum aperture size D which can be used. When the end of the fiber is a distance d from the lens, light rays outside the cross-hatched region enter the fiber at angles too great to allow total internal reflection. The quantity

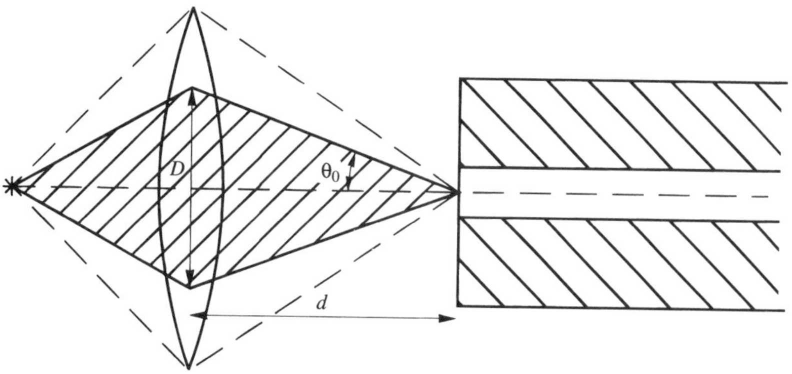

Fig. 17.3. Geometry for focusing light into fiber.

$\sin\theta_0 \simeq D/2d$ is called the numerical aperture NA of the system. So, from Eq. (17.4)

$$NA = \sin\theta_0 = \sqrt{n_1^2 - n_2^2} = n_1\sqrt{2\Delta}. \tag{17.5}$$

If the point source in Fig. (17.3) is placed a great distance from the lens, then $d = f$. In this case

$$NA = \frac{D}{2f}. \tag{17.6}$$

If the lens is chosen to be no larger than necessary, then the lens diameter will be D. The ratio f/D is a measure of the focusing/light collecting properties of the lens, called the f/number. So to match a distant source to the fiber $2NA = 1/(f/\text{number})$.

This analysis of a step-index cylindrical optical fiber by means of simple ray theory has so far considered only the paths of meridional rays, those rays whose trajectories lie within the plane containing the fiber axis. We have also tacitly assumed that Snell's law strictly delineates between those rays that totally internally reflect, and are *bound* within the fiber, and those that escape or *refract* into the cladding. However, if we consider the trajectories of skew rays we shall see that a third class of rays, so-called *tunnelling* or *leaky* rays exists. These are rays that appear at first glance to satisfy the Snell's law requirement for total internal reflection. However, because the refraction occurs at a curved surface they can in certain circumstances leak propagating energy into the cladding. In our description of the phenomenon we follow the essential arguments and notation given by Snyder and Love[17.1].

Fig. (17.4) shows the paths of skew rays in step-index and graded-index fibers projected onto a plane perpendicular to the fiber axis. The path of a particular skew ray is characterized by the angle θ_z it makes with the z axis and the angle θ_ϕ that it makes with the azimuthal direction. In a step-index fiber these angles are the same at all points along the ray path, in a graded-index fiber they vary with the radial distance of the ray from the fiber axis. The projected ray paths of skew rays are bounded, for a particular family of skew rays, by the inner and outer *caustic* surfaces.

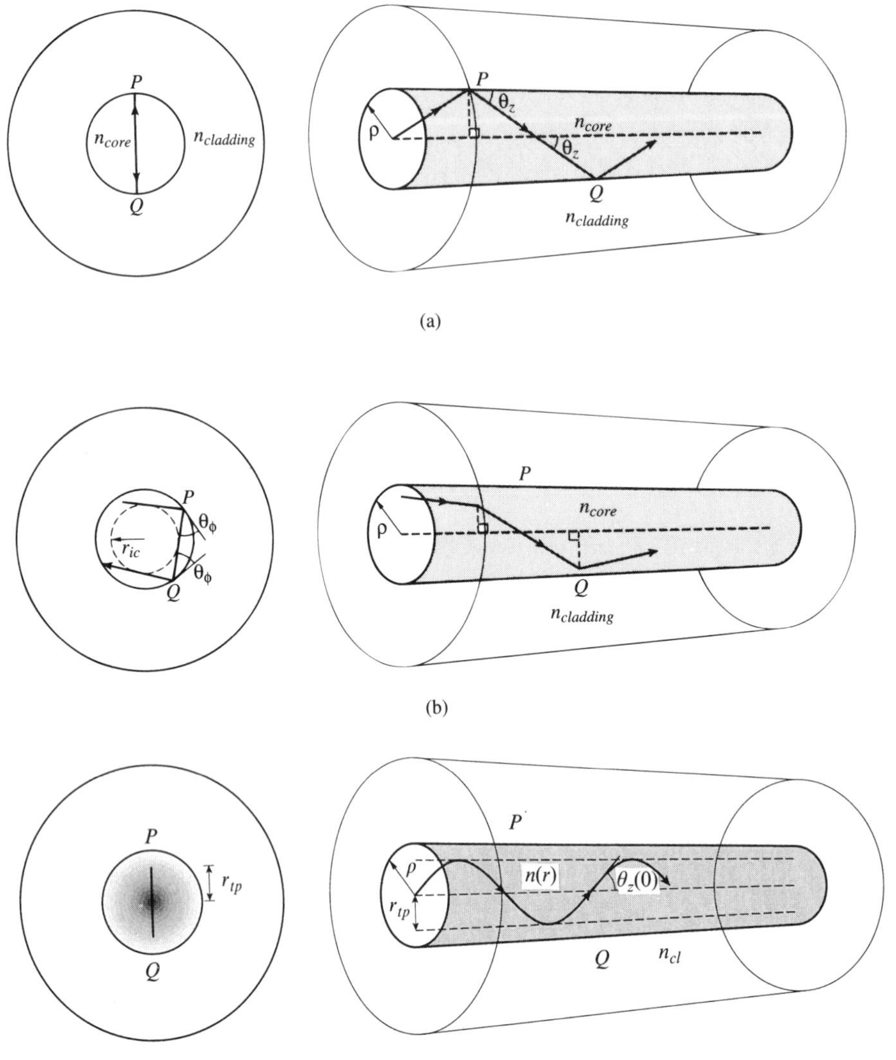

Fig. 17.4. (a) Projected path of a meridional ray in a step-index cylindrical fiber; (b) projected path of a skew ray in a step-index cylindrical fiber; (c) projected path of a meridional ray in a graded-index cylindrical fiber.

17.2.2 *Step-Index Fibers*

In a step-index fiber the outer caustic surface is the boundary between core and cladding. Fig. (17.4a) shows clearly that the radius of the inner caustic in a step-index fiber is

$$r_{ic} = \rho \cos \theta_\phi. \tag{17.7}$$

For meridional rays $\theta_\phi = \pi/2$ and $r_{ic} = 0$.

When a skew ray strikes the boundary between core and cladding it makes an angle α with the surface normal where

$$\cos \alpha = \sin \theta_z \sin \theta_\phi. \tag{17.8}$$

Simple application of Snell's law predicts that total internal reflection should occur if

$$\sin \alpha > \frac{n_{cladding}}{n_{core}}. \tag{17.9}$$

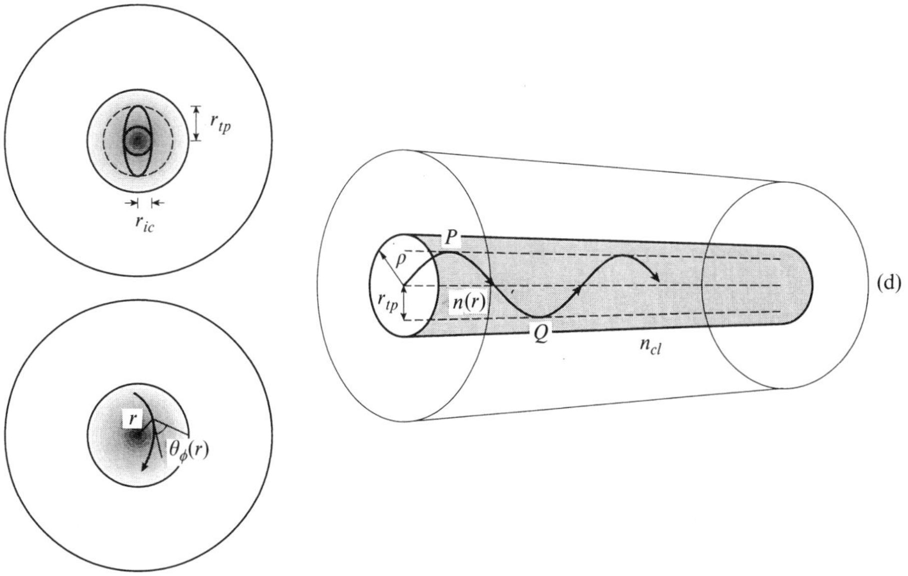

Fig. 17.4. (*cont.*) (d) projected path of a skew ray in a graded-index fiber showing the radial dependence of the azimuthal angle $\theta_\phi(r)$.

We introduce a new angle θ_c, related to the critical angle α_c by

$$\theta_c + \alpha_c = 90^\circ.$$

Rays can be classified according to the values of these various angles:

$$\theta \le \theta_z < \theta_c \qquad \text{bound rays;}$$
$$0 \le \alpha \le \alpha_c \qquad \text{refracting rays;}$$
$$\theta_c \le \theta_z \le 90^\circ; \alpha_c \le \alpha \le 90^\circ \qquad \text{tunnelling rays.}$$

It is the existence of this third class of rays that is somewhat surprising[17.2]. In order to explain the existence of such rays we need to consider various qualities that remain constant, or *invariant* for rays in a step-index fiber. These invariants follow for simple geometric reasons and are:

$$\overline{\beta} = n_{core} \cos\theta_z, \tag{17.10}$$
$$\overline{\ell} = n_{core} \sin\theta_z \cos\theta_\phi. \tag{17.11}$$

It follows from Eq. (17.8) that

$$\overline{\beta}^2 + \overline{\ell}^2 = n_{core}^2 \sin^2\alpha. \tag{17.12}$$

For meridional rays $\overline{\ell} = 0$, whereas for skew rays $\overline{\ell} > 0$. In terms of these ray invariants the classification of rays in a step-index fiber becomes:

$$n_{cladding} < \overline{\beta} \le n_{core} \qquad \text{bound rays;} \tag{17.13}$$

$$0 \le \overline{\beta}^2 + \overline{\ell}^2 < n_{cladding}^2 \qquad \text{refracting rays;} \tag{17.14}$$

$$n_{cladding}^2 < \overline{\beta}^2 + \overline{\ell}^2 \le n_{core}^2 \text{ and } 0 \le \overline{\beta} < n_{cladding} \qquad \text{tunnelling rays.} \tag{17.15}$$

An important consequence of the ray invariants is that the length of time it takes light to travel a distance L along the fiber is independent of skewness. From Fig. (17.4) it is easy to see that the distance from P to Q is

$$L_p = 2\rho\frac{\sin\theta_\phi}{\sin\theta_z} = \frac{2\rho N_{core}\sqrt{n_{core}^2 - \overline{\beta}^2 - \overline{\ell}^2}}{n_{core}^2 - \overline{\beta}^2}. \tag{17.16}$$

The corresponding distance travelled along in the axial direction is

$$L = \frac{L_p}{\cos\theta_z} \tag{17.17}$$

and the time taken is

$$\tau = \frac{L_p n_{core}}{c_0\cos\theta_z}. \tag{17.18}$$

The time taken per unit distance travelled is

$$\frac{\tau}{L} = \frac{n_{core}}{c_0\cos\theta_z}, \tag{17.19}$$

which clearly depends only on axial angle θ_z, and does not depend on skew angle θ_ϕ. This has important consequences for optical pulse propagation along a fiber. In this ray model an optical pulse injected at angle θ_z will take a single fixed time to travel along the fiber, independent of whether it travels as a meridional or skew ray.

17.2.3 Graded-Index Fibers

To discover the important ray invariants in a cylindrically symmetric graded index fiber we use Eq. (15.31)

$$\frac{d}{ds}\left[n(\mathbf{r})\frac{d\mathbf{r}}{ds}\right] = \text{grad } n(\mathbf{r}). \tag{17.20}$$

If we transform to cylindrical coordinates, for the case where n only varies radially

$$\frac{d}{ds}\left[n(r)\frac{dr}{ds}\right] - rn(r)\left(\frac{d\phi}{ds}\right)^2 = \frac{dn(r)}{dr}, \tag{17.21}$$

$$\frac{d}{ds}\left[n(r)\frac{d\phi}{ds}\right] + \frac{2n(r)}{r}\frac{d\phi}{ds}\frac{dr}{ds} = 0, \tag{17.22}$$

$$\frac{d}{ds}\left[n(r)\frac{dz}{ds}\right] = 0. \tag{17.23}$$

The path of a ray in projection in this case is shown in Fig. (17.4). At each point along the ray path the tangent to the ray path makes an angle $\theta_z(r)$ with the fiber axis.

In the projected path shown in Fig. (17.4) the projected tangent to the ray path makes an angle with the radius of 90-$\theta_\phi(r)$.

As before, from Eq. (17.8),

$$\cos\alpha(r) = \sin\theta_z(r)\sin\theta_\phi(r). \tag{17.24}$$

Various other relations follow:

$$\frac{dr}{ds} = \cos\alpha(r), \tag{17.25}$$

$$\frac{dz}{ds} = \cos\theta_z(r), \tag{17.26}$$

and because

$$ds^2 = dr^2 + r^2 d\phi^2 + dz^2 \tag{17.27}$$

it also follows that

$$\frac{d\phi}{ds} = \frac{1}{r}\sin\theta_z(r)\cos\theta_\phi(r). \tag{17.28}$$

Therefore, we can integrate Eq. (17.23) to give

$$\overline{\beta} = n(r)\frac{dz}{ds} = n(r)\cos\theta_z(r), \tag{17.29}$$

where $\overline{\beta}$ is a ray invariant.

Integration of Eq. (17.22) gives

$$\overline{\ell} = \frac{r^2}{\rho}n(r)\frac{d\phi}{ds} = \frac{r}{\rho}n(r)\sin\theta_z(r)\cos\theta_\phi(r), \tag{17.30}$$

where we have introduced a normalizing radius ρ so that $\overline{\ell}$ remains dimensionless. This could be a reference radius, or more usually the boundary between core and cladding when only the core has a graded index.

If we eliminate θ_z from Eqs. (17.29) and (17.30) we get

$$\cos\theta_\phi(r) = \frac{\rho}{r}\frac{\overline{\ell}}{\sqrt{n^2(r) - \overline{\beta}^2}}. \tag{17.31}$$

Now, as shown schematically in Fig. (17.4), skew rays take a helical path in the core, but not one whose cross-section is necessarily circular. A particular family of skew rays will not come closer to the axis than the inner caustic and will not go further from the axis than the outer caustic. Inspection of Fig. (17.4) reveals that on each caustic $\theta_\phi(r) = 0$. Consequently, the radii of the inner and outer caustic, r_{ic}, r_{tp}, respectively, are the roots of

$$g(r) = n^2(r) - \overline{\beta}^2 - \overline{\ell}^2\frac{\rho^2}{r^2} = 0. \tag{17.32}$$

17.2.4 Bound, Refracting, and Tunnelling Rays

We are now in a position to explain mathematically the distinction between these three classes of rays. From Eq. (17.29) we can write

$$\frac{dz}{ds} = \frac{\overline{\beta}}{n(r)} \tag{17.33}$$

and from Eq. (17.30)

$$\frac{d\phi}{ds} = \frac{\rho\overline{\ell}}{r^2 n(r)}. \tag{17.34}$$

Subsitution from Eq. (17.34) into Eq. (17.21), noting that

$$\frac{d}{ds} \equiv \frac{dz}{ds}\frac{d}{dz}, \tag{17.35}$$

gives

$$\overline{\beta}^2\frac{d^2r}{dz^2} - \overline{\ell}^2\frac{\rho^2}{r^3} = \frac{1}{2}\frac{dn^2(r)}{dr}. \tag{17.36}$$

Writing $d^2r/dz^2 = r'dr'/dr$ where $r' = dr/dz$ gives

$$\overline{\beta}^2 r' \frac{dr'}{dr} - \overline{\ell}^2 \frac{\rho^2}{r^3} = \frac{1}{2}\frac{dn^2(r)}{dr}, \tag{17.37}$$

which can be integrated to give

$$\frac{1}{2}\overline{\beta}^2 (r')^2 + \frac{1}{2}\frac{\overline{\ell}^2 \rho^2}{r^2} = \frac{1}{2}n^2(r) + \text{ constant.} \tag{17.38}$$

Using Eq. (17.32) allows Eq. (17.38) to be written as

$$\overline{\beta}^2 \left(\frac{dr}{dz}\right)^2 = g(r) + \overline{\beta}^2 + \text{ constant.} \tag{17.39}$$

At the outer caustic surface $(dr/dz)_{r_{tp}} = 0$, and also $g(r_{tp}) = 0$, so the constant of integration in Eq. (17.39) is $-\overline{\beta}^2$. Finally, therefore, we have

$$\overline{\beta}^2 \left(\frac{dr}{dz}\right)^2 = g(r) = n^2(r) - \overline{\beta}^2 - \overline{\ell}^2 \frac{\rho^2}{r^2}. \tag{17.40}$$

Ray paths can only exist if $g(r)$ is nonnegative.

The roots of $g(r)$ determine the location of the inner and outer caustic surfaces. For a refracting ray there is no outer caustic surface: the ray reaches the core boundary (or the outer surface of the fiber). If we assume for simplicity that $n(r)$ increases monotonically from the axis this requires $g(\rho) > 0$. From Eq. (17.40) this requires

$$n_{cladding}^2 > \overline{\beta}^2 + \overline{\ell}^2. \tag{17.41}$$

There is only an inner caustic in this case, as shown in Fig. (17.5a).

For a ray to be bound $g(r) < 0$ everywhere in the cladding, which requires

$$n_{cladding} < \overline{\beta}, \tag{17.42}$$

which additionally requires, not surprisingly, that

$$n_{cladding} < \overline{\beta} \leq n_{core}. \tag{17.43}$$

Both meridional and skew bound rays exist. For the meridional rays $\overline{\ell} = 0$ and $g(r)$ has only a single root, corresponding to r_{tp} where

$$n(r_{tp}) = \overline{\beta}, \tag{17.44}$$

as shown in Fig. (17.5b). For the skew bound rays there are two roots of $g(r)$ corresponding to the inner and outer caustic surfaces, as shown in Fig. (17.5c).

Tunneling rays fall into a different category where

$$\overline{\beta} < n_{cladding} \text{ and } n_{cladding}^2 < \overline{\beta}^2 + \overline{\ell}^2. \tag{17.45}$$

The behavior of $g(r)$ in this case is such that between r_{tp} and an outer radius r_{rad}, $g(r)$ is negative and no ray paths exist. However, for $r > r_{rad}$, $g(r)$ is nonnegative as shown in Fig. (17.5d). The ray 'tunnels' from radius r_{tp} to r_{rad},

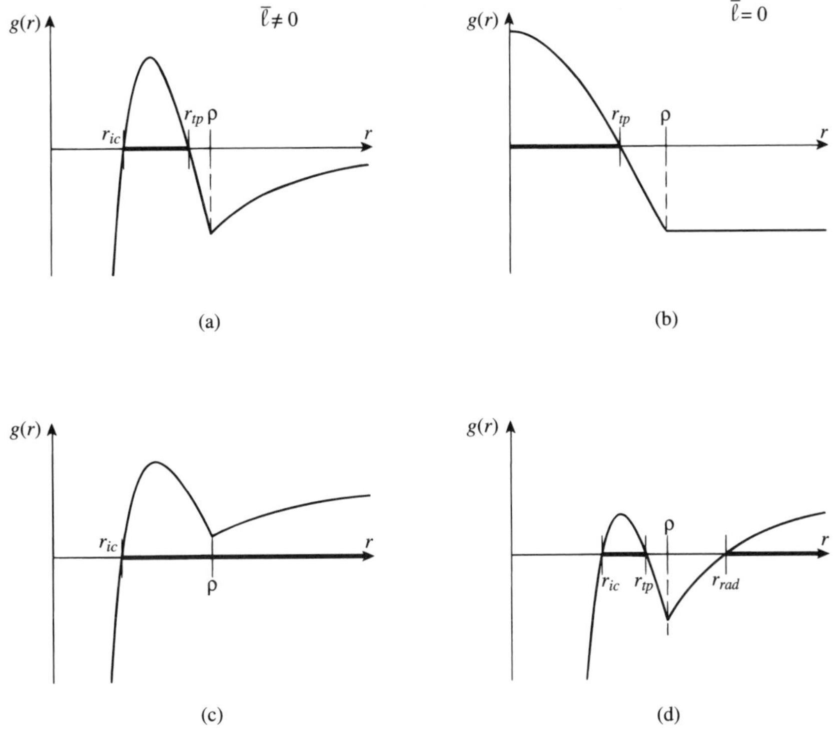

Fig. 17.5. Schematic form of the function $g(r)$, Eq. (17.40) for a graded-index fiber for different classes of ray; (a) skew bound ray; (b) meridional bound ray; (c) refracting ray; (d) tunnelling ray.

where it becomes a free ray. Since $g(r_{rad}) = 0$, it follows that

$$r_{rad} = \frac{\rho \bar{\ell}}{\sqrt{n^2_{\text{cladding}} - \bar{\beta}^2}}. \tag{17.46}$$

The mechanism by which energy leaks from the core to the outer part of the fiber is similar to the evanescent coupling of energy between two closely spaced surfaces, at one of which total internal reflection occurs, as was discussed in Chapter 14. Depending on the circumstances, tunnelling or 'leaky' rays may correspond to a very gradual loss of energy from the fiber and the fiber can guide energy in such rays over relatively long distances.

17.3 Ray Theory of a Dielectric Slab Guide

Before considering the wave theory of a cylindrical fiber in detail we can gain further insight into the propagation characteristics of meridional rays by analyzing the two-dimensional problem of a dielectric slab waveguide.

A dielectric slab of dielectric constant ϵ_1 will guide rays of light by total internal reflection provided the medium in which it is immersed has dielectric constant $\epsilon_2 < \epsilon_1$. For light rays making angle θ with the axis as shown in Fig. (17.6) total internal reflection will occur provided $\sin\theta_i > \sqrt{\epsilon_2/\epsilon_1}$.

Although there might appear to be an infinite number of such rays, this is not so. As the ray makes its zig-zag path down the guide the associated wavefronts must remain in phase or the wave amplitude will decay because of destructive

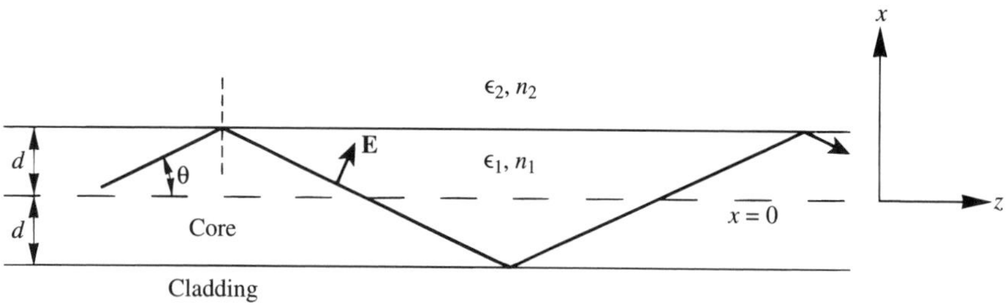

Fig. 17.6. Geometry describing meridional rays in a two-dimensional slab waveguide.

interference. We can analyze the slab as a Fabry–Perot resonator. Only those rays that satisfy the condition for maximum stored energy in the slab will propagate. This corresponds to the upward and downward components of the travelling wave constructively interfering. The component of the propagation constant perpendicular to the fiber axis for the ray shown in Fig. (17.1) is $k_1 \cos\theta_i$, so for constructive interference

$$4dk_1 \cos\theta_i = 2m\pi, \tag{17.47}$$

which gives the condition

$$\cos\theta_i = \frac{m\lambda_0}{4n_1 d}. \tag{17.48}$$

For simplicity we have neglected the phase shift that occurs when the wave totally internally reflects at the upper and lower core/cladding boundaries.

A given value of m in Eq. (17.47) corresponds to a phase shift of $2m\pi$ per round trip between the upper and lower boundaries of the slab: Such a *mode* will not propagate if θ is greater than the critical angle. In other words the cut-off condition for the mth mode is

$$\frac{d}{\lambda_0} = \frac{m}{4\sqrt{n_1^2 - n_2^2}}. \tag{17.49}$$

This condition is identical for waves polarized in the xz plane (P-waves) or polarized in the y direction (S-waves). Only the lowest mode, $m = 0$, has no cutoff frequency. A dielectric waveguide that is designed to allow the propagation of only the lowest mode is called a *single-mode* waveguide. The thickness of the guiding layer needs to be quite small to accomplish this. For example: at an operating wavelength of 1.55 μm and indices $n_1 = 1.5$, $n_2 = 1.485$ the maximum guide width that will support single-mode operation is 3.66 μm. Such waveguides are important because, in a simple sense, there is only one possible ray path in the guide. A short pulse of light injected into such a guide will emerge as a single pulse at its far end. In a multimode guide, a single pulse can travel along more than one path, and can emerge as multiple pulses. This is undesirable in a digital optical communication link. We can estimate when this would become a problem by using Eq. (17.48)

For the $m = 0$ mode in a guide of total length L the path length is L. For the $m = 1$ mode, the total path length along the guide is $L + \Delta L$, where

$$L + \Delta L = L/\sqrt{1 - (\lambda_0/4n_1 d)^2}. \tag{17.50}$$

A single, short optical pulse injected simultaneously into these two modes will emerge from the fiber as two pulses separated in time by an interval $\Delta\tau$ where

$$\Delta\tau = \frac{n_1 \Delta L}{c_0} = \frac{n_1 L}{c_0\sqrt{1-(\lambda_0/4n_1 d)^2}} - \frac{n_1 L}{c_0}. \tag{17.51}$$

This effect is called *group delay*. For example, if $L = 1$ km, $n_1 = 1.5$, and $d/\lambda_0 = 5$ then $\Delta\tau = 2.78$ ns. Clearly, communication rates in excess of about 200 MHz would be impossible in this case. As the number of modes increases, the bandwidth of the system decreases further.

17.4 The Goos–Hänchen Shift

We can examine the discussion of the previous section in a little more detail by including the phase shift that occurs when a wave undergoes total internal reflection. For an S-wave striking the core/cladding boundary the reflection coefficient is

$$\rho = \frac{n_1 \cos\theta_1 - n_2 \cos\theta_2}{n_1 \cos\theta_1 + n_2 \cos\theta_2}, \tag{17.52}$$

where we have made use of the effective impedance for S-waves at angle of incidence, θ_1, or refraction, θ_2, at the boundary, namely

$$Z'_{\text{core}} = \frac{Z_0}{n_1 \cos\theta_1}, \tag{17.53}$$

$$Z'_{\text{cladding}} = \frac{Z_0}{n_2 \cos\theta_2}. \tag{17.54}$$

For angles of incidence greater than the critical angle, $\cos\theta_2$ becomes imaginary and can be written as

$$\cos\theta_2 = i\sqrt{\left(\frac{n_1}{n_2}\right)^2 \sin^2\theta_1 - 1}. \tag{17.55}$$

In terms of the critical angle, θ_c,

$$\begin{aligned} \cos\theta_2 &= i\frac{n_1}{n_2}\sqrt{\sin^2\theta_1 - \sin^2\theta_c} \\ &= i\frac{n_1}{n_2}X. \end{aligned} \tag{17.56}$$

If we write the phase shift on reflection for the S-wave as ϕ_s, then from Eqs. (17.52) and (17.56)

$$\rho = |\rho| e^{i\phi_s} = \frac{\cos\theta_1 - iX}{\cos\theta_1 + iX}. \tag{17.57}$$

It is easy to show that $|\rho| = 1$: all the wave energy is reflected as a result of total internal reflection. Therefore, if we write

$$\cos\theta_1 - iX = e^{i\phi_s/2},$$

then

$$\tan(\phi_s/2) = -\frac{\sqrt{\sin^2\theta_1 - \sin^2\theta_c}}{\cos\theta_1}. \tag{17.58}$$

For P-waves the phase shift on reflection, ϕ_p, satisfies

$$\tan(\phi_p/2) = -\frac{\sqrt{\sin^2\theta_1 - \sin^2\theta_c}}{(n_2/n_1)\cos\theta_1}. \tag{17.59}$$

In practical optical waveguides usually $n_1 \simeq n_2$ so $\phi_s \simeq \phi_p = \phi$. We can put Eq. (17.58) in a more convenient form by using angles measured with respect to the fiber axis. We define

$$\theta_z = \pi/2 - \theta_1,$$
$$\theta_a = \pi/2 - \theta_c.$$

In a fiber with $n_1 \simeq n_2$, both θ_z and θ_a will be small angles. With these definitions

$$\phi = -2\frac{\sqrt{\theta_a^2 - \theta_z^2}}{\theta_z}. \tag{17.60}$$

If we represent the electric field of a guided wave incident on the slab boundary as

$$E_i(z) = E_1 e^{i\beta_1 z} \tag{17.61}$$

then after total internal reflection the wave becomes

$$E_r(z) = E_1 e^{\beta_1 z + \phi(\theta_1)}, \tag{17.62}$$

where $\beta_1 = k_1 \sin\theta_1$ is the propagation constant parallel to the z axis of the slab. Because this wave is laterally confined within the core we can include the effects of diffraction by considering the wave as a group of rays near the value θ_1. The phase shift $\phi(\beta_1)$ can be expanded as a Taylor series for propagation constants near β_1:

$$\phi(\beta) = \phi(\beta_1) + (\beta - \beta_1)\left(\frac{\partial\phi}{\partial\beta}\right)_{\beta_1}. \tag{17.63}$$

Therefore, the reflected wave can be written as

$$E_r(z) = E_1 \exp\left[i(\beta z + \phi(\beta_1) + (\beta - \beta_1)\left(\frac{\partial\phi}{\partial\beta}\right)_{\beta_1}\right]$$
$$= E_0 \exp\left[i(\beta z + \beta\left(\frac{\partial\phi}{\partial\beta}\right)_{\beta_1}\right] \tag{17.64}$$

where all the terms independent of angle have been incorporated into the new complex amplitude E_0. The additional phase factor $\beta\left(\partial\phi/\partial\beta\right)_{\beta_1}$ can be interpreted as a shift, Z_s, in the axial position of the wave as it reflects at the boundary, as shown in Fig. (17.7).

This shift in axial position Z_s is called the Goos–Hänchen shift[17.3]

$$Z_s = -\left(\frac{\partial\phi}{\partial\beta}\right)_{\beta_1}, \tag{17.65}$$

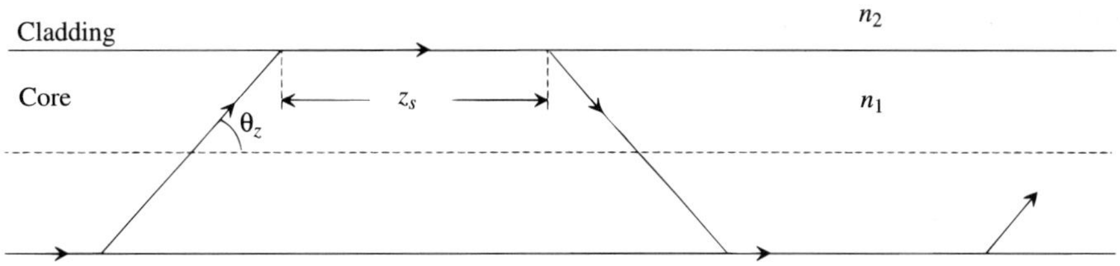

Fig. 17.7. The Goos–Hänchen shift.

which can be evaluated as

$$Z_s = -\left(\frac{\partial\phi}{\partial\theta_z}\frac{\partial\theta_z}{\partial\beta}\right)_{\beta_1}, \tag{17.66}$$

where

$$\frac{\partial\theta_z}{\partial\beta} = -\frac{1}{k\sin\theta_z}. \tag{17.67}$$

By differentiating Eq. (17.60) and assuming that the angles θ_z and θ_a are small, we get

$$Z_s \simeq \frac{\lambda}{\pi n_1\theta_z}\frac{1}{\sqrt{\theta_a^2-\theta_z^2}}. \tag{17.68}$$

This shift in axial position makes the distance travelled by the ray in propagating a distance ℓ along the fiber shorter than it would be without the shift. However, the shift is very small unless the ray angle is close to the critical angle, in which case the evanescent portion of the associated wave penetrates very far into the cladding.

17.5 Wave Theory of the Dielectric Slab Guide

When a given mode propagates in a slab, light energy is guided along in the slab medium or core. There are, however, finite field amplitudes that decay exponentially into the external medium. The propagating waves in this case, because they are guided along by the slab, but are not totally confined within it, are often called *surface waves*. This distinguishes them from the types of wave that propagate inside, and are totally confined by, hollow conducting structures – such as rectangular or cylindrical microwave waveguides. We can find the electromagnetic field profiles in the one-dimensional slab guide shown in Fig. (17.6) by using Maxwell's equations. We look for a propagating solution with fields that vary like $e^{i\omega t-\gamma z}$, where γ is the propagation constant.

There are both P- and S-wave solutions to the wave equation. Clearly, as can be seen from Fig. (17.6), the P-wave has an E_z component of its electric field so it could also be called a transverse magnetic field (TM) wave. Similarly the S-wave is a transverse electric field (TE) wave. From the curl equations, written in the form

$$\begin{aligned}\text{curl }\mathbf{E} &= -i\omega\mu\mathbf{H},\\ \text{curl }\mathbf{H} &= i\omega\epsilon\mathbf{E},\end{aligned} \tag{17.69}$$

the following equations result

$$\frac{\partial E_z}{\partial y} + \gamma E_y = -i\omega\mu H_x, \tag{17.70}$$

$$-\gamma H_x - \frac{\partial H_z}{\partial x} = i\omega\epsilon E_y, \tag{17.71}$$

$$-\gamma E_x - \frac{\partial E_z}{\partial x} = -i\omega\mu H_y, \tag{17.72}$$

$$\frac{\partial H_z}{\partial y} + \gamma H_y = i\omega\epsilon E_x, \tag{17.73}$$

$$\frac{\partial E_y}{\partial x} - \frac{\partial E_x}{\partial y} = -i\omega\mu H_z, \tag{17.74}$$

$$\frac{\partial H_y}{\partial x} - \frac{\partial H_x}{\partial y} = i\omega\epsilon E_z. \tag{17.75}$$

An expression for H_x can be found by eliminating E_y from Eqs. (17.70) and (17.71), or vice versa. E_x and H_y can be found from Eqs. (17.72) and (17.73).

$$E_x = \frac{-1}{\gamma^2 + k^2}\left(\gamma\frac{\partial E_z}{\partial x} + i\omega\mu\frac{\partial H_z}{\partial y}\right), \tag{17.76}$$

$$H_y = \frac{-1}{\gamma^2 + k^2}\left(i\omega\epsilon\frac{\partial E_z}{\partial x} + \gamma\frac{\partial H_z}{\partial y}\right), \tag{17.77}$$

$$E_y = \frac{1}{\gamma^2 + k^2}\left(-\gamma\frac{\partial E_z}{\partial y} + i\omega\mu\frac{\partial H_z}{\partial x}\right), \tag{17.78}$$

$$H_x = \frac{1}{\gamma^2 + k^2}\left(i\omega\epsilon\frac{\partial E_z}{\partial y} + \gamma\frac{\partial H_z}{\partial x}\right), \tag{17.79}$$

where $k = \omega\sqrt{\mu\epsilon}$.

From the Helmholtz equation (Eq. (16.4))

$$\nabla^2\mathbf{E} + k^2\mathbf{E} = 0; \quad \nabla^2\mathbf{H} + k^2\mathbf{H} = 0. \tag{17.80}$$

Since $\partial^2/\partial z^2$ is equivalent to multiplication by $-\gamma^2$, Eq. (17.80) can be written as

$$\frac{\partial^2\mathbf{E}}{\partial x^2} + \frac{\partial^2\mathbf{E}}{\partial y^2} = -(\gamma^2 + k^2)\mathbf{E}, \tag{17.81}$$

$$\frac{\partial^2\mathbf{H}}{\partial x^2} + \frac{\partial^2\mathbf{H}}{\partial y^2} = -(\gamma^2 + k^2)\mathbf{H}. \tag{17.82}$$

17.6 *P*-Waves in the Slab Guide

For P-waves in the slab guide $E_y = 0, \partial/\partial y \equiv 0$, so from Eq. (17.77)

$$H_y = -\frac{-i\omega\epsilon}{\gamma^2 + k^2}\frac{\partial E_z}{\partial x} \tag{17.83}$$

and from Eq. (17.81)

$$\frac{\partial^2 E_z}{\partial x^2} = -(\gamma^2 + k^2)E_z. \tag{17.84}$$

The solutions to Eq. (17.84) are sine or cosine functions, or exponentials. If we wish the waves to be confined to the slab we must choose the solution in medium 2 where the field amplitudes decay exponentially in the positive (or negative) x direction. So, for example, using Eqs. (17.83) and (17.84) the x dependence of E_z and H_y in the cladding will be

$$E_{z_2} = Be^{-\alpha_x x}; \quad H_{y_2} = -\frac{i\omega\epsilon_2}{\alpha_x} Be^{-\alpha_x x} \quad x \geq d, \tag{17.85}$$

$$E_{z_2} = Be^{\alpha_x x}; \quad H_{y_2} = \frac{i\omega\epsilon_2}{\alpha_x} Be^{\alpha_x x} \quad x \leq -d. \tag{17.86}$$

Inside the slab the fields are

$$E_{z_1} = A \sin k_x x; \quad H_{y_1} = \frac{-i\omega\epsilon_1}{k_x} A \cos k_x x \tag{17.87}$$

or

$$E_{z_1} = A \cos k_x x; \quad H_{y_1} = \frac{i\omega\epsilon_1}{k_x} A \sin k_x x, \tag{17.88}$$

where we have taken note that in medium 2

$$\alpha_x^2 = -(\gamma^2 + k_2^2), \tag{17.89}$$

where $\alpha_x > 0$, and in medium 1

$$k_x^2 = \gamma^2 + k_1^2, \tag{17.90}$$

where $k_x > 0$.

A propagating wave exists in the slab provided γ is imaginary. If γ were to be real then the $e^{-\gamma z}$ variation of the fields would represent an attenuated, nonpropagating *evanescent* wave. For γ to be imaginary, from Eq. (17.90), $k_1^2 > k_x^2$; therefore, $k_x^2 < \omega^2\mu_0\epsilon_1$.

To find the relation between α_x and k_x we use the boundary condition that at $x = d$ the E_z and H_y fields (which are the tangential fields at the boundary between the two media) must be continuous. Therefore,

$$(E_{z_1}/H_{y_1})_{x=d} = (E_{z_2}/H_{y_2})_{x=d}. \tag{17.91}$$

For the modes with odd E_z field symmetry in the slab, (the sine solution above)

$$\frac{-k_x}{i\omega\epsilon_1} \tan k_x d = \frac{-\alpha_x}{i\omega\epsilon_2}, \tag{17.92}$$

so

$$\alpha_x = \frac{\epsilon_2}{\epsilon_1} k_x \tan k_x d. \tag{17.93}$$

It is straightforward to show that for the even symmetry E_z-field solution in the slab (the cosine solution above)

$$\alpha_x = -\frac{\epsilon_2}{\epsilon_1} k_x \cot k_x d. \tag{17.94}$$

To determine the propagation constant γ for the odd-symmetry modes we must solve simultaneously Eqs. (17.89), (17.90), and (17.93). Elimination of γ from

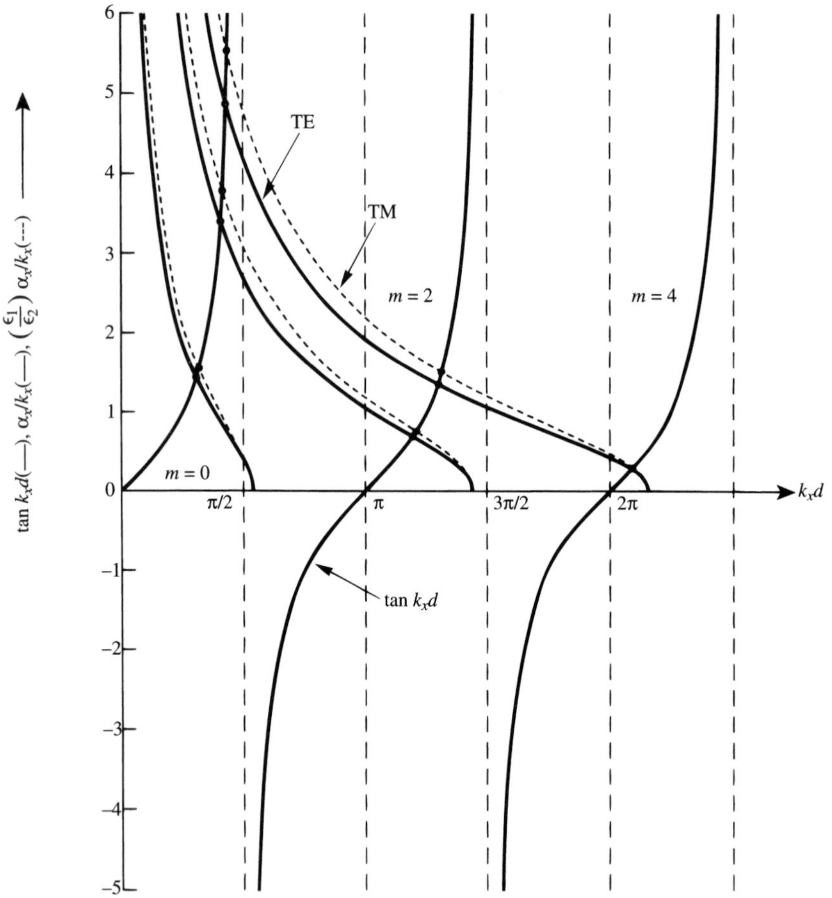

Fig. 17.8. Graphical solution of Eqs. (17.96) and (17.121) for the odd symmetry TM (P-wave) and TE (S-wave) modes in a slab waveguide. For these curves $n_1 = 1.5, n_2 = 1.4$. The three pairs of curves from left to right correspond to $\lambda_0 = 2\ \mu$m, 0.75 μm, and 0.5 μm, respectively.

Eqs. (17.89) and (17.90) gives

$$\alpha_x = \sqrt{k_1^2 - k_2^2 - k_x^2} = \sqrt{\omega^2\mu_0(\epsilon_1 - \epsilon_2) - k_x^2}. \tag{17.95}$$

We have assumed that $\mu_1 = \mu_2 = \mu_0$, as most optically transparent materials have relative permeability very close to unity.

From Eqs. (17.93) and (17.95):

$$\tan k_x d = \frac{\epsilon_1}{\epsilon_2}\sqrt{\frac{\omega^2\mu_0(\epsilon_1 - \epsilon_2)}{k_x^2} - 1}. \tag{17.96}$$

The solution to this equation can be illustrated graphically, as shown in Fig. (17.8). Whatever the value of ω there is always at least one solution to this equation. For small values of $k_x d$ Eq. (17.93) gives

$$\alpha_x = \frac{\epsilon_2}{\epsilon_1}k_x^2 d \tag{17.97}$$

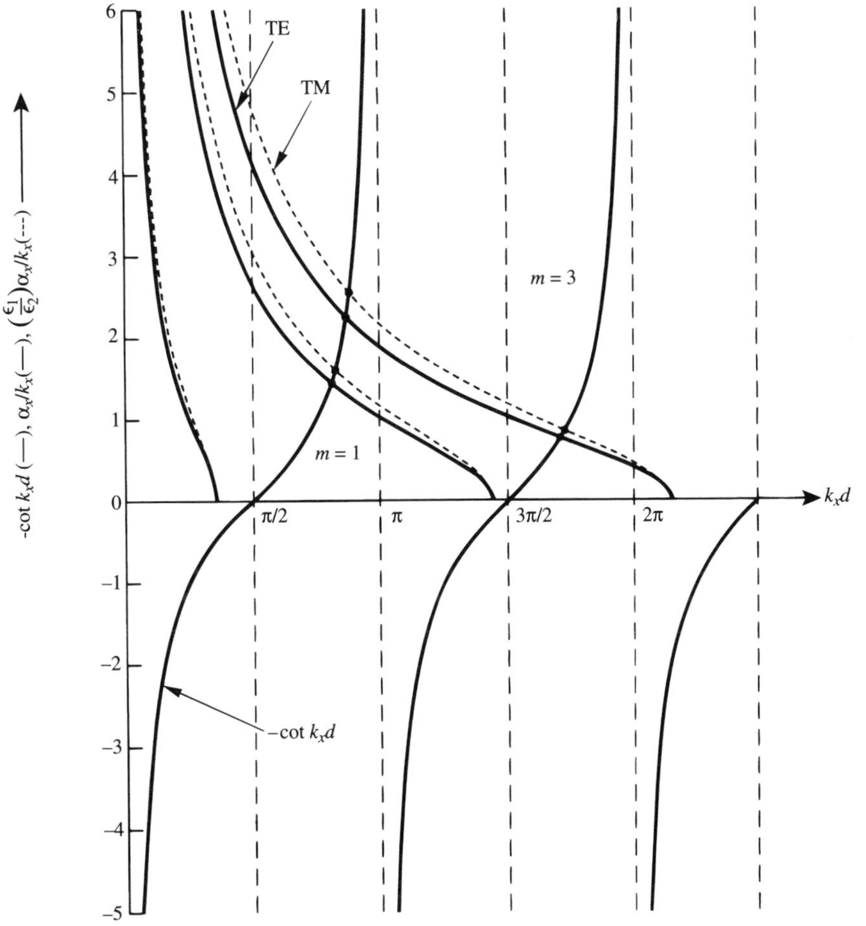

Fig. 17.9. Graphical solution of Eqs. (17.101) and ((17.128) for even-symmetry TM (P- and S-wave) and TE (S-wave) modes in a slab waveguide. For these curves $n_1 = 1.5, n_2 = 1.4$. The three pairs of curves from left to right correspond to $\lambda_0 = 3$ μm, 0.75 μm, and 0.5 μm, respectively. For the 3 μm wavelength the curves do not intersect and Eq. (17.101), for example, will have no solution.

and from Eq. (17.95)

$$k_x^4 d^4 + \left(\frac{\epsilon_1}{\epsilon_2}\right)^2 k_x^2 d^2 = \left(\frac{\epsilon_1}{\epsilon_2}\right)^2 \omega^2 \mu_0 (\epsilon_1 - \epsilon_2) d^2. \tag{17.98}$$

Neglecting the term in $k_x^4 d^4$ gives

$$k_x \simeq \omega^2 \mu_0 (\epsilon_1 - \epsilon_2) = k_1 - k_2 \tag{17.99}$$

and from Eq. (17.90), if $k_x \to 0$

$$\gamma \to \pm i k_1. \tag{17.100}$$

Therefore as either $\epsilon_1 \to \epsilon_2$, or d becomes very small, α_x also decreases and energy extends far into medium 2. The propagation constant of the wave approaches the value it would have in medium 2 alone.

The solution of Eq. (17.95) with the smallest value of $k_x d$ is called the *fundamental* P-wave mode of the guide. This mode can propagate at any frequency – it has no cut-off frequency. However, this is not true for the even-symmetry E_z field solution. In this case, from Eqs. (17.94) and (17.95)

$$\cot k_x d = -\frac{\epsilon_1}{\epsilon_2}\sqrt{\frac{\omega^2\mu_0(\epsilon_1-\epsilon_2)}{k_x^2}-1}. \tag{17.101}$$

The graphical solution of this equation is given in Fig. (17.9). In order for this equation to have a solution, $k_x d$ must be greater than $\pi/2$. This cannot be so while at the same time $\omega \to 0$.

17.7 Dispersion Curves and Field Distributions in a Slab Waveguide

For a propagating mode to exist, from either Eq. (17.96) or (17.101)

$$\omega^2\mu_0(\epsilon_1-\epsilon_2) \geq k_x^2. \tag{17.102}$$

In this case Eq. (17.90) implies

$$\omega^2\mu_0(\epsilon_1-\epsilon_2) \geq \gamma^2 + \omega^2\mu_0\epsilon_1, \tag{17.103}$$

which writing $\gamma = \pm i\beta$ implies $\beta \geq \omega^2\mu_0\epsilon_2$. By solving Eq. (17.96) or Eq. (17.101) for various values of ω we can plot the variation of β with ω for the S- and P-wave modes. These curves are called dispersion curves.

From Eq. (17.90)

$$\beta^2 = k_1^2 - k_x^2 = \omega^2\mu_0\epsilon_1 - k_x^2. \tag{17.104}$$

By solving for k_x from either Eq. (17.96) for the odd-symmetry E_z solution or Eq. (17.101) for the even-symmetry solution, we can produce the dispersion curves shown in Fig. (17.10). The curve labelled $m = 0$ is the fundamental mode, the curves labelled $m = 1, 2, 3$ etc. are alternate even-, odd-, even-symmetry E_z field higher modes. The E_x and H_y components of the P-wave solution are easily found from Eqs. (17.76) and (17.77). Some of their E_z field distributions within the slab are shown in Fig. (17.11). For the fundamental mode,

$$\begin{aligned} E_x(\text{in medium 1}) &= \frac{-\gamma}{\gamma^2+k_1^2}\frac{\partial E_z}{\partial x} = -\frac{A\gamma}{k_x}\cos k_x x, \\ E_x(\text{in medium 2}) &= -\frac{\gamma}{\gamma^2+k_2^2}\frac{\partial E_z}{\partial x} = -\frac{B\gamma}{\alpha_x}e^{-\alpha_x x}. \end{aligned} \tag{17.105}$$

Note that for each mode as the frequency increases the wave is confined more and more to the core and its propagation constant $\beta \to k_1$.

$$\begin{aligned} H_y(\text{in medium 1}) &= -\frac{i\omega\epsilon_1}{\gamma^2+k_1^2}\frac{\partial E_z}{\partial x} = -\frac{iA\omega\epsilon_1}{k_x}\cos k_x x, \\ H_y(\text{in medium 2}) &= -\frac{i\omega\epsilon_2}{\gamma^2+k_2^2}\frac{\partial E_z}{\partial x} = \frac{-iB\omega\epsilon_2}{\alpha_x}e^{-\alpha_x x}. \end{aligned} \tag{17.106}$$

The tangential component of magnetic field must be continuous at, say, $x = d$. So from Eq. (17.106)

$$\frac{A}{B} = \frac{\epsilon_2 k_x e^{-\alpha_x d}}{\epsilon_1 \alpha_x \cos k_x d}, \tag{17.107}$$

which from Eq. (17.28) gives

$$\frac{A}{B} = \frac{e^{-\alpha_x d}}{\sin k_x d}. \tag{17.108}$$

Fig. 17.10. Dispersion curves for P-wave modes in a slab waveguide. The solid dispersion curves are for the odd-symmetry E_z modes, the even-symmetry E_z modes. The curve has been calculated for $n_1 = 3$, $n_2 = \sqrt{3}$, $d = 0.6\ \mu$m, unrealistic values for a practical waveguide, but chosen to illustrate the dispersion curves clearly. The curves marked $m = 0, 2, 4, 6$ are calculated using Eq. (17.96), those for $m = 1, 3, 5$ from Eq. (17.101).

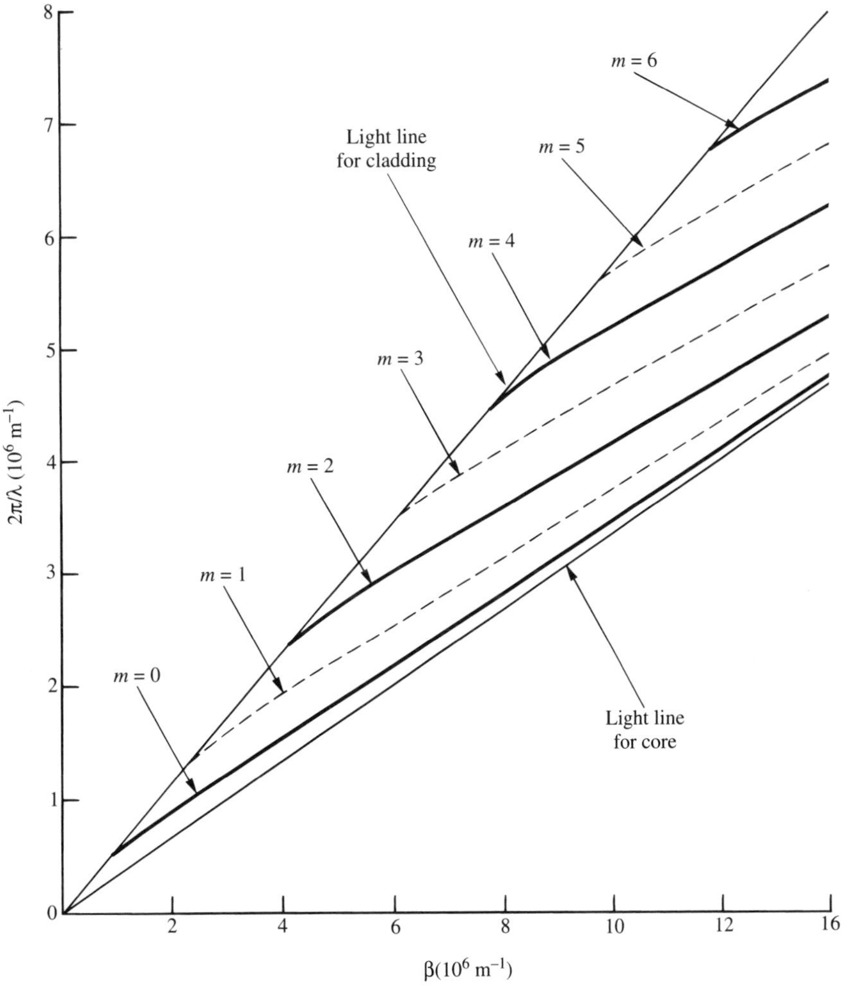

The actual value of A (or B) is determined from the energy flux in the dielectric slab. The transverse intensity distribution in the guide is

$$I = E_x H_y, \tag{17.109}$$

which from Eqs. (17.105) and (17.106) gives

$$I_1 \text{ (in medium 1)} = \frac{i\gamma w\epsilon_1 A^2}{k_x^2} \cos^2 k_x x, \tag{17.110}$$

$$I_2 \text{ (in medium 2)} = \frac{i\gamma w\epsilon_2 \beta^2}{\alpha_x^2} e^{-2\alpha_x x}. \tag{17.111}$$

The total energy flux per unit width is

$$\phi = 2\int_0^d I_1 dx + 2\int_d^\infty I_2 dx. \tag{17.112}$$

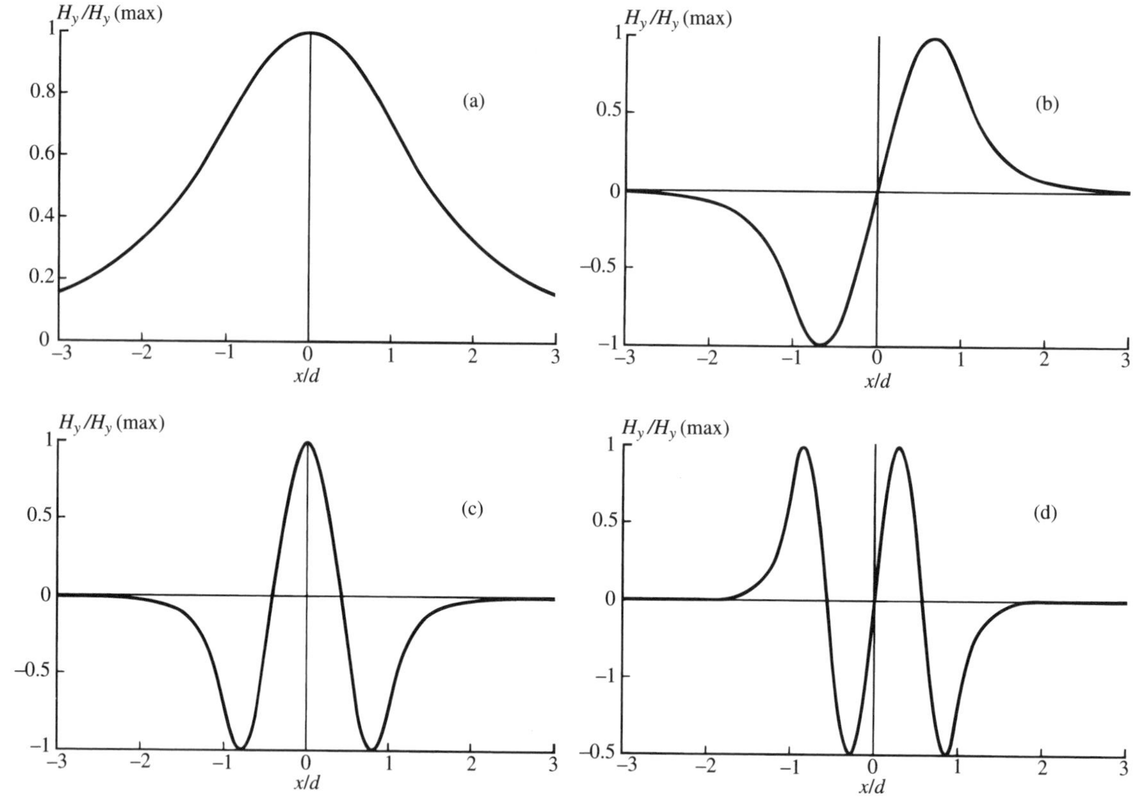

Fig. 17.11. H_y field variation for P-wave (TM wave) odd- and even-symmetry modes in a slab waveguide. The curves have been calculated for $n_1 = 1.5, n_2 = 1.485$: (a) $k_x d = \pi/4$, odd E_z symmetry, (b) $k_x d = 3\pi/4$, even E_z symmetry, (c) $k_x d = 5\pi/4$, odd E_z symmetry, (d) $k_x d = 7\pi/4$, even E_z symmetry.

17.8 S-Waves in the Slab Guide

For S-waves in the slab guide the electric field points only in the y direction, $\partial/\partial y \equiv 0$, and there are H_x and H_z components of magnetic field. In this case we start from Eq. (17.82) and write

$$\frac{\partial^2 H_z}{\partial x^2} = -(\gamma^2 + k^2)Hz. \tag{17.113}$$

For the modes with odd symmetry of the H_z field

$$H_{z_1} = C \sin k_x x \qquad | x | \leq d, \tag{17.114}$$

$$H_{x_1} = \frac{C\gamma k_x}{\gamma^2 + k_1^2} \cos k_x x = \frac{C\gamma}{k_x} \cos k_x x, \tag{17.115}$$

$$E_{y_1} = \frac{iC\omega\mu_1}{\gamma^2 + k_1^2} k_x \cos k_x x = \frac{iC\omega\mu_1}{k_x} \cos k_x x, \tag{17.116}$$

$$H_{z_2} = De^{-\alpha_x x}, \qquad x \geq d, \tag{17.117}$$

$$H_{x_2} = -\frac{C\gamma\alpha_x}{\gamma^2 + k_2^2} e^{-\alpha_x x} = \frac{C\gamma}{\alpha_x} e^{-\alpha_x x}, \tag{17.118}$$

$$E_{y_2} = -\frac{iD\omega\mu_2\alpha_x}{\gamma^2 + k_2^2} e^{-\alpha_x x} = \frac{iD\omega\mu_2}{\alpha_x} e^{-\alpha_x x}. \tag{17.119}$$

It is straightforward to show, in a similar manner to before, that

$$\alpha_x = \frac{\mu_2}{\mu_1} k_x \tan k_x d \tag{17.120}$$

and

$$\tan k_{xd} = \frac{\mu_1}{\mu_2}\sqrt{\frac{\omega^2 \mu_0(\epsilon_1 - \epsilon_2)}{k_x^2} - 1}. \tag{17.121}$$

Even as $\omega \to 0$ there is always at least one solution to the equation. This is the *fundamental*, $m = 0$, S-wave mode of the guide. The graphical solution of Eq. (17.121) in Fig. (17.8) illustrates this.

Modes which have even symmetry of the H_z field have field components

$$H_{z_1} = C \cos k_x x, \qquad |x| \leq d, \tag{17.122}$$

$$H_{x_1} = -\frac{C\gamma}{k_x} \sin k_x x, \tag{17.123}$$

$$E_{y_1} = -\frac{iC\omega\mu_1}{k_x} \sin k_x x, \tag{17.124}$$

$$H_{z_2} = De^{-\alpha_x x}, \qquad x \geq d, \tag{17.125}$$

$$H_{x_2} = \frac{C\gamma}{\alpha_x} e^{-\alpha_x}, \tag{17.126}$$

$$E_{y_2} = \frac{iD\omega\mu_2}{\alpha_x} e^{-\alpha_x x}. \tag{17.127}$$

The defining equation for the modes is

$$\alpha_x = -\frac{\mu_2}{\mu_1} k_x \cot k_x d. \tag{17.128}$$

Because most optical materials have $\mu \simeq \mu_0$ the cutoff frequencies of the S-wave modes above the fundamental are slightly lower, by a factor ϵ_2/ϵ_1, than the corresponding P-wave modes as shown in Fig. (17.9).

17.9 Practical Slab Guide Geometries

Slab guides that are infinite in one dimension are convenient mathematically but do not exist in practice. Practical slab guides utilize index changes or discontinuities in both the x and y directions (for a guide intended to propagate waves in the z direction). Some examples are shown in Fig. (17.12). Such structures can be fabricated by methods similar to those used in integrated circuit manufacture. In each case the fields outside the guiding layer decay exponentially. The fields are broadly similar to those discussed previously for the infinite slab guide. However, the modes that exist are no longer pure P- or S-wave modes, they possess all components of **E** and **H**, although they can still be predominantly P or S in character in some circumstances.

Slab guide geometries exist in most semiconductor diode lasers and in many LEDs as discussed in Chapter 13. Consequently it is possible to integrate the light source with a slab waveguide and then incorporate active components such as modulators, couplers, and switches into the structure. The resultant *integrated optic* package can serve as a complete transmitting module with full potential for amplitude or phase modulation, or multiplexing. We shall meet examples of

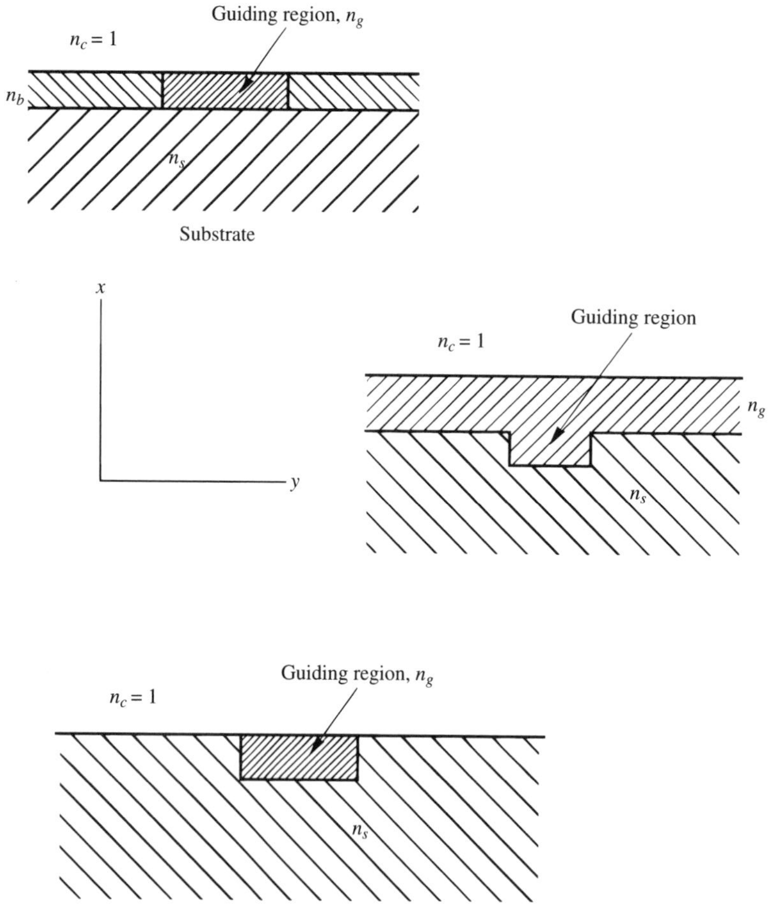

Fig. 17.12. Some practical slab waveguide geometrics.

these devices later, in Chapter 19. In order to realize a practical guided-wave optical communication link over long distances slab waveguides are not practical. The information-carrying light beam must be transmitted in an optical fiber – a cylindrical dielectric waveguide.

17.10 Cylindrical Dielectric Waveguides

The ray description of the slab waveguide is useful for describing the path of a meridional ray down a fiber. However, to obtain a more detailed understanding of the propagation characteristics of the fiber we again use Maxwell's equations. In cylindrical coordinates the curl equation (Eq. (17.56)) is:

$$\text{curl}\,\mathbf{E} = \left(\frac{1}{r}\frac{\partial E_z}{\partial \phi} - \frac{\partial E_\phi}{\partial z}\right)\hat{\mathbf{r}} + \left(\frac{\partial E_r}{\partial z} - \frac{\partial E_z}{\partial r}\right)\hat{\boldsymbol{\phi}} + \left[\frac{1}{r}\frac{\partial (rE_\phi)}{\partial r} - \frac{1}{r}\frac{\partial E_r}{\partial \phi}\right]\hat{\mathbf{z}} = i\omega\mu\mathbf{H} \qquad (17.129)$$

with a similar equation for curl $\mathbf{H}$. E_r, E_ϕ, E_z, respectively, are the radial, azimuthal, and axial components of the electric field, with associated unit vectors $\hat{\mathbf{r}}, \hat{\boldsymbol{\phi}}$, and $\hat{\mathbf{z}}$.

The Helmholtz equations Eqs. (17.81) and (17.82) become:

$$\begin{aligned}\frac{\partial^2 \mathbf{E}}{\partial r^2}+\frac{1}{r}\frac{\partial \mathbf{E}}{\partial r}+\frac{1}{r^2}\frac{\partial^2 \mathbf{E}}{\partial \phi^2}&=-(\gamma^2+k^2)\mathbf{E},\\ \frac{\partial^2 \mathbf{H}}{\partial r^2}+\frac{1}{r}\frac{\partial \mathbf{H}}{\partial r}+\frac{1}{r^2}\frac{\partial^2 \mathbf{H}}{\partial \phi^2}&=-(\gamma^2+k^2)\mathbf{H},\end{aligned} \tag{17.130}$$

where, as before, $k^2 = \omega^2\mu\epsilon$.

We start by examining the solution of Eq. (17.130) for the axial component of the field. We look for a separable solution of the form

$$E_z(r,\phi) = R(r)\Phi(\phi). \tag{17.131}$$

The z dependence of E_z is $e^{i\omega t-\gamma z}$, as before, but this factor will be omitted for simplicity. Substituting from (17.131) into (17.130) gives

$$r^2\frac{R''}{R}+\frac{rR'}{R}+k_c^2r^2=-\frac{\Phi''}{\Phi}, \tag{17.132}$$

where primes denote differentiation and we have written $k_c^2 = \gamma^2 + k^2$.

The only way for Eq. (17.132) to be satisfied, since the LHS is only a function of r, and the RHS is only a function of ϕ, is for each side to be independently equal to a constant. Therefore,

$$r^2\frac{R''}{R}+\frac{rR'}{R}+k_c^2r^2=\nu^2, \tag{17.133}$$

which is usually written as

$$R''+\frac{1}{r}\frac{\partial R}{\partial r}+\left(k_c^2-\frac{\nu^2}{r^2}\right)R=0 \tag{17.134}$$

and

$$-\frac{\Phi''}{\Phi}=\nu^2. \tag{17.135}$$

The general solution of Eq. (17.135) is

$$\Phi(\phi) = A\sin\nu\phi + B\cos\nu\phi. \tag{17.136}$$

ν must be a positive or negative integer, or zero, so that the function $\Phi(\phi)$ is single valued: i.e., $\Phi(\phi+2\pi) = \Phi(\phi)$.

Eq. (17.134) is called Bessel's equation. Its solutions are Bessel functions, which are the cylindrical geometry equivalents of the real or complex exponential solutions we encountered in the slab guide.

If $k_c^2 > 0$ the general solution of Eq. (17.134) is

$$R(r) = CJ_\nu(k_cr) + DY_\nu(k_cr), \tag{17.137}$$

where $J_\nu(k_cr)$ is the Bessel function of the first kind and $Y_\nu(k_cv)$ is the Bessel function of the second kind.† These are oscillatory functions of r, as shown in Fig. (17.13).

If $k_c^2 < 0$ the general solution of Eq. (17.134) is

$$R(r) = C'I_\nu(|k_c|r) + D'K_\nu(|k_c|r), \tag{17.138}$$

where $I(|k_c|r), K_\nu(|k_c|r)$ are the modified Bessel functions of the first and second

† The Y_ν Bessel function of the second kind is also written as N_ν by some authors.

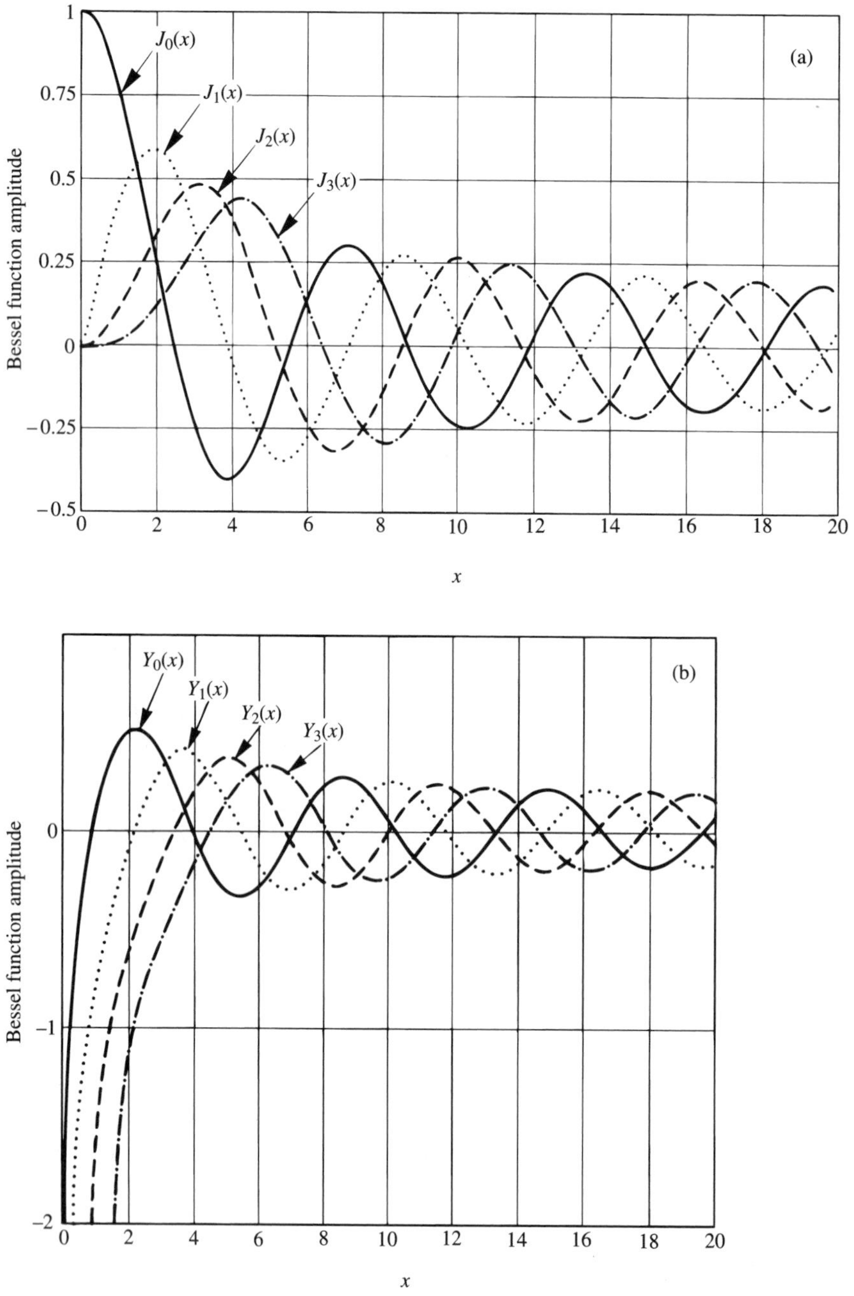

Fig. 17.13. Bessel functions of (a) the first and (b) second kinds.

kinds, respectively. These are monotonically increasing and decreasing functions of r as shown in Fig. (17.14).

Before proceeding further we can make several physical observations to help simplify our analysis.

(i) We cannot use the equations above to solve for wave propagation in a graded-index fiber – a fiber where the dielectric constant varies smoothly

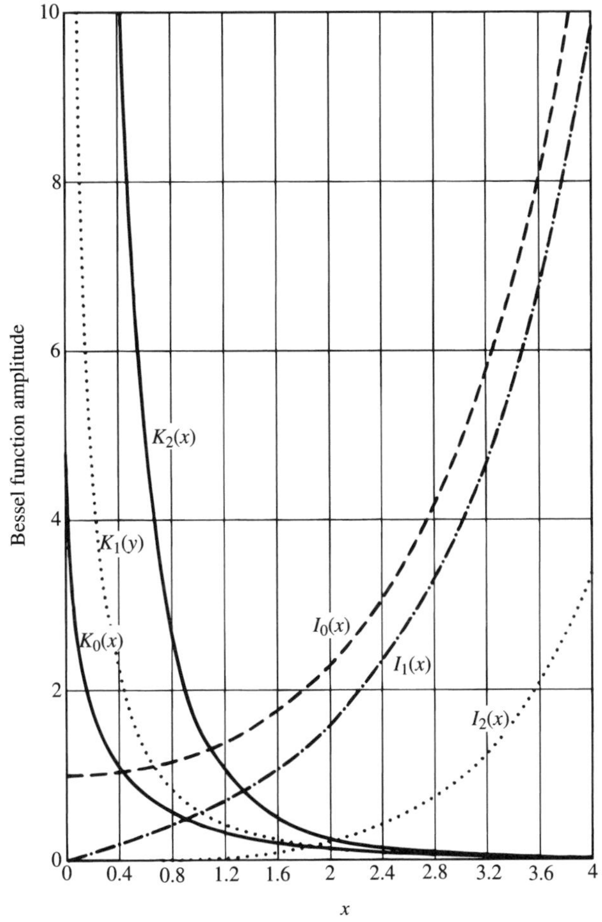

Fig. 17.14. Modified Bessel functions of the first and second kinds.

in the radial direction. In such fibers ϵ is a function of r (the quadratic index fiber is an example) and the wave equation analysis is much more complex. References [17.1] and [17.4] can be consulted for details.

(ii) We can use these equations to find the field components in a step-index fiber, whose radial refractive index profile is as shown in Fig. (17.15). We find a solution for the core, where the dielectric constant $\epsilon = n_1^2$ and match this solution to an appropriate solution for the cladding, where $\epsilon_2 = n_2^2$.

(iii) In the core we expect an oscillatory form of the axial field to be appropriate (as it was for the slab guide). Since the function $Y_\nu(k_c r)$ blows up as $r \to 0$ we reject it as physically unreasonable. Only the $J_\nu(k_c r)$ variation needs to be retained.

(iv) We are interested in fibers with $n_1 > n_2$ that guide energy. The fields in the cladding must decay as we go further away from the core. Since

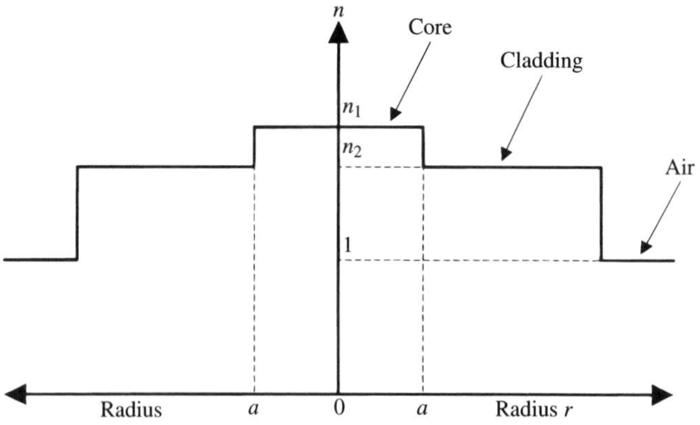

Fig. 17.15. Radial refractive index profile of a step-index cylindrical fiber.

$I_\nu(k_c r)$ blows up as $r \to \infty$ we reject it as physically unreasonable. Only the $K_\nu(k_c r)$ variation needs to be retained.

(v) In the core

$$k_{c_1}^2 > 0 \text{ implies } \gamma^2 + k_1^2 > 0 \tag{17.139}$$

and for a propagating solution $\gamma \equiv i\beta$; therefore

$$k_1^2 > \beta^2. \tag{17.140}$$

In the cladding $k_{c_2}^2 < 0$, and therefore $\gamma^2 + k_2^2 < 0$ which implies $k_2^2 < \beta^2$. In summary

$$k_2 < \beta < k_1. \tag{17.141}$$

The closer β approaches k_1, the more fields are confined to the core of the fiber, while for β approaching k_2, they penetrate far into the cladding. When β becomes equal to k_2 the fields are no longer guided by the core: they penetrate completely into the cladding.

(vi) If $k_c^2 > 0$, then the axial fields in the cladding would no longer be described by the K_ν Bessel function, instead they would be described by the J_ν Bessel function. These oscillatory fields no longer decay evanescently in the cladding and the wave is no longer guided by the core. We can take $k_c^2 = 0$ as a demarcation point between a wave being guided, or leaking from the core.

If we take the $\sin \nu\phi$ azimuthal variation for the axial electric field then in the core we have,

$$E_{z_1} = AJ_\nu(k_{c_1} r) \sin \nu\phi \quad (r \le a), \tag{17.142}$$

and in the cladding

$$E_{z_2} = CK_\nu(|\, k_{c_2} \,|\, r) \sin \nu\phi \quad (r \ge a), \tag{17.143}$$

where the subscripts 1 and 2 indicate core and cladding fields, respectively. These fields must be continuous at the boundary between core and cladding so

$$E_{z_1}(r = a) = E_{z_2}(r = a). \tag{17.144}$$

Far into the cladding the evanescent nature of the fields becomes apparent, since for large values of r

$$K_\nu(|k_{c_2}|r) \propto \frac{e^{-|k_{c_2}|r}}{\sqrt{|k_{c_2}|r}}. \tag{17.145}$$

The corresponding magnetic fields are

$$H_z = BJ_\nu(k_{c_1}) \cos \nu\phi \quad (r \le a), \tag{17.146}$$

$$H_z = DK_\nu(k_{c_2}) \cos \nu\phi \quad (r \ge a). \tag{17.147}$$

In this case, where we have assumed the $\sin \nu\phi$ for the electric field, we must have the $\cos \nu\phi$ variation for the H_z field to allow matching of the tangential fields (which include both z and ϕ components) at the core/cladding boundary. We must allow for the possibility of both axial electric and magnetic field components in the fiber: so-called *hybrid* modes. However, if $\nu = 0$ then Eq. (17.143) shows that $E_z = 0$: this is a TE mode. If we had chosen the $\cos \nu\phi$ variation for the axial electric field, then with $\nu = 0$ we would have a TM mode. Thus, TE and TM modes only exist for $\nu = 0$. To find the relationship between the field components orthogonal to the fiber axis and the E_z and H_z components we work from Eq. (17.129) and the corresponding equation for curl **H**. Some tedious algebra will yield the results

$$E_r = -\frac{i}{k_c^2}\left(\frac{\gamma}{i}\frac{E_z}{\partial r} + \omega\mu\frac{1}{r}\frac{\partial H}{\partial \phi}\right), \tag{17.148}$$

$$E_\phi = -\frac{i}{k_c^2}\left(\frac{\gamma}{i}\frac{1}{r}\frac{\partial E_z}{\partial \phi} - \omega\mu\frac{\partial H_z}{\partial r}\right), \tag{17.149}$$

$$H_r = -\frac{i}{k_c^2}\left(\frac{\gamma}{i}\frac{\partial H_z}{\partial r} - \omega\epsilon\frac{1}{r}\frac{\partial E_z}{\partial \phi}\right), \tag{17.150}$$

$$H_\phi = -\frac{i}{k_c^2}\left(\frac{\gamma}{i}\frac{1}{r}\frac{\partial H_z}{\partial \phi} + \omega\epsilon\frac{\partial E_z}{\partial r}\right). \tag{17.151}$$

Note that for a TE mode $(E_z = 0), E_r \sim \partial H_z/\partial\phi$ so that if H_z has a variation $\sim \cos \nu\phi$, where ν is a constant, then E_r must vary as $\sin \nu\phi$. For a TM mode $(H_z = 0)$, and $H_r \sim \partial E_z/\partial\phi$ with similar consequences. Bearing these facts in mind let us write down the full variation of the fields in the core and cladding as:

17.10.1 Fields in the Core

$$E_z = AJ_\nu\left(\frac{ur}{a}\right)\sin \nu\phi, \tag{17.152}$$

$$E_r = \left[-\frac{A\gamma}{(u/a)}J_\nu'\left(\frac{ur}{a}\right) + B\frac{i\omega\mu_0}{(u/a)^2}\frac{\nu}{r}J_\nu\left(\frac{ur}{a}\right)\right]\sin \nu\phi, \tag{17.153}$$

$$E_\phi = \left[-\frac{A\gamma}{(u/a)^2}\frac{\nu}{r}J_\nu\left(\frac{ur}{a}\right) + B\frac{i\omega\mu_0}{(u/a)}J_\nu'\left(\frac{ur}{a}\right)\right]\cos \nu\phi, \tag{17.154}$$

$$H_z = BJ_\nu\left(\frac{ur}{a}\right)\cos \nu\phi, \tag{17.155}$$

$$H_r = \left[A\frac{i\omega\epsilon_1}{(u/a)^2}\frac{\nu}{r}J_\nu\left(\frac{ur}{a}\right) - \frac{B\gamma}{(u/a)}J_\nu'\left(\frac{ur}{a}\right)\right]\cos \nu\phi, \tag{17.156}$$

$$H_\phi = \left[-A\frac{i\omega\epsilon_1}{(u/a)}J_\nu'\left(\frac{ur}{a}\right) + \frac{B\gamma}{(u/a)^2}\frac{\nu}{r}J_\nu\left(\frac{ur}{a}\right)\right]\sin \nu\phi, \tag{17.157}$$

where $u = k_{c_1} a = a\sqrt{k_1^2 + \gamma^2}$ is called the *normalized transverse phase constant* in the core region.

17.10.2 Fields in the Cladding

$$E_z = CK_\nu\left(\frac{wr}{a}\right)\sin\nu\phi, \tag{17.158}$$

$$E_r = \left[\frac{C\gamma}{(w/a)}K_\nu'\left(\frac{wr}{a}\right) - D\frac{i\omega\mu_0}{(w/a)^2}\frac{\nu}{r}K_\nu\left(\frac{wr}{a}\right)\right]\sin\nu\phi, \tag{17.159}$$

$$E_\phi = \left[\frac{C\gamma}{(w/a)^2}\frac{\nu}{r}K_\nu\left(\frac{wr}{a}\right) - D\frac{i\omega\mu_0}{(w/a)}K_\nu'\left(\frac{wr}{a}\right)\right]\cos\nu\phi, \tag{17.160}$$

$$H_z = DK_\nu\left(\frac{wr}{a}\right)\cos\nu\phi, \tag{17.161}$$

$$H_r = \left[-C\frac{i\omega\epsilon_2}{(w/a)^2}\frac{\nu}{r}K_\nu\left(\frac{wr}{a}\right) + \frac{D\gamma}{(w/a)}K_\nu'\left(\frac{wr}{a}\right)\right]\cos\nu\phi, \tag{17.162}$$

$$H_\phi = \left[C\frac{i\omega\epsilon_2}{(w/a)}K_\nu'\left(\frac{wr}{a}\right) - \frac{D\gamma}{(w/a)^2}\frac{\nu}{r}K_\nu\left(\frac{wr}{a}\right)\right]\sin\nu\phi, \tag{17.163}$$

where $w =\mid k_c \mid a = -a\sqrt{k_2^2 + \gamma^2})$ is called the *normalized transverse attenuation coefficient in the cladding.* If $w = 0$, this determines the *cut off* condition for the particular field distribution.

17.10.3 Boundary Conditions

To determine the relationship between the constants A, B, C, and D in Eqs. (17.152)–(17.163) we need to match appropriate tangential and radial components of the fields at the boundary. If we neglect magnetic behavior and put $\mu_1 = \mu_2 = \mu_0$, for the tangential fields:

$$\begin{aligned} E_{z_1} &= E_{z_2}, \\ E_{\phi_1} &= E_{\phi_2}, \\ H_{z_1} &= H_{z_2}, \\ H_{\phi_1} &= H_{\phi_2}, \end{aligned} \tag{17.164}$$

and for the radial fields

$$\begin{aligned} \epsilon_1 E_{r_1} &= \epsilon_2 E_{r_2}, \\ H_{r_1} &= H_{r_2}, \end{aligned} \tag{17.165}$$

where all fields are evaluated at $r = a$, and the subscripts 1 and 2 refer to core and cladding fields, respectively. It can be shown that use of these relationship in conjunction with Eqs. (17.152)–(17.163) leads to the condition

$$\left[\frac{J_\nu'(u)}{uJ_\nu(u)} + \frac{K_\nu'(w)}{wK_\nu(w)}\right]\left[\frac{\epsilon_1}{\epsilon_2}\frac{J_\nu'(u)}{uJ_\nu(u)} + \frac{K_\nu'(w)}{wK_\nu(w)}\right] = \nu^2\left(\frac{1}{u^2} + \frac{1}{w^2}\right)\left(\frac{\epsilon_2}{\epsilon_2}\frac{1}{u^2} + \frac{1}{w^2}\right). \tag{17.166}$$

Furthermore,

$$u^2 + w^2 = (k_1^2 + \gamma^2)a^2 - (k_2^2 + \gamma^2)a^2 = (k_1^2 - k_2^2)a^2. \tag{17.167}$$

Since $k_1 = 2\pi n_1/\lambda_0, k_2 = 2\pi n_2/\lambda_0$, Eq. (17.167) gives

$$u^2 + w^2 = 2k_1^2 a^2 \Delta. \tag{17.168}$$

Δ is the relative refractive index difference between core and cladding

$$\Delta = \frac{(n_1^2 - n_2^2)}{2n_1^2} \simeq \frac{(n_1 - n_2)}{n_1}. \tag{17.169}$$

A fiber is frequently characterized by its V number where

$$V = \sqrt{u^2 + w^2}, \tag{17.170}$$

which can be written as

$$\begin{aligned} V &= \frac{2\pi a}{\lambda_0}\sqrt{n_1^2 - n_2^2} = \frac{2\pi a}{\lambda_0}(NA) \\ &= kn_1 a\sqrt{2\Delta}. \end{aligned} \tag{17.171}$$

V is also called the normalized frequency.

Eqs. (17.166) and (17.168) can be solved for u and w to determine the specific field variations in the fiber.

17.11 Modes and Field Patterns

The modes in a cylindrical dielectric waveguide are described as either TE, TM or hybrid. The hybrid modes fall into two categories, HE modes where the axial electric field E_z is significant compared to E_r or E_ϕ, and EH modes where the axial magnetic field H_z is significant compared to H_r or H_ϕ. These modes are further characterized by two integers v and ℓ. A mode characterized as HE_{12}, for example, has azimuthal field variations like $\sin\phi$ or $\cos\phi$, and since $\ell = 2$, has two radial oscillations of the axial electric field within the core. For the TE and TM modes $v = 0$. For the E_z field variation given by Eq. (17.152)

$$E_z = AJ_v\left(\frac{ur}{a}\right)\sin v\phi \tag{17.172}$$

the number of radial oscillations of the axial electric field is determined by how many times $J_v(x)$ has crossed the axis as x varies from 0 to ur/a. Each time the Bessel function oscillates from positive to negative, or vice-versa, there is a new value of ur/a where the oscillatory function can merge smoothly with the radially decaying field distribution in the cladding.

The field distributions are easiest to write down for the $TE_{0\ell}$ and $TM_{0\ell}$ modes. Setting $v = 0$ in Eqs. (17.152)–(17.157) we get the fields of the $TE_{0\ell}$ mode, for example, in the core:

$$E_z = 0, \tag{17.173}$$

$$E_r = 0, \tag{17.174}$$

$$E_\phi = -B\frac{i\omega\mu_0}{(u/a)}J_1\left(\frac{ur}{a}\right), \tag{17.175}$$

$$H_z = BJ_0\left(\frac{ur}{a}\right), \tag{17.176}$$

$$H_r = \frac{B\gamma}{(u/a)}J_1\left(\frac{ur}{a}\right). \tag{17.177}$$

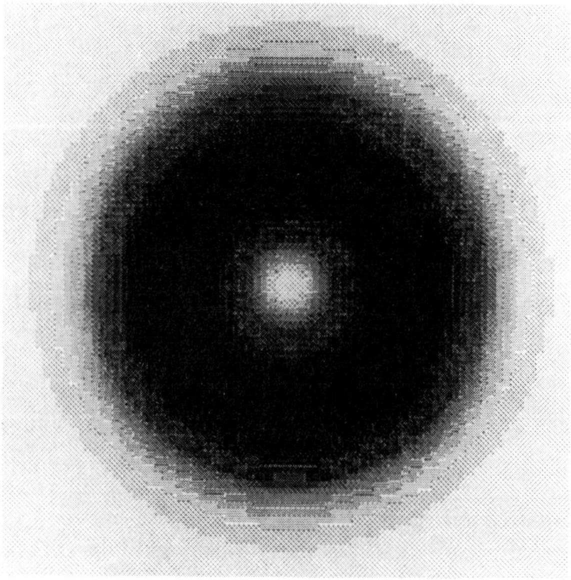

Fig. 17.16. Intensity pattern of axisymmetric TE_{01} mode.

We have made use of the relation $J_0'(x) = -J_1(x)$. The $TE_{0\ell}$ modes correspond to choosing the constants $A = C = 0$ in Eqs. (17.152)–(17.163).

We could have chosen the $\cos \nu\phi$ variation for the E_z field in which we would have got the fields of the $TM_{0\ell}$ mode by setting $\nu = 0$, which in this case implies $B = D = 0$.

The transverse fields of the mode give rise to a characteristic Poynting vector or intensity distribution, which can be called the *mode* pattern. For the $TE_{0\ell}$ mode this is

$$S(r) = E_\phi H_r \propto J_1^2\left(\frac{ur}{a}\right). \tag{17.178}$$

Because $J_1(0) = 0$, this is a donut-shaped intensity distribution as shown in Fig. (17.16). The intensity distribution for the TM_{01} mode looks the same. We shall see shortly that these are *not* the *fundamental* modes of the step-index fiber. These modes cannot propagate at a given frequency for an arbitrarily small core diameter – they have a *cutoff* frequency. The only mode that has no cutoff frequency is the HE_{11} mode. To discuss this further, and in order to find a way of grouping modes with similar mode patterns together we need to work in the *weakly-guiding approximation*[17.5].

17.12 The Weakly-Guiding Approximation

In the weakly-guiding approximation we neglect the difference in refractive index between core and cladding and set $\epsilon_1 = \epsilon_2$. In this case the boundary condition relation, Eq. (17.166) becomes

$$\frac{J_\nu'(u)}{uJ_\nu(w)} + \frac{K_\nu'(w)}{wK_\nu(w)} = \pm\nu\left(\frac{1}{u^2} + \frac{1}{w^2}\right). \tag{17.179}$$

For the $TE_{0\ell}$ and $TM_{0\ell}$ mode this gives

$$\frac{J_1(u)}{uJ_0(u)} + \frac{K_1(w)}{wK_0(w)} = 0, \tag{17.180}$$

where we have made use of the relation $K_0'(w) = -K_1(w)$.

By the use of a series of relationships between Bessel functions, given in Appendix 10, when the positive sign is taken on the right hand side of Eq. (17.179), the following equation results

$$\frac{J_{\nu+1}(u)}{uJ_\nu(n)} + \frac{K_{\nu+1}(w)}{wK_\nu(w)} = 0. \tag{17.181}$$

This equation describes the EH modes. When the negative sign is taken in Eq. (17.179) the following equation for the HE modes is obtained

$$\frac{J_{\nu-1}(u)}{uJ_\nu(u)} - \frac{K_{\nu-1}(w)}{wK_\nu(w)} = 0. \tag{17.182}$$

By the use of the Bessel function relations given in Appendix 10, Eq. (17.182) can be rearranged into the form

$$\frac{J_{\nu-1}(u)}{uJ_{\nu-2}(u)} + \frac{K_{\nu-1}(w)}{wK_{\nu-2}(w)} = 0. \tag{17.183}$$

If we introduce a new integer m, which takes the values

$$m = \begin{cases} 1 & \text{for TE and TM modes,} \\ \nu + 1 & \text{for EH modes,} \\ \nu - 1 & \text{for HE modes,} \end{cases} \tag{17.184}$$

then a unified boundary condition relation results that includes Eqs. (17.180), (17.181), and (17.183). It is

$$\frac{uJ_{m-1}(u)}{J_m(u)} = -\frac{wK_{m-1}(w)}{K_m(w)}. \tag{17.185}$$

If $m - 1 < 0$, as would be the case for $HE_{\nu\ell}$ modes with $\nu = 1$, then the Bessel function relations $J_{-\nu} = (-1)^\nu J_\nu$ and $K_{-\nu} = K_\nu$ would be useful in evaluating Eq. (17.185).

If the V-number of the fiber is known, given by Eqs. (17.170) and (17.171), then the values of u and w that satisfy Eq. (17.106) can be determined. Once these parameters are known the fields in the fiber can be calculated from Eqs. (17.152)–(17.163).

17.13 Mode Patterns

In the weakly-guiding approximation the Poynting vector variation within the fiber is again calculated from

$$S = E_t H_t, \tag{17.186}$$

where the transverse fields in the fiber are

$$\begin{aligned} \mathbf{E}_t &= E_r \hat{\mathbf{r}} + E_\phi \hat{\phi}, \\ \mathbf{H}_t &= H_r \hat{\mathbf{r}} + H_\phi \hat{\phi}. \end{aligned} \tag{17.187}$$

The transverse intensity distribution has a characteristic mode pattern. Similarities between these mode patterns allow groups of TE, TM, and hybrid modes to be grouped together as LP (linearly polarized) modes, as shown in Fig. (17.17). Each

LP mode is characterized by the two integers m and ℓ: ℓ describes the number of radial oscillations of the field that occur in the core; m describes the number of variations of E_x or E_y that occur as ϕ varies from 0 to 2π.

In the weakly-guiding approximation the transverse field distributions of these LP modes can be written in the form[17.5]

Core:

$$E_x = E_m \frac{J_m\left(ur/a\right)}{J_m(u)} \cos m\phi, \tag{17.188}$$

$$H_y = \frac{E_m}{Z_{core}}. \tag{17.189}$$

Cladding:

$$E_x = E_m \frac{K_m\left(ur/a\right)}{K_m(u)} \cos m\phi, \tag{17.190}$$

$$H_y = \frac{E_x}{Z_{cladding}}, \tag{17.191}$$

where E_x is a constant, and $Z_{core}, Z_{cladding}$ are the impedances of core and cladding, respectively. The designation of these modes as 'linearly polarized' does not imply that the transverse field points in a specific unique direction everywhere within the fiber. We have already seen that this was not the case for the $TE_{0\ell}$ mode, see Eqs. (17.173)–(17.177). However, if the Poynting vector obtained from E_xH_y or $-E_yH_x$ is examined, LP modes with the same m will have similar transverse mode patterns, as shown in Fig. (17.17). The field patterns associated with the various modes belonging to a particular $LP_{m\ell}$ mode are actually quite different[17.4],[17.6], as is also shown schematically in Fig. (17.17).

The schematic Poynting vector distributions for the LP modes shown in Fig. (17.17) are the patterns appropriate to either (E_xH_y) or $(-E_yH_x)$. For example, for the $TE_{0\ell}$ mode discussed earlier

$$E_x = -E_\phi \sin\phi, \tag{17.192}$$

$$E_y = E_\phi \cos\phi, \tag{17.193}$$

$$H_x = H_r \cos\phi, \tag{17.194}$$

$$H_y = H_r \sin\phi, \tag{17.195}$$

and the x-polarized intensity distribution in the core can be written as

$$I_x \propto J_1^2\left(\frac{ur}{a}\right)\sin^2\phi. \tag{17.196}$$

For $LP_{m\ell}$ modes with $\ell \geq 1$ there are four constituent transverse or hybrid modes. This results because each hybrid mode consists of two degenerate modes, corresponding to the choice of the $\cos \nu\phi$ or $\sin \nu\phi$ azimuthal variation of the fields. The TM or TE modes (for which $\nu = 0$) are singly degenerate and axisymmetric. The LP_{01} mode (HE_{11}) is doubly degenerate, corresponding to the choice of the origin for ϕ. For example, if $\phi = 0$ corresponds to the x direction the mode will be linearly polarized in the x direction. If $\phi = 0$ corresponds to the y direction the mode will be linearly polarized in the y direction.

17.14 Cutoff Frequencies

The field in the cladding decays with increasing distance from the core so long as $w > 0$. However, as $w \to 0$, the propagation constant β approaches the value k_2 appropriate to the cladding and the wave is no longer confined. The condition $w = 0$ determines the cutoff condition for the mode in question, which from Eq. (17.185) becomes

$$\frac{uJ_{m-1}(u)}{J_m(u)} = 0. \qquad (17.197)$$

The unique property of the HE_{11} mode can be illustrated with this equation. This is the LP_{01} mode with $m = 0$. Its Poynting vector distribution is axisymmetric.

Only the $J_0(u)$ Bessel function is nonzero at $u = 0$, so Eq. (17.197) has the solution $u = 0$ in this case. Therefore, the HE_{11} mode has no cutoff frequency – or viewed another way, it can propagate in a fiber of small core radius a. Since

$$u = a\sqrt{k_1^2 + \gamma^2}, \qquad (17.198)$$

and at cutoff

$$u = V = k_0 n_1 a\sqrt{2\Delta}, \qquad (17.199)$$

$u = 0$ implies ω (the wave frequency) $= 0$. All other modes in the fiber have a cutoff frequency below which they cannot propagate. The cutoff frequencies are determined by the zeros $J_{m-1,\ell}$ of the Bessel function J_{m-1}, a few values of which are given in Table (17.1). For the $TM_{0\ell}$ and $TE_{0\ell}$ modes the cutoff condition is

$$\frac{J_0(u)}{J_1(u)} = 0, \qquad (17.200)$$

so $u = j_{0\ell}$ at cutoff.

For the $HE_{\nu\ell}$ hybrid mode with $\nu \geq 2$ the cutoff condition is

$$\frac{J_\nu(u)}{J_{\nu+1}(u)} = 0, \qquad (17.201)$$

so $u = j_{\nu,\ell}$ at cutoff.

For the $EH_{\nu\ell}$ hybrid modes the cutoff condition is

$$\frac{J_{\nu-2}(u)}{J_{\nu-1}(u)} = 0, \qquad (17.202)$$

so $u = j_{\nu-2,\ell}$ at cutoff.

The smallest $j_{m-1,\ell}$ is j_{01}, so the LP_{11} has the lowest nonzero cutoff frequency. The cutoff wavelength for this mode is found from

$$k_0 n_1 a\sqrt{2\Delta} = 2.405, \qquad (17.203)$$

so the cutoff wavelength is

$$\lambda_c = \frac{2\pi n_1 a\sqrt{2\Delta}}{2.405}. \qquad (17.204)$$

If a given mode in the fiber is not below cutoff then its normalized phase constant u has to take values that are bounded by values at which the $J_{m-1}(u)$ and $J_m(u)$ Bessel functions in Eq. (17.106) are equal to zero. This is shown in Fig. (17.18)[17.5].

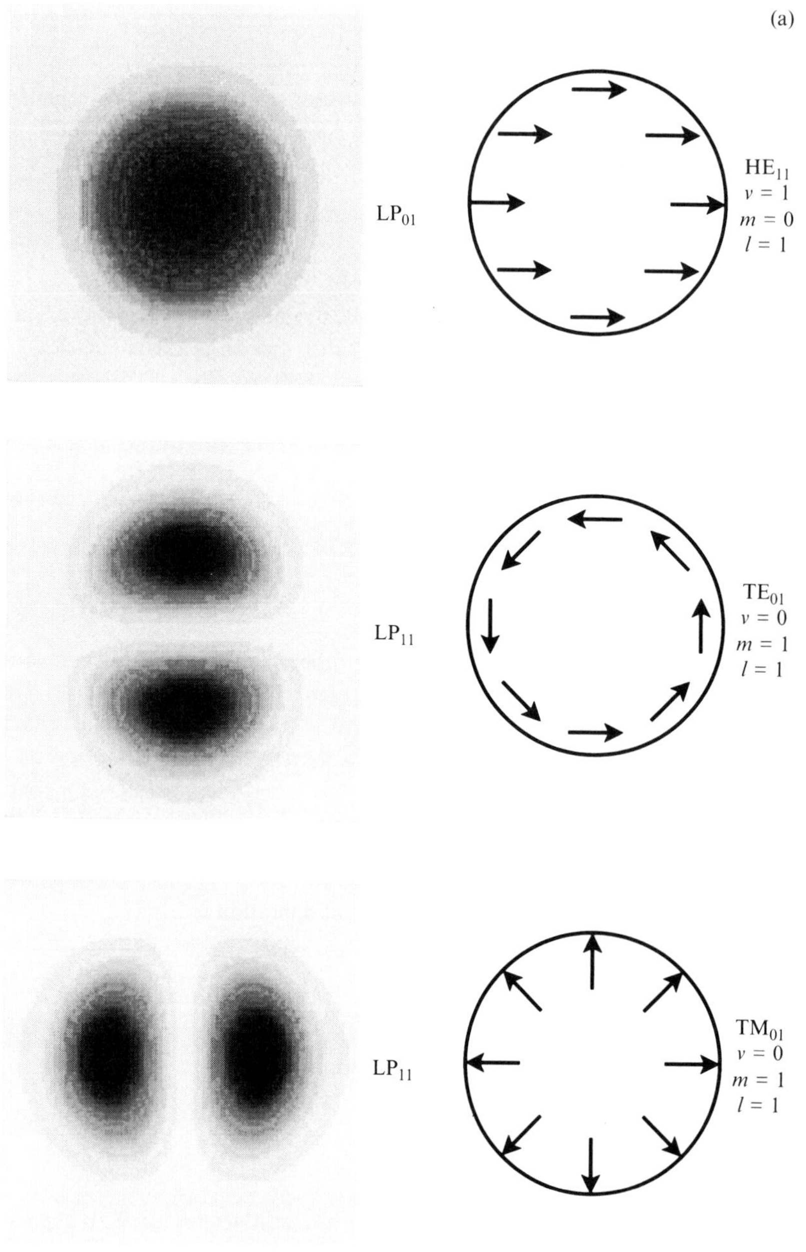

Fig. 17.17.
(a) Characteristic field and intensity patterns of low order linearly polarized modes. Although the transverse field distributions are different for different TE, TM, and hybrid modes that constitute a given LP mode, their x-polarized intensity distributions are qualitatively similar.

The value of u must be between the zeros of $J_{m-1}(u)$ and $J_m(u)$, that is for modes with $\ell = 0$, or 1:

$$J_{m-1,\ell} < u < J_{m,\ell}. \tag{17.205}$$

If this condition is not met then it is not possible to match the J Bessel function to the decaying K Bessel function at the core boundary. At the lower limit of the permitted values of u in Fig. (17.18), the cutoff frequency, $V = u$, while at the upper limit $V \to \infty$.

Fig. 17.17. (*cont.*)

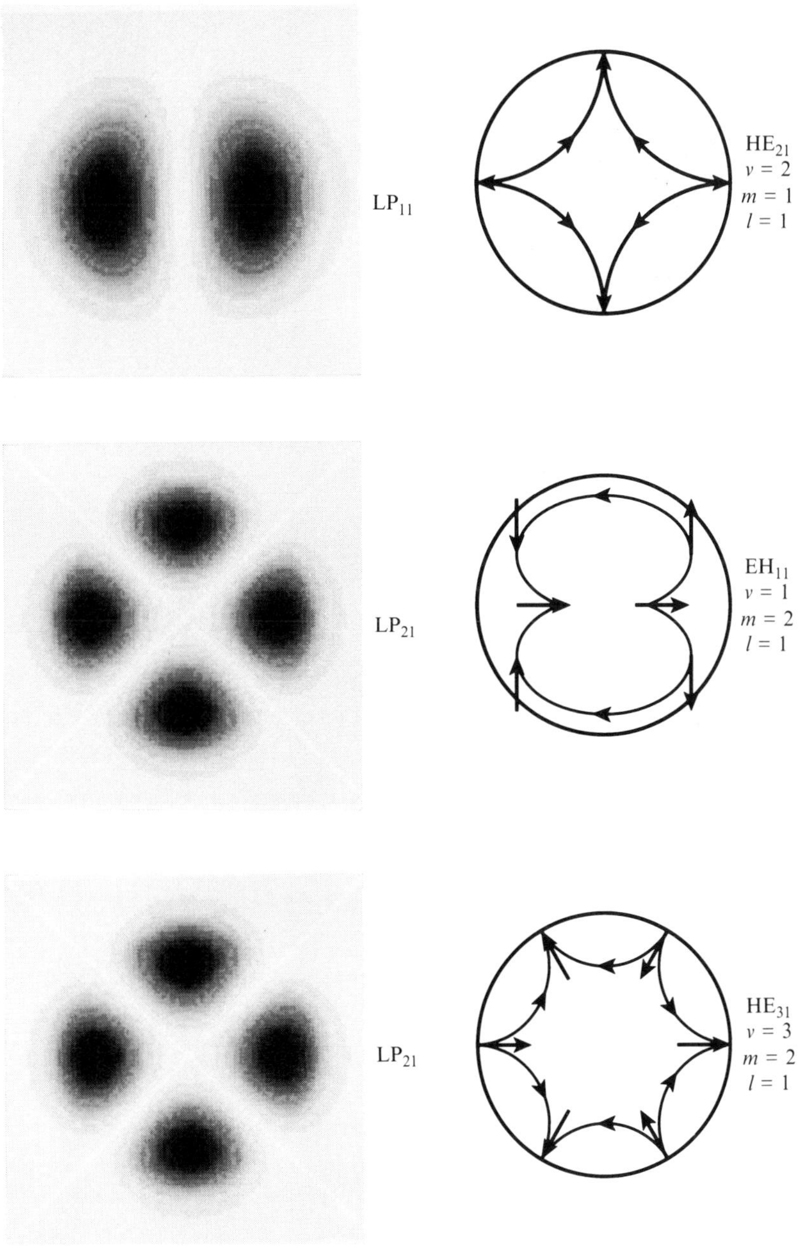

17.14.1 Example

For a fiber with $a = 3$ μm, $n_1 = 1.5$ and $\Delta = 0.002$ the cutoff wavelength is $\lambda_c =$ 744 nm. This fiber would support only a single mode for $\lambda > \lambda_c$.

Alternatively, to design a fiber for single frequency operation at $\lambda_0 = 1.55$ μm we want

$$a < \frac{2.405\lambda_0}{2\pi n_1 \sqrt{2\Delta}}. \tag{17.206}$$

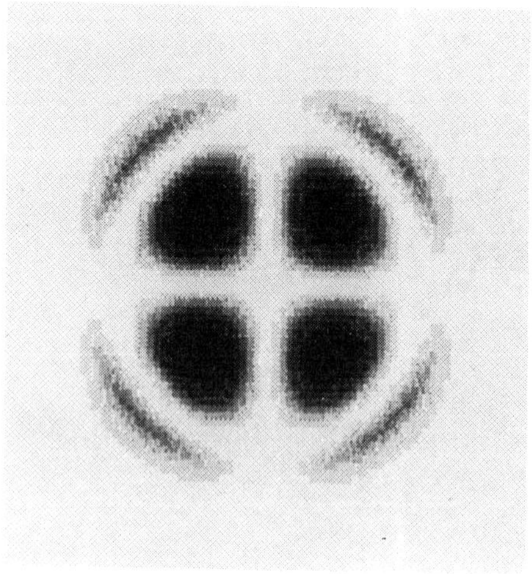

LP$_{12}$

(b)

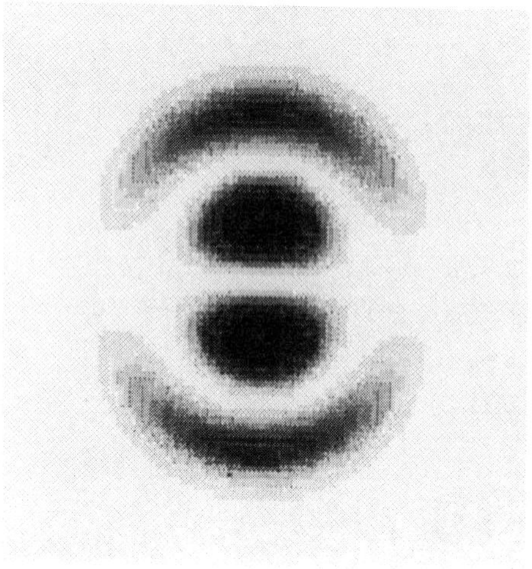

LP$_{22}$

Fig. 17.17. (*cont.*)
(b) Intensity patterns of the higher order LP modes LP$_{12}$ and LP$_{22}$ calculated for $u = 6$.

Table 17.1. *Low order zeros of Bessel functions.*

j_0	j_1	j_2	j_3
$j_{01} = 2.405$	$j_{11} = 3.832$	$j_{21} = 5.136$	$j_{31} = 6.380$
$j_{02} = 5.520$	$j_{12} = 7.106$	$j_{22} = 8.417$	$j_{32} = 9.761$
$j_{03} = 8.654$	$j_{13} = 10.173$	$j_{23} = 11.620$	$j_{33} = 13.015$

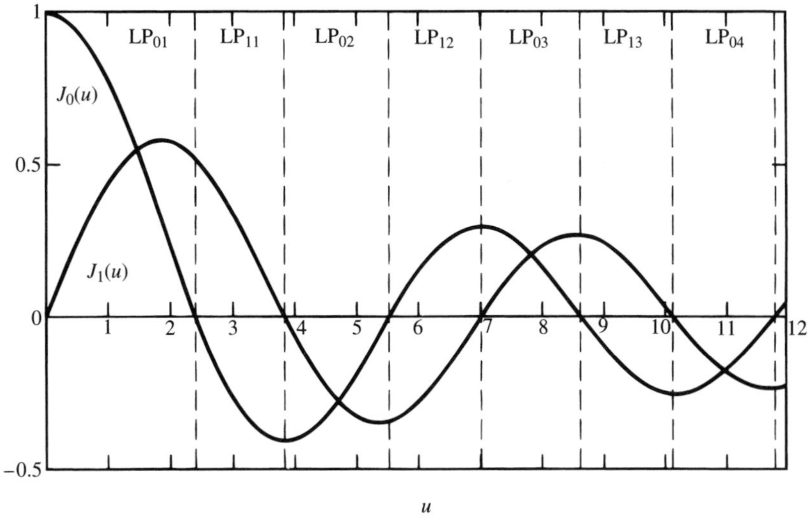

Fig. 17.18. Superimposed variation of the Bessel functions $J_0(u)$ and $J_1(u)$ showing how the zeroes of the Bessel functions bound the values of u where different LP modes operate[17.5].

For $n_1 = 1.5$ and $\Delta = 0.002$ this requires $a < 6.254$ μm.

A fiber satisfying this condition would be a single-mode fiber for 1.55 μm, it would not necessarily be a single-mode fiber for shorter wavelengths. Single-mode fibers for use in optical communication applications at the popular wavelengths of 810 nm, 1.3 μm or 1.55 μm all have small core diameters, although for mechanical strength and ease of handling their overall cladding diameter is typically 125 μm. The fiber will have additional protective layers of polymer, fiber, and even metal armoring (for rugged applications such as underground or overhead communications links, or for laying under the ocean). Fig. (17.19) shows the construction of various multifiber cables of this kind.

17.15 Multimode Fibers

If the core of a step-index fiber has a larger radius relative to the operating wavelength than the value given by Eq. (17.206) then more than one mode can propagate in the fiber. The number of modes that can propagate in a given fiber can be determined from its V-number. For $V < 2.405$ the fiber will allow only the HE_{11} mode to propagate. For relatively small V-numbers the number of modes that are permitted to propagate can be determined by counting the number of Bessel function zeroes $j_{\nu\ell} < V$. For large V-numbers a good estimate of the number M of permitted modes is $M = V^2/2$.

In a multimode fiber most of the modes will be well above cutoff and well confined to the core. The fraction of the mode power that propagates in the cladding decreases markedly for these modes as is shown in Fig. (17.20).

17.16 Fabrication of Optical Fibers

There are two principal groups of fabrication techniques for the manufacture of low attenuation optical fibers: methods involving the direct drawing of fiber from molten glass in a crucible; and various vapor deposition methods in which a

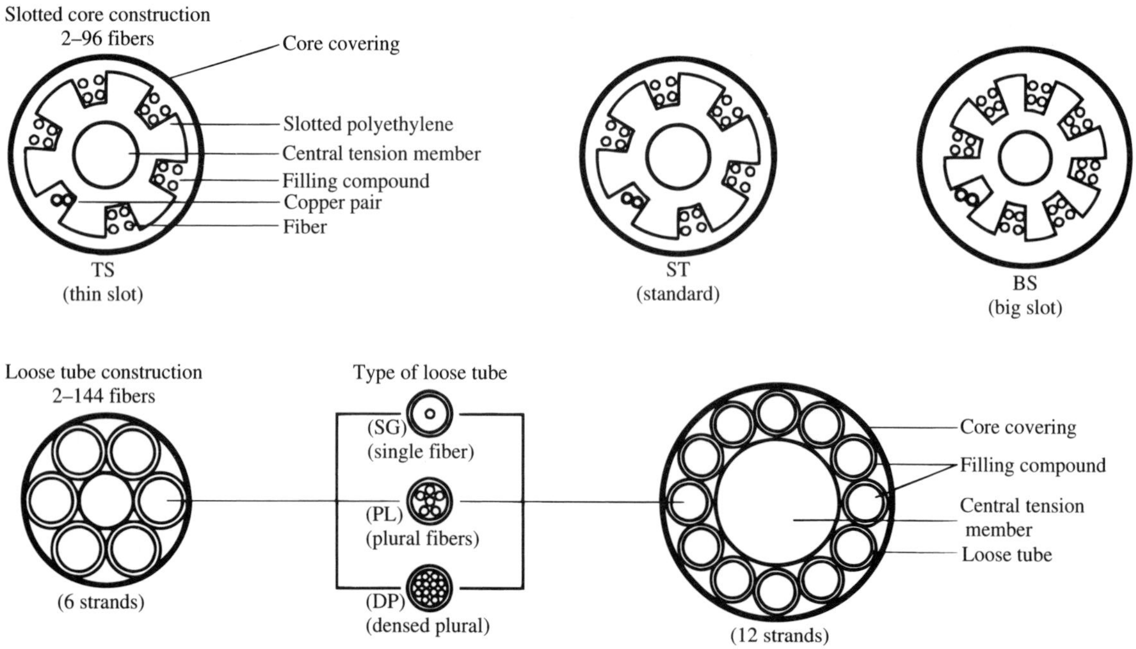

Fig. 17.19. Construction of multifiber optical cables.

'preform' with appropriate graded-index properties is produced and then drawn into a fiber. All these fabrication procedures rely on the use of highly pure starting material whether this be a borosilicate glass or silica, with added dopants. The brief discussion here can be supplemented with information from more specialized texts[17.6]–[17.9].

Fig. (17.21) shows schematically the double crucible method for the continuous production of clad fibers. The starting material can be fed in either in powder form or as preformed glass rods. This technique cannot be used for high melting point glasses, such as fused silica. The core and cladding materials can be doped to give the core a higher index than the cladding, for example, by doping with tantalum and sodium, respectively. By varying the drawing temperature the degree of interdiffusion between core and cladding can be controlled to give a weakly-guiding 'step-index' or 'graded-index' fiber. The addition of thallium to the core, followed by its diffusion into the cladding during the drawing process also gives a graded-index fiber. The double crucible method allows the continuous production of fiber with attenuation as low as 5 dB km^{-1}.

The production of single-mode versus multimode fiber is determined by the size of the drawing aperture from the inner crucible, the temperature of the molten glass, and the rate at which the fiber is drawn.

For the production of the lowest attenuation fibers based on silica doped with germanium, fluorine or phosphorus, particularly for the manufacture of single-mode fiber, a doped preform is fabricated by some form of vapor deposition. The three principal vapor deposition processes are shown schematically in Fig. (17.22).

In the outside chemical vapor deposition (OCVD) process materials such as $SiCl_4$, $GeCl_4$, $POCl_3$, BBr_3 or BCl_3 are oxidized in a burner and deposited on a

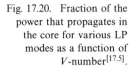

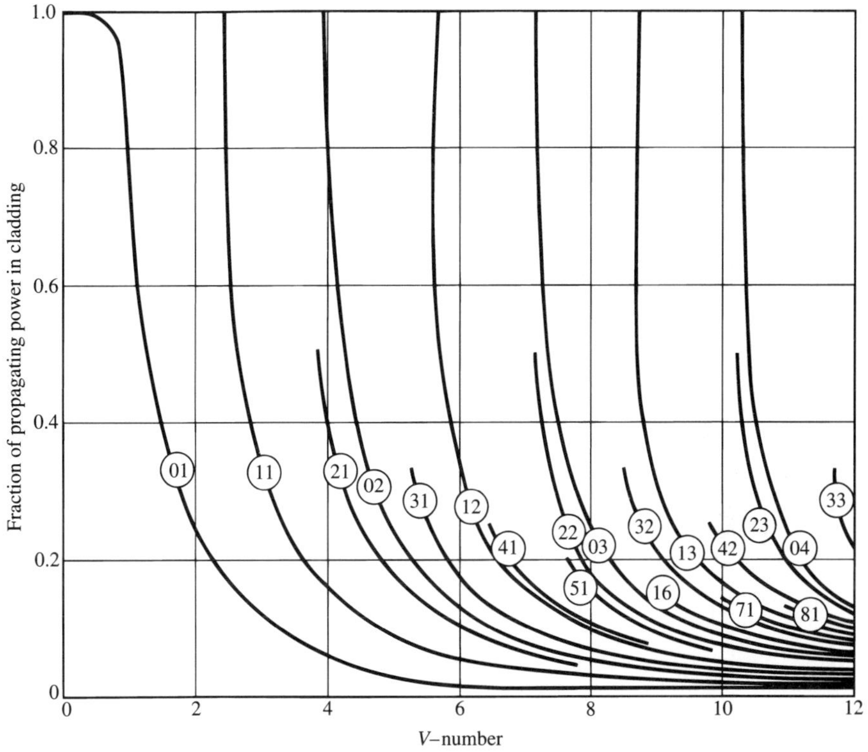

Fig. 17.20. Fraction of the power that propagates in the core for various LP modes as a function of V-number[17.5].

rotating silica mandrel as a 'soot'. By varying the vapor to be decomposed, layers of soot particles can be built up, layer upon layer, to generate the final refractive index profile desired. The mandrel, with soot over-layers, is then melted into a preform in a furnace. The preform can then be heated and drawn into a fiber.

In the inside chemical vapor deposition (ICVD) process the chemical vapors are decomposed inside a silica tube by heating them externally or with an internal RF-generated plasma. The tube with its inner layers of soot can then be collapsed and inserted into a second silica (cladding) tube to serve as the core of the final drawn fiber.

The vapor-phase axial deposition (VAD) process is a variant on the OCVD process in which the core and cladding glasses are deposited simultaneously onto the end of the preform, which is continuously drawn up into an electric furnace. The perform is rotated as deposition proceeds and as it is drawn up into the furnace its diameter decreases. The final preform is then taken away and heated for fiber drawing[17.6].

17.17 Dispersion in Optical Fibers

The use of optical fibers to transmit data at ever increasing rates, which in the laboratory have already exceeded 10 Gbit s^{-1} over kilometer long fibers, has required careful consideration of the effects of dispersion in the fiber. In its simplest sense dispersion is the spreading in width of a narrow optical pulse as it travels along the fiber. This spreading is greatest in multimode fibers, where in a ray description

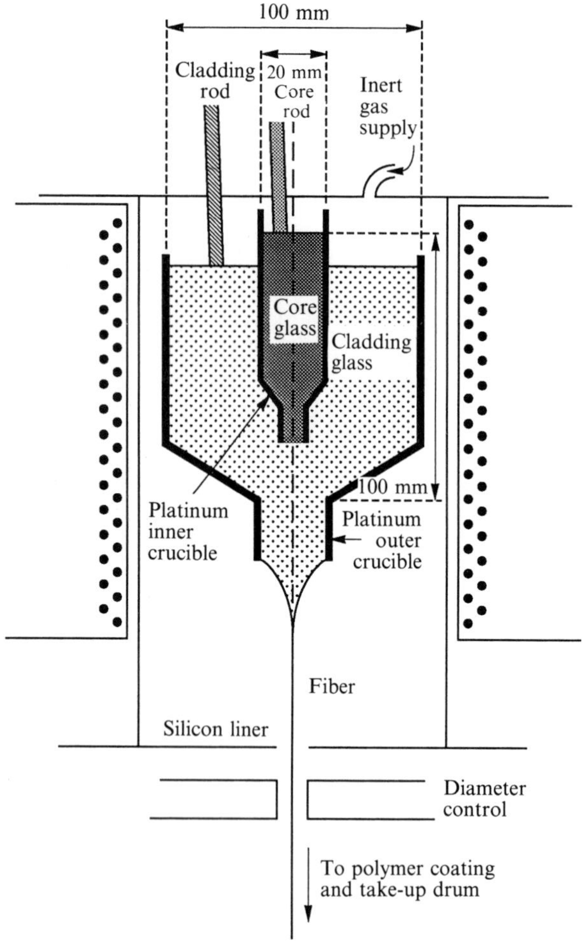

Fig. 17.21. Schematic illustration of the double crucible method for fabricating optical fibers.

each different mode corresponds to a different length ray trajectory inside the fiber. The different pulse transit times along these different ray paths lead to broadening or splitting of a narrow optical pulse. This is called *modal* dispersion. For example, for a step-index fiber the transit time difference between a ray that propagates along the axis and one that enters just within the *NA* is easily shown to be

$$\begin{aligned}\Delta T &= \frac{n_1^2}{n_2}\frac{\ell}{c_0}\Delta \\ &\simeq \frac{n_1 \ell \Delta}{c_0}.\end{aligned} \tag{17.207}$$

For $\Delta = 0.002, n_1 = 1.5$, this modal dispersion is 10 ns km^{-1}.

In single-mode fibers modal dispersion does not exist but dispersion still occurs. This residual dispersion comes from *material* dispersion $dn/d\lambda$, the variation of the refractive index of core and cladding with wavelength (or frequency), and *waveguide* dispersion, the variation of the waveguide propagation constant β with frequency. A detailed analysis of these effects has been given by Gowar[17.11], but a few general observations can be made.

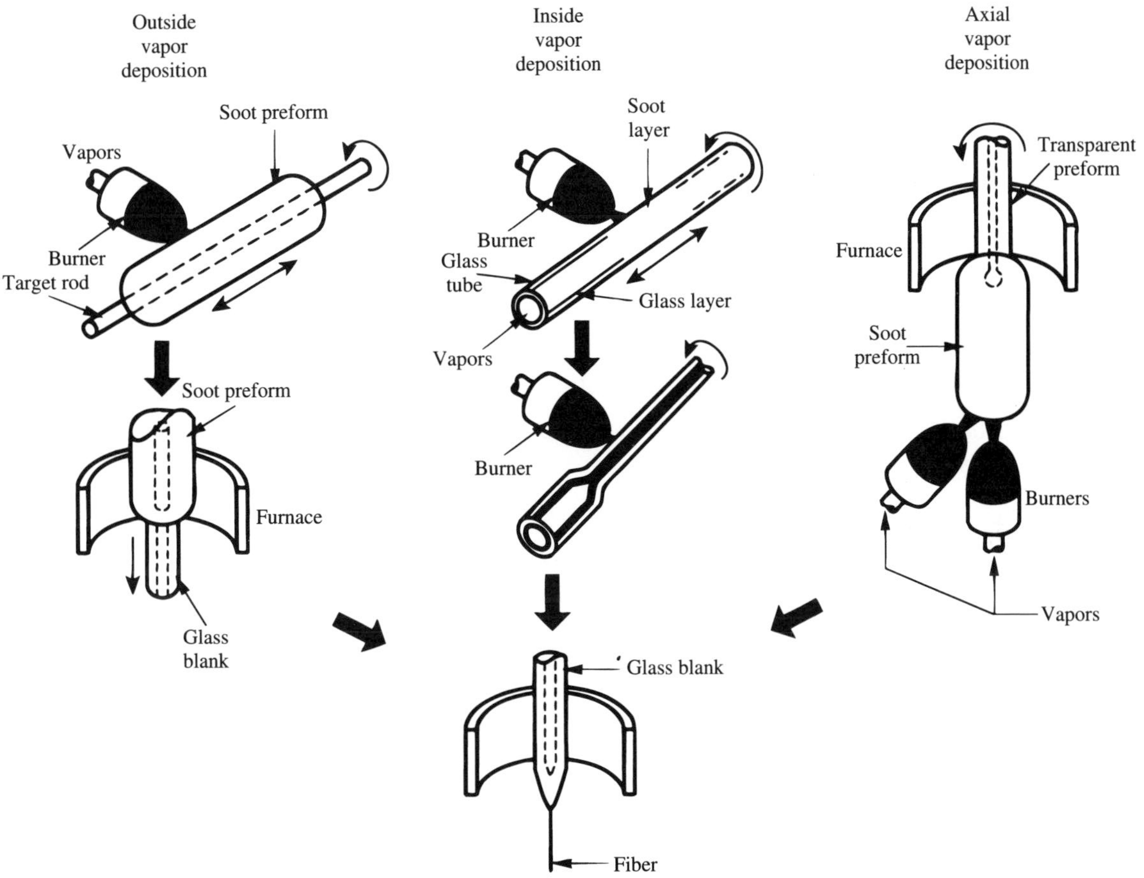

Fig. 17.22. Vapor deposition methods for fabricating doped fiber 'preforms'[17.10].

17.17.1 Material Dispersion

The transit time of a pulse in a single-mode fiber depends, among other factors, on its group velocity,† defined as

$$v_g = \frac{d\omega}{d\beta}. \tag{17.208}$$

In the simplest case of a pulse totally confined to the core

$$\beta = k_1 = \frac{\omega n_1}{c_0}. \tag{17.209}$$

It is easy to show that

$$v_g = \frac{c_0}{(n_1 - \lambda dn_1/d\lambda)} \tag{17.210}$$

We can define a *group refractive index*, n_{g_1}, where

$$v_g = \frac{c_0}{n_{g_1}}. \tag{17.211}$$

† The group velocity is a measure of the velocity with which the centroid of a pulse, which itself contains a range of frequencies, will travel. It is, therefore, an important parameter in characterizing the performance of an optical communication system that transmits information as a train of light pulses.

If a narrow optical pulse is travelling in the core, then its different spectral components will travel with different velocities. The spectral components cover a wavelength range $\Delta\lambda$ that depends on the monochromaticity of the source. For a light emitting diode (LED) $\Delta\lambda$ is typically $\simeq$ 40 nm at 800 nm; for a semiconductor laser the equivalent spectral frequency width will be in the range 1–100 MHz. The transit time of the pulse will be

$$T = \frac{\ell}{v_{g_1}} = \left(n_1 - \lambda\frac{dn_1}{d\lambda}\right)\frac{\ell}{c_0}, \tag{17.212}$$

and a narrow pulse (impulse) will spread to a width

$$\Delta T = -\frac{\ell}{c_0}\lambda^2\frac{d^2n}{d\lambda^2}\left(\frac{\Delta\lambda}{\lambda}\right). \tag{17.213}$$

This pulse broadening is minimized for a source with specific spectral purity $\Delta\lambda/\lambda$ when the quantity

$$|Y_m| = \left|\lambda^2\frac{d^2n}{d\lambda^2}\right| \tag{17.214}$$

is minimized. For doped silica fibers $Y_m = 0$ for wavelengths in the range 1.22 μm$< \lambda < 1.37$ μm, the minimum being at $\lambda_{md} = 1.276$ μm for a pure silica fiber. The spreading of the pulse will not actually be zero, even at this operating wavelength, because of the spectral width of the source. The residual pulse broadening that occurs for a source with $\lambda_0 = \lambda_{md}$ can be shown to be[17.11]

$$\frac{\Delta T}{\ell} = (-)\frac{\lambda_0^3}{8c_0}\left(\frac{d^3n}{d\lambda^3}\right)_{\lambda_0}\left(\frac{\Delta\lambda}{\lambda}\right)^2. \tag{17.215}$$

For pure silica at $\lambda_0 = 1.276$ μm, $\Delta T/\ell = 2\times 10^{-11}(\Delta\lambda/\lambda)^2$ s m.

For a high data rate optical communication link using a single mode semiconductor laser running at 10 Gbits s^{-1} the effective laser linewidth is broadened to $\simeq$ 20 GHz, so $\Delta\lambda/\lambda = 0.00009$ and $\Delta T/\ell \simeq 1.5\times 10^{-16}$ s km^{-1}.

Operation at a wavelength of 1.55 μm, where the attenuation of silica is a minimum, but $Y_m = -0.01$, leads to more broadening. Even so, this broadening is extremely small. For the optical link described above running at 1.55 μm, $\Delta T/\ell \simeq 3.5$ ps km^{-1}. In so-called *dispersion-shifted* fibers, doping, usually of germania, GeO_2, can move the minimum dispersion wavelength out to the minimum attenuation wavelength.

17.17.2 Waveguide Dispersion

When a mode propagates, its propagation constant is $\gamma = i\beta$ where β simultaneously satisfies both the equations

$$\beta = \sqrt{k_1^2 - (u/a)^2} \tag{17.216}$$

and

$$\beta = \sqrt{(w/a)^2 + k_2^2}. \tag{17.217}$$

At cutoff $w = 0$, $\beta = k_2$ and the wave propagates completely in the cladding. As the frequency ω of the wave rises far above the cut-off value $\beta \to k_1$. The

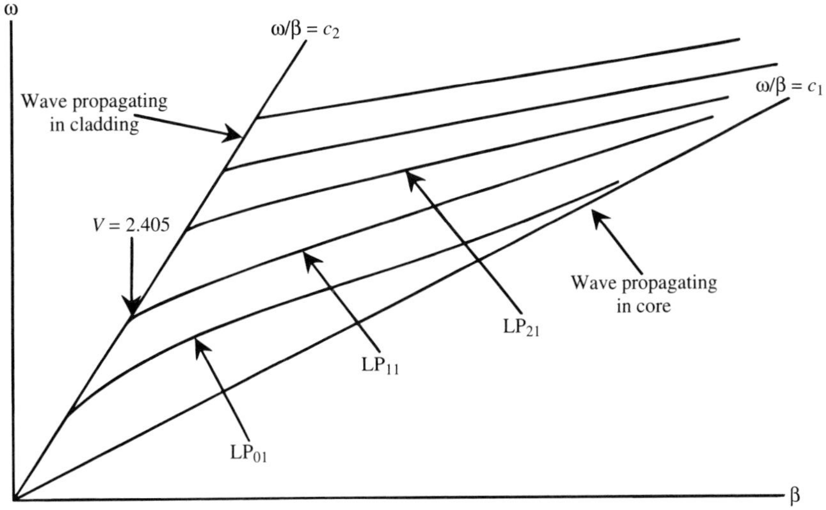

Fig. 17.23. Schematic variation of dispersion curves of different LP modes in a step-index fiber. The diagram is shown neglecting the variation of index with frequency and greatly exaggerates the difference between n_1 and n_2.

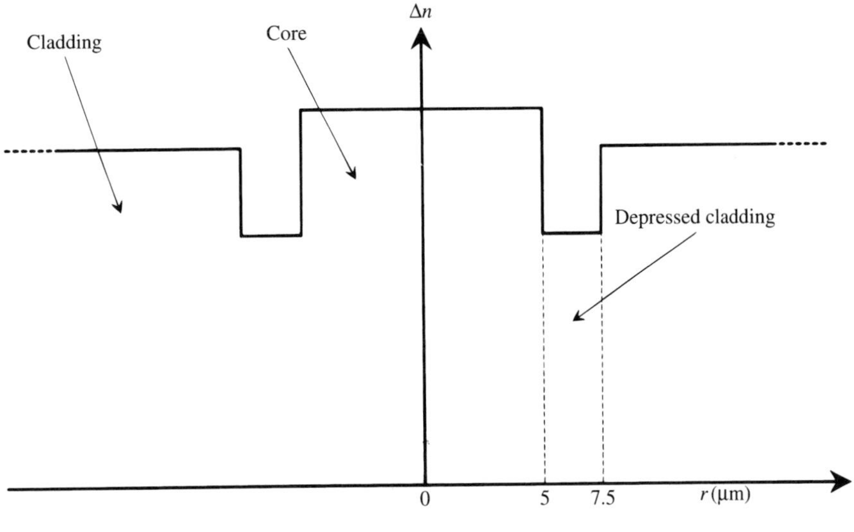

Fig. 17.24. Schematic radial refractive index variation of a W-fiber.

variation of β with ω is called a waveguide dispersion curve. At any point on a dispersion curve the *phase* velocity of the mode is ω/β, while the *group* velocity is $d\omega/d\beta$. The dispersion curve for a particular mode is bounded by the β values appropriate for core and cladding as shown in Fig. (17.23). The so-called *light lines* have ω/β slopes c_1 and c_2, where c_1 and c_2 are the velocities of light in core and cladding, respectively. Waveguide dispersion can be minimized by varying the radial refractive index distribution, for example, as in the so-called W-fiber, whose radial refractive index variation is shown schematically in Fig. (17.24).

It is also possible to design a single-mode fiber so that the waveguide dispersion and residual material dispersion cancel each other out at wavelengths longer than 1.276 μm, for example, to 1.6 μm in a fiber with core diameter of 4 μm[17.11].

17.18 Solitons

The material of an optical fiber is not only dispersive, it is also nonlinear. For sufficiently intense optical pulses propagating in the fiber the variation of the refractive index with intensity,

$$n = n_0 + n_2 I, \tag{17.218}$$

where I is the intensity within the fiber, affects the dispersion characteristics of the fiber. It can be shown[17.12] that, if the fiber is being operated at a wavelength where $dv_g/d\lambda < 0$ (or $d^2\beta/d\omega^2 < 0$), a so-called region of *anomalous*† dispersion, effects of nonlinearity and waveguide dispersion can cancel each other out. The consequence of this is the development of a stable pulse, called a *soliton*, that propagates along the fiber without broadening.

If the intensity profile of the pulse with time is written as

$$I(z,t) \propto |E(z,t)e^{i\omega_0 t}|^2, \tag{17.219}$$

where $E(z,t)$ is the electric field envelope, then the behavior of the pulse is dscribed by the *nonlinear Schrödinger* equation, which in normalized form can be written as

$$-i\frac{\partial u}{\partial q} + \frac{1}{2}\frac{\partial^2 u}{\partial^2 s} + |u|^2 u = 0, \tag{17.220}$$

where the relationship between the actual and normalized variables is:

$$u = \tau\sqrt{\frac{n_2\omega}{2Zc_0}\left|\frac{d^2\beta}{d\omega^2}\right|^2} E(z,t), \tag{17.221}$$

$$s = (t - z/v_g)/\tau, \tag{17.222}$$

$$q = \left|\frac{d^2\beta}{d\omega^2}\right| z/\tau^2. \tag{17.223}$$

s is a dimensionless variable that measures the relative position in time of points of the pulse envelope relative to the location of the point that travels with the group velocity v_g.

One solution of Eq. (17.220) is the so-called *fundamental* soliton, described by

$$u(q,\tau) = \text{sech}\ \tau e^{-iq/2}. \tag{17.224}$$

Apart from a varying phase factor this pulse propagates without changing its shape: the effects of dispersion and nonlinearity have cancelled each other out.

17.19 Erbium-Doped Fiber Amplifiers

The development of dispersion-shifted fiber and the use of soliton transmission offers solutions to the problem of very high data rate transmission along very long optical fibers. In dispersion-shifted fiber the combined effects of material and waveguide dispersion can be reduced to a very low value[17.13]

$$D_s = 0.1 \text{ ps km}^{-1} \text{ nm}^{-1}, \tag{17.225}$$

† Anomalous dispersion occurs on the long wavelength side of the minimum dispersion (material dispersion plus waveguide dispersion) in the fiber, which in silica implies $\lambda > 1.3\ \mu$m.

where the actual pulse spreading will be calculated as

$$\frac{\Delta T}{\ell} = D_s \Delta\lambda. \tag{17.226}$$

For example, with a bit rate of 2.5 Gbit s^{-1} the maximum transmission distance is about 40 000 km. Unfortunately, dispersion-shifted fibers operating at 1.55 μm have roughly 0.03 dB km^{-1} greater loss than single-mode fibers operating at the attenuation minimum near 1.3 μm in pure silica fibers. On the other hand, the dispersion in the low attenuation fibers at 1.3 μm is greater

$$D_s \simeq 17 \text{ ps km}^{-1} \text{ nm}^{-1}, \tag{17.227}$$

so the maximum transmission distance at 2.5 Gbit s^{-1} is 235 km. Most of the currently deployed long-haul optical fiber links were designed for operation at 1.3 μm, however, as the demand for higher data rates increases the use of dispersion-shifted fiber operating at 1.55 μm becomes very attractive. Because of the higher attenuation in these fibers the optical signal intensity must be periodically regenerated. This has been done traditionally with a *repeater*. This is a module incorporating an optical detector, pulse reshaping electronics, and a laser to retransmit the bit stream with renewed intensity. Such repeaters placed at intervals along a long fiber link periodically correct for the effects of dispersion, as pulse reshaping prevents these effects being cumulative. This is unnnecessary at 1.55 μm as dispersion is negligible – only pulse amplification is necessary. It is very attractive to perform this amplification without a traditional repeater. A semiconductor laser amplifier (a semiconductor laser structure with very low reflectivity or angled facets to prevent self-oscillation) can do the job, but the development of erbium-doped fiber amplifiers (EFDA) presents a better solution.

Very early in the development of solid-state lasers it was observed that erbium, Er^{3+}, doped into various hosts, including glass[17.14], would operate as an optically pumped laser, and it was demonstrated that laser oscillation could be obtained using a doped glass fiber[17.15]. It was some time later, in 1987, before Payne and his coworkers demonstrated the first erbium doped fiber with amplification properties that made it useful in telecommunication applications[17.16]. The attractive feature of EDFAs is that they can be pumped with the light from semiconductor lasers operating at 800 nm, 980 nm or 1480 nm. The energy level scheme and pumping process responsible for creating gain around 1.5 μm with a 980 nm pump is shown in Fig. (17.25). Because the upper level of the amplifying transition $^4I_{13/2}$ is metastable with a lifetime of around 11.4 ms it is possible to create significant gain with modest CW pump powers (tens of mW). The fluorescence peak at 1.54 μm dominates the spectrum of Er^{3+} doped silica glass when the material is pumped with shorter wavelength radiation[17.17]. The width of the emission is about 30 nm (FWHM) so semiconductor lasers operating at 1.55 μm are well within the gain curve. Gains in excess of 30 dB are readily obtained. Erbium concentrations typically range up to 1000 ppm: fibers with higher doping concentration will have a larger absorption coefficient for the pump laser so a shorter length of fiber is needed to give equivalent gain. Typical lengths of erbium doped fiber range from meters to hundreds of meters. For further details concerning the optimization of amplifier performance with respect to gain, bandwidth, fiber doping concentration, length, and noise behavior the interested reader should consult the specialized literature[17.13],[17.18].

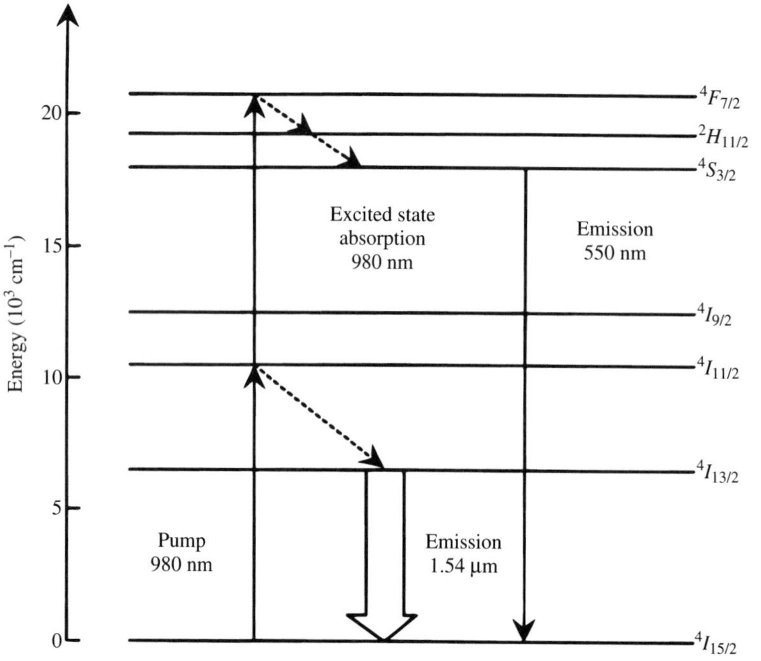

Fig. 17.25. Partial energy level diagram of Er^{3+} ions in silica glass showing the 980 nm pump process and the amplifying transition at 1.54 μm. Excited state absorption can also occur leading to green fluorescence and may reduce amplifier efficiency.

Fig. (17.26) shows schematically four configurations in which EFDAs can be operated. The use of isolators can enhance amplifier performance by discouraging feedback-induced oscillation from the couplers, or in between lengths of erbium doped fiber to reduce the possibility of oscillation or amplified spontaneous emission (ASE). If 980 nm pumps are used then provision must be made for the pump light not to pass through the isolator, since an in-fiber isolator optimized for 1.55 μm will pass 1.48 μm pump light efficiently but not 980 nm.

In summary, the advantages of EDFAs in advanced optical communication networks are many: high gain, low noise, high power conversion efficiency, high saturation intensity, insensitivity to polarization states and wide spectral bandwidth. Because of their wide spectral bandwidth they are particularly attractive in schemes involving *wavelength division multiplexing* (WDM). In such a scheme multiple channels of information are transmitted along the same fiber at different wavelengths. For example, with a bandwidth of 30 nm at 1.55 μm the frequency bandwidth of an EFDA is 3.74 THz. Therefore, for a series of lasers operating at wavelengths λ_1, λ_2, etc., near 1.55 μm, 375 2.5 Gbit s^{-1} channels spaced by 10 GHz could be amplified simultaneously. Furthermore, EFDAs can do this with low cross talk.†

Because many fibers that are already in service were designed for use at 1.3 μm intensive research efforts are continuing to produce doped fiber amplifiers with gain at this wavelength. Praesodymium, Pr^{3+}, and neodymium, Nd^{3+}, doped fluorozirconate (ZBLAN) glass is receiving a great deal of attention in this regard[17.13].

† The effect where modulation at one wavelength λ_i is transferred to another amplified laser beam at wavelength λ_j.

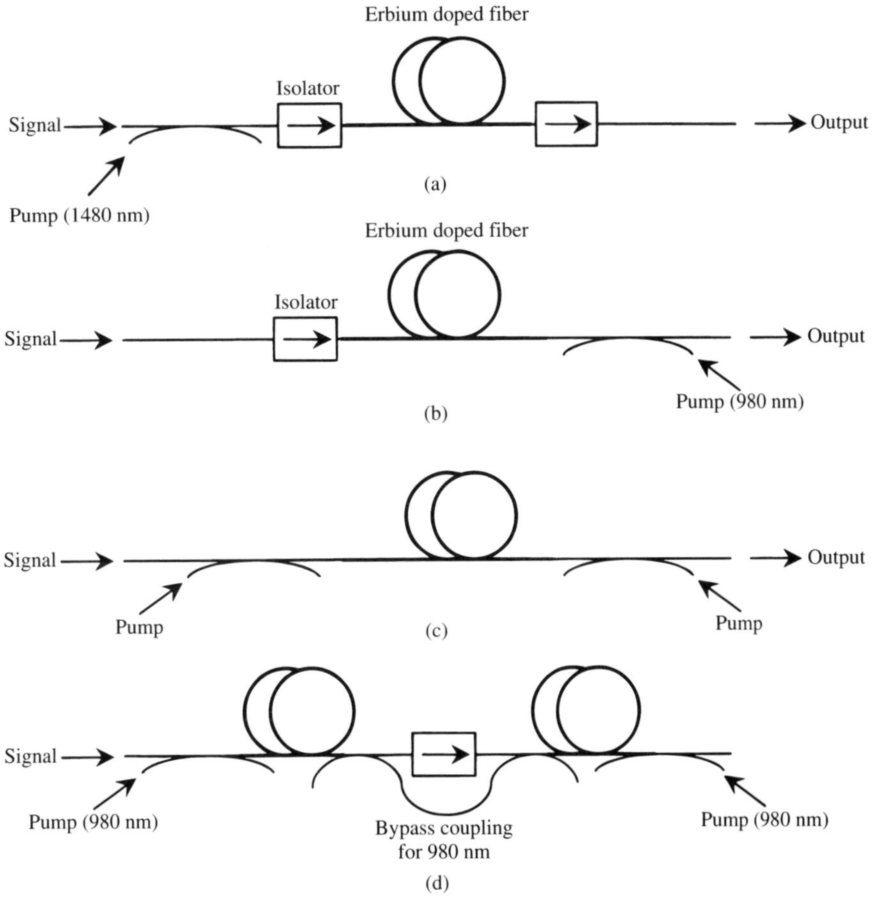

Fig. 17.26. Four configurations for operating erbium doped fiber amplifiers: (a) forward pumping, (b) backward pumping, (c) bidirect ional pumping, (d) bidirectional pumping with bypass coupler so 980 nm pump radiation can pass isolator.

17.20 Coupling Optical Sources and Detectors to Fibers

In order to couple light from a source into a fiber efficiently the light from the source must, in the first instance, be injected within the numerical aperture. In the case of multimode fibers this is generally sufficient to ensure efficient collection and efficient guiding of the injected source light. However, when step-index single-mode fibers are used with a laser source the injected light must additionally be matched as closely as possible to the lateral intensity profile of the mode that will propagate in the fiber. This cannot be done exactly because the injected light will typically have a Gaussian radial intensity profile, whereas the single mode lateral intensity profile is a zero order Bessel function with exponentially decaying wings. For optimum coupling a circularly symmetric Gaussian laser beam should be focused so that its beamwaist is on, and parallel to, the cleaved end face of the fiber, as shown in Fig. (17.27).

Optimum coupling of the Gaussian beam into the HE_{11} mode occurs when $w_0 = 0.61a$ and the V-number of the fiber is $V = 3.8$[17.1], in otherwords close to the value where the next highest mode HE_{12} will propagate. In this situation, if Fresnel losses at the fiber entrance face are ignored, the maximum coupling efficiency is 85%. The coupling efficiency from a Gaussian beam to a single-mode

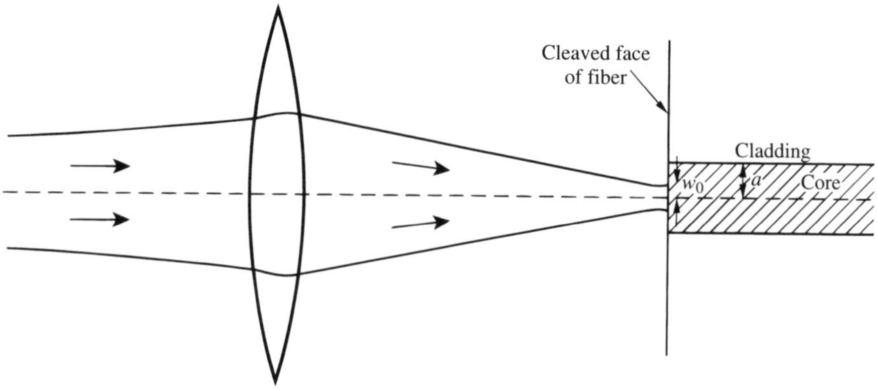

Fig. 17.27. Geometry for focusing a TEM_{00} Gaussian beam into a single-mode fiber.

or multimode fiber is somewhat complex to calculate in the general case[17.1]. To optimize coupling the input Gaussian must be matched to the equivalent Gaussian profile of the propagating mode. A useful approximation that can be used for a step-index fiber of core radius a is

$$w_{\text{mode}} = \frac{a}{\sqrt{2 \ln V}}. \tag{17.228}$$

If the end of the fiber is illuminated with too large a focused spot significant energy enters higher order modes, which leak out of the cladding. This is readily demonstrated in the laboratory. If a visible laser is focused into a fiber the first few centimeters of the fiber will glow brightly as nonpropagating modes are excited and then energy leaks from the cladding.

In the arrangement shown in Fig. (17.27) the lens can very conveniently be replaced by a GRIN lens, which allows butt coupling of source to GRIN. The parameters of the GRIN lens must be chosen commensurate with the input laser spot size to give an appropriate beamwaist at the output face of the GRIN.

If the laser source to be coupled to the fiber is not generating a circularly symmetric Gaussian beam, then anamorphic focusing optics must be used to render the beam circular and focus it onto the end face of the fiber. A typical semiconductor laser may generate an elliptical Gaussian beam, one in which the beam is characterized by different spot sizes in two orthogonal directions. In this case a combination of a cylindrical and spherical lens can be used to give improved focusing. If the emitting facet of the semiconductor laser is sufficiently small then a microfabricated spherical lens can be used to couple source and fiber.

The problem of coupling an optical detector to a fiber efficiently is an easy one. Most optical detectors are of larger area than the modal spot leaving the fiber so direct butt-coupling from fiber to detector is straightforward.

17.20.1 Fiber Connectors

If two identical fibers are to be joined together this can be done with either mechanical coaxial connectors or through a *fusion* splicer. In either case the flat, orthogonally cleaved ends of the two fibers must be brought together so that their two faces are in very close proximity, parallel, and the fibers are coaxial, as

Fig. 17.28. Alignment of optical fibers for connection: (a) optimum, (b) offset, (c) tilted.

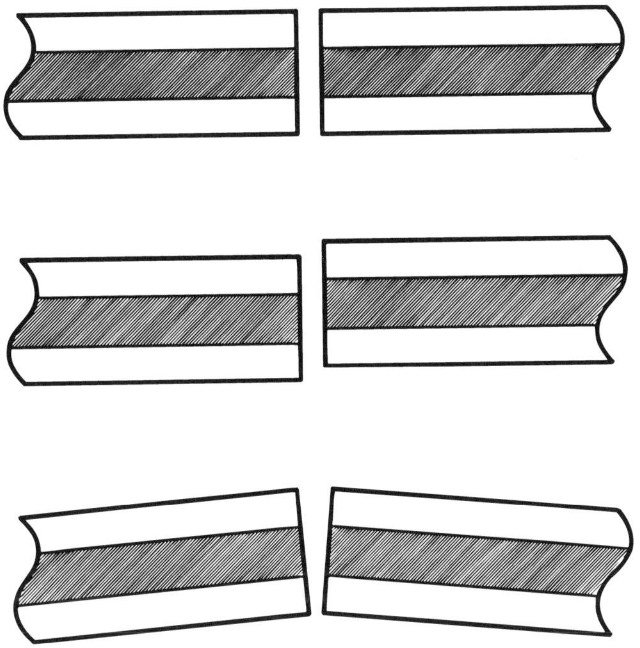

shown in Fig. (17.28a). Any tilt or offset of the fibers will reduce the coupling efficiency[17.19].

Precision, mechanical connectors are able to couple two fibers with typical losses of 0.1–0.2 dB. Generally, the connector will contain a small quantity of index-matching fluid that is squeezed in the small interface gap between the two fibers.

In a fusion splicer the two fiber ends in Fig. (17.28a) are brought into contact and external heat quickly applied to melt and fuse the two fibers together. Such splices will have a typical insertion loss below 0.1 dB.

17.21 Problems

(17.1) A lens of focal length 10 mm is being used to focus light from a point source into a multimode fiber with $n_{core} = 1.5$, $n_{cladding} = 1.48$. What is the practical maximum fraction of the light that can be injected into the fiber and be guided? How are source, lens, and fiber positioned to accomplish this? Use a simple ray model to obtain your answer.

(17.2) The curves in Fig. (17.5) have been calculated for a graded-core fiber with $n(r) = 3.5 - \frac{1}{2}n_2 r^2$ and $n_{cladding} = 1.625$. The core radius is 25 μm. From the curves (a) – (d) estimate the ray invariants $\overline{\beta}$ and $\overline{\ell}$ in each case.

(17.3) The Goos–Hänchen shift can be regarded as equivalent to the penetration of the totally internally reflected ray into the cladding so that the 'effective' thickness of the slabs is $2d_{eff}$. Derive an expression for d_{eff} using the parameters in Eq. (17.60).

(17.4) Prove that for an optical fiber being operated at a center wavelength λ_0 with a source of spectral width $\Delta\lambda$ the minimum pulse broadening as a function of distance caused by material dispersion is

$$\frac{\Delta T}{\ell} = (-)\frac{\lambda_0^3}{8c_0}\left(\frac{d^3n}{d\lambda^3}\right)_{\lambda_0}\left(\frac{\Delta\lambda}{\lambda}\right)^2.$$

(17.5) A fiber with a parabolic index profile has

$$n(r) = n_0 - \tfrac{1}{2}n_2r^2$$

with $n_0 = 1.5$ and $n_2 = 10^9\ \mathrm{m}^{-2}$. Design a single lens focusing arrangement that will optimally couple a laser with $\lambda_0 = 180$ nm and $w_0 = 2$ mm into the fiber. How would you modify the focusing arrangement if the laser emitted a Gaussian beam with $w_0 = 5\ \mu$m from the laser facet?

(17.6) Repeat Problem (17.3) but design a focusing system using a parabolic index GRIN lens that is 20 mm long. What parabolic profile parameter n_2 for the GRIN lens is needed?

(17.7) Use Eqs. (17.105) and (17.106) to plot transverse mode patterns of the P-wave and S-wave modes in a slab guide. Choose reasonable values for the wavelength and slab dimensions. Use $n_2 = 3, n_2 = 1$.

(17.8) Calculate the cutoff wavelengths for the modes LP_{01}, LP_{11}, and LP_{21} in a cylindrical fiber with $n_{core} = 1.5$, $\Delta = 0.005$, $r = 3\ \mu$m. What is the V-number of this fiber for the fundamental mode at a wavelength half its minimum wavelength for single mode operation?

(17.9) A planar slab guide has n_2=1.5, n_1=1.498, $2d = 10\ \mu$m. Calculate the maximum electric field in the guiding layer for a TM_{01} mode if the total guided flux is 1 W mm^{-1}. The wavelength of the injected radiation is 1.3 μm.

(17.10) Plot the transverse intensity distribution in a cylindrical guide with parameters similar to the slab guide described in Problem (17.7) for the HE_{22} and EH_{12} modes. Choose an appropriate wavelength so that these modes will propagate.

(17.11) Evaluate the number of modes, M, that will propagate in fibers of core diameters 10 μm and 50 μm, respectively, when they are excited by sources of wavelength 1.55 μm and 0.85 μm. Take n_{core} =1.465, $n_{cladding}$=1.460.

(17.12) What fraction of the power travels outside the cladding in a cylindrical single-mode step-index guide with $2a$ =5 μm, λ_0=1.55 μm, $n_{core} = 1.48$, $\Delta = 0.005$?

(17.13) Prove Eq. (17.210).

(17.14) Prove that Eq.(17.166) is the necessary condition for the field boundary conditions at the core/cladding boundary to be satisfied.

(17.15) Plot the normalized propagation constant $\beta\lambda/2\pi$ for the modes HE_{11}, (HE_{21}, TE_{01}, TM_{01}), and (HE_{12}, HE_{11} and HE_{31}) in a step-index fiber with $n_1 = 1.5$, $n_2 = 1.485$, $a = 2\ \mu$m for V-numbers from 0 to 10.

(17.16) Repeat Problem (17.15) in the weakly-guiding approximation and show that the seven dispersion curves in Problem (17.15) coalesce into three dispersion curves characteristic of the LP_{01}, LP_{11}, and LP_{21} modes[17.11].

(17.17) (a) A fiber with a parabolic index profile has

$$n(r) = n_0 - \tfrac{1}{2}n_2r^2$$

with $n_0 = 1.5$ and $n_2 = 10^9$ m^{-2}. Design a single lens focusing arrangement that will optimally couple a laser with $\lambda_0 = 810$ nm and $\omega_0 = 2$ mm into the fiber. How would you modify the focusing arrangement if the laser emitted a Gaussian beam with ω_0=5 μm from the laser facet?

(17.18) Repeat Problem (17.17) but design a focusing system using a parabolic index GRIN lens that is 20 mm long. What parabolic profile parameter n_2 for the GRIN lens is needed?

(17.19) Plot dispersion curves for the S-wave modes in a slab waveguide with the same parameters as Fig. (17.10).

References

[17.1] A.W. Snyder and J.D. Love, *Optical Waveguide Theory*, Chapman and Hall, London.

[17.2] A.W. Snyder, 'Leaky-ray theory of optical wavegides of circular cross section,' *Appl. Phys.*, **4**, 273–298, 1974.

[17.3] A.L. Snyder and J.D. Love, 'The Goos-Hänchen shift,' *Appl. Opt.*, **15**, 236–8, 1976.

[17.4] T. Okoshi, *Optical Fibers*, Academic Press, New York, 1982.

[17.5] D. Gloge, 'Weakly Guiding Fibers,' *Appl. Opt.*, **10**, 2252–2258, 1971.

[17.6] P.K. Cheo, *Fiber Optics and Optoelectronics*, 2nd Edition, Prentice Hall, Englewood Cliffs, New Jersey, 1990,

[17.7] J.E. Midwinter, *Optical Fibers for Transmission*, Wiley, New York, 1979.

[17.8] S.E. Miller and A.G. Chynoweth, Editors, *Optical Fiber Telecommunications*, Academic, New York, 1979.

[17.9] P.C. Schultz, 'Progress in optical waveguide process and materials,' *Appl. Opt.*, **18**, 3684–3693, 1979.

[17.10] D.B. Keck, 'Single-mode fibers out perform multimode cables,' *IEEE Spectrum*, March 1983, pp. 30–37.

[17.11] J. Gowar, *Optical Communication Systems*, Prentice Hall, Englewood Cliffs, New Jersey, 1984.

[17.12] H.A. Haus, *Waves and Fields in Optoelectronics*, Prentice-Hall, Englewood Cliffs, New Jersey, 1984.

[17.13] A. Bjarklev, *Optical Fiber Amplifiers: Design and System Applications*, Artech House, Boston, 1993.

[17.14] E. Snitzer and R. Woodcock, 'Yb^{3+}-Er^{3+} glass laser,' *Appl. Phys. Lett.*, **6**, 45–46, 1965.

[17.15] C.J. Koester and E. Snitzer, 'Amplification in a fiber laser,' *Appl. Opt.*, **3**, 1182–1186, 1964.

[17.16] R.J. Means, L. Reekie, I.M. Jauncy, and D.N. Payne, 'High-gain rare-earth-doped fiber amplifier operating at 1.54 μm,' *Proceedings of the Conference on Optical Fiber Communications, OFC '87*, Washington, DC, Vol. 3, p. 187, 1987.

[17.17] B.J Ainslie, 'A review of the fabrication and properties of erbium-doped fibers for optical amplifiers,' *J. Lightwave Tech.*, **9**, 220–227, 1991.

[17.18] *Journal of Lightwave Technology, Special Issue on Optical Amplifiers*, Vol. 9, No. 2, 1991.

[17.19] D. Marcuse, 'Loss analysis of single-mode fiber splices,' *Bell. Sys. Tech. J.*, **56**, 703–718, 1977.

18

Optics of Anisotropic Media

18.1 Introduction

In this chapter we discuss wave propagation in anisotropic media. We shall see that in such media the electric vector of a propagating wave is not in general parallel to its polarization direction – defined by the direction of its electric displacement vector. Further, for propagation of plane waves in a particular direction through an anisotropic medium two distinct possible polarization directions exist, and waves having these polarization directions propagate with different velocities. We shall discuss an ellipsoidal surface called the indicatrix and show how with its aid the allowed polarization directions and their corresponding refractive indices can be determined for wave propagation in a given direction. Other three-dimensional surfaces related to the indicatrix and their use in describing different optical properties of anisotropic media are also discussed. We shall concentrate our attention primarily on uniaxial crystals, which have optical properties that can be referred to an indicatrix with two equal axes, and will discuss how such crystals can be used to control the polarization characteristics of light.

Important anisotropic optical media are generally crystalline and their optical properties are closely related to various symmetry properties possessed by crystals. To assist the reader who is not familiar with basic ideas of crystal symmetry, Appendix 8 summarizes a number of aspects of this subject that should be helpful in reading this chapter and a number of those succeeding it.

18.2 The Dielectric Tensor

In an isotropic medium the propagation characteristics of electromagnetic waves are independent of their propagation direction. This generally implies that there is no direction within such a medium that is any different from any other. Clearly then, we can class gases and liquids, but not liquid crystals, as isotropic media, provided there are no externally applied fields present. Such a field would, of course, imply the existence of a unique direction in the medium – that of the field. As an example of a situation where an isotropic medium becomes anisotropic in an external field, we mention the case of a gas in a magnetic field, where the gas changes the polarization characteristics of a wave that propagates in the field direction. This phenomenon is called the Faraday effect. In most circumstances, cubic crystals of the highest symmetry, crystal classes m3m and 432, behave as isotropic media.

In an isotropic medium the electric displacement vector **D** and its associated electric field **E** are parallel, in other words we write

$$\mathbf{D} = \epsilon_r \epsilon_0 \mathbf{E}, \tag{18.1}$$

where ϵ_r is the scalar dielectric constant, which in the general case is a function of frequency. This is equivalent to saying that the polarization induced by the field and the field itself are parallel

$$\mathbf{P} = \epsilon_0 \chi \mathbf{E}, \tag{18.2}$$

where χ is the scalar susceptibility. We will restrict ourselves in what follows to materials that are neither absorbing nor amplifying, so that both ϵ_r and χ are real.

In an anisotropic medium **D** and **E** are no longer necessarily parallel and we write

$$\mathbf{D} = \overline{\overline{\epsilon}}_r \epsilon_0 \mathbf{E}. \tag{18.3}$$

where $\overline{\overline{\epsilon}}_r$ is the dielectric tensor,† which in matrix form referred to three arbitrary orthogonal axes is

$$\overline{\overline{\epsilon}}_r \equiv \begin{pmatrix} \epsilon_{xx} & \epsilon_{xy} & \epsilon_{xz} \\ \epsilon_{yx} & \epsilon_{yy} & \epsilon_{yz} \\ \epsilon_{zx} & \epsilon_{zy} & \epsilon_{zz} \end{pmatrix}. \tag{18.4}$$

So for example:

$$\begin{aligned} D_x &= \epsilon_0(\epsilon_{xx}E_x + \epsilon_{xy}E_y + \epsilon_{xz}E_z), \\ D_y &= \epsilon_0(\epsilon_{yx}E_x + \epsilon_{yy}E_y + \epsilon_{yz}E_z), \\ D_z &= \epsilon_0(\epsilon_{zx}E_x + \epsilon_{zy}E_y + \epsilon_{zz}E_z). \end{aligned} \tag{18.5}$$

By making the appropriate choice of axes the dielectric tensor can be diagonalized. With this choice of axes, called the principal axes of the material, Eq. (18.5) in matrix form becomes

$$\begin{pmatrix} D_x \\ D_y \\ D_z \end{pmatrix} = \epsilon_0 \begin{pmatrix} \epsilon_x & 0 & 0 \\ 0 & \epsilon_y & 0 \\ 0 & 0 & \epsilon_z \end{pmatrix} \begin{pmatrix} E_x \\ E_y \\ E_z \end{pmatrix}, \tag{18.6}$$

where ϵ_x, ϵ_y and ϵ_z are called the principal dielectric constants.

Alternatively, we can describe the anisotropic character of the medium with the aid of the susceptibility tensor $\overline{\overline{\chi}}$

$$\mathbf{P} = \epsilon_0 \, \overline{\overline{\chi}} \, \mathbf{E}, \tag{18.7}$$

where $\overline{\overline{\chi}}$ has the matrix form when referred to three arbitrary orthogonal axes

$$\overline{\overline{\chi}} \equiv \begin{pmatrix} \chi_{xx} & \chi_{xy} & \chi_{xz} \\ \chi_{yx} & \chi_{yy} & \chi_{yz} \\ \chi_{zx} & \chi_{zy} & \chi_{zz} \end{pmatrix}. \tag{18.8}$$

In the principal coordinate system

$$\overline{\overline{\chi}} \equiv \begin{pmatrix} \chi_x & 0 & 0 \\ 0 & \chi_y & 0 \\ 0 & 0 & \chi_z \end{pmatrix}. \tag{18.9}$$

Since $\mathbf{D} = \epsilon_0 \mathbf{E} + \mathbf{P}$, it is clear that in a principal coordinate system

$$\epsilon_x = 1 + \chi_x, \text{ etc.} \tag{18.10}$$

† For a brief review of some aspects of tensor algebra that should be of value see Appendix 9.

We can understand why crystals with less than cubic symmetry have susceptibilities that depend on the direction of the applied field by considering the physical relationship between polarization and applied field. When a field **E** is applied to a crystal, it displaces both electrons and nuclei from their equilibrium positions in the lattice and induces a net dipole moment per unit volume (polarization), which we can write in the form:

$$\mathbf{P} = \sum_j N_j e_j \Delta\mathbf{r}_j. \tag{18.11}$$

N_j is the density of species j with charge e_j in the crystal and $\Delta\mathbf{r}_j$ is the displacement of this charged species from its equilibrium position. We should note that when the applied field is at optical frequencies, only electrons make any significant contribution to this polarization, particularly the most loosely bound outer valence electrons of the ions within the lattice. The ions themselves are too heavy to follow the rapidly oscillating applied field. If the applied electric field has components, E_x, E_y and E_z, then in equilibrium we can write

$$E_x = -k_{jx}(\Delta\mathbf{r}_j)_x \tag{18.12}$$

and two other similar equations, where k_{jx} is a restoring force constant appropriate to the x component of the displacement of the charge j from its equilibrium position; thus

$$\Delta\mathbf{r}_j = \left(\frac{E_x}{k_{jx}}\hat{\mathbf{i}} + \frac{E_y}{k_{jy}}\hat{\mathbf{j}} + \frac{E_z}{k_{jz}}\hat{\mathbf{k}}\right), \tag{18.13}$$

which is a vector parallel to **E** only if all three force constants are equal. The equality of these force constants for displacement along three orthogonal axes would imply an arrangement with cubic symmetry of the ions in the lattice about the charged particle being considered.

18.3 Stored Electromagnetic Energy in Anisotropic Media

If we wish the stored energy density in an electromagnetic field to be the same in an anisotropic medium as it is in an isotropic one then we require

$$U = \frac{1}{2}(\mathbf{E}\cdot\mathbf{D} + \mathbf{B}\cdot\mathbf{H}), \tag{18.14}$$

which gives

$$U = \frac{1}{2}(\overline{\overline{\epsilon}}_r\, \epsilon_0 \mathbf{E}\cdot\mathbf{E} + \mathbf{B}\cdot\mathbf{H}). \tag{18.15}$$

Now, in any medium the net power flow into unit volume is†

$$\begin{aligned}\frac{\partial U}{\partial t} &= -\nabla\cdot(\mathbf{E}\times\mathbf{H}) = \mathbf{E}\cdot\frac{\partial \mathbf{D}}{\partial t} + \cdot\mathbf{H}\frac{\partial \mathbf{B}}{\partial t}\\ &= \epsilon_0\mathbf{E}\cdot\,\overline{\overline{\epsilon}}_r\,\frac{\partial \mathbf{E}}{\partial t} + \mathbf{H}\cdot\frac{\partial \mathbf{B}}{\partial t},\end{aligned} \tag{18.16}$$

where we have assumed that the medium is nonconductive so that $\mathbf{j} = 0$. Eq. (18.16) gives the rate of change of stored energy within unit volume, which must also be

† See Appendix 5.

given by the time derivative of Eq. (18.15)

$$\frac{\partial U}{\partial t} = \frac{1}{2}\left[\epsilon_0 \stackrel{=}{\epsilon_r} \frac{\partial}{\partial t}(\mathbf{E}\cdot\mathbf{E}) + \frac{\partial}{\partial t}(\mathbf{B}\cdot\mathbf{H})\right]. \tag{18.17}$$

Most optical crystals are not, or are only very slightly, magnetic so we can assume that $\mathbf{B} = \mu_0\mathbf{H}$ since $\mu_r \simeq 1$. In this case

$$\frac{1}{2}\frac{\partial}{\partial t}\mathbf{B}\cdot\mathbf{H} = \mathbf{H}\cdot\frac{\partial \mathbf{B}}{\partial t}. \tag{18.18}$$

Comparing the first term on the RHS of Eq. (18.16) with the first two terms on the RHS of Eq. (18.17)

$$\epsilon_0\mathbf{E}\cdot \stackrel{=}{\epsilon_r} \frac{\partial \mathbf{E}}{\partial t} = \frac{1}{2}\left(\epsilon_0 \stackrel{=}{\epsilon_r} \mathbf{E}\cdot\frac{\partial \mathbf{E}}{\partial t} + \epsilon_0 \stackrel{=}{\epsilon_r} \frac{\partial \mathbf{E}}{\partial t}\cdot\mathbf{E}\right), \tag{18.19}$$

which implies that

$$\stackrel{=}{\epsilon_r} \mathbf{E}\cdot\frac{\partial \mathbf{E}}{\partial t} = \stackrel{=}{\epsilon_r} \frac{\partial \mathbf{E}}{\partial t}\cdot\mathbf{E}. \tag{18.20}$$

Written out in full Eq. (18.20) is

$$\begin{aligned}
&\epsilon_{xx}E_x\dot{E}_x + \epsilon_{xy}E_y\dot{E}_x + \epsilon_{xz}E_z\dot{E}_x + \epsilon_{yx}E_x\dot{E}_y + \epsilon_{yy}E_y\dot{E}_y \\
&+\epsilon_{yz}E_z\dot{E}_y + \epsilon_{zx}E_x\dot{E}_z + \epsilon_{zy}E_y\dot{E}_z + \epsilon_{zz}E_z\dot{E}_z \\
&= \epsilon_{xx}\dot{E}_xE_x + E_{xy}\dot{E}_yE_x + \epsilon_{xz}\dot{E}_zE_x + \epsilon_{yx}\dot{E}_xE_y \\
&+\epsilon_{yy}\dot{E}_yE_y + \epsilon_{yz}\dot{E}_zE_y + \epsilon_{zx}\dot{E}_xE_z + \epsilon_{zy}\dot{E}_yE_z + \epsilon_{zz}\dot{E}_zE_z
\end{aligned} \tag{18.21}$$

Clearly $\epsilon_{xy} = \epsilon_{yx}; \epsilon_{xz} = \epsilon_{zx}$; etc. so the dielectric tensor only has six independent terms. By working in the principal coordinate system, all the off-diagonal terms of the dielectric tensor become zero, which greatly simplifies consideration of the wave propagation characteristics of anisotropic crystals. In this case we can write

$$D_x = \epsilon_0\epsilon_xE_x; \quad D_y = \epsilon_0\epsilon_yE_y; \quad D_z = \epsilon_0\epsilon_zE_z, \tag{18.22}$$

where we are writing $\epsilon_x = \epsilon_{xx}$, etc., for simplicity and the electrical energy density in the crystal becomes

$$U_E = \frac{1}{2}\mathbf{E}\cdot\mathbf{D} = \frac{1}{2\epsilon_0}\left(\frac{D_x^2}{\epsilon_x} + \frac{D_y^2}{\epsilon_y} + \frac{D_z^2}{\epsilon_z}\right). \tag{18.23}$$

Eq. (18.23) shows that the electric displacement vectors from a given point that correspond to a constant stored electrical energy describe an ellipsoid.

18.4 Propagation of Monochromatic Plane Waves in Anisotropic Media

Let us assume that a monochromatic plane wave of the form

$$\mathbf{D} = \mathbf{D_0}e^{i(\omega t - \mathbf{k}\cdot\mathbf{r})} = \mathbf{D}_0e^{i\phi} \tag{18.24)}$$

can propagate through an anisotropic medium. The direction of $\mathbf{D}$ specifies the direction of polarization of this wave. The wave vector $\mathbf{k}$ of this plane wave is normal to the wavefront (the plane where the phase of the wave is everywhere equal) and has magnitude $|\mathbf{k}| = \omega/c$. The phase velocity of the wave is c, which is related to the velocity of light *in vacuo*, c_0, by the refractive index n experienced by the wave according to $c = c_0/n$. The phase velocity of the wave in a particular wave vector direction in the crystal can also be written as $c = \omega/|\mathbf{k}|$.

We assume that Maxwell's equations still hold, so that in the absence of currents or free charges

$$\text{div } \mathbf{D} = 0, \tag{18.25}$$

$$\text{div } \mathbf{B} = 0, \tag{18.26}$$

$$\text{curl } \mathbf{E} = \frac{-\partial \mathbf{B}}{\partial t}, \tag{18.27}$$

$$\text{curl } \mathbf{H} = \frac{\partial \mathbf{D}}{\partial t}. \tag{18.28}$$

We stress that because $\mathbf{D}$ and $\mathbf{E}$ are now related by a tensor operation, div $\mathbf{E}$ is no longer zero.

Taking the curl of both sides of Eq. (18.27)

$$\text{curl curl}\mathbf{E} = -\frac{\partial}{\partial t}(\text{ curl } \mu_0 \mathbf{H}) = -\mu_0 \frac{\partial}{\partial t}\left(\frac{\partial \mathbf{D}}{\partial t}\right) = -\mu_0 \frac{\partial^2}{\partial t^2}(\epsilon_0 \stackrel{=}{\epsilon}_r \mathbf{E}) \tag{18.29}$$

and using the vector identity curl curl $\mathbf{E} =$ grad (div $\mathbf{E}$) $- \nabla^2 \mathbf{E}$ gives

$$\nabla^2 \mathbf{E} - \text{ grad (div}\mathbf{E}) = \epsilon_0 \mu_0 \stackrel{=}{\epsilon}_r \frac{\partial^2 \mathbf{E}}{\partial t^2}. \tag{18.30}$$

In the principal coordinate system the x component of Eq. (18.31) is, for example,

$$\frac{\partial^2 E_x}{\partial x^2} + \frac{\partial^2 E_x}{\partial y^2} + \frac{\partial^2 E_x}{\partial z^2} - \frac{\partial}{\partial x}\left(\frac{\partial E_x}{\partial x} + \frac{\partial E_y}{\partial y} + \frac{\partial E_z}{\partial z}\right) = \epsilon_0 \mu_0 \epsilon_x \frac{\partial^2 E_x}{\partial t^2}. \tag{18.31}$$

For a plane wave of the form above, again in the principal coordinate system,

$$\epsilon_0 \epsilon_x E_x = D_x = (\mathbf{D}_0)_x e^{i[wt-(k_x x + k_y y + k_z z)]}, \tag{18.32}$$

where k_x, k_y and k_z are the three orthogonal components of the wave vector and ϵ_x depends on the angular frequency ω; with similar equations for E_y and E_z. Substituting for $E_x = D_x/\epsilon_0 \epsilon_x$ in Eq. (18.32) gives

$$\frac{D_x}{\epsilon_x}(k_x^2 + k_y^2 + k_z^2) - k_x \left(\frac{k_x D_x}{\epsilon_x} + \frac{k_y D_y}{\epsilon_y} + \frac{k_z D_z}{\epsilon_z}\right) = \epsilon_0 \mu_0 \omega^2 D_x \tag{18.33}$$

and two other similar equations. Recognizing that

$$\epsilon_0 \mu_0 = \frac{1}{c_0^2} \tag{18.34}$$

and

$$(k_x^2 + k_y^2 + k_z^2) = |\mathbf{k}|^2, \tag{18.35}$$

and writing

$$\frac{\epsilon_0 c_0^2}{|\mathbf{k}|^2}(k_x E_x + k_y E_y + k_z E_z) = P^2, \tag{18.36}$$

Eq. (18.33) becomes

$$D_x = \frac{-k_x P^2}{c^2 - c_x^2}. \tag{18.37}$$

Similarly,

$$D_y = \frac{-k_y P^2}{c^2 - c_y^2}, \tag{18.38}$$

$$D_z = \frac{-k_z P^2}{c^2 - c_z^2}, \tag{18.39}$$

where $c_x = c_0/\sqrt{\epsilon_x}; c_y = c_0/\sqrt{\epsilon_y}$ and $c_z = c_0/\sqrt{\epsilon_z}$ are called the principal phase velocities of the crystal. Now, since div $\mathbf{D} = -i(k_x D_x + k_y D_y + k_z D_z) = 0$, we have

$$\frac{k_x^2}{c^2 - c_x^2} + \frac{k_y^2}{c^2 - c_y^2} + \frac{k_z^2}{c^2 - c_z^2} = 0. \tag{18.40}$$

Multiplying both sides of Eq. (18.41) by c^2 and rearranging gives

$$\frac{k_x^2 n_x^2}{n^2 - n_x^2} + \frac{k_y^2 n_y^2}{n^2 - n_y^2} + \frac{k_z^2 n_z^2}{n^2 - n_z^2} = 0, \tag{18.41}$$

where we have put $n_x = \sqrt{\epsilon_x}$, $n_y = \sqrt{\epsilon_y}$, and $n_z = \sqrt{\epsilon_z}$; n_x, n_y and n_z are called the principal refractive indices of the crystal. Eqs. (18.40) and (18.41), which is called Fresnel's equation, are quadratic in c^2 and n^2 respectively. Thus in general there are two possible solutions c_1, c_2 and n_1, n_2, respectively, for the phase velocity and refractive index of a monochromatic wave propagating through an anisotropic medium with wave vector **k**. However, when **k** lies in certain specific directions both roots of Eqs. (18.40) and (18.41) become equal. These special directions within the crystal are called *optic axes*. The fact that, for example, $\sqrt{n_1^2}$ has two roots $+n_1$ and $-n_1$ merely corresponds to each solution of Eq. (18.42) allowing a wave to propagate in a given direction either with wave vector **k** or $-$**k**.

18.5 The Two Possible Directions of D for a Given Wave Vector are Orthogonal

We can show that the two solutions of Eqs. (18.40) and (18.41) correspond to two different possible linear polarizations of a wave propagating with wave vector **k** and that these two solutions have mutually othogonal polarization. Let us first illustrate this by considering some special cases.

If we send a monochromatic wave into an anisotropic crystal travelling in the z direction, but linearly polarized in the x direction, then from Maxwell's Eqs. (18.27) and (18.28)

$$\frac{\partial E_x}{\partial z} = -\mu_r \mu_0 \frac{\partial H_y}{\partial t}, \tag{18.42}$$

$$\frac{\partial H_y}{\partial z} = -\epsilon_x \epsilon_0 \frac{\partial E_x}{\partial t}, \tag{18.43}$$

where we have taken the axes x, y, and z to correspond to the principal coordinate system of the crystal. Taking the z derivative of Eq. (18.43) and substituting for $\partial H_y/\partial z$ from (18.43) gives

$$\frac{\partial^2 E_x}{\partial z^2} = \mu_r \mu_0 \epsilon_x \epsilon_0 \frac{\partial^2 E_x}{\partial t^2}, \tag{18.44}$$

which is a one-dimensional wave equation with a solution of the form

$$E_x = |\mathbf{E_0}|_x e^{i(\omega t - k_x z)}, \tag{18.45}$$

where $k_x = \omega\sqrt{\mu_r \mu_0 \epsilon_x \epsilon_0}$.

Thus the wave propagates along the z axis with a phase velocity $c_1 = c_0/n_x$, where $n_x^2 = \epsilon_x$.

If we repeat this derivation, but for a wave propagating in the z direction and polarized in the y direction we find that the wave now propagates with a phase velocity $c_2 = c_0/n_y$ where $n_y^2 = \epsilon_y$. This illustrates, that, at least in this special case, a wave propagating in the z direction of the principal coordinate system has two possible orthogonal allowed linear polarizations, E_x and E_y, that travel with respective phase velocities $c_1 = c_0/n_x$ and $c_2 = c_0/n_y$. It is straightforward to show that for propagation in the x direction there are two possible orthogonal linear polarizations E_y and E_z with corresponding phase velocities c_0/n_y and c_0/n_z with similar behavior for a wave propagating in the y direction. Clearly, in these special cases where a wave propagates along one principal axis, and is polarized along a second, both **D** and **E** are parallel and the two possible orientations of **D** for a given wave vector are orthogonal. However, for propagation in an arbitrary direction **D** and **E** are no longer parallel. To find the angular relationship between the two directions of **D** which are allowed for a particular wave vector we use Eqs. (18.37), (18.38), and (18.39). If we designate quantities that refer to the two solutions by the subscripts 1 and 2, then

$$\left.\begin{aligned} D_{j_1} &= \frac{-k_j P_1^2}{c_1^2 - c_j^2} \\ D_{j_2} &= \frac{-k_j P_2^2}{c_2^2 - c_j^2} \end{aligned}\right\} \quad j = x, y, z \tag{18.46}$$

and

$$\begin{aligned} \mathbf{D}_1 \cdot \mathbf{D}_2 &= \sum_{j=x,y,z} \frac{k_j^2 P_1^2 P_2^2}{(c_1^2 - c_j^2)(c_2^2 - c_j^2)} \\ &= \frac{P_1^2 P_2^2}{(c_2^2 - c_1^2)} \sum_{j=x,y,z} \frac{k_j^2}{(c_1^2 - c_j^2)} - \frac{k_j^2}{(c_2^2 - c_j^2)}, \end{aligned} \tag{18.47}$$

which is zero by virtue of Eq. (18.40). Thus, the two possible directions of **D** for a given wave vector are orthogonal.

18.6 Angular Relationships between D, E, H, k, and the Poynting Vector S

The electric displacement vector **D** and wave vector **k** are, by definition and by virtue of Eq. (18.25), mutually perpendicular. Further, since the values of **D**, **E**, and **H** are constant over the phase front of a plane wave, the phase factor ϕ must be the same for all these three vectors. We can show that this is so provided certain angular relationships exist between them. For **D**, **E**, and **H** with a phase dependence of the form $e^{i(\omega t - \mathbf{k}\cdot\mathbf{r})}$ we can replace the operation $\partial/\partial t$ by multiplication by $i\omega$ and the operation $\partial/\partial x$ by multiplication by $-ik_x$. So, for

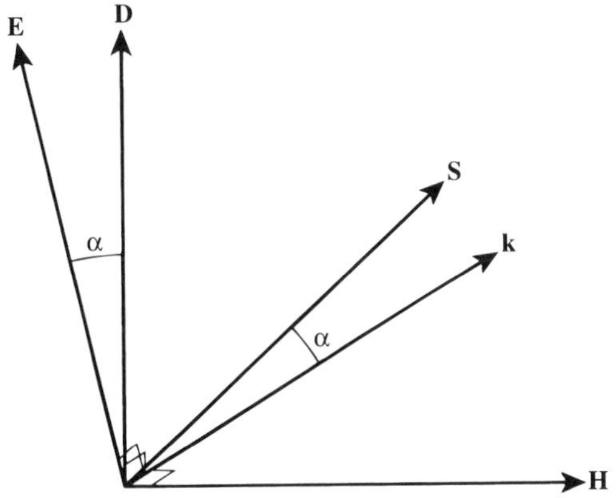

Fig. 18.1. Angular relationship between **D**, **E**, **H**, **k**, and **S** vectors in an anisotripic medium.

example,

$$\frac{\partial \mathbf{D}}{\partial t} = i\omega \mathbf{D}, \tag{18.48}$$

$$\text{curl } \mathbf{E} = i(\mathbf{E} \times \mathbf{k}), \tag{18.49}$$

which can be verified by writing out both sides in cartesian coordinates, and

$$\text{div } \mathbf{E} = -i(\mathbf{E} \cdot \mathbf{k}) \tag{18.50}$$

with similar relations for curl **H**, etc.

Thus from Eqs. (18.49) and (18.27)

$$\mathbf{E} \times \mathbf{k} = \omega \mu \mu_0 \mathbf{H}, \tag{18.51}$$

and similarly

$$\mathbf{H} \times \mathbf{k} = \omega \mathbf{D}. \tag{18.52}$$

Thus, **H** is normal to last **D**, **E**, and **k**, and the three are coplanar. **D** and **E** make an angle α with one another where

$$\alpha = \arccos\left(\frac{\mathbf{E} \cdot \mathbf{D}}{|\,\mathbf{E}\,||\,\mathbf{D}\,|}\right) = \arccos\left(\frac{\epsilon_0(\overline{\overline{\epsilon}}_r\ \mathbf{E}) \cdot \mathbf{E}}{|\,\mathbf{E}\,||\,\mathbf{D}\,|}\right). \tag{18.53}$$

These angular relationships are illustrated in Fig. (18.1). The direction of the Poynting vector $\mathbf{S} = \mathbf{E} \times \mathbf{H}$ is, by definition, perpendicular to both the **E** and **H** vectors. **S** defines the direction of energy flow within the medium and we identify it as the direction of the *ray* of familiar geometric optics. Whereas in isotropic media the ray is always parallel to the wave vector, and is therefore perpendicular to the wavefront, this is no longer so in anisotropic media, except for propagation along one of the principal axes.

To summarize, we have shown that transverse electromagnetic plane waves can propagate through anisotropic media, but for propagation in a general direction two distinct allowed linear polarizations specified by the direction of **D** can exist for the wave. We have shown that these two allowed polarizations are orthogonal and that the wave propagates with a phase velocity (the velocity of the surface of constant phase – the wavefront) which depends on which of these two polarizations it has. Clearly, a wave of arbitrary polarization that enters such an anisotropic medium will not in general correspond to one of the allowed polarizations, and will therefore be resolved into two linearly polarized components polarized along the allowed directions. Each component propagates with a different phase velocity. All that remains for us to do to characterize fully these allowed polarization directions is to specify their orientation with respect to the principal axes of the medium. This is done with the aid of a geometric figure called the *index ellipsoid* or *indicatrix*.

18.7 The Indicatrix

The indicatrix, wave-normal, or index ellipsoid is an ellipsoid with the equation

$$\frac{x^2}{n_x^2} + \frac{y^2}{n_y^2} + \frac{z^2}{n_z^2} = 1, \tag{18.54}$$

that allows us to determine the refractive index for monochromatic plane waves as a function of their direction of polarization. It is, apart from a scale factor, equivalent to the surface mapped out by the **D** vectors corresponding to a constant energy density at a given frequency. This ellipsoid can be visualized as oriented inside a crystal consistent with the symmetry axes of the crystal. For example, in any crystal with perpendicular symmetry axes, such as those belonging to the cubic, tetragonal, hexagonal, trigonal or orthorhombic crystal systems, the axes of the ellipsoid, which are the principal axes of the crystal, are parallel to the three axes of symmetry of the crystal. For the orientation of the indicatrix to be consistent with the symmetry of the crystal, planes of mirror symmetry within the crystal must coincide with planes of symmetry of the indicatrix: namely the xy, yz, and zx planes. In the monoclinic system, crystal symmetry is referred to three axes, two of which, the a and c axes of crystallographic terminology, intersect at acute and obtuse angles and a third, the b axis, which is perpendicular to the a and c axes. In such crystals, one of the axes of the indicatrix must coincide with the b axis, but the other two axes have any orientation, although this is fixed for a given crystal and wavelength. In triclinic crystals whose symmetry is referred to three unequal length, nonorthogonal axes, the indicatrix can take any orientation, although this is fixed for a given crystal and wavelength.

We use the geometrical properties of the indicatrix to determine the refractive indices and polarizations of the two monochromatic plane waves that can propagate through the crystal with a given wave vector. This is illustrated in Fig. (18.2), which shows the direction of the wave vector **k** for a monochromatic wave propagating through a crystal, drawn relative to the orientation of the indicatrix in the crystal. The plane surface that is orthogonal to the wave vector, and that passes through the center of the indicatrix, intersects this ellipsoid in an ellipse called the *intersection ellipse*. The semiaxes of this ellipse define the directions of the two allowed **D** polarizations that can propagate through the crystal with the given wave vector **k**, and the lengths of these semiaxes give the refractive indices experienced by these two polarizations. That this interpretation is consistent follows for a number of reasons. For a wave of given intensity propagating through the crystal, the stored energy density must be independent of polarization direction, otherwise we could either absorb or extract energy from the wave merely by rotating the crystal. This cannot be so as we have already specified that the crystal is transparent. Consequently, for a given wave vector the many possible **D** polarizations trace out an ellipse, equivalent

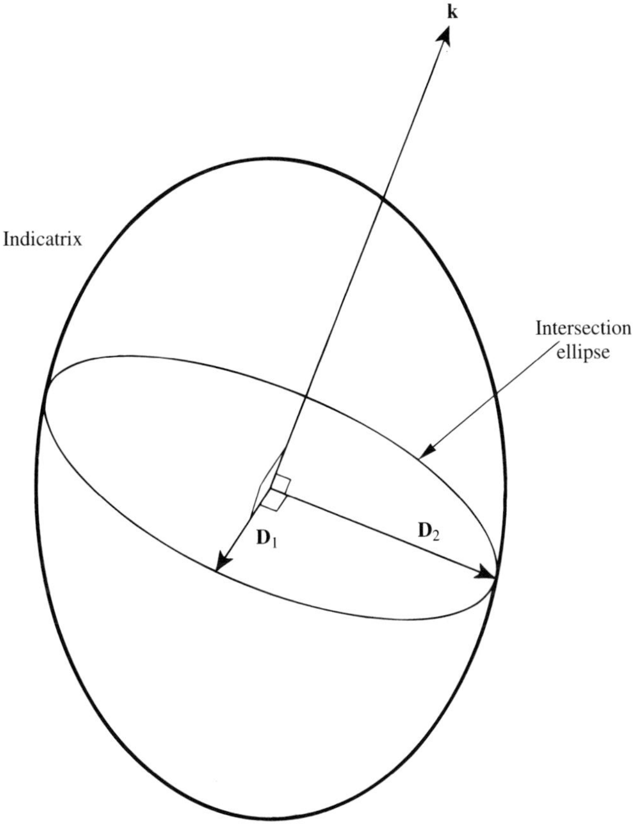

Fig. 18.2. The ellipsoidal indicatrix showing the wave vector direction **k** and the intersection ellipse that is perpendicular to **k**. The two permitted linear polarization directions $\mathbf{D}_1$ and $\mathbf{D}_2$ lie along the semi-axes of the intersection ellipse.

to the intersection ellipse. Further, it can be shown that the lengths of the semiaxes of the intersection ellipse corresponding to a given wave vector $\mathbf{k} = k_x\hat{\mathbf{i}} + k_y\hat{\mathbf{j}} + k_z\hat{\mathbf{k}}$ are the two roots n_1 and n_2 of Fresnel's Eq. (18.41). Thus, only the two **D** vectors parallel to these semiaxes simultaneously satisfy the condition of both being on the ellipse and having the appropriate refractive indices to satisfy Fresnel's equations.

In the general case there are two **k** vector directions through the center of the indicatrix for which the intersection ellipse is a circle. This is a fundamental geometric property of ellipsoids. These two directions are called the *principal optic axes* and are fixed for a given crystal and frequency of light. Waves can propagate along these optic axes with any arbitrary polarization, as in these directions the refractive index is not a function of polarization.

In cubic crystals the indicatrix is a sphere called the isotropic indicatrix, there are no specific optic axes as the indicatrix is anaxial and the propagation of waves is independent of both the directions of **k** and **D**.

In crystals belonging to the tetragonal, hexagonal, and trigonal crystal systems the crystal symmetry requires that $n_x = n_y$ and the indicatrix reduces to an ellipsoid of revolution. In this case there is only one optic axis, oriented along the axis of highest symmetry of the crystal, the z axis (or c axis). These crystals classes, listed in Table (18.1), are said to be uniaxial: a discussion of their properties is considerably simpler than for crystals belonging to the less symmetric orthorhombic, monoclinic, and triclinic crystal systems, which are *biaxial*.

Table 18.1. *Uniaxial crystal classes.*

Hexagonal	Trigonal	Tetragonal
$\bar{6}$2m	3m	$\bar{4}$2m
6mm	32	4mm
622	3	422
$\bar{6}$		$\bar{4}$
6		4

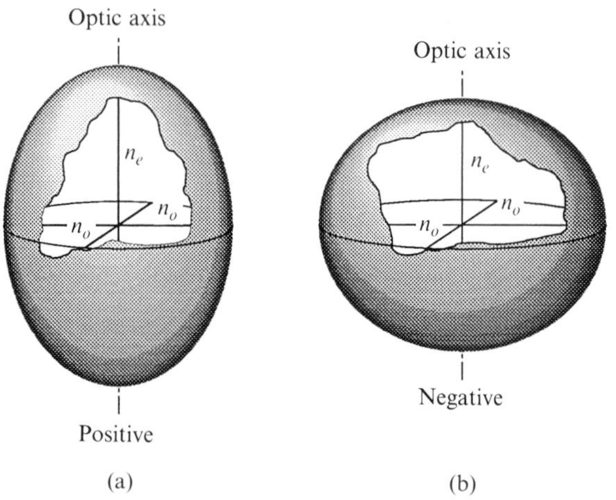

Fig. 18.3. (a) Indicatrix in a positive crystal, (b) indicatrix in a negative crystal.

18.8 Uniaxial Crystals

The equation of the uniaxial indicatrix is

$$\frac{x^2+y^2}{n_o^2}+\frac{z^2}{n_e^2}=1, \tag{18.55}$$

where n_o is the index of refraction experienced by waves polarized perpendicular to the optic axis, called *ordinary* or O-waves; n_e is the index of refraction experienced by waves polarized parallel to the optic axis, called *extraordinary* or E-waves. If $n_e > n_o$ the indicatrix is a prolate ellipsoid of revolution as shown in Fig. (18.3a) and such a crystal is said to be *positive uniaxial.*

If $n_e < n_o$ the indicatrix is an oblate ellipsoid of revolution as shown in Fig. (18.3b) and the crystal is said to be *negative uniaxial.* Fig. (18.4) shows the orientation of this ellipsoid inside a negative uniaxial crystal of calcite.

Because uniaxial crystals have indicatrices which are circularly symmetric about the z (optic) axis, their optical properties depend only on the polar angle θ that the wave vector **k** makes with the optic axis and not on the azimuthal orientation of **k** relative to the x and y axes. Thus, we can illustrate all their optical characteristics by considering propagation in any plane containing the optic axis, as shown in Fig. (18.5). For propagation with the wave vector in the direction ON, the two allowed polarizations are as indicated: perpendicular to the optic axis, and in the plane containing ON and the optic axis OS. The major semiaxis of the intersection ellipse is OQ in this case. The length OQ gives the refractive index for waves

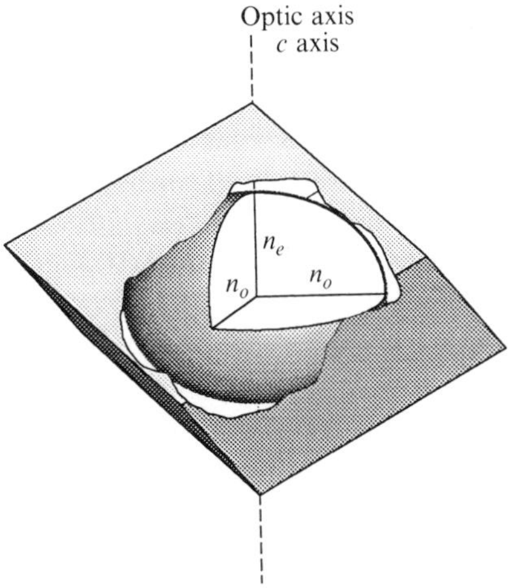

Fig. 18.4. Rhomboidal calcite crystal showing the orientation of the indicatrix.

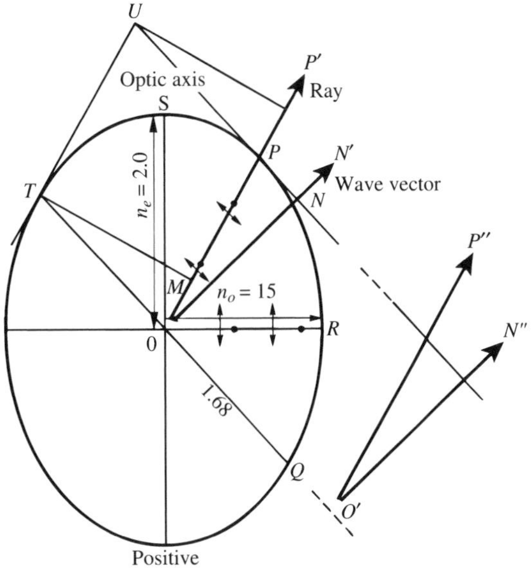

Fig. 18.5. Section of positive uniaxial indicatrix containing optic axis showing wave-normal and ray directions.

polarized parallel to OQ (E-waves). The refractive index of the O-wave is given by the minor semiaxis of the intersection ellipse, which in a uniaxial crystal is independent of direction and is equal to the length OR. In Fig. (18.5) the O-wave with wave vector in the direction ON propagates with phase velocity $c_0/n_o = c_0/1.5$, while the E-wave propagates with phase velocity $c_0/OQ = c_0/1.68$. Thus, in this positive crystal the O-wave travels faster than the E-wave.

In a uniaxial crystal the O-waves have no component of $\mathbf{D}$ along the z axis, thus from Eq. (18.7)

$$D_x = \epsilon_0 \epsilon_x E_x, \tag{18.56}$$

$$D_y = \epsilon_0 \epsilon_y E_y. \tag{18.57}$$

In uniaxial crystals, since $\epsilon_x = \epsilon_y$, $\mathbf{D}$ and $\mathbf{E}$ are parallel for the O-wave and the O-ray is parallel to the O-wave vector. However, the ray and wave vector of the E-wave are not parallel, except for propagation along, or perpendicular to, the optic axis. We can find the direction of the E-ray by a simple geometric construction shown in Fig. (18.5). The wavevector is in the direction ON, and TQ is the major axis of the intersection ellipse. The tangent to the ellipse at the point T is parallel to the ray direction OP. Note that the tangent to the ellipse at point P is orthogonal to the wave vector. The wave index of refraction of the extraordinary wave with wave vector **k** is given by the length OT. The length TM gives the ray index of refraction, which is a measure of the velocity with which energy flows along the ray OP.

Light propagating along the ray OP as an E-wave is not polarized perpendicular to the ray direction, however, the **E** vector of this wave is perpendicular to the ray, parallel to the direction MT in Fig. (18.5). Since light propagating along the ray OP with wave vector **k** travels a distance OP in the time the wave front (which is parallel to the direction PN) travels a distance ON, it is clear that the velocity of light along the ray is greater than the wave-normal velocity (the phase velocity). If the wave vector is in the direction OR in Fig. (18.5) then both wave-normal and ray are parallel for the E- and O-waves. As indicated by the dots, the O-wave propagates more quickly in this positive crystal.

18.9 Index Surfaces

The refractive index $n_e(\theta)$ of an E-wave propagating at angle θ to the optic axis in a uniaxial crystal can be calculated with reference to Fig. (18.6). For this wave the value of $n_e(\theta)$ is given by the length OX in the figure, the major semiaxis of the intersection ellipse. The cartesian coordinates of the point P relative to the origin O are

$$\begin{aligned} x &= -n_e(\theta)\cos\theta, \\ z &= n_e(\theta)\sin\theta, \\ y &= 0. \end{aligned} \tag{18.58}$$

This point lies on the indicatrix so

$$\frac{n_e^2(\theta)\cos^2(\theta)}{n_o^2} + \frac{n_e^2(\theta)\sin^2\theta}{n_e^2} = 1, \tag{18.59}$$

which gives

$$n_e(\theta) = \frac{n_o n_e}{\sqrt{n_e^2\cos^2\theta + n_o^2\sin^2\theta}}. \tag{18.60}$$

We can use this relationship to specify a surface called the extraordinary index surface, which shows geometrically the index of refraction of extraordinary waves in a uniaxial crystal as a function of their direction of propagation. The cartesian coordinates on this surface must satisfy

$$n_e^2(\theta) = x^2 + y^2 + z^2 \tag{18.61}$$

and also

$$\sin^2\theta = \frac{x^2 + y^2}{x^2 + y^2 + z^2}, \tag{18.62}$$

$$\cos^2\theta = \frac{z^2}{x^2 + y^2 + z^2}. \tag{18.63}$$

Fig. 18.6. Section of uniaxial indicatrix containing the optic axis.

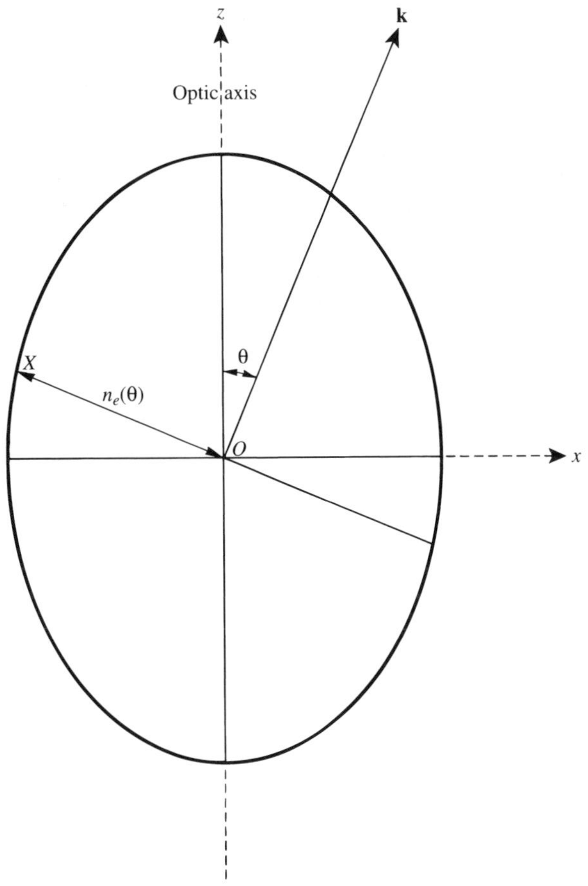

Substituting from Eqs. (18.61), (18.62), and (18.63) in Eq. (18.60) gives

$$\frac{x^2+y^2}{n_e^2}+\frac{z^2}{n_o^2}=1. \tag{18.64}$$

We might have expected this, since for an E-wave propagating perpendicular to the optic axis $z=0$ and $\sqrt{x^2+y^2}=n_e(0)=n_e$. For an E-wave propagating along the z axis (which in this case is actually also an O-wave) $x^2+y^2=0$ and $z=n_e(90)=n_o$.

The index surface for O-waves is, of course, a sphere since the index of refractive of such waves is independent of their propagation direction. The equation of this surface is

$$x^2+y^2+z^2=n_o^2. \tag{18.65}$$

Sections of the ordinary and extraordinary index surfaces for both positive and negative uniaxial crystals which contain the optic axis are shown in Fig. (18.7). Such sections are called principal sections.

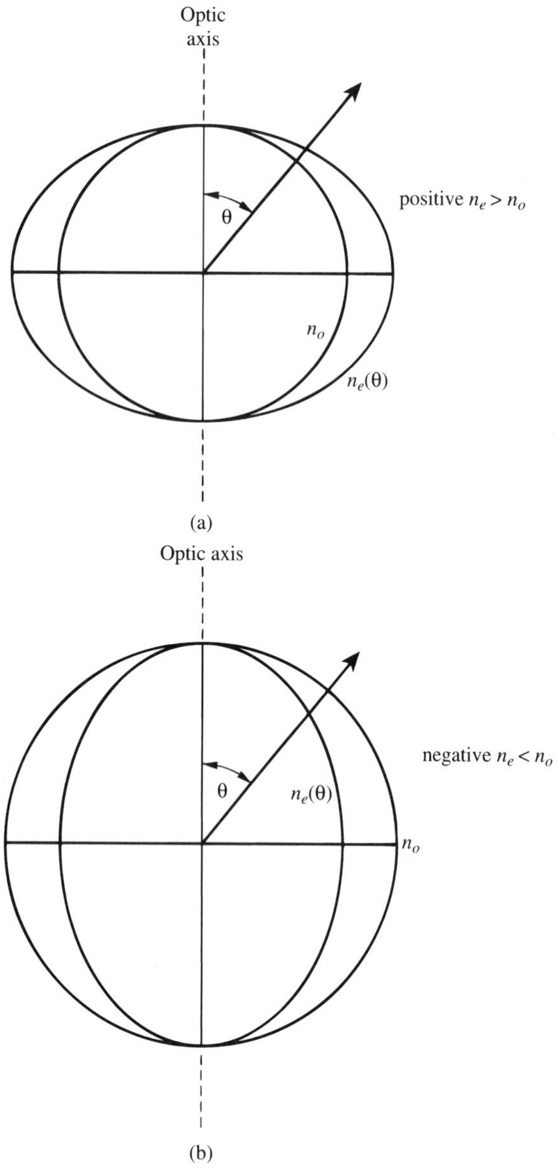

Fig. 18.7. Ordinary and extraordinatry index surfaces: (a) positive crystal; (b) negative crystal.

18.10 Other Surfaces Related to the Uniaxial Indicatrix

There are several other surfaces related to the uniaxial indicatrix whose geometrical properties describe various aspects of the propagation characteristics of uniaxial crystals.

The wave-velocity surface describes the velocity of waves in their direction of propagation, this surface is, like the index surface, a two-shelled surface. The wave-velocity surface for the O-waves is clearly

$$n_o^2(x^2 + y^2 + z^2) = c_0^2, \tag{18.66}$$

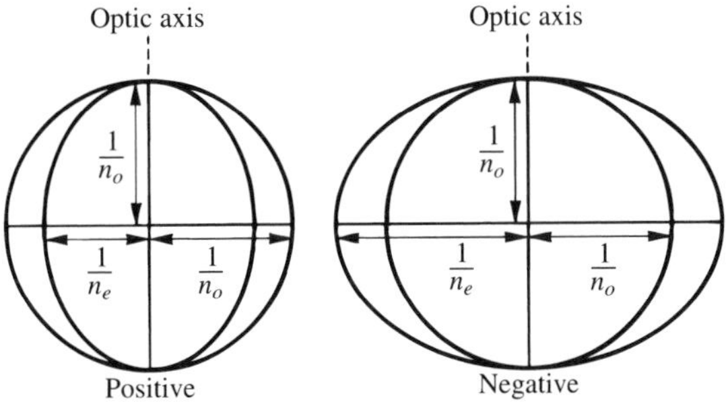

Fig. 18.8. Ray-velocity surfaces for positive and negative uniaxial crystals.

while for the E-waves

$$n_e^2(\theta)(x^2 + y^2 + z^2) = c_0^2 \tag{18.67}$$

which since

$$\sin\theta = \frac{x^2 + y^2}{x^2 + y^2 + z^2}; \quad \cos^2\theta = \frac{z^2}{x^2 + y^2 + z^2},$$

from Eq. (18.60) gives

$$\frac{x^2 + y^2}{n_e^2} + \frac{z^2}{n_o^2} = \frac{(x^2 + y^2 + z^2)^2}{c_0^2}. \tag{18.68}$$

This surface is not an ellipsoid but an ovaloid of revolution.

The ray-velocity surface describes the velocity of rays in their direction of propagation, which for E-rays is, as we have mentioned already, not the same as the direction of their wave vectors except for propagation along a principal axis. Along the x and y principal axes the extraordinary ray and wave velocities are both c_0/n_e, while along the z axis they both have the value c_0/n_o. It is left as an exercise to the reader to show that the ray-velocity surface for extraordinary rays is an ellipsoid of revolution with semiaxes c_0/n_o and c_0/n_e satisfying the equation

$$\frac{x^2 + y^2}{1/n_e^2} + \frac{z^2}{1/n_o^2} = c_0^2. \tag{18.69}$$

The ray-velocity surface for O-rays is clearly a sphere with equation

$$x^2 + y^2 + z^2 = \frac{c_0^2}{n_o^2}. \tag{18.70}$$

Fig. (18.8) shows principal sections of ray-velocity surfaces in both positive and negative crystals.

18.11 Huygenian Constructions

When a linearly polarized wave of arbitrary polarization direction enters an anisotropic medium, it will be resolved into two components polarized along the two allowed polarization directions determined by the direction of the wave vector relative to the axes of the indicatrix. In a uniaxial crystal these two components propagate as O- and E-waves respectively. At the entry surface of the anisotropic medium refraction of these waves, and of their corresponding rays, occurs. For normal incidence at the boundary of the anisotropic medium no refraction of either the O- or E-wave vectors occurs so that within the medium the wave vectors of both O- and E-waves remain parallel. However, except for propagation along a principal axis, the E-ray deviates from the common direction of the O- and E-wave vectors and the O-ray. Thus at the exit surface of the medium, where refraction renders O- and E-rays parallel once again, the E-ray will have been laterally displaced from the O-ray. In simple terms we can say that the E-ray is not refracted at the surface of the medium according to Snell's law. For other than normal incidence the O- and E-wave vectors refract separately according to Snell's law. For the O-wave calculation of the wave vector direction inside the medium is straightforward. For angle of incidence θ_i the angle of refraction obeys

$$\sin\theta_i = n_o \sin\theta_r \tag{18.71}$$

However, for the E-wave the refraction of the wave vector obeys

$$\sin\theta_i = n_e(\theta) \sin\theta_r, \tag{18.72}$$

where θ is the angle that the refracted wave vector makes with the optic axis. To illustrate these geometric optical properties of anisotropic media it is instructive to use *Huygenian constructions* using the ray-velocity surfaces for the O- and E-rays. In Huygenian constructions we treat each point on the entrance boundary of the anisotropic media as a secondary emitter of electromagnetic waves. In a given time the distance travelled by all the O- or E-rays leaving this point as a function of direction traces out the appropriate ray-velocity surface. The geometric paths of the O- and E-rays are perpendicular to the envelope of the ray-velocity surfaces that arise from these secondary emitters. To illustrate this, consider first the simple case shown in Fig. (18.9) where light enters a negative uniaxial crystal normally travelling along the optic axis. Secondary emitters at A and B give rise to the ray-velocity surfaces shown and contribute to the envelope PP', which in this case is both the wavefront and the ray front (the surface normal to the ray direction). Fig. (18.10) illustrates the situation that results when light is incident normally on a positive uniaxial crystal travelling in a plane perpendicular to the optic axis. In this case, the ray-velocity surfaces which arise from secondary emitters at A and B define two envelopes, the surfaces OO' and EE' which are respectively the ordinary wave and ray-front and the extraordinary wave and ray-front. The O-ray travels faster than the E-ray as indicated by the dots and arrows. In both Figs. (18.9) and (18.10) the wave vector and ray, which are perpendicular to their respective fronts, remain parallel. When the incident radiation does not travel along a principal axis in the crystal the O- and E-rays are no longer coincident, as shown in Fig. (18.11), which shows light travelling at an angle to the optic axis striking a planar uniaxial crystal slab cut normal to the optic axis. In this case the utility of the ray-velocity surface in determining the ray directions within the crystal is clear. The incident wavefronts are parallel to PP'; by the time the wavefront reaches point A the secondary emitter at B has given rise to the O- and E-ray-velocity surfaces shown. The O- and E-ray fronts are the tangents from A to the respective ray-velocity surfaces. The ray directions are perpendicular to these ray fronts. Once again, in this positive crystal the O-ray travels more quickly than the E-ray. On leaving the crystal the O- and E-rays become parallel to each other once more but the E-ray has been laterally displaced. This phenomenon is called *double refraction*. If the input to the crystal

Fig. 18.9. Huygenian construction for light entering a negative uniaxial crystal normally and travelling along the optic axis[18.1]. The number on the crystal face is an index that gives the orientation of the face[18.2].

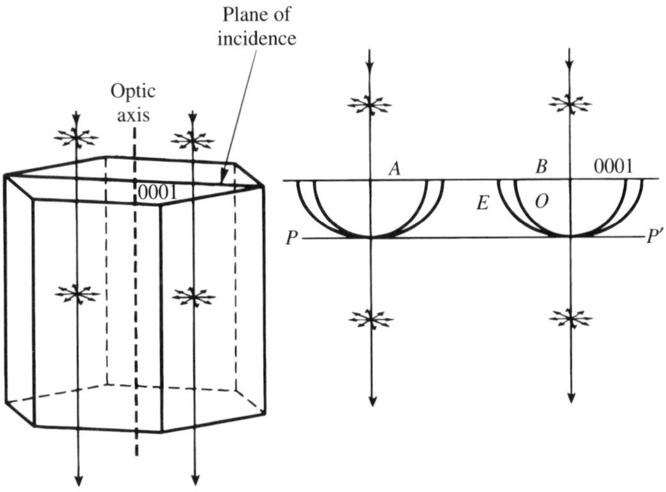

Fig. 18.10. Huygenian construction for light entering a positive uniaxial crystal normally and travelling perpendicular to the optic axis[18.1].

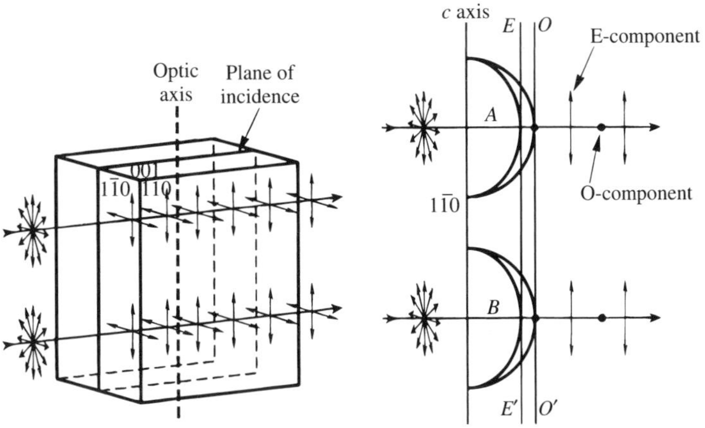

Fig. 18.11. Huygenian construction for light that does not enter the crystal normally and does not propagate along a principal axis direction[18.1].

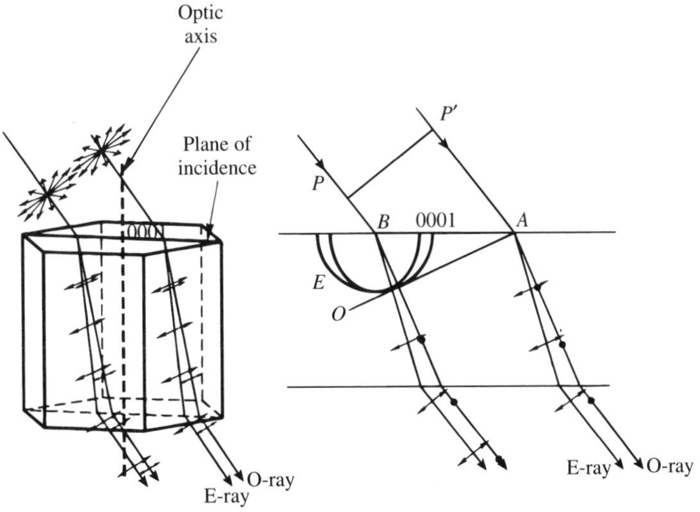

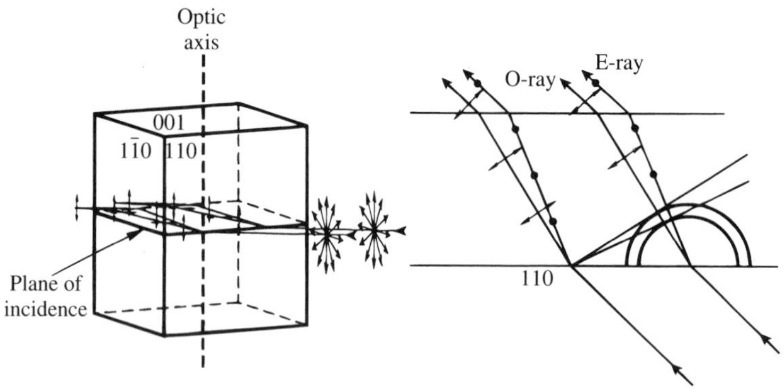

Fig. 18.12. Huygenian construction for light travelling perpendicular to optic axis that refracts on entering the crystal[18.1]. O- and E-waves (and rays) refract at different angles. The indices on the crystal faces specify their orientation[18.2].

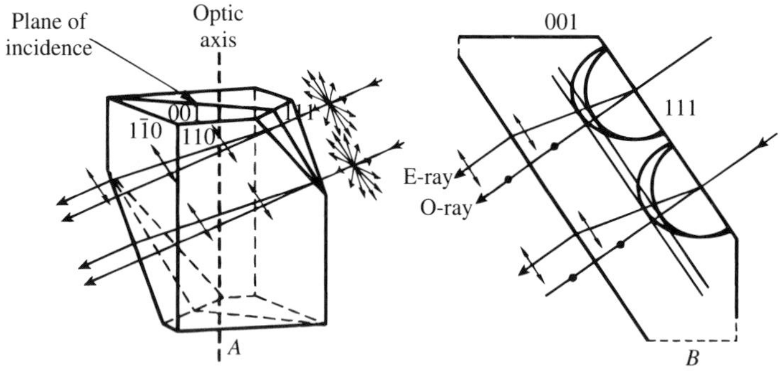

Fig. 18.13. Huygenian construction for a rather general case[18.1].

were a narrow laser beam, linearly polarized at some angle between the two allowed polarization directions within the crystal, it would be resolved into an O- and an E-ray and emerge from the crystal as two separate orthogonally polarized beams.

Further examples of double refraction are shown in Figs. (18.12) and (18.13) with the paths of the O- and E-rays determined by Huygenian constructions. Fig. (18.13) shows that, even in normal incidence, if the input wave vector is not along a principal axis the E-ray is displaced laterally from the O-ray. The refraction of the O-ray in every case follows the familiar refraction laws of geometric optics, with the angles of incidence θ_i and refraction θ_r at the crystal boundary obeying Snell's law:

$$n_o = \frac{\sin\theta_i}{\sin\theta_r}. \tag{18.73}$$

The path of the E-ray is controlled by the anisotropic character of the medium and must be determined from a nonspherical ray-velocity surface.

The separation of the O- and E-rays within an anisotropic medium is the basic phenomenon operative in the various polarizing devices used for controlling the direction of polarization of light beams, particularly laser beams. The construction and mode of operation of some common types of laser polarizer are illustrated in Fig. (18.14).

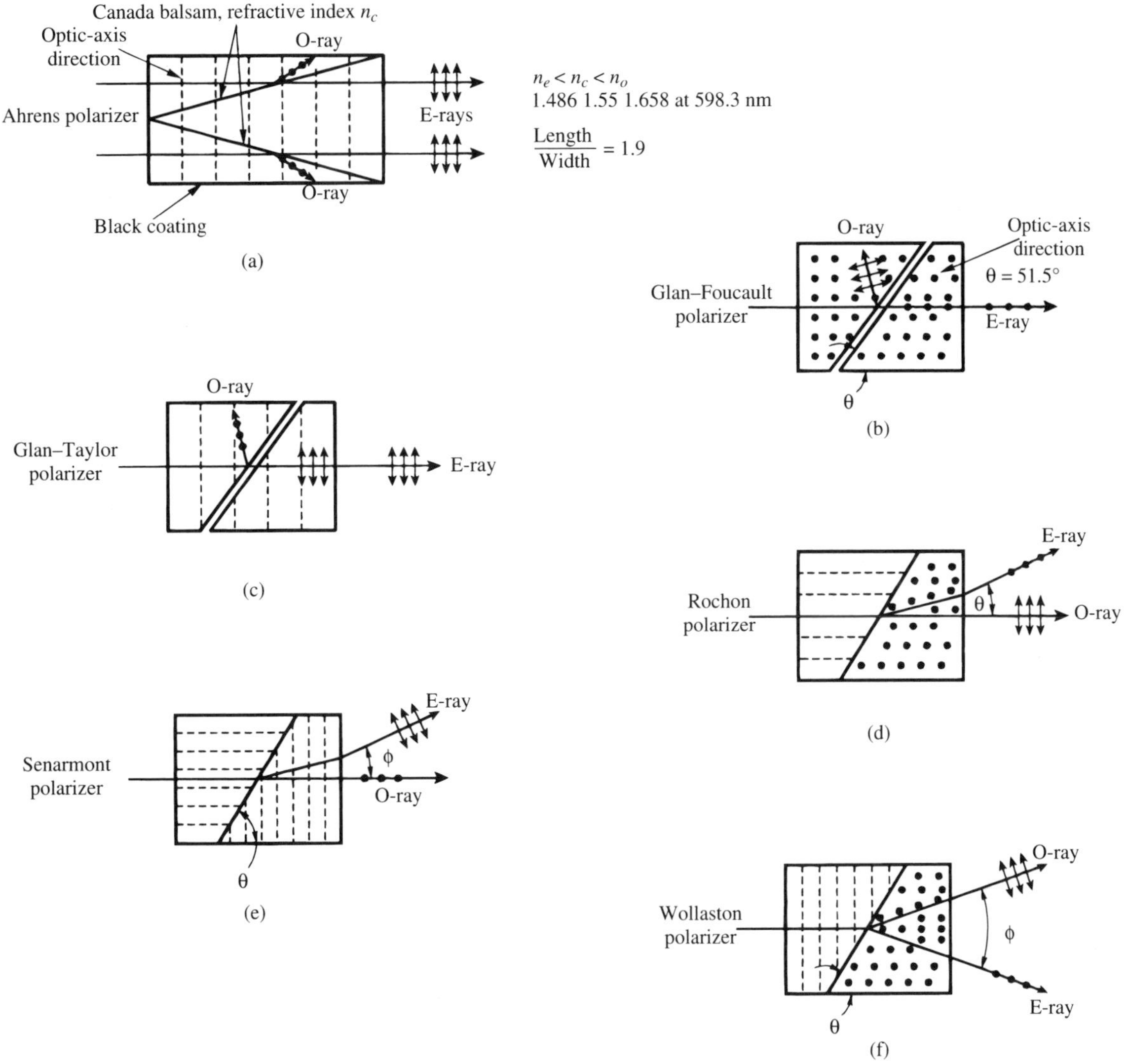

Fig. 18.14. Construction of various kinds of polarizer. (For $(d)-(f)$ the separation of the O- and E-rays is much exaggerated.)

18.12 Retardation

Unless a light beam is travelling in the direction of a principal axis, and is polarized parallel to a principal axis, when it is incident normally on a planar uniaxial crystal slab, it will be resolved within this material into O- and E-waves. If the direction of the wave vector is at angle θ to the optic axis then the wave velocities of these waves will be c_0/n_o and $c_0/n_e(\theta)$, respectively. In a positive crystal the O-wave will travel faster than the E-wave and in a negative crystal vice versa.

The wave vector of the ordinary wave is

$$|\mathbf{k}_o| = k_o = \frac{\omega n_o}{c_0} = \frac{2\pi n_o}{\lambda_0} \tag{18.74}$$

and of the extraordinary wave

$$|\,\mathbf{k}_e\,| = k_e = \frac{\omega n_e(\theta)}{c_0} = \frac{2\pi n_e(\theta)}{\lambda_0}, \tag{18.75}$$

where ω and λ_0 are the angular frequency and wavelength *in vacuo*, respectively, of the incident wave.

On passing through a crystal of thickness L the phase changes for the O- and E-waves, respectively, are

$$\phi_o = k_o L = \frac{2\pi n_o L}{\lambda_0}, \tag{18.76}$$

$$\phi_e = k_e L = \frac{2\pi n_e(\theta) L}{\lambda_0}. \tag{18.77}$$

The phase difference (*retardation*) introduced by the crystal is

$$\Delta\phi = \phi_e - \phi_o = \frac{2\pi L}{\lambda_0}\left[n_e(\theta) - n_o\right]. \tag{18.78}$$

For an incident wave of the form

$$D = A\cos(\omega t - kr) \tag{18.79}$$

linearly polarized at an angle β to the ordinary polarization direction, the ordinary and extraordinary waves will be

$$D_o = A\cos\beta\cos(\omega t - k_0 r), \tag{18.80}$$

$$D_e = A\sin\beta\cos(\omega t - k_e r), \tag{18.81}$$

and at the exit face of the crystal

$$D_o = A\cos\beta\cos(\omega t - \phi_o), \tag{18.82}$$

$$D_e = A\sin\beta\cos(\omega t - \phi_e). \tag{18.83}$$

If the input to the crystal is a narrow beam of light, these two orthogonally polarized output beams will in general be displaced laterally from one another and will not recombine to form a single output beam. However, if the input to the crystal is a plane wave, or a narrow beam travelling perpendicular to the optic axis, these two electric vectors recombine to form a resultant single displacement vector with magnitude

$$D_{out} = \sqrt{D_o^2 + D_e^2}, \tag{18.84}$$

which makes an angle α with the ordinary polarization direction, where

$$\tan\alpha = \frac{D_e}{D_o} = \tan\beta\frac{\cos(\omega t - \phi_e)}{\cos(\omega t - \phi_0)} \tag{18.85}$$

as illustrated in Fig. (18.15).

In the simplest case where $\phi_e - \phi_o = 2m\pi$, with m being any positive or negative integer or zero, $\tan\alpha = \tan\beta$ and the output wave is linearly polarized in the same direction as the input.

If $\phi_e - \phi_o = (2m+1)\pi$, then $\tan\alpha = -\tan\beta$; $\alpha = -\beta$ and the output wave is linearly polarized but rotated by 2β from its original polarization direction. This rotation occurs through a rotation by angle β towards the O-polarization direction followed by a further rotation through angle β.† Since a retardation $\phi = (2m+1)\pi$

† For this reason a wave plate alone cannot be used to make an optical isolator as can a Faraday rotator.

Fig. 18.15. Ordinary and extraordinary **D** vector directions at the output face of a crystal slab.

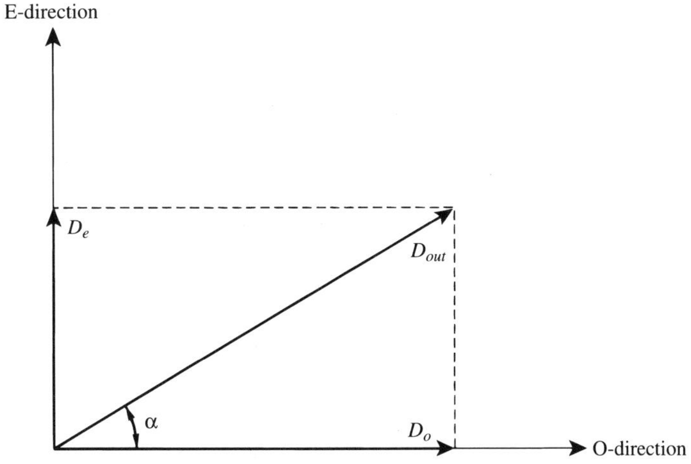

is equivalent to a path difference of $(2m+1)\lambda/2$ a crystal that rotates the plane of linear polarization by 2β is called a ((2m + 1)th order) half-wave (retardation) plate. It is most usual to cut such a crystal so its faces are parallel to the optic axis and to polarize the input at 45° to the optic axis, in which case the output is linearly polarized and rotated 90° from the input. When the input is polarized in this manner and $\phi = \phi_e - \phi_o = (2m+1)\pi/2$ then $\tan\alpha = \pm\tan(\omega t - \phi_o)$.

(i) For m even $\tan\alpha = \omega t - \phi_o$.

(ii) For m odd $\tan\alpha = -(\omega t - \phi_o)$.

In both these cases the resultant displacement vector has magnitude, from Eqs. (18.82), (18.83), and (18.84), with $\beta = 45°$

$$D_{out} = \frac{A}{\sqrt{2}}\sqrt{\cos^2(\omega t - \phi_o) + \cos^2(\omega t - \phi_o - \phi)}$$
$$= \frac{A}{\sqrt{2}}, \qquad (18.86)$$

so the output displacement vector has a constant magnitude but rotates about the direction of propagation with constant angular velocity ω. This is circularly polarized light. If the electric vector rotates in a clockwise direction when viewed in the direction of propagation, as when m is odd above, the light is said to be *left hand circularly polarized*, as illustrated in Fig. (18.16). When the electric vector rotates counter-clockwise the light is *right hand circularly polarized.* A crystal which introduces a retardation ϕ of $(2m+1)\pi/2$ is called a *quarter-wave plate* (strictly a (2m + 1)th order quarter-wave plate).

In the general case when ϕ is not an integral number of half-wavelengths, or if the input to a quarter-wave plate is not polarized at 45° to the optic axis, the resultant output will be elliptically polarized, as illustrated also in Fig. (18.16). In this case the displacement vector and the electric vector trace out an ellipse as they rotate in time as the wave propagates. To prove this we rewrite Eqs. (18.82) and (18.83) as

$$D_o = p = a\cos z, \qquad (18.87)$$

$$D_e = q = b\cos(z - \phi) = b\cos z\phi + b\sin z\sin\phi, \qquad (18.88)$$

(a)

(b)

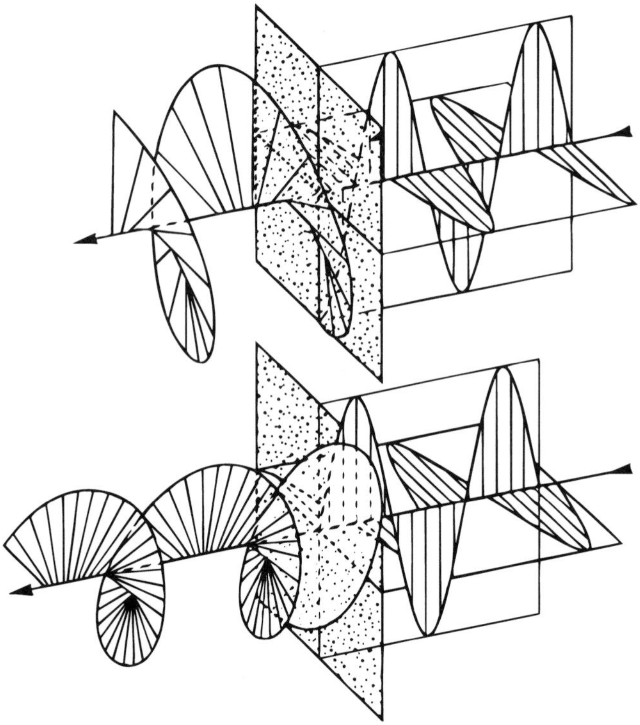

Fig. 18.16. Left-hand circularly polarized light (a) and (b) left hand elliptically polarized light[18.1].

where $a = A\cos\beta, b = A\sin\beta$, and $z = (\omega t - \phi_0)$.

From Eq. (18.88)

$$\frac{q^2}{b^2} - \frac{2q}{b}\cos z\cos\phi + \cos^2 z\cos^2\phi = \sin^2 z\sin^2\phi, \tag{18.89}$$

and substituting from Eq. (18.87)

$$\frac{q^2}{b^2} - \frac{2pq}{ab}\cos\phi + \frac{p^2}{a^2}\cos^2\phi = \left(1 - \frac{p^2}{a^2}\right)\sin^2\phi, \tag{18.90}$$

giving finally

$$\frac{p^2}{a^2} + \frac{q^2}{b^2} - \frac{2pq}{ab}\cos\phi - \sin^2\phi = 0. \tag{18.91}$$

This is the equation of an ellipse, one axis of which makes an angle ψ with the O-polarization direction, where

$$\tan 2\psi = \frac{2ab\cos\phi}{a^2 - b^2}. \tag{18.92}$$

This ellipse has its axes coincident with the O and E vibration directions if $\psi = (2m+1)\pi/2$. Thus, a quarter-wave plate produces elliptically polarized light if the incident light is not polarized exactly half-way between the O and E allowed polarization directions.

A first order quarter-wave plate is very thin; as can be seen from Eq. (18.78), its thickness L is

$$L = \frac{\lambda_0}{4(n_e - n_o)}. \tag{18.93}$$

For calcite (Iceland spar), a mineral form of calcium carbonate that is commonly used to make polarizing optics, $n_o = 1.658$, $n_e = 1.486$, so for a wavelength of 500 nm the thickness of a first order quarter-wave plate is only 726.7 nm $\sim$ 0.0007 mm. It is not practical to cut a crystalline slab as thin as this, except for a birefringent material such as mica which cleaves readily into very thin slices. $(2m+1)$th order quarter-wave plates can be of practical thickness for large values of m but suffer severely from the effects of temperature: the plate only has to expand or contract very slightly and it will cease to be a quarter-wave plate at the wavelength for which it was designed. To overcome this drawback, temperature compensated plates can be made. These consist of one $(2m+1)$th order quarter-wave plate stacked on top of a plate which produces a retardation of $m\pi$ but whose optic axis is perpendicular to the optic axis (E-wave polarization direction) of the first quarter-wave plate. The total retardation of the combination is

$$\phi = \phi_1 - \phi_2 = (2n+1)\pi/2 - n\pi = \pi/2 \tag{18.94}$$

so it is equivalent to a first order quarter-wave plate but is not sensitive to temperature changes.

18.13 Biaxial Crystals

The optical properties of biaxial crystals can be related to an ellipsoid with three unequal axes, the biaxial indicatrix, whose equation is

$$\frac{x^2}{n_x^2} + \frac{y^2}{n_y^2} + \frac{z^2}{n_z^2} = 1, \tag{18.95}$$

where n_x, n_y, and n_z are the three principal refractive indices of the material. It is the normal convention to label the axes so that $n_x < n_y < n_z$. When this is done the two optic axes, those directions through the crystal along which the direction of propagation of waves is independent of their polarization direction, lie in the xz plane. If the two optic axes are closer to the z axis then the x axis the crystal is said to be *positive*, otherwise it is *negative*. Some of the important features of positive and negative biaxial indicatrices are illustrated in Fig. (18.17). The acute angle between the two optic axes is labelled $2V$ – the *optic angle*. Whichever axis bisects this acute angle is called the *acute bisectrix* (*Bxa*), this is the z axis in positive crystals, the x axis in negative crystals. The other axis in each case is called the *obtuse bisectrix* (*Bxo*).

The angle between one of the optic axes and the z axis is V_z and is given by the expression

$$V_z = \tan^{-1}\sqrt{\frac{1/n_x^2 - 1/n_y^2}{1/n_y^2 - 1/n_z^2}}, \tag{18.96}$$

which is an angle $< 45°$ in a positive crystal.

Because of the lack of rotational symmetry of the biaxial indicatrix, the optical properties of biaxial crystals are more complicated than for uniaxial ones. For wave propagation along the three principal axes wave-normals and rays of both of the two allowed polarization directions are parallel. For propagation in a symmetry plane of the biaxial indicatrix, as shown in Fig. (18.18), the two allowed polarization directions are perpendicular to, and in, the plane of symmetry. The

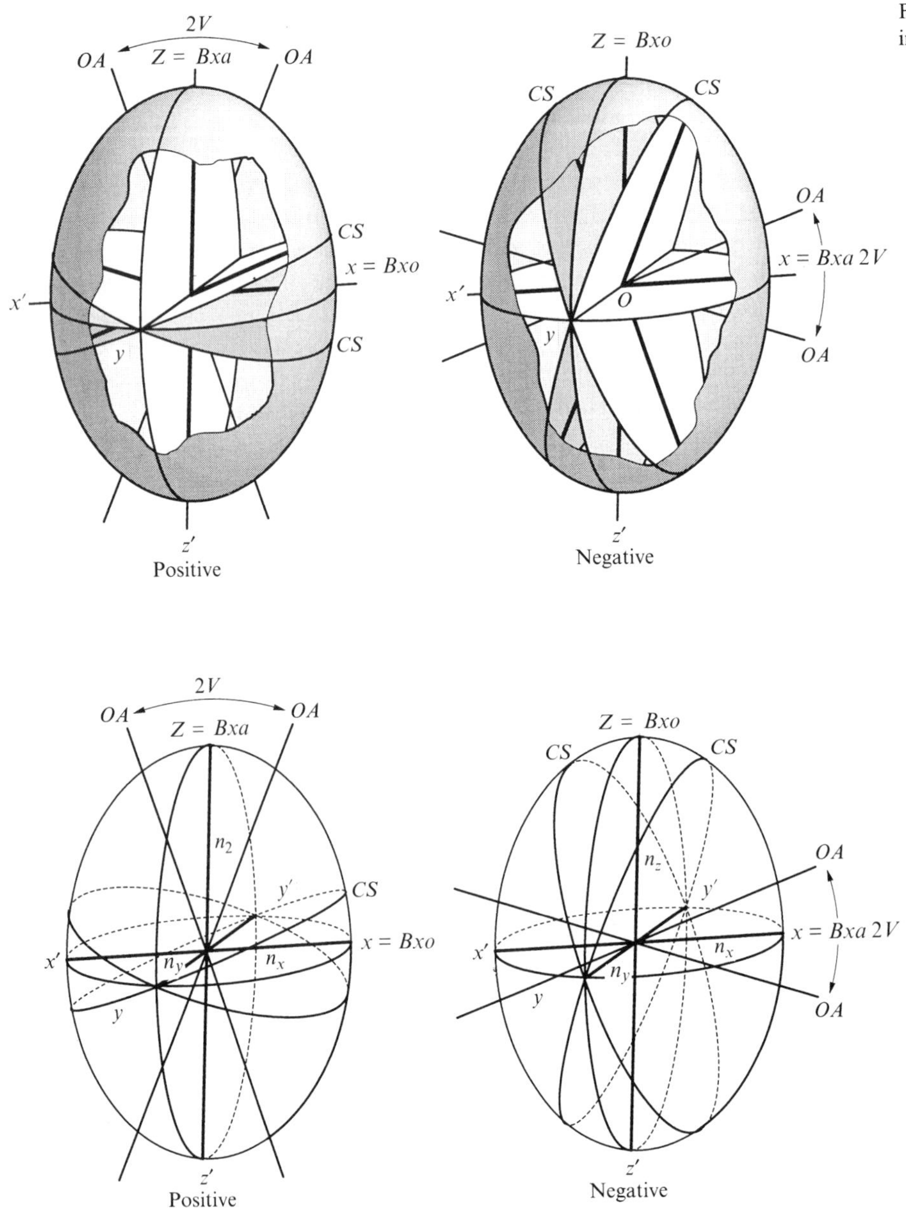

Fig. 18.17. Biaxial indicatrices.

component polarized perpendicular to the symmetry plane has its polarization direction parallel to a principal axis and therefore has its **E** direction parallel to **D**. Thus, for this polarization the wave-normal direction ON and the ray OS_1 are parallel. Both the wave and ray refractive indices for this polarization have the value n_y. The ray direction for the wave polarized in the symmetry plane is the direction OS, which is found by a construction similar to the one described for finding the direction of the E-ray in a uniaxial crystal. The wave refractive index for this polarization is given by the length OT and its ray refractive index by the length OM; OM is less than OT as the ray travels further than the wave in the

Fig. 18.18. A wave and ray in a symmetry plane of a biaxial indicatrix[18.1].

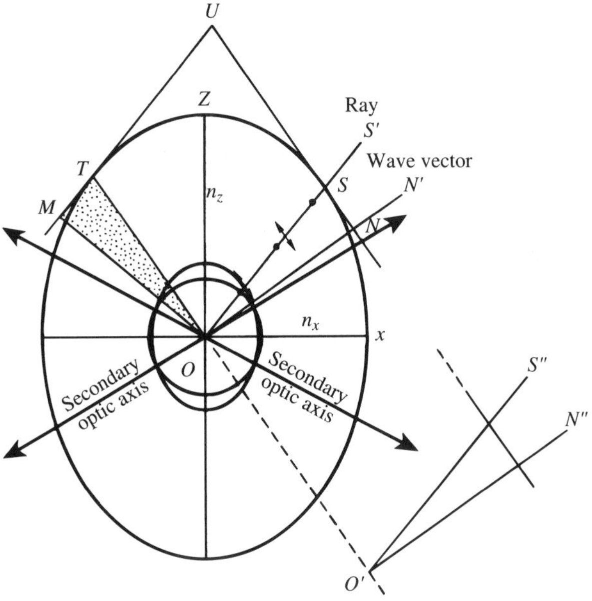

Fig. 18.19. Construction that indicates how wave vector and ray directions are related for the two permitted polarization directions in a biaxial crystal[18.1].

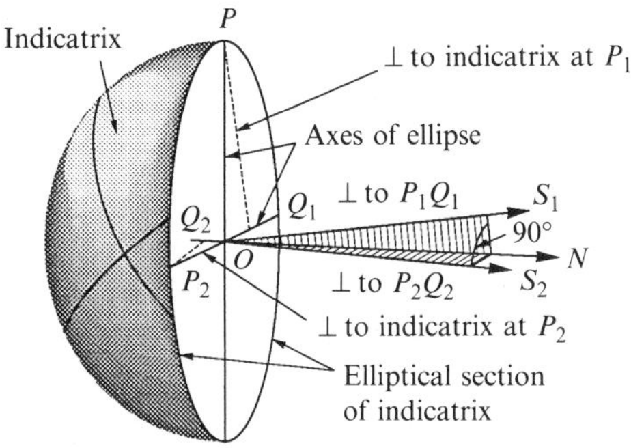

same period of time. There are two directions for which $OM = n_y$ in Fig. (18.18). These are ray directions for which the ray velocity is the same for both polarization directions; of course, the wave vectors of these two waves are not colinear with their rays. These directions of equal ray velocity are called the secondary optic axes and are generally within two degrees of the primary optic axes, which are directions of equal wave velocity. For propagation in an arbitrary direction not lying in a symmetry plane, both allowed polarization directions have rays which are not parallel to their wave normals. This is illustrated in Fig. (18.19) which shows the two ray directions OS_1, OS_2 that correspond to the two allowed polarization directions OP_1, OP_2 of a wave propagating in the general direction ON.

The surfaces which are related to the indicatrix of a biaxial crystal are more complicated than for uniaxial crystals, although for propagation in symmetry planes of the indicatrix they are somewhat like, but not quite the same as, sections of uniaxial surfaces. Fig. (18.20) illustrates the ray-velocity surface of a biaxial crystal. This is a two-shelled surface, one shell for each polarization direction corresponding

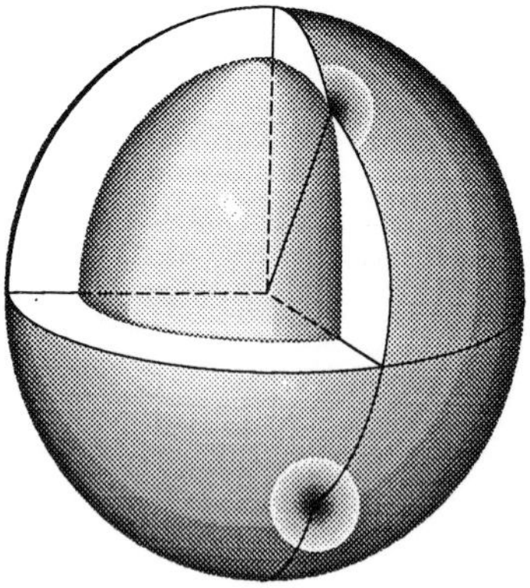

Fig. 18.20. Cut away view of the ray velocity surface of a biaxial crystal[18.1].

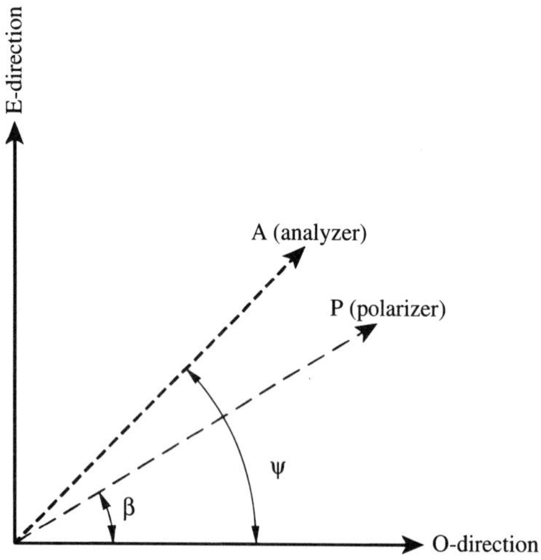

Fig. 18.21. Orientation directions for analysis of polarizer/waveplate/analyzer combinations.

to a given ray direction. The four dimples where one shell cuts through the other (two are visible in Fig. (18.20)) are the directions of the secondary optic axes.

18.14 Intensity Transmission Through Polarizer/Waveplate/Polarizer Combinations

A wave is passed through a linear polarizer (P) whose preferred direction is at angle β to the O-direction of a succeeding waveplate W, as shown in Fig. (18.21), and is then transmitted through a second linear polarizer (A) whose preferred axis makes an angle ψ with the O-direction of the wave plate. The second polarizer, usually called the analyzer, will not in general permit all the radiation emerging from the wave plate to pass. This PWA combination serves as an adjustable attenuator of a light beam.

From Eqs. (18.82) and (18.83) it is easy to see that the **D** vector transmitted through the analyzer has magnitude

$$D = A\cos\beta\cos(\omega t - \phi_o)\cos\psi + A\sin\beta\cos(\omega t - \phi_e)\sin\psi, \qquad (18.97)$$

which can be rearranged to give

$$D = D_0 \cos(\omega t - \phi_e + \chi), \tag{18.98}$$

where

$$D_0 = A\sqrt{(\cos\beta\cos\psi\cos\Delta\phi + \sin\beta\sin\psi)^2 + (\cos\beta\cos\psi\sin\Delta\phi)^2} \tag{18.99}$$

and

$$\tan\chi = \frac{\cos\beta\cos\psi\sin\Delta\phi}{\cos\beta\cos\psi\cos\Delta\phi + \sin\beta\sin\psi}. \tag{18.100}$$

$\Delta\phi = \phi_e - \phi_o$ is the retardation produced by the wave plate.

18.14.1 Examples

To illustrate the value of Eq. (18.100) some specific examples are in order:

(1) The input and output polarizers are parallel: $\psi = \beta$. In this case

$$D_0 = A\sqrt{1 - \tfrac{1}{2}\sin^2 2\beta(1 - \cos\Delta\phi)}. \tag{18.101}$$

If $\Delta\phi = 0$ (or if $\Delta\phi$ is any multiple of 2π) then $D_0 = A$: all the light is transmitted as we would have expected.

If $\beta = 45^\circ$ then

$$D_0 = A\sqrt{1 - \tfrac{1}{2}(1 - \cos\Delta\phi)}. \tag{18.102}$$

If $\Delta\phi = \pi$ (or any odd multiple of π) then $D_0 = 0$. If $\Delta\phi = \pi/2$ (or any odd multiple of $\pi/2$) then $D_0 = A/\sqrt{2}$. Since the transmitted intensity is proportional to D_0^2 half of the incident light is transmitted.

(2) The input and output polarizers are crossed: $\psi = \beta \pm \pi/2$. In this case

$$D_0 = \frac{1}{\sqrt{2}}\sin 2\beta\sqrt{1 - \cos\Delta\phi}. \tag{18.103}$$

If $\beta = 45^\circ$ then

$$D_0 = \frac{1}{\sqrt{2}}\sqrt{1 - \cos\Delta\phi}. \tag{18.104}$$

If $\Delta\phi = 0$ then $D_0 = 0$.

If $\Delta\phi = (2m + 1)\pi$, the wave plate is a half-wave plate, and $D_0 = A$. All the incident light is transmitted.

If $\Delta\phi = (2m + 1)\pi/2$, the waveplate is a quarter-wave plane and $D_0 = A/\sqrt{2}$. Half the incident light is transmitted.

These findings are summarized in Table (18.2).

18.15 The Jones Calculus

Over 50 years ago R. Clark Jones introduced a very useful technique for describing the change in polarization state of a light wave as it passed through an optical system containing various interfaces and polarizing elements.[18.3] The *Jones Calculus* treats the optical system as a linear system characterized by an appropriate Jones matrix that transforms vectors describing the polarization state of the wave. In this sense the Jones Calculus is analogous to paraxial ray analysis.

Table 18.2. *Transmittance of PWA combinations*

P/A	W	Transmittance
$\parallel$	0	1
$\parallel$	$\lambda/4$	1/2
$\parallel$	$\lambda/2$	0
$\perp$	0	0
$\perp$	$\lambda/4$	1/2
$\perp$	$\lambda/2$	1

18.15.1 The Jones Vector

The electric field of a light wave linearly polarized along the x axis can be written as

$$\mathbf{E}_x = A_x \sin(\omega t + \phi_x)\hat{\mathbf{i}}, \tag{18.105}$$

which in complex exponential notation is

$$\mathbf{E}_x = A_x e^{i\phi_x} e^{i\omega t}\hat{\mathbf{i}}. \tag{18.106}$$

On the other hand a light wave linearly polarized along the y axis can be written as

$$\mathbf{E}_y = A_y e^{i\phi_y} e^{i\omega t}\hat{\mathbf{j}}. \tag{18.107}$$

The superposition of the electric fields $\mathbf{E}_x$ and $\mathbf{E}_y$ leads in general to elliptically polarized light with electric field

$$\mathbf{E} = \left(A_x e^{i\phi_x}\hat{\mathbf{i}} + A_y e^{i\phi_y}\hat{\mathbf{j}}\right) e^{i\omega t}. \tag{18.108}$$

The complex amplitudes of the x and y components of this wave form the two elements of the Jones vector $\mathbf{J}$, where

$$\mathbf{J} = \begin{pmatrix} A_x e^{i\phi_x} \\ A_y e^{i\phi_y} \end{pmatrix}. \tag{18.109}$$

Clearly, light linearly polarized along the x axis has

$$\mathbf{J} = \begin{pmatrix} A_x e^{i\phi_x} \\ 0 \end{pmatrix}, \tag{18.110}$$

with a similar result for linearly polarized along y.

Light linearly polarized at angle β to the axis has

$$\mathbf{J}_\beta = \begin{pmatrix} E_0 \cos\beta \\ E_0 \sin\beta \end{pmatrix}. \tag{18.111}$$

Right hand circularly polarized (rcp) light has

$$\mathbf{J}_{rcp} = \begin{pmatrix} E_0 e^{i\phi_x} \\ E_0 e^{i(\phi_x+\pi/2)} \end{pmatrix} \tag{18.112}$$

and left hand circularly polarized (lcp) light has

$$\mathbf{J}_{rcp} = \begin{pmatrix} E_0 e^{i\phi_x} \\ E_0 e^{i(\phi_x-\pi/2)} \end{pmatrix}. \tag{18.113}$$

In an isotropic medium the intensity of the wave is

$$I \propto \mathbf{E}^* \cdot \mathbf{E}, \tag{18.114}$$

which in terms of the Jones vector gives

$$I \propto \mathbf{J}^* \cdot \mathbf{J} = A_x^2 + A_y^2. \tag{18.115}$$

It is generally simple and convenient to use the Jones vector in its normalized form, in which

$$\mathbf{J}^* \cdot \mathbf{J} = 1. \tag{18.116}$$

In this case the three vectors in Eqs. (18.111), (18.113), and (18.114) would become

$$\mathbf{J}_\beta = \begin{pmatrix} \cos\beta \\ \sin\beta \end{pmatrix}, \tag{18.117}$$

$$\mathbf{J}_{rcp} = \frac{1}{\sqrt{2}}\begin{pmatrix} 1 \\ e^{i\pi/2} \end{pmatrix} \text{ or } \frac{1}{\sqrt{2}}\begin{pmatrix} e^{-i\pi/4} \\ e^{i\pi/4} \end{pmatrix}, \tag{18.118}$$

$$\mathbf{J}_{lcp} = \frac{1}{\sqrt{2}}\begin{pmatrix} 1 \\ e^{-i\pi/2} \end{pmatrix} \text{ or } \frac{1}{\sqrt{2}}\begin{pmatrix} e^{i\pi/4} \\ e^{-i\pi/4} \end{pmatrix}. \tag{18.119}$$

The second description for circularly polarized light is an alternative symmetrical way of writing the column vector, since only the phase difference between the x and y component is significant.

In this notation a general elliptically polarized beam has a Jones vector that can be written as

$$\mathbf{J} = \begin{pmatrix} \cos\beta e^{-i\Delta/2} \\ \sin\beta e^{i\Delta/2} \end{pmatrix}. \tag{18.120}$$

18.15.2 The Jones Matrix

In a linear system description the output Jones vector of a light wave after it has interacted with an optical system has components that are linearly related to its input components.

If

$$\mathbf{J}_{\text{in}} = E_x\hat{\mathbf{i}} + Ey\hat{\mathbf{j}} \tag{18.121}$$

and

$$\mathbf{J}_{\text{out}} = E_x'\hat{\mathbf{i}} + E'y\hat{\mathbf{j}}, \tag{18.122}$$

where phase factors are included in the complex amplitudes, we can now write

$$\begin{aligned} E_x' &= m_{11}E_x + m_{12}E_y, \\ E_y' &= m_{21}E_x + m_{22}E_y \end{aligned} \tag{18.123}$$

or in matrix form

$$\begin{pmatrix} E_x' \\ E_y' \end{pmatrix} = \begin{pmatrix} m_{11} & m_{12} \\ m_{21} & m_{22} \end{pmatrix} \begin{pmatrix} E_x \\ E_y \end{pmatrix}. \tag{18.124}$$

Eq. (18.124) introduces the Jones matrix $\mathbf{M}$ with elements m_{ij}, it can be rewritten as

$$\mathbf{J}_{out} = \mathbf{M}\mathbf{J}_{in}. \tag{18.125}$$

The determination of the Jones matrix for common optical elements is straightforward. This can be demonstrated with a few examples. (a) Isotropic element: since $\mathbf{J}_{out} = \mathbf{J}_{in}$ clearly

$$\mathbf{M} = \begin{pmatrix} 1 & 0 \\ 0 & 1 \end{pmatrix}. \tag{18.126}$$

(b) Linear polarizer oriented along x axis:

$$\mathbf{M} = \begin{pmatrix} 1 & 0 \\ 0 & 0 \end{pmatrix}. \tag{18.127}$$

(c) Linear polarizer oriented at angle θ to the x axis:

$$\mathbf{M} = \begin{pmatrix} \cos^2\theta & \sin\theta\cos\theta \\ \sin\theta\cos\theta & \sin^2\theta \end{pmatrix}. \tag{18.128}$$

(d) A waveplate that produces a retardation $\phi_x - \phi_y = \Gamma$: this case is worthy of closer consideration. If the waveplate has its principal axes parallel to the x and y axes then the refractive indices seen by the x and y components of the wave are n_x, n_y respectively. In this case

$$\begin{aligned} E'_x &= E_x e^{-ik_0\ell n_x}, \\ E'_y &= E_y e^{-ik_0\ell n_y}, \end{aligned} \tag{18.129}$$

where $k_0 = 2\pi/\lambda_0$ and ℓ is the thickness of the retarder. Since $\Gamma = k_0\ell(n_y - n_x)$, Eq. (18.129) can be written as

$$\begin{aligned} E'_x &= E_x e^{-ik_0\ell n_y} e^{i\Gamma}, \\ E'_y &= E_y e^{-ik_0\ell n_y} \end{aligned} \tag{18.130}$$

or in symmetrical form

$$\begin{aligned} E'_x &= E_x e^{-i\phi} e^{i\Gamma/2}, \\ E'_y &= E_y e^{-i\phi} e^{-i\Gamma/2}, \end{aligned} \tag{18.131}$$

where $\phi = k_0\ell(n_x + n_y)/2$. Therefore, the Jones matrix of the waveplate is

$$M = e^{-i\phi} \begin{pmatrix} e^{i\Gamma/2} & 0 \\ 0 & e^{-i\Gamma/2} \end{pmatrix}. \tag{18.132}$$

The phase factor $e^{-i\phi}$ in Eq. (18.129) can be omitted for most practical purposes.

A quarter-wave plate has a Jones matrix

$$\mathbf{M} = \begin{pmatrix} e^{i\pi/4} & 0 \\ 0 & e^{-i\pi/4} \end{pmatrix}. \tag{18.133}$$

A linearly polarized wave with

$$\mathbf{J} = \begin{pmatrix} 1/\sqrt{2} \\ 1/\sqrt{2} \end{pmatrix} \tag{18.134}$$

is linearly polarized at 45° to the fast (and slow) axis of the wave plate. Its electric field is in the $\mathbf{E}_1$ direction in Fig. (18.22). The output Jones vector is

$$\begin{aligned} \mathbf{J}_{out} &= \begin{pmatrix} e^{i\pi/4} & 0 \\ 0 & e^{-i\pi/4} \end{pmatrix} \begin{pmatrix} 1/\sqrt{2} \\ 1/\sqrt{2} \end{pmatrix} \\ &= \frac{1}{\sqrt{2}} \begin{pmatrix} e^{i\pi/4} \\ e^{-i\pi/4} \end{pmatrix}, \end{aligned} \tag{18.135}$$

Fig. 18.22. Input electric field directions for producing left and right hand circularly polarized light from linearly polarized light oriented at 45° to the fast and slow axis of a quarter-wave plate.

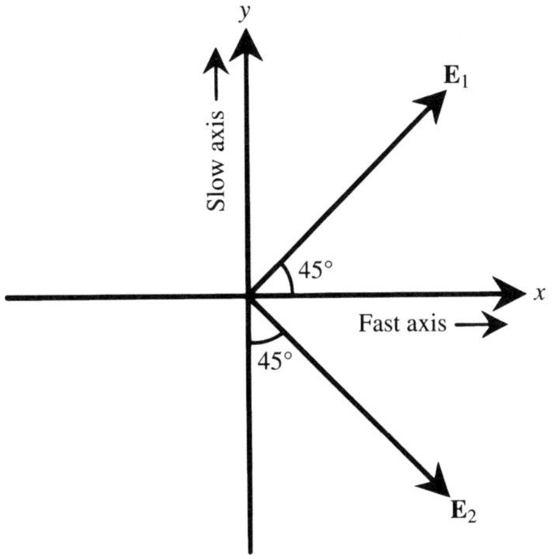

which has transformed the linearly polarized input light to left-hand circular polarization.

If the input linearly polarized light is rotated 90° with respect to the axis of the quarter-wave plate to the $\mathbf{E}_2$ direction in Fig. (18.22), then the output Jones vector will be

$$\mathbf{J}_{out} = \begin{pmatrix} e^{i\pi/4} & 0 \\ 0 & e^{-i\pi/4} \end{pmatrix} \begin{pmatrix} 1/\sqrt{2} \\ -1/\sqrt{2} \end{pmatrix}$$
$$= \frac{1}{\sqrt{2}} \begin{pmatrix} e^{i\pi/4} \\ e^{i3\pi/4} \end{pmatrix}$$
$$= \frac{1}{\sqrt{2}} e^{i\pi/2} \begin{pmatrix} e^{-i\pi/4} \\ e^{i\pi/4} \end{pmatrix}. \qquad (18.136)$$

This is right hand circularly polarized light, so a quarter-wave plate will convert linearly polarized light into left or right hand circularly polarized light depending on its orientation.

It is of value to view the action of the quarter-wave plate in a coordinate system that is rotated by 45°. What is now the Jones matrix for the plate for an input linearly polarized wave with Jones vector

$$\mathbf{J}_{in} = \begin{pmatrix} 1 \\ 0 \end{pmatrix}?$$

Such a wave is polarized in the x' direction in Fig. (18.22). Now the transformation between the xy and $x'y'$ coordinate system is

$$\begin{pmatrix} x' \\ y' \end{pmatrix} = \frac{1}{\sqrt{2}} \begin{pmatrix} 1 & -1 \\ 1 & 1 \end{pmatrix} \begin{pmatrix} x \\ y \end{pmatrix} = \mathbf{S} \begin{pmatrix} x \\ y \end{pmatrix} \qquad (18.137)$$

and

$$\begin{pmatrix} x \\ y \end{pmatrix} = \frac{1}{\sqrt{2}} \begin{pmatrix} 1 & 1 \\ -1 & 1 \end{pmatrix} \begin{pmatrix} x' \\ y' \end{pmatrix} = \mathbf{S}^{-1} \begin{pmatrix} x \\ y \end{pmatrix}. \qquad (18.138)$$

In terms of the rotation matrix $\mathbf{S}$, in the $x'y'$ coordinate system

$$\mathbf{J}'_{out} = \mathbf{S}\mathbf{J}_{out}. \qquad (18.139)$$

In the xy coordinate system

$$\mathbf{J}_{out} = \mathbf{M}\mathbf{J}_{in}, \tag{18.140}$$

where

$$\mathbf{M} = \begin{pmatrix} e^{i\pi/4} & 0 \\ 0 & e^{-i\pi/4} \end{pmatrix} \tag{18.141}$$

and

$$\mathbf{J}_{in} = \mathbf{S}^{-1}\begin{pmatrix} 1 \\ 0 \end{pmatrix} = \mathbf{S}^{-1}\mathbf{J}'_{in}. \tag{18.142}$$

Combining Eqs. (18.139), (18.140), and (18.142)

$$\mathbf{J}'_{out} = \mathbf{S}\mathbf{M}\mathbf{S}^{-1}\mathbf{J}'_{in}, \tag{18.143}$$

so the new Jones matrix in the $x'y'$ coordinate system is

$$\mathbf{M}' = \mathbf{S}\mathbf{M}\mathbf{S}^{-1}. \tag{18.144}$$

In this case

$$\begin{aligned} \mathbf{M}' &= \frac{1}{2}\begin{pmatrix} 1 & -1 \\ 1 & 1 \end{pmatrix}\begin{pmatrix} e^{i\pi/4} & 0 \\ 0 & e^{-i\pi/4} \end{pmatrix}\begin{pmatrix} 1 & 1 \\ -1 & 1 \end{pmatrix} \\ &= \frac{1}{\sqrt{2}}\begin{pmatrix} 1 & i \\ i & 1 \end{pmatrix}. \end{aligned} \tag{18.145}$$

We can generalize from this example: if coordinate system $x'y'$ is obtained by a counterclockwise rotation by angle α from coordinate system xy, then the transformation matrix is

$$\mathbf{S} = \begin{pmatrix} \cos\alpha & \sin\alpha \\ -\sin\alpha & \cos\alpha \end{pmatrix}. \tag{18.146}$$

The transformation of the Jones matrix from xy to $x'y'$ obeys Eq. (18.144).
(e) Faraday rotator: a Faraday rotator that rotates the plane of linear polarization in a counterclockwise direction by angle θ has a Jones matrix

$$\mathbf{M}_{\mathrm{fr}} = \begin{pmatrix} \cos\theta & -\sin\theta \\ \sin\theta & \cos\theta \end{pmatrix}. \tag{18.147}$$

This brief survey provides the essentials that are needed to describe the polarization change that occurs in a multielement optical system. There are additional matrix methods and graphical techniques that can also be used to provide similar information, in particular the Mueller Calculus and the Poincaré sphere. The interest reader is referred to the specialized literature[18.4]–[18.7].

18.16 Problems

(18.1) Prove that in a uniaxial crystal the maximum angular separation of O- and E-rays occurs when the wave vector makes an angle θ with the optic axis that satisfies

$$\theta = \mathrm{arc}\tan(n_e/n_o).$$

(18.2) A uniaxial crystal has $n_o = 1.5$, $n_e = 2.0$. Calculate the index of refraction for E-waves travelling at 30° and 45° to the optic axis.

(18.3) What is the wave velocity for the waves in Problem (18.2) travelling at 30° and 45° to the optic axis?

(18.4) What are the ray velocities in (18.2) for rays travelling at 30° and 45° to the optic axis?

(18.5) What is the angle between the optic axis and the ray in Problem (18.2) when the wave vector is at 45° to the optic axis?

(18.6) The indicatrix of a particular crystal is

$$ax^2 + by^2 + cz^2 + dyz = 1.$$

Where are the principal axes?

(18.7) A laser beam travelling at an angle of incidence of 45° strikes a uniaxial crystal slab 20 mm thick. The faces of the slab are parallel to each other and are perpendicular to the optic axis. $n_o = 3$, $n_e = 2$. Calculate the lateral separation of the exiting O- and E-rays.

(18.8) In a crystal with $n_e = 1.5$, $n_o = 2$, a *ray* is travelling at an angle of 45° to the optic axis. What is the direction of the wave vector and what value of $n_e(\theta)$ does this wave see?

References

[18.1] E. Wahlstrom, *Optical Crystallography*, 3rd Edition, Wiley, New York, 1960.

[18.2] C. Kittel, *Introduction to Solid State Physics*, 6th Edition, Wiley, New York, 1986.

[18.3] R. Clark Jones, 'A new calculus for the treatment of optical systems, I, Description and discussion of the calculus,' *J. Opt. Soc. Am.*, **31**, 488–493, 1941; see also R. Clark Jones, *J. Opt. Soc. Am.*, **31**, 493–499, 1941; **31**, 500–503, 1941; **32**, 486–493, 1942; **37**, 107–110, 1942; **38**, 671–685, 1948; **46**, 126–131, 1956.

[18.4] D.S. Kliger, J.W. Lewis, and C.E. Randall, *Polarized Light in Optics and Spectroscopy*, Academic Press, San Diego, 1990.

[18.5] W.A. Shercliff, *Polarized Light: Production and Use*, Harvard University Press, Cambridge, MA, 1962.

[18.6] R.M.A. Azzam and N.M. Bashara, *Ellipsometry and Polarized Light*, North-Holland, Amsterdam, 1977.

[18.7] A. Yariv and P. Yeh, *Optical Waves in Crystals*, Wiley, New York, 1984.

19

The Electro-Optic and Acousto-Optic Effects and Modulation of Light Beams

In this chapter we shall explain how the distortion produced in a crystal lattice by the application of an electric field or by the passage of a sound wave affects the propagation of light through the crystal. These effects – the electro-optic and acousto-optic effects are of considerable practical importance as they can be used to amplitude and phase modulate light beams, shift their frequencies, and alter the direction in which they travel.

19.1 Introduction to the Electro-Optic Effect

When an electric field is applied to a crystal, the ionic constituents move to new locations determined by the field strength, the charge on the ions, and the restoring force. As we saw in Chapter 18 unequal restoring forces along three mutually perpendicular axes in the crystal lead to anisotropy in the optical properties of the medium. When an electric field is applied to such a crystal, in general, it causes a change in the anisotropy. These changes can be described in terms of the modification of the indicatrix by the field – both in terms of the principal refractive indices of the medium and in the orientation of the indicatrix. If these effects can be described, to first order, as being linearly proportional to the applied field then the crystal exhibits the *linear* electro-optic or Kerr effect. We shall see that this results only if the crystal lattice lacks a center of symmetry.† So, some cubic crystals can exhibit the linear electro-optic effect. If the crystal possesses a center of symmetry (or is even an isotropic material such as a gas or liquid) a change in optical properties can result that depends, to first order, on the square of the applied field. This is the *quadratic* electro-optic or Kerr effect. Both the linear and quadratic electro-optic effects can be used effectively in various optical devices.

19.2 The Linear Electro-Optic Effect

For any cartesian coordinate system the equation of the indicatrix has the general form

$$\left(\frac{1}{n^2}\right)_1 x^2 + \left(\frac{1}{n^2}\right)_2 y^2 + \left(\frac{1}{n^2}\right)_3 z^2 + 2\left(\frac{1}{n^2}\right)_4 yz + 2\left(\frac{1}{n^2}\right)_5 xz + 2\left(\frac{1}{n^2}\right)_6 xy = 1. \tag{19.1}$$

† A crystal possesses a center of symmetry if there are identical particles in the lattice at vectors **r** and **-r** where **r** is a position vector measured from an appropriate origin.

If x, y, z are chosen to be principal axes the cross-terms in yz, xz and xy disappear and, in general, the equation of a biaxial indicatrix (Eq. (18.96)) results. For the linear electro-optic effect, an applied electric field changes each of the coefficients $(1/n^2)_i$ above in a way that is linearly dependent on the E_x, E_y, and E_z components of the field. This change in these coefficients is described by the electro-optic tensor $\overline{\overline{r}}$ through the relationship

$$\Delta\left(\frac{1}{n^2}\right)_i = \sum_{j=1}^{3} r_{ij}E_j, \tag{19.2}$$

where $E_1 = E_x, E_2 = E_y, E_3 = E_z$. So for example:

$$\Delta\left(\frac{1}{n^2}\right)_1 = r_{11}E_x + r_{12}E_y + r_{13}E_z \tag{19.3}$$

and

$$\begin{pmatrix} \Delta(1/n^2)_1 \\ \Delta(1/n^2)_2 \\ \Delta(1/n^2)_3 \\ \Delta(1/n^2)_4 \\ \Delta(1/n^2)_5 \\ \Delta(1/n^2)_6 \end{pmatrix} = \begin{pmatrix} r_{11} & r_{12} & r_{13} \\ r_{21} & r_{22} & r_{23} \\ r_{31} & r_{32} & r_{33} \\ r_{41} & r_{42} & r_{43} \\ r_{51} & r_{52} & r_{53} \\ r_{61} & r_{62} & r_{63} \end{pmatrix} \begin{pmatrix} E_x \\ E_y \\ E_z. \end{pmatrix}. \tag{19.4}$$

Typical magnitudes of the r_{ij} coefficients are 10^{-12} m V^{-1}. The form of the matrix describing $\overline{\overline{r}}$ depends on the symmetry of the crystal and is closely related to the symmetry of the piezoelectric† tensor $\overline{\overline{d}}$, which relates polarization produced in a medium to stress σ‡ The matrix describing $\overline{\overline{r}}$ is the transpose of the matrix describing $\overline{\overline{d}}$. Nye has tabulated in a convenient way the form of $\overline{\overline{d}}$. Table (19.1) is a slight variant on the format given by Nye.§

† *Piezoelectricity* is the phenomenon in which application of a force to a crystal causes a voltage to develop between the faces of the crystal. The voltage appears because the force has caused polarization – separation of positive and negative changes within the medium.

‡ *Stress* is force per unit area and can take the form of compressive or shear stress. For a more detailed discussion see Nye[19.1].

§ His piezoelectric coefficients d_{ij} describe polarization produced by stress in the form

$$\begin{pmatrix} P_x \\ P_y \\ P_z \end{pmatrix} = \begin{pmatrix} d_{11} & d_{12} & d_{13} & d_{14} & d_{15} & d_{16} \\ d_{21} & d_{22} & d_{23} & d_{24} & d_{25} & d_{26} \\ d_{31} & d_{32} & d_{33} & d_{34} & d_{35} & d_{36} \end{pmatrix} \begin{pmatrix} \sigma_{11} \\ \sigma_{22} \\ \sigma_{33} \\ \sigma_{23} \\ \sigma_{13} \\ \sigma_{13} \\ \sigma_{12} \end{pmatrix},$$

so, for example,

$$P_x = d_{11}\sigma_1 + d_{12}\sigma_{22} + d_{13}\sigma_{33} + d_{14}\sigma_{23} + d_{15}\sigma_{13} + d_{16}\sigma_{12}\sigma_{12}.$$

σ_{12} is a shear stress – it is the stress produced in the y direction by a force acting on the face of a volume perpendicular to x. To be consistent with our notation for $\overline{\overline{r}}$, where factors of 2 appear in the cross-terms in Eq. (19.1) we need to write

$$P_x = d_{11}\sigma_1 + d_{12}\sigma_{22} + d_{13}\sigma_{33} + (\tfrac{1}{2}d_{14})2\sigma_{23} + (\tfrac{1}{2}d_{15})2\sigma_{13} + (\tfrac{1}{2}d_{16})2\sigma_{12}.$$

So all Nye's d_{ij} coefficients where $i \neq j$ are double the value they have in our definition, where in symmetry terms $d_{ij} \equiv r_{ji}$.

Table 9.1. *The electro-optic* $\overline{\overline{r}}$ *for the 32 crystal symmetry classes.* · *zero element,* • *nonzero element,* •——• *equal nonzero elements,* •- - - - - -o *equal nonzero elements of opposite sign. Centrosymmetric classes – all elements of* $\overline{\overline{r}}$ *are zero.*

Triclinic

$$
\begin{pmatrix}
\bullet & \bullet & \bullet \\
\bullet & \bullet & \bullet \\
\bullet & \bullet & \bullet \\
\bullet & \bullet & \bullet \\
\bullet & \bullet & \bullet \\
\bullet & \bullet & \bullet
\end{pmatrix}
$$

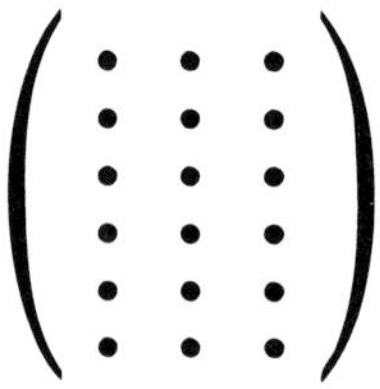

Example: Calcium thiosulphate ($CaS_2O_3.6H_2O$) strontium tartrate ($SrH_2 (C_4H_4O_6)_2.4H_2O$)

Monoclinic

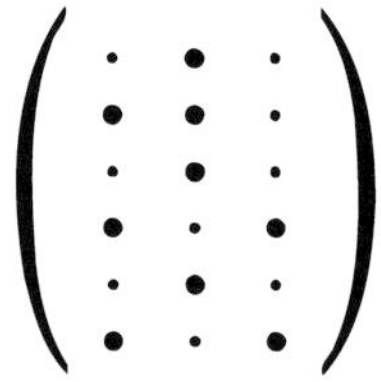

$$
\begin{pmatrix}
\cdot & \bullet & \cdot \\
\cdot & \bullet & \cdot \\
\cdot & \bullet & \cdot \\
\cdot & \cdot & \bullet \\
\cdot & \bullet & \cdot \\
\cdot & \cdot & \bullet
\end{pmatrix}
$$

2 (symmetry axis parallel to y)

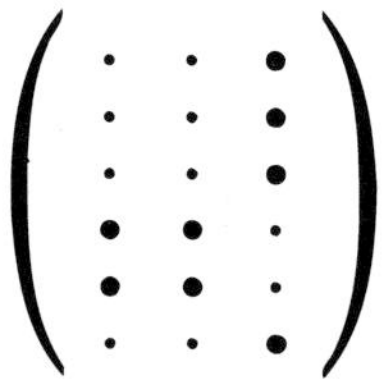

$$
\begin{pmatrix}
\cdot & \cdot & \bullet \\
\cdot & \cdot & \bullet \\
\cdot & \cdot & \bullet \\
\bullet & \cdot & \cdot \\
\bullet & \cdot & \cdot \\
\cdot & \bullet & \cdot
\end{pmatrix}
$$

2 (symmetry axis parallel to z)

Examples: lithium sulphate ($LiSO_4.H_2O$), tartaric acid triglycine sulphate

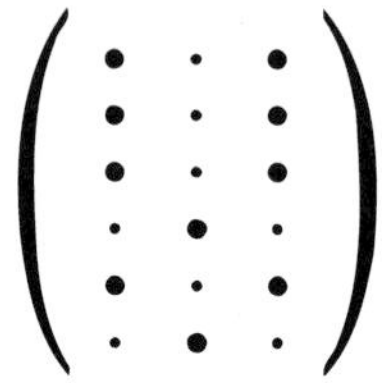

$$
\begin{pmatrix}
\bullet & \cdot & \bullet \\
\bullet & \cdot & \bullet \\
\bullet & \cdot & \bullet \\
\cdot & \bullet & \cdot \\
\bullet & \cdot & \bullet \\
\cdot & \bullet & \cdot
\end{pmatrix}
$$

m (perpendicular to y)

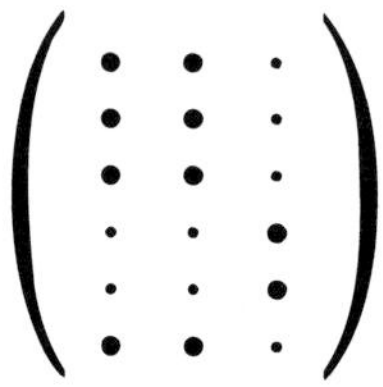

$$
\begin{pmatrix}
\bullet & \bullet & \cdot \\
\bullet & \bullet & \cdot \\
\bullet & \bullet & \cdot \\
\cdot & \cdot & \bullet \\
\cdot & \cdot & \bullet \\
\bullet & \bullet & \cdot
\end{pmatrix}
$$

m (perpendicular to z)

Example: potassium nitrite (KNO_2)

Orthorhombic

$$
\begin{pmatrix}
\cdot & \cdot & \cdot \\
\cdot & \cdot & \cdot \\
\cdot & \cdot & \cdot \\
\bullet & \cdot & \cdot \\
\cdot & \bullet & \cdot \\
\cdot & \cdot & \bullet
\end{pmatrix}
$$

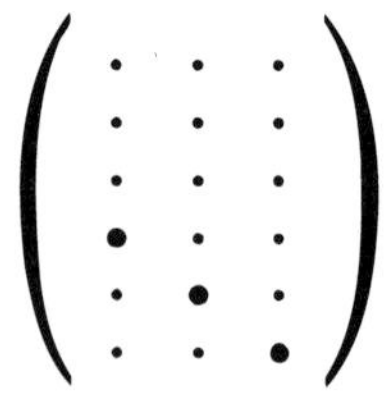

222

Examples: α–iodic acid (α–HIO_3), magnesium sulphate ($MgSO_4.7H_2O$), Rochelle salt ($KNaC_4H_4O_6.4H_2O$)

$$
\begin{pmatrix}
\cdot & \cdot & \bullet \\
\cdot & \cdot & \bullet \\
\cdot & \cdot & \bullet \\
\cdot & \bullet & \cdot \\
\bullet & \cdot & \cdot \\
\cdot & \cdot & \cdot
\end{pmatrix}
$$

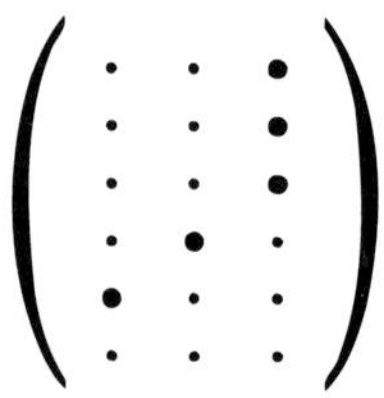

mm2

Examples: Barium sodium niobate ($Ba_2NaNb_5O_{15}$), polyvinylidine fluoride (PVF), ($CH_2CF_2)_n$

Table 9.1. (*cont.*)

Tetragonal

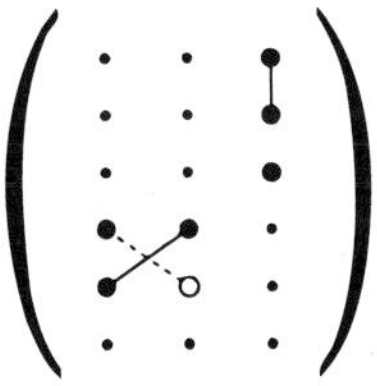

Example: iodo succinimide $(CH_2CO)_2NI$

4

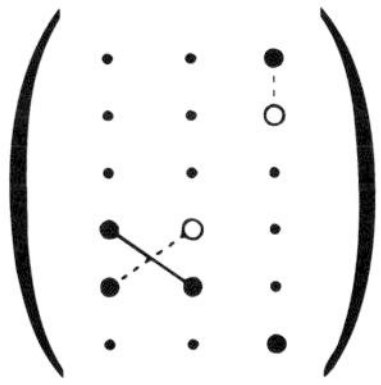

Examples: pentaerythritol $(C(CH_2OH)_4)$, cahnite $(Ca_4B_2As_2O_{12}.4H_2O)$

$\bar{4}$

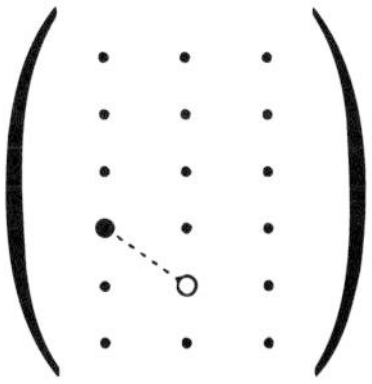

Example: nickel sulphate $NiSO_4.6H_2O$

422

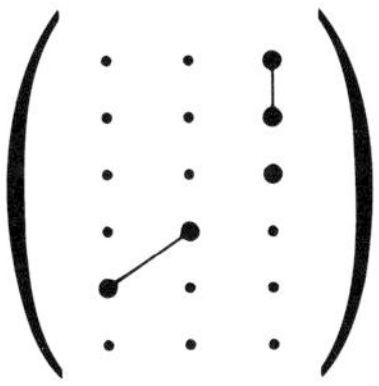

Example: $BaTiO_3$

4 mm

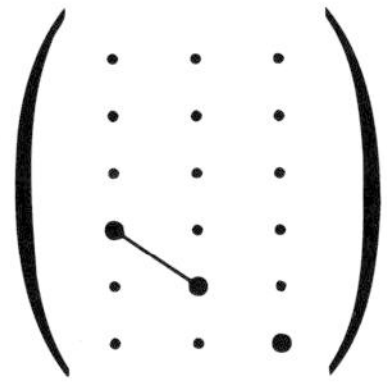

Examples: KDP, ADP, $CdGeAs_2$, $AgGaSe_2$, $AgGaS_2$

$\bar{4}2$ m (2-fold axis parallel to *x*)

Cubic

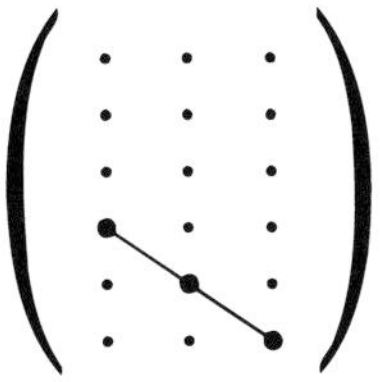

Examples: ($\bar{4}3$ m): GaAs, InAs, GaP, ZnSe, CdTe, InSb. (23): Sodium chlorate ($NaClO_3$), sodium bromate ($NaBrO_3$)

$\bar{4}3$ mm, 23

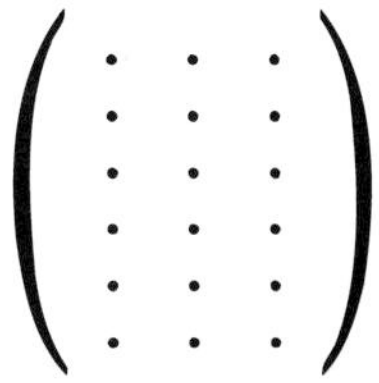

432 (lacks a center of symmetry, but other symmetry elements make all elements zero)

Trigonal

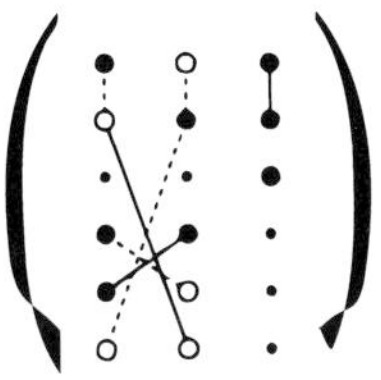

Example: sodium periodate $NaIO_4.3H_2O$

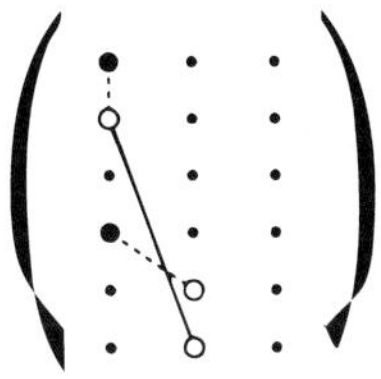

Examples: quartz, Te, mercury sulphide (α–HgS)

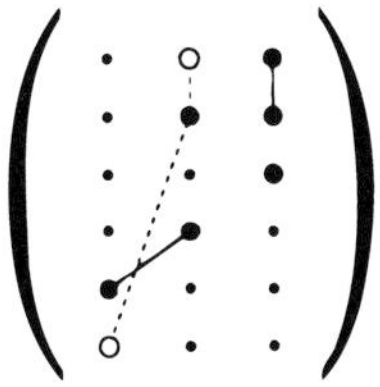

3m (m perpendicular to *x*) standard orientation

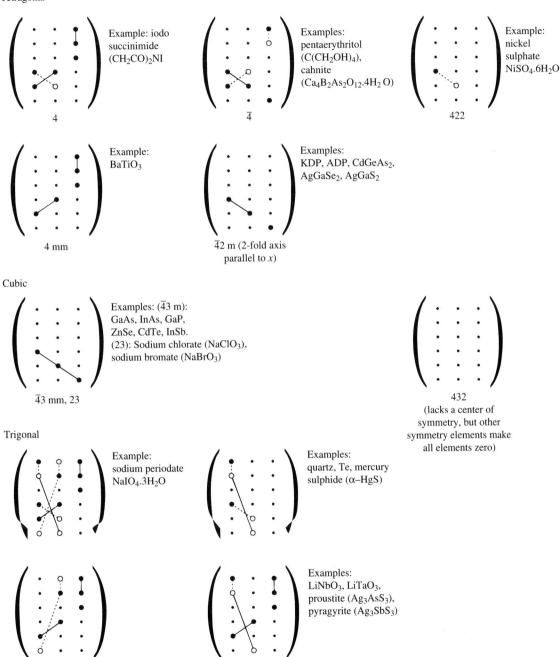

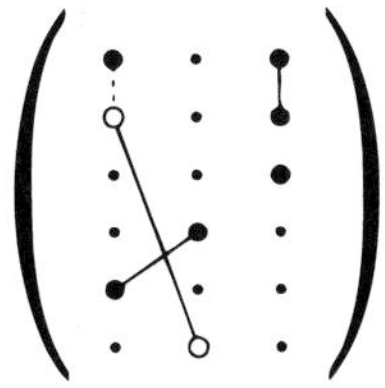

Examples: $LiNbO_3$, $LiTaO_3$, proustite (Ag_3AsS_3), pyragyrite (Ag_3SbS_3)

3m (m perpendicular to *y*)

Table 9.1. (*cont.*)

Hexagonal

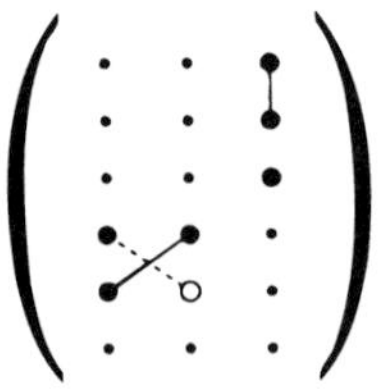

Examples: lithium iodate ($LiIO_3$), Iodoform (CHI_3)

6

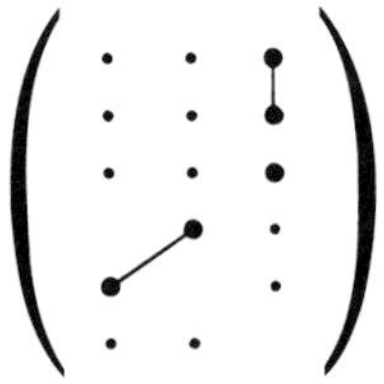

Examples: CdS, CdSe, ZnS, ZnO

6 mm

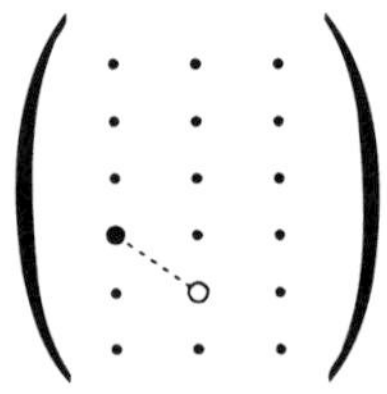

Examples: barium aluminate ($BaAl_2O_4$) kalsilite ($KAlSiO_4$)

622

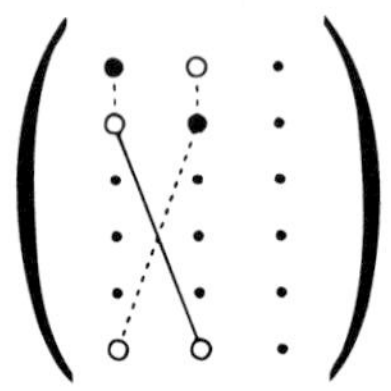

$\bar{6}$

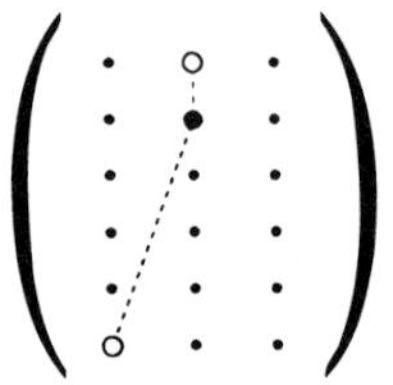

Examples: benitoite ($BaTiSi_3O_9$)

$\bar{6}$ m2 (m perpendicular to x) standard orientation

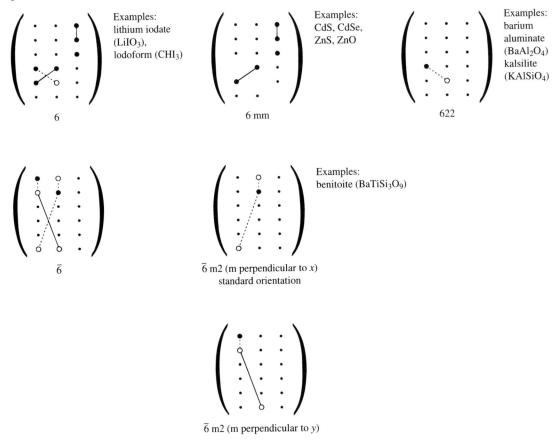

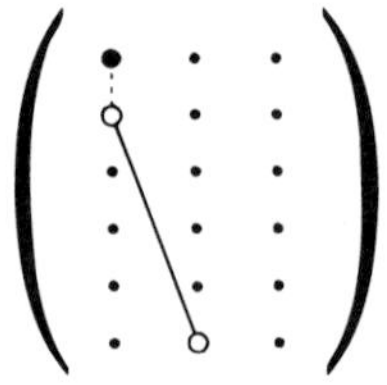

$\bar{6}$ m2 (m perpendicular to y)

With an electric field applied the equation of the indicatrix changes from Eq. (19.1) to

$$\begin{aligned}&\left[\frac{1}{n_1^2}+\Delta\left(\frac{1}{n^2}\right)_1\right]x^2+\left[\frac{1}{n_2^2}+\Delta\left(\frac{1}{n^2}\right)_2\right]y^2+\left[\frac{1}{n_3^2}+\Delta\left(\frac{1}{n^2}\right)_3\right]z^2\\&\quad+2\left[\frac{1}{n_4^2}+\Delta\left(\frac{1}{n^2}\right)_4\right]yz+2\left[\frac{1}{n_5^2}+\Delta\left(\frac{1}{n^2}\right)_5\right]xz\\&\quad+2\left[\frac{1}{n_6^2}+\Delta\left(\frac{1}{n^2}\right)_6\right]xy=1,\end{aligned}\tag{19.5}$$

The use of the index i on $\Delta(1/n^2)_i$ is an example of what is called *contracted* notation: i takes the values 1,2,3,4,5, and 6 where $1 \equiv xx, 2 \equiv yy, 3 \equiv zz, 4 \equiv yz, 5 \equiv xz, 6 \equiv xy$. Therefore the index i represents the specific quadratic term with which it is associated in the equation of the indicatrix. We shall have cause to use this notation further.

So, if our original axes were principal axes the indicatrix becomes

$$\left[\frac{1}{n_x^2}+\Delta\left(\frac{1}{n^2}\right)_1\right]x^2+\left[\frac{1}{n_y^2}+\Delta\left(\frac{1}{n^2}\right)_2\right]y^2$$
$$+\left[\frac{1}{n_z^2}+\Delta\left(\frac{1}{n^2}\right)_3\right]z^2+2\Delta\left(\frac{1}{n^2}\right)_4 yz+2\Delta\left(\frac{1}{n^2}\right)_5 xz$$
$$+2\Delta\left(\frac{1}{n^2}\right)_6 xy=1. \tag{19.6}$$

The application of the electric field has introduced cross-terms so x, y, z are no longer appropriate principal axes – the indicatrix has been changed in shape and rotated in space.

In a centrosymmetric crystal the change in shape of the indicatrix produced by a field **E** must be the same as for field –**E**, as these two fields are identical as far as the lattice is concerned. Therefore

$$\Delta\left(\frac{1}{n^2}\right)_i=\sum_{j=1}^{3} r_{ij}E_j=-\sum_{j=1}^{3} r_{ij}E_j, \tag{19.7}$$

which can only be true if $r_{ij}=0$. Consequently, there is no *linear* electro-optic effect (or any piezoelectric effect) in a centrosymmetric crystal.

The linear electro-optic effect is closely related to the inverse piezoelectric effect – where application of an electric field to a noncentrosymmetric crystal leads to an actual change in crystal shape[19.1]. The change in crystal shape leads to strain and the shape and orientation of the indicatrix are altered. With this additional contribution to the change in index coefficients we can write

$$\Delta\left(\frac{1}{n^2}\right)_i=\sum_{k=1}^{6} p_{ik}S_j+\sum_{j=1}^{3} r_{ij}E_j; \tag{19.8}$$

j takes the values $1, 2, 3; i$ and k take the values $1 \rightarrow 6$ (contracted notation). S_j is a component of the strain;† p_{ij} is the *elasto-optic* or *photoelastic* tensor. At high frequencies the inertia of the crystal prevents it straining macroscopically, so the first term on the right of Eq. (19.8) is zero. At high frequencies the residual change in indices is called the *Pockel's* effect. At low frequencies the elasto-optic effects cannot be ignored. However, since the deformation leading to strain is generally caused by the inverse piezoelectric effect‡ – and is therefore also related to applied field – it is possible to incorporate all changes in index into a single low frequency linear electro-optic tensor $\overset{=dc}{r}$ and

† Each strain component is a measure of the distortion of the crystal shape in a particular axial direction or of the rotation of the structure about a particular axial direction. For further details see Nye[19.1].

‡ If strain results solely from the inverse piezoelectric effect then

$$r_{ij}^{dc}=r_{ij}+\sum_{k=1}^{6} p_{ik}d_{jk}.$$

Table 19.2. *Low and high frequency values of elements of the electro-optic tensor for various crystals.*

Material	Symmetry	Electro-optic coefficients (10^{-12} m/V) at specific wavelength (μm) low frequency	high frequency
KDP	$\bar{4}2m$	r_{63}=-10.5 $r_{41} = 8.8$	r_{63}=8.15
ADP	$\bar{4}2m$	r_{63} = - 8.5 $r_{41} = 8.6$	$r_{63} = 5.5$
CDA (CsH_2AsO_4)	$\bar{4}2m$	$r_{63} = 8.6$	
$LiNbO_3$	3m	$r_{33} = 32.2(\lambda = 0.633)$ $r_{13} = 10(\lambda=0.633)$ $r_{22} = 7(\lambda=0.633)$ $r_{51} = 32(\lambda=0.633)$	$r_{33} = 30.8(\lambda=0.633)$ $r_{13} = 8.6(\lambda=0.633)$ $r_{22} = 3.4(\lambda=0.633)$ $r_{51} = 28(\lambda=0.633)$
$LiTaO_3$	3m		$r_{33} = 35.8(\lambda=0.633)$ $r_{13} = 7(\lambda = 0.633)$ $r_{22} \simeq 1\ (\lambda=0.633)$ $r_{51} = 20(\lambda=0.633)$
$BaTiO_3$	4mm	$r_{51}=1640(\lambda=0.546)$	$r_{51}=820(\lambda=0.546)$ $r_{33} = 28(\lambda=0.633)$ $r_{13}=8(\lambda=0.633)$
$PbTiO_3$	4mm		$r_{33} = 5.9(\lambda=0.633)$ $r_{13} = 13.8(\lambda=0.633)$
$Ba_2NaNb_5O_{15}$	mm2	$r_{33}=48(\lambda=0.633)$ $r_{13}=15(\lambda=0.633)$ $r_{23}=13(\lambda=0.633)$ $r_{42}=92(\lambda=0.633)$ $r_{51}=90(\lambda=0.633)$	$r_{33}=29(\lambda= 0.633)$ $r_{13}=7(\lambda=0.633)$ $r_{23}=8(\lambda=0.633)$ $r_{42} = 75(\lambda=0.633)$ $r_{51}=88(\lambda=0.633)$
GaAs	$\bar{4}3m$	$r_{41}=1.6(\lambda=10.6)$	$r_{41}=1.5(\lambda=10.6)$
ZnSe	$\bar{4}3m$	$r_{41} = 2.0(\lambda=0.546)$	$r_{41}=2.0(\lambda=0.633)$
ZnS	6mm		$r_{33}=1.8(\lambda=0.633)$ $r_{13}=0.9(\lambda=0.633)$
SiO_2(quartz)	32	r_{11}=-0.47(λ=0.5) $r_{41}=0.2$	$r_{11}=0.29(\lambda=0.633)$
CuCl	$\bar{4}3m$	$r_{41}=3.6(\lambda=0.633)$ $r_{41}=3.2(\lambda=10.6)$	$r_{41}=2.35(\lambda=0.633)$ $r_{41}=2.2(\lambda=3.39)$

write

$$\Delta\left(\frac{1}{n^2}\right)_i = \sum_{j=1}^{3} r_{ij}^{dc} E_j. \tag{19.9}$$

We shall not deal explicitly with this point further, but it should be noted that whenever a coefficient r_{ij} is written in what follows, this coefficient may be significantly different in magnitude between low and high frequencies. Table (19.2) gives some examples for important linear electro-optic materials.

One additional effect is worth noting – the photoelastic effect. This is the change in index coefficients produced directly by applied stress (no applied electric field is involved). For this effect, from Eq. (19.8) with no applied electric field

$$\Delta\left(\frac{1}{n^2}\right)_i = \sum_{k=1}^{6} p_{ik} S_j. \tag{19.10}$$

In a medium in which the photoelastic effect is isotropic, Eq. (19.10) can be written in the simple form

$$\frac{1}{n^2} - \frac{1}{n_0^2} = pS, \tag{19.11}$$

which gives

$$\Delta n = -\frac{n_0^3}{2} pS, \tag{19.12}$$

where p is the photoelastic or elasto-optic constant and S is the strain (deformation/length). Eq. (19.12) is particularly useful in describing the change in refractive index that occurs when a sound wave passes through an isotropic medium.

The photoelastic effect can also be described by the equation

$$\Delta\left(\frac{1}{n^2}\right)_i = \sum_{k=1}^{6} \pi_{ik}\sigma_k. \tag{19.13}$$

The π_{ik} are piezo-optical coefficients whose typical magnitude is 10^{-12} m^2 N^{-1}; the σ_k are the components of the stress (for further discussion see Nye[19.1]).

19.3 The Quadratic Electro-Optic Effect

In a crystal possessing a center of symmetry, or in an isotropic medium, there is no linear electro-optic effect. However, there is a change in the constants of the indicatrix that is related to second order products of applied electric field components.

For an isotropic medium the index of refraction of the medium changes according to

$$\frac{1}{n^2} = \frac{1}{n_0^2} + RE^2, \tag{19.14}$$

where R is a coefficient for the quadratic electro-optic or *Kerr* effect. Usually, the effect is characterized by the Kerr constant

$$B = \frac{\Delta n}{\lambda_0 E^2} = \frac{n_0^3 R}{2\lambda_0}. \tag{19.15}$$

In a centrosymmetric crystal whose indicatrix can be written as

$$\sum_{i=1}^{3} \frac{x_i^2}{n_i^2} = 1, \tag{19.16}$$

the quadratic electro-optic effect changes the indicatrix to[19.2]

$$\sum_{i=1}^{3}\sum_{j=1}^{3}\sum_{k=1}^{3}\sum_{\ell=1}^{3} \left(\frac{1}{n_{ij}^2} + R_{ijk\ell}E_k E_\ell\right) x_i x_j = 1, \tag{19.17}$$

where $n_{ij} = n_i$ for $i = j, n_{ij} = 0$ for $i \neq j$.

In contracted notation the changes in coefficients of the indicatrix are

$$\Delta\left(\frac{1}{n^2}\right)_i = \sum_{k=1}^{3}\sum_{\ell=1}^{3} R_{ik\ell}E_k E_\ell. \tag{19.18}$$

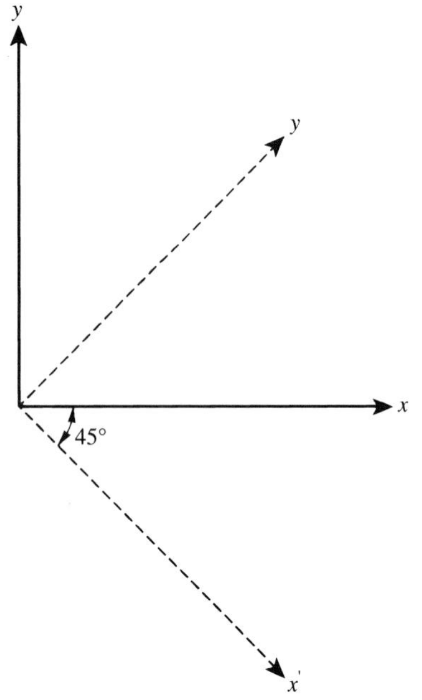

Fig. 19.1. Rotation of principal axes in a $\bar{4}2$m symmetry crystal when an electric field is applied in the z direction.

19.4 Longitudinal Electro-Optic Modulation

To illustrate how an applied electric field affects an electro-optic crystal it is instructive to consider the specific important material KDP (potassium dihydrogen phosphate, KH_2PO_4). The related materials KD*P (KD_2PO_4), ADP (ammonium dihydrogen phosphate, $(NH_4)H_2PO_4$), and AD*P, $(NH_4)D_2PO_4$) behave similarly. For these materials, symmetry $\bar{4}2$m, the electro-optic tensor has the form

$$\overline{\overline{r}} = \begin{pmatrix} 0 & 0 & 0 \\ 0 & 0 & 0 \\ 0 & 0 & 0 \\ r_{41} & 0 & 0 \\ 0 & r_{41} & 0 \\ 0 & 0 & r_{63} \end{pmatrix}. \tag{19.19}$$

In the presence of an applied field $\mathbf{E} = E_x\hat{\mathbf{i}} + E_y\hat{\mathbf{y}} + E_z\hat{\mathbf{k}}$ the indicatrix, written in its original principal axes coordinates is

$$\frac{x^2 + y^2}{n_o^2} + \frac{z^2}{n_e^2} + 2r_{41}E_x yz + 2r_{41}E_y xz + 2r_{63}E_z xy = 1. \tag{19.20}$$

For a field applied in the z direction only, Eq. (19.20) becomes

$$\frac{x^2 + y^2}{n_o^2} + \frac{z^2}{n_e^2} + 2r_{63}E_z xy = 1. \tag{19.21}$$

To find the new principal axes we note that Eq. (19.21) is symmetric in x and y, so the new x and y axes must be rotated by 45° from the original axes, as shown in Fig. (19.1). The z axis is unchanged.

The old axes are related to the new according to

$$x = x' \cos 45^\circ + y' \sin 45^\circ,$$
$$y = -x' \sin 45^\circ + y' \cos 45^\circ. \tag{19.22}$$

Substitution in Eq. (19.21) gives

$$\left(\frac{1}{n_o^2} - r_{63}E_z\right) x'^2 + \left(\frac{1}{n_o^2} + r_{63}E_z\right) y'^2 + \frac{z^2}{n_e^2} = 1. \tag{19.23}$$

The indicatrix now no longer has any cross-terms so our choice of new principal axes was correct. The crystal has now become biaxial (but only slightly so) with

$$n_{x'}^2 = \frac{n_o^2}{1 - n_o^2 r_{63} E_z}; \quad n_{y'}^2 = \frac{n_o^2}{1 + n_o^2 r_{63} E_z}; \quad n_{z'}^2 = n_e^2. \tag{19.24}$$

If $n_o^2 r_{63} E_z << 1$, which is true for any reasonable electric field, these equations become

$$n_{x'} = n_o(1 + \tfrac{1}{2} n_o^2 r_{63} E_z); n_{y'} = n_o(1 - \tfrac{1}{2} n_o^2 r_{63} E_z); n_{z'} = n_e. \tag{19.25}$$

If a wave propagates in the z direction through such a crystal the retardation is

$$\Delta\phi = \frac{2\pi L}{\lambda_0}(n_x' - n_y') = \frac{2\pi L n_o^3 r_{63} E_z}{\lambda_0}. \tag{19.26}$$

The crystal has become a wave plate whose retardation is linearly proportional to the field. It is common practice to characterize such a crystal by the voltage, V_π, necessary to produce a retardation of π. Since this applied voltage acts across a length L of crystal, $E_z = V_\pi / L$, so from Eq. (19.26)

$$V_\pi = \frac{\lambda_0}{2 n_o^3 r_{63}}. \tag{19.27}$$

For a crystal of KDP which has $r_{63} = -10.5 \times 10^{-12}$ m V^{-1}, n$_o$ = 1.51 and for an operating wavelength of 632.8 nm, Eq. (19.27) gives $V_\pi = 8752$ V. A crystal of KDP with this longitudinally applied voltage has been transformed into a half-wave plate and will, therefore, transform a linearly polarized input wave into a linearly polarized output beam whose polarization direction has been rotated from its original direction. If the input wave is polarized at 45° to the x' and y' directions, as shown if Fig. (19.2), that is polarized parallel to the original x or y directions of the undisturbed indicatrix, then the half-wave plate rotates the plane of linear polarization by 90°. When the crystal is operated in this way, as the applied voltage increases from 0 to V_π the output makes a transition from linear polarization → elliptical polarization with the major axis aligned along the original polarization direction → circular polarization → elliptical polarization with the minor axis aligned along the original polarization direction → 90° rotated linear polarization as shown in Fig. (19.3). We shall see shortly how this modulation of the polarization state can be used to make an electro-optic amplitude modulator.

Longitudinal electro-optic modulators of this sort, in which the field is applied along the light propagation direction, are somewhat inconvenient insofar as the light must pass through the electrodes used to apply the field: To minimize loss the electrodes are usually deposited as a fine metallic grid on the crystal surfaces shown in Fig. (19.2).

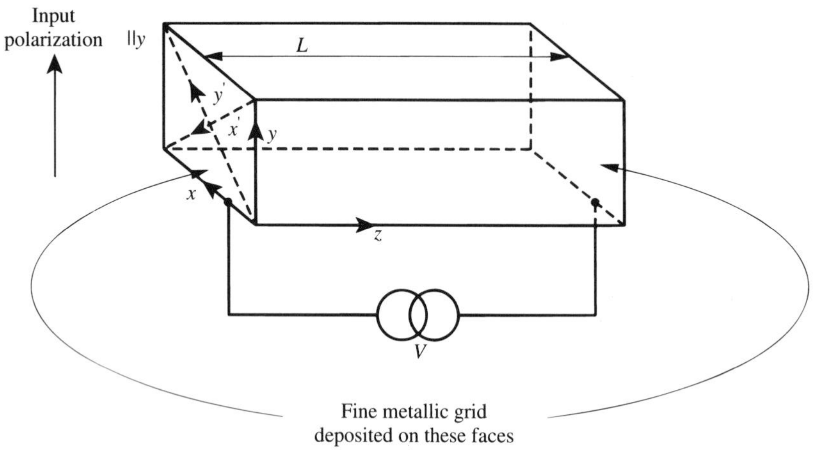

Fig. 19.2. KDP crystal orientation for longitudinal electro-optic modulation.

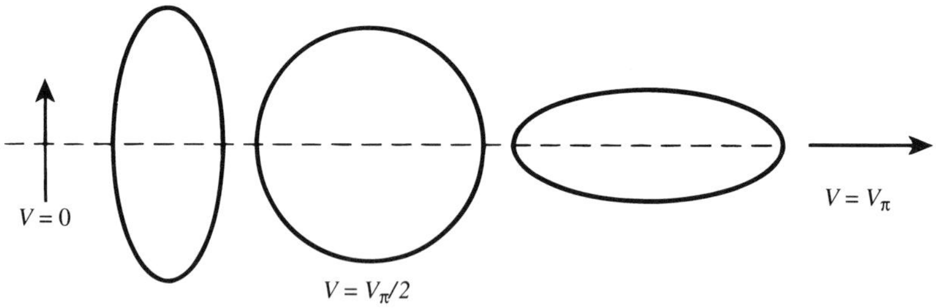

Fig. 19.3. Changes in polarization state resulting from different amounts of electro-optic retardation for a crystal operated as in Fig. (19.2).

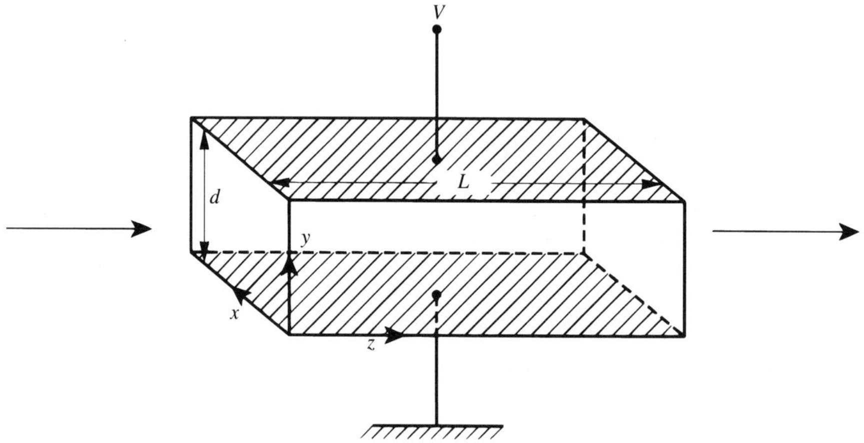

Fig. 19.4. A $LiNbO_3$ crystal setup for use as a transverse electro-optic amplitude modulator.

19.5 Transverse Electro-optic Modulation

It is more convenient to apply the electric field transversely between electrodes placed on side faces of the crystal as shown in Fig. (19.4). As an example of how this is done we can consider the important electro-optic material lithium niobate $LiNbO_3$.†

† This material is also strongly piezoelectric and is used to fabricate acousto-optic modulators.

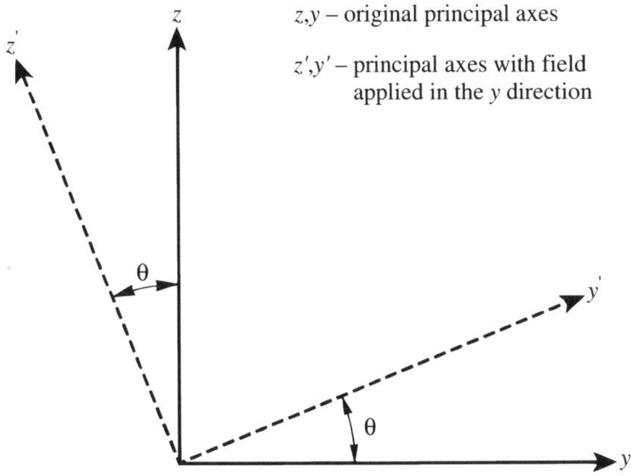

Fig. 19.5. Rotation of the principal axes in $LiNbO_3$ when a field is applied in the y principal axis direction.

Its crystal symmetry is 3m so the form of its linear electro-optic tensor is

$$\overline{\overline{r}}=\begin{pmatrix} 0 & -r_{22} & r_{13} \\ 0 & r_{22} & r_{13} \\ 0 & 0 & r_{33} \\ 0 & r_{51} & 0 \\ r_{51} & 0 & 0 \\ -r_{22} & 0 & 0 \end{pmatrix}. \tag{19.28}$$

Without an applied electric field the crystal has uniaxial symmetry with an indicatrix of the form

$$\frac{x^2}{n_o^2}+\frac{y^2}{n_o^2}+\frac{z^2}{n_e^2}=1. \tag{19.29}$$

With a field applied in the y principal axis direction the indicatrix becomes

$$\left(\frac{1}{n_o^2}-r_{22}E_y\right)x^2+\left(\frac{1}{n_o^2}+r_{22}E_y\right)y^2+\frac{z^2}{n_e^2}+2r_{51}E_y yz=1. \tag{19.30}$$

Clearly, because of the introduction of the cross-term in yz the original principal areas are no longer appropriate. The effect of the field has been to render the crystal biaxial and rotate the indicatrix in the yz plane. Let us assume that the indicatrix has been rotated so that its new principal axes y', z' are at an angle θ to the original principal axes, as shown in Fig. (19.5). The x direction remains a principal axis with the field E_y applied, as no cross-term involving x has been introduced into the equation of the indicatrix.

We can write the old coordinates in terms of the new as

$$\begin{aligned} x &= x', \\ y &= y'\cos\theta - z'\sin\theta, \\ z &= z'\cos\theta + y'\sin\theta. \end{aligned} \tag{19.31}$$

If we substitute from Eq. (19.31) into Eq. (19.30) the condition that x', y', z' are the new principal axes is that the cross-term in z', y' must vanish, that is,

$$\left(\frac{1}{n_e^2}-\frac{1}{n_o^2}-r_{22}E_y\right)\sin\theta\cos\theta+r_{51}E_y(2\cos^2\theta-1)=0. \tag{19.32}$$

We can assume that $r_{51}E_y$ is small enough that θ is a small angle and write $\sin\theta \simeq \theta$, $\cos\theta \simeq 1$. Eq. (19.32) then gives

$$\theta = \frac{-r_{51}E_y}{(1/n_e^2 - 1/n_o^2 - r_{22}E_y)}. \tag{19.33}$$

With $r_{51} = 28 \times 10^{-12}$ m V^{-1}, $r_{22} = 3.4 \times 10^{-12}$ m V^{-1}, $n_e = 2.21, n_o = 2.3$ and a voltage of 1 kV applied across a 1 mm thick crystal.

$$\theta = \frac{-28 \times 10^{-12} \times 10^6}{(\frac{1}{2.21})^2 - (\frac{1}{2.3})^2 - 3.4 \times 10^{-12} \times 10^6} = 1.78 \text{ mrad} = 0.1^\circ \tag{19.34}$$

Indeed, θ is a small angle. Since in $LiNbO_3$, $n_o > n_e$, θ is a negative angle, so in reality the rotation of the axes is clockwise, not counter-clockwise as shown in Fig. (19.5). With respect to the new principal axes the equation of the indicatrix is now

$$\begin{aligned}\left(\frac{1}{n_o^2} - r_{22}E_y\right)x'^2 + \left[\left(\frac{1}{n_o^2} + r_{22}E_y\right)\cos^2\theta + \frac{\sin^2\theta}{n_e^2} + r_{51}E_y\sin 2\theta\right]y'^2 \\ + \left[\left(\frac{1}{n_o^2} + r_{22}E_y\right)\sin^2\theta + \frac{\cos^2\theta}{n_e^2} - r_{51}E_y\sin 2\theta\right]z'^2 = 1.\end{aligned} \tag{19.35}$$

With $\cos\theta \simeq 1, \sin\theta \simeq \theta$

$$\begin{aligned}\left(\frac{1}{n_o^2} - r_{22}E_y\right)x'^{e} + \left(\frac{1}{n_o^2} + r_{22}E_y + 2r_{51}E_y\theta + \frac{\theta^2}{n_e^2}\right)y'^2 \\ + \left[\left(\frac{1}{n_o^2} + r_{22}E_y\right)\theta^2 - 2r_{51}E_y\theta + \frac{1}{n_e^2}\right]z'^2 = 1.\end{aligned} \tag{19.36}$$

The new principal refractive indices satisfy the equations

$$\begin{aligned}\frac{1}{n_{x'}^2} &= \frac{1}{n_o^2} - r_{22}E_y, \\ \frac{1}{n_{y'}^2} &= \frac{1}{n_o^2} + r_{22}E_y + 2r_{51}E_y\theta + \frac{\theta^2}{n_o^2}, \\ \frac{1}{n_{z'}^2} &= \left(\frac{1}{n_o^2} + r_{22}E_y\right)\theta^2 - 2r_{51}E_y\theta + \frac{1}{n_e^2}.\end{aligned} \tag{19.37}$$

Since θ is very small we can actually neglect the rotation of the indicatrix and write its equation as

$$\frac{x^2}{n_x^2} + \frac{y^2}{n_y^2} + \frac{z^2}{n_z^2} = 1, \tag{19.38}$$

where

$$\begin{aligned}n_x &= \frac{n_o}{\sqrt{1 - n_o^2 r_{22}E_y}} = n_o\left(1 + \tfrac{1}{2}n_o^2 r_{22}E_y\right), \\ n_y &= \frac{n_o}{\sqrt{1 + n_o^2 r_{22}E_y}} = n_o\left(1 - \tfrac{1}{2}n_o^2 r_{22}E_y\right), \\ n_z &= n_e.\end{aligned} \tag{19.39}$$

For waves travelling in the z direction the indices of refraction experienced for polarizations along x and y are n_x and n_y, respectively. For the transversely operated electro-optic modulator shown in Fig. (19.4) the retardation is

$$\Delta\phi = \frac{2\pi L}{\lambda_0}(n_x - n_y) = \frac{4\pi L n_o^3 r_{22} V}{\lambda_0 d}. \tag{19.40}$$

The half-wave voltage V_π of the modulator is the voltage required to make it act as a half-wave plate.

From Eq. (19.40), if $\Delta\phi = \pi$

$$V_\pi = \frac{\lambda_0 d}{4 L n_o^3 r_{22}} \tag{19.41}$$

For example: with $d = 5$ mm, $L = 10$ mm, $n_o = 2.3, r_{22} = 3.4 \times 10^{-12}$ m V^{-1} at an operating wavelength of 530 nm; $V_\pi = 1600$ V.

To illustrate further how a change in the orientation of the applied field and the light propagation direction affects the way in which an electro-optic crystal behaves, let us now consider what happens when light propagates through a $LiNbO_3$ crystal in the y direction, but with the field applied along the z direction.

In this case the indicatrix becomes

$$\left(\frac{1}{n_o^2} + r_{13}E_z\right)x^2 + \left(\frac{1}{n_o^2} + r_{13}E_z\right)y^2 + \left(\frac{1}{n_e^2} + r_{33}E_z\right)z^2 = 1. \tag{19.42}$$

The application of the field has left the crystal uniaxial, but the new ordinary and extraordinary refractive indices satisfy

$$\begin{aligned} \frac{1}{n_o'^2} &= \left(\frac{1}{n_0^2} + r_{13}E_z\right), \\ \frac{1}{n_e'^2} &= \left(\frac{1}{n_e^2} + r_{33}E_z\right). \end{aligned} \tag{19.43}$$

Since $r_{13}E_z << 1/n_0^2$ and $r_{33}E_z << 1/n_e^2$ for any reasonable applied field strengths,

$$\begin{aligned} n_o' &= \frac{n_o}{\sqrt{1 + r_{13}n_o^2 E_z}} = n_o\left(1 - \tfrac{1}{2} r_{13} n_o^2 E_z\right), \\ n_e' &= \frac{n_e}{\sqrt{1 + r_{33}n_e^2 E_z}} = n_e\left(1 - \tfrac{1}{2} r_{33} n_e^2 E_z\right). \end{aligned} \tag{19.44}$$

If a wave passes through such a crystal in the y direction it will, in general, split into an O-wave polarized in the x direction, and an E-wave polarized along the z direction. For the geometry shown in Fig. (19.6) the retardation is

$$\Delta\phi = \frac{2\pi L}{\lambda_0}(n_e' - n_o), \tag{19.45}$$

which gives

$$\Delta\phi = \frac{2\pi L}{\lambda_0}\left[n_e - n_o + \tfrac{1}{2}\left(r_{33}n_e^3 - r_{13}n_o^3\right)E_z\right]. \tag{19.46}$$

The half-wave voltage of the crystal used in this way is

$$V_\pi = \frac{\lambda_0 d}{L(r_{33}n_e^3 - r_{13}n_o^3)}. \tag{19.47}$$

With $d = 5$ mm, $L = 10$ mm, $r_{33}n_e^3 - r_{13}n_o^3 = 224 \times 10^{-12}$ m V^{-1}, and an operating wavelength of 530 nm; $V_\pi = 1183$ V. Although this is a slightly lower half-wave voltage than in the configuration corresponding to Eq. (19.41) discussed previously,

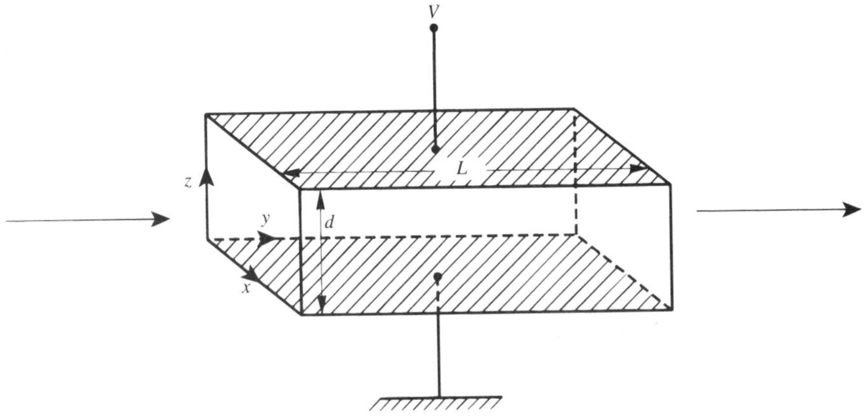

Fig. 19.6. $LiNbO_3$ crystal orientation for transverse electro-optic modulation.

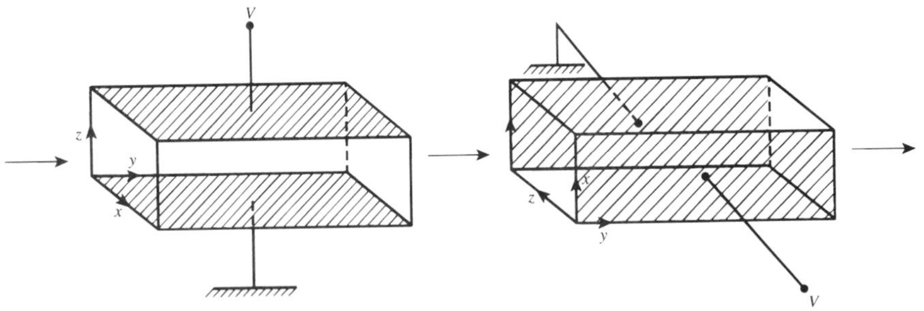

Fig. 19.7. Transverse electro-optic modulator using two $LiNbO_3$ crystals to cancel zero field birefringence.

there is a potential drawback to using the crystal in this way. Eq. (19.46) shows that the crystal operated in this orientation acts as a wave plate – because the wave is not propagating along the optic axis. The field-independent birefringent retardation $2\pi L(n_e - n_o)/\lambda_0$ will be temperature sensitive and can lead to practical stability problems. However, in practice this difficulty can be overcome by building an electro-optic device using pairs of crystals in series as shown in Fig. (19.7). The orientations of the x and z axes in the two crystals are arranged to be orthogonal so that the constant term in Eq. (19.46) cancels out. In order that the field-dependent retardation in Eq. (19.46) does not also cancel out the field polarity in the second crystal must be reversed.

19.6 Electro-Optic Amplitude Modulation

We have seen through the above example that an electro-optic crystal can be used to produce a retardation that is directly proportional to an applied voltage. To use this effect to make a useful amplitude modulator we must place the electro-optic retardation modulator in an experimental arrangement that incorporates a polarizer, wave plate, and analyzer, as shown in Fig. (19.8). This figure shows a $LiNbO_3$ crystal operated in the transverse field geometry in which the light propagates in the z direction and the field is applied in the y direction. We can analyze the performance of such an arrangement by using Eq. (18.100). A specific case will serve as a good example. If the input and output polarizers are

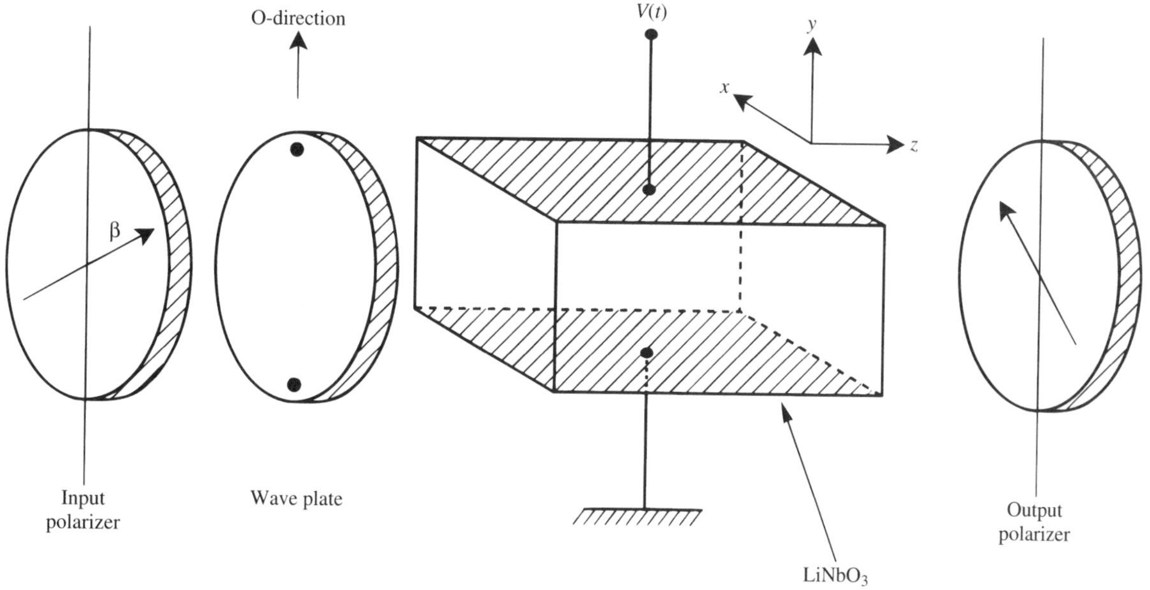

Fig. 19.8. Transverse electro-optic amplitude modulator using $LiNbO_3$.

crossed then the fractional transmitted intensity (neglecting reflection losses at any interfaces involved) is, without the electro-optic crystal

$$\frac{I_t}{I_0} = \tfrac{1}{2}\sin^2 2\beta(1 - \cos\Delta\phi_{WP}), \tag{19.48}$$

where $\Delta\phi_{WP}$ is the retardation produced by the wave plate and β is the angle between the O direction of the wave plate and the input polarization direction. It is usual to set $\beta = 45°$ and to insert the electro-optic crystal so that one of its perferred polarization directions is parallel to the O direction of the waveplate. In this case the overall retardation can be written as

$$\Delta\phi = \Delta\phi_{WP} + aV(t), \tag{19.49}$$

where $\Delta\phi_{WP}$ is the retardation produced by the wave plate and a is the constant of proportionality between applied voltage $V(t)$ and retardation in the electro-optic crystal. The constant a may be positive or negative depending on the orientation of the preferred directions in the crystal. In this case from Eq. (19.48) we can write

$$\begin{aligned}\frac{I_T}{I_0} &= \tfrac{1}{2}\{1 - \cos[\Delta\phi_{WP} + aV(t)]\} \\ &= \tfrac{1}{2}\{1 - \cos\Delta\phi_{WP}\cos[aV(t)]) + \sin\Delta\phi_{WP}\sin[aV(t)]\}.\end{aligned} \tag{19.50}$$

The amplitude modulation will only be linear if the time-dependent part of I_T/I_0 is proportional to $V(t)$. This will only be true if (i) $\Delta\phi_{WP} = \pi/2$, making the crystal a quarter-wave plate and (ii) $aV(t) << 1$ so that $\sin[aV(t)] = aV(t)$. If these conditions are satisfied

$$\frac{I_T}{I_0} = \tfrac{1}{2}[1 + aV(t)]. \tag{19.51}$$

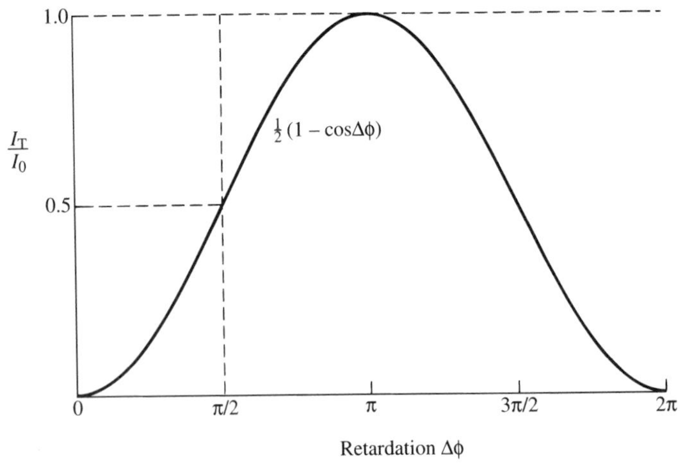

Fig. 19.9. Transmittance characteristic of the electro-optic modulator shown in Fig. (19.8) as a function of total retardation.

If the input modulating signal is

$$V(t) = V_0 \sin \omega_m t \tag{19.52}$$

then the intensity is

$$\frac{I_T}{I_0} = \tfrac{1}{2}(1 + aV_0 \sin \omega_{\mathrm{int}}). \tag{19.53}$$

The modulated output is a faithful representation of the original modulation as shown in Fig. (19.9), which plots Eq. (19.50) as a function of the total retardation. Modulation about the 'bias' point $\Delta\phi_{WP} = \pi/2$ produces linear modulation. For the geometry shown in Fig. (19.8) the retardation produced by the $LiNbO_3$ is given by Eq. (19.40) and

$$a = \frac{4\pi L n_o^3 r_{22}}{\lambda_0 d}. \tag{19.54}$$

Without the wave plate, the modulation would be nonlinear as, from Eq. (19.48)

$$\frac{I_T}{I_0} = \tfrac{1}{2}\,[1 - \cos(aV(t)]\,, \tag{19.55}$$

which for $aV(t) << 1$ gives

$$\frac{I_T}{I_0} = \tfrac{1}{2}a^2V^2(t). \tag{19.56}$$

This is clearly nonlinear modulation.

19.7 Electro-Optic Phase Modulation

If the input polarizer in Fig. (19.8) is rotated so that the light is linearly polarized parallel to the y direction and the wave plate and output polarizer are removed, the arrangement of linear polarizer and crystal becomes an electro-optic phase modulator. In this situation the electric field of the light wave emerging from the crystal can be written as

$$E_y^{\omega}(t) = E_0 \cos\,[\omega t + \phi(t)]\,. \tag{19.57}$$

$\phi(t)$ is the phase modulation, which from Eq. (19.39) is

$$\phi(t) = \left(\frac{n_o^3 r_{22}}{2d}\right) V(t). \tag{19.58}$$

If $V(t) = V_0 \sin \omega_m t$, then Eq. (19.57) can be written as

$$\begin{aligned} E_y^{(\omega)}(t) &= E_0 \cos(\omega t + m \sin \omega_m t) \\ &= E_0[\cos \omega t \cos(m \sin \omega_m t) - \sin \omega t \sin(m \sin \omega_m t)], \end{aligned} \tag{19.59}$$

where

$$m = \left(\frac{n_o^3 r_{22}}{2d}\right) V_0 \tag{19.60}$$

is called the *depth of modulation.*

The frequency spectrum of the phase-modulated wave is interesting. We use the following identities[19.3]:

$$\cos(m \sin \omega_m t) = J_0(m) + 2\sum_{k=1}^{\infty} J_{2k}(m) \cos(2k\omega_m t), \tag{19.61}$$

$$\sin(m \sin \omega_m t) = 2\sum_{k=0}^{\infty} J_{2k+1}(m) \sin[(2k+1)\omega_m t], \tag{19.62}$$

$$2 \sin A \sin B = \cos(A - B) - \cos(A + B), \tag{19.63}$$

$$2 \cos A \cos B = \cos(A - B) + \cos(A + B), \tag{19.64}$$

and write Eq. (19.59) as:

$$\begin{aligned} E_y^{(\omega)}(t) = E_0[&J_0(m) \cos \omega t + J_1(m) \cos(\omega + \omega_m)t - J_1(m) \cos(\omega + \omega_m)t \\ &+ J_2(m) \cos(\omega + 2\omega_m)t + J_2(m) \cos(\omega - 2\omega_m)t \\ &+ J_3(m) \cos(\omega + 3\omega_m)t - J_3(m) \cos(\omega - 3\omega_m)t \\ &+ \ldots]. \end{aligned} \tag{19.65}$$

The frequency spectrum of the phase-modulated light has acquired all the frequencies $(\omega \pm n\omega_n)$, where n is an integer, with amplitudes given by the Bessel function amplitudes $J_n(m)$. The center frequency ω is called the *carrier* frequency. Fig. (19.10) shows such a spectrum, calculated for $m = 1$.

Phase modulation has important communication applications, since the term $\phi(t)$ in Eq. (19.57) can be modulated in an arbitrary way to include information on the light beam. This is the optical analog of radio wave phase modulation (PM) or frequency modulation (FM)[19.4].†

19.8 High Frequency Waveguide Electro-Optic Modulators

In a high frequency electro-optic modulator the electrical drive to the modulator must be applied using some form of stripline configuration so that the induced phase or polarization modulation propagates at, or close to, the phase velocity of the light signal travelling through the modulator. If this is not done, the

† The concepts of PM and FM are closely related since the overall phase of the wave $\Phi(t) = \omega t + \phi(t)$ obeys $d\Phi/dt = \omega + d\phi/dt$, so $\omega + d\phi/dt$ can be regarded as the *instantaneous* frequency of the wave. Loosely speaking we could say that $FM = d/dt(PM)$.

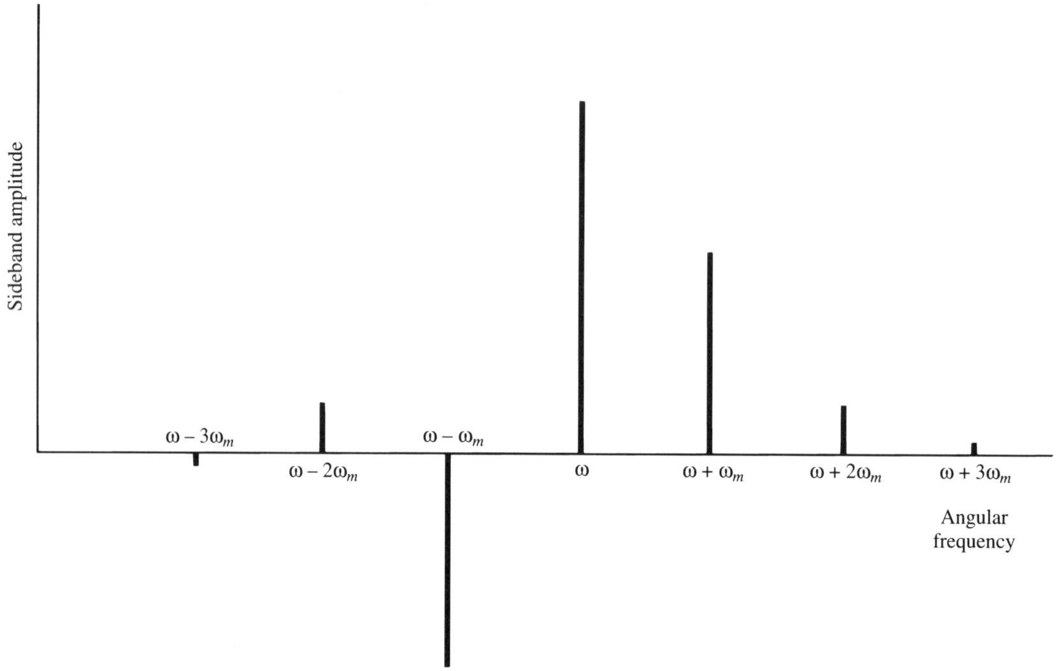

Fig. 19.10. Phase-modulated spectrum for a modulation depth $m = 1$.

light propagating through the crystal sees an electric field of varying phase as it propagates and the effects of the field on the phase or polarization state of the wave can cancel out.

19.8.1 Straight Electrode Modulator

In a stripline electro-optic modulator of the straight type shown in Fig. (19.11) high frequency response is limited by travelling wave mismatch effects. We can represent the drive voltage at position z along this structure as

$$V(z,t) = V_0 \sin\left(\omega_m t - \frac{\omega_m n_m}{c_0} z\right), \tag{19.66}$$

where c_0 is the velocity of light in vacuo, ω_m is the drive frequency, and n_m is the effective refractive index for the drive field (such that the phase velocity of the drive field is c_0/n_m).

The optical field propagates with phase velocity,

$$c = \frac{c_0}{n}, \tag{19.67}$$

where n is the effective refractive index for the waveguide mode.

If the optical field enters the crystal at $z = 0$ at time zero, then the voltage that is seen as it propagates can be found from Eq. (19.66) by writing $t = zn/c_0$, which is the time it takes the optical wave to reach location z:

$$V(z) = V_0 \sin\left[\frac{\omega_m z}{c_0}(n - n_m)\right]. \tag{19.68}$$

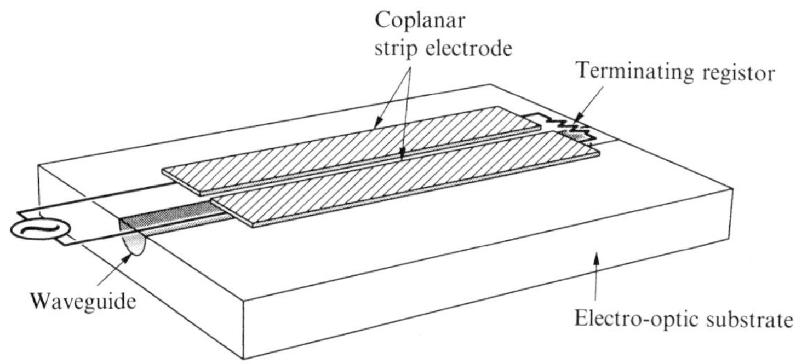

Fig. 19.11. Travelling wave electro-optic modulator (courtesy of Suzanne Hiser).

The electro-optically induced change in the index n can be written as

$$\Delta n(z) = aV(z), \tag{19.69}$$

where a is a constant that depends on the specific geometry and crystal being used.

The total phase shift experienced in propagating a length ℓ through the crystal is

$$\begin{aligned}\Delta\phi &= \int_0^\ell \frac{\omega \Delta n(z)}{c_0} dz \\ &= \frac{aV_0\omega}{c_0} \int_0^\ell \sin\left[\frac{\omega_m z}{c_0}(n - n_m)\right] dz,\end{aligned} \tag{19.70}$$

which gives

$$\Delta\phi = -\frac{aV_0}{(n - n_m)}\left[\cos\frac{\omega_m \ell}{c_0}(n - n_m) - 1\right]. \tag{19.71}$$

We make the substitution

$$\theta = \pi\omega_m/\omega_c, \tag{19.72}$$

where

$$\omega_c = \frac{\pi c_0}{(n_m - n)\ell} \tag{19.73}$$

is called the cutoff frequency.

The total phase shift can be written as

$$\begin{aligned}\Delta\phi &= \frac{aV_0\ell}{\pi c_0}\omega_c\left[\cos\theta - 1\right] \\ &= \frac{aV_0\ell\omega_m}{2c_0(\theta/2)}\left[\cos^2(\theta/2) - \sin^2(\theta/2) - 1\right] \\ &= \frac{aV_0\ell\omega_m}{2c_0(\theta/2)}\left[-2\sin^2(\theta/2)\right] \\ &= -\frac{aV_0\omega_m\ell}{c_0}\left[\frac{\sin(\theta/2)}{(\theta/2)}\right]\sin(\theta/2).\end{aligned} \tag{19.74}$$

Eq. (19.74) can be written as

$$\Delta\phi = -\frac{aV_0\omega_m^2\pi\ell}{2\omega_c c_0(\pi\omega_m/2\omega_c)}\frac{\sin^2(\theta/2)}{(\theta/2)}$$
$$= -\frac{a\pi V_0\omega_m^2\ell}{2c_0\omega_c}\text{sinc}^2(\theta/2), \tag{19.75}$$

giving finally†

$$\Delta\phi = -\frac{aV_0\omega_m^2\ell^2}{2c_0^2}(n_m - n)\text{sinc}^2(\theta/2). \tag{19.76}$$

The optimum length for the modulator satisfies

$$\frac{\omega_m\ell(n-n_m)}{c_0} = \pi; \text{ or } \omega_m = \omega_c, \theta = \pi. \tag{19.77}$$

For $LiNbO_3$, $n_m = 4.2, n = 2.146$; so, for example, with $\ell = 10$ mm the cutoff frequency is 7.36 GHz.

The electric field acts on the wave-guide region of the electro-optic crystal as shown in Fig. (19.12). In $LiNbO_3$, for example, the wave- guiding region is made by diffusing titanium into the crystal. The buffer layer is usually Al_2O_3 or SiO_2, and prevents the propagating wave, in this geometry a TM wave, from interacting with the metal electrodes. A practical device might have an indiffused channel 7 μm wide, a buffer layer 0.26 μm thick, and a 3 μm thick Cr/Au electrode on the top. The electrode thickness is chosen to be larger than the skin depth at the design frequency. The index difference between guiding region and substate is typically 0.01–0.02.

In the Z-cut, TM mode of operation, the electric field of the mode has x and z components.

For $LiNbO_3$, the linear electro-optic tensor is

$$\bar{\bar{r}} = \begin{pmatrix} 0 & -3.4 & 8.6 \\ 0 & 3.4 & 8.6 \\ 0 & 0 & 30.8 \\ 0 & 28 & 0 \\ 28 & 0 & 0 \\ -3.4 & 0 & 0 \end{pmatrix} \times 10^{-12} \text{ m V}^{-1}. \tag{19.78}$$

For this material typical optical losses are $\sim$ 1 dB cm^{-1} at $\lambda = 633$ nm and $\sim$ 0.3 dB cm^{-1} at $\lambda = 1.32$ μm

The column vector of index changes once again obeys

$$\Delta\left(\frac{1}{n^2}\right) = \bar{\bar{r}}\begin{pmatrix} E_x \\ E_y \\ E_z \end{pmatrix}, \tag{19.79}$$

where E_x, E_y, and E_z are the components of the externally applied electric field.

For the Z-cut the applied field has only a z component and therefore the only change in the indicatrix is

$$\Delta\left(\frac{1}{n^2}\right)_3 = r_{33}E_z. \tag{19.80}$$

† $\text{sinc}(\theta/2) = \sin(\theta/2)/(\theta/2)$.

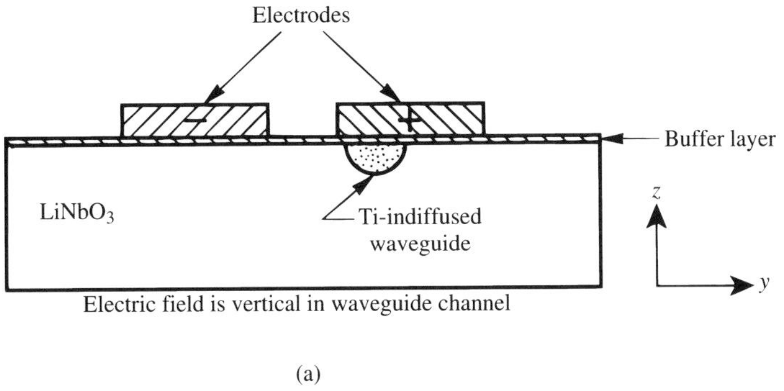

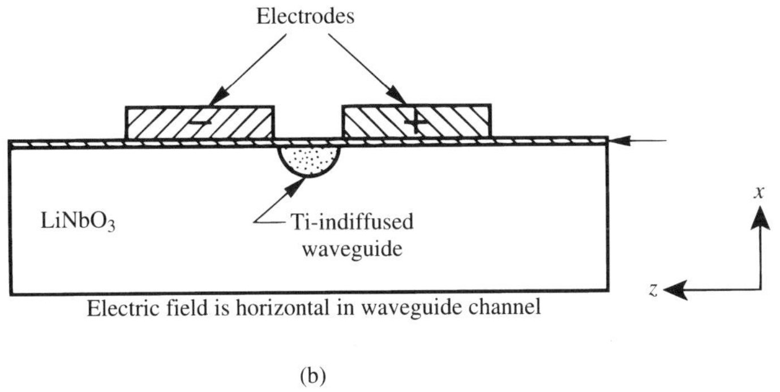

Fig. 19.12. Geometries of high frequency $LiNbO_3$ waveguide electro-optic modulators (courtesy of Suzanne Hiser): (a) Z-cut, TM mode, (b) X-cut, TE mode.

For the X-cut, the field is also applied in the z direction, and the wave propagates in the y direction. The only electric field component of the mode is in the z direction, which is a TE mode – it has x and y directed magnetic field components. With the X-cut, $\Delta(1/n^2)_3 = r_{33}E_z$ is still the only change in the indicatrix.

19.9 Other High Frequency Electro-Optic Devices

Various kinds of waveguide structure can be fabricated in $LiNbO_3$, as shown in Fig. (19.13).

In the Mach–Zehnder interferometer configuration the waves propagating in the two arms can experience different phase shifts, so can emerge in or out of phase. The device can be used as an amplitude modulator or switch. If the phase shift between the two arms is ϕ the output intensity I_{out} will be related to the output intensity I_{in} according to

$$\frac{I_{out}}{I_{in}} = \tfrac{1}{2}(1 + \cos\phi) = \cos^2(\phi/2). \tag{19.81}$$

Variation of ϕ in such a device will not produce linear modulation. However, if the device is biased by $\pi/2$, for example, by having one interferometer arm slightly longer than the other (as shown in Fig. (19.13d)) then we can write $\phi = \pi/2 + \Delta\phi$.

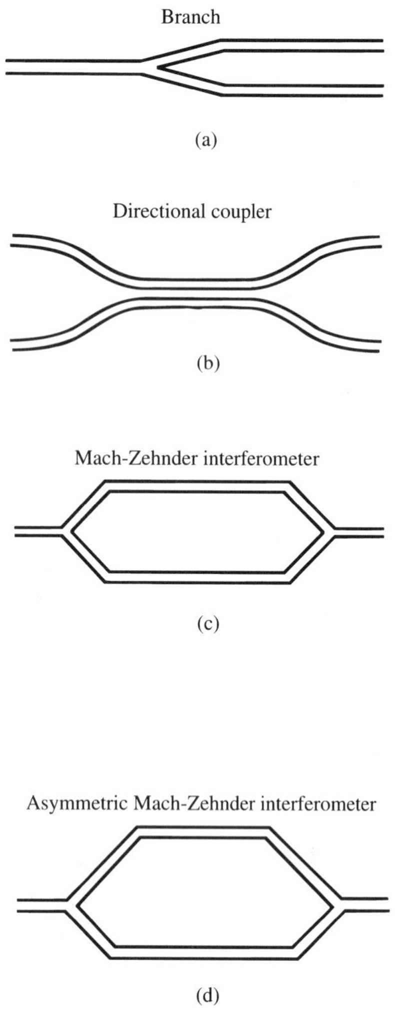

Fig. 19.13. Integrated waveguide structures.

In this case

$$\frac{I_{out}}{I_{in}} = \tfrac{1}{2}[1 + \cos[\pi/2 + \Delta\phi]]$$
$$= \tfrac{1}{2}[1 + \cos(\pi/2)\cos\Delta\phi - \sin(\pi/2)\sin\Delta\phi], \qquad (19.82)$$

which gives

$$\frac{I_{out}}{I_{in}} = \tfrac{1}{2} - \tfrac{1}{2}\Delta\phi \qquad (19.83)$$

for small phase modulation $\Delta\phi$.

This provides linear modulation of the intensity about a mean value $I_{out} = \frac{1}{2}I_{in}$. To achieve passive biasing of $LiNbO_3$, the path difference is $\Delta L = 152$ nm, this provides $\phi = \pi/2$. Typical devices are 10–20 mm long with the guiding regions 20 μm apart.

Because it is not possible to match the velocity of the excitation field and optical field over a long distance at high modulation frequencies, a phase-reversed scheme

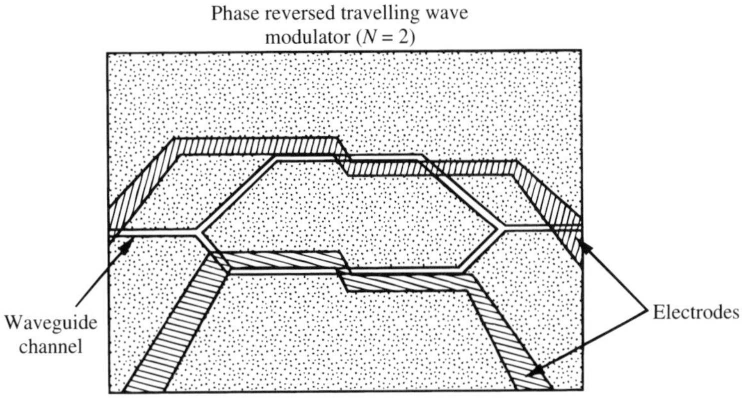

Fig. 19.14. Travelling wave electro-optic modulator with periodic phase-reversal electrodes (courtesy of Suzanne Hiser).

can be used to extend the bandwidth. After each length of the device over which the optical and drive fields get out of phase by π the polarity of the drive field is reversed as shown in Fig. (19.14). Such a device will in theory have a pass band response. Broad frequency response can be obtained by aperiodically varying the length between phase reversals of the drive field, as shown in Fig. (19.15a). By increasing the number of aperiodic phase reversal electrodes much broader response than with a periodic phase reversed structure with different numbers of sections N can be obtained, as shown in Fig. (19.15b).

19.10 Electro-Optic Beam Deflectors

The electro-optic effect can also be used to deflect or scan a laser beam. The simplest way to do this is to use a prism of electro-optic material in which an applied field varies the bulk index of an ordinary or extraordinary wave passing through the material, as shown in Fig. (19.16).

Fig. (19.17) shows a double prism electro-optic beam deflector that uses two KDP prisms with their optic axes reversed with respect to each other. By applying a voltage as shown, the index of refraction along the path *cd* can be made larger than that along the path *ab*. This causes the phase fronts passing through the crystal to tilt and the output beam deflects.† Electro-optic beam deflectors have not found as widespread use as have acousto-optic beam deflectors, which are our next topic in this chapter.

19.11 Acousto-Optic Modulators

When a sound wave propagates through a medium it produces regions of compression and rarefaction. The medium may support both longitudinal and shear sound waves depending on the sound wave propagation direction relative to the crystal symmetry axes. For a plane sound wave at frequency ω_s travelling in the z direction the change in density as a function of position can be written as

$$\Delta\rho = \Delta\rho_0 \cos(\omega_s t - k_s z), \tag{19.84}$$

† Problem(19.1) involves a quantitative analysis of this device.

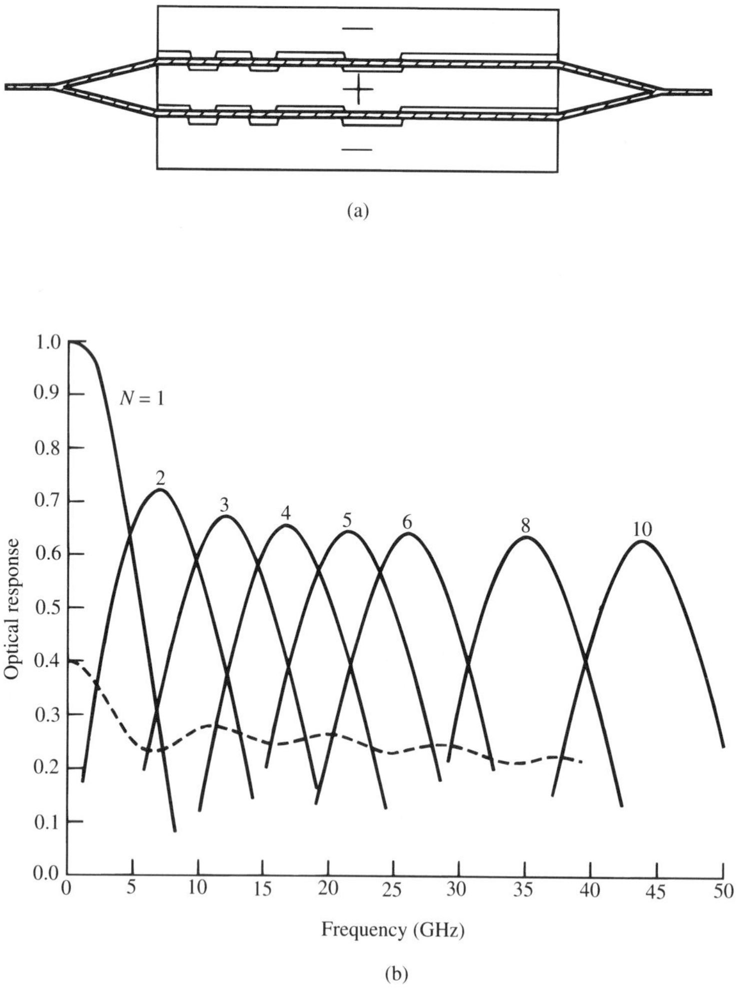

Fig. 19.15. (a) Travelling wave electro-optic modulator with S-bit Barker code, aperiodic, phase reversal electrodes, (b) frequency response of various periodic and aperiodic travelling wave electro-optic modulators. The dashed line is for a 13-bit Barker code aperiodic, phase reversal electrode configuration (courtesy of Suzanne Hiser).

where $k_s = 2\pi/\lambda_s$, with λ_s being the wavelength of the sound wave. For small relative changes in density we expect the change in refractive index to be $\Delta n \propto \Delta\rho$. From Eqs. (19.12) and (19.84) it is clear that

$$\Delta n = -\frac{n_0^3}{2} p S_0 \cos(\omega_s t - k_s z), \tag{19.85}$$

where S_0 is the peak value of the strain produced by the passage of the sound wave. Eq. (19.85) is most useful in describing the photoelastic effect in an isotropic medium, where the behavior is described by a single photoelastic coefficient p. In anisotropic media the change in index will depend on the direction of both the light wave and sound wave relative to the summetry axes. We will not deal with the details of acousto-optic effects in such anisotropic materials here; detailed information is available in the specialized literature[19.5]–[19.8].

If a light wave passes through a medium with a spatial variation in refractive index like Eq. (19.67) it is diffracted. This phenomenon is closely related to Brillouin

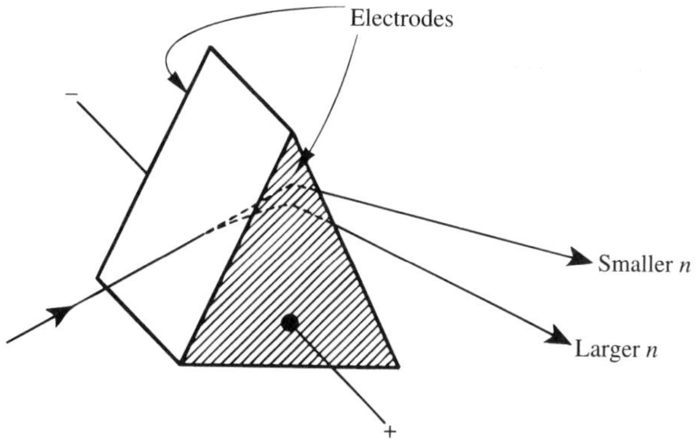

Fig. 19.16. Simple single prism electro-optic laser beam deflector.

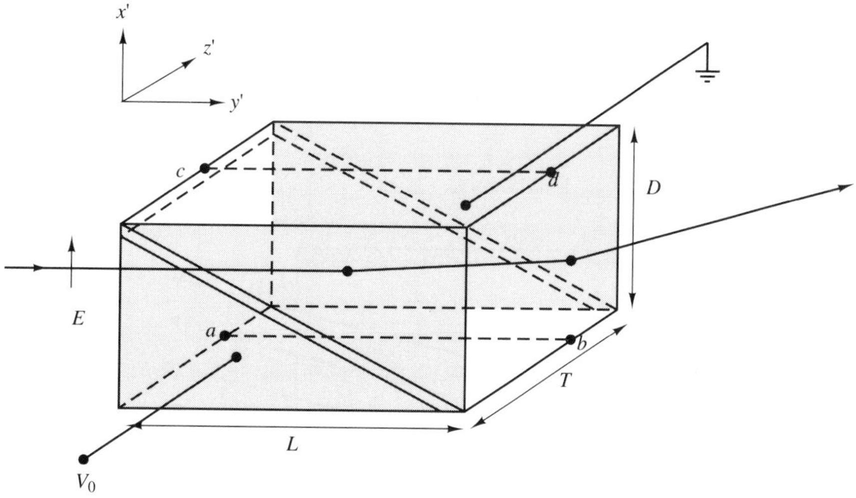

Fig. 19.17. Double prism KDP electro-optic beam deflector. The upper and lower prisms have their z (optic) axes reversed with respect to each other. The prisms are cut so that their orthogonal edges are parallel to the x', y', z' directions – the principal axis directions in KDP when an external electric field is applied in the z direction.

scattering, the interaction of light with naturally ocurring density fluctuations in media, first predicted in 1922[19.9].

Practical application of the diffraction of light by sound waves is accomplished in one of two geometries, shown in Figs. (19.18) and (19.19), in which an acoustical transducer (a piezoelectric transducer) is used to drive a sound wave through the acousto-optic medium. In the geometry shown in Fig. (19.18), where the interaction is called the *Debye–Sears* effect[19.10], the interaction between light wave and sound wave occurs over a relatively short length – in the so-called *Raman–Nath* regime[19.11]. In the geometry of Fig. (19.19) the light wave interacts with essentially plane sound waves and the phenomenon is described as acoustic *Bragg* diffraction: it is as if the light reflects and diffracts from the planes of density variation in the medium. We can delineate the difference between Raman–Nath and Bragg diffraction with the aid of Fig. (19.20). The regions of compression or rarefaction in the medium have width $\sim \lambda_s/2$. Light travelling in one of these layers spreads by diffraction at an angle

$$\theta_{diff} \sim \frac{2\lambda}{\lambda_s}. \tag{19.86}$$

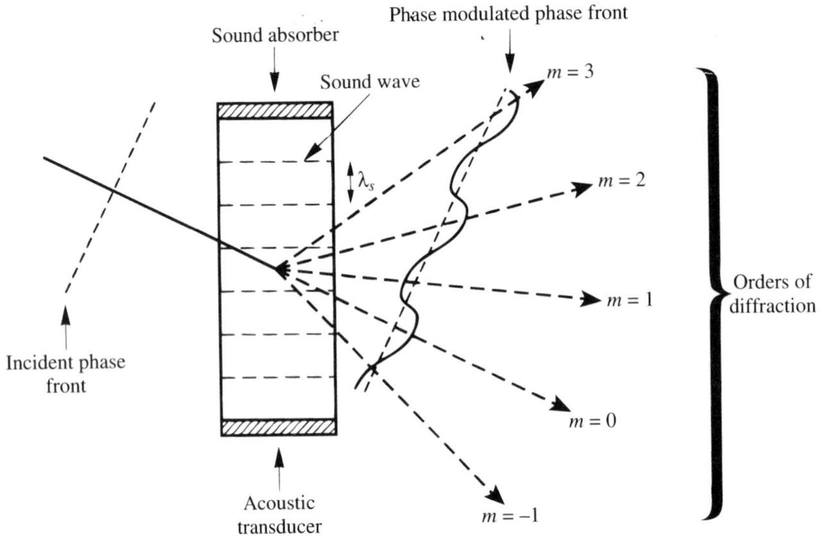

Fig. 19.18. Acousto-optic device using the Debye–Sears effect.

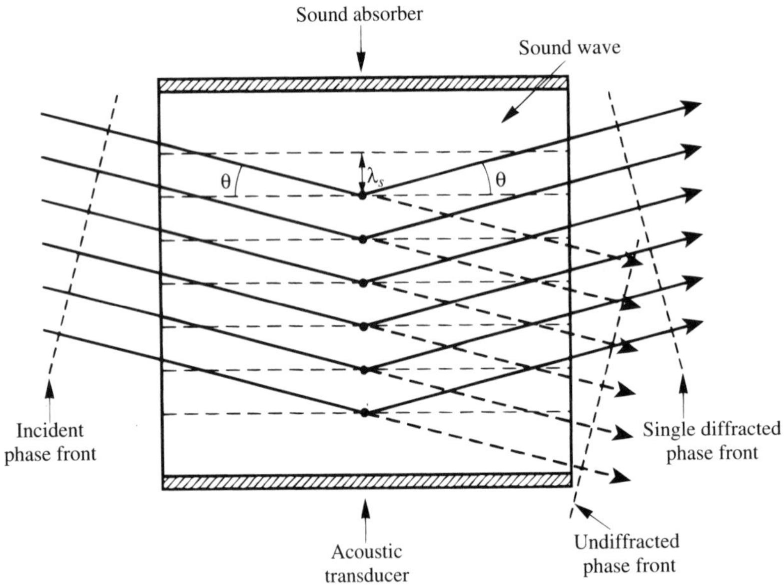

Fig. 19.19. Acousto-optic device using Bragg diffraction.

The maximum useful interaction length, ℓ_{max} over which the wave remains within a compressed (or rarefied layer) satisfies

$$\tan\theta_{diff} = \frac{\lambda_s}{2\ell_{max}}, \tag{19.87}$$

which for small angles gives

$$\ell_{max} = \frac{\lambda_s^2}{4\lambda}. \tag{19.88}$$

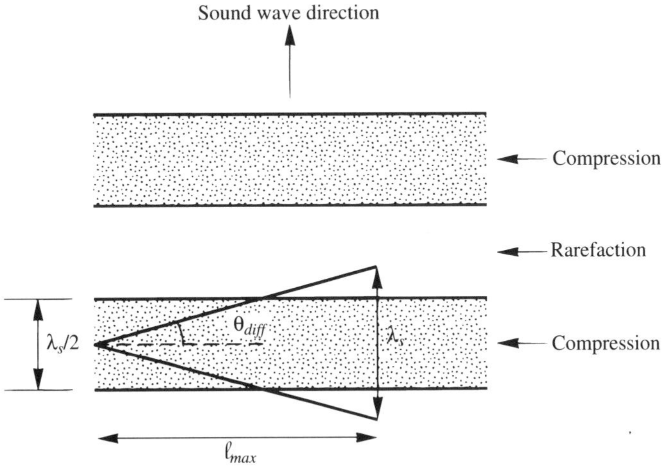

Fig. 19.20. Schematic distribution of density changes produced by passage of a sound wave through a medium used to show how the maximum interaction length in the Raman-Nath regime is calculated.

For example, with 1.3 μm laser radiation interacting with 40 MHz sound waves in TeO_2 (an important acousto-optic material) in a crystal orientation where v_s=4.2 km s^{-1} and λ_s=0.105 mm: ℓ_{max}=2 mm.

An alternative parameter that has been introduced to delineate the Raman–Nath and Bragg diffraction regimes is[19.12]

$$Q = \frac{2\pi\lambda_0 L}{n\lambda_s^2}. \tag{19.89}$$

For $Q < 1$, Raman–Nath diffraction is dominant, while for $Q > 7$ the device operates in the Bragg regime.

The light emerging from a Debye–Sears acousto-optic modulator contains several angular diffraction orders, each of which is Doppler shifted by virtue of the fact that the light has interacted with a *moving* sound wave. We can illustrate this with the simple model shown in Fig. (19.21). We model the moving sound wave as a series of small scatters travelling with velocity v_s. Light reflected from A and B will constructively interfere if

$$\sin\theta_1 + \sin\theta_2 = \frac{m\lambda}{\lambda_s}, \tag{19.90}$$

where m is an integer.†

We can calculate the Doppler shift of the light diffracted in Fig. (19.21) by considering the shortening of the ray path that occurs as the sound wave moves from point B to point A. In moving from B to A, the sound wave causes the distance travelled by a light ray to change from the path length of ray 1 to that of ray 2. This reduction in path length is $m\lambda$, and it occurs in a time $1/f_s$, where f_s is the sound frequency. Therefore, light observed along path 1 appears to come from a source whose velocity is

$$v_{source} = m\lambda f_s \tag{19.91}$$

† Eq. (19.90) should be slightly modified in an anisotropic material to account for the different refractive indices experienced by waves travelling in the θ_1 and θ_2 directions. However, this difference in index is generally small and we shall not worry about it.

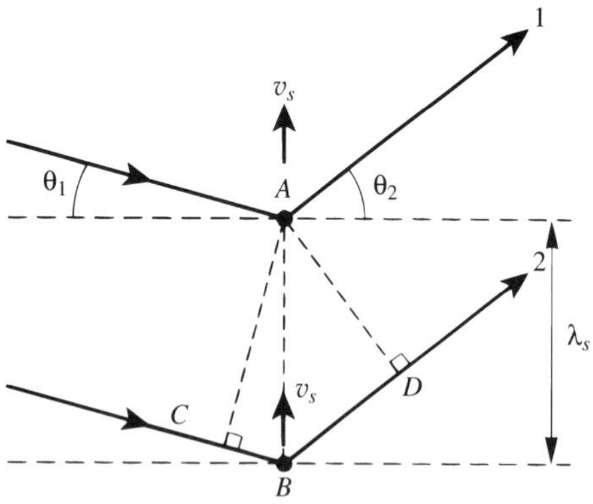

Fig. 19.21. Simple diagram for calculating diffraction from sound waves.

towards the observer. The resultant Doppler shift is

$$\Delta \nu = v_{source}\frac{\nu_0}{c}, \tag{19.92}$$

where ν_0 is the light frequency. Therefore,

$$\Delta \nu = m f_s. \tag{19.93}$$

Each diffracted order is Doppler shifted by the appropriate multiple of the sound frequency. The Doppler shift is negative for waves diffracted in the downward direction shown in Fig. (19.21).

When an acousto-optic device works in the Bragg regime, as shown in Fig. (19.19), the incident light wave interacts with broad sound-wave phase fronts. This imposes a more restrictive condition on the permitted directions for diffraction. Not only must the incident and diffracted angles satisfy Eq. (19.90) but the angles θ_1 and θ_2 must be equal – as if the acoustic phase fronts act as mirrors for the incident light. This situation is illustrated in Fig. (19.22). The Bragg diffraction condition is†

$$\sin\theta = \frac{m\lambda}{2\lambda_s}, \tag{19.94}$$

where $m = 1$. Brillouin was the first to point out that because the sound wave produces a sinusoidal disturbance in the medium, higher order diffraction effects, where $m > 1$, do not occur.

If the sound wave is moving in the upward direction in Fig. (19.22), with velocity v_s^+ then the diffracted light is up-shifted to frequency $\nu + f_s$. If the sound wave is travelling downwards with velocity v_s^-, then the diffracted wave is down-shifted to frequency $\nu - f_s$.

Acousto-optic interactions can be viewed in two additional ways. In the particle interaction model of the process each incident light photon collides with a quantized particle of acoustic energy (a phonon) creating a photon with different energy and momentum. This is shown schematically in Fig. (19.23). An incident photon with wave vector $\mathbf{k}_1$ collides with a phonon with wave vector $\mathbf{k}_s$ to create a photon with wave vector $\mathbf{k}_2$. Since the momentum of each of these particles is $\hbar\mathbf{k}$, momentum is conserved through the vector relationship:

$$\mathbf{k}_2 = \mathbf{k}_1 \pm \mathbf{k}_s, \tag{19.95}$$

† Once again we are neglecting the different refractive indices that would be experienced by the incident and diffracted waves in an anisotropic medium.

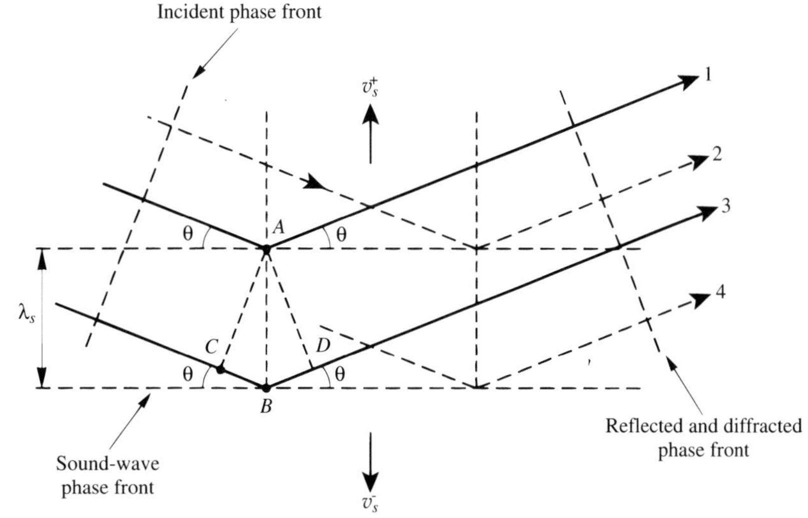

Fig. 19.22. Illustration of the Bragg condition for diffraction from a sound wave. If $CB+BD=m\lambda$ then $\sin\theta = m\lambda/2\lambda_s$ and the diffracted rays from all sound-wave phase fronts separated by λ_s remain in phase.

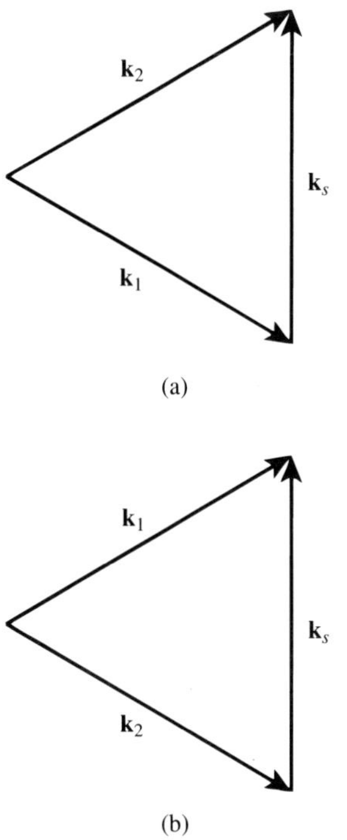

Fig. 19.23. Vector addition of wave vectors in the particle interaction model of an acousto-optic interaction: (a) Doppler frequency up-shift, (b) Doppler frequency down-shift.

and energy is conserved through the relationship

$$\omega_2 = \omega_1 \pm \omega_s. \tag{19.96}$$

The $\pm$ sign in Eqs. (19.95) and (19.96) corresponds to the two geometries shown in Fig. (19.23).

The interchange of energy among the three waves involved in Eq. (19.96) is also viewable as a general class of interaction called a *parametric* interaction. These are nonlinear interactions and will be discussed in greater detail in Chapters 20 and 21.

19.12 Applications of Acousto-Optic Modulators

Acousto-optic modulator can be used as frequency shifters, frequency modulators, optical beam deflectors and scanners, and optical switches. Used in one of these specific ways they find application in communication systems, sensors, optical information storage and retrieval, xerography, laser control, and spectrum analysis. They are generally used in the Bragg regime. Their performance efficiency is characterized by their diffraction efficiency η, which is[19.6]

$$\eta = \sin^2\left(\frac{\pi L}{\lambda\cos\theta}\sqrt{\frac{M_2 I_s}{2}}\right). \tag{19.97}$$

In this equation L is the interaction width of the light beam with the sound wave, I_s is the intensity of the sound wave, θ is the Bragg angle and M_2 is a *figure of merit* (FOM) for the acousto-optic material[19.13].† The FOM is a material constant that allows the relative efficiencies of different acousto-optic materials to be compared, given by

$$M_2 = \frac{n_0^6 p^2}{\rho v_s^3}, \tag{19.98}$$

where n_0 is the refractive index, p is the photoelastic constant, ρ is the density, and v_s is the sound velocity.

19.12.1 *Diffraction Efficiency of TeO_2*

For a longitudinal sound wave the M_2 FOM for TeO_2 is 34.5×10^{-15} s^3 kg^{-1} and the sound velocity is 4.2 km s^{-1}. Consequently, at 40 MHz the sound wavelength is 0.105 mm. For a laser wavelength of 1.06 μm with a refractive index of 2.26 the Bragg angle is

$$\theta = \text{arc}\ \sin(\lambda/2\lambda_s) = 0.128°.$$

For an acousto-optic modulator with a 1 mm×1 mm cross-section and an injected sound-wave power of 1 W, from Eq. (19.97) the diffraction efficiency is

$$\eta = \sin^2\left(\frac{\pi\times10^{-3}\times2.26}{1.06\times10^{-6}}\sqrt{\frac{34.5\times10^{-15}\times1}{2\times10^{-6}}}\right),$$

which gives $\eta = 59\%$.

19.12.2 *Acousto-Optic Modulators*

The Doppler shift of the Bragg diffracted light in Fig. (19.19) is the same as the frequency applied to the acoustic transducer. If the drive signal is of the form

$$V(t) = V_0\cos\{[\omega_s + \Delta\omega(t)]\,t\}, \tag{19.99}$$

† Other figures of merit M_1 and M_3 have also been proposed to include factors such as bandwidth[19.14] and an independence of beam dimensions[19.15], respectively.

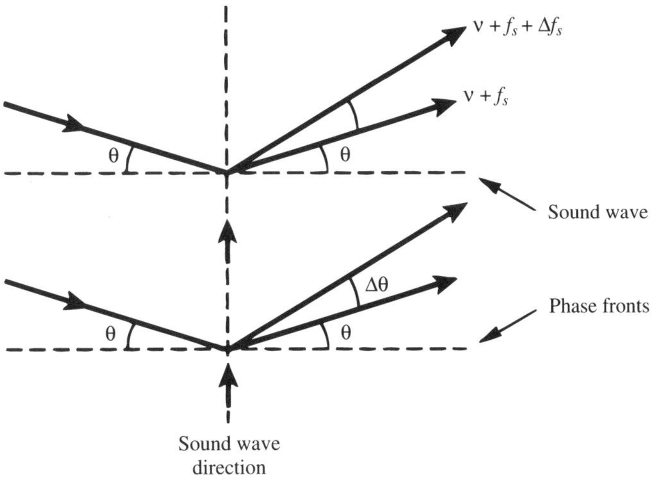

Fig. 19.24. Variation in diffraction angle that results from a change in sound-wave frequency.

which is a frequency modulated carrier wave at frequency ω_S, then provided $\Delta\omega \ll \omega_s$ the beam deflection angle is essentially constant but the frequency of the light within this beam will also be frequency modulated. If $\Delta\omega$ is not small with respect to ω_s then the deflection angle will be modulated. As far as the diffracted light in a particular direction is concerned, this corresponds to intensity modulation. However, the easiest way to achieve intensity modulation is to utilize the undiffracted beam and modulate its intensity by varying the drive signal so as to transfer incident light intensity into the diffracted beam. Intracavity acousto-optic modulators can be used in this way to both Q-switch and mode-lock lasers. The maximum intensity modulation frequency that can be achieved in this way depends on the transit time of the sound wave through the diameter of the incident light beams, so for high modulation frequencies the beam must be focused into the acousto-optic medium.

19.12.3 Acousto-Optic Beam Deflectors and Scanners

If an acousto-optic device is operated at the Bragg angle corresponding to a drive frequency ω_S, then variation of ω_S will lead to variation in the diffraction angle, thereby serving to deflect or scan the diffracted beam. Because the Bragg condition is only strictly satisfied at frequency ω_S, deviations from this value deflect the beam, but with concomitant reduction in diffracted intensity as the frequency moves further from ω_S. In Fig. (19.24) a variation in ω_s to $\omega_s + \Delta\omega_s$ changes θ to $\theta + \Delta\theta$. From Eq. (19.90) we can write

$$\sin\theta + \sin(\theta + \Delta\theta) = \frac{k_s + \Delta k_s}{k}, \tag{19.100}$$

where $k_s = 2\pi f_s/v_s$ and $k = 2\pi/\lambda$. Incorporating the Bragg condition (Eq. (19.94)) into Eq. (19.100) for small deflection angles gives

$$\Delta\theta = \frac{\Delta k_s}{k}. \tag{19.101}$$

The utility of a beam scanner is frequently determined by the number of resolvable deflected spots N that it can produce. For a Gaussian beam of minimum spot size

ω_0 focused at the Bragg angle

$$N = \frac{\Delta\theta_{max}}{\theta_{beam}}, \tag{19.102}$$

which gives

$$N = \frac{\pi\omega_0\Delta\theta_{max}}{\lambda}. \tag{19.103}$$

The maximum deflection angle is

$$\Delta\theta_{max} = \frac{\lambda\Delta f_{max}}{v_s} \tag{19.104}$$

and, since the sound-wave transit time through the beam is

$$\tau = \frac{2w_0}{v_s}, \tag{19.105}$$

we get

$$N = \left(\frac{\pi}{2}\right)\tau\Delta f_{max}. \tag{19.106}$$

A maximum reasonable sound wave frequency variation is $\Delta f_{max} = f_s/2$, where f_s is the sound-wave frequency corresponding to the Bragg angle. In this case

$$\tau = \frac{2\omega_0}{f_s\lambda_s} \tag{19.107}$$

and

$$N = \frac{\pi\omega_0}{2\lambda_s}. \tag{19.108}$$

The maximum number of resolvable spots is really the maximum number of distinguishable angles by which the incident beam is deflected. Each diffracted beam could be focused to a distinct spot with an appropriate lens.

19.12.4 RF Spectrum Analysis

If the RF drive signal applied to the transducer in Fig. (19.19) has a frequency spectrum $F(\omega_s)$, then each frequency component in this spectrum produces an angular deflection whose angle depends on ω_s, and whose intensity depends on the amplitude $F(\omega_s)$ of the drive signal applied to the transducer at frequency ω_s. If the overall diffracted beam is focused by a lens, the different angular components of this beam will be focused to spots in different lateral locations in the focal plane of the lens, as shown in Fig. (19.25). The overall arrangement acts like a Fourier transform device, input RF frequencies f_1, f_2, f_3 produce different angles of diffraction, which are focused by the lens at different spatial locations. If an array of light detectors is placed along the line AB then a rapid parallel computation of the input RF spectrum occurs – this is true *optical computation.*

19.13 Construction and Materials for Acousto-Optic Modulators

An acousto-optic modulator is fabricated from a material with a high acousto-optic figure of merit, such as tellurium dioxide (TeO_2) or lead molybdate ($PbMbO_4$) to which has been bonded a piezoelectric transducer material such as lithium niobate

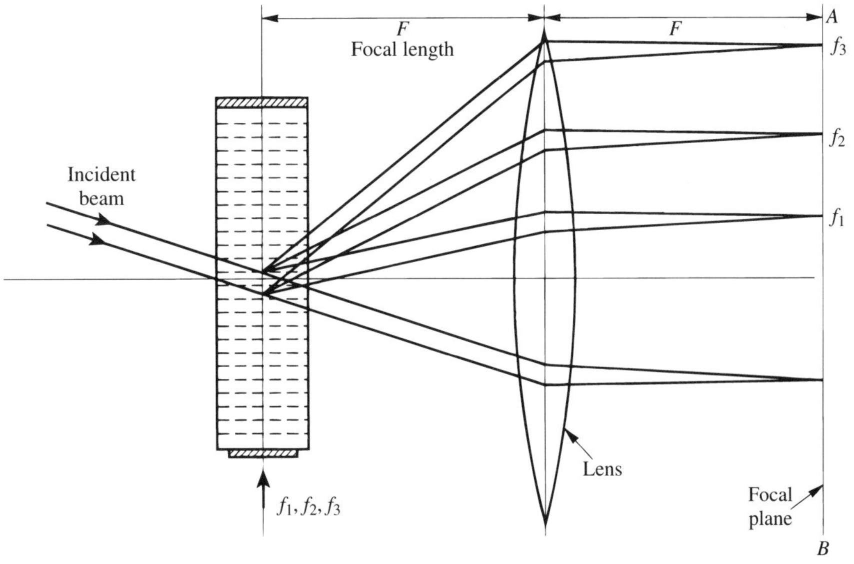

Fig. 19.25. Acousto-optic Bragg diffraction for RF spectrum analysis.

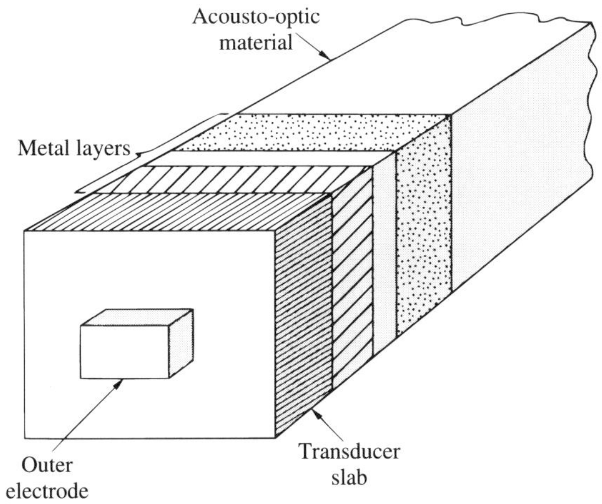

Fig. 19.26. Acousto-optic device showing the transducer and metal layers for acoustic matching.

($LiNbO_3$). The transducer is generally driven at an RF frequency: 40 MHz and 80 MHz are common values for fixed frequency devices. Consequently, it is also important that the acousto-optic material have low sound-wave attenuation at the sound-wave frequency being used. Since the acoustic attenuation in these materials varies $\sim f_s^2$ the attenuation coefficient is quoted in units of dB m^{-1} GHz^{-2}. Table (19.3) lists several important acousto-optic materials and their performance characteristics.

The transducer is a thin ceramic slab bonded to one or more metallized layers on the acousto-optic material. A second metallized layer is placed on the other side of the slab to provide electrical contact, as shown schematically in Fig. (19.26). Thus, the transducer presents an essentially capacitive load to the drive circuit. For high frequency operation the thickness of the transducer is reduced to a few micrometers, and in this way acousto-optic modulators operating at several GHz

Table 19.3. *Characteristics of acousto-optic materials (additional values can be found in References [19.6] and [19.8]).*

Material	Optical trans-mission (μm)	Wave-length (μm)	Density (kg m^{-3})	Mode and propagation direction	v_s (km s^{-1})	Polarization direction	n	M_2 (FOM) (10^{-15} s^{-3} kg^{-1})
Fused silica	0.2–4.5	0.633	2650	L	5.96	⊥	1.46	1.51
Diamond	0.2–5	0.589	3520	L[100]	17.5	‖	2.417	1.02
TeO_2	0.35–5	0.633	6210	L[001]	4.2	⊥	2.26	34.5
$PbMbO_4$	0.42–5.5	0.633	6950	L[001]	3.63	‖	2.262	36.3
						⊥	2.386	36.1
$LiNbO_3$	0.4–4.5	0.633	4640	L[100]	6.57	35° to y	2.2	7.0
				S[001]	3.59	⊥	2.29	2.92
$Pb_5(GeO_4)(VO_4)_2$	0.52–5.5	0.633	7150	L[001]	3.45	‖	2.275	50.6
				L[010]	3.11	‖	2.281	35.6
α-ZnS	0.4–12	0.633	4090	L[001]	5.82	‖	2.351	3.4
		0.633		S[001]	2.63	see [19.6]	2.348	8.4
β-ZnS	0.4–12	0.633	4100	L[110]	5.51	‖	2.35	3.41
α-HgS	0.62–16	0.633	8100	L[001]	2.45	⊥	2.887	960
		1.153						
Tl_3AsS_4	0.6–12	0.633	6200	L[001]	2.15	‖	2.825	800
		1.153						
Ge	2–20	10.6	5320	L[111]	5.50	‖	4.00	840
		10.6		S[100]	3.51	arbitrary	4.00	290
GaAs	1–11	1.153	5340	L[110]	5.15	‖	3.37	104
		1.153		S[100]	3.32	arbitrary	3.37	46.3
KDP	0.25–1.7	0.633	2340	L[100]	5.50	‖	1.51	1.91
ZnTe	0.55–20	0.633	5670	L[110]	3.87	‖	2.983	45.1
		1.153						
$LiIO_3$	0.3–6	0.633	4500	L[001]	4.13	⊥	1.88	8.0
		0.633		L[100]	4.3	[001]	1.74	13.0
GaP	0.6–10	0.633	4130	L[110]	6.32	‖	3.31	44.6
				S[100]	4.13	arbitrary	3.31	24.1
Si	1–5–10	10.6	2330	L[111]	9.85	‖	3.42	6.2

(a). L-longitudinal mode; S-shear modes. Directions are given in crystallographic notation: for example [100]≡ x direction, [001]≡ y direction; [110]≡ in xy plane at 45° to x. For further details on this notation consult Kittel[19.17].

(b). ‖ and ⊥, parallel and perpendicular to the acoustic-wave vector, respectively.

have been fabricated. The performance of the device is optimized by matching the transducer electrically to the device circuit and acoustically to the acousto-optic material[19.16]. The maximum frequency, f_{max}, at which an acousto-optic modulator can modulate a light beam is limited by the transit time of the sound wave across the beam diameter D

$$f_{max} = \frac{v_s}{D}. \qquad (19.109)$$

However, a beam focused to a small beamwaist will not be able to interact coherently along a long path through the sound wave. Generally speaking, except in mode-locking applications, the sound wave travels unidirectionally in the acousto-optic material, a reflected sound wave can be suppressed by an acoustic absorber or matched acoustic load on the end of the material remote from the transducer.

In high frequency devices the sound wave attenuates significantly as it leaves the transducer, so in a sufficiently long device there is negligible reflected sound wave.

19.14 Problems

(19.1) Fig. (19.17) shows a double prism electro-optic beam deflector described by Fowler and Schlafer[19.18]. Prove that the deflection angle of an input beam polarized in the extraordinary direction is, for small deflection angles,

$$\theta = \frac{L}{A} n_0^3 r_{63} V_0,$$

where A is the area of the input face, and V_0 is the applied field.

References

[19.1] J.F. Nye, *Physical Properties of Crystals*, Oxford University Press, Oxford, 1957.

[19.2] G.W.C. Kaye and T.H. Laby, *Tables of Physical and Chemical Constants*, 14th Edition, Longman, NY, 1973.

[19.3] M. Abramowitz and I.A. Stegun, *Handbook of Mathematical Function*, Dover, New York, 1965.

[19.4] H. Taub and D.L. Schilling, *Principles of Communication Systems*, McGraw-Hill, New York, 1971.

[19.5] E.G. Spencer, P.V. Lenzo, and A.A. Ballman, 'Dielectric materials for electro-optic, elastooptic and ultrasonic device applications,' *Proc. IEEE*, **55**, 2074–2108, 1967.

[19.6] N. Uchida and N. Niizeki, 'Acousto-optic deflection materials and techniques,' *Proc. IEEE*, **61**, 1073–1092, 1973.

[19.7] A. Korpel, *Acousto-Optics*, Marcel Dekker, New York, 1988.

[19.8] I.C. Chang, 'Acousto-optic devices and applications,' *IEEE Trans. Sonics Ultrasonics*, **SU-23**, 2–22, 1976.

[19.9] L. Brillouin, 'Diffusion de la lumiere et des rayons X par un corps transparent homogene in fluence de l'agitation thermique,' *Ann. Phys. (Paris)*, 9th sec., **17**, 88–122, 1922.

[19.10] P. Debye and F.W. Sears, 'On the scattering of light by supersonic waves,' *Proc. Nat'l. Acad. Sci*, **18**, 409–414, 1932.

[19.11] C.V. Raman and N.S.N. Nath, 'The diffraction of light by sound waves of high frequency, Part II,' *Proc. Ind. Acad. Sci.*, **A2**, 413–420, 1936.

[19.12] W.R. Klein and B.D. Cook, 'Unified approach to ultrasonic light diffraction,' *IEEE Trans. Sonics Ultrasonics*, **SU-14**, 123–124, 1967.

[19.13] T.M. Smith and A. Korpel, 'Measurement of light-sound interaction efficiency in solids,' *IEEE J. Quantum Electron.*, **QE-1**, 283–284, 1965.

[19.14] E. Gordon, 'A review of acousto-optical deflection and modulation devices,' *Proc. IEEE*, **54**, 1391–1401, 1966.

[19.15] R. Dixon, 'Photoelastic properties of selected materials and their relevance for application to acoustic light modulations and scanners,' *J. Appl. Phys.*, **38**, 5149–5153, 1967.

[19.16] E.H. Young, Jr., and S-K. Yao, 'Design considerations for acousto-optic devices,' *Proc. IEEE*, **69**, 54–64, 1981.

[19.17] Charles Kittel, *Introduction to Solid State Physics*, 6th Edition, Wiley, New York, 1986.

[19.18] V.J. Fowler and J. Schlafer, 'A survey of laser beam deflection techniques,' *Proc. IEEE*, **54**, 1437–1444, 1966.

20

Introduction to Nonlinear Processes

20.1 Introduction

In this chapter we shall begin our discussion of nonlinear phenomena that are important in optics. When one or more electromagnetic waves propagate through any medium they produce polarizations in the medium that, in principle, oscillate at all the possible sum and difference frequencies that can be generated from the incoming waves. These polarizations, which oscillate at these new frequencies, give rise to corresponding electromagnetic waves. Thus, we get phenomena such as harmonic generation: for example, when infrared light is converted into visible or ultraviolet light, and various other frequency mixing processes. These nonlinear processes can be described by a series of nonlinear susceptibilities or mixing coefficients. These coefficients will be defined and their origin traced to the anharmonic character of the potential that describes the interaction of particles in the medium.

20.2 Anharmonic Potentials and Nonlinear Polarization

When an electromagnetic wave, or waves, propagates through a medium a *total* electric field acts on each particle of the medium. This total field contains components at all the frequencies contained in the input waves. Each particle of the medium will be displaced from its equilibrium position by the action of this field. Positive ions, and nuclei will be displaced in the direction of the field, negative ions and electrons will be displaced in the opposite direction to that of the field. The resultant separation of centers of positive and negative charge creates dipoles in the medium. The resultant dipole moment per unit volume (polarization) can be written schematically as

$$\mathbf{P} = \epsilon_0(\chi^{(1)}\mathbf{E} + \chi^{(2)}\mathbf{E}^2 + \chi^{(3)}\mathbf{E}^3 + \ldots), \tag{20.1}$$

where $\chi^{(1)}$ is the already familiar linear susceptibility, $\chi^{(2)}$ is the second order nonlinear polarizability and so on.

We can explain the dependence of the polarization on the square and higher powers of the field in terms of the restoring force that acts on a particle of the medium when it is displaced from its equilibrium position. Consider the simple rectangular lattice structure shown in Fig. (20.1). If particle A is displaced towards particle B it will be pushed back to equilibrium by the repulsion of particle B and the attraction of particle C. We can describe the size and direction of the restoring force on particle A in terms of its potential energy as it moves along the line between B and C. This potential will be of the schematic form shown in Fig. (20.2).

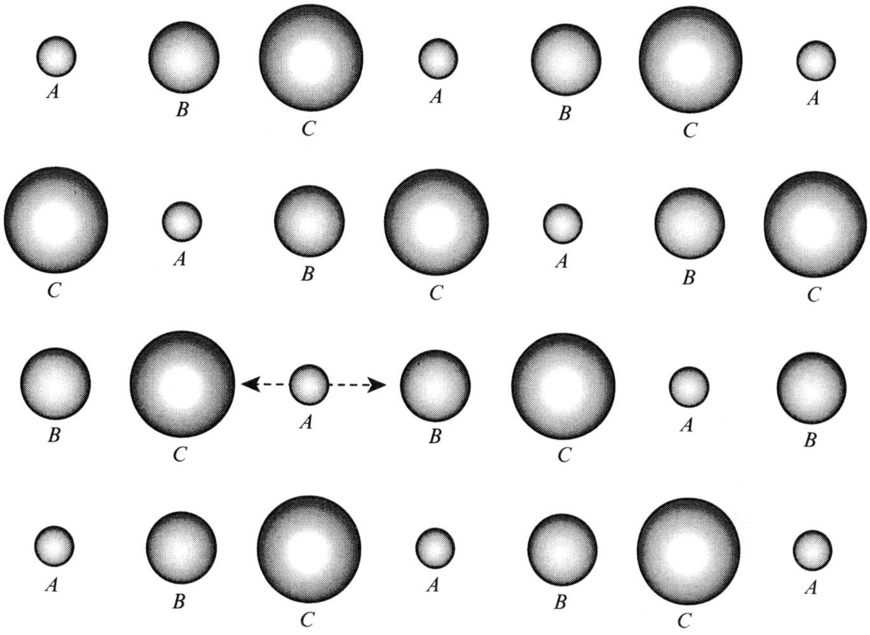

Fig. 20.1. Rectangular lattice structure that lacks a center of symmetry.

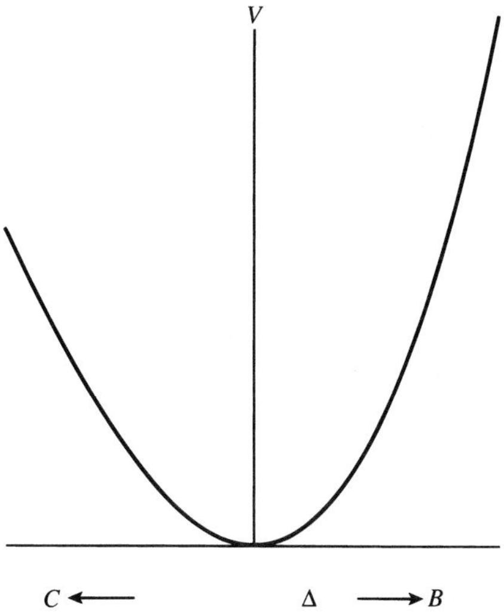

Fig. 20.2. Nonsymmetrical potential that results for displacement within a lattice that lacks a center of symmetry.

It is clear that this potential is not symmetric about $x = 0$. The restoring force acting on particle A as it moves a distance Δ towards particle B is clearly less than when it moves the same distance towards C. This occurs because the lattice structure lacks a center of symmetry. In this case $V(-x) \neq V(x)$.†

† A structure has a center of symmetry if a particle placed at position vector **r** sees exactly the same location and distribution of surrounding particles as a particle placed at −**r**. Crystalline structures can lack a center of symmetry but gases and liquids have a centrosymmetric structure.

We can always write the potential $V(x)$ as a power series

$$V(x) = \frac{1}{2}kx^2 + \frac{1}{3}bx^3 + \frac{1}{4}cx^4 + \frac{1}{5}dx^5 + \ldots, \tag{20.2}$$

where in general $b, c, \ll k$.

If $b, c \ldots = 0$ the potential is a *harmonic* potential, otherwise it is called an *anharmonic* potential. If the structure is centrosymmetric so that $V(-x) = V(x)$ then $b, d \ldots = 0$.

The restoring force that acts on the particle is

$$F(x) = -\frac{\partial V}{\partial x} = -kx - bx^2 \ldots \tag{20.3}$$

so, for example,

$$\begin{aligned} F(\Delta) &= -k\Delta - b\Delta^2, \\ F(-\Delta) &= k\Delta - b\Delta^2. \end{aligned} \tag{20.4}$$

These are unequal forces for equal magnitude displacements in opposite directions.

For very small displacements Δ the forces become equal and a particle mass m displaced from its equilibrium position will execute a simple harmonic oscillation with frequency $\omega_0 = \sqrt{k/m}$. When an electric field† $\frac{1}{2}[E(\omega)e^{i\omega t} + cc]$ is applied to the particles making up the medium it is mostly the outermost electrons on the atoms of the medium that move the most. This is particularly so at optical frequencies where the inertia of the much heavier nuclei and positive ions as a whole prevents them following the rapid oscillation of the applied field to any significant extent.

We can write down an equation of motion for one of the electrons in a very similar form to the one we used in the electron oscillator model for radiative transitions. (See Eq. (2.105)).

$$\frac{d^2x}{dt^2} + 2\Gamma\frac{dx}{dt} + \frac{k}{m}x + \frac{b}{m}x^2 = -\frac{e}{2m}\left[E(\omega)e^{i\omega t} + E^*(\omega)e^{-i\omega t}\right]. \tag{20.5}$$

For simplicity, we are ignoring the higher order nonlinear terms involving x^3, x^4, etc. We try as a solution to Eq. (20.5)

$$x(t) = \frac{x_1}{2}(e^{i\omega t} + e^{-i\omega t}) + \frac{x_2}{2}(e^{2i\omega t} + e^{-2i\omega t}). \tag{20.6}$$

Substituting, and comparing the coefficients of the terms containing $e^{i\omega t}$ gives

$$x_1\left[(\omega_0^2 - \omega^2) + 2i\omega\Gamma + \frac{b}{m}x_2\right] = -\frac{e}{m}E(\omega), \tag{20.7}$$

where, as before, $\omega_0 = \sqrt{k/m}$. If we assume that $bx_2/m \ll (\omega_0^2 - \omega^2) + 2i\omega\Gamma$, then we have

$$x_1 = -\frac{e}{m}E(\omega)\frac{1}{[(\omega_0^2 - \omega^2) + 2i\omega\Gamma]}. \tag{20.8}$$

For N identical electrons per unit volume the resultant polarization that oscillates

† This field is, strictly speaking, the *local* field within the medium that is seen by individual atoms or ions. We shall discuss at the conclusion of this chapter the distinction between this local field and the *externally applied field*.

at angular frequency ω is

$$P^{\omega}(t) = -\frac{Ne}{2}(x_1 e^{i\omega t} + cc), \tag{20.9}$$

which can be written in terms of the linear susceptibility as

$$P^{\omega}(t) = \frac{\epsilon_0 \chi^{(1)}(\omega)}{2}[E(\omega)e^{i\omega t} + cc]. \tag{20.10}$$

So

$$\chi^{(1)}(\omega) = \frac{Ne^2}{m\epsilon_0}\frac{1}{[(\omega_0^2 - \omega^2) + 2i\omega\Gamma]}. \tag{20.11}$$

Near resonance, $\omega \simeq \omega_0$, and Eq. (20.11) would reduce to Eq. (2.114). If we substitute from Eq. (20.6) in Eq. (20.5) and compare the coefficients of the terms containing $e^{2i\omega t}$ we get

$$x_2 = -\frac{be^2E^2(\omega)}{2m^3}\frac{1}{(\omega_0^2 - 4\omega^2 + 4i\omega\Gamma)[(\omega_0^2 - \omega^2) + 2i\omega\Gamma]^2}. \tag{20.12}$$

The resultant polarization at the second harmonic is

$$P^{2\omega}(t) = -\frac{Ne}{2}(x_2 e^{2i\omega t} + cc), \tag{20.13}$$

which can be written in terms of the Fourier amplitude $P(2\omega)$ as

$$P^{2\omega}(t) = \frac{1}{2}[P(2\omega)e^{2i\omega t} + cc]. \tag{20.14}$$

Clearly, $P(2\omega) = -Nex_2$. If we define a second harmonic generation (SHG) coefficient $d(2\omega)$ by the relationship

$$P(2\omega) = \epsilon_0 d(2\omega)E(\omega)E(\omega), \tag{20.15}$$

then

$$d(2\omega) = \frac{bNe^3}{2\epsilon_0 m^3}\frac{1}{(\omega_0^2 - 4\omega^2 + 4i\omega\Gamma)[(\omega_0^2 - \omega^2) + 2i\omega\Gamma]^2}, \tag{20.16}$$

which by the use of Eq. (20.11) for the linear susceptibility gives

$$d(2\omega) = \frac{b\epsilon_0^2}{2N^2e^3}[\chi^{(1)}(2\omega)][\chi^{(1)}(\omega)]^2. \tag{20.17}$$

$d(2\omega)$ has units of m V^{-1}. Eq. (20.17) shows that, within the framework of this simple model, the SHG coefficient can be determined from the *linear* susceptibilities at the fundamental and second harmonic frequencies. Estimates of SHG properties made in this way have met with some success[20.1],[20.2].

The SHG coefficient defined in Eq. (20.15) is a scalar quantity. In reality, in anisotropic media the polarization of the second harmonic frequency will depend on the direction of propagation of the fundamental wave, and its state of polarization, so a SHG tensor $\overline{\overline{r}}$ must be used. This will be discussed in more detail in the next section.

If the fundamental wave can be written as

$$E(t) = E(\omega)\cos(\omega t - \mathbf{k_1} \cdot \mathbf{r}) \tag{20.18}$$

then

$$P^{2\omega}(t) = \epsilon_0 d(2\omega) E^2(\omega) \cos(2\omega t - 2\mathbf{k}_1 \cdot \mathbf{r}), \tag{20.19}$$

so the $d(2\omega)$ coefficient is a *practical* coefficient that determines the amplitude of the *real* polarization at the second harmonic frequency.

We could continue this form of analysis to derive coefficients for third harmonic generation and higher terms. However, it is better in many ways to generalize our discussion to include all possible nonlinear iteractions that can occur when one or more waves at different frequencies interact within a medium. This is done by the introduction of a series of nonlinear susceptibilities.

20.3 Nonlinear Susceptibilities and Mixing Coefficients

If two waves, with angular frequencies ω_1, ω_2, respectively, interact within a nonlinear medium, then the second order polarization $\epsilon_0 \chi^{(2)} \mathbf{E}^2$ will clearly continue the frequencies $|\omega_1 \pm \omega_2|, 2\omega_1, 2\omega_2$ and dc. The magnitudes of the polarization produced at all these sum and difference frequencies are formally related to the electric field amplitudes involved by nonlinear susceptibility tensors. These tensors reveal the relationship between the cartesian components of the induced nonlinear polarization and the cartesian components of the interacting fields. Unfortunately, the way in which these nonlinear susceptibilities have been defined in the literature has not always been consistent, and has been the source of much confusion for many years. We will follow the approach of Byer[20.3] and Shen[20.4] and relate *real* polarizations to *real* fields. We take each electric field as

$$\mathbf{E}^{\omega_i}(t) = \mathbf{E}(\omega_i) \cos(\omega_i t - \mathbf{k}_i \cdot \mathbf{r}), \tag{20.20}$$

which can be written in terms of its Fourier amplitudes as

$$\mathbf{E}^{\omega_i}(t) = \frac{1}{2}[\mathbf{E}(\omega_i) e^{i(\omega_i t - \mathbf{k}_i \cdot \mathbf{r})} + cc]. \tag{20.21}$$

For a second order process involving the interaction of two fields at frequencies ω_1, ω_2 the nonlinear susceptibility tensor for sum-frequency generation is defined by the relation

$$\mathbf{P}^{\omega_3}(t) = \epsilon_0 \overline{\overline{\chi}}^{(2)}(-\omega_3, \omega_1, \omega_2) \mathbf{E}^{\omega_1}(t) \mathbf{E}^{\omega_2}(t). \tag{20.22}$$

The sum frequency is $\omega_3 = \omega_1 + \omega_2$, which is reflected in the formal notation $\overline{\overline{\chi}}^{(2)}(-\omega_3, \omega_1, \omega_2)$. In this specification of the nonlinear susceptibility, the algebraic sum of the frequencies contained within parentheses is zero. This reflects the conservation of energy that occurs during the nonlinear process. Specifically, in terms of the components of the vectors involved

$$P_i^{\omega_3}(t) = \epsilon_0 \sum_{jk} \chi_{ijk}^{(2)}(-\omega_3, \omega_1, \omega_2) E_j^{\omega_1}(t) E_k^{\omega_2}(t), \tag{20.23}$$

where i, j, k, can take the values x, y, or z.

The polarization at the frequency ω_3 is a *real* polarization, which can be written in terms of its Fourier amplitudes $P(\omega_3)$ as

$$\mathbf{P}^{\omega_3}(t) = \frac{1}{2}\left[\mathbf{P}(\omega_3) e^{i(\omega_3 t - \mathbf{k}_3 \cdot \mathbf{r})} + cc\right]. \tag{20.24}$$

Substitution of Eqs. (20.24) and (20.21) in Eq. (20.22) reveals the relationship between the Fourier amplitudes

$$\mathbf{P}(\omega_3) = \epsilon_0 \frac{\overset{=}{\chi}^{(2)}}{2}(-\omega_3, \omega_1, \omega_2)\mathbf{E}(\omega_1)\mathbf{E}(\omega_2)e^{-i[(\mathbf{k}_3-\mathbf{k}_1-\mathbf{k}_2)\cdot\mathbf{r}]}. \tag{20.25}$$

or in terms of the components of the vectors

$$P_i(\omega_3) = \frac{\epsilon_0}{2}\sum_{jk}\chi^{(2)}_{ijk}(-\omega_3, \omega_1, \omega_2)E_j(\omega_1)E_k(\omega_2)e^{-i[(\mathbf{k}_3-\mathbf{k}_1-\mathbf{k}_2)\cdot\mathbf{r}]}. \tag{20.26}$$

Eq. (20.26) can be written in the repeated subscript shorthand notation of tensor algebra as

$$P_i(\omega_3) = \frac{\epsilon_0}{2}\chi_{ijk}E_j(\omega_1)E_k(\omega_2). \tag{20.27}$$

The repeated subscripts j and k on the right hand side of this equation imply the summation over j and k. From now on, for simplicity, we will omit writing the exponential factor in Eq. (20.26) explicitly.† An alternative way of writing Eqs. (20.21) and (20.26) is also used, based on a nonlinear mixing tensor $\overset{=}{d}$. In terms of this tensor the relationship between the Fourier amplitudes is

$$\mathbf{P}(\omega_3) = \epsilon_0 \cdot \overset{=}{d}^{(2)}(-\omega_3, \omega_1, \omega_2)E(\omega_1)E(\omega_2). \tag{20.28}$$

The relationship between $\overset{=}{\chi}^{(2)}$ and $\overset{=}{d}^{(2)}$ is such that

$$\chi^{(2)}_{ijk}(-\omega_3, \omega_1, \omega_2) = 2d_{ijk}(-\omega_3, \omega_1, \omega_2). \tag{20.29}$$

It is instructive to examine in full the dependence of one of the polarization components in Eq. (20.26) on the various tensor components and frequencies involved. The summation covers all field components and frequencies that are *distinguishable* from each other. For example:

$$\begin{aligned} P_x(\omega_3) = \frac{\epsilon_0}{2}[&\chi_{xxx}E_x(\omega_1)E_x(\omega_2) + \chi_{xyy}E_y(\omega_1)E_y(\omega_2) + \chi_{xzz}E_z(\omega_1)E_z(\omega_2) \\ &+ \chi_{xyz}E_y(\omega_1)E_z(\omega_2) + \chi_{xzy}E_z(\omega_1)E_y(\omega_2) \\ &+ \chi_{xzx}E_z(\omega_1)E_x(\omega_2) + \chi_{xxz}E_x(\omega_1)E_z(\omega_2) \\ &+ \chi_{xxy}E_x(\omega_1)E_y(\omega_2) + \chi_{xyx}E_y(\omega_1)E_x(\omega_2)], \end{aligned} \tag{20.30}$$

where we have omitted the full frequency description of the χs. Note that the two terms, $\chi_{xyz}(-\omega_3, \omega_1, \omega_2)E_y(\omega_1)E_z(\omega_2)$ and $\chi_{xzy}(-\omega_3, \omega_1, \omega_2)E_z(\omega_1)E_y(\omega_2)$ are physically distinguishable. A unique frequency is associated with each field component.

The form of the tensor $\overset{=}{\chi}^{(2)}$ has to be consistent with the symmetry of the crystal. In particular, $\overset{=}{\chi}^{(2)}$ vanishes if the medium possesses inversion symmetry. To show why this must be so, consider the polarization component given by Eq. (20.26), which in repeated subscript notation is

$$P_i(\omega_3) = \frac{\epsilon_0}{2}\chi_{ijk}E_j(\omega_1)E_k(\omega_2). \tag{20.31}$$

If we reverse the direction of both electric fields then the polarization must also

† We shall see later that a most desirable situation for nonlinear mixing results when $\mathbf{k}_3 - \mathbf{k}_1 - \mathbf{k}_2 = 0$, the so-called *phase-matched* condition.

reverse direction because the fields see an identical arrangement of atoms in the medium when they are reversed. Consequently,

$$-P_i(\omega_3) = \frac{\epsilon_0}{2}\chi_{ijk}E_j(\omega_1)E_k(\omega_2). \tag{20.32}$$

Eqs. (20.31) and (20.32) are inconsistent unless $\chi_{ijk} = 0$.

For the 21 of the 32 crystal classes that lack inversion symmetry the number of nonzero, independent elements of χ_{ijk} depends on the symmetry. Only for the triclinic class 1 are all 27 elements of $\overset{=}{\chi}^{(2)}$ nonzero and independent.

Eq. (20.23) can be generalized to the mixing of n different waves so that

$$P_\ell^{\omega=\omega_1+\omega_2+\ldots\omega_n}(t) = \epsilon_0 \sum_{\ell_1,\ell_2\ldots\ell_n} \chi^{(n)}_{\ell\ell_1\ell_2\ldots\ell_n}(-\omega,\omega_1,\omega_2\ldots\omega_n) \times E_{\ell_1}^{\omega_1}(t)E_{\ell_2}^{\omega_2}(t)\ldots E_{\ell_n}^{\omega_n}(t). \tag{20.33}$$

In terms of the Fourier amplitudes involved

$$P_\ell(\omega) = \frac{\epsilon_0}{2^{(n-1)}}\sum_{\ell_1\ell_2\ldots\ell_n} \chi_{\ell\ell_1\ell_2\ldots\ell_n}^{(-\omega,\omega_1,\omega_2\ldots\omega_n)}E_{\ell_1}(\omega_1)E_{\ell_2}(\omega_2)\ldots E_{\ell_n}(\omega_n) \tag{20.34}$$

and in terms of a nonlinear mixing tensor

$$\mathbf{P}(\omega) = \epsilon_0 \overset{=}{d}^{(n)}(-\omega,\omega_1,\omega_2\ldots\omega_n)\mathbf{E}(\omega_1)\mathbf{E}(\omega_n). \tag{20.35}$$

Comparison of Eqs. (20.34) and (20.35) shows that

$$\overset{=}{d}^{(n)} = \overset{=}{\chi}^{(n)}/2^{(n-1)}. \tag{20.36}$$

The above results must be modified if one or more of the frequencies $\omega, \omega_1, \omega_2 \ldots \omega_n$ is dc.

For a dc field $E_i(t) = E(0)$. There is no factor of $\frac{1}{2}$ in this equation as there is in Eq. (20.21). Consequently if p of the interacting fields are at dc Eq. (20.36) becomes

$$\overset{=}{d}^{(n)} = \overset{=}{\chi}^{(n)}/2^{(n-1-p)}. \tag{20.37}$$

20.4 Second Harmonic Generation

An interesting, and potentially confusing, situation results if two or more of the frequencies in Eq. (20.34) become equal to each other. To show here that the definition of nonlinear susceptibilities remains consistent in this case let us consider the specific situation of second harmonic generation (SHG). If the frequencies ω_1 and ω_2 approach each other in Eq. (20.30), then when $\omega_1 = \omega_2$ we get

$$\begin{aligned} P_x(2\omega) = \frac{\epsilon_0}{2}\Big[&\chi_{xxx}E_x(\omega)E_x(\omega) + \chi_{xyy}E_y(\omega)E_y(\omega) + \chi_{xzz}E_z(\omega)E_z(\omega) \\ &+ \chi_{xyz}E_y(\omega)E_z(\omega) + \chi_{zy}E_z(\omega)E_y(\omega) \\ &+ \chi_{xzx}E_z(\omega)E_x(\omega) + \chi_{xxz}E_x(\omega)E_z(\omega) \\ &+ \chi_{xxy}E_x(\omega)E_y(\omega) + \chi_{xyx}E_y(\omega)E_x(\omega)\Big]. \end{aligned} \tag{20.38}$$

Each χ_{ijk} written above is shorthand for $\chi_{ijk}(-2\omega, \omega, \omega)$. Note that each of the nine terms is still physically distinguishable from the others. However, a fundamental question must be asked. Is the second harmonic being generated by the mixing together of two *distinct* waves, each at the same frequency ω, or are the fields of a single wave interacting nonlinearly with themselves to generate the second harmonic? The actual values of the χ_{ijk} coefficients must differ in these two circumstances, otherwise we would have an apparent discontinuity in the amount of second harmonic generated. This distinction can be clearly illustrated for the situation where the input electric fields only have an x directed component. For two distinct waves, each of intensity I, the electric field of each wave is $E_x(\omega) = \sqrt{2ZI}$ where Z is the impedance of the medium at frequency ω. Therefore, from Eq. (20.38)

$$P_x(2\omega) = \frac{\epsilon_0}{2}\chi_{xxx}^{mix}(-2\omega, \omega, \omega)(2ZI), \tag{20.39}$$

where the superscript *mix* stresses the independence of the two waves.

For SHG from a single input wave of intensity $2I$, whose field amplitude is $E_x(\omega) = \sqrt{4ZI}$, Eq. (20.38) gives

$$P_x(2\omega) = \frac{\epsilon_0}{2}\chi_{xxx}^{SHG}(-2\omega, \omega, \omega)(4ZI). \tag{20.40}$$

Eqs. (20.39) and (20.40) must correspond to the same induced polarization amplitude, therefore

$$\chi_{xxx}^{SHG}(-2\omega, \omega, \omega) = \tfrac{1}{2}\chi_{xxx}^{mix}(-2\omega, \omega, \omega). \tag{20.41}$$

SHG is such an important and useful nonlinear process that the above fine distinction is avoided through the use of the SHG tensor $\overset{=2\omega}{d}$ defined by

$$P_i(2\omega) = \epsilon_0 \sum_{jk} d_{ijk} E_j(\omega) E_k(\omega) e^{-i[(\mathbf{k}_{2\omega} - 2\mathbf{k}_\omega)\cdot\mathbf{r}]}. \tag{20.42}$$

In the use of the d_{ijk} coefficients no distinction is made between terms like $d_{xyz}E_y(\omega)E_z(\omega)$ and $d_{xzy}E_z(\omega)E_y(\omega)$, so in matrix notation

$$\begin{pmatrix} P_x(2\omega) \\ P_y(2\omega) \\ P_z(2\omega) \end{pmatrix} = \epsilon_0 \begin{pmatrix} d_{xxx} & d_{xyy} & d_{xzz} & d_{xyz} & d_{xzx} & d_{xxy} \\ d_{yxx} & d_{yyy} & d_{yzz} & d_{yyz} & d_{yzx} & d_{yxy} \\ d_{zxx} & d_{zyy} & d_{zzz} & d_{zyz} & d_{zzy} & d_{zxy} \end{pmatrix} \times \begin{pmatrix} E_x^2(\omega) \\ E_y^2(\omega) \\ E_z^2(\omega) \\ 2E_y(\omega)E_z(\omega) \\ 2E_z(\omega)E_x(\omega) \\ 2E_x(\omega)E_y(\omega) \end{pmatrix}. \tag{20.43}$$

Eq. (20.43) can be simplified further through the use of contracted notation, in which

$$\begin{aligned} x &= 1; y = 2; z = 3 \\ xx &= 1; yy = 2; zz = 3 \\ yz &= 4; zx = 5; xy = 6 \end{aligned} \tag{20.44}$$

and $\overset{=}{d}$ can be written as

$$\overset{=}{d} = \begin{pmatrix} d_{11} & d_{12} & d_{13} & d_{14} & d_{15} & d_{16} \\ d_{21} & d_{22} & d_{23} & d_{24} & d_{25} & d_{26} \\ d_{31} & d_{32} & d_{33} & d_{34} & d_{35} & d_{36} \end{pmatrix}. \tag{20.45}$$

20.4.1 Symmetries and Kleinman's Conjecture

The formally defined nonlinear susceptibility coefficients, for example, $\chi_{ijk}(-\omega_3, \omega_1, \omega_2)$, are related to each other in various ways. Common sense dictates that any permutation of the frequencies $\omega_1, \omega_2, \omega_3$ cannot affect the value of the coefficient, provided the axial indices i, j, k are also permuted in the same way. Therefore

$$\chi_{ijk}(-\omega_3, \omega_1, \omega_2) = \chi_{ikj}(-\omega_3, \omega_2, \omega_1) = \chi_{kij}(\omega_2, -\omega_3, \omega_1) \text{ etc.} \tag{20.46}$$

Furthermore, if the nonlinear polarization is to be a real quantity, then $[P_i(\omega_3)]^* = P_i(-\omega_3)$, which provides the additional relationship

$$[\chi_{ijk}(-\omega_3, \omega, \omega_2)]^* = \chi_{ijk}(\omega_3, -\omega_1, -\omega_2). \tag{20.47}$$

If the nonlinear material is transparent at all three frequencies ω_1, ω_2, and ω_3, then $\overline{\overline{\chi}}$ will be real and if $\overline{\overline{\chi}}$ is independent of frequency all permutations of axial index and associated frequency become irrelevant:

$$\chi_{ijk}(-\omega_3, \omega_1, \omega_2) = \chi_{ijk} = \chi_{jik} = \chi_{jki} \text{ etc.} \tag{20.48}$$

This is Kleinman's conjecture[20.5]. Because all materials are to some extent dispersive it cannot be absolutely true, but it has been verified experimentally to be a very good approximation in most practical nonlinear media. If Kleinman's conjecture is valid we have relationships such as $d_{14} = d_{25} = d_{36}; d_{15} = d_{31}; d_{23} = d_{34}$; etc. Independent of Kleinman's conjecture the elements of the SHG tensor are also in general not all independent, but are constrained by the symmetry of the crystal. For triclinic crystals of class 1 all elements of $\overline{\overline{d}}$ are nonzero and independent, for crystal classes with inversion symmetry all the elements of $\overline{\overline{d}}$ are zero. In other cases many elements of $\overline{\overline{d}}$ will either be zero or related to the elements of the matrix. In most nonlinear crystals that are widely used, such as ADP, KDP, $LiNbO_3$, there are only one or two important elements of the $\overline{\overline{d}}$ tensor, as can be seen from the values tabulated for these and other crystals in Table (20.1).

20.5 The Linear Electro-Optic Effect

If one of the input waves, say ω_2, is at dc then Eq. (20.36) becomes

$$\mathbf{P}(\omega_1) = \epsilon_0 \overline{\overline{d}}^{EO}(-\omega_1, \omega_1, 0)\mathbf{E}(\omega_1)\mathbf{E}(0), \tag{20.49}$$

where the tensor $\overline{\overline{d}}^{EO}$ describes the linear electro-optic effect, which we have already discussed, but in a phenomenologically different way. Symmetry requires that $d_{ijk}(-\omega, \omega, 0) = d_{kij}(0, -\omega, \omega)$, where $d_{kij}(0, -\omega, \omega)$ is a component of the optical rectification (OR) tensor $\overline{\overline{d}}^{OR}$, so the symmetries of the linear electro-optic mixing tensor and the tensor for optical rectification should be the same. Further, since SHG and OR occur simultaneously when a wave interacts nonlinearly with a medium, we expect the symmetry but not the values of the individual elements $\overline{\overline{d}}^{2\omega}$, $\overline{\overline{d}}^{EO}$ and $\overline{\overline{d}}^{OR}$ to be the same. The specific form of $\overline{\overline{d}}^{2\omega}$ for each of the 32 crystal classes has been tabulated by Byer[20.3], for example, for the crystal for class 3m

$$\overline{\overline{r}} = \begin{pmatrix} 0 & 0 & 0 & 0 & d_{15} & -d_{22} \\ -d_{22} & d_{22} & 0 & d_{15} & 0 & 0 \\ d_{31} & d_{31} & d_{33} & 0 & 0 & 0 \end{pmatrix}. \tag{20.50}$$

However, we do not need to tabulate all the forms of $\overline{\overline{d}}$ for the different crystal classes here. The symmetry of $\overline{\overline{d}}$ is the same as the symmetry of the piezoelectric tensor discussed in Chapter 19. There it is pointed out that the symmetry of $\overline{\overline{d}}$ is the same as the symmetry of the transpose of the linear electrooptic tensor $\overline{\overline{d}}$. Therefore, the symmetry of $\overline{\overline{d}}$ can be determined from tabulated symmetries given in Chapter 19. This is, perhaps, not too surprising in view of the fact that $\overline{\overline{d}}^{EO}$ and $\overline{\overline{d}}$ describe the same phenomenon.

20.6 Parametric and Other Nonlinear Processes

Parametric amplification is a process in which a weak electric field at frequency ω_s, called the *signal*, interacts nonlinearly with a strong electric field at frequency ω_p, called the *pump*, in such a way that the amplitude of the signal at frequency ω_s is increased. To conserve energy in this process, as photons are converted from frequency ω_p to ω_s, additional photons at frequency ω_i, called the *idler* frequency, must be generated simultaneously in such a way that

$$\omega_s + \omega_i = \omega_p. \tag{20.51}$$

The process is described by the nonlinear susceptibility tensor $\overline{\overline{\chi}}\,(\omega_p - \omega_s, -\omega_p, \omega_s)$. Parametric amplification has been studied in conventional nonlinear electric circuits for many years, for example, in resonant circuits containing voltage-dependent capacitors. The interested reader can find brief discussions of such circuits in [20.6] and [20.7].

Higher order nonlinear processes described by the tensor $\overline{\overline{\chi}}^{(3)}$ are of some importance because they are permitted to occur in media that possess a center of inversion symmetry, such as gases and liquids. We will not discuss any of these phenomena in detail here but will list some of the important ones. Third harmonic generation (THG) described by the tensor $\overline{\overline{\chi}}\,(-3\omega, \omega, \omega, \omega)$ has been studied quite widely as a means for generating shorter wavelengths from powerful laser sources[20.11]. The quadratic electro-optic effect is described by the tensor $\chi(-\omega, \omega, 0, 0)$. Coherent Anti-Stokes Raman Scattering (CARS) described by the tensor $\chi(-\omega_2, \omega_1, \omega_1, -\omega_2)$ is a useful spectroscopic technique for studying species in difficult situations – such as during combustion. If the nonlinear medium is dispersive, then the nonlinear tensors become complex and the real and imaginary parts of the appropriate tensor may describe different processes, for example, $\mathscr{R}\,\overline{\overline{\chi}}\,(-\omega, \omega, \omega, -\omega)$ describes intensity-induced birefringence, $\mathscr{I}(-\omega, \omega, \omega, -\omega)$ describes two-photon absorption.

There are also important nonlinear interactions that depend on *magnetic* field components of the nonlinearly interacting waves. For example the polarization effect described by

$$\mathbf{P}(\omega) = -i\epsilon_0\, \overline{\overline{\chi}}^{B}\,(-\omega, \omega, 0)\mathbf{E}(\omega)\mathbf{B}(0), \tag{20.52}$$

which depends linearly on a dc magnetic flux density $\mathbf{B}(0)$, describes the Faraday effect – the rotation of the plane of polarization of linearly polarized light as it propagates through a Faraday-active medium along an axial directed magnetic field.

The tensor $\overline{\overline{\chi}}^{BB}\,(-\omega, \omega, 0, 0)$ describes birefringence induced by a dc magnetic

Table 20.1. *Important elements of the* $\overline{\overline{d}}^{\dagger}$ *second harmonic generation tensor for various nonlinear materials.*

Crystal	Symmetry group	$\lambda(\mu m)$	$d_{ij}(10^{-12}\ m\ V^{-1})$
$Ba_2NaNb_5O_{15}$(banana)	mm2	1.06	d_{15}=7.4±2.3,d_{24}=6.5±0. 7
	3m	1.15	d_{33}=9±-0.7
Ag_3AsS_3(proustite)	3m	1.5	d_{15}=15.3,d_{22}=25.5
$LiNbO_3$	3m	1.15	d_{31}=-5.4±0.5,d_{22}=2.6± 1
			d_{33}=-49±9
HgS	32	10.6	d_{11}=77±26
SiO_2	32	1.06	d_{11}=0.32±0.02,d_{14}=0.003
Te	32	10.6	d_{11}=922±293
$BaTiO_3$	4mm	1.06	d_{15}=17.7±1.5,d_{31}=-18.8±1.5
			d_{33}=-7.1±0.5
ADP($NH_4H_2PO_4$	$\bar{4}$2m	1.06	d_{36}=0.57±0.07,d_{14}=0.6 3±0.03
AD*P($NH_4D_2PO_4$	$\bar{4}$2m	0.694	d_{36}=0.56
KDP(KH_2PO_4)	$\bar{4}$2m	1.06	d_{36}=0.51±0.02
KD*P(KD_2PO_4)	$\bar{4}$2m	1.06	d_{36}=0.46±0.02
$AgGaS_2$	$\bar{4}$2m	10.06	d_{36}=18±3
$CdGeAs_2$	$\bar{4}$2m	10.6	d_{36}=877±175
$LiIO_3$	6	1.06	d_{31}=6.03±0.51,d_{33}=6.28±0.51
CdTe	$\bar{4}$3m	10.6	d_{14}=168±67
GaAs	$\bar{4}$3m	1.06	d_{36}=137±27
		10.6	d_{36}=134±45
GaP	$\bar{4}$3m	10.6	d_{14}=100±18
InAs	$\bar{4}$3m	10.6	d_{14}=419±126
ZnS	$\bar{4}$3m	10.6	d_{14}=30.6±8.4
CdS	6mm	1.06	d_{15}=17.7±1.5, d_{31}=-16.2±1
			d_{33}=32±2
		10.6	d_{15}=29±7,d_{31}=26.4±6.3
		10.6	d_{15}=29±7,d_{31}=26.4±6.3
			d_{33}=44±13

† The d_{ij} coefficients are given in contracted notation, if Kleinman's conjecture were correct $d_{14} \equiv d_{xyz} \equiv d_{zxy} \equiv d_{36}$, for example. The values given represent a compilation of the apparently most reliable values listed by Zernike and Midwinter[20.6] and Kurtz[20.7]. There is some disagreement in the literature concerning the value of many of these coefficients[20.8], the errors listed come from the same literature, but are probably optimistic values.

flux density – the Cotton–Mouton effect[20.12]. For a fuller discussion of these various phenomena the interested reader should consult Tang and Rabin[20.13].

20.7 Macroscopic and Microscopic Susceptibilities

In our general discussion of nonlinear susceptibilities we have been dealing with applied fields **E** and resultant *macroscopic* polarizations **P**, which are related schematically by Eq. (20.1). However, when we attempt to relate these macroscopically measured quantities to an atomic model of the nonlinear process, such as in Eq. (20.5), we must be careful to make the distinction between the applied

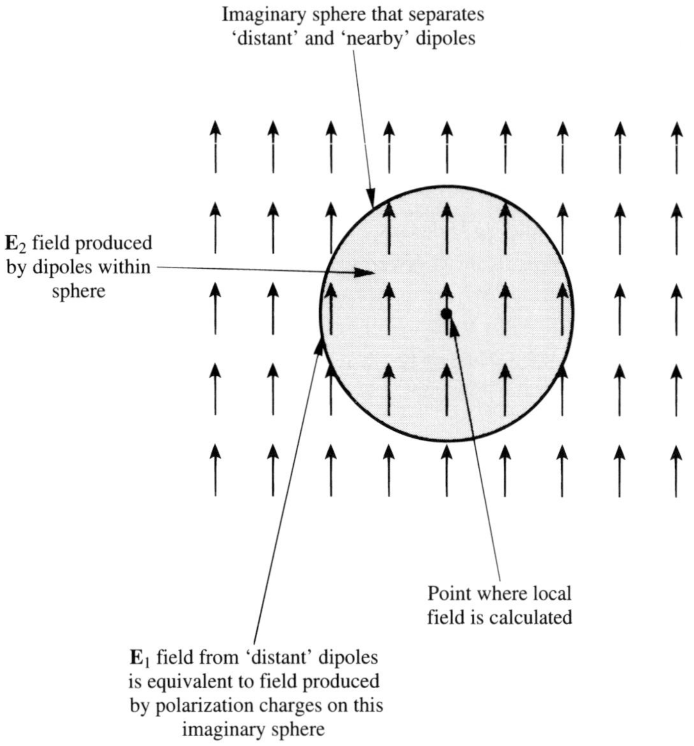

Fig. 20.3. Imaginary spherical region around a particle in a polarized medium that delineates nearby and distant dipoles.

macroscopic field, and the field actually experienced by the individual particles of the medium. This *local* field is different from the macroscopic field because of the polarization of the medium. We follow the arguments presented by Kittel[20.14] in relating the macroscopic and local fields.

The macroscopic field $\mathbf{E}$ generates an induced linear polarization in the medium according to

$$\mathbf{P} = \epsilon_0 \chi^{(1)} \mathbf{E}. \tag{20.53}$$

We will neglect the tensor character of $\chi^{(1)}$ to make our analysis a little more straightforward. Each particle in the medium sees a local electric field $\mathbf{E}_{loc}$ where

$$\mathbf{E}_{loc} = \mathbf{E} + \mathbf{E}_1 + \mathbf{E}_2. \tag{20.54}$$

The induced dipoles, which give rise to the macroscopic polarization $\mathbf{P}$, macroscopically contribute to the fields $\mathbf{E}_1$ and $\mathbf{E}_2$. The field $\mathbf{E}_1$ is produced, at the point of interest, by all the dipoles that are so far away as to form a continuous polarization. $\mathbf{E}_2$ is the field of dipoles sufficiently close to the observation point that their discrete nature is apparent. The field $\mathbf{E}_1 + \mathbf{E}_2$ is the total field acting at the observation point that results from all the dipoles in the medium.†

$$\mathbf{E}_1 + \mathbf{E}_2 = \frac{1}{4\pi\epsilon_0} \sum_i \frac{3(\boldsymbol{\mu}_i \cdot \mathbf{r}_i)}{r_i^5} \mathbf{r}_i - \frac{\boldsymbol{\mu}_i}{r_i^3} \tag{20.55}$$

† We are not including in our analysis the so-called depolarization field, which results from dipoles on the outside surface of a finite object of specific shape[20.14]. This field is zero for long thin crystals.

when $\boldsymbol{\mu}_i$ is the dipole moment (vector) of the ith particle of the medium. We can separate the dipoles, near the observation point from the distant dipoles by an imaginary sphere centered at the observation point as shown in Fig. (20.3). The field $\mathbf{E}_1$, which comes from the distant dipole can be shown to be[20.14]

$$\mathbf{E}_1 = \frac{\mathbf{P}}{3\epsilon_0}, \tag{20.56}$$

where $\mathbf{P}$ is the total macroscopic polarization defined in Eq. (20.53), which is related to the $\mathbf{D}$ vector that appears in Maxwell's equations by

$$\mathbf{D} = \epsilon_0\mathbf{E} + \mathbf{P}. \tag{20.57}$$

In an isotropic nonmetallic medium it can be shown that $\mathbf{E}_2 = 0$. Therefore in an isotropic medium

$$\mathbf{E}_{loc} = \mathbf{E} + \frac{\mathbf{P}}{3\epsilon_0}. \tag{20.58}$$

In nonisotropic media $\mathbf{E}_{loc}$ is determined by a Lorentz correction factor $\mathscr{L}$ such that

$$\mathbf{E}_{loc} = \mathbf{E} + \frac{\mathscr{L}}{\epsilon_0}P \tag{20.59}$$

On a microscopic scale, each individual dipole moment $\boldsymbol{\mu}_i$ is related to the local field by

$$\boldsymbol{\mu}_i = \overline{\alpha}_i\mathbf{E}_{\text{loc}} \tag{20.60}$$

where $\overline{\alpha}_i$ is the spatially averaged polarizability of the ith particle. Clearly,

$$\mathbf{P} = \sum_i N_i\overline{\alpha}_i\mathbf{E}_{\text{loc}} \tag{20.61}$$

where N_i is the number of particles of type i per unit volume of the medium. Combining Eqs. (20.58) and (20.61) gives

$$\mathbf{P} = \sum_i N_i\overline{\alpha}_i\left(\mathbf{E} + \frac{\mathbf{P}}{3\epsilon_0}\right), \tag{20.62}$$

and by rearranging

$$\mathbf{P} = \frac{\sum_i N_i\overline{\alpha}_i\mathbf{E}}{1 - \sum_i N_i\overline{\alpha}_i/3\epsilon_0}. \tag{20.63}$$

Also,

$$\mathbf{D} = \left(\epsilon_0 + \frac{\sum_i N_i\overline{\alpha}_i}{1 - \sum_i N_i\overline{\alpha}_i/3\epsilon_0}\right)\mathbf{E} = \epsilon_0\epsilon_r\mathbf{E}, \tag{20.64}$$

where ϵ_r is the dielectric constant. Remembering that $n = \sqrt{\epsilon}$, Eq. (20.64) gives

$$\frac{\epsilon_r - 1}{\epsilon_r + 2} = \frac{n^2 - 1}{n^2 + 2} = \frac{1}{3\epsilon_0}\sum_i N_i\overline{\alpha}_i, \tag{20.65}$$

which is called the Clausius–Mossotti equation. It relates a macroscopic quantity

$\epsilon_r = 1 + \chi^{(1)}$ to a microscopic one α. For linear phenomena, $\mathbf{P}$ is related to $\mathbf{E}$ by Eq. (20.53). Combining Eqs. (20.58) and (20.53) gives

$$\mathbf{E}_{loc} = \frac{\epsilon_r + 2}{3}\mathbf{E} = f\mathbf{E}, \tag{20.66}$$

where $\epsilon_r = 1 + \chi$ is the dielectric constant of the medium. It is easy to show that

$$f = 1 + \chi^{(1)}\mathscr{L}. \tag{20.67}$$

The value of f predicted by Eq. (20.67) is satisfactory for ionic crystals of cubic symmetry, but in crystals of other symmetry its exact calculation can be complicated. We will treat it as an empirical factor that relates the local field to the externally applied field.

Eq. (20.61) can be written in the form

$$\mathbf{P} = \epsilon_0\chi_m^{(1)}\mathbf{E}_{loc}, \tag{20.68}$$

where the linear microscopic susceptibility $\chi_m^{(1)}$ is clearly

$$\chi_m^{(1)} = \frac{1}{\epsilon_0}\sum_i N_i\bar{\alpha}_i. \tag{20.69}$$

Combining Eqs. (20.68) and (20.66) gives

$$\mathbf{P} = \epsilon_0 f\chi_m^{(1)}\mathbf{E}_{loc}. \tag{20.70}$$

Clearly,

$$\chi^{(1)} = f\chi_m^{(1)}. \tag{20.71}$$

When nonlinear phenomena occur in a medium the connection between microscopic and macroscopic susceptibilities becomes more subtle. Ducuing[20.15] has presented a simple analysis of this situation. Consider the case of second harmonic generation where a macroscopic nonlinear polarization $P_{NLS}(2\omega)$ is generated according to

$$P_{NLS}(2\omega) = \frac{\epsilon_0}{2}\chi^{(2)}E^2(\omega), \tag{20.72}$$

where $\chi^{(2)}(2\omega)$ is a shorthand notation for $\chi^{(2)}(-2\omega, \omega, \omega)$. For simplicity we will ignore the vector or tensor qualities of the variables in Eq. (20.72). The subscript NLS, which stands for 'nonlinear source,' is included to stress that the origin of the polarization is distinctly nonlinear, and that this polarization can serve as a source of waves in Maxwell's equations. We shall examine this point in detail in the next chapter. The subtlety that arises is that the electric field $E(2\omega)$ generated as a result of $P_{NLS}(2\omega)$ makes an additional contribution to the *total* polarization at frequency $2\omega, P(2\omega)$, through the linear susceptibility at frequency 2ω. On a microscopic scale each particle of the medium experiences a local second harmonic field $E_{loc}(2\omega)$ and the microscopic nonlinear polarization is

$$P_m(2\omega) = \frac{\epsilon_0}{2}\chi_m^{(2)}(2\omega)[E_{loc}(\omega)]^2 = \frac{\epsilon_0}{2}\chi_m^{(2)}(2\omega)f^2(\omega)E^2(\omega). \tag{20.73}$$

Eq. (20.73) effectively defines the microscopic nonlinear susceptibility $\chi_m^{(2)}(2\omega)$.

The total polarization at frequency 2ω is

$$P(2\omega) = \epsilon_0\chi_m^{(1)}(2\omega)E_{loc}(2\omega) + \frac{\epsilon_0}{2}\chi_m^{(2)}(2\omega)f^2(\omega)E^2(\omega). \tag{20.74}$$

The local field at frequency 2ω is

$$E_{loc}(2\omega) = E(2\omega) + \frac{\mathscr{L}(2\omega)}{\epsilon_0} P(2\omega). \tag{20.75}$$

Combining Eqs. (20.74) and (20.75) gives

$$P(2\omega) = \frac{\epsilon_0 \chi_m^{(1)}(2\omega)E(2\omega) + (\epsilon_0/2)\chi_m^{(2)}(2\omega)f^2(\omega)E^2(\omega)}{1 - \chi_m^{(1)}(2\omega)\mathscr{L}(2\omega)} \tag{20.76}$$

and from Eqs. (20.67) and (20.71)

$$\mathscr{L}(2\omega) = \frac{f(2\omega) - 1}{\chi^{(1)}(2\omega)} = \frac{f(2\omega) - 1}{f(2\omega)\chi_m^{(1)}(2\omega)}. \tag{20.77}$$

Combining Eqs. (20.76) and (20.77) gives

$$P(2\omega) = \epsilon_0 \chi_m^{(1)}(2\omega) f(2\omega) E(2\omega) + \frac{\epsilon_0}{2} f(2\omega) f^2(\omega) \chi_m^{(2)}(2\omega) E^2(\omega). \tag{20.78}$$

On close examination, the first term on the right hand side of Eq. (20.78) should be obvious: this linear term would be expected to occur if any second harmonic field is present. The second term in Eq. (20.78) clearly must correspond to the nonlinear source polarization defined in Eq. (20.72) as it is the only term present that depends on $E^2(\omega)$. Therefore, comparing Eqs. (20.72) and (20.78)

$$\chi^{(2)}(2\omega) = f(2\omega) f^2(\omega) \chi_m^{(2)}(2\omega). \tag{20.79}$$

This concludes our rather formal discussion of nonlinear processes and how they are quantified. In the next chapter we will pursue this discussion further but on a more practical basis.

20.8 Problems

(20.1) By substituting from Eq. (20.6) into Eq. (20.5) and examining the dc terms, derive an expression for the coefficient for optical rectification $d(0)$ defined by the relationship

$$P(\omega = 0) = \epsilon_0 d(0) E^*(\omega) E(\omega).$$

References

[20.1] R.C. Miller, 'Optical second harmonic generation in piezoelectric crystals,' *Appl. Phys. Lett.*, **5**, 17–19, 1964.

[20.2] C.G.B. Garrett and F.N.H. Robinson, 'Miller's phenomenological rule for computing nonlinear susceptibilities,' *IEEE J. Quant. Electron.*, **QE-2**, 328–329, 1966.

[20.3] R.L. Byer, 'Parametric Oscillation and nonlinear materials,' in *Nonlinear Optics*, P.G. Harper and B.S. Wherrett, Eds., Academic Press, London, 1977.

[20.4] Y.R. Shen, *The Principles of Nonlinear Optics* Wiley, New York, 1984.

[20.5] D.A. Kleinman, 'Nonlinear Dielectric Polorization in Optical Media,' *Phys. Rev.*, **126**, 1977–1979, 1962.

[20.6] F. Zernike and J.E. Midwinter, *Applied Nonlinear Optics*, Wiley, New York, 1973.

[20.7] S.K. Kurtz, 'Measurement of nonlinear optical susceptibilities,' in *Quantum Electronics*, H. Rabin and C.L. Tang, Eds., Vol. 1A, Academic Press, New York, 1975.

[20.8] S. Singh, 'Nonlinear Optical Materials,' in *Handbook of Lasers*, R.J. Pressley, Ed., CRC Press, Cleveland, 1971.

[20.9] A. Yariv, *Optical Electronics*, 3rd Edition, Holt, Rinehold & Winston, New York, 1985.

[20.10] S.Y. Liao, *Microwave Devices and Circuits*, 2nd Edition, Prentice-Hall, Englewood Cliffs, 1985.

[20.11] C.R. Vidal, 'Four-wave frequency mixing in gases,' in *Tunable Lasers*, L.R. Mollenauer and J.C. White, Eds., Topics in Applied Physics, Vol. 59, Springer-Verlag, Berlin, 1987.

[20.12] R.W. Ditchburn, *Light*, 3rd Edition, Academic Press, London, 1976.

[20.13] *Quantum Electronics*, H. Rabin and C.L. Tang, Eds., Vols. 1A and 1B, Academic, New York, 1975.

[20.14] C. Kittel, *Introduction to Solid State Physics*, 6th Edition, Wiley, New York, 1986.

[20.15] J. Ducuing, 'Microscopic and Macroscopic Nonlinear Optics,' in *Nonlinear Optics*, P.G. Harper and B.S. Wherrett, Eds., Academic Press, London, 1977.

21

Wave Propagation in Nonlinear Media

21.1 Introduction

In this chapter we will follow up our rather formal discussion of nonlinear phenomena begun in the last chapter. We will describe through the use of Maxwell's equations how nonlinear interactions actually lead to the production of waves at new frequencies in a medium. Specifically, we will examine the practically useful phenomena of second harmonic generation and parametric amplification and oscillation. For each phenomenon, specific examples of how these processes are optimized in practice will be given.

21.2 Electromagnetic Waves and Nonlinear Polarization

To show how a nonlinearly generated polarization leads to the production of waves at new frequencies the following equations of electromagnetic theory will be used:

$$\text{Ampère's law: curl } \mathbf{H} = \mathbf{j} + \frac{\partial \mathbf{D}}{\partial t}, \tag{21.1}$$

where $\mathbf{j}$(A m^{-2}) is current density.

$$\text{Faraday's law: curl } \mathbf{E} = -\mu \frac{\partial \mathbf{H}}{\partial t}, \tag{21.2}$$

$$\text{Ohm's law: } \mathbf{j} = \sigma \mathbf{E}, \tag{21.3}$$

where σ (S m^{-1}) is the conductivity of the medium, and

$$\mathbf{D} = \epsilon_0 \mathbf{E} + \mathbf{P}. \tag{21.4}$$

When a second order nonlinear process occurs in the medium the total polarization can be written as

$$\mathbf{P} = \epsilon_0 \overset{=(1)}{\chi} \mathbf{E} + \mathbf{P}_{NLS}, \tag{21.5}$$

where the nonlinear source (NLS) polarization has cartesian components

$$(P_{NLS})_i = \epsilon_0 \sum_{jk} \chi_{ijk} E_j E_k. \tag{21.6}$$

From Eqs. (21.1) and (21.4) we have

$$\begin{aligned} \text{curl } \mathbf{H} &= \sigma \mathbf{E} + \epsilon_0 \frac{\partial \mathbf{E}}{\partial t} + \frac{\partial \mathbf{P}}{\partial t} \\ &= \sigma \mathbf{E} + \epsilon_0 \overset{=}{\epsilon}_r \frac{\partial \mathbf{E}}{\partial t} + \frac{\partial}{\partial t} \mathbf{P}_{NLS}. \end{aligned} \tag{21.7}$$

In Eq. (21.7) we have included all the *linear* polarization effects in the term containing the dielectric tensor $\overline{\overline{\epsilon}}_r$. Use of the vector identity curl curl $\mathbf{E}$ = grad div $\mathbf{E} - \nabla^2\mathbf{E}$ in conjunction with Eqs. (21.2) and (21.7) gives

$$\nabla^2\mathbf{E} = \mu\sigma\frac{\partial\mathbf{E}}{\partial t} + \mu\epsilon_0\epsilon_r\frac{\partial^2\mathbf{E}}{\partial t^2} + \mu\frac{\partial^2}{\partial t^2}\mathbf{P}_{NLS}. \tag{21.8}$$

For simplicity we are neglecting the anisotropy of the medium: in other words we are assuming that ϵ_r is a scalar and div $\mathbf{E} = 0$. Eq. (21.8) is a modified version of the wave equation with the addition of a loss term (because of nonzero conductivity σ) and a new source term.

If we are dealing with plane waves (and laser beams come close to being treatable this way unless we are interested in fine detail), then without loss of generality we can analyze a situation in which all the waves propagate in the z direction and the only field components present lie in the x or y directions. In this case Eq. (21.8) becomes

$$\frac{\partial^2 E_k}{\partial z^2} = \mu\sigma_k\frac{\partial E_k}{\partial t} + \mu\epsilon_0\epsilon_r\frac{\partial^2 E_k}{\partial t^2} + \mu\epsilon_0\frac{\partial^2}{\partial t^2}\chi_{kij}E_iE_j, \tag{21.9}$$

where ϵ_r is the dielectric constant appropriate to the frequency of wave E_k, and σ_k is the appropriate conductivity. The subscripts i, j, k, can only take the values x and y.† Eq. (21.9) shows clearly the consistency relationship between the various field components present. If the field components E_i, E_j are at frequencies ω_1, ω_2, respectively, then clearly E_k must contain a component of frequency $\omega_3 = \omega_1 + \omega_2$ for both sides of Eq. (21.9) to balance.

We can write the plane waves present as

$$\begin{aligned} E_i^{\omega_1}(z,t) &= \frac{1}{2}\left[E_{1i}(z)e^{i(\omega_1 t - k_1 z)} + cc\right], \\ E_j^{\omega_2}(z,t) &= \frac{1}{2}\left[E_{2j}(z)e^{i(\omega_2 t - k_2 z)} + cc\right], \\ E_k^{\omega_3}(z,t) &= \frac{1}{2}\left[E_{3k}(z)e^{i(\omega_3 t - k_3 z)} + cc\right]. \end{aligned} \tag{21.10}$$

The NLS polarization at frequency ω_3, which drives the nonlinearly generated wave at frequency ω_3 is, cf. Eq. (20.23)

$$\left[P_{NLS}^{\omega_3}(z,t)\right]_k = \frac{\epsilon_0\chi_{kij}}{4}E_{1i}(z)E_{2j}(z)e^{i(\omega_1+\omega_2)t-(k_1+k_2)z} + cc \tag{21.11}$$

The left hand side of Eq. (21.9) can be written as

$$\begin{aligned} \frac{\partial^2 E_k^{\omega_3}}{\partial z^2}(z,t) =& \frac{1}{2}\frac{\partial^2}{\partial z^2}\left[E_{3k}(z)e^{i(\omega_3 t - k_3 z)} + cc\right] \\ =& -\frac{1}{2}\left[k_3^2 E_{3k}(z)e^{i(\omega_3 t - k_3 z)} + 2ik_3\frac{dE_{3k}(z)}{dz}e^{i(\omega_3 t - k_3 z)}\right. \\ &\left. - \frac{d^2E_{3k}}{dz^2}(z)e^{i(\omega_3 t - k_3 z)}\right] + cc \end{aligned} \tag{21.12}$$

If E_{3k} is sufficiently slowly varying in the z direction, then we can assume that

† The repeated indices in the product $\chi_{kij}E_iE_j$ in Eq. (21.9) and in following equations imply the summation $\sum_{ij}\chi_{kij}E_iE_j$.

$2k_3 dE_{3k}/dz \gg d^2E_{3k}/dz^2$, and neglect the term containing d^2E_{3k}/dz^2. With this approximation we have

$$\frac{\partial^2 E_k^{\omega_3}}{dz^2}(z,t) = -\frac{1}{2}\left[k_3^2 E_{3k}(z)e^{i(\omega_3 t - k_3 z)} + 2ik_3\frac{dE_{3k}(z)}{dz}e^{i(\omega_3 t - k_3 z)}\right] + cc \quad (21.13)$$

Similar relations exist for $\partial^2 E_i^{\omega_1}/\partial z^2(z,t)$ and $\partial^2 E_j^{\omega_2}/\partial z^2(z,t)$.

If we substitute in Eq. (21.9) and compare the terms that oscillate at frequency $\omega_3 = \omega_1 + \omega_2$, we have

$$\left[-\frac{k_3^2}{2}E_{3k}(z) - ik_3\frac{dE_{3k}(z)}{dz}\right]e^{i(\omega_3 t - k_3 z)} = \frac{1}{2}(i\omega_3\mu\sigma_3 - \omega_3^2\mu\epsilon_0\epsilon_3)E_{3k}e^{i(\omega_3 t - k_3 z)} + \frac{\omega_3^2\mu\epsilon_0\chi_{kij}}{4}E_{1i}E_{2j}e^{i(\omega_1+\omega_2)t-(k_1+k_2)z}. \quad (21.14)$$

Noting that $\omega_3^2\mu\epsilon_0\epsilon_3 = k_3^2$ in this case, where ϵ_3 is the dielectric constant at frequency ω_3, Eq. (21.14) can be simplified as

$$ik_3\frac{dE_{3k}}{dz}e^{-ik_3 z} = -\frac{i\omega_3\mu\sigma_3}{2}E_{3k}e^{-ik_3 z} + \frac{\omega_3^2\mu\epsilon_0\chi_{kij}}{4}E_{1i}E_{2j}e^{-i(k_1+k_2)z} \quad (21.15)$$

giving

$$\frac{dE_{3k}}{dz} = -\frac{\sigma_3}{2}\sqrt{\frac{\mu}{\epsilon_3\epsilon_0}}E_{3k} - \frac{i\omega_3}{4}\sqrt{\frac{\mu}{\epsilon_3\epsilon_0}}\epsilon_0\chi_{kij}E_{1i}E_{2j}e^{-i(k_1+k_2-k_3)z}. \quad (21.16)$$

Bearing in mind that NLS polarization like Eq. (21.11) can also be written for nonlinear generation of frequencies ω_1 and ω_2, for example,

$$\left[P_{NLS}^{\omega_2}(z,t)\right]_j = \frac{\epsilon_0\chi_{jki}}{4}E_{3k}(z)E_{1i}^*(z)e^{i[(\omega_3-\omega_1)t-(k_3-k_1)z]} + cc, \quad (21.17)$$

it is straightforward to show that

$$\frac{dE_{1i}}{dz} = -\frac{\sigma_1}{2}\sqrt{\frac{\mu}{\epsilon_1\epsilon_0}}E_{1i} - \frac{i\omega_1}{4}\sqrt{\frac{\mu}{\epsilon_1\epsilon_0}}\epsilon_0\chi_{ikj}E_{3k}E_{2j}^*e^{-i(k_3-k_2-k_1)z} \quad (21.18)$$

and

$$\frac{dE_{2j}}{dz} = -\frac{\sigma_2}{2}\sqrt{\frac{\mu}{\epsilon_2\epsilon_0}}E_{2j} - \frac{i\omega_2}{4}\sqrt{\frac{\mu}{\epsilon_2\epsilon_0}}\epsilon_0\chi_{jki}E_{3k}E_{1i}^*e^{-i(k_3-k_1-k_2)z}. \quad (21.19)$$

We have omitted the precise frequency dependency of, for example, $\chi_{kij} = \chi_{kij}(-\omega_3, \omega_1, \omega_2)$, but this is important and must not be completely ignored – as we shall see later. If all three frequencies ω_1, ω_2 and ω_3 are different and Kleinman's conjecture is valid, we can ignore index permutations on χ_{ijk} and write them all as χ.

Eqs. (21.16), (21.18), and (21.19) are the fundamental equations that couple three waves through a second order nonlinearity. We can use them to describe various important nonlinear processes.

Eqs. (21.16), (21.18), and (21.19) can be written in a simple, compact form as

$$\frac{dE_{1i}}{dz} = -\alpha_1 E_{1i} - i\kappa_1 E_{3k} E_{2j}^* e^{-i\Delta kz}, \tag{21.20}$$

$$\frac{dE_{2j}}{dz} = -\alpha_2 E_{2j} - i\kappa_2 E_{3k} E_{1i}^* e^{-i\Delta kz}, \tag{21.21}$$

$$\frac{dE_{3k}}{dz} = -\alpha_3 E_{3k} - i\kappa_3 E_{1i} E_{2j} e^{i\Delta kz}, \tag{21.22}$$

where

$$\Delta k = k_3 - k_2 - k_1, \tag{21.23}$$

$$\alpha_\ell = \frac{\sigma_\ell}{2}\sqrt{\frac{\mu}{\epsilon_\ell \epsilon_0}}; \ell = 1, 2, 3, \tag{21.24}$$

and

$$\kappa_\ell = \frac{\omega_\ell}{4}\sqrt{\frac{\mu}{\epsilon_\ell \epsilon_0}}\epsilon_0 \chi; \ell = 1, 2, 3, \tag{21.25}$$

which can be written as

$$\kappa_\ell = \frac{\omega_\ell \chi}{4 n_\ell c_0}; \ell = 1, 2, 3. \tag{21.26}$$

Note that α_ℓ is the electric field absorption coefficient for each wave. In the absence of any nonlinearity $\kappa_\ell = 0$ and each of the Eqs. (21.20)–(21.22) has the general solution

$$E_\ell = E_0 e^{-\alpha_\ell z}. \tag{21.27}$$

If the nonlinear crystal is transparent at all three frequencies, ω_1, ω_2, and ω_3, then the loss terms in Eqs. (21.20)–(21.22) can be neglected. Furthermore, if $k_3 = k_1 + k_2$, the so-called *phase-matched* condition, then we have three simple coupled equations.

$$\frac{dE_{1i}}{dz} = -i\kappa_1 E_{2j}^* E_{3k}, \tag{21.28}$$

$$\frac{dE_{2j}}{dz} = -i\kappa_2 E_{1i}^* E_{3k}, \tag{21.29}$$

and

$$\frac{dE_{3k}}{dz} = -i\kappa_3 E_{1i} E_{2j}. \tag{21.30}$$

Because the crystal is transparent at all three frequencies $\kappa_\ell = \kappa_\ell^*$.

The intensity of each wave is

$$I_\ell = \frac{E_\ell E_\ell^*}{2 Z_\ell}; \ell = 1, 2, 3, \tag{21.31}$$

where Z_ℓ is the characteristic impedance for the frequency and polarization direction involved. The rate of intensity change of each of the waves is

$$\frac{dI_\ell}{dz} = \frac{1}{2Z_\ell}\left(E_\ell \frac{dE_\ell^*}{dz} + E_\ell^* \frac{dE_\ell}{dz}\right). \tag{21.32}$$

So, for example, for the wave at ω_1, from Eq. (21.28)

$$\frac{dI_1}{dz} = \frac{i\kappa_1}{2Z_1}\left(E_{1i} E_{2j} E_{3k}^* - E_{1i}^* E_{2j}^* E_{3k}\right). \tag{21.33}$$

Similarly,

$$\frac{dI_2}{dz} = \frac{i\kappa_2}{2Z_2}\left(E_{1i}E_{2j}E^*_{3k} - E^*_{1i}E^*_{2j}E_{3k}\right) \tag{21.34}$$

and

$$\frac{dI_3}{dz} = -\frac{i\kappa_3}{2Z_3}\left(E_{1i}E_{2j}E^*_{3k} - E^*_{1i}E^*_{2j}E_{3k}\right). \tag{21.35}$$

Each impedance factor can be written as

$$Z_\ell = \frac{1}{n_\ell}\sqrt{\frac{\mu}{\epsilon_0}} = \frac{\mu c_0}{n_\ell} \tag{21.36}$$

so

$$\frac{\kappa_\ell}{Z_\ell} = \frac{\omega_\ell \chi}{4\mu}. \tag{21.37}$$

Therefore, Eqs. (21.33)–(21.35) can be combined to read

$$\frac{1}{\omega_1}\frac{dI_1}{dz} = \frac{1}{\omega_2}\frac{dI_2}{dz} = -\frac{1}{\omega_3}\frac{dI_3}{dz}, \tag{21.38}$$

or in terms of the photon fluxes N_ℓ at each of the frequencies

$$\frac{dN_1}{dz} = \frac{dN_2}{dz} = -\frac{dN_3}{dz}. \tag{21.39}$$

Eqs. (21.38) and (21.39) are alternative statements of the *Manley–Rowe* relations.[21.1] They are essentially statements of the principle of energy conservation. They hold for any nonlinear interaction involving three frequencies. For a source frequency generation experiment giving $\omega_1+\omega_2=\omega_3$ each photon generated at ω_3 is accompanied by the disappearance of one photon at ω_1 and one at ω_2. In a difference frequency generation experiment giving $\omega_3-\omega_1=\omega_2$ each photon generated at ω_2 is accompanied by the disappearance of one photon at ω_3 and the *appearance* of one photon at ω_1. For a parametric process of the form $\omega_3=\omega_1+\omega_2$, each photon at ω_3 that disappears leads to the simultaneous production of two photons – one at ω_1 and one at ω_2.

21.3 Second Harmonic Generation

The field of modern nonlinear optics was heralded by the observation of radiation at the second harmonic wavelength of 347 nm when a ruby laser, with a fundamental wavelength of 694 nm, was focused into a quartz crystal[21.2]. This process of second harmonic generation (SHG) is described by setting $\omega_1=\omega_2=\omega$, and $\omega_3=2\omega$, in Eqs. (21.17)–(21.19).

If we assume that very little fundamental radiation at frequency ω is converted into the second harmonic then we can set $dE_{1i}/dz=0$, and $E_{1i}=E_{1i}(0)$. In other words we ignore *depletion*. If, furthermore, we neglect losses by setting $\alpha_3=0$, then Eq. (21.16) becomes

$$\frac{dE_k(2\omega)}{dz} = -\frac{i\omega}{2}\sqrt{\frac{\mu}{\epsilon_3\epsilon_0}}\epsilon_0\chi_{kij}E_{1i}(0)E_{1j}(0)e^{i\Delta kz}, \tag{21.40}$$

where $\Delta k = k_3 - 2k_1$ is the wave vector mismatch.

In an anisotropic material the magnitude of Δk will depend on the direction of propagation relation to the principal axes of the material (see Chapter 18), as will the value of ϵ_3. Introduction of the experimental SHG coefficient $d_{kij}^{2\omega}$ allows Eq. (21.40) to be written as

$$\frac{dE_k(2\omega)}{dz} = -i\omega\sqrt{\frac{\mu}{\epsilon_3\epsilon_0}}\epsilon_0 d_{kij}^{2\omega} E_{1i}(0)E_{1j}(0)e^{i\Delta kz}. \tag{21.41}$$

Strictly speaking, Eq. (21.41) involves a summation over the components E_{1i}, E_{1j}, of the input beam. However, in any given situation of propagation direction and input beam polarization we can use the *effective* SHG coefficient d_{eff} that describes the situation. The value of d_{eff} for some specific crystals and geometries is discussed in the next section.

If SHG takes place in a crystal of length L and there is no input at 2ω, then the solution of Eq. (21.41) with the boundary condition $E_k(2\omega)(z=0)=0$ is

$$E_k(2\omega)(z=L) = -\omega\sqrt{\frac{\mu}{\epsilon_3\epsilon_0}}\epsilon_0 d_{eff} E_{1i}(0)E_{1j}(0)\frac{e^{i\Delta kL}-1}{\Delta k}. \tag{21.42}$$

In an anisotropic medium with a propagation direction **k**, chosen as the z direction in this case, E_{1i}, E_{1j} will correspond to the permitted polarization directions $\mathbf{D}_1, \mathbf{D}_2$ in the crystal (see Section 18.7). Therefore, it is logical to choose these two directions as our x and y axes. The fundamental will have an ordinary and/or extraordinary component, and depending on the direction of $E_k(2\omega)$ (x or y direction) the second harmonic will also have ordinary and extraordinary components. The simplest case arises when the fundamental is *either* an ordinary or an extraordinary wave, in which case the second harmonic intensity is

$$I(2\omega) = \frac{E_k^*(2\omega)E_k(2\omega)}{2Z(2\omega)}, \tag{21.43}$$

where $Z(2\omega)$ is the impedance of the medium at the second harmonic frequency

$$Z(2\omega) = \sqrt{\frac{\mu}{\epsilon_3\epsilon_0}} \tag{21.44}$$

and the value of ϵ_3 will depend on the polarization direction of the second harmonic.

From Eqs. (21.42), (21.43), and (21.44)

$$I(2\omega) = \frac{1}{2}\sqrt{\frac{\mu}{\epsilon_3\epsilon_0}}\omega^2\epsilon_0^2 d_{eff}^2 |E_{1i}|^4 L^2 \frac{\sin^2(\Delta kL/2)}{(\Delta kL/2)^2}. \tag{21.45}$$

Introducing the intensity of the fundamental beam

$$I(\omega) = \frac{|E_{1i}|^2}{2Z(\omega)} \tag{21.46}$$

gives

$$I(2\omega) = \frac{2\omega^2 d_{\text{eff}}^2}{c_0^3 n^{2\omega}(n^\omega)^2} I^2(\omega)L^2\frac{\sin^2(\Delta kL/2)}{(\Delta kL/2)^2}, \tag{21.47}$$

where $c_0 = 1/\sqrt{\mu_0\epsilon_0}$ is the velocity of light *in vacuo*, $n^{2\omega} = \sqrt{\epsilon_3}$ is the refractive index at the second harmonic frequency, and $n^\omega = \sqrt{\epsilon_1}$ is the refractive index at the fundamental frequency. Eq. (21.47) can be written as†

$$I(2\omega) = AI^2(\omega)L^2\text{sinc}^2(\Delta kL/2) \tag{21.48}$$

† sinc $(x) = \sin(x)/x$.

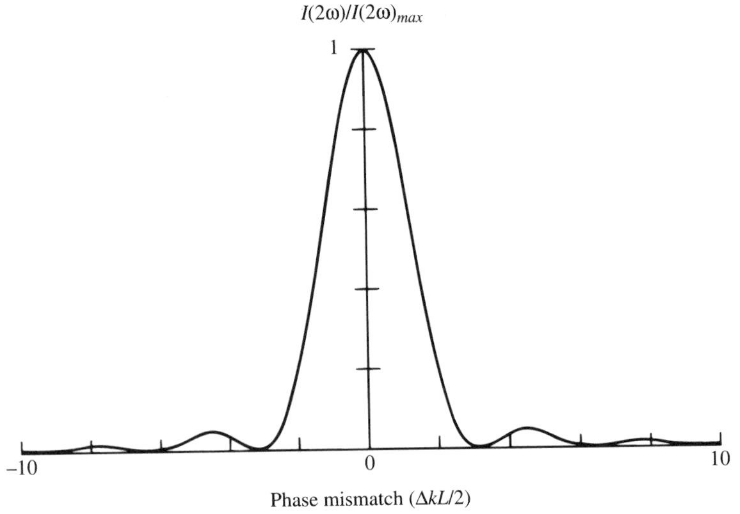

Fig. 21.1. Variation of second harmonic conversion efficiency as a function of wave vector mismatch Δk and crystal length L given by the relation $\mathrm{sinc}^2(\Delta kL/2)$.

where

$$A = \frac{2\omega^2 d_{eff}^2}{c_0^3 n^{2\omega}(n^\omega)^2}. \tag{21.49}$$

Eq. (21.48) shows clearly that the second harmonic intensity is proportional to the square of the input intensity and is maximized if $\Delta k = 0$. If $\Delta k = 0$, the process is phase-matched. The relative variation of the second harmonic conversion efficiency described by the function $\sin^2(\Delta kL/2)/(\Delta kL/2)^2$ is shown in Fig. (21.1). There is a rapid reduction in SHG with detuning from $\Delta k = 0$. Because of dispersion, the variation of refractive index with wavelength, the refractive indices $n^{2\omega}$ and n^ω will be different if both correspond to waves polarized in the same direction. In this case the second harmonic radiation generated at each point within the medium will get out of step with second harmonic radiation generated at later points leading to alternate constructive and destructive interference of the second harmonic waves. In the phase-matched condition this does not happen. To achieve the condition $\Delta k = 0$ requires $n^{2\omega} = n^\omega$, which can be accomplished by using the birefringence of an anisotropic material and obtaining SHG that is orthogonally polarized to the fundamental.

21.4 The Effective Nonlinear Coefficient d_{eff}

For a given crystal there will be specific propagation directions and polarizations for optimizing a particular nonlinear process. The generated polarization that results in a particular geometry involves summations over input field components and elements of the appropriate nonlinear tensor $\overline{\overline{d}}$, for example, as given by Eq. (20.44). However, it is possible to write a simple equation for the various amplitudes involved. For example, for a second order process

$$P(\omega_1 + \omega_2) = d_{eff} E(\omega_1) E(\omega_2), \tag{21.50}$$

where d_{eff} is an effective nonlinear coefficient that includes all the summations that apply in the particular geometry in question. This concept is best illustrated

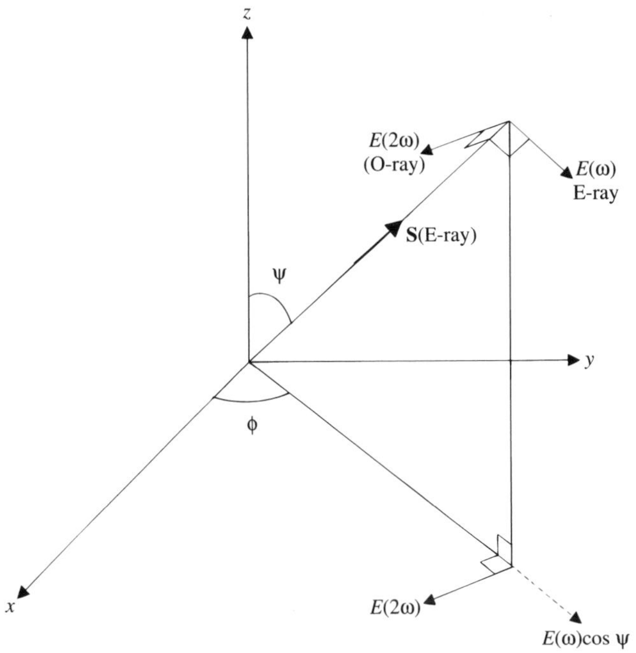

Fig. 21.2. Geometry for type-I phase matching in a positive uniaxial crystal showing fundamental field and second harmonic polarization directions.

with an example. Consider the case of second harmonic generation in tellurium, which belongs to point group 32 and is positive uniaxial ($n_e > n_o$). For this crystal, if we assume Kleinman's conjecture

$$\begin{aligned} P_x(2\omega) &= d_{11}E_x^2(\omega) - d_{11}E_y^2(\omega), \\ P_y(2\omega) &= -2d_{11}E_x(\omega)E_y(\omega), \\ P_z(2\omega) &= 0. \end{aligned} \tag{21.51}$$

For type-I phase matching in this crystal we have a fundamental E-wave and a second harmonic O-wave. The geometry for phase matching in this case is shown in Fig. (21.2). The fundamental ray direction **S** is shown travelling at an angle ψ to the optic axis: this is not exactly equal to the phase-matching angle because of the angle between **S** and **k**. For a ray at the azimuthal angle shown the input electric field components are:

$$\begin{aligned} E_x &= \cos\psi\cos\phi|\mathbf{E}(\omega)|, \\ E_y &= \cos\psi\sin\phi|\mathbf{E}(\omega)|, \\ E_z &= -\sin\psi|\mathbf{E}(\omega)|. \end{aligned} \tag{21.52}$$

From Eq. (21.51) the components of the second harmonic polarization are

$$\begin{aligned} P_x(2\omega) &= d_{11}\cos^2\psi\cos 2\phi|\mathbf{E}(\omega)|^2, \\ P_y(2\omega) &= -d_{11}\cos^2\psi\sin 2\phi|\mathbf{E}(\omega)|^2. \end{aligned} \tag{21.53}$$

The *effective* polarization that points in the correct direction to generate a second harmonic O-wave must point in the direction shown in Fig. (21.2). This effective

Table 21.1. d_{eff} *coefficients when Kleinman's conjecture is valid.*

Crystal class	Two O-rays and one E-ray	Two E-rays and one O-ray
622	0	0
6mm	$-d_{15}\sin\psi$	0
$\bar{6}m2$	$-d_{22}\cos\psi\sin 3\phi$	$d_{22}\cos^2\psi\cos 3\phi$
$\bar{6}$	$\cos\psi(d_{11}\cos 3\phi-$ $d_{22}\sin 3\phi)$	$\cos^2\psi(d_{11}\sin 3\phi+$ $d_{22}\cos 3\phi)$
6	$d_{15}\sin\psi$	0
$\bar{4}2m$	$-d_{14}\sin\psi\sin 2\phi$	$d_{14}\sin 2\psi\cos 2\phi$
$\bar{4}$	$-\sin\psi(d_{14}\sin 2\phi+$ $d_{15}\cos 2\phi)$	$\sin 2\psi(d_{14}\cos 2\phi-$ $d_{15}\sin 2\phi)$
422	0	0
4mm	$d_{15}\sin\psi$	0
4	$d_{15}\sin\psi$	0
32	$d_{11}\cos\psi\cos 3\phi$	$d_{11}\cos^2\psi\sin 3\phi$
3m	$d_{15}\sin\psi - d_{22}\cos\psi\sin 3\phi$	$d_{22}\cos^2\psi\cos 3\phi$
3	$d_{15}\sin\psi+\cos\psi(d_{11}\cos 3\phi-$ $d_{22}\sin 3\phi)$	$\cos^2\psi(d_{11}\sin 3\phi+$ $d_{22}\cos 3\phi)$

polarization is

$$\begin{aligned} P_{eff}(2\omega) &= d_{11}\cos^2\psi\cos 2\phi\sin\phi|\mathbf{E}(\omega)|^2 + d_{11}\cos^2\psi\sin 2\phi\cos\phi|\mathbf{E}(\omega)|^2 \\ &= d_{11}\cos^2\psi\sin 3\phi(\mathbf{E}(\omega)|^2. \end{aligned} \tag{21.54}$$

Comparison with Eq. (21.50) shows that

$$d_{eff} = d_{11}\cos^2\psi\sin 3\phi. \tag{21.55}$$

Therefore, for maximum second harmonic conversion efficiency in this case the fundamental should propagate at the phase-matching angle, and at an azimuthal angle of either 30° or 90°. Table (21.1) lists the d_{eff} coefficients for the 13 uniaxial crystal classes that lack inversion symmetry, assuming the validity of Kleinman's conjecture. These coefficients can be used to describe both second harmonic generation and frequency mixing.

21.5 Phase Matching

Collinear phase matching occurs when the waves involved have a common wave vector direction. In this case, the vector condition

$$\Delta\mathbf{k} = \mathbf{k}_3 - \mathbf{k}_1 - \mathbf{k}_2 = 0 \tag{21.56}$$

reduces, in the case of SHG, to

$$\Delta k = 2\pi\left[\frac{n(2\omega)}{\lambda(2\omega)} - \frac{2n(\omega)}{\lambda(\omega)}\right] = 0, \tag{21.57}$$

so $n(2\omega) = n(\omega)$.

In the most general case involving noncollinear waves and many frequencies through a susceptibility tensor $\chi(-\omega, \omega_1, \omega_2 \ldots \omega_n)$ the phase matching condition is

$$\mathbf{k}(\omega) - \mathbf{k}(\omega_1) - \mathbf{k}(\omega_2) \ldots - \mathbf{k}(\omega_n) = 0. \tag{21.58}$$

21.5.1 Second Harmonic Generation

For SHG we can achieve phase matching in a sufficiently birefringent crystal by having a fundamental *ordinary* wave and a second harmonic *extraordinary wave*, or vice versa. So, for example in a positive uniaxial crystal we seek the condition $n_o^{2\omega}(\theta) = n_e^{\omega}$. How this is done is illustrated by the phase-matching diagram shown in Fig. (21.3a). This diagram shows principal sections of the ordinary and extraordinary index surfaces of a positive uniaxial crystal for both the fundamental and second harmonic frequencies. Normal dispersion in such a crystal leads to $n_o^{2\omega} > n_o^{\omega}$ and $n_e^{2\omega}(\theta) > n_e^{\omega}(\theta)$, as shown.

If the birefringence in a negative crystal is large enough, then the ordinary index surface at the fundamental frequency will intersect the extraordinary index surface at the second harmonic at the *phase-matching angle*, θ_m. From Eq. (18.58) the condition for phase matching becomes for a negative uniaxial crystal

$$n_e^{2\omega}(\theta_m) = \frac{n_o^{2\omega} n_e^{2\omega}}{\sqrt{(n_e^{2\omega})^2 \cos^2\theta_m + (n_o^{2\omega})^2 \sin^2\theta_m}} = n_o^{\omega}, \qquad (21.59)$$

which can be rearranged to give

$$\sin^2\theta_m = \frac{(n_o^{\omega})^{-2} - (n_o^{2\omega})^{-2}}{(n_e^{2\omega})^{-2} - (n_o^{2\omega})^{-2}}. \qquad (21.60)$$

In a positive crystal phase matching is achieved with an extraordinary wave at the fundamental and an ordinary wave at the second harmonic. It is easy to show in this case that

$$\sin^2\theta_m = \frac{(n_o^{2\omega})^{-2} - (n_o^{\omega})^{-2}}{(n_e^{\omega})^{-2} - (n_o^{\omega})^{-2}}. \qquad (21.61)$$

In practical applications of phase matching a crystal would be cut with end faces perpendicular to the phase-matching direction as shown in Fig. (21.3b).

21.5.2 Example

Tellurium, a positive uniaxial crystal belonging to class 32 has been used for SHG with a CO_2 laser. Interpolation from the refractive index data of Caldwell and Fan[21.3] gives at 10.6 μm and 5.3 μm

$$n_o^{\omega} = 4.794; n_o^{2\omega} = 4.856; n_e^{\omega} = 6.243$$

Substitution into Eq. (21.61) gives $\theta_m = 14.4°$.

Phase-matching angles are frequently calculated through the use of *Sellmaier* equations, which are semiempirical formulae that give the variation of the ordinary and extraordinary refractive indices of a material. A typical pair of Sellmaier equations would be of the form[21.4]

$$n_o^2(\lambda) = a_o + \sum_k \frac{b_{ok}}{\lambda^2 - \lambda_k^2}, \qquad (21.62)$$

$$n_e^2(\lambda) = a_e + \sum_k \frac{b_{ek}}{\lambda^2 - \lambda_k^2}, \qquad (21.63)$$

where the summation covers the various absorption resonances, λ_k, that contribute to dispersion. Usually only one or two such resonances are important, so

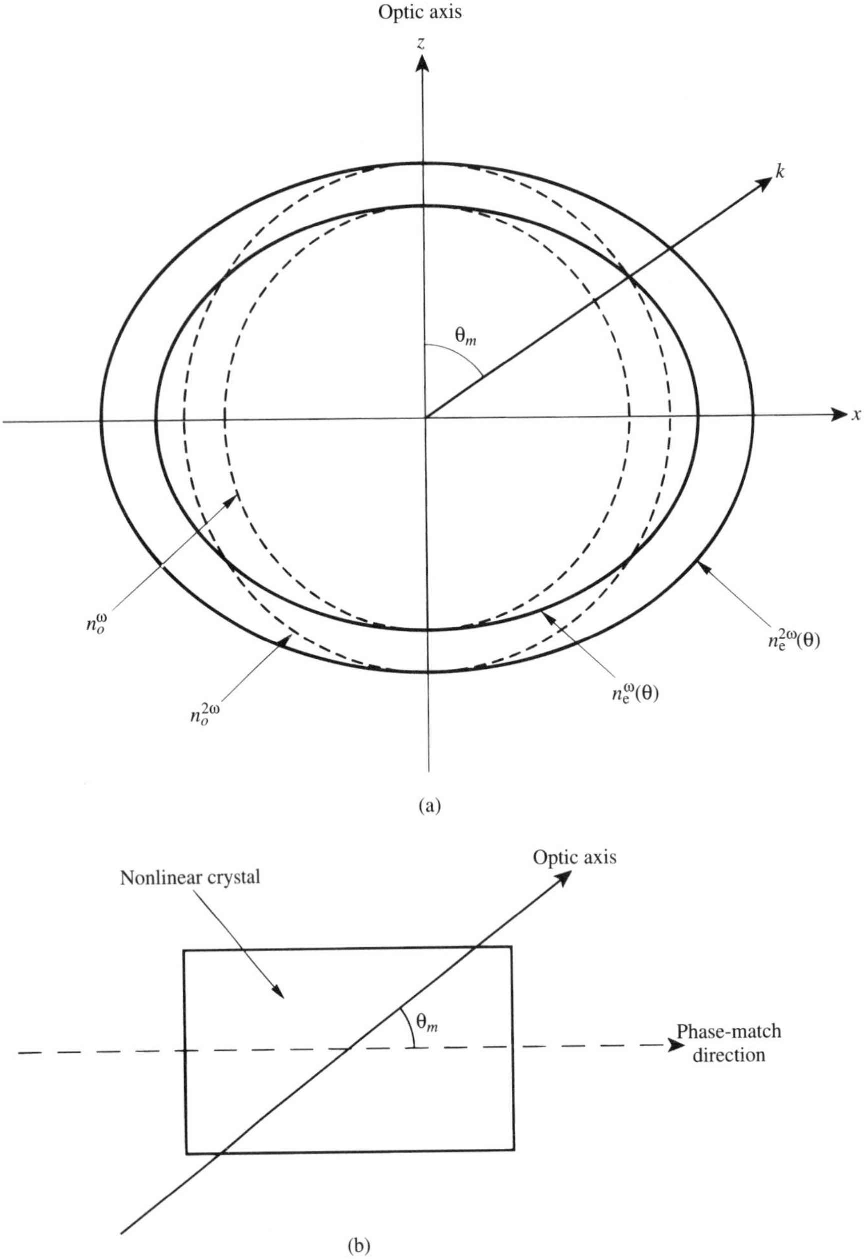

Fig. 21.3. Geometry for type-I phase matching in a positive uniaxial crystal: (a) indicatrices, (b) crystal cut with faces perpendicular to phase-matching direction.

Eqs. (21.62) and (21.63) often only have one or two terms in the summation. For example, for the nonlinear material proustite Ag_3AsS_3[21.5]:

$$n_o^2(\lambda) = 9.220 + \frac{0.4454}{\lambda^2 - 0.1264} - \frac{1733}{1000 - \lambda^2}, \tag{21.64}$$

$$n_e^2(\lambda) = 7.007 + \frac{0.3230}{\lambda^2 - 0.1192} - \frac{660}{100 - \lambda^2}. \tag{21.65}$$

where μ is measured in μm.

21.5.3 Phase Matching in Sum-Frequency Generation

In the examples of phase-matched SHG we have considered so far the fundamental waves are polarized in a unique direction, while the second harmonic waves are polarized in an orthogonal direction. This is a special case of *type-I* phase matching. Generalized to the mixing of two waves at frequencies ω_1 and ω_2 to generate a sum frequency $\omega_3 = \omega_1 + \omega_2$, type-I phase matching has the waves at ω_1 and ω_2 polarized in the same direction.

Type-II phase matching occurs in this case when the waves at ω_1 and ω_2 are orthogonally polarized. For collinear phase matching we want

$$\mathbf{k}_3 = \mathbf{k}_1 + \mathbf{k}_2, \tag{21.66}$$

that is,

$$\frac{n_3}{\lambda_3} = \frac{n_1}{\lambda_1} + \frac{n_2}{\lambda_2}. \tag{21.67}$$

Therefore, we can summarize the phase-matching conditions for collinear sum-frequency generation as follows:

Positive Uniaxial Crystal

Type I: $\omega_1 \equiv$ E-wave; $\omega_2 \equiv$ E-wave; $\omega_3 \equiv$ O-wave

$$n_o^{\omega_3} = \frac{\lambda_3}{\lambda_1} n_e^{\omega_1} + \frac{\lambda_3}{\lambda_2} n_e^{\omega_2} \tag{21.68}$$

Type II: $\omega_1 \equiv$ O-wave; $\omega_2 \equiv$ E-wave; $\omega_3 \equiv$ O-wave

$$n_o^{\omega_3} = \frac{\lambda_3}{\lambda_1} n_o^{\omega_1} + \frac{\lambda_3}{\lambda_2} n_e^{\omega_2} \tag{21.69}$$

Negative Uniaxial Crystal

Type I: $\omega_1 \equiv$ O-wave; $\omega_2 \equiv$ O-wave; $\omega_3 \equiv$ E-wave

$$n_e^{\omega_3} = \frac{\lambda_3}{\lambda_1} n_o^{\omega_1} + \frac{\lambda_3}{\lambda_2} n_o^{\omega_2} \tag{21.70}$$

Type II: $\omega_1 \equiv$ E-wave; $\omega_2 \equiv$ O-wave; $\omega_3 \equiv$ E-wave

$$n_e^{\omega_3} = \frac{\lambda_3}{\lambda_1} n_e^{\omega_1} + \frac{\lambda_3}{\lambda_2} n_o^{\omega_2} \tag{21.71}$$

21.6 Beam Walk-Off and 90° Phase Matching

The refractive indices of crystals depend on temperature so the phase-matching angle can be temperature tuned. A highly desirable geometry results if the phase-matching angle can be tuned to 90° in a uniaxial crystal. If this can be accomplished, and it is not feasible in every case, then beam walk-off caused by double refraction can be eliminated. This beam walk-off results because in either type-I or type-II phase matching both ordinary and extraordinary waves are participating in the nonlinear interactions. For collinear phase matching all the wave normals involved point in the same direction, but the ray directions (Poynting vectors) do not. Nonlinear generation usually involves narrow focused laser beams. Such focused beams generate large electric fields and lead to efficient nonlinear conversion. However, along the path of such a beam in the crystal, which will follow the ray direction, a nonlinearly generated wave of orthogonal polarization will travel at a different angle. The input and nonlinearly generated rays do not remain collinear

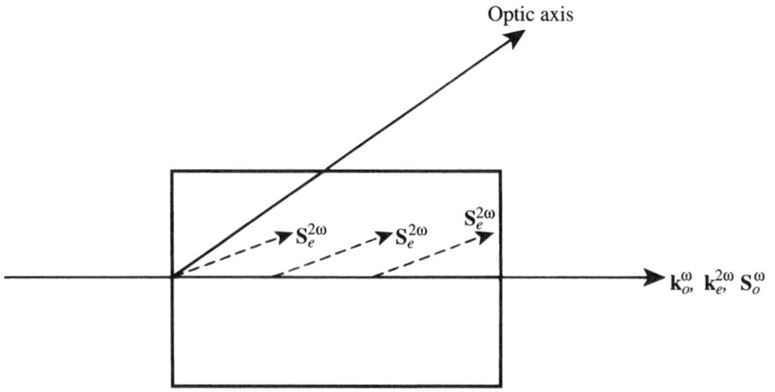

Fig. 21.4. Beam walk-off in second harmonic generation in a negative uniaxial crystal that results when the phase-matching direction is not a principal axis direction. The wave vectors of both the fundamental and second harmonic and the ray direction of the fundamental all propagate collinearly through the crystal.

Table 21.2. *Conditions for 90° phase matching (PM) in selected crystals.*

Crystal	Wavelength (μm)	90° PM temperature (°C)
KDP	0.5145	-11.0
ADP	0.5145	∼-9.2
$LiNbO_3$	1.1523	210
Ag_3AsS_3	1.152	12
CD*A (CsH_2AsO_4)	1.06	103

along the path. For example, for SHG in a negative uniaxial crystal, the second harmonic ray continually 'walks-off' from the fundamental as shown in Fig. (21.4). The problem of beam walk-off is more serious for mixing processes using type-II phase matching. In this case the input waves are orthogonally polarized and walk-off from each other until no nonlinear mixing occurs. To first order, if phase matching occurs at 90° to the optic axis then no beam walk-off occurs. Table (21.2) lists some of the crystals, wavelengths, and temperatures for which this can be accomplished.

When a focused Gaussian beam is used to maximize the second harmonic or mixing efficiency in a nonlinear crystal there is a second order contribution to beam walk-off because the focused beam contains a range of effective wave vector directions[21.6].

21.7 Second Harmonic Generation with Gaussian Beams

To maximize the conversion efficiency in SHG it is usual to focus a fundamental Gaussian mode laser beam into the nonlinear crystal. For high energy pulsed lasers this is generally done by placing the nonlinear crystal outside the fundamental laser cavity and focusing the fundamental beam with a lens. The second harmonic pulse will be of shorter time duration than the fundamental because its time profile will be

$$I(2\omega, t) \propto I^2(\omega, t). \qquad (21.72)$$

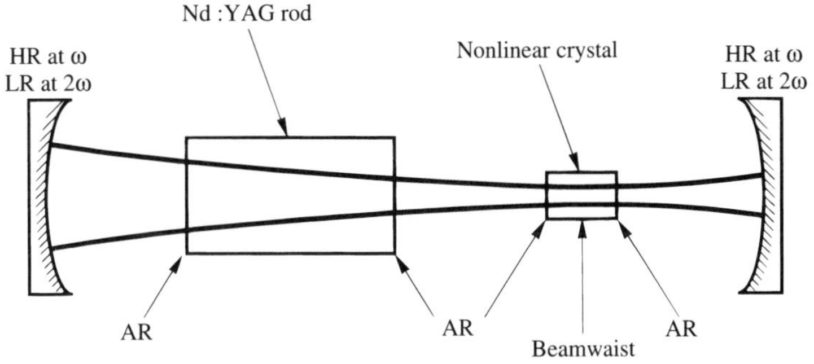

Fig. 21.5. Schematic arrangement for intracavity SHG of a Nd:YAG laser. HR – high reflectance; LR – low reflectance.

For CW SHG an efficient way to achieve maximum fundamental intensity in the crystal is to place the nonlinear crystal inside the laser resonator. If the nonlinear crystal is not too long, we can treat the phase fronts of the input Gaussian beam within the crystal as plane. SHG results therefore from a plane wave with a Gaussian transverse intensity distribution

$$I(\omega, r) = I_0 e^{-2r^2/w_0^2}, \tag{21.73}$$

with total power

$$P(\omega) = \frac{1}{2}\pi w_0^2 I_0. \tag{21.74}$$

If we assume phase matching then the radial variation of the second harmonic intensity is, from Eq. (21.48),

$$I(2\omega, r) = AL^2 I_0^2 e^{-4r^2/w_0^2}, \tag{21.75}$$

so the spot size of the second harmonic beam in the crystal is $w_0/\sqrt{2}$.

The total second harmonic power is

$$P(2\omega) = 2\pi AL^2 I_0^2 \int_0^\infty r e^{-4r^2/w_0^2} dr, \tag{21.76}$$

which gives

$$P(2\omega) = \frac{AL^2}{\pi w^2} P^2(\omega). \tag{21.77}$$

21.7.1 *Intracavity SHG*

The crystal is placed at the beamwaist of the intracavity mode as shown in Fig. (21.5). To minimize undesirable losses, which would prevent the threshold for oscillation at the fundamental being reached, the nonlinear crystal should have negligible loss at the fundamental wavelength. In addition, to prevent intracavity reflection losses or alignment problems, all the intracavity faces should be antireflection coated (AR) at the fundamental wavelength. Intracavity SHG provides a high fundamental intensity, but there may be instability of the generated second harmonic light if the laser operates on more than one axial mode.

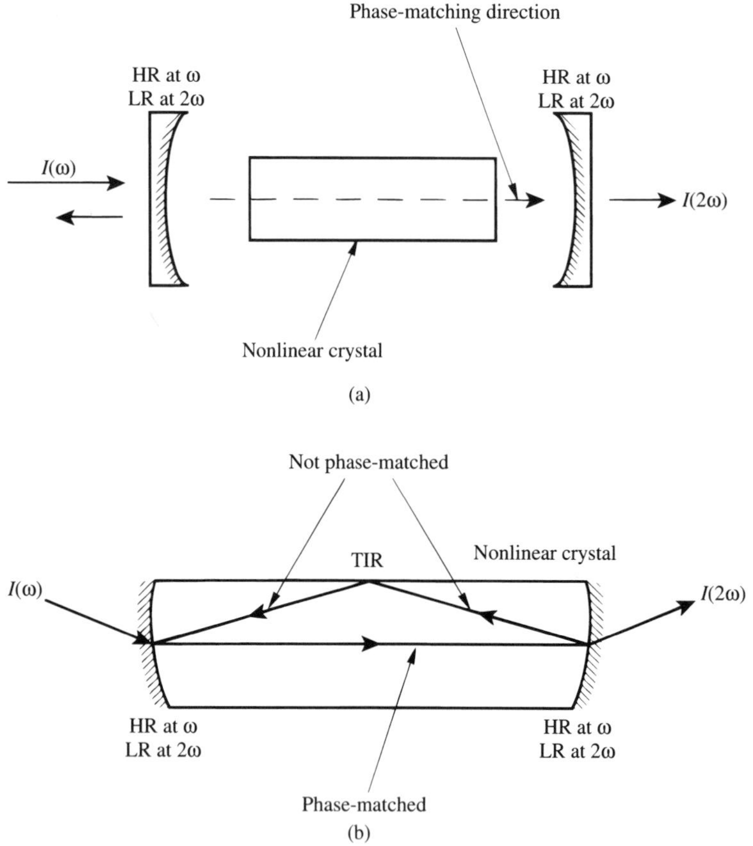

Fig. 21.6. Arrangements for SHG outside the laser cavity: (a) conventional standing-wave, (b) monolithic ring resonator (TIR–total internal reflection).

21.7.2 External SHG

The efficiency of SHG outside the cavity can be enhanced by placing the nonlinear crystal in an external resonator, which has mirrors that are highly reflective at the fundamental wavelength. This can be done in two ways, as shown in Fig. (21.6).

The reflectors can be separate from the nonlinear crystal, but it is better to incorporate them directly on the polished crystal faces. In the geometry of Fig. (21.6a) the second harmonic is phase-matched in both directions so emerges through both mirrors. Unidirectional SHG can be obtained through the use of a ring resonator, shown in Fig. (21.6b). The path of the second harmonic is only phase-matched in the direction shown and emerges predominantly through the right hand mirror. Although it is possible to make the left hand mirror in Fig. (21.6a) highly reflective at the second harmonic, there is then the possibility of destructive interference between the counterpropagating second harmonic beams in the crystal.

21.7.3 The Effects of Depletion on Second Harmonic Generation

If a high conversion efficiency to the second harmonic is obtained, then it is no longer reasonable to ignore depletion. If type-I phase matching is being used we can rewrite Eqs. (21.28) and (21.30) as

$$\frac{dE(\omega)}{dz} = -i\kappa_1 E^*(\omega)E(2\omega), \tag{21.78}$$

$$\frac{dE(2\omega)}{dz} = -i\kappa_3 E(\omega)E(\omega). \tag{21.79}$$

However, some care must be taken at this point in determining the values of the κs. Written out in full

$$\kappa_1 = \frac{\omega}{4n_1c_0}\chi(-\omega,-\omega,2\omega), \tag{21.80}$$

$$\kappa_3 = \frac{2\omega}{4n_3c_0}\chi(-2\omega,\omega,\omega). \tag{21.81}$$

$\chi(-\omega,-\omega,2\omega)$ is a coefficient that describes difference frequency generation – it enters because when depletion occurs second harmonic radiation can be reconverted back to the fundamental by the process. $\chi(-2\omega,\omega,\omega)$ describes second harmonic generation. Therefore, as discussed in Chapter 20

$$\chi(-2\omega,\omega,\omega) = \frac{1}{2}\chi(-\omega,-\omega,2\omega). \tag{21.82}$$

Therefore, in terms of the d_{eff} coefficient for SHG and with $n_1 = n_3 = n$

$$\kappa_1 = \kappa = \frac{\omega d_{eff}}{n_2 c_0}, \tag{21.83}$$

$$\kappa_3 = \kappa = \frac{\omega d_{eff}}{n c_0}. \tag{21.84}$$

The Manley–Rowe relations require that energy be conserved in the conversion of fundamental to second harmonic, so from Eq. (21.38),

$$\frac{1}{\omega}\frac{d(I(\omega)/2)}{dz} = -\frac{1}{2\omega}\frac{dI(2\omega)}{dz}, \tag{21.85}$$

where we are imagining the SHG process to be equivalent to the mixing together of two beams each of intensity $I(\omega)/2$.

Therefore, following the treatment given by Byer[21.7]

$$\frac{d}{dz}[I(\omega)+I(2\omega)] = 0 \tag{21.86}$$

and

$$I(\omega)+I(2\omega) = \text{ constant } = I_0, \tag{21.87}$$

which can be written in terms of the electric fields as

$$|E(\omega)^2| + |E(2\omega)|^2 = 2ZI_0, \tag{21.88}$$

where Z is the impedance of the medium, identical for both frequencies ω and 2ω under phase-matched conditions. If we write the complex amplitudes $E(\omega), E(2\omega)$ in terms of amplitude and phase factors we have

$$E(\omega) = \mathscr{E}(\omega)e^{i\phi_\omega},$$
$$E(2\omega) = \mathscr{E}(2\omega)e^{i\phi_{2\omega}}, \tag{21.89}$$

and Eqs. (21.78) and (21.79) become

$$\frac{d\mathscr{E}(\omega)}{dz} = -i\kappa\mathscr{E}(\omega)\mathscr{E}(2\omega)e^{-i(2\phi_\omega-\phi_{2\omega})}, \tag{21.90}$$

$$\frac{d\mathscr{E}(2\omega)}{dz} = -i\kappa\mathscr{E}(\omega)\mathscr{E}(\omega)e^{i(2\phi_\omega-\phi_{2\omega})}. \tag{21.91}$$

For the amplitudes $\mathscr{E}(\omega), \mathscr{E}(2\omega)$ to be real it is clearly necessary for the factor $ie^{-i(2\phi_\omega-\phi_{2\omega})}$ to be real. We choose the negative sign for Eq. (21.90) because we expect the fundamental to decrease in amplitude as it is converted to the second harmonic. Therefore, we have

$$\frac{d\mathscr{E}(\omega)}{dz} = -\kappa\mathscr{E}(\omega)\mathscr{E}(2\omega), \tag{21.92}$$

$$\frac{d\mathscr{E}(2\omega)}{dz} = \kappa\mathscr{E}(\omega)\mathscr{E}(\omega). \tag{21.93}$$

Substituting from Eq. (21.88) into Eq. (21.93) gives

$$\frac{d\mathscr{E}(2\omega)}{dz} = \kappa[2ZI_0 - \mathscr{E}^2(2\omega)], \tag{21.94}$$

and integrating

$$\int_0^{\mathscr{E}(2\omega,L)} \frac{d\mathscr{E}(2\omega)}{2ZI_0 - \mathscr{E}^2(2\omega)} = \kappa\int_0^L dz, \tag{21.95}$$

which gives

$$\frac{1}{\sqrt{2ZI_0}}\tanh^{-1}\left[\frac{\mathscr{E}(2\omega,L)}{\sqrt{2ZI_0}}\right] = \kappa L, \tag{21.96}$$

so,

$$\mathscr{E}(2\omega, L) = \sqrt{2ZI_0}\tanh\left(\kappa L\sqrt{2ZI_0}\right) \tag{21.97}$$

and finally,

$$I(2\omega, L) = I_0\tanh^2(\Gamma L), \tag{21.98}$$

where

$$\Gamma = \kappa\sqrt{2ZI_0}. \tag{21.99}$$

From Eq. (21.88)

$$\begin{aligned} I(\omega, L) &= I_0\left[1 - \tanh^2(\Gamma L)\right] \\ &= I_0\,\mathrm{sech}^2(\Gamma L). \end{aligned} \tag{21.100}$$

Fig. (21.7) shows the growth of the second harmonic towards 100% conversion efficiency, and the decay of the fundamental to zero predicted by Eqs. (21.99) and (21.100) for plane waves. For Gaussian beams the conversion efficiency is actually limited to about 60%.

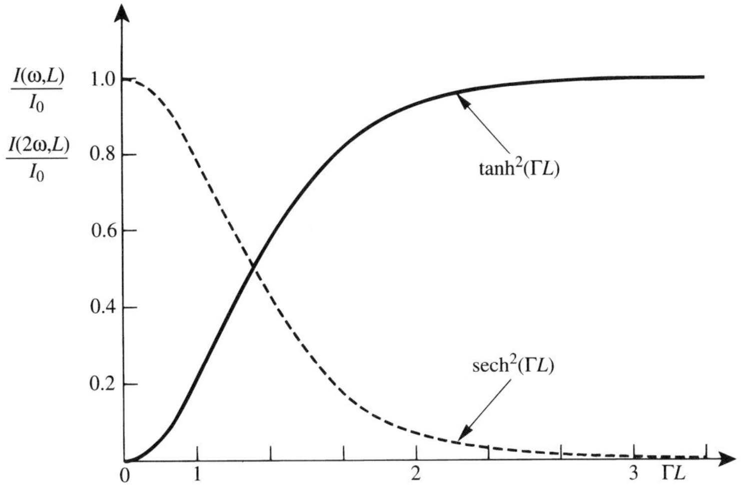

Fig. 21.7. Growth of second harmonic and decay of fundamental as a function of crystal length during efficient second harmonic generation.

21.8 Up-Conversion and Difference-Frequency Generation

These are two processes that can be analyzed in a similar way to SHG. In up-conversion, or sum-frequency generation, two lower energy photons, ω_1, ω_2, combine in a nonlinear medium to generate a higher energy photon, ω_3. One application of the process $\omega_1 + \omega_2 = \omega_3$ is for a strong laser beam at frequency, ω_1, in the visible to be mixed with a weak infrared signal, ω_2. The up-converted photons, ω_3, will then be capable of detection with very sensitive photodetectors. This provides a means for the detection of very weak infrared light signals[21.8]. The up-conversion process can also be used to provide visual imaging of an infrared illuminated scene[21.9]. The efficiency, and practical utility, of these approaches depends critically on phase matching. Incoming radiation will be up-converted most efficiently at angles that vary with wavelength[21.6].

Difference-frequency generation can be used to obtain coherent infrared radiation at wavelengths that are difficult to generate by other means[21.10],[21.11]. The process $\omega_3 = \omega_1 - \omega_2$ will only work efficiently if the nonlinear medium is transparent at all three wavelengths involved. GaAs has been successfully used for the generation of tunable far-infrared by the mixing together of different lines from two CO_2 lasers[21.12],[21.13]. $LiNbO_3$ has been used to generate far-infrared radiation with a wavelength as long as 1.23 mm by the mixing of two ruby laser beams. However, tunable infrared radiation can also be generated by other means, such as stimulated Raman scattering[21.14]–[21.16] and with a spin-flip Raman laser[21.17], a discussion of which is outside the scope of this book.

An important application of sum-frequency generation is in the generation of blue–green light by mixing the output of a semiconductor diode laser with the output of a diode-pumped Nd:YAG laser. An 810 nm diode laser mixed with a 1.06 μm Nd:YAG laser provides radiation at 459 nm, which is transmitted efficiently through water. Fig. (21.8) shows one of the configurations that has been used experimentally to do this.

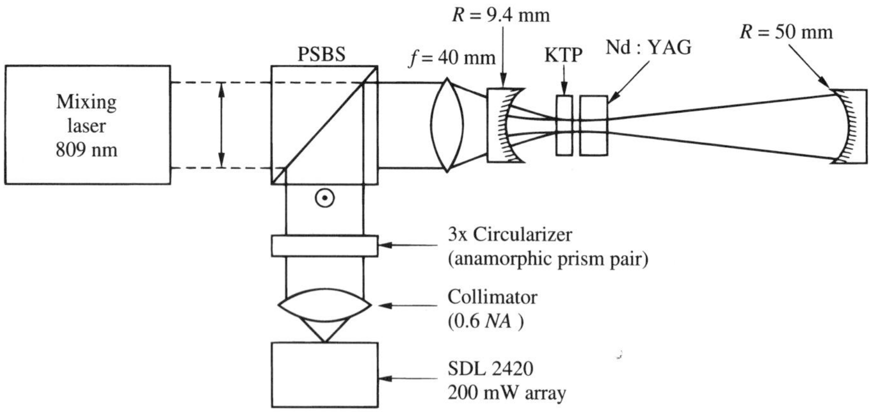

Fig. 21.8. Schematic arrangement for generating blue–green light by mixing the output of a semiconductor laser with a 1.06 μm diode-pumped Nd:YAG laser[21.17].

21.9 Optical Parametric Amplification

This is a process in which a weak laser beam at frequency ω_1, the *Signal*, is amplified by a nonlinear interaction involving a strong laser beam at frequency ω_3, the *Pump*. To conserve energy additional photons at a frequency ω_2, the *Idler* must be simultaneously generated. The process is represented as

$$\omega_3 = \omega_1 + \omega_2.$$

In a sense this process is the reverse of sum-frequency generation. Its attractiveness comes from its tunability, since in theory the pump photons at ω_3 can be divided in infinitely many ways into pairs of photons, ω_1, ω_2. To describe this process mathematically we again use Eqs. (21.20)–(21.22).

If we neglect depletion of the pump we can assign the pump field a constant value $E_3 = E_P$. Further, by assuming a particular set of polarizations for the fields involved we can drop the subscripts on the electric field components and write these equations as

$$\begin{aligned} \frac{dE_1}{dz} + \alpha_1 E_1 &= -i\kappa_1 E_P E_2^* e^{-i\Delta kz}, \\ \frac{dE_2^*}{dz} + \alpha_2 E_2^* &= i\kappa_2 E_P^* E_1 e^{i\Delta kz}. \end{aligned} \tag{21.101}$$

We look for a solution in which both the signal, E_1, and idler, E_2, fields grow exponentially, and try as our solution

$$\begin{aligned} E_1(z) &= E_S e^{(\Gamma' - i\Delta k/2)z}, \\ E_2(z) &= E_I e^{(\Gamma' - i\Delta k/2)z}. \end{aligned} \tag{21.102}$$

Substitution into Eq. (21.101) gives

$$\begin{aligned} E_S(\Gamma' + \alpha_1 - i\Delta k/2) + E_I^*(i\kappa_1 E_P) &= 0, \\ E_S(-i\kappa_2 E_P^*) + E_I^*(\Gamma' + \alpha_2 + i\Delta k/2) &= 0. \end{aligned} \tag{21.103}$$

In order for Eq. (21.103) to represent the solution for arbitrary initial values of the signal and idler fields, the determinant of the coefficients must be zero, therefore

$$(\Gamma' + \alpha_1 - i\Delta k/2)(\Gamma' + \alpha_2 + i\Delta k/2) - \kappa_1\kappa_2|E_P|^2 = 0. \tag{21.104}$$

Solving for Γ' gives

$$\Gamma' = -\left(\frac{\alpha_1+\alpha_2}{2}\right) \pm \frac{1}{2}\left[(\alpha_1-\alpha_2)^2 - 2i\Delta k(\alpha_1-\alpha_2) - 4\left(\frac{\Delta k}{2}\right)^2 + 4\kappa_1\kappa_2|E_P|^2\right]^{1/2}. \quad (21.105)$$

Eq. (21.105) can be put in a simpler form if we assume that $\alpha_1 = \alpha_2 = \alpha$ and introduce a parametric gain coefficient Γ defined by

$$\Gamma^2 = \kappa_1\kappa_2|E_P|^2 = \frac{\omega_1\omega_2 d_{eff}^2|E_P|^2}{n_1 n_2 c_0^2}, \quad (21.106)$$

so that

$$\Gamma' = -\alpha \pm \left[\Gamma^2 - \left(\frac{\Delta k}{2}\right)^2\right]^{1/2}. \quad (21.107)$$

If we introduce the effective gain coefficient g defined by

$$g = \left[\Gamma^2 - \left(\frac{\Delta k}{2}\right)^2\right]^{1/2}, \quad (21.108)$$

then we can write the general solution to Eq. (21.101) as

$$\begin{aligned} E_1(z) &= \left(E_{1A}e^{gz} + E_{1B}e^{-gz}\right)e^{-\alpha z}e^{-i\Delta k z/2}, \\ E_2(z) &= \left(E_{2A}e^{gz} + E_{2B}e^{-gz}\right)e^{-\alpha z}e^{-i\Delta k z/2}. \end{aligned} \quad (21.109)$$

If the parametric amplification process occurs in a nonlinear medium that stretches from $z = 0$ to $z = \ell$ and the input signal and idler amplitudes satisfy $E_1(z=0) = E_S(0)$ and $E_2(z=0) = E_I(0)$, then Eqs. (21.109) and (21.101) can be solved to determine the constants E_{1A}, E_{1B}, E_{2A}, and E_{2B}. The solution is

$$E_S(\ell)e^{\alpha\ell} = E_S(0)e^{-i\frac{\Delta k\ell}{2}}\left[\cosh g\ell + \frac{i\Delta k}{2g}\sinh g\ell\right] - \frac{i\kappa_1}{g}E_P E_I^*(0)e^{-i\frac{\Delta k\ell}{2}}\sinh g\ell, \quad (21.110)$$

$$E_I(\ell)e^{\alpha\ell} = E_I(0)e^{-i\frac{\Delta k\ell}{2}}\left[\cosh g\ell + \frac{i\Delta k}{2g}\sinh g\ell\right] - \frac{i\kappa_2}{g}E_P E_S^*(0)e^{-i\frac{\Delta k\ell}{2}}\sinh g\ell, \quad (21.111)$$

where $\kappa_i = \omega_i d_{eff}/n_i c_0$, as before.

Note that Eqs. (21.110) and (21.111) are symmetric under interchange of the designation S (signal) and I (idler), as expected.

A few interesting features of the above analysis can be pointed out. Unless $\Gamma > \Delta k/2$ no continuous growth of signal or idler occurs. In this case g becomes pure imaginary and the exponential growth predicted by Eqs. (21.110) and (21.111) becomes only oscillatory since $\cosh ix = \cos x$ and $\sinh ix = i\sin x$.

Optimum signal and idler generation results if $\Delta k = 0$. This is the phase-matching condition already discussed and the same relationships between the refractive indices and wavelengths involved hold true as were given in Eqs. (21.68)–(21.71).† If no idler input is supplied, then the parameter amplifier can still generate an idler wave by amplifying its own noise. In this case with phase matching Eqs. (21.110) and (21.111) would be

$$E_S(\ell)e^{\alpha\ell} = E_S(0)\cosh g\ell \tag{21.112}$$

and

$$E_I(\ell)e^{\alpha\ell} = -\frac{i\kappa_2}{g}E_P E_S^*(0)\sinh g\ell. \tag{21.113}$$

The fractional single pass intensity increase corresponding to Eq. (21.112) is

$$\frac{I_S(\ell)}{I_S(0)} = e^{-2\alpha\ell}\cosh^2 g\ell, \tag{21.114}$$

which for $g\ell >> 1$ can be written

$$I_S(\ell) = I_S(0)e^{2(g-\alpha)\ell}. \tag{21.115}$$

In a phase-matched situation

$$g^2 = \frac{\omega_1\omega_2 d_{eff}^2|E_p|^2}{n_1 n_2 c_0^2}. \tag{21.116}$$

21.9.1 Example

$LiNbO_3$ is a popular nonlinear material for use in parametric amplifiers and oscillators. For a pump wavelength of 532 nm (obtained from a frequency-doubled Nd:YAG laser) and for signal and idler wavelengths of 647 nm and 3 μm, we can use $n_1 = n_2 = n_o = 2.24$. This is a negative uniaxial crystal with $n_e < n_o$ and phase matching can be achieved with signal and idler waves propagating as ordinary waves. The pump at ω_3 propagates as an extraordinary wave. The phase-matching condition is given by Eq. (21.70). This crystal can be temperature tuned to achieve 90° phase matching, as shown in Fig. (21.9)[21.7]. We can use $n_3 = n_e = 2.16$, and $d_{eff} = d_{31} = 6.25 \times 10^{-12}$ m V^{-1}.

For a pump intensity of 10 GW m^{-2}, well below the crystal damage threshold of 500–1400 GW m^{-2}, the pump field is

$$|E_P|^2 = \frac{2Z_0 I_p}{n_e}. \tag{21.117}$$

Substitution of numerical values in Eq. (21.116) gives $g = 23.5$ m^{-1}. In practice the crystal in such a system might be 10 mm long so $g\ell \simeq 0.235$.

It is clear from this example that substantial gains in such a parametric amplifier are not obtained without large pump intensities. However, by placing the pumped crystal in a resonant cavity to provide feedback at the signal and/or the idler wavelengths, the device can be converted into a parametric oscillator.

† The additional complexities introduced by the use of Gaussian beams and the effects of beam walk-off are not dealt with here. For further details the interested reader should consult the specialized literature[21.17],[21.18].

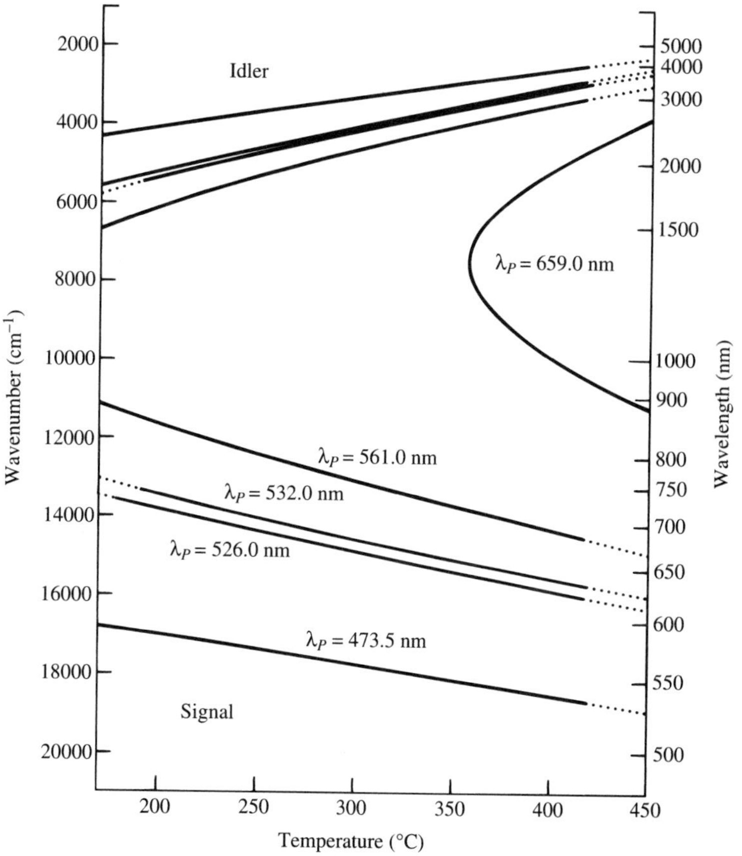

Fig. 21.9. Temperature tuning curve for $LiNbO_3$.

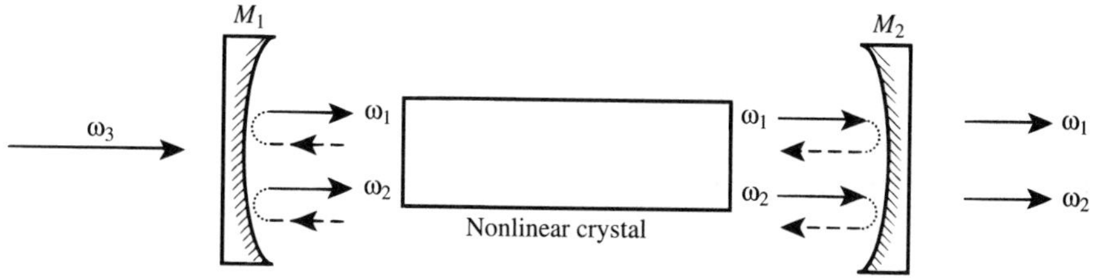

Fig. 21.10. Schematic arrangement of parametric oscillator. Mirror M_1 has high transmittance for the pump wavelength.

21.10 Parametric Oscillators

Allthough Eq. (21.115) suggests that for the possibility of oscillation $g > \alpha$, namely that gain must exceed loss, we can better examine the conditions necessary for the onset of parametric oscillation by using Eqs. (21.110) and (21.111).

In the schematic parametric oscillator shown in Fig. (21.10) the pump wave generally only makes a single pass through the crystal so the signal and idler waves are only amplified when they travel in the same direction as the pump. A more stable, lower loss device can be fabricated if the mirrors are deposited directly on the ends of the crystal.

The resonant cavity formed by the two mirrors M_1 and M_2 may provide significant feedback at both signal and idler wavelengths, in which case it is referred to as a doubly resonant optical parametric oscillator (DRO). If the cavity provides feedback at only signal or idler wavelengths it is a singly resonant optical parametric oscillator (SRO). We can incorporate both these situations into our analysis by enlarging our definition of the loss coefficients α_1 and α_2. So far, these coefficients have referred to intracavity losses within the nonlinear medium. However, the leakage of an oscillating field through M_1 or M_2 is equivalent to additional intracavity loss.

As we have seen in Chapter 4 this loss is related to the passive cavity lifetime τ. If the reflectances of the cavity mirrors at the signal and idler wavelengths are $R_1(\omega_1)$ and $R_2(\omega_2)$ then

$$\tau_1 \simeq \frac{L_1}{c_0(1-R_1)}, \quad \tau_2 \simeq \frac{L_2}{c_0(1-R_2)}, \tag{21.118}$$

where L_1, L_2 are the optical lengths of the cavity at the signal and idler wavelengths.† In a passive cavity a field at frequency ω_1, would decay according to

$$E_1(t) = E_1(0)e^{-t/2\tau_1}. \tag{21.119}$$

It is straightforward to show that if the effect of the intracavity loss coefficients α_i is included, then

$$\tau_i \simeq \frac{L_i}{c_0(1+2\alpha_i\ell - R_i)}, \tag{21.120}$$

provided both $\alpha_i\ell, R_i << 1$.

In the phase-matched case Eqs. (21.110) and (21.111) become

$$E_S(\ell) = e^{-\alpha\ell}\left[E_S(0)\cosh\Gamma\ell - \frac{i\kappa_1}{\Gamma}E_P E_I^*(0)\sinh\Gamma\ell\right], \tag{21.121}$$

$$E_I^*(\ell) = e^{-\alpha\ell}\left[E_I^*(0)\cosh\Gamma\ell + \frac{i\kappa_2}{\Gamma}E_P^* E_S(0)\sinh\Gamma\ell\right]. \tag{21.122}$$

With an additional pass through the nonlinear medium and reflections at the cavity mirrors, Eqs. (21.121) and (21.122) become, at the threshold of oscillation when changes in field amplitudes occur on a round trip in the cavity,

$$E_S(0) = R_1 e^{-2\alpha\ell}\left[E_S(0)\cosh\Gamma\ell - \frac{i\kappa_1}{\Gamma}E_P E_I^*(0)\sinh\Gamma\ell\right], \tag{21.123}$$

$$E_I^*(0) = R_2 e^{-2\alpha\ell}\left[E_I^*(0)\cosh\Gamma\ell + \frac{i\kappa_2}{\Gamma}E_P^* E_S(0)\sinh\Gamma\ell\right]. \tag{21.124}$$

If we incorporate mirror reflectance and intracavity losses into a single loss coefficient we can write

$$R_1 e^{-2\alpha\ell} = e^{-\alpha_S\ell}, \tag{21.125}$$

$$R_2 e^{-2\alpha\ell} = e^{-\alpha_I\ell}. \tag{21.126}$$

† For example, for a cavity of geometric length ℓ and crystal length L : $L_1 = (\ell - L) + n_1 L$.

Note that in terms of α_S and α_I

$$\tau_1 \simeq \frac{L_1}{c_0 \alpha_S \ell}; \tau_2 \simeq \frac{L_2}{c_o \alpha_I \ell}. \tag{21.127}$$

A nontrivial solution of the simultaneous equations requires once again that the determinant of coefficients be zero. Noting that $\Gamma^2 = \kappa_1 \kappa_2 |E_P|^2$, it is straightforward to show that

$$\cosh \Gamma \ell = 1 + \frac{\alpha_S \alpha_I \ell^2}{2 + \alpha_S \ell + \alpha_I \ell}. \tag{21.128}$$

In the DRO case $\alpha_S \ell$ and $\alpha_I \ell$ will both be small, so the threshold condition can be written as

$$\Gamma^2 \ell^2 \simeq \alpha_S \alpha_I \ell^2. \tag{21.129}$$

For the SRO threshold with $\alpha_I \ell >> \alpha_S \ell$

$$\Gamma^2 \ell^2 \sim 2\alpha_S \ell. \tag{21.130}$$

Thus, the SRO threshold is $2/\alpha_I$ greater than the DRO. We can express Eq. (21.129) in terms of the threshold pump intensity (in the crystal) by using Eqs. (21.106), (21.117), (21.125), and (21.126). If the mirror reflectance R_1 (signal) and R_2 (idler) are $\simeq 1$ and intracavity losses are negligible, then we can write Eqs. (21.125) and (21.126) as

$$\alpha_S \ell = 1 - R_1, \tag{21.131}$$

$$\alpha_I \ell = 1 - R_2. \tag{21.132}$$

The threshold pump intensity is

$$(I_P)_{threshold} = \frac{n_1 n_2 n_3 c_0^2 (1 - R_1)(1 - R_2)}{2\omega_1 \omega_2 d_{eff}^2 Z_0 \ell^2}. \tag{21.133}$$

21.10.1 Example

Let us consider a $LiNbO_3$ crystal with $\ell = 50$ mm; $n_1 = n_2 = 2.24, n_3 = 2.16, \lambda_1 = 647$ nm; $\lambda_2 = 3$ μm; $\lambda_3 = 532$ nm. We consider a DRO with 99% reflectance mirrors, so $(1 - R_1), (1 - R_2)$ are both 0.01. Substitution of these values in Eq. (21.133) gives $(I_P)_{threshold} \simeq 7 \times 10^5$ W m^{-2} or 0.7 W mm^{-2}: a modest pump intensity.

This illustrates the relative ease (*in principle*) with which a DRO can be operated. Indeed, many investigators have operated such devices using crystals such as $LiNbO_3$, Ba_2NaNbO_{15}† with pump powers of only a few milliwatts focused into the crystal. However, the practical difficulties of actually operating parametric oscillators should not be minimized. In order to achieve stable DRO operation much care must be given to the phase-matching condition (90° phase matching prevents beam walk-off), and the actual resonant conditions for signal and idler wavelengths. In practice this requires a stable single-frequency pump laser and careful cavity length control to achieve resonance for both signal and idler waves. These stringent requirements generally mean that pulsed parametric oscillators are generally operated in an SRO configuration, CW systems may be operated in a

† Colloquially referred to as 'BANANA'.

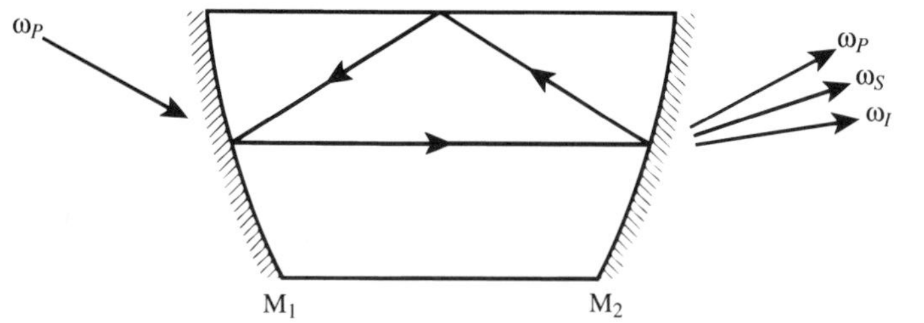

Fig. 21.11. Monolithic ring doubly resonant optical parametric oscillator. Mirror M_1 is transparent to the pump wavelength, while both mirrors M_1 and M_2 are reflective at signal and idler wavelengths. The angle between the emerging (depleted) pump, signal, and idler is exaggerated, but results from the different refractive indices of the nonlinear crystal at the three frequencies ω_P, ω_S, and ω_I.

DRO configuration. The signal chooses an appropriate cavity resonance and then the idler adjusts itself to satisfy energy conservation. Byer and coworkers[21.20] have achieved considerable success in the operation of CW $LiNbO_3$ DROs that use a monolithic ring configuration, like that shown in Fig. (21.11). In this ring geometry the parametric amplification process occurs unidirectionally along the single-pass pump direction.

21.11 Parametric Oscillator Tuning

Although noncollinear phase matching has been used in parametric oscillators, collinear phase matching is generally used. For example, the pump may be an extraordinary wave, travelling at angle θ to the optic axis, and signal and idler are ordinary waves. The collinear phase-matching condition can be restated as

$$\omega_3 n_e^{\omega_3}(\theta) = \omega_1 n_o^{\omega_1} + \omega_2 n_o^{\omega_2} \tag{21.134}$$

and for energy conservation

$$\omega_3 = \omega_1 + \omega_2. \tag{21.135}$$

Eqs. (21.134) and (21.135) can be solved to determine ω_1 and ω_2, for fixed ω_3 (say) provided the variation of $n_e(0)$ with the tuning parameter is known.

The use of angle, temperature, and electro-optic tuning is discussed in much more detail than here by Byer and his coworkers[21.20] and Shen[21.14].

In temperature tuning, which allows the use of 90° phase matching, we can write for the type-I phase matching previously mentioned

$$\omega_3 n_e^{\omega_3}(T) = \omega_1 n_o^{\omega_1}(T) + \omega_2 n_o^{\omega_2}(T). \tag{21.136}$$

At temperature $T + \Delta T$, the output frequencies change to $\omega_1 + \Delta\omega$ and $\omega_2 - \Delta\omega$. Provided we are not too close to the *degenerate* operating point,† where $\omega_1 = \omega_2$, we can write

$$\begin{aligned}\omega_3\left[n_e^{\omega_3}(T) + \frac{\partial n_e^{\omega_3}}{\partial T}\Delta T\right] =& (\omega_1 + \Delta\omega)\left[n_o^{\omega_1}(T) + \frac{\partial n_o^{\omega_1}}{\partial T}\Delta T + \frac{\partial n_o^{\omega_1}}{\partial \omega}\Delta\omega\right]\\ &+ (\omega_2 - \Delta\omega)\left[n_o^{\omega_2}(T) + \frac{\partial n_o^{\omega_2}}{\partial T}\Delta T - \frac{\partial n_o^{\omega_2}}{\partial \omega}\Delta\omega\right],\end{aligned} \tag{21.137}$$

† At the degenerate operating points where $\omega_1 = \omega_2 = \omega_3/2$, the properties of the emitted degenerate radiation can become very interesting. The device becomes a subharmonic generator – providing two identical photons for each pump photon. In the ideal case this is called 'squeezed light.' The interested reader should consult Yariv[21.21].

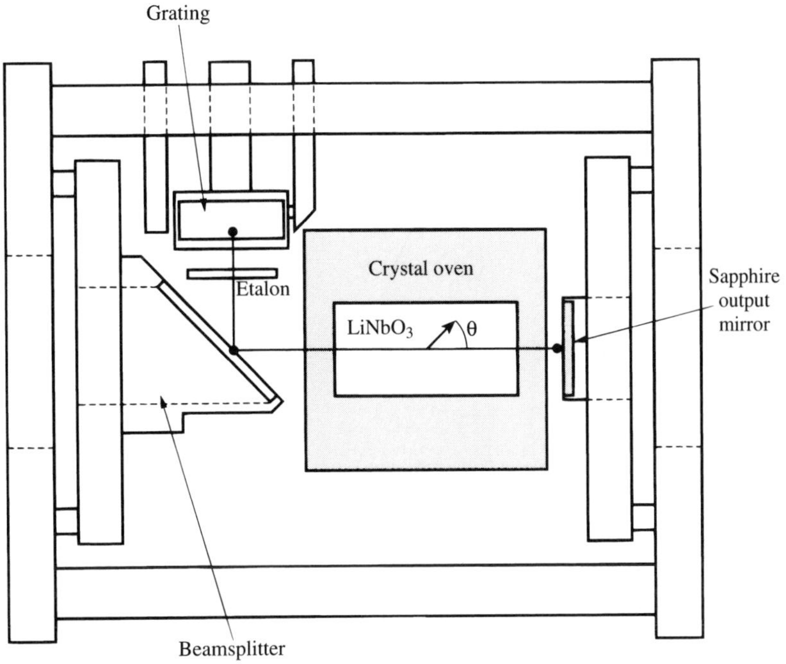

Fig. 21.12. Angle-tuned $LiNbO_3$ singly resonant[21.7] parametric oscillator.

which gives

$$\Delta\omega = \frac{[\omega_3(\partial n_e^{\omega_3}/\partial T) - \omega_1(\partial n_o^{\omega_1}/\partial T) - \omega_2(\partial n_o^{\omega_2}/\partial T)]\Delta T}{n_o^{\omega_1} - n_o^{\omega_2} + \omega_1(\partial n_o^{\omega_1}/\partial\omega) - \omega_2(\partial n_o^{\omega_2}/\partial\omega)} \tag{21.138}$$

Near the degenerate operating condition $T = T_0$ and $\omega_1 = \omega_3/2$. In this case the calculation of the tuning equation (21.138) cannot neglect higher order terms in the Taylor sense expansions of the variation of $n_o^{\omega_1}, n_o^{\omega_2}$ with frequency. In this case

$$\Delta\omega = \left\{\frac{\omega_3\left[\partial n_e^{\omega_3}/\partial T - \partial n_e^{\omega_1}/\partial T\right]_{T=T_0}}{\left[2\partial n_o^{\omega_1}/\partial\omega + \omega_3(\partial^2 n_o^{\omega_1}/\partial\omega^2)/2\right]_{\omega_1=\omega_3/2}}\right\}(\Delta T)^{1/2}. \tag{21.139}$$

Fig. (21.9) shows temperature tuning curves calculated in this way for $LiNbO_3$ parametric oscillators.

In angle tuning, which can be accomplished in an arrangement shown in Fig. (21.12), a change in angle from θ to $\theta + \Delta\theta$ changes the operating frequencies to $\omega_1 + \Delta\omega$ and $\omega_2 - \Delta\omega$. Not too close to the degeneracy point

$$\Delta\omega \simeq \omega_3 \frac{\partial n_e^{\omega_3}}{\partial\theta}\Delta\theta \Bigg/ \left[\left(\omega_1\frac{\partial n_o^{\omega_1}}{\partial\omega_1} - \omega_2\frac{\partial n_o^{\omega_2}}{\partial\omega_2}\right) + n_o^{\omega_1} - n_o^{\omega_2}\right]. \tag{21.140}$$

Near the degenerate operating point

$$\Delta\omega = \left[\omega_3\left(\frac{\partial n_e^{\omega_3}}{\partial\theta}\right)_{\theta_0} \Bigg/ \left(2\frac{\partial n_o}{\partial\omega} + \frac{\omega_3}{2}\frac{\partial^2 n_o}{\partial\omega^2}\right)_{\omega=\omega_3/2}\right]^{1/2}(\Delta\theta)^{1/2}. \tag{21.141}$$

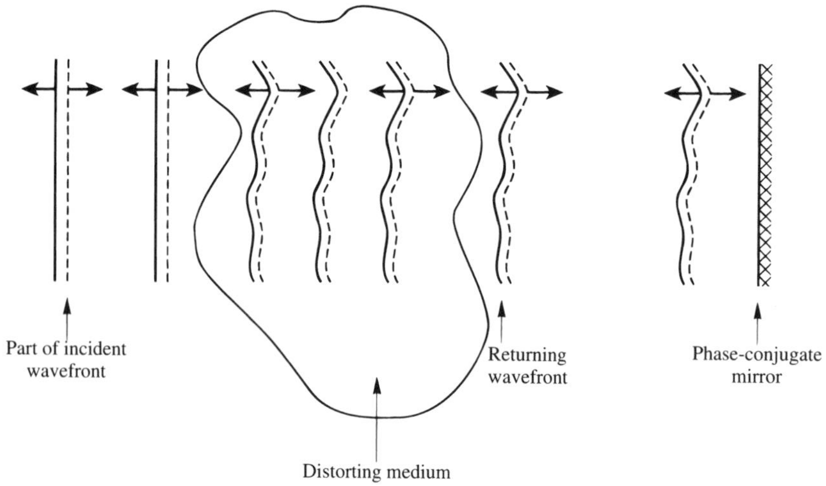

Fig. 21.13. Schematic illustration of wave front correction of a distorted wave front by reflection from a conjugate mirror.

21.12 Phase Conjugation

Before completing this chapter two additional nonlinear optical phenomena are worthy of brief mention. The first of these, optical phase conjugation, uses a nonlinear mixing process to generate the perfect counter-propagating wave, both in direction and phase, of an incident wave.

For example, an incident wave of the form

$$E_1 = E_0 e^{i(\omega t - kz)} \tag{21.142}$$

will, after reflection at a phase-conjugate mirror become

$$E_2 = aE_0^* e^{i(\omega t + kz)}, \tag{21.143}$$

where a is a constant that indicates the reflection efficiency of the 'mirror.' The reflected wave retraces the path of the incident wave and its phase variation in space will replicate exactly the phase variation of the input wave. If one could observe the variation in time of the moving phase fronts of the reflected wave, it would appear exactly as if an equivalent 'movie' of the incident phase fronts were being run backwards. The interesting consequence of this is that if a wave travels through a medium that distorts its optical phase fronts, such as an aberrated optical system, an optical fiber, or along an atmospheric path, then after a reflection at a phase conjugate mirror the wave will retrace its path and the phase distortions produced on its outward path to the phase conjugate mirror will be exactly unfolded on the return path as shown schematically in Fig. (21.13). The wave returns to its starting point undistorted.

Of particular interest is the ability of phase conjugation to compensate for the random fluctuations imposed on the phase front of an optical wave as it traverses a path through the atmosphere. This is of interest in connection with line-of-sight coherent optical communication through the atmosphere, in optical astronomy, and in imaging through the turbulent atmosphere. The atmosphere behaves as a random medium in the sense that its refractive index varies randomly in time according to

$$n(\mathbf{r}, t) = n_0 + n_1(\mathbf{r}, t).$$

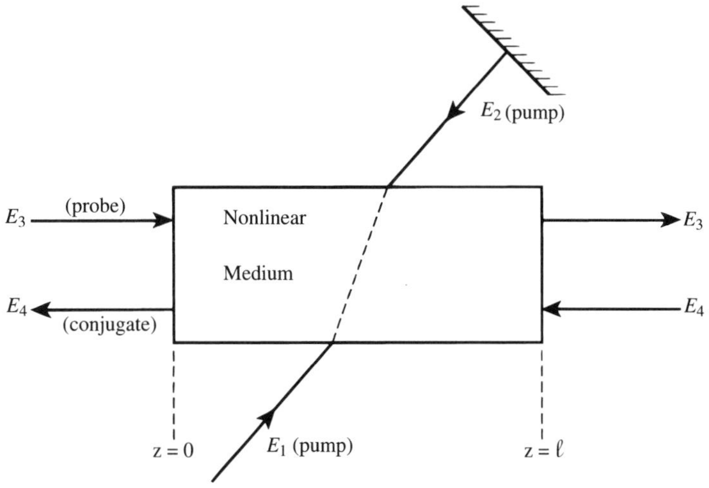

Fig. 21.14. Schematic arrangement for phase conjugation by four wave mixing.

The fluctuating contribution to the index, $n_1(\mathbf{r}, t)$, contributes to the haze seen over a hot highway, and to the twinkling of distant lights or stars at night. These fluctuations occur on a time scale up to about 1 kHz.

To illustrate the attractiveness of phase conjugation in correcting for optical properties of the atmosphere a futuristic example can be given. Suppose that it is desired to focus an intense laser beam onto a target through the turbulent atmosphere, for example, in a 'Star Wars' application. It is possible, in principle, to generate the phase conjugate of a weak signal that has travelled back from target to source location and then amplify this weak phase conjugate signal and project it back to a small focal spot on the target.

How do we make a phase-conjugate mirror? We use a nonlinear mixing process involving a third order nonlinear susceptibility tensor $\chi^{(3)}(-\omega_4, \omega_1, \omega_2, -\omega_3)$. Because four frequencies are involved this process is called four-wave mixing. In practice all the frequencies $\omega_1, \omega_2, \omega_3$, and ω_4 may be the same. A common geometry for doing this is shown in Fig. (21.14). In this arrangement a pump laser beam passes through the nonlinear medium and is then reflected back on itself. The probe beam is incident on the nonlinear medium at an arbitrary angle. In a phase-matched situation where all waves have the same frequency, if $\mathbf{k}_1 = -\mathbf{k}_2$; then automatically $\mathbf{k}_4 = -\mathbf{k}_3$. In and of itself, this explains how the phase-conjugate wave is generated in the counter-propagating direction to the probe.

If we write the electric fields of the waves involved in the form

$$\mathbf{E}_i(t) = \tfrac{1}{2} E_i e^{i(\omega_i t - \mathbf{k}_i \cdot \mathbf{r})} + cc \tag{21.144}$$

then the polarization that results at frequency $\omega_4 = \omega_1 + \omega_2 - \omega_3$ can be written (see Chapter 20) as

$$P_i^{\omega_4} = 6\epsilon_0 \chi_{ijk\ell}(-\omega_4, \omega_1, \omega_2, -\omega_3) E_{1j} E_{2k} E_{3\ell}^* e^{-i(\mathbf{k}_1 + \mathbf{k}_2 - \mathbf{k}_3)\cdot\mathbf{r}}. \tag{21.145}$$

If specific polarizations are chosen for all the waves involved, then a simplified version of Eq. (21.145) is

$$P^{\omega_4} = \epsilon_0 \chi^{(3)} E_1 E_2 E_3^* e^{-i(\mathbf{k}_1 + \mathbf{k}_2 - \mathbf{k}_3)\cdot\mathbf{r}}. \tag{21.146}$$

If all four waves have the same frequency ω, then the polarization $P(\omega)$ can arise in five ways – from the combinations $E_1E_2E_3^*$, $E_1E_1^*E_3$, $E_2E_2^*E_3$, $E_3E_3^*E_3$, and $E_4E_4^*E_3$. Other combinations would not generate frequency ω, or be phase matched. For example the combination $E_1E_2E_3$ would generate a nonlinear polarization at frequency 3ω. If we substitute the nonlinear polarization P^{ω_4} in the wave equation, in an analogous manner to obtaining Eqs. (21.18) and (21.19), we get in a lossless medium

$$\frac{dE_4}{dz} = \frac{i\omega}{2}\sqrt{\frac{\mu}{\epsilon}}\epsilon_0\chi^{(3)}\left(|E_1|^2+|E_2|^2+|E_3|^2+|E_4|^2\right)E_4 + \frac{i\omega}{2}\sqrt{\frac{\mu}{\epsilon}}\epsilon_0\chi^{(3)}E_1E_2E_3^*, \tag{21.147}$$

where we have assumed that the process is phased match with $\mathbf{k}_1 = -\mathbf{k}_2$, and $\mathbf{k}_4 = -\mathbf{k}_3$ the + signs on the RHS of Eq. (21.147) result because E_4 is propagating in the $-z$ direction. In a similar way

$$\frac{dE_3}{dz} = -\frac{i\omega}{2}\sqrt{\frac{\mu}{\epsilon}}\epsilon_0\chi^{(3)}\left(|E_1|^2+|E_2|^2+|E_3|^2+|E_4|^2\right)E_3 - \frac{i\omega}{2}\sqrt{\frac{\mu}{\epsilon}}\epsilon_0\chi^{(3)}E_1E_2E_4^*. \tag{21.148}$$

If the pump beams are intense relative to the probe and conjugate beams, then we can neglect terms in $|E_3|^2$ and $|E_4|^2$ to get

$$\frac{dE_4}{dz} = \frac{i\omega}{2}\sqrt{\frac{\mu}{\epsilon}}\epsilon_0\chi^{(3)}\left(|E_1|^2+|E_2|^2\right)E_4 + \frac{i\omega}{2}\sqrt{\frac{\mu}{\epsilon}}\epsilon_0\chi^{(3)}E_1E_2E_3^* \tag{21.149}$$

and

$$\frac{dE_3}{dz} = -\frac{i\omega}{2}\sqrt{\frac{\mu}{\epsilon}}\epsilon_0\chi^{(3)}\left(|E_1|^2+|E_2|^2\right)E_3 - \frac{i\omega}{2}\sqrt{\frac{\mu}{\epsilon}}\epsilon_0\chi^{(3)}E_1E_2E_4^*. \tag{21.150}$$

If we treat $|E_1|^2$ and $|E_2|^2$ as constants (so we are neglecting any depletion effects) then we can redefine E_3 and E_4 as

$$E_3' = E_3\exp\left[-i\omega\sqrt{\frac{\mu}{\epsilon}}\epsilon_0\chi^{(3)}(|E_1|^2+|E_2|^2)\right], \tag{21.151}$$

$$E_4' = E_4\exp\left[i\omega\sqrt{\frac{\mu}{\epsilon}}\epsilon_0\chi^{(3)}(|E_1|^2+|E_2|^2)\right]. \tag{21.152}$$

All this really means is that we have multiplied the fields E_3 and E_4 by a constant phase factor. Eqs. (21.149) and (21.150) now reduce to the simple forms

$$\frac{dE_4'}{dz} = \frac{i\omega}{2}\sqrt{\frac{\mu}{\epsilon}}\epsilon_0\chi^{(3)}E_1E_2E_3'^*, \tag{21.153}$$

$$\frac{dE_3'^*}{dz} = \frac{i\omega}{2}\sqrt{\frac{\mu}{\epsilon}}\epsilon_0\chi^{(3)}E_1^*E_2^*E_4'. \tag{21.154}$$

Writing $\kappa^* = (\omega/2)\sqrt{\mu/\epsilon}\epsilon_0\chi^{(3)}E_1E_2$ gives us the final coupled equations describing phase conjugation

$$\frac{dE_3}{dz} = -i\kappa^* E_4^*, \tag{21.155}$$

$$\frac{dE_4}{dz} = i\kappa^* E_3. \tag{21.156}$$

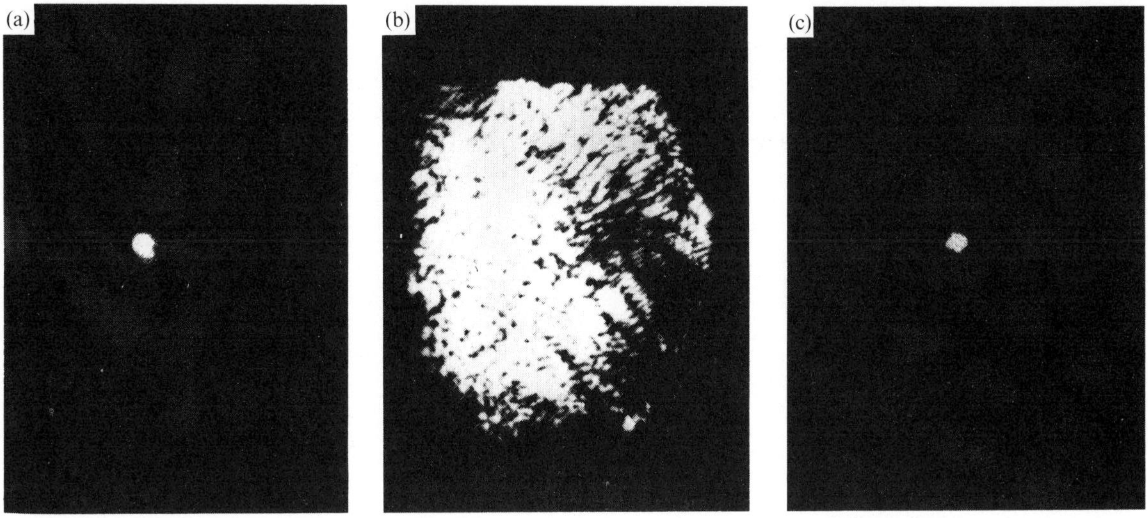

Fig. 21.15. Illustration of how a phase-conjugate mirror can recreate an undistorted wavefront from one that has been distorted in propagating through a medium (or aberrated optical system). Photograph (a) shows an unperturbed focused laser beam; (b) shows this beam after it has been phase aberrated on passage through an etched glass plate; and (c) shows the beam after it has reflected from a plane-conjugate mirror and repassed through the same etched glass plate. From reference [21.24]. Reproduced with the permission of the publisher.

If we take $E_3(z = 0) = E_P$ and $E_4(z = \ell) = 0$, which represents the practical situation of a finite pump entering the nonlinear medium, and zero initial conjugate wave entering the nonlinear medium at $z = \ell$, then the solution of Eqs. (21.155) and (21.156) is[21.22],[21.23]

$$E_4(z = 0) = -i \left[\frac{\kappa^*}{|\kappa|} \tan(|\kappa|\ell) \right] E_P^* \tag{21.157}$$

and

$$E_3(z = \ell) = E_P / \cos(|\kappa|\ell). \tag{21.158}$$

Eq. (21.157) shows that the nonlinearly reflected wave at $z = 0$ is the conjugate of the pump multiplied by a 'gain' factor $-i(\kappa^*/|\kappa|)\tan(|\kappa|\ell)$. This 'gain' factor will not be significant in general without sufficient pump intensity. We can illustrate this by considering the use of carbon disulphide, CS_2, a material with a large $\chi^{(3)}$.

21.12.1 Phase Conjugation in CS_2

For this medium $\epsilon_0\chi^{(3)} \equiv 6\epsilon_0\chi_{yxxy}(-\omega, \omega, \omega, -\omega) \simeq 2.025 \times 10^{31}$ (C m V^{-3}) and n=1.62. For this element of the nonlinear susceptibility tensor the pump beams ω_1, ω_2 are polarized orthogonally to the probe and conjugate beams. If $E_1 = E_2 = E$, then $E^2 = 2ZI_P$ and

$$|\kappa| = \left(\frac{\pi c_0}{\lambda_0 n^2} \right) (\epsilon_0\chi^{(3)})(2Z_0^2 I_P) \tag{21.159}$$

At a pump intensity of 1 MW mm^2, $|\kappa| = 38.8$ m^{-1} so for a 'gain' of 1 we need a length of $\simeq$ 26 mm. Intensities of this magnitude require the use of pulsed lasers, but media with much larger effective $\epsilon_0\chi^{(3)}$ are available that allow the observation of phase conjugation with CW lasers.

The practical ability of a phase-conjugate mirror to remove distortions from a laser beam is illustrated in Fig. (21.15).

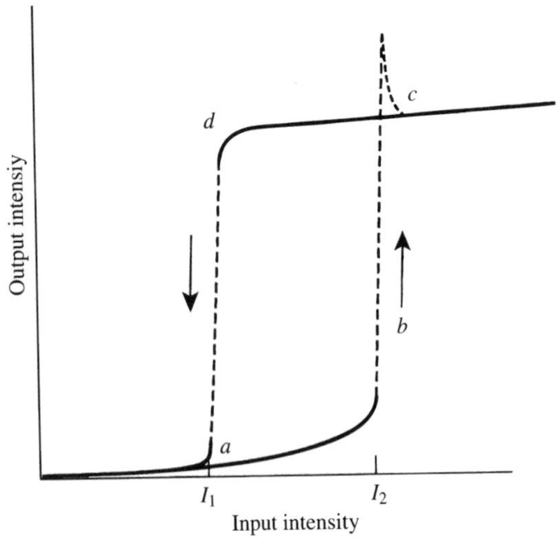

Fig. 21.16. Input intensity–output intensity characteristic curves for a bistable medium.

21.13 Optical Bistability

Optical bistability is a phenomenon in which the output intensity from an optical system can have two potential values for the same value of input intensity for some range of value of these parameters. In other words its $I_{out} - I_{in}$ curve appears as shown schematically in Fig. (21.16). This is just like the hysteresis curve familiar in an analysis of the magnetization of ferromagnetic materials[21.25].

Just as the magnetic recording medium on a disk or tape has the ability to store information because the magnetization exhibits hysteresis, optical bistable devices can perform a memory function. For example, if the input intensity is increased from zero to I_2 to point b in Fig. (21.16) then the output intensity makes a rapid transition to point c. The output intensity could at this point be regarded as representing a '1', and this value will be sustained as long as the input intensity is maintained above value I_1 – point d on the curve. The input intensity must be reduced below I_1 (after it has been above I_2) for the output intensity to return to a low value point a, which can be regarded as representing a '0'.

Many bistable systems consist of a nonlinear medium placed between the mirrors of an optical cavity, and we will use such a system to give a brief introduction to the phenomenon. The model we will discuss is the one first described by Szöke, Danen, Goldhar, and Kurnit[21.26]. A Fabry–Perot interferometer containing an absorbing medium, shown schematically in Fig. (21.17), is illuminated at the peak of one of its transmission resonances. The input wavelength therefore satisfies $m\lambda/2 = \ell$, where ℓ is the spacing of the two mirrors, each assumed to have reflectance R. The intensity absorption coefficient of the medium is α.

We can model the system in a fairly simple way. If the input electric field amplitude is E_i, then the transmitted electric field amplitude is (see Section 4.5)

$$E_t = E_i T e^{-\alpha\ell/2}\left(1 + Re^{-\alpha\ell}e^{-i\delta} + R^2 e^{-2\alpha\ell}e^{-2i\delta} + \ldots\right), \qquad (21.160)$$

where T and R are the transmittance and reflectance, respectively, of the mirrors, which are assumed to be identical. the parameter δ is the phase shift experienced on a single pass from one mirror to the other. Summing the geometric series in

Fig. 21.17. Schematic arrangement for illustration of absorptive optical bistability. A Fabry–Perot interferometer having mirrors of reflectance R contains an absorbing medium. The various travelling wave directions involved are shown.

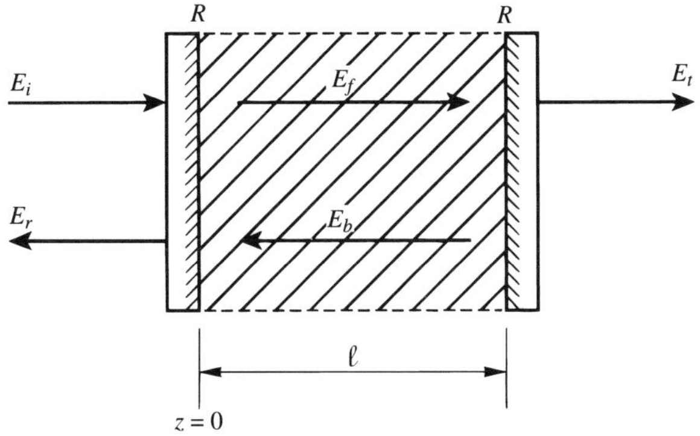

Eq. (21.160) gives

$$E_t = \frac{T E_i e^{-\alpha\ell/2}}{1 - Re^{-i\delta}e^{-\alpha\ell}}. \tag{21.161}$$

At resonance $\delta = 0$ and since intensity $\propto$ |electric field|^2 Eq. (21.161) gives

$$\frac{I_t}{I_i} \simeq \frac{e^{-\alpha\ell}}{\left[\left(1 - Re^{-\alpha\ell}\right)/(1-R)\right]^2}. \tag{21.162}$$

If $\alpha\ell << 1$, we can expand the exponential terms in Eq. (21.162) to give

$$\frac{I_t}{I_i} \simeq \frac{1}{(1+\alpha\ell)\left[1 + \left(\frac{R\alpha\ell}{1-R}\right)\right]^2} \simeq \frac{1}{(1+k)^2}, \tag{21.163}$$

provided $1 - R >> \alpha\ell$ and where

$$k = \frac{R\alpha\ell}{1-R}. \tag{21.164}$$

Interesting things happen if α is intensity-dependent (as it actually always is). If the absorbing medium is treated as a simple two-level homogeneously broadened system, then its absorption coefficient will be

$$\alpha = \frac{\alpha_0}{1 + I_e/I_s}, \tag{21.165}$$

where α_0 is its absorption coefficient at zero intensity, I_e is the effective intracavity intensity, and I_s is the saturation intensity. When this saturation of absorption is included we can write

$$k = \frac{k_0}{1 + I_e/I_s}, \tag{21.166}$$

where $k_0 = R\alpha_0\ell/(1-R)$. We need to relate the effective intracavity intensity to the incident intensity, bearing in mind that the intracavity field E_{cavity} is a standing wave, which for $\alpha\ell, T << 1$, will be

$$E_{cavity} = E_f \cos(\omega t - kz) + E_b \cos(\omega t + kz), \tag{21.167}$$

where E_f, E_b are the forward and backward travelling waves amplitudes, respectively, and in this case $E_f \simeq E_b$. Rearrangement of Eq. (21.167) gives

$$E_{cavity} = 2E_b \cos\omega t \cos kz. \tag{21.168}$$

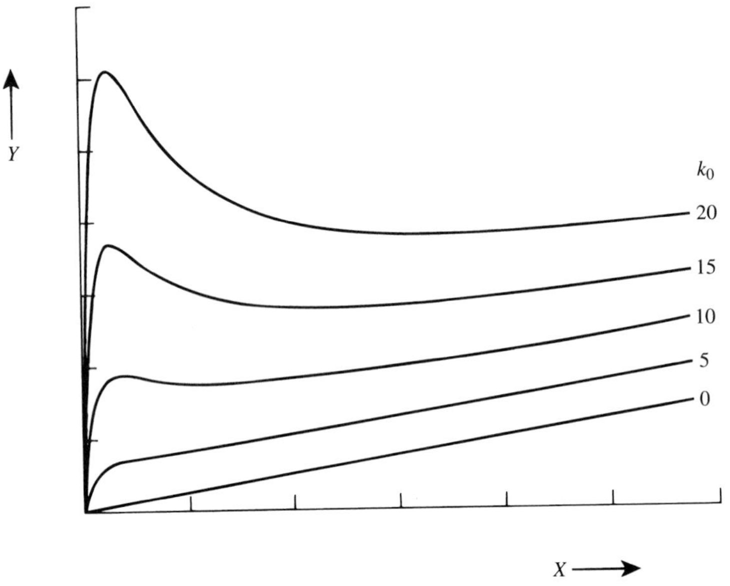

Fig. 21.18. Bistability curves calculated from Eq. (21.172)

Since $E_b = E_t/\sqrt{T}$, Eq. (21.168) can be rewritten as

$$\frac{I_{cavity}}{I_t} = \frac{2}{T}\cos^2 \omega t \cos^2 kz. \tag{21.169}$$

The average of $\cos^2 kz$ over the whole cavity is $\frac{1}{2}$, so

$$\overline{I_{cavity}} = \frac{I_t}{T}\cos^2 \omega t. \tag{21.170}$$

Therefore, we can replace the effective intracavity intensity by $I_e = I_t/T$, and from Eq. (21.163)

$$\frac{I_i}{I_t} = \left[1 + \left(\frac{k_0}{1 + I_t/TI_s}\right)\right]^2. \tag{21.171}$$

If we write $Y = I_i/TI_s$ and $X = I_t/TI_s$, where Y and X now represent normalized input and output intensities, respectively, then

$$Y = X\left[1 + \frac{k_0}{1 + X}\right]^2. \tag{21.172}$$

Eq. (21.172) is a fundamental equation of bistability. For a given value of Y, there are two possible values for X. However, for true bistable behavior the curve of Y versus X must pass through a minimum between the two values of X. Fig. (21.18) shows curves calculated from Eq. (21.172) for various values of k_0. Note, for example, that the curve for $k_0 = 15$ will permit bistability, where $k_0 = 5$ will not. We can identify when bistable behavior can occur by noting that it requires $dY/dX < 0$.

The condition $dY/dX = 0$ gives, from Eq. (21.172),

$$X = \frac{k_0 - 2 \pm \sqrt{k_0(k_0 - 8)}}{2}. \tag{21.173}$$

For a physically meaningful value of X this requires $k_0 \geq 8$: that is,

$$\frac{\alpha_0 \ell R}{1-R} > 8, \tag{21.174}$$

which for $1 - R \simeq 0$ gives

$$\frac{\alpha_0 \ell}{T} > 4. \tag{21.175}$$

The actual behavior of a real bistable system will deviate from the predictions of this simple steady-state absorptive optical bistable model because it neglects simple features of real systems such as linear background absorption, transverse intensity profiles, and the effect of standing waves in the cavity. Bistability can also result not only from absorptive effects but from dispersive effects in the nonlinear medium. A full review of the whole subject has been given by Gibbs[21.27]. In potentially important applications of bistability the state of the bistable system is switched from one of its two stable states to the other by the action of a light signal. Thus, the transient behavior of the system is important.

The ability of 'light to control light' in an optical bistable system opens up the possibility of using such devices in all-optical computers, and as all-optical switches in optical communication systems. The challenge is to make these devices smaller, faster, and practical economically.

21.14 Practical Details of the Use of Crystals for Nonlinear Applications

One of the commonest methods for growing nonlinear crystals is the Czochralski method, discussed in Chapter 8. The resultant boule will be oriented according to the orientation of the seed crystal – quite often along the c axis of the crystal, which is identical with the optic axis in uniaxial crystals. After growth the orientation of the boule can be confirmed by X-ray crystallographic techniques. For angle-tuned phase-matching a section must be cut from the boule so its faces are perpendicular to the phase-matching direction, as shown in Fig. (21.3b). To minimize Fresnel reflection losses it may be desirable to antireflection coat the input and output faces of the nonlinear crystal. Some crystals, like ADP, KDP, are hygroscopic and must be protected from atmospheric moisture. Such crystals can be placed in a hermetically sealed container with antireflection coated input and output windows. If the windows have a refractive index close to that of the crystal and an appropriate *index-matching* fluid is placed around the crystal then Fresnel losses can be minimized.

Efficient nonlinear generation frequently requires the use of very intense pulsed lasers. Therefore, of particular importance in the selection of materials for nonlinear applications is the resistance of those materials to optical damage. Damage can result from thermal heating, multiphoton absorption, self-focusing and dielection breakdown, among other mechanisms[21.28]. The intensities necessary to induce damage depend on the material, wavelength, pulse duration of the intense source, and whether the nonlinear material is being operated in a phase-matched arrangement[21.29]. Semiconductor materials damage in the 0.5–1 TW m^{-2} range† for 100 ns pulses. Sapphire damages around 250 TW m^{-2} (250 MW mm^{-2}) for

† 1 TW (terawatt) is 10^{12} W.

pulses ~10 ns long. Most nonlinear materials have damage thresholds in the 0.1–10 TW m^{-2} range, which for a Q-switched laser pulse 10 ns long limits maximum pulse energies to 0.1 J mm^{-2}. It is easy to see from these numbers how relatively small energy laser pulses focused to small spot sizes can lead to damage. A 1 J laser pulse, 10 ns long focused to a Gaussian spot size of 1 μm has a peak intensity of 6.4 TW m^{-2}, and generates a peak electric field at the focus of almost 70 MV m^{-1}. This is well in excess of the value necessary to produce dielectric breakdown, which is ~ 3 MV m^{-1}.

The phenomenon of self-focusing, which can lead to optical damage is an interesting one in its own right. This is a nonlinear phenomen in which the refractive index becomes a function of the laser intensity, namely:

$$n = n_0 + n_1 I. \tag{21.176}$$

If a fundamental Gaussian-mode laser beam enters such a medium it is focused because the radial intensity variation in the Gaussian beam leads to the production of a positive lens. the radial index profile in the medium will be

$$n(r) = n_0 + n_1 I_0 e^{-r^2/2w^2}, \tag{21.177}$$

which leads to focusing. However, the effect can be negated by diffraction, which makes the Gaussian beam want to diverge more rapidly as its spot size gets smaller. A more detailed discussion can be found elsewhere[21.14],[21.21]. An interesting consequence of self-focusing is for any spatial nonuniformity of a high intensity laser beam to be amplified so that the beam may break up into a collection of tiny filaments[21.30]. In very high energy pulsed laser systems, such as are used for laser-induced fusion studies (see Chapter 24), the laser profile must be continuously smoothed, for example, by spatial filtering (see Chapter 16), to prevent self-focusing and internal damage.

21.15 Problems

(21.1) Use Eqs. (21.64) and (21.65) to calculate the phase-matching conditions for second harmonic generation of a 10.6 μm laser.

(21.2) Prove Eq. (21.139).

(21.3) Prove that in a uniaxial crystal

$$\frac{dn_e^{\omega_3}(\theta)}{d\theta} = -\frac{n_e^{\omega_3}(\theta)^3}{2}\sin 2\theta\left[\left(\frac{1}{n_e^{\omega_3}}\right)^2 - \left(\frac{1}{n_o^{\omega_3}}\right)^3\right],$$

where θ is the angle of wave propagation with respect to the optic axis.

(21.4) Prove Eq. (21.140).

(21.5) Prove Eq. (21.141).

(21.6) Prove that in a phase-matched up-conversion process, the conversion efficiency of the detected frequency ω_2 to the up-converted frequency ω_3 is

$$\frac{I_3(z)}{I_2(0)} = \frac{\omega_3}{\omega_2}\sin^2\Gamma(z),$$

where

$$\Gamma = \sqrt{(\kappa_3^2|E_1|^2)},$$

where E_1 is the field amplitude of the pump beam at frequency ω, and $K = \omega d_{eff}/nc$. Neglect any depletion of the strong pump beam. Explain why the z dependence of the conversion is different from the second harmonic generation conversion efficiency given by Eq. (21.98).

(21.7) $BeSO_4.4H_2O$ is a crystal used for generation of ultraviolet light by second harmonic generation. Its refractive indices can be described by the Sellmaier equations

$$n_o^2 = 2.1545 + \frac{0.00835}{\lambda^2 - 0.01606} - 0.03573\lambda^2,$$

$$n_e^2 = 2.0335 + \frac{0.00806}{\lambda^2 - 0.01354} - 0.01970\lambda^2$$

with λ given in μm.

Calculate the phase matching angles for both type-I and type-II second harmonic generation with a fundamental input at 530 nm.

References

[21.1] J.M. Manley and H.E. Rowe, 'General energy in nonlinear resistances,' *Proc. IRE*, **47**, 2115, 1959.

[21.2] P.A. Franken, A.E. Hill, C.L. Peters, and G. Weinrich, 'Generation of optical harmonics,' *Phys. Rev. Lett.*, **7**, 118–119 1961.

[21.3] R.W. Caldwell and H.Y. Fan, 'Optical properties of tellurium and selenium,' *Phys. Rev.*, **114**, 664–675, 1959.

[21.4] M. Born and E. Wolf, *Principles of Optics*, 6th Edition, Pergamon Press, Oxford, 1980.

[21.5] M.V. Hobden, 'The dispersion of the refractive indices of proustite (Ag_3AsS_3),' *Opto-Electronics*, **1**, 159, 1969.

[21.6] F. Zernike and J.E. Midwinter, *Applied Nonlinear Optics*, Wiley, New York, 1973.

[21.7] R.L. Byer, 'Parametric oscillators and nonlinear materials,' in *Nonlinear Optics*, P.F. Harper and B.S. Wherrett, Eds., Academic, London, 1977.

[21.8] H.A. Smith, and H. Mahr, 'An infrared detector for astronomy using up-conversion techniques,' paper presented at the International Quantum Electronics Conference, Kyoto, Japan, September 1970.

[21.9] J.E. Midwinter, 'Image conversion from 10.6 μm to the visible in lithium niobate,' *Appl. Phys. Lett.*, **12**, 68–70, 1968.

[21.10] R.L. Byer in *Nonlinear Infrared Generation*, Y.R. Shen, Ed., Springer-Verlag, Berlin, 1977.

[21.11] F. Zernike, in 'Methods of experimental physics,' Vol. XV *Quantum Electronics*, Part B, C.L. Tang, Ed., Academic, New York, 1979.

[21.12] B. Lax, R.L. Aggarwal, and G. Favrot, 'Far-infrared step-tunable coherent radiation source: 70 μm to 2 mm,' *Appl. Phys. Lett.*, **23**, 679–681, 1973.

[21.13] B. Lax and R.L. Aggarwal, 'Far-infrared generation and applications', *J. Opt. Soc. Am.*, **64**, 533, 1974.

[21.14] Y.R. Shen, *The Principles of Nonlinear Optics*, Wiley, New York, 1984.

[21.15] R. Frey and F. Pradere, 'High-efficiency narrow-linewidth Raman amplification and spectral compression,' *Opt. Lett.*, **5**, 374–376, 1980.

[21.16] J.J. Tiee and C. Wittig, 'CF_4 and NOCL molecular lasers operating in the 16-μm region,' *Appl. Phys. Lett.*, **30**, 420–422, 1977.

[21.17] W. Lenth, R.M. Macfarlane and W.P. Risk, 'Generation of blue-green laser radiation using nonlinear frequency up conversion,' paper VL2.3 presented at the

1988 meeting of the Lasers and Electro-optics Society, Santa Clara, CA, November 2-4, 1988.
[21.18] M.J. Colles and C.R. Pidgeon, 'Tunable lasers,' *Rept. Prog. Phys.*, **38**, 329, 1975.
[21.19] R.L. Byer, 'Optical parametric oscillators,' in *Quantum Electronics*, Vol. 1B, H. Rabin and C.C. Tang, Eds., Academic Press, New York, 1975.
[21.20] R.C. Eckardt, C.D. Nabors, W.J. Kozlovsky, and R.L. Byer, 'Optical parametric oscillator frequency tuning and control,' *J. Opt. Soc. Am.*, **B8**, 646–667, 1991.
[21.21] A. Yariv, *Quantum Electronics*, 3rd Edition, Wiley, New York, 1989.
[21.22] A. Yariv and D.M. Pepper, 'Amplified reflection, phase conjugation and oscillation in degenerate four-wave mixing,' *Opt. Lett.*, **1**, 16–18, 1977.
[21.23] G.C. Bjorklund, 'Conjugate wave-front generation and image reconstruction by four-wave mixing,' *Appl. Phys. Lett.*, **31**, 592–594, 1977.
[21.24] C.R. Guiliano, 'Applications of optical phase conjugation,' *Phys. Today*, April 1981, p. 27.
[21.25] I.D. Mayergoyz, *Mathematical Models of Hysteresis*, Springer-Verlag, New York, 1991.
[21.26] A. Szöke, V. Danen, J. Goldhar, and N.A. Kurnit, 'Bistable optical element and its applications,' *Appl. Phys. Lett.*, **15**, 376–379, 1969.
[21.27] H.M. Gibbs, *Optical Bistability: Controlling Light with Light*, Academic Press, Orlando, FL, 1985.
[21.28] J.F. Ready, *The Effects of High Power Laser Radiation*, Academic Press, New York, 1971.
[21.29] A.G. Glass and A.H. Guenther, 'Damage in laser materials,' *Appl. Opt.*, **11**, 832–840, 1972.
[21.30] S.C. Abbi and H. Malin, 'Correlation of filaments in nitrobenzene with laser spikes,' *Phys. Rev. Lett.*, **26**, 604–606, 1971.

22

Detection of Optical Radiation

22.1 Introduction

In this chapter we will review some of the fundamentals of the optical detection process. This will include a discussion of the randomly fluctuating signals, or *noise* that appear at the output of any detector. We will then examine some of the practical characteristics of various types of optical detector. The chapter will conclude with a discussion of the limiting detection sensitivities of important detectors used in various ways.

Photon detectors operate by absorbing the photons coming from a source and using the absorbed energy to produce a change in the electrical characteristics of their active element(s). This can occur in many ways. In a photomultiplier or vacuum photodiode the incoming photons are absorbed in a photoemissive surface and through the photoelectric effect free electrons are produced. These electrons can be accelerated and detected as an electrical current. In a semiconductor photodiode or photovoltaic detector, absorption of a photon at a *p–n* or *p–i–n* junction creates an electron–hole pair. The electron and hole separate because of the energy barrier at the junction – each carrier moves to the region where it can reduce its potential energy, as shown in Fig. (22.1).

Thermal radiation detectors use the heating effect produced by absorbed photons to change some characteristic of the detector element. In a bolometer, the heating of carriers changes their mobility and consequently the resistivity of the detector element. In a thermopile the heating effect is used to generate a voltage through the thermoelectric (Seebeck) effect.

Pyroelectric detectors utilize the change in surface charge that results when certain crystals (ones that can possess an internal electric dipole moment) are heated.

22.2 Noise

22.2.1 Shot Noise

The ability of a photodetector to detect an incoming light signal is limited by the intrinsic fluctuations, or noise, both of the incoming light itself and of the background electrical current fluctuations generated by the detector. In the photon description of a monochromatic light beam an incoming beam of intensity $< I >$ has an average photon flux associated with it of $< N >$ photons m^{-2} where

$$< N >= \frac{< I >}{h\nu}. \tag{22.1}$$

If the rate of arrival of photons at the detector is examined closely it will be observed to fluctuate about this average value. Strictly speaking, all we can ever

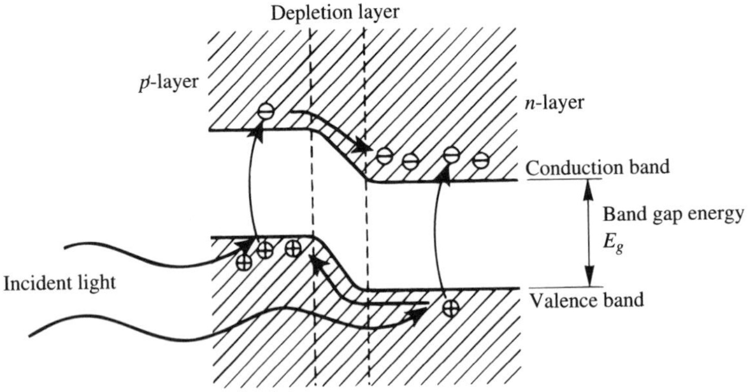

Fig. 22.1. Photoexcitation at a p–n junction.

really observe is the rate of appearance of photoelectrons or carriers produced by the disappearance of photons at the active surface of the detector. We assume that these events are directly related by a quantum efficiency factor η, where

$$\eta = \frac{\text{number of carriers produced}}{\text{number of absorbed photons}}. \tag{22.2}$$

The fluctuations in photon flux, N, give rise to a fluctuation in the rate of production of photoproduced carriers. These fluctuations appear as a randomly varying current superimposed on the average current. The average current from the photodetector is

$$\bar{i} = \frac{\eta e}{h\nu} < I > A, \tag{22.3}$$

which can be written as

$$\bar{i} = \mathscr{R} < I > A, \tag{22.4}$$

where A is the detector area, and $\mathscr{R} = e\eta/h\nu$ is called the responsivity, which has units of A W^{-1}. The fluctuations in current resulting from the fluctuating appearance of photoproduced carriers is called *shot*, or *photon* noise. The frequency spectrum of these current fluctuations can be calculated by considering the frequency components in the current produced by the elemental current of a single photoproduced carrier. If we take one of these elemental current pulses to be of a Gaussian shape we can write

$$i(t) = \frac{e}{\sigma\sqrt{2\pi}} \exp(-t^2/2\sigma^2), \tag{22.5}$$

where the shape of the pulse is normalized so that

$$\int_{-\infty}^{\infty} i(t)dt = e. \tag{22.6}$$

The charge on an electron is $-e$.

The frequency spectrum of this current is

$$F(\omega) = \frac{1}{2\pi} \int_{-\infty}^{\infty} i(t) \exp(-i\omega t)dt, \tag{22.7}$$

which gives

$$F(\omega) = \frac{e}{2\pi}\exp\left(-\frac{\sigma^2\omega^2}{2}\right). \tag{22.8}$$

For sufficiently low frequencies, for example, such that $\sigma^2\omega^2/2 < 0.01$, the exponential factor is very close to unity, and the Fourier transform is flat. Simply stated, this result says that randomly occurring narrow pulses contribute broadband 'white' noise.

In terms of its Fourier transform the current can be written as

$$i(t) = \int_{-\infty}^{\infty} F(\omega)\exp(i\omega t)d\omega. \tag{22.9}$$

If this current flows into a resistance R, then the average power dissipated over time T, where T is greater than the pulse duration, is

$$P = \frac{1}{T}\int_{-T/2}^{T/2} \frac{i^2(t)}{R}dt = \frac{1}{RT}\int_{-T/2}^{T/2} i(t)\int_{-\infty}^{\infty} F(\omega)\exp(i\omega t)d\omega, \tag{22.10}$$

which can be rearranged as

$$P = \frac{1}{RT}\int_{-\infty}^{\infty} F(\omega)\left[\int_{-\infty}^{\infty} i(t)\exp(i\omega t)dt\right]d\omega. \tag{22.11}$$

The limits on the inner integral have been set to $\pm\infty$ without significant error since the pulse is localized within the time interval T. Since $i(t)$ is real, it follows from Eq. (22.7) that

$$F(\omega) = \frac{1}{2\pi}\int_{-\infty}^{\infty} i(t)\exp(i\omega t)dt = F(-\omega). \tag{22.12}$$

Substitution in Eq. (22.9) gives

$$P = \frac{2\pi}{RT}\int_{-\infty}^{\infty} |F(\omega)|^2 d\omega, \tag{22.13}$$

which can be written as

$$P = \frac{4\pi}{RT}\int_{0}^{\infty} |F(\omega)|^2 d\omega. \tag{22.14}$$

The fraction of the power spectrum between ω and $\omega + d\omega$ is $P(\omega)d\omega$, where the power spectral density function is clearly

$$P(\omega) = \frac{4\pi(|F(\omega)|^2}{RT}. \tag{22.15}$$

For a flow of $\eta < N >$ charge carriers per second, which are assumed to be uncorrelated, the total power spectral density is $\eta < N > P(\omega)$. The spectral energy density supplied per second is

$$U(\omega) = \eta < N > P(\omega)T, \tag{22.16}$$

where $T = 1$ s, which can be written as

$$U(\omega) = 4\pi\frac{\eta < N > |F(\omega)|^2}{R}. \tag{22.17}$$

This energy can be considered to result from a noise current generator whose mean squared magnitude is

$$< i_N^2(\omega) >= 4\pi\eta < N > |F(\omega)|^2 d\omega. \tag{22.18}$$

For frequencies where $F(\omega)$ is flat

$$< i_N^2(\omega) >= \frac{\eta < N > e^2}{\pi} d\omega. \tag{22.19}$$

Since $\eta < N >= \bar{i}/e$ and $d\omega = 2\pi df$ we can write the shot noise spectrum in terms of conventional frequency as

$$< i_N^2(f) >= 2e\bar{i}df. \tag{22.20}$$

Even if all other sources of noise in the detector and its following electronics can be eliminated, shot noise is inescapable. It is produced by the signal. The best signal to noise (power) ratio to be expected is the shot-noise-limited (SNL) value

$$\left(\frac{S}{N}\right)_{SNL} = \frac{\bar{i}^2 R}{i_N^2 R} = \frac{(e\eta < I > A/h\nu)^2}{2e^2\eta < I > A/h\nu}, \tag{22.21}$$

giving

$$\left(\frac{S}{N}\right)_{SNL} = \frac{\eta < I > A}{2h\nu\Delta f} = \frac{\eta P}{2h\nu\Delta f}, \tag{22.22}$$

where $P =< I > A$ is the total power (W) falling on the detector.

This signal-to-noise ratio is frequently called the limiting *video* or *direct* detection limit.† It is the best that can be done with a detector directly receiving only the incoming light it is desired to detect. For a detector with unity quantum efficiency the minimum detectable light signal ($S/N = 1$) is

$$P_{min} = 2h\nu\Delta f. \tag{22.23}$$

The sampling time for digital data, by Nyquist's theorem, is generally set so that data is sampled at twice the maximum frequency being observed, i.e., $\Delta T = 1/(2\Delta f)$. The minimum detectable signal therefore corresponds to one detected photon per sampling interval. In practice, additional sources of noise are present in a detection system involving a detector and other electronic devices. Principal among these are $1/f$ noise, Johnson noise (also called thermal, Nyquist or resistance noise), and in semiconductor detectors generation–recombination (gr noise).

22.2.2 Johnson Noise

At temperatures above absolute zero the thermal energy of the charge carriers in any resistor leads to fluctuations in local charge density. These fluctuating charges cause local voltage gradients that can drive a corresponding current into the rest of the circuit. The quantitative treatment of this thermal noise can be carried out in several different ways[22.1]. Nyquist[22.2] proposed a theorem that stated that the noise power generated by a circuit element did not depend specifically on the nature of the circuit element, but only on the temperature of the component and the frequency band being examined. He proved this result thermodynamically and derived a quantitative expression for the noise power by considering the energy

† Also called the photon-noise or quantum-limited signal-to-noise ratio.

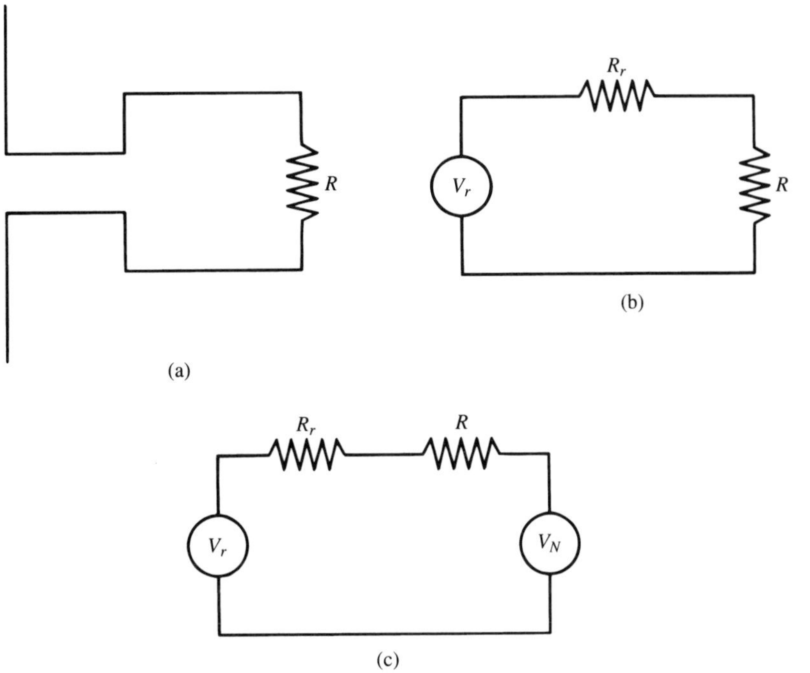

Fig. 22.2. Circuits used in an analysis of Johnson noise: (a) antenna connected to a resistor; (b) including the radiation resistance of the antenna; (c) showing the noise source associated with the resistor.

flow between two resistors of equal value connected together by a cable with characteristic impedance R. The noise power can also be calculated by considering the voltages and currents associated with a RLC tuned circuit. However, we choose here to calculate the noise power by relating the current and voltage fluctuations to the available radiation energy at temperature T described by black-body radiation. We do this by considering the circuit shown in Fig. (22.2a) in which an antenna is connected to a resistance R. The antenna can be modelled as a voltage source with an associated series resistance R_r – called its radiation resistance. If the antenna is a wire of length ℓ, then the electric field of the black-body radiation will induce a voltage in the antenna. For radio and microwave frequencies, we can use the Rayleigh–Jeans approximation to Planck's radiation formula to calculate this induced voltage. The energy density of the radiation interacting with the antenna within a frequency band of width Δf at frequency f is

$$\Delta\rho = \frac{8\pi f^2}{c^3} kT\Delta f. \tag{22.24}$$

On average, because the black-body radiation is unpolarized, only one-third of this energy actually interacts with a linear antenna, oriented (say) in the x direction. The mean-squared electric field in the antenna direction within the frequency band Δf is†

$$\Delta(E_x^2) = \frac{cZ_0\Delta\rho}{3}. \tag{22.25}$$

† For a plane linearly polarized wave the intensity within Δf can be written as $\Delta < I >= \Delta(E_x^2)/2Z_0 = c\Delta\rho/6$. The factor of 6 comes from the three orthogonal propagation directions in space and the two orthogonal polarization directions associated with each of them.

The mean-square voltage induced in the antenna within the band Δf is

$$\Delta < V_r^2 >= \ell^2 \Delta(E_x)^2 = \frac{8\pi f^2}{3c^2} \ell^2 kTZ_0 \Delta f. \tag{22.26}$$

This voltage induces a current i_a in the antenna, which can flow into an external short-circuit.

Since an ideal antenna cannot dissipate any energy it must reradiate the energy it receives. We can model this reradiation process by associating a radiation resistance R_r with the antenna as shown in Fig. (22.2b) such that the *apparent* dissipation in R_r balances the received power. For this to be so

$$\frac{\Delta(V_r^2)}{R_r} = i_r^2 R_r. \tag{22.27}$$

Now the power radiated by an antenna of length ℓ carrying an oscillating current of magnitude i_r is[22.3],[22.4]

$$W = \frac{\pi Z_0 i_r^2 \ell^2 f^2}{3c^2}. \tag{22.28}$$

Equating W with $\frac{1}{2} i_r^2 R_r$ gives

$$R_r = \frac{2\pi f^2 \ell^2 Z_0}{3c^2}. \tag{22.29}$$

If the antenna is connected to a resistor R, as shown in Fig. (22.2b), then the fluctuating current in the antenna will dissipate a power $V_r^2 R/(R_r + R)^2$ in the resistor R. If the entire circuit shown in Fig. (22.2b) is in thermal equilibrium, then for the resistor R to heat would violate the second law of thermodynamics. Consequently, the resistor R must drive power back into the antenna to balance the power it receives. Consequently, the mean-square voltage generated by R within the band Δf must satisfy

$$\frac{< V_N^2 >}{R} = \frac{\Delta < V_r^2 >}{R_r} = \frac{8\pi f^2 \ell^2}{3c^2} kTZ_0 \Delta f \times \frac{3c^2}{2\pi f^2 \ell^2 Z_0} = 4kT\Delta f, \tag{22.30}$$

which gives

$$< V_N^2 >= 4kTR\Delta f. \tag{22.31}$$

The overall situation can be represented as shown in Fig. (22.2c). Alternatively, the noisy resistor can be treated as a resistor R with a parallel current source whose mean-square value within the band Δf is

$$< i_N^2 >= \frac{4kT}{R} \Delta f. \tag{22.32}$$

Thus, this source of noise can be reduced by cooling the offending component to a low temperature.

If a shot-noise-limited detector drives a circuit with an input resistance R the equivalent circuit involves both the shot noise current source and the Johnson noise current source, as shown in Fig. (22.3).

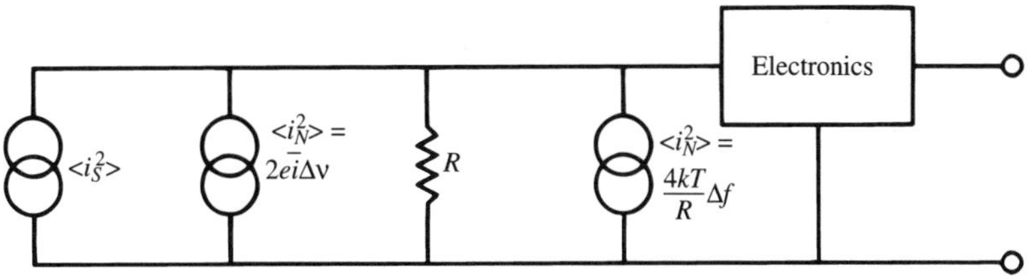

Fig. 22.3. Equivalent circuit of a shot-noise-limited detector driving a resistive load.

22.2.3 Generation–Recombination Noise and 1/f Noise

These two types of noise are important in semiconductor detectors. $1/f$ noise, or current noise, has a power spectrum that depends inversely on the frequency. It can be described by a noise current generator

$$< i_N^2 >= \frac{K < i >^\alpha \Delta f}{f^\beta}, \tag{22.33}$$

where $< i >$ is the average current through the detector, K is a constant, and typically $\alpha = 2$ and $\beta = 1$. The noise that depends inversely on the frequency, the $1/f$ noise, originates from many causes, such as the diffusion of charge carriers, the presence of impurity atoms and lattice defects in the material, and interaction of charge carriers with the surface of the semiconductor. To achieve high S/N ratios low frequency detection should be avoided because of $1/f$ noise – which may increase in magnitude down to frequencies of 10^{-4} Hz. The trend towards higher and higher data rates in optical communication systems moves system operation well away from $1/f$ noise, although at high and intermediate frequencies generation–recombination (gr) noise will still be a factor in determining the S/N ratio. Generation–recombination noise arises from statistical fluctuations in the number of carriers in the detector. In this sense it is closely related to photon noise, the fluctuation in the number of generated carriers, but the gr noise results from the secondary carrier density fluctuations arising from random electron–hole recombinations. This recombination process has a characteristic lifetime τ_0, which can be very short, 1 ns or less, so we expect this source of noise to contribute up to frequencies $\sim 1/\tau_0$. Indeed, this is the case, the gr noise spectrum is flat up to a frequency $\sim 1/\tau_0$ and can be shown to correspond to an equivalent noise power generator[22.5]

$$< i_N^2 >= \frac{4 < i >^2 \tau_0}{< N > (1 + 4\pi^2 f^2 \tau_0^2)}, \tag{22.34}$$

where $< N >$ is the average number of charge carriers. If the time it takes a charge carrier to travel through the detector into the external current is τ_d then we can write

$$\frac{< N >}{\tau_d} = \frac{< i >}{e} \tag{22.35}$$

and the gr noise current generator can be written as

$$< i_N^2 >= \frac{4e < i > (\tau_0/\tau_d)}{(1 + 4\pi^2 f^2 \tau_0^2)}. \tag{22.36}$$

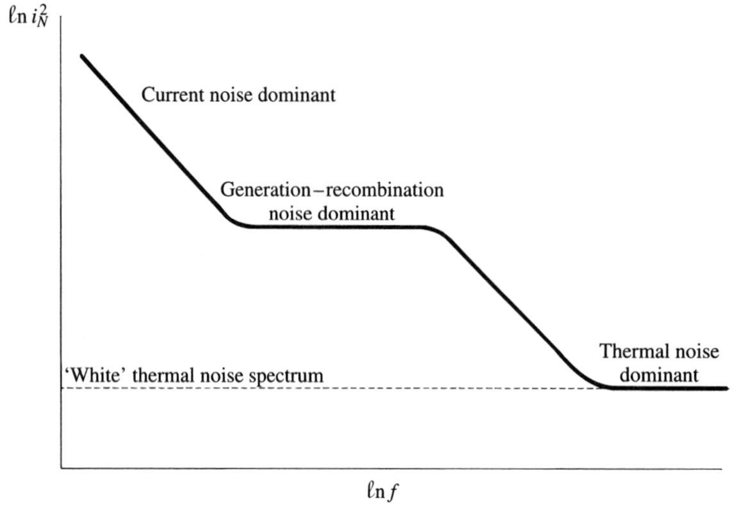

Fig. 22.4. Schematic variation of noise with frequency in a semiconductor detector.

Thus, we expect gr noise to be less in a detector with a rapid recombination time and also less in heavily doped semiconductors where the number of charge carriers will be greater at a given average current $< i >$ than in an intrinsic material.

Thus, in a semiconductor detector there are contributions from $1/f$ noise, gr noise, and thermal noise. The relative contributions of these noise sources vary with frequency in the schematic way shown in Fig. (22.4).

22.3 Detector Performance Parameters

22.3.1 Noise Equivalent Power

It is important to consider under what conditions of operation the performance of a detector will be primarily limited by generation–recombination noise, since for all but low frequencies gr noise generally dominates over $1/f$ and Johnson noise. This can be carried out by considering the *noise equivalent power* (NEP) of the detector, which for a good detector is a practical measure of the magnitude of the gr noise. The NEP is the rms value of a sinusoidally modulated light signal falling on the detector that gives rise to an rms electrical signal equal to the rms noise voltage. The NEP is usually specified in terms of a black-body source, a reference bandwidth, usually 1 Hz, and the modulation frequency of the radiation.

For example NEP (500 K, 900, 1) implies a black-body illuminator whose temperature is 500 K, a 900 Hz modulation frequency, and a reference bandwidth of 1 Hz. If the illuminating signal has intensity I (W m^{-2}) and falls on a detector of area A then we can write

$$NEP = \frac{IA}{\sqrt{\Delta f}}\frac{V_N}{V_S}, \tag{22.37}$$

where V_S, V_N are the signal and noise voltages, respectively, measured with a bandwidth Δf.

We can rearrange Eq. (22.37) to give the equivalent intensity needed to generate a S/N ratio of 1 in a 1 Hz bandwidth as

$$I(S/N = 1, \Delta f = 1 \text{ Hz}) = \frac{NEP}{A}. \tag{22.38}$$

For example a detector with an *NEP* of 10^{-12} W $\text{Hz}^{-1/2}$ needs a total power of 10^{-12} W of black-body power to fall onto its sensitive surface to generate a signal equal to the *detector noise*. The photon noise limit for such a detector would correspond to a received intensity of

$$I \text{ (photon-noise-limited, } S/N = 1, \Delta f = 1 \text{ Hz}) = 2h\nu.$$

So the relative magnitudes of the detector-noise-limited minimum detectable light signal and that set by photon noise is

$$\frac{I \text{ (detector-noise-limited)}}{I \text{ (photon-noise-limited)}} \simeq \frac{NEP}{2Ah\bar{\nu}}. \tag{22.39}$$

Strict equality does not hold in Eq. (22.39) as the *NEP* is generally defined for black-body radiation and the quantum $h\bar{\nu}$ implies an appropriate average frequency for the incoming radiation.

22.3.2 *Detectivity*

Many detectors exhibit an *NEP* that is proportional to the square root of the detector area; so a detector-area-independent parameter, the *detectivity* D^* is frequently used, specified by

$$D^* = \frac{\sqrt{A}}{NEP}, \tag{22.40}$$

where A is the area of the detector. D^* is specified in the same way as *NEP*: for example, D^* (500 K, 900, 1). To specify the variation in response of a detector with wavelength, the *spectral detectivity* is often used. Thus, the symbol $D^*(\lambda, 900, 1)$ would specify the response of the detector to radiation of wavelength λ, modulated at 900 Hz and detected with a 1 Hz bandwidth. D^* is measured in units cm $\text{Hz}^{1/2}$ W^{-1}.

The performance of photodiodes used in optical communication systems is frequently specified in terms of their responsivity $\mathscr{R}$, usually specified in A W^{-1}, which characterizes the response of the detector to unit irradiance. At high data rates the actual S/N performance of these detectors will depend on the wide-band amplification electronics with which they are used. Specification of their performance in terms of *NEP* or D^* becomes less relevant.

22.3.3 *Frequency Response and Time Constant*

The frequency response of a detector is the variation of responsivity or radiant sensitivity as a function of the modulation frequency of the incident radiation. The frequency variation of $\mathscr{R}$ and the time constant of the detector are generally related according to

$$\mathscr{R}(f) = \frac{\mathscr{R}(0)}{(1 + 4\pi^2 f^2 \tau^2)^{1/2}}. \tag{22.41}$$

For frequencies above $1/2\pi\tau$ the response is falling off significantly and at high enough modulation frequencies the detector will provide a dc output proportional to the average intensity.

22.4 Practical Characteristics of Optical Detectors

The development of optical detectors has occurred, in common with various branches of electronics, by a series of advances from the use of gas-filled tubes and vacuum tubes to various semiconductor devices. However, whereas in general electronics the vacuum tube is now reserved for specialized applications, vacuum-tube optical detectors such as the photomultiplier are still in widespread use.

22.4.1 *Photoemissive Detectors*

Photoemissive detectors include gas-filled and vacuum photodiodes, photomultiplier tubes, and photochanneltrons. These are all photon detectors that utilize the photoelectric effect. When radiation of frequency v falls upon a metal surface, electrons are emitted, provided the photon energy hv is greater than a minimum critical value ϕ, called the *work function*, which is characteristic of the material being irradiated. A simplified energy level diagram illustrating this effect for a metal–vacuum interface is shown in Fig. (22.5a). For most metals, ϕ is in a range from 4–5 eV(1 eV$\equiv \lambda = 1.24\ \mu$m), although for the alkali metals it is lower, for example, 2.4 eV for sodium and 1.8 eV for cesium. Pure metals or alloys, particularly beryllium–copper, are used as photoemissive surfaces in ultraviolet and vacuum-ultraviolet detectors.

Lower work functions, and consequently sensitivities that can be extended into the infrared, can be obtained with special semiconductor materials. Fig. (22.5b) shows a schematic energy level diagram of a semiconductor–vacuum interface. In this case the work function is defined as $\phi = E_{vac} - E_F$, where μ is the energy of the Fermi level. In a pure semiconductor, the Fermi level is in the middle of the band gap, as illustrated in Fig. (22.5b). In a *p*-type doped semiconductor, μ moves down toward the top of the valence band, while in *n*-type material it moves up toward the bottom of the conduction band. Consequently, ϕ is not as useful a measure of the minimum photon energy for photoemission as it is for a metal. The electron affinity χ is a more useful measure of this minimum energy for a semiconductor. Except at absolute zero, photons with energy $hv > \chi$ cause photoemission. Photons with energy $hv > E_g$ lead to the production of carriers in the conduction band; this leads to intrinsic photoconductivity, which is the operative detection mechanism in various infrared detectors, such as InSb.

(a) Vacuum Photodiodes. Once electrons are liberated from a photoemissive surface (a photocathode), they can be accelerated to an electrode positively charged with respect to the cathode – the anode – and generate a signal current. If the acceleration of photoelectrons is directly from cathode to anode through a vacuum, the device is a vacuum photodiode. Because the electrons in such a device take a very direct path from anode to cathode and can be accelerated by high voltages – up to several kilovolts in a small device – vacuum photodiodes have the fastest response of all photoemissive detectors. Risetimes of 100 ps or less can be achieved. External connections and electronics are generally the limiting factors in obtaining short risetimes from such devices. However, vacuum photodiodes are not very

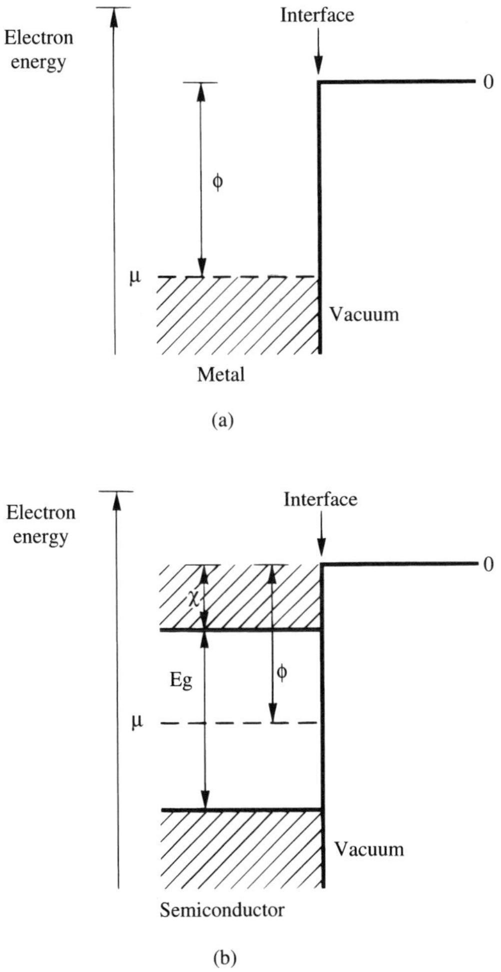

Fig. 22.5. Band structure of (a) a metal-vacuum interface and (b) a pure-semiconductor–vacuum interface (ϕ = work function; χ = electron affinity; E_g = band-gap energy; μ = Fermi level).

sensitive, since at most one electron can be obtained for each photon absorbed at the photocathode. In principle, of course, the limiting sensitivity of the device is set by its quantum efficiency. Practical quantum efficiencies for photoemissive materials range up to about 0.4.

If the space between photocathode and anode is filled with a noble gas, photoelectrons will collide with gas atoms and ionize them, yielding secondary electrons. Thus, an electron multiplication effect occurs. However, because the mobility of the electrons moving from cathode to anode through the gas is slow, these devices have a long response time, typically about 1 ms. Gas-filled photocells are no longer competitive with solid-state detectors in practical applications.

(b) Photomultipliers. If photoelectrons are accelerated *in vacuo* from the photocathode and allowed to strike a series of secondary electron emitting surfaces, called *dynodes*, held at progressively more positive voltages, a considerable electron multiplication can be achieved and a substantial current can be collected at the anode. Such devices are called *photomultipliers* (PMTS). Practical gains of 10^9 (anode electrons per photoelectron) can be achieved from these devices for

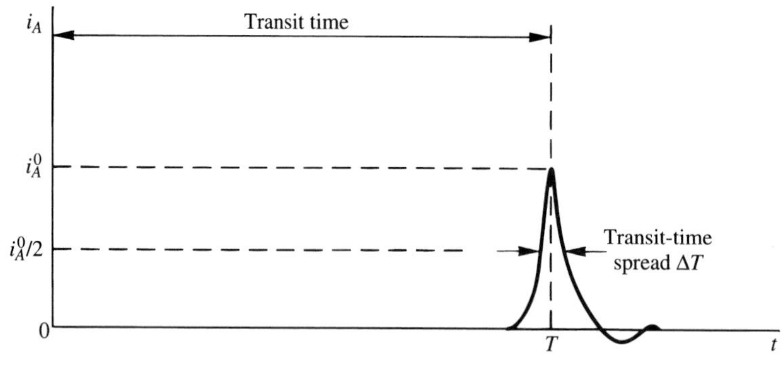

Fig. 22.6. Typical anode pulse produced by a single photoelectron emission at the cathode of a photomultiplier tube, t is the time following photoemission. The transit time T, the transit-time spread ΔT, and the peak anode current i_A^0 all fluctuate from pulse to pulse.

short light pulses. Because of their very high gain, photomultipliers can generate substantial signals when only a single photon is detected: for example, an anode pulse of 2 ns duration containing 10^9 secondary photoelectrons produced from a single photoelectron will generate a voltage pulse of 4 V across 50 Ω. This, coupled with their low noise, makes photomultipliers effective single-photon detectors. Photomultiplier D^*-values can range up to 10^{16} cm Hz$^{1/2}$ W^{-1}. The dark-adapted human eye, which can detect bursts of about 10 photons in the blue, comes close to this sensitivity.

The time response characteristics of photomultiplier tubes depend to a considerable degree on their internal dynode arrangement. The response of a given device can be specified in terms of the output signal at the anode that results from a single photoelectron emission at the photocathode. This is illustrated in Fig. (22.6). Because electrons passing through the dynode structure can generally take slightly different paths, secondary electrons arrive at the anode at different times. The anode pulse has a characteristic width called the *transit-time spread*, which usually ranges from 0.1 to 20 ns. The time interval between photoemission at the cathode and the appearance of an anode pulse is called the *transit time*, and is usually a few tens of nanoseconds. The transit time and transit-time spread also fluctuate slightly from one single-photoelectron-produced anode pulse to the next.

There are four main types of dynode structure in common usage in photomultiplier tubes; these are illustrated in Fig. (22.7). The circular cage structure is compact and can be designed for good electron collection efficiency and small transit-time spread. This dynode structure works well with opaque photocathodes, but is not very suitable for high amplification requirements where a larger number of dynodes is required. The box-and-grid and Venetian-blind structures offer very good electron-collection efficiency. Because they collect multiplied electrons independently of their path through the dynode structure, a wide range of secondary-electron trajectories is possible, leading to a large transit-time spread and slow response. Typical response times of these types of tube are 10–20 ns.

The focused dynode structure is designed so that electrons follow paths of similar length through the dynode structure. To accomplish this, electrons that deviate too much from a specified range of trajectories are not collected at the next dynode. These types of tube offer short response times, typically 1–2 ns.

Venetian-blind tubes can easily be extended to many dynode stages and have a very stable gain in the presence of small power supply fluctuations. These tubes also

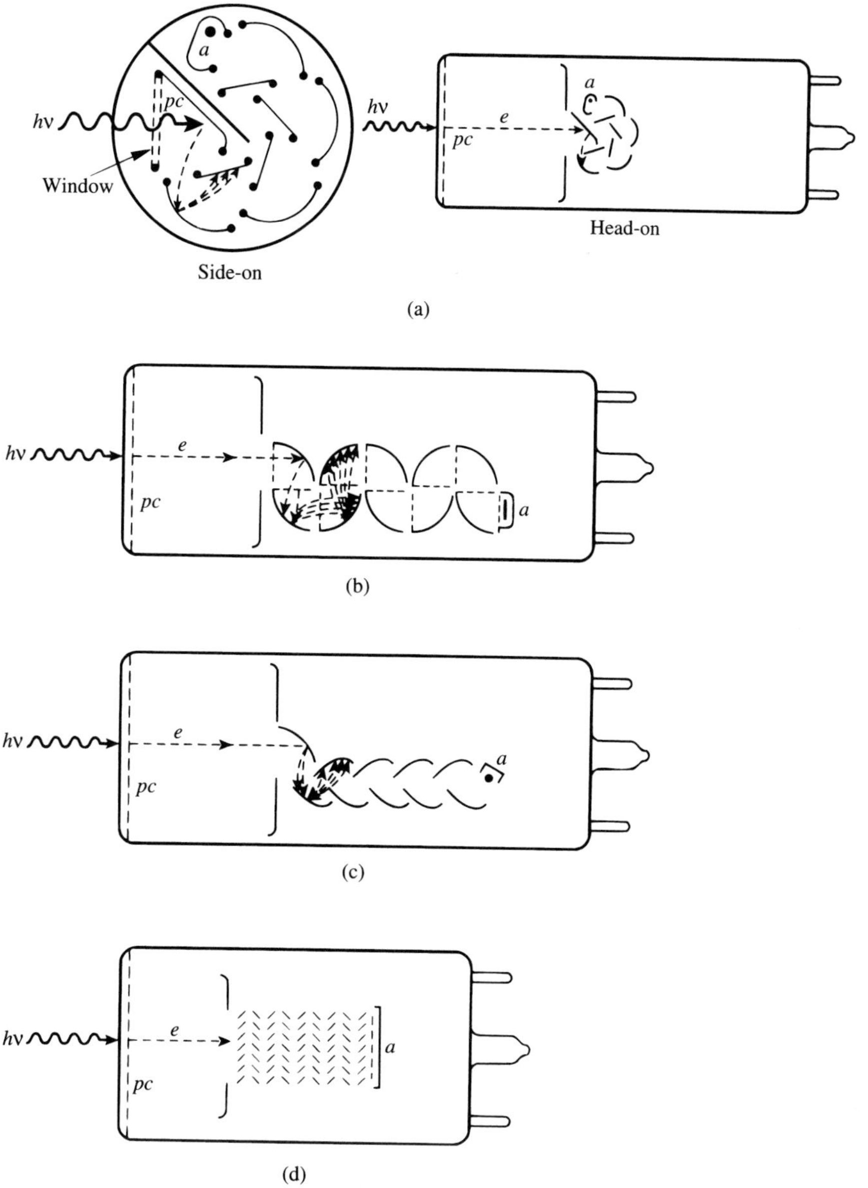

Fig. 22.7. Schematic diagram of the internal structure of various types of photomultiplier tube: (a) circular cage; (b) box and grid; (c) focused dynode; (d) Venetian blind. *a* – anode; *pc* – photocathode.

have an optically opaque dynode structure, which contributes to their exhibiting very low dark-current noise when operated under appropriate conditions.

(c) Photocathode and Dynode Materials. The performance of a photomultiplier depends not only on its internal structure, but also on the photoemissive material of its photocathode and the secondary-electron-emitting material of its dynodes. The wavelength dependence of various commercially available photocathode materials is shown in Fig. (22.8); some radiant sensitivities and quantum efficiencies are given in Table (22.1). The short-wavelength cutoff of a given material depends on its work function. This cutoff is not sharp because, except at absolute zero,

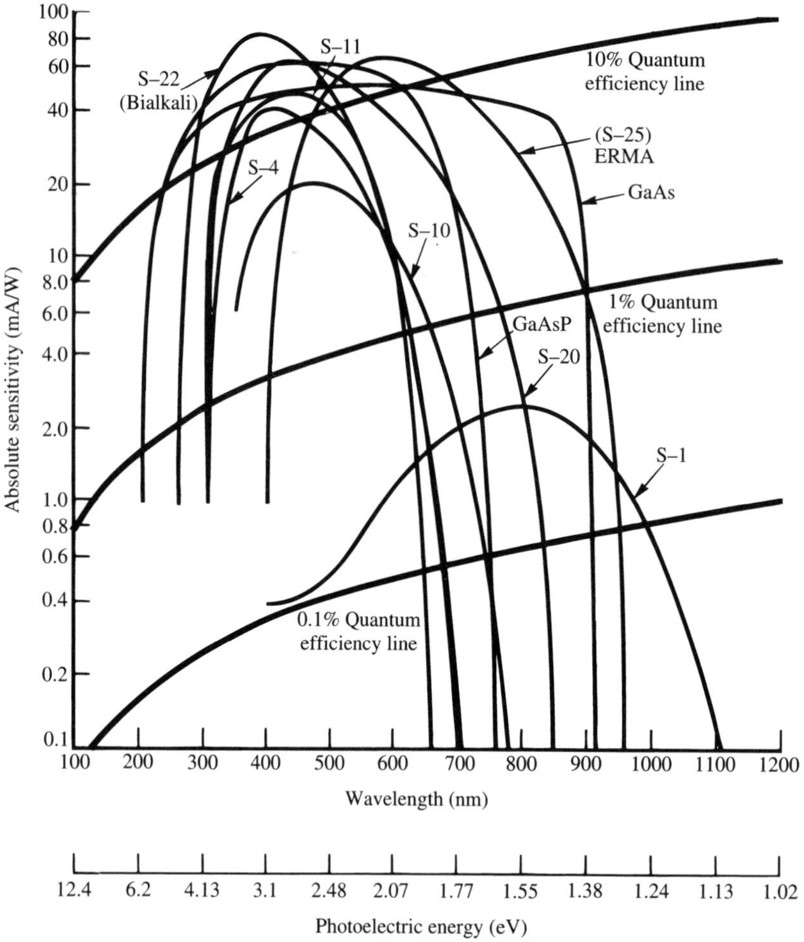

Fig. 22.8. Wavelength dependence of radiant sensitivity of several commercially available photocathode materials.

there are always a few electrons high up in the conduction band available for photoexcitation by low energy photons. Some few of these electrons, because of their thermal excitation, will be emitted without any photostimulation. This contributes the major part of the dark current observed from the photocathode. Materials that have low work functions, and are consequently more red- and infrared-sensitive, have higher – often much higher – dark currents than materials that are optimized for visible and ultraviolet sensitivity.

Photocathodes are available in both opaque and semitransparent forms, depending on the model of phototube. In the semitransparent form the photoelectrons are ejected from the thin photoemissive layer on the opposite side from the incident light. In both types the photocathode has to perform two important functions: it must absorb incident photons and allow the emitted photoelectron to escape. The latter event is inhibited if photons are absorbed too deep in a thick photoemissive layer, or if the photoelectron suffers energy loss from scattering in the layer. If the emitted photoelectron has too much energy, it can excite a further electron across the band gap. This pair-production phenomenon inhibits the release of

Table 22.1. *Characteristics of photoemissive surfaces.*

Cathode	Radiant sensitivity (mA W^{-1}) 515 nm	694 nm	1.06 μm	Peak quantum efficiency (%)	λ_{peak} (μm)	Dark current[a] (A cm^{-2})
S-1 Cs–O–Ag	0.6	2	0.4	0.08	800	9×10^{-13}
S-10 Cs–O–Ag–Bi[b]	20	2.7	0	5	470	9×10^{-16}
S-11 Cs_3Sb on MnO[c]	39	0.2	0	13	440	10^{-16}
S-20 (Cs)Na_2KSb (trialkali)	53	20	0	18	470	3×10^{-16}
S-22(bialkali)	42	0	0	26	390	$1\text{-}6\times10^{-18}$
GaAs[b]	48	28	0	14	560[e]	10^{-16}
GaAsP[d]	60	30	0	19	400[e]	3×10^{-15}
InGaAs[d]	–	–	4.3	–	–[e]	3×10^{-14}
InGaAsP[d]	–	–	–	47	300[e]	2.5×10^{-13}
S-25 (ERMA)[f]	53	26	0	25	430	1×10^{-15}
Cs–Te (solar blind)	–	–	–	15	254	2.5×10^{-17}

Note: Table shows typical values, but these can vary greatly from one manufacturer to another.
[a] At room temperature.
[b] Cathode designated S-3 is similar.
[c] Several types of CsSb photocathodes exist where the CsSb is deposited on different opaque and semitransparent substrates and various window materials are used. These photocathodes have the designations S-4, 5, 13, 17, and 19 as well as S-11.
[d] Negative-electron-affinity (NEA) photoemitters.
[e] May show no wavelength of maximum quantum efficiency; quantum efficiency falls with increasing wavelength. However, exact spectral characteristics will depend on thickness of photoemitter and whether it is used in transmission or not.
[f] Extended-red S-20.

photoelectrons from the photoemissive layer, and accounts for the ultraviolet cut-off characteristics of the different materials shown in Fig. (22.8). With reference to Fig. (22.5) it can be shown that for pair production to occur the incident photon energy must be greater than $2E_g$. Photoelectrons have the best chance of escaping, and the photocathode its highest quantum efficiency, for materials where $\chi < E_g$.

Practical photoemissive materials fall into two main categories: classical photoemitters and negative-electron-affinity (NEA) materials. Classical photoemitters generally involve an alkali metal or metals, a group-V element such as phosphorus, arsenic, antimony, or bismuth, and sometimes silver and/or oxygen. Examples are the Ag–O–Cs(S-1) photoemitter, which has the highest quantum efficiency beyond about 800 nm of any classical photoemitter, and Na_2KSbCs, the so-called trialkali (S-20) cathode.

NEA photoemitters utilize a photoconductive single-crystal semiconductor substrate with a very thin surface coating of cesium and usually a small amount of oxygen. The cesium (oxide) layer lowers the electron affinity below the value it would have in the pure semiconductor, achieving an effectively negative value. Examples of such NEA photoemitters are GaAs (CsO) and InP (CsO). NEA emitters can offer very high quantum efficiency and extended infrared response. GaAs (CsO), for example, has higher quantum efficiency in the near-infrared then an S-1 photocathode.

The performance of the dynode material in photomultiplier tubes is specified in terms of the secondary-emission ratio δ as a function of energy. For a phototube with n dynodes the gain is δ^n. In the past, the commonest dynode materials were CsSb, AgMgO, and BeCuO. The last is also used as the primary photoemitter in windowless photomultipliers that are operated *in vacuo* for the detection of vacuum-ultraviolet radiation. BeCuO has the desirable property that it can be reactivated after exposure to air. The above materials have δ-values of 3–4. Newer NEA dynode materials have much higher δ-values: in particular, that of GaP can range up to 40 for an incident-electron input energy of 800 eV. With such high δ-values a photomultiplier tube needs fewer dynodes for a given gain, which means a more compact and faster-response tube can be built. In very many commercial photomultipliers, the first dynode at least is now frequently made of GaP. This offers improved characterization of the single-photoelectron response of the tube, which is important in designing a system for optimum signal-to-noise ratio.

The accelerating voltages are supplied to the dynodes of a photomultiplier by a resistive voltage divider called a *dynode chain.* The relative resistance values in the chain determine the distribution of voltages applied to the dynodes. The total chain resistance R determines the chain current at a given total photocathode–anode applied voltage. Many photomultiplier tubes have one or more focusing electrodes between the photocathode and the first dynode. The voltage on these electrodes can be adjusted to optimize the collection of photoelectrons from the photocathode.

The actual response-time behavior of the photomultiplier can be determined by observing its single-photoelectron response. This is done by looking at the anode pulses with a fast oscilloscope. The photocathode does not need to be illuminated for this to be done; sufficient noise pulses will usually be observed. The pulses should appear as in Fig. (22.6). These anode pulses reflect the time distribution and number of secondary electrons reaching the anode following single (or multiple) photoelectron emissions from the photocathode. If the height distribution of anode pulses is measured, a distribution such as is shown in Fig. (22.9a) will probably be seen. The idealized distribution shown in Fig. (22.9b) may be seen from newer tubes with GaP dynodes, which have a very high δ-value.

22.4.2 *Photoconductive Detectors*

Photoconductive detectors can operate through either intrinsic or extrinsic photoconductivity. The physics of intrinsic photoconductivity is illustrated in Fig. (22.10a). Photons with energy $h\nu > E_g$ excite electrons across the band gap. The electron–hole pair that is thereby created for each photon absorbed leads to an increase in conductivity – which comes mostly from the electrons. Semiconductors with small band gaps respond to long wavelength radiation but must be cooled accordingly: otherwise thermally excited electrons swamp any small photoconductivity effects. Table (22.2) lists commonly available intrinsic photoconductive detectors together with their usual operating temperature and the limit of their long wavelength response, λ_0, together with some representative figures for detectivities and time

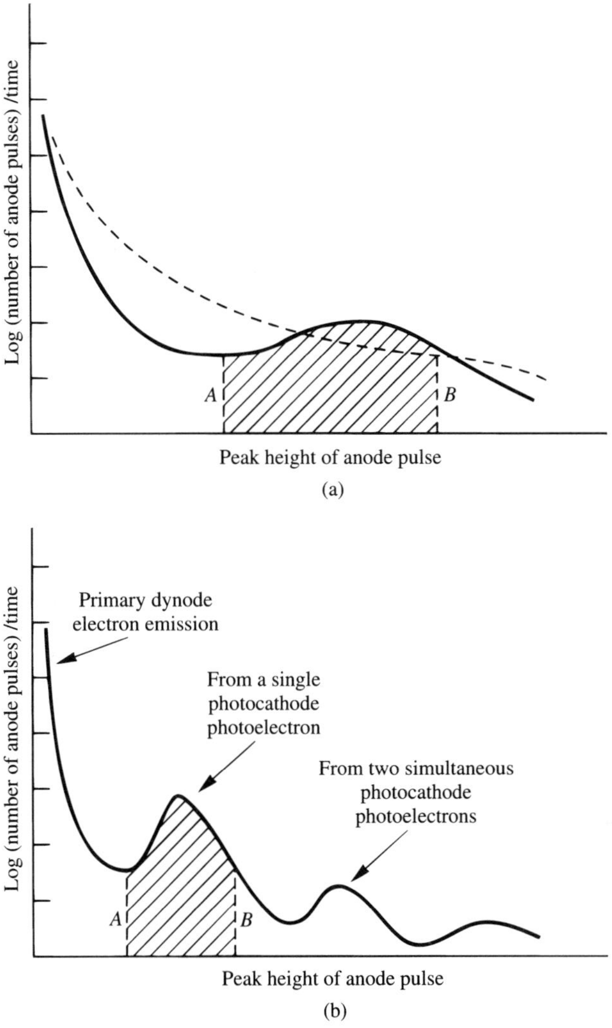

Fig. 22.9. Schematic photomultiplier anode pulse-height distributions: (a) two forms likely to be observed in practice; (b) idealized form from tube with dynodes having a high and well defined secondary emission coefficient. The best signal-to noise-ratio would be obtained in a photo-counting experiment by collecting only anode pulses in a height range roughly indicated by the shaded region *AB*.

constants. Note that silicon and germanium are also operated in both photovoltaic and avalanche modes.

If a semiconductor is doped with an appropriate material, impurity levels are produced between the valence and conduction bands as shown in Fig. (22.9b). Impurity levels that are able to accept an electron excited from the conduction band are called *acceptor levels*, whereas impurity levels that can have an electron excited from them into the conduction band are called *donor levels*. Thus, in Fig. (22.10b) photons with energy $h\nu > E_A$ excite an electron to the impurity level, leaving a hole in the valence band and thereby giving rise to *p*-type extrinsic photoconductivity. Photons with energy $h\nu > E_D$ will excite an electron into the conduction band, giving *n*-type extrinsic photoconductivity. For example, gold doped germanium has an acceptor level 0.15 eV above the valence band and is an extrinsic *p*-type photoconductor, as is copper doped germanium, which has an acceptor level 0.041 eV above the valence band. These are the two most commonly

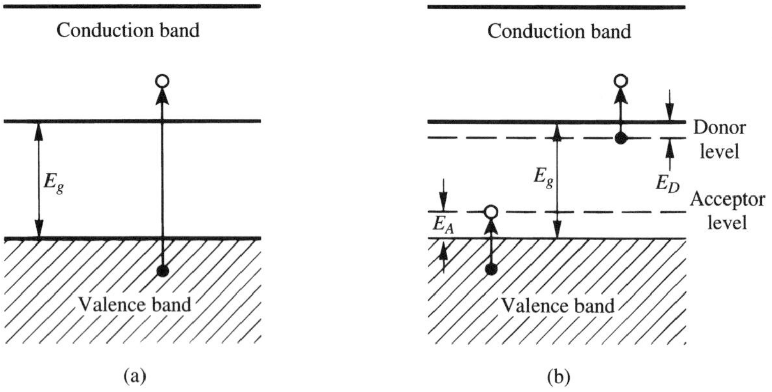

Fig. 22.10. Mechanism for: (a) intrinsic photoconductivity; (b) extrinsic photoconductivity.

Table 22.2. *Intrinsic photoconductive detectors.*

Semiconductor	T (K)	E_g (eV)	λ_0 (μm)	D^*(max) (cm $Hz^{1/2}$ W^{-1})	τ
CdS	295	2.4	0.52	3.5×10^{14}	50 ms
CdSe	295	1.8	0.69	2.1×10^{11}	$\simeq$10 ms
Si	295[a]	1.12	1.1	$\leq2\times10^{12}$	50 ps
Ge	295[a]	0.67	1.8	$\leq10^{11}$	10 ns
PbS	295	0.42	2.5	$\leq2\times10^{11}$	0.1–10 ms
	195	0.35	3.0	$\leq4\times10^{11}$	0.1–10 ms
	77	0.32	3.3	$\leq8\times10^{11}$	0.1–10 ms
PbSe	295	0.25	4.2	1×10^9–5×10^9	1 μs
	195	0.22	5.4	1.5–4×10^{10}	30–50 μs
	77	0.21	5.8	$\leq3\times10^{10}$	50 μs
InSb[b]	77	$\simeq$0.22	5.5–7.0	$\leq3\times10^{10}$	0.1–1 μs
$Hg_{0.8}Cd_{0.2}Te$	77	$\simeq$0.1	12–25	10^9–10^{11}	>1 ns

[a] Increased sensitivity can be obtained by cooling.
[b] More commonly operated in a photovoltaic mode.

used extrinsic photoconductive detectors, responding out to about 9 μm and 30 μm respectively. Curves showing the variation of their D^* with wavelength are given in Fig. (22.11).

All photoconductive detectors, whether intrinsic or extrinsic, are operated in essentially the same way, although there are wide differences in packaging geometry. These differences arise from differing operating temperatures and speed-of-response considerations. A schematic diagram which shows the main construction features of a liquid-nitrogen-cooled photoconductive or photovoltaic detector is given in Fig. (22.12). Uncooled detectors can be of much simpler construction – for example, in a transistor or flat solar cell package.

One feature of the cooled detector design shown in Fig. (22.12) is worthy of note. The field of view of the detector is generally restricted by an aperture, which is kept at the temperature of the detection element. This shields the detector from ambient infrared radiation, which peaks at 9.6 μm. For detection of low level narrow-band infrared radiation the influence of background radiation can

Fig. 22.11. $D^*(\lambda)$ as a function of wavelength for various photoconductive detectors (courtesy of Hughes Aircraft Company).

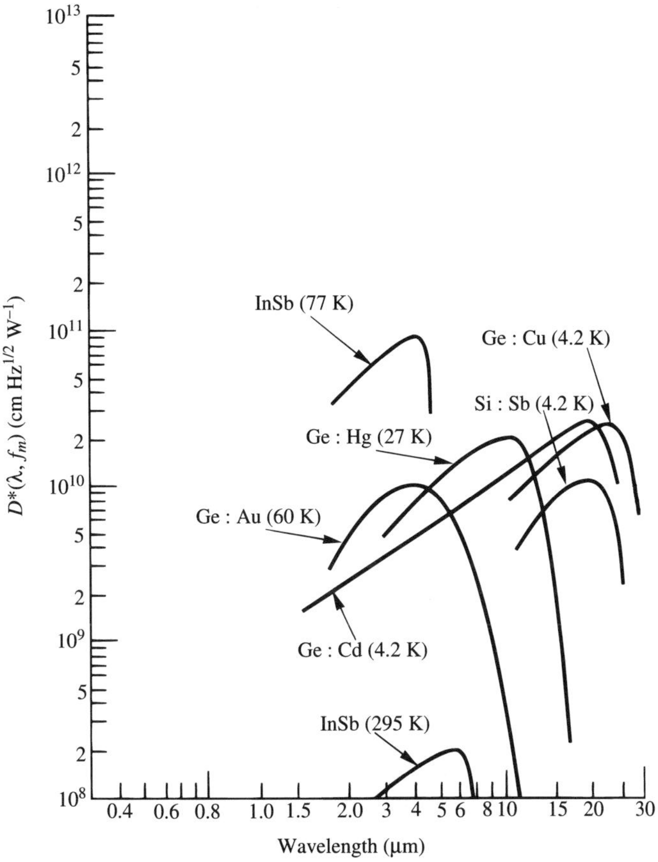

Fig. 22.12. Radiation-shielded liquid-nitrogen-cooled photoconductive or photovoltaic infrared detector assembly.

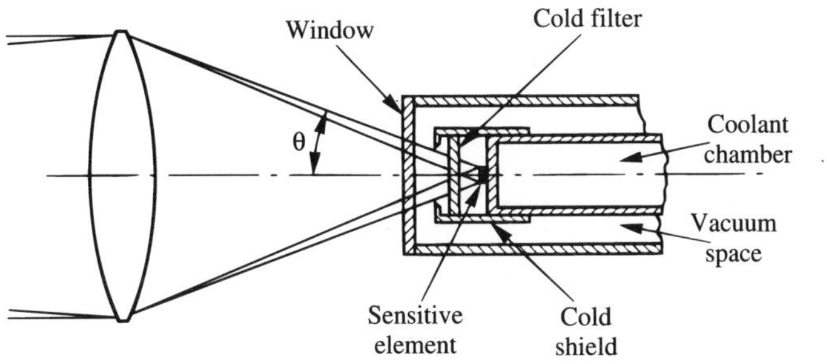

be further reduced by incorporating a cooled narrow-band filter in front of the detector element. The filter will only radiate beyond the cutoff wavelength of the detector, and it restricts transmitted ambient radiation to a narrow band.

Fig. (22.13) shows a basic biasing circuit commonly used for operating photoconductive detectors. R_d is the detector dark resistance. It is easy to see that the change in voltage, ΔV, that appears across the load resistor R_L, for a small change ΔR in the resistance of the detector is

$$\Delta V = \frac{-V_0 R_L \Delta R}{(R_d + R_L)^2}. \tag{22.42}$$

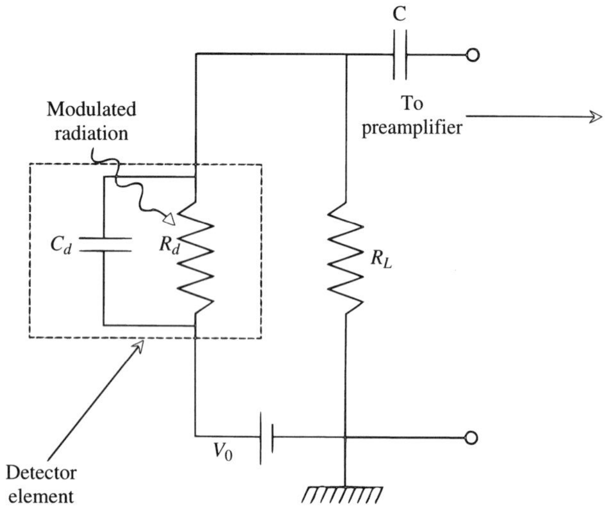

Fig. 22.13. Simple biasing circuit for operating a photoconductive detector with modulated radiation.

This is at a maximum when $R_L = R_d$. Thus it is common practice to bias the detector with a load resistance equal to the detector's dark resistance. The bias voltage is selected to give a bias current through the detector that gives optimum detectivity.

A few photoconductive detectors are worthy of brief extra comment. Silicon and germanium are much more commonly used for photodiodes, frequently in an avalanche mode. These devices are discussed in the next section. Lead sulfide detectors have high impedance, 0.5–100 MΩ, and slow response, but are sensitive detectors in the spectral region between 1.2 and 3 μm and can be used uncooled. $D^*(\lambda)$ curves for these detectors and lead selenide are shown in Fig. (22.14).

Table (22.3) lists the characteristics of some extrinsic photoconductive detectors. Because of the small energy gaps involved in these detectors, they all operate at cryogenic temperatures. The use of extrinsic photoconductivity for the detection of far-infrared radiation requires the introduction of appropriate doping material into a semiconductor in order to generate an acceptor or donor impurity level extremely close to the valence or conduction bands, respectively. Well characterized impurity levels can be generated in germanium in this way using gallium, indium, boron, or beryllium doping, but the long wavelength sensitivity limit is restricted to about 124 μm. Longer wavelength sensitive, extrinsic photoconductivity can be observed in appropriately doped InSb in a magnetic field. However, bulk InSb can be used more efficiently for infrared detection in a different photoconductive mode entirely[22.6]–[22.8]. Even at the low temperature at which far-infrared photoconductive detectors operate ($\leq$4 K), there are carriers in the conduction band. These free electrons can absorb far-infrared radiation efficiently and move into higher energy states within the conduction band. This change in energy results in a change of mobility of these free electrons, which can be detected as a change in conductivity. These *hot-carrier-effect* photoconductive detectors can be used successfully over a wavelength range extending from 50 to 10,000 μm; they have detectivities up to 2×10^{12} cm $\text{Hz}^{1/2}$ W^{-1} and response times down to 10 ns or less. They are frequently operated in a large magnetic field (several hundred kA m^{-1} or more).

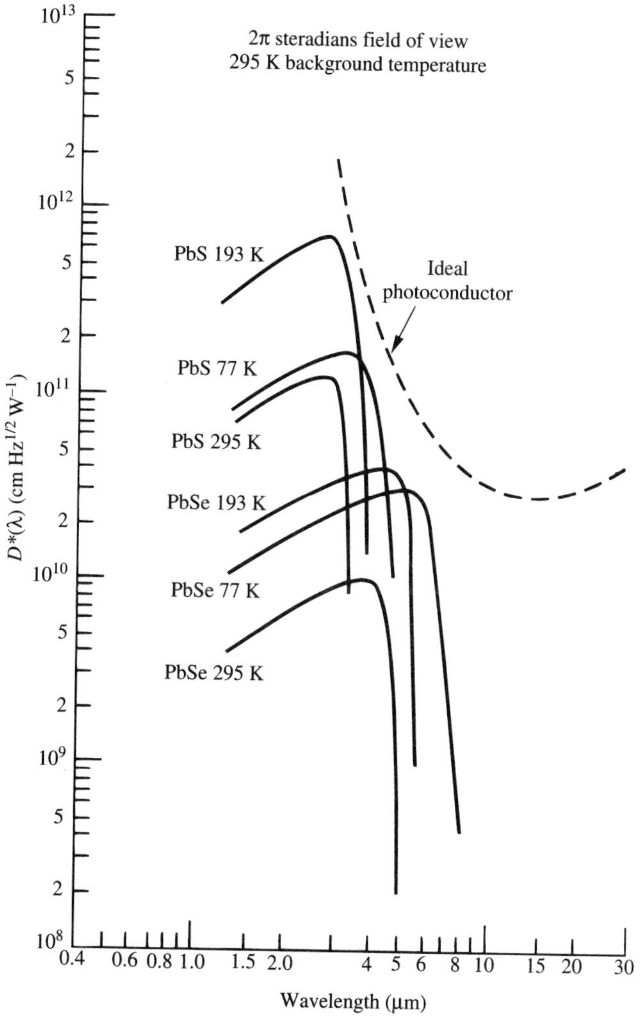

Fig. 22.14. D^* as a function of wavelength for various lead-salt photoconductive detectors (courtesy of Hughes Aircraft Company).

Table 22.3. *Extrinsic photoconductive detectors.*

Semiconductor	Impurity	T(K)	$\lambda_0(\mu m)$	D^*	τ
Ge:Au[a]	p-type	77	8.3	3×10^9–10^{10}	30 ns
Ge:Hg[b]	p-type	<28	14	$(1$–$2)\times10^{10}$	>0.3 ns
Ge:Cd	p-type	<21	21	$(2$–$3)\times10^{10}$	10 ns
Ge:Cu[b]	p-type	<15	30	$(1$–$3)\times10^{10}$	>0.4 ns
Ge:Zn[a,b]	p-type	<12	38	$(1$–$2)\times10^{10}$	10 ns
Ge:Ga	p-type	<3	115	2×10^{10}	>1 μs
Ge:In	p-type	4	111	—	<1 μs
Si:Ga	p-type	4	17	10^9–10^{10}	>1 μs
Si:As	n-type	<20	22	$(1$–$3.5)\times10^{10}$	0.1 μs

[a] Sometimes also contain silicon.
[b] Sometimes also contain antimony.

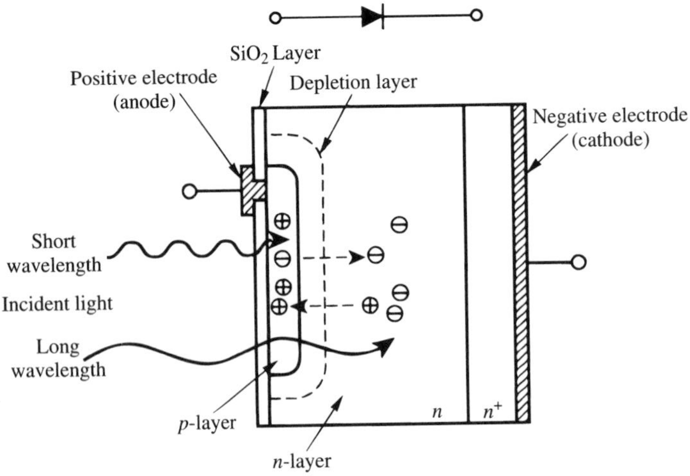

Fig. 22.15. Construction of a p–n junction photodiode (courtesy of Hamamatsu Corporation).

22.4.3 *Photovoltaic Detectors (Photodiodes)*

In a photovoltaic detector photoexcitation of electron–hole pairs occurs near a junction when radiation of energy greater than the band gap is incident on the junction region. Extrinsic photoexcitation is rarely used in photovoltaic photodetectors. The internal energy barrier of the junction causes the electron and hole to separate, creating a potential difference across the junction. This effect is illustrated for a p–n junction in Fig. (22.1). Fig. (21.15) shows how an actual p–n photodiode is constructed. Other types of structure are also used as illustrated in Fig. (22.16), such as p–i–n, Schottky-barrier (a metal deposited onto a semiconductor surface) and heterojunctions. The p–n and p-i-n structures are the most commonly used. All these devices are commonly called photodiodes. The characteristics of some important photodiodes are listed in Table (22.4). Important photodiodes include silicon for detection of radiation between 0.1 and 1.1 μm and detectors based on the InGaAs(P) system for the region between 0.9 and 1.7 μm, which encompasses the important fiber optical communication wavelengths of 1.3 μm and 1.55 μm (see Chapter 24). Typical spectral responsivities for some of the materials are shown in Fig. (22.17). Other photodiodes with more specialized applications include germanium between 0.4 and 1.8 μm, indium arsenide between 1 and 3.8 μm, indium antimonide between 1 and 7 μm, lead–tin telluride between 2 and 18 μm, and mercury–cadmium telluride between 1 and 12 μm. Some typical curves of $D^*(\lambda)$ for these photodiodes are shown in Fig. (22.18). These spectral response regions are not all necessarily covered by a detector operating at the same temperature; for example, InSb responds to 7 μm at 300 K but to wavelengths no longer than 5.6 μm at 77 K. The wavelength response of PbSnTe and HgCdTe depends also on the stoichiometric composition of the crystal. All these photodiodes have very high quantum efficiency, defined in this case as the ratio of photons absorbed to mobile electron–hole pairs produced in the junction region. Values in excess of 90% have been observed in the case of silicon.

When a photodiode detector is illuminated with radiation of energy greater than the band gap, it will generate a voltage and can be operated in the very simple circuit illustrated in Fig. (22.19a). However, it is much better to operate

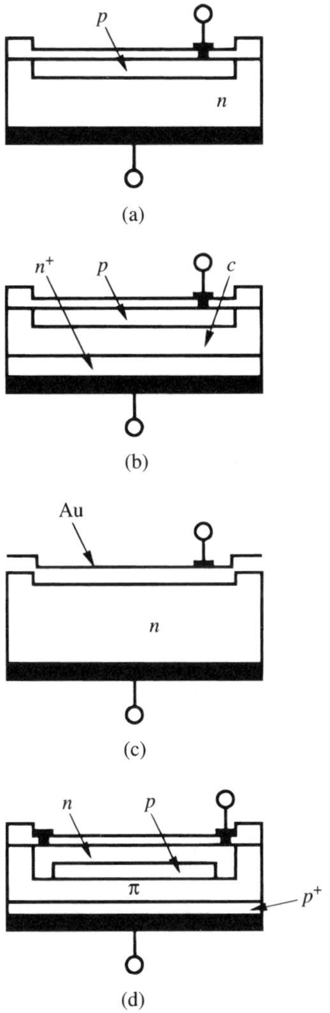

Fig. 22.16. Schematic construction of various photodiode types: (a) planar diffusion type; (b) *p–i–n*; (c) Schottky type; (d) avalanche (courtesy of Hamamatsu Corporation).

a photodiode detector in a reverse-biased mode, as shown in Fig. (22.19b), where a positive voltage is applied to the *n*-type side of the junction and negative to the *p*-type. In this case, the observed photosignal is seen as a change in current through the load resistor. The difference between the two modes of operation can be easily seen from Fig. (22.20a), which shows the *I–V* characteristic of a photodiode in the dark and in the presence of illumination. In the forward bias direction the light response of the photodiode is nonlinear. A photodiode responds much more linearly to changes in light intensity and has greater detectivity when operated in the reverse-biased mode. At increasing levels of reverse bias the photodiode first enters an *avalanche* region before it breaks down at sufficiently large reverse bias. Ideal operation is obtained when the diode is operated in the current mode with an operational amplifier that effectively holds the photodiode voltage at zero – its optimum bias point. A simple practical circuit which can be used to operate a photodiode in this way is shown in Fig. (22.20b). In this circuit, the bias voltage V_B is not necessary, but for many photodiodes will improve the

Table 22.4. *Photovoltaic detectors (photodiodes).*

Semiconductor	T (K)	Wavelength range (μm)	D^*(max) (or NEP)[b]	τ
Si	300	0.2–1.1	$\leq 2\times10^{13}$	>6ps
InGaAs[a]	300	0.9–1.7	6×10^{-14}(NEP, W $\mathrm{Hz}^{-1/2}$)	20ps
GaAsP[a]	300	0.3–0.76	2×10^{-15}(NEP, W $\mathrm{Hz}^{-1/2}$)	
Ge	300	0.4–1.8	10^{11}	0.3 ns
InAs	300	1–3.8	$\leq 4\times10^{9}$	5 ns–1 μs
InAs	77	1–3.2	4×10^{11}	0.7 μs
InSb	300	1–7	1.5×10^{8}	0.1 μs
InSb	77	1–5.6	$\leq 2\times10^{11}$	>25 ns
PbSnTe	77	2–18	$\leq 10^{11}$	20 ns–1 μs
HgCdTe	77	1–25	10^{9}–10^{11}	>1.6 ns

[a] Precise performance depends on stoichiometry. Quaternary versions of these detectors based on InGaAsP are also used.
[b] The D^* or NEP of these detectors are typical values. The actual S/N performance of these detectors will in practical applications depend on the inevitable additional noise added by amplification electronics[22.9].

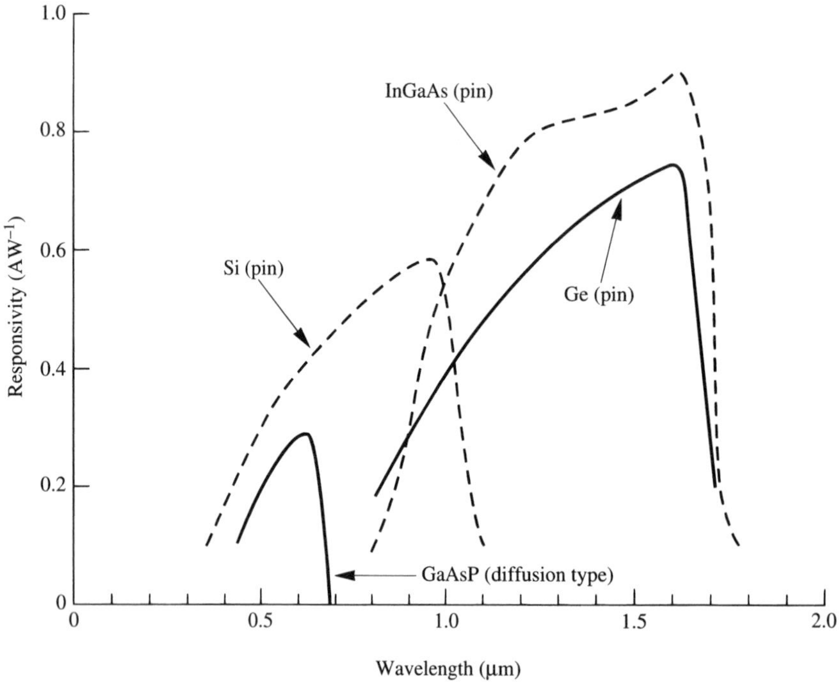

Fig. 22.17. Spectral responsivities of some important near-infrared photodiodes.

speed of response, albeit at the expense of an increase in noise. The $p-i-n$ structure is most commonly used in these devices because its performance, in terms of quantum efficiency (number of useful carriers generated per photon absorbed) and frequency response, can be readily optimized. These devices have very low

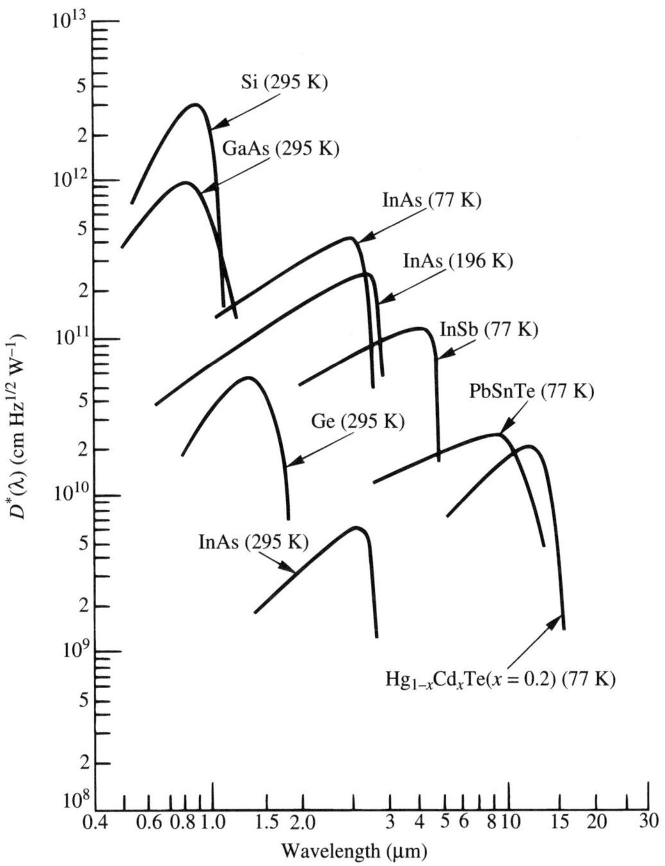

Fig. 22.18. D^* as a function of wavelength for various photovoltaic detectors (courtesy of Hughes Aircraft Company).

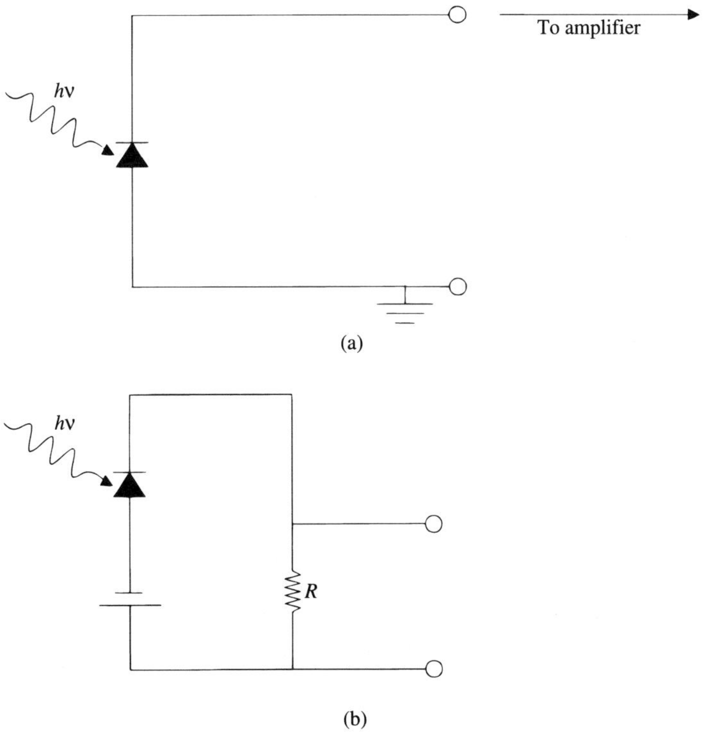

Fig. 22.19. Photovoltaic detector operated in: (a) open-circuit mode; (b) reversed-biased mode.

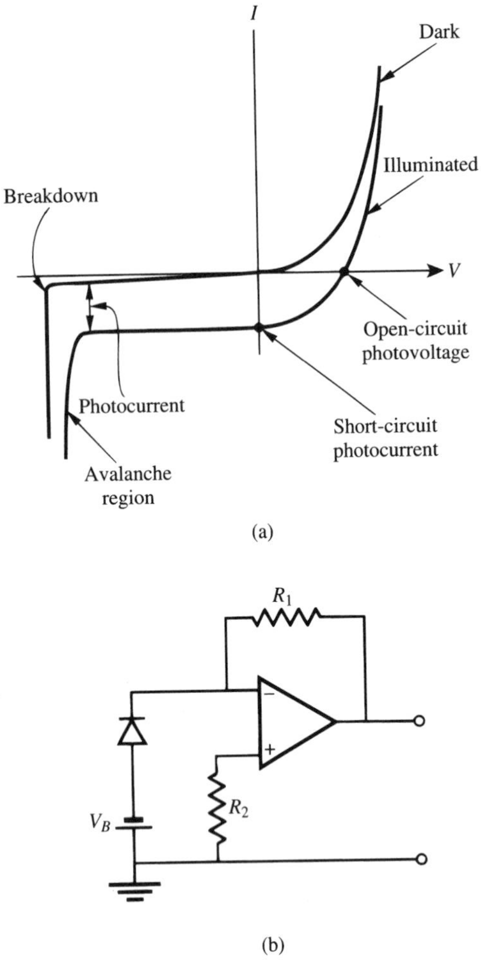

Fig. 22.20. (a) Current voltage characteristics of a photodiode;
(b) Current-mode-operated photodiode.

noise and fast response. In practice, the limiting sensitivity that can be obtained with them will be determined by the noise of the associated amplifier circuitry.

If the reverse-bias voltage on a photodiode is increased, photoinduced charge carriers can acquire sufficient energy traversing the junction region to produce additional electron–hole pairs. Such a photodiode exhibits current gain and is called an *avalanche photodiode* (APD). It is in some respects the solid-state analog of the photomultiplier. Avalanche photodiodes are noisier than *p–i–n* photodiodes, but because they have internal gain, the practical sensitivity that can be achieved with them is greater. Because of their importance *p–i–n* photodiodes and APDs are worthy of more detailed discussion.

22.4.4 *p–i–n* Photodiodes

In a simple *p–n* junction photodiode using the structure shown in Fig. (22.1) reverse bias leads to a current that increases linearly with incident optical power over up to nine orders of magnitude, say from 1 pW to 1 mW. In order for electron–hole pairs created by photon absorption to appear as useful current in

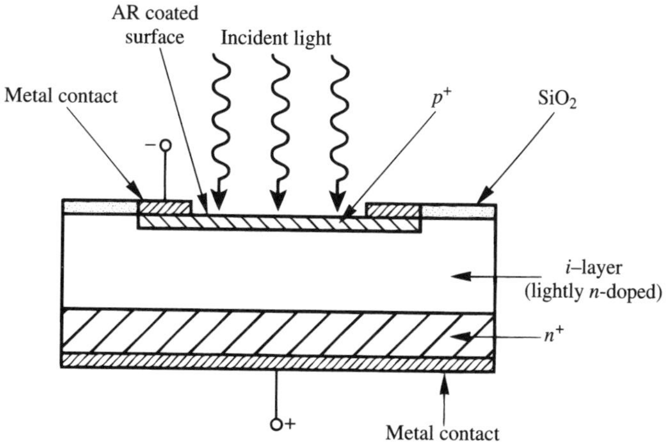

Fig. 22.21. Schematic construction of a p–i–n photodiode. The lightly n-doped intrinsic layer is frequently designated as a ν-layer.

the external circuit they must be swept out of the junction region and collected at the electrodes before they have a chance to recombine. This is best accomplished if as much of the photon absorption as possible occurs in a thick depletion layer that is close to the electrodes, as shown in Fig. (22.21). Under reverse bias there are very few mobile charge carriers in the depleted i-layer. There is a build-up of electrons on the heavily doped p^+ side of the device and of holes on the heavily doped n^+ side. Thus, the static electrical state is that of a parallel plate capacitor of capacitance

$$C = \frac{\epsilon_0 \epsilon_r A}{d}. \tag{22.43}$$

For a typical device with $d = 20\ \mu$m, $A \simeq 10^{-8}$ m^2, $\epsilon_v = 12; C = 0.05$ pF. The intrinsic time constant of the diode driving a 50Ω load is 2.7 ps. These are high speed devices. There are very many variations in the detailed construction of $p - i - n$ photodiodes[22.9]–[22.11]. It is possible to use heterostructures in these devices, for example, the p^+–i–n^+ layers can be GaAlAs/GaAs/GaAs or $(InGaAsP)_1/(InGaAsP)_2/InP$.† If the layer through which incident radiation enters has a larger band gap than the absorbing (intrinsic) layer, then long wavelength photons will not be absorbed in the surface layer. As is common in these layered semiconductor structures (see also Chapter 13), additional highly doped layers may be included adjacent to contact (metal) electrodes. If a metal contact layer is placed on a layer that is not sufficiently heavily doped a rectifying Schottky diode results. Such structures can actually be used as semiconductor detectors themselves, particularly for ultraviolet detection. A thin, transparent gold layer is placed on a substrate of GaAsP or GaP, as shown as the n-layer in Fig. (22.16c).

22.4.5 *Avalanche Photodiodes*

When the reverse bias of a photodiode is increased sufficiently the internal electric field can accelerate photo-generated charge carriers to sufficient energy that they can excite additional electron–hole pairs. Both electrons and holes can contribute to this process, which is shown schematically for an electron in Fig. (22.22). A

† The subscripts indicate different stoichiometries. For example $(InGaAsP)_1$ could be InP or InGaAs.

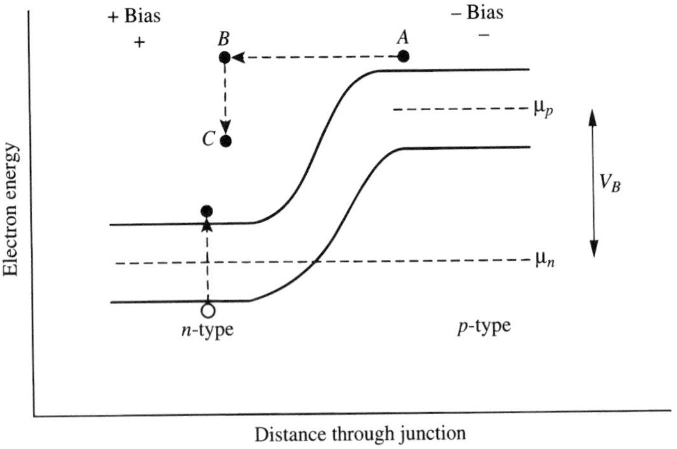

Fig. 22.22. Illustration of avalanche electron–hole pair production by an electron crossing a reverse-biased p–n junction. The Fermi energies are separated by the bias voltage V_B across the junction. An electron crossing the junction from A to B has sufficient energy to excite an electron-hole pair and simultaneously fall to energy C.

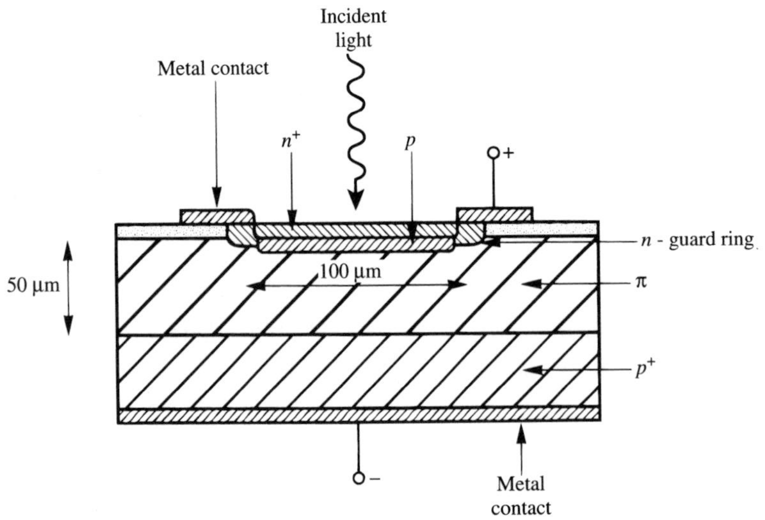

Fig. 22.23. Schematic construction of a planar n^+-p-π-p^+ APD with a guard ring and showing typical dimensions.

schematic layout of a typical avalanche photodiode (APD) is shown in Fig. (22.23). In this structure electron-hole pairs are created initially mostly in the lightly p-doped intrinsic layer (called a π-layer), because the n^+–p junction region is very thin. There is sufficient voltage across the π layer for photogenerated electrons and holes to drift rapidly across it. At the n^+–p junction near the positive electrode there is a large field gradient and efficient avalanche multiplication occurs. The n-doped region at the edge of the p-doped region is called a *guard ring*. It prevents edge effects at the boundary between the n^+-, p-, and π-layers. This keeps the avalanche region electric field uniform: there are no high fields at the edge where an avalanche could become destructive breakdown.

The performance of an APD is characterized by its multiplication factor M, which is the number of electron–hole pairs generated by the dominant carrier in the detector – electrons in silicon, holes in InGaAs/InP devices. In the latter we see once again the use of a heterostructure, which allows control of where incident radiation is absorbed. Incident radiation can pass through the larger band gap InP and then be absorbed in the InGaAs. It is important in both APDs, and *p–i–n* photodiodes for the highest response speed that electron-hole pairs be produced in a fully depleted region of the device. If the charge carriers are generated in a region where electrons or holes are present in significant concentration, the motion of the photogenerated carriers is slowed by diffusion through these undepleted regions. The fastest response is obtained if the photogenerated carriers are pulled out of the depleted region by the bias field at the maximum possible speed. This occurs at the so-called *saturation velocity*, v_s, which is typically on the order of 10^5 m s^{-1}.

To achieve saturation velocity requires an applied field in the depleted region on the order of 1 MV m^{-1}. In a 50 μm thick depleted π-layer, as shown in Fig. (22.23), this requires a reverse bias of 50 V. A rough estimate for the speed of response will then be 0.5 ns. Of course, to determine the precise response of the detector it would be necessary to include the actual spatial distribution of photogenerated carriers in the device[22.9]. The lateral dimensions of the device must also be kept small so as to minimize capacitance effects on the speed of response.

22.5 Thermal Detectors

Thermal detectors, in principle, have a detectivity that is independent of wavelength from the vacuum-ultraviolet upward as shown in Fig. (22.24), which also shows representative $D^*(\lambda)$ values for various types. In practice, the absorbing properties of the 'black' surface of the detector will, in general, show some wavelength dependence, and the necessity for a protective window on some detector elements may limit the useful spectral bandwidth of the device. Most commonly available thermal detectors, although by no means as sensitive as various types of photodetectors, achieve spectral response very far into the infrared – to the microwave region in fact – while conveniently operating at room temperature. Each of these detectors is discussed briefly below. Putley[22.7] gives a more detailed discussion.

(a) Thermopile. Thermopiles, although they are one of the earliest forms of infrared detector, are still widely used. Their operation is based on the Seebeck effect, where heating the junction between two dissimilar conductors generates a potential difference across the junction. An ideal device should have a large Seebeck coefficient, low resistance (to minimize ohmic heating), and a low thermal conductivity (to minimize heat loss between the hot and cold junctions of the thermopile). These devices are usually operated with an equal number of hot (irradiated) and cold (dark) junctions, the latter serving as a reference to compensate for drifts in ambient temperature. Both metal (copper–constantan, bismuth–silver, antimony–bismuth) and semiconductor junctions are used as the active elements. The junctions can take the form of evaporated films, which improves the robustness of the devices and reduces their time constant, although this is still slow (0.1 ms at best). Because a thermopile has very low output impedance, it must be used with a specially designed low noise amplifier, or with a step-up transformer[22.12].

(b) Pyroelectric Detectors. These are detectors that utilize the change in surface charge that results when certain asymmetric crystals (ones that can possess an internal electric dipole moment) are heated. The crystalline material is fabricated as the dielectric in a small capacitor, and the change in charge is measured when the element is irradiated. Thus, these devices are inherently ac detectors. If

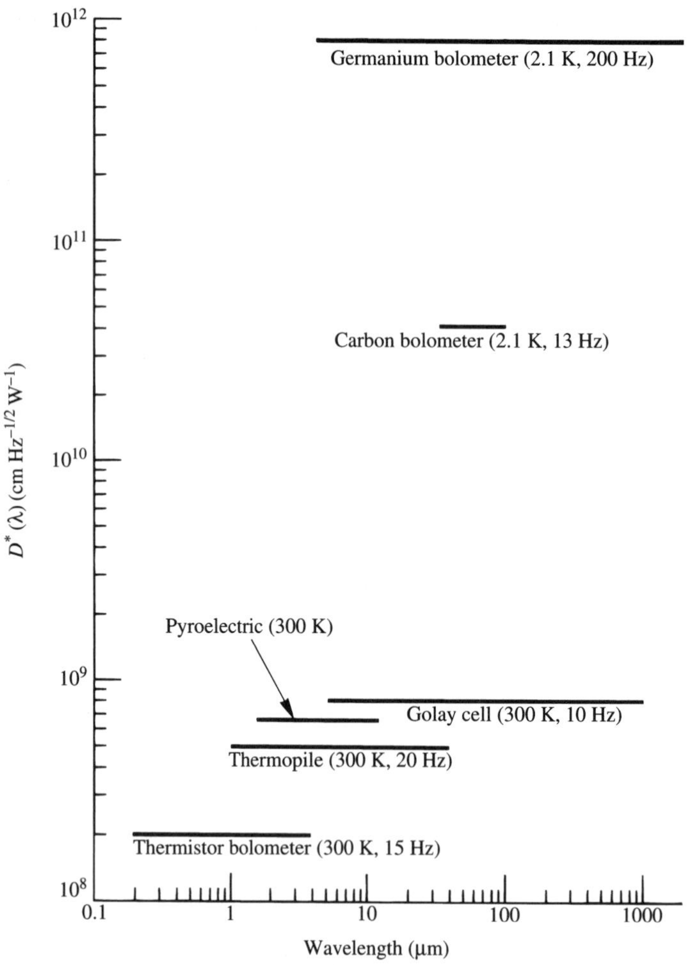

Fig. 22.24. Representative $D^*(\lambda)$ response curves for various thermal detectors assuming total absorption of incident radiation. The operating temperature, modulation frequency, and typical useful wavelength range are shown for each detector. In each case the detector is assumed to view a hemispherical enclosure at a temperature of 300 K.

the chopping frequency of the input radiation is slow compared to the thermal relaxation time of the crystal, the crystal remains close to thermal equilibrium and the current response is small. When the chopping period becomes shorter than the thermal relaxation time, much greater heating and current response results. The responsivity of the detector in this case can be written as

$$R = \frac{p(T)}{\rho C_p d} \quad (\mathrm{A\ W^{-1}}), \tag{22.44}$$

where $p(T)$ is the pyroelectric coefficient at temperature T, d is the spacing of the capacitor electrodes, and ρ and C_p are the density and specific heat of the crystal, respectively.

The equivalent circuit of a pyroelectric detector is a current source in parallel with a capacitance, which can range from a few to several hundred picofarads. For optimum performance, the resultant high impedance must be matched to a high-input-impedance, low-output-impedance amplifier. These detectors can have response times as short as 2 ps. Their detectivities are comparable with those of thermopiles, and they also have flat spectral response. Consequently, they can replace the thermopile in many applications as a convenient, room temperature, wide-spectral-sensitivity detector of infrared and visible light.

(c) Bolometer. The resistance of a solid changes with temperature according to a relation of the form

$$R(T) = R_0[1 + \gamma(T - T_0)], \tag{22.45}$$

where γ is the temperature coefficient of resistance, typically about 0.05 K^{-1} for a metal, and R_0 is the resistance at temperature T_0.

A *bolometer* is constructed from a material with a large temperature coefficient of resistance. Absorbed radiation heats the bolometer element and changes its resistance. Bolometers utilize metal, semiconductor, or almost superconducting elements. Metal bolometers utilize fine wires (platinum or nickel) or metal films. The mass of the element must be kept small in order to maximize its temperature rise. Even so, the response time is fairly long ($\geq$1 ms). Semiconductor bolometer elements (thermistors) have larger absolute values of γ and have largely replaced metals except where very long-term stability is required.

It is usual to operate bolometer elements in pairs in a bridge circuit, as shown in Fig. (22.25). One element is irradiated, while the second serves as a reference and compensates for changes in ambient temperature. Thermistors have a negative *I–V* characteristic above a certain current and will exhibit destructive thermal runaway unless operated with a bias resistor. It is therefore usually best to operate the thermistor at currents below the negative-resistance part of its *I–V* characteristic.

(d) The Golay Cell. In a Golay cell (named for its inventor, M.J.E. Golay)[22.13], radiation is absorbed by a metal film that forms one side of a small sealed chamber containing xenon (used because of its low thermal conductivity). Another wall of the chamber is a flexible membrane, which moves as the xenon is heated. The motion of the membrane is used to change the amount of light reflected to a photodetector. The operating principle and essential design features of a modern Golay cell are shown in Fig. (22.26). Although these detectors are fragile, they are very sensitive and are still widely used for far-infrared spectroscopy.

22.6 Detection Limits for Optical Detector Systems

In a practical detection system the detector itself is coupled to various electronic devices such as amplifiers, filters, pulse counters, limiters, discriminators, phase-locked loops, etc. It is beyond our scope here to deal with all the consequent realities of how the optical detection limit is influenced by the additional noise contributions of these devices. Nor can we deal in detail with how various ways of encoding information onto a light signal can be used to enhance the signal-to-noise ratio of the overall detection system. For additional details the reader should consult the more specialized literature[22.9],[22.14]–[22.18]. However, in a well designed optical detection system the performance should be primarily limited only by the characteristics of the detector itself. Therefore, in our discussion of fundamental detector limits we will deal with only the unavoidable noise sources of the detector and its associated resistors, as shown schematically in Fig. (22.3).

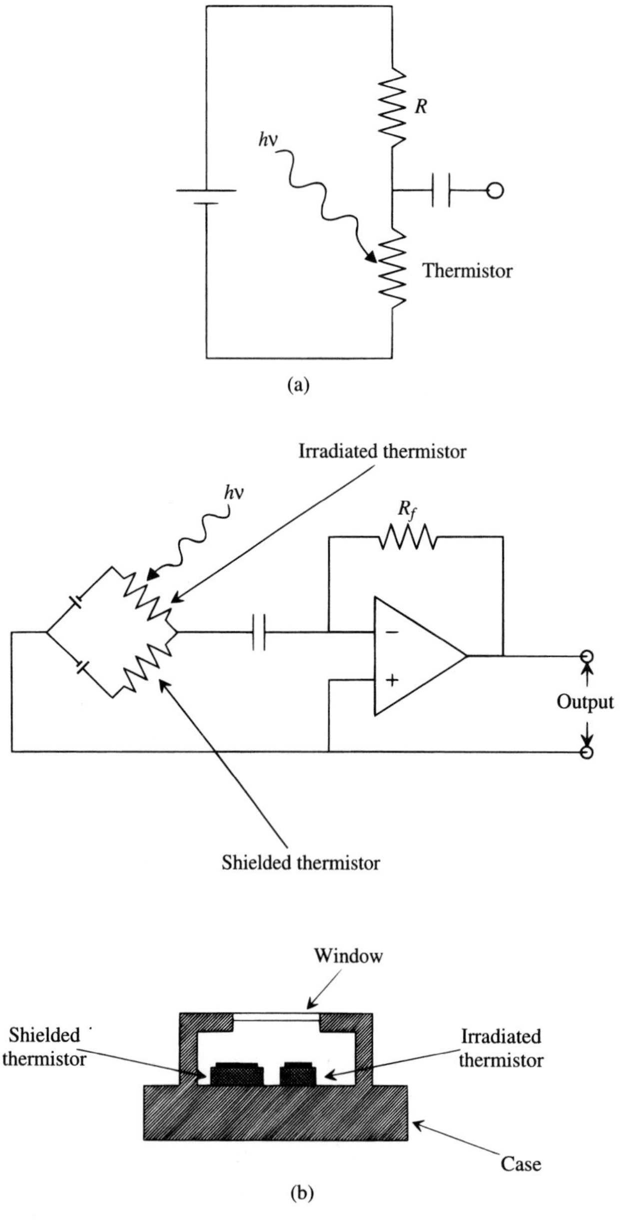

Fig. 22.25. Bridge operating circuit for thermistor bolometers using compensating shielded thermistor, with device construction shown.

22.6.1 Noise in Photomultipliers

Noise in photomultipliers comes from several sources:

(a) Thermionic emission from the photocathode.

(b) Thermionic emission from the dynodes.

(c) Field emission from dynodes (and photocathode) at high interdynode voltages. In this phenomenon the potential gradient near the emitting surface is sufficiently great to liberate electrons from the material.

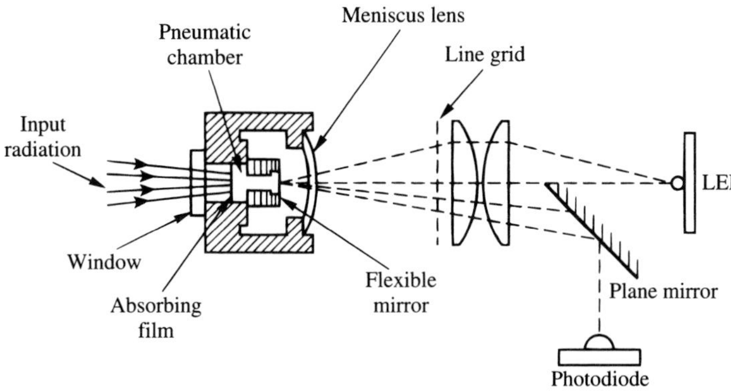

Fig. 22.26. Schematic design of the Golay detector. The top half of the line grid is illuminated by the LED and imaged back on the lower half of the grid by the flexible mirror and meniscus lens. Any radiation-induced deformation of the flexible mirror moves the image of the line grid and changes the illumination reaching the photodiode.

(d) Radioactive materials in the tube envelope, for example ^{40}K in glass.

(e) Electrons striking the tube envelope and causing fluorescence.

(f) Electrons striking the dynodes and causing fluorescence.

(g) Electrons colliding with residual atoms of vapor in the tube, cesium, for example, and causing fluorescence.

(h) Cosmic rays.

Of these sources thermionic emission from the photocathode is generally the most important. It increases with the area of the photocathode. Thermionic emission from the dynodes leads to smaller anode pulses than for electrons originating at the photocathode. Therefore, when a photomultiplier is used in *photon counting*, where individual anode pulses are counted, discrimination against some noise can be effected by counting only within a specific range of anode pulse lengths, as shown in Fig. (22.9). Small anode pulses are likely to originate with noise source (b). Large anode pulses are likely to originate with noise sources (c)–(h).

22.6.2 Photon Counting

Photomultipliers are *extremely* sensitive detectors of ultraviolet, visible and near-infrared radiation. They have the unique ability to provide a macroscopic signal output from a single photoelectron liberated at their photocathode. In photon counting the anode pulses are counted individually and the pulse count rate in the case of illumination is compared with the dark count rate. The signal-to-noise ratio of this process can be further enhanced if the arrival time of signal photons can be isolated to a time window of width τ, for example by synchronizing the pulse counting electronics to the excitation of the phenomenon leading to the light.

Let us suppose that the (weak) light source to be detected emits $\overline{N}_1/\eta$ photons per second that are absorbed in the photocathode. Consequently, the average number of signal anode photoelectron counts per second is $\overline{N}_1$. The number of signal counts in a time interval, which can be assumed to occur randomly within this time period, is† $\overline{N}_1\tau$, with a variance

$$\overline{(N_1 - \overline{N}_1)^2\tau} = \overline{N}_1\tau. \tag{22.46}$$

† For a more detailed discussion of this point see Chapter 23.

If there are zero dark counts the relative accuracy with which $\overline{N}_1$ can be determined is

$$\frac{\sqrt{\overline{(N_1 - \overline{N}_1)^2\tau}}}{\overline{N}_1\tau} = \frac{1}{\sqrt{\overline{N}_1\tau}}. \tag{22.47}$$

Therefore, by counting for a long time good accuracy can be obtained: for $\overline{N}_1\tau = 10^6$ counts the accuracy is 0.1%.

If in addition there are dark counts that are detected randomly at a rate $\overline{N}_2$ per second, then the variance in the number of dark counts in a time τ is

$$\overline{(N_2 - \overline{N}_2)^2\tau} = \overline{N}_2\tau. \tag{22.48}$$

The relative accuracy with which the number of signal counts can be determined is

$$\frac{\sqrt{\overline{N}_1\tau + \overline{N}_2\tau}}{\overline{N}_1\tau} = \sqrt{\frac{1 + \overline{N}_2/\overline{N}_1}{\overline{N}_1\tau}}. \tag{22.49}$$

An estimate of the limiting sensitivity of photon counting can be obtained by setting the relative accuracy to unity and choosing $\tau = 1$. The minimum sensitivity is then

$$(\overline{N}_1)_{min} = \frac{1}{2}\left[1 + \sqrt{1 + 4\overline{N}_2}\right], \tag{22.50}$$

and since generally $\overline{N}_2 >> 1$.

$$(\overline{N}_1)_{min} = \sqrt{\overline{N}_2}. \tag{22.51}$$

So, for example, with a 500 nm source and a photomultiplier with quantum efficiency of 20% and 10^2 dark counts per second the minimum detectable optical power is

$$P_{min} = 50h\nu = 2 \times 10^{-17}\ \text{W}. \tag{22.52}$$

22.6.3 *Signal-to-Noise Ratio in Direct Detection*

In many applications of photomultiplier tubes the anode pulses are integrated to give a fluctuating analog current. The shot noise originating at the photocathode is, from Eq. (22.20)

$$< i_N^2 >_c = 2e(\bar{i}_c + \bar{i}_d)\Delta f, \tag{22.53}$$

where $\bar{i}_c$ is the average photocathode current produced by a light source and $\bar{i}_d$ is the average photocathode dark current. This noise is multiplied by the amplification of the electron number by interaction with the N dynodes of the tube. If each dynode has a secondary emission multiplication efficiency δ the overall gain of the tube is†

$$G = \delta^N. \tag{22.54}$$

Because of statistical fluctuations in the secondary emission process and, in addition, because electrons can originate thermionically from the dynodes, the noise at

† The first dynode often has a larger δ value than the others, but we will assume that δ is an average value.

the anode is further increased by a noise factor F. The overall noise appearing at the anode is

$$< i_N^2 >_A = 2eG^2F(\bar{i}_c + \bar{i}_d)\Delta f. \tag{22.55}$$

For $\delta = 4$, and a 14-stage tube $G = 2.6 \times 10^8$, and typically $F \simeq \delta/(\delta - 1) = 1.33$.

A photomultiplier is a current source: to convert this current to a detected voltage the amplified photoelectron current passes through an anode resistor of value R. This anode resistor may be part of the photomultiplier circuit, or may be provided in whole or in part by the input impedance of a following amplifier stage. The Johnson noise from the resistor is

$$< i_N^2 >_R = \frac{4kT\Delta f}{R}. \tag{22.56}$$

To optimize detection of a signal it is common practice to amplitude modulate the signal at some angular frequency ω_m so as to permit synchronous detection at this frequency[22.12]. The optical power reaching the photocathode can be represented in this case as

$$P = P_0(1 + m \sin \omega_m t), \tag{22.57}$$

where m is a modulation parameter.

The average photocathode current is

$$\bar{i}_c = \frac{e\eta P_0}{h\nu}, \tag{22.58}$$

and

$$i_c(t) = \bar{i}_c(1 + m \sin \omega_m t). \tag{22.59}$$

At the anode the time-varying part of this current is

$$i_s(t) = \bar{i}_c G m \sin \omega_m t. \tag{22.60}$$

The signal-to-noise ratio (S/N) at the input to the electronics in Fig. (22.3) is therefore

$$\frac{< i_s^2 >}{< i_N^2 >_A + < i_N^2 >_R} = \frac{\bar{i}_c^2 G^2 m^2/2}{2eG^2F(\bar{i}_c + \bar{i}_d)\Delta f + 4kT\Delta f/R}. \tag{22.61}$$

The noise from the photomultipler tube is usually sufficiently large that the Johnson noise can be neglected. If it is assumed that $\bar{i}_d >> \bar{i}_c$, then for $m = 1$ and $S/N = 1$ we have, from Eqs. (22.58) and (22.61)

$$(P_0)_{min} = \frac{2h\nu}{\eta}\sqrt{\frac{F\bar{i}_d\Delta f}{e}}. \tag{22.62}$$

Example: Typical values for a good photomultiplier will be $\eta = 0.2, F \simeq 1$, $\bar{i}_d \simeq 10^{-15}$ A. For a 530 nm source Eq. (22.62) gives $(P_0)_{min} = 3 \times 10^{-16}$ W.

22.6.4 Direct Detection with *p–i–n* Photodiodes

The shot noise from a *p–i–n* photodiode is

$$< i_N^2 >_1 = 2e(\bar{i}_s + i_d)\Delta f, \tag{22.63}$$

where $\bar{i}_s$ is the average signal current, and i_d is the dark current (usually specified for a low noise photodiode in units of nA Hz$^{-1/2}$ or pA Hz$^{-1/2}$).

The average signal current is, in a similar way to Eq. (22.58),

$$\bar{i}_s = \frac{e\eta P_0}{h\nu}, \tag{22.64}$$

where η is the quantum efficiency, which is much larger than for a photomultiplier: values of 0.7–0.8 are common. For the simple equivalent current discussed previously (Fig. (22.3)) there is an additional Johnson noise contribution of magnitude

$$(i_N^2)_2 = \frac{4kT\Delta f}{R}. \tag{22.65}$$

The overall S/N ratio for direct detection of an unmodulated signal is

$$\frac{< i_s^2 >}{< i_n^2 >_1 + < i_N^2 1 >_R} = \frac{(e\eta P_0/h\nu)^2 R}{2eR(e\eta P_0/h\nu + i_d)\Delta f + 4kT\Delta f}. \tag{22.66}$$

This S/N ratio would be reduced by a factor $m^2/2$ for a modulated signal (cf. Eq.(22.61)).

If shot noise dominates over dark current and Johnson noise, then the shot-noise-limited S/N ratio is

$$\frac{S}{N} = \frac{\eta P_0}{2h\nu\Delta f}. \tag{22.67}$$

In a practical application using a photodiode there will be additional stages of electronic amplification that add noise. It is common to characterize the effect of an electronic circuit on the noise by its *noise figure*, F_N.

The noise figure can be defined conveniently as

$$F_N = \frac{\text{noise power at output of circuit}}{\text{amplified noise power at the input}}. \tag{22.68}$$

For input Johnson noise the mean-square, amplified, output noise current is

$$< i_N^2 >_1 = \frac{4kTG\Delta f}{R}, \tag{22.69}$$

where G is the power gain of the amplifier within the frequency band being considered. The actual output mean-square noise current is

$$< i_N^2 >_2 = \frac{4kTGF_N\Delta f}{R}. \tag{22.70}$$

It is as if the amplifier were noiseless but the input Johnson noise is increased by the noise figure.

The noise figure is frequently quoted in dB, a noise figure of 3 dB would represent a doubling of the output noise over the value expected from a noiseless amplifier. The term *noise temperature*, T_N, is also used defined by

$$F_N = 1 + \frac{T_N}{T_{ambient}}. \tag{22.71}$$

Example: We illuminate an InGaAs photodiode with a responsivity of 0.8 A W^{-1} at 1.3 μm in an optical communication link in which there is 30 dB of loss between a 10 mW source and receiver. The system band width is 100 MHz, the dark current is 5 nA (equivalent to 0.5 pA Hz$^{-1/2}$). We assume that the amplification electronics has a noise figure of 6 dB (relative to a 50 Ω input). We note the following:

(a) Received power $P_0 = 10^{-3} \times 10$ mW $= 10$ μW;

(b) Signal current $\bar{i}_s = 8$ μA;

(c) Dark current $\bar{i}_d = 5$ nA (dominated by $\bar{i}_s$);

(d) Signal Power $= (\bar{i}_s)^2 R = 3.2$ nW;

(e) Shot noise power $= 2e(\bar{i}_s + \bar{i}_d)\Delta f R = 1.28 \times 10^{-14}$ W;

(f) Johnson noise power $= 4kT\Delta f = 4(1.38 \times 10^{-22})(300)(10^8) = 1.66$ pW;

(g) Effective Johnson noise power including noise figure $= 10^{0.6} \times 1.66$ pW $=$ 6.6 pW.

In this case the Johnson noise is dominant, the effective S/N ratio is

$$\frac{S}{N} = \frac{3.2 \times 10^{-9}}{6.6 \times 10^{-12}} = 485.$$

22.6.5 Direct Detection with APDs

In an APD there is a multiplication of the number of charge carriers by a factor M. This multiplication can result from secondary ionizations produced by both electrons and holes. It is desirable that one or other of these charge carriers should have a significantly greater secondary ionization coefficient† than the other[22.19]. The current in an APD increases by the factor M but the associated shot noise increases further because in a given avalanche process M will fluctuate, taking values $M, M \pm 1, M \pm 2$, etc. The mean-square noise current thereby increases, not by a factor M^2, but by a factor FM^2, where F is called the *noise factor*. In silicon APDs F typically lies in the range 2–20. Therefore, the shot noise becomes

$$< i_N^2 >_1 = 2eFM^2(\bar{i}_s + i_d)\Delta f. \tag{22.72}$$

Both photo- and dark-generated carriers contribute to the shot noise. The overall S/N ratio is modified from Eq. (22.66) and becomes

$$\frac{< i_s^2 >}{< i_N^2 >_1 + < i_N^2 >_2} = \frac{M^2(e\eta P_0/h\nu)^2 R}{2eRFM^2(e\eta P_0/h\nu + i_d)\Delta f + 4kT\Delta f}. \tag{22.73}$$

Eq. (22.73) shows that the S/N ratio improves with increasing multiplication as the Johnson noise contribution becomes less important until the avalanche shot noise becomes dominant. The avalanche-limited S/N ratio is

$$\frac{S}{N} = \frac{\eta P_0}{2Fh\nu\Delta f}, \tag{22.74}$$

a reduction of the signal-noise-ratio from the quantum-limited value by the noise factor.

The NEP can be computed for both *p–i–n* photodiodes and APDs by setting $S/N = 1$ in equations like (22.66) and (22.73) and thereby the minimum input

† The secondary ionization coefficient is the number of secondary electron–hole pairs produced per unit length by a specific charge carrier (electron or hole) in travelling through the material.

power for $S/N = 1$ is determined. For $P_0 = P_{min}$, we expect the dark current to be larger than the signal current so Eq. (22.73) becomes

$$\frac{S}{N} = \frac{(Me\eta P_0/h\nu)^2}{(2eFM^2 i_d + 4kT/R)\Delta f}. \tag{22.75}$$

For $S/N = 1$ and a 1 Hz bandwidth $P_0 = P_{min} = NEP$, which gives for an APD

$$NEP(\mathrm{W\ Hz^{-1/2}}) = \frac{h\nu}{Me\eta}\sqrt{2eFM^2 i_d + 4kT/R}. \tag{22.76}$$

The equivalent result for a *p–i–n* diode can be obtained by setting $M = F = 1$.

Example. For an InGaAs APD with F=10, M=100, R=50Ω, $i_d = 2$ nA, and responsivity 0.8 A W^{-1}, the quantum efficiency is

$$\eta = \frac{0.8(6.626 \times 10^{-34})(3 \times 10^8)}{(1.6 \times 10^{-19})(1.55 \times 10^{-6})} = 0.64. \tag{22.77}$$

From Eq. (22.73) the NEP is

$$\begin{aligned} &NEP(\mathrm{W\ Hz^{-1/2}}) \\ &= \frac{(2 \times 1.6 \times 10^{-19} \times 10^5 \times 2 \times 10^{-9} + 4 \times 1.38 \times 10^{-22} \times 300/50)^{1/2}}{0.8 \times 100} \\ &= 4.7 \times 10^{-13}\ \mathrm{W\ Hz^{-1/2}}. \end{aligned}$$

Note that in this case the thermal noise is dominant.

22.7 Coherent Detection

We have seen several examples in this chapter of how the signal-to-noise ratio of an optical detection system is limited by noise, in particular by Johnson noise and the electronic noise of the amplification stages that follow an optical detector. Even if these sources of noise were not present the signal-to-noise ratio would be limited by shot noise. In the direct detection schemes discussed so far the quantum noise limit set by shot noise is rarely attainable. However, it is possible to achieve the quantum limit for detection by the use of *coherent detection.* In this scheme the detector is illuminated simultaneously by the *signal* light and by a second source of light called the *local oscillator* (lo), which must be phase coherent with the signal. The degree of phase coherency that is required to make this scheme work well has been discussed in detail by Salz[22.20].

A useful rule of thumb is that the local oscillator must be phase coherent over a time long enough to receive the information being transmitted. In an optical communication system in which binary information is being detected the required phase coherence time is on the order of the pulse duration being detected.

The schematic way in which coherent detection is carried out is shown in Fig. (22.27). For optimum performance not only must the signal and lo be phase coherent, but their phase fronts and polarization states must be matched at the detector surface. This corresponds, for example, to a situation in which two linearly polarized TEM_{00} Gaussian laser beams, which are coaxial and of equal spot size, coincide in a beamwaist at the detector. Deviations from this ideal geometry, either through spot size mismatch or angular or lateral misalignment, decrease the efficiency of the detection process.

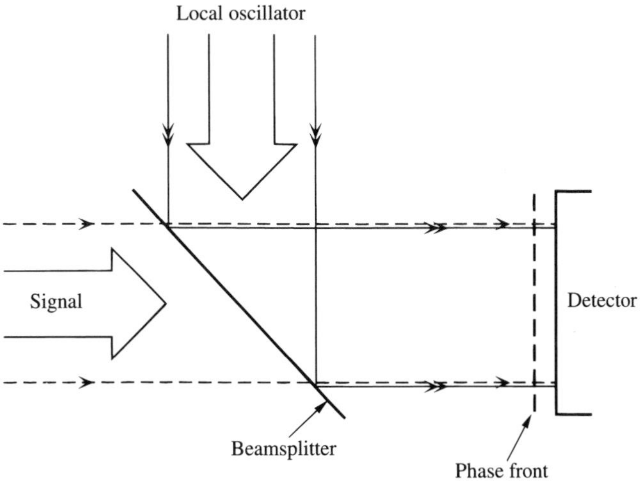

Fig. 22.27. Schematic arrangement for optical mixing.

We represent the electric field of the signal beam at the detector surface as

$$E_s(t) = E_1 \cos(w_1 t + \phi_1) \tag{22.78}$$

and of the local oscillator as

$$E_{lo}(t) = E_2 \cos(w_2 t + \phi_2). \tag{22.79}$$

The detector responds to the intensity of the light falling on it. Therefore, the detector current is

$$i(t) = \frac{\mathscr{R}A}{Z}[E_s(t) + E_{lo}(t)]^2, \tag{22.80}$$

where $\mathscr{R}$ is the detector responsivity, A is the effective detector area, and Z is the characteristic impedance of the medium in front of the detector. Substitution from Eqs. (22.78) and (22.79) into (22.80) gives

$$\begin{aligned} i(t) = \frac{\mathscr{R}A}{Z}\, &\left[E_1^2 \cos^2(\omega_1 + \phi_1) + E_2^2 \cos^2(\omega_2 t + \phi_2)\right. \\ &\left. + 2E_1E_2 \cos(\omega_1 t + \phi_1)\cos(\omega_2 t + \phi_2)\,\right]. \end{aligned} \tag{22.81}$$

By the use of well known trigenometrical identities† the detector current can be written as

$$\begin{aligned} i(t) = \frac{\mathscr{R}A}{Z}\Bigg\{ &\frac{E_1^2}{2} + \frac{E_1^2}{2}\cos 2(\omega_1 t + \phi_1) + \frac{E_2^2}{2} \\ &+ \frac{E_2^2}{2}\cos 2(\omega_2 t + \phi_2) + E_1E_2 \cos[(\omega_1 + \omega_2)t + \phi_1 + \phi_2] \\ &+ E_1E_2\cos[(\omega_1 - \omega_2)t + \phi_1 - \phi_2]\Bigg\}. \end{aligned} \tag{22.82}$$

The detector does not respond to the rapidly oscillating terms at frequencies ω_1, ω_2,

† $\cos^2 x = \frac{1}{2}(1 + \cos 2x)$; $\cos X + \cos Y = 2\cos\left(\frac{X+Y}{2}\right)\cos\left(\frac{X-Y}{2}\right)$.

and $(\omega_1 + \omega_2)$ in Eq. (22.82),† so the detector current is

$$i(t) = \frac{\mathscr{R}A}{Z}\left\{\frac{E_1^2}{2} + \frac{E_2^2}{2} + E_1E_2\cos[(\omega_1 - \omega_2)t + \phi_1 - \phi_2]\right\}. \tag{22.83}$$

We are assuming that the difference frequency $(\omega_1 - \omega_2)$ is within the response range of the detector. The generation of this difference frequency is called *optical mixing*. If $\omega_1 = \omega_2$ it is a *homodyne* mixing process, otherwise it is a *heterodyne* process. Similar processes occur in the operation of FM radio receivers. Note that the first two terms in Eq. (22.83) correspond to the detector current resulting from signal and local oscillator independently.

The detector shot noise is

$$< i_N^2 >_1 = \left[\frac{2e\mathscr{R}A}{Z}\left(\frac{E_1^2}{2} + \frac{E_2^2}{2}\right) + i_d\right]\Delta f, \tag{22.84}$$

where i_d is the detector dark current. In addition we expect a Johnson noise contribution of

$$< i_N^2 >_2 = \frac{4kT\Delta f}{R}. \tag{22.85}$$

It is usual to operate a coherent detection scheme in which $E_2 >> E_1$: a weak signal is mixed with a strong local oscillator. Therefore, the effective signal current is

$$i_s(t) = \frac{\mathscr{R}AE_1E_2}{Z}\cos[(\omega_1 - \omega_2)t + \phi_1 - \phi_2]. \tag{22.86}$$

Eq. (22.86) provides essential insight into the desirability of coherent detection for detection of a *weak* coherent optical field E_1. The signal current generated when this weak optical field is mixed with a local oscillator field E_2 is multiplied by the magnitude of the local oscillator field. If $\omega_1 \neq \omega_2$ Eqs. (22.84)–(22.86) give a signal-to-noise ratio

$$\frac{< i_s^2(t) >}{< i_N^2 >_1 + < i_N^2 >_2} = \frac{\mathscr{R}^2A^2E_1^2E_2^2}{2Z^2}\bigg/\left[\frac{e\mathscr{R}A}{Z}(E_1^2 + E_2^2) + i_d + \frac{4kT}{R}\right]\Delta f. \tag{22.87}$$

With a sufficiently powerful local oscillator its shot noise dominates and the dark current contribution and Johnson noise can be neglected. In this case, for $E_2 >> E_1$ Eq. (22.87) reduces to

$$\frac{S}{N} = \frac{\mathscr{R}AE_1^2}{2eZ\Delta f}. \tag{22.88}$$

Noting that the average signal power is

$$P_s = \frac{AE_1^2}{2Z} \tag{22.89}$$

and that $\mathscr{R} = \eta e/h\nu$, Eq. (22.88) becomes

$$\frac{S}{N} = \frac{\eta P_s}{h\nu\Delta f}. \tag{22.90}$$

† In a more advanced analysis that reflects the fact that optical detectors respond to a light signal by absorbing photons these high frequency terms do not even appear. See Chapter 23 for a further discussion.

It might appear from a comparison of Eqs. (22.67) and (22.90) that ideal heterodyne detection has quantum-limited performance twice that of direct detection. However, to make such a comparison one must assume that it is possible to use the same signal detection bandwidth Δf in both schemes. If an unmodulated CW laser beam of power P_s is to be detected then Eq. (22.67) would predict that the direct detection (dd) photon-noise-limited signal-to-noise ratio would be

$$\left(\frac{S}{N}\right)_{dd} = \frac{\eta P_s}{2h\nu\Delta f}. \tag{22.91}$$

However, in practice one could not hope to achieve this S/N ratio because the resultant dc detector signal would be overwhelmed by $1/f$ noise.

It is always necessary to modulate the signal in some way. For example, if the signal power is

$$P = P_0(1 + m\sin\omega_m t) \tag{22.92}$$

the detected signal at frequency ω_m could be band-pass filtered. In this case the *effective* signal to noise ratio is

$$\left(\frac{S}{N}\right)_{dd} = \frac{m^2\eta P_s}{4h\nu\Delta f}. \tag{22.93}$$

Particular care must be taken in comparing a homodyne receiver with a heterodyne receiver. If both signal and local oscillator are CW signals at the same frequency it is meaningless to talk of detection of the signal, since signal and local oscillator are essentially indistinguishable. To discuss the signal-to-noise ratio it is essential to consider how the signal beam is being modulated, since it is through modulation that information is transferred. To illustrate this we consider a situation in which the signal beam is phase modulated so that

$$E_s(t) = E_1\cos(\omega_1 t + m\sin\omega_m t). \tag{22.94}$$

In this case the information being transferred is the sinusoidal phase modulation at frequency ω_m. For small values of m, which is called the *modulation depth*, Eq. (22.94) can be rewritten as†

$$E_s(t) = E_1[J_0(m)\cos\omega_1 t + J_1(m)\cos(\omega_1+\omega_m)t - J_1(m)\cos(\omega_1-\omega_m)t]. \tag{22.95}$$

The signal beam has acquired *sidebands* at frequencies separated by $\pm\omega_m$ from the carrier frequency ω_1.

The mixing process, in this case, can be written as

$$i(t) = \frac{\mathcal{R}A}{Z}[E_1\cos(\omega_1 + m\sin\omega_m t) + E_2\cos(\omega_2 + \phi_2)]^2. \tag{22.96}$$

By the use of Eq. (22.95), and with the assumption that $m << 1$, so that $[J_1(m)]^2 << J_1(m)$, Eq. (22.95) reduces to

$$i(t) = \frac{\mathcal{R}A}{Z}\left\{\frac{E_2^2}{2} + \frac{E_1^2 J_0^2(m)}{2} - 2J_1(m)E_1E_2\sin[(\omega_1 - \omega_2 - \phi_2)t]\sin\omega_m t\right\} \tag{22.97}$$

† See Section 19.7.

after the (nonexistent) high frequency terms are eliminated. The useful signal current contained in Eq. (22.97) is

$$i_s(t) = -\frac{2\mathscr{R}AJ_1(m)}{Z}E_1E_2\sin[(\omega_1 - \omega_2 - \phi_2)t]\sin\omega_m t. \tag{22.98}$$

The information term $\sin\omega_m t$ is present as a low frequency modulation of the intermediate frequency (if) $(\omega_1 - \omega_2)$. In a similar way to Eq. (22.87) the signal-to-noise ratio is

$$\frac{< i_s^2(t) >}{< i_N^2 >_1 + < i_N^2 >_2} = \frac{\mathscr{R}^2A^2J_1^2(m)E_1^2E_2^2}{Z^2\left[e\mathscr{R}AE_2^2/Z + i_d + 4kT/R\right]\Delta f}, \tag{22.99}$$

where we have made the assumption that $E_1 << E_2$, as is usually the case in a heterodyne receiver. For a sufficiently powerful local oscillator Eq. (22.99) gives

$$\frac{S}{N} = \frac{\mathscr{R}AJ_1^2(m)E_1^2}{eZ\Delta f}. \tag{22.100}$$

For homodyne detection, we would optimize performance by setting $\phi_2 = \pi/2$ in Eq. (22.98). In this case the signal-to-noise ratio is

$$\left(\frac{S}{N}\right)_{homodyne} = \frac{2\mathscr{R}AJ_1^2(m)E_1^2}{eZ\Delta f}. \tag{22.101}$$

In this situation the homodyne S/N is twice the value achieved by heterodyne detection. It should be stressed, however, that the S/N in coherent detection schemes depends on the precise modulation scheme, and also on the contribution to the signal-to-noise ratio of the demodulation scheme used to extract the information term ($\sin\omega_m t$ in Eq. (22.98)), from the if signal at $\omega_1 - \omega_2$[22.17],[22.21]. A discussion of the merits of various demodulators, involving band-pass filters, discriminators, limiters, and phase-locked loops is beyond our scope here.

We can note that the signal power in the two sidebands in Eq. (22.95) can be written as

$$P_s = \frac{E_1^2J_1^2(m)}{Z}, \tag{22.102}$$

so that in terms of the sideband power the heterodyne S/N is, as before,

$$S/N = \frac{\eta P_s}{h\nu\Delta f}. \tag{22.103}$$

Example: An interesting example of the coherent detection process is provided under conditions of *suppressed carrier* operation. If the modulation depth in Eq. (22.94) is set to a value $m = 2.405$, then $J_0(m) = 0$ and the carrier is suppressed. For a detector with $\eta = 0.6$, and bandwidth 100 MHz operating at 1.55 μm the minimum detectable signal power – defined to correspond to $S/N = 1$ – is

$$P_s = \frac{h\nu\Delta f}{1}, \tag{22.104}$$

which in this case gives

$$P_s = \frac{(6.626 \times 10^{-34})(3 \times 10^8)(10^8)}{0.6(1.55 \times 10^{-6})} = 21 \text{ pW}.$$

22.8 Bit-Error Rate

We conclude this chapter with a discussion of a final performance parameter that is widely used to characterize the performance of optical communication systems. This is the *bit-error rate* (BER) used to describe the performance of an optical detection system receiving binary-encoded signals. In its simplest form such a scheme has a binary 'one' represented by the detection of an optical pulse by the detector. A binary 'zero' is represented by the failure to detect such a pulse. The 'ones' and 'zeroes' constitute a bit stream at a prescribed clock rate, as shown schematically in Figs. (22.28a) and (22.28b). The detection of a 'one' is characterized by the detector signal raising above a threshold level ℓ_1. The detection of a 'zero' is characterized by the detector signal failing to rise above a lower threshold level ℓ_2. In the presence of additive noise, represented schematically by Fig. (22.28c), the overall detector signal will appear as shown schematically in Fig. (22.28d). The questions that then arise are: when will the noise reduce the detector signal below ℓ_1 when a 'one' is present, and when will the noise be above level ℓ_2 when a 'zero' is being detected. Both of these possibilities lead to a bit detection error, as shown schematically in Fig. (22.28e). One simple way to model this process is to assume that the detector gives rise to a noise current that fluctuates in a Gaussian fashion – more sophisticated analyses would examine more closely the statistical nature of both the detection and noise processes and the different ways in which the binary signal is encoded[22.17],[22.21].

For a detector noise current with variance $< i_N^2 >$ that is Gaussian distributed about zero the probability that the noise is *below* a set level is

$$< i_N \leq \ell) = \frac{1}{\sigma\sqrt{2\pi}} \int_{-\infty}^{\ell} e^{-x^2/2\sigma^2} dx, \tag{22.105}$$

where $\sigma^2 =< i_N^2 >$.

On average a simple binary stream of 'ones' and 'zeroes' will have equal numbers of ones and zeroes. Therefore, for a large number N of transmitted bits there will be $N/2$ 'ones' and $N/2$ 'zeroes'. There will be a bit error whenever a 'one' is present but the noise current is $< -\ell_1$; there will also be a bit error whenever a 'zero' is present but the noise is *above* level ℓ_2. Therefore, we can write the probability of error as

$$p_e = \frac{1}{2} \left[P(i_N < -\ell_1) + P(i_N > \ell_2) \right]. \tag{22.106}$$

A simple result can be obtained if $\ell_1 = \ell_2 = i_s/2$, where i_s is the detector current corresponding to a 'one'. In this case

$$p_e = \frac{1}{2\sigma\sqrt{2\pi}} \left(\int_{-\infty}^{-i_s/2} e^{-x^2/2a^2} dx + \int_{i_s/2}^{\infty} e^{-x^2/2\sigma^2} dx \right), \tag{22.107}$$

which becomes

$$p_e = \frac{1}{\sigma\sqrt{2\pi}} \int_{i_s/2}^{\infty} e^{-x^2/2a^2} dx = \frac{1}{\sqrt{\pi}} \int_{i_s/(2\sqrt{2}\sigma)}^{\infty} e^{-t^2} dt. \tag{22.108}$$

Eq. (22.108) can be written in terms of the error function, erf(z), defined by the relation[22.22]

$$\operatorname{erf}(z) = \frac{2}{\sqrt{\pi}} \int_0^z e^{-t^2/dt} \tag{22.109}$$

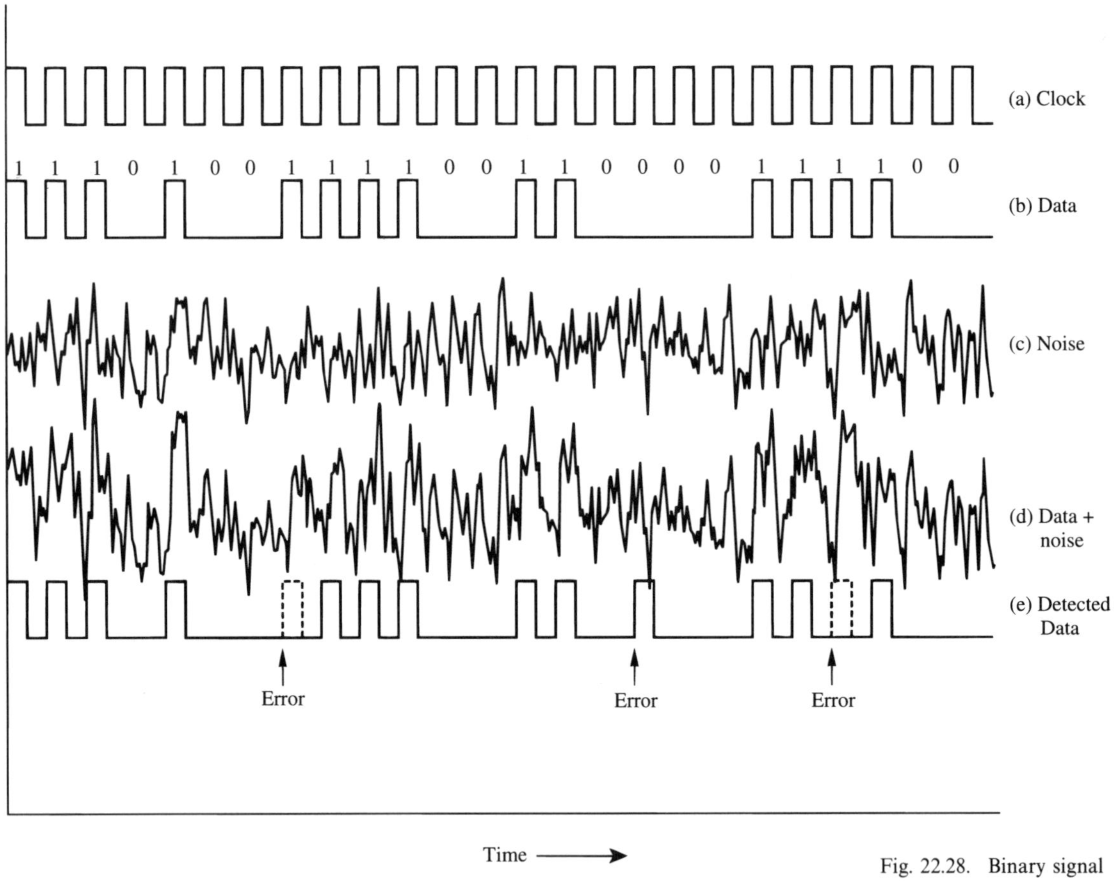

Fig. 22.28. Binary signal transmission showing clock, data, the effect of added noise and the resultant detection of 'ones' and 'zeroes'.

where erf $(z) \rightarrow 1$ as $z \rightarrow \infty$. Therefore,

$$p_e = \frac{1}{2}\left[1 - \text{erf}\left(\frac{i_s}{2\sqrt{2}\sigma}\right)\right] = \frac{1}{2}\text{erfc}\left(\frac{i_s}{2\sqrt{2}a}\right) \tag{22.110}$$

and finally

$$p_e = \frac{1}{2}\text{erfc}\left(\frac{i_s}{2\sqrt{2}\sqrt{< i_N^2 >}}\right). \tag{22.111}$$

The signal-to-noise ratio can be characterized by

$$\frac{S}{N} = \left[\frac{i_s}{(\text{RMS noise current})}\right]^2, \tag{22.112}$$

so

$$p_e = \frac{1}{2}\text{erfc}\left[\frac{1}{2\sqrt{2}}\sqrt{\frac{S}{N}}\right] \tag{22.113}$$

Fig. (22.29) shows a plot of this probability of error, or bit-error rate, as a function of S/N. A frequently used performance measure is $P_e < 10^{-9}$, which implies $S/N = 141$, or 21.5 dB. This corresponds to i_s/RMS noise current $= 11.89$, so

Fig. 22.29. Bit-error rate as a function of signal-to-noise ratio in a simple binary detection system.

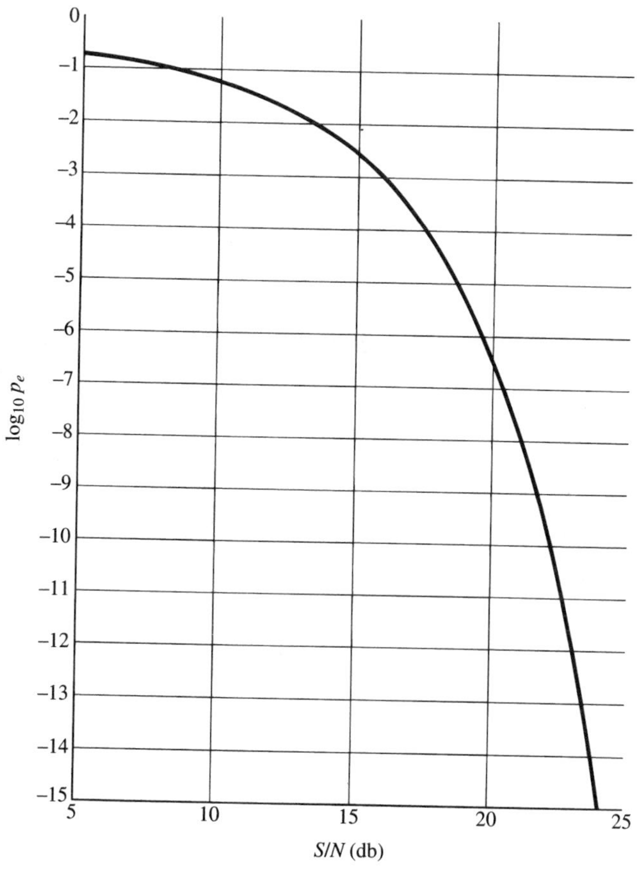

an optical signal only 11.89 times greater than the dark current level is needed to achieve a bit-error rate of 10^{-9}.

In a practical test of bit-error rate a random, long sequence of binary pulses (10^{15}–10^{21}) would be transmitted from source to receiver and the received bit sequence compared with a record of the transmitted sequence.

References

[22.1] A. van der Ziel, *Noise*, Prentice-Hall, New York, 1954.

[22.2] H. Nyquist, 'Thermal agitation of electric charge in conductors,' *Phys. Rev.* **32**, 110–113, 1928.

[22.3] S. Ramo, J.R. Whinney, and T. Van Duzer, *Fields and Waves in Communication Electronics*, 2nd Edition, Wiley, New York, 1989.

[22.4] B.I. Bleaney and B. Bleaney, *Electricity and Magnetism*, Oxford, University Press, Oxford, 3rd Edition, 1976.

[22.5] P.W. Kruse, L.D. McGlanchlin, and R.B. McQuistan, *Elements of Infrared Technology: Generation, Transmission, and Detection*, Wiley, New York, 1962.

[22.6] E.H. Putley, 'Indium Antimonide Submillimeter Photoconductive Detectors,' *Appl. Opt.*, **4**, 649–656, 1969.

[22.7] E.H. Putley, 'Thermal Detectors,' in *Optical and Infrared Detectors*, R.J. Keyes, Ed., 2nd Edition, Springer, Berlin, 1980.

[22.8] E.H. Putley and D.H. Martin, 'Detectors,' in *Spectroscopic Techniques for Infrared, Submillimeter and Millimeter Waves*, D.H. Martin, Ed., North-Holland, Amsterdam, 1967.

[22.9] J. Gowar, *Optical Communication Systems*, Prentice-Hall, Englewood Cliffs, New Jersey, 1984.

[22.10] H. Kressel, Editor, *Semiconductor Devices for Optical Communication*, Topics in Applied Physics, Vol. 39, Springer-Verlag, Berlin, 1982.

[22.11] H. Melchior, M.B. Fisher, and F.R. Arams, 'Photodetectors for Optical Communication Systems,' *Proc. IEEE*, **58**, 1466–1486, 1970.

[22.12] J.H. Moore, C.C. Davis, and M.A. Coplan, *Building Scientific Apparatus*, 2nd Edition, Addison Wesley, Reading, Massachusetts, 1989.

[22.13] M.J.E. Golay, 'A Pneumatic Infra-Red Detector,' *Rev. Sci. Instrum.*, **18**, 357–362, 1947.

[22.14] H. Taub, and D.L. Schilling, *Principles of Communication Systems*, McGraw-Hill, New York, 1971.

[22.15] H. Kressel, Editor, *Semiconductor Devices for Optical Communication*, Topics in Applied Physics, Vol. 39, Springer-Verlag, Berlin, 1982.

[22.16] P.K. Cheo, *Fiber Optics and Optoelectronics*, 2nd Edition, Prentice-Hall, Englewood Cliffs, New Jersey, 1990.

[22.17] R.M. Gagliardi and S. Karp, *Optical Communications*, Wiley, New York, 1976.

[22.18] H.B. Killen, *Fiber Optic Communications*, Prentice-Hall, Englewood Cliffs, 1991.

[22.19] R.J. McIntyre, 'Multiplication noise in uniform avalanche diodes,' *IEEE Trans. Electron. Devices*, **ED-13**, 164–168, 1966.

[22.20] J. Salz, 'Coherent lightwave communications,' *AT&T Tech. J.*, **64**, 2153–2209, 1985.

[22.21] Paul E. Green, Jr., *Fiber Optic Networks*, Prentice-Hall Englewood Cliffs, 1993.

[22.22] M. Abramowitz and I.A. Stegun, *Handbook of Mathematical Functions*, Dover, New York, 1965.

23

Coherence Theory

23.1 Introduction

In this chapter we will put the concept of *coherence* on a more mathematical basis. This will involve the formal definition of a number of functions that are used to describe the coherence properties of optical fields. These include the analytic signal, various correlation functions, and the degree of coherence for describing both temporal and spatial coherence. We shall see that the degree of temporal coherence is quantitatively related to the lineshape function and that the degree of spatial coherence between two points is determined by the size, intensity distribution, and location of an illuminating source. We will use the wave equation, and special solutions to the wave equation called Green's functions, to show how spatial coherence varies from point to point.

The chapter will conclude with a discussion of how intensity fluctuations of a source depend on its coherence properties and we will examine a specific scheme, the Hansbury-Brown–Twiss experiment by which this relationship is studied. This discussion will involve a discussion of 'photon statistics,' the time variation of the 'detection' of *photons* from a source. In a quantum mechanical context, square-law detectors respond to these quantized excitations of the optical field, which we call photons.

In classical coherence theory it is advantageous to represent the real electromagnetic field by a complex quantity, both for its mathematical simplicity and because it serves to emphasize that coherence theory deals with phenomena that are sensitive to the 'envelope' or average intensity of the field. In spite of the fact that such a complex representation is rather artificial in the classical theory it provides the best correspondence with the approach of quantum theory, where an equivalent representation is intimately connected with the process by which an electromagnetic field is detected.

23.2 Square-Law Detectors

Suppose we illuminate a photodetector with a plane, monochromatic light wave containing two closely spaced frequencies, for example one for which the electric field can be written as

$$\mathbf{E} = \mathbf{A}\cos\omega t + \mathbf{B}\cos(\omega + \Delta\omega)t, \tag{23.1}$$

where $\mathbf{A}$ and $\mathbf{B}$ are parallel.

The detector responds to the intensity of this field and therefore to the square of the electric field, its output is

$$i \propto A^2\cos^2\omega t + B^2\cos^2(\omega+\Delta\omega)t + 2AB\cos\omega t\cos(\omega+\Delta\omega)t, \tag{23.2}$$

which can be rewritten as

$$i \propto \frac{A^2}{2}(1+\cos 2\omega t) + \frac{B^2}{2}\,[1+\cos 2(\omega+\Delta\omega)t] + AB\cos(2\omega+\Delta\omega)t + AB\cos\Delta\omega t. \tag{23.3}$$

If we make the practical assumption that typical optical detectors cannot follow the high frequency terms at $2\omega, 2(\omega+\Delta\omega)$, and $(2\omega+\Delta\omega)$ in the above expression, then the actual observed response can be calculated, as we have seen previously,

$$i \propto \frac{A^2}{2} + \frac{B^2}{2} + AB\cos\Delta\omega t, \tag{23.4}$$

where we have assumed that the detector can respond to the beat signal at frequency $\Delta\omega$. If we use complex notation and write Eq. (23.1) as

$$E = Ae^{i\omega t} + Be^{i(\omega+\Delta\omega)t} \tag{23.5}$$

then the detector response is $i \propto EE^* = \mid E\mid^2$, which gives

$$\begin{aligned} i &\propto [Ae^{i\omega t} + Be^{i(\omega+\Delta\omega)t}][Ae^{-i\omega t} + Be^{-i(\omega+\Delta\omega)t}] \\ &\propto A^2 + B^2 + ABe^{i\Delta\omega t} + ABe^{-i\Delta\omega t} \\ &\propto A^2 + B^2 + 2AB\cos\Delta\omega t \end{aligned} \tag{23.6}$$

and the use of the complex notation gives the same frequency dependence as Eq. (23.4). In other words the apparent averaging effect of the detector is automatically included by writing the fields in complex form.† This emphasizes the critical physical reality of the detection process. The detector does *not* average the high frequency terms that appear in Eq. (23.3). These terms do not, in fact, exist for a photon detector. The detector responds to quanta of the electromagnetic field – photons. The conversion of these photons to charge carriers in the photodetector is well represented by the use of the complex notation in Eq. (23.6).

23.3 The Analytic Signal

We can generalize the use of complex fields to describe real fields by introducing the *complex analytic signal.*

If we have a real scalar quantity $V^{(r)}(t)$, where the superscript (r) indicates the reality of the quantity, which may, for example, represent a cartesian component of the electric field **E**, then $V^{(r)}(t)$ possesses a Fourier transform,

$$V^{(r)}(t) = \int_{-\infty}^{\infty} V^{(r)}(\nu)e^{-2\pi i\nu t}d\nu, \tag{23.7}$$

where

$$V^{(r)}(\nu) = \int_{-\infty}^{\infty} e^{2\pi i\nu t}V^{(r)}(t)dt. \tag{23.8}$$

For positive t this integral contains terms

$$(\cos 2\pi i\nu t + i\sin 2\pi i\nu t)\,V^{(r)}(t).$$

† In general if $a(t), b(t)$ are two real fields with *complex* amplitudes A and B then

$$\overline{a(t)b(t)} = \frac{1}{2}\mathcal{R}(AB^*)$$

For negative t the integral contains terms

$$(\cos 2\pi i v t - i \sin 2\pi i v t)\, V^{(r)}(-t).$$

So we can write

$$V^{(r)}(-v) = \int_{-\infty}^{\infty} e^{-2\pi i v t} V^{(r)}(t) dt, \tag{23.9}$$

and

$$[V^{(r)}(-v)]^* = \int_{-\infty}^{\infty} e^{2\pi i v t} V^{(r)}(t) dt. \tag{23.10}$$

Therefore, $[V^{(r)}(-v)]^* = V^{(r)}(v)$, and $[V^{(r)}(v)]^* = V^{(r)}(-v)$.

Thus all the information in $V^{(r)}(v)$ is contained in the positive frequencies.

The complex analytic signal corresponding to $V^{(r)}(t)$ is defined by

$$V(t) = \int_{0}^{\infty} e^{-2\pi i v t} V^{(r)}(v) dv \tag{23.11}$$

and includes only contributions from positive frequencies. The positive frequency Fourier amplitudes can be written as

$$\begin{aligned} V^{(r)}(v) &= \int_{-\infty}^{\infty} V(t) e^{2\pi i v t} dt \text{ for } v \geq 0, \\ &= 0 \qquad\qquad \text{for } v < 0. \end{aligned} \tag{23.12}$$

A real field, $V^{(r)}(t)$ for example, could be

$$A \cos 2\pi v t = \frac{A}{2}(e^{i2\pi v t} + e^{-i2\pi v t}). \tag{23.13}$$

For such a field $V(t)$ can be written in terms of its real and imaginary parts as

$$V(t) = \frac{1}{2}[V^{(r)}(t) + i V^{(i)}(t)]. \tag{23.14}$$

A simple example of such an analytic signal could be $Ae^{-2\pi i v t}$.

If $V^{(r)}(t)$ satisfies the wave equation

$$\nabla^2 V^{(r)}(t) - \frac{1}{c^2}\frac{\partial^2 V^{(r)}(t)}{\partial t^2} = 0,$$

then both $V^{(i)}(t)$ and the analytic signal $V(t)$ do so also.

From Eq. (23.11)

$$\begin{aligned} V(t + i\tau) &= \int_{0}^{\infty} e^{-2\pi v(it-\tau)} V^{(r)}(v) dv \\ &= \int_{0}^{\infty} e^{2\pi v \tau} e^{-2\pi i v t} V^{(r)}(v) dv. \end{aligned} \tag{23.15}$$

The integrand $\to \infty$ for $\tau > 0$ so the derivative of $V(t + i\tau) = V(z)$ is not single-valued. Thus $V(t+i\tau)$ is only analytic† in the lower half plane, $V(t)$ is the boundary value as $\tau \to 0^-$.

† Possesses a finite derivative.

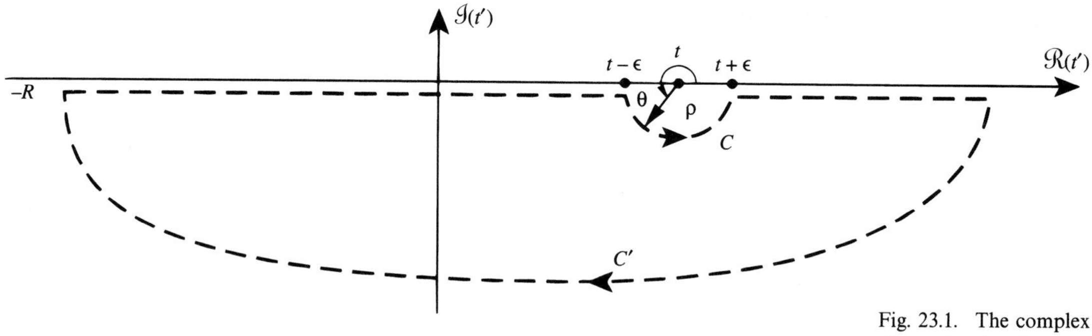

Fig. 23.1. The complex plane, showing the contour over which the integral in Eq. (23.19) is evaluated.

23.3.1 *Hilbert Transforms*

A fundamental theorem of complex variables states that a function $f(z)$ which has no poles in the lower (or upper) z plane has real and imaginary parts that are related by Hilbert transforms[23.1],[23.2].

When

$$V(t) = \frac{1}{2}[V^{(r)}(t) + iV^{(i)}(t)], \tag{23.16}$$

and

$$V^{(i)}(t) = \frac{1}{\pi} PV \int_{-\infty}^{\infty} \frac{V^{(r)}(t')}{t'-t} dt' \tag{23.17}$$

$$V^{(r)}(t) = -\frac{1}{\pi} PV \int_{-\infty}^{\infty} \frac{V^{(i)}(t')}{t'-t} dt'. \tag{23.18}$$

PV indicates the Cauchy principal value† of the integral at the singular point $t' = t$.

Let us integrate the integral below over the closed contour shown in Fig. (23.1).

$$\int_{contour} \frac{V(t')}{t'-t} dt' = \int_{C'} \frac{V(t')}{t'-t} dt' + \int_{-R}^{t-\epsilon} \frac{V(t')}{t'-t} dt' + \int_{t+\epsilon}^{R} \frac{V(t')}{t'-t} dt' + \int_{C} \frac{V(t')}{t'-t} dt' = 0. \tag{23.19}$$

Now, the integral

$$\int_{C'} \frac{V(t')}{t'-t} dt' \text{ vanishes as } |\, t' \,| \to \infty.$$

We integrate $\int_C \left[V(t')/(t'-t)\right] dt'$ over the infinitesimally small half circle of radius

† Cauchy Principal Value. Let $q(x) \to \infty$ as $x \to a$, then the principal value of

$$\int_b^c q(x)dx \text{ written } PV \int_b^c q(x)dx (b < a < c)$$

is

$$\lim_{\delta \to 0} \left\{ \int_b^{a-\delta} q(x)dx + \int_{a+\delta}^{c} q(x)dx \right\}.$$

ρ surrounding the point t.

$$\int_C \frac{V(t')}{t'-t}dt' = V(t)\int_C \frac{dt'}{t'-t} = V(t)\int_C \frac{d(\rho e^{i\theta})}{\rho e^{i\theta}}. \tag{23.20}$$

Over this half circle ρ is constant, θ goes from π to 2π, and the integral becomes

$$V(t)i\int_\pi^{2\pi} d\theta = \pi i V(t). \tag{23.21}$$

Hence,

$$\int_C \frac{V(t')}{t'-t}dt' = \pi i V(t). \tag{23.22}$$

So,

$$V(t) = \frac{i}{\pi}\left[\int_{-R}^{t-\epsilon} \frac{V(t')}{t'-t}dt' + \int_{t+\epsilon}^{R} \frac{V(t')}{t'-t}dt'\right]. \tag{23.23}$$

As $R \to \infty, \epsilon \to 0$ the quantity in the square bracket becomes the Cauchy principal value, i.e.,

$$V(t) = \frac{i}{\pi}PV\int_{-\infty}^{\infty} \frac{V(t')}{t'-t}dt'. \tag{23.24}$$

Thus,

$$V^{(r)}(t) + iV^{(i)}(t) = \frac{i}{\pi}PV\int_{-\infty}^{\infty}\left\{\frac{V^{(r)}(t')}{t'-t} + \frac{iV^{(i)}(t')}{t'-t}\right\}dt'. \tag{23.25}$$

Comparing real and imaginary parts of Eq. (23.25)

$$V^{(r)}(t) = -\frac{1}{\pi}PV\int_{-\infty}^{\infty} \frac{V^{(i)}(t')}{t'-t}dt', \tag{23.26}$$

$$V^{(i)}(t) = \frac{1}{\pi}PV\int_{-\infty}^{\infty} \frac{V^{(r)}(t')}{t'-t}dt'. \tag{23.27}$$

23.4 Correlation Functions

Let the (complex) electromagnetic field with polarization μ at a point $\mathbf{r}$ at time t be $V_\mu(\mathbf{r}, t)$. The $(m+n)$th order correlation function is an ensemble average of the products of these field functions considered at different space-time points and polarizations, defined as follows:

$$\Gamma^{(m,n)}_{\mu_1 \ldots \mu_{m+n}}(\mathbf{r}_1 \ldots \mathbf{r}_{m+n}, t_1 \ldots t_{m+n}) = < \prod_{j=1}^{m} V^*_{\mu_j}(\mathbf{r}_j, t_j) \prod_{k=m+1}^{m+n} V_{\mu_k}(\mathbf{r}_k, t_k) > \tag{23.28}$$

where the <> brackets indicate an ensemble average.†

An important class of optical fields are those that are *stationary* and *ergodic*. For a *stationary* field the correlation functions are independent of the choice of

† An ensemble average would be obtained by measuring the fields $V_{\mu_i}(\mathbf{r}_i, t_i)$ in a large number of systems that as closely as possible are identical. Such systems would contain identical sources, boundaries, temperatures, and detector configurations, but statistical fluctuations from one system to the other would cause individual fields to be different.

time origin. For an *ergodic* field the ensemble average can be replaced by the time average, i.e.,

$$< \ldots >= \lim_{T\to\infty} \frac{1}{2T} \int_{-T}^{T} \ldots dt. \tag{23.29}$$

So, leaving the polarization indices to be understood,

$$\Gamma^{(m,n)}(\mathbf{r}_1 \ldots \mathbf{r}_{m+n}; t_1 \ldots t_{m+n}) \equiv \Gamma^{m,n}(\mathbf{r}_1 \ldots \mathbf{r}_{m+n}, \tau_2 \ldots \tau_{m+n})$$

$$= \lim_{T\to\infty} \frac{1}{2T} \int_{-T}^{T} \prod_{j=1}^{m} V^*(\mathbf{r}_j, t+\tau_j) \prod_{k=m+1}^{m+n} V(\mathbf{r}_k, t+\tau_k) dt, \tag{23.30}$$

where τ_1 has been set to zero, and $\tau_j = t_j - t_1$ (the stationarity condition).

It is to be understood that $\mathbf{r}_j$ implies polarization μ_j, although we are primarily concerned with linearly polarized fields where all the μ_j are the same. In all that follows we will be assuming linearly polarized fields all polarized in the same direction. If this were not the case we could resolve the field into components along orthogonal axes.

The correlation function of second order $m = n = 1$ for stationary and ergodic fields is called the *mutual coherence function* and is written

$$\Gamma(\mathbf{r}_1, \mathbf{r}_2, \tau) = \lim_{T\to\infty} \frac{1}{2T} \int_{-T}^{T} V^*(\mathbf{r}_1, t) V(\mathbf{r}_2, t+\tau) dt. \tag{23.31}$$

$\Gamma(\mathbf{r}_1, \mathbf{r}_2, \tau)$ is frequently written $\Gamma_{12}(\tau)$. Clearly

$$\Gamma^*_{12}(\tau) = \Gamma_{21}(-\tau). \tag{23.32}$$

We can also write

$$\Gamma(\mathbf{r}_1, \mathbf{r}_2, \tau) = \Gamma(\mathbf{r}_1, t_1, \mathbf{r}_2, t_2) = \overline{V^*(\mathbf{r}_1, t) V(\mathbf{r}_2, t+\tau)}. \tag{23.33}$$

This function is important in the mathematical description of effects such as interference and diffraction. When $\mathbf{r}_1 = \mathbf{r}_2 = \mathbf{r}$

$$\Gamma(\mathbf{r}, \mathbf{r}, \tau) = \overline{V^*(\mathbf{r}, t) V(\mathbf{r}, t+\tau)} = \Gamma_{11}(\tau). \tag{23.34}$$

This is the *autocorrelation function.* We define the *intensity* at point $\mathbf{r}$ at time t as

$$I(\mathbf{r}, t) = V^*(\mathbf{r}, t) V(\mathbf{r}, t). \tag{23.35}$$

We have already illustrated this in a simple case for two different frequencies illuminating a photodetector.

It can be shown that[23.3],[23.4]

$$\overline{V_1^{(r)}(t) V_2^{(i)}(t+\tau)} = -\overline{V_1^{(i)}(t) V_2^{(r)}(t+\tau)}, \tag{23.36}$$

$$\overline{V_1^{(r)}(t) V_2^{(r)}(t+\tau)} = \overline{V_1^{(i)}(t) V_2^{(i)}(t+\tau)}, \tag{23.37}$$

and of course

$$\Gamma_{12}(\tau) = \overline{V_1^*(t) V_2(t+\tau)} = \frac{1}{4} \left[\overline{V_1^{(r)}(t) V_2^{(r)}(t+\tau)} + \overline{V_1^{(i)}(t) V_2^{(i)}(t+\tau)} \right]$$
$$- \frac{1}{4} \left[\overline{V_1^{(i)}(t) V_2^{(r)}(t+\tau)} - \overline{V_1^{(r)}(t) V_2^{(i)}(t+\tau)} \right], \tag{23.38}$$

where for convenience we are writing $V(\mathbf{r}_1, t_1) = V_1(t_1)$.

Since

$$V(t) = \int_0^\infty e^{-2\pi i\nu t} V^{(r)}(\nu) d\nu, \tag{23.39}$$

we can write

$$V(t) = \int_{-\infty}^\infty e^{-2\pi i\nu t} V(\nu) d\nu, \tag{23.40}$$

since for $\nu \geq 0, V(\nu) = V^{(r)}(\nu)$ and for $\nu < 0, V(\nu) = 0$.

Note that by definition $V(t)$ contains only positive frequencies. Similarly for the correlation function

$$\Gamma(\mathbf{r}_1, \mathbf{r}_2, \tau) = \int_0^\infty \Gamma(\mathbf{r}_1, \mathbf{r}_2, \nu) e^{-2\pi i\nu t} d\nu. \tag{23.41}$$

This is equivalent to the Wiener–Khintchine theorem[23.5],[23.6] in the theory of stationary stochastic processes.

The *mutual spectral density is*

$$\begin{aligned} \Gamma(\mathbf{r}_1, \mathbf{r}_2, \nu) &= \int_{-\infty}^\infty \Gamma(\mathbf{r}_1, \mathbf{r}_2, \tau) e^{2\pi i\nu\tau} d\tau \text{ for } \nu \geq 0 \\ &= 0. \qquad\qquad\qquad\qquad \text{for } \nu < 0. \end{aligned} \tag{23.42}$$

$V(\mathbf{r}, t)$ obeys the wave equation

$$\left(\nabla^2 - \frac{1}{c^2} \frac{\partial^2}{\partial t^2} \right) V(\mathbf{r}, t) = 0, \tag{23.43}$$

where

$$\nabla^2 = \frac{\partial^2}{\partial x^2} + \frac{\partial^2}{\partial y^2} + \frac{\partial^2}{\partial z^2} \tag{23.44}$$

is an operator that acts in the vector space of the vector $\mathbf{r} = x\hat{\mathbf{i}} + y\hat{\mathbf{j}} + z\hat{\mathbf{x}}$. Multiplying Eq. (23.43) by the complex conjugate of the analytic field at the space-time point $(\mathbf{r}', t')$

$$\left(\nabla^2 - \frac{1}{c^2} \frac{\partial^2}{\partial t^2} \right) V(\mathbf{r}, t) V^*(\mathbf{r}', t') = 0. \tag{23.45}$$

This equation holds for all the waves $V(\mathbf{r}, t)$ in the ensemble, so it holds for the ensemble average.

$$\left(\nabla^2 - \frac{1}{c^2} \frac{\partial^2}{\partial t^2} \right) \Gamma(\mathbf{r}, t, \mathbf{r}', t') = 0. \tag{23.46}$$

In a similar way, since $V(\mathbf{r}, t) V^*(\mathbf{r}', t') = [V(\mathbf{r}, t)^* V(\mathbf{r}', t')]^*$

$$\left(\nabla'^2 - \frac{1}{c^2} \frac{\partial^2}{\partial t'^2} \right) \Gamma(\mathbf{r}, t, \mathbf{r}', t') = 0, \tag{23.47}$$

where

$$\nabla'^2 = \frac{\partial^2}{\partial x'^2} + \frac{\partial^2}{\partial y'^2} + \frac{\partial^2}{\partial z'^2} \tag{23.48}$$

is an operator that acts in the vector space of the vector $\mathbf{r}' = x'\hat{\mathbf{i}} + y'\hat{\mathbf{j}} + z'\hat{\mathbf{k}}$.

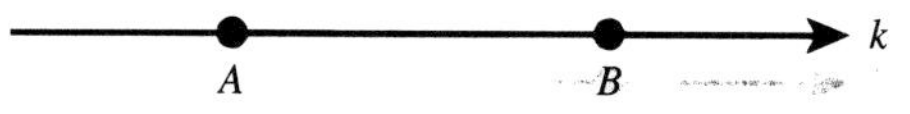

Fig. 23.2. Diagram for explaining temporal coherence.

Now since

$$\Gamma(\mathbf{r}, t, \mathbf{r}', t') = \int_0^\infty \Gamma(\mathbf{r}, \mathbf{r}', \nu) e^{-2\pi i \nu \tau} d\nu, \tag{23.49}$$

where $\tau = t' - t$,

$$\begin{aligned} \frac{\partial^2}{\partial t^2}(\Gamma(\mathbf{r}, t, \mathbf{r}', t')) &= \frac{\partial^2}{\partial t^2} \int_0^\infty \Gamma(\mathbf{r}, \mathbf{r}', \nu) e^{-2\pi i \nu (t'-t)} d\nu \\ &= -4\pi \nu^2 \int_0^\infty \Gamma(\mathbf{r}, \mathbf{r}', \nu) e^{-2\pi i \nu (t'-t)} d\nu. \end{aligned} \tag{23.50}$$

Hence,

$$\nabla^2 \int_0^\infty \Gamma(\mathbf{r}, \mathbf{r}', \nu) e^{-2\pi i \nu \tau} d\nu + \frac{4\pi}{c^2} \int_0^\infty \nu^2 \Gamma(\mathbf{r}, \mathbf{r}', \nu) e^{-2\pi i \nu (t'-t)} d\nu = 0, \tag{23.51}$$

so

$$\nabla^2 \Gamma(\mathbf{r}, \mathbf{r}', \nu) + k^2 \Gamma(\mathbf{r}, \mathbf{r}', \nu) = 0, \tag{23.52}$$

where

$$k = \frac{2\pi \nu}{c} = \frac{\omega}{c}. \tag{23.53}$$

Similarly

$$\nabla'^2 \Gamma(\mathbf{r}, \mathbf{r}', \nu) + k^2 \Gamma(\mathbf{r}, \mathbf{r}', \nu) = 0 \tag{23.54}$$

and $\Gamma(\mathbf{r}, \mathbf{r}', \nu)$ obeys the homogeneous Helmholtz equation.

23.5 Temporal and Spatial Coherence

Temporal coherence implies a fixed phase relationship between the field amplitudes at two points A and B spaced a distance apart in the direction of the wave vector as shown in Fig. (23.2). Temporal coherence can be studied with a Michelson interferometer as shown in Fig. (23.3).

The path difference between the two arms of the interferometer is Δs. Interference will be observed if $\Delta s < \ell_c$, the coherence length. The coherence length is $\ell_c = c\tau_c$ where τ_c is the coherence time. If the bandwidth of the source is $\Delta \nu$, then $\tau_c \sim 1/\Delta \nu$

Examples: For a low pressure mercury lamp $\tau_c \approx$ few ns, therefore the coherence length is a few cm.

For a stable laser, $\Delta \nu \simeq 1$ kHz, $\tau_c \simeq 1$ ms and the coherence length $\simeq 3 \times 10^5$ m.

If the analytic signals coming from the two arms of the interferometer in Fig. (23.3) are V_1, V_2, then the total analytic signal at some observation point $\mathbf{x}$ is

$$V(\mathbf{x}, t) = V_1(\mathbf{x}, t) + V_2(\mathbf{x}, t + \tau) = V_1(t) + V_2(t + \tau), \tag{23.55}$$

where τ is the time difference caused by the path difference $\Delta s : \tau = \Delta s/c$.

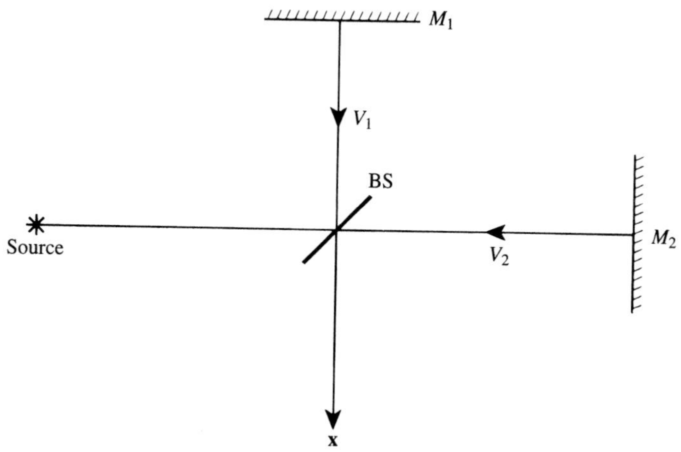

Fig. 23.3. Simple Michelson interferometer used in a discussion of the autocorrelation function of a light source. BS – beamsplitter.

The average intensity observed at **x** is

$$\begin{aligned} I(\mathbf{x}) &= \overline{V^*(\mathbf{x},t)V(\mathbf{x},t)} \\ &= \overline{V_1^*(t)V_1(t)} + \overline{V_2^*(t+\tau)V_2(t+\tau)} \\ &\quad + \overline{V_1^*(t)V_2(t+\tau)} + \overline{V_2^*(t+\tau)V_1(t)}. \end{aligned} \tag{23.56}$$

We can write $V_1 = a_1 V_0; V_2 = a_2 V_0$ where $V_0(t)$ is related to the analytic field at the beamsplitter and the factors a_1, a_2 take into account parameters such as different reflectivities at mirrors M_1 and M_2.

We can put

$$\overline{V_1^*(t)V_1(t)} = a_1^* a_1 \mid V_0(t) \mid^2 = \mid a_1 \mid^2 \mid V_0(t) \mid^2 = \mid a_1 \mid^2 I_0, \tag{23.57}$$

$$\overline{V_2^*(t+\tau)V_2(t+\tau)} = \mid a_2 \mid^2 \mid V_0(t) \mid^2 = \mid a_2 \mid^2 I_0, \tag{23.58}$$

and

$$\overline{V_1^*(t)V_2(t+\tau)} = a_1^* a_2 \Gamma_{11}(\tau), \tag{23.59}$$

where

$$\Gamma_{11}(\tau) = V_0(\mathbf{x},t)V_0(\mathbf{x},t+\tau) \tag{23.60}$$

and

$$\overline{V_2^*(t+\tau)V_1(t)} = a_2^* a_1 \Gamma_{11}(-\tau) = a_2^* a_1 \Gamma_{11}^*(\tau). \tag{23.61}$$

Thus,

$$\begin{aligned} I(\mathbf{x}) &= \mid a_1 \mid^2 I_0 + \mid a_2 \mid^2 I_0 + a_1^* a_2 \Gamma_{11}(\tau) + a_2^* a_1 \Gamma_{11}^*(\tau) \\ &= \mid a_1 \mid^2 I_0 + \mid a_2 \mid^2 I_0 + 2\,\mathscr{R}\,[a_1^* a_2 \Gamma_{11}(\tau)] \end{aligned} \tag{23.62}$$

In the simplest case $\mid a_1 \mid^2 = \mid a_2 \mid^2 = K^2$ and

$$I(\mathbf{x}) = 2K^2[I_0 + \mathscr{R}\,\Gamma_{11}(\tau)] \tag{23.63}$$

and the variation in average intensity at **x** with time τ is contained in the autocorrelation function.

If $\tau > \tau_c$ then $\Gamma_{11}(\tau) = 0$: temporal coherence is measured by the autocorrelation function.

We write

$$| a_1 |^2 I_0 = I_1, \tag{23.64}$$

$$| a_2 |^2 I_0 = I_2. \tag{23.65}$$

The product $a^* a_2$ is generally real so we can write

$$a_1^* a_2 = \sqrt{I_1 I_2}/I_0. \tag{23.66}$$

Introducing a normalized coherence function

$$\gamma(\mathbf{x}, \mathbf{x}, \tau) = \frac{\Gamma(\mathbf{x}, \mathbf{x}, \tau)}{\sqrt{I_1 I_2}} = \frac{\Gamma_{11}(\tau)}{\sqrt{I_1 I_2}}, \tag{23.67}$$

we can write

$$I(\mathbf{x}) = I_1 + I_2 + 2\sqrt{I_1 I_2} \mathscr{R}\, \gamma(\mathbf{x}, \mathbf{x}, \tau) \tag{23.68}$$

$\gamma(\mathbf{x}, \mathbf{x}, \tau)$ is called the *degree of temporal coherence.* In the case where $I_1 = I_2 = I$

$$\gamma(x, x, \tau) = \Gamma_{11}(\tau)/I. \tag{23.69}$$

If the source is quasi-monochromatic we use the slowly varying envelope (SVE) approximation and write

$$\gamma(\mathbf{x}, \mathbf{x}, \tau) = | \gamma(\mathbf{x}, \mathbf{x}, \tau) | \, e^{-2\pi i \nu_0 \tau}, \tag{23.70}$$

where $|\gamma(\mathbf{x}, \mathbf{x}, \tau)|$ is a much more slowly varying function of time than $e^{-2\pi i \nu_0 \tau}$. This separates a physically observable time variation $|\gamma(\mathbf{x}, \mathbf{x}, \tau)|$ from the rapid oscillation at the optical frequency ν.

In this case

$$I(\mathbf{x}) = I_1 + I_2 + 2\sqrt{I_1 I_2} \, | \gamma(\mathbf{x}, \mathbf{x}, \tau) | \cos 2\pi \nu_0 \tau. \tag{23.71}$$

For perfectly monochromatic light $|\gamma(\mathbf{x}, \mathbf{x}, \tau)| = 1$; $\gamma(\mathbf{x}, \mathbf{x}, \tau)$ is frequently written $\gamma(\tau)$.

We can define the coherence time exactly as

$$\tau_c^2 = \frac{1}{N} \int_{-\infty}^{\infty} \tau^2 \, | \gamma(\tau) |^2 \, d\tau = \frac{2}{N} \int_0^{\infty} \tau^2 \, | \gamma(\tau) |^2 \, d\tau, \tag{23.72}$$

where

$$N = \int_{-\infty}^{\infty} | \gamma(\tau) |^2 \, d\tau. \tag{23.73}$$

We can show that

$$\int_{-\infty}^{\infty} | \gamma(\tau) |^2 \, d\tau = \int_0^{\infty} g^2(\nu) d\nu, \tag{23.74}$$

where $g(\nu)$ is the normalized spectral density of the light or lineshape function. This is essentially Parsevaal's theorem[23.6].

Remember that

$$\Gamma(\mathbf{x},\mathbf{x},\nu) = \int_{-\infty}^{\infty} \Gamma(\mathbf{x},\mathbf{x},\tau)e^{2\pi i\nu\tau}d\tau \text{ for } \nu \geq 0$$
$$= 0 \text{ for } \nu < 0, \tag{23.75}$$

$$\Gamma(\mathbf{x},\mathbf{x},\tau) = \int_{0}^{\infty} \Gamma(\mathbf{x},\mathbf{x},\nu)e^{-2\pi i\nu\tau}d\nu, \tag{23.76}$$

and

$$I(\mathbf{x}) = \Gamma(\mathbf{x},\mathbf{x},0) = \int_{0}^{\infty} \Gamma(\mathbf{x},\mathbf{x},\nu)d\nu. \tag{23.77}$$

The lineshape function $g(\nu)$ is defined formally as

$$g(\nu) = \frac{\Gamma(\mathbf{x},\mathbf{x},\nu)}{\int_0^\infty \Gamma(\mathbf{x},\mathbf{x},\nu)d\nu} = \frac{\Gamma(\nu)}{\int_0^\infty \Gamma(\nu)d\nu}. \tag{23.78}$$

Clearly

$$\int_0^\infty g(\nu)d\nu = 1.$$

It we remember that $\Gamma(\mathbf{x},\mathbf{x},\nu)$ is 0 for $\nu < 0$ we can just as easily write

$$g(\nu) = \frac{\Gamma(\nu)}{\int_{-\infty}^\infty \Gamma(\nu)d\nu} = \frac{\Gamma(\nu)}{I(\mathbf{x})}, \tag{23.79}$$

in which case

$$\int_{-\infty}^\infty g(\nu)d\nu = 1, \tag{23.80}$$

where $g(\nu) = 0$ for $\nu < 0$.

Now

$$\gamma(\mathbf{x},\mathbf{x},\tau) = \frac{\Gamma(\mathbf{x},\mathbf{x},\tau)}{I(\mathbf{x})} = \int_0^\infty \frac{\Gamma(\mathbf{x},\mathbf{x},\nu)}{I(x)}e^{-2\pi i\nu\tau}d\nu$$
$$= \int_0^\infty g(\nu)e^{-2\pi i\nu\tau}d\nu. \tag{23.81}$$

So, $\gamma(\tau)$ and $g(\nu)$ are Fourier transforms of each other. Since $g(\nu)$ is real,

$$\gamma^*(\mathbf{x},\mathbf{x},\tau) = \int_0^\infty g(\nu)e^{2\pi i\nu\tau}d\nu. \tag{23.82}$$

So

$$\int_{-\infty}^\infty |\gamma(\tau)|^2 d\tau = \int_{-\infty}^\infty \int_0^\infty \frac{\Gamma(\mathbf{x},\mathbf{x},\tau)}{I(\mathbf{x})}\frac{\Gamma^*(\mathbf{x},\mathbf{x},\nu)}{I(x)}e^{2\pi i\nu\tau}d\nu d\tau$$
$$= \int_0^\infty \frac{|\Gamma(\mathbf{x},\mathbf{x},\nu)|^2}{[I(x)]^2}d\nu = \int_0^\infty [g(\nu)]^2 d\nu. \tag{23.83}$$

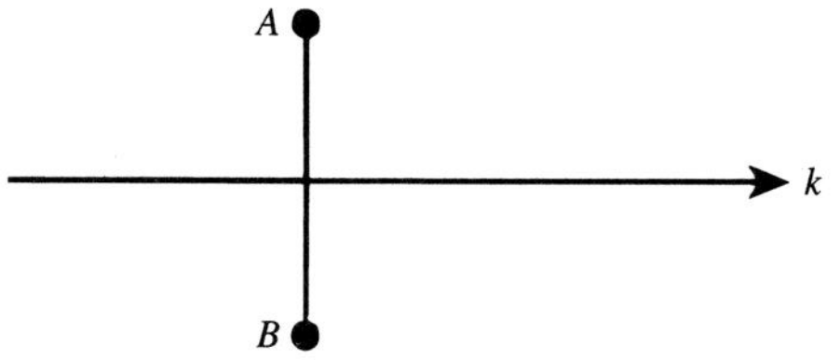

Fig. 23.4. Diagram for explaining spatial coherence.

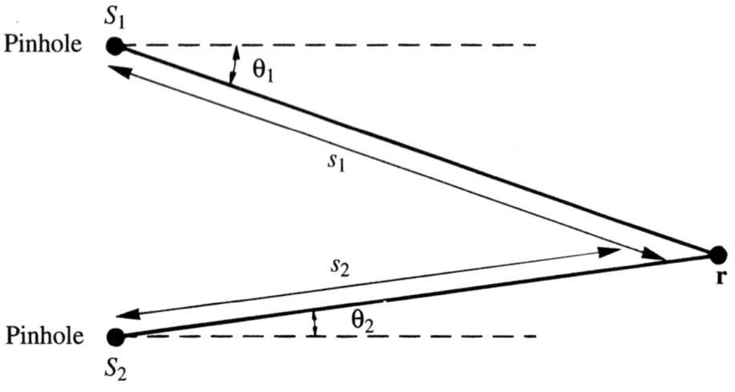

Fig. 23.5. Young's slits interference experiment with two pinholes S_1 and S_2.

23.6 Spatial Coherence

Spatial coherence in the simplest case tells us about the phase relationship between the field amplitudes at two points in a plane normal to the wave vector, as shown in Fig. (23.4).

The existence, or not, of spatial coherence between the fields at two points can be demonstrated in a Young's slits experiment, shown schematically in Fig. (23.5).

The intensity observed at point **r** is

$$I(\mathbf{r}) = \overline{I(\mathbf{r}, t)} = \overline{V^*(\mathbf{r}, t)V(\mathbf{r}, t)}, \tag{23.84}$$

where

$$V(\mathbf{r}, t) = a_1 V(\mathbf{r}_1, t - t_1) + a_2 V(\mathbf{r}_2, t - t_2). \tag{23.85}$$

$V(\mathbf{r}_1, t - t_1)$ is the analytic field at pinhole S_1, etc., a_1 and a_2 are complex, time-independent transmission factors that depend on the mean frequency of the wave, the size of the pinholes and the geometry of the experiment, for example the distances s_1, s_2 from each of the pinholes to point **r**.

In a similar way to our previous analysis we get

$$I(\mathbf{r}) = \overline{I(\mathbf{r}, t)} = | a_1 |^2 I(\mathbf{r}_1) + | a_2 |^2 I(\mathbf{r}_2) + 2\mathscr{R}[a_1^* a_2 \Gamma(\mathbf{r}_1, \mathbf{r}_2, \tau)], \tag{23.86}$$

where

$$\Gamma_{12}(\tau) = \Gamma(\mathbf{r}_1, \mathbf{r}_2, \tau) = \overline{V^*(\mathbf{r}_1, t)V(\mathbf{r}_2, t + \tau)} \tag{23.87}$$

is the mutual correlation function, and $\tau = t_1 - t_2$.

We have used the stationarity condition

$$\overline{V^*(\mathbf{r}_1, t - t_1)V(\mathbf{r}_2, t - t_2)} = \overline{V^*(\mathbf{r}_1, t)V[\mathbf{r}_2, t + (t_1 - t_2)]}. \tag{23.88}$$

$|a_1|^2 I(\mathbf{r}_1)$ is the intensity arising at point **r** from pinhole S_1 alone, if pinhole S_2 is closed.

It can be shown that $a_1^* a_2$ is real. In fact

$$a_1 = \frac{ik\cos\theta_1}{2\pi s_1} = \frac{iks_1}{2\pi s_1^2}\cos\theta_1, \tag{23.89}$$

where θ_1 is the angle shown in Fig. (23.5).

We can identify the $2\pi s_1^2$ factor in the denominator as the surface area of the half sphere illuminated through the pinhole, iks_1 is a complex amplitude and $\cos\theta_1$ is an obliquity factor. The intensity at pinhole 1 is

$$I(\mathbf{r}_1) = \Gamma(\mathbf{r}_1, \mathbf{r}_1, 0) = I_1. \tag{23.90}$$

We can introduce the *degree of coherence*

$$\gamma(\mathbf{r}_1, \mathbf{r}_2, \tau) = \gamma_{12}(\tau) = \frac{\Gamma(\mathbf{r}_1, \mathbf{r}_2, \tau)}{\sqrt{\Gamma(\mathbf{r}_1, \mathbf{r}_1, 0)\Gamma(\mathbf{r}_2, \mathbf{r}_2, 0)}} = \frac{\Gamma_{12}(\tau)}{\sqrt{I_1 I_2}}. \tag{23.91}$$

If we write

$$| a_1 |^2 I(\mathbf{r}_1) = I_1(\mathbf{r}), \tag{23.92}$$

$$| a_2 |^2 I(\mathbf{r}_2) = I_2(\mathbf{r}), \tag{23.93}$$

then

$$I(\mathbf{r}) = I_1(\mathbf{r}) + I_2(\mathbf{r}) + 2\sqrt{I_1(\mathbf{r}) I_2(\mathbf{r})}\,\mathscr{R}\left[\gamma(\mathbf{r}_1, \mathbf{r}_2, \tau)\right]. \tag{23.94}$$

If the light illuminating both pinholes is monochromatic, then we can write

$$V(r_1, t - t_1) = \sqrt{I_1} e^{-2\pi i v(t - t_1)}, \tag{23.95}$$

which gives

$$I(\mathbf{r}) = I_1(\mathbf{r}) + I_2(\mathbf{r}) + 2\sqrt{I_1(\mathbf{r}) I_2(\mathbf{r})}\cos[2\pi v(t_2 - t_1)]. \tag{23.96}$$

Eq. (23.96) describes an ideal two-pinhole interference pattern.

From Schwarz's inequality† it follows that

$$\begin{aligned}\mathscr{R}\,\Gamma_{12}(\tau) \leq | \Gamma_{12}(\tau) \leq \sqrt{\Gamma_{11}(\tau)\Gamma_{22}(\tau)} \\ \leq \sqrt{I_1 I_2}\end{aligned} \tag{23.97}$$

and therefore,

$$\mathscr{R}\gamma_{12}(\tau) \leq | \gamma_{12}(\tau) | \leq 1. \tag{23.98}$$

If we are dealing with light that is not quite monochromatic, but is quasi-monochromatic, contained within a frequency band Δv much less than its center frequency v_0, it is reasonable to write

$$\gamma_{12}(\tau) = |\gamma_{12}(\tau)| e^{-2\pi i v_0 \tau}. \tag{23.99}$$

This is the SVE approximation again, $| \gamma_{12}(\tau) |$ is a slowly varying function that remains constant over many oscillations of the phase factor.

† Schwarz's inequality states that for two functions, $f(x), g(x)$

$$\int f f^* dx \int g g^* dx \geq \left| \int f^* g dx \right|^2$$

In this case

$$I(\mathbf{r}) = I_1(\mathbf{r}) + I_2(\mathbf{r}) + 2\sqrt{I_1(\mathbf{r})I_2(\mathbf{r})]} \mid \gamma_{12}(\tau) \mid \cos 2\pi\nu_0\tau. \tag{23.100}$$

Such an interference pattern would show the characteristic light and dark regions of a Young's slits interference pattern in the center of the pattern, but the fringes would grow less distinct moving away from the center. This is the effect of temporal coherence showing up. As we move away from the center of the interference pattern the time delay difference, τ, for propagation from the pinholes grows larger. The time delay at the edges of the pattern approaches the value $\tau = 2d/c$, where $2d$ is the pinhole spacing.

Defining the visibility, V, of the fringes as

$$V = \frac{I_{max} - I_{min}}{I_{max} + I_{min}} \tag{23.101}$$

gives

$$V = \frac{2\sqrt{I_1 I_2} \mid \gamma_{12}(\tau) \mid}{I_1 + I_2}, \tag{23.102}$$

which when $I_1(\mathbf{r}) = I_2(\mathbf{r})$ gives

$$V = \mid \gamma_{12}(t_1 - t_2) \mid . \tag{23.103}$$

For incoherent illumination of the two pinholes, $|\gamma_{12}(\tau)| = 0$ and no fringes are observed.

Perfect fringes are observed in the case of completely coherent illumination,

$$\mid \gamma_{12}(\tau) \mid = 1. \tag{23.104}$$

The intermediate case $0 < \mid \gamma_{12}(\tau) \mid < 1$ represents partial coherence.

To distinguish between the factors $\gamma_{12}(\tau)$ is often called the *complex degree of coherence* and $\mid \gamma_{12}(\tau) \mid$ the *degree of coherence.*

23.7 Spatial Coherence with an Extended Source

Consider the case of two pinholes illuminated by an extended source which takes the form of a square with sides of length Δa as shown in Fig. (23.6).

If fringes are seen, then at least some spatial coherence exists between light arriving at Q from P_1 and P_2. For this to happen the overall field at P_1, which results from a superposition of the disturbances originating from all points on the source, must be phase related to the overall field at P_2. If each point on the source is an independent emitter the overall fields will have a stable phase relationship provided the extreme ends of the source, A_1, A_2, are equidistant from P_1 (or P_2) within a half wavelength. It is easy to show that this leads to the condition $\theta\Delta a < \lambda_0$ (for quasi-monochromatic light of center wavelength λ_0), so the pinholes must be arranged within an area around O of size

$$\Delta A \approx (r\theta)^2 \simeq \frac{r^2\lambda_0^2}{\Delta a^2} = \frac{c^2 r^2}{\nu_0^2 S}, \tag{23.105}$$

where S is the area of the source. ΔA is called the coherence area in plane A. The volume of coherence is

$$\Delta V = \Delta A \ell_c = \frac{c}{\Delta\nu}\frac{c^2}{\nu_0^2}\frac{r^2}{S} = \frac{\lambda_0}{\Delta\lambda}\left(\frac{r}{\Delta a}\right)^2 \lambda_0^3. \tag{23.106}$$

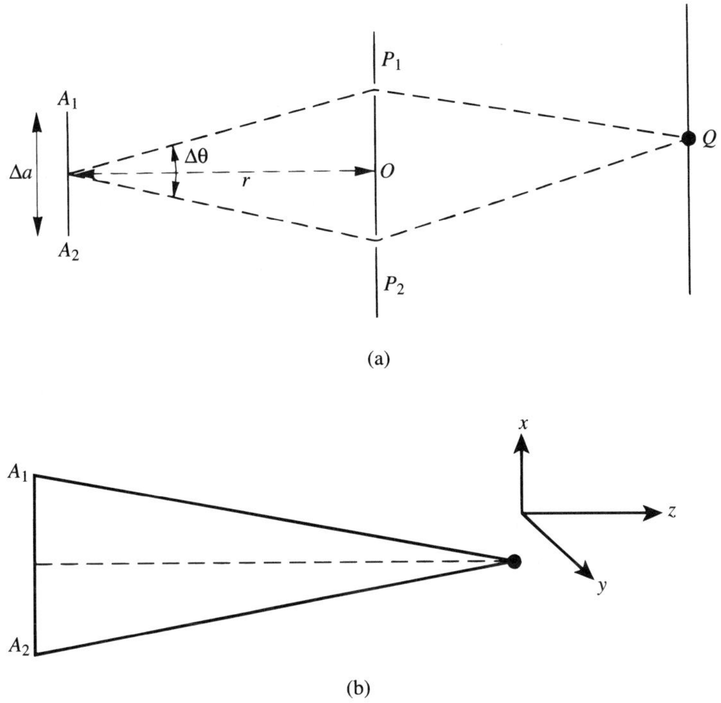

Fig. 23.6. (a) Geometry of a Young's slits experiment where an extended source is used, (b) coordinate system used in discussing uncertainties in photon momentum.

The volume of coherence defined in this way has a length in the direction of propagation of the light equal to the coherence length, and a cross-section equal to the coherence area.

We can show that this volume is also the volume corresponding to one cell of phase space of the photons from the source.

The volume of an elementary cell of phase space is given by[23.7],[23.8]

$$\Delta p_x \Delta p_y \Delta p_z \Delta q_x \Delta q_y \Delta q_z = h^3, \tag{23.107}$$

where Δp_x..., Δq_x... are the uncertainties in the components of the momentum and position, respectively, of the photon. We can estimate these uncertainties from the geometry of the double pinholes shown in Fig. (23.6b).

The momentum of a photon is h/λ_0. The momentum in the x direction for photons from A_1 is $-(h/\lambda_0)\sin(\theta/2) \simeq -h\theta/2\lambda_0$ and for photons from A_2 is $h\theta/2\lambda_0$. Thus,

$$\Delta p_x = \Delta p_y = \frac{h\theta}{\lambda_0} = \frac{h}{\lambda_0}\frac{\Delta a}{r}. \tag{23.108}$$

Δp_z comes mainly from uncertainties in the frequency of the source, i.e.,

$$\Delta p_z = \frac{h}{\lambda_0^2}\Delta\lambda, \tag{23.109}$$

so

$$\frac{h^2\Delta_a^2}{\lambda_0^2 r^2} \cdot \frac{h\Delta\lambda}{\lambda_0^2}\Delta q_x \Delta q_y \Delta q_z = h^3, \tag{23.110}$$

which gives

$$\Delta q_x \Delta q_y \Delta q_z = \frac{\lambda_0}{\Delta\lambda}\left(\frac{r}{\Delta a}\right)^2 \lambda_0^3 : \tag{23.111}$$

identical with the volume of coherence.

We can introduce the degeneracy parameter δ that represents the average number of photons in the same state of polarization in the volume of coherence. Thus, it represents the average number of photons in the same polarization state that cross the area of coherence per coherence time. If N_ν is the average number of photons emitted from unit area of the source, per unit frequency interval, per unit solid angle around the direction normal to the source, then it is easy to show that

$$\delta = \frac{1}{2} N_\nu S \Delta\nu \Delta\Omega \tau_c. \tag{23.112}$$

S is again the area of the source, $\Delta\Omega$ is the solid angle subtended by the coherence area at the source, $\Delta\nu$ is the bandwidth of the source, and τ_c is its coherence time. The factor $\frac{1}{2}$ arises because the light coming from the source is assumed to be unpolarized. The solid angle is

$$\Delta\Omega = \frac{\Delta A}{r^2} = \frac{c^2}{\nu_0^2 S}, \tag{23.113}$$

and because $\Delta\nu\tau_c \simeq 1$

$$\delta \simeq \frac{c^2}{2\nu_0^2} N_\nu. \tag{23.114}$$

For black-body radiation the energy density $\rho(v)$(J m^{-3} Hz^{-1}) is

$$\rho(\nu) = \frac{8\pi h\nu^3}{c^3} \frac{1}{(e^{h\nu/kt} - 1)}, \tag{23.115}$$

so the photon flux N_ν is

$$\begin{aligned} N_\nu &= \frac{\rho(\nu)c}{4\pi h\nu} = \frac{c}{4\pi} \frac{8\pi h\nu^3}{c^3 h\nu} \frac{1}{(e^{h\nu/kT} - 1)} \\ &= \frac{2\nu^2}{c^2} \frac{1}{(e^{h\nu/kT} - 1)}. \end{aligned} \tag{23.116}$$

Therefore, from Eq. (23.114)

$$\delta \simeq 1/(e^{h\nu/kT} - 1). \tag{23.117}$$

This is identical with the mean number of photons in the mode – the occupation number $\overline{n}$, introduced in Chapter 1, which is generally very small except at low frequencies and/or high temperatures.

23.8 Propagation Laws of Partial Coherence

The classic Young's slits interference experiment is carried out with a point source illuminating the two pinholes. This ensures that the two pinholes will be within the area of coherence. One could imagine, in principle, this experiment being done using a star as the point source, with a suitable narrow band filter placed over the pinholes to ensure quasi-monochromaticity. It is clear, however, that if

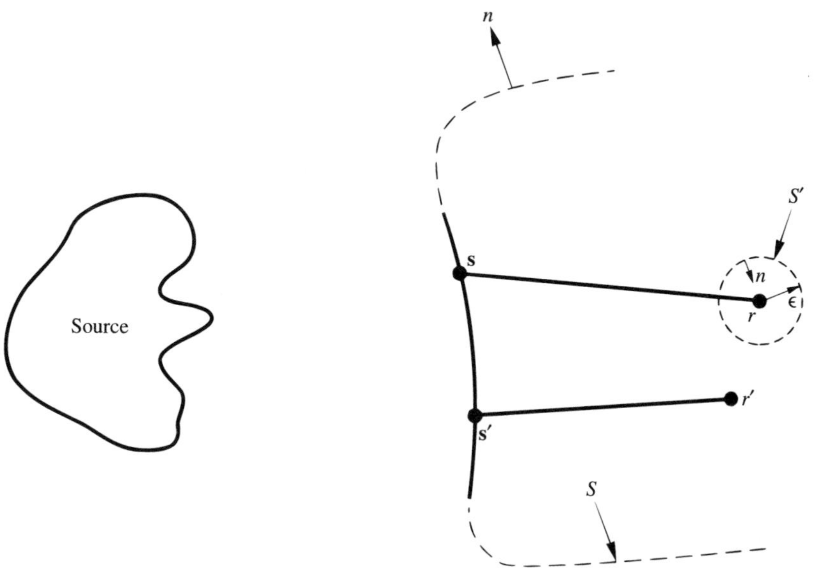

Fig. 23.7. Diagram used in discussion of the propagation of the mutual coherence function from a surface S to a surface S'.

it were possible to place the two pinholes very close to the star, no fringes would be observed because each pinhole would be illuminated by light from different points on the surface of the star – the pinholes would no longer be within the area of coherence. Therefore, the placement of the pinholes relative to an *extended* source is crucial in determining whether fringes will be seen. We can analyze this phenomenon mathematically by showing that the mutual coherence function varies as it propagates, so that the fields at laterally-separated points go from no coherence very near the source to a high degree of coherence far away.

We assume that we know the mutual coherence function $\Gamma_{12}(\tau) = \Gamma(\mathbf{r}, \mathbf{r}', \tau)$ over a certain surface S of space. The problem is to determine $\Gamma_{12}(\tau) = \Gamma(\mathbf{r}, \mathbf{r}', \tau)$ in terms of its value $\Gamma(\mathbf{s}, \mathbf{s}', \tau)$ for points $\mathbf{s}, \mathbf{s}'$ on the surface S, as shown in Fig. (23.7).

Now, we know that the mutual coherence function $\Gamma(\mathbf{r}, \mathbf{r}', \nu)$ satisfies the homogeneous Helmholtz equation

$$(\nabla^2 + k^2)\Gamma(\mathbf{r}, \mathbf{r}', \nu) = 0, \tag{23.118}$$

and

$$(\nabla'^2 + k^2)\Gamma(\mathbf{r}, \mathbf{r}', \nu) = 0. \tag{23.119}$$

We choose a Green's function $G(\mathbf{r}, \mathbf{s}, \nu)$ which satisfies

$$(\nabla^2 + k^2)G(\mathbf{r}, \mathbf{s}, \nu) = -\delta(\mathbf{r} - \mathbf{s}). \tag{23.120}$$

This Green's function is like a source of waves placed at $\mathbf{r}$. We have

$$G\nabla^2\Gamma + Gk^2\Gamma = 0,$$

where

$$\Gamma = \Gamma(\mathbf{s}, \mathbf{r}', \nu) \text{ and } G = G(\mathbf{r}, \mathbf{s}, \nu) \tag{23.121}$$

and also

$$\Gamma\nabla^2 G + \Gamma k^2 G = -\delta(\mathbf{r} - \mathbf{s})\Gamma(\mathbf{s}, \mathbf{r}', \nu). \tag{23.122}$$

Subtracting Eq. (23.122) from Eq. (23.121) gives

$$G\nabla^2\Gamma - \Gamma\nabla^2 G = \delta(\mathbf{r}-\mathbf{s})\Gamma(\mathbf{s},\mathbf{r}',\nu). \tag{23.123}$$

Using Green's theorem[23.7]

$$\int_V (G\nabla^2\Gamma - \Gamma\nabla^2 G)dV = \int_S (G\nabla\Gamma - \Gamma\nabla \mathbf{G})\cdot d\mathbf{S}, \tag{23.124}$$

where V is the volume contained within the closed surface S, gives

$$\int_S (G\nabla\Gamma - \Gamma\nabla \mathbf{G})\cdot d\mathbf{S} = \int_S \left(G\frac{\partial\Gamma}{\partial n} - \frac{\partial G}{\partial n}\Gamma\right) dS$$

$$= \int_V \delta(\mathbf{r}-\mathbf{s})\Gamma(\mathbf{s},\mathbf{r}',\nu)dV = \Gamma(\mathbf{r},\mathbf{r}',\nu), \tag{23.125}$$

where $\partial/\partial n$ is a derivative in the positive normal direction (outward pointing) on the surface S.

We must be careful about including the singular point $\mathbf{r}$ inside the integration volume in Eq. (23.125), but in this case it can be shown to be justified. If we use $G = e^{ik|\mathbf{r}-\mathbf{s}|}/4\pi|\mathbf{r}-\mathbf{s}|$ and exclude the singular point at $\mathbf{s}=\mathbf{r}$ by a surface S' of radius ϵ as shown in Fig. (23.7), then in this case the volume integral goes to zero and

$$\int_S \left(G\frac{\partial\Gamma}{\partial n} - \frac{\partial G}{\partial n}\Gamma\right) dS = -\int_{S'} \left(G\frac{\partial\Gamma}{\partial n} - \frac{\partial G}{\partial n}\Gamma\right) dS. \tag{23.126}$$

Writing $G = e^{ik\epsilon}/\epsilon$ gives

$$\int_S \left(G\frac{\partial\Gamma}{\partial n} - \frac{\partial G}{\partial n}\Gamma\right) dS = -\int_{S'} \frac{e^{ik\epsilon}}{4\pi\epsilon}\frac{\partial\Gamma}{\partial n}(\mathbf{r}) - \Gamma(\mathbf{r})\left(\frac{e^{ik\epsilon}}{4\pi\epsilon^2} - \frac{ike^{ik\epsilon}}{4\pi\epsilon}\right) dS. \tag{23.127}$$

It should be noted that we have used $\partial/\partial n = -\partial/\partial\epsilon$ because the normal n points away from surface S' into the small volume surrounding point $\mathbf{r}$. Thus

$$\int_S \left(G\frac{\partial\Gamma}{\partial n} - \frac{\partial G}{\partial n}\Gamma\right) dS = -4\pi\epsilon^2 \left[\frac{e^{ik\epsilon}}{4\pi\epsilon}\frac{\partial\Gamma}{\partial n}(\mathbf{r}) - \Gamma(\mathbf{r})\left(\frac{e^{ik\epsilon}}{4\pi\epsilon}\right)\left(\frac{1}{\epsilon} - ik\right)\right]$$
$$= \Gamma(\mathbf{r}) \text{ in the limit as } \epsilon \to 0. \tag{23.128}$$

This justifies Eq. (23.125). In summary,

$$\Gamma(\mathbf{r},\mathbf{r}',\nu) = \int_S \left\{ G(\mathbf{r},\mathbf{s},\nu)\frac{\partial\Gamma}{\partial n}(\mathbf{s},\mathbf{r}',\nu) - \frac{\partial G}{\partial n}(\mathbf{r},\mathbf{s},\nu)\Gamma(\mathbf{s},\mathbf{r}',\nu)\right\} dS. \tag{23.129}$$

It is particularly convenient if we choose a Green's function such that $G(\mathbf{r},\mathbf{s},\nu)=0$ on the surface S. Then

$$\Gamma(\mathbf{r},\mathbf{r}',\nu) = \int_S -\frac{\partial G}{\partial n}(\mathbf{r},\mathbf{s},\nu)\Gamma(\mathbf{s},\mathbf{r}',\nu)dS, \tag{23.130}$$

where $\partial/\partial n$ is a normal derivative at point $\mathbf{s}$. We can now use a second Green's function that satisfies

$$(\nabla'^2 + k^2)G'(\mathbf{r}',\mathbf{s}',\nu) = -\delta(\mathbf{r}'-\mathbf{s}'), \tag{23.131}$$

and which also vanishes on the surface S to get in a similar way

$$\Gamma(\mathbf{s}, \mathbf{r}', \nu) = \int_S -\frac{\partial G'}{\partial n'}(\mathbf{r}', \mathbf{s}', \nu)\Gamma(\mathbf{s}, \mathbf{s}', \nu)(\mathbf{s}, \mathbf{s}', \nu)dS', \tag{23.132}$$

where $\partial/\partial n'$ is a normal derivative at point $\mathbf{s}'$. Combining Eqs. (23.130) and (23.132)

$$\Gamma(\mathbf{r}, \mathbf{r}', \nu) = \int_S \int_S \frac{\partial G}{\partial n}(\mathbf{r}, \mathbf{s}, \nu)\frac{\partial G'}{\partial n'}(\mathbf{r}', \mathbf{s}', \nu)\Gamma(\mathbf{s}, \mathbf{s}', \nu)dSdS' \tag{23.133}$$

and $\Gamma(\mathbf{r}, \mathbf{r}', \nu)$ can be determined from its value $\Gamma(\mathbf{s}, \mathbf{s}', \nu)$ on a surface.

Since we have already shown that $\Gamma(\mathbf{r}, \mathbf{r}, \nu) = \Gamma^*(\mathbf{r}, \mathbf{r}, \nu)$, it is straightforward to show that

$$\Gamma(\mathbf{r}, \mathbf{r}', \nu) = \Gamma^*(\mathbf{r}', \mathbf{r}, \nu). \tag{23.134}$$

Hence,

$$G(\mathbf{r}, \mathbf{s}, \nu) = G'^*(\mathbf{r}, \mathbf{s}, \nu). \tag{23.135}$$

If we define

$$K(\mathbf{r}', \mathbf{s}', \nu) = \frac{\partial G'}{\partial n'}(\mathbf{r}', \mathbf{s}', \nu) \tag{23.136}$$

then

$$\Gamma(\mathbf{r}, \mathbf{r}', \nu) = \int_S \int_S K^*(\mathbf{r}, \mathbf{s}, \nu)K(\mathbf{r}', \mathbf{s}', \nu)\Gamma(\mathbf{s}, \mathbf{s}', \nu)dSdS'. \tag{23.137}$$

$K(\mathbf{r}, \mathbf{s}, \nu)$ is called the diffraction function – this is the complex amplitude at $\mathbf{r}$ caused by a unit point source of frequency ν situated at $\mathbf{s}$. The mutual coherence function between the two points $\mathbf{r}, \mathbf{r}'$ is

$$\Gamma(\mathbf{r}, \mathbf{r}', \tau) = \int_0^\infty \Gamma(\mathbf{r}, \mathbf{r}', \nu)e^{-2\pi i\nu\tau}d\nu. \tag{23.138}$$

23.9 Propagation from a Finite Plane Surface

When we used Green's theorem in the preceding section we were converting a volume integral to an integral over a closed surface. However, if we are concerned with finite area surfaces, which are not closed, we have to justify neglecting the integral over the rest of the closed surface that surrounds the finite region in which we are interested.

Now we have already shown that

$$\Gamma(\mathbf{r}, \mathbf{r}', \nu) = \int_S \left\{ G(\mathbf{r}, \mathbf{s}, \nu)\frac{\partial \Gamma}{\partial n}(\mathbf{s}, \mathbf{r}', \nu) - \frac{\partial G}{\partial n}(\mathbf{r}, \mathbf{s}, \nu)\Gamma(\mathbf{s}, \mathbf{r}', \nu) \right\} dS. \tag{23.139}$$

A typical basic Green's function that could be used in this equation is

$$G(\mathbf{r}, \mathbf{s}, \nu) = \frac{e^{ik|\mathbf{r}-\mathbf{s}|}}{4\pi \mid \mathbf{r} - \mathbf{s} \mid}. \tag{23.140}$$

It is straightforward to show that this function satisfies the scalar Helmholtz equation

$$(\nabla_s^2 + k^2)G(\mathbf{r}, \mathbf{s}, \nu) = -\delta(\mathbf{r} - \mathbf{s}), \tag{23.141}$$

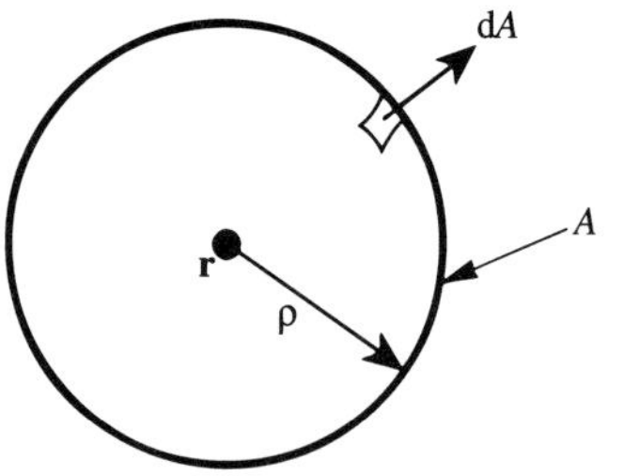

Fig. 23.8. Spherical volume surrounding a point $\mathbf{r}$ where a Green's function such as given in Eq. (23.140) becomes singular.

the s subscript on ∇_s indicates differentiation with respect to the coordinates of point $\mathbf{s}$. Except at the point $\mathbf{r}$ the Green's function satisfies

$$(\nabla_s^2 + k^2)G(\mathbf{r},\mathbf{s},\nu) = 0. \tag{23.142}$$

It follows that

$$(\nabla_s^2 + k^2)\frac{e^{ik|\mathbf{r}-\mathbf{s}|}}{4\pi \mid \mathbf{r}-\mathbf{s}\mid} = \frac{e^{ik(|\mathbf{r}-\mathbf{s}|)}}{4\pi}\nabla_s^2\left(\frac{1}{\mid \mathbf{r}-\mathbf{s}\mid}\right). \tag{23.143}$$

We can show that the RHS of Eq. (23.143) is a δ-function. $\nabla_s^2(1/\mid \mathbf{r}-\mathbf{s}\mid)$ is zero except when $\mid \mathbf{r}-\mathbf{s}\mid = 0$, and consequently

$$\int_{all\ space}\frac{e^{ik|\mathbf{r}-\mathbf{s}|}}{4\pi}\nabla_s^2\left(\frac{1}{\mid \mathbf{r}-\mathbf{s}\mid}\right) = \frac{1}{4\pi}\int_{all\ space}\nabla_s^2\left(\frac{1}{\mathbf{r}-\mathbf{s}}\right). \tag{23.144}$$

We can evaluate this integral by surrounding the point $\mathbf{r}$ by a small sphere V of radius ρ and surface A as shown in Fig. (23.8). Writing $\mid \mathbf{r}-\mathbf{s}\mid = R$

$$\frac{1}{4\pi}\int_{all\ space}\nabla_s^2\left(\frac{1}{R}\right) = \frac{1}{4\pi}\int_V \nabla_s^2\left(\frac{1}{R}\right) = \frac{1}{4\pi}\int A\nabla_S\left(\frac{1}{R}\right)\cdot d\mathbf{A}, \tag{23.145}$$

where $d\mathbf{A}$ is outward pointing. Finally,

$$\frac{1}{4\pi}\int_A \nabla_S\left(\frac{1}{R}\right)\cdot d\mathbf{A} = \frac{1}{4\pi}\int_A \frac{\rho\hat{\mathbf{r}}}{\rho^3}\cdot\rho^3\hat{\mathbf{r}}d\Omega = -1, \tag{23.146}$$

where $\hat{\mathbf{r}}$ is a unit vector in the direction of ρ and $d\Omega$ is an elementary solid angle. This verifies Eq. (23.141).

Consider the finite aperture shown in Fig. (23.9) where the rest of the plane is, or can be imagined to be, blocked off with an opaque screen. The closed surface S_0 under consideration consists of three parts, the aperture S, the infinite screen S_1, and the infinite half sphere S_2

Now,

$$\begin{aligned}\Gamma(\mathbf{r},\mathbf{r}',\nu) &= \int_{S_0}\left[G(\mathbf{r},\mathbf{s},\nu)\frac{\partial\Gamma(\mathbf{s},\mathbf{r}',\nu)}{\partial n} - \frac{\partial G(\mathbf{r},\mathbf{s},\nu)}{\partial n}\Gamma(\mathbf{s},\mathbf{r}',\nu)\right]dS \\ &= \int_{S+S_1+S_2}\left[G(\mathbf{r},\mathbf{s},\nu)\frac{\partial\Gamma(\mathbf{s},\mathbf{r}',\nu)}{\partial n} - \frac{\partial G(\mathbf{r},\mathbf{s},\nu)}{\partial n}\Gamma(\mathbf{s},\mathbf{r}',\nu)\right]dS.\end{aligned} \tag{23.147}$$

Now, if $G = (1/4\pi)e^{-ikR}/R$ then on S_2, $\mathbf{r}$ is essentially at the center of the sphere since on S_2, R is large

$$\frac{\partial G}{\partial n} = \frac{1}{4\pi}\left(ik - \frac{1}{R}\right)\frac{e^{ikR}}{R} \approx ikG \text{ as } R \to \infty. \tag{23.148}$$

Fig. 23.9. A closed surface S_0 consisting of three parts, an illuminated aperture S, an infinite opaque screen S_1 and an infinite half 'sphere', S_2. In the limit the distance R of a point **r** becomes infinite.

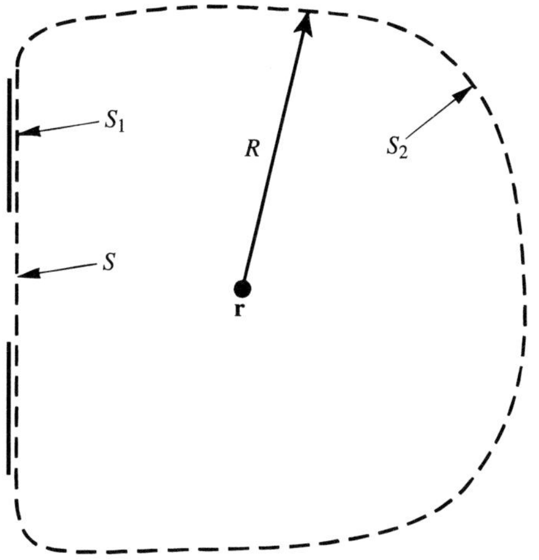

Therefore, the surface integral over S_2 becomes

$$\int_{S_2} \left(G\frac{\partial \Gamma}{\partial n} - ikG\Gamma \right) dS = \int_{\Omega} G\left(\frac{\partial \Gamma}{\partial n} - ik\Gamma \right) R^2 d\omega, \tag{23.149}$$

where Ω is the solid angle subtended by S_2 at $\mathbf{r}$ and $\Gamma = \Gamma(\mathbf{s}, \mathbf{r}', \nu)$. Now $|RG|$ is finite on S_2, therefore the integral in Eq. (23.147) will vanish provided Γ has the property.

$$\lim R \to \infty \quad R\left(\frac{\partial \Gamma}{\partial n} - ik\Gamma \right) = 0 \tag{23.150}$$

over the whole solid angle. This requirement is, in scalar diffraction theory, which involves similar considerations, known as the *Sommerfeld radiation condition*[23.9],[23.10] and is satisfied if Γ vanishes at least as fast as a diverging spherical wave.

The rest of the surface integral can be reduced to an integral only over the illuminated aperture provided Γ fulfills the *Kirchoff boundary conditions*: (1) Across the aperture S the field distribution Γ and its derivative $\partial\Gamma/\partial n$ are exactly the same as they would be in the absence of the screen. (2) Over the portion of the surface obscured by the screen, S_1, the field distribution Γ and $\partial\Gamma/\partial n$ are identically zero.

Thus, we have finally

$$\Gamma(\mathbf{r}, \mathbf{r}', \nu) = \int_{S(aperture\ only)} \left(G\frac{\partial \Gamma}{\partial n} - \frac{\partial G}{\partial n}\Gamma \right) dS. \tag{23.151}$$

Unfortunately, the Kirchoff boundary conditions are unnecessarily restrictive. The requirement that $\partial\Gamma/\partial n$ vanishes at the boundary between the aperture and screen is unreasonable. This restriction can be avoided by choosing a Green's function that vanishes over the plane surface comprising $S + S_1$. In studies of diffraction this is called the Rayleigh–Sommerfeld theory[23.2]. In this case

$$\Gamma(\mathbf{r}, \mathbf{r}', \nu) = \int_S -\frac{\partial G}{\partial n}\Gamma dS. \tag{23.152}$$

Green's functions that vanish over a plane do exist. For example, suppose G is generated by two identical point sources, one at $\mathbf{r}$ and one in the mirror image

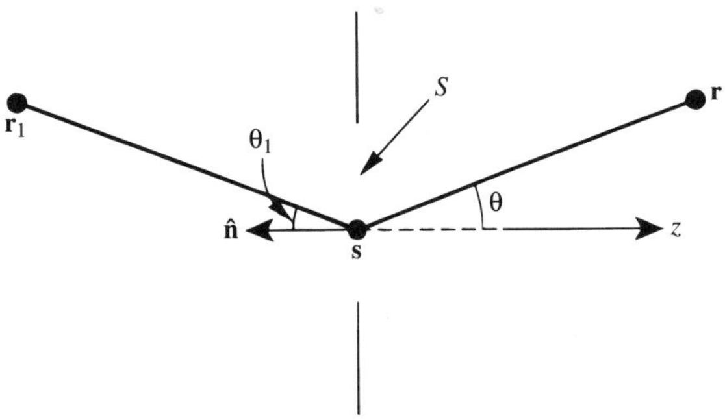

Fig. 23.10. Geometry used in discussing a Green's function that vanishes over a plane. Point $\mathbf{r}_1$ is a mirror image of $\mathbf{r}$ in the plane S.

position $\mathbf{r}_1$ on the opposite side of the aperture, but as shown schematically in Fig. (23.10) oscillating in antiphase. In this case

$$G(\mathbf{r},\mathbf{s},\nu) = \frac{1}{4\pi}\left(\frac{e^{ik|\mathbf{r}-\mathbf{s}|}}{|\,\mathbf{r}-\mathbf{s}\,|} - \frac{e^{ik|\mathbf{r}_1-\mathbf{s}|}}{|\,\mathbf{r}_1-\mathbf{s}\,|}\right). \tag{23.153}$$

Clearly if $\mathbf{r}_1$ lies on S this function vanishes, as then $\mathbf{r}$ lies on S also. The Green's function $G^*(\mathbf{r},\mathbf{s},\nu)$ would be equally satisfactory in this regard.

If $\mathbf{s}$ is the point $(x_s, y_s, z_s), \mathbf{r}$ is the point (x, y, z), and $\mathbf{r}_1$ the point (x_1, y_1, z_1) where

$$z_1 = z_s - (z - z_s) = 2z_s - z, \tag{23.154}$$

as shown in Fig. (23.10), then

$$|\,\mathbf{r}-\mathbf{s}\,| = \sqrt{(x-x_s)^2 + (y-y_s)^2 + (z-z_s)^2} \tag{23.155}$$

and

$$|\,\mathbf{r}_1-\mathbf{s}\,| = \sqrt{(x-x_s)^2 + (y-y_s)^2 + (z+z_s)^2}. \tag{23.156}$$

$\hat{\mathbf{n}}$ is an outward pointing unit vector in Fig. (23.10).

The Green's function above can be written as

$$G(\mathbf{r},\mathbf{s},\nu) = \frac{1}{4\pi}\left(\frac{e^{ikR}}{R} - \frac{e^{ikR_1}}{R_1}\right) \tag{23.157}$$

and the normal derivative as

$$\frac{\partial}{\partial n} = +\frac{\partial}{\partial R}\cos\theta; \quad \frac{\partial}{\partial n} = -\frac{\partial}{\partial R_1}\cos\theta_1 = -\frac{\partial}{\partial R_1}\cos\theta. \tag{23.158}$$

Hence, for the above Green's function

$$\frac{\partial G}{\partial n} = \frac{1}{4\pi}\left\{\cos\theta\left(ik - \frac{1}{R}\right)\frac{e^{ikR}}{R} + \cos\theta\left(ik - \frac{1}{R_1}\right)\frac{e^{ikR_1}}{R_1}\right\}. \tag{23.159}$$

Moreover, $R_1 = R$ and we have finally

$$\frac{\partial G}{\partial n} = 2\left(ik - \frac{1}{R}\right)\frac{e^{ikR}}{4\pi R}\cos\theta \tag{23.160}$$

where $R = |\,\mathbf{r}-\mathbf{s}\,|$.

A corresponding Green's function for the variables $\mathbf{r}', \mathbf{s}'$ is

$$G(\mathbf{r}',\mathbf{s}',\nu) = \frac{1}{4\pi}\left(\frac{e^{ikR'}}{R'} - \frac{e^{ikR_1'}}{R_1'}\right) \tag{23.161}$$

with $R'_1 = R'$. So

$$\frac{\partial G'}{\partial n'} = 2\left(ik - \frac{1}{R'}\right)\frac{e^{ikR'}}{4\pi R'}\cos\theta'. \tag{23.162}$$

The angle between $\mathbf{r} - \mathbf{s}$ and the normal is θ, and between $\mathbf{r}' - \mathbf{s}'$ and the normal is θ'. The primes on the variables in Eqs. (23.161) and (23.162) are not intended to indicate that the Green's functions G' and G are fundamentally different, but to distinguish the points over which the integration is being performed.

Restating Eq. (23.133), using $G = G'^*$

$$\Gamma(\mathbf{r}, \mathbf{r}', \nu) = \int_S \int_S \frac{\partial G^*}{\partial n}(\mathbf{r}, \mathbf{s}, \nu)\frac{\partial G}{\partial n'}(\mathbf{r}', \mathbf{s}', \nu)\Gamma(\mathbf{s}, \mathbf{s}', \nu)dSdS', \tag{23.163}$$

which gives

$$\Gamma(\mathbf{r}, \mathbf{r}', \nu) = \frac{1}{4\pi^2}\int_S \int_S (1 + ikR)(1 - ikR') \times \frac{e^{-ik(R-R')}}{R^2R'^2}\cos\theta\cos\theta'\Gamma(\mathbf{s}, \mathbf{s}', \nu)dSdS'. \tag{23.164}$$

The mutual coherence function can be found by taking the Fourier transform of Eq. (23.164)

$$\Gamma(\mathbf{r}, \mathbf{r}', \tau) = \frac{1}{4\pi^2}\int_0^\infty \int_S \int_S (1 + ikR)(1 - ikR') \times \frac{e^{-ik(R-R')-2\pi i\nu\tau}}{R^2R'^2} \times \cos\theta\cos\theta'\Gamma(\mathbf{s}, \mathbf{s}', \nu)dSdS'd\nu. \tag{23.165}$$

If the observation points $\mathbf{r}, \mathbf{r}'$ are not too close to the aperture, we can assume that $k = 2\pi/\lambda = 2\pi\nu/c >> 1/R, 1/R'$. Consequently,

$$\Gamma(\mathbf{r}, \mathbf{r}', \tau) = \frac{1}{4\pi^2}\int_0^\infty \int_S \int_S \frac{(-ikR)(ikR')\cos\theta\cos\theta'}{R^2R'^2}\Gamma(\mathbf{s}, \mathbf{s}', \nu) \times e^{-2\pi i\nu[\tau+(R-R')/c]}dSdS'd\nu. \tag{23.166}$$

Writing $ik\cos\theta/2\pi R = K$; $ik\cos\theta'/2\pi R' = K'$ gives

$$\Gamma(\mathbf{r}, \mathbf{r}', \tau) = \int_S \int_S \Gamma\left[\mathbf{s}, \mathbf{s}', \left(\tau + \frac{R - R'}{c}\right)\right] K^*K'dSdS'. \tag{23.167}$$

The intensity at the point $\mathbf{r}$ is $\Gamma(\mathbf{r}, \mathbf{r}, 0)$, which from Eq. (23.167) gives

$$I(\mathbf{r}) = \int_S \int_S \sqrt{I(\mathbf{s})I(\mathbf{s}')}\gamma\left(\mathbf{s}, \mathbf{s}', \frac{R - R'}{c}\right) K^*K'dSdS'. \tag{23.168}$$

Note that all the quantities on the RHS are measurable.

If the illuminated surface consists of two pinholes 1 and 2, then from Eq. (23.168) performing the first integral over the surface S gives

$$I(\mathbf{r}) = \int_S \sqrt{I(\mathbf{s})I_1}\gamma\left(\mathbf{s}, 1, \frac{R - R_1}{c}\right) K^*K_1 + \sqrt{I(\mathbf{s})I_2}\gamma\left(\mathbf{s}, 2, \frac{R - R_2}{c}\right) K^*K_2, \tag{23.169}$$

and integrating again

$$
\begin{aligned}
I(\mathbf{r}) =& I_1 K_1^* K_2 \gamma(1,1,0) + \sqrt{I_2 I_1}\gamma\left[2,1,\frac{R_2 - R_1}{c}\right] K_2^* K_1 \\
&+ I_2 K_2^* K_2 \gamma(2,2,0) + \sqrt{I_2 I_2}\gamma\left[1,2,\frac{R_1 - R_2}{c}\right] K_1^* K_2.
\end{aligned}
\tag{23.170}
$$

Now $\gamma(1,1,0) = \gamma(2,2,0) = 1$ and $\gamma(1,2,(R_1 - R_2)/c) \equiv \gamma(1,2,\tau) = \gamma_{12}(\tau)$, where $\tau = (R_1 - R_2)/c$ and $\gamma(2,1,(R_2 - R_1/c) \equiv \gamma(2,1,-\tau) = \gamma_{12}^*(\tau)$. Hence

$$
I(\mathbf{r}) = | K_1 |^2 I_1 + | K_2 |^2 I_2 + 2\sqrt{I_1 I_2}\mathscr{R}[K_1^* K_2 \gamma_{12}(\tau)],
\tag{23.171}
$$

a similar result to Eq. (23.62). We can identify K_1, K_2 with the complex propagation factors a_1, a_2 previously introduced.

23.10 van Cittert–Zernike Theorem

An important feature of the mutual coherence function is the acquisition of partial coherence by the very process of propagation.

If the surface S is that of the source itself, for example a star, then clearly $\Gamma(\mathbf{s},\mathbf{s}',\nu)$ vanishes unless $\mathbf{s}$ and $\mathbf{s}'$ are sufficiently close together, say no more than a meter apart. Physically this simply means that the light sources are statistically independent over any macroscopic interval.

In the integral

$$
\int_S \int_S \frac{\partial G}{\partial n}(\mathbf{r},\mathbf{s},\nu)\frac{\partial G^*}{\partial n'}(\mathbf{r}',\mathbf{s}',\nu)\Gamma(\mathbf{s},\mathbf{s}',\nu)dSdS'
$$

we can use

$$
\Gamma(\mathbf{s},\mathbf{s}',\nu) = \delta_s(\mathbf{s}-\mathbf{s}')\,[\Gamma(\mathbf{s},\nu)]\,,
\tag{23.172}
$$

where δ_s denotes a surface δ function. Thus the integral becomes

$$
\Gamma(\mathbf{r},\mathbf{r}',\nu) = \int_S \frac{\partial G}{\partial n}(\mathbf{r},\mathbf{s},\nu)\frac{\partial G^*}{\partial n}(\mathbf{r}',\mathbf{s},\nu)\Gamma(\mathbf{s},\nu)dS,
\tag{23.173}
$$

which need not vanish for $\mathbf{r} \neq \mathbf{r}'$, so the field is now partially coherent.

With our previous assumptions, i.e., that $k \gg 1/R$, and using the same 'mirror image' Green's function as before, Eq. (23.166) gives

$$
\begin{aligned}
\Gamma(\mathbf{r},\mathbf{r}',\tau) = & \int_0^\infty d\nu e^{-2\pi i\nu\tau} \int_S \int_S \Gamma(\mathbf{s},\mathbf{s}',\nu) \\
& \times e^{-ik(R-R')} K^* K' dSdS'.
\end{aligned}
\tag{23.174}
$$

The intensity at the point $\mathbf{r}$ is $\Gamma(\mathbf{r},\mathbf{r},0)$, which from Eq. (23.170) gives

$$
I(\mathbf{r}) = \int_0^\infty d\nu \int_S \int_S \Gamma(\mathbf{s},\mathbf{s}',\nu) \mid K \mid^2 dSdS'
\tag{23.175}
$$

and if $\Gamma(\mathbf{s},\mathbf{s}',\nu)$ is a surface δ-function.

$$
I(\mathbf{r}) = \int_0^\infty d\nu \int_S \Gamma(\mathbf{s},\mathbf{s},\nu) \mid K \mid^2 dS = \int_S I(\mathbf{s}) \mid K \mid^2 dS,
\tag{23.176}
$$

where we have used

$$\Gamma(\mathbf{s}, \mathbf{s}, \tau = 0) = I(\mathbf{s}) = \int_0^\infty \Gamma(\mathbf{s}, \mathbf{s}, \nu) d\nu. \tag{23.177}$$

Eq. (23.176) is on examination an obvious result: the intensity observed from an extended, incoherent source is found by integating the intensities produced by each infinitesimal surface element of the source over the area of the source.

If the extended, spatially incoherent source is quasi-monochromatic, then provided $[\tau + (R - R')/c] \gg 1/\Delta\nu$, Eq. (23.174) gives

$$\Gamma(\mathbf{r}, \mathbf{r}', \tau) = \int_0^\infty e^{-2\pi i \nu \tau} d\nu \int_S e^{-ik(R-R')} K^* K dS. \tag{23.178}$$

If the source is monochromatic $I(\mathbf{s}, \nu)$ is only significant for frequencies ν near ν_0, therefore

$$\Gamma(\mathbf{r}, \mathbf{r}', \tau) = e^{-2\pi i \nu_0 \tau} |\overline{K}|^2 \int_S I(\mathbf{s}) e^{-i\bar{k}(R-R')} dS, \tag{23.179}$$

where we have assumed small angles, such that $\cos\theta, \cos\theta' \approx 1$ and $\bar{k}$ is the value of k appropriate to ν_0, similarly for $\overline{K}$.

The degree of coherence is

$$\gamma(\mathbf{r}, \mathbf{r}', \tau) = \gamma_{12}(\tau) = \frac{\Gamma(\mathbf{r}, \mathbf{r}', \tau)}{\sqrt{I(\mathbf{r}) I(\mathbf{r}')}}. \tag{23.180}$$

Writing $\gamma_{12}(\tau) = \gamma_{12}(0) \exp(-2\pi i \nu_0 \tau)$, where $\gamma_{12}(0)$ is the *degree of spatial coherence*

$$\gamma_{12}(0) = \frac{|\overline{K}|^2}{\sqrt{I(\mathbf{r}) I(\mathbf{r}')}} \int_S I(\mathbf{s}) e^{-i\bar{k}(R-R')} dS. \tag{23.181}$$

So, the degree of spatial coherence of an extended monochromatic source can be calculated from the intensity distribution over the source.

We can rewrite Eq. (23.179) for the mutual coherence function $\Gamma(\mathbf{r}, \mathbf{r}', \tau)$ as

$$\Gamma(\mathbf{r}, \mathbf{r}', \tau) = \Gamma(\mathbf{s}, \mathbf{s}, \tau) \int_S e^{-i\bar{k}(R-R')} K^* K' dS \tag{23.182}$$

provided $I(\mathbf{s}, \nu)$ does not depend on $\mathbf{s}$, where $\Gamma(\mathbf{s}, \mathbf{s}, \tau)$ is the Fourier transform of $I(\mathbf{s}, \nu)$.

For such a uniform source

$$\gamma(\mathbf{r}, \mathbf{r}', \tau) = \gamma(\tau) \gamma(\mathbf{r}, \mathbf{r}', 0), \tag{23.183}$$

$$\gamma(\tau) = \Gamma(\mathbf{s}, \mathbf{s}, \tau)/I(\mathbf{s}), \tag{23.184}$$

and

$$\gamma(\mathbf{r}, \mathbf{r}', 0) = \gamma_{12}(0), \tag{23.185}$$

where $\gamma(\tau)$ is the degree of temporal coherence of the source and $\gamma(\mathbf{r}, \mathbf{r}', 0)$ is the degree of spatial coherence. $\gamma(\tau)$ is a function of $\Delta\nu$, whilst $\gamma(\mathbf{r}, \mathbf{r}', 0)$ is a function of the geometric properties of the source.

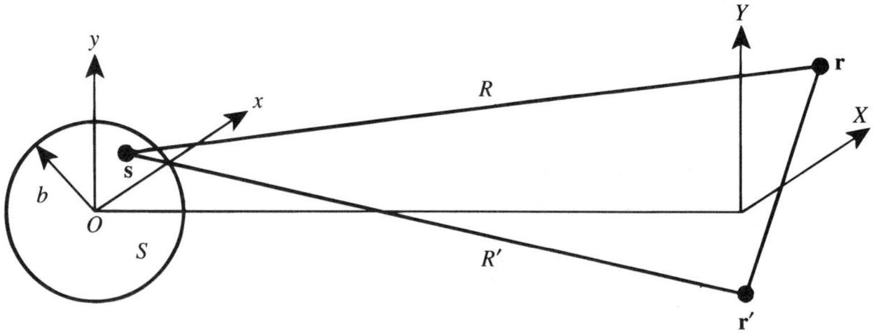

Fig. 23.11. Geometry of a circular source centered at O relative to two observation points $\mathbf{r}$ and $\mathbf{r}'$.

23.11 Spatial Coherence of a Quasi-Monochromatic, Uniform, Spatially Incoherent Circular Source

The geometry of the source and its location relative to the observation points $\mathbf{r}$ and $\mathbf{r}'$ is shown in Fig. (23.11).

The degree of spatial coherence between the points $\mathbf{r}, \mathbf{r}'$ is

$$\gamma(\mathbf{r}, \mathbf{r}', 0) = \gamma_{12}(0) = \frac{1}{\sqrt{I(\mathbf{r})I(\mathbf{r}'}} \int_S I(\mathbf{s}) e^{-i\bar{k}(R-R')} K^* K' dS, \tag{23.186}$$

where

$$I(\mathbf{r}) = \int_S I(\mathbf{s}) \mid K \mid^2 dS. \tag{23.187}$$

For a uniform source

$$I(\mathbf{r}) = I_S \int_S \mid K \mid^2 dS. \tag{23.188}$$

If we make the far field approximation, namely $R^2 = R'^2 \gg S$ (area of source) and $R^2 \gg (R - R')^2$, then $I(\mathbf{r}) = I(\mathbf{r}') = I|K|^2$, where $I = \pi b^2 I_S$.

Further, in the far field approximation, for a quasi-monochromatic source K^*K' is constant, $=\mid \overline{K} \mid^2$, and

$$\gamma_{12}(0) = \frac{1}{\pi b^2} \int_S e^{-i\bar{k}(R-R')} dS. \tag{23.189}$$

It is left as an exercise for the reader to prove that

$$\bar{k}(R - R') = \bar{k}(\mid \mathbf{r} - \mathbf{s} \mid - \mid \mathbf{r}' - \mathbf{s} \mid) \simeq -\frac{\bar{k}}{R} \mathbf{s} \cdot (\mathbf{r} - \mathbf{r}'). \tag{23.190}$$

Integrating over the circular source

$$\gamma_{12}(0) = \frac{\mid K \mid^2 \int_0^b \int_0^{2\pi} I_S e^{(-i\bar{k}/R) sr \cos\phi} s ds d\phi}{I_S \pi b^2 \mid K \mid^2}. \tag{23.191}$$

The inner integral in Eq. (23.191) can be related to the Bessel function of zero order[23.11] giving

$$\gamma_{12}(0) = \frac{2}{I_S b^2} \int_0^b I_S J_0 \left(\frac{\bar{k} s r}{R} \right) s ds \tag{23.192}$$

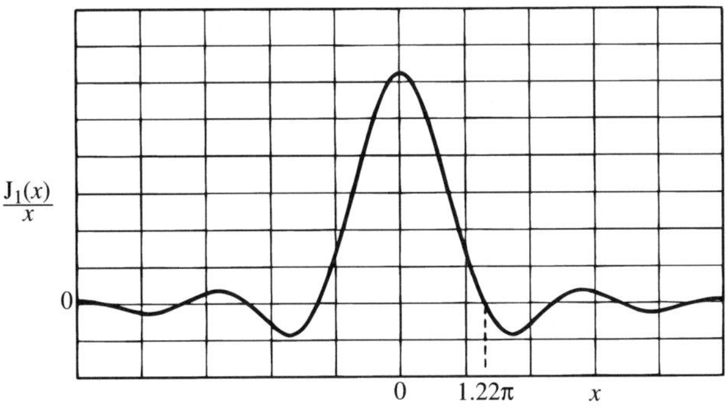

Fig. 23.12. Spatial variation of the coherence of a circular source as predicted by Eq. (23.194) (the Airy function).

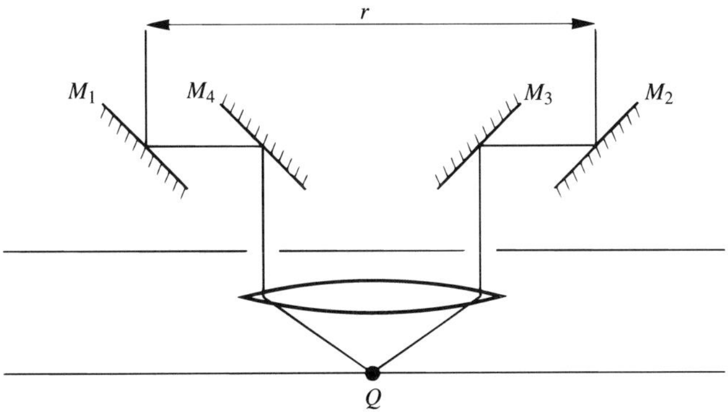

Fig. 23.13. Schematic of Michelson stellar interferometer used to study the spatial coherence between two points separated by a distance r.

and[23.11]

$$\gamma_{12}(0) = \frac{2}{b^2}\left(\frac{Rb}{\bar{k}r}\right) J_1\left(\frac{\bar{k}br}{R}\right) \tag{23.193}$$

which can be written as

$$\gamma_{12}(0) = \frac{2J_1(x)}{x} \tag{23.194}$$

where $x = \bar{k}br/R$ and $r = |\mathbf{r} - \mathbf{r}'|$. This function is shown in Fig. (23.12).

Eq. (23.194) occurs in a similar form in the calculation of the far field (Fraunhofer) diffraction pattern of a circular aperture – the Airy† diffraction pattern.

Eq. (23.194) can be used to determine the angular diameter of distant objects, such as stars, in the Michelson stellar interferometer, which allows the visibility of interference fringes produced from laterally separated points in an incoming phase front from the star to be studied as a function of lateral separation, as shown in Fig. (23.13).

The visibility of the interference fringes first vanishes when $J_1(x) = 0$, i.e., when $x = 3.83 = 1.22\pi$. This occurs when

$$\frac{b}{R} = \alpha = \frac{0.61\bar{\lambda}}{r}, \tag{23.195}$$

† Named for the astronomer Sir George Airy.

where α is the angular diameter of the star. The drawback of this technique is that the visibility of the fringes is also dependent on temporal coherence, which is affected by atmospheric turbulence, limiting the method to large stars.

For example, for a star the size of the sun $b \approx 10^9$ m at a distance of 1 light year ($R \approx 10^{16}$ m) and for a mean $\bar{k} \approx 10^7$ m^{-1}($\bar{\nu}_0 \approx 10^{15}$ Hz), the distance r for the visibility first to vanish is somewhat less than 4 m.

23.12 Intensity Correlation Interferometry

In the Michelson stellar interferometer the fundamental quantity that determines the visibility of the interference fringes is, just as it was in the case of the Young's interference experiment previously considered, $\mathscr{R}\Gamma(\mathbf{r}, \mathbf{r}', \tau)$. If the optical paths are equal, then in theory the visibility is determined by spatial coherence alone, in other words by

$$\gamma_{12}(0) = \frac{\Gamma(\mathbf{r}, \mathbf{r}', 0)}{\sqrt{I(\mathbf{r})I(\mathbf{r}')}}. \tag{23.196}$$

However, in practice atmospheric turbulence degrades the degree of coherence between the two points $\mathbf{r}$, $\mathbf{r}'$ below the value that would be expected if the waves propagated through vacuum the entire distance from the source to $\mathbf{r}$ and $\mathbf{r}'$.

Now $\Gamma(\mathbf{r}, \mathbf{r}', \tau) = \overline{V^*(\mathbf{r}, t)V(\mathbf{r}', t')}$, where $t' - t = \tau$. The effect of atmospheric turbulence can be considered to modify the analytic signal at point $\mathbf{r}$ by a random, rapidly varying, phase factor, so $V(\mathbf{r}, t)$ changes according to

$$V(\mathbf{r}, t) \rightarrow V(\mathbf{r}, t)e^{i\phi(t)}, \tag{23.197}$$

where $\phi(t) = \phi(\mathbf{r}, t)$ is the phase factor at point $\mathbf{r}$. Consequently

$$\Gamma_{obs}(\mathbf{r}, \mathbf{r}', \tau) = e^{i[\phi(t) - \phi'(t')]}\Gamma_{source}(\mathbf{r}, \mathbf{r}', \tau). \tag{23.198}$$

Even when $t = t'$, the rapidly varying phase factor $e^{i[\phi(t) - \phi'(t']}$ need not vanish as the functions ϕ and ϕ' could be different and largely independent, as happens for a spatial interval greater than a few meters. Thus, amplitude correlation interferometry breaks down because $\mathscr{R}\Gamma_{obs}$ is no longer equal to $\mathscr{R}\Gamma_{source}$.

The effect of random atmospheric-induced fluctuations on the analytic signals can be largely eliminated by using *intensity correlation interferometry*. In such experiments the quantity that is measured is $\mid \gamma(\mathbf{r}, \mathbf{r}', \tau) \mid^2$, generally for $\tau = 0$.

Since $\gamma(\mathbf{r}, \mathbf{r}', \tau) = \Gamma(\mathbf{r}, \mathbf{r}', \tau)/\sqrt{I(\mathbf{r})}$, when

$$\Gamma_{obs}(\mathbf{r}, \mathbf{r}', \tau) = e^{i[\phi(t) - \phi'(t')]}\Gamma_{source}(\mathbf{r}, \mathbf{r}', \tau), \tag{23.199}$$

it is clear that

$$\mid \Gamma_{obs}(\mathbf{r}, \mathbf{r}', \tau) \mid = \mid \Gamma_{source}(\mathbf{r}, \mathbf{r}', \tau) \mid, \tag{23.200}$$

so that the phase distortions that plague amplitude correlation interferometry are rendered harmless. The technique of intensity correlation interferometry was developed by R. Hanbury-Brown and R.Q. Twiss[23.11] and demonstrated in several interesting experiments beginning in the mid-1950s. Their first experiments were performed with radio waves and later ones used visible light. Using these techniques they were able to extend the baseline of their stellar interferometer to 200 m and thus were able to measure stars of much smaller angular diameter than with amplitude correlation interferometry.

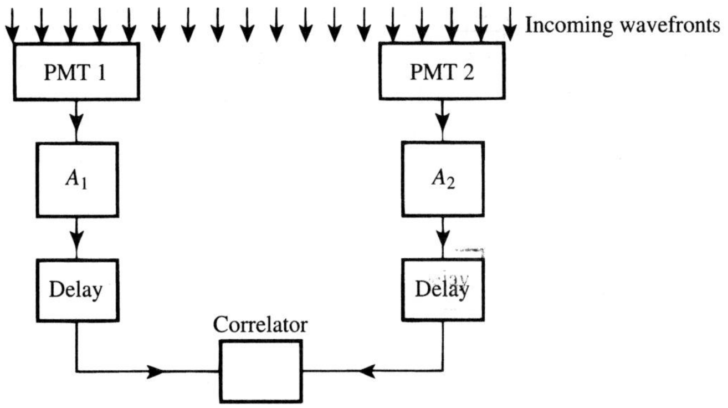

Fig. 23.14. Intensity correlation (Hanbury-Brown–Twiss) interferometer for use with visible light.

The Hanbury-Brown–Twiss (HBT) interferometer measures the angular diameter of stars by measuring $| \gamma_{12}(0) |^2$. As we have already shown $| \gamma_{12} |$ first vanishes for a $| \mathbf{r} - \mathbf{r}' |$ separation of $\bar{k}b \mid \mathbf{r} - \mathbf{r}' \mid /R = 3.83$, i.e., $\alpha = b/R = 0.61\bar{\lambda}/|\mathbf{r} - \mathbf{r}'|$, so by determining the first vanishing point of $| \gamma_{12}(0)|^2$ the angular diameter can be determined.

The principle of the HBT interferometer is illustrated in Fig. (23.14). The two time delays allow compensation to be made for differences in the time of arrival of the wavetrains at the two phototubes. In Hanbury–Brown and Twiss's first optical experiments the correlations were determined from the analog signals from the photomultipliers, although in later experiments the ability of photomultipliers to detect individual photons was used to allow correlations to be measured from photoelectron count fluctuations. We shall examine this situation in detail in Section 23.16.

23.13 Intensity Fluctuations

In an experiment in which an optical signal is being detected by a photodetector, we expect the intensity falling on the photodetector to fluctuate. Fast photodectors, such as photomultipliers, avalanche photodiodes and *p–i–n* photodiodes, can follow such intensity fluctuations on quite a fast time scale, but not as fast as an optical period. If the source is quasi-monochromatic it emits many wavetrains of different frequencies lying within some frequency range $\Delta\nu$. Because there is normally no definite phase relationship between these different frequency components, the resultant real wave amplitude $V^{(r)}(t)$ must be regarded as a stochastic time function, which can be expanded as a Fourier series in a long interval T in the form

$$V^{(r)}(t) = \sum_n a_n \cos\left(\frac{2\pi nt}{T}\right) + b_n \sin\left(\frac{2\pi nt}{T}\right). \tag{23.201}$$

The Fourier coefficients a_n and b_n can be regarded as statistically independent Gaussian variables with the same variance, provided $V^{(r)}(t)$ is stationary and T is sufficiently long. Such assumptions apply mainly to spontaneous emission from sources in thermal equilibrium and not to lasers. We call the light from such sources *chaotic light*. If the field $V^{(r)}(t)$ is quasi-monochromatic, then in the slowly varying envelope approximation we can write

$$V(t) = U(t)e^{2\pi i\nu_0 t}, \tag{23.202}$$

where ν_0 is the center frequency and $U(t)$ is the envelope function, which is nearly constant over any interval short compared with $1/\Delta\nu(\approx \tau_c)$; note that $|V(t)| = |U(t)|$. In other words, the intensity fluctuations of a quasi-monochromatic source occur on a time scale greater than the coherence time of the source.

We had previously, from Eq. (23.41),

$$\begin{aligned}\Gamma(\mathbf{r},\mathbf{r}',\tau) &= \int_0^\infty \Gamma(\mathbf{r},\mathbf{r}',\nu)e^{-2\pi i\nu\tau}d\nu \\ &= e^{-2\pi i\nu_0\tau}\int_{-\nu_0}^\infty \Gamma(\mathbf{r},\mathbf{r}',\nu_0+\nu')e^{-2\pi i\nu'\tau}d\nu'\end{aligned} \tag{23.203}$$

where we have put $\nu = \nu_0 + \nu'$.

If the field is quasi-monochromatic, the integral in Eq. (23.203) vanishes for any ν' that is much outside the range $|\nu'| \lesssim \Delta\nu$. In this frequency range $2\pi\nu'\tau << 1$ provided $\tau << \tau_c$.

Hence,

$$\begin{aligned}\Gamma(\mathbf{r},\mathbf{r}',\tau) &= e^{2\pi i\nu_0\tau}\int_{-\nu_0}^\infty \Gamma(\mathbf{r},\mathbf{r}',\nu_0+\nu')d\nu' \\ &= e^{2\pi i\nu_0\tau}\Gamma(\mathbf{r},\mathbf{r}',0) \text{ for } \tau \ll \tau_c.\end{aligned} \tag{23.204}$$

Now, the autocorrelation function of $U(t)$ is

$$\begin{aligned}\overline{U^*(t)U(t+\tau)} &= \overline{V^*(t)V(t+\tau)e^{2\pi i\nu_0\tau}} \\ &= e^{2\pi i\nu_0\tau}\Gamma(\mathbf{r},\mathbf{r},\tau) \\ &= \Gamma(\mathbf{r},\mathbf{r},0) = \Gamma_{11}(0) \text{ for } \tau \ll \frac{1}{\Delta\nu}.\end{aligned} \tag{23.205}$$

In other words, $U(t)$ is a slowly varying function over time intervals $\approx \tau \ll 1/\Delta\nu$. We can identify $|U(t)|^2$ as the instantaneous value of the intensity $I(t)$, a slowly varying function of t for $t \ll \tau_c$.

Now, we previously quoted the result†

$$\overline{V_1^{(r)}(t)V_2^{(i)}(t+\tau)} = -\overline{V_1^{(i)}(t)V_2^{(r)}(t+\tau)} \tag{23.206}$$

if we let $V_1 = V_2$ and $\tau = 0$

$$\overline{V_1^{(r)}(t)V_1^{(i)}(t)} = -\overline{V_1^{(i)}(t)V_1^{(r)}(t)} \tag{23.207}$$

and the values of $V_1^{(r)}(t)$ and $V_1^{(i)}(t)$ at the same time t have no correlation.

If $V_1^{(r)}(t)$ is Gaussian then $V_1^{(i)}(t)$ is Gaussian also and both fluctuating quantities have the same variance σ^2. Thus, we can write the joint probability distribution of $V^{(r)}(t)$ and $V^{(i)}(t)$ as

$$\begin{aligned}&p\left[V^{(r)}(t), V^{(i)}(t)\right] dV^{(r)}dV^{(i)}(t) \\ &\quad = \frac{1}{2\pi\sigma^2}\exp\frac{-[V^{(r)^2}(t)+V^{(i)^2}(t)]}{2\sigma^2}dV^{(r)}(t)dV^{(i)}(t).\end{aligned} \tag{23.208}$$

† Results such as this are easily verified by, for example, setting $V^{(r)}(t) = \cos\omega t$; $V^{(i)}(t) = \sin\omega t$.

Fig. 23.15. Coordinate system for transforming the probability density $V(t)$ into polar coordinates.

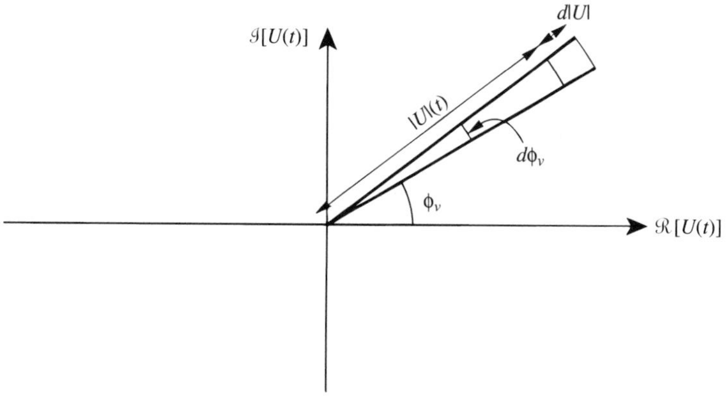

Now

$$| U(t) |^2 = | V(t) |^2 = \frac{1}{4}\left[V^{(r)^2}(t) + V^{(i)^2}(t)\right]. \tag{23.209}$$

If we write $U(t)$ in the form

$$U(t) = | U(t) | e^{i\phi_u} \tag{23.210}$$

then

$$V(t) = U(t)e^{2\pi i\nu_0 t} = | U(t) | e^{i(\phi_u + 2\pi\nu_0 t)}. \tag{23.211}$$

Hence, $\phi_u = \phi_v - 2\pi\nu_0 t$, where ϕ_v is defined from

$$V(t) = | V(t) | e^{i\phi_v}. \tag{23.212}$$

We can transform the joint probability distribution Eq. (23.208) into polar coordinates as shown in Fig. (23.15) and get

$$\begin{aligned} &p(| V(t) |, \phi_v)d | V(t) | d\phi_v \\ &= \frac{1}{2\pi\sigma^2} | V(t) | \exp\left[\frac{- | V(t) |^2}{2\sigma^2}\right] d | V(t) | d\phi_v, \end{aligned} \tag{23.213}$$

which gives

$$\begin{aligned} p(| U(t) |, \phi_u + 2\pi\nu_0 t)d | U(t) | d\phi_u &= \frac{1}{2\pi\sigma^2} | U(t) | \\ \times \exp\left[\frac{- | U(t) |^2}{2\sigma^2}\right] d | U(t) | d\phi_u. \end{aligned} \tag{23.214}$$

Note that the argument $\phi_u + 2\pi\nu_0 t$ does not appear as a variable in this probability distribution. Hence, all arguments are equally probable and we can write

$$p(| U |, \phi_u)d|u|d\phi_u = \frac{1}{2\pi\sigma^2} | U | \left(\exp \frac{- | U |^2}{2\sigma^2}\right) d | U | d\phi_u \tag{23.215}$$

and integrating over all values of the argument $(0 \to 2\pi)$

$$p|U|d|U| = \frac{| U |}{\sigma^2} \exp\left(\frac{- | U |^2}{2\sigma^2}\right) d | U |, \tag{23.216}$$

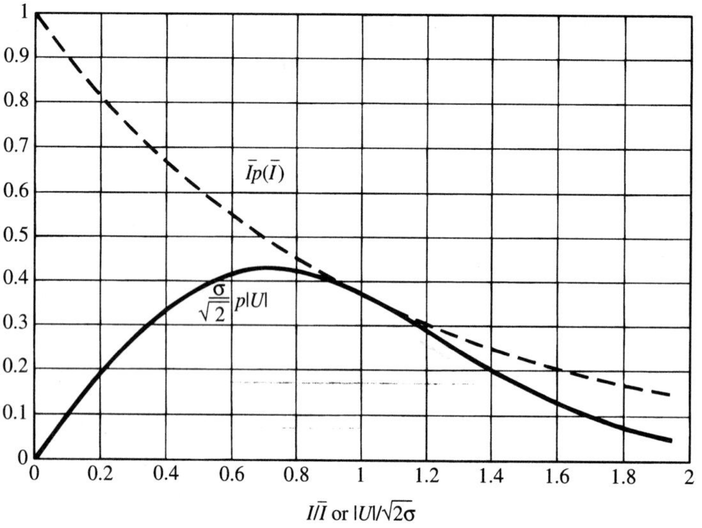

Fig. 23.16. Probability distribution for the analytic signal amplitude $|u|$, called the Rayleigh distribution, and for the instantaneous intensity I, of chaotic light.

which is called the Rayleigh distribution, shown in Fig. (23.16).

The distribution of $| U |^2$ is, from Eq. (23.216)

$$p|U|^2 d|U|^2 = \frac{1}{2\sigma^2} \exp\left(\frac{-| U |^2}{2\sigma^2}\right) d \mid U \mid^2 . \tag{23.217}$$

The average value of $|U|^2$ is

$$\overline{U^2} = \int_0^\infty \frac{| U |^2}{2\sigma^2} \exp\left(\frac{-| U |^2}{2\sigma^2}\right) d \mid U^2 \mid = 2\sigma^2. \tag{23.218}$$

Consequently, since $|U|^2 = I$, and writing $\bar{I} = 2\sigma^2$ the distribution of the instantaneous intensity is

$$p(I)dI = \frac{1}{\bar{I}} \exp\left(-\frac{I}{\bar{I}}\right) dI, \tag{23.219}$$

also shown in Fig. (23.16).

23.14 Photon Statistics

Since photomultiplier tubes have the ability to detect individual photons, albeit not with 100% quantum efficiency, it is important to determine how intensity fluctuations $I(t)$ influence the detection of photoelectron counts from an illuminated phototube.

Although the phenomenon of photoemission is fundamentally quantum mechanical in nature, we can still consider the statistics of the photoelectrons classically by making the plausible assumption that photoelectron counts are directly proportional to the incident intensity. Consequently, the statistical distribution of photoelectron counts reflects the statistics of the electromagnetic field illuminating the photocathode. As far as the full quantum mechanical treatment of this problem is considered correlations between classical analytic fields become coincidences between detected photons at different points or delayed coincidences in time between photons detected at the same point.

The simplest type of photon statistics results when the source is of constant intensity over the observation period. For relatively long observation periods T this implies temporal coherence of the source such that $\tau_c > T$.

23.14.1 Constant Intensity Source

If we make the plausible assumption that the rate of observed photoelectron counts is directly proportional to the instantaneous intensity, then the differential probability dp that a count is observed in a time interval dt when a phototube is illuminated with radiation intensity $I(t)$ is

$$dp(t) = \alpha I(t)dt = \alpha V^*(t)V(t)dt, \tag{23.220}$$

where the parameter α is a measure of the quantum efficiency of the detector and also depends on the area of the detector, the average spectral characteristics of the incident radiation, the operating conditions of the phototube and its associated pulse counting electronics, etc.

If there are no random fluctuations in the intensity $I(t)$ then it is reasonable to assume that the probabilities of detecting photons in distinct time intervals are statistically independent. The probability that no count occurs in the time interval dt' at time t' is $1 - dp(t')$. Thus the joint probability that no count occurs in an entire interval from t to $t+T$ is

$$\prod_{k=0}^{n}\left[1 - dp\left(t + \frac{kT}{n}\right)\right] = P(0,T,t). \tag{23.221}$$

In the limit as $dp \to 0$, and $n \to \infty$ we can write

$$\ln P = \sum \ln[1 - dp(t')] = \sum_{\lim dp \to 0} -dp(t') = -\int_t^{t+T} dp(t'), \tag{23.222}$$

giving

$$P(0,T,t) = \exp\left[-\int_t^{t+T} dp(t')\right] = \exp\left[-\alpha\int_t^{t+T} I(t')dt'\right], \tag{23.223}$$

where it must be stressed that we have assumed that there are no intensity fluctuations.

In a similar way the probability that one count occurs between t and $t+T$ is

$$\begin{aligned} P(1,T,t) &= \sum_{k''=0}^{\infty} dp\left(\frac{k''T}{n} + t\right)\prod_{k'=0}^{\infty}\left[1 - dp\left(t + \frac{k'T}{n}\right)\right] \\ &= \int_t^{t+T} dp(t'')\exp\left[-\int_t^{t+T} dp(t')\right] \end{aligned} \tag{23.224}$$

in the limit as $k'', k' \to \infty, dp \to 0, n \to \infty$. We have used the fact that the probability of obtaining a count in a time interval dt'' at time t'' and at no other times within the interval $t \to t+T$ is the same as the probability of a count at t'' in dt'' multiplied by the probability of no counts in the rest of the interval $t \to t+T$. Summing this probability over all infinitesimal time intervals dt'' within the range $t \to t+T$ gives the integral in Eq. (23.224).

Thus,

$$P(1, T, t) = \alpha \int_t^{t+T} I(t'')dt'' \exp\left[-\alpha \int_t^{t+T} I(t')dt'\right]. \qquad (23.225)$$

In a similar way we can show that the probability of obtaining n counts in the interval $t \to t + T$ is

$$P(n, T, t) = \frac{1}{n!}\left(\alpha \int_t^{t+T} I(t'')dt''\right)^n \exp\left[-\alpha \int_t^{t+T} I(t')dt'\right], \qquad (23.226)$$

where we have made use of the fact that n counts observed at times $t_1, t_2 ... t_n$ in intervals $dt_1, dt_2 ... dt_n$ can be arranged among themselves in $n!$ different ways.

If we put $\mu = \alpha \int_t^{t+T} I(t')dt'$, then

$$P(n, T, t) = \frac{\mu^n}{n!} e^{-\mu}. \qquad (23.227)$$

This is a Poisson distribution, which we write as $P(n, \mu)$.

The mean number of counts observed in the interval $t \to t + T$ is

$$\overline{n} = \sum_{n=0}^{\infty} nP(n, \mu) = \sum_{n=0}^{\infty} \frac{\mu^n}{(n-1)!} e^{-\mu} = \mu. \qquad (23.228)$$

The mean-square number is

$$\begin{aligned}\overline{n^2} &= \sum_{n=0}^{\infty} n^2 P(n, \mu) \\ &= \sum_{n=0}^{\infty} \frac{n\mu^n}{(n-1)!} e^{-\mu} \\ &= \sum_{n=0}^{\infty} [(n-1)+1] \frac{\mu^n}{(n-1)!} e^{-\mu} = \mu^2 + \mu. \end{aligned} \qquad (23.229)$$

The variance of the distribution of n is

$$\sigma^2 = \overline{(n - \overline{n})^2} = \overline{n^2} - (\overline{n})^2. \qquad (23.230)$$

This quantity is a measure of the fluctuations in n about its mean value. For a Poisson distribution, from Eqs. (23.228) and (23.229)

$$\sigma^2 = \mu^2 + \mu - \mu^2 = \mu = \overline{n}. \qquad (23.231)$$

23.14.2 Random Intensities

We have shown that when the light intensity has no random fluctuations

$$P(n, T, t) = \frac{1}{n!}\left(\alpha \int_t^{t+T} I(t')dt'\right)^n \exp\left[-\alpha \int_t^{t+T} I(t')dt'\right]. \qquad (23.232)$$

If $I(t)$ is constant $= I$ then

$$P(n, T, t) = \frac{1}{n!}(\alpha I T)^n e^{-\alpha I T}, \qquad (23.233)$$

and it is clear that the choice of time t is unimportant.

However, when $I(t)$ is fluctuating, it is apparent that $P(n, T, t)$ will depend on the choice of t. It can be shown, by dividing the time interval T into small intervals, that even in this case $P(n, T, t)$ has the same form as Eq. (23.233). In the case of fluctuating intensity,

$$P(n, T, t) = \frac{1}{n!}(\alpha U_I)^n e^{-\alpha U_I}, \tag{23.234}$$

where

$$U_I = \int_t^{t+T} I(t')dt'. \tag{23.235}$$

In practice, the quantity $P(n, T, t)$ is not likely to be measured, as this would require many identical detectors simultaneously irradiated coherently with the same radiation field at time t. What is usually measured is $P(n, T)$ which is an ensemble or time average of $P(n, T, t)$ that takes into account the distribution of $I(t)$, and consequently of U_I.

If $I(t)$ is a random variable then U_I will also be a random variable with some distribution $p(U)$. Thus, if we average over the ensemble of U_I,

$$P(n, T) = \int_0^\infty \frac{(\alpha U_I)^n}{n!} e^{-\alpha U_I} p(U_I)du. \tag{23.236}$$

Even when $I(t)$ fluctuates, it remains essentially constant for time intervals $T \ll \tau_c$. In this case

$$U_I = \int_t^{t+T} I(t')dt' = I(t)T = IT \tag{23.237}$$

and $p(U_I) = p(I)$, which we previously derived. Therefore,

$$P(n, T) = \int_0^\infty \frac{(\alpha I T)^n}{n!} e^{-\alpha IP} p(I)dI, \tag{23.238}$$

where we stress that this applies to $T \ll \tau_c$.

If we substitute $p(I)$ from Eq. (23.219) in Eq. (23.238) and integrate we get

$$P(n, T) = \frac{(\alpha \bar{I} T)^n}{(1 + \alpha \bar{I} T)^{n+1}}, \tag{23.239}$$

where

$$\bar{I} = \int_0^\infty I p(I)dI. \tag{23.240}$$

We can recognize $\alpha \bar{I} T$ as $\bar{n}$, the average number of counts during the period T, so

$$p(n, T) = \frac{\bar{n}^n}{(1 + \bar{n})^{n+1}}. \tag{23.241}$$

This is the Bose–Einstein distribution for n identical particles in one quantum state[23.12].

Note that

$$\bar{n} = \sum_{n=0}^\infty nP(n, T) = \int_0^\infty \sum_{n=0}^\infty \frac{n(\alpha I T)^n}{n!} e^{-\alpha I T} p(I)dI$$

$$= \int_0^\infty \alpha I T p(I)dI = \alpha T \bar{I}. \tag{23.242}$$

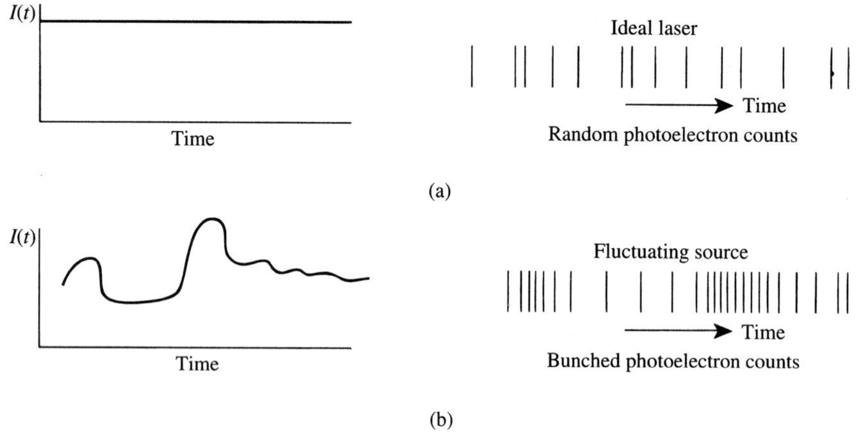

Fig. 23.17. Schematic variation of photoelectron counts from (a) a constant intensity source, such as an ideal laser, where the distribution of counts in Poissonian, and (b) a source whose intensity is fluctuating, which gives rise to photoelectron 'bunching.'

The mean-square number of counts is

$$\overline{n^2} = \sum_{n=0}^{\infty} n^2 P(n,T) = \int_0^{\infty} \left(\alpha^2 T^2 I^2 + \alpha T I\right) p(I) dI$$

$$= \alpha^2 T^2 \overline{I^2} + \alpha T \bar{I}, \qquad (23.243)$$

and the variance is

$$\sigma^2 = \overline{n^2} - (\bar{n})^2 = \alpha T \bar{I} + \alpha^2 T^2 [(\overline{I^2}) - (\bar{I})^2]. \qquad (23.244)$$

This always exceeds $\bar{n}$ unless $p(I)$ is a δ-function, $\delta(\bar{I} - I)$. For the thermal distribution $p(I) = e^{-I/\bar{I}}/\bar{I}$ and

$$\overline{I^n} = n!(\bar{I})^n. \qquad (23.245)$$

Consequently,

$$\sigma^2 = \alpha T \bar{I} + \alpha^2 T^2 [2(\bar{I})^2] = \bar{n}(1 + \bar{n}), \qquad (23.246)$$

which exceeds the value for a Poisson distribution. This implies that the fluctuations of the counts recorded by a phototube illuminated by Gaussian light depart from classical statistics in such a way that the variance always exceeds that given by a Poisson distribution. It follows that counts do not occur at random, but exhibit a bunching effect. Although this bunching effect is characteristic of Bose particles it is not for this reason that it occurs. The bunching is correlated with intensity fluctuations of the source as illustrated in Fig. (23.17). Indeed, no bunching is observed in the output of an ideal laser, which is still a source of Bose particles, as shown in Fig. (23.17). If two light beams are partially coherent, the bunching in the two beams will be correlated and it is this correlation that is responsible for the intensity correlations found by Hanbury–Brown and Twiss.

When bunching occurs, photocounts are no longer independent, once a count is detected there is an increased conditional probability of obtaining a second count within a short time $\tau \approx \tau_c$.

23.15 The Hanbury-Brown–Twiss Interferometer

This interferometer measures the intensity correlations between the waves illuminating two detectors. The quantity measured is

$$\overline{I_1(t)I_2(t+\tau)} = \frac{1}{16}\left\{\overline{V_1^{(r)^2}(t)V_2^{(r)^2}(t+\tau)} + \overline{V_1^{(i)^2}(t)V_2^{(r)^2}(t+\tau)} + \overline{V_1^{(r)^2}(t)V_2^{(i)2}(t+\tau)} + \overline{V_1^{(i)^2}(t)V_2^{(i)2}(t+\tau)}\right\} xi \tag{23.247}$$

Now, it can be shown that for two complex analytic signals $V_1(t)$ and $V_2(t)$

$$\overline{V_1(t)V_2(t+\tau)} = 0, \tag{23.248}$$

$$\overline{V_1^{(r)}(t)V_2^{(r)}(t+\tau)} = \overline{V_1^{(i)}(t)V_2^{(i)}(t+\tau)}, \tag{23.249}$$

$$\overline{V_1^{(i)}(t)V_2^{(r)}(t+\tau)} = -\overline{V_1^{(r)}(t)V_2^{(i)}(t+\tau)}, \tag{23.250}$$

and as we have mentioned previously

$$\overline{V_1^{(i)}(t)V_2^{(r)}(t)} = 0. \tag{23.251}$$

From these relations, if $\tau = 0$ and $V_1(t) = V_2(t)$

$$\overline{V_1^{(r)^2}(t)} = \overline{V_1^{(i)^2}(t)} = 2\overline{|\, V_1(t)\,|^2} = 2\overline{I_1}, \tag{23.252}$$

where $\overline{I_1}$ is the mean value of $I_1(t)$ (averaged over a time long compared with τ_c).

Now,

$$\begin{aligned}\Gamma_{12}(\tau) &= \overline{V_1^*(t)V_2(t+\tau)} \\ &= \frac{1}{4}\left\{\overline{V_1^{(r)}(t)V_2^{(r)}(t+\tau)} + \overline{V_1^{(i)}(t)V_2^{(i)}(t+\tau)} + i\overline{V_1^{(r)}(t)V_2^{(i)}(t+\tau)} - i\overline{V_1^{(i)}(t)V_2^{(r)}(t+\tau)}\right\},\end{aligned} \tag{23.253}$$

which using Eq. (23.250) gives

$$\Gamma_{12}(\tau) = \frac{1}{2}\overline{V_1^{(r)}(t)V_2^{(r)}(t+\tau)} - \frac{1}{2}i\overline{V_1^{(i)}(t)V_2^{(r)}(t+\tau)}. \tag{23.254}$$

Clearly,

$$\frac{1}{2}\overline{V_1^{(r)}(t)V_2^{(r)}(t+\tau)} = \mathscr{R}\Gamma_{12}(\tau) = \frac{1}{2}\overline{V_1^{(i)}(t)V_2^{(i)}(t+\tau)}, \tag{23.255}$$

$$-\frac{1}{2}\overline{V_1^{(i)}(t)V_2^{(r)}(t+\tau)} = \mathscr{I}\Gamma_{12}(\tau) = \frac{1}{2}\overline{V_1^{(r)}(t)V_2^{(i)}(t+\tau)}. \tag{23.256}$$

We now use the following relation that applies to Gaussian functions $X(t)$ and $Y(t)$, namely

$$\overline{X^2Y^2} = \overline{X^2}\;\overline{Y^2} + 2\overline{XY}^2. \tag{23.257}$$

If X and Y are independent, $\overline{XY} = 0$.

Applying this relation to the first of the four terms in Eq. (23.247) for

$\overline{I_1(t)I_2(t+\tau)}$

$$\overline{V_1^{(r)^2}(t)V_2^{(r)^2}(t+\tau)} = \overline{V_1^{(r)^2}(t)}\ \overline{V_2^{(r)^2}(t+\tau)} + 2[\overline{V_2^{(r)}(t)V_2^{r}(t+\tau)}]^2$$
$$= 4\bar{I}_1\bar{I}_2 + 2[\overline{V_1^{(r)}(t)V_2^{(r)}(t+\tau)}]^2$$
$$= 4\bar{I}_1\bar{I}_2 + 8\{\mathscr{R}[\Gamma_{12}(\tau)]\}^2. \quad (23.258)$$

Similarly, for the other three terms

$$\overline{V_1^{(i)^2}(t)V_2^{(r)^2}(t+\tau)} = 4\bar{I}_1\bar{I}_2 + 8\{\mathscr{I}\Gamma_{12}(\tau)\}^2, \quad (23.259)$$
$$\overline{V_1^{(r)^2}(t)V_2^{(i)^2}(t+\tau)} = 4\bar{I}_1 I_2 + 8\{\mathscr{I}\Gamma_{12}(\tau)\}^2, \quad (23.260)$$
$$\overline{V_1^{(i)^2}(t)V_2^{(i)^2}(t+\tau)} = 4\bar{I}_1\bar{I}_2 + 8\{\mathscr{R}\Gamma_{12}(\tau)\}^2. \quad (23.261)$$

Therefore,

$$\overline{I_1(t)I_2(t+\tau)} = \bar{I}_1\bar{I}_2 + |\ \Gamma_{12}(\tau)\ |^2. \quad (23.262)$$

We can normalize Eq. (23.262) to give

$$\frac{\overline{I_1(t)I_2(t+\tau)}}{\bar{I}_1\bar{I}_2} = 1 + |\gamma_{12}(\tau)|^2. \quad (23.263)$$

The left hand side of Eq. (23.263) defines the degree of second order coherence $\gamma_{12}^{(2)}(\tau)$, so

$$\gamma_{12}^{(2)}(\tau) = 1 + |\gamma_{12}(\tau)|^2. \quad (23.264)$$

Clearly, for $\tau \gg \tau_c, \gamma_{12}^{(2)} \to 1$, and $\gamma_{12}^{(2)}(0) = 2$. Eq. (23.264) holds true for all kinds of chaotic light, even if the fields are not Gaussian variates. For an ideal laser $\tau_c \to \infty$ and the degree of second order coherence $\gamma_{12}^{(2)}(\tau) = 2$. For special kinds of so-called *nonclassical* light, Eq. (23.264) does not hold and all that can be proven is that[23.13]

$$0 \le \gamma_{12}^{(2)}(\tau) < \infty. \quad (23.265)$$

If we measure intensity fluctuations $\Delta I_1(t)$ and $\Delta I_2(t)$ from the mean values $\bar{I}_1$ and $\bar{I}_2$ then

$$\Delta I_1(t) = I_1(t) - \bar{I}_1; \qquad \Delta I_2(t) = I_2(t) - \bar{I}_2; \quad (23.266)$$

$$\overline{\Delta I_1(t)} = 0; \qquad \overline{\Delta I_2(t) = 0}. \quad (23.267)$$

Hence,

$$\overline{I_1(t)I_2(t+\tau)} = \overline{[\bar{I}_1 + \Delta I_1(t)][\bar{I}_2(t+\tau) + \Delta I_2(t+\tau)]}$$
$$= \bar{I}_1\bar{I}_2 + \overline{\Delta I_1(t)\Delta I_2(t+\tau)}. \quad (23.268)$$

Consequently

$$\overline{\Delta I_1(t)\Delta I_2(t+\tau)} = |\ \Gamma_{12}(\tau)\ |^2. \quad (23.269)$$

By normalizing

$$\overline{\Delta I_1(t)\Delta I_2(t+\tau)} = \bar{I}_1\bar{I}_2\ |\ \gamma_{12}(\tau)\ |^2, \quad (23.270)$$

which therefore yields $|\ \gamma_{12}(\tau)\ |^2$, or $|\ \gamma_{12}(0)\ |^2$ if τ is set equal to zero. These fluctuations will be a function of the geometry and of the coherence time of the source.

The validity of the method of intensity correlation interferometry was first demonstrated for radio waves[23.11] and there was some doubt originally as to whether the method would still work in the optical region where the best available detectors (photomultiplier tubes) were inherently quantum mechanical in nature and detected photons rather than classical waves.

23.16 Hanbury-Brown–Twiss Experiment with Photon Count Correlations

When this experiment is performed by counting photons then the measured quantity can be written in terms of the mean number of electron counts as

$$\overline{I_1(t')I_2(t'+\tau)} \propto \overline{n_1 n_2}, \tag{23.271}$$

where n_1, n_2 are the number of counts recorded when the two phototubes are illuminated for a time T at the times $t', t'+\tau$, respectively.

$$\overline{n_1 n_2} = \sum_{n_1=0}^{\infty}\sum_{n_2=0}^{\infty} n_1 n_2 \overline{P_1(n, T, t')P_2(n_2, T, t'')}, \tag{23.272}$$

where $t'' = t' + \tau$ and $P(n, T, t)$ is given by Eq. (23.226).

Written out in full

$$\overline{n_1 n_2} = \sum_{n_1=0}^{\infty}\sum_{n_2=0}^{\infty} \frac{n_1 n_2}{n_1! n_2!}\left[\int_0^T \alpha_1 I_1(t')dt'\right]^{n_1}\left[\exp\left(-\int_0^T \alpha_1 I_1(t')dt'\right)\right]$$
$$\times\left[\int_0^T \alpha_2 I_2(t'')dt''\right]^{n_2}\left[\exp\left(-\int_0^T \alpha_2 I_2(t'')dt''\right)\right], \tag{23.273}$$

which after some algebraic manipulation gives

$$\begin{aligned}\overline{n_1 n_2} &= \alpha_1\alpha_2 \int_0^T\int_0^T \overline{I_1(t')I_2(t'')}dt'dt'' \\ &= \alpha_1\alpha_2 \int_0^T\int_0^T \left[\overline{I}_1\overline{I}_2 + |\Gamma_{12}(t''-t')|^2\right]dt'dt'' \\ &= \alpha_1\alpha_2 \int_0^T\int_0^T \overline{I}_1\overline{I}_2\left[1+|\gamma_{12}(\tau)|^2\right]dt'dt''.\end{aligned} \tag{23.274}$$

If we can write $\gamma_{12}(\tau) = \gamma_{12}(0)\gamma_{11}(\tau)$ – a situation of *cross-spectral purity*, then

$$\overline{n_1 n_2} = \alpha_1\alpha_2\overline{I}_1\overline{I}_2T^2 + \alpha_1\alpha_2\overline{I}_1\overline{I}_2|\gamma_{12}(0)|^2\int_0^T\int_0^T |\gamma_{11}(t''-t')|^2 dt'dt'', \tag{23.275}$$

which can be rewritten as

$$\overline{n_1 n_2} = \overline{n}_1\overline{n}_2 + \overline{n}_1\overline{n}_2[\xi(T)/T]|\gamma_{12}(0)|^2, \tag{23.276}$$

where we have written

$$[\xi(T)/T] = \frac{1}{T^2}\int_0^T\int_0^T |\gamma_{11}(t''-t')|^2 dt'dt''. \tag{23.277}$$

It follows that the fluctuations of n_1 and n_2 are correlated and that the correlation is given by

$$\overline{\Delta n_1 \Delta n_2} = \overline{n_1 n_2} - \overline{n}_1\overline{n}_2 = \overline{n}_1\overline{n}_2[\xi(T)/T]|\gamma_{12}(0)|^2. \tag{23.278}$$

This formula is analogous to Eq. (23.270) that we derived previously by considering classical waves.

It turns out that the correlations predicted by Eq. (23.278) are difficult to observe in practice unless $[\xi(T)/T]$, which is related to the coherence time of the source and the observation time T, is sufficiently large. This is equivalent also to the degeneracy parameter of the light being sufficiently large. If $\delta \ll 1$, bunching and correlation effects are not easily observed.

Note that

$$\begin{aligned}\xi(T) &= \frac{1}{T}\int_0^T\int_0^T |\gamma_{11}(t'-t'')|^2 dt'dt'' \\ &= \frac{2}{T}\int_0^T (T-\tau)|\gamma_{11}(\tau)|^2 d\tau. \end{aligned} \tag{23.279}$$

If $\gamma_{11}(\tau) = 1$ for the time interval T, then $\xi(T) = T$. The limiting value of $\xi(T)$ as $T \to \infty$ is

$$\xi(T) = 2\int_0^\infty |\gamma_{11}(\tau)|^2 d\tau, \tag{23.280}$$

which, from Eq. (23.67) is a finite quantity. Therefore as $T \to \infty, \xi(T)/T \to 0$, and all correlations in the photoelectron counts disappear.

References

[23.1] P.M. Morse and H. Feshbach, *Methods of Theoretical Physics*, McGraw-Hill, New York, 1953.

[23.2] W.R. LePage, *Complex Variables and the Laplace Transform for Engineers*, Dover, New York, 1961.

[23.3] J. Perina, *Coherence of Light*, Van Nostrand Reinhold, London, 1972.

[23.4] M. Françon, *Optical Interferometry*, Academic Press, New York, 1966.

[23.5] B.P. Lathi, *Linear Systems and Signals*, Berkeley-Cambridge Carmichael, CA, 1992.

[23.6] J.G. Proakis and M. Salehi, *Communication Systems Engineering*, Prentice-Hall, Englewood Cliffs, 1994.

[23.7] H. Margenau and G.M. Murphy, *The Mathematics of Physics and Chemistry*, 2nd Edition, Van Nostrand, Princeton, 1956.

[23.8] L.D. Landau and E.M. Lifshitz, *Statistical Physics*, 2nd Revised Edition, Pergamon, Oxford, 1969.

[23.9] M. Born and E. Wolf, *Principles of Optics*, 6th Edition, Pergamon Press, Oxford, 1980.

[23.10] J.W. Goodman, *Introduction to Fourier Optics*, McGraw-Hill, New York, 1968.

[23.11] R. Hanbury-Brown and R.Q. Twiss, 'A new type of interferometer for use in radio astronomy,' *Phil Mag.*, **45**, 663–682, 1954; see also 'A test of a new type of stellar interferometer on Sirius,' *Nature* **178**, 1046–1048, 1956 and 'Interferometry of the intensity fluctuations in light III, Applications to astronomy,' *Proc. Roy. Soc.*, **A248**, 199–221, 1958, 'Interferometry of the intensity fluctuations in light IV, a test of an intensity interferometer on Sirius A,' *Proc. Roy. Soc.*, **A248**, 222–221, 1958.

[23.12] F. Reif, *Fundamentals of Statistical and Thermal Physics*, McGraw-Hill, New York, 1965.

[23.13] R. London, *The Quantum Theory of Light*, 2nd Edition, Oxford University Press, Oxford, 1983.

24
Laser Applications

24.1 Optical Communication Systems

24.1.1 Introduction

Optical communication systems have a long history. Ancient man signalled with smoke and fire, often relaying messages from mountain top to mountain top. However, this optical communication scheme had limited transmission capacity. They could serve as a warning, as Queen Elizabeth the First of England planned when she had a network of bonfires erected to be set in the event of a seaborne invasion from Spain. The smoke signals transmitted by native Americans had the capacity to transmit various messages. Since the end of the eighteenth century messages have been passed by semaphore – the use of flags to indicate the transmission of one letter at a time. This form of communication could transmit information at a rate of about one letter per second over a direct line of sight, although messages could be relayed over long distances. Such means of communication were not very secure: anyone in the line of sight to the message sender could read the information (if he knew the code). The message could also be intercepted and altered during the relay process as the Count of Monte Cristo did to his advantage[24.1].

Another historical use of optical communications involved the heliograph – a device to reflect the sun's rays from a transmitting to receiving station using a code. This technique was widely used by the US Cavalry in the desert south-west of the United States up until the early part of the twentieth century. For optical communication to progress past these early efforts, an information carrying channel had to be developed that was reliable, inexpensive, and that could be used over long distances, preferably at high rates of data transmission. It was demonstrated as early as 1854 that light could be guided inside a transparent medium with a refractive index discontinuity with its surroundings: John Tyndall[24.2] showed light being transmitted along a stream of water flowing from a container.

The development of optical fibers possessing extremely low attenuation has provided optical communication systems with the information carrying channel that they need. But this did not come all at once. In the early 1960s the attenuation of glass fibers made from conventional glass was about 1000 dB km^{-1} in the visible† so messages could realistically be transmitted only a distance of a few meters. In 1966, Kao and Hockham[24.3] pointed out that the attenuation of glass was largely caused by the presence of impurity metal ions, such as iron, copper, vanadium, and chromium, and that if a glass could be developed with

† If the power emerging at the far end of a fiber of length L km is P_2, for a power input P_1, the attenuation in dB/km is $(10/L)\log(P_1/P_2)$.

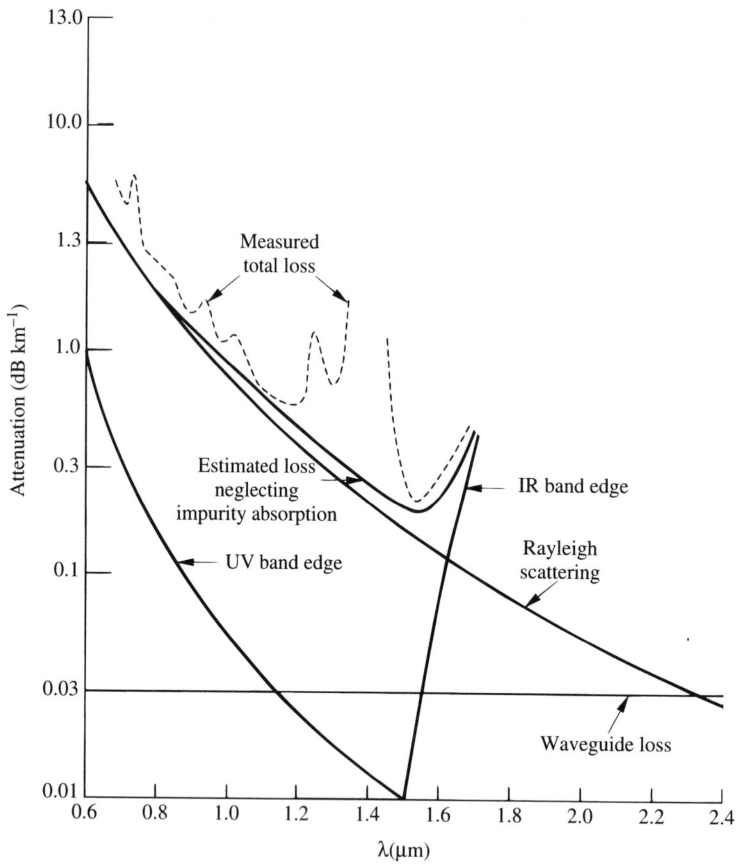

Fig. 24.1. Variation of attenuation with wavelength of a modern, low loss, germania doped optical fiber.

attenuation of only 20 dB km^{-1}, then optical communication could become a reality. By developing ways for eliminating absorbing impurities from glass used to make optical fibers, Corning achieved this figure in 1970, and by 1975 the attenuation was down to 4 dB km^{-1}. Attenuation in state-of-the-art fibers has declined steadily since then, reaching 0.5 dB km^{-1} in 1976 and 0.2 dB km^{-1} in 1979.

These reductions in attenuation have been accomplished by developing optical fiber fabrication techniques that eliminate impurities, particularly hydroxyl ions when the lowest attenutions are required. It should be acknowledged that the lowest attentuation levels have not been achieved in the visible but at two near-infrared wavelengths near 1.3 μm and 1.55 μm. Fig. (24.1) shows the variation of attentuation with wavelength of a typical modern fiber[24.4]. The development of low loss, low dispersion, single-mode fibers, with the possibility of soliton transmission over long distances, may serve to satisfy the world's insatiable appetite for ever-increasing data rates. This is evidenced by Fig. (24.2), which shows how the data rates of communication channels have grown since the early days of the telegraph. As recently as 1990, Hitachi Ltd announced that they had developed a 40 Gbit s^{-1} link operating over a 40 km long single-mode fiber.

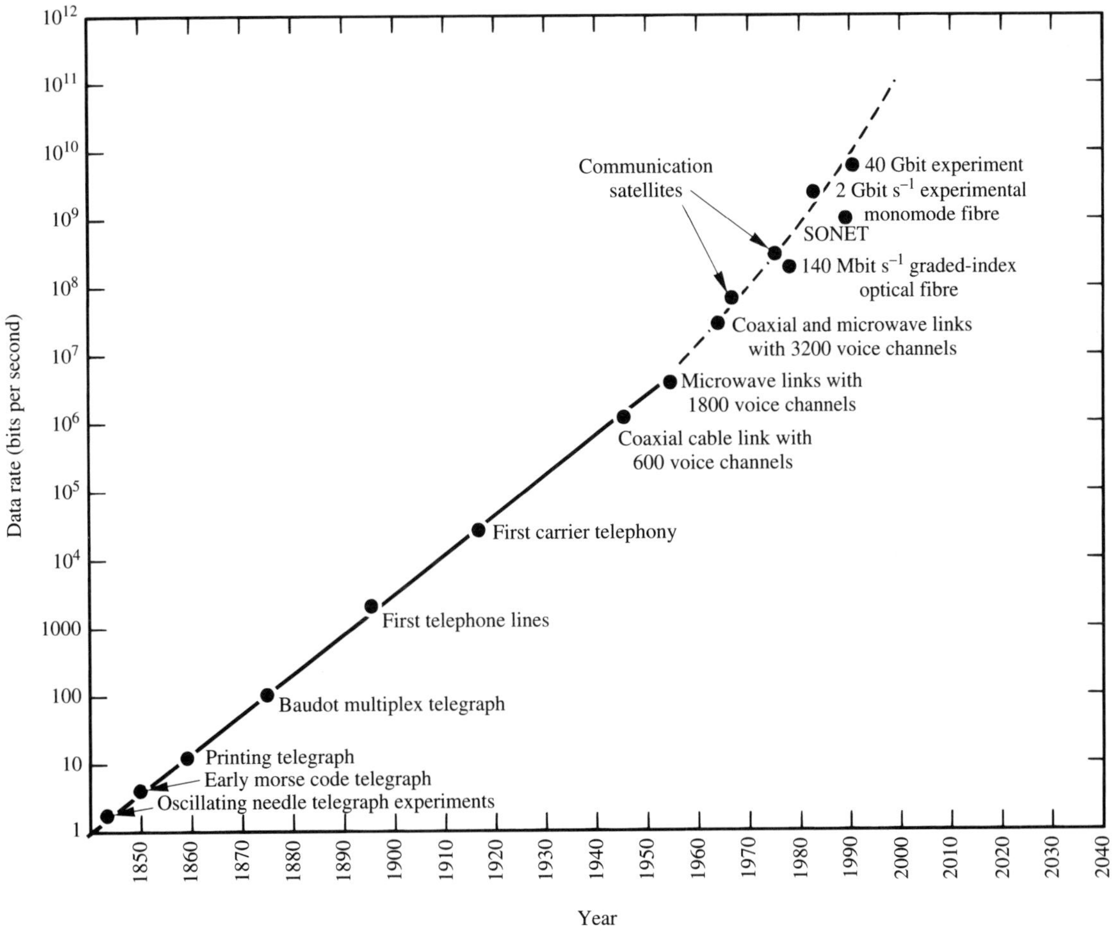

Fig. 24.2. Historical variation of data transmission rates for various communication schemes since the early days of the telegraph.

24.1.2 Absorption in Optical Fibers

Residual absorption in low loss modern optical fibers comes from several sources: residual ultraviolet absorption, residual infrared absorption, and Rayleigh scattering. Hydroxyl ions (OH^-) have a fundamental vibrational absorption at 2.73 μm, with overtone absorptions at 1.39, 0.95 and 0.72 μm. The hydroxyl content must be kept below 1 ppm in order that attenuation at 0.9 μm be less than 1 dB km^{-1}.

Even if the water content of the fiber is minimized, residual absorption remains from the infrared absorption of the fundamental vibrations of the bonds that make up the glass, which are 7.3 μm for B–O, 8.0 μm for P–O, 9.0 μm for Si–O, and 11.0 μm for Ge–O. Because these absorption features are very broad, their 'wings' extend down to short wavelengths. Thus, the farther the fundamental absorption is from the desired operating wavelength, the better. For this reason a germania doped fiber achieves very low attenuation, as shown in Fig. (24.2)[24.4]. By the same token the 'wings' of the ultraviolet absorption of the constituents of the fiber extend upwards in wavelength and contribute a small, but limiting amount of absorption, as can also be seen in Fig. (24.1). The remaining significant, unavoidable attenuation mechanism in the fiber is Rayleigh scattering – the natural tendency for any atom or module to reradiate in all directions a part of the electromagnetic radiation incident on it. The magnitude of this scattering is $\propto 1/\lambda^4$, and is enhanced by the random structural character of glass. Microscopic variations in material density is a glass fiber lead to concomitant variations in refractive index that enhance Rayleigh scattering. This loss can be minimized by drawing the fiber during production so as to minimize compositional variations.

24.1.3 Optical Communication Networks

Fiber optic links have been extensively installed worldwide. Most long distance (trunk) lines use single-mode fiber and increasingly high data rates. An emerging standard is the 2.488 Gbit s^{-1} maximum data rate of SONET (synchronous optical network). A typical SONET optical fiber link will carry 32,000 simultaneous two-way voice or data channels on one pair of fibers. SONET is designed to interface with and replace existing communication networks that operate at various data rates. The lowest level SONET signal is called the Synchronous Transport Signal Level 1 (STS-1), which has a data rate of 51.84 Mbit s^{-1}. Higher data rates are achieved by multiplexing N of these data streams up to a maximum of N=48. SONET uses single-mode fiber and different types of light source depending on the ranges involved. Short reach links (up to 2 km) use LEDs or multimode lasers operating at 1.31 μm; intermediate reach links (up to 15 km) use 50 μW single- or multi-mode laser transmitters operating at 1.31 μm or 1.55 μm. Long reach links (up to 40 km) use 500 μW laser transmitters at 1.31 μm or 1.55 μm. For local area networks (LANs) that use LED sources and multimode fiber the Ethernet standard (3 Mbit s^{-1}) is widely used, although when higher data rates are required the Fiber Distributed Data Interface (FDDI) standard is becoming established. This is a 100 Mbit s^{-1} channel that through technological developments allows interfacing of a FDDI optical network with twisted-pair cable for final connection to transmit/receive users of the network. In these hybrid systems the electrical connection is typically only the last 100 m or so to the user from the fiber-based part of the system. The standard being currently developed for interfacing FDDI networks via twisted-pair cables over these typically 100 m connections is called the Twisted-Pair Distributed Data Interface (TPDDI). At 100 Mbit s^{-1} it is possible to use graded-index fibers, with typical modal dispersion of 0.2 ns km^{-1} and either LED or laser sources. GaAlAs/GaAs($\sim$850 nm) based lasers are being superseded by 1.3 μm and 1.55 μm InGaAsP-based lasers for connections in excess of a few kilometers.

In long distance fiber links the signal decays with distance so a repeater must be installed periodically to regenerate the signal. The repeater consists essentially of an optical detector (usually an avalanche photodiode), pulse reshaping electronics, and a laser transmitter. The ability to reshape and synchronize the bits in the data stream means that the effects of dispersion are not cumulative from one repeater to the next. Erbium doped fiber amplifiers (EDFAs) are emerging as a practical means for doing this without any optical$\rightarrow$ electronic $\rightarrow$ optical conversion. In the EDFA an optical pulse is amplified directly for transmission to the next repeater. However, in such a scheme the effects of dispersion must be compensated for in the EFDA as dispersion effects will be cumulative from one repeater to the next. Two laser wavelengths are in use in long distance, high-data-rate systems: 1.3 μm sources operate at the material dispersion minimum; 1.55 μm sources operate at the attenuation minimum in the fiber. In *dispersion-shifted* fiber the waveguide dispersion is modified by reducing the core diameter until the overall (material + waveguide) dispersion minimum coincides with the attenuation minimum. In selecting repeater spacing the power transmitted into the fiber and system losses, including connectors and fiber splices, must be considered in order to calculate the acceptable distance.

Example. A fiber link has a laser transmitter that injects –7 dBm† into the fiber, the fiber has attenuation of 1 dB km^{-1} and an additional loss of 6 dB from connectors and splices. The minimum received power for the required bit error rate (BER) is –30 dB m. The *power margin* for this system is defined as

$$P_m = -7 \text{ dBm } - (-30 \text{ dB m}) \equiv 23 \text{ dB}$$

with 6 dB additional loss the acceptable receiver spacing is 17 km.

The longest fiber links are those that span the world's oceans, including two across the Atlantic from North America to Europe and one from the United States to Japan. Many additional long distance fiber links are planned: by 1995 there should be nine across the Atlanta, four from the United States to Japan, one from Japan to China, with additional links for Australia, Korea, India, South and Central America, South Africa, and New Zealand[24.5]. The transatlantic cable laid in 1990 contains six fibers and can carry 40,000 telephone calls at one time; the next generation of cable will carry double this number of calls. These fiber cables are heavily armored against shark bites and in the shallow part of the ocean are buried under the ocean floor to prevent their being snagged by fishing boats. Repeaters are placed every 55–70 km. The repeaters have built-in redundancy, incorporating extra detectors and lasers so that if one device fails the channel remains functional. Because it is prohibitively expensive to dredge up cables to modify or repair repeaters a long term goal is to eliminate them in favor of erbium doped fiber amplifiers. The amplifying sections of fiber perform the repeater functions of maintaining signal strength, but dispersion effects can be cumulative along the entire length of the link. Serious consideration is being given to installing a trans-Pacific cable that will overcome this dispersion problem by utilizing soliton transmission along the fiber.

24.1.4 Optical Fiber Network Architectures

Fig. (24.3) shows three popular topologies of optical networks. In these arrangements of transmitters and receivers, each transmitter/receiver pair on the network, which is called a *node* or *subscriber* can communicate with every other. The construction of such networks relies on the use of various passive optical components for their implementation. The most important of these is the fiber coupler, one form of which is shown schematically in Fig. (24.4), that either splits or combines optical signals. This is a four-port device. In the simplest sense such a device is fabricated by fusing together two single mode fibers so that their cores come close together, allowing light to leak from one core to the other. Depending on the splitting ratio between the ports the device can be described as either a *tap* or *splitter*. For example, in Fig. (24.4b), if intensity I_1, enters port 1 it is split on output between ports 3 and 4. Port 2 functions in the same way, although usually

†0 dB m ≡ 1 mW, so –7 dB m is 7 dB below 1 mW, namely 0.2 mW.

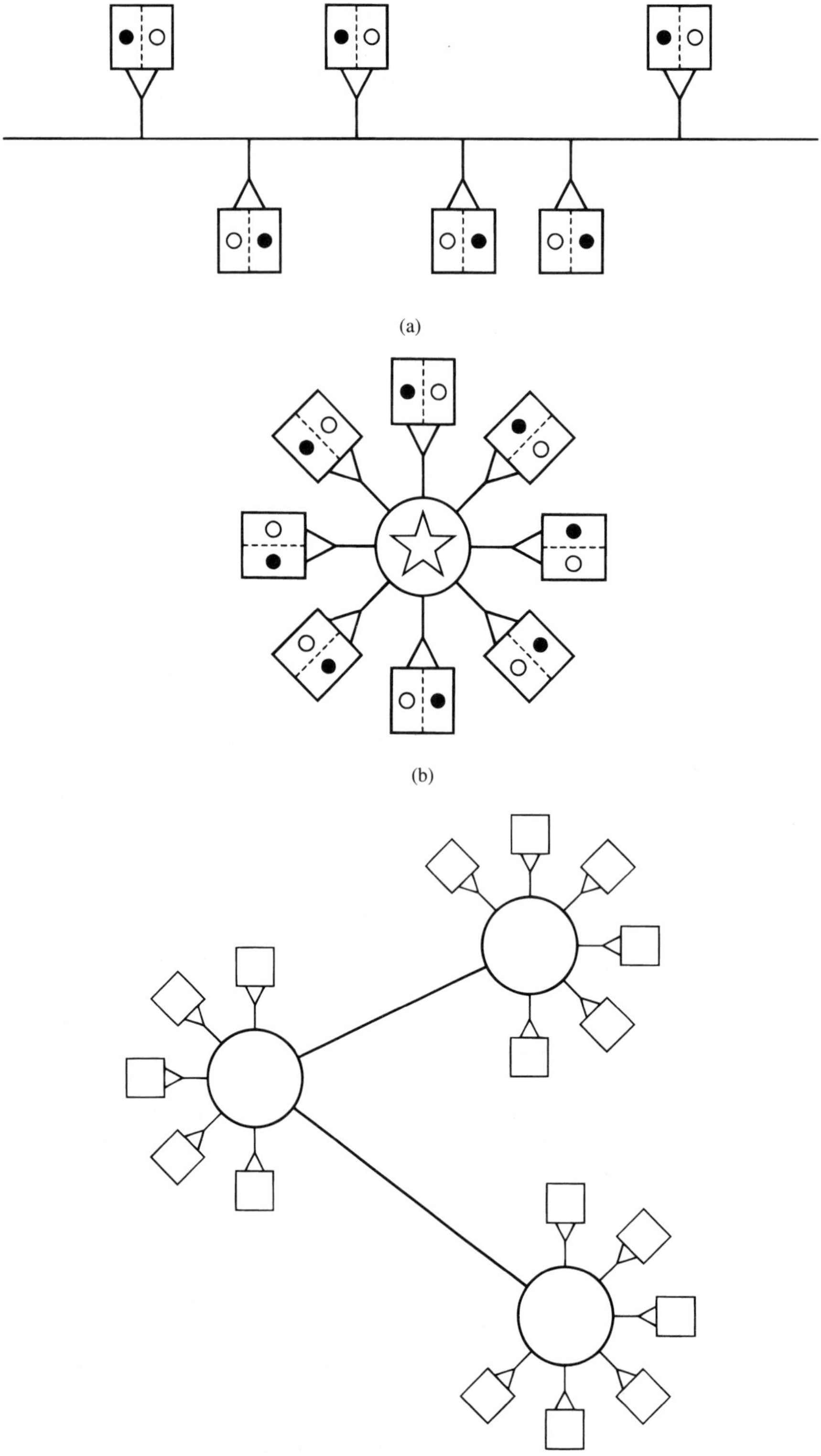

Fig. 24.3. Optical network topologies • – transmitter, o – receiver: (a) bus, (b) star, (c) multilevel star.

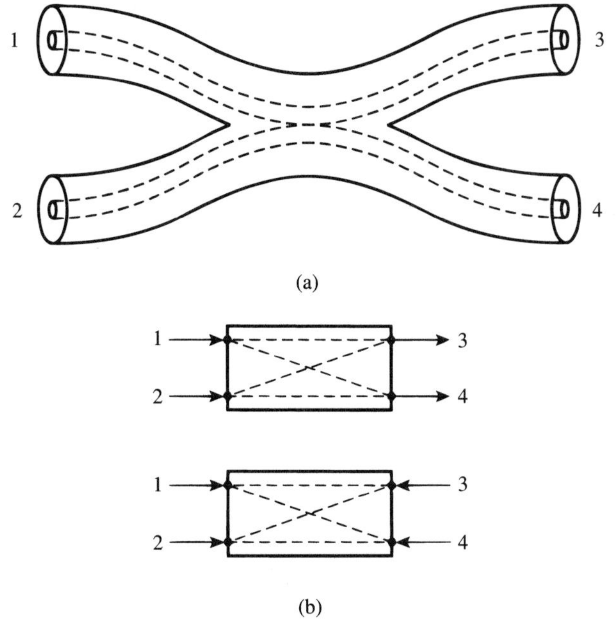

Fig. 24.4. Optical fiber couplers: (a) schematic of a fused fiber optic coupler, (b) bidirectional coupling pathways of coupler.

one of ports 1 or 2 will be unused if it is desired to act as a Y (or T) coupler. If the intensities I_3, I_4 leaving ports 3 and 4, respectively, on injection of I_1 are equal the coupler is referred to as a 3 dB coupler or splitter, on the other hand if 99% of the input power passes from port 1 to port 3, with only 1% leaving at port 4 the device is described as a 20 dB coupler or tap, since $I_4 = I_1/100$, which is 20 dB below I_1.

24.1.5 Coding Schemes in Optical Networks

As we have seen in previous chapters a laser (or LED) source can be modulated in various ways, either by direct modulation of the source, for example by modulating the drive current of a semiconductor laser, or with an external amplitude or phase modulator. Almost all the practical implementations of optical communication networks utilize some form of binary modulation, which can amount simply to switching the laser on to transmit a '1' and off to transmit a '0'. It is also possible to switch the laser between two discrete frequencies (FM) or two distinct polarization states (polarization modulation). It is beyond our scope here to go into detail regarding the precise implementation, advantages and disadvantages of various schemes, for details the reader should consult the specialized literature[24.6]–[24.8]. However, a few details of how the binary information is encoded for transmission in the optical channel are worthy of note, and can be explained by reference to Fig. (24.5). These schemes are generally called pulse code modulation (PCM).

The transmission of the digital data is synchronous with a clock rate. In the return-to-zero (RZ) format each '1' is emitted as a pulse during the first half of each clock cycle. In the non-return-to-zero (NRZ) format the pulse train stays 'high' for an entire clock cycle when a '1' is to be transmitted, and stays 'low' for the entire clock cycle when a '0' is to be transmitted. In the NRZI format the

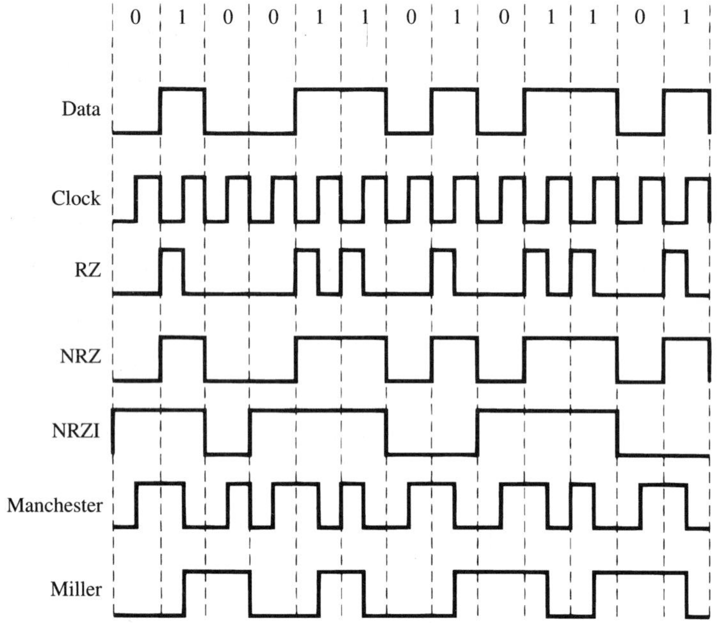

Fig. 24.5. Pulse code modulation formats.

pulse train only makes a transition from 'high' or 'low' when the succeeding bit is a '0'. The drawback of all these schemes is that the received data stream does not allow recovery of the clock rate: for example in the RZ scheme if a long stream of zeroes is transmitted. However, other schemes such as the Manchester and Miller encoding schemes allow recovery of the clock rate from the received data stream. The Manchester code is essentially a pulse-position modulation. For a '0' the signal makes a transition to the 'high' state in the middle of each clock cycle, each '1' is transmitted as a return from high to low in the middle of the clock cycle.† In the Miller scheme the transitions from high to low occur in the center of the clock cycle when a '1' is transmitted and at the beginning of the clock cycle when a '0' is transmitted after a '0'.

24.1.6 Line-of-Sight Optical Links

Line-of-sight optical links, which involve the free space propagation of the encoded light signal from source to receiver, have only limited, short range application in ground-based situations. Turbulence, scattering, and absorption by the atmosphere introduce amplitude and phase noise into the transmitted signal and cause significant signal amplitude decay over moderate distances. However, ground-to-satellite, satellite-to-satellite, and deep-space communication links represent important application areas for line-of-sight optical communications. When the communication channel must be guaranteed, as in a commercial satellite ⟷ ground link, microwaves are generally preferred because their long wavelengths can penetrate clouds. An optical downlink, even to a generally cloudless location such as the mountains of Arizona or Hawaii, or the deserts of Australia cannot provide 100% reliable

† In logic terms the Manchester-encoded output is obtained as an exclusive OR operation between data and clock.

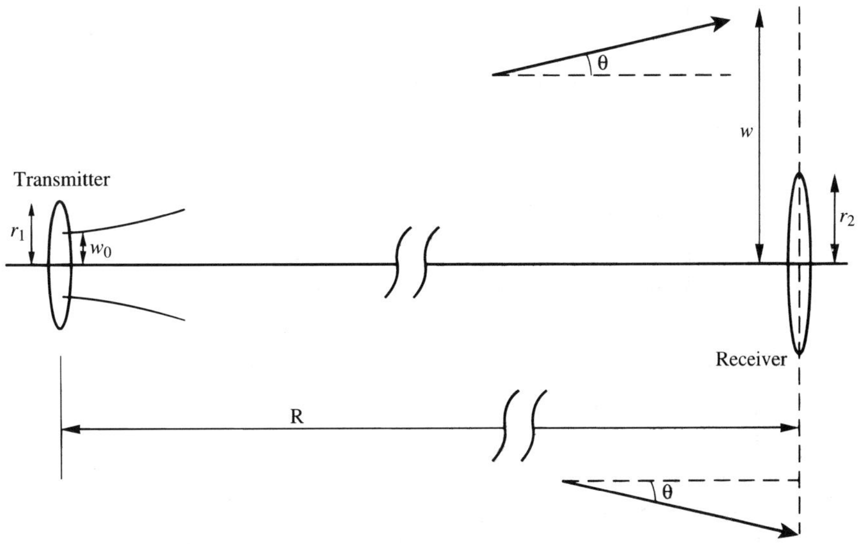

Fig. 24.6. Geometry used in describing the performance of a line-of-sight optical link.

visibility conditions. Nonetheless, there are specialized applications of such links: a blue-green laser can be transmitted from an aircraft or satellite and provide a secure high data rate communication channel to a submerged submarine. Optical links between satellites in near-earth orbit will be of increasing importance in the future for global communications. A series of satellites in synchronous earth orbit (at a height of 38 600 km) can provide global communications coverage and can relay information from closer-in satellites used for earth observation. The NASA Earth Observing System (EOS) program envisages the use of such satellites for continuous, long term monitoring of earth resources, climatic conditions, and atmospheric composition. The primary advantage of optical links for space communications is that a much more directional beam can be sent from source to receiver, so transmitter requirements for long distance transmission are reduced. The most challenging application of such a link is in deep-space communications. In explorations of the outer planets in programs such as Pioneer and Voyager, data is relayed to and from the spacecraft with microwaves. Because the transmission distances are so long, high transmitter powers are required, which presents a problem for the spacecraft end of the system, which must rely on solar power. If a short wavelength laser is used as the transmitter on the spacecraft, a substantial reduction in transmitted power is possible because a more directional beam is possible.

The transmit/receive power analysis of such a link, and of any reasonably long range line-of-sight link, can be carried out with reference to Fig. (24.6). The transmitter and receiver apertures are drawn schematically as lenses, although the transmitter and receiver will likely use reflective telescopes. We assume that the transmitter telescope sends a collimated Gaussian TEM_{00} mode beam of power P_1 towards the receiver situated at a distant range R. The diffraction angle of the transmitted beam is

$$\theta = \frac{\lambda}{\pi w_0}, \tag{24.1}$$

where we can safely assume θ is a small angle. The transmitter aperture is chosen sufficiently large so as not to truncate this beam significantly: a commonly chosen condition for this is $r_1 = 1.4w_0$. At long range, R, the spot size is

$$w = R\theta, \tag{24.2}$$

and the power collected by the receiver is

$$P_2 = 2P_1 \left(\frac{r_2}{w}\right)^2. \tag{24.3}$$

Combining Eqs. (24.1), (24.2), and (24.3) gives

$$P_2 = 2\left(\frac{\pi r_2 w_0}{R\lambda}\right)^2 P_1. \tag{24.4}$$

So, all other parameters being equal, received power varies as $1/\lambda^2$. This highlights the diffraction advantage of an optical link. This advantage between 405 nm (a frequency-doubled GaAs laser) and 30 GHz microwaves is 6×10^8. Even allowing for the fact that larger microwave transmit/receive antennas than optical telescopes can be built, the advantage of the optical link is clear.

24.1.6.1 Example

We assume a spacecraft transmitter power of 0.1 W at 405 nm at a distance corresponding to the orbit of Jupiter, 780×10^6 km; a transmitted spot size $w_0 = 150$ mm, and a receiver aperture of 1.27 m (corresponding to a 100 inch telescope). The received power calculated from Eq. (24.4) is 1.9×10^{-10} W. If we are using an APD as our receiver that has an excess noise factor, F, the minimum detectable received power in a bandwidth Δf for an S/N ratio of unity is

$$P_{\text{min}} = \frac{2eF\Delta f}{\mathcal{R}}, \tag{24.5}$$

where $\mathcal{R}$ is the responsivity.† With F=6, and $\mathcal{R} = 0.5$ A W^{-1},

$$P_{min} = 3.84 \times 10^{-18} \Delta f \text{ W}.$$

For a BER of 10^{-6} (which would be acceptable in such a long range link) the actual required S/N ratio is $\simeq 19.6$ dB, or $S/N = 10$. Therefore, for a 200 kHz bandwidth (100 kbit s^{-1}), the minimum required power is $P_{min} = 3.84 \times 10^{-11}$ W: our proposed link would function at this data rate.

24.2 Holography

24.2.1 Wavefront Reconstruction

Holography is lensless three-dimensional photography. A conventional photograph is only a flat record of a real image projected onto a photographic film. Information about the three-dimensional character of the object is almost entirely lost during the photographic recording process. Only the blurring of the images of objects that are not within the *depth of field* of the camera preserves any record of the location of the object relative to the camera. A *hologram* on the other hand is a special 'photograph' of an object that retains information about the phase of waves coming from the object. Holography has a much longer history than that

† See Chapter 22.

of the laser, although it was the invention of the laser that made the production of high quality holograms a reality[24.9],[24.10].

In 1947, Denis Gabor was looking for a way to improve the quality of images that could be obtained with an electron microscope[24.11]. Although the electron microscopes of that day should have been capable of resolving down to the atomic scale, aberrations of the electron optics prevented them from doing so. Gabor conceived the idea of recording an electron hologram, which would contain amplitude and phase information, and then reconstructing a corrected image of the object by optical means. He did succeed in producing holograms and performing image reconstruction, but the optical sources available to him were not sufficiently temporally coherent to allow high quality images to be achieved. For his pioneering work he was awarded the 1971 Nobel prize for physics[24.12]–[24.14].

The high degree of coherence obtainable with a laser has made the production of holograms and applications of holography into a large subject in its own right[24.15]–[24.20]. The principles underlying holography share much in common with the phenomena of interference and diffraction that we discussed in Chapter 6 and the concept of optical mixing discussed in Chapters 19 and 22. Taken together, these phenomena belong to the subject of *Fourier optics*[24.21]–[24.24]

In conventional photography the photographic emulsion becomes dark according to the local intensity of the real image falling on it. The emulsion is a mixture of silver halide crystals of different sizes, predominantly silver bromide, in a gelatin matrix. Incident light frees electrons from the halide, but the electrons become trapped. Addition of developer reduces the silver ions in an exposed crystal to silver atoms, leading to a strongly absorbing region where exposure levels were high. This is how a photographic negative is produced. If the incident field acting on the emulsion is

$$O(x, y) = |O(x, y)|e^{i\phi(x,y)}, \tag{24.6}$$

where $\phi(x, y)$ is the phase of the light at point (x, y) in the image then the incident intensity is

$$I(x, y) \propto |O(x, y)|^2 \tag{24.7}$$

and all phase information is lost. After exposure the field transmission coefficient of the film can be written as

$$t(x, y) = t_0 - \beta' I(x, y). \tag{24.8}$$

This linear relationship holds true provided the intensity at the location (x, y) is not too high. Therefore, the field transmission coefficient of the developed film is

$$t(x, y) = t_0 - \beta|O(x, y)|^2. \tag{24.9}$$

If we illuminate the developed film uniformly the transmitted intensity is

$$I_T(x, y) \propto \left[t_0 - \beta|O(x, y)|^2\right]^2. \tag{24.10}$$

If the film is not overexposed, the dominant term of interest in Eq. (24.10) is

$$I_T(x, y) \propto -2\beta t_0|O(x, y)|^2, \tag{24.11}$$

which is a *negative* reconstruction of the original image intensity distribution but without any phase information.

We make a hologram, and preserve phase information in the photographic

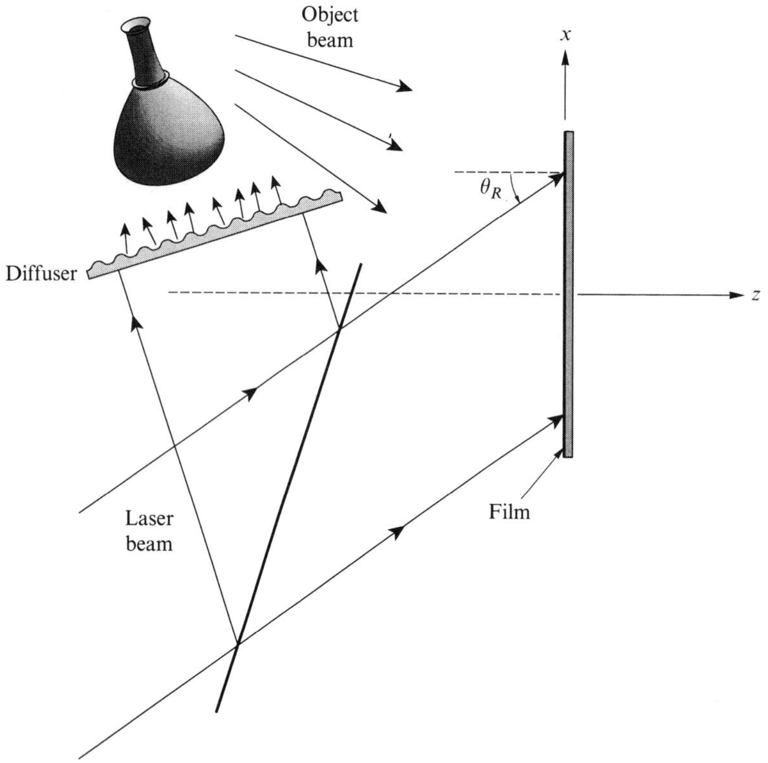

Fig. 24.7. Schematic arrangement for making a simple hologram.

record, by allowing temporally and spatially coherent light to illuminate the object. Light reflected and scattered from the object falls on a photographic film together with a reference wave supplied directly to the film with the same laser as is used to illuminate the object. The schematic way in which this is done is shown in Fig. (24.7). No lenses are used to image the object onto the film. The relative location of all the components involved must be kept constant to much less than a wavelength of the light being used during the exposure. Holograms of moving objects can be taken with pulsed lasers. The reference wave is incident on the film at angle θ_R.

The field from the object is

$$O(x, y) = |O(x, y)|e^{i\phi_O(x,y)}, \tag{24.12}$$

where the amplitude $|O(x, y)|$ depends on the combined reflection plus scattering efficiency from each point on the object and $\phi_O(x, y)$ is a composite phase factor that contains information about the relative location of each point on the object relative to the film. If an observer were to place his or her eye in the location of the film then he or she would *see* the object in full three-dimensionality. It does not matter precisely how the object is illuminated, whether this be directly by a collimated or expanding laser beam, or after sending a laser beam through a diffuser to illuminate the object uniformly as shown in Fig. (24.7). The illuminated object would be clearly visible to the observer in any case.

The field sent directly to the film from the same laser that illuminates the object

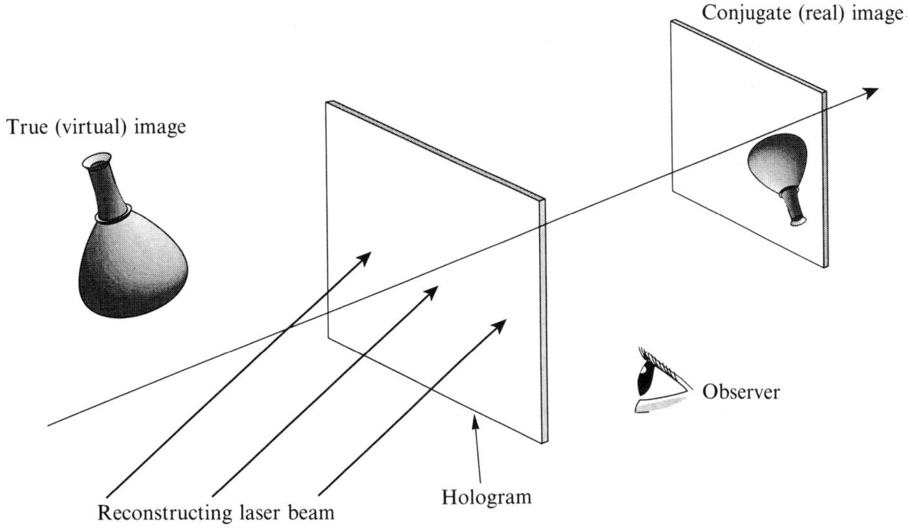

Fig. 24.8. Reconstruction of the image stored in a hologram showing the location of the real and virtual images.

is called the reference field, which is written as

$$R(x, y) = |R(x, y)|e^{i\phi_R(x,y)}. \tag{24.13}$$

In an actual exposure $|R(x, y)|$ might vary very little from point to point on the film. Furthermore, the phase factor $\phi_R(x, y)$ will only have some smooth variation across the film depending on whether the reference wave is a plane or expanding spherical wave.

The transmission of the developed film in this case will depend on the total illuminating field $O(x, y) + R(x, y)$, and will be

$$t(x, y) = t_0 - \beta|O(x, y) + R(x, y)|^2, \tag{24.14}$$

which in a simplified notation can be written as

$$t(x, y) = t_0 - \beta OO^* - \beta(RR^*) - \beta(OR^*) - \beta(O^*R). \tag{24.15}$$

If the developed film is now illuminated by a laser that generates a reference wave just like the original R, then the transmitted field is

$$E_t = t_0R - \beta(OO^*)R - \beta(RR^*)R - \beta O(R^*R) - \beta O^*RR. \tag{24.16}$$

If the field from the object is much weaker than the reference field we can drop the second term in (24.16). The interesting remaining terms can be written out in full as

$$\begin{aligned} &- \beta|R(x, y)|^2R(x, y)e^{i\phi_R(x,y)} - \beta O(x, y)e^{i\phi_O(x,y)}|R(x, y)|^2 \\ &- \beta O^*(x, y)e^{-i\phi_O(x,y)}|R(x, y)|^2e^{2i\phi_R(x,y)}. \end{aligned} \tag{24.17}$$

The first term in (24.17) is just the reference wave. The second term is the *primary reconstruction* of the original object. Apart from the (almost) spatially uniform illumination factor $|R(x, y)|^2$ this field is $\propto O(x, y)e^{i\phi_O(x,y)}$ – identical to the original field scattered by the illuminated object. This wave, if received in a geometry identical to the original exposure geometry, will appear exactly as it did when coming from the original object. Since the waves were diverging from the original object, the third term in (24.17) also corresponds to a diverging wave so will be viewed as a virtual image by an observer, as shown in Fig. (24.8). The final term in (24.17) is the complex conjugate wave of the original object, except for the phase factor $e^{2i\phi_R(x,y)}$. Therefore, this corresponds to a set of waves that converge to form a real image of the original object, as shown in Fig. (24.8). Although this simple analysis does not show it, this image is inverted.

The developed hologram does not contain a recognizable image of the object as a conventional photograph does. The hologram generally appears as a collection of bright and dark lines with bright and dark patterns of concentric circles. The stored image can be regarded as a representation of the diffraction pattern of the object. We saw in Chapter 6 that the Fraunhofer diffraction pattern of an obstacle (or aperture) can be represented in terms of the Fourier transform of the amplitude transmission function of the obstacle. If the diffraction pattern is viewed close to the object so that the Fresnel diffraction pattern is being observed then the concept of the diffraction pattern as a simple Fourier transform must be modified. The object must be viewed as a series of point sources, each of which gives rise to a Huygenian secondary wavelet. The contributions of these secondary wavelets depend on a factor e^{ikr}/r, because each is a source of an expanding spherical wave. A detailed analysis has been given by Smith[24.15].

The hologram formed by exposure of a film and the production of a complex pattern of transmission is called as *amplitude hologram.* However, it is possible to process the film chemically by *bleaching* so that the dark silver atoms are converted into silver salts that are transparent but have a different refractive index than the surrounding film. This converts the recorded hologram into a pattern of phase information and is called a *phase hologram.* As we shall see in the next section the production of an image in holography can be viewed as a diffraction process. Diffraction will occur for structures that have variations in either amplitude or phase transmission. However, the greater transmission through a bleached hologram provides much brighter reconstructions of the object[24.25],[24.26]

24.2.2 *The Hologram as a Diffraction Grating*

To reinforce the interpretation of a hologram as a diffraction pattern consider the simple situation shown in Fig. (24.9a). The simplest kind of object is a distant point source, the waves from which can be treated as plane. The reference beam is also a plane wave. In the geometry of Fig. (24.9a) the wave from the object is

$$O(x) = E_1 e^{i(\omega t + kx\sin\theta_1 - kz\cos\theta_1)}. \tag{24.18}$$

The reference wave is

$$R(x) = E_2 e^{i(\omega t - kx\sin\theta_2 - kz\cos\theta_2)}. \tag{24.19}$$

The transmission of the film placed in the plane $z = 0$ is

$$t(x) = t_0 - \beta|(E_1 e^{ikx\sin\theta_1} + E_2 e^{-ikx\sin\theta_2})|^2, \tag{24.20}$$

which gives

$$E(x) = t_0 - \beta E_1^2 - \beta E_2^2 - 2\beta E_1 E_2 \cos[k(\sin\theta_1 + \sin\theta_2)x]. \tag{24.21}$$

$E(x)$ clearly corresponds to a regular series of bright/dark bands on the film. The spacing d of these bands is determined by the condition

$$k(\sin\theta_1 + \sin\theta_2)d = 2\pi, \tag{24.22}$$

which gives

$$d = \frac{\lambda}{\sin\theta_1 + \sin\theta_2}. \tag{24.23}$$

The interfering incident and reference waves have *written* a diffraction grating into the film.

In the language of Fourier theory $E(x)$ is a monochromatic function in k-space. As such it is a representation of a wave travelling at a well defined angle – the incident wave from the object.

If we illuminate the film pattern corresponding to Eq. (24.21) with the reference wave, the transmitted field, in the plane $z = 0$, is

$$\begin{aligned} E_t(x) =& t_0 R(x) - \beta E_1^2 R(x) - \beta E_2^2 R(x) \\ &- 2\beta E_1 E_2^2 \cos[k(\sin\theta_1 + \sin\theta_2)x]\cos(\omega t - kx\sin\theta_2). \end{aligned} \tag{24.24}$$

The first three terms on the RHS of Eq. (24.24) correspond to the reference wave being transmitted through the film. However, the final term is the interesting one. It can be written as

$$E_t(x) = -\beta E_2^2[E_1\cos(\omega t + kx\sin\theta_1) + E_1\cos(\omega t - 2kx\sin\theta_2 - kx\sin\theta_1)]. \tag{24.25}$$

The first term in Eq. (24.25) is a reconstruction of the field of the incident wave *measured at the plane z=0.* This field propagates away from the film as

$$E_1(x) = -\beta E_2^2 E_1\cos(\omega t + kx\sin\theta_1 - kz\cos\theta_1), \tag{24.26}$$

which apart from a proportionality constant is equivalent to Eq. (24.18). The second term in Eq. (24.25) is a new plane wave reconstruction of the original incident wave that propagates away from the film at angle θ_3 as

$$E_3(x) = -\beta E_2^2 E_1\cos(\omega t - kx\sin\theta_3 - kz\cos\theta_3), \tag{24.27}$$

where

$$\sin\theta_3 = 2\sin\theta_2 + \sin\theta_1. \tag{24.28}$$

Now the condition for diffraction at angle θ_3 from a grating of spacing d is

$$d(\sin\theta_3 - \sin\theta_2) = \lambda. \tag{24.29}$$

Substituting for d from Eq. (24.23) gives $\sin\theta_3 = 2\sin\theta_2 + \sin\theta_1$, identical to Eq. (24.28).

We conclude, therefore, that the reference wave recreates two versions of the original incident wave: one travels in the same direction as the original wave, for a more complex set of input waves this corresponds to the true or virtual image seen by the observer in Fig. (24.9b). The second wave corresponds to the reconstruction of the conjugate or real image in Fig. (24.9b).

24.2.3 Volume Holograms

The holograms that we have been considering so far are called *plane holograms* because we have been tacitly assuming that interference of object and reference waves writes the hologram in a particular plane. We are neglecting the thickness of the photographic emulsion. However, since actual emulsions can be up to 20 μm

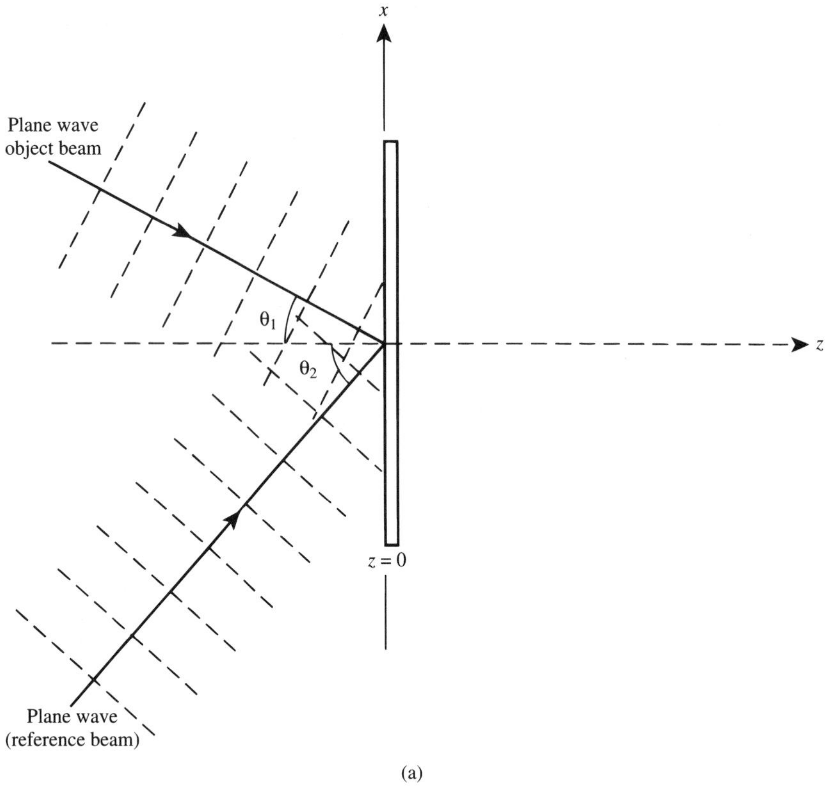

Fig. 24.9. (a) Production of a hologram from a plane wave object beam and a plane wave reference beam.

thick it is possible, and in many cases desirable, to write a *volume hologram* in which interference fringes are written throughout the emulsion. The hologram can be thought of as a continuous distribution of plane holograms at different values of z through the emulsion thickness. Just as the simple plane hologram produced by the interference of two plane waves was a planar diffraction grating the volume hologram produced in this way will have a structure where the regions of bright and dark are distributed in a series of parallel planes in the recording medium. Fig. (24.10) shows schematically how the transition from plane hologram to volume hologram changes the pattern of recorded information. Whether the volume hologram is a transmission or reflection hologram is determined by the way in which the hologram is recorded. A reflection hologram is recorded with the reference beam incident on the film from the opposite side from the object as shown in Fig. (24.11a) so the resulting interference patterns are almost parallel to the film. A transmission hologram is recorded as shown in Fig. (24.12).

Because the interference patterns that constitute the volume hologram are families of planes, during reconstruction of the wavefronts not only must a diffraction condition like Eq. (24.29) be satisfied, but so also must the Bragg condition for reflection from the periodic planes in the hologram. Consequently during wavefront reconstruction only one of the two images, the true or conjugate will be reconstructed, depending on the direction of the reference beam used for reconstruction, as shown in Figs. (24.12b) and (24.12c). The diffraction and Bragg conditions for both cannot be satisfied at the same time. Reflection holograms are of particular interest, both aesthetic and practical, because they can be reconstructed with white

Fig. 24.9. (*cont.*) (b) reconstruction of a simple plane wave hologram showing diffracted waves corresponding to production of real and virtual images.

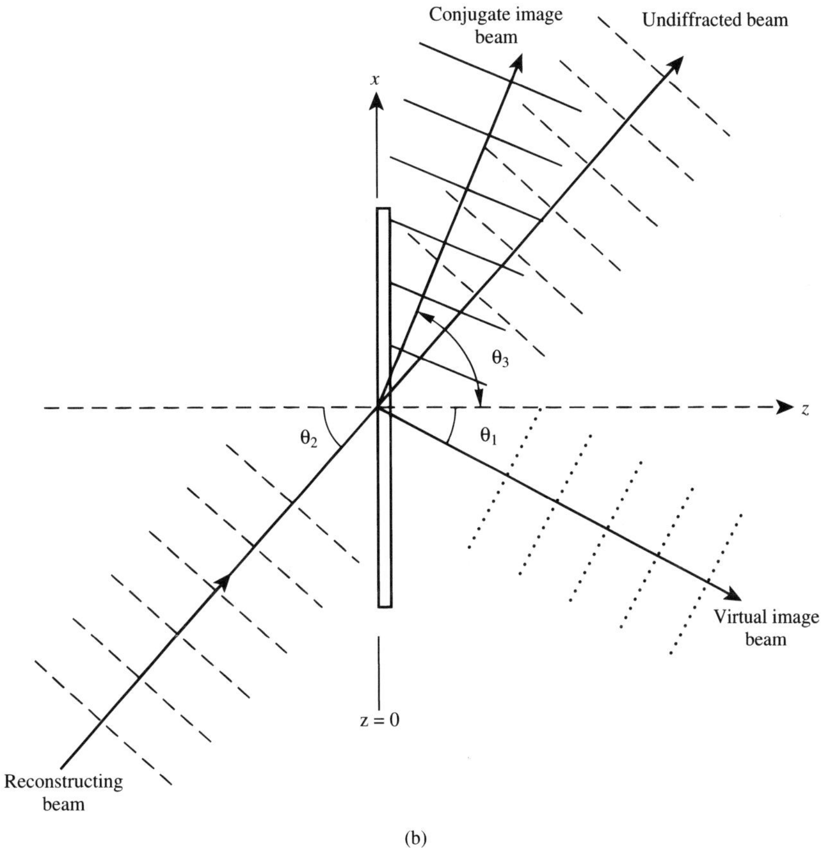

light. These holograms are generally bleached because of the increased diffraction efficiency that this provides. The hologram picks out the appropriate narrow band of wavelengths that satisfy both the diffraction and Bragg reflection conditions. The reconstructed image will be colored, and the color depends on the viewing angle. Full color holograms can be produced by recording the hologram with the light from red, green, and blue lasers and then using these same colors during reconstruction.

White light holograms reconstruct images that lack the granularity, or *speckle* that always appears when highly coherent light is used to either illuminate an object or reconstruct an image. The primary cause of speckle is the microscopically rough surface of the illuminated object, or the diffuser if one is being used during hologram recording. In the case of a microscopically rough surface, each surface feature acts as a source of an expanding spherical wavelet. The many wavelets from the whole object add vectorially at each point in space and there will inevitably be a pattern of interference maxima and minima produced. This is the speckle pattern. It is, for example, the term $-\beta OO^*$ that we neglected in Eq. (24.15). When a speckle pattern is viewed with the naked eye the granularity appears to fluctuate. The speckle pattern itself is stationary in space, but the eye in its constant small motions moves through the various maxima and minima and leads to a subjective twinkling. The speckle pattern produced by an object contains

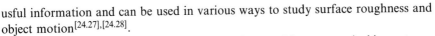

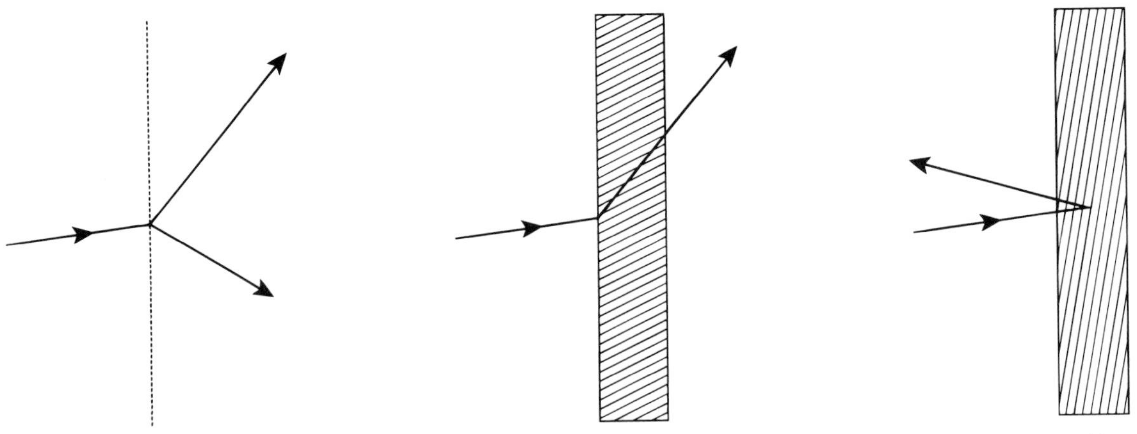

(a) (b) (c)

Fig. 24.10. Schematic illustration of the transition from plane to volume hologram: (a) plane hologram in which the interference pattern written as the hologram is located in a plane. The real and virtual image waves are shown schematically resulting from a reconstructing wave; (b) volume hologram in which the interference of object and reference waves has written fringes almost perpendicular to the plane of the film. Only a single reconstructed wave results; (c) volume hologram in which the interference fringes are almost parallel to the film. The direction of the reconstructed wave that results from illumination is shown.

usful information and can be used in various ways to study surface roughness and object motion[24.27],[24.28].

Holography has found application in several areas. Of most practical importance is nondestructive testing. Extremely small deformations of an object can be detected by superimposing the reconstructed images of the object in its two different states: for example, before and after a stress is applied[24.29]. Holography is also used in optical memories[24.20], and in this context is intimately related to the whole field of optical information processing[24.15],[24.17],[24.31],[24.32], which is beyond our scope here. Practical details of recording holograms have been described by Iizuka[24.20] and Smith[24.15].

24.3 Laser Isotope Separation

The different isotopes of an element are chemically virtually indistinguishable and physically different to only a small extent. Consequently, it has always been extremely challenging to remove separate pure isotopes from the mixture of isotopes found for each element in nature. If very small quantities of the pure isotope are required, then it is possible to use a mass spectrometer to separate the ions of different isotopes. However, to produce larger quantities of commercially important isotopes such as heavy hydrogen (deuterium) or to enrich uranium from its naturally occuring 0.7% of uranium-235 to the 3.2% required for light-water nuclear reactors requires different schemes to be used. This problem is easiest to solve for the separation of hydrogen (H_2) and deuterium (D_2), since the deuterium is twice as heavy as the hydrogen. Electrolysis of water, which naturally contains about 0.015% of its hydrogen as heavy hydrogen, produces hydrogen and deuterium at different rates at the cathode, so sequential repeated electrolysis can enrich water in D_2O.†

The use of lasers for isotope separation is based on the spectroscopic difference between isotopes of an element[24.34],[24.35]. The wavelengths of absorption differ for atoms or molecules that are chemically identical but contain different isotopes. Much of the stimulus for the development of laser isotope separation arose from a

† An added complexity is presented by the presence of mixed water HDO.

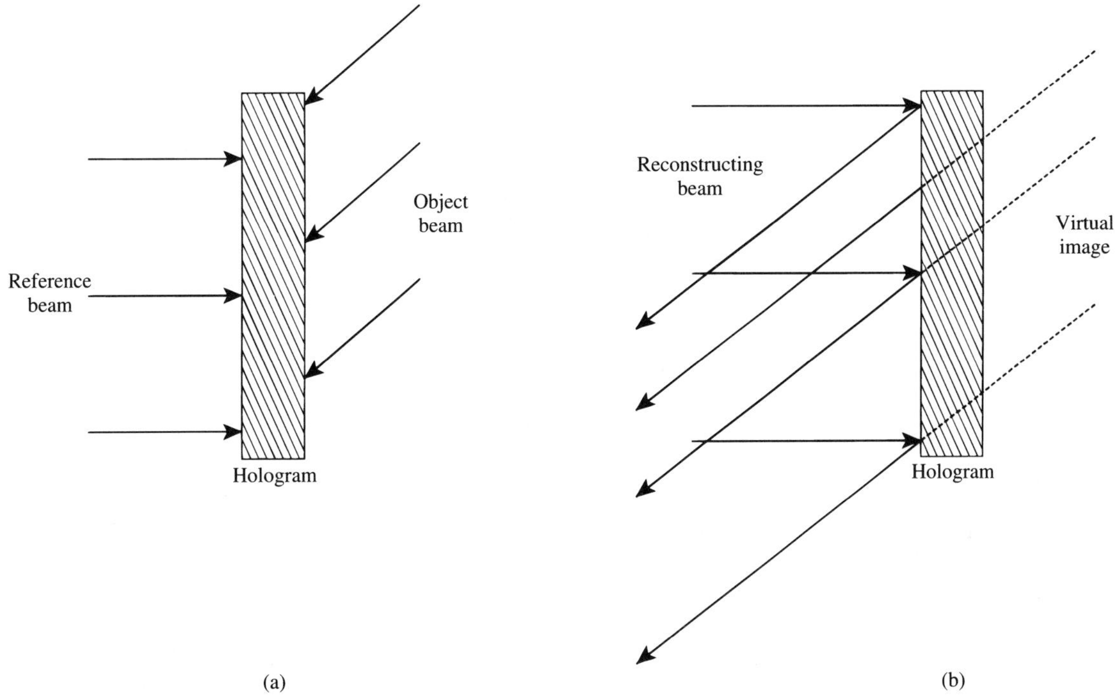

Fig. 24.11.
(a) Arrangement for writing a reflection volume hologram,
(b) reconstruction of image.

need to improve the efficiency (and lower the cost) of increasing the concentration of ^{235}U relative to ^{238}U. The traditional method for uranium enrichment is gaseous diffusion. The gas uranium hexafluoride (UF_6) is allowed to diffuse through a series of porous barriers. The $^{235}UF_6$ diffuses slightly faster so a large number of successive diffusions, in which the slightly $^{235}UF_6$ enriched gas is passed to each diffusion stage, leads to the desired enrichment Approximately 1240 stages of diffusion are required to produce 3.2% $^{235}UF_6$. It is also possible to produce this enrichment in a gas centrifuge where the central forces acting on the different isotope species of UF_6 are used to produce enrichment. Both these techniques require gigantic amounts of electric power for their operation.

The way in which lasers can be used to produce isotopic enrichment is well illustrated by one of the schemes that has been used to enrich atomic uranium. Fig. (24.13) shows in a schematic way the energy levels that are involved in the separation of ^{235}U and ^{238}U. In one implementation of this scheme at the LLNL in California a beam of uranium atoms emerges from an oven at 2600 K. At this temperature 27% of the uranium atoms are in a metastable sub-level of the ground state. ^{235}U atoms are excited from this sub-level of energy $E_1(620\ \mathrm{cm}^{-1})$ by a xenon ion laser with $\lambda_1 = 378.1$ nm to an intermediate level of energy E_2. The wavelength of the transition from E_1' to E_2' in the ^{238}U is at a slightly longer wavelength, so these atoms are not excited. Excited ^{235}U atoms that reach level E_2 are ionized with a krypton laser operating at 350.7 nm or 246.4 nm. By this two-step photodissociation process $^{235}U^+$ ions are selectively produced. The ions are collected by deflecting them to a collector with a strong electric field.† This isotope

† The photoionizing processes in the currently used uranium isotope separation at LLNL have not been revealed in detail, but the photons needed for the process are generated by copper-vapor-laser-pumped tunable dye lasers.

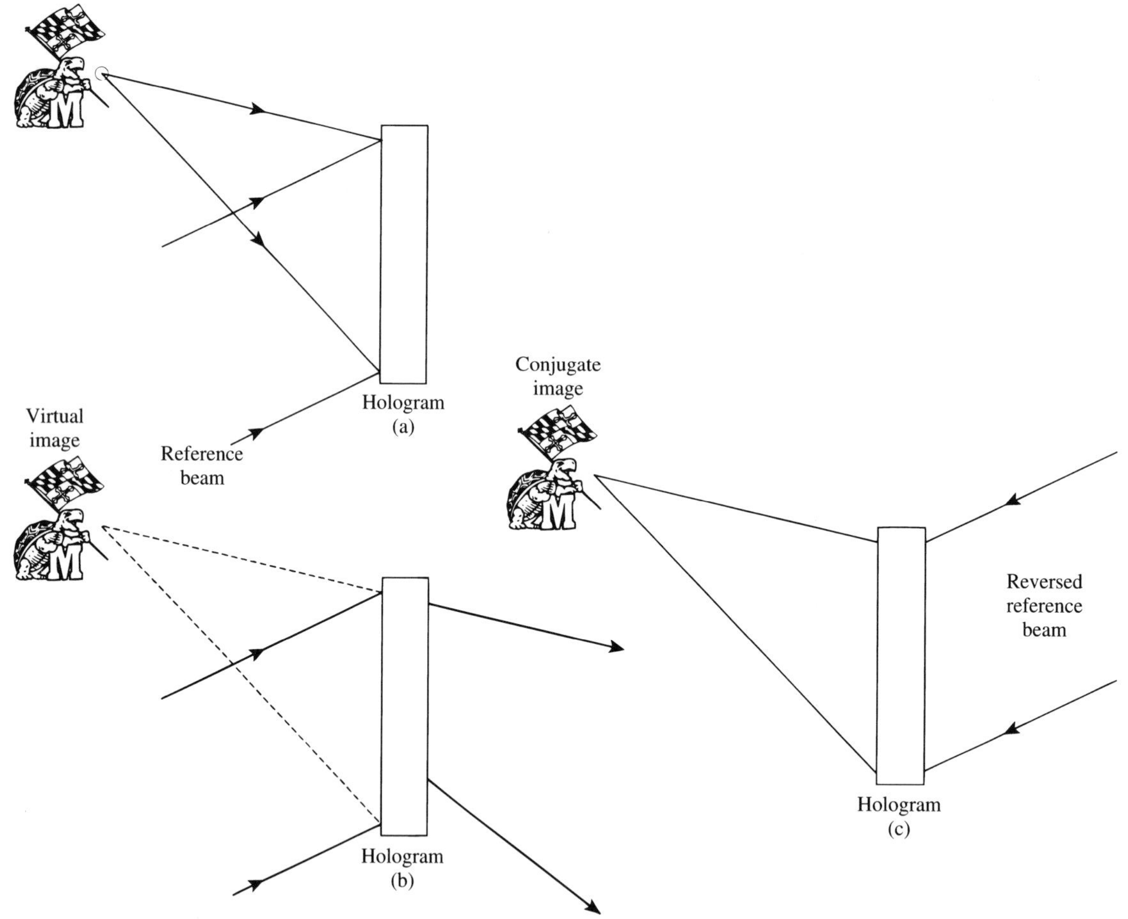

Fig. 24.12.
(a) Arrangement for writing a transmission volume hologram,
(b) reconstruction of virtual image, (c) reconstruction of conjugate (real) image. The happy terrapin shown is Testudo, the mascot of the University of Maryland, College Park.

separation does not, in reality, proceed quite so straightforwardly as described. Excited uranium atoms can collide and exchange energy, thereby scrambling the process. However, this scheme has been adapted by the United States Department of Energy as the principal means for meeting the future need for enriched uranium fuel after the current generation of gaseous diffusion plants ceases to operate.

There are also some interesting laser isotope separation processes that use molecular species. Two schemes that have been used are shown in Fig. (24.14). A tunable ultraviolet laser selectively excites one isotopic species of a molecule to a new electronic state where it dissociates. The fragments are thereby enriched in the isotopic species whose absorption line is coincident with the excitation laser. This scheme has been used to separate carbon isotopes through the predissociation of formaldehyde.

$$H_2^{13}CO \xrightarrow{h\nu} H_2 + {}^{13}CO_2. \tag{24.30}$$

Different isotopes of hydrogen have been separated from formaldehyde in a similar way

$$HD^{12}CO \xrightarrow{h\nu} HD + {}^{12}CO. \tag{24.31}$$

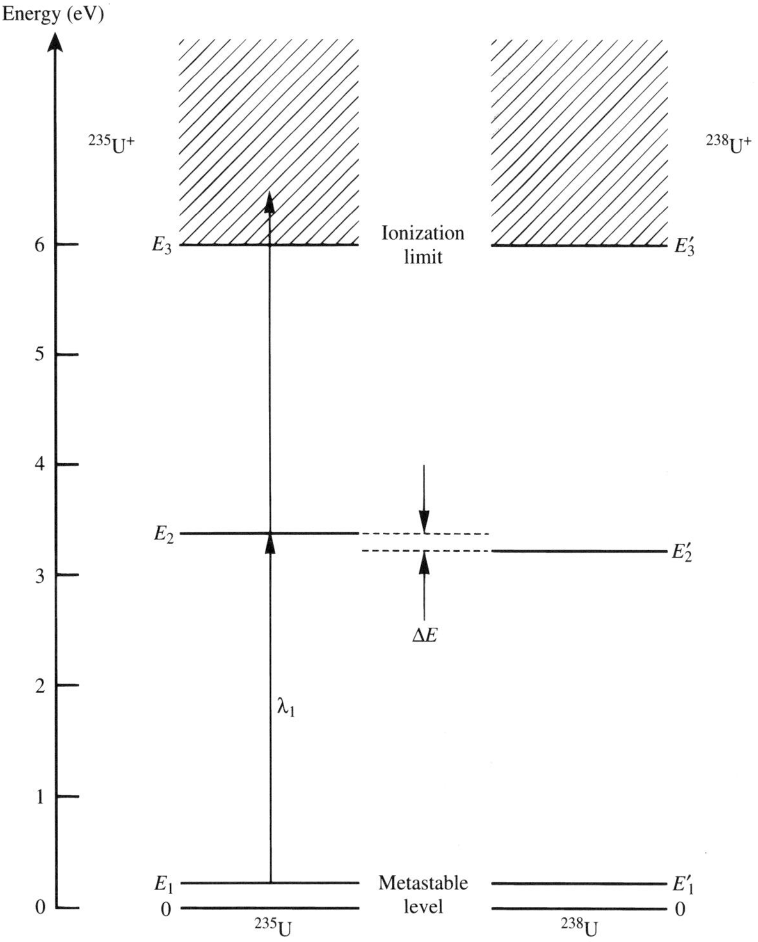

Fig. 24.13. Schematic energy level scheme of ^{235}U and ^{238}U showing how laser isotope separation can be performed.

The second scheme shown in Fig. (24.14) is one that has been actively studies as a means for separating ^{235}U-rich fragments from UF_6. A laser that is resonant with the v_3 vibration–vibration absorption in the ground electronic state of UF_6 selectively excites $^{235}UF_6$. This vibrationally excited molecule is then dissociated with a second energetic photon to produce a $^{235}UF_5$ fragment and a fluorine atom. The first selective excitation in UF_6 is provided with a 16 μm photon, the second with a 200–300 nm photon from an excimer laser. The $^{235}UF_5$ fragments combine chemically and can be separated. In order that the 16 μm absorption occurs in $^{235}UF_6$ and not $^{238}UF_6$, the UF_6 has to be cooled by expansion through a supersonic nozzle. This ensures that the absorption lines are sharp and higher vibrational levels are not populated thermally[24.36].

There has also been considerable success in separating isotopes through a process of multiphoton absorptions. In this process a molecule sequentially absorbs many identical photons and climbs up its ladder of vibrational energies to a sufficiently high energy that dissociation takes place. Although this process should not be efficient, because the vibrational energy levels of molecules are not equally spaced,

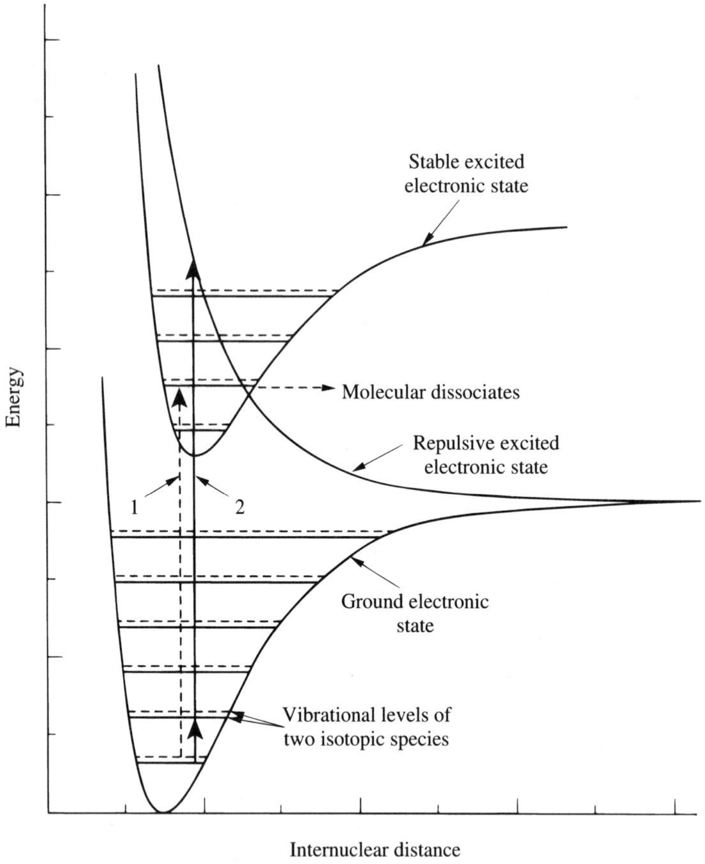

Fig. 24.14. Laser isotope schemes for molecular species. In process 1 a single energetic photon excites a particular isotope to a level where the molecule dissociates. In scheme 2 an infrared photon selectively excites a vibrational state of one isotope, which is then dissociated with a second, energetic photon.

it is in fact surprisingly so. For example, the various isotopes of sulfur in SF_6 have been separated by processes such as

$$^{32}SF_6 + n(h\nu) \rightarrow \; ^{32}SF_4 + 2F, \qquad (24.32)$$

$$SF_4 + H_2O \rightarrow SOF_2 + 2HF. \qquad (24.33)$$

The SOF_2 is chemically inert and can be separated from the other species[24.37].

One final isotope separation scheme uses laser enhanced chemical reactions. These schemes rely for their operation on the fact that excited states of atoms or molecules are frequently more reactive than those in the ground state. For example, the isotopes of chlorine can be separated by irradiating a mixture of iodine chloride and bromobenzene with red light from a dye laser. In order to obtain a sufficiently great photon flux the reaction cell is placed inside the cavity of the dye laser

$$I^{37}Cl + h\nu(\lambda = 605 \text{ nm}) \rightarrow (I^{37}Cl)^*, \qquad (24.34)$$

$$(I^{37}Cl)^* + C_6H_5Br \rightarrow C_6H_5\,^{37}Cl + I + Br. \qquad (24.35)$$

The resultant chlorobenzene ($C_6H_5{}^{37}Cl$) is enriched in ^{37}Cl.

24.4 Laser Plasma Generation and Fusion

Short pulse, high energy laser pulses when focused produce extremely high energy flux, electric and magnetic fields. For example, a 1 kJ 1 ns rectangular laser pulse focused to a spotsize of 1 μm produces an energy flux of 6.4×10^{23} W m^{-2}, an electric field of 2.2×10^{13} V m^{-2}, and a magnetic flux density in free space of 73 kT. Fields substantially smaller than this can strip outer and inner shell electrons from the atoms of a target material. Consequently, intense focused lasers have found widespread use in the production of plasmas from the surface of target materials. These laser produced plasmas can be copious producers of X-rays, which can be relatively coherent. Such X-rays have significant potential for high resolution holography and lithography. Because diffraction limits resolvable feature size in a microscope or lithography system, the shorter the wavelength being used the finer the resolution. Most laser-induced plasma generated X-radiation is coherent because of amplified spontaneous emission (ASE) effects. The rapid ionization of a target material, frequently a thin metal foil, irradiated with an intense laser pulse[24.38]–[24.40], can lead to a short lived population inversion on energy levels of the resultant multiply ionized ions. In many ways the process is an energetic analog of the excitation processes in self-terminating pulsed gas lasers such as copper vapor and nitrogen.

The irradiation of various materials at very high laser fluences (J m^{-2}) has, in and of itself, provided a fruitful means for studying matter at very high temperatures and pressures. However, much of the impetus for such studies has been provided by the desire to demonstrate laser-driven, inertial confinement, nuclear fusion. Nuclear fusion occurs in stars and hydrogen bomb explosions, and there has been work for several decades on the controlled production of such thermonuclear reactions on a laboratory scale for nuclear power generation.

Most controlled thermonuclear reactions (CTRs) are initiated in magnetically contained plasmas heated to high temperatures in reactors such as Tokamaks and Stellerators[24.41]. In these reactors a plasma, an electrically neutral mixture of gaseous ions and free electrons, is confined by a complex arrangement of magnetic fields and heated. Mixtures of deuterium, D, and tritium, T, are used as fuels. At sufficiently high temperatures the following fusion reactions occur:

$$\mathrm{D+D} \rightarrow\ ^3\mathrm{He}(0.82\ \mathrm{MeV}) + \mathrm{n}(2.45\ \mathrm{MeV}), \tag{24.36}$$

$$\mathrm{D+D} \rightarrow\ \mathrm{T}(1.01\ \mathrm{MeV}) + \mathrm{H}(3.02\ \mathrm{MeV}), \tag{24.37}$$

$$\mathrm{D+D} \rightarrow\ ^4\mathrm{He}(3.5\ \mathrm{MeV}) + \mathrm{n}(14.1\ \mathrm{MeV}), \tag{24.38}$$

$$\mathrm{D+{}^3He} \rightarrow\ ^4\mathrm{He}(3.6\ \mathrm{MeV})+\mathrm{H}(14.7\ \mathrm{MeV}). \tag{24.39}$$

At temperatures of 100 million degrees the first two reactions occur at roughly equivalent rates. The sustainability of a nuclear fusion reactor is frequently characterized in terms of the *Lawson criterion*, which says that at a temperature of 5 keV (1 eV=11 600 K), the product of ion number density, n_i, and containment time, τ, must satisfy

$$n_i\tau > 10^{20}\ \mathrm{m}^{-3}\ \mathrm{s}. \tag{24.40}$$

In a magnetically-confined reactor containment times are typically in the millisecond to second range.

The approach taken in laser fusion research is that of *inertial* confinement. The idea is to heat a small pellet of nuclear fuel, D–T mixtures are the most likely to be used, to high temperatures so fast that the mass of the fuel pellet prevents it from dispersing because of its own inertia. The Lawson criterion is thereby satisfied with a larger ion density and smaller containment time than for magnetically confined CTR. For a fuel pellet of radius a the natural inertial containment time is

$$\tau_i \simeq a/v_s, \tag{24.41}$$

where v_s is the velocity of sound in the fuel material at its elevated temperature. At a temperature of 10 keV the sound speed is $\sim 10^6$ m s^{-1}, so for a fuel pellet of radius 1 mm, the inertial containment time is only 1 ns. To accomplish the heating of a fuel pellet in a short time, while simultaneously confining it and enhancing the achievement of the Lawson criterion, the fuel pellet is irradiated simultaneously by a large number of high energy, short pulse laser beams. The irradiation should be as spherically symmetric as possible. In the Lawrence Livermore National Laboratory (LLNL) NOVA laser ten beams provide up to 24 kJ of energy in 1 ns pulses at a wavelength of 351 nm. This wavelength is obtained by third harmonic generation of the beams from oscillator/amplifier chains of Nd:phosphate-glass lasers operating at 1.053 μm. The irradiation of the fuel pellet vaporizes its outer surface to form a plasma. The plasma allows the laser beam to penetrate a certain distance, to the *critical-density* surface, where substantial energy is absorbed. The hot plasma conducts thermal energy inwards to the solid boundary of the fuel pellet. The rapid heating of the solid boundary ablates material, which explodes away from the rest of the pellet like a rocket engine. The equal and opposite reaction force of Newton's third law exerts an enormous pressure, perhaps 10^{12} atm, and compresses the remainder of the fuel pellet inwards. The core of the pellet is compressed to as much as 100 times solid density. The culmination of this compression and heating process is that perhaps 10% of the original fuel pellet is heated and compressed sufficiently that it undergoes a fusion reaction. The aim is to generate 100 MJ of fusion energy (equivalent to the explosion energy of 48 lb of TNT)† in each explosion and heating process. Fig. (24.15) shows schematically some of the processes that occur during the laser fusion process. Although in early experiments the fuel pellets were cooled solid (DT) spheres, or small glass microspheres containing (DT) gas at high pressure, current work is concentrating on the production of foam spheres that will hold the liquid DT[24.42]. The foam should contain only the light elements carbon, hydrogen, and oxygen and have a small pore size (<1 μm) to hold the liquid DT stably.

To heat the fuel pellet 'smoothly', so that its core is not heated too rapidly until it is sufficiently compressed, a special shape of laser pulse appears to be necessary. Theoretically the ideal laser pulse is

$$I(t) = I_0(1 - t/t_c)^{-2}, \tag{24.42}$$

where t_c is a characteristic time that depends on the precise characteristics of the fuel pellet. This pulse shape rises slowly at first and ultimately deposits a significant fraction of its total energy in a very short period at the end of the pulse.

Unfortunately, although the production of commercial quantities of electrical power by laser fusion can be made to appear feasible[24.43]–[24.45] and complete

† The explosive force of the microfusion 'bomb' is not as destructive as TNT since 80% of the energy is carried by neutrons. These neutrons can be absorbed in a surrounding blanket of ^{6}Li to regenerate the tritium $^6\mathrm{Li}+\mathrm{n}\rightarrow{}^4\mathrm{He}+\mathrm{T}$($\simeq$ 4.6 MeV).

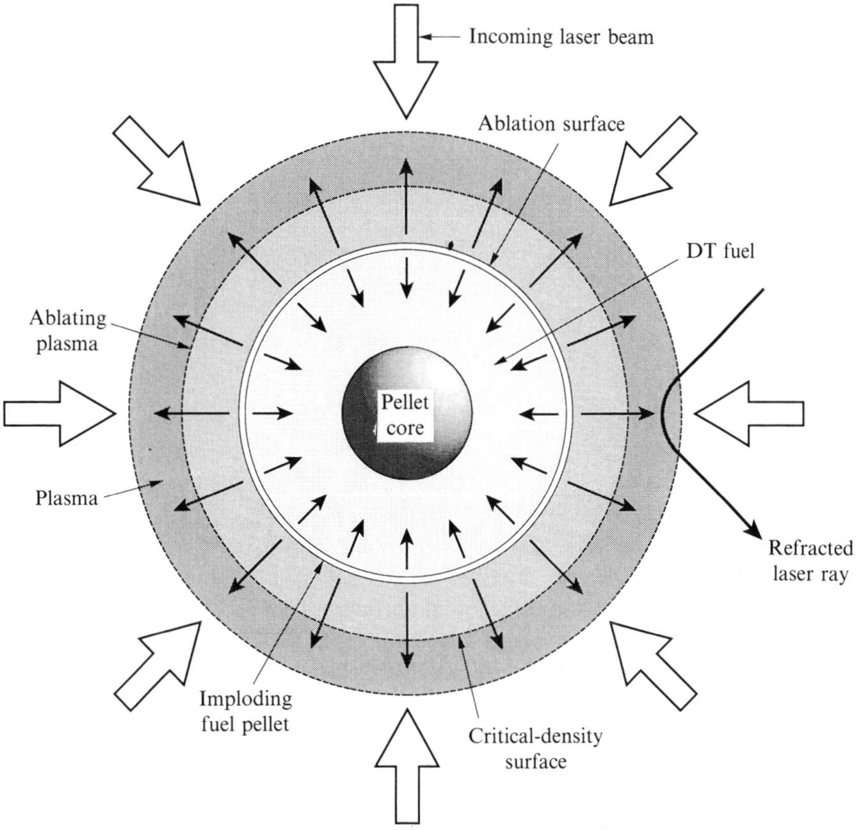

Fig. 24.15. Schematic illustration of the laser fusion process where a spherical pellet of nuclear fuel is irradiated symmetrically with intense laser beams.

reactor concepts have been developed, the 'break-even' point, where the fusion reaction generates as much energy as is injected into the drive lasers, seems far away. However, this research has led to a substantial understanding of laser–matter interactions and significant development in materials research.

24.5 Medical Applications of Lasers

Almost from its inception the laser has found applications in medicine. Sometimes the laser has emerged as a means for carrying out a new surgical procedure, in other cases the laser presents an alternative means for carrying out an existing procedure. Perhaps the best known of these surgical applications of the laser has been in the reattachment of detached retinas. If the detached retina will settle back into the correct position it can literally be 'welded' back into place by focusing a CW argon ion or a pulsed or CW Nd:YAG laser through the lens of the eye onto the back of the eye. Bleeding inside the eye can also be stopped by laser heating the site of bleeding – a process called *photocoagulation*. The same lasers that present a serious eye-hazard, because they can penetrate to the back of the eye, can be used for these surgical procedures inside the eye. As can be seen from Fig. (24.16) wavelengths beyond about 1.5 μm are absorbed before reaching the retina. Lasers of these longer wavelengths can still be hazardous to the eye, but primarily as a means of causing burns to the outer surface of the cornea. The

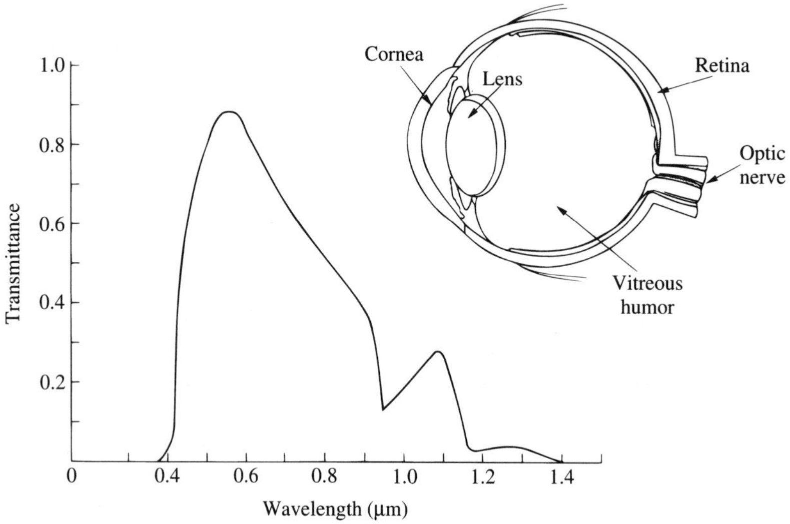

Fig. 24.16. Simplified cross-section of the human eye showing the spectral transmittance to the retina from the outside of the eye.

ability of lasers to burn, or cut, the outer surfaces of the cornea has led to new, and still controversial, procedures in which a laser is used to sculpt the cornea to correct vision defects. This procedure is called radial keratotomy and is generally carried out with excimer lasers.

Ultraviolet, visible, and infrared lasers are readily absorbed by body tissue so they have an ability to act as scalpels in specialized treatments. For treatment of certain skin conditions, including some cancers, the ability of the laser to burn, or photoablate,† and selectively remove thin layers near the surface without damaging underlying tissue is very valuable. Much of the ability of the laser to burn locally, and selectively, is based on the absorption of the laser wavelength by water. Consequently CO_2 lasers are more effective than argon ion or Nd:YAG lasers in these scalpel-like applications. However, the absorption of water in the body is largest near 3 μm so lasers near this wavelength, such as erbium:YAG, are attracting great interest.

A clearly beneficial application of pulsed blue-green dye lasers is in the fragmentation of kidney stones that have left the kidney and become trapped in the urinary tract. Pulses of intense visible light directed along an optical fiber at the kidney stone successively break the stone into several pieces that are then able to leave the body. It is preferable to use a tunable dye laser for this treatment so as to maximize the energy absorption by the offending stones while at the same time not heating up any blood that is in the vicinity of the stone.

For general application as scalpels lasers are attractive when used to cut very vascular tissue, which bleeds a great deal when cut. The laser cuts and cauterizes at the same time, the result being an incision with a layer of burnt tissue on both sides of the cut. Lasers are even finding application in 'painless' dentistry. Lasers have many additional non-surgical uses in diagnostics and monitoring, for example in analytical procedures, and as optical sources in various types of sensors for temperature, pressure, and blood flow.

† The rapid absorption by tissue of energy from a pulsed laser that leads to an ejection of material from the surface by shock without significant burning.

For performing surgery inside the body the laser beam must be directed inside an optical fiber system that at the same time provides the surgeon with a TV image of the region at the end of the fiber. An exciting application of lasers in this way involves the burning away or ablation, of the plaque inside blocked arteries, thereby opening them up once again to blood flow. This procedure is called laser angioplasty and is worthy of further discussion.

24.5.1 Laser Angioplasty

Arteriosclerosis is the constriction, or blocking, of blood flow through arteries. The material that leads to this problem is called plaque, a composite material containing layers of fat, fibrous tissue, muscle tissue, and varying amounts of calcium (as $Ca_3(PO_4)_2$). There has been great interest, and some success, in using both CW and pulsed lasers to ablate plaque away. A laser beam is directed at the blockage in an artery along an optical fiber inserted through a catheter into the artery. The catheter is threaded over a previously placed guide wire. The use of lasers in this way is called *laser angioplasty*, in contrast with the earlier (and still more widely used and successful) technique of balloon angioplasty. In this technique a small balloon is threaded into the occluded region of the artery and then filled with fluid, thereby compressing the plaque against the wall of the artery.

The problems that have been faced in the use of lasers to remove plaque have involved:

(i) Avoidance of damage to the wall of the artery, which is only about 1 mm thick. This problem has been alleviated to a large extent by placing a ball lens on the end of the fiber[24.46].

(ii) The opened region can reocclude within a few months.

(iii) Thermal damage to healthy tissue can occur. The use of pulsed lasers seems to avoid the thermal damage problem while short intense pulses are effective in photoablating the plaque. Sub-microsecond ultraviolet pulses from XeCl lasers at 308 nm are being investigated in this application[24.47] as is the use of holmium lasers operating at 2.1 μm. These lasers are used in a long pulse mode delivering a few joules per pulse in 1/4 second long pulses at 1-10 pulses per second. Their wavelength of 2.1 μm is strongly absorbed by water. Both these ultraviolet and near-infrared wavelengths can be delivered to the occluded site along silica fibers. The treatment of the occluded artery is frequently followed up by an application of balloon angioplasty.

References

[24.1] Alexandre Dumas, *The Count of Monte Cristo*, Dodd Mead, New York, 1979.

[24.2] J. Tyndall, *Roy. Inst. Great Britain Proc.*, **6**, 189, 1870–1872.

[24.3] K.C. Kao and G.A. Hockham, 'Dielectric-fiber surface waveguides for optical frequencies,' *Proc. IEEE*, **113**, 1151–1158, 1966.

[24.4] T. Miya, Y. Terunuma, T. Hosaka, and T. Miyashita, 'Ultimate low-loss single-mode fiber at 1.55 μm,' *Electron. Lett.*, **15**, 106–108, 1979.

[24.5] P.E. Green, Jr, *Fiber Optic Networks*, Prentice-Hall, Englewood Cliffs, 1993.

[24.6] David J. Morris, *Pulse Code Formats for Fiber Optical Data Communication*, Marcel Dekker, New York, 1983.

[24.7] Harold B. Killen, *Fiber Optic Communications,* Prentice-Hall, Englewood Cliffs, 1991.

[24.8] Joseph C. Palais, *Fiber Optic Communications*, 3rd Edition, Prentice Hall, Englewood Cliffs, 1992.

[24.9] E.N. Leith and J. Upatnieks, 'Wavefronts reconstruction with continuous-tone objects,' *J. Opt. Soc. Am.*, **53**, 1377–1381, 1963.

[24.10] E.N.Leith and J. Upatnieks, 'Wavefront reconstruction with diffused illumination and three-dimensional objects,' *J. Opt. Soc. Am.*, **54,** 1295–1301, 1969.

[24.11] D. Gabor, 'Holography, 1948–71,' *Proc. IEEE*, **60**, 655–668, 1972.

[24.12] D. Gabor, 'A new microscopic principle,' *Nature*, **161**, 777–778, 1948.

[24.13] D. Gabor, 'Microscopy by reconstructed wavefronts,' *Proc. Roy. Soc.*, **A197**, 454–487, 1949.

[24.14] D. Gabor, 'Microscopy by reconstructed wavefronts, II,' *Proc. Roy. Soc.*, **A197**, 454–487, 1949.

[24.15] H.M. Smith, *Principles of Holography*, 2nd Edition, Wiley Interscience, New York, 1975.

[24.16] G.W. Stroke, *An Introduction to Coherent Optics and Holography.*

[24.17] W.T. Cathey, *Optical Information Processing and Holography*, Wiley, New York, 1974.

[24.18] H.J. Caulfield and S. Lu, *The Applications of Holography*, Wiley-Interscience, New York, 1970.

[24.19] B.J. Thompson, 'Principles and Applications of Holography,' in *Applied Optics and Optical Engineering*, Vol. 6, R. Kingslake and B.J. Thompson, Eds., Academic, New York, 1980.

[24.20] K. Iizuka, *Engineering Optics*, 2nd Edition, Springer-Verlay, Berlin, 1987.

[24.21] J.W. Goodman, *Introduction to Fourier Optics*,, McGraw-Hill, New York, 1968.

[24.22] J.D. Gaskill, *Linear Systems, Fourier Transforms, and Optics*, 2nd Edition, Wiley, New York, 1978.

[24.23] P.M. Duffieux, *The Fourier Transform and Its Applications to Optics*, 2nd Edition, Wiley, New York, 1983.

[24.24] A. Papoulis, *Systems and Transforms with Applications in Optics*, McGraw-Hill, New York, 1968.

[24.25] J. Upatnieks and C. Leonard, 'Diffraction efficiency of bleached photographically recorded interference patterns,' *Appl. Opt.*, **8**, 85–89, 1969.

[24.26] M. Young and Kittridge, 'Amplitude and phase holograms exposed on Agfa-Gevaert 10E75 planes,' *Appl. Opt.*, **8**, 2453–2454, 1969.

[24.27] M. Francon, *Laser Speckle and Applications in Optics*, Academic Press, New York, 1979.

[24.28] J.C. Dainty, Ed., *Laser Speckle and Related Phenomena*, 2nd Edition, Topics in Applied Physics, Vol. 9, Springer-Verlag, Berlin, 1984.

[24.29] W. Schumann and M. Dubas, *Holographic Interferometry*, Springer Series on Optical Science, Vol. 16, Springer-Verlag, Berlin, 1979.

[24.30] Y.I. Ostrovsky, M.M. Butusov, and G.V. Ostrovskaya, *Interferometry by Holography*, Springer Series on Optical Science, Vol. 20, Springer-Verlag, Berlin, 1980.

[24.31] E. Marom, A.A. Friesem, and W. Wiener-Arenear, *Applications of Holography and Optical Data Processing*, Pergamon, New York, 1977.

[24.32] F.T.S. Yu, *Optical Information Processing*, Wiley, New York, 1983.

[24.33] S.H. Lee, Ed., *Optical Information Processing*, Topics in Applied Physics, Vol. 48, Springer-Verlag, Berlin, 1981.

[24.34] V.S. Letokhov and C.B. Moore, 'Laser isotope separation,' *Sov. J. Quant. Electron*, **6**, 129–150, 1976 and **6**, 249–276, 1976.

[24.35] C.D. Cantrell, S.M. Freund, and J.L. Lyman, 'Laser-induced Chemical Reactions and Isotope Separation,' in *Laser Handbook*, Vol. III, M.L. Stitch Ed., North Holland, Amsterdam, 1979.

[24.36] R.J. Jensen, O.P. Judd, and J.A. Sullivan, 'Separating isotopes with lasers,' *Los Alamos Science*, **3**, 2–33, 1982.

[24.37] R.V. Ambartsumyan, Yu A. Gorokhov, V.S. Letokhov, G.N. Makarov, and A.A. Puretskii, 'Explanation of the selective dissociation of the SF_6 molecule in a strong IR laser field,' *JETP Lett.*, **23**, 22–25, 1976.

[24.38] D.L. Matthews *et al.*, 'Demonstration of a soft X-ray amplifier,' *Phys. Rev. Lett.*, **54**, 110–113, 1985.

[24.39] M.D. Rosen *et al.*, 'The exploding foil technique for achieving a soft X-ray laser,' *Phys. Rev. Lett.*, **54**, 106–109, 1985.

[24.40] S. Suckewer *et al.*, 'Amplified spontaneous emission at 18.2 nm,' *Phys. Rev. Lett.*, **55**, 1753–1756, 1985.

[24.41] Robert W. Conn, 'The engineering of magnetic fusion reactors,' *Sci. Am.*, October 1983.

[24.42] R.A. Sacks and D.H. Darling, 'Direct drive cryogenic ICF capsules employing DT wetted foam,' *Nucl. Fusion*, **27**, 447–452, 1987.

[24.43] J. Nuckolls, L. Wood, A. Thiessen, and G. Zimmerman, 'Laser compression of matter to super-high densities: thermonuclear (CTR) applications,' *Nature*, **249**, 139–142, 1972.

[24.44] L.A. Booth, D.A. Freiwald, T.G. Frank, and F.T. Finch, 'Prospects of generating power with laser-driven fusion,' *Proc. IEEE*, **64**, 1460–1482, 1976.

[24.45] R.S. Craxton, R.C. McCrory, and J.M. Soures, 'Progress in laser fusion,' *Sci. Am.*, **255**, 68–79, August 1986.

[24.46] G.S. Abela and G.R. Barbeau, 'Laser angioplasty: potential effects and current limitations,' in *Textbook of Interventional Cardiology*, E.J. Topol, Ed., W.B. Saunders, Philadelphia, PA, 1990, chapter 37, pp. 724–737.

[24.47] D.L. Singleton, G. Paraskevopoulos, G.S. Jolly, R.S. Irwin, and D.J. McKenny, 'Excimer lasers in cardiovascular surgery: Ablation products and photoacoustic spectrum of arterial wall,' *Appl. Phys. Lett.*, **48**, 878–880, 1986.

APPENDIX 1
Optical Terminology

Light is electromagnetic radiation with a wavelength between 0.1 nm and 100 μm. Of course, this selection of wavelength range is somewhat arbitrary; short wavelength vacuum-ultraviolet light between 0.1 nm and 10 nm (so called because these wavelengths – and a range above up to about 200 nm – are absorbed by air and most gases) might as easily be called soft X-radiation. At the other end of the wavelength scale, between 100 μm and 1000 μm, the sub-millimeter wave region of the spectrum, lies a spectral region where conventional optical methods become difficult, as do the extension of microwave techniques which are easily applied to the centimeter and millimeter wave regions. The use of optical techniques in this region, and even in the millimeter region, is often called quasi-optics.

Table (A1.1) summarises the important parameters that are used to characterize light and the media through which it passes. A few comments on the table are appropriate. Although the velocity of light in a medium depends both on the relative magnetic permeability μ_r and dielectric constant ϵ_r of the medium, for all practical optical materials $\mu_r \simeq 1$ so the refractive index and dielectric constant are related by

$$n = \sqrt{\epsilon_r}. \tag{A1.1}$$

When light propagates in an anisotropic medium, such as a crystal of lower than cubic symmetry, n and ϵ_r will, in general, depend on the direction of propagation of the wave and its polarization state. The velocity of light *in vacuo* is currently the most precisely known of all the physical constants – its value is known[A1.1] within 4 cms^{-1}. A redefinition of the meter has been adopted[A1.2] based on the cesium atomic clock frequency standard[A1.3] and a velocity of light of exactly 2.99792458×10^8 ms^{-1}. The velocity, wavelength, and wavenumber of light travelling in the air have slightly different values than they have in *vacuo*. Tables are available[A1.4] that tabulate corresponding values of $\overline{v}$ *in vacuo* and in standard air.

A plane electromagnetic wave travelling in the z direction can, in general, be decomposed into two independent, linearly polarized components.† The electric and magnetic field pairs associated with each of these components are themselves mutually orthogonal and transverse to the direction of propagation. They can be written as (E_x, H_y) and (E_y, H_x). The ratio of the mutually orthogonal **E** and **H** components is called the *impedance* Z of the medium.

$$\frac{E_x}{H_y} = \frac{-E_y}{H_x} = Z = \sqrt{\frac{\mu_r\mu_0}{\epsilon_r\epsilon_0}}. \tag{A1.2}$$

† The wave could also be decomposed into left and right hand circularly polarized components.

Table A1.1. *Fundamental parameters of electromagnetic radiation and optical media.*

Parameter	Symbol	Value	Units
Velocity of light in free space *in vacuo*	$c_0 = \sqrt{\mu_0 \epsilon_0}$	2.99792458 $\times 10^{8a}$	m s^{-1}
Permeability of free space	μ_0	$4\pi \times 10^{-7}$	H m^{-1}
Permittivity of free space	ϵ_0	8.85416×10^{-12}	F m^{-1}
Velocity of light in a medium	$c = 1/\sqrt{\mu_r \mu_0 \epsilon_r \epsilon_0} = c_0/n$		m s^{-1}
Refractive index	$n = \sqrt{\mu_r \epsilon_r}$		dimensionles
Relative permeability of a medium	μ_r	usually 1	dimensionless
Dielectric constant of a medium	ϵ_r		dimensionless
Frequency	$\nu = c/\lambda$		Hz 10^9 Hz = 1 GHz (gigahertz) 10^{12} Hz = 1 THz (terahertz)
Wavelength *in vacuo*	$\lambda_0 = c_0/\nu$		m
Wavelength in a medium	$\lambda = c/\nu = \lambda_0/n$		m 10^{-3} m = 1 mm (millimeter) 10^{-6} m = 1 μm (micrometer) 10^{-9} m = 1 nm (nanometer) 10^{-12} m = 1 pm (picometer) 1 Å(angstrom) = 10^{-10} m 1 μ (micron) = 1 μm 1 mμ (millimicron) = 1 nm
Wavenumber	$\bar{\nu} = 1/\lambda$		cm^{-1} (Kayser)
Wave vector	$\mathbf{k}, \lvert \mathbf{k} \rvert = 2\pi/\lambda$		m^{-1}
Photon energy	$E = h\nu$		J 1 eV (electron volt) = 1.60202×10^{-19} J
Electric field of wave	$\mathbf{E}$		V m^{-1}
Magnetic field of wave	$\mathbf{H}$		A m^{-1}
Poynting vector	$\mathbf{P} = \mathbf{E} \times \mathbf{H}$		W m^{-2}
Intensity	$I = \langle \lvert \mathbf{P} \rvert \rangle_{AV} = \lvert \mathbf{E} \rvert^2/2Z$		W m^{-2}
Impedance of medium	$Z = E_x/H_y = -E_y/H_x = \sqrt{\mu_r \mu_0/\epsilon_r \epsilon_0}$		Ω
Impedance of free space	$Z_0 = \sqrt{\mu_0/\epsilon_0}$	376.7	Ω

[a]See text.

The negative sign in Eq. (A1.2) arises because (y, x, z) is not a right handed coordinate system.

The Poynting vector, **S**, where

$$\mathbf{S} = \mathbf{E} \times \mathbf{H}, \tag{A1.3}$$

is a vector which points in the direction of energy propagation of the wave. The average magnitude of the Poynting vector is called the intensity I

$$I = \langle | \mathbf{S} | \rangle_{AV} = \frac{|\mathbf{E}|^2}{2Z}. \tag{A1.4}$$

The factor of 2 comes from time-averaging the square of the sinusoidally varying electric field.

The wave vector **k** points in the direction perpendicular to the phase front of the wave (the surface of constant phase). In an isotropic medium **k** and **S** are always parallel.

The photon flux corresponding to an electromagnetic wave of average intensity, I, and frequency, ν, is

$$N = I/h\nu = I\lambda/hc, \tag{A1.5}$$

where h is Planck's constant, 6.626×10^{-34} J s.

For a wave of intensity 1 W m^{-2} and wavelength 1 μm *in vacuo* $N = 5.04\times 10^{18}$ photons s^{-1} m^{-2}. Photon energy is sometimes measured in electron volts (eV): 1 eV $= 1.60202\times 10^{-19}$ J. A photon of wavelength 1 μm has an energy of 1.24 eV. It is often important, particularly in the infrared, to know the correspondence between photon and thermal energies. The characteristic thermal energy at absolute temperature is kT. At 300 K, $kT = 4.14\times 10^{-21}$ J $= 208.6$ cm^{-1} $= 0.026$ eV.

References

[A1.1] K.M. Baird, D.S. Smith, and B.G. Whitford, "Confirmation of the currently accepted value 299792458 meters per second for the speed of light", *Opt. Commun.*, **31**, 367–368, 1979.

[A1.2] E.R. Cohen andB.N. Taylor, Committee on Data for Science and Technology, CODATA Bulletin, **No. 63**, 1986.

[A1.3] G.W.C. Kaye and T.H. Laby, *Tables of Physical and Chemical Constants*, 14th Edition, Longman, New York, 1972.

[A1.4] C.D. Coleman, W.R. Bozman, and W.F. Meggers, *Table of Wavenumbers*, Vols. 1–2, 1960.

APPENDIX 2

The δ-Function

The δ-function† is a *pathological*‡ mathematical function with the following properties:

$$\begin{aligned} \delta(k - k_0) &= 0, \qquad k \neq k_0, \\ \delta(k - k_0) &= \infty \qquad k = k_0, \\ \int_{-\infty}^{\infty} \delta(k - k_0) dk &= 1. \end{aligned} \tag{A2.1}$$

The δ-function is the Fourier transform of a monochromatic complex exponential function, $f(x)$, where

$$f(x) = e^{-ik_0 x}, \tag{A2.2}$$

and therefore,

$$\delta(k - k_0) = \frac{1}{2\pi} \int_{-\infty}^{\infty} e^{i(k-k_0)x} dx. \tag{A2.3}$$

We can simplify our notation by writing $\delta(k - k_0)$ as $\delta(k)$ when the function is centered at the origin, $k_0 = 0$. Additional properties of the δ-function are

$$\delta(t) = \delta(-t), \tag{A2.4}$$

$$\int f(t)\delta(t - t_0) dt = f(t_0), \tag{A2.5}$$

$$\delta(at) = \frac{1}{a}\delta(t), \qquad a > 0, \tag{A2.6}$$

$$\int \delta(t - t')\delta(t' - t'')dt' = \delta(t - t''), \tag{A2.7}$$

$$f(t)\delta(t - t_0) = f(t_0)\delta(t - t_0). \tag{A2.8}$$

The derivative of the δ-function can also be defined:

$$\delta'(k) = \frac{1}{2\pi} \int_{-\infty}^{\infty} ike^{ikx} dx, \tag{A2.9}$$

$$\delta'(k) = -\delta'(-k), \tag{A2.10}$$

$$\int f(k)\delta'(k - k_0) dk = -f'(k_0). \tag{A2.11}$$

† Originally introduced in quantum theory, by P.A.M. Dirac, see P.A.M. Dirac, *Principles of Quantum Mechanics*, Oxford University Press, Oxford, 3rd Edition 1947. In engineering applications it is often called the *unit impulse function*.

‡ Strictly speaking it is no function at all.

Care must be taken in assigning values to the integrals above because of the singular behavior of $\delta(k)$. If, however, $\delta(k)$ is written as the limit of a rational function then such problems can be avoided.

For example, a normalized Gaussian function can be written as

$$g(x - x_0) = \frac{1}{\sigma\sqrt{2\pi}} e^{-(x-x_0)^2/2\sigma^2}, \tag{A2.12}$$

so that

$$\int_{-\infty}^{\infty} g(x - x_0) dx = \sigma \frac{1}{a\sqrt{2\pi}} \int_{\infty}^{\infty} e^{-(x-x_0)^2/2\sigma^2} dx = 1. \tag{A2.13}$$

This allows us to write the δ-function as the limiting case of a normalized Gaussian function whose width shrinks to zero, that is,

$$\delta(x - x_0) = \lim_{\sigma \to 0} \frac{1}{a\sqrt{\sigma\pi}} e^{-(x-x_0)^2/2\sigma^2}. \tag{A2.14}$$

The δ-function can be written in many different ways as the limit of an appropriately normalized function. For example, in terms of a normalized Lorentzian function.

$$\delta(v - v_0) = \lim_{\Delta v \to 0} \frac{(2/\pi\Delta v)}{1 + [2(v - v_0)/\Delta v]^2}. \tag{A2.15}$$

We can extend the definition of the δ-function to two or more dimensions. The three-dimensional δ-function is

$$\begin{aligned} \delta(\mathbf{k}) &= \delta(k_x)\delta(k_y)\delta(k_x) \\ &= \frac{1}{(2\pi)^3} \int_{-\infty}^{\infty} \int_{-\infty}^{\infty} \int_{-\infty}^{\infty} e^{i(k_x x + k_y y + k_z z)} dx dy dz \\ &= \left(\frac{1}{2\pi}\right)^3 \int_{-\infty}^{\infty} e^{i\mathbf{k}\cdot\mathbf{r}} d\mathbf{r}. \end{aligned} \tag{A2.16}$$

The δ-function is closely related to the unit step or *Heaviside* function $u(t)$ defined by

$$\begin{aligned} u(t) &= 0, \qquad t < 0, \\ &= 1, \qquad t > 0, \end{aligned} \tag{A2.17}$$

such that

$$\delta(t) = \frac{d}{dt} u(t). \tag{A2.18}$$

Appendix 3

Black-Body Radiation Formulas

The energy density distribution with frequency of black-body radiation is

$$\rho(\nu) = \frac{8\pi h\nu^3}{c^3} \frac{1}{e^{h\nu/kT} - 1}, \tag{A3.1}$$

where $\rho(\nu)$ is measured in units J m^{-3} Hz^{-1}.

The corresponding energy distribution with wavelength is

$$\rho(\lambda) = \frac{8\pi hc}{\lambda^5} \frac{1}{e^{hc/\lambda kT} - 1}, \tag{A3.2}$$

where $\rho(\lambda)d\lambda$ is the energy stored per unit volume that lies within the region $\lambda \to \lambda + d\lambda$, measured in units J m^{-3} m^{-1}. From $\rho(\lambda)$ can be derived an expression for the *spectral emittance*, $M_{e\lambda}$, the total power emitted per unit wavelength interval into a solid angle 2π by unit area of the black body, where

$$M_{e\lambda} = \frac{c_1}{\lambda^5(e^{c_2/\lambda T} - 1)}, \tag{A3.3}$$

$c_1 = 2\pi hc^2$ is called the *first radiation constant*, it has the value 3.7418×10^{-16} W m^2; $c_2 = ch/k$ is called the *second radiation constant*, it has the value 1.43877×10^{-2} m K.

The *radiance*, L_e, of an extended source is the radiant flux that it emits per unit solid angle per unit area of the source:

$$L_e = \frac{\delta I_e}{\delta S_n}, \tag{A3.4}$$

where the area δS_n is the projection of the surface element of the source in the direction being considered and δI_e is the *radiant intensity*, the amount of power emitted by this surface element in a particular direction per solid angle, measured in watts per steradian. A black-body has a radiance that is independent of the viewing angle – it is said to be a perfectly diffuse or *Lambertian* radiator. Clearly for such a source the radiant intensity at an angle θ to the normal to its surface is

$$I_e(\theta) = I_e(0)\cos\theta. \tag{A3.5}$$

For a surface element δS, the projected area at angle θ to the normal is $\delta S\cos\theta$, so (A3.5) in conjunction with (A3.4) gives $L_e = I_e(0)$. The total flux emitted per unit area is its *radiant emittance*, M_e, where

$$M_e = \pi I_e(0). \tag{A3.6}$$

The relationship between spectral emittance, $M_{e\lambda}$, and spectral radiance, $L_{e\lambda}$ (W sr^{-1} m^{-2} m^{-1}), is

$$M_{e\lambda} = \pi L_{e\lambda}. \tag{A3.7}$$

For a black-body source the wavelength of maximum emittance, λ_m, obeys Wien's displacement law,

$$\lambda_m T = 2.898 \times 10^{-3}\text{m K}. \tag{A3.8}$$

The total radiant emittance of a black body at temperature T is

$$M_e = \int_0^\infty M_{e\lambda} d\lambda = \frac{2\pi^5 k^4}{15c^2h^3} T^4 = \sigma T^4. \tag{A3.9}$$

This is a statement of the Stefan–Boltzmann law. The coefficient σ, called the *Stefan–Boltzmann constant*, has a value of 5.6705×10^{-8} W m^3 K^{-4}. The spectral emittance in terms of photons, N_λ, is

$$N_\lambda = \frac{M_{e\lambda}}{hc/\lambda}. \tag{A3.10}$$

APPENDIX 4
RLC Circuit

A4.1 Analysis of a Driven *RLC* Circuit

For an *RLC* circuit such as shown in Fig. (2.14) with a driving voltage $v(t)$, application of Kirchoff's law to the circuit gives

$$L\frac{dI}{dt} + RI + \frac{Q}{C} = V(t), \tag{A4.1}$$

where I is the current in the circuit and Q is the change on the capacitor. This can be more conveniently written as

$$L\frac{d^2I}{dt^2} + R\frac{dI}{dt} + \frac{I}{C} = \frac{dV}{dt}. \tag{A4.2}$$

Before considering the case of a sinusoidal driving voltage we will consider the behavior of the circuit if the voltage source is shorted out. In this case

$$L\frac{d^2I}{dt^2} + R\frac{dI}{dt} + \frac{I}{C} = 0. \tag{A4.3}$$

For this linear, second order differential equation we try the solution $I = I_0 e^{-at}$, which gives

$$La^2 - Ra + \frac{1}{C} = 0 \tag{A4.4}$$

and, therefore

$$a = \frac{R \pm \sqrt{R^2 - 4L/C}}{2L}. \tag{A4.5}$$

If $R > 4L/C, a$ is real and the current decays purely exponentially. In this case

$$a = \frac{R \pm R\sqrt{1 - 4L/R^2C}}{2L} \simeq \frac{R \pm R(1 - 2L/R^2C)}{2L}, \tag{A4.6}$$

which can be written

$$a \simeq \frac{R \pm R}{2L} \mp \frac{1}{RC}. \tag{A4.7}$$

Clearly, we must choose the lower sign in order that the current decays as

$$I_0 = I_0 e^{-t/RC}, \tag{A4.8}$$

which is the familiar result for decay of a capacitor through a resistor.

If $R < 4L/C$ then

$$a = \frac{R - i\sqrt{4L/C - R^2}}{2L} = \alpha - i\omega', \tag{A4.9}$$

where $\alpha = R/2L$ and ω' is the natural oscillation frequency of the circuit

$$\omega' = \sqrt{\frac{1}{LC} - \frac{R^2}{4L^2}}. \quad (A4.10)$$

The decay of the current is then given by the damped oscillation

$$I = I_0 e^{-R/2L} e^{i\omega' t}. \quad (A4.11)$$

If the circuit is driven with a sinusoidal voltage of the form $V_0 e^{i\omega t}$, then it is reasonable to assume that the current will be at the frequency ω. Therefore, as a solution to Eq. (A4.2) we try $I = I_0 e^{i\omega t}$ where I_0 may be complex (includes a phase factor). Substitution in Eq. (A4.2) gives

$$\left(-L\omega^2 + i\omega R + \frac{1}{C}\right) I_0 = i\omega V_0, \quad (A4.12)$$

which gives

$$I_0 = \frac{iV_0}{(1/\omega C - L\omega) + iR} = \frac{i(1/\omega C - L\omega - iR)}{(1/\omega C - L\omega)^2 + R^2} V_0. \quad (A4.13)$$

This can be written in the form

$$I_0 = \frac{V_0 e^{i\phi}}{\sqrt{(1/\omega C - L\omega)^2 + R^2}}, \quad (A4.14)$$

where $\tan\phi = (1/\omega C - L\omega)/R$.

The amplitude of the current varies with frequency according to

$$| I_0 | = \frac{| V_0 |}{\sqrt{(1/\omega C - L\omega)^2 + R^2}} = \frac{| V_0 | /R}{\sqrt{1 + (1 - \omega^2 LC/\omega RC)^2}}. \quad (A4.15)$$

Maximum current results at the resonance frequency of the circuit, where $(1/\omega_0 C - L\omega_0) = 0$. Therefore

$$\omega_0 = \frac{1}{\sqrt{LC}}. \quad (A4.16)$$

Eq. (A4.15) can be written in the form

$$| I_0 | = \frac{| V_0 | /R}{\sqrt{1 + (1 - \omega\sqrt{LC})^2 (1 + \omega\sqrt{LC})^2 / \omega^2 R^2 C^2}}, \quad (A4.17)$$

which near resonance, $\omega \simeq \omega_0$, becomes

$$| I_0 | = \frac{| V_0 | /R}{\sqrt{1 + [2(\omega - \omega_0)/\omega_0^2 RC]^2}}. \quad (A4.18)$$

The power $W(\omega)$ dissipated in the circuit is $R \mid I_0 \mid^2$. Therefore

$$W(\omega) = \frac{| V_0 |^2 /R}{1 + [2(\omega - \omega_0)/\omega_0^2 RC]^2}, \quad (A4.19)$$

which is a Lorentzian-shaped resonance curve with FWHM

$$\Delta\omega = \omega_0^2 RC. \quad (A4.20)$$

The *quality factor*, Q of the circuit is defined as

$$Q = \frac{\omega_0}{\Delta\omega} = \frac{1}{\omega_0 RC}. \tag{A4.21}$$

In terms of Q the power spectrum of the circuit is

$$W(\omega) = \frac{|V_0|^2}{1 + [2(\omega - \omega_0)Q/\omega_0]^2}. \tag{A4.22}$$

APPENDIX 5

Storage and Transport of Energy by Electromagnetic Fields

The total energy (per unit volume) stored by a system of electric and magnetic fields is,† in an isotropic medium,

$$\rho = \frac{1}{2}(\mathbf{D}\cdot\mathbf{E}+\mathbf{B}\cdot\mathbf{H}) = \frac{1}{2}(\epsilon\epsilon_0 E^2 + \mu\mu_0 H^2). \tag{A5.1}$$

In an electromagnetic wave this energy is transported along in the direction of energy flow, which in an isotropic medium is in the same direction as the wave vector (the normal to the wavefront). The rate of energy flow across unit area per second is given by a vector **S** called the Poynting vector. If we consider a plane-polarized wave with fields E_x and H_y, then **S** is in the direction of the z axis and has magnitude

$$\begin{aligned} |\,\mathbf{S}\,| = c\rho &= \frac{1}{2\sqrt{\epsilon\epsilon_0\mu\mu_0}}(\epsilon\epsilon_0 E_x^2 + \mu\mu_0 H_y^2) \\ &= \frac{1}{2}\left(\sqrt{\frac{\epsilon\epsilon_0}{\mu\mu_0}}E_x^2 + \sqrt{\frac{\mu\mu_0}{\epsilon\epsilon_0}}H_y^2\right) = \frac{1}{2}\left(\frac{E_x^2}{Z_0} + Z_0 H_y^2\right) = E_x H_y \\ &= \frac{E_x^2}{Z_0} = Z_0 H_y^2, \end{aligned} \tag{A5.2}$$

so

$$S_z = E_x H_y. \tag{A5.3}$$

In general

$$\mathbf{S} = \mathbf{E}\times\mathbf{H}. \tag{A5.4}$$

The total energy per unit time that is carried by the wave into a volume V bounded by a surface A, as shown in Fig. (A5.1), is $\int_A -(\mathbf{S}\cdot\hat{\mathbf{A}})dA$; the minus sign is present because $\hat{\mathbf{A}}$ is a unit vector directed away from the surface. Now from Gauss's theorem

$$\int_A -(\mathbf{S}\cdot\hat{\mathbf{A}})dA = \int_V -\operatorname{div}\mathbf{S}\,dV \tag{A5.5}$$

and

$$\int_V -\operatorname{div}\mathbf{S}\,dV = \int_V -\operatorname{div}(\mathbf{E}\times\mathbf{H})dV. \tag{A5.6}$$

† W.H. Hayt, Jr., *Engineering Electromagnetics*, 5th Edition, McGraw-Hill, New York, 1989.

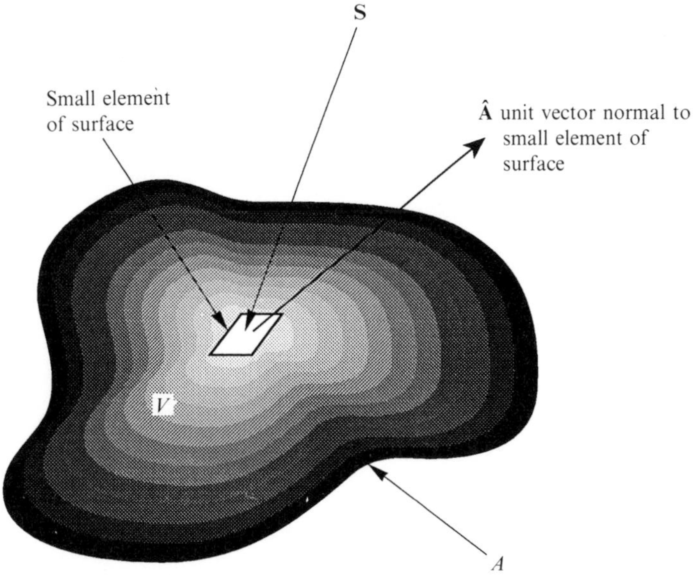

Fig. A5.1. Transport of electromagnetic energy into a region of space of volume V and bounded by a closed surface S.

Since div($\mathbf{E} \times \mathbf{H}$) = $\mathbf{H} \cdot$ curl$\mathbf{E}$ − $\mathbf{E} \cdot$ curl$\mathbf{H}$, using Maxwell's equations we get

$$\text{div}(\mathbf{E} \times \mathbf{H}) = \mathbf{H} \cdot \left(\frac{\partial \mathbf{B}}{\partial t}\right) - \mathbf{E} \cdot \left(\mathbf{j} + \frac{\partial \mathbf{D}}{\partial t}\right), \tag{A5.7}$$

where in Eq. (A5.7) $\mathbf{j}$ is the current density flowing in the medium. Now, remember that

$$\mathbf{D} = \epsilon_0 \mathbf{E} + \mathbf{P} = \epsilon\epsilon_0 \mathbf{E}, \tag{A5.8}$$

$$\mathbf{B} = \mu_0(\mathbf{H} + \mathbf{M}) = \mu\mu_0 \mathbf{H}, \tag{A5.9}$$

where $\mathbf{M}$ is the magnetization (magnetic dipole moment vol. $^{-1}$) in the medium. Therefore,

$$\int_V -\text{div}(\mathbf{E} \times \mathbf{H}) dV = \int_V \left[\mathbf{H} \cdot \left(\mu_0 \frac{\partial \mathbf{H}}{\partial t}\right) + \mu_0 \mathbf{H} \cdot \frac{\partial \mathbf{M}}{\partial t} + \mathbf{E} \cdot \mathbf{j} + \mathbf{E} \cdot \left(\epsilon_0 \frac{\partial \mathbf{E}}{\partial t}\right) + \mathbf{E} \cdot \frac{\partial \mathbf{P}}{\partial t}\right] dV. \tag{A5.10}$$

Using the equality

$$\frac{1}{2}\frac{\partial}{\partial t}(\mathbf{E} \cdot \mathbf{E}) = \mathbf{E} \cdot \frac{\partial \mathbf{E}}{\partial t} \tag{A5.11}$$

from Eq. (A5.10), the flow of energy into the volume V becomes

$$\int_V -\text{div}(\mathbf{E} \times \mathbf{H}) dV = \int_V \left[\mathbf{E} \cdot \mathbf{j} + \frac{\partial}{\partial t}\left(\frac{\epsilon_0}{2}\mathbf{E} \cdot \mathbf{E}\right) + \frac{\partial}{\partial t}\left(\frac{\mu_0}{2}\mathbf{H} \cdot \mathbf{H}\right) + \mu_0 \mathbf{H} \cdot \frac{\partial \mathbf{M}}{\partial t}\right] + \mathbf{E} \cdot \frac{\partial \mathbf{P}}{\partial t} dV, \tag{A5.12}$$

where the terms on the RHS correspond, from left to right, to: (i) the work done by the field in causing conventional current flow (ohmic heating), (ii) and (iii) the rate of increase of vacuum electromagnetic energy, (iv) the work done on magnetic dipoles in the volume, and (v) the final term on the RHS of Eq. (A5.12) the term $\mathbf{E} \cdot \partial \mathbf{P}/\partial t$ represents work done by the electromagnetic field in polarizing the medium.

Appendix 6

The Reflection and Refraction of a Plane Electromagnetic Wave at the Boundary Between Two Isotropic Media of Different Refractive Index

Fig. (A6.1) shows the directions of propagation and field directions for the incident, reflected, and transmitted electromagnetic waves at a planar boundary. The discussion is restricted to P-waves, but a similar discussion is readily developed for S-waves. For the incident wave the electric field varies with position as

$$\begin{aligned} E_1 &= E_i e^{i(\omega t - k_1 s_1)} \\ &= E_i e^{i(\omega t - k_1 x \sin\theta_1 - k_1 z \cos\theta_1)}, \end{aligned} \tag{A6.1}$$

where s_1 is distance measured along the direction of propagation.

For the reflected wave

$$\begin{aligned} E_3 &= E_r e^{i(\omega t - k_3 s_3)} \\ &= E_r e^{i(\omega t - k_3 x \sin\theta_3 + k_3 z \cos\theta_3)}, \end{aligned} \tag{A6.2}$$

and for the transmitted wave

$$E_2 = E_t e^{i(\omega t - k_2 x \sin\theta_2 - k_2 z \cos\theta_2)}. \tag{A6.3}$$

Similar expressions exist for the magnetic fields. At the boundary, $z = 0$, the tangential electric field must be continuous.† Therefore,

$$E_1 \cos\theta_1 + E_3 \cos\theta_3 = E_2 \cos\theta_3, \tag{A6.4}$$

which gives

$$E_i \cos\theta_1 e^{-ik_1 x \sin\theta_1} + E_r \cos\theta_3 e^{-ik_3 x \sin\theta_3} = E_t \cos\theta_2 e^{-ik_2 x \sin\theta_2}. \tag{A6.5}$$

Eq. (A6.5) must be true at all points along the boundary – all values of x. This can only be true if all the exponents are equal, which since $k_1 = k_3 = 2\pi n_1/\lambda_0$, and $k_2 = 2\pi n_2/\lambda_0$, gives

$$\theta_1 = \theta_3. \tag{A6.6}$$

The angle of reflection equals the angle of incidence. Furthermore,

$$n_1 \sin\theta_1 = n_2 \sin\theta_2, \tag{A6.7}$$

† S. Ramo, J.R. Whinnery and T. Van Duzer, *Fields and Waves in Communication Electronics*, 2nd Edition, Wiley, New York, 1984.

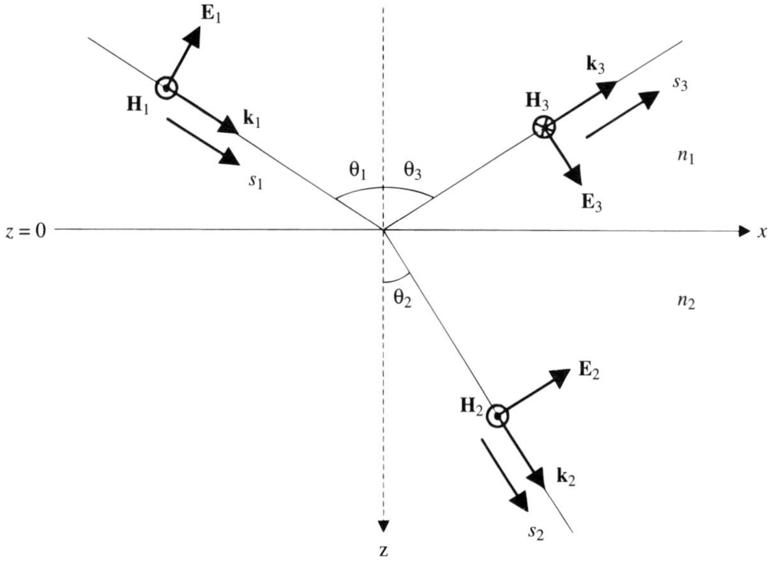

Fig. A6.1. Field and propagation directions for incident, reflected, and transmitted waves at a planar boundary.

which is Snell's law.

Tangential magnetic fields must also be continuous at the boundary so:

$$H_i - H_r = H_t, \tag{A6.8}$$

which can be written in terms of impedances as

$$\frac{E_i}{Z_1} - \frac{E_r}{Z_1} = \frac{E_t}{Z_2}. \tag{A6.9}$$

Eq. (A6.5) can also be restated as

$$E_i \cos\theta_1 + E_r \cos\theta_1 = E_t \cos\theta_2. \tag{A6.10}$$

Combination of Eqs. (A6.9) and (A6.10) gives

$$\frac{E_r}{E_i} = \frac{Z_2 \cos\theta_2 - Z_1 \cos\theta_1}{Z_2 \cos\theta_2 + Z_1 \cos\theta_1}, \tag{A6.11}$$

$$\frac{E_t}{E_i} = \frac{2Z_2 \cos\theta_1}{Z_2 \cos\theta_2 + Z_1 \cos\theta_1}. \tag{A6.12}$$

In terms of field components parallel to the boundary

$$\frac{E_{rx}}{E_{ix}} = \rho = \frac{Z_2 \cos\theta_2 - Z_1 \cos\theta_1}{Z_2 \cos\theta_2 + Z_1 \cos\theta_1}, \tag{A6.13}$$

$$\frac{E_{tx}}{E_{ix}} = \frac{E_t \cos\theta_2}{E_i \cos\theta_1} = \tau = \frac{2Z_2 \cos\theta_2}{Z_2 \cos\theta_2 + Z_1 \cos\theta_1}. \tag{A6.14}$$

with the introduction of the effective impedance $Z' \equiv Z \cos\theta$ (for P-waves),† we

† For S-waves the effective impendance is $Z/\cos\theta$.

have the universal formulas

$$\rho = \frac{Z_2' - Z_1'}{Z_1' + Z_2'}, \tag{A6.15}$$

$$\tau = \frac{2Z_2'}{Z_1' + Z_2'}. \tag{A6.16}$$

For transmission in the opposite direction across the boundary we would have

$$\rho' = \frac{Z_1' - Z_2'}{Z_1' + Z_2'}, \tag{A6.17}$$

$$\tau' = \frac{2Z_1'}{Z_1' + Z_2'}. \tag{A6.18}$$

Note that $\rho' = -\rho, |\rho'| = |\rho|$.

The fractional energy transmission across the boundary, the transmittance is

$$T = 1 - |\rho|^2, \tag{A6.19}$$

which is independent of direction across the boundary. For lossless media, all the impedances are real and, from Eqs. (A6.15), (A6.16), and (A6.18),

$$T = 1 - |\rho|^2 = \frac{4Z_1'Z_2'}{(Z_1' + Z_2')^2}. \tag{A6.20}$$

Clearly then, from Eqs. (A6.16) and (A6.18)

$$T = \tau\tau'. \tag{A6.21}$$

Appendix 7

The Vector Differential Equation for Light Rays

Proof that $(d/ds)\left(n d\mathbf{r}/ds\right) = \mathbf{grad}\ n$

The optical length L in a uniform homogeneous medium is $n\ell$, where n is the (constant) refractive index along a straight line of geometric path length ℓ. In a situation where the refractive index varies from point to point the optical path from point P_1 to point P_2 is

$$L = \int_{P_1}^{P_2} n(\mathbf{r}) ds, \tag{A7.1}$$

where s is distance measured along the path of the ray. *Fermat's principle* states that the actual path taken by a ray will be the one for which L is a minimum.

We can generalize the idea of the optical length by introducing a quantity called the *optical path* or *eikonal*, ζ, defined as

$$\zeta = n\mathbf{r} \cdot \hat{\mathbf{s}}. \tag{A7.2}$$

$\mathbf{r}$ is a point on the light ray measured with respect to an arbitrary origin, as shown in Fig. (A7.1), $\hat{\mathbf{s}}$ is a unit vector in the direction of the ray path at point $\mathbf{r}$. In an isotropic medium (one where the wave vector $\mathbf{k}$ and Poynting vector are parallel) a typical monochromatic wave has field amplitudes that vary from point to point as $e^{i(\omega t - \mathbf{k}\cdot\mathbf{r})}$. The phase variation with distance depends on the factor $\mathbf{k} \cdot \mathbf{r}$, which can be written as

$$\mathbf{k} \cdot \mathbf{r} = \left(\frac{2\pi}{\lambda_0}\right) n\mathbf{r} \cdot \hat{\mathbf{s}} = \left(\frac{2\pi}{\lambda_0}\right) \zeta, \tag{A7.3}$$

so we can see that the eikonal is closely related to the spatial variation of the phase of the wave. Points where $\zeta =$ constant correspond to phase fronts.

From Eq. (A7.2)

$$\text{grad}\ \zeta = \text{grad}\ (n\mathbf{r} \cdot \hat{\mathbf{s}}). \tag{A7.4}$$

Provided n does not vary too rapidly with position we can write Eq. (A7.4) as†

$$\text{grad}\ \zeta = n\ \text{grad}(\mathbf{r} \cdot \hat{\mathbf{s}}). \tag{A7.5}$$

† For an alternative discussion of this point see M. Born and E. Wolf, *Principles of Optics*, 6th Edition, Pergamon Press, Oxford, 1980.

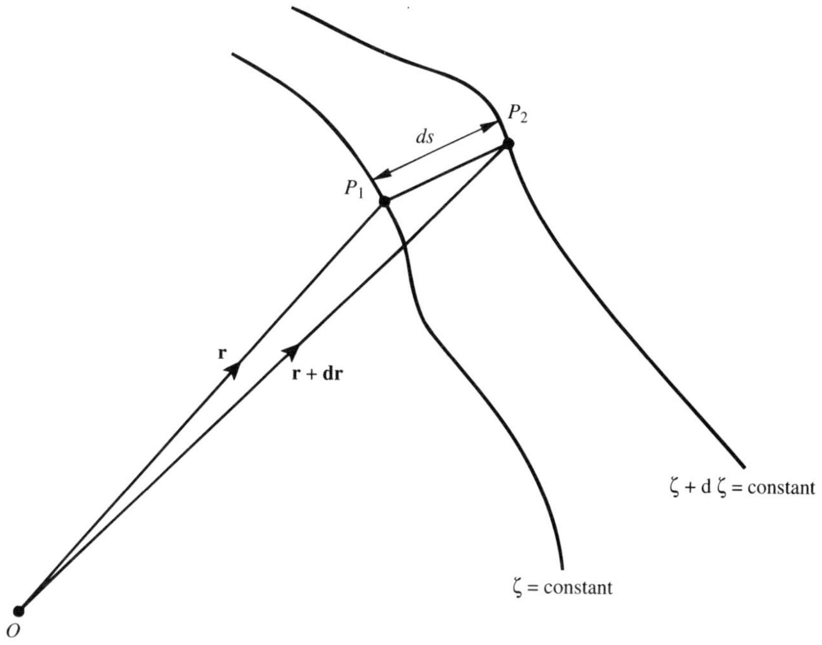

Fig. A7.1. Geometry of phase front, ray path, and position vectors used in discussing the ray path in a medium where the refractive index varies with position.

By the use of a vector identity

$$\text{grad }(\mathbf{u}\cdot\mathbf{v}) = (\mathbf{v}\cdot\nabla)\mathbf{u} + (\mathbf{u}\cdot\nabla)\mathbf{v} + \mathbf{v}\times\text{ curl }\mathbf{u} + \mathbf{u}\times\text{ curl }\mathbf{v} \tag{A7.6}$$

we get

$$\text{grad }\zeta = n[(\hat{\mathbf{s}}\cdot\nabla)\mathbf{r} + (\mathbf{r}\cdot\nabla)\hat{\mathbf{s}}] + n[\mathbf{r}\times\text{curl }\hat{\mathbf{s}} + \hat{\mathbf{s}}\times\text{ curl }\mathbf{r}]. \tag{A7.7}$$

Now, curl $\mathbf{r} = 0$; curl $\hat{\mathbf{s}} = 0$; and because $\hat{\mathbf{s}}$ is not explicitly a function of x, y, or z, $(\mathbf{r}\cdot\nabla)\,\hat{\mathbf{s}} = 0$. Furthermore,

$$(\hat{\mathbf{s}}\cdot\nabla)\mathbf{r} = \hat{\mathbf{s}}. \tag{A7.8}$$

Eq. (A7.8) can be proved as follows: If $\hat{\mathbf{s}} \equiv a\hat{\mathbf{i}} + b\hat{\mathbf{j}} + c\hat{\mathbf{k}}$, and $\mathbf{r} = x\hat{\mathbf{i}} + y\hat{\mathbf{j}} + z\hat{\mathbf{k}}$, then

$$\begin{aligned}(\mathbf{s}\cdot\nabla)\mathbf{r} &= \left(a\frac{\partial}{\partial x} + b\frac{\partial}{\partial y} + c\frac{\partial}{\partial z}\right)(x\hat{\mathbf{i}} + y\hat{\mathbf{j}} + z\hat{\mathbf{k}})\\ &= a\hat{\mathbf{i}} + b\hat{\mathbf{j}} + c\hat{\mathbf{k}}\\ &= \hat{\mathbf{s}}.\end{aligned}$$

Consequently, from Eqs. (A7.6) and (A7.7),

$$\text{grad }\zeta = n\hat{\mathbf{s}}. \tag{A7.9}$$

In physical terms this is equivalent to saying that the ray path is perpendicular to the phase front.

In Fig. (A7.1) an incremental ray path ds between two closely spaced phase fronts is shown. Clearly,

$$\mathbf{r} + d\mathbf{r} = \mathbf{r} + ds\hat{\mathbf{s}}, \tag{A7.10}$$

so

$$\frac{d\mathbf{r}}{ds} = \hat{\mathbf{s}}. \tag{A7.11}$$

Therefore,

$$n\frac{d\mathbf{r}}{ds} = n\hat{\mathbf{s}} = \text{grad } \zeta \tag{A7.12}$$

and consequently

$$\frac{d}{ds}\left(n\frac{d\mathbf{r}}{ds}\right) = \frac{d}{ds}(\text{grad } \zeta). \tag{A7.13}$$

Now,

$$\frac{d}{ds}(\text{grad } \zeta) \equiv \left(\frac{dx}{ds}\frac{\partial}{\partial x} + \frac{dy}{ds}\frac{\partial}{\partial y} + \frac{dz}{ds}\frac{\partial}{\partial z}\right) \text{grad } \zeta. \tag{A7.14}$$

Therefore,

$$\frac{d}{ds}(\text{grad } \zeta) = \left(\frac{d\mathbf{r}}{ds} \cdot \nabla\right) \text{grad } \zeta, \tag{A7.15}$$

which from Eq. (A7.12) can be written as

$$\frac{d}{ds}(\text{grad } \zeta) = \frac{1}{n}(\text{grad } \zeta \cdot \nabla) \text{ grad } \zeta. \tag{A7.16}$$

Use of Eq. (A7.6) with $\mathbf{u} \equiv \mathbf{v} \equiv$ grad ζ gives

$$\begin{aligned} \text{grad (grad } \zeta \cdot \text{ grad } \zeta) &= (\text{grad } \zeta \cdot \nabla) \text{ grad } \zeta \\ &\quad + (\text{grad } \zeta \cdot \nabla)\text{grad } \zeta + 2\text{grad } \zeta \times \text{curl grad } \zeta. \end{aligned} \tag{A7.17}$$

The last term on the RHS of Eq. (A7.17) is zero because curl grad $\zeta = 0$. Therefore,

$$(\text{grad } \zeta \cdot \nabla)\text{grad } \zeta = \frac{1}{2} \text{ grad}(\text{grad } \zeta)^2. \tag{A7.18}$$

Combining Eqs. (A7.12), (A7.13), (A7.16), and (A7.18) gives

$$\frac{d}{ds}\left(n\frac{d\mathbf{r}}{ds}\right) = \frac{1}{2n} \text{ grad } (n^2) \tag{A7.19}$$

and since

$$\begin{aligned} \text{grad } (n^2) &= n \text{ grad } n + (\text{grad } n)n \\ &= 2n \text{ grad } n, \end{aligned} \tag{A7.20}$$

we have the final desired result

$$\frac{d}{ds}\left(n\frac{d\mathbf{r}}{ds}\right) = \text{grad } n. \tag{A7.21}$$

APPENDIX 8

Symmetry Properties of Crystals and the 32 Crystal Classes

The external form that crystals of a particular material adopt is a reflection of the symmetry of the internal arrangement of the atoms that make up the structure. The symmetry of the crystal is determined by observing what *symmetry operations* apply to it. These symmetry operations include rotational symmetry, the existence of planes of mirror symmetry and inversion symmetry. The highest order of rotational symmetry possessed by a crystal determines to which of the seven *crystal systems* it belongs. Rotational symmetries that are two-fold, three-fold, four-fold, and six-fold are readily observed in nature.† If a crystal possesses a six-fold axis of symmetry, for example, then rotation of the crystal by $360/6 = 60°$ brings the crystal into a configuration that is identical to the one in which it started. A mirror plane m is a self-explanatory symmetry element: if half of a crystal structure is a mirror image of the other half relative to a particular plane through the crystal then that plane is a mirror plane. The existence of a center of inversion symmetry implies that if a face exists on the crystal perpendicular to a particular vector direction **r** from the origin, then a crystal face will also exist perpendicular to the vector **-r** from the origin. In terms of the internal structure an atom at coordinate (x, y, z) has an identical counterpart at $(-x, -y, -z)$. Crystals can possess an axis of n-fold *inversion* symmetry: for example in a crystal of six-fold inversion symmetry a rotation of 60° brings a crystal face on the bottom of the crystal into a position directly inverted from where an original face on the top of the crystal was in the 0° orientation. A crystal possessing a 6-fold axis of inversion symmetry would be characterized as having a $\bar{6}$ axis.

The precise interplay between symmetry elements is best illustrated by giving specific examples of the 32 crystal classes that exist within the seven crystal systems. These seven systems are characterized by the highest order of rotational symmetry that they possess as shown in Table (A8.1)

The 32 crystal crystals are tabulated in Table (A8.2), which gives their designation and lists briefly the symmetry elements that they possess. The highest symmetry cubic class, designated m3m, provides a good example of how the various symmetry operations are related. This crystal class has three orthogonal four-fold axes, four three-fold axes (along the cube body diagonals), six two-fold axes, nine mirror planes and a center of inversion symmetry, as shown schematically in Fig. (A8.1). For details, and excellent drawings, of the crystal forms that occur in

† Some examples of five-fold symmetry have been observed on a microscopic scale, but we will not deal with them further here.

Table A8.1. *The seven crystal systems.*

System	Highest order of rotational symmetry
Triclinic	No axes of symmetry (1-fold axes)
Monoclinic	One 2-fold axis
Orthorhombic	Three orthogonal 2-fold axes
Trigonal	One 3-fold axis
Hexagonal	One 6-fold axis
Tetragonal	One 4-fold axis
Cubic	Four 3-fold axes

Table A8.2. *The 32 crystal crystals.*

System	Designations of classes
Triclinic	1, $\bar{1}$
Monoclinic	2,m,2/m
Orthorhombic	mm,222,mmm
Trigonal	3,$\bar{3}$, 3m,$\bar{3}$m,32
Hexagonal	6,$\bar{6}$,6/m,6mm,$\bar{6}m$2,62,6/mmm
Tetragonal	4,$\bar{4}$,4/m,4mm,$\bar{4}$2m,42,4/mmm
Cubic	23,m3,$\bar{4}$3m,43,m3m

$\bar{2}$ indicates an axis of two-fold inversion symmetry.
m indicates a plane of mirror symmetry.
6/m indicates a six-fold axis orthogonal to a mirror plane.
4/mmm indicates a four-fold axis orthogonal to a mirror plane and additional mirror planes within which the four-fold axis lies.

the 32 crystal classes the interested reader should consult F.C. Phillips.‡ However, a few examples might be helpful.

A8.1 Class 6mm

This class has a six-fold symmetry axis at the intersection of two sets of three planes of mirror symmetry. Fig. (A8.2) gives an example of a crystal of zinc oxide that has this symmetry.

A8.2 Class $\bar{4}$2m

This class, represented by the important nonlinear crystals ADP and KDP, has an axis of four-fold inversion symmetry along the intersection of two orthogonal mirror planes, with two two-fold axes at 45° to these planes. Fig. (A8.3) shows these symmetry elements in a crystal of urea ($CO(NH_2)_2$).

‡ F.C. Phillips, *An Introduction to Crystallography*, 2nd Edition, Longmans, Green and Co., London, 1956.

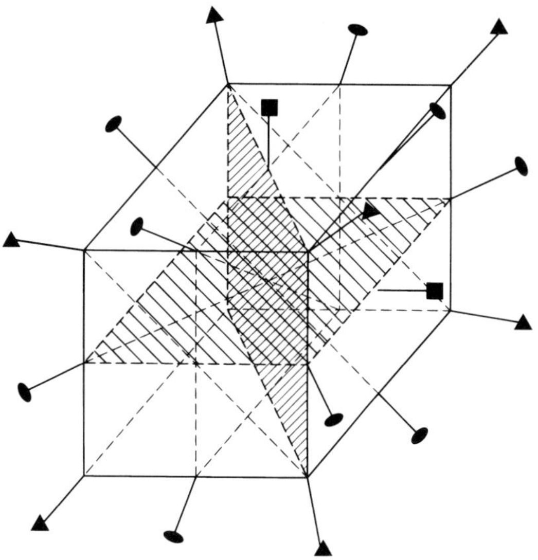

Fig. A8.1. The symmetry elements of a cubic crystal of the highest symmetry, class m3m. The black squares indicate axes of four-fold symmetry, the triangles indicate axes of three-fold symmetry (triad axes) and the lenticular shapes indicate axes of two-fold symmetry (dyad axes). Various planes of mirror symmetry are also shown.

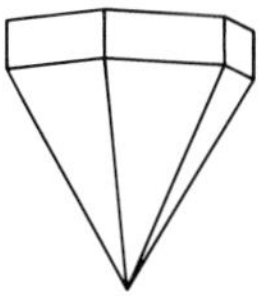

Fig. A8.2. A zinc oxide crystal form with symmetry 6mm.

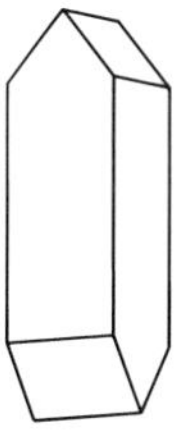

Fig. A8.3. An urea crystal form with symmetry $\bar{4}2$m.

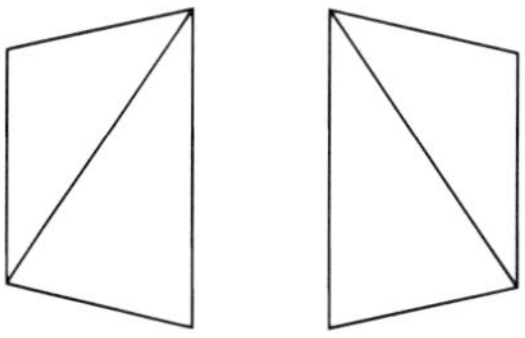

Fig. A8.4. Sphenoidal crystal form with symmetry 222.

A8.3 Class 222

These crystals have three mutually orthogonal two-fold axes, but no mirror planes, the orthorhombic sphenoidal forms shown in Fig. (A8.4) illustrate this symmetry. These crystals do not possess any planes of mirror symmetry.

Appendix 9

Tensors

One kind of tensor that is of particular interest to us here is a so-called tensor of the *second rank*,† denoted for example by $\overline{\overline{T}}$, which transforms one vector into another, for example,

$$\mathbf{B} = \overline{\overline{T}}\ \mathbf{A}. \tag{A9.1}$$

The tensor transforms the vector **A** into a new vector **B**, which is different in both magnitude *and direction* from **A**.

For example, the susceptibility tensor $\overline{\overline{\chi}}$ relates polarization and electric field

$$\mathbf{P} = \overline{\overline{\chi}}\ (\epsilon_0 \mathbf{E}). \tag{A9.2}$$

Clearly, the tensor can be written as an array of coefficients

$$\overline{\overline{\chi}} = \begin{pmatrix} \chi_{11} & \chi_{12} & \chi_{13} \\ \chi_{21} & \chi_{22} & \chi_{23} \\ \chi_{31} & \chi_{32} & \chi_{33} \end{pmatrix}. \tag{A9.3}$$

Physically, it does not matter what set of coordinate axes we choose, an electric field in a particular direction relative to specific crystal directions always gives the same polarization vector. However, the specific value of the coefficients T_{ij} will depend on the coordinate system chosen. Thus, these coefficients must change, or *transform*, with a change in coordinate axes so as to preserve the specific relationship between **P** and **E**. Tensors are mathematically described as quantities that transform in a specific way under coordinate axis rotations.

In a principal axis system (x, y, z) the tensor $\overline{\overline{\chi}}$ is diagonal

$$\overline{\overline{\chi}} = \begin{pmatrix} \chi_{11} & 0 & 0 \\ 0 & \chi_{22} & 0 \\ 0 & 0 & \chi_{33} \end{pmatrix}. \tag{A9.4}$$

Let us choose a different set of axes (x', y', z') in the following way:

(a) rotate by angle ϕ about the z axis: this produces axes x'', y', z where

$$\begin{aligned} x'' &= x\cos\phi + y\sin\phi, \\ y' &= -x\sin\phi + y\cos\phi, \end{aligned} \tag{A9.5}$$

(b) rotate by angle θ about the new (y') axis: this gives axes x', y', z', where

$$\begin{pmatrix} x' \\ y' \\ z' \end{pmatrix} = \begin{pmatrix} \cos\theta\cos\phi & \sin\phi\cos\theta & -\sin\theta \\ -\sin\phi & \cos\phi & 0 \\ \cos\phi\sin\theta & -\sin\phi\sin\theta & \cos\theta \end{pmatrix} \begin{pmatrix} x \\ y \\ z \end{pmatrix}. \tag{A9.6}$$

† A scalar is a tensor of rank zero, a vector is a tensor of rank 1.

Eq. (A9.6) can be used to demonstrate the transformation of a vector

$$\mathbf{r}' = \mathbf{R}\mathbf{r}, \tag{A9.7}$$

where $\mathbf{r}, \mathbf{r}'$, are the representations of a vector in the two coordinate systems and $\mathbf{R}$ is the *rotation* matrix.

If we write Eq. (A9.4) in a nonprincipal coordinate system that results from a rotation $\mathbf{R}$ of the principal coordinate system then

$$\mathbf{P}' = \epsilon_0 \,\overline{\overline{\chi}}\,'\mathbf{E}', \tag{A9.8}$$

where $\overline{\overline{\chi}}\,'$ is the representation of the $\overline{\overline{\chi}}$ tensor in the primed coordinate system. Since $\mathbf{P}' = \mathbf{RP}$, $\mathbf{E}' = \mathbf{RE}$, we have

$$\mathbf{RP} = \epsilon_0 \overline{\overline{\chi}}\,'\mathbf{RE} \tag{A9.9}$$

and

$$\mathbf{P} = \epsilon_0 \mathbf{R}^{-1} \overline{\overline{\chi}}\,'\mathbf{RE} = \epsilon_0 \,\overline{\overline{\chi}}\, \mathbf{E}. \tag{A9.10}$$

Therefore, in terms of the tensor representation in the principal axis system, $\overline{\overline{\chi}}$,

$$\mathbf{R}^{-1} \overline{\overline{\chi}}\,'\mathbf{R} = \overline{\overline{\chi}}\,' \tag{A9.11}$$

and

$$\overline{\overline{\chi}}\,' = \mathbf{R}\,\overline{\overline{\chi}}\,'\mathbf{R}^{-1}. \tag{A9.12}$$

Let us consider a simple case in which $\phi = 90°, \theta = 90°$. This transforms (x, y, z) to (x', y', z'), where from the standpoint of directions $x' \equiv -z; y' \equiv -x; z' \equiv y$.

The transformation matrix is

$$\mathbf{R} = \begin{pmatrix} 0 & 0 & -1 \\ -1 & 0 & 0 \\ 0 & -1 & 0 \end{pmatrix}. \tag{A9.13}$$

It is easy to show that

$$\mathbf{R}^{-1} = \begin{pmatrix} 0 & -1 & 0 \\ 0 & 0 & -1 \\ -1 & 0 & 0 \end{pmatrix}, \tag{A9.14}$$

so a susceptibility tensor in a principal axis system, such as Eq. (A9.4), would become

$$\overline{\overline{\chi}}\,' = \mathbf{R}\,\overline{\overline{\chi}}\,\mathbf{R}^{-1} = \begin{pmatrix} \chi_3 & 0 & 0 \\ 0 & \chi_1 & 0 \\ 0 & 0 & \chi_2 \end{pmatrix}, \tag{A9.15}$$

which correctly reflects the axis rearrangement that has occurred.

Note that $\mathbf{R}^{-1} = \tilde{\mathbf{R}}$, where $\tilde{\mathbf{R}}$ is the transpose of matrix $\mathbf{R}$,† so we can write $(\mathbf{R}^{-1})_{ik} = \mathbf{R}_{ki}$. Therefore, Eq. (A9.12) can be written

$$\chi'_{ij} = \sum_k \sum_\ell R_{ik} R_{j\ell} \chi_{k\ell} \tag{A9.16}$$

or in the repeated index convention that is commonly used in tensor algebra

$$\chi'_{ij} = R_{ik} R_{j\ell} \chi_{k\ell}. \tag{A9.17}$$

† If all the elements of a matrix are real and $\mathbf{R}^{-1} = \tilde{\mathbf{R}}$, the matrix is said to be *orthogonal*.

We can generalize this discussion to tensors of the *third rank*, such as the electro-optic tensor $\overline{\overline{d}}$, so that the coefficients of this tensor change under a coordinate transformation as

$$d'_{ijk} = R_{i\ell} R_{jm} R_{kn} d_{\ell mm}. \tag{A9.18}$$

The form of the $\overline{\overline{d}}$ tensor that is tabulated in Table (19.1) reflects the fact that d'_{ijk} must be identical to $d_{\ell mn}$ if the coordinate transformation, or general symmetry transformation, takes the crystal to an identical form – such as a 60° rotation about the six-fold axis in a class 6 crystal.

Appendix 10
Bessel Function Relations

The Bessel function relations used in deriving the unified boundary condition Eqs. (17.102)–(17.104) are:

$$J_{-\nu}(u) = (-1)^{\nu} J_{\nu}(u), \tag{A10.1}$$

$$K_{-\nu}(w) = K_{\nu}(w), \tag{A10.2}$$

$$2J'_{\nu} = J_{\nu-1} - J_{\nu+1}, \tag{A10.3}$$

$$-2K'_{\nu} = K_{\nu-1} + K_{\nu+1}, \tag{A10.4}$$

$$J_{\nu+1}(u) + J_{\nu-1}(u) = 2\left(\frac{\nu}{u}\right) J_{\nu}(u) \tag{A10.5}$$

$$K_{\nu+1}(w) - K_{\nu-1}(w) = 2\left(\frac{\nu}{w}\right) K_{\nu}(w). \tag{A10.6}$$

The following equations result from (A10.1)–(A10.6)

$$\frac{J'_{\nu}(u)}{uJ_{\nu}(u)} = \frac{J_{\nu-1}(u)}{uJ_{\nu}(u)} - \frac{\nu}{u^2} = -\frac{J_{\nu+1}}{uJ_{\nu}} + \frac{\nu}{u^2}, \tag{A10.7}$$

$$\frac{K'_{\nu}(w)}{wK_{\nu}(w)} = -\frac{K_{\nu-1}(w)}{wK_{\nu}(w)} - \frac{\nu}{w^2} = -\frac{K_{\nu+1}(w)}{wK_{\nu}(w)} + \frac{\nu}{w^2}. \tag{A10.8}$$

Appendix 11
Green's Functions

In the most general sense, the use of Green's functions is a method for solving certain differential equations *including* their boundary conditions.† For example, suppose it is desired to find the solution $g(x)$ for some inhomogeneous differential equation of the general form

$$L[g(x)] = f(x), \tag{A11.1}$$

where $L[g(x)]$ includes various differential operators acting on the function $g(x)$. If the solution to this equation, including its boundary conditions $G(x, x')$, can be determined for a function $f(x)$ that corresponds to a δ-function at $x = x'$, then we could write

$$L[G(x, x')] = \delta(x - x'). \tag{A11.2}$$

A general function $f(x)$ can be imagined as made up of a series of δ-functions since

$$f(x) = \int_{-\infty}^{\infty} f(x')\delta(x - x')dx'. \tag{A11.3}$$

Consequently, the solution to Eq. (A11.1) can be found in a similar way from the Green's function

$$g(x) = \int_{-\infty}^{\infty} G(x, x')g(x')dx'. \tag{A11.4}$$

To make this discussion more specific we could choose examples from many different areas: the motion of damped vibrating systems subject to arbitrary forces, diffusion, the conduction of heat, or electromagnetic theory. Perhaps the simplest example comes from the theory of the electrostatic potential.

When a distribution of charge of charge density $\rho(\mathbf{r})$ is present in an unbounded region, the scalar potential ϕ obeys Poisson's equation

$$\nabla^2\phi = -\frac{\rho}{\epsilon}, \tag{A11.5}$$

where ϵ is the permittivity of the region being considered. The Green's function for this situation is the potential that results from a unit point charge situated at the origin. This is equivalent to choosing $x' = 0$ in our definition of the Green's function. For this function

$$\nabla^2 G(r, 0) = -\frac{\delta(r)}{\epsilon}. \tag{A11.6}$$

† H. Margenau and G.M. Murphy, *The Mathematics of Physics and Chemistry*, 2nd Edition, Van Nostrand, Princeton, New Jersey, 1956.

If we imagine replacing the point charge at the origin with a small spherical region of radius a, then the charge density within this region is

$$\rho_0 = \frac{3}{4\pi a^3}. \tag{A11.7}$$

Clearly as $a \to 0$, $\rho_0 \to \infty$ and

$$\int \rho_0 dV = \int_0^a 4\pi r^2 \left(\frac{3}{4\pi a^3}\right) dv = 1. \tag{A11.8}$$

So as $a \to 0$, ρ_0 is an appropriately normalized δ-function.

Now, the potential produced by a unit point charge at the origin is well known. This is the Green's function for the problem

$$G(r, 0) = \frac{1}{4\pi\epsilon r}. \tag{A11.9}$$

Writing Eq. (A11.6) in spherical coordinates

$$\frac{1}{r^2}\frac{\partial}{\partial r}\left(r^2\frac{\partial G}{\partial r}\right) = -\frac{\delta(r)}{\epsilon} \tag{A11.10}$$

and substituting from Eq. (A11.9) gives

$$\frac{1}{r^2}\frac{\partial}{\partial r}\left(r^2\frac{\partial G}{\partial r}\right) = \frac{1}{r^2}\frac{\partial}{\partial r}\left(-\frac{1}{4\pi\epsilon}\right). \tag{A11.11}$$

Integrating the right hand side of Eq. (A11.11) gives

$$\int \frac{1}{r^2}\frac{\partial}{\partial r}\left(-\frac{1}{4\pi\epsilon}\right) dV = 4\pi \int \frac{\partial}{\partial r}\left(-\frac{1}{4\pi\epsilon}\right) dv = -\frac{1}{\epsilon}, \tag{A11.12}$$

the same result as integrating the right hand side of Eq. (A11.10). So, indeed, the Green's function is a solution of Eq. (A11.6). Therefore, from Eq. (A11.4) we can see that the potential distribution of a distribution of charges is

$$\phi(\mathbf{r}) = \int \frac{\rho(\mathbf{r}')dV}{4\pi\epsilon|\mathbf{r} - \mathbf{r}'|}. \tag{A11.13}$$

In other words the Green's function has allowed us to solve the inhomogeneous differential equation (Poisson's equation) by integrating over the solutions corresponding to individual point charges – δ-functions – since their charge densities are infinite yet their total charge integrated over all space is finite.

If we solve the slightly more general problem of a point charge inside a region where a series of boundaries are at specified potentials we thereby determine the appropriate Green's function $G(\mathbf{r}, \mathbf{r}')$. The solution to this boundary value problem for a distribution of charge is

$$\phi(\mathbf{r}) = \int G(\mathbf{r}, \mathbf{r}')\rho(r')dV. \tag{A11.14}$$

Let us consider one more example. For monochromatic waves the wave equation reduces to the Helmholtz equation

$$\nabla^2\psi + k^2\psi = 0. \tag{A11.15}$$

If we are considering spherical waves originating from a point source, then in spherical coordinates Eq. (A11.15) can be written as

$$\frac{1}{r^2}\frac{\partial}{\partial r}\left(r^2\frac{\partial\psi}{\partial r}\right)a + k^2\psi = 0, \qquad \text{(A11.16)}$$

which clearly has the solution, except at $r = 0$,

$$\psi(\mathbf{r}) = \mathbf{G}(\mathbf{r}, 0) = \frac{1}{4\pi r}e^{\pm ikr}. \qquad \text{(A11.17)}$$

This is the appropriate Green's function for the problem, which satisfies the equation

$$\nabla^2 G(\mathbf{r}, 0) + k^2 G(\mathbf{r}, 0) = -\delta(\mathbf{r}). \qquad \text{(A11.18)}$$

If we have a distribution of point sources of complex amplitude $g(\mathbf{r}')$ then the full solution to the Helmholtz equation can be obtained as an integral in a similar way to Eq. (A11.13)

$$\psi_{Total}(\mathbf{r}) = \frac{1}{4\pi}\int \frac{g(\mathbf{r}')}{|\mathbf{r} - \mathbf{r}'|}e^{\pm ik|\mathbf{r}-\mathbf{r}'|}dV. \qquad \text{(A11.19)}$$

For further details of how this approach is extended to non-monochromatic sources the reader should consult Panofsky and Phillips.†.

† W.K.H. Panofsky and M. Philips, *Classical Electricity and Magnetism*, Addison-Wesley, Reading, MA, 1955.

APPENDIX 12

Recommended Values of Some Physical Constants

Quantity	Symbol	Value (with SD uncertainty)
Speed of light in a vacuum	c_0	2.99792458×10^8 ms^{-1} (exact)
Permeability of a vacuum	μ_0	$4\pi \times 10^{-7}$ H m^{-1}
Permitivity of a vacuum, $1/\mu_0 c^2$	ϵ_0	$8.854187817... \times 10^{-12}$ F m^{-1}
Elementary charge (of photon)	e	$1.60217733(49) \times 10^{-19}$ C
Unified atomic mass constant	m_u	$1.6605402(10) \times 10^{-27}$ kg
Rest mass:		
of electron	m_e	$9.1093897(54) \times 10^{-31}$ kg
	m_e/m_u	$5.48579903(13) \times 10^{-4}$
of proton	m_p	$1.6726231(10) \times 10^{-27}$ kg
	m_p/m_u	1.007276470(12)
	m_p/m_e	1836.152701(37)
of neutron	m_n	$1.6749286(10) \times 10^{-27}$ kg
	m_n/m_u	1.008664904(14)
Planck's constant	h	$6.6260755(40) \times 10^{-34}$ J s
$h/2\pi$	$\hbar$	$1.05457266(63) \times 10^{-34}$ J s
Avogadro's constant	N_A	$6.0221367(36) \times 10^{23}$ mol^{-1}
Boltzmann's constant	k	$1.380658(12) \times 10^{-23}$ J K^{-1}
Stefan–Boltzmann constant	σ	$5.67051(19) \times 10^{-8}$ W m^{-2} K^{-4}
First radiation constant, $2\pi hc^2$	c_1	$3.7417749(22) \times 10^{-16}$ J m^2 s^{-1}
Second radiation constant, hc/k	c_2	$1.438769(12) \times 10^{-2}$ m K

Index